Advanced Machine Learning Applications in Big Data Analytics

Advanced Machine Learning Applications in Big Data Analytics

Editors

Taiyong Li
Wu Deng
Jiang Wu

MDPI

Basel • Beijing • Wuhan • Barcelona • Belgrade • Novi Sad • Cluj • Manchester

Editors
Taiyong Li
Southwestern University of Finance and Economics
China

Wu Deng
Civil Aviation University of China
China

Jiang Wu
Southwestern University of Finance and Economics
China

Editorial Office
MDPI
St. Alban-Anlage 66
4052 Basel, Switzerland

This is a reprint of articles from the Special Issue published online in the open access journal *Electronics* (ISSN 2079-9292) (available at: https://www.mdpi.com/journal/electronics/special_issues/ML_Big_Data).

For citation purposes, cite each article independently as indicated on the article page online and as indicated below:

Lastname, A.A.; Lastname, B.B. Article Title. *Journal Name* **Year**, *Volume Number*, Page Range.

ISBN 978-3-0365-8486-7 (Hbk)
ISBN 978-3-0365-8487-4 (PDF)
doi.org/10.3390/books978-3-0365-8487-4

Contents

About the Editors

Taiyong Li

Taiyong Li received his Ph.D. from Sichuan University, Chengdu, China, in 2009, and he is currently a Full Professor at the School of Computing and Artificial Intelligence, Southwestern University of Finance and Economics. His research expertise lies in machine learning, computer vision, image processing, and evolutionary computation, focusing on clustering, image security, and time series analysis. He has published over 80 papers in journals and conferences, including ASOC, NEUROCOM, TVCJ, ECM, MTAP, CVPR, etc. Eight of his papers have been selected as highly cited in ESI, his Google Scholar H-index is 24, and he has led or participated in multiple national-level and industry projects. He is an Electronics and Frontiers Artificial Intelligence in Finance guest editor or review editor, and a reviewer of more than 30 journals, including AI Med, AI Rev, EAAI, ENTROPY, ESWA, FIN, IJBC, IJIST, SUPERCOM, PR, PLOS ONE, SWARM EVOL COMPUT, and so on.

Wu Deng

Wu Deng received a Ph.D. in computer application technology from Dalian Maritime University, Dalian, China, in 2012. He is currently a Professor at the College of Electronic Information and Automation, Civil Aviation University of China, Tianjin, China. His research interests include artificial intelligence, optimization method, and fault diagnosis. He has published over 120 papers in journals and conferences, including IEEE T-SMCA, IEEE T-ITS, IEEE TIM, IEEE TR, INS, KBS, etc. His Google Scholar H-index is 36.

Jiang Wu

Jiang Wu received his Ph.D. from Sichuan University, Chengdu, China, in 2008, and he is currently a Full Professor at the School of Computing and Artificial Intelligence, Southwestern University of Finance and Economics. His primary research interests include machine learning and image processing. He has published more than 60 papers in journals and conferences. He has led or participated in multiple national-level and industry projects. He reviews some journals and conferences, such as ENTROPY, J INF SCI, and PACIS.

electronics

Editorial

Advanced Machine Learning Applications in Big Data Analytics

Taiyong Li [1,*], Wu Deng [2] and Jiang Wu [1]

1 School of Computing and Artificial Intelligence, Southwestern University of Finance and Economics, Chengdu 611130, China; wuj_t@swufe.edu.cn
2 College of Electronic Information and Automation, Civil Aviation University of China, Tianjin 300300, China; wdeng@cauc.edu.cn
* Correspondence: litaiyong@gmail.com

1. Introduction

We are currently living in the era of big data. Discovering valuable patterns from big data has become a very hot research topic, which holds immense benefits for governments, businesses, and even individuals. Advanced machine learning models and algorithms have emerged as effective approaches to analyze such data. At the same time, these methods and algorithms are prompting applications in the field of big data.

Considering advanced machine learning and big data together, we have selected a series of relevant works in this special issue to showcase the latest research advancements in this field. Specifically, a total of thirty-three articles are included in this special issue, which can be roughly categorized into six groups: time series analysis, evolutionary computation, pattern recognition, computer vision, image encryption, and others.

2. Brief Description of the Published Articles

2.1. Time Series Analysis

Li et al. [1] proposed an integrated model combining bagging and stacking for short-time traffic-flow prediction. The model incorporates vacation and peak time features, as well as occupancy and speed information. A stacking model with ridge regression as the meta-learner was established and optimized using the bagging model to obtain the Ba-Stacking model. The base learners' information structure was modified by weighting the error coefficients to improve utilization, resulting in a DW-Ba-Stacking model. Experiment results showed that the DW-Ba-Stacking model had the highest prediction accuracy for short-term traffic flow compared with traditional models.

Li et al. [2] proposed a nonlinear integrated forecasting model combining auto-regressive and moving average (ARMA), grey system theory model (GM), and back-propagation (BP) model optimized by genetic algorithms (GA) to improve the forecasting accuracy of China coastal bulk coal freight index (CBCFI). The predicted values of ARMA and GM were used as input training samples for the neural network. A genetic algorithm was used to optimize the BP network to better exploit the prediction accuracy of the combined model. The combined ARMA-GM-GABP model was shown to have improved prediction accuracy and can effectively solve the CBCFI forecasting problem.

Wang et al. [3] proposed a new time series classification method called CEEMD-MultiRocket. It combined complementary ensemble empirical mode decomposition (CEEMD) with an improved MultiRocket algorithm to increase classification accuracy. The raw time series was first decomposed into three sub-series using CEEMD. The improved MultiRocket was applied to the raw time series, the selected decomposed sub-series and the first-order difference of the raw time series to generate the final classification results. Experimental results showed that CEEMD-MultiRocket ranked second in classification accuracy on the 109 datasets from the UCR repository against a spread of state-of-the-art TSC models, only behind HIVE-COTE 2.0, but with only 1.4% of the latter's computing load.

Citation: Li, T.; Deng, W.; Wu, J. Advanced Machine Learning Applications in Big Data Analytics. *Electronics* **2023**, *12*, 2940. https://doi.org/10.3390/electronics12132940

Received: 29 June 2023
Accepted: 3 July 2023
Published: 4 July 2023

Bousbaa et al. [4] proposed an incremental and adaptive strategy using the online stochastic gradient descent algorithm (SGD) and particle swarm optimization metaheuristic (PSO). Two techniques were involved in data stream mining (DSM): adaptive sliding windows and change detection. The study focused on forecasting the value of the Euro in relation to the US dollar. Results showed that the flexible sliding window proved its ability to forecast the price direction with better accuracy compared to using a fixed sliding window.

Han et al. [5] proposed a model named LST-GCN to improve the accuracy of traffic flow predictions. They simulated spatiotemporal correlations by optimizing GCN parameters using an LSTM network. This method improved the traditional method of combining recurrent neural networks and graph neural networks in spatiotemporal traffic flow prediction. Experiments conducted on the PEMS dataset showed that their proposed method was more effective and outperformed other state-of-the-art methods.

2.2. Evolutionary Computation

Gao et al. [6] introduced an enhanced slime mould algorithm (MSMA) with a multi-population strategy and proposed a prediction model based on the modified algorithm and the support vector machine (SVM) algorithm called MSMA-SVM to provide a reference for postgraduate employment decision and policy formulation. The multi-population strategy improved the solution accuracy of the algorithm and the proposed model enhanced the ability to optimize the SVM. Experiments showed that the modified slime mould algorithm had better performance compared to other algorithms and the optimal SVM model had better classification ability and more stable performance for predicting employment stability.

Bao et al. [7] introduced two strategies to address the shortcomings of the butterfly optimization algorithm (BOA): the random replacement strategy and the crisscross search strategy. These strategies were combined to create the random replacement crisscross BOA (RCCBOA). In order to evaluate the performance of RCCBOA, the author conducted comparative experiments with nine other advanced algorithms on the IEEE CEC2014 functional test set, and founded it is effective when combining RCCBOA with support vector machine (SVM) and feature selection (FS).

Zhang et al. [8] proposed an improved matrix particle swarm optimization algorithm (IMPSO) to optimize DNA sequence design. The algorithm incorporated centroid opposition-based learning and a dynamic update based on signal-to-noise ratio distance to search for high-quality solutions. The results showed that the proposed method achieved satisfactory outcomes and higher computational efficiency.

Song et al. [9] developed a multi-strategy adaptive particle swarm optimization (APSO/DU) to accelerate the solving speed of the mean-semivariance (MSV) model . A constraint factor was introduced to control velocity weight and reduce blindness in the search process. A dual-update (DU) strategy was designed based on new speed and position update strategies. The experiment results showed that the APSO/DU algorithm had better convergence accuracy and speed.

Li et al. [10] designed an intelligent prediction model for talent stability in higher education using a kernel extreme learning machine (KELM) and proposed a differential evolution crisscross whale optimization algorithm (DECCWOA) for optimizing the model parameters. The DECCWOA was shown to achieve high accuracy and fast convergence in solving both unimodal and multimodal functions. The DECCWOA was combined with KELM and feature selection (DECCWOA-KELM-FS) to achieve efficient talent stability intelligence prediction for universities or colleges in Wenzhou. The results showed that the performance of the proposed model outperformed other comparative algorithms. The created system can serve as a reliable way to predict higher education talent flows.

Wang et al. [11] proposed a new algorithm called SEGDE to solve the capacitated vehicle routing problem (CVRP). It combined the saving mileage algorithm (SMA), sequential encoding (SE), and gravitational search algorithm (GSA) to address the problems of the differential evolution (DE) algorithm. The SMA was used to initialize the population of the DE. The SE approach was used to adjust the differential mutation strategy. The GSA was applied to adjust the evolutionary search direction and improve search efficiency. Four CVRPs were tested with SEGDE and the results showed that SEGDE effectively solved CVRPs with better performance.

2.3. Pattern Recognition

Miu et al. [12] proposed a two-step method to more finely classify the event type of stock announcement news. First, candidate event trigger words and co-occurrence words were extracted and arranged in order of common expressions. Then, final event types were determined using three proposed criteria. Based on the real data of the Chinese stock market, this method constructed 54 event types (p = 0.927, f = 0.946), and included some types not discussed in previous studies.

Jia et al. [13] proposed a new hybrid graph network recommendation model called the user multi-behavior graph network (UMBGN) to make full use of multi-behavior user-interaction information. This model used a joint learning mechanism to integrate user–item multi-behavior interaction sequences and a user multi-behavior information-aware layer was designed to focus on the long-term multi-behavior features of users and learn temporally ordered user–item interaction information through BiGRU and AUGRU units. Experiments on three public datasets showed that this model outperformed the best baselines.

Fatehi et al. [14] investigated the effectiveness of adversarial attacks on clinical document classification and proposed a defense mechanism to develop a robust neural network (CNN) model and counteract these attacks. Various black-box attacks based on concatenation and editing adversaries were applied on unstructured clinical text. A defense technique based on feature selection and filtering was proposed to improve the robustness of the models. Experimental results showed that small perturbations caused a significant drop in performance and the proposed defense mechanism avoided this drop and enhanced the robustness of the CNN model for clinical document classification.

Yin et al. [15] proposed an improved hierarchical clustering algorithm called PRI-MFC to solve the problems of traditional hierarchical clustering algorithms. The algorithm was tested on artificial and real datasets and the experimental results showed superiority in clustering effect, quality, and time consumption.

Yang et al. [16] proposed an intelligent fault diagnosis method for bearings based on variational mode decomposition (VMD), composite multi-scale dispersion entropy (CMDE), and deep belief network (DBN) with particle swarm optimization (PSO) algorithm. The number of modal components decomposed by VMD was determined by the observation center frequency and reconstructed according to the kurtosis. The CMDE of the reconstructed signal was calculated to form training and test samples for pattern recognition. PSO was used to optimize the parameters of the DBN model for fault identification. Through experiment comparison, it was proved that the VMD-CMDE-PSO-DBN method had application value in intelligent fault diagnosis.

Chen et al. [17] proposed an improved least squares support vector machines method to solve the problem of the abnormality or loss of quick access recorder (QAR) data. This method used the entropy weight method to obtain index weights, principal component analysis for dimensionality reduction, and LS-SVM for data fitting and repair. The method was tested using QAR data from multiple real plateau flights and showed high accuracy and fit degree. This proved that the improved least squares support vector machines machine learning model could effectively fit and supplement missing QAR data in the plateau area through historical flight data.

Yu et al. [18] proposed a novel hierarchical heterogeneous graph attention network to model global semantic relations among nodes for emotion-cause pair extraction (ECPE). This method introduced all types of semantic elements involved in ECPE. A pair-level subgraph was constructed to explore the correlation between pair nodes and their different neighboring nodes. Two-level heterogeneous graph attention networks were used to achieve representation learning of clauses and clause pairs. Experiments on benchmark datasets showed that this proposed model achieved significant improvement over 13 compared methods.

2.4. Computer Vision

Fan et al. [19] proposed an infrared vehicle target detection algorithm based on an improved version of YOLOv5. The algorithm used the DenseBlock module to increase shallow feature extraction ability, and the Ghost convolution layer replaced the ordinary convolution layer to improve network feature extraction ability. The detection accuracy of the whole network was enhanced by adding a channel attention mechanism and modifying the loss function. Experimental results showed that the addition of DenseBlock and EIOU modules alone improved detection accuracy by 2.5% and 3%, respectively, compared to the original YOLOv5 algorithm. The combination of DenseBlock and Ghost convolution had the best effect, and when adding three modules at the same time, the mAP fluctuation was smaller, reaching 73.1%, which was 4.6% higher than the original YOLOv5 algorithm.

Guerrero-Ibañez et al. [20] proposed a model based on convolutional neural networks to identify and classify tomato leaf diseases using a public dataset and photographs taken in the fields to improve crop yields. Generative adversarial networks were used to avoid overfitting. The proposed model achieved an accuracy greater than 99% in detecting and classifying diseases in tomato leaves.

Zhang et al. [21] proposed a Hemerocallis citrina Baroni maturity detection method based on a deep learning algorithm, called the GGSC YOLOv5 algorithm. This method integrated a lightweight neural network and dual attention mechanism. The improved GGSC YOLOv5 algorithm reduced the number of parameters and Flops by 63.58% and 68.95%, respectively, and reduced the number of network layers by about 33.12% in terms of model structure. The detection precision was up to 84.9%, an improvement of about 2.55%, and the real-time detection speed increased from 64.16 FPS to 96.96 FPS.

Chen et al. [22] proposed a method for detecting abnormal pilot behavior during flight based on an improved YOLOv4 deep learning algorithm and an attention mechanism. The CBAM attention mechanism was introduced to improve the feature extraction capability of the deep neural network. The improved YOLOv4 recognition rate was significantly higher than the unimproved algorithm. The experimental results showed that the improved YOLOv4 had a high mAP, accuracy, and recall rate.

Jin et al. [23] proposed a quantum dynamic optimization algorithm called quantum dynamic neural architecture search (QDNAS) to find the optimal structure for a candidate network. The proposed QDNAS viewed the iterative evolution of the optimization over time as a quantum dynamic process. Experiments on four benchmarks showed that QDNAS was consistently better than all baseline methods in image classification tasks.

Yue et al. [24] designed a detection algorithm called TP-ODA for border patrol object detection. This algorithm improved the detection frame imbalance problem and optimized the feature fusion module of the algorithm with the PDOEM structure. The TP-ODA algorithm was tested on the Border Patrol object dataset BDP and showed improvement in mAP, GFLOPs, model volume, and FPS compared to the baseline model.

Ye et al. [25] proposed an innovative classification method for hyperspectral remote sensing images (HRSIs) called IPCEHRIC, which utilized the advantages of enhanced PSO algorithm, convolutional neural network (CNN), and extreme learning machine (ELM). Experiment conducted on Pavia University data and actual HRSIs after Jiuzhaigou 7.0 earthquake, and results showed that IPCEHRIC could accurately classify these data with stronger generalization, faster learning ability, and higher classification accuracy.

2.5. Image Encryption

Huang et al. [26] proposed a polymorphic mapping-coupled map lattice with information entropy for encrypting color images, improving the traditional one-dimensional-mapping coupled lattice.The original 4x4 matrix was extended and a new pixel-level substitution method was proposed using the huffman idea. The idea of polymorphism was employed and the pseudo-random sequence was diversified and homogenized. Experiments were conducted on three plaintext color images, "Lena", "Peppers" and "Mandrill", and the results showed that the algorithm had a large key space, better sensitivity to keys and plaintext images, and a better encryption effect.

Chen et al. [27] proposed a new digital image encryption algorithm based on the splicing model and 1D secondary chaotic system. The algorithm divided the plain image into four sub-parts using quaternary coding, which could be coded separately. The key space was big enough to resist exhaustive attacks due to the use of a 1D quadratic chaotic system. Experimental results showed that the algorithm had high security and a good encryption effect.

2.6. Others

Muntean et al. [28] proposed a methodological framework based on design science research for designing and developing data and information artifacts in data analysis projects. They applied several classification algorithms to previously labeled datasets through clustering and introduced a set of metrics to evaluate the performance of classifiers. Their proposed framework can be used for any data analysis problem that involves machine learning techniques.

Zheng et al. [29] proposed a novel KNN-based consensus algorithm that classified transactions based on their priority. The KNN algorithm calculated the distance between transactions based on factors that impacted their priority. Experimental results obtained by adopting the enhanced consensus algorithm showed that the service level agreement(SLA) was better satisfied in the BaaS systems.

Liu et al. [30] proposed a coordinated output strategy for peak shaving and frequency regulation using existing energy storage to improve its economic development and benefits in industrial parks. The strategy included profit and cost models, an economic optimization model for dividing peak shaving and frequency regulation capacity, and an intra-day model predictive control method for rolling optimization. The experimental results showed a 10.96% reduction in daily electricity costs using this strategy.

Hussain et al. [31] presented a COVID-19 warning system based on a machine learning time series model using confirmed, detected, recovered, and death case data. The author compared the performanceof long short-term memory (LSTM), auto-regressive (AR), PROPHET and autoregressive integrated moving average (ARIMA) models for predicting patients' confirmed, and found the PROPHET and AR models had low error rates in predicting positive cases.

Xie et al. [32] presented an effective solution for the problem of confidentiality management of digital archives on the cloud. The basic concept involved setting up a local server between the cloud and each client of an archive system to run a confidentiality management model of digital archives on the cloud. This model included an archive release model and an archive search model.The archive release model encrypted archive files and generated feature data for the archive data. The archive search model transformed query operations on the archive data submitted by a searcher. Both theoretical analysis and experimental evaluation demonstrated the good performance of the proposed solution.

Providence et al. [33] discussed the influence of temporal and spatial normalization modules on multi-variate time series forecasts. The study encompassed various neural networks and their applications. Extensive experimental work on three datasets showed that adding more normalization components could greatly improve the effectiveness of canonical frameworks.

3. Future Directions

We believe that advanced machine learning and big data will continue to develop. On one hand, advanced machine learning algorithms will discover more valuable patterns from big data, thereby fueling the emergence of new applications for big data. On the other hand, the constantly increasing volume of big data has raised higher demands for advanced machine learning, leading to the development of more effective and efficient machine learning algorithms. Therefore, developing new machine learning algorithms for big data analysis and expanding the application scenarios of big data are important research directions in the future.

Acknowledgments: We would like to thank all the authors for their papers submitted to this special issue. We would also like to acknowledge all the reviewers for their careful and timely reviews to help improve the quality of this special issue. Finally, we would like to thank the editorial team of the *Electronics* journal for all the support provided in the publication of this special issue.

Conflicts of Interest: The authors declare no conflict of interest.

References

1. Li, Z.; Wang, L.; Wang, D.; Yin, M.; Huang, Y. Short-Term Traffic-Flow Forecasting Based on an Integrated Model Combining Bagging and Stacking Considering Weight Coefficient. *Electronics* **2022**, *11*, 1467. [CrossRef]
2. Li, Z.; Piao, W.; Wang, L.; Wang, X.; Fu, R.; Fang, Y. China coastal bulk (Coal) freight index forecasting based on an integrated model combining ARMA, GM and BP model optimized by GA. *Electronics* **2022**, *11*, 2732. [CrossRef]
3. Wang, P.; Wu, J.; Wei, Y.; Li, T. CEEMD-MultiRocket: Integrating CEEMD with Improved MultiRocket for Time Series Classification. *Electronics* **2023**, *12*, 1188. [CrossRef]
4. Bousbaa, Z.; Sanchez-Medina, J.; Bencharef, O. Financial Time Series Forecasting: A Data Stream Mining-Based System. *Electronics* **2023**, *12*, 2039. [CrossRef]
5. Han, X.; Gong, S. LST-GCN: Long Short-Term Memory Embedded Graph Convolution Network for Traffic Flow Forecasting. *Electronics* **2022**, *11*, 2230. [CrossRef]
6. Gao, H.; Liang, G.; Chen, H. Multi-population enhanced slime mould algorithm and with application to postgraduate employment stability prediction. *Electronics* **2022**, *11*, 209. [CrossRef]
7. Bao, H.; Liang, G.; Cai, Z.; Chen, H. Random replacement crisscross butterfly optimization algorithm for standard evaluation of overseas Chinese associations. *Electronics* **2022**, *11*, 1080. [CrossRef]
8. Zhang, W.; Zhu, D.; Huang, Z.; Zhou, C. Improved Multi-Strategy Matrix Particle Swarm Optimization for DNA Sequence Design. *Electronics* **2023**, *12*, 547. [CrossRef]
9. Song, Y.; Liu, Y.; Chen, H.; Deng, W. A Multi-Strategy Adaptive Particle Swarm Optimization Algorithm for Solving Optimization Problem. *Electronics* **2023**, *12*, 491. [CrossRef]
10. Li, H.; Ke, S.; Rao, X.; Li, C.; Chen, D.; Kuang, F.; Chen, H.; Liang, G.; Liu, L. An Improved Whale Optimizer with Multiple Strategies for Intelligent Prediction of Talent Stability. *Electronics* **2022**, *11*, 4224. [CrossRef]
11. Wang, J.; Shang, S.; Jing, H.; Zhu, J.; Song, Y.; Li, Y.; Deng, W. A Novel Multistrategy-Based Differential Evolution Algorithm and Its Application. *Electronics* **2022**, *11*, 3476. [CrossRef]
12. Miu, F.; Wang, P.; Xiong, Y.; Jia, H.; Liu, W. Fine-Grained Classification of Announcement News Events in the Chinese Stock Market. *Electronics* **2022**, *11*, 2058. [CrossRef]
13. Jia, M.; Liu, F.; Li, X.; Zhuang, X. Hybrid Graph Neural Network Recommendation Based on Multi-Behavior Interaction and Time Sequence Awareness. *Electronics* **2023**, *12*, 1223. [CrossRef]
14. Fatehi, N.; Alasad, Q.; Alawad, M. Towards Adversarial Attacks for Clinical Document Classification. *Electronics* **2023**, *12*, 129. [CrossRef]
15. Yin, L.; Li, M.; Chen, H.; Deng, W. An Improved Hierarchical Clustering Algorithm Based on the Idea of Population Reproduction and Fusion. *Electronics* **2022**, *11*, 2735. [CrossRef]
16. Yang, E.; Wang, Y.; Wang, P.; Guan, Z.; Deng, W. An intelligent identification approach using VMD-CMDE and PSO-DBN for bearing faults. *Electronics* **2022**, *11*, 2582. [CrossRef]
17. Chen, N.; Sun, Y.; Wang, Z.; Peng, C. Improved LS-SVM Method for Flight Data Fitting of Civil Aircraft Flying at High Plateau. *Electronics* **2022**, *11*, 1558. [CrossRef]
18. Yu, J.; Liu, W.; He, Y.; Zhong, B. A Hierarchical Heterogeneous Graph Attention Network for Emotion-Cause Pair Extraction. *Electronics* **2022**, *11*, 2884. [CrossRef]
19. Fan, Y.; Qiu, Q.; Hou, S.; Li, Y.; Xie, J.; Qin, M.; Chu, F. Application of Improved YOLOv5 in Aerial Photographing Infrared Vehicle Detection. *Electronics* **2022**, *11*, 2344. [CrossRef]
20. Guerrero-Ibañez, A.; Reyes-Muñoz, A. Monitoring Tomato Leaf Disease through Convolutional Neural Networks. *Electronics* **2023**, *12*, 229. [CrossRef]
21. Zhang, L.; Wu, L.; Liu, Y. Hemerocallis citrina Baroni Maturity Detection Method Integrating Lightweight Neural Network and Dual Attention Mechanism. *Electronics* **2022**, *11*, 2743. [CrossRef]
22. Chen, N.; Man, Y.; Sun, Y. Abnormal Cockpit Pilot Driving Behavior Detection Using YOLOv4 Fused Attention Mechanism. *Electronics* **2022**, *11*, 2538. [CrossRef]

23. Jin, J.; Zhang, Q.; He, J.; Yu, H. Quantum Dynamic Optimization Algorithm for Neural Architecture Search on Image Classification. *Electronics* **2022**, *11*, 3969. [CrossRef]
24. Yue, L.; Ling, H.; Yuan, J.; Bai, L. A Lightweight Border Patrol Object Detection Network for Edge Devices. *Electronics* **2022**, *11*, 3828. [CrossRef]
25. Ye, A.; Zhou, X.; Miao, F. Innovative Hyperspectral Image Classification Approach Using Optimized CNN and ELM. *Electronics* **2022**, *11*, 775. [CrossRef]
26. Huang, P.; Li, D.; Wang, Y.; Zhao, H.; Deng, W. A Novel Color Image Encryption Algorithm Using Coupled Map Lattice with Polymorphic Mapping. *Electronics* **2022**, *11*, 3436. [CrossRef]
27. Chen, C.; Zhu, D.; Wang, X.; Zeng, L. One-Dimensional Quadratic Chaotic System and Splicing Model for Image Encryption. *Electronics* **2023**, *12*, 1325. [CrossRef]
28. Muntean, M.; Militaru, F.D. Design Science Research Framework for Performance Analysis Using Machine Learning Techniques. *Electronics* **2022**, *11*, 2504. [CrossRef]
29. Zheng, Q.; Wang, L.; He, J.; Li, T. KNN-Based Consensus Algorithm for Better Service Level Agreement in Blockchain as a Service (BaaS) Systems. *Electronics* **2023**, *12*, 1429. [CrossRef]
30. Liu, D.; Jin, Z.; Chen, H.; Cao, H.; Yuan, Y.; Fan, Y.; Song, Y. Peak Shaving and Frequency Regulation Coordinated Output Optimization Based on Improving Economy of Energy Storage. *Electronics* **2021**, *11*, 29. [CrossRef]
31. Hussain, M.; Islam, A.; Turi, J.A.; Nabi, S.; Hamdi, M.; Hamam, H.; Ibrahim, M.; Cifci, M.A.; Sehar, T. Machine Learning-Driven Approach for a COVID-19 Warning System. *Electronics* **2022**, *11*, 3875. [CrossRef]
32. Xie, J.; Xuan, S.; You, W.; Wu, Z.; Chen, H. An Effective Model of Confidentiality Management of Digital Archives in a Cloud Environment. *Electronics* **2022**, *11*, 2831. [CrossRef]
33. Providence, A.M.; Yang, C.; Orphe, T.B.; Mabaire, A.; Agordzo, G.K. Spatial and Temporal Normalization for Multi-Variate Time Series Prediction Using Machine Learning Algorithms. *Electronics* **2022**, *11*, 3167. [CrossRef]

electronics

MDPI

Article

Short-Term Traffic-Flow Forecasting Based on an Integrated Model Combining Bagging and Stacking Considering Weight Coefficient

Zhaohui Li [1,*], Lin Wang [1], Deyao Wang [1,*], Ming Yin [2] and Yujin Huang [1]

[1] School of Maritime Economics and Management, Dalian Maritime University, Dalian 116026, China; wang_lin@dlmu.edu.cn (L.W.); hyjin29@dlmu.edu.cn (Y.H.)
[2] Xuzhou Xugong Materials Supply Co., Ltd., Xuzhou 221000, China; y18004094826@163.com
* Correspondence: leezhaohui@dlmu.edu.cn (Z.L.); dream@dlmu.edu.cn (D.W.)

Abstract: This work proposed an integrated model combining bagging and stacking considering the weight coefficient for short-time traffic-flow prediction, which incorporates vacation and peak time features, as well as occupancy and speed information, in order to improve prediction accuracy and accomplish deeper traffic flow data feature mining. To address the limitations of a single prediction model in traffic forecasting, a stacking model with ridge regression as the meta-learner is first established, then the stacking model is optimized from the perspective of the learner using the bagging model, and lastly the optimized learner is embedded into the stacking model as the new base learner to obtain the Ba-Stacking model. Finally, to address the Ba-Stacking model's shortcomings in terms of low base learner utilization, the information structure of the base learners is modified by weighting the error coefficients while taking into account the model's external features, resulting in a DW-Ba-Stacking model that can change the weights of the base learners to adjust the feature distribution and thus improve utilization. Using 76,896 data from the I5NB highway as the empirical study object, the DW-Ba-Stacking model is compared and assessed with the traditional model in this paper. The empirical results show that the DW-Ba-Stacking model has the highest prediction accuracy, demonstrating that the model is successful in predicting short-term traffic flows and can effectively solve traffic-congestion problems.

Keywords: short-term traffic-flow forecasting; bagging model; stacking model; ridge regression; error coefficient

Citation: Li, Z.; Wang, L.; Wang, D.; Yin, M.; Huang, Y. Short-Term Traffic-Flow Forecasting Based on an Integrated Model Combining Bagging and Stacking Considering Weight Coefficient. *Electronics* **2022**, *11*, 1467. https://doi.org/10.3390/electronics11091467

Academic Editor: Stefano Ferilli

Received: 23 March 2022
Accepted: 28 April 2022
Published: 3 May 2022

1. Introduction

In recent years, as the economy has grown and people's quality of life has improved, people's demands for transportation has increased, and vehicles have progressively become the preferred mode of transportation. However, this has caused an increase in traffic congestion, and a contradiction and an intensification between the supply and demand of road traffic. As a result, comprehensive technologies and methods are urgently needed to properly control and monitor traffic flow, as well as to alleviate traffic congestion and other issues.

Traffic-flow prediction is fundamental in traffic management and dredging, and its accuracy is critical in resolving traffic-congestion issues. A vast number of experts have done extensive research on this in recent years, primarily utilizing a linear or nonlinear model to predict the following:

(1) Linear model

The historical average forecasting methods, the time series forecasting methods, and the Kalman wave forecasting methods were all used in the early days of traffic flow research. Some scholars use simple linear models to predict traffic flow, such as the autoregressive moving average model (ARIMA) model, which is suitable for predicting data with time rules, but the traffic flow has a strong non-linear trend, and its prediction accuracy for traffic flow is not high and has limitations [1–3]. D. Cvetek et al. used the collected data to compare some common time series methods such as ARIMA and SARIMA, showing tha the ARIMA model provides better performance in predicting traffic demand [4].

The Kalman wave is also used as a linear theory-prediction method by many scholars. Okutani firstly applied the Kalman wave to traffic-flow forecasting [5]. According to the inherent shortcomings of Kalman wave variance, Guo et al. proposed an adaptive Kalman wave energy update variance, which improved the prediction performance of the original model [6]. Israr Ullah et al. developed an artificial neural network (ANN)-based learning module to improve the accuracy of the Kalman filter algorithm [7]. Additionally, in the experiment of the indoor environment prediction in the greenhouse, good prediction results were obtained. Therefore, the Kalman wave model can effectively reduce the uncertainty and noise in the flow change in the prediction process, but it is difficult to predict the nonlinear change trend of the traffic flow.

(2) Non-linear model

With the recent development of technology, the widespread use of powerful computer and mathematical models is applied to this field [8]. Among them, the wavelet neural network, as a representative of the nonlinear theoretical model, has a better traffic-flow prediction effect. Gao et al. used the network model to predict short-term traffic flow and achieved good results [9]. Although the wavelet neural network converges faster and the prediction accuracy is higher, the existence of the wavelet basis function increases the complexity of the model.

Machine learning models have become research hotspots that have been widely used in many fields. They are best applied to the field of traffic flow. Qin et al. proposed a new SoC estimation method on the impact of temperature on SoC estimation, and the use of limited data to rapidly adjust the estimation model to new temperatures, which not only reduces the prediction error at a fixed temperature but also improves the prediction accuracy at a new temperature [10]. Xiong Ting et al. used the random forest model to predict the traffic flow and achieved high prediction accuracy, based on the combination of spatio-temporal features [11]. Lu et al. used the XGBoost model to predict the traffic flow at public intersections in Victoria and achieved high prediction accuracy [12]. Alajali et al. used the GBDT model to analyze the lane-level traffic flow data on the Third Ring Road in Beijing on the basis of feature processing and proved that the model has a good prediction effect and is suitable for the traffic prediction of different lanes [13]. On the basis of extracting features, Yu et al. used the KNN model to complete the prediction of the traffic node and route traffic flow, which achieved good prediction results [14].

Therefore, it can be concluded that the integrated model based on the decision tree is widely used and has high prediction accuracy, while the KNN model can eliminate the sensitivity to abnormal traffic flow in the prediction. Qin et al. proposed a slow-varying dynamics-assisted temporal CapsNet (SD-TemCapsNet) that introduced a long short-term memory (LSTM) mechanism to simultaneously learn slow-varying dynamics and temporal dynamics from measurements, achieving an accurate RUL estimation [15]. Although LSTM has been used by many scholars as a network model with high accuracy in terms of time series prediction, the complexity of the network itself is difficult to avoid. The gate recurrent unit model (GRU) can effectively solve this problem, which can complete the prediction of traffic with fewer parameters under the premise of meeting a certain prediction accuracy. Dai et al. used the GRU model to predict the traffic flow under the condition of making full use of the features and verified the effectiveness of the model through comparative analysis with the convolutional neural network [16]. As an evolutionary model of LSTM, the GRU can predict traffic flow with fewer parameters, under the premise of satisfying a certain prediction accuracy.

Although machine learning models perform well in traffic-flow prediction, the prediction performance of the single model is limited. Therefore, a model combining multiple single models has gradually become a trend [17]. Pengfei Zhu et al. integrated the GRU and BP to predict the frequency shift of unknown monitoring points, which effectively improved the prediction accuracy of a single model [18]. Although the above combined models can improve the accuracy to a certain extent, they are limited by the number of single models. The integrated model that mixes multiple models is gradually becoming favored by scholars and has been applied to various fields [19]. Shuai Wang et al. proposed a probabilistic approach using stacked ensemble learning that integrates random forests, long short-term memory networks, linear regression, and Gaussian process regression, for predicting cloud resources required for CSS applications [20]. Common ensemble models include bagging [21], boosting [22], and stacking [23]. Compared with other ensemble models, the stacking model has a high degree of flexibility, which can effectively integrate the changing characteristics of heterogeneous models to make the prediction results better.

In summary, the single prediction model has limitations, and the combined forecasting model has gradually become a trend. Common models that can integrate a single model include the entropy combination method, the inverse error group method, the ensemble learning method, and other combination methods. [24,25]. Among them, the comprehensive model is more practical. The bagging integration model and the boosting integration model, generally used for a homogeneous single model, are limited to a single model, while the stacking integration model is more commonly used for the fusion of heterogeneous models. Therefore, the first use of the bagging model is to optimize the base learner model and then optimize the stacking model, to improve the overall performance of the model.

2. Establishment of the DW-Ba-Stacking Model

In this section, a DW-Ba-Stacking model was put forwarded in detail. The DW-Ba-Stacking model consists of three parts in total, the stacking model (stacking), the bagging model (Ba), and the dynamic weighting adjustment (DW).

2.1. Stacking Model

Traffic flow trends are complex, and there are various models used in this field, among which machine learning models are widely used in traffic-flow prediction due to their good non-linear fitting. In order to obtain a stacking model with high accuracy, machine learning models with different merits and good applications in this field are selected for fusion: the random forest model, which is less prone to overfitting; the KNN model, which is insensitive to outliers; the decision-tree model; the XGBoost and GBDT models; the GRU model, which can effectively use temporal features; and the K-fold cross validation to prevent overfitting.

2.1.1. Principle of the Stacking Model

The stacking model obtains the final prediction by linear or non-linear processing of the sub-learners. The main principle is that the original data are first predicted by the base learner, and then the prediction is passed to the meta-learner to obtain the final result. To prevent overfitting, the data are usually trained by K fold cross-validation, as follows.

Let the original data set $M = \{(y_n, x_n)\}$, x_n be the feature variables of the n sample, y_n be the predictor variables of the n sample, and the number of base learners be L The data from the original dataset $1/K$ are used as the validation set, $M_{1/k}$; the rest of the data $M^{-1/k} = M - M_{1/k}$ are used as the training set; the divided data are fed into the base learner $A_L^{-1/k}$ for training, and the prediction results from K are obtained N_{kL}. The predictions from the base learner and y_n are then used as the training set for the meta-learner, which trains the model and makes predictions.

2.1.2. Machine Learning Models

Random Forest and KNN Models

The Random Forest model is a modified bagging algorithm. When the model is used for regression, the single model that is integrated is the CART regression tree. First, samples are drawn by bootstrap sampling with replacement; then, the corresponding regression trees are modelled for the m different samples drawn to form the forest; and, finally, the average of the predictions from the different regression trees is taken as the final prediction. The samples and features of the regression trees in the model are chosen randomly. Each regression tree built through bootstrap sampling is independent and uncorrelated. This feature increases the variation between models and enhances the generalization ability of the model. At the same time, the random nature of feature selection reduces the variability of the models. As the number of regression trees increases, the model error gradually converges, which reduces the occurrence of overfitting. This is why the model was selected as one of the base learners.

When the KNN model is used for classification, it determines the k sample types by searching for k samples in the historical data that are similar to the samples to be classified. The principle can be expressed as follows:

$$S = (X_1, Y_1), (X_2, Y_2) \ldots (X_N, Y_N) \tag{1}$$

where X is the feature vector, Y is the category of the example sample, and $i = (1, 2, 3 \ldots, N)$. The Euclidean distance is used to express the similarity between the sample to be classified and the feature sample in S. The Euclidean distance between the observed sample and the feature is calculated. Based on the calculated distances, find the closest K points to the object to be classified in S and determine the X category. The principle is shown in Figure 1. There are N samples with the categories $Q_1, Q_2, \ldots, Q_N$, which are N different categories. By testing the Euclidean distance between sample X_i and the N training sets, M samples that are closer to sample X_i are obtained, and if most of the M samples belong to a certain type, then sample X_i also belongs to that type. The model can be applied to both discrete and continuous features and is insensitive to outliers, so it is used as a base learner.

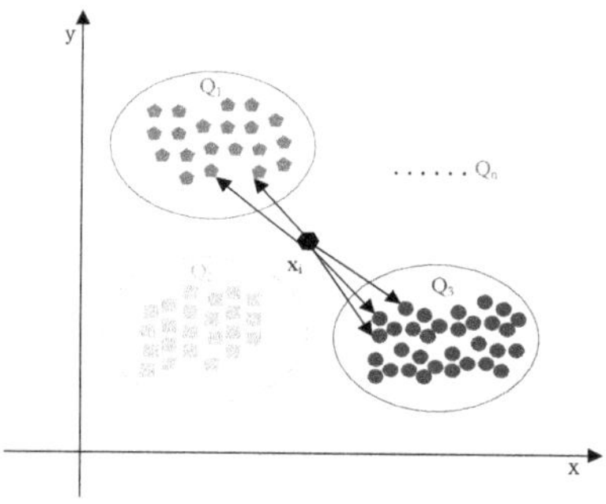

Figure 1. The KNN schematic diagram.

Decision Trees, and the GBDT and XGBoost Models

A decision tree is a model consisting of nodes and directed edges that allow predictions to be made by correspondence between attributes and objects. The internal nodes are the features of the object and the leaf nodes are the classes of the object. The model has a wide range of applications, and it is efficient and suitable for high-dimensional feature processing, which is why it has been chosen as one of the traffic-flow prediction models. It aims to summarize certain rules from the training dataset and eventually achieve the correct result. The essence is to find the optimal decision tree. The three more important features in the search process are attribute selection, decision tree generation, and decision tree pruning. The key to their generation is the division of the optimal attributes. Purity is a measure based on the assignment of attributes. The evaluation metrics for measuring purity include information gain, gain rate, and Gini index. The principle is shown in Figure 2.

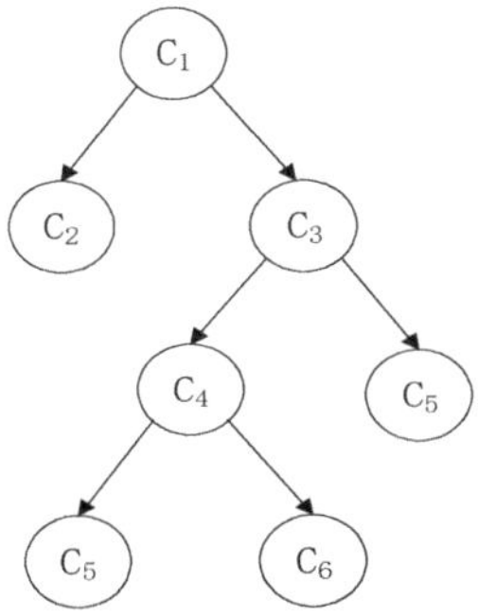

Figure 2. The decision tree model.

Both GBDT and XGBoost are algorithms that evolve by boosting. GBDT is formed by continuously fitting the residual error by updating the learners on the gradient. When the residual error reaches a certain limit, the model stops iterating and forms the final learner. The model can be very good at fitting non-linear data. However, the computational complexity will increase when the dimensionality is high and the traffic flow has fewer characteristic dimensions, so the model is suitable for prediction in this area. The regulator model is a linearly weighted combination of different regulators.

$$F_n(x) = \sum_{n=1}^{N} R(x;\theta_n) \tag{2}$$

where $T(x;\theta_n)$ is a weak regressor. The loss function of the weak regressor is

$$\hat{R}_n = argmin \sum_{i=1}^{M} L(y_i, F_{n-1}(x_i) + T(x;\theta_n)) \tag{3}$$

where $L(\cdot)$ is the loss function.

XGBoost and GBDT share the same principles and integrated model, with a process of continuously fitting the residuals and gradually reducing them. During the fitting process, the learner is updated with first-order derivatives and second-order derivatives. Specifically, the second-order Taylor expansion of the loss function and the positive term of the error component are used as the objective function during each round of iterations. It updates the parameters through the solution of the least significant graph. The positive term in the objective function controls the complexity of the model, reduces the variance of the model, and makes the learning process of the model easier, so this model is chosen as the base learner. The loss function L is

$$L = \sum_{i=1}^{M} l(y_i, \hat{y}_i) + \sum_{n=1}^{N} \Omega(f_k) \tag{4}$$

In the formula, the first half is the error between the predicted and actual values; the second half is the conventional term.

$$\Omega(f) = \gamma T + \frac{1}{2}\lambda \|\omega\|^2 \tag{5}$$

The Equations γ and λ are the penalty coefficients for the model.

GRU Model

A deep-learning model is one of the machine learning models. It can adapt well to the changing characteristics of data when the amount of data is appropriate. It has gradually been applied to various fields with good results. Zheng Jianhu et al. relied on deep learning (DL) to predict traffic flow through a time series analysis and carried out long-term traffic-flow prediction experiments based on the LSTM network-based traffic-flow prediction model, the ARIMA model, and the BPNN model [26]. It can be seen that regular sequences have won the favor of various scholars and that GRU is a more mature network for processing time series in recent years. Additionally, the earliest proposed network to deal with time series is RNN, but it is prone to gradient disappearance, leading to network performance degradation. Zhao et al. used long short-time memory (LSTM) to predict traffic flow under the premise of considering spatial factors in the actual prediction process and achieved high prediction accuracy [27], but the network model also has the disadvantage of poor robustness. In order to solve this problem, Li Yuelong et al. realized the optimization of the prediction performance of the network through the network space feature fusion rights protection unit [28]. It can be seen that although LSTM is used by many scholars as a network model with high time series prediction accuracy, the complexity of the network itself is difficult to avoid. The GRU model, on the other hand, can effectively reduce the network parameters while ensuring the performance of the model itself. Its structure is shown in Figure 3.

$$r_t = \sigma(W_r x_t \times U_r h_{t-1} + b_r) \tag{6}$$

$$z_t = \sigma(W_z x_t \times U_z h_{t-1} + b_z) \tag{7}$$

$$\widetilde{h}_t = \tanh(W_h x_t + U_h(h_{t-1} \otimes r_t) + b_r) \tag{8}$$

$$\widetilde{h}_t = (1 - z_t) \otimes h_{t-1} + z_t \otimes \widetilde{h}_t \tag{9}$$

where $\otimes$ is the product of the corresponding positions of the two matrices, σ is the activation function, W and U are the weight parameters of the network, and b is the bias parameter of the network, which is the state value of the hidden layer at different moments. The reset gate r_t determines the input ratio of the previous state information h_{t-1} to the current network cell; the update gate Z_t determines the deletion ratio of the previous state information. The entire network cell is filtered by the two gates to determine the valid information of the network cell. Compared with the LSTM model, the GRU model reduces one gate unit and only sets the reset gate and update gate to control the input and output information of the network unit, which reduces the complexity of the network and improves the network training speed.

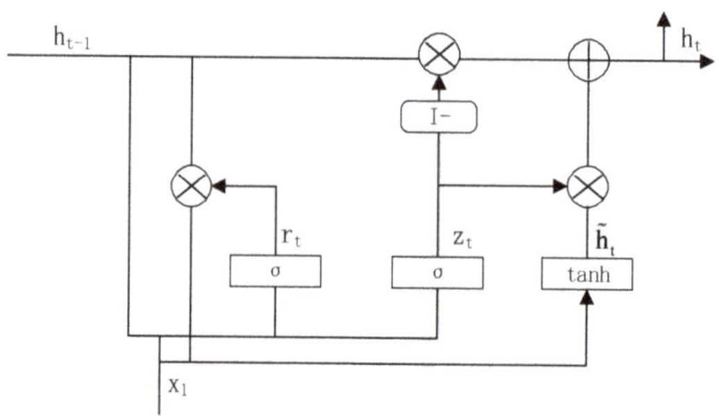

Figure 3. The GRU structure diagram.

2.2. *Bagging Model*

The overall architecture of the Ba-Stacking model included the bagging model processing stage and the stacking model processing stage. Because the bagging was only embedded as part of the stacking model, the stacking model architecture plays a big role. The more important processing phases are: the base learner processing phase and the meta learner processing phase. The base learner processing stage requires different base learners to obtain the prediction results, so the choice of the base learner plays an important role. The meta-learner processing stage is more important because it includes a large amount of raw data information, so it is important that the effect of using the base learner information affects the final prediction results. However, the output information of different base learners is duplicated, and the data variability is not strong enough to extract the effective information of the output data. Therefore, to address the problem that the output information of base learners cannot be fully utilized, it is necessary to consider how to effectively utilize its output information and reflect its importance and variability.

To further improve the stacking model, this paper considers the use of the bagging algorithm to further optimize the base learner and reduce the base learner variance, as two ways to improve the potential performance of the meta-learner model in the stacking model.

Considering that the prediction effect of the base learner directly affects the final effect of the integrated model, the prediction effect of the base learner of the stacking-integrated model is optimized by the bagging algorithm. To better extract the base learner features, a ridge regression with linearity is used as the meta-learner, and the overall construction principle is shown in Figure 4.

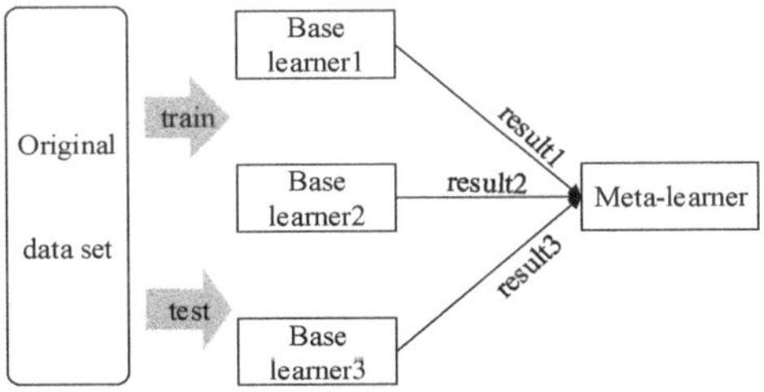

Figure 4. The Ba-Stacking model architecture diagram.

The process of this model is to optimize the data features of the stacking base learner based on its output information through the bagging algorithm and then further input this optimized data into the meta-learner in the stacking-integrated model for traffic prediction. The process consists of three parts: the first part builds the stacking base learner model by comparing and analyzing different features to obtain the optimal base learner model; the second part builds the stacking model and obtains the optimal stacking model by comparing and analyzing different base learner models and meta-learner models; finally, the bagging model is combined into the stacking model to build the Ba-Stacking model.

2.3. DW Model

The entropy value can be expressed as the uncertainty of each value. The entropy weighting method in the tradition weights the fixed coefficients of each model, but the certainty degree of different positions of the base learner can be deduced from the certainty degree of a specific position in each model.

Where the single model $Y_{ij}(i = 1, 2, \ldots, m; j = 1, 2, \cdots, n)$ is the base learner prediction and $L_i(i = 1, 2, \ldots, m)$ is the actual value, the entropy value is

$$h_{ij} = -\frac{e_{ij} ln(e_{ij} + 0.5)}{ln(N + 0.5)} \tag{10}$$

The addition of 0.5 to the Ln function in Equation (10) is to accommodate the calculation of zeros in the original series. h_{ij} is the entropy value derived from the error value e_{ij}, where e_{ij} is the absolute error indicator value. Because the characteristics of the meta-learner in the stacking-integrated model are the strong information characteristics of the base learner output, and the uncertainty of the base learner can be known according to its entropy value at different positions, the variability of the base learner model output information can be enhanced after the introduction of weights, which in turn improves the overall performance of the model. The degree of uncertainty of different models is determined by introducing the entropy value after the MSE is calculated, which is used when the dynamic parameters are calculated.

2.4. Model Construction

2.4.1. Dynamic Weighting Adjustment Model Process

In the stacking model, the degree of data deviation at different locations in the base learner output information varies, and fixed weighting cannot capture its dynamic change pattern, so dynamic weighting coefficients are designed in the model.

The coefficient is designed outside the meta-learners, and the dynamic weight coefficients are first solved according to the degree of deviation at different positions, and then the dynamic weight coefficients are weighted to adjust the base learner output information to achieve the extraction of dynamic change patterns. The weighting coefficients here include error weighting and entropy weighting.

$Y_{ij}(i = 1, 2, \cdots, m; j = 1, 2, \cdots, n)$ is the predicted value of the base learner, $L_i(i = 1, 2, \cdots, m)$ is the actual value, m is the number of elements, n is the number of base learners, and u_j is the predicted mean value of each base learner. The adjustment process of the output information of the base learner is

$$\begin{pmatrix} y_{11} & y_{12} & \cdots & y_{1n} \\ y_{21} & y_{22} & \cdots & y_{2n} \\ \cdots & \cdots & \cdots & \cdots \\ y_{m1} & y_{m2} & \cdots & y_{mn} \end{pmatrix}_{m\times n} \xrightarrow{\text{transform}} \begin{pmatrix} y_{11}x_{11} & y_{12}x_{12} & \cdots & y_{1n}x_{1n} \\ y_{21}x_{21} & y_{22}x_{22} & \cdots & y_{2n}x_{2n} \\ \cdots & \cdots & \cdots & \cdots \\ y_{m1}x_{m1} & y_{m2}x_{m2} & \cdots & y_{mn}x_{mn} \end{pmatrix}_{m\times n} \xrightarrow{\text{prediction}} \begin{pmatrix} l_1 \\ l_2 \\ \cdots \\ l_m \end{pmatrix}_{m\times 1} \tag{11}$$

In the process of adjustment, the key lies in the solution of dynamic weight coefficients x_{ij}.The solution process is as follows:

$$e_{ij} = \begin{pmatrix} |y_{11} - l_1| & |y_{12} - l_1| & \cdots & |y_{1n} - l_1| \\ |y_{21} - l_2| & |y_{22} - l_2| & \cdots & |y_{2n} - l_2| \\ \cdots & \cdots & \cdots & \cdots \\ |y_{m1} - l_{\mathrm{m}}| & |y_{m2} - l_{\mathrm{m}}| & \cdots & |y_{mn} - l_{\mathrm{m}}| \end{pmatrix}_{m\times n} \tag{12}$$

(1) Calculate the absolute error of each element e_{ij}, that is, the degree of deviation of each element: the absolute value of the difference between the predicted value y_{ij} and the actual value l_i of the base learner;

$$E_{ij} = \begin{pmatrix} \frac{|y_{11}-l_1|-u_1}{u_1'-u_1} & \frac{|y_{12}-l_1|-u_2}{u_2'-u_2} & \cdots & \frac{|y_{1n}-l_1|-u_n}{u_n'-u_n} \\ \frac{|y_{21}-l_2|-u_1}{u_1'-u_1} & \frac{|y_{22}-l_2|-u_2}{u_2'-u_2} & \cdots & \frac{|y_{2n}-l_2|-u_n}{u_n'-u_n} \\ \cdots & \cdots & \cdots & \cdots \\ \frac{|y_{m1}-l_m|-u_1}{u_1'-u_1} & \frac{|y_{m2}-l_m|-u_2}{u_2'-u_2} & \cdots & \frac{|y_{mn}-l_m|-u_n}{u_n'-u_n} \end{pmatrix}_{m\times n} \tag{13}$$

$$\overline{E}_{ij} = \begin{pmatrix} \frac{\sum\limits_{i=1}^{m} |y_{i1}-l_i|-mu_1}{m(u_1'-u_1)} & \frac{\sum\limits_{i=1}^{m} |y_{i2}-l_i|-mu_2}{m(u_2'-u_2)} & \cdots & \frac{\sum\limits_{i=1}^{m} |y_{in}-l_i|-mu_n}{m(u_n'-u_n)} \end{pmatrix}_{1\times n} \tag{14}$$

(2) Calculate the deviation rate E_{ij} and average deviation rate of each element E_{ij}, the normalized value of absolute error e_{ij}, and the normalized mean value of absolute error of each column n e_{ij}, respectively;

$$C_{ij} = \begin{pmatrix} \frac{u_1'-|y_{11}-l_1|}{u_1'-u_1} & \frac{u_2'-|y_{12}-l_1|}{u_2'-u_2} & \cdots & \frac{u_n'-|y_{1n}-l_1|}{u_n'-u_n} \\ \frac{u_1'-|y_{21}-l_2|}{u_1'-u_1} & \frac{u_2'-|y_{22}-l_2|}{u_2'-u_2} & \cdots & \frac{u_n'-|y_{2n}-l_2|}{u_n'-u_n} \\ \cdots & \cdots & \cdots & \cdots \\ \frac{u_1'-|y_{m1}-l_m|}{u_1'-u_1} & \frac{u_2'-|y_{m2}-l_m|}{u_2'-u_2} & \cdots & \frac{u_n'-|y_{mn}-l_m|}{u_n'-u_n} \end{pmatrix}_{m\times n} \tag{15}$$

$$\overline{C}_{ij} = \begin{pmatrix} \frac{mu'_1 - \sum_{i=1}^{m} |y_{i1} - l_i|}{m(u_1' - u_1)} & \frac{mu'_2 - \sum_{i=1}^{m} |y_{i2} - l_i|}{m(u_2' - u_2)} & \cdots & \frac{mu'_n - \sum_{i=1}^{m} |y_{in} - l_i|}{m(u_n' - u_n)} \end{pmatrix}_{1\times n} \tag{16}$$

(3) Calculate the contribution rate C_{ij} and the average contribution rate of each element C_{ij}, the value of 1 minus the deviation rate, and the value of 1 minus the average deviation rate, respectively.

The contribution rate calculated in Equation (14) is the dynamic weight coefficient C_{ij}. The adjusted output information reduces the prediction results influenced by errors or deviation information, making the information characteristics more representative. The coefficient matrices are used to adjust the training set and test set. The specific process is as follows:

- Training set

Adjust the change rule of the predicted value of the base learner: use the product of the predicted value of different positions and the dynamic weight coefficient as the new data. The specific process is shown in Figure 5.

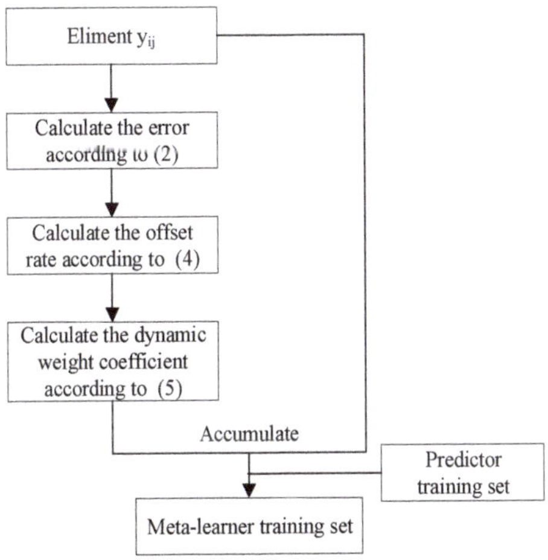

Figure 5. The meta learner training set adjustment process.

- Test set

Adjust the overall change law of the predicted value of the training set of the base learner: use the product of the predicted value of different positions and the average dynamic weight coefficient in the training set as the new data. The specific process is shown in Figure 6.

2.4.2. Ba-Stacking Model Optimization Process

The principle of the improved stacking ensemble model is shown in Figure 7. Assuming that the traffic flow data sequence has X records of data, N is the number of characteristic variables, the original data set is $\{(Y_{0X}, Q_{iX})\}$, $Y_{0X}(X = 1, 2, \ldots, N)$ is the predictor variable, and Q_{iX} is the characteristic variable. The specific steps of the model are as follows:

(1) Divide the original data into the training set and test set;
(2) Construct the corresponding prediction models, including random forest, XGBoost, the GBDT, and the decision-tree model;
(3) Use Q_{iX} and Y_{0X} to obtain the corresponding predicted values of different models through the bagging algorithm, denoted as $Y_{1X}, Y_{2X}, Y_{3X}, Y_{4X}, Y_{5X}, Y_{6X}$;

(4) Using $Y_{1X}, Y_{2X}, Y_{3X}, Y_{4X}, Y_{5X}, Y_{0X}$, obtain the weight coefficients by different adjustment methods, followed by the flow data of the adjusted base learner model, noted as $Y'_{1X}, Y'_{2X}, Y'_{3X}, Y'_{4X}, Y'_{5X}$;
(5) Using $Y'_{1X}, Y'_{2X}, Y'_{3X}, Y'_{4X}, Y'_{5X}, Y_{0X}$, build a meta-learner ridge regression mode to obtain the final traffic prediction values of the improved stacking integration model;
(6) Train the model with the training set. Once trained, the model will be tested using the test set.

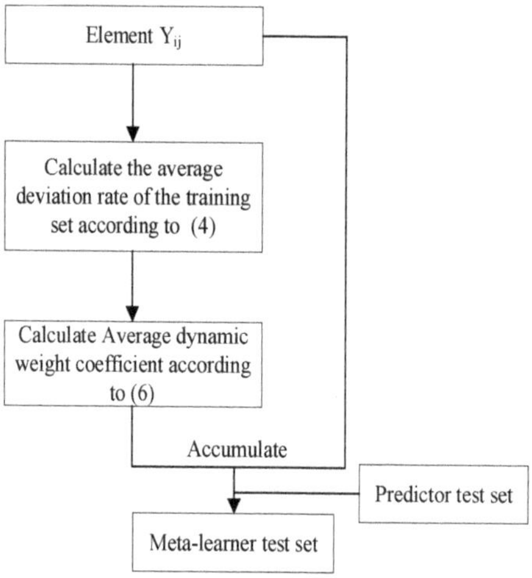

Figure 6. The meta learner test set adjustment process.

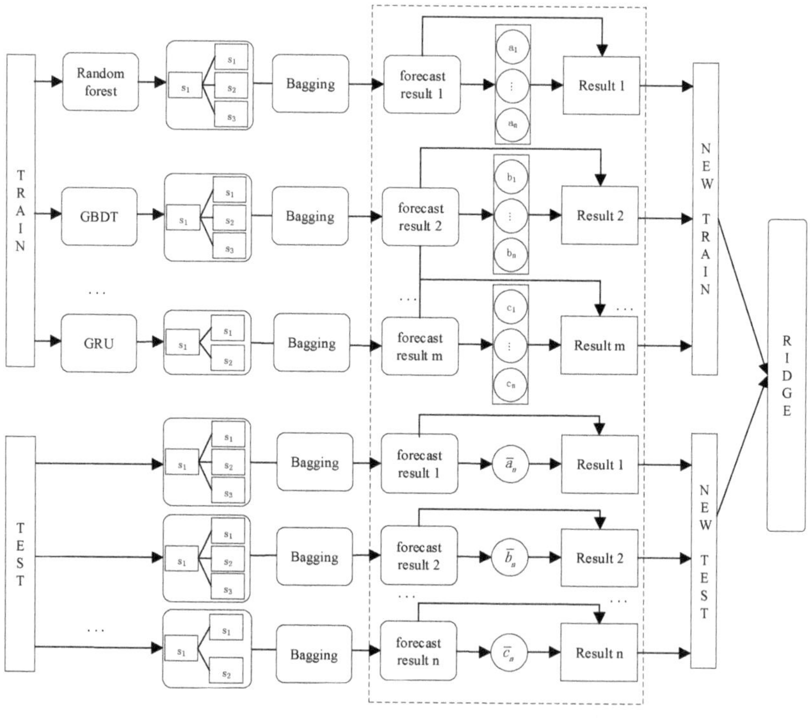

Figure 7. The DW-Ba-Stacking model principle diagram (ridge regression).

3. Problem Description and Data Progress

3.1. Overview of Short-Term Traffic Flows

Traffic flow is the volume of traffic formed by vehicles on a roadway. The factors influencing traffic flow include flow, speed, and occupancy. Traffic flow is an important indicator to determine traffic congestion, its prediction results are an important parameter to grasp the city traffic situation, and the selection of the prediction time period is an important step in the prediction process. There are various types of traffic flow data recorded by monitoring point detectors, including 30s, 5 min, 15 min, 30 min, and 60 min, depending on the collection time, and highways and urban roads depending on the type of road collected. Traffic-flow prediction results are usually obtained from the historical data function, that is, the historical information within the time of (t_{i-j}, t_{i-1}) predicts the current information of t_i.

$$t_i = f(t_{i-j}) \tag{17}$$

t_i stands for the current time of the indicator value t_{i-j} stands for the historical time period of the indicator value, and the function indicates the historical time period of the traffic value to predict the future time period of the traffic value. When the adjacent time interval Δt is less than or equal to 15 min, the formula represents the short time traffic flow forecast, so this paper chooses 15 min as the time interval for traffic-flow prediction and analysis.

In short-term traffic-flow forecasting, occupancy and speed are the important impact indicators of road traffic; this paper will focus on the two indicators as the flow of the constraint characteristics, and its historical trend to join the traffic-flow forecasting; the function relationship is shown as follows:

$$l_i = f(z_{i-j}, v_{i-j}, l_{i-j}) \tag{18}$$

$z_{i-j}, v_{i-j}, l_{i-j}$ refers to the values of occupancy, speed, and traffic flow indicators, respectively, at the historical moment; i is the current time; and j is the historical time period used. $j = 4$ is chosen for the prediction analysis in this paper.

3.2. Data Sources and Pre-Processing

3.2.1. Data Sources

The data selected for this paper is from the PORTAL dataset, which provides official traffic data for Portland, USA and Vancouver, Canada, with monitors recording traffic at five intervals: 30 s, 5 min, 15 min, 30 min, and 60 min. Traffic data for 15 min intervals on the I5NB highway in Portland, USA are selected for analysis, and the main data tables studied are the monitor data tables.

The data set occupancy and speed are data features that can be directly utilized in the data tables. These two indicators are also the actual indicators that affect the traffic flow, so these two indicators are used as input features to the model. The timestamp feature is the time recorded by the detector, which can provide some regular reference for the trend change of the traffic flow; the specific feature construction analysis is in Section 3.2.3, so this indicator is also used as an input feature, and the traffic flow is input into the model as an output feature for prediction. The data analyzed for the example in this paper is 100703 detector data, collected at the specific time of 1 February 2018 00:00:00–12 April 2020 0:00:00, with 96 data per day, comprising a total of 76,896 pieces of data. Seventy percent of the data set was used as the training set and 30% of the data set was used as the test set.

3.2.2. Feature Construction

According to the analysis in Section 3.1, occupancy and speed can effectively influence traffic flows, so these two indicators are entered into the forecasting model as intrinsic characteristics. In addition to the intrinsic characteristics, the trend of traffic flow can be influenced by certain external characteristics that can affect the accuracy of the forecast, especially the temporal characteristics: traffic flow has obvious cyclical characteristics, so the cyclical temporal characteristics are important characteristics affecting the traffic flow; for example, there is more traffic flow during peak hours or holidays, so the extraction of the temporal characteristics plays an important role. To explore the temporal characteristics of traffic flow in depth, the trend of traffic flow changes over a period of time is randomly selected for analysis, as shown in Figure 8.

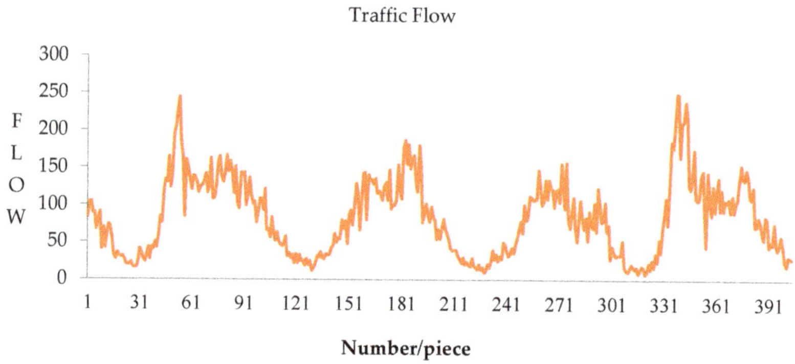

Figure 8. The traffic flow trend.

It can be seen that the same characteristics of variation occur each day, and it is obvious that there are two peaks, the peak commuting period and the peak leaving period, which are in line with the characteristics of real-life variation. This work makes full use of the historical data of the traffic flow and adds the relevant historical data of occupancy and speed as features to the prediction of the model as well. The specific construction process is as follows.

(1) Structured rest day features

Holidays and weekends are days off, and people can choose to stay at home or travel depending on the situation; therefore, the traffic flow situation is different between rest days and weekdays, so this feature is used as an important feature for predicting traffic flow. This work extracts holiday data and weekday data from the temporal features of the traffic flow collection.

(2) Construction work peak characteristics

The peak information is also used as an important indicator for predicting traffic flow, considering people's daily life habits, i.e., there will be normal commuting in the morning and evening, so there is more traffic flow at this time, which will also affect the prediction results. In this paper, 6:00–8:00 am and 17:00–19:00 pm are taken as the peak time periods. If this time is the peak hour, it is set to 1; otherwise, it is set to 0.

(3) Constructing historical indicator characteristics

Speed is the distance travelled by vehicles per unit of time, and occupancy is time occupancy and space occupancy, respectively, indicating the density of vehicles; these two indicators have a strong correlation with traffic flow, and this paper sets the sliding window to 4, i.e., occupancy and speed in 4 time periods as historical indicator features, aiming to extend the feature structure of the traffic-flow prediction model and improve the overall performance of the model.

(4) One-hot encoding processing

One-hot encoding, also known as one-hot encoding or one-valid encoding, is a method of encoding N states using N-bit state registers, each of which has its own register and only one of which is valid at any given time. The method uses N-bit status registers to encode N states, each of which has its own independent register bits and only one of which is valid at any given time. One-hot is a method for processing discrete data and converting different discrete data into continuous data, and this paper uses this method to convert temporal features into continuous temporal features.

Occupancy, speed, and traffic flow are all features of the original data table, while holidays, weekends, and peaks are expanded features of the original data table and are discrete data features. Therefore, this paper uses one-hot to process this discrete data and uses this data and the historical occupancy, speed, and traffic flow as features to input into the model. The time features are interpreted in detail as follows: a holiday feature of 0 means this time is not a holiday; a weekend feature of 1 means the time is a weekend; and a peak information feature of 0 means this time is not a peak time period.

3.2.3. Data Pre-Processing

In the process of traffic flow detection, the recorder of the detection data may be affected by some random factors to produce missing data, such as weather and climate, road driving conditions, the recorder itself, etc., and these data are important parts of the model prediction. The way the data are processed plays a key role in the accuracy of the prediction, so it is necessary to effectively deal with the missing part of the data. The difference in the data size will also cause an error in the prediction, and as the data size required for each monitoring point is different, some processing needs to be done to eliminate the error.

Missing Value Handling

There are two types of data loss: the first is the loss of an entire record, which can be caused by the failure of a logger, but this is uncommon; the second is the loss of part of a record, where a value is not recorded during the monitoring of the logger for reasons external to the logger, and thus part of the data is missing.

The traffic flow data in this paper have a low missing rate, and for the continuous variation characteristic of the missing values, this paper uses mean filling, specifically with the mean of the last five values of the same time attribute in history.

Data Normalization

Data normalization is an important step in data processing, where a certain amount of data is scaled down to a certain range so that the input features of the model vary within a smaller range, thereby eliminating the error generated in the model by the variability of the feature magnitudes. In this paper, we use the maximum-minimum normalization method to vary the original data features to within [0, 1], as a function of

$$x' = \frac{x - min}{max - min} \tag{19}$$

where *min* is the minimum value of each feature and *max* is the maximum value of each feature; the larger the value of the metric in each feature, the closer to 1 it will be after the change.

4. Experiment

4.1. Evaluation Indicators

This work selects the mean squared error (MSE) and mean absolute error (MAE) to evaluate the prediction effect of each model. The formula is shown as follows:

$$MSE = \frac{1}{n}\sum_{i=1}^{n}[Y\prime(i) - Y(i)]^2 \tag{20}$$

$$MAE = \frac{1}{n}\sum_{i=1}^{n}|Y\prime(i) - Y(i)| \tag{21}$$

where $Y\prime(i)$ is the predictor variable, $Y(i)$ is the actual variable, and n is the number of records of the data.

4.2. Model Prediction

In order to verify the effectiveness of the algorithm, this paper adds a comparative analysis with other models, including single models such as random forest, GBDT, other single models before and after feature optimization, stacking ensemble models before and after improvements, and other combined models.

4.2.1. Analysis of Feature Prediction Effect

In this paper, considering the correlation between historical data and future data, the first four periods of the time data of the occupancy rate and the speed are added to the characteristics of the model. In the actual model prediction, the addition of features has a more obvious optimization effect on the random forest. In order to analyze the effects of historical related characteristics, some single models such as XGBoost, DBDT, and decision tree are selected for comparative analysis, shown as Table 1.

Table 1. A comparative analysis of the prediction effects of different characteristics.

Base Learner	Historical Characteristics	Time Characteristics	MSE	MAE
Random forest	+	+	662.11	17.38
	+	-	745.40	18.08
	-	-	761.44	18.26
XGBoost	+	+	649.15	17.27
	+	-	762.76	18.40
	-	-	773.18	18.40
GBDT	+	+	648.21	17.25
	+	-	760.87	18.32
	-	-	778.18	18.44
Decision tree	+	+	754.31	18.45
	+	-	778.73	18.70
	-	-	789.33	18.81
KNN	+	+	754.37	18.63
	+	-	776.90	18.36
	-	-	789.40	18.50
GRU	+	-	744.73	18.54
	-	-	768.01	18.72

Note: + having this characteristic; - not having this characteristic.

It can be clearly seen from Table 1 that the selection of features has improved the overall model prediction performance. From the perspective of MSE and MAE, for all models, the structure of time features has different degrees of improvement for different models and determines whether the deep learning or the representative machine learning model is used. The more obvious are the GBDT model and the random forest model of the integrated tree model. The MSE has been improved by more than 20, followed by the XGBoost model and the GRU model, and the last is a relatively single KNN model and the decision-tree model. This conclusion shows that the single model is not as sensitive as the integrated mode.

For learners other than deep learning, after adding historical features and time features, each machine learning model experiences a greater degree of improvement: the accuracy of a single model is limited and the improved MSE is within 50. For the integrated model, the addition of this feature makes a greater contribution to the improvement: the MSE's improvement space is about 100, of which the boosting integrated model constitutes the largest improvement and the GBDT accuracy improvement is the largest, followed by

XGBoost. Therefore, from the analysis of the fusion of the two features, it can be analyzed that the integrated model is more sensitive to the model features.

In order to analyze part of the effect of model prediction, add Figures 9–11 for a more detailed analysis, i.e., to select one day's traffic flow data for analysis randomly, with the aim to analyze the prediction effects of different characteristics. It can be seen from the figure that the change trend of different models after adding features is roughly the same, and the prediction effect is better than that without adding features. The more features are integrated, the closer the prediction curve is to the original data line.

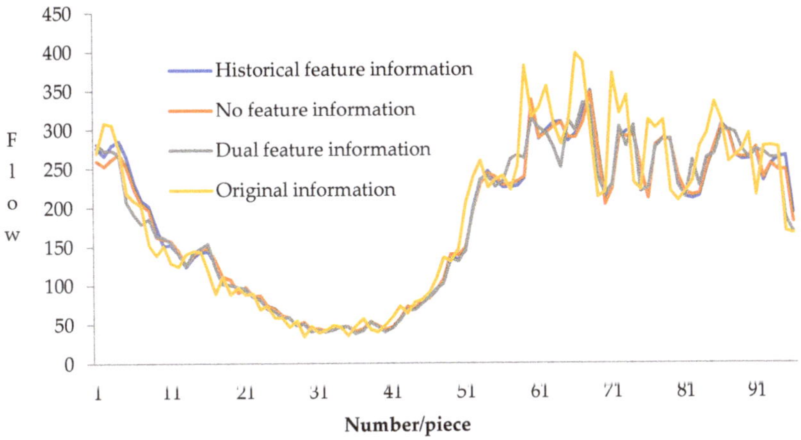

Figure 9. The GBDT feature-analysis diagram.

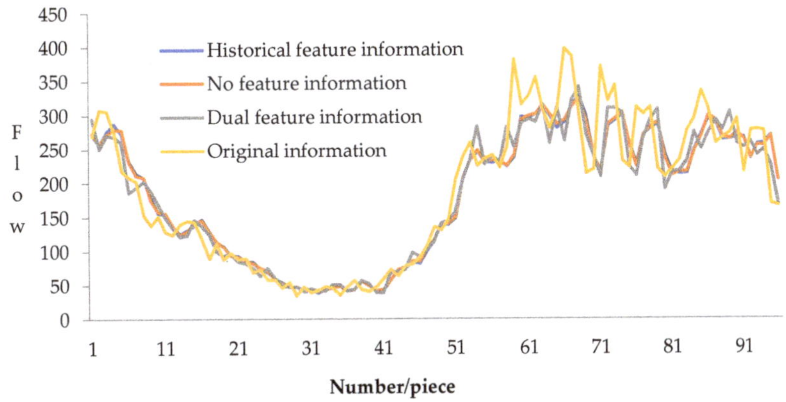

Figure 10. The decision tree feature-analysis diagram.

4.2.2. Single Model Parameter Setting

The grid search is a method of adjusting parameters. First, set a set of candidate values for the parameters you want to adjust, and then the grid search will exhaust various parameter combinations and find the best set of settings according to the set scoring mechanism. In the actual machine model, there are many parameters, so it is impossible to manually adjust the parameters in a timely and effective manner. Therefore, the parameters of different learning models can be automatically adjusted through the grid search method to obtain the parameters with the highest prediction accuracy. In this paper, the grid search is applied to the base learner, with the aim to find the parameter features when the accuracy is optimal.

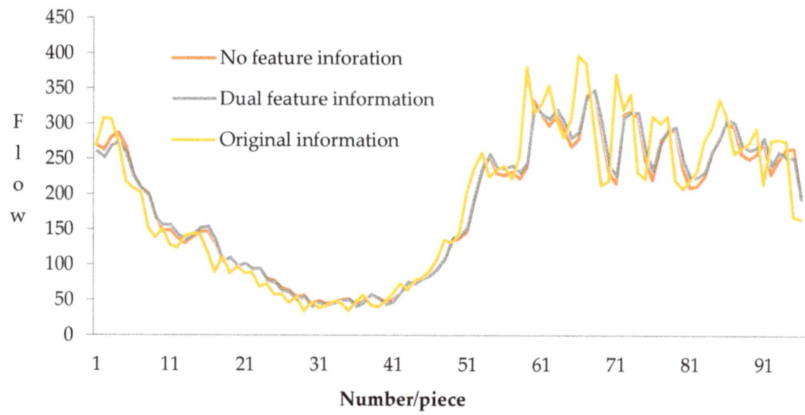

Figure 11. The GRU feature-analysis diagram.

In the model building process, other variables in the data table except the volume variable are used as the input variables, and the volume variable is used as the dependent variable to construct the following single predictive model. Among them, random forest, XGBoost, GBDT, KNN, and the decision tree use the network search method to adjust the parameters, and the GRU model adopts manual parameter adjustment. The parameter settings of each single model and the error after parameter adjustment are shown in Table 2. The prediction effects of different models are shown in the Figure 12.

Table 2. The single model prediction.

Base Learner	Parameter Setting	MSE	MAE
Random forest	The tree depth is 10, the number of trees is 160, the minimum number of samples for leaf nodes is 2, and the minimum number of samples for the node division is 5	662.11	17.38
XGBoost	The number of trees is 390, the minimum leaf node sample weight is 8, the random sampling ratio is 0.9, the number of columns randomly sampled in each tree accounts for 0.8, and the learning rate is 0.12	649.15	17.27
GBDT	The tree depth is 3, the number of trees is 470, the minimum number of samples for leaf nodes is 6, and the minimum number of samples for node division is 10	648.21	17.25
KNN	Take the number of adjacent points: 48	754.37	18.63
Decision tree	The tree depth is 7, the minimum number of samples for leaf nodes is 2, and the minimum number of samples for node division is 7	754.31	18.45
GRU	GRU has two layers of neurons, of which the number in the first layer is 64 and that in the second layer is 32. The dropout is 0.1	744.73	18.54

It can be seen that among many models, the integrated model performs well in this traffic-flow prediction. The GBDT model performs best, followed by the bagging algorithm, represented by random forest model. The deep-learning model GRU performs moderately well. The single-model KNN and decision tree perform poorly. It can be seen that, compared to the single model, the integrated model is more suitable for traffic-flow prediction, and the boosting integrated tree model performs better.

Figure 12 shows an error map of selected different models in a day. It can be seen that the error variation characteristics of six single models are the same. Among them, the fluctuation error of the KNN model and the decision-tree model is larger; the error fluctuation of the other models are smaller, indicating that the prediction stability of these four models is better. From Table 2, it can be seen that the prediction effects of the six models are distributed in two sets, of which GBDT has the best prediction effect, and its MSE is 648.21, which is 7.8% less than the MSE of KNN, with a larger error, while the prediction effect of Random forest, GDBT, and XGBoost is better. Therefore, from the perspective of overall or partial predictive analysis results, the stability and accuracy of integrated model prediction are higher than that of a single model.

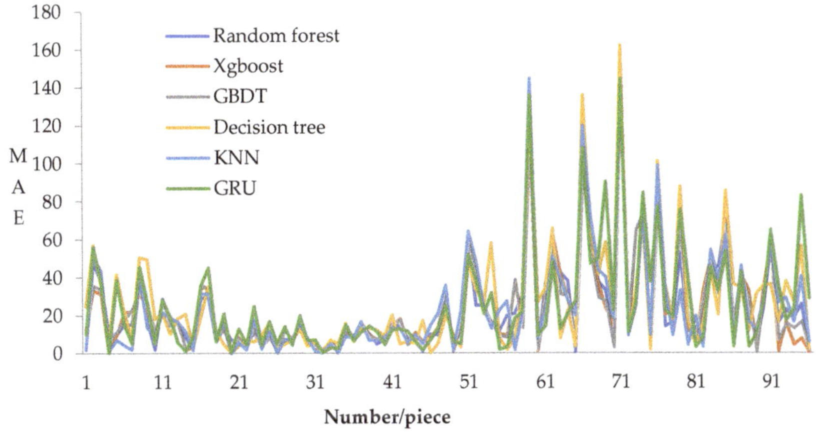

Figure 12. The error graphs of different models.

4.2.3. Pearson Characteristic Coefficient Analysis

The coefficients that measure the degree of correlation between variables include the Pearson correlation coefficient, the Spearman's correlation coefficient, and Kendall's correlation coefficient. Among them, the Pearson correlation coefficient can represent the linear coefficient value between variables. In recent years, it has been used by major models to screen the features of competitions, and it has good applicability. In the overall architecture of the stacking model, the output information of the base learner model is used as the important feature information of the prediction information, and the degree of its association with the prediction information affects the final prediction result. In this paper, the Pearson correlation coefficient is used to measure the correlation and screening process between the output information and prediction information of the base learner model. The coefficients obtained are shown in the Table 3.

Table 3. The Pearson coefficient analysis table.

	R	X	GB	D	K	G	Y
R	1	0.9962	0.9964	0.9890	0.9932	0.9895	0.9441
X	0.9962	1	0.9988	0.9869	0.9898	0.9891	0.9444
GB	0.9964	0.9988	1	0.9870	0.9900	0.9891	0.9442
D	0.9890	0.9869	0.9870	1	0.9829	0.9849	0.9354
K	0.9932	0.9898	0.9901	0.9829	1	0.9877	0.9355
G	0.9895	0.9891	0.989123	0.9849	0.9877	1	0.9357
Y	0.9441	0.9444	0.9442	0.9354	0.9354	0.9357	1

Note: R is the random forest model, X is the XGBoost model, GB is the GBDT model, D is the decision-tree model, K is the KNN model, G is the GRU model, and Y is the actual traffic flow variable.

In Table 3, the fourth column is the correlation degree between the features of the corresponding base learner and the predictor variables. The closer it is to 1, the greater the correlation. The correlation coefficients of all base learner variables and predictor variables are bigger than 0.9, indicating that the degree of correlation is greater, and its use effect will affect the final result. Under the premise that the base model is known, knowing how to choose an effective model plays a key role in the accuracy of the prediction results. Next, the selection of the model is analyzed in detail.

In order to analyze the effects of different base learners in the stacking model, this paper takes the ridge regression meta-learner as an example to establish the final prediction effect under different base learner combinations. The prediction results are shown in Table 4.

Table 4. The model selection analysis table.

Model Selection	Y/N	Y/N	Y/N	Y/N	Y/N	Y/N
Random forest	1	1	1	1	1	1
GBDT	1	1	0	1	1	1
XGBoost	1	0	1	1	1	1
KNN	1	1	0	0	0	0
Decision tree	1	1	1	0	1	0
GRU	1	1	1	1	1	1
MSE	638.15	638.92	638.71	643.67	638.63	643.01
MAE	17.07	17.06	17.07	17.17	17.08	17.15

Note: This applies to situations in which yes is 1; in other situations, it is 0.

The base learner in the stacking model selected in this paper has different characteristics, and knowing how to combine effective models has a greater impact on the final result. The above table is the MSE and MAE index values that combine different models. It can be seen that the smallest values of MSE and MAE indicators are achieved when the six models are combined. From Table 3, it can be seen that the correlation coefficient of each model is greater than 0.92, so the output information of the model-based learner and the actual information have a great correlation. After removing the models with small or large correlations, their accuracy is reduced to varying degrees. Therefore, the stacking model requires a certain degree of difference. When the integrated model represents all models with better base learner accuracy, its accuracy is not the highest, and after removing part of the model information in this table, its accuracy is reduced. Therefore, the stacking-integrated model of the six machine models proposed in this paper can make predictions more effectively.

4.2.4. Ba-Stacking Model Prediction

In order to analyze the improvement effect of the bagging algorithm integrated with different base learner models on the stacking integration algorithm, the Ba-Stacking model of different base learner models is established, and the final MSE and MAE are used to specifically evaluate the prediction effect, as shown in Table 5.

Table 5. The Ba-Stacking model prediction effect of different meta-learners.

Meta-Learner	Bagging					Evaluation Index	
	Random Forest	XGBoost	GBDT	Decision Tree	KNN	MSE	MAE
Ridge regression	-	-	-	-	-	638.15	17.07
	+	-	-	-	-	641.15	17.11
	-	+	-	-	-	637.33	17.06
	-	-	+	-	-	637.51	17.06
	-	-	-	+	-	638.14	17.07
	-	-	-	-	+	633.83	17.00
	+	+	+	+	+	634.95	17.00
	-	+	+	+	+	632.85	16.99

Note: + using this model; - not using this model

It can be seen from Table 5 that the prediction accuracy of the overall stacking model has decreased after the integration of the random forest optimized by bagging. The integration of other machine learning models optimized by bagging has improved the overall stacking model. The random forest model optimized by the bagging algorithm is not as good as the original random forest model, which affects the performance of the ensemble model. After bagging with the integration of other optimized models, the prediction accuracy of the stacking ensemble model has been improved compared to the original stacking ensemble model, and the base learner model that has been optimized by the bagging algorithm is integrated, namely, the optimized XGBoost, GBDT, decision-tree, and the stacking-integrated models after the KNN-based learner model make more accurate predictions. Therefore, whether from the horizontal or vertical angle of the table, it can be seen that the accuracy of the stacking model optimized by the bagging algorithm has improved the accuracy of the original model to varying degrees. We can know that this method optimizes the overall performance under the premise of optimizing the base learner.

4.2.5. DW-Ba-Stacking Model Prediction

In order to verify the effectiveness of this model, take the mentioned optimal single model prediction result as input and actual traffic flow as output; ridge regression is established as the original stacking ensemble model and the DW-Ba-Stacking model of the meta-learner. The prediction effect of each single model and each combination model is shown in Table 6, and the prediction effect is shown in the Figure 13.

Table 6. A performance analysis of each combination model.

Method	MSE	MAE
Random forest	662.11	17.38
XGBoost	649.15	17.27
GBDT	648.21	17.25
KNN	754.37	18.63
Decision tree	754.31	18.45
GRU	744.73	18.54
Stacking model	638.15	17.07
Ba-Stacking model	632.85	16.99
DW-Ba-Stacking model	619.59	16.87
Reciprocal error method combination	659.31	17.28

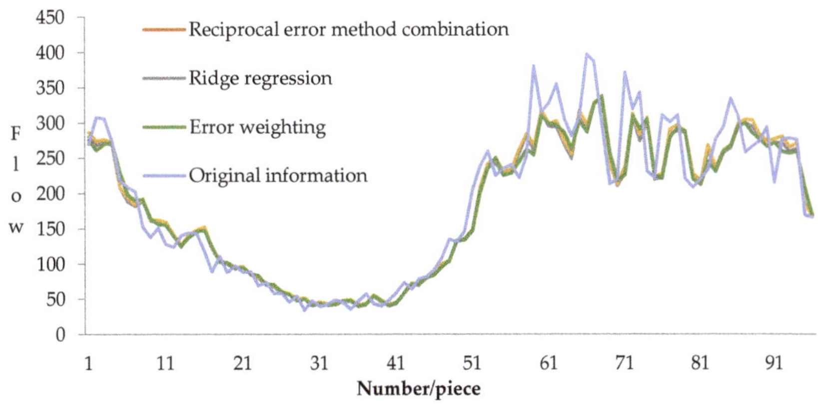

Figure 13. The single model error diagram.

Table 6 shows the prediction results of different combination models. From the prediction results, it can be seen that the prediction effect of other combination models is poor. Because there are many single models in this paper, the advantages of the single models cannot be well integrated; the stacking ensemble model has better prediction results than other combination models, among which is stacking. The base learners of the ensemble model are XGBoost, GBDT, decision tree, random forest, and GRU. The stacking model, whose meta-learner is ridge regression, is weighted by entropy; the MSE of the original model is reduced; and the MAE index value is reduced. The improved stacking model after error weighting is less than the MSE of the original model, and the MAE index value is reduced. Compared with the improved stacking ensemble model of the GRU meta-learner, the improved effect of the meta-learner is the ridge regression; obviously, it can be seen that the stacking ensemble models improved by different weights have optimization effects, and the stacking ensemble model of error-weighted ridge regression has the best optimization effect.

4.2.6. Comparative Analysis of Experimental Results

The model comparison analysis includes the basic learner model under different characteristics in the literature [9–12,15], the Ba-Stacking model optimized by the bagging algorithm, and the DW-Ba-Stacking model; the prediction results analysis as shown in the Table 7. The comparative analysis further verifies that the model proposed in this paper has higher prediction accuracy and stronger applicability.

Table 7. The comparison table of different models.

Models		MSE	MAE
Single model	Random forest [9]	706.68	18.72
	XGBoost [10]	694.99	18.54
	GBRT [11]	691.63	18.51
	KNN [12]	830.83	20.40
	GRU [15]	773.54	19.89
Meta-learner for ridge regression	Stacking	689.79	18.46
	Ba-Stacking	688.35	18.45
	DW-Ba-Stacking	681.39	18.22

5. Conclusions

With socio-economic improvements, traffic congestion will occur more frequently. Traffic-flow prediction can effectively manage and monitor traffic flow, and its prediction accuracy plays a crucial role in solving traffic-congestion problems. Machine learning algorithms have long been applied to the field of traffic-flow prediction, but individual models are greatly limited in terms of their predictive powers. Therefore, this paper applies the stacking-integrated learning model, which has been widely used in various fields in recent years, to traffic-flow prediction and provides a new idea for its prediction. A series of improvement measures are carried out to address the shortcomings of the traditional stacking-integrated learning model. The main objectives of this paper are as follows:

(1) In order to improve the shortcomings of the traffic prediction model with a single feature, temporal features such as holidays and historical features such as speed are constructed. Traffic flow is always recorded in the detector, so the time for the recorded parameters is clearer. In this paper, different time-feature information is extracted according to the specific time of the record: holiday information, weekend information features, and peak information; historical speed and occupancy features related to traffic flow are constructed according to the original data features, and the rationality of the introduced features is verified through the comparative analysis of different features. Thus, the best effect is obtained.

(2) The stacking integration model with the highest accuracy is obtained by filtering and optimizing the learners. First, we build machine learning models with different merits; then, we analyze the correlation coefficient between each model and the actual information by using the Pearson correlation coefficient; next, we select the stacking-integrated model with the highest prediction accuracy based on the weight of each model; and, finally, we embed the bagging model in this model to further improve the prediction accuracy of the model.

(3) According to the shortcomings of the stacking-integrated model, the stacking model two-layer is used as the object of improvement. With the goal of enhancing the variability between models and the correlation between predicted and actual information, the weights of different base learner models are adjusted so that the prediction accuracy is higher.

The main innovative work of this paper is to achieve the following:

(1) Realize the effective combination of the stacking model and bagging model, i.e., the construction of Ba-Stacking. The bagging model is used to optimize the output information features of the base learner in the stacking model, and the construction of the Ba-Stacking model is completed.
(2) Based on the Ba-Stacking model, the DW-Ba-Stacking model is constructed by weighting coefficients. The Ba-Stacking model with the meta-learners as ridge regression optimizes the base learner feature information by error coefficient.

In summary, this paper not only introduces the stacking-integrated model, which can effectively improve the accuracy of traffic-flow prediction, but also proposes an improved DW-Ba-Stacking model, which further improves the prediction accuracy of traffic flow while adjusting the internal structure, and provides a reference for the development of traffic-management strategies and implementation plans. In the future, the improved method can be applied to other fields with practical significance. However, in the process of improving the stacking ensemble model, this paper only pays attention to the prediction accuracy and does not consider the time efficiency, so there are some limitations in its level of improvement. In the future, the improved method can be applied to other fields with practical significance.

Author Contributions: Conceptualization, Z.L. and M.Y.; Data curation, M.Y. and D.W.; Formal analysis, Y.H.; Investigation, D.W.; Methodology, Z.L. and L.W.; Project administration, Z.L.; Validation, L.W. and D.W.; Writing – original draft, M.Y.; Writing – review & editing, Z.L., L.W., D.W. and Y.H. All authors have read and agreed to the published version of the manuscript.

Funding: This research was funded by the National Key R&D Program of China (no.2019YFD1101104).

Data Availability Statement: The data were obtained from portal (https://new.portal.its.pdx.edu/downloads/, 22 March 2022).

Conflicts of Interest: The authors declare that they have no conflict of interest.

References

1. Alghamdi, T.; Elgazzar, K.; Bayoumi, M.; Sharaf, T.; Shah, S. Forecasting Traffic Congestion Using ARIMA Modeling. In Proceedings of the 2019 15th International Wireless Communications & Mobile Computing Conference (IWCMC), Tangier, Morocco, 24–28 June 2019; pp. 1227–1232.
2. Min, X.; Hu, J.; Zhang, Z. Urban traffic network modeling and short-term traffic flow forecasting based on GSTARIMA model. In Proceedings of the 13th International IEEE Conference on Intelligent Transportation Systems, Funchal, Portugal, 19–22 September 2010; pp. 1535–1540.
3. Liu, X.W. Research on Highway Traffic Flow Prediction and Comparison based on ARIMA and Long-Short-Term Memory Neural Network. Master's Thesis, Southwest Jiaotong University, Chengdu, China, 2018. (In Chinese).
4. Cvetek, D.; Muštra, M.; Jelušić, N.; Abramović, B. Traffic Flow Forecasting at Micro-Locations in Urban Network using Bluetooth Detector. In Proceedings of the 2020 International Symposium ELMAR, Zadar, Croatia, 14–15 September 2020; pp. 57–60.
5. Iwao, J.; Stepphanedes Yorgos, J. Dynamic prediction of traffic volume through Kalman filtering theory. *Pergamon* **1984**, *18*, 1–11.
6. Guo, J.; Huang, W.; Williams, B.M. Adaptive Kalman filter approach for stochastic short-term traffic flow rate prediction and uncertainty quantification. *Transp. Res. Part C Emerg. Technol.* **2014**, *43*, 50–64. [CrossRef]
7. Ullah, I.; Fayaz, M.; Naveed, N.; Kim, D. ANN Based Learning to Kalman Filter Algorithm for Indoor Environment Prediction in Smart Greenhouse. *IEEE Access* **2020**, *8*, 159371–159388. [CrossRef]
8. Vlahogianni, E.I.; Karlaftis, M.G.; Golias, J.C. Short-term traffic forecasting: Where we are and where we're going. *Transp. Res. Part C Emerg. Technol.* **2014**, *43*, 3–19. [CrossRef]
9. Gao, J.; Leng, Z.; Qin, Y.; Ma, Z.; Liu, X. Short-term traffic flow forecasting model based on wavelet neural network. In Proceedings of the Control and Decision Conference (CCDC), Guiyang, China, 25–27 May 2013; pp. 5081–5084.
10. Qin, Y.; Adams, S.; Yuen, C. Transfer Learning-Based State of Charge Estimation for Lithium-Ion Battery at Varying Ambient Temperatures. *IEEE Trans. Ind. Inform.* **2021**, *17*, 7304–7315. [CrossRef]

11. Xiong, T.; Qi, Y.; Zhang, W.B.; Li, Q.M. Short-term traffic flow prediction model based on spatiotemporal correlation. *Comput. Eng. Des.* **2019**, *40*, 501–507. (In Chinese)
12. Lu, W.; Rui, Y.; Yi, Z.; Ran, B.; Gu, Y.A. Hybrid Model for Lane-Level Traffic Flow Forecasting Based on Complete Ensemble Empirical Mode Decomposition and Extreme Gradient Boosting. *IEEE Access* **2020**, *8*, 42042–42054. [CrossRef]
13. Alajali, W.; Zhou, W.; Wen, S.; Wang, Y. Intersection Traffic Prediction Using Decision Tree Models. *Symmetry* **2018**, *10*, 386. [CrossRef]
14. Yu, S.; Li, Y.; Sheng, G.; Lv, J. Research on Short-Term Traffic Flow Forecasting Based on KNN and Discrete Event Simulation. In Proceedings of the 15th International Conference on Advanced Data Mining and Applications, Foshan, China, 12–15 November 2019; pp. 853–862.
15. Qin, Y.; Yuen, S.C.; Qin, M.B.; Li, X.L. Slow-varying Dynamics Assisted Temporal Capsule Network for Machinery Remaining Useful Life Estimation. *arXiv* **2022**, arXiv:2203.16373. [CrossRef] [PubMed]
16. Dai, G.W.; Ma, C.X.; Xu, X.C. Short-Term Traffic Flow Prediction Method for Urban Road Sections Based on Space Time Analysis and GRU. *IEEE Access* **2019**, *7*, 143025–143035.
17. Hu, H.; Yan, W.; Li, H.M. Short-term traffic flow prediction of urban roads based on combined forecasting method. *Ind. Eng. Manag.* **2019**, *24*, 107–115.
18. Zhu, P.F.; Liu, Y. Prediction of distributed optical fiber monitoring data based on GRU-BP. In Proceedings of the 2021 International Conference on Intelligent Transportation, Big Data & Smart City (ICITBS), Xi'an, China, 27–28 March 2021; pp. 222–224.
19. Barboza, F.; Kimura, H.; Altman, E. Machine learning models and bankruptcy prediction. *Expert Syst. Appl.* **2017**, *83*, 405–417. [CrossRef]
20. Wang, S.; Yao, Y.; Xiao, Y.; Chen, H. Dynamic Resource Prediction in Cloud Computing for Complex System Simulatiuon: A Probabilistic Approach Using Stacking Ensemble Learning. In Proceedings of the 2020 International Conference on Intelligent Computing and Human-Computer Interaction (ICHCI), Sanya, China, 4–6 December 2020; pp. 198–201.
21. Liu, Y.; Yang, C.; Gao, Z.; Yao, Y. Ensemble deep kernel learning with application to quality prediction in industrial polymerization processes. *Chemom. Intell. Lab. Syst.* **2018**, *174*, 15–21. [CrossRef]
22. Zhang, X.M.; Wang, Z.J.; Liang, L.P. A Stacking Algorithm for Convolutional Neural Networks. *Comput. Eng.* **2018**, *44*, 243–247.
23. Li, B.S.; Zhao, H.Y.; Chen, Q.K.; Cao, J. Prediction of remaining execution time of process based on Stacking strategy. *Small Microcomput. Syst.* **2019**, *40*, 2481–2486. (In Chinese)
24. Sun, X.J.; Lu, X.X.; Liu, S.F. Research on combined traffic flow forecasting model based on entropy weight method. *J. Shandong Univ. Sci. Technol. (Nat. Sci. Ed.)* **2018**, *37*, 111–117. (In Chinese)
25. Gong, Z.H.; Wang, J.N.; Su, C. A weighted deep forest algorithm. *Comput. Appl. Softw.* **2019**, *36*, 274–278. (In Chinese)
26. Zheng, Z.H.; Huang, M.F. Traffic Flow Forecast Through Time Series Analysis Based on Deep Learning. *IEEE Access* **2020**, *8*, 82562–82570. [CrossRef]
27. Zhao, Z.; Chen, W.; Wu, X.; Chen, P.C.; Liu, J. LSTM network: A deep learning approach for short-term traffic forecast. *IET Intell. Transp. Syst.* **2017**, *11*, 68–75. [CrossRef]
28. Li, Y.L.; Tang, D.H.; Jiang, G.Y.; Xiao, Z.T.; Geng, L.; Zhang, F.; Wu, J. Residual LSTM Short-Term Traffic Flow Prediction Based on Dimension Weighting. *Comput. Eng.* **2019**, *45*, 1–5. (In Chinese)

electronics

MDPI

Article

China Coastal Bulk (Coal) Freight Index Forecasting Based on an Integrated Model Combining ARMA, GM and BP Model Optimized by GA

Zhaohui Li [1,*], Wenjia Piao [1,*], Lin Wang [1], Xiaoqian Wang [2], Rui Fu [3] and Yan Fang [1]

1 School of Maritime Economics and Management, Dalian Maritime University, Dalian 116026, China
2 Zhejiang Provincial Military Command, Hangzhou 310002, China
3 ZCCE, Faculty of Sciences and Engineering, Swansea University, Bay Campus, Fabian Way, Swansea SA1 8EN, UK
* Correspondence: leezhaohui@dlmu.edu.cn (Z.L.); windpwj@dlmu.edu.cn (W.P.)

Abstract: The China Coastal Bulk Coal Freight Index (CBCFI) is the main indicator tracking the coal shipping price volatility in the Chinese market. This index indicates the variable performance of current status and trends in the coastal coal shipping sector. It is critical for the government and shipping companies to formulate timely policies and measures. After investigating the fluctuation patterns of the shipping index and the external factors in light of forecasting accuracy requirements of CBCFI, this paper proposes a nonlinear integrated forecasting model combining ARMA (Auto-Regressive and Moving Average), GM (Grey System Theory Model) and BP (Back-Propagation) Model Optimized by GA (Genetic Algorithms). This integrated model uses the predicted values of ARMA and GM as the input training samples of the neural network. Considering the shortcomings of the BP network in terms of slow convergence and the tendency to fall into local optimum, it innovatively uses a genetic algorithm to optimize the BP network, which can better exploit the prediction accuracy of the combined model. Thus, establishing the combined ARMA-GM-GABP prediction model. This work compares the short-term forecasting effects of the above three models on CBCFI. The results of the forecast fitting and error analysis show that the predicted values of the combined ARMA-GM-GABP model are fully consistent with the change trend of the actual values. The prediction accuracy has been improved to a certain extent during the observation period, which can better fit the CBCFI historical time series and can effectively solve the CBCFI forecasting problem.

Keywords: CBCFI; combined prediction model; ARMA; GM; GA; BP

Citation: Li, Z.; Piao, W.; Wang, L.; Wang, X.; Fu, R.; Fang, Y. China Coastal Bulk (Coal) Freight Index Forecasting Based on an Integrated Model Combining ARMA, GM and BP Model Optimized by GA. *Electronics* **2022**, *11*, 2732. https://doi.org/10.3390/electronics11172732

Academic Editor: Alberto Fernandez Hilario

Received: 31 July 2022
Accepted: 25 August 2022
Published: 30 August 2022

1. Introduction

The China (Coastal) Bulk Coal Freight Index (CBCFI) published by the Shanghai Shipping Exchange reflects the pricing of coastal coal shipping in China [1]. It includes the daily complex index and spot ratios relating to various routes/kinds of vessels in the coastal coal service market [2]. CBCFI is used to reflect the changes in the level of bulk freight in China's coastal bulk transport market [3]. It can not only reflect the changes in the level of shipping rates in the coastal bulk market but also objectively reflect the degree of fluctuations in the transport market. So, it can, to a certain extent, reflect the economic development of China and the trend of coastal bulk trade. The release of CBCFI helps the development of the shipping index system in the China coastal coal transportation market [4]. As the "barometer" of the coastal coal transportation market, the index can accurately and timely reflect the dramatic and frequent price fluctuations in the coastal coal transportation market [5]. So, it is essential for shipping operators and investors to use an effective model to forecast CBCFI accurately when developing relevant strategies. However, there is a scarcity of analytical studies on the volatility of China's coastal bulk cargo market. At the same time, although the existing studies can provide guidance for

CBCFI prediction, the prediction accuracy of the models is not high enough to accurately predict CBCFI data with volatility. All of these make it worthwhile and meaningful to propose an effective method for the accurate prediction of CBCFI.

Scholars worldwide have conducted studies on the potential volatility patterns and trend forecasting of the shipping index. For example, Chen [6] developed a grey system theory based on the Baltic dry bulk shipping index forecasting model; Liang et al. [7] presented a neural network based on the export container shipping index estimation model; Lian et al. [8,9] constructed the ARMA model to forecast the shipping index, and demonstrated the applicability of the time series model in index forecasting; Zhou et al. [10] developed a GARCH model to analyze the seasonality, cyclicality, persistence and asymmetry patterns of the fluctuations of the coastal container shipping index; Adland et al. [11] used a nonlinear randomness model to explore the trend of the international market shipping index; Shan et al. [12] used wavelet analysis and ARIMA model to forecast China's export container shipping index. In addition, Li et al. [13] created a prediction model with BP neural network improved by genetic algorithm and verified that the improved BP neural network gets higher prediction accuracy and faster convergence speed than the traditional BP neural network. By analyzing the research methods of the above scholars, we find that the forecasting methods for CBCFI mainly including: the ARIMA model, GARCH model, neural network, SVM model, wavelet analysis and so on. The above models have good guiding significance for CBCFI forecasting research, but there are also certain shortcomings. First, the GARCH model is based on statistics and theory. Before building these forecasting models, the non-linear and non-stationary shipping index need to be smoothed, which will inevitably destroy the intrinsic characteristics of the shipping index to a certain extent. Second, wavelet analysis is not free from the constraint of pre-selected basis functions, there is too much subjectivity in the selection of parameters, and the selection of different parameters produces results that vary greatly and lack adaptability. Third, the above scholars mostly use a single linear or a nonlinear model to forecast the shipping index. However, because of the complexity of CBCFI series, these models are easily influenced by their own characteristics, which can result in a decrease in forecast credibility.

Considering the shortcomings mentioned above, this paper proposes a combined ARMA-GM-BP model based on GA optimization for short-term forecasting of CBCFI, and then presents a BP network optimized by genetic algorithm to simulate nonlinear combination functions and creates an ARMA-GM-GABP combined forecasting model. First, we use the ARMA model and GM (1,1) model to take the prediction value of CBCFI for the given time respectively. Then, these two values are two-dimensionally input into a GA-BP neural network, the GA-BP neural network model then combines these two predicted values nonlinearly, predicts its fitting error and further corrects the predicted values, finally outputs the predicted CBCFI value. In this paper, we innovatively use a genetic algorithm to optimize the BP network, allowing it to avoid the defects of the BP model and improve the combined model's prediction accuracy. The combined model compensates well for the slow error correction of the ARMA model and the fact that the GM model is only suitable for predicting sequences that grow monotonically at an approximately exponential rate. Furthermore, the combined model has a high prediction accuracy and fits the CBCFI historical time series better.

The paper is organized as follows. Section 1 introduces the research background, literature review and the significance of this work, including research motivation, knowledge gap, problem statement and a brief introduction to combinatorial model building. Section 2 provides a theoretical introduction of the models used in this paper and introduces the construction principle of the combined forecasting model. Section 3 builds the models and uses the ARMA model, the GM (1,1) model and the ARMA-GM-GABP combined forecasting model to predict the CBCFI values respectively, and then compares the predicting results of these three models through the error evaluation index. Section 4 reviews the whole paper and proposes directions for future improvement.

2. Combination Construction of the Forecasting Model

The CBCFI sequence's extreme volatility proves the coal transportation market's considerable risk. According to data volatility research, CBCFI has complicated nonlinear properties. And the typical single prediction model has limits in data prediction. Therefore, on the basis of in-depth research on ARMA, GM and BP models, this paper establishes the ARMA-GM-GABP nonlinear combination model to predict CBCFI.

2.1. ARMA Model

American statisticians Box and Jenkins proposed the Autoregressive Moving Average Model as a time series forecasting method [14]. A time series is a collection of continuous observations of a single variable over a period of time, organized in a time sequence. We mainly use the autoregressive model (AR model) and moving average model (MA model) to statistically describe the stochastic nature of time series [15]. Although both the ARMA and ARIMA models are a hybrid of AR and MA models, their application objects are different. The ARMA model is used to model stationary time series, while the ARIMA model is used to model non-stationary time series [16,17]. The term "stationarity of time series" refers to the fact that the statistical law of time series does not vary over time [18]. It means that the statistical properties of the random process of time series data that generate variables do not change. A stationary time series can be regarded as a curve moving up and down around its mean value [19]. In practical applications, it is necessary to check whether the time series is stable first. If it is a non-stationary series, it must be smoothed. After stationary processing, ARMA can analyze these data. The most common expression of ARMA (p,q) is:

$$\chi_t = c + \varphi_1\chi_{t-1} + \varphi_2\chi_{t-2} + \varphi_3\chi_{t-3} + \ldots + \varphi_p\chi_{t-p} + \varepsilon_t + \theta_1\varepsilon_{t-1} + \theta_2\varepsilon_{t-2} + \ldots + \theta_q\varepsilon_{t-q} \tag{1}$$

In the formula, the first half is the autoregressive part and the non-negative integer p is the autoregressive order, $\varphi_1, \ldots, \varphi_p$ is the autoregressive coefficient, the second half is the moving average part, the non-negative integer q is the moving average order and $\theta_1, \ldots, \theta_q$ is the moving average coefficient.

In summary, the ARMA model is built on a smooth time series. Before using the ARMA model for forecasting, the data used should be pre-processed first to eliminate periodicity and trendiness and make it meet the smoothness requirements. Then, the truncated and trailing tails of the autocorrelation and partial correlation functions are judged to make pattern recognition [20]. The model structure that is most similar to the change process of the pre-processed data series is selected. After determining the order of the model by the fixed-order method, use the least squares estimation method to find the model parameters φ and θ. Finally, the model is tested for suitability by determining whether the residual series of the model is a white noise series. If it passes the test, we can use this model to predict the value.

2.2. GM Model

In grey system theory, the GM (1,1) model is the most widely used grey dynamic prediction model. The grey model accumulates the original data to generate a new series in order to weaken the random terms and increase their regularity. It is mainly used to fit and estimate the eigenvalues of a single principal element in a complex system [21]. CBCFI has obvious dynamic characteristics and uncertainties, which is consistent with the characteristics of the gray system [22]. The GM (1,1) model typically uses newly generated data sequences. Taking the cumulative generation as an example:

1. Suppose the original time series data: $X^{(0)} = \left(X_1{}^{(0)}, X_2{}^{(0)}, \ldots, X_j{}^{(0)}, \ldots, X_n{}^{(0)}\right)$;
2. Data accumulation generates a new sequence: $X^{(1)} = \left(X_1{}^{(1)}, X_2{}^{(1)}, \ldots, X_j{}^{(1)}, \ldots, X_n{}^{(1)}\right)$, where $x_{(j)}{}^{(1)} = \sum_{i=1}^{j} x_{(i)}{}^{(0)}$;

3. Grade ratio test: Generally, the level ratio of ${\sigma_k}^{(0)}$ to $X^{(0)}$ and its range are used to judge whether a high precision GM (1,1) model can be established for a given sequence. Grade ratio definition: ${\sigma_k}^{(0)} = \frac{{x_{(k-1)}}^{(0)}}{{x_{(k)}}^{(0)}}$, If ${\sigma_k}^{(0)} \in (e^{-\frac{2}{n+1}}, e^{\frac{2}{n+1}})$ is satisfied, $X^{(0)}$ can be regarded as the modeling object of GM (1,1);
4. The change trend of the new series is approximately described by the following differential equation, where a is the development gray level and u is the endogenous control gray level:

$$\frac{dx^{(1)}}{dt} + ax^{(1)} = u \tag{2}$$

5. Using the ratio of mean square error and the probability of small error to test the prediction accuracy of the GM (1,1) model.

As a very important grey forecasting model, the GM (1,1) model has a number of significant modeling advantages [23]. For example, the theoretical principles of the model are relatively simple, the model requires fewer sample data and does not require the sample data to meet specific probability distribution characteristics. At the same time, the parameter solution of the model is relatively simple, the prediction precision is relatively high, and the prediction test of the model is relatively simple. As a result, the GM (1,1) model has now been applied with some success in a number of areas.

2.3. BP Model Improved by GA

There are many unknown variables in the change process of CBCFI, and the neural network does not need to consider the relationship among variables. As long as the nodes of the input layer and the output layer are defined, the network system can be trained continuously until the test accuracy reaches the set value [24].

BP neural networks, also known as back propagation neural networks, are trained with sample data to continuously modify the network weights and thresholds so that the error function decreases in the negative gradient direction, approximating the desired output [25]. The BP neural network model topology consists of an input layer, a hidden layer and an output layer. The input layer receives the sample and calculates it through the hidden layer and outputs it through the output layer. When the output value differs significantly from the expected value, the error is propagated backward and the weights of each layer are modified by the output layer through the hidden layer. This process repeats alternately until the error is reduced to an acceptable range or a predetermined number of training periods are performed. The main components of a predictive model using the BP neural network algorithm include: the determination of the input samples, the number of input and output layers and the number of hidden layers [26].

The initial weights and biases of a single BP neural network are completely random. Although the BP network corrects the initial weights and biases during the training process, they have a significant impact on the outcome. The basic idea of genetic algorithms is to simulate the process of biological evolution [27]. It starts from a population that represents a potential set of solutions to a problem and uses fitness as a basis for evaluating the merits of individuals. Then, it repeatedly uses selection, crossover and variation operators on the population so that the population gradually approaches the optimal solution. Genetic algorithms have strong environmental self-adaptation and self-learning capabilities, and their highly parallel global search algorithms can overcome the shortcomings of BP neural networks [28,29]. The combination of genetic algorithms and BP neural network not only helps to avoid BP neural networks from falling into local minima but also accelerates the convergence speed of the network and enhances the learning ability and the generalization ability of the model [30]. Therefore, we use a genetic algorithm to optimize the BP network to achieve the purpose of efficient solution and global optimization search.

The GA-BP neural network model uses a genetic algorithm to perform a global search on the range of weights to find the optimal initial weight values and thresholds for the BP neural network model first. Then the BP neural network model begins the training process with the optimal initial weight values and thresholds provided by the genetic algorithm and approximates the optimal solution to the prediction problem. Finally, this model outputs the prediction values that achieve the desired prediction accuracy of the initial setting.

Steps of improving the BP neural network by GA.

- Initialize the population.
- We encode all the weights and thresholds in the network as real numbers. Each individual is represented by a set of chromosomes with the following chromosome form: $w_{11}, w_{12} \ldots, w_{ij}, a_1, a_2 \ldots, a_l, w_{11}, w_{12} \ldots, w_{jk}, b_1, b_2 \ldots, b_m$. w_{ij} is the connection weight between the input and hidden layers; $a = \{a_1, a_2, \ldots, a_l\}$ is the hidden layer threshold; w_{jk} is the hidden layer and output layer connection weight; $b = \{b_1, b_2, \ldots, b_m\}$ is the output layer threshold. This experiment started with a population of 100 persons.
- Choose a fitness function.
- The less the absolute value of the error in the BP neural network, the better. The higher the fitness score in the genetic algorithm, the better. As a result, the fitness function is the inverse of the BP neural network goal function.

$$F(x) = \left(\sum_{k=1}^{m} \sqrt{\sum_{k=1}^{m} \left(y_k^q - V_k^q \right)^2} \right)^{-1} \tag{3}$$

- Select genes.
- Using the roulette method to choose individuals in the population. Those with high fitness are chosen to be passed down to the next generation.
- Operation of crossover mutation.
- The basic action of a genetic algorithm is to generate new individuals. The goal is to improve the coding structure of the individual. The mutation process includes changing the gene value at some sites of an individual string in the population, which can lead to the generation of new individuals and allow the genetic algorithm to perform a local random search.
- Operation on a cyclic basis.
- If the fitness of an individual reaches a certain threshold, or if the fitness of the individual and group no longer rises, the algorithm can terminate. Otherwise, the loop will restart at the second stage. We use the connection weights and thresholds optimized by the genetic algorithm as the initial weights and thresholds. The GA-BP neural network is trained until the error requirements are met or the maximum number of training times is reached.

2.4. ARMA-GM-GABP Combined Model Construction

In recent years, combined forecasting models have been increasingly used in forecasting problems because of their general advantage of higher forecasting accuracy compared to single forecasting models. Normally, there are limitations to the practical application of a single forecasting model. For example, although the GM (1,1) model is very good at reducing the volatility of the original modeled data series, there are certain disadvantages relative to other models in terms of portraying the periodicity and trend of the original modeled data series. Combined models can be a good way to overcome the shortcomings of individual prediction models.

In this paper, we consider the construction of the combination model from the following aspects. First of all, the success of the combination model depends largely on the choice of the model. Considering the volatility and periodicity of CBCFI, our work chooses the ARMA model, which is suitable for linear prediction and has high short-term prediction accuracy and the GM (1,1) model, which can effectively reduce the volatility of the data. These two models provide better prediction results than other machine learning models. The ARMA model captures the periodicity and trend information in the original modeling data series, and the GM (1,1) model can effectively reduce the volatility of the original modeling data series. Then, considering the defects of slow convergence speed and the easy falling into a local optimum of the BP network, the genetic algorithm is used to optimize the BP network. This operation significantly improves the convergence speed and convergence performance of the model and at the same time, it greatly reduces the prediction error of the model and better exploits the prediction accuracy of the model. Considering the characteristics of the three models above, we finally choose to combine these three models, the combined ARMA-GM-GABP prediction model is obtained.

The principle of the nonlinear combination forecasting model refers to the nonlinear combination of different forecasting methods. The nonlinear function $f(x)$ is:

$$\hat{y} = f(x) = f(t_1, t_2, \ldots, t_n) \quad (4)$$

where $t(x)$ (i = 1, 2, ... , n) represents the prediction results of i-th prediction methods. The combined forecasting model makes comprehensive use of the advantages of each single model, so the forecast accuracy of $f(x)$ is higher than that of $t_i(x)$. Since a single hidden layer BP network can arbitrarily approximate a continuous nonlinear function, this paper attempts to use BP neural networks to model the nonlinear combinatorial prediction function $f(x)$, so as to achieve the purpose of nonlinear combinatorial modeling and prediction using the ARMA model and GM (1,1) model. The basic idea of the ARMA-GM-BP combined forecasting model: First, we obtain the CBCFI prediction values of the ARMA and grey GM (1,1) models for the given date. Then, we two-dimensionally input the predicted values into the BP neural network model optimized by the genetic algorithm, the GA-BP neural network model then combines these two predicted values nonlinearly, predicts its fitting error and further corrects the predicted values, finally outputs the predicted CBCFI value. Figure 1 shows the specific process of the combined ARMA-GM-BP forecasting model.

The specific implementation steps of the combined forecasting model are as follows:

1. Establish an ARMA single-term forecast model to obtain the forecast value $t_1 = (t_{11}, t_{12}, \ldots, t_{1n})$ (t_{1i} is the predicted value in the i-th day).
2. Establish a GM (1,1) single prediction model to obtain the predicted value $t_2 = (t_{21}, t_{22}, \ldots, t_{2n})$ (t_{2i} is the predicted value in the i-th day).
3. Determine the BP neural network structure according to the number of input and output parameters of the fitting function: Use the predicted value (t_{11}, t_{21}), (t_{12}, t_{22}) ... , (t_{1m}, t_{2m}) of ARMA model and GM (1,1) model as the two-dimensional input training sample of the BP neural network ($m < n$), the number of input nodes is 2. Use $r_i = (r_1, r_2, \ldots, r_m)$ as an output target (where r_i is the true CBCFI value at day i), the number of the output node is 1.
4. Attempt to use genetic algorithms to improve the weight and threshold of the BP neural network. First, we use the fitness function to calculate the individual fitness value, then use selection, crossover and mutation operations to determine the optimal fitness value corresponding to persons.
5. Initial weight and threshold assignment to the network using a genetic algorithm to obtain the optimal individual, and then we use the trained combination model to predict the test sample $(t_{1m+1}, t_{2m+1}) \ldots, (t_{1n}, t_{2n})$. Predict the CBCFI for the test date and compare it with the true value to test the prediction ability of the network.

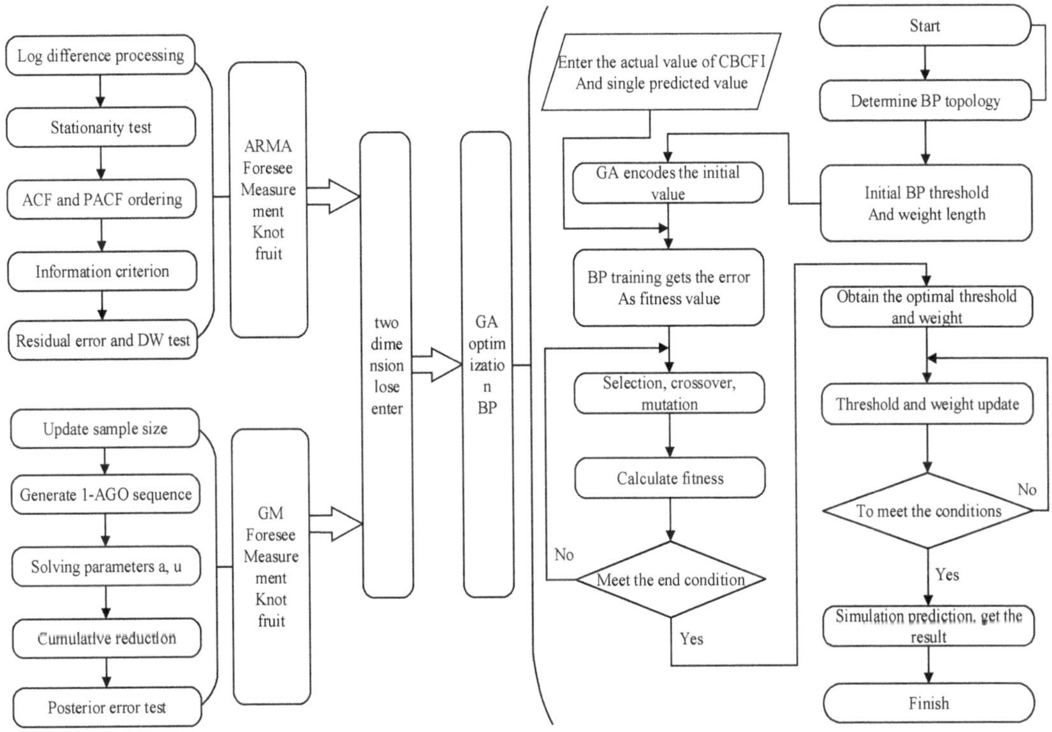

Figure 1. Structure of the ARMA-GM-GABP combined model.

3. Empirical Analysis

In this paper, we use the China Coastal Bulk Coal Freight Index from January 2014 to November 2019 as the sample. This work set the data from January 2014 to July 2019 as the training set and the data from August to October 2019 as the test set. The training set contains 2038 data and the test set contains 61 data. Then, we use November 2019 data as the forecast set to conduct a comparative analysis of forecast accuracy based on three models, the ARMA model, GM model and ARMA-GM-GABP combination model.

3.1. Data Volatility Analysis

Figure 2 depicts the trend of CBCFI in the sample range. The figure shows that the CBCFI data fluctuate greatly and there is a phenomenon of sharp rise and fall. In addition, the data show obvious fluctuation clustering characteristics, the larger changes are relatively concentrated in one period, while the smaller changes are relatively concentrated in another period.

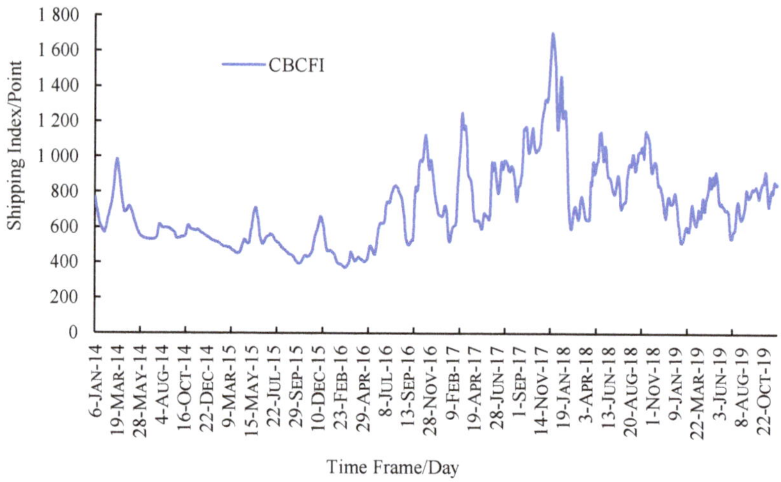

Figure 2. Historical data of CBFI.

The large volatility of CBCFI data is mainly caused by comprehensive changes in the shipping market's capacity and turnover in different periods. Since 2014, the growth rate of coal demand has slowed down, leading to oversupply in the charter market. The overall trend of the domestic coastal coal shipping market is sluggish, and CBCFI continues to bottom out. Coal transport prices remained low in 2015, shipping rates were able to rebound sharply in May due to a significant reduction in coal imports and a significant reduction in domestic coal prices. In December 2017, due to the accelerated pace of coal reserves in power plants in winter and the obstacles to shipping capacity in the northern region, shipping rates have jumped, reaching their highest value in recent years, which is 1706.2. The increase in hydropower squeezed coastal thermal power production in July 2019, prompting a reduction in coal consumption by high-energy-consuming enterprises, making coal pulling less active, but then, affected by typhoons and a lack of capacity supply in the market, supporting higher freight prices.

Table 1 shows the descriptive statistical characteristics of the mean, standard deviation, kurtosis and JB statistic of the CBCFI data within the sample interval. The skewness is $1.046825 > 0$, the kurtosis is $4.349061 > 3$ and the probability p corresponding to the JB statistic is 0, which shows that the data have a clear spike-right skew, a large deviation from the normal distribution, and a spike and fall situation.

Table 1. Statistics characteristics of sample.

Parameter	Eigenvalues
Mean	719.7643
Standard deviation	231.9314
Skewness	1.046825
Kurtosis	4.349061
JB statistics	371.9415
p value	0.000000

3.2. *Predicting from ARMA Model*

First, we establish the ARMA prediction model. It can be seen from Figure 2 that the CBCFI data fluctuate greatly in the selected sample interval, and the changing trend shows a non-stationary state. In order to eliminate the instability of the coal shipping index, we choose the daily return rate of the index as the research object. The daily return rate of the index adopts the calculation formula of the logarithmic return rate, and the daily return rate of CBCFI is expressed as:

$$\text{RCBCFI} = \ln\text{CBCFI}_t - \ln\text{CBCFI}_{t-1} \quad (5)$$

In the above formula, RCBCFI is the daily return on CBCFI after first order logarithmic differencing, CBCFI_t is the daily coal freight index corresponding to day t, and CBCFI_{t-1} is the daily coal shipping index corresponding to day $t-1$. After the first-order logarithmic difference processing, the change trend of CBCFI's daily return is shown in Figure 3.

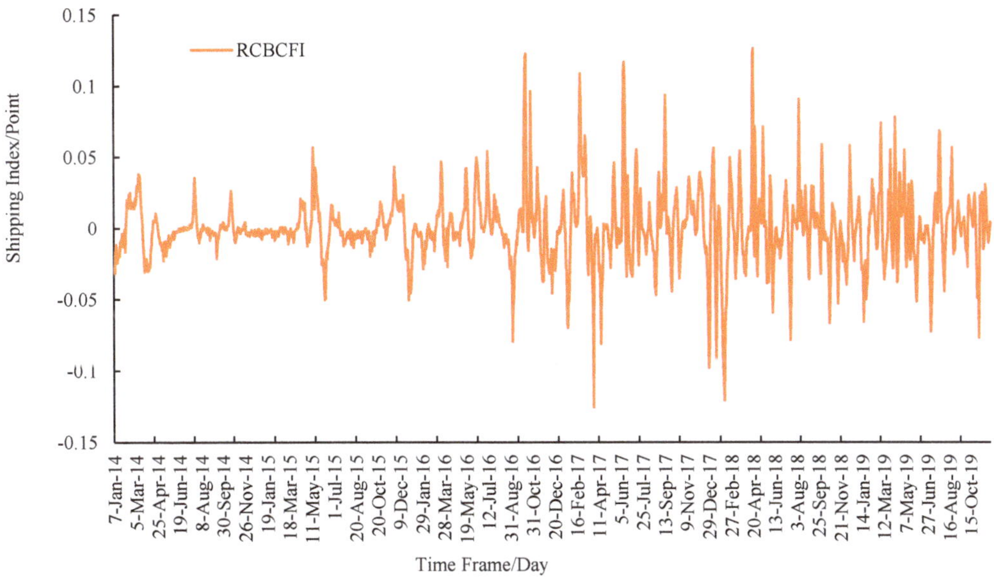

Figure 3. Historical data of CBFI return series.

We use the ADF method to test whether RCBCFI is stationary. The T statistic is −15.97711 less than the critical value of 1% of the significance level −2.566702. The concomitant probability is 0.0000, indicating that the RCBCFI sequence does not have a unit root and is a stationary sequence, which is suitable for constructing the ARMA forecast model. By analyzing the statistical characteristics of the autocorrelation function and partial autocorrelation function, we preliminarily determine that there are two preselected models, ARMA (1,1) and ARMA (1,2).

Our work sequentially tests the two preselected models from the low level. It can be seen from the comparison of model statistics and T test results in Table 2 that the ARMA (1,2) model has passed the T test, and all indicators are overall better than the ARMA (1,1). Therefore, the ARMA (1,2) model is determined as the optimal prediction model. Then we perform an autocorrelation test on the estimated ARMA (1,2) model residuals. It is found that the autocorrelation functions of the samples are all within the 95% confidence interval, and the corresponding probability p values of the Q statistic are far greater than the test level of 0.05. Therefore, it is considered that there is no autocorrelation in the residual sequence of the model ARMA (1,2) estimation results, that is, the model construction is reasonable.

Table 2. Parameter estimation results of ARMA model.

Preselected Model	Variable	Coefficient	Standard Error	t Value	p Value	R2	AIC Value	SC Value
ARMA (1,1)	AR(1) MA(1)	0.708 0.389	0.034 0.041	31.298 13.056	0.000 0.000	0.708	−5.796	−5.788
ARMA (1,2)	AR(1) MA(1) MA(2)	0.632 0.489 0.141	0.033 0.040 0.037	18.396 11.952 3.786	0.000 0.000 0.000	0.711	−5.801	−5.793

3.3. Predicting from GM Model

According to the modeling steps of the GM model, we first test the original CBCFI sequence for extreme ratios. The test results find that the extreme ratios of the sequence are included in the required range and meet the modeling requirements. Then we use MATLAB software to write a program and train the model to obtain the GM (1,1) model parameters $a = 0.0393$, $\mu = 758.8001$. The model prediction formula is:

$$x^{(1)}{}_{k+1} = -18605.2106e^{-0.0393k} + 19307.8906 \tag{6}$$

To test the prediction accuracy of the above model, we use the formula $C = \frac{S_2}{S_1}$, $S^2 = \frac{1}{n}\sum_{k=1}^{n}\left(x_k{}^{(0)} - \overline{x}\right)^2$ and $S_2{}^2 = \frac{1}{n-1}\sum_{k=2}^{n}(\varepsilon(k) - \overline{\varepsilon})^2$ to calculate the mean square error ratio C of the model, where $\varepsilon(k)$ is the difference between the original sequence and the predicted sequence; $\overline{\varepsilon}$ is the average of the residual sequence $\varepsilon(k)$; S_1 and S_2 are the standard deviations of the original series and the residual series respectively. The calculation result shows the mean squared error ratio $C = 0.02097 < 0.35$. Then we continue to use the formula $P = \{|\varepsilon(k) - \overline{\varepsilon}| < 0.6745S_1\}$ to calculate the probability of small error $p = 1 > 0.95$. Finally, we refer to the gray prediction accuracy test grade standard table to know that the above model has passed the test, which is the better model with level 1.

3.4. ARMA-GM-GABP Combined Model Prediction

According to the previous analysis, we use the BP neural network optimized by GA to simulate the nonlinear function. The number of nodes in the input layer is 2 and the number of nodes in the output layer is 1. In this paper, the number of hidden layer neurons affects the model fitting effect and calculation time. According to the empirical formula $h = \sqrt{m+n} + \alpha$, the number of hidden layer neurons h is preliminarily determined, where m is the number of input layer nodes, n is the number of output layer nodes and α is an adjustment constant between 1–10. After calculation, it is found that h is between 2 and 11. After investigating the effect of BP training with different hidden layers, we check the effect for each BP prediction result with the test group data and calculate the difference between the fitted value and the corresponding actual value. The results show that when the number of neurons is 5, the fitting effect of the neural network is better and the calculation time is shorter. Therefore, our work selects the 2-5-1 BP network prediction model. The activation function of the hidden layer is the tansig function, and the activation function of the output layer is the purelin function. The number of training times is 1000, and the error setting is 0.0001.

GA uses real number coding. Based on Equation $l = n_1 \times n_2 + n_2 \times m + n_2 + m$, where n_1 is the number of neurons in the input layer, m is the number of neurons in the output layer, and n_2 is the number of neurons in the hidden layer, we can calculate the length of each individual code l is 21.The initial population size of the experiment is 100. We select the inverse of the BP neural network objective function as the fitness function. The selection operation uses the roulette method, the crossover operation uses the two-point arithmetic crossover method, the mutation operation uses the basic bit mutation method, and the number of iterations is 600. Finally, the 2-5-1 BP neural network with the best fit is found through iterative training and is used to predict the CBCFI values for November 2019.

3.5. Analysis of the Predicting Results

Figure 4 shows the prediction results of the ARMA model, Figure 5 shows the GM model prediction results and Figure 6 shows the prediction results of the combined model. It can be seen from the prediction fitting that the prediction result of the ARMA model can reflect the changing trend of the actual CBCFI value to a certain extent, but when the data change greatly, a big error will occur. At the same time, after an error occurs, it takes at least two units of time to correct it. For data with large fluctuations, the ARMA model will cause the predicted data to be too big or too small. The prediction of the GM model shows that it is suitable for approximating the prediction of exponential growth, so the prediction value is credible under the premise of no large data fluctuations. Therefore, the GM model is suitable for data prediction of wavelet motion. For data with large fluctuations, there is a big error between the predicted results and actual values in most cases. Through the ARMA-GM-GABP combined model, the trend of the predicted sequence is close to the actual sequence, and the correction time after an error in the forecast is no more than 1 time unit. Compared with the ARMA model and the GM model, the prediction accuracy is greatly improved.

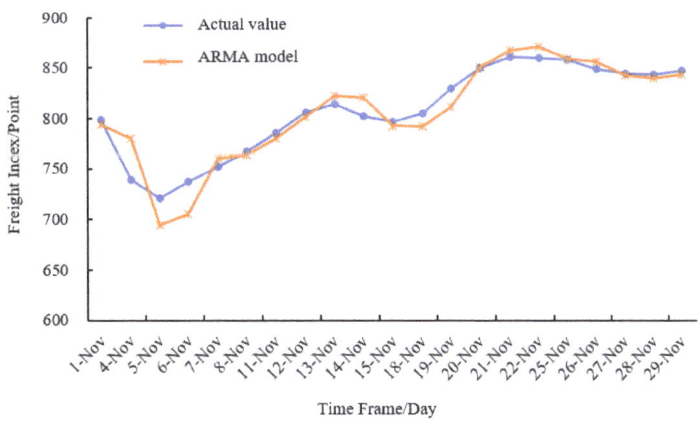

Figure 4. Comparison between forecasting result of ARMA model and real value.

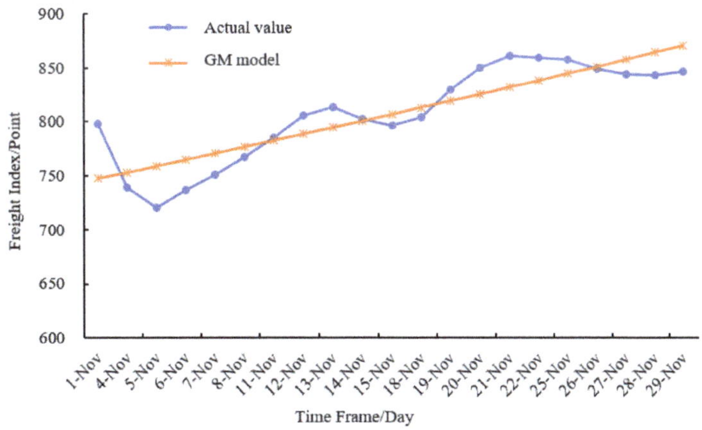

Figure 5. Comparison between forecasting result of GM model and real value.

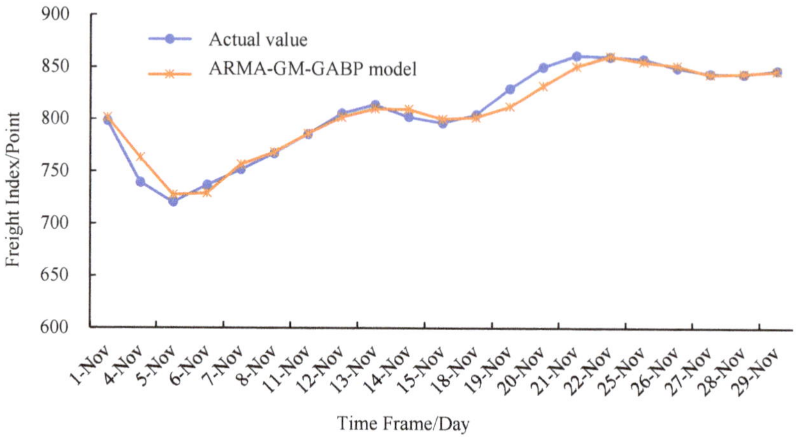

Figure 6. Comparison between forecasting result of ARMA-GM-GABP model and real value.

In order to comprehensively assess the forecasting performance of the ARMA-GM-GABP combined model, we use the following four indicators as the assessment criteria: Mean Absolute Error (MAE), Root Mean Square Error (RMSE), Hill Inequality Coefficient (TIC) and Absolute Error (AE). MAE is used to evaluate the predictive effect of the smooth part of the CBCFI, RMES is used to evaluate the predictive effect of the high values in the CBCFI, and TIC and AE are used to evaluate the predictive power of the model and the degree of model fit. These four indicators are used to test the prediction accuracy of each model. The lower these four values, the lower the prediction error. Compared to other evaluation indicators, these four indicators can provide a more accurate assessment of the model's prediction of the high part, the smooth part and the overall trends of the CBCFI. Therefore, it is very suitable to use these assessment indicators for evaluating the predictive effectiveness of the volatile CBCFI. The calculation formula is as follows:

$$E_{AE} = |\hat{x}_t - x_t| \tag{7}$$

$$E_{MAE} = \frac{1}{n}\sum_{t=N}|\hat{x}_t - x_t| \tag{8}$$

$$E_{RMSE} = \sqrt{\frac{\sum\limits_{t=N}(\hat{x}_t - x_t)^2}{N}} \tag{9}$$

$$E_{TIC} = \frac{\sqrt{\frac{\sum\limits_{t=N}(\hat{x}_t - x_t)^2}{N}}}{\sqrt{\frac{\sum\limits_{t=N}\hat{x}_t{}^2}{N}} + \sqrt{\frac{\sum\limits_{t=N}x_t{}^2}{N}}} \tag{10}$$

In the above formula: $\hat{x}_t$ is the predicted value of CBCFI; x_t is the actual value of CBCFI; N is the data sample size.

The comparative analysis of predictive indicators of these three models ARMA, GM and ARMA-GM-GABP is shown in Table 3 and Figure 7. The test results show that:

- The MAE of the combined ARMA-GM-GABP forecasting model is 5.8780, a decrease of 44.16% compared to the ARMA model and 67.37% compared to the GM (1,1) model. This suggests that the ARMA-GM-GABP model has improved the prediction of the smooth part of the CBCFI series.
- The RMSE of the combined ARMA-GM-GABP forecasting model is 8.5889, a decrease of 42.25% compared to the ARMA model and 60.1% compared to the GM (1,1) model.

It shows that the prediction accuracy of the high-value part of the model is then significantly improved.
- The combined ARMA-GM-GABP forecasting model has improved the predictive ability and fits of the CBCFI with a TIC of 0.0053, a decrease of 42.4% compared to the ARMA model and 60.15% compared to the GM (1,1) model. It shows that the combined ARMA-GM-GABP forecasting model has better forecasting ability compared to other models.
- From Figure 7, we can find that the AE curve of the combined model is the lowest in the vast majority of cases. There are only four instances where it is not the lowest and will correct to be the lowest within a unit of time.

Table 3. Forecasting index of three models.

Predictive Model	E_{MAE}	E_{RMSE}	E_{TIC}
ARMA	10.5260	14.8729	0.0092
GM	18.0132	21.5242	0.0133
Combined	5.8780	8.5889	0.0053

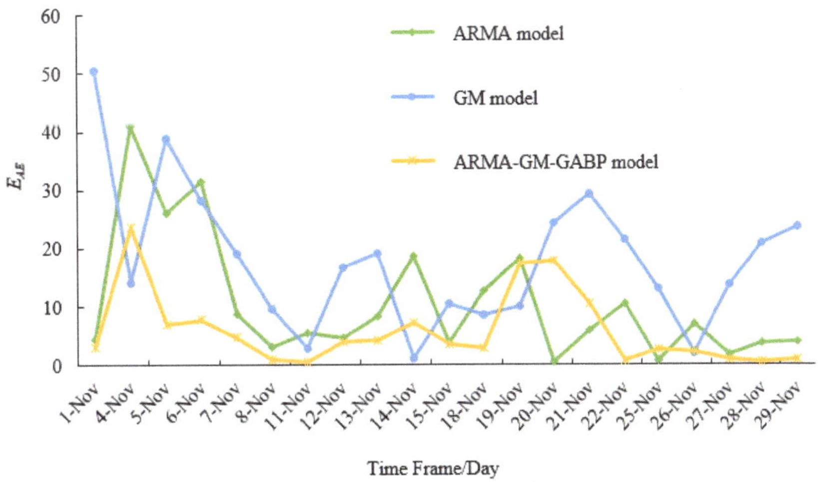

Figure 7. EAE of three prediction models.

All of the above suggests that the ARMA-GM-GABP combined model is more suitable for CBCFI forecasting than the ARMA model and the GM (1,1) model.

4. Conclusions

In this paper, we select CBCFI as the research object. First of all, our work uses ARMA model prediction. ARMA model is the most commonly used model to deal with time series. By fitting the linear characteristics of the time series, it can often get good results. However, for the CBCFI, which is a noisy and non-smooth series, the linear analysis alone does not give a good result. Second, we use the GM (1,1) model, which is the most widely used grey dynamic prediction model in grey system theory. Only a few prediction values of this model have relatively small errors, but the other prediction values can only reflect the growth trend of the data series to a certain extent, and the prediction accuracy is relatively low.

In response to the large fluctuations in the CBCFI, which contains noise and the series itself is non-linear and non-stationary. This paper establishes a combined ARMA-GM-GABP forecasting model to forecast the CBCFI. The empirical analysis results show that the

ARMA-GM-GABP combined model has the following advantages compared to traditional forecasting models:

- Effective capture of trend information.
- ARMA model can capture the periodicity and trend information of the original data series well. Therefore, using the predicted data from the ARMA model as input to the GABP model allows the combined model to more accurately predict trends of the CBCFI.
- Noise reduction.
- The GM (1,1) model allows for effective noise reduction of the original CBCFI series, providing stable input data for the GA-BP prediction model.
- Better prediction and accuracy.
- Compare to the ARMA model and GM (1,1) model, the MAE and RMESE of the combined ARMA-GM-GABP forecasting model decreased by 67.37% and 60.09%, respectively. Its prediction error is significantly smaller than that of the ARMA model and GM (1,1) model.
- General predictive applicability.
- The combined ARMA-GM-GABP forecasting model is constructed based on the CBCFI series in the years since 2014. Due to the extreme volatility of CBCFI over this time period, the ARMA-GM-GABP combined forecasting model is trained to have general predictive applicability to CBCFI over different time periods.

Above all, the ARMA-GM-GABP combined model can make up for the deficiency of the single prediction models. It has good modeling and prediction advantages for dealing with the original modeling data with volatility, periodicity and trend, so the model has good prediction performance. The ARMA-GM-GABP combined model provides scientifically accurate forecasts of the CBCFI, which can support the government and relevant departments in better macroeconomic regulation and control and enable relevant enterprises and participants in the coastal shipping market to better obtain market information and grasp market dynamics. The areas for improvement of the combined forecasting model include: (1) Optimization of single-term models ARMA and GM; (2) Considering the selection of models with better forecasting effects as single-term forecasting models. In the future, as our research progresses, we will try to improve the combined model and extend it to other shipping indices, in order to further verify the validity and practicality of the model in practical applications, and provide support for shipping market operators and investors to better grasp market trends and formulate strategies.

Author Contributions: Conceptualization, W.P. and L.W.; methodology, Z.L.; software, W.P. and L.W. and Y.F.; validation, W.P. and L.W.; formal analysis, X.W.; investigation, X.W. and Y.F.; resources, Z.L.; data curation, X.W and W.P.; writing—original draft preparation, W.P. and L.W.; writing—review and editing, Z.L. and R.F.; project administration, Y.F. All authors have read and agreed to the published version of the manuscript.

Funding: This work was funded by the National Natural Science Foundation of China under Grant 71801028, the Social Science Planning Fund of Liaoning Province Grant L18CTQ004, and China Postdoctoral Science Foundation Grant 2015M571292.

Institutional Review Board Statement: Not applicable.

Informed Consent Statement: Not applicable.

Data Availability Statement: The data presented in this study are available upon request.

Conflicts of Interest: The authors declare no conflict of interest.

References

1. Tan, H.; Yang, J. An Empirical Analysis of the Correlation between China's Coastal and Yangtze River Coal Freight Price Volatility Based on VAR. *J. Wuhan Univ. Technol.* **2021**, *45*, 161–165.
2. Wang, S.; Chen, J.; Yu, S. China's coastal and international dry bulk freight rates linkage. *China Navig.* **2016**, *39*, 114–118.

3. Liu, C.; Liu, J.; Yang, J. Evaluation of Volatility of Coastal Coal Freight Index Based on ARCH Family Models. *J. Wuhan Univ. Technol.* **2012**, *3*, 445–449.
4. Xiao, W.; Xu, C.; Liu, A. Hybrid LSTM-Based Ensemble Learning Approach for China Coastal Bulk Coal Freight Index Prediction. *J. Adv. Transp.* **2021**, *2021*, 5573650. [CrossRef]
5. Wang, S. Analyzing the Influence of Each Influencing Factor on the Freight Rate of Coastal Coal Based on Analytic Hierarchy Process. *Int. Core J. Eng.* **2020**, *6*, 256–261.
6. Chen, Y. Forecasting Baltic dry index with unequal-interval grey wave forecasting model. *J. Dalian Marit. Univ.* **2015**, *41*, 96–101.
7. Liang, W.; Lu, C. Export Containerized Freight Index Estimation Model Based on Neural Network. *Comput. Simul.* **2013**, *30*, 421–425.
8. Lian, Y. *Crude Oil Tanker Freight Rate Forecasting Based on ARMA and Artificial Neural Network*; Shanghai Jiao Tong University: Shanghai, China, 2015.
9. Yuan, Y.; Wang, B. Forecast on highway logistics freight price index of China by ARIMA model. *Math. Pract. Theory* **2017**, *47*, 52–57.
10. Zhou, Y.; Yang, J. A study on the fluctuation characteristics of China's coastal container shipping index. *J. Wuhan Univ. Technol.* **2022**, *44*, 32–39.
11. Adland, R.; Cullinane, K. The nonlinear dynamics of spot freight rates in tanker markets. *Transp. Res. Part E Logist. Transp. Rev.* **2006**, *42*, 211–224. [CrossRef]
12. Shan, F. China Export Container Freight Index Forecast Based on Wavelet Analysis and ARIMA Model. Master's thesis, Dalian Maritime University, Dalian, China, 2013.
13. Li, C. Price Forecasting Analysis of BP Neural Network Based on Improved Genetic Algorithm. *Comput. Technol. Dev.* **2018**, *28*, 144–151.
14. Wang, Y. Analysis and Forecast of Stock Price Based on ARMA Model. *Product. Res.* **2021**, *09*, 124–127.
15. Xu, D.; Zhou, C.; Guan, C. Prediction method of equipment failure rate based on ARMA-BP combined model. *Agriculture* **2022**, *12*, 793.
16. Wu, D.; Wu, C. Research on the Time-Dependent Split Delivery Green Vehicle Routing Problem for Fresh Agricultural Products with Multiple Time Windows. *Agriculture* **2022**, *12*, 793. [CrossRef]
17. Xu, Y.; Chen, X. Comparison of Seasonal ARIMA Model and LSTM Neural Network Forecast. *Stat. Decis.* **2021**, *37*, 46–50.
18. Liu, Z.; Ding, Y.; Yan, J. Frequency prediction of SVR-ARMA combined model based on particle swarm optimization. *Vib. Test. Diagn.* **2020**, *40*, 374–380.
19. Wang, H. Multi-objective optimization design of ARMA control chart considering both efficiency and cost. *Oper. Res. Manag.* **2021**, *30*, 80–86.
20. Chen, H.Y.; Miao, F.; Chen, Y.J.; Xiong, Y.J.; Chen, T.A. Hyperspectral image classification method using multifeature vectors and optimized KELM. *IEEE J. Sel. Top. Appl. Earth Obs. Remote Sens.* **2021**, *14*, 2781–2795. [CrossRef]
21. Gao, C.; Kong, X.; Shen, X. Evaluation of Frost Resistance of Stress-damaged Lightweight Aggregate Concrete Based on GM (1,1). *Eng. Sci. Technol.* **2021**, *53*, 184–190.
22. Wang, H.; Jing, W.; Zhao, G. Fatigue life prediction of hydraulic support base based on gray system model GM (1, 1) improved Miner criterion. *J. Shanghai Jiao Tong Univ.* **2020**, *54*, 106–110.
23. Kang, C.; Gong, L.; Wang, Z. Predicting the deterioration of hydraulic concrete by using grey residual GM (1,1)-Markov model. *J. Water Resour. Water Transp. Eng.* **2021**, *1*, 95–103.
24. Zhou, X.B.; Ma, H.J.; Gu, J.G.; Chen, H.L.; Deng, W. Parameter adaptation-based ant colony optimization with dynamic hybrid mechanism. *Eng. Appl. Intell.* **2022**, *114*, 105139. [CrossRef]
25. Qian, K.; Hou, Z.; Sun, D. Sound Quality Estimation of Electric Vehicles Based on GA-BP Artificial Neural Networks. *Appl. Sci.* **2020**, *10*, 5567. [CrossRef]
26. Wu, D.; Hong, N.; Yi, L.; Hui, C.; Hui, Z. An adaptive different evolution algorithm based on belief space and generalized opposition=based learning for resource allocation. *Appl. Soft Comput.* **2022**, *127*, 1568–4946.
27. Chen, Y.; Hu, Y.; Zhang, S.; Mei, X.; Shi, Q. Optimized Erosion Prediction with MAGA Algorithm Based on BP Neural Network for Submerged Low-Pressure Water Jet. *Appl. Sci.* **2020**, *10*, 2926. [CrossRef]
28. Zhang, G.; Zheng, Y.; Liao, K. Research on Ink Color Matching Based on GABP Algorithm. *J. Xi'an Univ. Technol.* **2019**, *35*, 113–119.
29. Yao, R.; Guo, C.; Deng, W.; Zhao, H.M. A novel mathematical morphology spectrum entropy based on scale-adaptive techniques. *ISA Trans.* **2022**, *126*, 691–702. [CrossRef]
30. Yu, Q.; Zhang, Z.; Qu, Y. Prediction of Lost Load of Power Grid Blackout Based on ARMA-GABP Combined Model. *China Power* **2018**, *51*, 38–44.

electronics

MDPI

Article

CEEMD-MultiRocket: Integrating CEEMD with Improved MultiRocket for Time Series Classification

Panjie Wang, Jiang Wu, Yuan Wei and Taiyong Li *

School of Computing and Artificial Intelligence, Southwestern University of Finance and Economics, Chengdu 611130, China
* Correspondence: litaiyong@gmail.com

Abstract: Time series classification (TSC) is always a very important research topic in many real-world application domains. MultiRocket has been shown to be an efficient approach for TSC, by adding multiple pooling operators and a first-order difference transformation. To classify time series with higher accuracy, this study proposes a hybrid ensemble learning algorithm combining Complementary Ensemble Empirical Mode Decomposition (CEEMD) with improved MultiRocket, namely CEEMD-MultiRocket. Firstly, we utilize the decomposition method CEEMD to decompose raw time series into three sub-series: two Intrinsic Mode Functions (IMFs) and one residue. Then, the selection of these decomposed sub-series is executed on the known training set by comparing the classification accuracy of each IMF with that of raw time series using a given threshold. Finally, we optimize convolution kernels and pooling operators, and apply our improved MultiRocket to the raw time series, the selected decomposed sub-series and the first-order difference of the raw time series to generate the final classification results. Experiments were conducted on 109 datasets from the UCR time series repository to assess the classification performance of our CEEMD-MultiRocket. The extensive experimental results demonstrate that our CEEMD-MultiRocket has the second-best average rank on classification accuracy against a spread of the state-of-the-art (SOTA) TSC models. Specifically, CEEMD-MultiRocket is significantly more accurate than MultiRocket even though it requires a relatively long time, and is competitive with the currently most accurate model, HIVE-COTE 2.0, only with 1.4% of the computing load of the latter.

Keywords: time series classification; complementary ensemble empirical mode decomposition (CEEMD); MultiRocket; feature selection; hybrid model

Citation: Wang, P.; Wu, J.; Wei, Y.; Li, T. CEEMD-MultiRocket: Integrating CEEMD with Improved MultiRocket for Time Series Classification CEEMD-MultiRocket: Integrating CEEMD with Improved MultiRocket for Time Series Classification. *Electronics* **2023**, *12*, 1188. https://doi.org/10.3390/electronics12051188

Academic Editor: Daniel Gutiérrez Reina

Received: 30 January 2023
Revised: 23 February 2023
Accepted: 28 February 2023
Published: 1 March 2023

1. Introduction

A time series is a set of data arranged in chronological order, which is widely applied in different domains in real life. With the fast advancement of information acquisition equipments and improvement of acquisition methods, time series have gotten more sophisticated, and their application involves a wide variety of fields, such as traffic [1], energy [2,3], finance [4], medical diagnosis [5–7] and social media [8]. By classifying time series into groups based on their underlying stochastic process, we can gain insights into the underlying phenomenon being measured and potentially make predictions. This involves identifying features in the time series data that are indicative of the underlying process, such as the autocorrelation structure, the distribution of values, or the frequency spectrum. Therefore, time series classification (TSC), as a task of characterizing a series of values observed at a continuous time as belonging to one of two or more categories, has always been the focus of research [9].

Several TSC algorithms have been presented over the years. These algorithms are generally separated into traditional approaches and deep learning approaches. The main groups of traditional TSC algorithms are introduced as follows: (1) Distance-based classifiers use distance metrics to determine class membership, and their representatives include a combination of K-Nearest Neighbors (KNN) and Dynamic Time Warping (DTW) [10] and Proximity Forest [11]. (2) Frequency-based classifiers are based on frequency data extracted from time series, and their representative is Random Interval Spectral Ensemble (RISE) [12], which is viewed as a popular Time Series Forest (TSF) [13] variation. (3) Interval-based classifiers rely their classification on information contained in distinct series intervals, and their representatives include TSF and Diverse representation Canonical Internal Forest (DrCIF) [14]. DrCIF builds on RISE and TSF, and uses the catch22 [15] to expand the original features. (4) Dictionary-based classifiers first convert discrete "words" from real-valued time series. The distribution of the retrieved symbolic terms is used as the basis for classification. Their representatives include Bag of Symbolic-Fourier-Approximation Symbols (BOSS) [16] and Temporal Dictionary Ensemble (TDE) [17]. (5) Shapelets are short subsequences of time series that are typical of their class. It is possible to utilize them to discover the similarity between two time series belonging to the same class [18]. Their representatives include Shapelet Transformation (ST) [19] and Shapelet Transform Classifier (STC) [20].

An ensemble classifier is a meta ensemble based on the previously described classifiers, and the typical representatives include HIVE-COTE [21], HIVE-COTE 2.0 [22], InceptionTime [23] and Time Series Combination of Heterogeneous and Integrated Embedding Forest (TS-CHIEF) [24]. HIVE-COTE 2.0 is a meta ensemble consisting of four components: STC, TDE, DrCIF and Arsenal [22]. InceptionTime is a collection of five TSC deep learning models generated by cascading numerous inception modules [23]. Each model has the same design but distinct random initialization weight values. TS-CHIEF builds on an ensemble tree-structured classifier that incorporates the most efficient time series embeddings created in the previous ten years of study [24].

On the other hand, deep learning methods for TSC are generally classified into two types: generative models and discriminative models [25].

The most common generative models include Stacked Denoising Auto-Encoders (SDAE) [26,27] and Echo State Networks (ESN) [28]. To model the time series, SDAE is preceded by an unsupervised pre-training stage [26,27]. As Recurrent Neural Networks (RNN) frequently experience the vanishing gradient problem as a result of training on lengthy time series [29], ESNs were created to ameliorate the difficulties of RNNs [30]. Discriminative models are classifiers that can quickly figure out how to transfer a time series' original input to a dataset's output, which is a probability distribution over the class variables. These models may be further classified into two types: (1) deep learning models using hand-engineered features and (2) end-to-end deep learning models [30]. The translation of series into images utilizing specialized imaging approaches is the most common feature extraction algorithm for hand-engineered approaches, such as recurrence plots [31,32] and Gramian fields [33]. In contrast, end-to-end deep learning tries to include the feature learning procedure while optimizing the discriminative classifier [34]. Convolutional Neural Networks (CNN) are the most extensively used for the TSC issue due to their robustness and training efficiency [30].

Overall, the state-of-the-art (SOTA) TSC models in terms of classification accuracy mainly include HIVE-COTE and its variants, TS-CHIEF, InceptionTime, Rocket, MiniRocket, MultiRocket, etc. [35]. Among them, Rocket, MiniRocket, and MultiRocket are not only accurate, but also ensure scalability. Rocket employs lots of randomly initialized convolution kernels for feature extraction, and uses a linear classifier for classification, without training the kernels [36]. MiniRocket is about 75 times faster than Rocket, and it employs a limited number of kernels and just one pooling operation [37]. MultiRocket is built on MiniRocket and uses the same set of convolution kernels that are used in MiniRocket [35]. MultiRocket differs in two ways from MiniRocket. On one hand, MultiRocket uses the first-order difference of raw time series, along with the raw time series, as the inputs to the classification model. On the other hand, MultiRocket includes three extra pooling operators in addition to PPV to derive more discriminative features.

Although Rocket and its improved versions MiniRocket and MultiRocket have achieved satisfactory classification performance, there is certainly room for improvement in series transform, the design of convolution kernels and feature extraction. To solve the existing defects and enhance classification performance, this study proposes a novel hybrid ensemble leaning model incorporating Complementary Ensemble Empirical Mode Decomposition (CEEMD) and improved MultiRocket, namely CEEMD-MultiRocket, to enhance the classification performance of time series. Raw time series is firstly divided into three sub-series utilizing CEEMD [38–40]. The sub-series refer to the individual Intrinsic Mode Functions (IMFs) that make up the decomposition of the raw time series into its oscillatory components. Since the decomposition is performed using a sifting process that extracts the highest frequency component first and continues with lower frequency components until the residual is obtained, these three sub-series represent high-, medium- and low-frequency portions of the original time series, respectively. Since not every decomposed sub-series as the input has a positive contribution to the performance of the classification model, the selection of the more crucial sub-series and pruning the redundant and less important ones are necessary to enhance the final classification performance and reduce computational complexity. The selection of these decomposed sub-series is executed on the known training set by comparing the classification accuracy of each sub-series with that of the raw time series using a given threshold. Finally, we improve the original MultiRocket and apply it to the raw time series and the selected decomposed sub-series to derive features and generate the final classification results. In improved MultiRocket, the convolution kernels are modified, and one additional pooling operator is applied to convolution outputs. CEEMD-MultiRocket has been empirically tested with 109 datasets from the UCR time series repository. Compared with some SOTA classification models, the experiments demonstrate that our proposed CEEMD-MultiRocket achieves promising classification performance. Specifically, our proposed CEEMD-MultiRocket is more accurate than MultiRocket even though it takes a relatively long time, and is competitive with the HIVE-COTE 2.0 which ranks the best at present in terms of classification accuracy, only with a small fraction of the training time of the latter. One of the main theoretical and technical implications of CEEMD-MultiRocket is that it is the first time that CEEMD has been integrated with convolution kernel transform for the feature extraction of time series, making it outperform almost all of the previous SOTA methods. Furthermore, CEEMD-MultiRocket improves convolution kernel and pooling operator design and is demonstrated to be a fast, effective and scalable method for time series classification tasks, showing that the optimization of convolution kernels and pooling operator is a promising field worth studying for improving classification performance. The main contributions of this research lie in five aspects:

(1) A novel hybrid TSC model that integrates CEEMD and improved MultiRocket is proposed. Raw time series is decomposed into high-, medium- and low-frequency portions, and convolution kernel transform is utilized to derive features from the raw time series, the decomposed sub-series and the first-order difference of raw time series. This kind of transformation is able to obtain more detailed and discriminative information of time series from various aspects.

(2) A sub-series selection method is proposed based on the whole known training data. This method selects the more crucial sub-series and prunes the redundant and less important ones, which helps to further enhance classification performance and also reduce computational complexity.

(3) The length and number of convolution kernels are modified, and one additional pooling operator is applied to convolutional outputs in our improved MultiRocket. These improvements contribute to the enhancement of classification accuracy.

(4) Extensive experiments demonstrate that the proposed classification algorithm is more accurate than most SOTA algorithms for TSC.

(5) We further analyze some characteristics of the proposed CEEMD-MultiRocket for TSC, including the CEEMD parameter settings, the selection of decomposed sub-series, the design of convolution kernel and pooling operators.

The rest of this paper is organized as follows. Section 2 briefly introduces CEEMD and MultiRocket. Section 3 gives the description of the proposed CEEMD-MultiRocket algorithm in detail, including CEEMD and sub-series selection, improved MultiRocket and feature extraction. Section 4 reports experimental results and assesses the proposed algorithm in terms of accuracy and training time. Section 5 discusses the impact of the CEEMD parameters, the threshold setting for sub-series selection, the convolution kernel length and an additional pooling operator on the classification performance of CEEMD-MultiRocket, followed by conclusions in Section 6.

2. Related Works

2.1. Complementary Ensemble Empirical Mode Decomposition

CEEMD [38] is an extension built on Ensemble Empirical Mode Decomposition (EEMD) [41] and Empirical Mode Decomposition (EMD) [42]. EMD is a time-frequency analysis approach which is created for nonlinear and nonstationary signals or time series [43]. EMD applies local extreme points of the raw time series to form the envelope step by step, separates fluctuations or trends at diverse scales and generates a group of relatively stable components, including IMFs and one residue. Specifically, the EMD algorithm involves iteratively extracting local oscillations from the signal by means of a sifting process. The extracted oscillations are called IMFs, and they represent the underlying oscillatory modes that make up the signal. The remaining signal after extracting the IMFs is called the residue, which contains the trends and other non-oscillatory components. The main disadvantage of mode mixing in EMD is that the significantly diverse scales may appear in the same IMF component [44]. To reduce mode mixing, EEMD was proposed [41]. In EEMD, IMFs are defined as a combination of time series and white noise with a limited amplitude, which can significantly reduce mode mixing. Despite the fact that EEMD has effectively handled the mode mixing problem, the residual noise in signal reconstruction has increased. Therefore, CEEMD was proposed, where a specific type of white noise was introduced at each stage of the decomposition [45]. It not only suppresses the mode mixing but also reduces the reconstruction signal errors caused by residue noise. The CEEMD is described as follows:

(1) Add two equal-amplitude, opposite-phase white noises to the signal $x(t)$, to obtain the following sequences.

$$\begin{cases} P_i(t) = x(t) + n_i(t) \\ N_i(t) = x(t) - n_i(t) \end{cases} \tag{1}$$

where $n_i(t)$ is the white noise superimposed in the ith stage, $P_i(t)$ and $N_i(t)$ denote the sequence after adding noise in the ith stage .

(2) CEEMD firstly breaks down the sequence's noise to generate the components IMF, C_{1_j} and the trend surplus r_1.

(3) In the same way, process the white noise with opposing symbols in step (1) to generate the components C_{-1_j} and r_{-1}.

(4) Repeat steps (1)~(3) n times to obtain n sets.

(5) The ultimate result is chosen as the average of the components of two sets of residual positive and negative white noise acquired by repeated decomposition, i.e.,

$$\begin{cases} C_i(t) = \frac{1}{2n}\sum_{j=1}^{n}(C_{n_j} + C_{-n_j}) \\ r_n(t) = \frac{1}{2n}\sum_{j=1}^{n}(r_j + r_{-j}) \end{cases} \tag{2}$$

2.2. MultiRocket

Rocket employs lots of randomly initialized convolution kernels for transform, applies pooling operators to convolutional outputs and uses a linear classifier, without training the kernels [36]. For Rocket, a time series is convolved using 10 k random convolution kernels, whose weights are sampled from $N(-1,1)$; length is selected from $\{7,9,11\}$ with equal probability; padding is alternating; dilation is exponentially scaled; and bias is sampled from $U(-1,1)$. Additionally, the Proportion of Positive Values (PPV) and global max pooling (Max) pooling operators are applied to each convolutional output to generate two features, and to generate 20 K features in total for each input series. Finally, for a larger dataset, the derived features are employed to train a logistic regression classifier, while for a relatively small dataset, a ridge regression classifier is trained. Rocket has been proved to be a efficient, fast and novel algorithm for the feature extraction of time series [36].

MiniRocket is built on Rocket and becomes further deterministic by pre-defining a set of convolution kernels with fixed lengths and weights. MiniRocket retains the dilation and PPV, while it discards the max pooling which is of no benefit for enhancing the classification accuracy [37]. It performs a convolution operation on the input series using a fixed group of 84 kernels with each kernel generating multi-dilation (74 by default) and using different bias which are obtained by sampling on the convolutional output from a randomly selected instance in the training set. Since only PPV is used in MiniRocket, the number of features (84 × 119 = 9996 by default) generated by MiniRocket is only about half of the number of features generated by Rocket.

The kernels used in MultiRocket are the same as MiniRocket. Unlike MiniRocket, MultiRocket injects the diversity of features by adding the first-order difference of raw time series and three additional pool operators to enhance the performance of MiniRocket. Inspired by DrCIF, MultiRocket uses the first-order difference of raw time series as the input to offer more diverse information related to the transformation of raw time series. MultiRocket has 84 fixed convolution kernels and each convolutional kernel will produce 74 kinds of dilation. Firstly, MultiRocket performs a convolution operation on the input series and the first-order difference of the input series using the kernels with dilations to obtain the convolutional outputs. Next, four features (PPV and an additional three pooling operators) are calculated for each convolutional output and then about 50 k (more accurately, 84 × 74 × 2 × 4 = 49,728) features are generated. Finally, a linear regression classifier is trained on the features. MultiRocket is faster than all TSC algorithms (except for MiniRocket) and more accurate than all TSC algorithms (except for HIVE-COTE 2.0) [35].

In summary, Rocket, MiniRocket and MultiRocket are representations of the scalable and most accurate algorithms on the UCR time series repository. As a series of algorithms, their differences can be seen in Table 1.

Table 1. Summary of changes from Rocket to MiniRocket and then to MultiRocket.

	Rocket	MiniRocket	MultiRocket
kernel length	$7,9,11$	9	9
weights	$N(0,1)$	$-1,2$	$-1,2$
bias	$U(-1,1)$	from convolutional output	from convolutional output
dilation	random	fixed (range $\lfloor 2^0 \rfloor, \cdots, \lfloor 2^{max} \rfloor$)	fixed (range $\lfloor 2^0 \rfloor, \cdots, \lfloor 2^{max} \rfloor$)
padding	random	fixed	fixed
pooling operators	PPV, MAX	PPV	PPV, MPV, MIPV, LSPV
num. features	20 K	10 K	50 K

3. The Proposed CEEMD-MultiRocket

This research proposes a hybrid ensemble model that combines CEEMD and improved MultiRocket, termed CEEMD-MultiRocket, for TSC. The proposed model includes three steps which are decomposition, sub-series selection and feature extraction and classification, as demonstrated in Figure 1.

Step 1: Decomposition. Each time series in a dataset is decomposed into three sub-series using CEEMD: IMF_i ($i = 1, 2$) and one residual.

Step 2: Sub-series selection. In order to enhance the final classification accuracy and decrease computational load, the selection of these decomposed sub-series is executed on the whole known training dataset by comparing the classification accuracy of each decomposed sub-series with that of the raw time series using a pre-set threshold.

Step 3: Feature extraction and classification.The convolution kernel transform is applied to the raw time series, the selected sub-series and the first-order difference of raw series. Then, five pooling operators are designed to extract features from the convolutional output. Finally, a ridge regression classifier is trained using these extracted features. In our improved MultiRocket, the length and number of convolution kernels are modified, and one additional pooling operator is applied to the convolutional output.

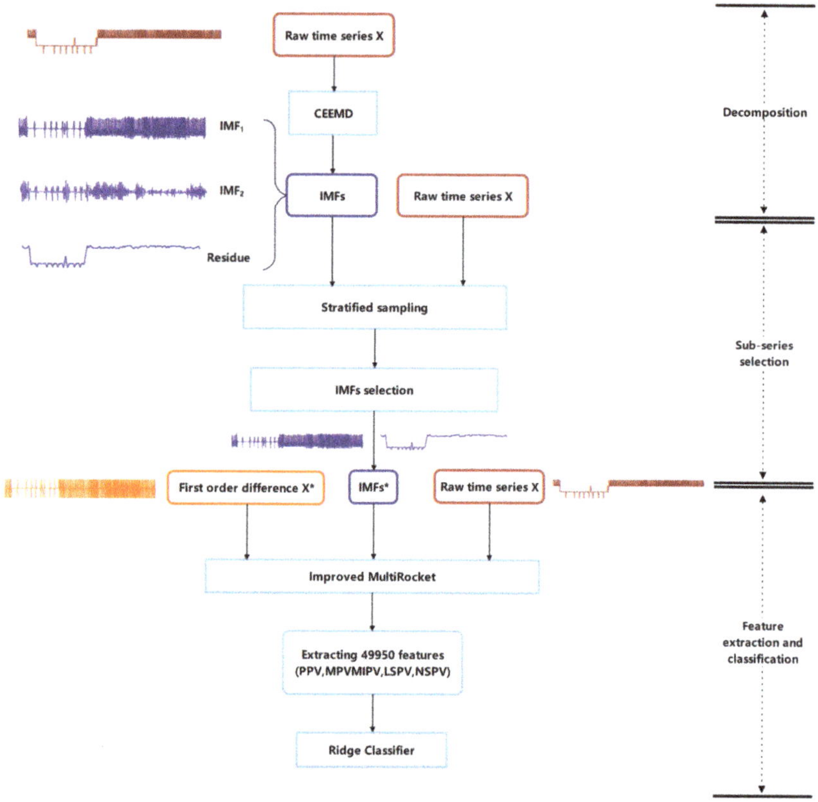

Figure 1. The flowchart of the proposed CEEMD-MultiRocket.

Firstly, the proposed CEEMD-MultiRocket applies CEEMD to decompose raw time series into three sub-series (two IMFs and one residue), each of which contains information about the different frequency of raw time series. In general, the first and second IMF represent the high- and medium-frequency portions and the residue represents the low-frequency portion of raw time series. Secondly, in order to enhance the final classification performance and decrease computational complexity, it is necessary to select the most crucial sub-series and discard less important ones. The selection of these sub-series is executed on the whole known training set which is further subdivided into training and testing sets using stratified sampling. Improved MultiRocket is used for the raw time series and each sub-series on the newly generated training and testing sets, and the appropriate sub-series is selected when its testing accuracy is higher than a given threshold, which is set to a percentage of the testing accuracy of raw time series. Finally, convolution operation is performed on the raw time series, the selected sub-series and the first-order difference of raw time series, respectively. It should be specially noted that the transform is only applied to the raw time series and its first-order difference when there is a dataset without any selected sub-series. Feature extraction is conducted on each convolutional output, and these extracted features are eventually applied to train a ridge regression classifier. In improved MultiRocket, the length and number of convolution kernels are modified, and five pooling operators are used in each convolutional output to derive features. The combination of these modifications has the potential to enhance the classification performance of MultiRocket. Overall, this hybrid ensemble learning paradigm, CEEMD-MultiRocket, can diversify the input series and comprehensively extract more extensive features from the raw series and the decomposed sub-series for classification, which makes it possible to enhance classification performance.

3.1. CEEMD and Sub-Series Selection

The CEEMD algorithm is usually applied in the field of signal processing, which decomposes raw time series into several components to obtain better classification performance [46]. The proposed CEEMD-MultiRocket firstly uses the CEEMD decomposition algorithm to decompose raw time series into two IMFs and a residue. Figure 2 illustrates a decomposition of a time series from the electricity consumption dataset ScreenType from the UCR repository [47] using CEEMD. The length of each series in the ScreenType dataset is 720 (24 h of readings taken every 2 min). The x-axis represents the time (every 2 min) and the y-axis represents the electricity consumption in Figure 2.

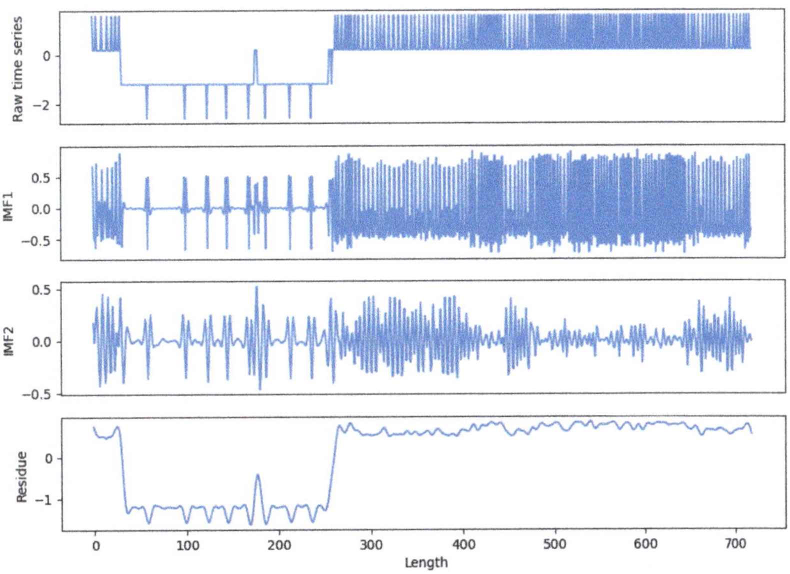

Figure 2. A raw time series and its corresponding sub-series decomposed by CEEMD in the ScreenType dataset.

It is a challenging issue to select the appropriate sub-series for extracting discriminative characteristics of the raw time series for time series analysis correctly [48]. To select the appropriate sub-series (two IMFs or one residue) as the inputs to our classification model, we propose a novel approach to select the appropriate sub-series generated by CEEMD using the known training data. The main idea is to subdivide the original training dataset into two parts including a new training dataset and a new testing dataset, then train a classification model using improved MultiRocket and obtain the testing accuracy, and finally select the sub-series with satisfactory testing accuracies. This kind of selection approach is based on the inference that the sub-series with relatively high testing accuracy may contain more potentially useful characteristics as the input to improve MultiRocket. The decomposition and sub-series selection are described as follows:

(1) Utilize CEEMD to decompose raw time series into two IMFs and a residue. For convenience, we refer to three of them as IMFs.
(2) Perform stratified sampling to subdivide the original training set and its corresponding three IMFs into new training sets and testing sets, respectively, and ensure that the new training and testing sets contain all labels (the split ratio is 1:1).
(3) Apply improved MultiRocket to the newly generated training set of original training time series, and then obtain the testing classification accuracy $acc_{original}$ on the newly generated testing set. Perform the corresponding operations for each IMF and obtain the testing classification accuracy $acc_i, i = 1, 2, 3$.
(4) Select the IMFs whose testing accuracies acc_i are more than $acc_{original} \times threshold$ as the inputs of improved MultiRocket. We refer to the selected IMFs as $IMFs^*$ throughout the paper, which may contain 0–3 sub-series generated by CEEMD. The threshold is set to 0.9 by default.

By comparing the testing accuracy of each sub-series with that of raw time series, we can select the most crucial sub-series and discard the redundant and less important ones as the inputs to classification model, thereby enhancing classification performance and reducing computational cost.

3.2. Improved MultiRocket

This section provides a comprehensive explanation of improved MultiRocket which retains the basic architecture as the original MultiRocket [35]. The main difference between the improved MultiRocket and the original MultiRocket lies in two aspects. The first is the modification of convolution kernel length, and the second involves an addition pooling operator for feature extraction. We expect that these modifications are able to significantly enhance classification ability. The comparison of original MultiRocket and improved MultiRocket is listed in Table 2.

Table 2. Comparison of MultiRocket and improved MultiRocket.

	MultiRocket	Improved MultiRocket
kernel length	9	6
num. kernels	84	15
weights	$-1, 2$	$-1, 2$
bias	from convolutional output	from convolutional output
dilation	fixed (range $\lfloor 2^0 \rfloor, \cdots, \lfloor 2^{max} \rfloor$)	fixed (range $\lfloor 2^0 \rfloor, \cdots, \lfloor 2^{max} \rfloor$)
padding	fixed	fixed
pooling operators	PPV, MPV, MIPV, LSPV	PPV, MPV, MIPV, LSPV, NSPV
num. features	50 K	50 K

3.2.1. Convolution Kernels

Improved MultiRocket employs 15 fixed convolution kernels with length 6 and has fixed weights. Except for the length of convolution kernel, the dilation, the padding and bias of the improved MultiRocket are the same as those of MultiRocket. The detailed kernel design, dilation, bias and padding are described as follows.

- Kernel length and weight setting: To simplify the computation complexity as much as possible, the number of convolution kernels ought to be as small as possible [37]. Therefore, our proposed CEEMD-MultiRocket tries to employ 15 convolution kernels with length 6 instead of the 84 kernels with length 9 in the original MultiRocket. The convolution kernel weights are restricted to two values, α and β, and there are 2^6 = 64 possible dual-valued kernels with a length of 6. Improved MultiRocket employs the subset of convolution kernels that have two values of β, and this provides a total of $C_6^2 = 15$ fixed kernels, which strikes a good balance between computing efficiency and classification accuracy. In the improved MultiRocket, we set the weight $\alpha = -1$ and $\beta = 2$. As long as α and β increase by multiples, equivalently, that is $\beta = -2\alpha$, it has no effect on the results, because bias and features are extracted from the output of convolution [37]. Since the original MultiRocket uses 84 kernels with length 9, the number of kernels used in our improved MultiRocket is less than a fifth of the number of kernels in the original MultiRocket, effectively decreasing computing load.

- Dilation: The dilations used by each kernel are the same and fixed. The total number of dilations of each kernel n depends on the number of features, where $n = f/3/15$ and f represent the total number of features (50 k by default). Dilations are specified in range $\{\lfloor 2^0 \rfloor, \ldots, \lfloor 2^{\max} \rfloor\}$, where the exponent obeys a uniform distribution between 0 and max = min $(\frac{l_{input}-1}{l_{kernel}-1}, 64)$, where l_{kernel} is the length of the kernel and l_{input} is the length of input time series.
- Bias: Bias values are determined by the convolutional outputs for each kernel/dilation combination. For a kernel/dilation combination, we randomly select a training example and calculate its convolutional output, then sample from the convolutional output based on many quantiles to obtain bias, in which the quantiles are drawn from a low-discrepancy sequence.
- Padding: Each combination of kernel and dilation alternates between using and not using padding, with half of the combinations using padding.

We refer to the feature vector extracted by the convolution operation as Z and the length of the input time series as l. According to [36], the result of applying a kernel to a time series, X, from index i in X can be obtained using Equation (3):

$$Z = X_i \times w = b + (\sum_{j=0}^{l_{kernel}-1} X_{i+(j\times d)} \times w_j) \tag{3}$$

where ω is weights, d is dilation and b is bias of the kernel.

3.2.2. Pooling Operators

MultiRocket injects diversity through two main aspects: the first-order difference of raw time series and an additional three pooling operators. In order to enhance the diversity of derived features, we propose an additional pooling operator, Number of Stretch of Positive Values (NSPV), to extract more comprehensive features from the convolutional output. Thus, we employ five pooling operators together to derive features, including the four existing pooling operators used in the original MultiRocket [35]. Table 3 summarizes the pooling operators in the improved MultiRocket, including Proportion of Positive Values (PPV), Mean of Positive Values (MPV), Mean of Indices of Positive Values (MIPV), Longest Stretch of Positive Values (LSPV) and NSPV.

Table 3. The summary of pooling operators in improved MultiRocket uses a virtual example to illustrate that the four pooling operators in original MultiRocket cannot distinguish different scenarios with different convolutional outputs. Each convolutional output contains 6 zeros and 6 ones, MPV = 1, PPV = 0.5, MIPV = 5.5, LSPV = 2.

Convolutional Outputs	PPV	MPV	MIPV	LSPV	NSPV
A = [1, 1, 0, 0, 0, 1, 1, 0, 0, 0, 1, 1]	0.5	1	5.5	2	3
B = [1, 0, 1, 0, 0, 1, 1, 0, 0, 1, 0, 1]	0.5	1	5.5	2	1
C = [1, 1, 0, 0, 1, 0, 0, 1, 0, 0, 1, 1]	0.5	1	5.5	2	2

PPV was first used in Rocket. It uses Equation (4) to compute the proportion of positive value of Z.

$$\mathrm{PPV}(Z) = \frac{1}{l}\Sigma_{i=1}^{l}[z_i > 0] \tag{4}$$

MPV, MIPV and LSPV were used in MultiRocket. The MPV value is calculated using Equation (5), where m is the number of positive values in Z.

$$\text{MPV}(Z) = \frac{\sum_{l}^{i=1} z_i[z_i > 0]}{m} \tag{5}$$

MIPV is calculated by Equation (6), where i^+ is the index of positive value and m is the number of positive values in Z.

$$\text{MIPV}(Z) = \begin{cases} -1 & \text{otherwise} \\ \frac{1}{m}\sum_{j=1}^{m} i_j^+ & \text{if } m > 0 \end{cases} \tag{6}$$

LSPV is calculated using Equation (7) and represents the maximum length of any subsequence of positive values in Z.

$$\text{LSPV}(Z) = \max\left[j - i \mid \forall_{i \le k \le j}\, z_k > 0\right] \tag{7}$$

We propose NSPV to calculate the number of continuous subsequences with positive values in Z, as defined in Equation (8). It can offer a distinctive kind of information compared with the other four pooling operators provided in the original MultiRocket. As shown in Table 3, NSPV is the key to distinguishing between three time series, A to C.

$$\text{NSPV}(Z) = \sum_{k=1}^{n}\left[j - i > 1 \mid \forall_{i \le k \le j}\, z_k > 0\right] \tag{8}$$

3.3. Feature Extraction

Original MultiRocket produces 50 k features by default. For a fair comparison, the improved MultiRocket extracts five aggregate features from each convolutional output and also generates about 50 k (more accurately, 15 × 222 × 3 × 5 = 49,950) features for each time series by default. Specifically, it has 15 fixed convolution kernels and each convolution kernel produces 222 kinds of dilation, making the length of the convolution kernel from 6 to the input time series' length. The input data consists of three parts: (1) the raw time series; (2) the selected $IMFs^*$; and (3) the first-order difference of raw time series. Firstly, the three parts as the inputs are convolved by each combination of kernel and dilation in turn to obtain the convolutional output. Next, a total of 49,950 features are eventually derived by calculating five features for each convolutional output using five pooling operators. Finally, a ridge regression classifier is trained on the extracted features.

4. Experimental Results

4.1. Datasets

To better assess the performance of the proposed CEEMD-MultiRocket, the experiments were conducted on 109 univariate time series classification datasets from the UCR time series repository [47], which includes datasets from many different fields and has been used to evaluate various TSC models.

4.2. Experimental Settings

From the perspective of classification accuracy and runtime analysis, the proposed CEEMD-MultiRocket was contrasted with some SOTA algorithms for classifying time series, including MultiRocket, HIVE-COTE 2.0, InceptionTime, MiniRocket, Arsenal, STC, TS-CHIEF, DrCIF, TDE and ProximityForest. In order to directly compare the classification accuracy with other most accurate algorithms as mentioned above, we assessed the proposed CEEMD-MultiRocket on 30 resamples of 109 datasets from the UCR univariate time series repository used in [22,35] and adopted exactly the same resampling method as used in MultiRocket [35]. Thus, on the basis of the same data sets and stratified split, we examined the effectiveness of our proposed CEEMD-MultiRocket in enhancing classification accuracy, and compared the runtime of CEEMD-MultiRocket with that of the above time series classification algorithms.

In addition, the noise standard deviation was set to 0.4 and the number of realizations was set to 30 in CEEMD. The threshold of sub-series selection was set to 0.9 of the testing accuracy of raw time series. We performed CEEMD using MATLAB R2016a and improved MultiRocket using pycharm IDE on a cluster with an Intel Xeon Gold 5218 CPU @2.30 GHz using a single thread.

4.3. Results and Analysis

4.3.1. Classification Results

We compared CEEMD-MultiRocket with the 10 SOTA TSC algorithms mentioned above to examine the effectiveness of our proposed method. Figure 3 illustrates the mean rank of CEEMD-MultiRocket in comparison to the 10 TSC algorithms. According to a two-sided Wilcoxon signed-rank test with Holm correction (as a post hoc test to the Friedman test), algorithms connected with a black line have no pairwise statistical difference in their accuracy [49]. The Wilcoxon signed-rank test is a nonparametric statistical hypothesis test used to determine if two related samples have the same distribution, which is often used to compare the significance of the differences between two related samples. From Figure 3, we can see that CEEMD-MultiRocket is significantly more accurate than MultiRocket and other TSC algorithms, and only marginally less accurate than HIVE-COTE 2.0. Note that the accuracy difference between CEEMD-MultiRocket and HIVE-COTE 2.0 is statistically insignificant, showing that CEEMD-MultiRocket achieves almost the same level of classification performance as the latter.

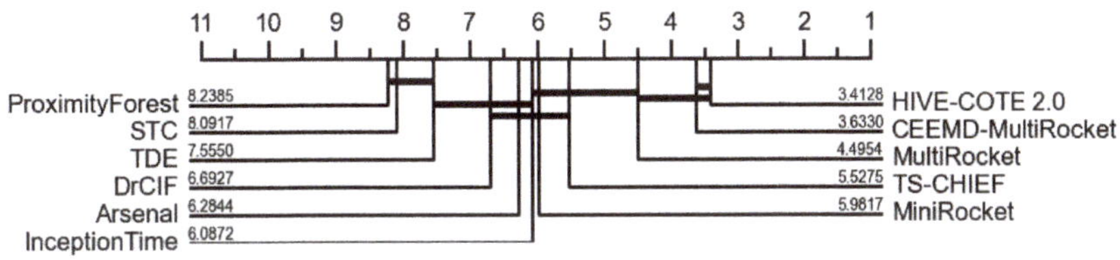

Figure 3. Mean rank of CEEMD-MultiRocket in terms of accuracy over 30 resamples of 109 datasets from the UCR time series repository, against 10 other SOTA algorithms.

Figure 4 shows the pairwise difference of the CEEMD-MultiRocket and 10 other SOTA TSC algorithms in terms of statistical significance. The first row of each cell in the matrix indicates the wins, draws and losses of the algorithm in the Y-axis versus the algorithm in the X-axis, and the second row shows the p-value of the Holm-corrected two-sided Wilcoxon signed-rank test between the pairwise algorithms. The bold numbers in the cells represent that significant differences do not exist in the classification accuracy of the pairwise algorithms after applying the Holm correction. As shown in Figure 4, our proposed CEEMD-MultiRocket is significantly more accurate than all SOTA classification algorithms except for HIVE-COTE 2.0, where the p-values for most of algorithms are close to 0. CEEMD-MultiRocket outperforms MultiRocket with 70 wins and only 31 losses out of 109 datasets. Compared with HIVE-COTE 2.0, CEEMD-MultiRocket achieves higher accuracy on 51 datasets, lower on 50 datasets and is the only algorithm with a p-value close to 1 after applying Holm correction.

As we can see, there are many different SOTA algorithms that can be used for time series classification, and the suitability of a particular algorithm depends on various factors such as the length of the time series, the sampling frequency, the number of classes and the complexity of underlying patterns. In general, if the time series are very short, with only a few data points, then simpler algorithms, such as nearest neighbor or decision trees, may be more appropriate. These algorithms can be effective for small datasets and can quickly classify time series based on their similarity to other time series in the training set.

On the other hand, if the time series are very long and have a high sampling frequency, then more complex algorithms, such as RNN or CNNs, may be more suitable. Through the experiments on 109 datasets, we find that our CEEMD-MultiRocket algorithm performs well in classification and outperforms the vast majority of existing classification algorithms. Among these 109 datasets, the shortest time series length is 15 (SmoothSubspace), and the longest is 2844 (Rock), indicating that our algorithm is effective for both short and long time series.

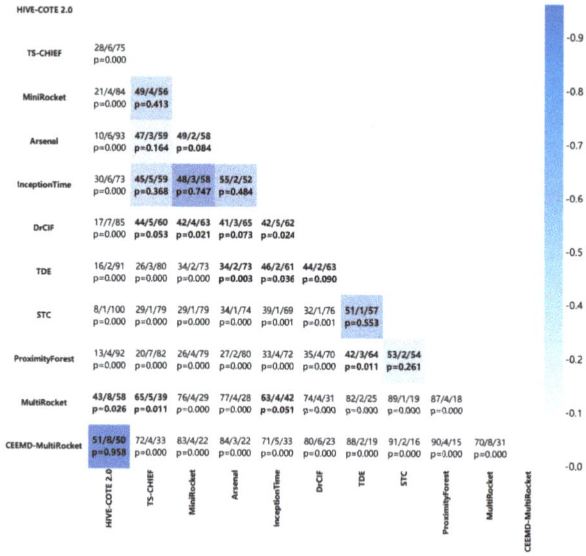

Figure 4. Pairwise difference between CEEMD-MultiRocket and 10 other SOTA algorithms in terms of statistical significance.

Figure 5 illustrates the pairwise accuracy of CEEMD-MultiRocket versus MultiRocket over 30 resamples of the 109 datasets from the UCR time series repository. Overall, we can find that most points are scattered above the dotted line, showing that CEEMD-MultiRocket achieves significantly better classification accuracy than MultiRocket.

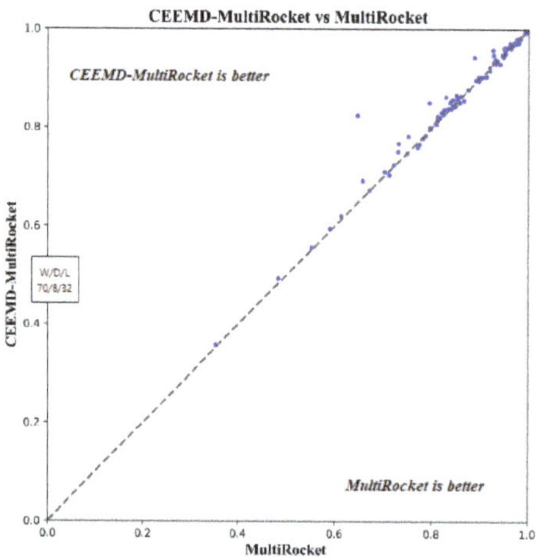

Figure 5. Pairwise accuracy of CEEMD-MultiRocket versus MultiRocket over 30 resamples of 109 datasets from the UCR time series repository.

4.3.2. Runtime Analysis

Although CEEMD-MultiRocket significantly outperforms MultiRocket in terms of accuracy, the additional decomposition operation, the sub-series selection and one extra pooling operator increase computational complexity. Fortunately, the reduction in the number of convolution kernels decreases the runtime of our proposed CEEMD-MultiRocket algorithm. We evaluated the overall training time of 10 other SOTA algorithms to train a single resample on 112 UCR datasets and compared the training time of these algorithms with our proposed CEEMD-MultiRocket, as shown in Table 4. From Table 4, we can find that CEEMD-MultiRocket is obviously faster than most TSC algorithms. All the SOTA algorithms, except the Rocket family, take lots of time to train, as reported by [22]. As for the Rocket family, MiniRocket, unsurprisingly, is the fastest, with a training time of under 4 min. After that is MultiRocket, which can be trained in under 24 min. Next is Rocket, with a training time of over 4 h, followed by our CEEMD-MultiRocket, taking under 5 h for training. Arsenal is an ensemble of Rocket, with a training time of about 28 h. The training time of six non-Rocket algorithms running on a high-performance computing (HPC) cluster with a single thread was reported in [22], which used higher-level hardware than ours. DrCIF takes about 2 days and the ensemble algorithms, including STC, HIVE-COTE 1.0/2.0 and TS-CHIEF, take at least 4 days. By comparison, the Rocket family algorithms use lots of randomly initialized convolution kernels for feature extraction, and only use one linear classifier for classification, without training the kernels, while non-Rocket algorithms, such as HIVE-COTE 2.0, TS-CHIEF, etc., integrate many classifiers with a large number of parameters and therefore require a lot of time for training. As a result, we can find that these non-Rocket algorithms are clearly more time-consuming than our algorithm. Specifically, the training time of CEEMD-MultiRocket consists of three parts: CEEMD (4.11 h), the IMFs selection (26.3 min) and model training using the improved MultiRocket (20.3 min). Due to the decomposition cost of raw time series, the proposed CEEMD-MultiRocket is slower than original MultiRocket, but it significantly outperforms MultiRocket in classification accuracy. Compared with HIVE-COTE 2.0, our proposed CEEMD-MultiRocket achieves almost the same level of classification accuracy but only costs 1.4% of the training time, showing that our proposed CEEMD-MultiRocket is an effective algorithm in TSC.

Table 4. Runtime to train single resample of 112 UCR datasets. The runtime of Rocket family and CEEMD-MultiRocket algorithm is calculated by running with a single thread on Intel Xeon Gold 5218 CPU. The runtime of the others is cited from [22].

TSC Algorithm	Total Training Time
MiniRocket (10 k features)	3.61 min
MultiRocket (50 k features)	23.65 min
Rocket (20 k features)	4.25 h
CEEMD-MultiRocket (50 k features)	4.88 h
Arsenal	27.91 h
DrCIF	45.40 h
TDE	75.41 h
STC	115.88 h
HIVE-COTE 2.0	340.21 h
HIVE-COTE 1.0	427.18 h
TS-CHIEF	1016.87 h

5. Discussion

For a more comprehensive evaluation of CEEMD-MultiRocket, we continue to discuss several characteristics of the proposed algorithm on 109 datasets from the UCR time series repository in detail, including the parameter setting of CEEMD, the sub-series selection, the convolution kernel design and pooling operators.

5.1. CEEMD Parameter Settings

During the procedure of decomposing raw time series, CEEMD adds a specific white noise to the time series. The addition of white noise is an important step in the CEEMD method that can eliminate the mode mixing problem and help to improve the accuracy and reliability of the decomposition results. The decomposition contains two main parameters: the number of realizations R and the noise signal intensity N. Different numbers of realizations and noise signal intensities may produce different IMFs. Experiments were conducted on 109 datasets from the UCR time series repository to assess the impact of these two parameters for classification performance. Figure 6 shows the mean rank of different noise intensities (N = 0.1, 0.2, 0.3, 0.4, 0.5, 0.6, 0.7) applied on CEEMD-MultiRocket with a fixed number of realizations R = 30. Figure 7 shows the mean rank of different numbers of realizations (R = 10, 20, 30, 40, 50, 60, 70) applied on CEEMD-MultiRocket with a fixed noise intensity N = 0.4.

As shown in Figure 6, CEEMD-MultiRocket obtains the best mean rank in classification accuracy on the 109 datasets when N = 0.4. This indicates that a too high or too low noise intensity leads to a reduction in classification accuracy, though there is not a statistically significant difference in classification accuracy. The findings of the experiment reveal that the intensity of noise has only a marginal impact on the classification performance and an optimum value for the intensity of added noise is somewhere around 0.4.

Figure 7 shows that CEEMD-MultiRocket achieves the best classification accuracy when the number of realizations R = 70, but the difference of classification performance between R = 70 and R = 30 is relatively small and statistically insignificant. As shown in Section 4.3.2, the decomposition of raw time series takes up most of the runtime (about 84%) of CEEMD-MultiRocket, it is necessary to reduce the decomposition time as much as possible. Since the number of realizations R = 30 takes less than half of the time of R = 70, we adopt 30 as the number of realizations in our CEEMD-MultiRocket to achieve a balance between the classification accuracy and time cost.

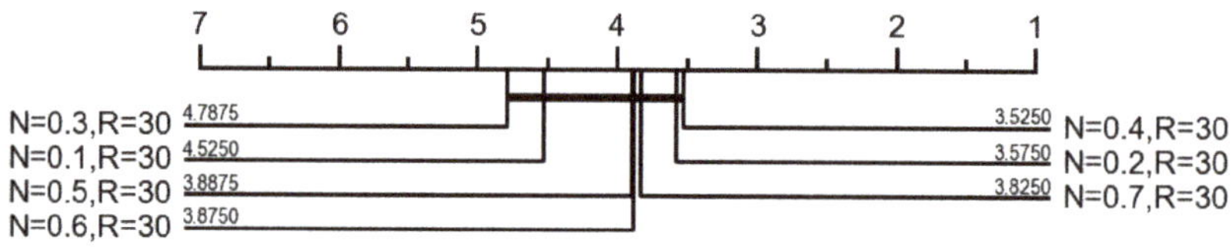

Figure 6. Mean rank of different noise intensities applied on CEEMD-MultiRocket with a fixed number of realizations.

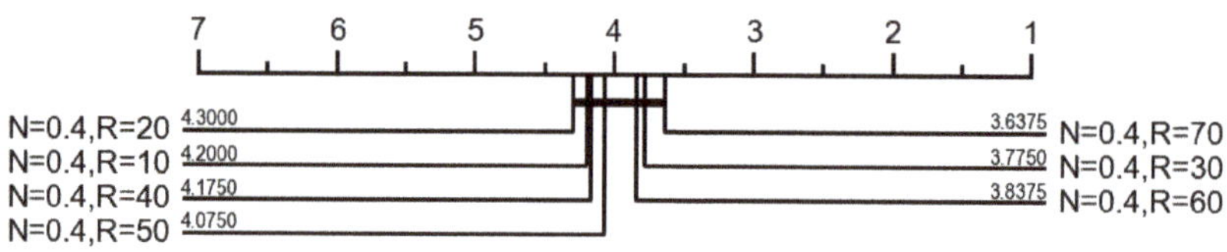

Figure 7. Mean rank of different numbers of realizations applied on CEEMD-MultiRocket with a fixed noise intensity.

5.2. Sub-Series Selection

Integrating all decomposed sub-series as the input to the classification model is not necessarily a guarantee of improvement of classification performance. Therefore, it is necessary to select the most important sub-series and discard less important ones to enhance the overall classification performance and decrease the computational complexity. The selection is executed on the whole known training set which is further divided into training and testing sets by stratified sampling, by comparing the testing accuracy (using improved MultiRocket) of each IMF with that of the raw time series, respectively, using a predetermined threshold. When the ratio of the testing accuracy of the IMF to that of the raw time series is more than the given threshold, this IMF is selected as the input to the classification model. We set different thresholds and the corresponding classification results are shown in Figure 8.

From Figure 8, we can find that CEEMD-MultiRocket achieves the best classification accuracy when the value of the threshold is 0.9, although the difference by setting different thresholds is negligible and statistically insignificant in terms of the classification accuracy. Table 5 shows the number of datasets with 0, 1, 2 or 3 IMFs which are selected using different thresholds on 109 datasets. When the threshold is set to 0, all three IMFs are unconditionally selected for each dataset, and the classification accuracy is the worst because some of these IMFs may produce negative impacts on the performance of the classification algorithm. When the threshold is set to 1, more than half of the datasets do not have any IMFs selected. Although this can decrease the computational complexity, it may also lose many crucial sub-series and reduce the classification performance. We find that when the threshold is set to 0.9, 93 datasets out of all 109 datasets select at least one decomposed IMF, and our model achieves the best classification accuracy due to the addition of appropriate $IMFs^*$.

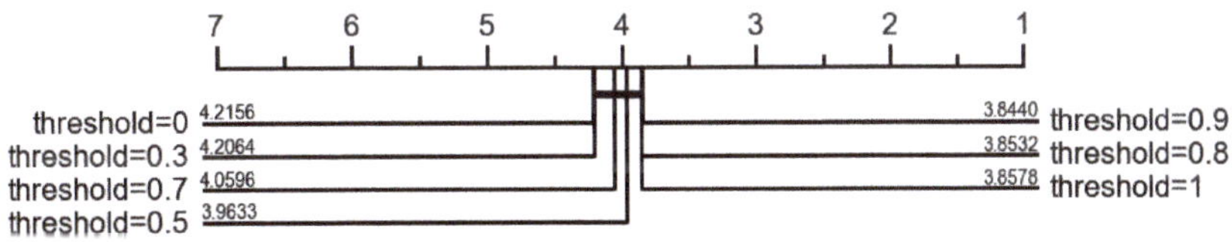

Figure 8. Mean rank of CEEMD-MultiRocket using different threshold.

Table 5. Number of datasets of selecting different number of IMFs using different thresholds on 109 datasets from the UCR time series repository.

Threshold	Number of IMFs Selected			
	0	1	2	3
0	0	0	0	109
0.3	0	3	3	103
0.5	0	14	7	88
0.7	0	31	13	65
0.8	1	40	22	46
0.9	16	48	17	28
1	60	31	11	7

5.3. Convolution Kernel Design

In CEEMD-MultiRocket, we decrease the length of the convolution kernel to 6. Figure 9 demonstrates the effectiveness of different kernel lengths on classification accuracy. In Figure 9, 6_2 means the kernel length is 6, in which two weights are one value and the remaining weights are another value. Thus, it gives $C_6^2 = 15$ fixed kernels in total. As can be seen from Figure 9, convolution kernels of length 5, 6 or 7 significantly outperform other convolution kernels with a length of 8, 9 or 11 in terms of classification accuracy.

It is also worth mentioning that the entire set of kernels of length 6 produces higher accuracy than the 6_2 kernel subset, but the classification accuracy is statistically insignificant. Since the 6_2 kernel subset only has about a quarter of the number of convolution kernels of the entire set of kernels with length 6, it is particularly suitable for the optimizations of avoiding multiplications, which can significantly shorten training time but achieve almost the same level of classification accuracy. Therefore, the convolution kernel of length 6_2 is applied in the proposed CEEMD-MultiRocket.

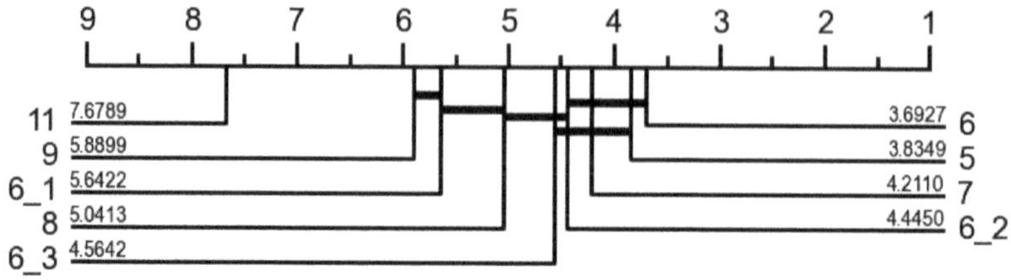

Figure 9. Mean rank of CEEMD-MultiRocket using different convolution kernel lengths.

5.4. Pooling Operators

In CEEMD-MultiRocket, we add an extra pooling operator NSPV (Number of Stretch of Positive Values) to enrich the discriminatory power of derived features. Figure 10 compares the effectiveness of using all five pooling operators (PPV, MPV, MIPV, LSPV, NSPV), four pooling operators (PPV, MPV, MIPV, LSPV), two pooling operators (PPV and NSPV) and only PPV in CEEMD-MultiRocket with 50k features. The experimental result shows that compared with four pooling operators, an additional NSPV pooling operator is able to significantly increase classification accuracy, indicating that NSPV contributes to the improvement of classification performance in CEEMD-MultiRocket. Furthermore, we also find that using four pooling operators (PPV + MPV + MIPV + LSPV) is not significantly better than only using two pooling operators (PPV + NSPV).

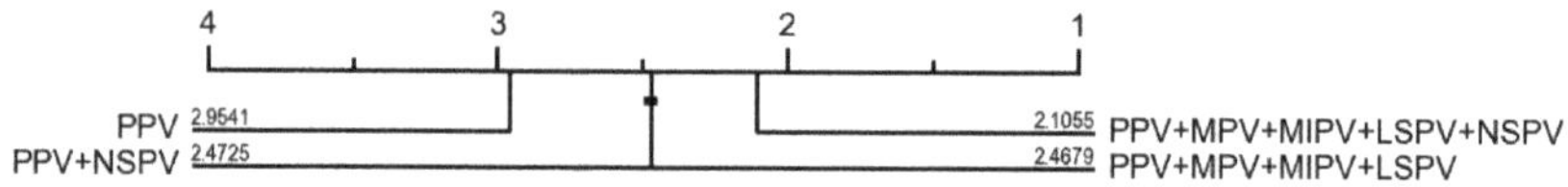

Figure 10. Mean rank of CEEMD-MultiRocket using different combinations of pooling operators.

5.5. Summary

From the above results and analysis, some findings can be summarized as follows:

(1) Decomposing raw time series into sub-series and extracting features from them can obtain more detailed and discriminative information from various aspects, which significantly contributes to the enhancement of classification accuracy.
(2) Selecting the more crucial sub-series and pruning the redundant and less important ones can both enhance classification performance and reduce computational complexity.
(3) The optimization in convolution kernel design can generate more efficient transform, which helps to improve the overall classification accuracy.
(4) The additional pooling operator NSPV enriches the discriminatory power of derived features.

6. Conclusions

To enhance the classification performance of the original MultiRocket, this study proposes a hybrid classification model CEEMD-MultiRocket which integrates CEEMD and improved MultiRocket. Firstly, the CEEMD algorithm is employed to decompose raw time series into two IMFs and one residue, which represent the high-, medium- and low-frequency portions of raw time series, respectively. Then, the selection of these decomposed sub-series is conducted on the whole known training set which is further divided into new training and testing sets using stratified sampling, by comparing the classification accuracy of each sub-series with that of the raw time series using a given threshold. Finally, we improve the convolutional kernel and pooling operators of the original MultiRocket, apply the improved MultiRocket to the raw time series, the selected decomposed sub-series and the first-order difference of raw time series to extract features, and build a ridge regression classifier. The experimental results demonstrate that: (1) in comparison to all SOTA classification algorithms except for HIVE-COTE 2.0, the proposed algorithm can significantly enhance the classification accuracy on 109 datasets from the UCR time series repository; (2) CEEMD-MultiRocket achieves almost the same level of classification accuracy as HIVE-COTE 2.0, with a fraction of the computing cost of the latter; (3) the CEEMD algorithm has the ability to generate a variety of representations of raw time series as the inputs of the algorithm, which contributes to the improvement of classification accuracy; (4) the improvement of convolution kernel length and the reduction in the number of convolution kernels can enhance classification performance while reducing computational load; and (5) the additional pooling operator contributes to enhancing the classification accuracy.

There are two main limitations in our work: (1) CEEMD is a relatively time-consuming decomposition method; (2) the values of weights in the convolution kernel are pre-defined and cannot be dynamically adjusted in line with increases in the dilation. The main directions for future research could be extended in two aspects: (1) continuing to improve MultiRocket to build the hybrid ensemble classification algorithm for time series; (2) considering faster decomposition algorithms and sub-series selection algorithms to improve the runtime and classification accuracy of the algorithm.

Author Contributions: Conceptualization, P.W., J.W. and Y.W.; Formal analysis, P.W. and J.W.; Investigation, P.W. and Y.W.; Methodology, P.W. and J.W.; Project administration, J.W.; Resources, J.W. and T.L.; Software, P.W. and J.W.; Supervision, T.L.; Validation, P.W. and Y.W.; Writing—original draft, P.W., J.W. and Y.W.; Writing—review and editing, J.W., Y.W. and T.L. All authors have read and agreed to the published version of the manuscript.

Funding: This work was supported by the Ministry of Education of Humanities and Social Science Project (grant no. 19YJAZH047) and the Social Practice Research for Teachers of Southwestern University of Finance and Economics (grant no. 2022JSSHSJ11).

Institutional Review Board Statement: Not applicable.

Informed Consent Statement: Not applicable.

Data Availability Statement: All the data in this paper are publicly available. They can be accessed at https://www.cs.ucr.edu/~eamonn/time_series_data/ (all accessed on 20 October 2022).

Conflicts of Interest: The authors declare no conflict of interest.

References

1. Rezaei, S.; Liu, X. Deep learning for encrypted traffic classification: An overview. *IEEE Commun. Mag.* **2019**, *57*, 76–81. [CrossRef]
2. Susto, G.A.; Cenedese, A.; Terzi, M. Time-series classification methods: Review and applications to power systems data. *Big Data Appl. Power Syst.* **2018**, 179–220. [CrossRef]
3. Li, T.; Qian, Z.; Deng, W.; Zhang, D.; Lu, H.; Wang, S. Forecasting crude oil prices based on variational mode decomposition and random sparse Bayesian learning. *Appl. Soft Comput.* **2021**, *113*, 108032. [CrossRef]
4. Chao, L.; Zhipeng, J.; Yuanjie, Z. A novel reconstructed training-set SVM with roulette cooperative coevolution for financial time series classification. *Expert Syst. Appl.* **2019**, *123*, 283–298. [CrossRef]
5. Ebrahimi, Z.; Loni, M.; Daneshtalab, M.; Gharehbaghi, A. A review on deep learning methods for ECG arrhythmia classification. *Expert Syst. Appl. X* **2020**, *7*, 100033. [CrossRef]
6. Wu, J.; Zhou, T.; Li, T. Detecting epileptic seizures in EEG signals with complementary ensemble empirical mode decomposition and extreme gradient boosting. *Entropy* **2020**, *22*, 140. [CrossRef]
7. Craik, A.; He, Y.; Contreras-Vidal, J.L. Deep learning for electroencephalogram (EEG) classification tasks: A review. *J. Neural Eng.* **2019**, *16*, 031001. [CrossRef]
8. Liu, Y.; Wu, Y.F. Early detection of fake news on social media through propagation path classification with recurrent and convolutional networks. In Proceedings of the AAAI Conference on Artificial Intelligence, New Orleans, LA, USA, 2–7 February 2018; Volume 32.
9. Pantiskas, L.; Verstoep, K.; Hoogendoorn, M.; Bal, H. Taking ROCKET on an efficiency mission: Multivariate time series classification with LightWaves. *arXiv* **2022**, arXiv:2204.01379.
10. Nishikawa, Y.; Sannomiya, N.; Itakura, H. A method for suboptimal design of nonlinear feedback systems. *Automatica* **1971**, *7*, 703–712.
11. Lucas, B.; Shifaz, A.; Pelletier, C.; O'Neill, L.; Zaidi, N.; Goethals, B.; Petitjean, F.; Webb, G.I. Proximity forest: An effective and scalable distance-based classifier for time series. *Data Min. Knowl. Discov.* **2019**, *33*, 607–635. [CrossRef]

12. Flynn, M.; Large, J.; Bagnall, T. The contract random interval spectral ensemble (c-RISE): The effect of contracting a classifier on accuracy. In *Proceedings of the International Conference on Hybrid Artificial Intelligence Systems*; Springer: Berlin/Heidelberg, Germany, 2019; pp. 381–392.
13. Deng, H.; Runger, G.; Tuv, E.; Vladimir, M. A time series forest for classification and feature extraction. *Inf. Sci.* **2013**, *239*, 142–153. [CrossRef]
14. Middlehurst, M.; Large, J.; Bagnall, A. The canonical interval forest (CIF) classifier for time series classification. In Proceedings of the 2020 IEEE International Conference on Big Data (Big Data), Atlanta, GA, USA, 10–13 December 2020; pp. 188–195.
15. Lubba, C.H.; Sethi, S.S.; Knaute, P.; Schultz, S.R.; Fulcher, B.D.; Jones, N.S. catch22: CAnonical Time-series CHaracteristics: Selected through highly comparative time-series analysis. *Data Min. Knowl. Discov.* **2019**, *33*, 1821–1852. [CrossRef]
16. Schäfer, P. The BOSS is concerned with time series classification in the presence of noise. *Data Min. Knowl. Discov.* **2015**, *29*, 1505–1530. [CrossRef]
17. Middlehurst, M.; Large, J.; Cawley, G.; Bagnall, A. The temporal dictionary ensemble (TDE) classifier for time series classification. In Proceedings of the Machine Learning and Knowledge Discovery in Databases: European Conference, ECML PKDD 2020, Ghent, Belgium, 14–18 September 2020; Part I; Springer: Berlin/Heidelberg, Germany, 2021; pp. 660–676.
18. Bagnall, A.; Lines, J.; Bostrom, A.; Large, J.; Keogh, E. The great time series classification bake off: A review and experimental evaluation of recent algorithmic advances. *Data Min. Knowl. Discov.* **2017**, *31*, 606–660. [CrossRef]
19. Lines, J.; Davis, L.M.; Hills, J.; Bagnall, A. A shapelet transform for time series classification. In Proceedings of the 18th ACM SIGKDD International Conference on Knowledge Discovery and Data Mining, Beijing, China, 12–16 August 2012; pp. 289–297.
20. Bostrom, A.; Bagnall, A. Binary shapelet transform for multiclass time series classification. In *Transactions on Large-Scale Data- and Knowledge-Centered Systems XXXII: Special Issue on Big Data Analytics and Knowledge Discovery*; Springer: Berlin/Heidelberg, Germany, 2017; pp. 24–46.
21. Bagnall, A.; Flynn, M.; Large, J.; Lines, J.; Middlehurst, M. On the usage and performance of the hierarchical vote collective of transformation-based ensembles version 1.0 (hive-cote v1.0). In Proceedings of the Advanced Analytics and Learning on Temporal Data: 5th ECML PKDD Workshop, AALTD 2020, Ghent, Belgium, 18 September 2020; Revised Selected Papers 6; Springer: Berlin/Heidelberg, Germany, 2020; pp. 3–18.
22. Middlehurst, M.; Large, J.; Flynn, M.; Lines, J.; Bostrom, A.; Bagnall, A. HIVE-COTE 2.0: A new meta ensemble for time series classification. *Mach. Learn.* **2021**, *110*, 3211–3243. [CrossRef]
23. Ismail Fawaz, H.; Lucas, B.; Forestier, G.; Pelletier, C.; Schmidt, D.F.; Weber, J.; Webb, G.I.; Idoumghar, L.; Muller, P.A.; Petitjean, F. Inceptiontime: Finding alexnet for time series classification. *Data Min. Knowl. Discov.* **2020**, *34*, 1936–1962. [CrossRef]
24. Shifaz, A.; Pelletier, C.; Petitjean, F.; Webb, G.I. TS-CHIEF: A scalable and accurate forest algorithm for time series classification. *Data Min. Knowl. Discov.* **2020**, *34*, 742–775. [CrossRef]
25. Längkvist, M.; Karlsson, L.; Loutfi, A. A review of unsupervised feature learning and deep learning for time-series modeling. *Pattern Recognit. Lett.* **2014**, *42*, 11–24. [CrossRef]
26. Bengio, Y.; Yao, L.; Alain, G.; Vincent, P. Generalized denoising auto-encoders as generative models. *Adv. Neural Inf. Process. Syst.* **2013**, *26*.
27. Hu, Q.; Zhang, R.; Zhou, Y. Transfer learning for short-term wind speed prediction with deep neural networks. *Renew. Energy* **2016**, *85*, 83–95. [CrossRef]
28. Gallicchio, C.; Micheli, A. Deep echo state network (deepesn): A brief survey. *arXiv* **2017**, arXiv:1712.04323.
29. Pascanu, R.; Mikolov, T.; Bengio, Y. Understanding the exploding gradient problem. *arXiv* **2012**, arXiv:1211.5063.
30. Ismail Fawaz, H.; Forestier, G.; Weber, J.; Idoumghar, L.; Muller, P.A. Deep learning for time series classification: A review. *Data Min. Knowl. Discov.* **2019**, *33*, 917–963. [CrossRef]
31. Hatami, N.; Gavet, Y.; Debayle, J. Classification of time-series images using deep convolutional neural networks. In Proceedings of the Tenth International Conference on Machine Vision (ICMV 2017); Vienna, Austria, 13–15 November 2017; SPIE: Bellingham, WA, USA, 2018; Volume 10696, pp. 242–249.
32. Tripathy, R.; Acharya, U.R. Use of features from RR-time series and EEG signals for automated classification of sleep stages in deep neural network framework. *Biocybern. Biomed. Eng.* **2018**, *38*, 890–902. [CrossRef]
33. Wang, Z.; Oates, T. Imaging time-series to improve classification and imputation. In Proceedings of the Twenty-Fourth International Joint Conference on Artificial Intelligence, Buenos Aires, Argentina, 25–31 July 2015.
34. Nweke, H.F.; Teh, Y.W.; Al-Garadi, M.A.; Alo, U.R. Deep learning algorithms for human activity recognition using mobile and wearable sensor networks: State of the art and research challenges. *Expert Syst. Appl.* **2018**, *105*, 233–261. [CrossRef]
35. Tan, C.W.; Dempster, A.; Bergmeir, C.; Webb, G.I. MultiRocket: Multiple pooling operators and transformations for fast and effective time series classification. *Data Min. Knowl. Discov.* **2022**, *36*, 1623–1646. [CrossRef]
36. Dempster, A.; Petitjean, F.; Webb, G.I. ROCKET: Exceptionally fast and accurate time series classification using random convolutional kernels. *Data Min. Knowl. Discov.* **2020**, *34*, 1454–1495. [CrossRef]
37. Dempster, A.; Schmidt, D.F.; Webb, G.I. Minirocket: A very fast (almost) deterministic transform for time series classification. In Proceedings of the 27th ACM SIGKDD Conference on Knowledge Discovery & Data Mining, Singapore, 14–18 August 2021; pp. 248–257.

38. Torres, M.E.; Colominas, M.A.; Schlotthauer, G.; Flandrin, P. A complete ensemble empirical mode decomposition with adaptive noise. In Proceedings of the 2011 IEEE International Conference on Acoustics, Speech and Signal Processing (ICASSP), Prague, Czech Republic, 22–27 May 2011; pp. 4144–4147.
39. Zhou, Y.; Li, T.; Shi, J.; Qian, Z. A CEEMDAN and XGBOOST-based approach to forecast crude oil prices. *Complexity* **2019**, *2019*, 1–15. [CrossRef]
40. Li, T.; Qian, Z.; He, T. Short-term load forecasting with improved CEEMDAN and GWO-based multiple kernel ELM. *Complexity* **2020**, *2020*, 1–20. [CrossRef]
41. Wu, Z.; Huang, N.E. Ensemble empirical mode decomposition: A noise-assisted data analysis method. *Adv. Adapt. Data Anal.* **2009**, *1*, 1–41. [CrossRef]
42. Huang, N.E.; Shen, Z.; Long, S.R.; Wu, M.C.; Shih, H.H.; Zheng, Q.; Yen, N.C.; Tung, C.C.; Liu, H.H. The empirical mode decomposition and the Hilbert spectrum for nonlinear and non-stationary time series analysis. *Proc. R. Soc. London. Ser. A Math. Phys. Eng. Sci.* **1998**, *454*, 903–995. [CrossRef]
43. Li, S.; Zhou, W.; Yuan, Q.; Geng, S.; Cai, D. Feature extraction and recognition of ictal EEG using EMD and SVM. *Comput. Biol. Med.* **2013**, *43*, 807–816. [CrossRef] [PubMed]
44. Tang, B.; Dong, S.; Song, T. Method for eliminating mode mixing of empirical mode decomposition based on the revised blind source separation. *Signal Process.* **2012**, *92*, 248–258. [CrossRef]
45. Wu, J.; Chen, Y.; Zhou, T.; Li, T. An adaptive hybrid learning paradigm integrating CEEMD, ARIMA and SBL for crude oil price forecasting. *Energies* **2019**, *12*, 1239. [CrossRef]
46. Li, T.; Zhou, M. ECG classification using wavelet packet entropy and random forests. *Entropy* **2016**, *18*, 285. [CrossRef]
47. Dau, H.A.; Bagnall, A.; Kamgar, K.; Yeh, C.C.M.; Zhu, Y.; Gharghabi, S.; Ratanamahatana, C.A.; Keogh, E. The UCR time series archive. *IEEE/CAA J. Autom. Sin.* **2019**, *6*, 1293–1305. [CrossRef]
48. Chai, J.; Wang, Y.; Wang, S.; Wang, Y. A decomposition–integration model with dynamic fuzzy reconstruction for crude oil price prediction and the implications for sustainable development. *J. Clean. Prod.* **2019**, *229*, 775–786. [CrossRef]
49. Demšar, J. Statistical comparisons of classifiers over multiple data sets. *J. Mach. Learn. Res.* **2006**, *7*, 1–30.

electronics

MDPI

Article

Financial Time Series Forecasting: A Data Stream Mining-Based System

Zineb Bousbaa [1,*,†,‡], Javier Sanchez-Medina [2,‡] and Omar Bencharef [1,‡]

1 Computer and System Engineering Laboratory, Faculty of Science and Technology, Cadi Ayyad University, Marrakech 40000, Morocco; o.bencharef@uca.ma

2 Innovation Center for the Information Society (CICEI), Campus of Tafira, University of Las Palmas de Gran Canaria, 35017 Las Palmas de Gran Canaria, Spain; javier.sanchez@ulpgc.es

* Correspondence: zineb.bousbaa@ced.uca.ma

† Current address: Faculty of Sciences and Technology, Cadi Ayyad University, Marrakesh 40000, Morocco.

‡ These authors contributed equally to this work.

Abstract: Data stream mining (DSM) represents a promising process to forecast financial time series exchange rate. Financial historical data generate several types of cyclical patterns that evolve, grow, decrease, and end up dying. Within historical data, we can notice long-term, seasonal, and irregular trends. All these changes make traditional static machine learning models not relevant to those study cases. The statistically unstable evolution of financial market behavior yields a progressive deterioration in any trained static model. Those models do not provide the required characteristics to evolve continuously and sustain good forecasting performance as the data distribution changes. Online learning without DSM mechanisms can also miss sudden or quick changes. In this paper, we propose a possible DSM methodology, trying to cope with that instability by implementing an incremental and adaptive strategy. The proposed algorithm includes the online Stochastic Gradient Descent algorithm (SGD), whose weights are optimized using the Particle Swarm Optimization Metaheuristic (PSO) to identify repetitive chart patterns in the FOREX historical data by forecasting the EUR/USD pair's future values. The data trend change is detected using a statistical technique that studies if the received time series instances are stationary or not. Therefore, the sliding window size is minimized as changes are detected and maximized as the distribution becomes more stable. Results, though preliminary, show that the model prediction is better using flexible sliding windows that adapt according to the detected distribution changes using stationarity compared to learning using a fixed window size that does not incorporate any techniques for detecting and responding to pattern shifts.

Keywords: data stream mining; forex; online learning; adaptive learning; incremental learning; sliding window; concept drift; financial time series forecasting

Citation: Bousbaa, Z.; Sanchez-Medina, J.; Bencharef, O. Financial Time Series Forecasting: A Data Stream Mining-Based System. *Electronics* **2023**, *12*, 2039. https://doi.org/10.3390/electronics12092039

Academic Editors: Taiyong Li, Wu Deng and Jiang Wu

Received: 19 February 2023
Revised: 22 April 2023
Accepted: 23 April 2023
Published: 28 April 2023

1. Introduction

Financial Time Series Exchange Rate Forecasting (FTSERF) is a growing field because many investors are interested in it. Artificial intelligence, as a computer science sub-field, helped in this development by being part of trading decision systems. Machine learning models became important parts of these systems because they were able to accurately predict the exchange rate and, as a result, increase the chances of making good profits.

When we look at the work done on the algorithms used for FTSERF, we can see that researchers in both public and private institutions have tried out all the machine learning tools that are available. Those tools may be supervised, such as classification [1], regression [2], recommender systems [3], and reinforement learning [4]. They also include unsupervised algorithms, such as clustering and association analysis, ranking, and anomaly detection techniques [5]. The research field of FTSERF using machine learning is wide.

Some studies limit their focus to forecasting future trends, while others go beyond that to implement trading strategies that work on maximizing profit.

Speaking about the FTSERF difficulties, dealing with the data's volatile and chaotic nature is the biggest one. To learn from historical financial time series datasets, you have to be able to adapt to new patterns all the time. This adaptivity is called reacting to the detected changes within the data stream mining (DSM) context. This task is challenging as it requires both recognizing real changes and avoiding false alerts. There are many change detection techniques. Ref. [6] cites some techniques, including the CUSUM test, the geometric moving average test, statistical tests, and drift detection methods.

Our paper's contribution is to show how integrating DSM techniques into the online learning process can increase FTSERF performance. In our experimental study, we propose the SGD algorithm optimized using the PSO. We managed the changes that occurred in the financial time series trends by implementing a sliding window mechanism. The sliding window size is flexible. It is minimized when a high fluctuation is detected and maximized when the time series pattern is more or less stable. As a change detection mechanism, we test for each sliding window the stationarity of the data stream within, and based on the results, we decide to maintain or adapt the window size that will be passed as input to our forecasting model. We have compared going through the learning process using a traditional algorithm vs. integrating DSM techniques. The traditional algorithm combines the SGD, whose parameters are optimized using the PSO metaheuristic periodically. The DSM version involves the adaptive sliding window mechanism and the statistical stationarity test.

The remainder of this paper is structured as follows. The following section carries out a literature review that we have performed concerning data mining, in addition to DSM's application to FTSERF. For further illustration, Section 3 is devoted to describing our dataset's components, its preprocessing, analysis, and input selection processes. Section 4 represents the various proposed forecasting system components and illustrates its architecture and algorithm. Section 4 also shows the various experimental studies, analysis of their results, and further discussions. Finally, some concluding remarks are made in Section 5.

2. Literature Review

2.1. Machine Learning Application to Financial Forecasting

2.1.1. Overview

The financial forecasting research field is very dynamic. The value forecasting of financial assets is a field that attracts the interest of multiple profiles, from accountants, mathematicians, and statisticians to computer scientists. In addition, when it comes to investing, we find that investors with no scientific background easily integrate themselves into the field. Tools like technical analysis are highly used as they are easy to apply and have shown effectiveness in trading. Despite the fact that financial asset markets first appeared in the 16th century (see Figure 1), the first structured approaches to economic forecasting of these assets date from the last century [7–10].

Research employs methods ranging from mathematical models to statistical models like ARMA, suggested by Wold in 1938 by combining both AR and MA schemes [11], as well as to macroeconomic models like Tinbergen's (1939) [12]. By this time, various accounting, mathematical, statistical, and machine learning models were constructed to predict different kinds of financial assets, which led to their exponential increase. One of the main apparent reasons for this increase is that this research field is one of the most highly funded, since financial investment generates profit. Funding comes from many sources; we can mainly mention big asset management companies and investment banks.

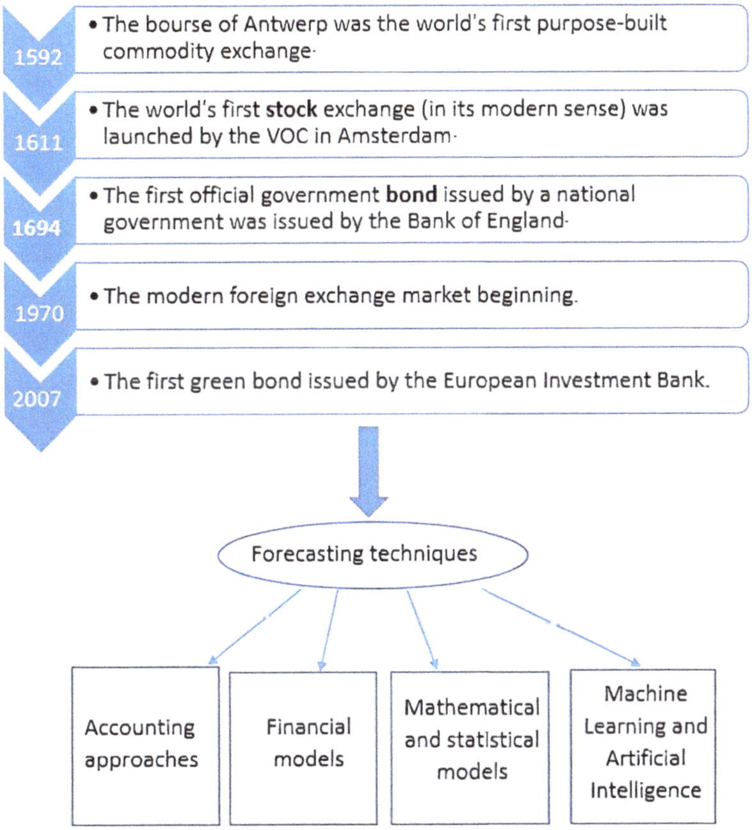

Figure 1. The chronology of financial asset appearance and the existing forecasting approaches for their valuation [7–10].

Back in the 1980s, exploring historical data was more difficult since the amount of data had grown, but computational power was still weak [13]. By the 1990s, more economic and statistical models had appeared, and they showed better performance than a random walk. Still, these models do not perform the same way over all the financial assets [14].

However, by the beginning of the 21st century, machine learning models were all the rage because computers had become faster and were capable of more work. During this time, many hybrid algorithms, such as moving averages that take into account regressivity (ARIMA) or algorithms that combine neural networks and traditional time series, have been proposed. We also noticed the rapid, exponential growth in the number of papers in this research field from 1998 to 2016. There has been a wide range of topics ranging from credit rating to inflation forecasting and risk management, which has saturated this research field and made finding innovative ideas more challenging [13].

On the other hand, scientists became more aware in the 1980s of how important it was to process textual data. There were attempts to import other predictors developed from linguistics by Frazier in 1984 [15]. More progress has been achieved, such as word spotting using naïve statistical methods (Brachman and Khabaza 1996) [16]. Sentiment analysis resources were proposed at the beginning of the 21st century (Hu and Liu 2004) [17]. Sentiment analysis is a critical component that employs both machine learning algorithms and knowledge-based methodologies, particularly for analyzing the sentiments of social media users [18]. From 2010 on, social media data increased exponentially, which attracted the interest of the news analytics community in processing real-time data (Cambria et al. 2014) [13,19].

While reviewing some surveys that shed light on research in financial forecasting, we find some surveys that suggest categorizing the proposed models in different ways. The study in [20] distinguished between singular and hybrid models, which include non-linear models such as artificial neural networks (ANN), support vector machines (SVMs), particle swarm optimization (PSO), as well as linear models like autoregressive integrated moving average (ARIMA), etc. Meanwhile, the study in [21] revealed fundamental and technical analysis as the two approaches that are commonly used to analyze and predict financial market behaviors. It also distinguished between statistical and machine learning approaches, assuming that machine learning approaches deal better with complex, dynamic, and chaotic financial time series. In addition, Ref. [21] points out that the profit analysis of the suggested forecasting techniques in real-world applications is generally neglected. As a solution, they suggest a detailed process for creating a smart trading system. As a solution, it is composed of data preparation, algorithm definition, training, forecasting evaluation, trading strategies, and money evaluation [21]. Papers can also be summarized based on their primary goal, which could be preprocessing, forecasting, or text mining. They can likewise be classified based on the nature of the dataset, whether qualitatively derived from technical analysis or quantitatively retrieved from financial news, business financial reports, or other sources [21]. In addition, Ref. [22] distinguished between parametric statistical methods like Discriminant Analysis and Logistic Regression, Non-Parametric Statistical Methods like Decision Tree and Nearest Neighbor, or Soft Computing techniques like Fuzzy Logic, Support Vector Machine, and Genetic Algorithm.

Essential conclusions are extracted from the survey [21], where it is considered that no approach is better in every way than another, as there is no well-established methodology to guide the construction of a successful, intelligent trading system. Another issue is highlighted in [22] concerning the use of metrics like RMSE and MSE that depend on the scale vs. those that do not, such as the MAPE metric. On the other hand, some papers consider that once a forecasting system is trained, it is expected well in advance to forecast future values. However, this is not possible in the financial time series study case, as they change over time, making retraining with updated data necessary to collect the most recent knowledge about the state of the market. This final issue has been our motivation to work on the DSM application to FTSERF.

2.1.2. Optimization Techniques

In this paper, we combined two optimization techniques, SGD and PSO, as a form of hybridization in order to improve exchange rate forecasting performance.

The gradient descent has been suggested for non-linear problems by Haskell Curry back in 1944 [23]. In the gradient descent, we work on finding the optimal weights for the regression function by minimizing the error loss function. In our study case, the regression function weights depend on the number of explanatory variables used to forecast our target variable. More details about our dataset structure will be shared in a later section. The different types of gradient descent showed great results in dealing with fluctuating patterns, and even with a limited training dataset [24], this makes it an adequate choice for dealing with the chaotic and volatile nature of financial time series. Having an adaptive system is also a must when dealing with change. Gradient descent is flexible enough to be combined with adaptive mechanisms. For example, Ref. [25] propose an adaptive gradient descent willing to deal with a time delay in receiving the input data. SGD can address various types of problems. For example, in [26], a financial continuous-time problem is solved using SGD in continuous time (SGDCT), which showed better results compared to the classic SGD dealing with high-dimensional continuous-time problems. The possibility to combine the SGD with other techniques in order to adapt to the target study's challenges is a strong point of this algorithm. Another example is the study [27] where a functional gradient descent is proposed in order to forecast the historical interest rate progress possibilities.

The PSO optimization technique is an iterative algorithm that works and converges closer and faster to the solution search space. It is based on a population of solutions collaborating together to reach the optimum results [28]. Despite its limitations, such as the inability to guarantee a good convergence and its computational cost, researchers still use it in financial optimization problems in particular and in other study fields as well, as it helps exceed the convergence limits in many cases. PSO is proposed for the first time by [29,30] as a solution to deal with problems presented in the form of nonlinear continuous functions. In the literature, we find a lot of studies showing how the PSO helped achieve a new score for forecasting time series or any other type of data. In [31], authors have experimented with how PSO can help optimize the results given by neural networks compared to the classic backpropagation algorithm for time series forecasting. On the other hand, the article [32] shows how currency exchange rate forecasting capacity can be polished by adjusting the model function parameters and using the PSO as a booster to the generalized regression neural network algorithm's performance. Another work combining PSO and neural networks is [33], which worked on predicting the Singapore stock market index. Results demonstrate the effectiveness of using the particle swarm optimization technique to train neural network weights. The performance of the PSO FFNN is assessed by optimizing the PSO settings. In addition, recurrent neural network results predicting stock market behavior have been optimized using PSO in [34]. The study's findings demonstrate that the model used in this work has a noticeable impact on performance compared to before the hybridization. Finally, we cite [35] where a competitive swarm optimizer that is a PSO variant proved its efficiency dealing with datasets having a large number of features. The experiment shows how the proposed technique can show a fast convergence to greater accuracy.

2.2. Data Stream Mining Application to Financial Forecasting

2.2.1. Data Stream Mining

DSM is the sequence for learning the evolving data stream's behavior. Data streams could be formed in different data structures, such as trees, particularly ordered and unordered ones. Data streams are present in several study cases, such as financial tickers, network monitoring, traffic management performance metrics, log records, clickstreams for web tracking, etc.

Data stream models need to assume that the sequence of the data that are coming and that are already here is potentially infinite. Thus, it is impossible to keep the data stored. In some circumstances, these models must also handle data in real time since the stream of data might be significant. The distribution of data could change in the future. Thus, historical information could be detrimental to the present situation. The algorithms of data streams have been the subject of extensive research, leading to computer paradigms that optimize memory and time-per-item consumption. The nature of the distribution change has also been studied, where the sliding window approach is used to manage this issue. The oldest element in a window is deleted to receive the newest one. Fixing the window size cannot be prioritized. The change rate needs to be continuously studied since it may itself vary over time [6].

2.2.2. Online Learning

Online learning is one of the main techniques for dealing with data streams. It is a fundamental paradigm of computational learning theory. The algorithms that only employ a small amount of prior data storage are covered under the field of online machine learning in big data streams. After each prediction, an online learner receives the feedback. Online learners are often compared to the top predictors in terms of their excess loss [36,37]. The learning process is called online when it is incremental and adaptive.

For example, effective reinforcement learning is necessary for adaptive real-time machine learning. Due to the nature of online learning, which involves continuous streams of real-time data and adaptive learning from a limited sample size, the algorithm should constantly interact with its environment to optimize the reward [38]. Prequential learning is also one of the efficient techniques used for online learning. It serves as an alternative to the standard holdout evaluation that was carried over from batch-setting issues. The prequential analysis is specifically created for stream environments. Each sample has two distinct functions; it is analyzed sequentially in the order of receipt and then rendered inaccessible. This approach makes predictions for each instance, tests the model, and then trains it using the same sample (partial fit). The model is continually put to the test on fresh samples. Validation techniques are also frequently used to help models be adaptive and incremental. We can mention the following methods: data division into training, test, and holdout groups. We can also cite cross-validation techniques including k-fold, leave-one-out, leave-one-group, and nested. The problem with validation techniques is the risk of overfitting. Techniques for adapting a model are many. We can mention, for example, computing the area under the ROC Curve (AUC) using constant time and memory [39].

2.2.3. Incremental Learning

Incremental learning is a real-time learning approach that builds a similar model as a batch learning algorithm. In theory, the stream of observations could go on indefinitely, making it impossible to wait until all observations are received. Instead of accumulating and storing all inputs and applying batch learning to the full series of received instances, one should apply a batch learning algorithm to each new input [40]. A machine learning paradigm known as incremental learning changes what has already been learned whenever new instances appear. The biggest distinction between incremental learning and classical machine learning is that the former does not presuppose the existence of an appropriate training set before the learning process [41]. Many algorithms, such as neural networks, use epochs to make the model incremental, which helps optimize computational power consumption, especially when dealing with large datasets.

2.2.4. Adaptive Learning

On the other hand, adaptive learning techniques aid in employing a continuously enhanced learning strategy that keeps the system current and maintains its excellent performance. Input and output values, as well as the associated attributes, are continuously monitored and learned through the adaptive learning process. Additionally, it continuously improves its accuracy by learning from events that could change market behavior in real time. Adaptive artificial intelligence takes into account the feedback from the operational environment and responds to it to produce data-driven predictions [42].

Concept drift or changes in the way data are spread out are problems that make it necessary to use adaptive learning techniques. It is a situation where the statistical characteristics of the class variable, or the target we wish to forecast, change with time [43]. Concept drift in machine learning and data mining describes how relationships between the input and output data in the underlying problem vary over time. The concept drift issue is especially prevalent in some areas where forecasts are ordered by time, such as time series forecasting and predictions on streaming data, and it needs to be specifically checked for and addressed [44]. Concept drift can be sudden, gradual, or recurrent. To effectively handle it, a system must be able to quickly adapt, be resilient to noise, be able to separate the noise from it, notice and respond to severe drifts in the model's performance, and capture up-to-date data trends [45]. For the provided models, techniques, and libraries for data stream classification, we can mention [5], which is a Java-based open-source DSM framework. It also provides regression models. However, more resources are provided when dealing with classification problems.

Machine learning models use different techniques to detect concept drift in classification. We can mention AUC, which detects and adapts to concept drift for classification models. AUC is computed with memory and constant time using a sliding window [46]. AUC evaluates the ranking abilities of a classification. The accuracy metric can be a good choice in the high-class imbalance ratio case, but it may poorly display the concept drift and be biased in identifying the principal class. The foundation for drift identification in unbalanced streams should be AUC. The study in [46] includes Page–Hinkey (PH) statistical test with some updates, include the best outcomes, or very nearly so, except that ADWIN might require more time and memory when change is constant or there is no change.

Regarding concept drift detection for regression problems, there is a technique that consists of studying the eigenvalues and eigenvectors. It allows the characterization of the distribution using an orthogonal basis. This approach is the object of the principal component analysis. A second technique is to monitor covariance. In probability theory and statistics, covariance measures the joint variability of two random variables, or how far two random variables differ when combined. It is also used for two sets of numerical data, calculating deviations from the mean. The covariance between two random variables, X and Y, is 0 if they are independent. The opposite, however, is untrue [47]. For further details, Ref. [48] shows the covariance matrix types. The cointegration study is another technique that detects concept drift, identifying the sensitivity degree of two variables to the same average price over a specified period. The use of it in econometrics is common. It can be used to discover mean-reversion trading techniques in finance [49]. In our study, we compared the use of a fixed versus a flexible window size. We are also involved in studying the stationary process thing the AUC in the process. We conclude that class inequality has an impact on both prequential accuracy and AUC. However, AUC is statistically more discriminant. While accuracy can only reveal genuine drifts, it shows both real and virtual drifts. The authors used post hoc analysis, and the results confirm that AUC performs the best but has issues with highly imbalanced streams. Another family of methods is adaptive decision trees. It can adaptively learn from the data stream, and it is not necessary to understand how frequently or quickly the stream will change [6]. The ADWIN window is also a great technique for adaptive learning. We can either fix its size or make it variable. ADWIN is a parameter-free adaptive size sliding window. When two large sub-windows are too different, the window's older section is dropped. ADWIN reacts when sudden, infrequent or slow, gradual changes occur. To evaluate ADWIN's performance, we can follow the accuracy evolution of the false alarm probability, the probability of accurate detection, and the average delay time in detection. Some hybrid models, such as ADWIN with the Kalman filter, have demonstrated that they producat is related to regression problems; more details about it will be explained in the preliminaries section.

2.2.5. Implications of Econometric Methods

For a complete and successful real-world investment strategy in financial market assets, financial economics theory employment is recommended. It is needed to study factors such as time, risks, and investment costs and how they can play a major role in encouraging or discouraging a certain decision. Having agents that are economy-oriented in an investment strategy decision system can prevent the limitations that machine learning would present. Data science and financial economics-oriented agents can work together for more profitable approaches and better forecasting systems. In this context, we cite [50], where authors used machine learning to predict intra-day realized volatility. The study considered stock market crashes such as the European debt crisis, the China–United States trade war, and COVID-19. They tested multiple machine learning algorithms: seasonal autoregressive integrated moving averages (SARIMA), heterogeneous autoregressive with diurnal effects (HAR-D), ordinary least squares (OLS), least absolute shrinkage and selection operator (LASSO), XGBoost, multilayer perceptron (MLP), and long short-term memory (LSTM). They used three training schemes, where in the singular scheme they built distinct models for each stock. The universal scheme consists of constructing models with all stock data. The third training scheme, which is called the augmented scheme, also works on constructing models with all stock data except that predictors take into account market volatility. Training, validation, and testing sets are used in the training process. General results showed the advantage of incorporating volatility, which enhanced the forecasting ability. This study can show how the combination of machine learning and finance knowledge can lead to better predictions and more efficient decision systems.

The paper in [51] has also conducted an interesting study where forecasting simulations are made using econometric methods and other simulations are carried out using machine learning methods. The experiment reveals the importance of considering financial factors such as market maturity in addition to technical factors such as the used forecasting methods and evaluation metrics for good forecasting performance. Results show how Support Vector Machines (SVMs), which is a machine learning method, have given better results than the autoregressive model (AR), which is an econometric method. Advanced machine learning techniques show efficiency in detecting market anomalies in numerous significant financial markets. Authors also criticize studies that judge machine learning efficiency based on experiments applying traditional models instead of advanced ones like sliding windows and optimization mechanisms. They also refer to the fact that forecasting results do not necessarily lead to good returns. In addition to that, many researchers do not consider the transaction cost in their trading simulations.

In the literature, we can find multiple research studies showing how financial and machine learning methods can both contribute to efficient financial forecasting and investment systems. The study in [52] shows how combining statistical and machine learning metrics enhances the forecasting system's evaluation performance. They compare the forecasting abilities of well-known machine learning techniques: multilayer perceptrons (MLP) and support vector machine (SVM) models, the deep learning algorithm, and long short-term memory (LSTM), in order to predict the opening and closing stock prices for the Istanbul Stock Exchange National 100 Index (ISE-100). The evaluation metrics used are MSE, RMSE, and R^2. In addition, statistical tests are made using IBM SPSS statistics software in order to evaluate the different machine learning models' results. The findings of this study demonstrate how favorable MLP and LSTM machine learning models are for estimating opening and closing stock prices. Authors of the study in [53] also recommended combining fundamental economic knowledge with machine learning systems, as experts' judgment strengthens ultimate risk assessment. Their experiment compared machine learning algorithms to a statistical model in risk modeling. The study shows how extreme gradient boosting (XGBoost) succeeds in generating stress-testing scenarios surpassing the classical method. However, the lack of balance complicates class detection for machine learning models. Another challenge is that their dataset was limited to the Portuguese environment and needs to be expanded to other markets in order to improve the system's validation.

We also find finance and economy-oriented papers that have explored machine learning algorithms and proved their efficiency to forecast financial market patterns and generate good returns for investors. For example, authors in [7] demonstrated how asset pricing with machine learning algorithms is promising in finance. They implemented linear, tree, and neural network-based models. They used machine learning portfolios as metastrategies, where the first metastrategy combines all the models they have built and the second one selects the best-performing models. Results show how high-dimensional machine learning approaches can approximate unknown and potentially complex data-generating processes better than traditional economic models.

We also cite the study in [54], which shows how institutional investors use machine learning to estimate stock returns and analyze systemic financial risks for better investment decisions. The authors concluded that big data analysis can efficiently contribute to detecting outliers or unusual patterns in the market. They also recommend data-driven or data science-based research as a promising avenue for the finance industry.

Despite its efficiency in forecasting, machine learning applications in financial markets have some limitations. For example, authors in [55] concluded that machine learning systems' performance can vary depending on the studied market conditions. They evaluated the following machine learning algorithms: elastic net (Enet), gradient-boosted regression trees (GBRTs), random forest (RF), variable subsample aggregation (VASA), and neural networks with one to five layers (NN1–NN5). In addition, they tested ordinary least squares (OLS), regression LASSO, an efficient neural network, and gradient-boosted regression trees equipped with a Huber loss function. They encountered difficulties when using data from the US market but achieved good results as they worked on the Chinese stock market time series. However, their experiment did not involve advanced optimization techniques for hyperparameter selection and adaptation to each time series nature.

An interesting book that compares econometric models to machine learning models is [56]. The study shows how econometrics leans more toward statistical significance, while machine learning models focus more on the data's behavior over time. Machine learning advantages include the fact that they do not skip important information related to data interaction, unlike econometric models. In addition, machine learning models have the capacity to break down complex trends into simple patterns. They also better prevent overfitting using validation datasets. On the other hand, econometric models' advantage is the fact that their results are explicable, unlike those of machine learning methods, whose learning process includes black boxes. Overall, the references provide valuable insights into the advantages and limitations of machine learning and econometric models in financial forecasting and highlight the need for careful evaluation and interpretation of their results. In our current work, we focused on showing how DSM techniques can boost online algorithms to adapt to financial time series data with time-varying characteristics. Statistical methods play a major role in our system, as we use the stationarity test to detect if there is a change in the data stream distribution.

2.2.6. Data Stream Mining Application to Financial Forecasting

Data mining involves machine learning, statistics, information retrieval, and pattern recognition. The DSM handles data mining tasks as well as the volume, speed, and shifting patterns of data streams. The DSM technique improves and maintains the model's performance using pattern mining. As older data stream distribution can become irrelevant or damaging to the forecasting accuracy, fitting to newer data using adaptive learning is a must. The DSM is a recent machine learning domain that appeared as a result of multiple domains that generate data streams.

Mining changing data streams employs multiple methods. They must save excellent information, forget useless information, and enhance the model. We can mention as an example the sliding window technique, which retains W data items in a window. New elements replace the oldest ones, where a time t element expires at the time t + W for the sake of memory optimization. The window size W may adjust with the data distribution pace, either externally or during the learning process [6].

The scientists in the FTSERF field using machine learning models have benefited from various incremental and adaptive approaches. They have also proposed novel approaches and hybridized the existing ones. Most of these studies involve adaptive or incremental techniques that are not among the DSM approaches. Few works concretely involved them compared to the total. We can cite [57,58], where sliding windows are used. In those works, adaptivity has been ensured using optimization techniques. For [57], the used technique is called ELM-Jaya, where the final solution is based on the most effective and the weakest solutions. The parameters are adapted in the study in [58] thanks to the PSO metaheuristic, Genetic Algorithm (GA), and neural networks, which are adaptive algorithms by nature.

Incremental approaches include, for example, the uni-iterative approach, where the model receives one instance at each iteration, as in [59,60]. Secondly, multi-iterative systems have multiple instances instead of one. Refs. [61,62] are examples of this type. The third technique is the sliding window. As examples of studies we can cite [63,64]. The sliding window algorithm has a maintained window that keeps instances that have been read most recently, and according to specified rules, older examples are removed [6]. The fourth technique deals with real-time learning, which is efficient but could face difficulties due to the time factor's strictness. Online learning is the process of making predictions about a series of instances, one after another, and being rewarded or penalized for each one. Before making a prediction, the learner often receives a description of the situation. The learner's objective is to maximize the cumulative benefit or, conversely, reduce the cumulative loss [65]. Finally, we find no incremental studies in the literature that can be used for batch learning, such as [66,67]. A set or a series of observations are accepted as a single input by a batch learning algorithm. The algorithm creates its model, and it does not continue to learn. In contrast to online learning, batch learning is standing [65].

Adaptive approaches include several categories; we can mention concept drift for change detection to update decision systems, as in [68,69]. Secondly, forgetting factors are highly used in the FTSERF field, and they are especially dedicated for models that rely on weight updates for model tuning. Refs. [70,71] are two examples from this category. The third technique we mention is the order selection technique. It analyzes statistics, makes decisions based on sentimental analysis modules, and votes, such as in [72,73]. The fourth technique is pattern selection, which entails identifying profitable patterns and testing them, as in [74,75]. Last, the weight update is a very common technique. It consists of using new data to adjust the system parameters, or weights. It is proposed in many studies, such as in [76,77]. More information concerning the state of the art of DSM application for FTSERF will be presented in another work, our global survey.

3. Preliminaries

3.1. Dataset Description

In our dataset, we have chosen to include three currency pairs. The EUR/USD pair represents our target to forecast, while the GBP/USD and the JPY/USD pairs are included because of their significant impact and correlation to our target pair. Each pair's historical data have open, high, low, and close prices. The dataset we used ranges from 30 May 2000 to 28 February 2017, later expanded to 30 November 2022. More information can be found in the data availability statement.

Our dataset also integrated 12 technical indicators calculated for each one of the three used pairs: the stochastic oscillator, the Relative Strength Index (RSI), the StochRSI oscillator, the Moving Average Convergence and Divergence (MACD), the average directive index (ADX), the Williams% R, the Commodity Channel Index (CCI), the true mean range (ATR), the High-Low index, the ultimate oscillator, the Price Rate Of Change Indicator (ROC), the Bull power, and the Bear power. In addition, we used historical gold price data in Euro, US Dollar, British Pound, and Japanese Yen.

3.2. Dataset Preprocessing

Data preprocessing is a significant step between the data collection and the algorithm learning phases. It can use techniques like data transformation, encoding, and feature engineering to make the dataset easy for the algorithm to understand and work with. We may exclude at this phase unnecessary or redundant features. The exclusion can also be because of the weak correlation and impact on the target variable. Speaking about data preprocessing for DSM, the study in [78] provides a detailed survey. The authors cited how they converted unprocessed information into high-quality input. Techniques like integration, normalization, purification, and transformation are involved. In addition, the study integrated data reduction techniques, discretizing complex continuous feature spaces by choosing and removing unnecessary and distracting features.

In our experimental study, we first started with the input data granularity unification on a daily basis. Then we calculated for each one of the three pairs of time series from j-1 (1 day before the current value) to j-6 (6 days before the current value). The next step was calculating the technical indicators using their mathematical formulas. The process has been performed using a Python code we developed to make the process easy to repeat every time a new input comes. Further details will be shared in the dataset analysis and the experimental sections.

3.3. The Dataset Analysis and the Input Selection

The next step that follows the data preprocessing is the data analysis. It allows an understanding of the relevance of the data, and it also facilitates the choice of the algorithm to be used in the learning process. The data analysis includes three main steps: the univariate analysis, the bivariate analysis, and the multivariate analysis.

Whether we are looking at a qualitative or a quantitative feature changes how the univariate analysis is performed. In this analysis, we can study the following characteristics: the first quartile Q1, the minimum, the third quartile Q3, the median, the deciles, the 95th percentile, the 5th percentile, the maximum, the variance, the standard deviation, the range, the variable dispersion, and the symmetry index (skewness).

The bivariate analysis leads to linear links by restriction, transformation, or by studying the linear correlation. We test the strength of the correlation using the Pearson correlation coefficient, the H0 test, multiple linear regression, covariance, the assumptions of the simple linear model, the variance analysis table (ANOVA), the estimators table, and model validation methods like Anscombe.

The multivariate analysis techniques help choose the best combination of features and carry out the data dimension reduction. One of the most commonly used techniques is principal component analysis (PCA) [79].

Tests have been made in our previous work [2]. In Figure 2, we can see that dispersed residuals along a horizontal line without clear patterns are equally distributed on the upper and lower sides of the line. It is also noticeable that residuals have non-linear patterns. As we do not have any non-linear correlations, linear regression is a reasonable option for our study case.

Figure 3 shows that the residuals follow a straight line in the middle and do not deviate in a severe way, indicating that our quantile sets do in fact originate from normal distributions. However, residuals at the extremities curve off. This behavior typically indicates that our historical data have more extreme instances than would be expected if they really came from a normal distribution.

The spread-location plot is depicted in Figure 4. The residuals are distributed equally over the predictor ranges, and the horizontal line with evenly dispersed points supports the assumption of an equal variance.

Figure 5 displays the residual in relation to the leverage. Extreme values that might have an impact on a regression line can be seen in the lower right corner, outside of a dashed line, or at very few of Cook's distance points. These examples have an impact on the regression results, so excluding them will change the results.

The FactorMiner module in RStudio's PCA method is used for feature selection. It condensed the 147 columns to 30 variables with 99 percent of the information. The number of columns is depicted in Figure 6 together with the percentage of data that each one can carry. It aids in the optimization of memory usage during learning.

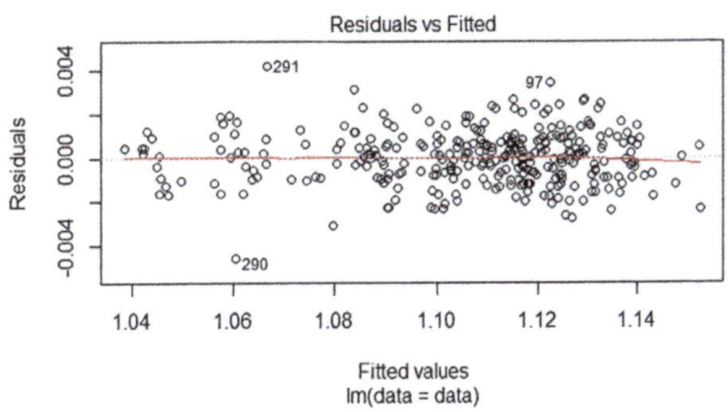

Figure 2. The residuals measure versus the fitted values.

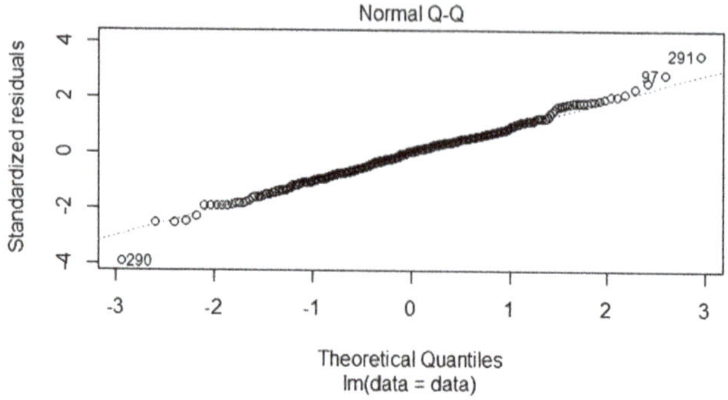

Figure 3. The normal Q-Q plot.

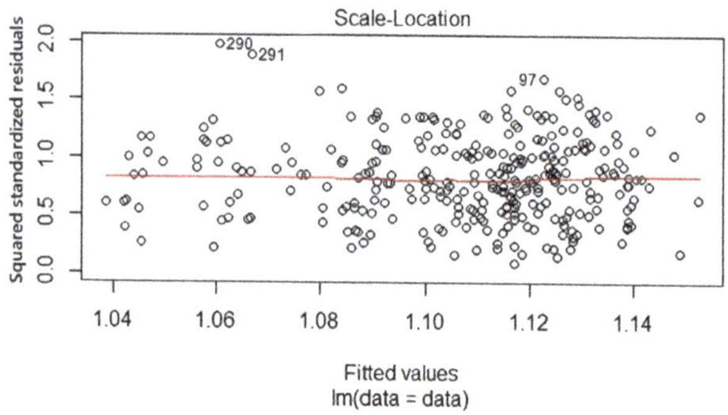

Figure 4. The location scale.

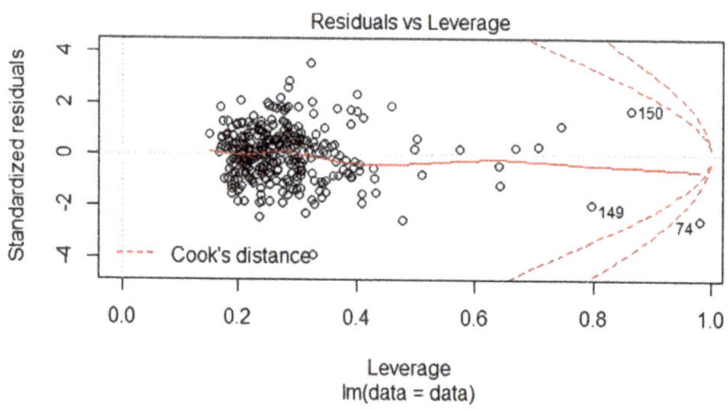

Figure 5. The leverage versus the residuals.

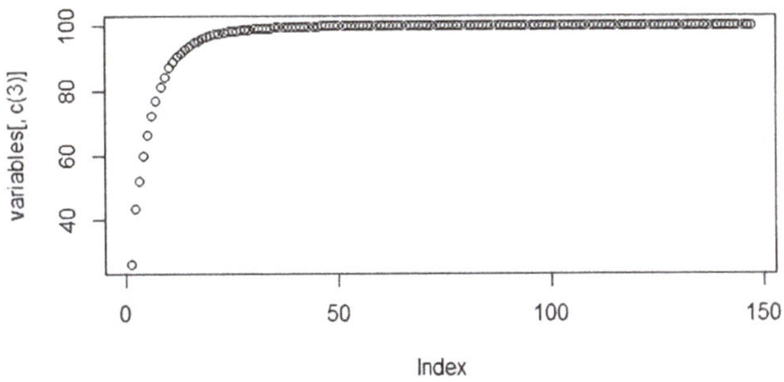

Figure 6. The data distribution analysis.

3.4. Methodologies Adopted

3.4.1. The Stationary Process

The stationary process is the technique we used in our proposed architecture. A stochastic process that is stationary in mathematics and statistics is one whose characteristics or probability distribution do not change as time passes. As a result, variables like the mean and the variance also remain constant over time [80]. Mathematically, a family of random variables is the typical definition of a stochastic or random process. Time series can be used to represent a variety of stochastic processes. A time series, on the other hand, is a collection of observations with integer indexes, but a stochastic process is continuous. The stochastic process is a process in which the characteristic variables undergo random fluctuations [81].

3.4.2. Stochastic Gradient Descent

The gradient descent algorithm reduces a function to its smallest value iteratively. The gradient descent algorithm is summarized in the formula below in a single line.

$$f(x) = \theta_0 + \sum_{n=1}^{p} \theta_i x_i = y \tag{1}$$

As we draw a random line through some of these data points in the space, this straight line's equation would be Y = mX + b, where m is the slope and b is the Y-axis intercept. A machine-learning model tries to predict what will happen with a new set of inputs based on what happened with a known set of inputs. The discrepancy between the expected and actual values would be the error:

$$\text{Error} = \text{Y (predicted)} - \text{Y (Actual)}$$

The concept of a cost function or a loss function is relevant here. The loss function calculates the error for a single training example in order to assess the performance of the machine learning algorithm.

The cost function, on the other hand, is the average of all the loss functions from the training samples. If the dataset has N total points and we want to minimize the error for each of those N points, the total squared error would be the cost function. Any machine learning algorithm's goal is to lower the cost function. To do this, we identify the value of X that results in the value of Y that is most similar to actual values. To locate the cost function minima, we devise the gradient descent algorithm formula [82].

The gradient descent algorithm looks like this:

Repeat until convergence

$$\theta_j := \theta_j - \frac{1}{m} \sum_{i=1}^{m} (h_\theta(x^i) - y^i) x_j^i \tag{2}$$

The SGD is similar to the gradient descent algorithm structure, with the difference that it processes one training sample at each iteration instead of using the whole dataset. SGD is widely used for training large datasets because it is computationally faster and can be processed in a distributed way. The fundamental idea is that we can arrive at a location that is quite near the actual minimum by focusing our analysis on just one sample at a time and following its slope. SGD has the drawback that, despite being substantially faster than gradient descent, its convergence route is noisier. Since the gradient is only roughly calculated at each step, there are frequent changes in the cost. Even so, it is the best option for online learning and big datasets.

3.4.3. PSO Metaheuristic Optimization Technique

Metaheuristics apply to discrete problems, and they can also adapt to continuous problems. These methods are stochastic, dealing with the combinatorial explosion of possibilities. They are inspired by physics and biology (such as evolutionary algorithms or ethology). They also share the same drawbacks: the difficulties of adjusting the method's parameters and the lengthy computation [83].

From the point of view of a particle, the PSO idea works by spreading a fleet of randomly made particles around a search space. Each particle has its own random speed and can evaluate its position to determine its best performance. This model is a good tool for solving linear and mixed-number problems and for situations where the numbers are mixed or continuous. More details about the PSO algorithm we used in our experimental study can be found in [2,83].

4. Experimental Setup and Analysis

4.1. The Proposed Architecture

During our survey, we noticed several techniques ensuring that the proposed system is adaptive, incremental, and consequently learning online. We can mention among adaptive approaches penalization for wrong predictions, higher weighting to more recent data, rewarding techniques, forgetting factors to ancient data, retraining, metaheuristics for parameter adapting, learning rate optimization, sliding windows, and iterations until the performance metric is optimized.

During our experimentation, we have chosen to study the paper in [57]. Their architecture shows an online model that processes financial data streams. They employ a sliding window of size 12, the model is incremental, and the system is adaptive by selecting a solution based on their best and worst performances. The model is tested in various cases: containing the dataset's statistical metrics, containing technical indicators, and both. In the experimental studies, their model predicts the next day, the following 3, 5, 7 and 15 days, and the following month. Additionally, they employ a variety of learning models, including Teaching Learning-Based Optimization (TLBO), the Jaya optimization method, Neural Networks (NNs), Functional Link Artificial Neural Networks (FLANNs) based on the PSO, and the Differential Evolution (DE) algorithm for weights optimization. Among the performance evaluation metrics they utilized were Theil's U, Annual Rental Value (ARV), Mean Absolute Error (MAE), and the Mean Absolute Percentage Error (MAPE).

Our idea consists of keeping the window size flexible depending on the concept drift instead of using a fixed window size. We minimize the window size when the concept drift occurs and negatively impacts our current model performance. We maximize the window size if the data trend is stable and no concept drift is detected. Our architecture in Figure 7 is composed of four parts. The first part consists of collecting and scraping the data from the sources we use. The second part focuses on the data preprocessing. In this stage, we first unify the granularity of our historical data on a daily basis, then compute 12 technical indicators for each of the three currency pairs. Online learning starts with the concept of drift detection; this part is responsible for maintaining the stability of our model's performance by continuously adapting it to new trends. We have tested several techniques and mainly chose the study that tests if the data is stochastic. The presence or absence of concept drift leads to the choice of the window size. The next step will be to fit our model. Since the PSO usage requires more calculations than the SGD alone, we only launch the PSO every 60 days. Using the PSO helps prevent falling into the local minima. Last but not least, we visualize our prediction results and performance metrics progress in several plots and tables.

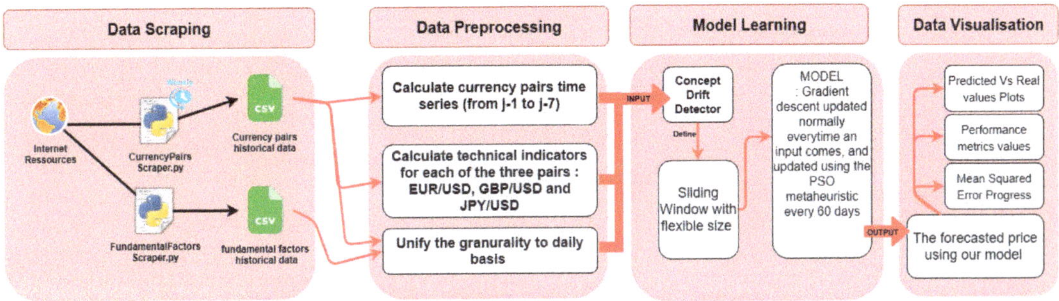

Figure 7. Our proposed architecture.

4.2. The Experiment Environment

The model has been developed using the Anaconda distribution of the Python and R programming languages for scientific computing. Specifically, we used Jupyter Notebook 6.4.12 for coding with Python version 3.9.13. Windows is the operating system where the experiment was carried out.

4.3. The Parameters' Description

This section explains the different parameters used in our algorithm. We initialized the sliding window size with the value 15, which is likely to be minimized when a change in the time series pattern is detected. We chose 15 because, from our preliminary tests, we saw that, in general, patterns vary from 15 days to another. The variable numRows is our index variable that we use to process the data stream. It is continuously updated by the last instance index that we processed in our model training. The variables rangeMin and rangeMax help us display the instances index from the dataset that our model is being trained with. The PSOApplication parameter helps us count how many days have been processed, so that as soon as we reach 60 days, we launch the PSO optimization technique to help the SGD boost its results.

The PSO parameters are mainly c1 and cmax and are used while updating the weights. xmin and xmax are the range where we try to find our solution. It can be challenging to fix it every time we use a new dataset, and limiting it can take a lot of tests until it is properly fixed. However, once fixed for a particular time series, we continue using it for training and forecasting future values. The number of particles depends on our time and computational power limitations. In this study, we fixed it at 20 particles after several tests on different values. The parameters vmin and vmax determine the speed of movement in the search space for each particle.

The gradient descent parameters are the tolerance that we use as a condition to stop the optimization. It represents the margin of error that we tolerate once it is reached. xmin and xmax are also given as parameters to the SGD.

4.4. The Proposed Search Algorithm

The model-learning phase is composed of many parts. When new instances are available, the stationarity test is performed to specify the next window size, and then the SGD model receives the new input to update the model parameters. Every 60 days, we update the model using the PSO optimization metaheuristic. Our proposed search Algorithm 1 steps are:

Algorithm 1: The SGD algorithm optimized using the PSO metaheuristic with an adaptive sliding window

```
Input: The window is initialized with the first 15 elements, and the SGD and the
       PSO metaheuristic parameters are initialized by learning from the first
       window instances.
Initialization;
windowsize = 15;
numrows = 15;
rangeMin = 0;
rangeMax = 15;
PSOApplication = 1;
while We have a new input with 15 instances or higher do
    Save the previous window in a variable previousWindow;
    Save the next 15 instances in a variable currentWindow;
    Create the target variable combining our target variable from the
      previousWindow and currentWindow datasets;
    if The target variable is stationary then
        windowsize = 15;
    else
        windowsize = 1;
        /* Put in the variable currentwindow the instances ranging from
           numrows+1-15 to numrows+1                                    */
    numrows=numrows+windowsize;
    Validate the current SGD model using the current window;
    Give as input the current window instance to the SGD to obtain new weights
      for our online model;
    if numrows>= 60*PSOApplication then
        Give as input the last 60 days to the PSO optimizer and receive as output
          the new weight for our SGD model;
        PSOApplication+=1;
```

The implementation of the SGD and PSO is inspired by our previous work, Ref. [2], where we developed the classic gradient descent optimized using Particle Swarm Optimization. We used almost the same gradient descent algorithm for the SGD, as the difference between them is limited to the used training data. For the gradient descent, all the training data are received as an input to the Algorithm 2, and then a backpropagation is applied to adjust the model weights. The SGD algorithm receives a new part of the training data at each iteration and adjusts its weights to that subset of data using a backpropagation.

Figure 8 shows our proposed method flowchart, which consists of adapting the sliding window size based on the stationarity statistical test. After this, the forecasting model receives the sliding window instances as input for validation and training.

Algorithm 2: Gradient algorithm descent optimized with Particle Swarm Optimization

Data: Algorithm input, c1, and cmax are used while updating the weights. xmin and xmax determine the search space of weights, determined thanks to the numerous tests carried out:

initialization;
Tolerance = 0.001;
$c_1 = 0.1$;
$c_{max} = (2 / 0.97725) * c_1$;
$x_{min} = -0.3$;
$x_{max} = 0.3$;
numberOfParticles = 20;
foreach *particle* $p_i \in P$ **do**
    initialization;
    Regression formula weights with random values:
    $w_i = \text{random}(x_{min}, x_{max})$;
    Particles' movement speed for each particle by a real random value:
    $v_{min} = (x_{min} - x_{max})/2$;
    $v_{max} = (x_{max} - x_{min})/2$;
end
while *MeanSquaredError > tolerance* **do**
    foreach *particle* $p_i \in P$ **do**
        Calculate the predicted value by the model using the multiple linear regression function:

$$y_i = w_0 + w_1 x_1 + x_2 x_2 + \cdots + w_d x_d = w_0 + \sum_{i=1}^{d} w_i x_i$$

        Calculate the value of the function that we are optimizing and that is presented by the derivative of the squared difference between the predicted values and the actual values:

$$\frac{\partial J_n(w)}{\partial w} = \frac{2}{n} \sum_{i=1}^{n} (y_i - w_0 - w_1 x_1 - w_2 x_2 - \cdots - w_d x_d) x_{i,j}$$

        Speeds vd and weights wd update with d the number of weights:
        $v_d \leftarrow c_1 v_d + random(0, c_{max})(p_{dxd}) + random(0, c_{mmax})(p_d - x_d) + random(, c_{max})(g_d - x_d)$
        $x_d \leftarrow x_d + v_d$
        with :

        - gd : The list of weights having realized the best results in the whole swarm.
        - pd : The list of weights having realized the best results in the current particle.
          Calculate the value of the error function derivative and store it in a variable called error;

        if *the error of the current particle is higher than the stored one in error variable* **then**
            Assign the error of the current particle to error variable;
        else
            Go to the next particle without doing anything;
        end
    end
end

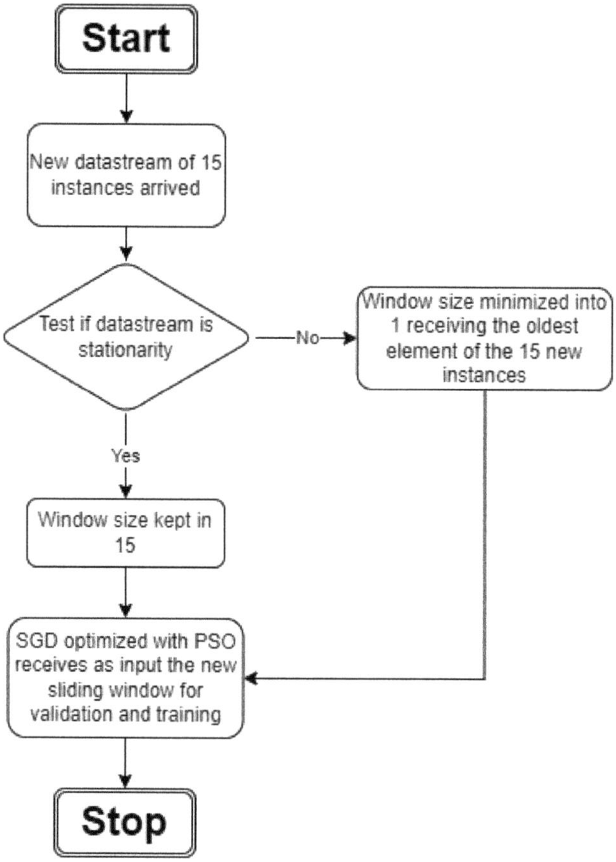

Figure 8. The proposed method flowchart.

4.5. The Performance Evaluation Metrics

The model quality measurement metrics are many, depending on the algorithm type and also on the study case. In our experiment, we evaluated our model using two techniques. The first one is the Mean Square Error (MSE), which represents the loss function we wish to optimize:

$$f(x) = \frac{1}{n}\sum_{i=1}^{n}(y_i - f(x_i))^2 \tag{3}$$

The function $f(x)$ is the gradient descent function for predicting our target variable, which is the EUR/USD exchange rate:

$$f(x) = w_0 + w_1x_1 + w_2x_2 + \cdots + w_dx_d = w_0 + \sum_{i=1}^{d} w_ix_i \tag{4}$$

Our goal is to find the optimal weights of the regression function in our iterative process by optimizing our loss function. Weights are optimized in our SGD using the following formula, where the tolerance is fixed to 0.001 in our experimental part and the error is simply the difference between the real and predicted value of the target variable, the EUR/USD exchange rate in our case:

$$w_i = w_i - (tolerance * error) \tag{5}$$

Our second metric is a classification metric, where we study the accuracy of predicting if the exchange rate will rise or fall. It is presented in the form of a percent, and its formula is the following:

$$Accuracy = \frac{TP + TN}{TP + TN + FP + FN} \quad (6)$$

The third metric we used is average rectified value *ARV*, which shows the average variation for a group of data points. The forecasting model performs identically if we calculate the mean over the series with *ARV* = 1. The model is considered worse than just taking the mean if *ARV* > 1.

$$ARV = \frac{\sum_{i=1}^{N}(y_i - f(x_i))^2}{\sum_{i=1}^{N}|(y_i - f(x_i))|} \quad (7)$$

4.6. Analysis of Results

This section will discuss the various steps and the evolution of our experimental study. In the first tests, we studied the variance and the MSE of the SGD algorithm during its learning process. As shown in Figure 7, in the model learning phase, the concept drift detector, the stationarity statistical test in this case, is applied, right after the sliding window size is determined accordingly, and then we work on optimizing the exchange rate forecasting margin error of the multilinear regression function that we optimize in our SGD algorithm.

For optimizing the regression model weights, both SGD and PSO every 60 days are involved in our experimental study. In SGD, weights are optimized using the gradient descent weights update, Formula (5), in the part of Algorithm 1 where we give as input the new window to the SGD. The second way weights are updated is based on the PSO formula in Algorithm 2 where the particle speed is added to the weight value:

$$x_d \leftarrow x_d + v_d \quad (8)$$

As illustrated in Algorithm 2, the speed is then calculated based on the the search space range, the best particle weight found, the best swarm weight found, and the current weight.

Figures 9 and 10 show that the SGD itself is making good progress. After receiving 1000 instances, the mean squared error (MSE) became more stable, and the variance reached its best stability after receiving 3000 instances.

We updated the learning rate by adding or subtracting 20% of its value and multiplying it by 0.99 or 1.01. We remarked that the learning rate has no impact on the model performance improvement in the case of our architecture. The results did not change and stayed similar to those obtained with the default parameters. As we noted, even when making the previously mentioned updates on the learning rate, the error still does not stabilize until the algorithm reaches around 1000 processed instances.

Figure 11 shows the EUR/USD close price historical data from 1 January 2001 to 1 January 2004. We notice that the value range changed completely comparing 2001 and 2002 to 2003 and 2004, revealing the importance of online learning for financial time series processing.

Figures 12 and 13 show the predicted values in orange versus the actual values in blue using the SGD alone. On the other hand, Figure 14 shows the results as we integrate the PSO metaheuristic every 60 days into the learning process. The accuracy for all the plots is good and reaches 82%. This means that the models correctly predict the price direction in 82% of the cases. The added value of the PSO metaheuristic is noticeable in terms of the margin error, which decreases significantly as the price decreases. The PSO helped minimize the margin error between the predicted and actual values as the price crashed between instances 20 and 30.

Figures 15 and 16 show, the EUR/USD daily close price time series and histogram, respectively, from 30/05/2000 to 28/07/2000. The price values show the volatility of the time series data stream that we need to deal with using concept drift detection techniques.

Tables 1–3 summarize statistical values such as the mean and the variance. They also contain a *p*-value that indicates whether the data is stationary or not. If the *p*-value is higher than 0.05, the null hypothesis (H0) cannot be rejected, and the data are non-stationary. The results show that in the case of this two-month time series, we have a stationary trend every 15 days, but as we study a whole month or two months, the trend is non-stationary.

We made tests to compare the fixed and flexible window sizes. For the fixed-size case, the chosen size is 15 instances at each iteration because, according to our statistical studies, the data tend to have the same pattern every two weeks. For the flexible window size, we study the next 15 days' stationarity. If the data are stationary, the algorithm receives 15 new instances. If the data are not stationary, the algorithm receives only one new instance.

Table 4 shows the prediction results for year 2000 EUR/USD historical data as it represents the first data received by the system. In most intervals except [75:90], [90:105], and [120:135], the mean squared error for the flexible-size window case exceeds the fixed-size window case. Meanwhile, for all intervals, we notice that the accuracy using the flexible-size window exceeds or equals the accuracy given using a fixed-size window. To illustrate the predicted vs. the real values, Figures 17 and 18 show the interval [60:74]. We can see that at each point, the real and predicted values are closer in the flexible approach compared to the fixed window approach. The ARV results are all way smaller than 1, which means that our model predicts way better than simply taking the mean. ARV also shows the data points' variation, and from the obtained values, we can see that the instances are not too correlated to one another.

Table 1. The EUR/USD statistics from 30 May 2000 to 28 July 2000.

Mean	0.9819729708789546
Variance	0.008933902258522159
ADF Statistic	−1.535025
p-value	0.516119
Critical Values	1%: −3.563; 5%: −2.919; 10%: −2.597

Table 2. The EUR/USD statistics from 30 May 2000 to 28 July 2000 split into two equal parts.

Data	1st 30 Days	2nd 30 Days
Mean	0.984582	0.979364
Variance	0.008592	0.008965
ADF Statistic	−1.084344	−1.193593
p-value	0.721275	0.676346
Critical Values	1%: −3.770; 5%: −3.005; 10%: −2.643	1%: −3.809; 5%: −3.022; 10%: −2.651

Table 3. The EUR/USD statistics from 30 May 2000 to 28 July 2000 split into four equal parts.

Data	1st 15days	2nd 15days	3rd 15days	4th 15days
Mean	0.983657	0.985506	0.986441	0.972288
Variance	0.008653	0.008529	0.008389	0.009440
ADF Statistic	−4.419853	−3.628841	−17.540606	−16.313688
p-value	0.000274	0.005233	0.0	0.0
Critical Values	1%: −4.473 5%: −3.290 10%: −2.772	1%: −4.473 5%: −3.290 10%: −2.772	1%: −4.473 5%: −3.290 10%: −2.772	1%: −4.473 5%: −3.290 10%: −2.772

Table 4. The flexible vs. the fixed sliding window results from year 2000 EUR/USD historical data.

Flexible Window Size				**Fixed Window Size**			
Range	**MSE**	**ARV**	**Accuracy**	**Range**	**MSE**	**ARV**	**Accuracy**
[0:15]	0.0027	0.0519	66.6	[0:15]	0.002902	0.0538	66.6
[1:16]	0.002534	0.0503	66.6	[15:30]	0.00528	0.0726	80.0
[2:17]	0.00237	0.0486	73.3				
[3:18]	0.00217	0.0465	73.3				
[4:19]	0.0028	0.0529	73.3				
[5:20]	0.002	0.0447	73.3				
[6:21]	0.00192	0.0438	80.0				
[7:22]	0.00201	0.0448	86.6				
[8:23]	0.0031	0.0556	86.6				
[9:24]	0.00404	0.0635	80.0				
[10:25]	0.00528	0.0726	80.0				
[11:26]	0.00608	0.0779	80.0				
[26:41]	0.00727	0.0852	66.6	[30:45]	0.01416	0,1189	66.6
[27:42]	0.00658	0.0811	66.6				
[28:43]	0.00586	0.0765	66.6				
[29:44]	0.00495	0.0703	66.6				
[30:45]	0.00494	0.0702	66.6				
[31:46]	0.00546	0.0738	66.6	[45:60]	0.00607	0.0779	93.3
[32:47]	0.00686	0.0828	66.6				
[33:48]	0.00623	0.0789	66.6				
[34:49]	0.00699	0.0836	66.6				
[35:50]	0.00788	0.0887	73.3				
[36:51]	0.00874	0.0934	73.3				
[37:52]	0.00917	0.0957	73.3				
[38:53]	0.00895	0.0946	73.3				
[39:54]	0.01016	0,1007	80.0				
[40:55]	0.00887	0.0941	86.6				
[41:56]	0.00856	0.0925	93.3				
[42:57]	0.00857	0.0925	93.3				
[43:58]	0.00874	0.0934	93.3				
[44:59]	0.00895	0.0946	93.3				
[45:60]	0.00849	0.0921	93.3				
[46:61]	0.00922	0.096	93.3	[60:75]	0.00936	0.0967	73.3

Table 4. *Cont.*

Flexible Window Size				Fixed Window Size			
Range	MSE	ARV	Accuracy	Range	MSE	ARV	Accuracy
[47:62]	0.00782	0.0884	93.3				
[48:63]	0.0071	0.0842	86.6				
[49:64]	0.00649	0.0805	86.6				
[50:65]	0.00635	0.0796	86.6				
[51:66]	0.00657	0.081	86.6				
[52:67]	0.00709	0.0842	80.0				
[53:68]	0.00843	0.0918	80.0				
[54:69]	0.00836	0.0914	80.0				
[55:70]	0.00908	0.0952	73.3				
[56:71]	0.01043	0,1021	73.3				
[57:72]	0.01237	0,1112	73.3				
[58:73]	0.00573	0.0756	73.3				
[59:74]	0.00464	0.0681	73.3				
[74:89]	0.00688	0.0829	80.0	[75:90]	0.0047	0.0685	80.0
[75:90]	0.00781	0.0883	80.0				
[76:91]	0.00917	0.0957	80.0	[90:105]	0.00672	0.0819	60.0
[77:92]	0.01095	0.1046	80.0				
[78:93]	0.0127	0.1126	86.6				
[79:94]	0.01469	0.1212	80.0				
[80:95]	0.01651	0.1284	73.3				
[81:96]	0.01672	0.1293	73.3				
[82:97]	0.01695	0.1301	73.3				
[83:98]	0.01727	0.1314	66.6				
[84:99]	0.01819	0.1348	66.6				
[85:100]	0.01864	0.4317	66.6				
[86:101]	0.0186	0.1363	66.6				
[87:102]	0.01881	0.1371	66.6				
[88:103]	0.01814	0.1346	66.6				
[89:104]	0.01837	0.1355	66.6				
[90:105]	0.01754	0.1324	60.0				
[91:106]	0.01636	0.1279	60.0	[105:120]	0.03948	0,1986	53.3
[92:107]	0.01427	0.1194	60.0				
[93:108]	0.01176	0.1084	66.6				
[94:109]	0.00956	0.0977	73.3				
[95:110]	0.00766	0.0875	73.3				
[96:111]	0.0074	0.086	73.3				
[97:112]	0.00712	0.0843	73.3				
[98:113]	0.00684	0.0827	66.6				
[99:114]	0.00586	0.0765	60.0				
[100:115]	0.00486	0.0697	60.0				
[101:116]	0.00624	0.0789	66.6				
[102:117]	0.00562	0.0749	60.0				
[103:118]	0.00836	0.0914	53.3				
[104:119]	0.01005	0.1002	60.0				
[105:120]	0.01068	0.1033	53.3				
[120:135]	0.01108	0.10526	60.0	[120:135]	0.00885	0.094	60.0

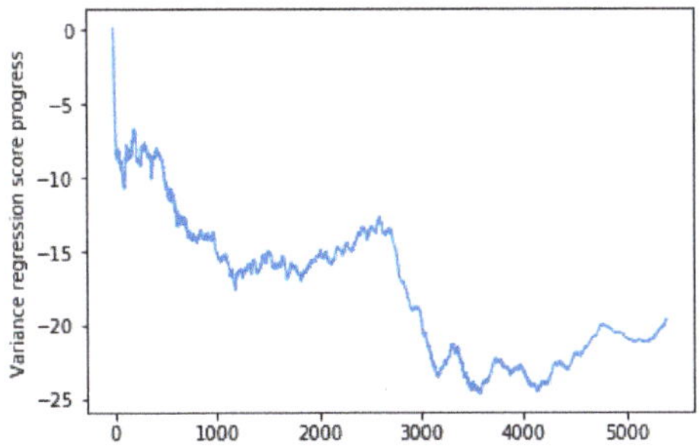

Figure 9. The variance regression score progress.

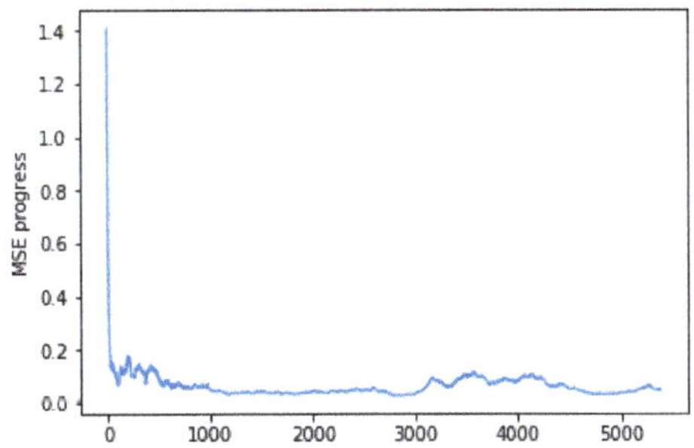

Figure 10. The MSE regression score progress.

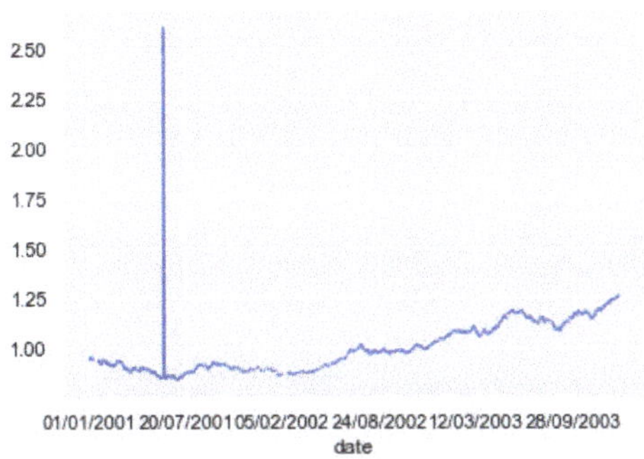

Figure 11. The EUR/USD close price historical data from 1 January 2001 to 1 January 2004.

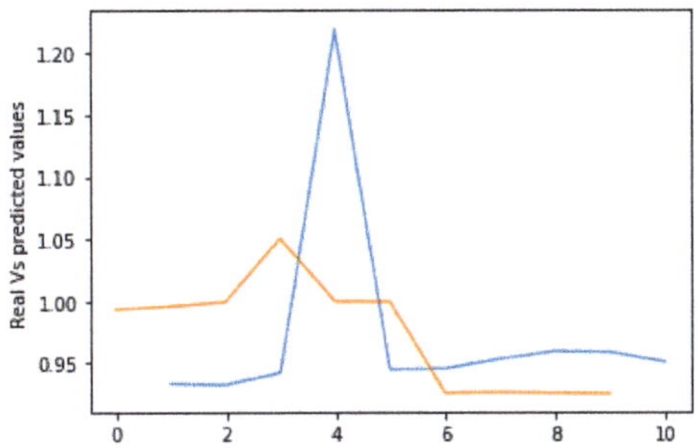

Figure 12. The real vs. the predicted values using the SGD algorithm.

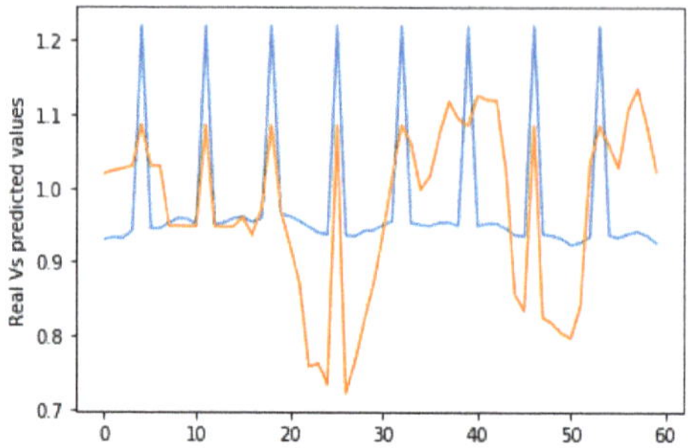

Figure 13. The real vs. the predicted values using the SGD algorithm on a bigger test dataset.

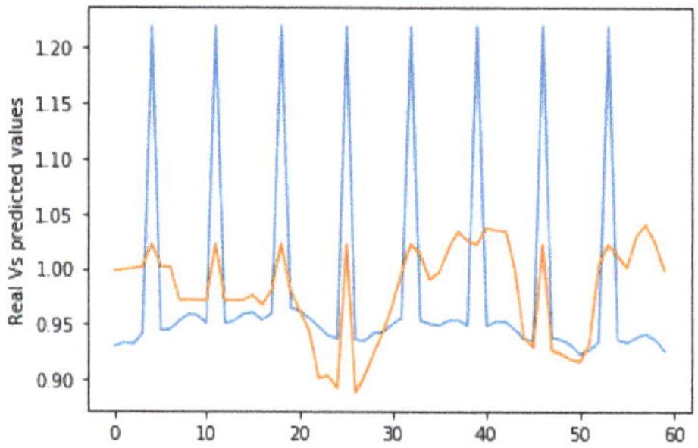

Figure 14. The real vs. the predicted values using the SGD algorithm optimized using the PSO metaheuristic every 60 days.

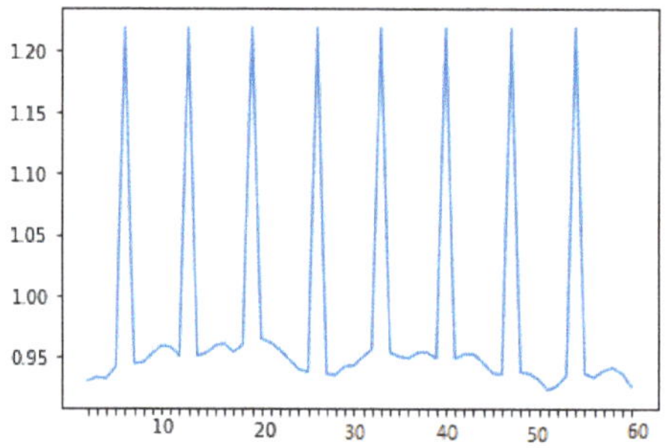

Figure 15. The EUR/USD daily close price time series from 30 May 2000 to 28 July 2000.

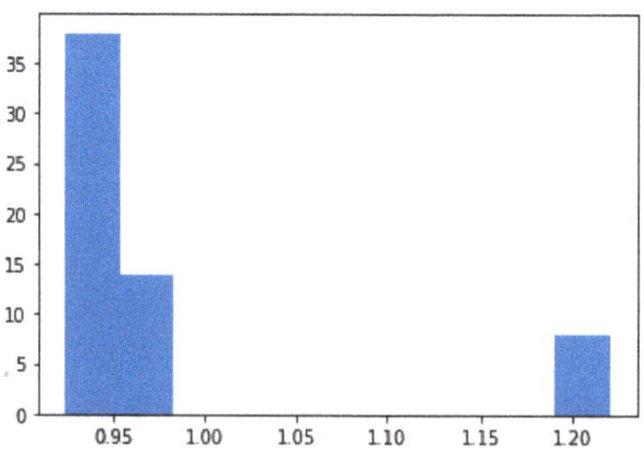

Figure 16. The EUR/USD daily close price histogram from 30 May 2000 to 28 July 2000.

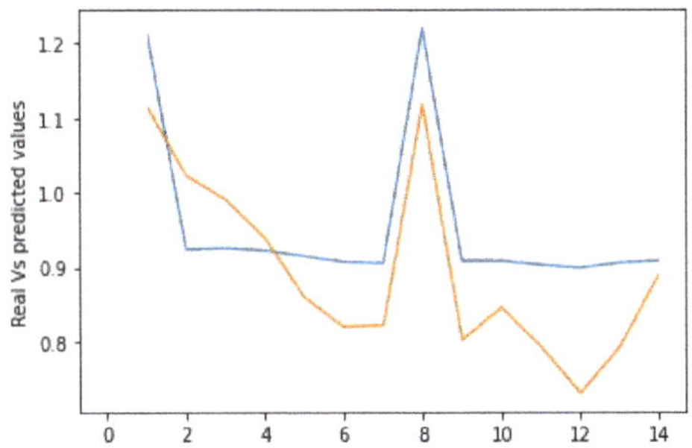

Figure 17. The flexible window: the predicted vs. the real value for the interval [60:74].

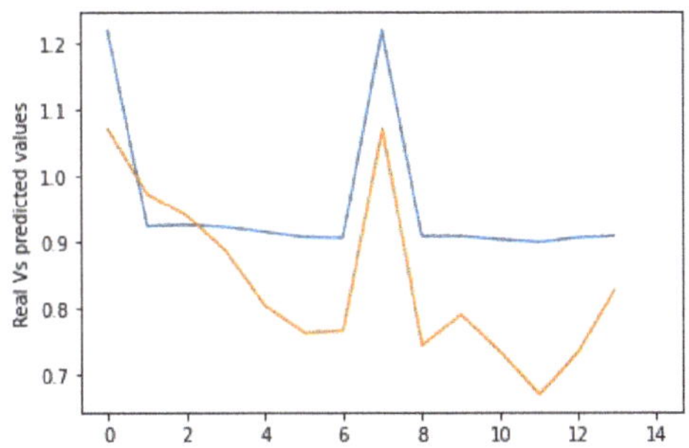

Figure 18. The fixed window: the predicted vs. the real value for the interval [60:74].

4.7. Discussions

Figure 9 shows the regression score variance. We see that the model should perform the learning through multiple sliding windows and receive a certain number of instances to reach the point where we can rely on the proposed algorithm results for decision making.

The same is true for Figure 10, where the mean squared error convergence reached its limit starting from receiving approximately 1000 instances. One of the biggest challenges of using gradient descent algorithms is building a model that converges as much as possible. In addition, the best convergence is not guaranteed with the first algorithm execution. Since the weights are often primarily initialized randomly, little by little we limit the search space of the optimal weights to a smaller range.

The learning rate speeds up as the gradient moves while descending. If you set it too high, your path will become unstable, and if you set it too low, the convergence will be slow. If you set it to zero, your model is not picking up any new information from the gradients. As we worked on updating the learning rate alpha by decreasing or increasing its value, we did not notice a difference, and we still obtained the best convergence beyond receiving 1000 instances. The fact that reducing the error to some extent only requires receiving a certain amount of data may help to explain those results.

On the other hand, Figure 11 reveals the importance of DSM to erase the old irrelevant models and build a newer one that fits the new data trends. However, keeping the irrelevant models aside for potential future use can be a good idea. As for some study cases, patterns can reappear occasionally or periodically.

Figures 12–14 compared integrating the PSO metaheuristic to online learning vs. not using it. The positive impact is noticed as the price crashes. The margin error was significantly reduced when the PSO was used. Even though the computational and time costs of using the PSO are higher, integrating it periodically to enhance the forecasting quality is promising.

The volatility illustrated in Figures 15 and 16 is one of the biggest challenges encountered in the FTSERF. It has to be managed by minimizing the risks that it reveals. In cases of high volatility, using flexible sliding windows becomes a must. By doing this, we can guarantee that the windows are the right size to see emerging trends and make wise choices.

As noticed from Figures 17 and 18, flexible sliding windows ensured the suggested algorithm had an optimal duration, accuracy, and error margin. The PSO periodic integration and the adaptive sliding windows achieved the fastest convergence. The training and forecasting performances of the algorithm with a flexible window size are better when we compare them to those of the learning algorithm with a fixed window size.

In traditional machine learning, the future fluctuations are adjusted based on previous expectation errors. It consists of investing historical knowledge about past fluctuations, and the model is making decisions or forecasts based on the training it went through. However, as we integrate DSM techniques, adaptive expectations are also ensured by calculating the statistical distribution for every new data stream. The model receives at each iteration fifteen instances, which are minimized to one instance at each iteration as soon as a high level of volatility is detected in the fifteen instances of the new sliding window, which makes the model more adaptive compared to real-time approaches without data stream mining techniques that work on detecting the change and reacting to it.

5. Conclusions and Perspectives

Our study aims to explore the DSM techniques' efficiency in financial time series forecasting. We mainly used the SGD, and for weight optimization, we integrated the PSO metaheuristic periodically every 60 days. Our target variable was the Euro's value relative to the US dollar. The first technique involved in DSM is adaptive sliding windows. We tested the cases of using a flexible window whose size changes depending on the data volatility versus using a fixed sliding window. The second technique involved in DSM is change detection. It is the stationarity statistical study where we test if the time series has a constant variance. The flexible sliding window proved its ability to forecast the price direction, as it achieved better accuracy compared to using a fixed sliding window. The adaptivity with the changes in the dataset patterns also assured better price and value forecasting with less margin error, especially as the PSO is involved. Future work will focus on testing more online models and concept drift techniques for financial time series and comparing the strengths and weaknesses of each one. Further experimental tests can also be performed by including other periods of crisis and testing other financial time series.

Author Contributions: Conceptualization, Z.B.; methodology, Z.B.; software, Z.B.; validation, Z.B.; formal analysis, Z.B.; investigation, Z.B.; resources, Z.B.; data curation, Z.B.; writing—original draft preparation, Z.B.; writing—review and editing, Z.B.; visualization, Z.B.; supervision, O.B. and J.S.-M.; project administration, O.B. and J.S.-M.; funding acquisition, O.B. and J.S.-M. All authors have read and agreed to the published version of the manuscript.

Funding: This work was supported by a scholarship received from Erasmus+ exchange program and funding from Centro de Innovación para la Sociedad de la Información, University of Las Palmas de Gran Canaria (CICEI-ULPGC).

Data Availability Statement: We published the dataset we used in this research at the following link: https://github.com/zinebbousbaa/eurusdtimeseries accessed on 27 March 2023.

Conflicts of Interest: All authors have no conflict of interest to disclose.

Abbreviations

The following abbreviations are used in this manuscript:

DSM	Data Stream Mining
FTSERF	Financial Time Series Exchange Rate Forecasting
PSO	Particle Swarm Optimization Metaheuristic
SGD	Stochastic Gradient Descent

References

1. Gerlein, E.A.; McGinnity, M.; Belatreche, A.; Coleman, S. Evaluating machine learning classification for financial trading: An empirical approach. *Expert Syst. Appl.* **2016**, *54*, 193–207. [CrossRef]
2. Bousbaa, Z.; Bencharef, O.; Nabaji, A. Stock Market Speculation System Development Based on Technico Temporal Indicators and Data Mining Tools. In *Heuristics for Optimization and Learning*; Springer: Berlin/Heidelberg, Germany, 2021; pp. 239–251.
3. Stitini, O.; Kaloun, S.; Bencharef, O. An Improved Recommender System Solution to Mitigate the Over-Specialization Problem Using Genetic Algorithms. *Electronics* **2022**, *11*, 242. [CrossRef]
4. Jamali, H.; Chihab, Y.; García-Magariño, I.; Bencharef, O. Hybrid Forex prediction model using multiple regression, simulated annealing, reinforcement learning and technical analysis. *Int. J. Artif. Intell. ISSN* **2023**, *2252*, 8938. [CrossRef]
5. Bifet, A.; Holmes, G.; Pfahringer, B.; Kranen, P.; Kremer, H.; Jansen, T.; Seidl, T. Moa: Massive online analysis, a framework for stream classification and clustering. In Proceedings of the First Workshop on Applications of Pattern Analysis, PMLR, Windsor, UK, 1–3 September 2010; pp. 44–50.
6. Bifet, A. *Adaptive Stream Mining: Pattern Learning and Mining from Evolving Data Streams*; Ios Press: Amsterdam, The Netherlands, 2010; Volume 207.
7. Thornbury, W.; Walford, E. *Old and New London: a Narrative of Its History, Its People and Its Places*; Cassell Publisher: London, UK, 1878; Volume 6.
8. Cummans, J. A Brief History of Bond Investing. 2014. Available online: http://bondfunds.com/ (accessed on 24 February 2018).
9. BIS site development project. Triennial central bank survey: Foreign exchange turnover in April 2016. *Bank Int. Settl.* **2016**. Available online: https://www.bis.org/publ/rpfx16.htm (accessed on 24 February 2018).
10. Lange, G.M.; Wodon, Q.; Carey, K. *The Changing Wealth of Nations 2018: Building a Sustainable Future*; Copyright: International Bank for Reconstruction and Development, The World Bank 2018, License type: CC BY, Access Rights Type: open, Post date: 19 March 2018; World Bank Publications: Washington, DC, USA, 2018; ISBN 978-1-4648-1047-3.
11. Makridakis, S.; Hibon, M. ARMA models and the Box–Jenkins methodology. *J. Forecast.* **1997**, *16*, 147–163. [CrossRef]
12. Tinbergen, J. Statistical testing of business cycle theories: Part i: A method and its application to investment activity. In *Statistical Testing of Business Cycle Theories*; Agaton Press: New York, NY, USA, 1939; pp. 34–89.
13. Xing, F.Z.; Cambria, E.; Welsch, R.E. Natural language based financial forecasting: a survey. *Artif. Intell. Rev.* **2018**, *50*, 49–73. [CrossRef]
14. Cheung, Y.W.; Chinn, M.D.; Pascual, A.G. Empirical exchange rate models of the nineties: Are any fit to survive? *J. Int. Money Financ.* **2005**, *24*, 1150–1175. [CrossRef]
15. Clifton, C., Jr.; Frazier, L.; Connine, C. Lexical expectations in sentence comprehension. *J. Verbal Learn. Verbal Behav.* **1984**, *23*, 696–708. [CrossRef]
16. Brachman, R.J.; Khabaza, T.; Kloesgen, W.; Piatetsky-Shapiro, G.; Simoudis, E. Mining business databases. *Commun. ACM* **1996**, *39*, 42–48. [CrossRef]
17. Hu, M.; Liu, B. Mining and summarizing customer reviews. In Proceedings of the Tenth ACM SIGKDD International Conference on Knowledge Discovery and Data Mining, Seattle, WA, USA, 22–25 August 2004; pp. 168–177.
18. Ali, T.; Omar, B.; Soulaimane, K. Analyzing tourism reviews using an LDA topic-based sentiment analysis approach. *MethodsX* **2022**, *9*, 101894. [CrossRef]
19. Cambria, E.; White, B. Jumping NLP curves: A review of natural language processing research. *IEEE Comput. Intell. Mag.* **2014**, *9*, 48–57. [CrossRef]
20. Rather, A.M.; Sastry, V.; Agarwal, A. Stock market prediction and Portfolio selection models: A survey. *Opsearch* **2017**, *54*, 558–579. [CrossRef]
21. Cavalcante, R.C.; Brasileiro, R.C.; Souza, V.L.; Nobrega, J.P.; Oliveira, A.L. Computational intelligence and financial markets: A survey and future directions. *Expert Syst. Appl.* **2016**, *55*, 194–211. [CrossRef]
22. Gadre-Patwardhan, S.; Katdare, V.V.; Joshi, M.R. A Review of Artificially Intelligent Applications in the Financial Domain. In *Artificial Intelligence in Financial Markets*; Springer: Berlin/Heidelberg, Germany, 2016; pp. 3–44.
23. Curry, H.B. The method of steepest descent for non-linear minimization problems. *Q. Appl. Math.* **1944**, *2*, 258–261. [CrossRef]
24. Shao, H.; Li, W.; Cai, B.; Wan, J.; Xiao, Y.; Yan, S. Dual-Threshold Attention-Guided Gan and Limited Infrared Thermal Images for Rotating Machinery Fault Diagnosis Under Speed Fluctuation. *IEEE Trans. Ind. Inform.* **2023**, 1–10. [CrossRef]
25. Lv, L.; Zhang, J. Adaptive Gradient Descent Algorithm for Networked Control Systems Using Redundant Rule. *IEEE Access* **2021**, *9*, 41669–41675. [CrossRef]
26. Sirignano, J.; Spiliopoulos, K. Stochastic gradient descent in continuous time. *Siam J. Financ. Math.* **2017**, *8*, 933–961. [CrossRef]
27. Audrino, F.; Trojani, F. Accurate short-term yield curve forecasting using functional gradient descent. *J. Financ. Econ.* **2007**, *5*, 591–623.
28. Bonyadi, M.R.; Michalewicz, Z. Particle swarm optimization for single objective continuous space problems: A review. *Evol. Comput.* **2017**, *25*, 1–54. [CrossRef]
29. Kennedy, J.; Eberhart, R. Particle swarm optimization. In Proceedings of the ICNN'95-International Conference on Neural Networks, Perth, WA, Australia, 27 November–1 December 1995; IEEE: Piscataway, NJ, USA, 1995; Volume 4, pp. 1942–1948.

30. Shi, Y.; Eberhart, R. A modified particle swarm optimizer. In Proceedings of the 1998 IEEE International Conference on Evolutionary Computation Proceedings. IEEE World Congress on Computational Intelligence (Cat. No. 98TH8360), Anchorage, AK, USA, 4–9 May 1998; IEEE: Piscataway, NJ, USA, 1998; pp. 69–73.
31. Jha, G.K.; Thulasiraman, P.; Thulasiram, R.K. PSO based neural network for time series forecasting. In Proceedings of the 2009 International Joint Conference on Neural Networks, Atlanta, GA, USA, 14–19 June 2009; IEEE: Piscataway, NJ, USA, 2009; pp. 1422–1427.
32. Wang, K.; Chang, M.; Wang, W.; Wang, G.; Pan, W. Predictions models of Taiwan dollar to US dollar and RMB exchange rate based on modified PSO and GRNN. *Clust. Comput.* **2019**, *22*, 10993–11004. [CrossRef]
33. Junyou, B. Stock Price forecasting using PSO-trained neural networks. In Proceedings of the 2007 IEEE Congress on Evolutionary Computation, Singapore, 25–28 September 2007; IEEE: Piscataway, NJ, USA, 2007; pp. 2879–2885.
34. Yang, F.; Chen, J.; Liu, Y. Improved and optimized recurrent neural network based on PSO and its application in stock price prediction. *Soft Comput.* **2021**, *27*, 3461–3476. [CrossRef]
35. Huang, C.; Zhou, X.; Ran, X.; Liu, Y.; Deng, W.; Deng, W. Co-evolutionary competitive swarm optimizer with three-phase for large-scale complex optimization problem. *Inf. Sci.* **2023**, *619*, 2–18. [CrossRef]
36. Auer, P. Online Learning. In *Encyclopedia of Machine Learning and Data Mining*; Sammut, C., Webb, G.I., Eds.; Springer: Boston, MA, USA, 2016; pp. 1–9.
37. Benczúr, A.A.; Kocsis, L.; Pálovics, R. Online machine learning algorithms over data streams. *J. Encycl. Big Data Technol.* **2018**, 1207–1218.
38. Julie, A.; McCann, C.Z. Adaptive Machine Learning for Changing Environments. 2018. Available online: https://www.turing.ac.uk/research/research-projects/adaptive-machine-learning-changing-environments (accessed on 1 September 2018).
39. Grootendorst, M. Validating your Machine Learning Model. 2019. Available online: https://towardsdatascience.com/validating-your-machine-learning-model-25b4c8643fb7 (accessed on 26 September 2018).
40. Gepperth, A.; Hammer, B. Incremental learning algorithms and applications. In *European Symposium on Artificial Neural Networks (ESANN)*; HAL: Bruges, Belgium, 2016.
41. Li, S.Z. *Encyclopedia of Biometrics: I-Z*; Springer Science & Business Media: Berlin/Heidelberg, Germany, 2009; Volume 2.
42. Vishal Nigam, M.J. Advantages of Adaptive AI Over Traditional Machine Learning Models. 2019. Available online: https://insidebigdata.com/2019/12/15/advantages-of-adaptive-ai-over-traditional-machine-learning-models/ (accessed on 15 December 2018).
43. Santos, J.D.D. Understanding and Handling Data and Concept Drift. 2020. Available online: https://www.explorium.ai/blog/understanding-and-handling-data-and-concept-drift/ (accessed on 24 February 2018).
44. Brownlee, J. A Gentle Introduction to Concept Drift in Machine Learning. 2020. Available online: https://machinelearningmastery.com/gentle-introduction-concept-drift-machine-learning/ (accessed on 10 December 2018).
45. Das, S. Best Practices for Dealing With Concept Drift. 2021. Available online: https://neptune.ai/blog/concept-drift-best-practices (accessed on 8 November 2018).
46. Brzezinski, D.; Stefanowski, J. Prequential AUC for classifier evaluation and drift detection in evolving data streams. In *Proceedings of the International Workshop on New Frontiers in Mining Complex Patterns*; Springer: Berlin/Heidelberg, Germany, 2014; pp. 87–101.
47. Dodge, Y. *The Concise Encyclopedia of Statistics*; Springer Science & Business Media: Berlin/Heidelberg, Germany, 2008.
48. Chan, J.; Choy, S. Analysis of covariance structures in time series. *J. Data Sci.* **2008**, *6*, 573–589. [CrossRef]
49. Ruppert, D.; Matteson, D.S. *Statistics and Data Analysis for Financial Engineering*; Springer: Berlin/Heidelberg, Germany, 2011; Volume 13.
50. Zhang, C.; Zhang, Y.; Cucuringu, M.; Qian, Z. Volatility forecasting with machine learning and intraday commonality. *arXiv* **2022**, arXiv:2202.08962.
51. Hsu, M.W.; Lessmann, S.; Sung, M.C.; Ma, T.; Johnson, J.E. Bridging the divide in financial market forecasting: machine learners vs. financial economists. *Expert Syst. Appl.* **2016**, *61*, 215–234. [CrossRef]
52. DEMİREL, U.; Handan, Ç.; Ramazan, Ü. Predicting stock prices using machine learning methods and deep learning algorithms: The sample of the Istanbul Stock Exchange. *Gazi Univ. J. Sci.* **2021**, *34*, 63–82. [CrossRef]
53. Guerra, P.; Castelli, M.; Côrte-Real, N. Machine learning for liquidity risk modelling: A supervisory perspective. *Econ. Anal. Policy* **2022**, *74*, 175–187. [CrossRef]
54. Kou, G.; Chao, X.; Peng, Y.; Alsaadi, F.E.; Herrera-Viedma, E. Machine learning methods for systemic risk analysis in financial sectors. *Technol. Econ. Dev. Econ.* **2019**, *25*, 716–742. [CrossRef]
55. Leippold, M.; Wang, Q.; Zhou, W. Machine learning in the Chinese stock market. *J. Financ. Econ.* **2022**, *145*, 64–82. [CrossRef]
56. Shivarova, A.; Matthew, F. Dixon, Igor Halperin, and Paul Bilokon: Machine learning in Finance from Theory to Practice. 2021. Available online: https://rdcu.be/daRTw (accessed on 8 November 2018).
57. Das, S.R.; Mishra, D.; Rout, M. A hybridized ELM-Jaya forecasting model for currency exchange prediction. *J. King Saud-Univ.-Comput. Inf. Sci.* **2020**, *32*, 345–366. [CrossRef]
58. Nayak, S.C. Development and performance evaluation of adaptive hybrid higher order neural networks for exchange rate prediction. *Int. J. Intell. Syst. Appl.* **2017**, *9*, 71. [CrossRef]

59. Yu, L.; Wang, S.; Lai, K.K. An Online BP Learning Algorithm with Adaptive Forgetting Factors for Foreign Exchange Rates Forecasting. In *Foreign-Exchange-Rate Forecasting with Artificial Neural Networks*; Springer: Boston, MA, USA, 2007; pp. 87–100. [CrossRef]
60. Soares, S.G.; Araújo, R. An on-line weighted ensemble of regressor models to handle concept drifts. *Eng. Appl. Artif. Intell.* **2015**, *37*, 392–406. [CrossRef]
61. Carmona, J.; Gavalda, R. Online techniques for dealing with concept drift in process mining. In *Proceedings of the International Symposium on Intelligent Data Analysis*; Springer: Berlin/Heidelberg, Germany, 2012; pp. 90–102.
62. Yan, H.; Ouyang, H. Financial time series prediction based on deep learning. *Wirel. Pers. Commun.* **2018**, *102*, 683–700. [CrossRef]
63. Barddal, J.P.; Gomes, H.M.; Enembreck, F. Advances on concept drift detection in regression tasks using social networks theory. *Int. J. Nat. Comput. Res. (IJNCR)* **2015**, *5*, 26–41. [CrossRef]
64. Chen, J.F.; Chen, W.L.; Huang, C.P.; Huang, S.H.; Chen, A.P. Financial time-series data analysis using deep convolutional neural networks. In Proceedings of the 2016 7th International Conference on Cloud Computing and Big Data (CCBD), Macau, China, 16–18 November 2016; IEEE: Piscataway, NJ, USA, 2016; pp. 87–92.
65. Sammut, C.; Webb, G.I. *Encyclopedia of Machine Learning*; Springer Science & Business Media: Berlin/Heidelberg, Germany, 2011.
66. Kumar Chandar, S. Fusion model of wavelet transform and adaptive neuro fuzzy inference system for stock market prediction. *J. Ambient. Intell. Humaniz. Comput.* **2019**, 1–9. [CrossRef]
67. Pradeepkumar, D.; Ravi, V. Forex rate prediction: A hybrid approach using chaos theory and multivariate adaptive regression splines. In *Proceedings of the 5th International Conference on Frontiers in Intelligent Computing: Theory and Applications*; Springer: Berlin/Heidelberg, Germany, 2017; pp. 219–227.
68. Wang, L.Y.; Park, C.; Yeon, K.; Choi, H. Tracking concept drift using a constrained penalized regression combiner. *Comput. Stat. Data Anal.* **2017**, *108*, 52–69. [CrossRef]
69. Baier, L.; Hofmann, M.; Kühl, N.; Mohr, M.; Satzger, G. Handling Concept Drifts in Regression Problems–the Error Intersection Approach. *arXiv* **2020**, arXiv:2004.00438.
70. Maneesilp, K.; Kruatrachue, B.; Sooraksa, P. Adaptive parameter forecasting for forex automatic trading system using fuzzy time series. In Proceedings of the 2011 International Conference on Machine Learning and Cybernetics, Guilin, China, 10–13 July 2011; IEEE: Piscataway, NJ, USA, 2011; Volume 1, pp. 189–194.
71. Yu, L.; Wang, S.; Lai, K.K. An online learning algorithm with adaptive forgetting factors for feedforward neural networks in financial time series forecasting. *Nonlinear Dyn. Syst. Theory* **2007**, *7*, 51–66.
72. Ilieva, G. *Fuzzy Supervised Multi-Period Time Series Forecasting*; Sciendo: Warszawa, Poland, 2019.
73. Bahrepour, M.; Akbarzadeh-T, M.R.; Yaghoobi, M.; Naghibi-S, M.B. An adaptive ordered fuzzy time series with application to FOREX. *Expert Syst. Appl.* **2011**, *38*, 475–485. [CrossRef]
74. Martín, C.; Quintana, D.; Isasi, P. Grammatical Evolution-based ensembles for algorithmic trading. *Appl. Soft Comput.* **2019**, *84*, 105713. [CrossRef]
75. Hoan, M.V.; Mai, L.C.; Hui, D.T. Pattern discovery in the financial time series based on local trend. In *Proceedings of the International Conference on Advances in Information and Communication Technology*; Springer: Berlin/Heidelberg, Germany, 2016; pp. 442–451.
76. Yu, L.; Wang, S.; Lai, K.K. Forecasting Foreign Exchange Rates Using an Adaptive Back-Propagation Algorithm with Optimal Learning Rates and Momentum Factors. In *Foreign-Exchange-Rate Forecasting with Artificial Neural Networks*; Springer: Boston, MA, USA, 2007; pp. 65–85.
77. Castillo, G.; Gama, J. An adaptive prequential learning framework for Bayesian network classifiers. In *Proceedings of the European Conference on Principles of Data Mining and Knowledge Discovery*; Springer: Berlin/Heidelberg, Germany, 2006; pp. 67–78.
78. Ramírez-Gallego, S.; Krawczyk, B.; García, S.; Woźniak, M.; Herrera, F. A survey on data preprocessing for data stream mining: Current status and future directions. *Neurocomputing* **2017**, *239*, 39–57. [CrossRef]
79. Husson, F.; Lê, S.; Pagès, J. *Analyse de Données avec R*; Presses universitaires de Rennes: Rennes, France, 2016.
80. Brockwell, P.J.; Davis, R.A. *Introduction to Time Series and Forecasting*; Springer: Berlin/Heidelberg, Germany, 2002.
81. Binder, M.D.; Hirokawa, N.; Windhorst, U. *Encyclopedia of Neuroscience*; Springer: Berlin/Heidelberg, Germany, 2009; Volume 3166.
82. Pandey, P. Understanding the Mathematics behind Gradient Descent. 2019. Available online: https://towardsdatascience.com/understanding-the-mathematics-behind-gradient-descent-dde5dc9be06e (accessed on 18 March 2019).
83. Clerc, M.; Siarry, P. Une nouvelle métaheuristique pour l'optimisation difficile: La méthode des essaims particulaires. *J3eA* **2004**, 3, 007. [CrossRef]

electronics

Article

LST-GCN: Long Short-Term Memory Embedded Graph Convolution Network for Traffic Flow Forecasting

Xu Han and Shicai Gong *

College of Science, Zhejiang University of Science and Technology, Hangzhou 310023, China; 222009252009@zust.edu.cn
* Correspondence: scgong@zafu.edu.cn; Tel.: +86-137-7758-5486

Abstract: Traffic flow prediction is an important part of the intelligent transportation system. Accurate traffic flow prediction is of great significance for strengthening urban management and facilitating people's travel. In this paper, we propose a model named LST-GCN to improve the accuracy of current traffic flow predictions. We simulate the spatiotemporal correlations present in traffic flow prediction by optimizing GCN (graph convolutional network) parameters using an LSTM (long short-term memory) network. Specifically, we capture spatial correlations by learning topology through GCN networks and temporal correlations by embedding LSTM networks into the training process of GCN networks. This method improves the traditional method of combining the recurrent neural network and graph neural network in the original spatiotemporal traffic flow prediction, so it can better capture the spatiotemporal features existing in the traffic flow. Extensive experiments conducted on the PEMS dataset illustrate the effectiveness and outperformance of our method compared with other state-of-the-art methods.

Keywords: traffic flow forecasting; long short-term memory network; graph convolutional network

Citation: Han, X.; Gong, S. LST-GCN: Long Short-Term Memory Embedded Graph Convolution Network for Traffic Flow Forecasting. *Electronics* **2022**, *11*, 2230. https://doi.org/10.3390/electronics11142230

Academic Editor: Wojciech Mazurczyk

Received: 16 June 2022
Accepted: 16 July 2022
Published: 17 July 2022

1. Introduction

In recent years, with the increase in the utilization rate of automobiles, the traffic flow on the road is increasing day by day. When the road is insufficient to accommodate vehicles, problems such as traffic congestion and traffic accidents will emerge. In this situation, traffic flow prediction is of great significance [1,2]. Traffic flow prediction refers to an analysis using traffic flow, speed and other information obtained by sensors in a certain road section for future prediction. It provides effective assistance in planning driving routes, thereby avoiding potential traffic jams.

Traffic flow prediction is inseparable from the temporal and spatial information in the road network. Individually considering any aspect of the information in the prediction will lead to a lack of information, and hence affect the accuracy of prediction. We need to predict outcomes from both a temporal and spatial perspective. Traffic data are recorded at fixed time points and fixed locations in space. Observations at adjacent locations and adjacent timestamps are not independent of each other, but are dynamically related. The key to such tasks is to explore dynamic correlations in data space and time to make accurate predictions.

With the advancement of technology, it has become easier to obtain data about the transportation networks, which also makes it more convenient for us to predict the traffic flow. Using cameras, sensors and other equipment on the highway, people can collect a large amount of time-series data, including traffic flow, speed, occupancy, and other information, which provides a solid data foundation for traffic forecasting, thus giving birth to a series of traffic forecast methods [3]. These include statistical methods and machine-learning methods. These methods either rely on feature engineering or cannot consider both the time and space information of the data and have certain limitations in the prediction of traffic flow. With the development of deep learning, some researchers tried to use graph convolutional networks to predict traffic flow or combine graph convolutional networks with recurrent neural networks to capture spatial and temporal features in traffic flow. Although much progress has been made in the prediction of traffic flow, most studies do not consider the periodicity of traffic flow, so the prediction of traffic flow still does not achieve the desired accuracy. To improve the accuracy of model predictions, we take into account the weekly and daily periodicity of traffic flow.

To make a more accurate traffic flow prediction, the LST-GCN model is proposed in this paper, and the LSTM model [4] is embedded into the parameter training of the GCN model [5], to capture the time and space information more synchronously. Further, we explore the internal relation of time and space, and reduce the number of parameter training, so as to make more accurate prediction.

The original combined model is relatively simple in processing data sets, such as the combined model of LSTM model and GCN model. For traffic flow data, the GCN model is used to update the node flow information at each moment separately to obtain data space information, and then using the LSTM model further combines the node traffic information at all times to obtain information about the time of the data. The disadvantage of this method is that the number of model parameters and calculations are large. In response to this problem, we propose a new LST-GCN embedded structure. Different from previous models, we directly embed the LSTM model into the update process of GCN parameters, which greatly reduces the number of parameters and the amount of computation. At the same time, the model can make good use of the temporal and spatial information of the data.

The remainder of this paper is organized as follows. The related works on traffic flow forecasting are discussed in Section 2. In Section 3, we propose some definitions about traffic flow and introduce the structure of the GCN model and LSTM models. Section 4 proposes the LST-GCN model to capture spatial correlations by learning topology through GCN networks and temporal correlations by embedding LSTM networks into the training process of GCN networks. In Section 5, a comprehensive assessment of the model performance is conducted using real road-traffic datasets. At the same time, the experimental results are discussed. Section 6 concludes the paper and provides an outlook on future work.

2. Related Work

2.1. Traffic Forecasting

There are two main types of methods for traffic flow forecasting: one is the statistical method and the other is the machine-learning method. The statistical methods mainly include ARIMA (autoregressive integrated moving average model) [6–8], HA (history average model) [3], ES (exponential smoothing model) [9] and KF (Kalman filter model) [10–13]. ARIMA models analyze time-series data and use them to make predictions about future traffic flows. The ARIMA model [6–8] assumes that the change in traffic flow is linear. The HA model [2] uses the least-squares method to evaluate the parameters of the model to further predict the traffic flow. The ES model [9] and the KF model [10–13] are suitable for making predictions on traffic flow with a smaller amount of data. The assumptions of these models are relatively strict. Once random interference occurs, the accuracy of the models will decrease. They rely on the assumption of stability. At the same time, these models cannot reflect the nonlinearity of traffic conditions. Therefore, the use of these models has certain limitations.

There are many machine-learning methods for traffic flow prediction, which are mainly divided into two categories: the traditional machine-learning method and the deep-learning method. The SVR (support vector regression) model [14], KNN (K-nearest neighbor) model [15], Bayesian model [16], fuzzy logic model [17], neural-network model [18], etc., as traditional machine-learning methods, are often used to predict traffic flow. The SVR model [14] introduces a supervised machine-learning method called regressive online support vector machines, which can make short-term traffic flow predictions for both typical and atypical conditions. The KNN model [15] takes the k value and d_m value of the nearest neighbors as the input parameters of the model, and combines the prediction range of multiple intervals to optimize the parameter values of the model, and then predict the value of traffic flow. The Bayesian model [16] first searches the manifold neighborhood, and then obtains a higher accuracy of the manifold neighborhood, and then proposes a traffic-state prediction method based on the expansion strategy of adaptive neighborhood selection. Fuzzy logic models [17] use fuzzy methods to classify input data into clusters, which in turn specify input–output relationships. The neural-network model [18] is the first attempt to build an artificial neural network based on historical traffic data, aiming to predict traffic volume based on historical data at major urban intersections. This type of model has strong nonlinear mapping ability, and the data requirements are not as strict as statistical methods, so it can better adapt to the uncertainty of traffic flow and effectively improve the prediction effect. However, the spatial structure of observation points is unstructured, and the above methods do not use the spatial structure information of the data, and only analyzing from the time dimension has certain limitations in improving the prediction accuracy.

The deep-learning models originally used for traffic flow prediction mainly include the GRU (gated recurrent unit) model [19] and LSTM model. The GRU model and LSTM model are important recursive neural-network models that are used to integrate and analyze temporal information to make predictions. Compared with the prediction models based on statistical learning and machine-learning methods, deep learning can model multidimensional features and realize the approximation of complex functions by learning the deep nonlinear network structures, which can better learn the abundant changes inherent in traffic flow. It can simulate its complex nonlinear relationship and greatly improve the accuracy of traffic flow prediction. However, these models also did not consider the influence of the spatial structure of the data on the prediction results, and did not fully mine the spatiotemporal characteristics of the traffic data. There are also certain limitations in predicting traffic flow.

Recently, models that consider spatiotemporal information have sparked a lot of research. Wu et al. [20] designed a feature fusion framework for short-term traffic flow prediction by combining the CNN (convolutional neural network) model with the LSTM model. This framework uses a one-dimensional CNN to describe the spatial features of traffic flow data. For the time-varying periodicity and temporal variation of the traffic flow, this framework utilizes two LSTM models. DCRNN, proposed by Li et al. [21], uses a bidirectional random walk to capture spatial dependencies and an encoder-decoder with predetermined sampling to capture temporal dependencies. Sun et al. [22] constructed a multibranch framework called TFPNet (traffic flow prediction network), a deep-learning framework for short-term traffic flow prediction. TPFNet uses a multi-layer fully convolutional network structure to extract the relationship from local to global hierarchical space. Zhao et al. [23] proposed the T-GCN model, which combines gated recurrent units with graph convolutional networks for short-term traffic flow prediction. Geng et al. [24] designed a spatiotemporal multigraph convolutional network that first encodes the non-Euclidean pairwise correlations between regions into multiple graphs, and then uses multigraph convolution to explicitly map these correlations. Diao et al. [25] used a dynamic Laplacian matrix estimator to discover changes in the Laplacian matrix, which in turn made predictions about traffic flow. Huang et al. [26] proposed the cosAtt model, a graph-attention network that integrates cosAtt and GCN into a spatial gating block. Lv et al. [27] modeled various global features in road networks, including spatial, temporal, and semantic correlations, and proposed a temporal multigraph convolutional network. Guo et al. [28] used the attention mechanism for traffic flow prediction and proposed an AST-GCN model. The attention mechanism has been applied in both time and space and achieved better prediction results.

2.2. Convolutions on Graphs

In order to solve the irregularity of the spatial neighborhood, Bruna et al. [29] made a breakthrough from the spectral space and proposed a spectral network on the graph. According to the knowledge of graph theory, they decompose the Laplacian matrix spectrally and use the obtained eigenvalues and eigenvectors to define the convolution operation in the spectral space. To simplify the problem of complexity, Defferrard et al. [30] proposed a Chebyshev network, which defined the convolution kernel as a polynomial form, and used Chebyshev expansion to approximate the calculation of the convolution kernel, which greatly improved the computational efficiency. After that, Kipf and Welling [5] simplified the Chebyshev network, using only a first-order approximate convolution kernel, and made a little sign change, resulting in the well-known graph-convolution network.

2.3. Long Short-Term Memory Network

Bengio et al. [31] proposed the RNN (recurrent neural network) model. Using the RNN model can help people process sequence data more efficiently. In the RNN model, people can reinput the output of a neuron at a certain time as the input to the neuron. For the dependencies between time-series data, the network structure of the RNN model can adequately maintain them. However, this model suffers from vanishing gradients and exploding gradients. To solve the problems of gradient disappearance and gradient explosion in the traditional RNN model, Hochreiter et al. [4] proposed the LSTM network. The LSTM network is improved from the traditional RNN model. Compared with the RNN model, the hidden unit of the LSTM model has more complexity. At the same time, the LSTM model has a wider range of applications than RNN and is a more effective sequence model. During the run of the model, the LSTM model can selectively add or subtract information by adding linear interventions.

3. Preliminaries

3.1. Traffic Networks

Definition 1. *Road network G. We use $G = (V, E, A)$ to denote a spatial network, as shown in Figure 1, where $|V| = N$ is the set of vertices and N is the number of vertices. E is the set of edges, which reflects the connections between road sections. $A \in \mathbb{R}^{N \times N}$ is the adjacency matrix of the network G. The value of each element represents the connectivity between the corresponding road segments. An element value of 1 indicates connectivity, and an element value of 0 indicates disconnection.*

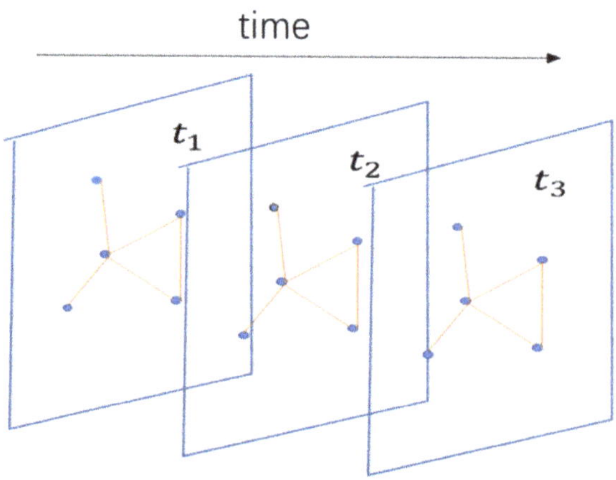

Figure 1. The spatial-temporal structure of traffic data, where the data at each time slice form a graph.

Definition 2. *The graph feature matrix* $X_G^{(t)} \in \mathbb{R}^{N \times C}$, *where C is the number of attribute features and t represents the time step. The graph signal matrix represents the observations of the spatial network G at the time step t.*

The problem of traffic flow data prediction can be described as learning a mapping function, f, which maps the historical spatiotemporal network sequence $\left(X_G^{(t-T+1)}, X_G^{(t-T+2)}, \ldots, X_G^{(t)}\right)$ into future observations of this spatiotemporal network $\left(X_G^{(t+1)}, X_G^{(t+2)}, \ldots, X_G^{(t+T')}\right)$, where T represents the length of the historical spatiotemporal network sequence and T' denotes the length of the target spatiotemporal network sequence to be predicted.

3.2. GCN Model

Based on the Chebyshev network, Kipf and Welling proposed the GCN model. The updated convolution formula of each layer of the GCN model node is as follows:

$$H^{(l+1)} = \sigma\left(\widetilde{D}^{-\frac{1}{2}}\widetilde{A}\widetilde{D}^{-\frac{1}{2}}H^{(l)}W^{(l)}\right), \tag{1}$$

$$\widetilde{A} = A + I_N \tag{2}$$

$$\text{and } \widetilde{D} = D + I_N \tag{3}$$

Among them, $H^{(l+1)}$ represents the node representation of the $l+1$-th layer, $H^{(l)}$ represents the node representation of the $l+1$-th layer, and $W^{(l)}$ represents the learnable parameters of the l-th layer. A represents the adjacency matrix, I_N represents the identity matrix, and D represents the degree matrix.

By determining the topological relationship between the central node and the surrounding nodes, the GCN model can simultaneously encode the topological structure of the road network and the attributes of the nodes, so that spatial dependencies can be captured on this basis.

3.3. LSTM Model

The LSTM model is a typical RNN (recurrent neural network) model, which is proposed to solve the problems of gradient disappearance and gradient explosion existing in the traditional RNN model. The structure diagram of LSTM is shown in Figure 2, and the Equations are shown in (4)~(9).

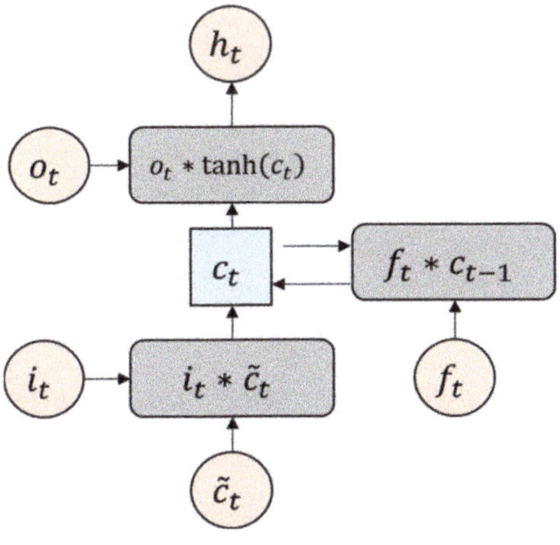

Figure 2. LSTM model diagram.

$$i_t = \sigma(W_i x_t + U_i h_{t-1} + b_i), \quad (4)$$

$$f_t = \sigma\left(W_f x_t + U_f h_{t-1} + b_f\right), \quad (5)$$

$$o_t = \sigma(W_o x_t + U_o h_{t-1} + b_o), \quad (6)$$

$$\widetilde{c}_t = \tanh(W_c x_t + U_c h_{t-1} + b_c), \quad (7)$$

$$c_t = i_t * \widetilde{c}_t + f_t * c_{t-1} \quad (8)$$

$$\text{and } h_t = o_t * \tanh(c_t). \quad (9)$$

where i_t controls the input of the input gate to $\widetilde{c}_t$, f_t controls the memory level of the forget gate for c_{t-1}, and o_t controls the output of $\tanh(c_t)$. Since the activation function is a sigmoid function, the values of i_t, f_t, and o_t are in between 0 and 1.

The LSTM model uses the hidden state of the previous moment and the parameter information of the current moment as input to determine the parameter state of the current moment. Due to the gating mechanism, the LSTM model retains the changing trend of historical parameter information when capturing the parameter information at the current moment. Therefore, the model can capture the time-varying features of traffic dynamics from parametric data. In this paper, we apply the LSTM model to learn the temporal-varying trend of traffic states.

4. Method

Figure 3 shows the general framework of the LST-GCN model. The model consists of three parts with the same structure, and the model is established by representing data from three perspectives: adjacent time, daily cycle, and weekly cycle. As shown in Figure 3, this paper takes χ_h, χ_d, and χ_w as input, respectively. We consider each sensor as a node, and the sensor information about the three dimensions of traffic flow, vehicle speed, and occupancy rate is regarded as the vector representation of the node. χ_h, χ_d, and χ_w represent the node representation of all nodes at the adjacent time, the daily cycle, and the weekly cycle, respectively.

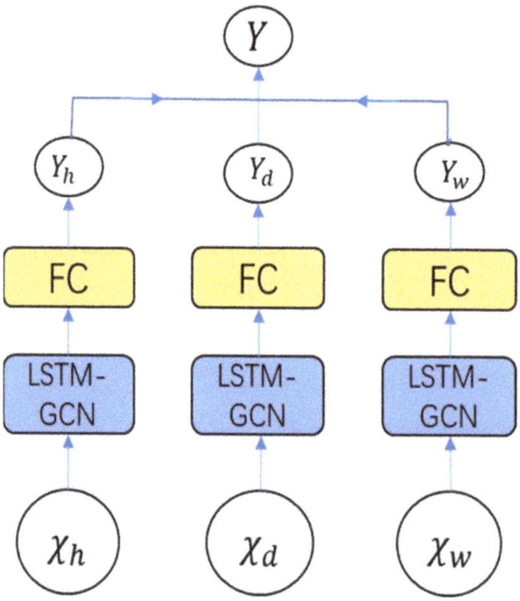

Figure 3. LST-GCN model frame diagram.

$\mathcal{X}_h \in \mathbb{R}^{N \times F \times T}$, N represents the number of nodes; the value of F is 3, which represents the three dimensions of traffic flow, vehicle speed, and occupancy; and T represents the length of the adjacent time slice.

We update the node representation through the LSTM-GCN block, and then use a fully connected layer to make predictions, and the results are denoted by Y_h, Y_d, and Y_w, respectively. Afterwards, the prediction results of the three series of proximity correlation, daily correlation, and weekly correlation are weighted and combined to obtain the final result, which is represented by Y.

Figure 4 shows the general framework of the LSTM-GCN block. Taking χ_h as an example, we take $X_{t_0-h+1}, X_{t_0-h+2}, \ldots, X_{t_0}$ as input. $\mathcal{X}_h \in \mathbb{R}^{N \times F \times T}$, N represents the number of nodes; the value of F is 3, which represents the three dimensions of traffic flow, vehicle speed, and occupancy; and T represents the length of the adjacent time slice. $X_{t_0-h+1}, X_{t_0-h+2}, \ldots, X_{t_0}$ represents the representation of each moment of χ_h. Through the LSTM-GCN block, we can update the node representation to obtain $X^1_{t_0-h+1}$, $X^1_{t_0-h+2}, \ldots, X^1_{t_0}$. Through the connection between the parameters, all GCN models are combined together and the representation of each vector is updated in time and space.

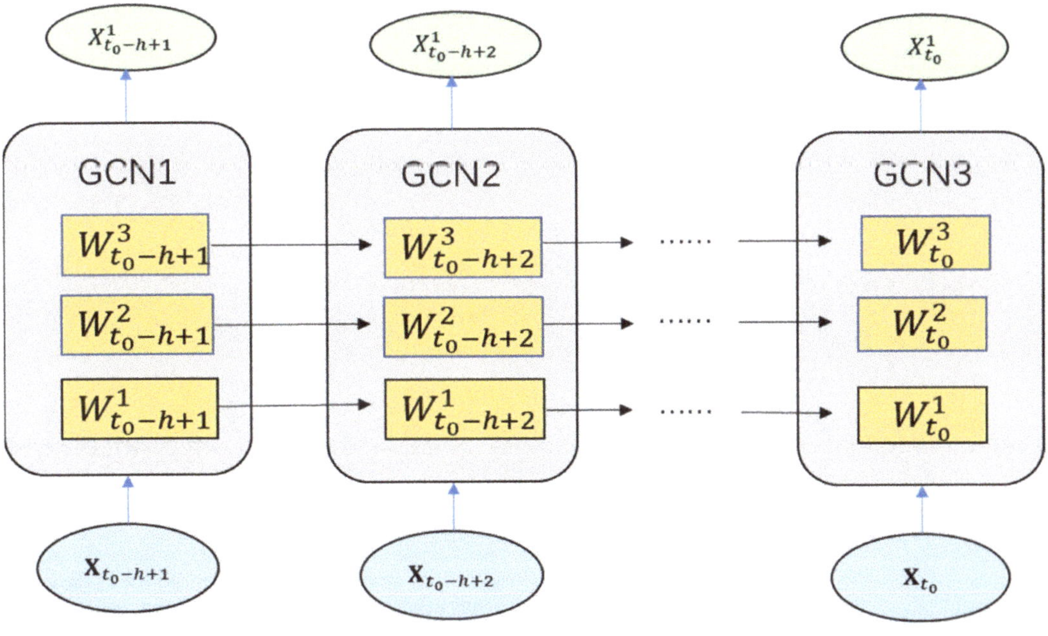

Figure 4. LSTM-GCN block diagram.

To explore the distribution of data from the perspective of space and time simultaneously, we introduce the LSTM model into the parameter update process of the GCN model. For the parameter $W^{(l)}$, we connect the $W^{(l)}$ at each moment through the LSTM model, as shown in Equation (10).

$$W_t^{(l)} = \mathrm{LSTM}\left(W_{t-1}^{(l)}\right) \tag{10}$$

Meanwhile, at time t, the convolution operation from the lth layer to the $l+1$-th layer is the same as that of the GCN model, as shown in Equation (11).

$$H_t^{(l+1)} = \mathrm{GCONV}\left(\widetilde{D}^{-\frac{1}{2}}\widetilde{A}\widetilde{D}^{-\frac{1}{2}}, H_t^{(l)}, W_t^{(l)}\right) \tag{11}$$

Combining Equations (10) and (11), we can obtain the update rule of node representation at $l+1$-th layer, as shown in Equation (12).

$$\left[H_t^{(l+1)}, W_t^{(l)}\right] = \mathrm{LST} - \mathrm{GCN}\left(\widetilde{D}^{-\frac{1}{2}}\widetilde{A}\widetilde{D}^{-\frac{1}{2}}, H_t^{(l)}, W_{t-1}^{(l)}\right) \tag{12}$$

Figure 5 illustrates the update of the node. At time t, the representation of the node at $l+1$-th layer is determined by the node and the parameters at l-th layer through convolution. Similarly, we can calculate the node representation of any layer. The node at the zeroth layer at time t is represented X_t corresponding to time t, that is, the vector representation of each sensor in the three dimensions of traffic flow, vehicle speed, and occupancy at time t. For the parameter W of each layer, we can update it through the LSTM model.

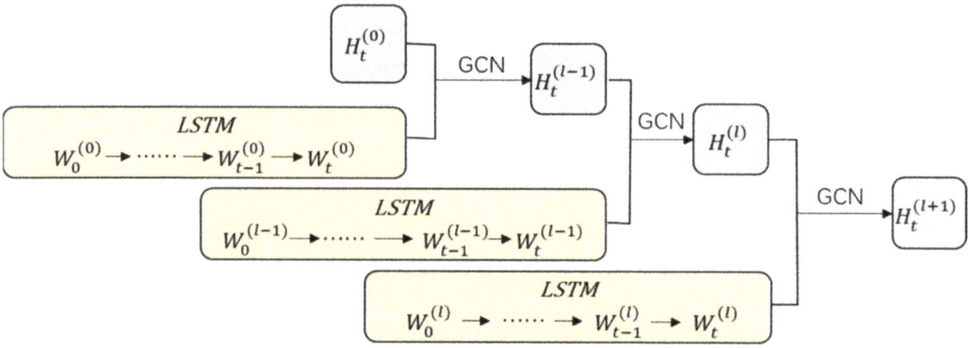

Figure 5. Node update.

5. Experiment

5.1. Data Set and Processing

To verify the effectiveness of our model, we used the California highway dataset. PEMS uses sensors to acquire real-world traffic data from more than 8100 locations on California highways and highway systems, which are integrated into multiple time intervals. We selected the PEMS04 dataset and the PEMS08 dataset. The PEMS04 dataset contains the traffic data of San Francisco Bay from 1 January 2018 to 28 February 2018 collected by 3848 sensors, including three aspects of traffic, speed, and occupancy, where we selected data from 307 of these sensors for verification. The PEMS08 dataset contains the traffic data of San Bernardino from 1 July 2016 to 31 August 2016 collected by 1979 sensors, including three aspects of traffic, speed, and occupancy, where we selected data from 170 of these sensors for verification.

We first removed redundant sensors with distances of less than 3.5 miles; some data were missing from the original traffic speed dataset due to equipment failures, etc. Considering the spatiotemporal characteristics of traffic data, we used linear interpolation for missing values.

The traffic information in both datasets was updated every 5 min. In chronological order, we selected the first 60% of the data as the training set, the middle 20% of the data as the validation set, and the last 20% of the data as the test set.

Since the distance between each sensor was different, we chose the inverse of the distance as the element value of the adjacency matrix, thereby constructing the adjacency matrix. Because of the different dimensions, we normalized all the data, as shown in Equation (13).

$$X_{norm} = \frac{X - X_{min}}{X_{max} - X_{min}}. \tag{13}$$

5.2. Experimental Setup

Considering the influence of periodicity on the experimental results, we divided the experimental data into adjacent time series, daily period series, and weekly period series. They are represented by χ_h, χ_d, and χ_w, respectively. We fed χ_h, χ_d, and χ_w as inputs to the three LSTM-GCN subnetworks for training, respectively, and combined the outputs of the three subnetworks into the final output. We conducted experiments on a server configured with a Xeon Platinum 8163 processor clocked at 2.7 GHz and an NVIDIA Tesla P100 graphics card with 16 GB of VRAM. When training on the PMES04 dataset, the number of iterations was 100, the batch size was 16, and the Adam optimizer was used to update the parameters with a learning rate of 0.01. When training on the PMES08 dataset, the number of iterations was 200, the batch size was 32, the Adam optimizer was used for parameter update, and the learning rate was 0.01.

5.3. Evaluation Indicators

The experiment tests the model performance through *RMSE* (root-mean-square error), *MAE* ((mean absolute error) and *MAPE* (mean absolute percentage error); the formulas are defined as follows:

$$MAE = \frac{1}{n}\sum_{i=1}^{n}|\hat{y}_i - y_i|, \tag{14}$$

$$MAPE = 100\% \times \frac{1}{n}\sum_{i=1}^{n}\left|\frac{\hat{y}_i - y_i}{y_i}\right| \tag{15}$$

$$\text{and } RMSE = \sqrt{\frac{1}{n}\sum_{i=1}^{n}(\hat{y}_i - y_i)^2} \tag{16}$$

where n is the number of predicted values, $\hat{y}_i$ is the predicted value, and y_i is the true value.

5.4. Results

As shown in Table 1, our model outperforms other models on both datasets. Since the HA model and the ARIMA model are linear models and only consider the information of the time dimension, the prediction effect of the models is relatively poor. The SVR model and the GRU model use machine-learning methods to analyze data, and have better nonlinear mapping capabilities than the HA model and the ARIMA model. However, the SVR model and the GRU model also only analyze the data from the time dimension, without considering the spatial dimension, so the prediction effect of the model is only better than the HA model and the ARIMA model. The ASTGCN model uses an attention mechanism from the temporal and spatial dimensions, respectively. Compared with the ARIMA model, the LSTM model, and the GRU model, the model considers the information of the spatial dimension, thereby significantly improving the prediction effect of the data. The LST-GCN model uses the LSTM model to update the parameters of the GCN model, which avoids the problem of too many parameters caused by separating the two models. It also considers the information of the time dimension and the space dimension. At the same time, the model also combines adjacent sequences and daily sequences. Three sequences of weekly sequence are used to predict the traffic flow. Considering the influence of periodicity on the prediction results, the data information is greatly utilized. Therefore, the model in this paper has achieved better prediction results than other models. For example, for the PEMS04 dataset, using RMSE, MAE, and MAPE as evaluation metrics, respectively, LST-GCN has an average improvement of 0.9%, 2.2%, and 1.3% compared with ASTGCN. For the PEMS08 dataset, using RMSE, MAE, and MAPE as evaluation metrics, respectively, LST-GCN achieves an average improvement of 2.5%, 3.7%, and 1.8% compared to ASTGCN.

Table 1. Average performance comparison of different approaches on PEMS04 and PEMS08.

Model	PMES04			PMES08		
	RMSE	MAE	MAPE(%)	RMSE	MAE	MAPE(%)
HA	54.16	36.68	19.69	44.06	29.46	15.25
ARIMA	68.16	32.01	19.17	43.31	24.05	14.34
SVR	45.75	29.45	17.09	36.98	23.13	13.81
GRU	45.16	28.64	16.27	35.96	22.25	13.03
ASTGCN	35.23	22.93	16.58	28.16	18.61	13.05
LST-GCN	**34.93**	**22.43**	**16.37**	**27.47**	**17.93**	**12.81**

To confirm the spatiotemporal prediction ability of the LST-GCN model, we respectively compared the LST-GCN model with the LSTM model and the GCN model. As shown in Figure 6, our LST-GCN model has a strong spatiotemporal prediction ability. Since the LSTM model only considers the impact of time factors on traffic flow, while the GCN model only considers the impact of spatial factors on traffic flow, these two models cannot fully consider the information of the data. Therefore, the prediction accuracy of the LSTM model and GCN model is relatively poor. For example, using RMSE as the evaluation metric, on the PEMS04 dataset, LST-GCN has an average improvement of 9.2% compared with GCN, and an improvement of 3.4% compared to LSTM. On the PEMS08 dataset, LST-GCN has an average improvement of 13.9% compared to GCN and 3.5% compared to LSTM.

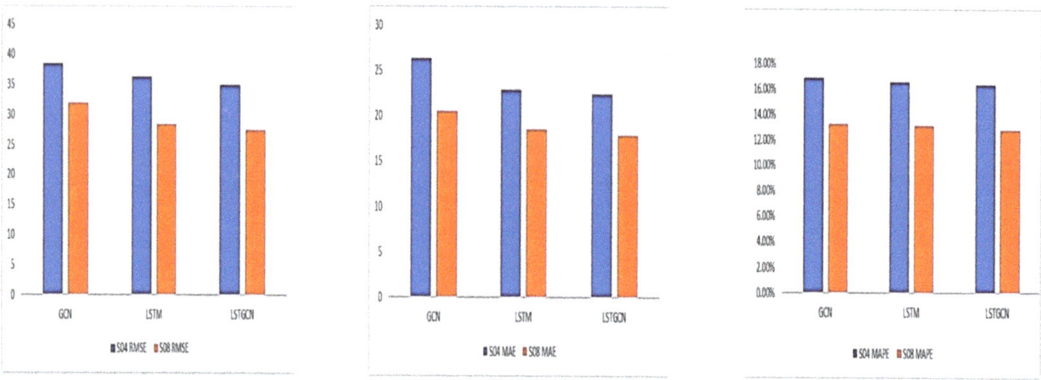

(a) (b) (c)

Figure 6. Average performance comparison of LST-GCN and GCN and LSTM on PEMS04 and PEMS08. (**a**) RMSE comparison of LST-GCN and GCN and LSTM on PEMS04 and PEMS08. (**b**) MAE comparison of LST-GCN and GCN and LSTM on PEMS04 and PEMS08. (**c**) MAPE comparison of LST-GCN and GCN and LSTM on PEMS04 and PEMS08.

Figure 7 shows how the prediction performance of the model varies with the range of prediction. With the increase in the prediction interval, the prediction error of the model will gradually increase, and the prediction effect will inevitably deteriorate. The RMSE, MAE, and MAPE values of the four models, HA, ARIMA, SVR, and GRU, increase continuously with the increase in prediction time, and the variation range is large. Compared with these four models, the ASTGCN model and the LST-GCN model continue to increase with the prediction time, but the variation range is relatively small. This is because the first four models only consider the impact of variation of time on the prediction results. With the increase in prediction interval, the time dimension information between roads on future traffic will have less and less impact, resulting in a lower and lower prediction accuracy of the model. In the long-term prediction, the spatiotemporal correlation is a more important predictor, so ASTGCN model and LST-GCN model are far superior to the other four models in the longer-term prediction. It can also be seen from the figure that the overall prediction effect of our LST-GCN model is better than that of ASTGCN model, which indicates that our LST-GCN model can better mine the spatiotemporal correlation of traffic data, to make more accurate predictions.

To better understand the LST-GCN model, we selected a road segment on the PEMS04 dataset and PEMS08 dataset, respectively, and visualized the prediction results on the test set. Figure 8a,b show the visualization results on two datasets, PEMS04 and PEMS08, respectively. It can be seen that the simulation effect of the model is better. It can be seen from the results that the prediction results of the LST-GCN model are relatively smooth. We speculate that it may be because the GCN model adds a smoothing filter to the Fourier domain and moves the filter to capture spatial features. This results in smoother experimental results.

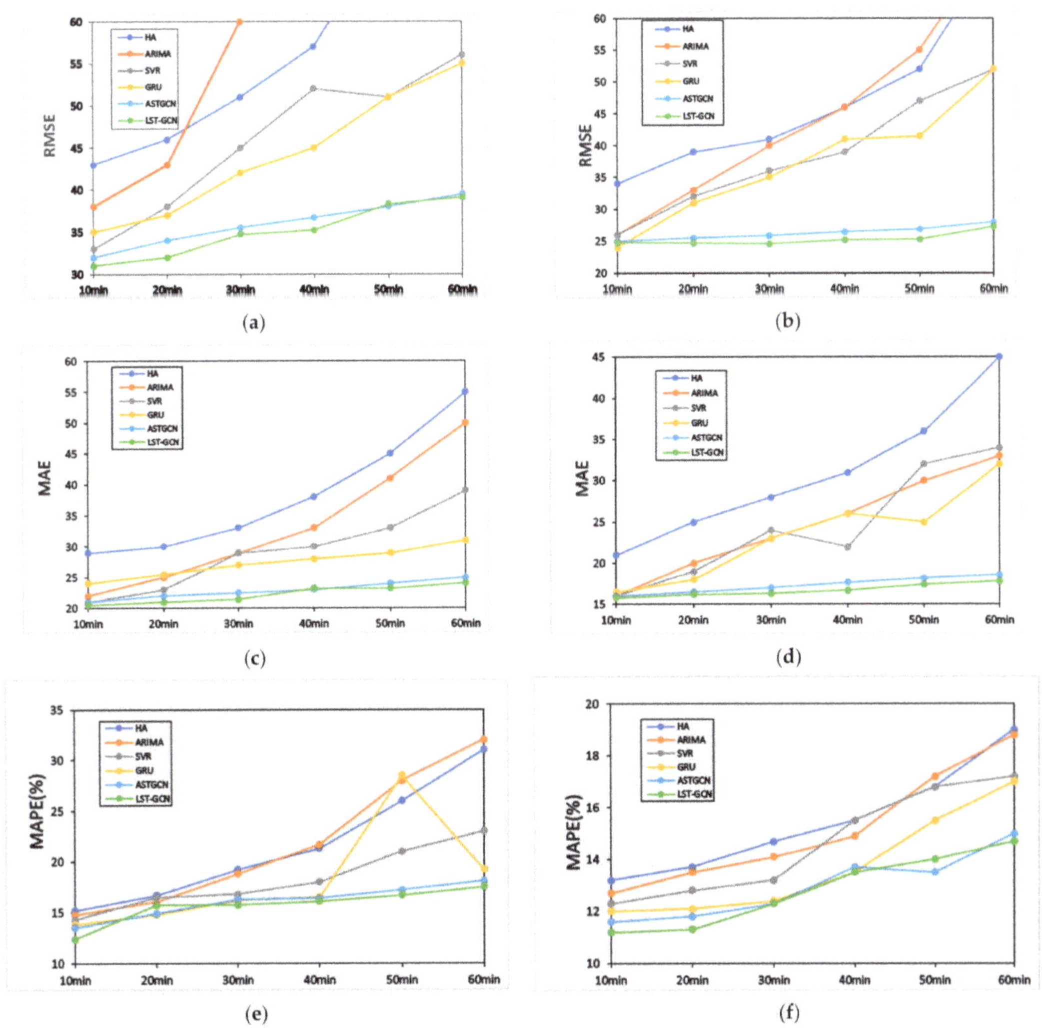

Figure 7. Performance changes of different methods as the forecasting interval increases. (**a**) Changes on PEMS04 dataset, based on RMSE. (**b**) Changes on PEMS08 dataset, based on RMSE. (**c**) Changes on PEMS04 dataset, based on MAE. (**d**) Changes on PEMS08 dataset, based on MAE. (**e**) Changes on PEMS04 dataset, based on MAPE. (**f**) Changes on PEMS08 dataset, based on MAPE.

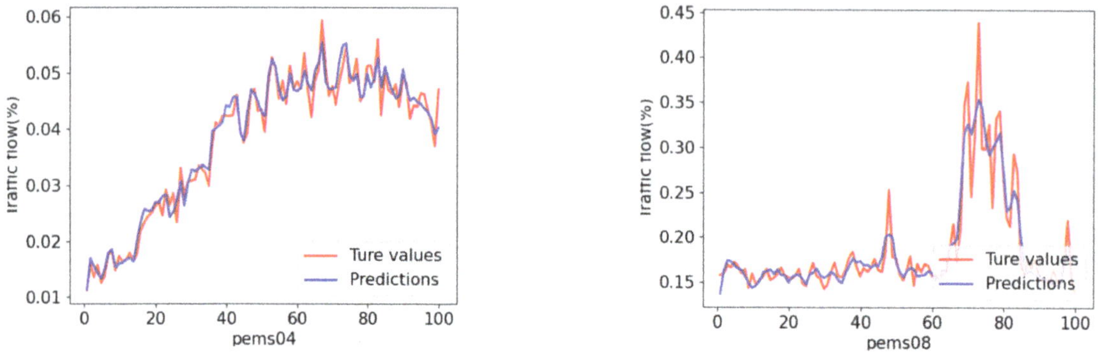

(a) (b)

Figure 8. The visualization results for prediction. (**a**) Results on PEMS04 dataset. (**b**) Results on PEMS08 dataset.

6. Discussion

Accurate and rapid traffic flow prediction is an important issue affecting the development of intelligent transportation. The original traffic prediction model basically has the problem of large-parameter data or an inability to make full use of the data information. The reason why our model results are better than other models is mainly because of the following advantages: (1) We propose a new LST-GCN structure, which directly embeds the LSTM model into the updating process of GCN parameters, reducing the number of parameters; (2) compared with the model with a single model structure, our model considers both time and space factors, and makes full use of data information.

Our model improves the performance of short-term traffic flow, but there are still some issues to consider. Considering the "memory" capability introduced by the LSTM model may have a negative impact on the time complexity. [32–34] This effect exists in many cyclic structures. This needs further research in future work.

7. Conclusions

According to the traffic flow prediction problem, this paper proposes a method to update the model parameters of the graph convolutional network model using the long short-term memory neural-network model. By embedding the long short-term memory neural network into the graph convolutional network and modeling from the perspective of time and space at the same time, we further explore the internal connection of time and space. At the same time, three sequences of adjacent sequence, daily sequence, and weekly sequence are combined to predict traffic flow, and the influence of periodicity on the prediction result is considered. Finally, the method in this paper is compared with several common methods for predicting traffic flow through three evaluation indicators—RMSE, MAE, and MAPE—and it is concluded that the model proposed in this paper is better than other models on the PEMS dataset.

In the future, the main directions that need to be studied are: (1) applying the LST-GCN model to more road segments and increasing the prediction period of the model; (2) considering more complex road conditions, and improving our model by taking into account other factors such as weather and traffic accidents; (2) applying the LST-GCN model to other scenarios such as air quality prediction, energy prediction, etc.

Author Contributions: Conceptualization, X.H. and S.G.; Methodology, X.H.; Formal Analysis, S.G.; Writing—Original Draft Preparation, X.H.; Writing—Review & Editing, S.G. All authors have read and agreed to the published version of the manuscript.

Funding: This research received no external funding.

Conflicts of Interest: The authors declare no conflict of interest.

Abbreviations

GCN	Graph convolutional network
LSTM	Long short-term memory network
ARIMA	Autoregressive integrated moving average model
HA	History average model
ES	Exponential smoothing model
KF	Kalman filter model
SVR	Support vector regression model
KNN	K-nearest neighbor model
GRU	Gated recurrent unit model
CNN	Convolutional neural network
RNN	Recurrent neural network

References

1. Hani, S.M. Traveler behavior and intelligent transportation systems. *Transp. Res. Part C Emerg. Technol.* **1999**, *7*, 73–74.
2. Li, Y.; Lin, Y.; Zhang, F. Research on geographic information system intelligent transportation systems. *Chung-Kuo K. Lu Hsueh Pao China J. Highw. Transp.* **2000**, *13*, 97–100.
3. Liu, J.; Guan, W. A summary of traffic flow forecasting methods. *J. Highw. Transp. Res. Dev.* **2004**, *3*, 82–85.
4. Ma, X.; Tao, Z.; Wang, Y. Long short-term memory neural network for traffic speed prediction using remote microwave sensor data. *Transp. Res. C Emerg. Technol.* **2015**, *54*, 187–197. [CrossRef]
5. Kipf, T.N.; Welling, M. Semi-Supervised Classification with Graph Convolutional Networks. *arXiv* **2016**, arXiv:1609.02907.
6. Levin, M.; Tsao, Y.-D. On forecasting freeway occupancies and volumes. *Transp. Res. Rec.* **1980**, *773*, 47–49.
7. Guo, J.; Huang, W.; Williams, B.M. Adaptive Kalman filter approach for stochastic short-term traffic flow rate prediction and uncertainty quantification. *Transp. Res. C Emerg. Technol.* **2014**, *43*, 50–64. [CrossRef]
8. Shi, G.; Guo, J.; Huang, W.; Williams, B.M. Modeling seasonal heteroscedasticity in vehicular traffic condition series using a seasonal adjustment approach. *J. Transp. Eng.* **2014**, *140*, 5. [CrossRef]
9. Chan, K.Y.; Dillon, T.S.; Singh, J.; Chang, E. Neural-network-based models for short-term traffic flow forecasting using a hybrid exponential smoothing and Levenberg–Marquardt algorithm. *IEEE Trans. Intell. Transp. Syst.* **2012**, *13*, 644–654. [CrossRef]
10. Kumar, S.V. Traffic flow prediction using Kalman filtering technique. *Procedia Eng.* **2017**, *187*, 582. [CrossRef]
11. Zhou, T.; Jiang, D.; Lin, Z.; Han, G.; Xu, X.; Qin, J. Hybrid dual Kalman filtering model for short-term traffic flow forecasting. *IET Intell. Transp. Syst.* **2019**, *13*, 1023–1032. [CrossRef]
12. Cai, L.; Zhang, Z.; Yang, J.; Yu, Y.; Zhou, T.; Qin, J. A noise-immune Kalman filter for short-term traffic flow forecasting. *Phys. A Stat. Mech. Appl.* **2019**, *536*, 122601. [CrossRef]
13. Zhang, S.; Song, Y.; Jiang, D.; Zhou, T.; Qin, J. Noise-identified Kalman filter for short-term traffic flow forecasting. In Proceedings of the IEEE 15th International Conference on Mobile Ad-Hoc Sensor Networks, Shenzhen, China, 11–13 December 2019; pp. 1–5.
14. Castro-Netoa, M.; Jeong, Y.S.; Jeong, M.K.; Hana, L. Online-svr for short-term traffic flow prediction under typical and atypical traffic conditions. *Expert Syst. Appl.* **2009**, *36*, 6164–6173. [CrossRef]
15. Chang, H.; Lee, Y.; Yoon, B.; Baek, S. Dynamic near-term traffic flow prediction: System-oriented approach based on past experiences. *IET Intell. Transp. Syst.* **2012**, *6*, 292–305. [CrossRef]
16. Su, Z.; Liu, Q.; Lu, J.; Cai, Y.; Jiang, H.; Wahab, L. Short-time traffic state forecasting using adaptive neighborhood selection based on expansion strategy. *IEEE Access* **2018**, *6*, 48210–48223. [CrossRef]
17. Yin, H.; Wong, S.C.; Xu, J.; Wong, C.K. Urban traffic flow prediction using a fuzzy-neural approach. *Transp. Res. Part C* **2002**, *10*, 85–98. [CrossRef]
18. Çetiner, B.G.; Sari, M.; Borat, O. A Neural Network Based Traffic-Flow Prediction Model. *Math. Comput. Appl.* **2010**, *15*, 269–278. [CrossRef]
19. Fu, R.; Zhang, Z.; Li, L. Using LSTM and GRU neural network methods for traffic flow prediction. In Proceedings of the 2016 31st Youth Academic Annual Conference of Chinese Association of Automation, Wuhan, China, 11–13 November 2016.
20. Wu, Y.; Tan, H. Short-term traffic flow forecasting with spatial-temporal correlation in a hybrid deep learning framework. *arXiv* **2016**, arXiv:1612.01022.
21. Li, Y.; Yu, R.; Shahabi, C.; Liu, Y. Diffusion convolutional recurrent neural network: Data-driven traffic forecasting. In Proceedings of the 6th International Conference on Learning Representations (ICLR), Vancouver, BC, Canada, 30 Apr–3 May 3 2018.
22. Sun, S.; Wu, H.; Xiang, L. City-Wide Traffic Flow Forecasting Using a Deep Convolutional Neural Network. *Sensors* **2020**, *20*, 421. [CrossRef]
23. Zhao, L.; Song, Y.; Zhang, C. T-GCN: A temporal graph convolutional network for traffic prediction. *IEEE Trans. Intell. Transp.* **2020**, *21*, 3848–3858. [CrossRef]

24. Geng, X.; Li, Y.; Wang, L. Spatiotemporal multi-graph convolution network for ride-hailing demand forecasting. *AAAI Conf. Artif. Intell.* **2019**, *33*, 3656–3663. [CrossRef]
25. Diao, Z.; Wang, X.; Zhang, D. Dynamic spatial-temporal graph convolutional neural networks for traffic forecasting. *AAAI Conf. Artif. Intell.* **2019**, *33*, 890–897. [CrossRef]
26. Huang, R.; Huang, C.; Liu, Y. Lsgcn: Long short-term traffic prediction with graph convolutional networks. *Int. Joint Conf. Artif. Intell.* **2020**, 2355–2361.
27. Lv, M.; Hong, Z.; Chen, L. Temporal multi-graph convolutional network for traffic flow prediction. *IEEE Trans. Intell. Transp. Syst.* **2020**, *22*, 3337–3348. [CrossRef]
28. Guo, S.; Lin, Y.; Feng, N.; Song, C.; Wan, H. Attention Based Spatial-Temporal Graph Convolutional Networks for Traffic Flow Forecasting. *AAAI Conf. Artif. Intell.* **2019**, *33*, 922–929. [CrossRef]
29. Bruna, J.; Zaremba, W.; Szlam, A.; LeCun, Y. Spectral networks and locally connected networks on graphs. *arXiv* **2014**, arXiv:1312.6203.
30. Defferrard, M.; Bresson, X.; Vandergheynst, P. Convolutional neural networks on graphs with fast localized spectral filtering. In *Advances in Neural Information Processing Systems*; Morgan Kaufmann Publishers Inc.: San Francisco, CA, USA, 2016.
31. Bengio, Y.; Simard, P.; Frasconi, P. Learning long-term dependencies with gradient descent is difficult. *IEEE Trans. Neural Netw.* **1994**, *5*, 157–166. [CrossRef]
32. Mauro, M.D.; Galatro, G.; Liotta, A. Experimental Review of Neural-Based Approaches for Network Intrusion Management. *IEEE Trans. Netw. Service Manag.* **2020**, *17*, 2480–2495. [CrossRef]
33. Dong, S.; Xia, Y.; Peng, T. Network Abnormal Traffic Detection Model Based on Semi-Supervised Deep Reinforcement Learning. *IEEE Trans. Netw. Service Manag.* **2021**, *18*, 4197–4212. [CrossRef]
34. Pelletier, C.; Webb, G.I.; Petitjean, F. Deep Learning for the Classification of Sentinel-2 Image Time Series. In Proceedings of the IGARSS 2019, Yokohama, Japan, 31 July 2019.

electronics

MDPI

Article

Multi-Population Enhanced Slime Mould Algorithm and with Application to Postgraduate Employment Stability Prediction

Hongxing Gao [1,2], Guoxi Liang [3,*] and Huiling Chen [4,*]

1 Graduate School, Wenzhou University, Wenzhou 325035, China; ghx@wzu.edu.cn
2 Wenzhounese Economy Research Institute, Wenzhou University, Wenzhou 325035, China
3 Department of Information Technology, Wenzhou Polytechnic, Wenzhou 325035, China
4 College of Computer Science and Artificial Intelligence, Wenzhou University, Wenzhou 325035, China
* Correspondence: guoxiliang2017@gmail.com (G.L.); chenhuiling.jlu@gmail.com (H.C.)

Abstract: In this study, the authors aimed to study an effective intelligent method for employment stability prediction in order to provide a reasonable reference for postgraduate employment decision and for policy formulation in related departments. First, this paper introduces an enhanced slime mould algorithm (MSMA) with a multi-population strategy. Moreover, this paper proposes a prediction model based on the modified algorithm and the support vector machine (SVM) algorithm called MSMA-SVM. Among them, the multi-population strategy balances the exploitation and exploration ability of the algorithm and improves the solution accuracy of the algorithm. Additionally, the proposed model enhances the ability to optimize the support vector machine for parameter tuning and for identifying compact feature subsets to obtain more appropriate parameters and feature subsets. Then, the proposed modified slime mould algorithm is compared against various other famous algorithms in experiments on the 30 IEEE CEC2017 benchmark functions. The experimental results indicate that the established modified slime mould algorithm has an observably better performance compared to the algorithms on most functions. Meanwhile, a comparison between the optimal support vector machine model and other several machine learning methods on their ability to predict employment stability was conducted, and the results showed that the suggested the optimal support vector machine model has better classification ability and more stable performance. Therefore, it is possible to infer that the optimal support vector machine model is likely to be an effective tool that can be used to predict employment stability.

Keywords: global optimization; meta-heuristic; support vector machine swarm intelligence

Citation: Gao, H.; Liang, G.; Chen, H. Multi-Population Enhanced Slime Mould Algorithm and with Application to Postgraduate Employment Stability Prediction. *Electronics* **2022**, *11*, 209. https://doi.org/10.3390/electronics11020209

Academic Editor: Marco Mussetta

Received: 5 December 2021
Accepted: 7 January 2022
Published: 10 January 2022

1. Introduction

In China, postgraduates are valuable talent resources. The employment quality of postgraduates is not only related to their own sense of social belonging and security, but it also affects social stability and sustainable development, where employment stability is an important measure of postgraduate employment quality. Employment stability not only affects the career development of individual graduate students, but it is also a focal issue of educational equity and social stability. Moreover, employment stability not only reflects practitioners' psychological satisfaction with the employment unit, employment environment, remuneration, and career development, but it is also an important indicator of employment quality. When the skill level and the salary level of the job match, employment stability is high. On the contrary, the practitioner will actively seek to change jobs if there is a disparity between those factors, and especially in cases where the salary level is extremely mismatched with the skill level, the practitioner will face the risk of being fired and will passively change jobs, and employment stability will be low. It can be seen that employment stability also determines the employment quality of graduate students to a large extent. In addition, for enterprises, if they can retain talent and maintain the job stability of new graduate students, they can not only reduce labor costs, but these enterprises can also achieve sustainable development. Therefore, it is necessary to analyze the employment stability of graduate students through the effective mining of big data related to post-graduation graduate employment and to construct an intelligent prediction model using a fusion of intelligent optimization algorithms and machine learning methods to verify the hypothesis of relevant relationships. At the same time, in order to provide a reference for postgraduate employment decision making and policy formulation by relevant departments, it is also necessary to dig into the key factors affecting the stable employment of postgraduates, conduct in-depth analyses of key influencing factors, and explore the main factors affecting the stability of postgraduate employment.

At present, many studies have been conducted by many researchers on employment and employment stability. Yogesh et al. [1] applied artificial intelligence algorithms to enrich the student employability assessment process. Li et al. [2] made full use of the C4.5 algorithm to generate a type of employment data mining model for graduates. Liu et al. [3] proposed a weight-based decision tree to help students improve their employability. Mahdi et al. [4] proposed a novel method based on support vector machines, which was applied to predicting cryptocurrency returns. Tu et al. [5] developed an adaptive SVM framework to predict whether students would choose to start a business or find a job after graduation. Additionally, there also have been many studies on swarm intelligence algorithms. Cuong-Le et al. [6] presented an improved version of the Cuckoo search algorithm (NMS-CS) using the random walk strategy. Abualigah et al. [7] presented a novel nature-inspired meta-heuristic optimizer called the reptile search algorithm (RSA). Nadimi-Shahraki et al. [8] introduced an enhanced version of the whale optimization algorithm (EWOA-OPF), which combines the Levy motion strategy and Brownian motion. Gandomi et al. [9] proposed an evolutionary framework for the seismic response formulation of self-centering concentrically braced frame systems.

Therefore, in order to better predict the employment stability of graduate students, this paper first proposes a modified slime mould algorithm (MSMA), the core of which is the use of a multi-population mechanism to further balance the exploration and development of the slime mould algorithm, effectively improving the accuracy of the solution of the original slime mould algorithm. Further, a MSMA-based SVM model (MSMA-SVM) is proposed, in which MSMA effectively enhances the accuracy of the classification prediction of the original SVM. To demonstrate the performance of MSMA, MSMA and the slime mould algorithm were first subjected to analytical experiments to obtain careful balance and diversity using the 30 benchmark functions in the IEEE CEC2017 as a basis. In addition, this paper not only compares MSMA with other traditional basic algorithms, including differential evolution (DE) [10], the slime mould algorithm (SMA) [11], the grey wolf optimizer (GWO) [12,13], the bat-inspired algorithm (BA) [14], the firefly algorithm (FA) [15], the whale optimizer (WOA) [16,17], moth–flame optimization (MFO) [18–20], and the sine cosine algorithm (SCA) [21], but it also compares MSMA with some algorithm variants that have previously demonstrated very good performance, including boosted GWO (OBLGWO) [22], the balanced whale optimization algorithm (BWOA) [17], the chaotic mutative moth–flame-inspired optimizer (CLSGMFO) [20], PSO with an aging leader and challengers (ALCPSO) [23], the differential evolution algorithm based on chaotic local search (DECLS) [24], the double adaptive random spare reinforced whale optimization algorithm (RDWOA) [25], the chaos-enhanced bat algorithm (CEBA) [26], and the chaos-induced sine cosine algorithm (CESCA) [27]. Ultimately, the comparative experimental results that were obtained for the benchmark functions effectively illustrate that MSMA not only provides better performance than the initial SMA, but that it is also offers greater superiority than many common similar algorithms. To make better predictions and judgments about the employment stability of graduate students, the comparative MSMA-SVM experiments and experiments for other machine learning approaches were conducted. The results of the experiments indicate that, among all the comparison methods, MSMA-SVM can obtain more accurate classification results and better stability using the four indicators.

The rest of this paper is structures as follows: Section 2 provides a brief introduction to SVM and SMA. In Sections 3 and 4, the proposed MSMA and the MSMA-SVM model are described in detail, respectively. Section 5 mainly introduces the data source and simulation settings. The experimental outcomes of MSMA on the benchmark functions and the MSMA-SVM on the real-life dataset are analyzed in Section 6. A discussion of the improved algorithm is provided in Section 7. Additionally, the last section provides summaries and advice as they pertain to the present research.

In conclusion, the present research contributes the following major innovations:

(a) This paper proposes a novel version of SMA that combines a multi-population strategy called MSMA.
(b) Experiments comparing MSMA with other algorithms are conducted on a benchmark function set. The experimental results demonstrate that the proposed algorithm can better balance the exploitation and exploration capabilities and has better accuracy.
(c) The MSMA algorithm is combined with the support vector machine algorithm to construct a prediction model for the first time, which is called MSMA-SVM. Additionally, the MSMA-SVM model is employed in entrepreneurial intention prediction experiments.
(d) The proposed MSMA in the benchmark function experiment and the MSMA-SVM in entrepreneurial intention prediction demonstrate better performance than their counterparts.

2. Background

2.1. Support Vector Machine

The core principle of SVMs is the development of a plane that is best able to divide two kinds of data in such a way where the distance between the two is maximized and where the classification has the greatest generalization power. Support-vector data are the closest data to the boundary. The SVM is often a supervised learning approach that is used to process classification data for the purpose of finding the best hyperplane that can properly separate positive and negative samples.

With the given data set $G = (x_i, y_i), i = 1, \ldots, N, x \in R^d, y \in \{\pm 1\}$, the hyperplane can be expressed as:

$$g(x) = \omega' x + b \tag{1}$$

In terms of the geometric understanding of the hyperplane, the maximization of the geometric spacing is equal to the minimization of $||\omega||$. The concept of a "soft interval" is introduced, and the slack variable $\xi_i > 0$ is applied in cases where there are few outliers. One of the key parameters that can influence the ability of SVM classification is the disciplinary factor c, which represents the ability to accommodate outliers. A standard SVM model is shown below:

$$\begin{cases} \min(\omega) = \frac{1}{2}||\omega||^2 + c \sum_{i=1}^{N} \xi_i{}^2 \\ \text{s.t } y_i(\omega^{\mathrm{T}} x_i + b) \geq 1 - \xi_i, i = 1, 2, \ldots, N \end{cases} \tag{2}$$

where ω is an inertia weight, and b is a constant.

In this way, the initial low dimensional sample set is mapped to the high dimensional space H, allowing the best classification surface to be established in a linear method. Meanwhile, the SVM non-linearly transforms the linearly inseparable sample set $\Phi : R^d \to H$. For the purposes of keeping the computed results of the sample set in the low dimensional space corresponding to the results of the inner product that is mapped to the high dimensional part, a suitable $k(x_i, x_j)$ is constructed using generalized function theory to denote the kernel function, with α_i denoting the Lagrange multiplier, and Equation (3) being converted to as it is seen below:

$$\begin{cases} Q(\alpha) = \frac{1}{2}\sum_{i=1}^{N} \alpha_i \alpha_j y_i y_j k(x_i, x_j) - \sum_{i=1}^{N} \alpha_i \\ \text{s.t} \sum_{i=1}^{N} a_i y_i = 0, 0 \le a_i \le C, i = 1, 2, \ldots, N \end{cases} \tag{3}$$

This paper adopts the generalized radial basis kernel function as the function model of the support vector machine, and its expression is as follows:

$$k(x,y) = e^{-\gamma||x_i - x_j||} \tag{4}$$

where γ is a kernel parameter, another element that is quite important to the classification performance of an SVM, and it represents the interaction's kernel function width.

2.2. Slime Mould Algorithm

Similar to many other recently proposed optimization algorithms, including Harris hawks optimization (HHO) [28], the Runge Kutta optimizer (RUN) [29], the colony predation algorithm (CPA) [30], and hunger games search (HGS) [31], SMA is a novel and high-performing swarm intelligence optimization algorithm that was developed by Li et al. [11], who were motivated by the slime mould's foraging behavior. Since its introduction, SMA has been applied to many problems such as image segmentation [32,33], engineering design [34], parameter identification in photovoltaic models [35], medical decision-making [36], and multi-objective problems [37]. In this section, some mathematical models related to the mechanisms and characteristics of SMA are presented.

During its approach to food, the slime mould can be approached by odors in the environment. To show its actions mathematically in terms of convergence, the expressions below can be used to simulate its shrinkage pattern:

$$\vec{X}_{(t+1)} = \begin{cases} \vec{X_b}(t) + \vec{v_b} \cdot \left(\vec{W} \cdot \vec{X_A}(t) - \vec{X_B}(t)\right), & r < p \\ \vec{v_c} \cdot \vec{X}_{(t)}, & r \ge p \end{cases} \tag{5}$$

where $\vec{v_b}$ is a parameter in $[-a,\ a]$, $\vec{v_C}$ that takes values in the range $[-1, 1]$, t acts as the the quantity of current iterations, $\vec{X_b}$ represents the location of the individual found to have the best fitness value, $\vec{X}$ represents the location of the slime mould, $\vec{X_A}$ and $\vec{X_B}$ represent two individuals chosen from the slime mould in a random way, $\vec{W}$ represents the weight of the slime mould, and r is a stochastic number in the range $[0, 1]$. In addition, p, $\vec{v_b}$, $\vec{v_c}$, and $\vec{W}$ are computed as follows:

$$p = \tanh\left(|S(i) - BF|\right) \tag{6}$$

$$\vec{v_b} = [-a, a] \tag{7}$$

$$\vec{v_c} = \text{rand}(-b, b) \tag{8}$$

$$a = \text{arctanh}\left(-\left(\frac{FEs}{Max\_FEs}\right) + 1\right) \tag{9}$$

$$b = 1 - \left(\frac{FEs}{Max\_FEs}\right) \tag{10}$$

$$\vec{W}((SI(FEs))) = \begin{cases} 1 + r \cdot log\left(\frac{BF - S(i)}{BF - WF} + 1\right), & \text{condition} \\ 1 - r \cdot log\left(\frac{BF - S(i)}{BF - WF} + 1\right), & \text{others} \end{cases} \tag{11}$$

$$smellIndex = \text{sort}(S) \tag{12}$$

where $i \in 1, 2, 3, \cdots, n$, $S(i)$ represents the fitness of $\vec{X}$, BF and WF are the currently gained best fitness and worst fitness, *Fes* is the current quantity of the evaluations, Max_FEs is the maximum quantity of the evaluations, $condition$ refers to the top half of the $S(i)$ ranking in the population, and SI represents the sequence of the fitness values arranged in ascending order.

When food is being wrapped and as the concentration of food exposed to the vein increases, the more powerful the propagation wave produced by the bio-oscillator and the quicker the cytoplasmic flow are, resulting in thicker veins. Equation (11) models the positive and negative feedback relationship between vein width and food concentration in slime moulds. If the food concentration is higher, the weight of the nearby area will increase, and at lower food concentrations, the weight of the area will decline, causing the slime mould to move on to explore other areas. Therefore, the motility behavior of slime moulds can be simulated using Equation (13).

$$\vec{X^*} = \begin{cases} rand \cdot (UB - LB) + LB & (1) \quad rand < z \\ \begin{cases} \vec{X_b}(t) + \vec{v_b} \cdot \left(W \cdot \vec{X_A}(t) - \vec{X_B}(t) \right) & r < p \quad (2), \\ \vec{v_c} \cdot \vec{X}_{(t)} & , r \geq p \quad (3) \end{cases} & rand \geq z \end{cases} \tag{13}$$

where the upper and lower bounds are expressed by UB and LB in the search range, and $rand$ and r are random values in $[0, 1]$. According to the original version, parameter z is set to 0.03.

While grasping food, the way in which the slime moulds change the cytoplasmic flux is mainly through the propagation wave of the biological oscillator, putting it in a more favorable position for food concentration. W, vb, and vc were used to imitate the changes observed in the venous width of slime moulds. The value of $\vec{vb}$ oscillates randomly between $[-a, a]$ and approaches 0 with increasing iterations of the primary key. The value of $\vec{vc}$ varies between $[-1, 1]$ and eventually converges to 0. The drifts of the two are monitored in Figure 1, and these drifts are also specific to the task considered in this work.

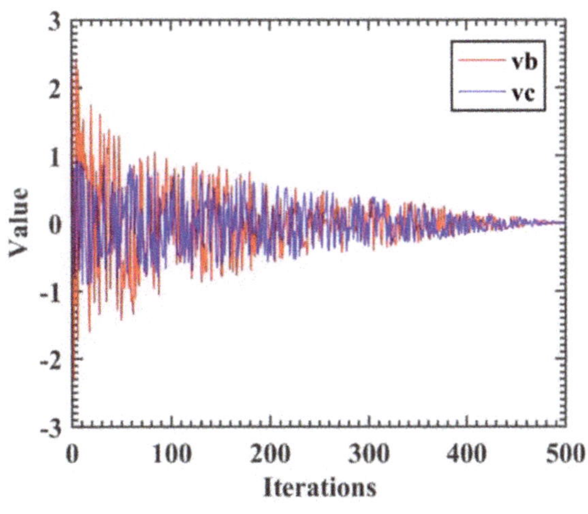

Figure 1. Variations in $\vec{v_b}$ and $\vec{v_c}$ trends.

The pseudo-code of the SMA is displayed in Algorithm 1.

Algorithm 1 Pseudo-code of SMA

```
Initialize the parameters popsize, Max_FEs;
Initialize the population of slime mould X_i (i = 1, 2, 3, ... n);
Initialize control parameters z, a;
While (t ≤ Max_FEs)
Calculate the fitness of slime mould;
Sorted in ascending order by fitness;
Update bestFitness, X_b;
Calculate the W by Equation (12);
For i = 1 to popsize
Update p by Equation(6);
Update vb by Equations (7) and (9);
Update vb by Equations (8) and (10);
      If rand < z
        update positions by Equation (13)(1);
      Else
        update p, vb, vc;
        If r < p
           update positions by Equation (13)(2);
        Else
           update positions by Equation (13)(3);
        End If
     End If
  End For
                          t = t + 1;
End While
Return bestFitness, X_b
```

3. Suggested MSMA

3.1. Multi-Population Structure

As an important factor that affects the information exchange between populations, the topological structure of the population also has a great impact on the balancing of the exploration and development processes. In the multi-population topological structure, the structure is mainly composed of three parts, which are the dynamic sub-population number strategy (DNS), the purposeful detecting strategy (PDS), the sub-populations regrouping strategy (SRS).

DNS means that the whole population is separated into many sub-populations after the first iteration. Usually, a sub-population is composed of two search individuals, and as the quantity of iterations increases, the quantity of the sub-populations gradually decreases, and the scale of the sub-populations increases. Additionally, only one sub-population is left in the search space, which represents the aggregation of all of the sub-populations at the ending of the iteration process. Smaller sub-populations can better help the swarm maintain its diversity. With the iteration process, the population change characteristics mainly show that the number of sub-populations gradually decreases and that the size of sub-populations expands. The strategy enables individuals in the population to exchange information more quickly and widely. In addition, the DNS implementation is decided by the feedback of the changing principle of the subgroup quantity and the cycle. To resolve the first problem, a set of integers $N = \{n_1, n_2, \cdots, n_{k-1}, n_k\}, n_1 > n_2 > \cdots > n_{k-1} > n_k$ are used, where the integer indicates the subgroup quantity. To ensure the implementation of the DNS strategy, the size of each sub-population remains unchanged in one iteration, that is, the whole number of individuals can be evenly divided by the quantity of the sub-populations. For that changing period, a fixed stage is used to adapt the structure of the whole population. The stage length is calculated by $Cgen = MaxFEs/\|N\|$, where $\|N\|$ is the quantity of the integers in N, and $MaxFEs$ delegates the preset number of evaluation times to ensure that the efficient variation of sub-population quantity is efficient.

In SRS, the proposed method uses the same sub-population reorganization strategy as the published enhanced particle swarm optimization [38], where $Stag_{best}$ represents the quantity of the best individual stagnations. The sub-population reorganization strategy will be executed when the whole population stagnates in the suggested approach, and the execution timing of the sub-population reorganization scheme is determined in this way. Additionally, the scale of the sub-population impacts the frequency with which this strategy is executed. As the scale of the sub-population increases, individuals need more iterations to obtain useful guidelines. Because of the above points, the $Stag_{best}$ calculation method is shown below: $Stag_{best} = S_{sub}/2$.

PDS enhances the capability of the presented method to get rid of the local optima, particularly in multi-modal problems. The collected population information is used to guide the swarm to energetically search rooms with a higher search value, and many researches have proven the superiority of the scheme [39,40]. To provide convenience for PDS execution, it is stipulated that each dimension of the search room be equal in size. The function of the segmentation mechanism is to help the search individuals collect information. For PDS, the segments are classified. When the best search agent and when the current individual are in the best exploration interval of the dimension, the best search individual will select a search segment in the worst exploration interval of the same dimension. If the fitness of that newly searched-for new candidate solution is superior to the current optimal record, the optimal single position will be substituted by the new solution. The underexplored intervals will be more fully explored because of the benefits imparted by PDS. Meanwhile, a taboo scheme was attached to the PDS to avoid repeatedly exploring the same area. When a segment s_j^i is searched, the variable tab_j^i that delegates the segment is set to 1. Additionally, segment s_j^i can only be found again when tab_j^i is reset to 0. All flag variables will be recorded as 0 when each segment of each dimension has been fully explored.

3.2. Proposed MSMA

The MSMA improvement principle is the addition of the dynamic multi-population structure to the original SMA. The whole population is divided into many subgroups with the same swarm scale at the start of the multi-population strategy. The equal scale of the subgroups not only simplifies the general population structure, but it also simplifies the process complication of adjusting and fusing the population structure. The multi-population structure is employed to lead the whole population's exploration tendency in the direction of the improved search methodology by updating the SMA function. With the continuation of the iterative process, the DNS strategy is to increase the scale of the sub-populations while reducing their number in order to guide this method to the exploitation stage. In addition, during the searching process, the PDS scheme is implemented to realize information sharing among sub-populations and enhances algorithm exploration capabilities as well. The SRS strategy will be executed to make the population jump out of the local optima when the population is located in the local optima. The pseudocode of Algorithm 2 below expresses those details of the MSMA framework.

Algorithm 2 Pseudo-code of MSMA

Initialize the parameters *popsize, Max_FEs;*
Initialize the population of slime mould X_i (i = 1, 2, 3, ... n);
Initialize control parameters z, a;
While ($t \leq Max\_FEs$)
Calculate the fitness of slime mould;
Sorted in ascending order by fitness;
Update $bestFitness,\ X_b$;
Calculate the W by Equation (12);
For $i = 1\ to\ popsize$
Update p by Equation (6);
Update vb by Equations (7) and (9);
Update vb by Equations (8) and (10);
  If rand < z
    $update\ positions\ by$ Equation (13)(1);
  Else
    $update\ p, vb,\ vc$;
    If r < p
      $update\ positions\ by$ Equation (13)(2);
    Else
      $update\ positions\ by$ Equation (13)(3);
    End If
  End If
End For
Perform DNS, SRS, and PDS from multi-population topological structure;
$t = t + 1$;
End While
Return $bestFitness,\ X_b$

The complexity of MSMA is mainly related to slime mould initialization, fitness calculation, weight calculation, position updating, and the complexity of DNS, SRS, and PDS. n represents the quantity of the slime mould, T represents the number of iterations, and dim represents the dimension of the objective function. Thus, the complexity of slime mould initialization is $O(n)$, the fitness calculation and ordering complexity is $O(T \times 3 \times (n + nlog\ n))$, the weight calculation complexity is $O(T \times n \times dim)$, and the position updating complexity is $O(T \times n \times dim)$. The DNS complexity is $O\ (T \times (n + T \times n))$. The SRS complexity is $O\ (T \times n)$. The PDS complexity is $O\ (T \times dim \times Rn)$, where Rn represents the quantity of segments in the dimension. Thus, the overall MSMA complexity is $O(n \times (1 + T \times n \times ((5 + T) + 3 \times log\ n + 3 \times dim)))$.

3.3. Proposed MSMA-SVM Method

Penalty factor C, the kernel parameter γ, and the optimal feature set are two important factors that determine the classification results and algorithm complexity of the SVM classification model. Usually, these two parameters are selected based on experience, resulting in poor efficiency and accuracy. The feature subset also uses the whole set or randomly selected variables, which also leads to poor efficiency and accuracy. Therefore, a new solution model MSMA-SVM was proposed, in which MSMA is used to optimize two vital parameters in SVM and in the feature subset. Then, the model will be applied to two special situations in the actual world: medical diagnosis situations and financial forecasting situations. The framework of the MSMA-SVM is displayed in Figure 2. The model mainly contains two important components. The left two columns use MSMA to optimize the two parameters and feature subset in the SVM model. In the right half, this optimized SVM obtained the classification accuracy (ACC) through 10-fold cross-validation (CV), nine of which were utilized for the training, and the rest was employed for test applications.

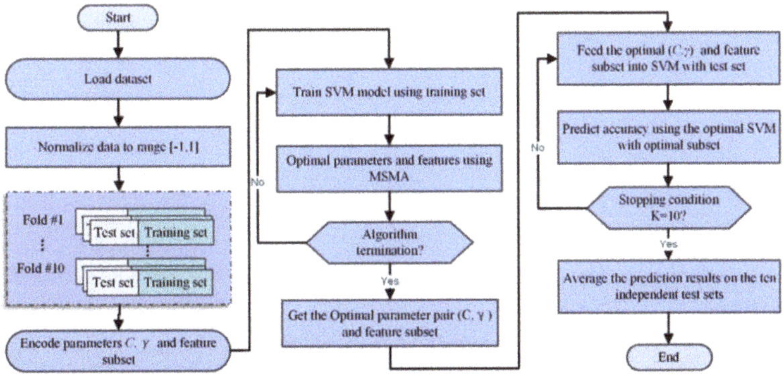

Figure 2. Flowchart of the suggested MSMA-SVM model.

4. Experiments

4.1. Collection of Data

The population studied in this article comprised (a total of 331) full-time postgraduate students from the class of 2016 at Wenzhou University. According to the comparison of the employment status of the 2016 postgraduate graduates after three years with the initial postgraduate graduate employment program in September 2019, it was found that 153 postgraduates (46.22%) had not changed workplaces in three years, and 178 postgraduates (53.78%) demonstrated separation behavior.

Through data mining and analyses gender, political outlook, professional attributes, academic system, situations where the student experienced difficulty, student origin, academic performance (average course grades, teaching practice grades, social practice grades, academic report grades, thesis grades), graduation destination, nature of initial employment unit, location of initial employment unit, initial employment position, degree of initial employment and its relevance to the student's major, monthly salary level during initial employment, employment variation, current employment status, nature of current employment unit, location of current employment unit, variation in employment location, current employment position, degree of current employment and its relevance to the student's major, current monthly salary level, and monthly salary difference (see Table 1), the authors explored the importance and intrinsic connection of each index and built an intelligent prediction model based on this information.

Table 1. Description of the total 26 attributes.

ID	Attribute	Description
F1	gender	Male and female students are marked as 1 and 2, respectively.
F2	political status (PS)	There are four categories: Communist Party members, reserve party members, Communist Youth League members, and the masses, denoted by 1, 2, 3, and 13, respectively.
F3	division of liberal arts and science (DLS)	Liberal arts and sciences are indicated by 1 and 2.
F4	years of schooling (YS)	The 3-year and 4-year academic terms are indicated by 3 and 4.
F5	students with difficulties (SWD)	There are four categories: non-difficult students, employment difficulties, family financial difficulties, and dual employment and family financial difficulties, which are indicated by 0, 1, 2, and 3, respectively.

Table 1. *Cont.*

ID	Attribute	Description
F6	student origin (OS)	There are three categories: urban, township, and rural, denoted by 1, 2, and 3, respectively.
F7	career development after graduation (CDG)	There are three categories of direct employment, pending employment, and further education, which are indicated by 1, 2, and 3, respectively.
F8	unit of first employment (UFE)	Employment pending is indicated by 0. State organizations are indicated by 10, scientific research institutions are indicated by 20, higher education institutions are indicated by 21, middle and junior high education institutions are indicated by 22, health and medical institutions are indicated by 23, other institutions are indicated by 29, state-owned enterprises are indicated by 31, foreign-funded enterprises are indicated by 32, private enterprises are indicated by 39, troops are indicated by 40, rural organizations are indicated by 55, and self-employment is indicated by 99.
F9	location of first employment (LFE)	Employment pending is indicated by 0, sub-provincial and above large cities by 1, prefecture-level cities by 2, and counties and villages by 3.
F10	position of first employment (PFE)	Employment pending is represented by 0, civil servants by 10, doctoral students and researchers by 11, engineers and technicians by 13, teaching staff by 24, professional and technical staff by 29, commercial service staff and clerks by 30, and military personnel by 80.
F11	degree of specialty relevance of first employment (DSRFE)	The correlation between major and job is measured, and the higher the percentage, the higher the correlation.
F12	monthly salary of first employment (MSFE)	Used to measure the average monthly salary earned, with higher values indicating higher salary levels.
F13	status of current employment (SCE)	Three years after graduation, the employment status is represented by 1, 2, and 3 for the categories of employment, pending employment, and further education, respectively.
F14	employment change (EC)	When comparing the employment units three years after graduation with initial employment units, no change is indicated by 0 and any change is indicated by 1.
F15	unit of current employment (UCE)	The nature of the employment unit three years after graduation is expressed in the same way as the nature of initial employment unit in F8.
F16	location of current employment (LCE)	The type employment location three years after graduation is expressed in the same way as the initial employment location in F9.
F17	change in place of employment (CPE)	Used to measure the changes in employment location from the initial employment location three years after graduation and is expressed as the difference between F16 current employment location type and F9 initial employment location type, and the larger the absolute value of the difference, the larger the change in employment location.
F18	position of current employment (PCE)	The job type three years after graduation is expressed in the same way as the initial employment job type in F10.
F19	specialty relevance of current employment (SRCE)	The professional relevance of employment three years after graduation is expressed in the same way as the initial employment job type in F11.
F20	monthly salary of current employment (MSCE)	The monthly salary level three years after graduation is expressed in the same way as the monthly salary level during initial employment in F12.

Table 1. *Cont.*

ID	Attribute	Description
F21	salary difference (SD)	Used to measure the changes in the graduates' monthly salary in their current employment and initial employment, i.e., the difference between F20 monthly salary level in current employment and F12 monthly salary level in initial employment, with a larger value indicating a larger increase in monthly salary.
F22	grade point average (GPA)	Used to assess the how much the postgraduate students learned while they were in school and is the average of the final grades of courses taken by graduate students, with higher averages indicating higher quality learning.
F23	scores of teaching practice (STP)	A method used to assess the quality of learning in postgraduate teaching practice sessions, with excellent, good, moderate, pass, and fail expressed as 1, 2, 3, 4, and 5, respectively.
F24	scores of social practices (SSP)	A method used to assess how much the postgraduate students learned in social practice sessions, with excellent, good, moderate, pass, and fail expressed as 1, 2, 3, 4, and 5, respectively.
F25	scores of academic reports (SAR)	A method used to assess how the must the postgraduate students learned during academic reporting sessions, with excellent, good, moderate, pass, and fail expressed as 1, 2, 3, 4, and 5, respectively.
F26	scores of graduation thesis (SGT)	A method used to assess the how much the postgraduate students learned during the thesis sessions, with excellent, good, moderate, pass, and fail expressed as 1, 2, 3, 4, and 5, respectively.

4.2. Experimental Setup

MATLAB R2018 software was utilized to conduct the experiment. The data were scaled to $[-1, 1]$ before classification. The k-fold cross-validation (CV) was used to split the data, where k was set to 10.

In addition, to ensure the same environment for all experiments, the experiments were conducted on a Windows 10 with Intel(R) Core (TM) i5^{-4200}H CPU @ 2.80 GHz and 8 GB of RAM. Coding was completed by using Matlab R2018.

5. Experimental Result

5.1. The Qualitative Analysis of MSMA

Swarm intelligence algorithms are good at solving many optimization problems, such as traveling salesman problems [41], feature selection [42–46], object tracking [47,48], wind speed prediction [49], PID optimization control [50–52], image segmentation [53,54], the hard maximum satisfiability problem [55,56], parameter optimization [22,57–59], gate resource allocation [60,61], fault diagnosis of rolling bearings [62,63], the detection of foreign fibers in cotton [64,65], large-scale supply chain network design [66], cloud workflow scheduling [67,68], neural network training [69], airline crew rostering problems [70], and energy vehicle dispatch [71]. This section conducts a qualitative analysis of MSMA.

Original SMA was selected for comparison with MSMA. Figure 3 displays the feasibility outcomes of the study comparing MSMA and SMA. There are five columns in the figure. The first column (a) is the position distribution for the MSMA search history on the three-dimensional plane. The second column (b) is the position distribution for the MSMA search history on the two-dimensional plane. In Figure 3b, the red dot represents the location of the optimal solution, and the black dot represents the MSMA search location. In the figure, the black dots are scattered everywhere on the entire search flat, which shows that MSMA performs a global search on the solution space. The black dots are significantly denser in the area around the red dots, which shows that MSMA has exploited the area to a greater extent in the areas where the best solution is situated. The third column (c) is the trajectory of the first dimension of the MSMA during the iteration. In Figure 3c, it is easy to see that the one-MSMA dimensional trajectory has large fluctuations. The amplitude of the trajectory fluctuation reflects the search range of the algorithm to a certain extent. The large fluctuation range of the trajectory indicates that the algorithm has performed a large-scale search. The fourth column (d) displays changes in the average MSMA fitness during the iteration. In Figure 3d, the average fitness of the algorithm shows huge fluctuations, but the overall fitness is decreasing. The fifth column (e) describes the MSMA and SMA convergence curves. In Figure 3e, the authors can clearly see that the MSMA convergence is lower than that of SMA, which shows that MSMA has better convergence performance.

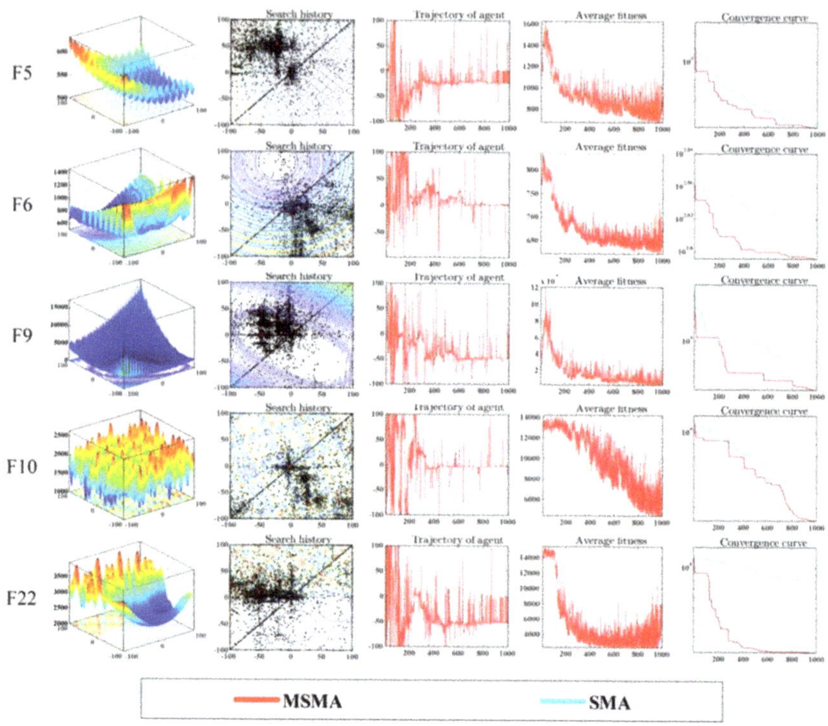

Figure 3. (**a**) Three-dimensional location distribution of MSMA; (**b**) two-dimensional location distribution of MSMA; (**c**) MSMA trajectory in the first dimension; (**d**) mean fitness of MSMA; (**e**) c MSMA and SMA convergence graphs.

Balance analysis and diversity analysis were carried out on the same functions. Figure 4 shows the outcomes of the balance study on MSMA and SMA. The three curves in each picture represent three different behaviors. As indicated in the legend, the red curve and blue curve represent exploration and exploitation, respectively. The large value of the curve indicates that this corresponding behavior is prominent in this algorithm. The green curve is an incremental–decremental curve. This curve can more intuitively reflect the changing trends in the two behaviors of the algorithm. When the curve increases, it means that exploration activities are currently dominant. The exploitation behavior is dominant in the opposite circumstances. Additionally, if these two are at the same stage, the increment–decrement curve has the best performance.

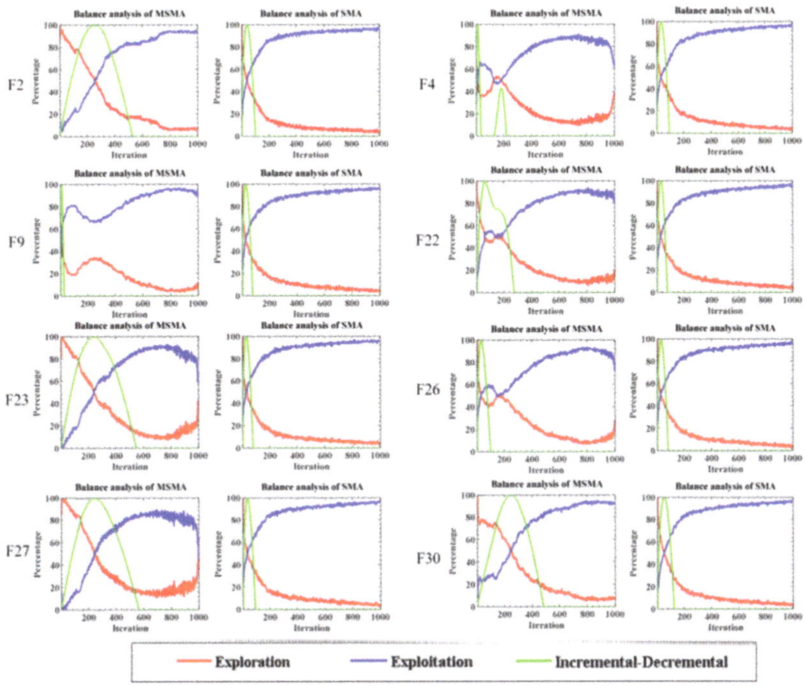

Figure 4. Balance analysis of MSMA and SMA.

The swarm intelligence algorithm will first perform a global search when solving optimization problems. After determining the position of the optimal solution, the area will be locally developed. Therefore, the authors see that exploration activities are dominant in MSMA and SMA at the beginning. MSMA spends more time on exploration than the original SMA, which can be clearly seen in F2, F23, F27, and F30. However, the proportion of MSMA exploration behavior on F4, F9, F22, and F26 is also higher than that of SMA. The authors can see that the exploration curves and exploitation curves of MSMA on F4, F9, F22, and F26 are not monotonous, but instead fluctuate. This fluctuation can be clearly observed when the MSMA exploration curve drops rapidly in the early phase. Because the fluctuation guarantees the proportion of exploration behavior, MSMA will not end the global exploration phase too quickly. This is a big difference in the balance between MSMA and SMA.

Figure 5 is the result of diversity analysis of MSMA and SMA. In Figure 5, the abscissa stands for the iteration quantity, and the ordinate represents the population diversity. At the beginning, the swarm is randomly generated, so the population diversity is very high. As the iteration progresses, the algorithm continues to narrow the search range, and the population diversity will decrease. The SMA diversity curve is a monotonically decreasing curve, which can be seen in Figure 5. However, MSMA is different. The fluctuations in the balance analysis are also reflected in the diversity curve. The authors can see that the F1, F3, F12, and F15 curves all have a reverse increase period in terms of diversity, while other functions are not obvious. This fluctuation period becomes more obvious when the MSMA diversity decreases rapidly in the early stage. Obviously, this ensures that MSMA can maintain high population diversity and wide search capabilities in the early and mid-term. In the later period, the MSMA diversity dropped to a low level and demonstrated good convergence ability.

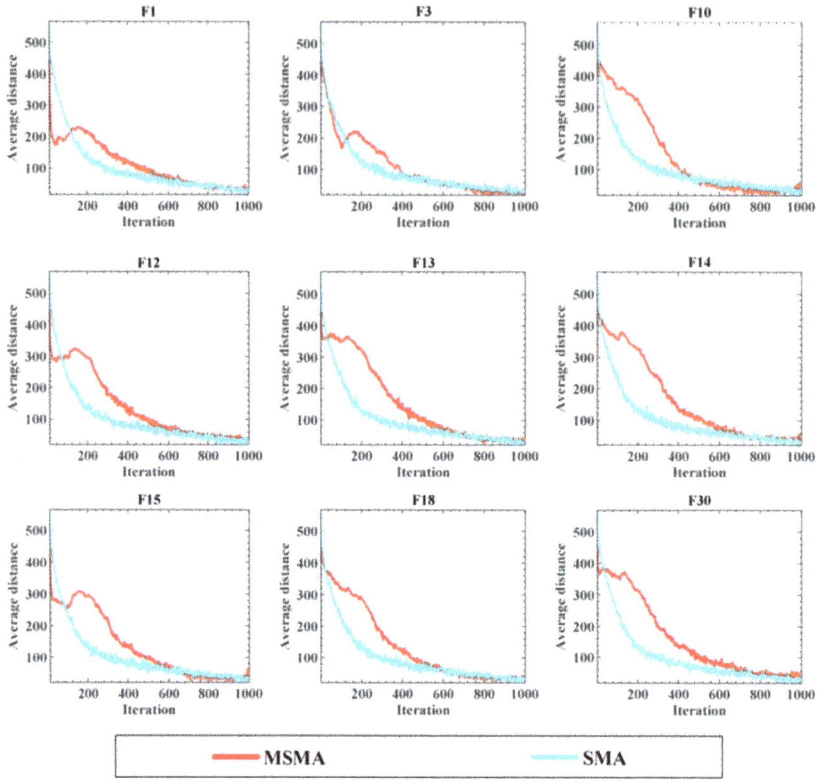

Figure 5. Diversity analysis of MSMA and SMA.

5.2. Comparison with Original Methods

In this section, the MSMA is compared with eight original swarm intelligence algorithms: SMA [11], DE [10], GWO [12,13], BA [14], FA [15], WOA [16,17], MFO [18–20], and SCA [21], to prove the performance of the MSMA. These comparison algorithms are classic and representative original algorithms that have been cited by many researchers for the sake of estimating the superiority of their own developed algorithms. In this experiment, the authors selected the CEC2017 [72] test function to judge the excellence of the involved algorithms and set the number of search agents to 30, the search agent dimension to 30, and the maximum evaluation times to 150,000. Every algorithm was run individually 30 times to obtain the mean value. In Table 2, the average and standard deviation that these algorithms searched for on different test functions is displayed, respectively. Obviously, the mean and standard deviations of the presented algorithm are lower than those of other compared ones for most functions. The Friedman [73] test is a non-parametric test method that can test whether multiple population distributions are significantly different. The Friedman test calculates the average performance differences for the chosen approaches and then compares them statistically to determine the ARV values (average ranking values) for the different methods. In Table 2, the MSMA algorithm ranks first in 22 benchmark functions, such as F1 and F3, proving that this paper's enhanced algorithm has numerous advantages over other algorithms that were compared using the CEC2017 benchmark functions. The Wilcoxon [74] symbolic rank test was used to test whether the algorithm that was improved in the article was significantly better than the others. In the Wilcoxon symbolic rank test, when the p value is lower than 0.05, the MSMA algorithm is obviously better than others in the present test functions. In Table 2, most of the p-values that were calculated by the MSMA and the comparison algorithm on the test function are less than 0.05. Therefore, the MSMA algorithm is more capable of searching for the optimal solution using the CEC2017 test function than other competitors.

Table 2. Comparison results of different original algorithms best scores obtained so far.

		MSMA	SMA	DE	GWO	BA	FA	WOA	MFO	SCA
F1	Avg	1.31×10^{2}	8.03×10^{3}	1.79×10^{3}	2.24×10^{9}	5.72×10^{5}	$1.55 \times 10^{+10}$	2.36×10^{7}	9.38×10^{9}	$1.41 \times 10^{+10}$
	Std	1.68×10^{2}	7.18×10^{3}	2.96×10^{3}	1.68×10^{9}	3.78×10^{5}	1.58×10^{9}	1.86×10^{7}	7.26×10^{9}	1.89×10^{9}
	Rank	1	3	2	6	4	9	5	7	8
	p-value		1.73×10^{-6}	3.88×10^{-6}	1.73×10^{-6}	1.73×10^{-6}	1.73×10^{-6}	1.73×10^{-6}	1.73×10^{-6}	1.73×10^{-6}
F2	Avg	8.31×10^{4}	6.49×10^{2}	$1.26 \times 10^{+24}$	$2.36 \times 10^{+32}$	1.75×10^{3}	$6.49 \times 10^{+34}$	$2.84 \times 10^{+26}$	$1.31 \times 10^{+38}$	$7.01 \times 10^{+36}$
	Std	2.55×10^{5}	9.69×10^{2}	$3.56 \times 10^{+24}$	$9.82 \times 10^{+32}$	8.51×10^{3}	$1.54 \times 10^{+35}$	$1.11 \times 10^{+27}$	$6.86 \times 10^{+38}$	$3.82 \times 10^{+37}$
	Rank	3	1	4	6	2	7	5	9	8
	p-value		7.51×10^{-5}	1.73×10^{-6}	1.73×10^{-6}	2.37×10^{-5}	1.73×10^{-6}	1.73×10^{-6}	1.73×10^{-6}	1.73×10^{-6}
F3	Avg	3.00×10^{2}	3.00×10^{2}	6.26×10^{4}	3.85×10^{4}	3.00×10^{2}	6.85×10^{4}	2.18×10^{5}	1.09×10^{5}	4.37×10^{4}
	Std	3.11×10^{-5}	2.80×10^{-1}	1.13×10^{4}	1.15×10^{4}	1.39×10^{-1}	7.95×10^{3}	6.96×10^{4}	5.83×10^{4}	7.84×10^{3}
	Rank	1	3	6	4	2	7	9	8	5
	p-value		1.73×10^{-6}	1.73×10^{-6}	1.73×10^{-6}	1.73×10^{-6}	1.73×10^{-6}	1.73×10^{-6}	1.73×10^{-6}	1.73×10^{-6}
F4	Avg	4.01×10^{2}	4.92×10^{2}	4.91×10^{2}	5.90×10^{2}	4.81×10^{2}	1.49×10^{3}	5.69×10^{2}	1.60×10^{3}	1.53×10^{3}
	Std	1.62×10^{0}	2.69×10^{1}	7.26×10^{0}	8.24×10^{1}	3.02×10^{1}	1.92×10^{2}	3.31×10^{1}	8.13×10^{2}	2.85×10^{2}
	Rank	1	4	3	6	2	7	5	9	8
	p-value		1.73×10^{-6}	1.73×10^{-6}	1.73×10^{-6}	1.73×10^{-6}	1.73×10^{-6}	1.73×10^{-6}	1.73×10^{-6}	1.73×10^{-6}
F5	Avg	5.90×10^{2}	5.94×10^{2}	6.26×10^{2}	6.09×10^{2}	7.94×10^{2}	7.66×10^{2}	7.81×10^{2}	7.13×10^{2}	7.96×10^{2}
	Std	2.31×10^{1}	2.55×10^{1}	8.69×10^{0}	2.52×10^{1}	5.42×10^{1}	1.12×10^{1}	6.49×10^{1}	5.43×10^{1}	1.89×10^{1}
	Rank	1	2	4	3	8	6	7	5	9
	p-value		5.86×10^{-1}	6.98×10^{-6}	6.04×10^{-3}	1.73×10^{-6}	1.73×10^{-6}	1.73×10^{-6}	1.73×10^{-6}	1.73×10^{-6}
F6	Avg	6.03×10^{2}	6.02×10^{2}	6.00×10^{2}	6.08×10^{2}	6.73×10^{2}	6.16×10^{2}	6.71×10^{2}	6.38×10^{2}	6.52×10^{2}
	Std	1.28×10^{0}	1.30×10^{0}	5.59×10^{-14}	3.70×10^{0}	1.16×10^{1}	2.57×10^{0}	1.02×10^{1}	1.20×10^{1}	4.36×10^{0}
	Rank	3	2	1	4	9	6	8	5	7
	p-value		2.85×10^{-2}	1.73×10^{-6}	2.35×10^{-6}	1.73×10^{-6}	1.73×10^{-6}	1.73×10^{-6}	1.73×10^{-6}	1.73×10^{-6}
F7	Avg	8.25×10^{2}	8.35×10^{2}	8.61×10^{2}	8.74×10^{2}	1.73×10^{3}	1.42×10^{3}	1.25×10^{3}	1.14×10^{3}	1.15×10^{3}
	Std	1.92×10^{1}	2.31×10^{1}	1.18×10^{1}	4.93×10^{1}	2.24×10^{2}	3.68×10^{1}	8.20×10^{1}	1.51×10^{2}	3.99×10^{1}
	Rank	1	2	3	4	9	8	7	5	6
	p-value		6.87×10^{-2}	4.29×10^{-6}	1.36×10^{-5}	1.73×10^{-6}	1.73×10^{-6}	1.73×10^{-6}	1.73×10^{-6}	1.73×10^{-6}
F8	Avg	8.82×10^{2}	9.04×10^{2}	9.24×10^{2}	8.96×10^{2}	1.02×10^{3}	1.06×10^{3}	1.01×10^{3}	1.02×10^{3}	1.06×10^{3}
	Std	2.07×10^{1}	3.00×10^{1}	9.78×10^{0}	2.54×10^{1}	4.55×10^{1}	1.20×10^{1}	5.04×10^{1}	5.39×10^{1}	2.19×10^{1}
	Rank	1	3	4	2	6	9	5	7	8
	p-value		5.67×10^{-3}	2.35×10^{-6}	1.85×10^{-2}	1.73×10^{-6}	1.73×10^{-6}	1.73×10^{-6}	1.73×10^{-6}	1.73×10^{-6}
F9	Avg	1.08×10^{3}	2.84×10^{3}	9.00×10^{2}	2.17×10^{3}	1.42×10^{4}	5.49×10^{3}	8.24×10^{3}	7.75×10^{3}	5.95×10^{3}
	Std	1.44×10^{2}	1.55×10^{3}	2.11×10^{-14}	1.05×10^{3}	5.31×10^{3}	6.66×10^{2}	2.81×10^{3}	1.97×10^{3}	1.11×10^{3}
	Rank	2	4	1	3	9	5	8	7	6
	p-value		8.47×10^{-6}	1.73×10^{-6}	2.35×10^{-6}	1.73×10^{-6}	1.73×10^{-6}	1.73×10^{-6}	1.73×10^{-6}	1.73×10^{-6}
F10	Avg	3.86×10^{3}	4.23×10^{3}	6.29×10^{3}	3.92×10^{3}	5.54×10^{3}	8.21×10^{3}	6.32×10^{3}	5.24×10^{3}	8.32×10^{3}
	Std	6.45×10^{2}	6.50×10^{2}	2.26×10^{2}	6.84×10^{2}	6.85×10^{2}	3.30×10^{2}	9.04×10^{2}	6.51×10^{2}	3.21×10^{2}
	Rank	1	3	6	2	5	8	7	4	9
	p-value		3.85×10^{-3}	1.73×10^{-6}	9.59×10^{-1}	1.73×10^{-6}	1.73×10^{-6}	1.73×10^{-6}	1.73×10^{-6}	1.73×10^{-6}
F11	Avg	1.19×10^{3}	1.27×10^{3}	1.18×10^{3}	2.17×10^{3}	1.29×10^{3}	3.95×10^{3}	2.11×10^{3}	4.85×10^{3}	2.49×10^{3}
	Std	3.49×10^{1}	5.92×10^{1}	2.27×10^{1}	1.00×10^{3}	6.85×10^{1}	6.09×10^{2}	7.63×10^{2}	4.69×10^{3}	5.58×10^{2}
	Rank	2	3	1	6	4	8	5	9	7
	p-value		1.24×10^{-5}	7.81×10^{-1}	1.73×10^{-6}	2.60×10^{-6}	1.73×10^{-6}	1.73×10^{-6}	1.73×10^{-6}	1.73×10^{-6}
F12	Avg	2.99×10^{3}	1.39×10^{6}	3.25×10^{6}	1.15×10^{8}	2.80×10^{6}	1.71×10^{9}	8.42×10^{7}	3.92×10^{8}	1.37×10^{9}
	Std	5.18×10^{2}	1.27×10^{6}	1.84×10^{6}	3.27×10^{8}	1.65×10^{6}	4.40×10^{8}	7.66×10^{7}	6.94×10^{8}	4.14×10^{8}
	Rank	1	2	4	6	3	9	5	7	8
	p-value		1.73×10^{-6}	1.73×10^{-6}	1.73×10^{-6}	1.73×10^{-6}	1.73×10^{-6}	1.73×10^{-6}	1.73×10^{-6}	1.73×10^{-6}
F13	Avg	4.43×10^{3}	2.80×10^{4}	8.88×10^{4}	3.05×10^{7}	3.75×10^{5}	7.12×10^{8}	2.29×10^{5}	4.63×10^{7}	5.09×10^{8}
	Std	1.76×10^{3}	2.43×10^{4}	5.01×10^{4}	8.51×10^{7}	1.76×10^{5}	1.95×10^{8}	3.25×10^{5}	1.93×10^{8}	1.46×10^{8}
	Rank	1	2	3	6	5	9	4	7	8
	p-value		1.73×10^{-6}	1.73×10^{-6}	1.73×10^{-6}	1.73×10^{-6}	1.73×10^{-6}	1.73×10^{-6}	1.73×10^{-6}	1.73×10^{-6}
F14	Avg	1.69×10^{3}	6.08×10^{4}	8.10×10^{4}	2.12×10^{5}	8.68×10^{3}	2.78×10^{5}	1.05×10^{6}	1.07×10^{5}	2.32×10^{5}
	Std	1.85×10^{2}	2.74×10^{4}	4.73×10^{4}	3.41×10^{5}	5.59×10^{3}	1.38×10^{5}	1.15×10^{6}	1.62×10^{5}	1.36×10^{5}
	Rank	1	3	4	6	2	8	9	5	7
	p-value		1.73×10^{-6}	1.73×10^{-6}	1.73×10^{-6}	1.73×10^{-6}	1.73×10^{-6}	1.73×10^{-6}	1.73×10^{-6}	1.73×10^{-6}
F15	Avg	2.04×10^{3}	2.97×10^{4}	1.54×10^{4}	2.44×10^{5}	1.36×10^{5}	7.90×10^{7}	7.87×10^{4}	6.23×10^{4}	2.07×10^{7}
	Std	2.08×10^{2}	1.46×10^{4}	1.08×10^{4}	6.85×10^{5}	6.24×10^{4}	3.57×10^{7}	4.96×10^{4}	5.54×10^{4}	1.60×10^{7}
	Rank	1	3	2	7	6	9	5	4	8
	p-value		2.35×10^{-6}	1.73×10^{-6}	1.73×10^{-6}	1.73×10^{-6}	1.73×10^{-6}	1.73×10^{-6}	1.73×10^{-6}	1.73×10^{-6}

Table 2. *Cont.*

		MSMA	SMA	DE	GWO	BA	FA	WOA	MFO	SCA
F16	Avg	2.22×10^{3}	2.52×10^{3}	2.17×10^{3}	2.47×10^{3}	3.61×10^{3}	3.56×10^{3}	3.67×10^{3}	3.14×10^{3}	3.74×10^{3}
	Std	2.05×10^{2}	3.15×10^{2}	1.40×10^{2}	2.68×10^{2}	4.46×10^{2}	1.68×10^{2}	6.32×10^{2}	3.36×10^{2}	1.80×10^{2}
	Rank	2	4	1	3	7	6	8	5	9
	p-value		5.29×10^{-4}	2.80×10^{-1}	1.29×10^{-3}	1.73×10^{-6}	1.73×10^{-6}	1.73×10^{-6}	1.73×10^{-6}	1.73×10^{-6}
F17	Avg	1.93×10^{3}	2.28×10^{3}	1.89×10^{3}	2.03×10^{3}	2.79×10^{3}	2.62×10^{3}	2.53×10^{3}	2.47×10^{3}	2.47×10^{3}
	Std	1.41×10^{2}	2.33×10^{2}	7.80×10^{1}	1.66×10^{2}	3.14×10^{2}	1.13×10^{2}	2.60×10^{2}	2.56×10^{2}	1.80×10^{2}
	Rank	2	4	1	3	9	8	7	5	6
	p-value		1.13×10^{-5}	4.53×10^{-1}	2.18×10^{-2}	1.73×10^{-6}	1.73×10^{-6}	1.73×10^{-6}	2.35×10^{-6}	1.73×10^{-6}
F18	Avg	2.19×10^{3}	3.49×10^{5}	4.56×10^{5}	7.66×10^{5}	2.28×10^{5}	5.26×10^{6}	3.15×10^{6}	6.37×10^{6}	4.05×10^{6}
	Std	1.67×10^{2}	3.22×10^{5}	2.64×10^{5}	9.10×10^{5}	2.49×10^{5}	2.44×10^{6}	3.44×10^{6}	9.45×10^{6}	2.28×10^{6}
	Rank	1	3	4	5	2	8	6	9	7
	p-value		1.73×10^{-6}	1.73×10^{-6}	1.73×10^{-6}	1.73×10^{-6}	1.73×10^{-6}	1.73×10^{-6}	1.73×10^{-6}	1.73×10^{-6}
F19	Avg	2.61×10^{3}	2.68×10^{4}	1.54×10^{4}	5.05×10^{6}	9.81×10^{5}	1.21×10^{8}	4.14×10^{6}	5.36×10^{6}	3.37×10^{7}
	Std	4.59×10^{2}	2.26×10^{4}	1.12×10^{4}	2.49×10^{7}	4.02×10^{5}	5.76×10^{7}	3.06×10^{6}	1.87×10^{7}	1.94×10^{7}
	Rank	1	3	2	6	4	9	5	7	8
	p-value		5.22×10^{-6}	1.92×10^{-6}	1.73×10^{-6}	1.73×10^{-6}	1.73×10^{-6}	1.73×10^{-6}	2.13×10^{-6}	1.73×10^{-6}
F20	Avg	2.28×10^{3}	2.40×10^{3}	2.20×10^{3}	2.38×10^{3}	3.03×10^{3}	2.65×10^{3}	2.73×10^{3}	2.71×10^{3}	2.70×10^{3}
	Std	1.07×10^{2}	1.89×10^{2}	8.56×10^{1}	1.26×10^{2}	2.14×10^{2}	8.76×10^{1}	1.96×10^{2}	2.28×10^{2}	1.17×10^{2}
	Rank	2	4	1	3	9	5	8	7	6
	p-value		4.99×10^{-3}	1.83×10^{-3}	4.99×10^{-3}	1.73×10^{-6}	1.73×10^{-6}	1.73×10^{-6}	2.35×10^{-6}	1.73×10^{-6}
F21	Avg	2.37×10^{3}	2.40×10^{3}	2.42×10^{3}	2.40×10^{3}	2.64×10^{3}	2.55×10^{3}	2.57×10^{3}	2.50×10^{3}	2.56×10^{3}
	Std	3.69×10^{1}	2.52×10^{1}	1.09×10^{1}	3.07×10^{1}	8.23×10^{1}	1.38×10^{1}	7.61×10^{1}	4.54×10^{1}	2.18×10^{1}
	Rank	1	3	4	2	9	6	8	5	7
	p-value		1.60×10^{-4}	1.73×10^{-6}	3.38×10^{-3}	1.73×10^{-6}	1.73×10^{-6}	1.73×10^{-6}	1.73×10^{-6}	1.73×10^{-6}
F22	Avg	2.30×10^{3}	5.31×10^{3}	4.57×10^{3}	5.36×10^{3}	7.27×10^{3}	3.95×10^{3}	6.36×10^{3}	6.40×10^{3}	8.74×10^{3}
	Std	8.02×10^{-1}	1.17×10^{3}	2.14×10^{3}	1.69×10^{3}	1.27×10^{3}	1.59×10^{2}	2.11×10^{3}	1.56×10^{3}	2.09×10^{3}
	Rank	1	4	3	5	8	2	6	7	9
	p-value		1.73×10^{-6}	1.73×10^{-6}	1.73×10^{-6}	1.73×10^{-6}	1.73×10^{-6}	1.73×10^{-6}	1.73×10^{-6}	1.73×10^{-6}
F23	Avg	2.74×10^{3}	2.75×10^{3}	2.78×10^{3}	2.77×10^{3}	3.31×10^{3}	2.92×10^{3}	3.08×10^{3}	2.85×10^{3}	3.01×10^{3}
	Std	2.31×10^{1}	2.77×10^{1}	1.29×10^{1}	3.96×10^{1}	1.70×10^{2}	1.25×10^{1}	1.06×10^{2}	4.02×10^{1}	3.07×10^{1}
	Rank	1	2	4	3	9	6	8	5	7
	p-value		8.59×10^{-2}	1.73×10^{-6}	3.61×10^{-3}	1.73×10^{-6}	1.73×10^{-6}	1.73×10^{-6}	1.73×10^{-6}	1.73×10^{-6}
F24	Avg	2.90×10^{3}	2.93×10^{3}	2.98×10^{3}	2.94×10^{3}	3.37×10^{3}	3.07×10^{3}	3.17×10^{3}	2.99×10^{3}	3.18×10^{3}
	Std	2.31×10^{1}	2.80×10^{1}	1.13×10^{1}	6.09×10^{1}	1.25×10^{2}	1.13×10^{1}	7.61×10^{1}	4.48×10^{1}	3.44×10^{1}
	Rank	1	2	4	3	9	6	7	5	8
	p-value		1.60×10^{-4}	1.73×10^{-6}	8.73×10^{-3}	1.73×10^{-6}	1.73×10^{-6}	1.73×10^{-6}	1.92×10^{-6}	1.73×10^{-6}
F25	Avg	2.88×10^{3}	2.89×10^{3}	2.89×10^{3}	3.00×10^{3}	2.92×10^{3}	3.64×10^{3}	2.98×10^{3}	3.23×10^{3}	3.24×10^{3}
	Std	2.14×10^{0}	1.42×10^{0}	2.86×10^{-1}	6.92×10^{1}	2.39×10^{1}	1.10×10^{2}	3.13×10^{1}	3.69×10^{2}	9.54×10^{1}
	Rank	1	2	3	6	4	9	5	7	8
	p-value		1.73×10^{-6}	1.92×10^{-6}	1.73×10^{-6}	1.92×10^{-6}	1.73×10^{-6}	1.73×10^{-6}	1.73×10^{-6}	1.73×10^{-6}
F26	Avg	4.41×10^{3}	4.63×10^{3}	4.86×10^{3}	4.81×10^{3}	9.93×10^{3}	6.65×10^{3}	7.13×10^{3}	5.97×10^{3}	7.07×10^{3}
	Std	2.73×10^{2}	2.29×10^{2}	9.40×10^{1}	4.56×10^{2}	1.04×10^{3}	1.49×10^{2}	1.34×10^{3}	5.01×10^{2}	2.27×10^{2}
	Rank	1	2	4	3	9	6	8	5	7
	p-value		2.77×10^{-3}	2.35×10^{-6}	5.71×10^{-4}	1.73×10^{-6}	1.73×10^{-6}	2.35×10^{-6}	1.73×10^{-6}	1.73×10^{-6}
F27	Avg	3.18×10^{3}	3.22×10^{3}	3.21×10^{3}	3.26×10^{3}	3.44×10^{3}	3.34×10^{3}	3.37×10^{3}	3.25×10^{3}	3.44×10^{3}
	Std	2.30×10^{1}	1.23×10^{1}	4.56×10^{0}	3.15×10^{1}	1.26×10^{2}	1.68×10^{1}	8.05×10^{1}	3.12×10^{1}	6.15×10^{1}
	Rank	1	3	2	5	8	6	7	4	9
	p-value		1.92×10^{-6}	2.60×10^{-6}	1.73×10^{-6}	1.73×10^{-6}	1.73×10^{-6}	1.73×10^{-6}	1.73×10^{-6}	1.73×10^{-6}
F28	Avg	3.15×10^{3}	3.24×10^{3}	3.22×10^{3}	3.47×10^{3}	3.14×10^{3}	3.98×10^{3}	3.36×10^{3}	4.40×10^{3}	3.88×10^{3}
	Std	5.70×10^{1}	3.12×10^{1}	1.96×10^{1}	1.37×10^{2}	6.20×10^{1}	9.78×10^{1}	4.44×10^{1}	1.02×10^{3}	1.46×10^{2}
	Rank	2	4	3	6	1	8	5	9	7
	p-value		7.69×10^{-6}	3.11×10^{-5}	1.73×10^{-6}	4.17×10^{-1}	1.73×10^{-6}	1.73×10^{-6}	1.73×10^{-6}	1.73×10^{-6}
F29	Avg	3.58×10^{3}	3.79×10^{3}	3.63×10^{3}	3.73×10^{3}	5.11×10^{3}	4.80×10^{3}	4.88×10^{3}	4.19×10^{3}	4.81×10^{3}
	Std	1.27×10^{2}	2.25×10^{2}	7.18×10^{1}	1.83×10^{2}	4.23×10^{2}	1.48×10^{2}	4.28×10^{2}	2.98×10^{2}	2.71×10^{2}
	Rank	1	4	2	3	9	6	8	5	7
	p-value		5.29×10^{-4}	7.86×10^{-2}	2.77×10^{-3}	1.73×10^{-6}	1.73×10^{-6}	1.73×10^{-6}	1.73×10^{-6}	1.73×10^{-6}
F30	Avg	8.37×10^{3}	1.75×10^{4}	1.67×10^{4}	7.79×10^{6}	1.67×10^{6}	1.03×10^{8}	1.81×10^{7}	3.54×10^{6}	7.72×10^{7}
	Std	2.08×10^{3}	4.73×10^{3}	5.47×10^{3}	9.38×10^{6}	9.70×10^{5}	4.04×10^{7}	1.32×10^{7}	7.66×10^{6}	3.40×10^{7}
	Rank	1	3	2	6	4	9	7	5	8
	p-value		1.92×10^{-6}	3.18×10^{-6}	1.73×10^{-6}	1.73×10^{-6}	1.73×10^{-6}	1.73×10^{-6}	1.92×10^{-6}	1.73×10^{-6}

The authors can more clearly understand the convergence speed and precision of the algorithm through the algorithm convergence graph. The authors have selected six representative algorithm convergence graphs from the CEC2017 test function. As shown in Figure 6, six convergence trend graphs are listed, namely F1, F12, F15, F18, F22, and F30. In the trends observed in the six convergence graphs, the MSMA algorithm converges quickly before approaching 5000 evaluations, but the convergence speed becomes slower at around 5000 to 20,000 evaluations, and then the convergence speed increases. Consequently, the MASA algorithm demonstrates a strong ability to remove the local optimal solution well. Furthermore, the optimal solutions that are searched for by the MSMA algorithm on these six test functions are better than those determined by the other algorithms that were compared.

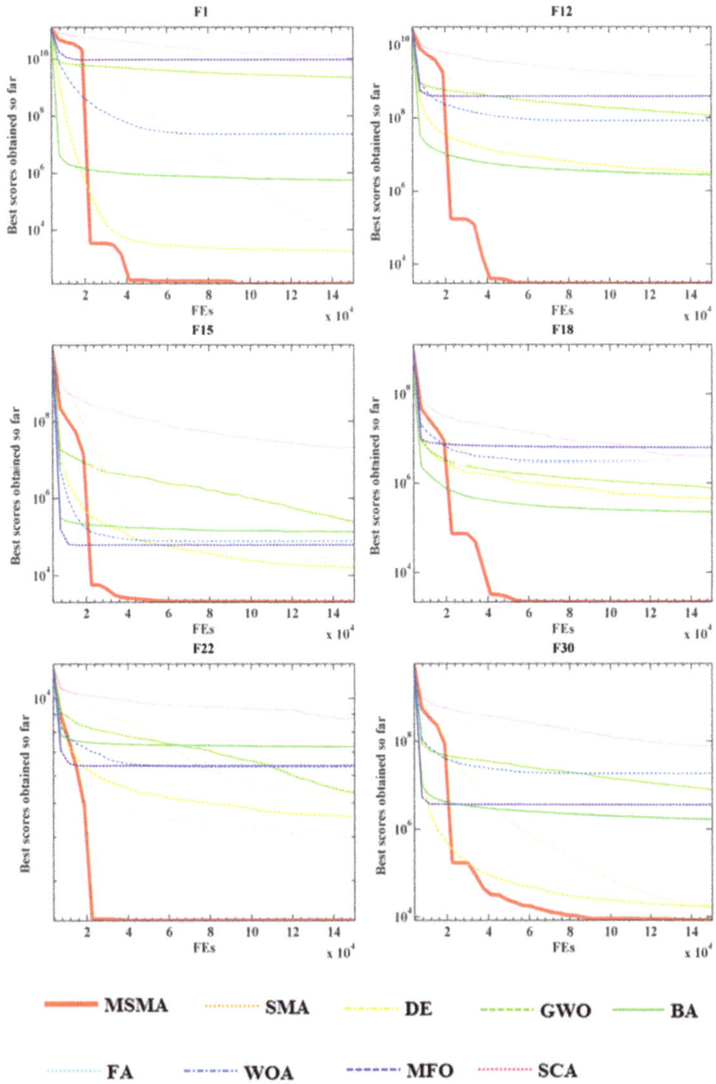

Figure 6. Convergence tendency of MSMA and original algorithms.

5.3. Comparison against Well-Established Algorithms

To prove the superiority of the MSMA algorithm, this section compares the MSMA algorithm with eight improved swarm intelligence algorithms, including OBLGWO [22], CLSGMFO [20], BWOA [17], RDWOA [25], CEBA [26], DECLS [24], ALCPSO [23], and CESCA [27]. Those comparison algorithms are improved by some classic original algorithms and have a strong ability to find optimal solutions. This section uses these algorithms to evaluate the superiority of the MSMA algorithm more precisely. The authors chose the CEC2017 test function as the test function and set the number of search agents to 30, the dimension of search agents to 30, and the maximum quantity of evaluations to 150,000. Every algorithm was run individually 30 times to obtain the average value. Table 3 shows the average fitness value and standard deviation for every algorithm on various test functions. The smaller the average fitness value and standard deviation, the better the algorithm performed on the current test function. As seen from the table, the average value and standard deviation of the MSMA on a few test functions are larger than some comparison algorithms, which m proves that the MSMA has great advantages over the other algorithms. This research uses Friedman's test to rank the algorithm's efficiency and to obtain the ARV value (average ranking value) of different algorithms. Observing Table 3, the authors can see that the MSMA algorithm ranks first in most test functions. This proves that the MSMA also has a relatively strong advantage compared to the other peers on the CEC2017 test functions. Additionally, the Wilcoxon signed-rank test was used to assess whether the MSMA algorithm performs significantly better than other advanced and improved algorithms in this experiment. Table 3 presents that the p values calculated on most test functions, and all of them are lower than 0.05. This proves that the MSMA algorithm has a big advantage over the remaining algorithms on most test functions.

The convergence diagram was employed to clearly understand the convergence trends of the algorithms on the test functions. The authors selected six representative con-vergence graphs from the CEC2017 test functions. As shown in Figure 7, when the con-vergence trend of the MSMA algorithm slows down, the algorithm convergence speed be-comes faster after a certain number of evaluations, which proves that it able to skip be-tween local optimal solution well. The MSMA algorithm searches for the optimal solution on these six test functions better than the other advanced and improved algorithms.

Table 3. Comparison results of different well-established algorithms.

		MSMA	OBLGWO	CLSGMFO	BWOA	RDWOA	CEBA	DECLS	ALCPSO	CESCA
F1	Avg	1.22×10^{2}	3.26×10^{7}	5.45×10^{3}	1.10×10^{9}	4.48×10^{7}	3.89×10^{3}	2.80×10^{3}	5.48×10^{3}	$5.71 \times 10^{+10}$
	Std	8.09×10^{1}	1.94×10^{7}	6.08×10^{3}	1.04×10^{9}	4.02×10^{7}	3.77×10^{3}	3.85×10^{3}	6.14×10^{3}	4.49×10^{9}
	Rank	1	6	4	8	7	3	2	5	9
	p-value		1.73×10^{-6}	3.52×10^{-6}	1.73×10^{-6}	1.73×10^{-6}	1.73×10^{-6}	1.92×10^{-6}	1.73×10^{-6}	1.73×10^{-6}
F2	Avg	1.05×10^{5}	$1.16 \times 10^{+18}$	$5.25 \times 10^{+13}$	$1.88 \times 10^{+30}$	$1.53 \times 10^{+17}$	8.48×10^{2}	$1.07 \times 10^{+26}$	$1.18 \times 10^{+17}$	$5.51 \times 10^{+45}$
	Std	4.32×10^{5}	$1.45 \times 10^{+18}$	$1.49 \times 10^{+14}$	$8.16 \times 10^{+30}$	$2.34 \times 10^{+17}$	3.30×10^{3}	$2.66 \times 10^{+26}$	$4.62 \times 10^{+17}$	$1.54 \times 10^{+46}$
	Rank	2	6	3	8	5	1	7	4	9
	p-value		1.73×10^{-6}	2.13×10^{-6}	1.73×10^{-6}	1.73×10^{-6}	4.90×10^{-4}	1.73×10^{-6}	1.92×10^{-6}	1.73×10^{-6}
F3	Avg	3.00×10^{2}	2.96×10^{4}	1.70×10^{4}	6.53×10^{4}	3.17×10^{4}	3.00×10^{2}	8.43×10^{4}	3.97×10^{4}	1.09×10^{5}
	Std	9.43×10^{-6}	6.71×10^{3}	4.55×10^{3}	1.11×10^{4}	8.77×10^{3}	2.07×10^{-2}	1.42×10^{4}	6.83×10^{3}	1.55×10^{4}
	Rank	1	4	3	7	5	2	8	6	9
	p-value		1.73×10^{-6}	1.73×10^{-6}	1.73×10^{-6}	1.73×10^{-6}	1.73×10^{-6}	1.73×10^{-6}	1.73×10^{-6}	1.73×10^{-6}
F4	Avg	4.01×10^{2}	5.35×10^{2}	4.96×10^{2}	7.18×10^{2}	5.27×10^{2}	4.50×10^{2}	4.95×10^{2}	5.06×10^{2}	1.57×10^{4}
	Std	1.95×10^{0}	3.64×10^{1}	2.43×10^{1}	9.64×10^{1}	3.15×10^{1}	3.74×10^{1}	1.04×10^{1}	4.45×10^{1}	2.38×10^{3}
	Rank	1	7	4	8	6	2	3	5	9
	p-value		1.73×10^{-6}	1.73×10^{-6}	1.73×10^{-6}	1.73×10^{-6}	3.52×10^{-6}	1.73×10^{-6}	1.73×10^{-6}	1.73×10^{-6}
F5	Avg	5.93×10^{2}	6.68×10^{2}	6.59×10^{2}	7.85×10^{2}	7.10×10^{2}	7.61×10^{2}	6.41×10^{2}	6.14×10^{2}	9.64×10^{2}
	Std	2.58×10^{1}	5.27×10^{1}	3.67×10^{1}	3.55×10^{1}	5.15×10^{1}	3.20×10^{1}	1.23×10^{1}	3.21×10^{1}	1.71×10^{1}
	Rank	1	5	4	8	6	7	3	2	9
	p-value		5.75×10^{-6}	4.73×10^{-6}	1.73×10^{-6}	1.92×10^{-6}	1.73×10^{-6}	4.29×10^{-6}	8.73×10^{-3}	1.73×10^{-6}
F6	Avg	6.03×10^{2}	6.20×10^{2}	6.25×10^{2}	6.68×10^{2}	6.19×10^{2}	6.61×10^{2}	6.00×10^{2}	6.08×10^{2}	7.03×10^{2}
	Std	1.84×10^{0}	1.36×10^{1}	1.14×10^{1}	5.47×10^{0}	6.09×10^{0}	4.07×10^{0}	1.12×10^{-13}	5.98×10^{0}	4.67×10^{0}
	Rank	2	5	6	8	4	7	1	3	9
	p-value		1.73×10^{-6}	1.73×10^{-6}	1.73×10^{-6}	1.73×10^{-6}	1.73×10^{-6}	1.73×10^{-6}	7.51×10^{-5}	1.73×10^{-6}
F7	Avg	8.27×10^{2}	9.54×10^{2}	9.09×10^{2}	1.28×10^{3}	9.72×10^{2}	1.27×10^{3}	8.75×10^{2}	8.55×10^{2}	1.54×10^{3}
	Std	2.12×10^{1}	6.76×10^{1}	5.79×10^{1}	6.67×10^{1}	6.66×10^{1}	4.55×10^{1}	1.07×10^{1}	3.20×10^{1}	4.64×10^{1}
	Rank	1	5	4	8	6	7	3	2	9
	p-value		1.73×10^{-6}	3.52×10^{-6}	1.73×10^{-6}	1.73×10^{-6}	1.73×10^{-6}	1.92×10^{-6}	8.31×10^{-4}	1.73×10^{-6}

Table 3. *Cont.*

		MSMA	OBLGWO	CLSGMFO	BWOA	RDWOA	CEBA	DECLS	ALCPSO	CESCA
F8	Avg	8.83×10^{2}	9.61×10^{2}	9.28×10^{2}	9.89×10^{2}	9.93×10^{2}	9.90×10^{2}	9.41×10^{2}	9.10×10^{2}	1.18×10^{3}
	Std	1.74×10^{1}	3.84×10^{1}	2.49×10^{1}	2.73×10^{1}	4.43×10^{1}	1.94×10^{1}	8.93×10^{0}	2.41×10^{1}	1.95×10^{1}
	Rank	1	5	3	6	8	7	4	2	9
	p-value		2.35×10^{-6}	2.88×10^{-6}	1.73×10^{-6}	1.92×10^{-6}	1.73×10^{-6}	1.73×10^{-6}	3.06×10^{-4}	1.73×10^{-6}
F9	Avg	1.03×10^{3}	4.25×10^{3}	3.26×10^{3}	6.66×10^{3}	5.35×10^{3}	5.29×10^{3}	9.00×10^{2}	1.94×10^{3}	1.45×10^{4}
	Std	1.32×10^{2}	2.71×10^{3}	9.16×10^{2}	9.50×10^{2}	1.90×10^{3}	2.58×10^{2}	8.94×10^{-2}	1.08×10^{3}	1.47×10^{3}
	Rank	2	5	4	8	7	6	1	3	9
	p-value		2.88×10^{-6}	1.73×10^{-6}	1.73×10^{-6}	1.73×10^{-6}	1.73×10^{-6}	1.73×10^{-6}	5.75×10^{-6}	1.73×10^{-6}
F10	Avg	3.93×10^{3}	5.48×10^{3}	5.05×10^{3}	6.68×10^{3}	4.99×10^{3}	5.31×10^{3}	6.71×10^{3}	4.38×10^{3}	8.65×10^{3}
	Std	5.84×10^{2}	1.11×10^{3}	6.26×10^{2}	8.24×10^{2}	6.41×10^{2}	5.86×10^{2}	2.77×10^{2}	8.41×10^{2}	2.46×10^{2}
	Rank	1	6	4	7	3	5	8	2	9
	p-value		1.64×10^{-5}	2.35×10^{-6}	1.73×10^{-6}	1.24×10^{-5}	3.18×10^{-6}	1.73×10^{-6}	3.16×10^{-2}	1.73×10^{-6}
F11	Avg	1.18×10^{3}	1.29×10^{3}	1.26×10^{3}	2.51×10^{3}	1.29×10^{3}	1.25×10^{3}	1.22×10^{3}	1.28×10^{3}	1.06×10^{4}
	Std	2.81×10^{1}	5.14×10^{1}	5.10×10^{1}	5.13×10^{2}	4.38×10^{1}	6.13×10^{1}	1.25×10^{1}	7.34×10^{1}	1.61×10^{3}
	Rank	1	7	4	8	6	3	2	5	9
	p-value		2.35×10^{-6}	5.75×10^{-6}	1.73×10^{-6}	2.35×10^{-6}	4.45×10^{-5}	6.34×10^{-6}	1.02×10^{-5}	1.73×10^{-6}
F12	Avg	2.82×10^{3}	2.09×10^{7}	1.68×10^{6}	1.49×10^{8}	4.00×10^{6}	1.46×10^{5}	5.04×10^{6}	3.46×10^{5}	$1.54 \times 10^{+10}$
	Std	4.40×10^{2}	2.14×10^{7}	1.81×10^{6}	1.00×10^{8}	2.27×10^{6}	2.53×10^{5}	2.16×10^{6}	5.30×10^{5}	1.82×10^{9}
	Rank	1	7	4	8	5	2	6	3	9
	p-value		1.73×10^{-6}	1.73×10^{-6}	1.73×10^{-6}	1.73×10^{-6}	1.73×10^{-6}	1.73×10^{-6}	1.73×10^{-6}	1.73×10^{-6}
F13	Avg	4.69×10^{3}	3.08×10^{5}	1.95×10^{5}	9.78×10^{5}	1.24×10^{4}	1.70×10^{4}	2.23×10^{5}	1.97×10^{4}	$1.39 \times 10^{+10}$
	Std	1.83×10^{3}	5.16×10^{5}	8.05×10^{5}	9.89×10^{5}	1.26×10^{4}	1.73×10^{4}	1.79×10^{5}	1.94×10^{4}	4.05×10^{9}
	Rank	1	7	5	8	2	3	6	4	9
	p-value		1.73×10^{-6}	1.92×10^{-6}	1.73×10^{-6}	9.63×10^{-4}	4.20×10^{-4}	1.73×10^{-6}	4.53×10^{-4}	1.73×10^{-6}
F14	Avg	1.95×10^{3}	8.01×10^{4}	6.88×10^{4}	1.44×10^{6}	2.35×10^{5}	3.62×10^{3}	1.13×10^{5}	3.53×10^{4}	5.46×10^{6}
	Std	1.16×10^{3}	6.51×10^{4}	6.79×10^{4}	1.58×10^{6}	1.94×10^{5}	2.18×10^{3}	7.96×10^{4}	8.56×10^{4}	2.62×10^{6}
	Rank	1	5	4	8	7	2	6	3	9
	p-value		1.73×10^{-6}	1.92×10^{-6}	1.73×10^{-6}	1.73×10^{-6}	1.80×10^{-5}	1.73×10^{-6}	1.73×10^{-6}	1.73×10^{-6}
F15	Avg	2.01×10^{3}	1.17×10^{5}	9.54×10^{3}	7.57×10^{5}	1.22×10^{4}	3.96×10^{3}	5.15×10^{4}	1.47×10^{4}	5.02×10^{8}
	Std	2.01×10^{2}	1.14×10^{5}	7.76×10^{3}	1.16×10^{6}	1.07×10^{4}	3.52×10^{3}	3.34×10^{4}	1.36×10^{4}	1.44×10^{8}
	Rank	1	7	3	8	4	2	6	5	9
	p-value		1.73×10^{-6}	2.88×10^{-6}	1.73×10^{-6}	2.60×10^{-6}	8.31×10^{-4}	1.73×10^{-6}	4.73×10^{-6}	1.73×10^{-6}
F16	Avg	2.21×10^{3}	2.94×10^{3}	2.87×10^{3}	3.87×10^{3}	2.82×10^{3}	3.14×10^{3}	2.34×10^{3}	2.62×10^{3}	6.02×10^{3}
	Std	2.73×10^{2}	3.06×10^{2}	3.66×10^{2}	5.28×10^{2}	3.71×10^{2}	3.48×10^{2}	1.55×10^{2}	3.36×10^{2}	5.57×10^{2}
	Rank	1	6	5	8	4	7	2	3	9
	p-value		1.73×10^{-6}	1.49×10^{-5}	1.73×10^{-6}	4.29×10^{-6}	1.73×10^{-6}	2.70×10^{-2}	1.60×10^{-4}	1.73×10^{-6}
F17	Avg	1.97×10^{3}	2.28×10^{3}	2.36×10^{3}	2.65×10^{3}	2.36×10^{3}	2.65×10^{3}	1.95×10^{3}	2.15×10^{3}	4.75×10^{3}
	Std	1.23×10^{2}	1.96×10^{2}	3.11×10^{2}	2.93×10^{2}	2.46×10^{2}	3.11×10^{2}	6.19×10^{1}	1.83×10^{2}	8.76×10^{2}
	Rank	2	4	6	7	5	8	1	3	9
	p-value		8.47×10^{-6}	1.97×10^{-5}	1.73×10^{-6}	7.69×10^{-6}	1.92×10^{-6}	5.04×10^{-1}	1.36×10^{-4}	1.73×10^{-6}
F18	Avg	2.20×10^{3}	1.75×10^{6}	3.78×10^{5}	5.38×10^{6}	7.65×10^{5}	9.68×10^{4}	7.13×10^{5}	5.27×10^{5}	5.57×10^{7}
	Std	1.57×10^{2}	1.81×10^{6}	3.12×10^{5}	4.77×10^{6}	8.51×10^{5}	7.27×10^{4}	3.24×10^{5}	1.09×10^{6}	2.69×10^{7}
	Rank	1	7	3	8	6	2	5	4	9
	p-value		1.73×10^{-6}	1.73×10^{-6}	1.73×10^{-6}	1.73×10^{-6}	1.73×10^{-6}	1.73×10^{-6}	1.73×10^{-6}	1.73×10^{-6}
F19	Avg	2.63×10^{3}	8.18×10^{5}	5.75×10^{3}	7.67×10^{6}	1.56×10^{4}	5.61×10^{3}	4.83×10^{4}	1.47×10^{4}	1.26×10^{9}
	Std	4.31×10^{2}	7.17×10^{5}	4.29×10^{3}	7.54×10^{6}	1.38×10^{4}	3.26×10^{3}	3.47×10^{4}	1.46×10^{4}	2.75×10^{8}
	Rank	1	7	3	8	5	2	6	4	9
	p-value		1.73×10^{-6}	4.86×10^{-5}	1.73×10^{-6}	2.35×10^{-6}	6.32×10^{-5}	1.73×10^{-6}	2.16×10^{-5}	1.73×10^{-6}
F20	Avg	2.32×10^{3}	2.49×10^{3}	2.49×10^{3}	2.75×10^{3}	2.54×10^{3}	2.90×10^{3}	2.22×10^{3}	2.44×10^{3}	3.23×10^{3}
	Std	1.40×10^{2}	1.15×10^{2}	2.23×10^{2}	1.96×10^{2}	2.00×10^{2}	1.81×10^{2}	8.02×10^{1}	1.86×10^{2}	1.12×10^{2}
	Rank	2	5	4	7	6	8	1	3	9
	p-value		4.20×10^{-4}	1.04×10^{-2}	3.18×10^{-6}	4.45×10^{-5}	1.73×10^{-6}	6.64×10^{-4}	6.84×10^{-3}	1.73×10^{-6}
F21	Avg	2.38×10^{3}	2.45×10^{3}	2.43×10^{3}	2.59×10^{3}	2.50×10^{3}	2.60×10^{3}	2.44×10^{3}	2.42×10^{3}	2.76×10^{3}
	Std	1.82×10^{1}	3.94×10^{1}	3.33×10^{1}	4.95×10^{1}	3.49×10^{1}	5.17×10^{1}	1.26×10^{1}	3.37×10^{1}	3.19×10^{1}
	Rank	1	5	3	7	6	8	4	2	9
	p-value		2.13×10^{-6}	1.02×10^{-5}	1.73×10^{-6}	1.73×10^{-6}	1.73×10^{-6}	1.73×10^{-6}	2.16×10^{-5}	1.73×10^{-6}
F22	Avg	2.30×10^{3}	2.90×10^{3}	2.30×10^{3}	7.18×10^{3}	6.06×10^{3}	7.16×10^{3}	4.39×10^{3}	4.73×10^{3}	9.35×10^{3}
	Std	7.47×10^{-1}	1.51×10^{3}	1.43×10^{0}	1.96×10^{3}	1.81×10^{3}	1.41×10^{3}	1.99×10^{3}	1.94×10^{3}	6.80×10^{2}
	Rank	1	3	2	8	6	7	4	5	9
	p-value		1.73×10^{-6}	1.04×10^{-3}	1.73×10^{-6}	1.73×10^{-6}	1.92×10^{-6}	1.73×10^{-6}	5.31×10^{-5}	1.73×10^{-6}
F23	Avg	2.73×10^{3}	2.82×10^{3}	2.79×10^{3}	3.10×10^{3}	2.89×10^{3}	3.39×10^{3}	2.79×10^{3}	2.80×10^{3}	3.46×10^{3}
	Std	2.69×10^{1}	4.28×10^{1}	3.48×10^{1}	1.20×10^{2}	7.39×10^{1}	2.00×10^{2}	1.23×10^{1}	6.07×10^{1}	5.09×10^{1}
	Rank	1	5	3	7	6	8	2	4	9
	p-value		1.92×10^{-6}	5.22×10^{-6}	1.73×10^{-6}	1.73×10^{-6}	1.73×10^{-6}	1.73×10^{-6}	3.11×10^{-5}	1.73×10^{-6}
F24	Avg	2.91×10^{3}	2.98×10^{3}	2.96×10^{3}	3.23×10^{3}	3.09×10^{3}	3.48×10^{3}	3.00×10^{3}	2.99×10^{3}	3.49×10^{3}
	Std	2.15×10^{1}	4.97×10^{1}	4.75×10^{1}	9.77×10^{1}	8.74×10^{1}	1.48×10^{2}	1.14×10^{1}	7.20×10^{1}	3.88×10^{1}
	Rank	1	3	2	7	6	8	5	4	9
	p-value		2.37×10^{-5}	8.47×10^{-6}	1.73×10^{-6}	1.73×10^{-6}	1.73×10^{-6}	1.73×10^{-6}	1.64×10^{-5}	1.73×10^{-6}

Table 3. *Cont.*

		MSMA	OBLGWO	CLSGMFO	BWOA	RDWOA	CEBA	DECLS	ALCPSO	CESCA
F25	Avg	2.88×10^3	2.93×10^3	2.90×10^3	3.08×10^3	2.92×10^3	2.90×10^3	2.89×10^3	2.90×10^3	5.53×10^3
	Std	1.78×10^0	2.33×10^1	1.89×10^1	5.01×10^1	2.15×10^1	1.74×10^1	3.67×10^{-1}	1.91×10^1	4.63×10^2
	Rank	1	7	3	8	6	4	2	5	9
	p-value		1.73×10^{-6}	1.92×10^{-6}	1.73×10^{-6}	1.73×10^{-6}	1.73×10^{-6}	1.73×10^{-6}	1.73×10^{-6}	1.73×10^{-6}
F26	Avg	4.50×10^3	5.55×10^3	3.84×10^3	7.90×10^3	5.61×10^3	6.08×10^3	5.00×10^3	4.99×10^3	1.11×10^4
	Std	2.57×10^2	4.03×10^2	1.32×10^3	1.04×10^3	1.27×10^3	2.40×10^3	9.28×10^1	5.57×10^2	5.86×10^2
	Rank	2	5	1	8	6	7	4	3	9
	p-value		1.73×10^{-6}	2.07×10^{-2}	1.92×10^{-6}	3.59×10^{-4}	3.32×10^{-4}	2.60×10^{-6}	1.25×10^{-4}	1.73×10^{-6}
F27	Avg	3.19×10^3	3.25×10^3	3.31×10^3	3.41×10^3	3.25×10^3	3.69×10^3	3.21×10^3	3.25×10^3	3.72×10^3
	Std	2.16×10^1	2.11×10^1	7.31×10^1	1.09×10^2	2.48×10^1	3.83×10^2	3.69×10^0	2.38×10^1	6.97×10^1
	Rank	1	3	6	7	5	8	2	4	9
	p-value		1.73×10^{-6}	1.73×10^{-6}	1.73×10^{-6}	1.73×10^{-6}	1.73×10^{-6}	4.29×10^{-6}	1.73×10^{-6}	1.73×10^{-6}
F28	Avg	3.13×10^3	3.30×10^3	3.23×10^3	3.49×10^3	3.28×10^3	3.14×10^3	3.23×10^3	3.23×10^3	7.09×10^3
	Std	5.03×10^1	3.61×10^1	1.88×10^1	1.01×10^2	2.96×10^1	5.78×10^1	2.15×10^1	3.55×10^1	4.95×10^2
	Rank	1	7	4	8	6	2	5	3	9
	p-value		1.73×10^{-6}	1.92×10^{-6}	1.73×10^{-6}	1.73×10^{-6}	5.44×10^{-1}	1.73×10^{-6}	1.73×10^{-6}	1.73×10^{-6}
F29	Avg	3.57×10^3	4.11×10^3	4.02×10^3	5.13×10^3	4.02×10^3	4.48×10^3	3.73×10^3	3.84×10^3	6.05×10^3
	Std	1.20×10^2	3.17×10^2	2.20×10^2	5.98×10^2	2.53×10^2	3.27×10^2	1.04×10^2	1.92×10^2	1.49×10^2
	Rank	1	6	5	8	4	7	2	3	9
	p-value		1.73×10^{-6}	2.88×10^{-6}	1.73×10^{-6}	1.73×10^{-6}	1.73×10^{-6}	9.32×10^{-6}	1.36×10^{-5}	1.73×10^{-6}
F30	Avg	9.82×10^3	4.24×10^6	1.19×10^5	3.50×10^7	2.74×10^4	9.74×10^3	3.89×10^4	1.84×10^4	2.74×10^9
	Std	3.02×10^3	2.92×10^6	1.69×10^5	2.91×10^7	1.92×10^4	4.48×10^3	2.49×10^4	1.44×10^4	7.83×10^8
	Rank	2	7	6	8	4	1	5	3	9
	p-value		1.73×10^{-6}	1.97×10^{-5}	1.73×10^{-6}	3.18×10^{-6}	5.44×10^{-1}	1.92×10^{-6}	2.11×10^{-3}	1.73×10^{-6}

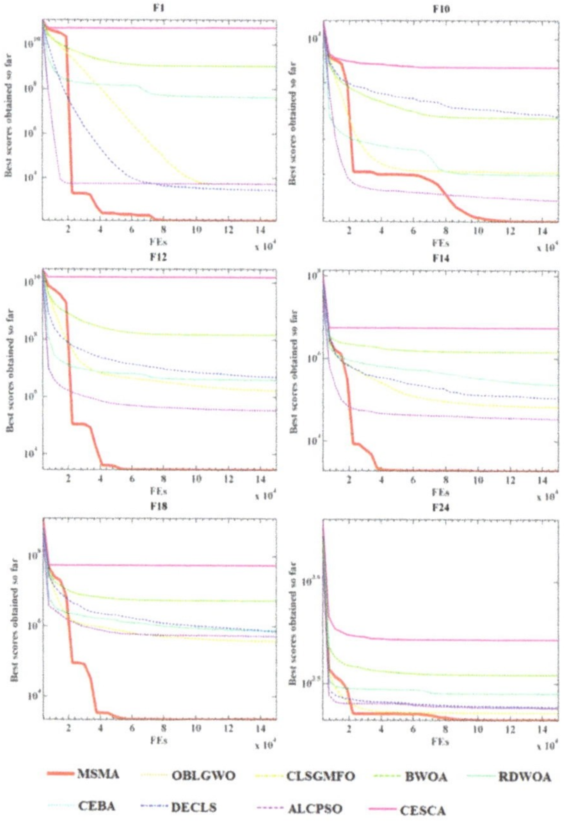

Figure 7. Convergence trends of MSMA and well-established algorithms.

5.4. Predicting Results of Employment Stability

During this experiment, the authors evaluated the validity of the MSMA-SVM with a feature selection (MSMA-SVM-FS) model relative to its peers, the detailed results of which are presented in Table 4. From the results that were obtained, the authors can conclude that the ACC obtained from MSMA-SVM-FS was 86.4%, the MCC was 72.9%, the sensitiv-ity was 82.3%, the specificity was 89.9%, and the standard deviations (STD) were 0.040, 0.081, 0.064, and 0.057, respectively. In addition, the optimal parameters and feature sub-sets were acquired directly by the MSMA method in our experiments, which means that the introduction of the multi-population structure mechanism results in the SMA having a stronger search capability and better accuracy.

Table 4. Classification results of MSMA-SVM-FS in the light of four metrics.

Fold	ACC	MCC	Sensitivity	Specificity
Num.1	0.848	0.702	0.733	0.944
Num.2	0.824	0.646	0.813	0.833
Num.3	0.909	0.819	0.875	0.941
Num.4	0.909	0.820	0.938	0.882
Num.5	0.909	0.817	0.867	0.944
Num.6	0.848	0.702	0.733	0.944
Num.7	0.879	0.756	0.867	0.889
Num.8	0.879	0.759	0.800	0.944
Num.9	0.788	0.576	0.800	0.778
Num.10	0.848	0.694	0.800	0.889
AVG	0.864	0.729	0.823	0.899
STD	0.040	0.081	0.064	0.057

With the aim of determining the efficiency of the approach, the authors compared it with five other successful machine learning models containing MSMA-SVM, SMA-SVM, ANN, RF, and KELM, is the results of which are displayed in Figure 8. The results show that MSMA-SVM-FS outperforms SMA-SVM, ANN, RF, and KELM in four evaluation metrics and that MSMA-SVM only outperforms MSMA-SVM-FS in sensitivity, but not in the other three metrics. Further, the STD is smaller than that of MSMA-SVM, SMA-SVM, ANN, RF, and KELM, indicating that the introduction of the multi-population structure strategy makes MSMA-SVM-FS perform better and results in it being more stable. On the ACC evaluation metric, the best performance was achieved by MSMA-SVM-FS with MSMA-SVM, which was 2.4% higher than the second ranked MSMA-SVM. This was closely followed by SMA-SVM and RF, with ANN achieving the worst result, which was 6.6% lower than that of MSMA-SVM-FS. The STD of MSMA-SVM-FS is smaller than that of MSMA-SVM and SMA-SVM, indicating that the MSMA-SVM and SMA-SVM models are less stable than MSMA-SVM-FS in coping with the situation but that the enhanced MSMA-SVM-FS model has much better results. On the MCC evaluation metric, the best results were still achieved with MSMA-SVM-FS followed by MSMA-SVM. MSMA-SVM was 4.6% lower than MSMA-SVM-FS accompanied by SMA-SVM and RF, and ANN had the worst effects, with values that were 12.5% lower than MSMA-SVM-FS, where MSMA-SVM-FS had the smallest STD of 0.081. In terms of sensitivity evaluation metrics, MSMA-SVM had the best effects along with MSMA-SVM-FS, only demonstrating a difference of 0.7%, accompanied by RF and SMA-SVM. The ANN model owns the worst effects, but concerning STD, MSMA-SVM-FS is the smallest at 0.064, and MSMA-SVM is the largest at 0.113. In terms of specificity metrics, MSMA-SVM-FS ranked first, accompanied by ANN, RF, KELM, MSMA-SVM, and SMA-SVM. MSMA-SVM-FS only differed from ANN by 2.4% and from MSMA-SVM by 5%; the worst was SMA-SVM at 84.9%. However, regarding STD, MSMA-SVM-FS was still the smallest at 0.057.

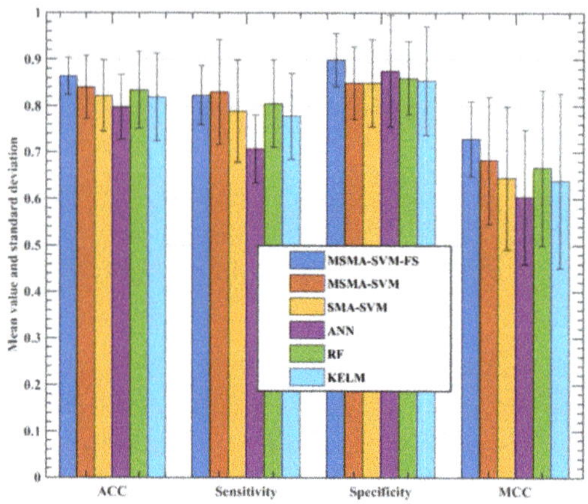

Figure 8. Classification results of five models in terms of four metrics.

During the process, the suggested MSMA not only achieved the optimal SVM super parameters settings, but it also achieved the best feature set. The authors took advantage of a 10-fold CV technique. Figure 9 illustrates the frequency of the major characteristics identified by the MSMA-SVM through the 10-fold CV procedure.

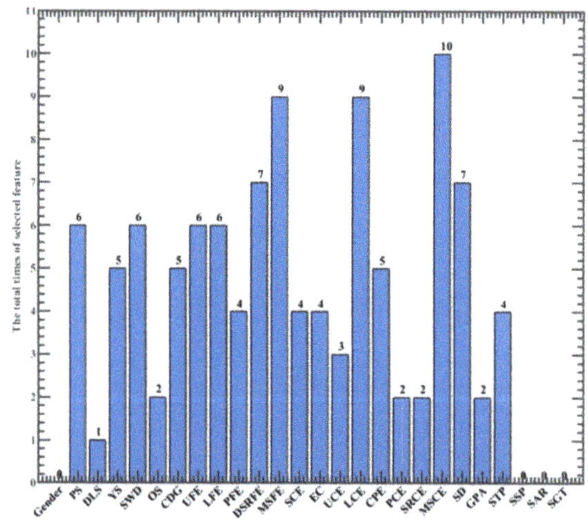

Figure 9. Frequency of the features chosen from MSMA-SVM through the 10-fold CV procedure.

As displayed in the chart, the monthly salary of current employment (F20), monthly salary of first employment (F12), change in place of employment (F17), degree of specialty relevance of first employment (F11), and salary difference (F21) were the five most frequent characteristics, which appeared 10, 9, 9, 7, and 7 times, respectively. Consequently, it was concluded that those characteristics may play a central part in forecasting graduate employment.

6. Discussion

The simulation results reveal the postgraduate student employment stability is influenced by the constraints of many factors, showing corresponding patterns in specific aspects and showing some inevitable links with most of the factors involved. Among them, the monthly salary of current employment (F20), the monthly salary of first employment (F12), change in place of employment (F17), degree of specialty relevance of first employment (F11), and salary difference (F21) have a great deal of influence on student employment stability. This section analyzes and predicts graduate student employment stability based on these five characteristic factors while further demonstrating the practical significance and validity of the MSMA-SVM model.

Among them, the monthly salary of current employment, the monthly salary of first employment, and salary difference can be unified into a wage category for analysis. First, in terms of employment area, graduate student employment is mainly concentrated in large and medium-sized cities with higher costs of living, and the monthly employment salary (F12, F20) is closely related to the costs associated with daily life in those environments; in addition, compared to undergraduates, graduate students have higher employment expectations, and they have higher salary requirements in terms of them being able to support themselves well. Secondly, the salary difference (F21) indicates the difference between the current monthly salary and the first monthly salary, and the salary difference can, to a certain extent, infer future salary packages. Graduate students do not choose employment immediately after their bachelor's degree, often because they believe that a higher level of education offers broader employment prospects. If the gap between the higher expectations that graduate students have and the real salary level is large, then graduate students will feel that the salary cannot does not reflect their human resource value and labor contribution, which will reduce their confidence in their current jobs and affect their job satisfaction, which will lead to separation behavior, and the probability of separation is higher for graduates at lower salary and benefit levels. Finally, from a comprehensive point of view, postgraduate employment looks at the current employment monthly salary, the first employment monthly salary, and salary difference in order to seek better career development and a more favorable working environment, improve quality of life, and achieve more sustainable and stable employment.

The degree of specialty relevance of first employment (F11) represents the relevance between the field of study and the work performed. According to the theory of person–job matching, it is only possible to obtain stable and harmonious career development when personal traits and career traits are consistent. On the one hand, graduate students choose their graduate majors independently based on their undergraduate professional knowledge and ability, which is reflective in their subjective future career aspirations. On the other hand, the disciplinary strength of graduate students, the influence of supervisors, academic ability and professionalism, and the demand of the job market all directly or indirectly affect the choice of graduate employment positions. If there is inconsistency between the professional structure and economic development structure in postgraduate training, or if there is a distance between academic goal of cultivation and real social and economic development, the deviation phenomenon between study major and the employment industry will appear, which will be specifically manifested as a low-relevance employment position and a job that is less relevant to the student's field of study. Therefore, graduate students are prone to making the decision to find another job that reflects their own values. Therefore, it can be seen that the degree of relevance that a student's major has on their first employment position can greatly affect the employment stability of graduate students.

Among them, changes in the place of employment (F17) represent the difference in location type between initial employment location and current employment location. First, in recent years, major cities have realized that talent is an important resource for urban development and frequently introduce unprecedented policies to attract talent. By virtue of developed economic conditions, perfect infrastructure, quality public services, and wide development space, large cities attract a continuous inflow of talent. Therefore, in order to squeeze into big cities, some postgraduates give up their majors and engage in jobs with a relatively low professional match; other postgraduates accumulate certain working experience in small and medium cities before rushing to the job market of big cities. Secondly, changes in employment location often follow changes in occupation. In our re-study sample, the authors found that among the 128 graduate students employed in non-staff positions, such as at private enterprises, 82 of them found their jobs with those establishments within three years of graduation, accounting for 64.06% of the students involved in the study, which is 10.28 percentage points higher than the average separation rate of the sample. On the one hand, the reason for this is that postgraduates working in established jobs have higher security in terms of social security, social reputation and occupational safety, and higher job stability. On the other hand and in contrast, non-established positions are a two-way selection market that is characterized by competition, and although employees can enjoy good income and security, the competition is fierce and stressful, so the probability of leaving is higher.

This subsection provides a detailed analysis of graduate student employment stability through MSMA-SVM model simulation experiments and actual survey sampling. From the monthly salary of current employment, monthly salary of first employment, and salary difference, it can be seen that graduate students first care about their salary because it represents the guarantee of current and future quality of life; The degree at which a student's specialization relevant to their first job indicates that when employment is consistent with the field of study, it is easier students to realize their own value and thus find a long-term and stable job. Changes in employment location indicate that graduate students are more likely to be employed in big cities with rich resources or in stable and established positions where they are able to realize their value. In summary, the MSMA-SVM model can reasonably analyze and predict the current employment situation of postgraduates, which will hopefully act as an effective reference for related postgraduate employment.

Due to its strong optimization capability, the developed MSMA can also be applied to other optimization problems, such as multi-objective or many optimization problems [75–77], big data optimization problems [78], and combination optimization problems [79]. Moreover, it can be applied to tackle the practical problems such as medical diagnosis [80–83], location-based service [84,85], service ecosystem [86], communication system conversion [87–89], kayak cycle phase segmentation [90], image dehazing and retrieval [91,92], information retrieval service [93–95], multi-view learning [96], human motion capture [97], green supplier selection [98], scheduling [99–101], and microgrid planning [102] problems.

7. Conclusions, Limitations, and Future Research

In this study, the authors developed an effective hybrid MSMA-SVM model that could be used to predict the employment stability of graduate students. This method's main innovation is the introduction a multi-population mechanism into the SMA, which further balances its exploration and exploitation abilities. The proposed MSMA can provide better solutions with better stability for the 30 CEC2017 benchmark functions when compared to several comparison algorithms. Meanwhile, it is possible to acquire better parameter combinations and feature subsets than other methods when using MSMA to optimize SVM. According to the employment stability prediction model for graduate students, it was found that the career stability of graduate students within three years of graduation is low, and the monthly salary level of initial employment, the relevance of initial employment, the location of the initial employment unit, and the nature of the initial employment unit are significant in predicting the exit behavior of graduates. The proposed method has more accurate and stable prediction and realization abilities when dealing with the problem of graduate employment stability prediction compared to other machine learning methods.

This article has some limitations. First of all, there were not enough research samples, and if more data samples are collected, better prediction performance with prediction accuracy can be obtained. Second, the incomplete sample attributes of the study create factors that affect the employment stability of graduate students, and these factors need to be discussed further. In addition, due to the fact that the study sample is only from one university, both the applicability of the model and the reliability of its prediction of postgraduate employment stability need to be proven further.

In future research, the authors will address the limitations for future work expansion, such as expanding the number of samples to enhance the prediction performance and accuracy of the model, expanding the number of employment attribute samples to enhance the precision of the model, and collecting samples from different regions to enhance the adaptability of the model. On the other hand, MSMA-SVM models will be applied to predict other problems such as disease diagnosis and financial risk prediction. In addition, it is expected that the MSMA algorithm can be extended to address different application areas such as photovoltaic cell optimization [103], resource requirement prediction [104,105], and the optimization of deep learning network nodes [106,107].

Author Contributions: Conceptualization, H.G. and H.C.; Methodology, H.C. and G.L.; software, G.L.; validation, H.C., H.G. and G.L.; formal analysis, H.G.; investigation, G.L. and G.L.; resources, H.C.; data curation, G.L.; writing—original draft preparation, G.L.; writing—review and editing, H.C., G.L. and H.G.; visualization, G.L. and H.G.; supervision, H.G.; project administration, G.L.; funding acquisition, H.C., H.G. and G.L. All authors have read and agreed to the published version of the manuscript.

Funding: This work was supported in part by The WenZhou Philosophy and Social Science Planning (21wsk205).

Data Availability Statement: The data involved in this study are all public data, which can be downloaded through public channels.

Conflicts of Interest: The authors declare no conflict of interest.

References

1. Bharambe, Y.; Mored, N.; Mulchandani, M.; Shankarmani, R.; Shinde, S.G. Assessing employability of students using data mining techniques. In Proceedings of the 2017 International Conference on Advances in Computing, Communications and Informatics (ICACCI), Manipal, Karnataka, India, 13–16 September 2017; pp. 2110–2114.
2. Li, L.; Zheng, Y.; Sun, X.H.; Wang, F.S. The Application of Decision Tree Algorithm in the Employment Management System. *Appl. Mech. Mater.* **2014**, *543-547*, 1639–1642. [CrossRef]
3. Liu, Y.; Hu, L.; Yan, F.; Zhang, B. Information Gain with Weight Based Decision Tree for the Employment Forecasting of Undergraduates. In Proceedings of the 2013 IEEE International Conference on Green Computing and Communications and IEEE Internet of Things and IEEE Cyber, Physical and Social Computing, Washington, DC, USA, 20–23 August 2013; pp. 2210–2213.
4. Mahdi, E.; Leiva, V.; Mara'Beh, S.; Martin-Barreiro, C. A New Approach to Predicting Cryptocurrency Returns Based on the Gold Prices with Support Vector Machines during the COVID-19 Pandemic Using Sensor-Related Data. *Sensors* **2021**, *21*, 6319. [CrossRef] [PubMed]
5. Tu, J.; Lin, A.; Chen, H.; Li, Y.; Li, C. Predict the Entrepreneurial Intention of Fresh Graduate Students Based on an Adaptive Support Vector Machine Framework. *Math. Probl. Eng.* **2019**, *2019*, 1–16. [CrossRef]
6. Cuong-Le, T.; Minh, H.-L.; Khatir, S.; Wahab, M.A.; Tran, M.T.; Mirjalili, S. A novel version of Cuckoo search algorithm for solving optimization problems. *Expert Syst. Appl.* **2021**, *186*, 115669. [CrossRef]
7. Abualigah, L.; Elaziz, M.A.; Sumari, P.; Geem, Z.W.; Gandomi, A.H. Reptile Search Algorithm (RSA): A nature-inspired meta-heuristic optimizer. *Expert Syst. Appl.* **2021**, *191*, 116158. [CrossRef]
8. Nadimi-Shahraki, M.H.; Taghian, S.; Mirjalili, S.; Abualigah, L.; Elaziz, M.A.; Oliva, D. EWOA-OPF: Effective Whale Optimization Algorithm to Solve Optimal Power Flow Problem. *Electronics* **2021**, *10*, 2975. [CrossRef]
9. Gandomi, A.H.; Roke, D. A Multi-Objective Evolutionary Framework for Formulation of Nonlinear Structural Systems. *IEEE Trans. Ind. Inform.* **2021**, 1. [CrossRef]
10. Storn, R.; Price, K. Differential evolution—A simple and efficient heuristic for global optimization over continuous spaces. *J. Glob. Optim.* **1997**, *11*, 341–359. [CrossRef]
11. Li, S.; Chen, H.; Wang, M.; Heidari, A.A.; Mirjalili, S. Slime mould algorithm: A new method for stochastic optimization. *Future Gener. Comput. Syst.* **2020**, *111*, 300–323. [CrossRef]
12. Mirjalili, S.; Mirjalili, S.M.; Lewis, A. Grey Wolf Optimizer. *Adv. Eng. Softw.* **2014**, *69*, 46–61. [CrossRef]
13. Zhao, X.; Zhang, X.; Cai, Z.; Tian, X.; Wang, X.; Huang, Y.; Chen, H.; Hu, L. Chaos enhanced grey wolf optimization wrapped ELM for diagnosis of paraquat-poisoned patients. *Comput. Biol. Chem.* **2019**, *78*, 481–490. [CrossRef]
14. Yang, X.-S. A New Metaheuristic Bat-Inspired Algorithm. In *Nature Inspired Cooperative Strategies for Optimization (NICSO 2010). Studies in Computational Intelligence*; González, J.R., Pelta, D.A., Cruz, C., Terrazas, G., Krasnogor, N., Eds.; Springer: Berlin/Heidelberg, Germany, 2010; pp. 65–74.
15. Yang, X.-S. Firefly Algorithms for Multimodal Optimization. In *International Symposium on Stochastic Algorithms*; Springer: Berlin/Heidelberg, Germany, 2009; pp. 169–178.
16. Mirjalili, S.; Lewis, A. The Whale Optimization Algorithm. *Adv. Eng. Softw.* **2016**, *95*, 51–67. [CrossRef]

17. Chen, H.; Xu, Y.; Wang, M.; Zhao, X. A balanced whale optimization algorithm for constrained engineering design problems. *Appl. Math. Model.* **2019**, *71*, 45–59. [CrossRef]
18. Mirjalili, S. Moth-flame optimization algorithm: A novel nature-inspired heuristic paradigm. *Knowl. Based Syst.* **2015**, *89*, 228–249. [CrossRef]
19. Xu, Y.; Chen, H.; Luo, J.; Zhang, Q.; Jiao, S.; Zhang, X. Enhanced Moth-flame optimizer with mutation strategy for global optimization. *Inf. Sci.* **2019**, *492*, 181–203. [CrossRef]
20. Xu, Y.; Chen, H.; Heidari, A.A.; Luo, J.; Zhang, Q.; Zhao, X.; Li, C. An efficient chaotic mutative moth-flame-inspired optimizer for global optimization tasks. *Expert Syst. Appl.* **2019**, *129*, 135–155. [CrossRef]
21. Mirjalili, S. SCA: A Sine Cosine Algorithm for solving optimization problems. *Knowl. Based Syst.* **2016**, *96*, 120–133. [CrossRef]
22. Heidari, A.A.; Abbaspour, R.A.; Chen, H. Efficient boosted grey wolf optimizers for global search and kernel extreme learning machine training. *Appl. Soft Comput.* **2019**, *81*, 105521. [CrossRef]
23. Chen, W.-N.; Zhang, J.; Lin, Y.; Chen, N.; Zhan, Z.-H.; Chung, H.; Li, Y.; Shi, Y.-H. Particle Swarm Optimization with an Aging Leader and Challengers. *IEEE Trans. Evol. Comput.* **2012**, *17*, 241–258. [CrossRef]
24. Jia, D.; Zheng, G.; Khan, M.K. An effective memetic differential evolution algorithm based on chaotic local search. *Inf. Sci.* **2011**, *181*, 3175–3187. [CrossRef]
25. Chen, H.; Yang, C.; Heidari, A.A.; Zhao, X. An efficient double adaptive random spare reinforced whale optimization algorithm. *Expert Syst. Appl.* **2020**, *154*, 113018. [CrossRef]
26. Yu, H.; Zhao, N.; Wang, P.; Chen, H.; Li, C. Chaos-enhanced synchronized bat optimizer. *Appl. Math. Model.* **2020**, *77*, 1201–1215. [CrossRef]
27. Lin, A.; Wu, Q.; Heidari, A.A.; Xu, Y.; Chen, H.; Geng, W.; Li, Y.; Li, C. Predicting Intentions of Students for Master Programs Using a Chaos-Induced Sine Cosine-Based Fuzzy K-Nearest Neighbor Classifier. *IEEE Access* **2019**, *7*, 67235–67248. [CrossRef]
28. Heidari, A.A.; Mirjalili, S.; Faris, H.; Aljarah, I.; Mafarja, M.; Chen, H. Harris hawks optimization: Algorithm and applications. *Futur. Gener. Comput. Syst.* **2019**, *97*, 849–872. [CrossRef]
29. Ahmadianfar, I.; Heidari, A.A.; Gandomi, A.H.; Chu, X.; Chen, H. RUN beyond the metaphor: An efficient optimization algorithm based on Runge Kutta method. *Expert Syst. Appl.* **2021**, *181*, 115079. [CrossRef]
30. Tu, J.; Chen, H.; Wang, M.; Gandomi, A.H. The Colony Predation Algorithm. *J. Bionic Eng.* **2021**, *18*, 674–710. [CrossRef]
31. Yang, Y.; Chen, H.; Heidari, A.A.; Gandomi, A.H. Hunger games search: Visions, conception, implementation, deep analysis, perspectives, and towards performance shifts. *Expert Syst. Appl.* **2021**, *177*, 114864. [CrossRef]
32. Zhao, S.; Wang, P.; Heidari, A.A.; Chen, H.; Turabieh, H.; Mafarja, M.; Li, C. Multilevel threshold image segmentation with diffusion association slime mould algorithm and Renyi's entropy for chronic obstructive pulmonary disease. *Comput. Biol. Med.* **2021**, *134*, 104427. [CrossRef] [PubMed]
33. Liu, L.; Zhao, D.; Yu, F.; Heidari, A.A.; Ru, J.; Chen, H.; Mafarja, M.; Turabieh, H.; Pan, Z. Performance optimization of differential evolution with slime mould algorithm for multilevel breast cancer image segmentation. *Comput. Biol. Med.* **2021**, *138*, 104910. [CrossRef]
34. Yu, C.; Heidari, A.A.; Xue, X.; Zhang, L.; Chen, H.; Chen, W. Boosting quantum rotation gate embedded slime mould algorithm. *Expert Syst. Appl.* **2021**, *181*, 115082. [CrossRef]
35. Liu, Y.; Heidari, A.A.; Ye, X.; Liang, G.; Chen, H.; He, C. Boosting slime mould algorithm for parameter identification of photovoltaic models. *Energy* **2021**, *234*, 121164. [CrossRef]
36. Shi, B.; Ye, H.; Zheng, J.; Zhu, Y.; Heidari, A.A.; Zheng, L.; Chen, H.; Wang, L.; Wu, P. Early Recognition and Discrimination of COVID-19 Severity Using Slime Mould Support Vector Machine for Medical Decision-Making. *IEEE Access* **2021**, *9*, 121996–122015. [CrossRef]
37. Premkumar, M.; Jangir, P.; Sowmya, R.; Alhelou, H.H.; Heidari, A.A.; Chen, H. MOSMA: Multi-Objective Slime Mould Algorithm Based on Elitist Non-Dominated Sorting. *IEEE Access* **2020**, *9*, 3229–3248. [CrossRef]
38. Xia, X.; Gui, L.; Zhan, Z.-H. A multi-swarm particle swarm optimization algorithm based on dynamical topology and purposeful detecting. *Appl. Soft Comput.* **2018**, *67*, 126–140. [CrossRef]
39. Zhang, L.; Zhang, C. Hopf bifurcation analysis of some hyperchaotic systems with time-delay controllers. *Kybernetika* **2008**, *44*, 35–42.
40. Geyer, C.J. *Markov Chain Monte Carlo Maximum Likelihood*; Interface Foundation of North America: Fairfax Sta, VA, USA, 1991.
41. Lai, X.; Zhou, Y. Analysis of multiobjective evolutionary algorithms on the biobjective traveling salesman problem (1,2). *Multimedia Tools Appl.* **2020**, *79*, 30839–30860. [CrossRef]
42. Zhang, Y.; Liu, R.; Wang, X.; Chen, H.; Li, C. Boosted binary Harris hawks optimizer and feature selection. *Eng. Comput.* **2021**, *37*, 3741–3770. [CrossRef]
43. Hu, J.; Chen, H.; Heidari, A.A.; Wang, M.; Zhang, X.; Chen, Y.; Pan, Z. Orthogonal learning covariance matrix for defects of grey wolf optimizer: Insights, balance, diversity, and feature selection. *Knowl. Based Syst.* **2020**, *213*, 106684. [CrossRef]
44. Zhang, X.; Xu, Y.; Yu, C.; Heidari, A.A.; Li, S.; Chen, H.; Li, C. Gaussian mutational chaotic fruit fly-built optimization and feature selection. *Expert Syst. Appl.* **2020**, *141*, 112976. [CrossRef]
45. Li, Q.; Chen, H.; Huang, H.; Zhao, X.; Cai, Z.-N.; Tong, C.; Liu, W.; Tian, X. An Enhanced Grey Wolf Optimization Based Feature Selection Wrapped Kernel Extreme Learning Machine for Medical Diagnosis. *Comput. Math. Methods Med.* **2017**, *2017*, 1–15. [CrossRef]

46. Liu, T.; Hu, L.; Ma, C.; Wang, Z.-Y.; Chen, H.-L. A fast approach for detection of erythemato-squamous diseases based on extreme learning machine with maximum relevance minimum redundancy feature selection. *Int. J. Syst. Sci.* **2013**, *46*, 919–931. [CrossRef]
47. Hu, K.; Ye, J.; Fan, E.; Shen, S.; Huang, L.; Pi, J. A novel object tracking algorithm by fusing color and depth information based on single valued neutrosophic cross-entropy. *J. Intell. Fuzzy Syst.* **2017**, *32*, 1775–1786. [CrossRef]
48. Hu, K.; He, W.; Ye, J.; Zhao, L.; Peng, H.; Pi, J. Online Visual Tracking of Weighted Multiple Instance Learning via Neutrosophic Similarity-Based Objectness Estimation. *Symmetry* **2019**, *11*, 832. [CrossRef]
49. Chen, M.-R.; Zeng, G.-Q.; Lu, K.-D.; Weng, J. A Two-Layer Nonlinear Combination Method for Short-Term Wind Speed Prediction Based on ELM, ENN, and LSTM. *IEEE Internet Things J.* **2019**, *6*, 6997–7010. [CrossRef]
50. Zeng, G.-Q.; Lu, K.; Dai, Y.-X.; Zhang, Z.; Chen, M.-R.; Zheng, C.-W.; Wu, D.; Peng, W.-W. Binary-coded extremal optimization for the design of PID controllers. *Neurocomputing* **2014**, *138*, 180–188. [CrossRef]
51. Zeng, G.-Q.; Chen, J.; Dai, Y.-X.; Li, L.-M.; Zheng, C.-W.; Chen, M.-R. Design of fractional order PID controller for automatic regulator voltage system based on multi-objective extremal optimization. *Neurocomputing* **2015**, *160*, 173–184. [CrossRef]
52. Zeng, G.-Q.; Xie, X.-Q.; Chen, M.-R.; Weng, J. Adaptive population extremal optimization-based PID neural network for multivariable nonlinear control systems. *Swarm Evol. Comput.* **2019**, *44*, 320–334. [CrossRef]
53. Zhao, D.; Liu, L.; Yu, F.; Heidari, A.A.; Wang, M.; Liang, G.; Muhammad, K.; Chen, H. Chaotic random spare ant colony optimization for multi-threshold image segmentation of 2D Kapur entropy. *Knowl. Based Syst.* **2021**, *216*, 106510. [CrossRef]
54. Zhao, D.; Liu, L.; Yu, F.; Heidari, A.A.; Wang, M.; Oliva, D.; Muhammad, K.; Chen, H. Ant colony optimization with horizontal and vertical crossover search: Fundamental visions for multi-threshold image segmentation. *Expert Syst. Appl.* **2020**, *167*, 114122. [CrossRef]
55. Zeng, G.-Q.; Lu, Y.-Z.; Mao, W.-J. Modified extremal optimization for the hard maximum satisfiability problem. *J. Zhejiang Univ. Sci. C* **2011**, *12*, 589–596. [CrossRef]
56. Zeng, G.; Zheng, C.; Zhang, Z.; Lu, Y. An Backbone Guided Extremal Optimization Method for Solving the Hard Maximum Satisfiability Problem. *Int. J. Innov. Comput. Inf. Control.* **2012**, *8*, 8355–8366. [CrossRef]
57. Shen, L.; Chen, H.; Yu, Z.; Kang, W.; Zhang, B.; Li, H.; Yang, B.; Liu, D. Evolving support vector machines using fruit fly optimization for medical data classification. *Knowl. Based Syst.* **2016**, *96*, 61–75. [CrossRef]
58. Wang, M.; Chen, H.; Yang, B.; Zhao, X.; Hu, L.; Cai, Z.; Huang, H.; Tong, C. Toward an optimal kernel extreme learning machine using a chaotic moth-flame optimization strategy with applications in medical diagnoses. *Neurocomputing* **2017**, *267*, 69–84. [CrossRef]
59. Wang, M.; Chen, H. Chaotic multi-swarm whale optimizer boosted support vector machine for medical diagnosis. *Appl. Soft Comput.* **2020**, *88*, 105946. [CrossRef]
60. Deng, W.; Xu, J.; Zhao, H.; Song, Y. A Novel Gate Resource Allocation Method Using Improved PSO-Based QEA. *IEEE Trans. Intell. Transp. Syst.* **2020**, *PP*, 1–9. [CrossRef]
61. Deng, W.; Xu, J.; Song, Y.; Zhao, H. An Effective Improved Co-evolution Ant Colony Optimization Algorithm with Multi-Strategies and Its Application. *Int. J. Bio-Inspired Comput.* **2020**, *16*, 158–170. [CrossRef]
62. Deng, W.; Liu, H.; Xu, J.; Zhao, H.; Song, Y. An Improved Quantum-Inspired Differential Evolution Algorithm for Deep Belief Network. *IEEE Trans. Instrum. Meas.* **2020**, *69*, 7319–7327. [CrossRef]
63. Zhao, H.; Liu, H.; Xu, J.; Deng, W. Performance Prediction Using High-Order Differential Mathematical Morphology Gradient Spectrum Entropy and Extreme Learning Machine. *IEEE Trans. Instrum. Meas.* **2020**, *69*, 4165–4172. [CrossRef]
64. Zhao, X.; Li, D.; Yang, B.; Ma, C.; Zhu, Y.; Chen, H. Feature selection based on improved ant colony optimization for online detection of foreign fiber in cotton. *Appl. Soft Comput.* **2014**, *24*, 585–596. [CrossRef]
65. Zhao, X.; Li, D.; Yang, B.; Chen, H.; Yang, X.; Yu, C.; Liu, S. A two-stage feature selection method with its application. *Comput. Electr. Eng.* **2015**, *47*, 114–125. [CrossRef]
66. Zhang, X.; Du, K.-J.; Zhan, Z.-H.; Kwong, S.; Gu, T.-L.; Zhang, J. Cooperative Coevolutionary Bare-Bones Particle Swarm Optimization With Function Independent Decomposition for Large-Scale Supply Chain Network Design With Uncertainties. *IEEE Trans. Cybern.* **2019**, *50*, 4454–4468. [CrossRef]
67. Chen, Z.-G.; Zhan, Z.-H.; Lin, Y.; Gong, Y.-J.; Gu, T.-L.; Zhao, F.; Yuan, H.-Q.; Chen, X.; Li, Q.; Zhang, J. Multiobjective Cloud Workflow Scheduling: A Multiple Populations Ant Colony System Approach. *IEEE Trans. Cybern.* **2019**, *49*, 2912–2926. [CrossRef]
68. Wang, Z.-J.; Zhan, Z.-H.; Yu, W.-J.; Lin, Y.; Zhang, J.; Gu, T.-L.; Zhang, J. Dynamic Group Learning Distributed Particle Swarm Optimization for Large-Scale Optimization and Its Application in Cloud Workflow Scheduling. *IEEE Trans. Cybern.* **2020**, *50*, 2715–2729. [CrossRef]
69. Yang, Z.; Li, K.; Guo, Y.; Ma, H.; Zheng, M. Compact real-valued teaching-learning based optimization with the applications to neural network training. *Knowl. Based Syst.* **2018**, *159*, 51–62. [CrossRef]
70. Zhou, S.-Z.; Zhan, Z.-H.; Chen, Z.-G.; Kwong, S.; Zhang, J. A Multi-Objective Ant Colony System Algorithm for Airline Crew Rostering Problem with Fairness and Satisfaction. *IEEE Trans. Intell. Transp. Syst.* **2020**, *22*, 6784–6798. [CrossRef]
71. Liang, D.; Zhan, Z.-H.; Zhang, Y.; Zhang, J. An Efficient Ant Colony System Approach for New Energy Vehicle Dispatch Problem. *IEEE Trans. Intell. Transp. Syst.* **2019**, *21*, 4784–4797. [CrossRef]
72. Liang, J.J.; Qu, B.Y.; Suganthan, P.N. Problem definitions and evaluation criteria for the CEC 2017 special session and competition on single objective real-parameter numerical optimization. *Tech. Rep.* **2016**, *635*, 490.

73. Derrac, J.; García, S.; Molina, D.; Herrera, F. A practical tutorial on the use of nonparametric statistical tests as a methodology for comparing evolutionary and swarm intelligence algorithms. *Swarm Evol. Comput.* **2011**, *1*, 3–18. [CrossRef]
74. García, S.; Fernández, A.; Luengo, J.; Herrera, F. Advanced nonparametric tests for multiple comparisons in the design of experiments in computational intelligence and data mining: Experimental analysis of power. *Inf. Sci.* **2010**, *180*, 2044–2064. [CrossRef]
75. Hua, Y.; Liu, Q.; Hao, K.; Jin, Y. A Survey of Evolutionary Algorithms for Multi-Objective Optimization Problems with Irregular Pareto Fronts. *IEEE/CAA J. Autom. Sin.* **2021**, *8*, 303–318. [CrossRef]
76. Zhang, W.; Hou, W.; Li, C.; Yang, W.; Gen, M. Multidirection Update-Based Multiobjective Particle Swarm Optimization for Mixed No-Idle Flow-Shop Scheduling Problem. *Complex Syst. Model. Simul.* **2021**, *1*, 176–197. [CrossRef]
77. Gu, Z.-M.; Wang, G.-G. Improving NSGA-III algorithms with information feedback models for large-scale many-objective optimization. *Futur. Gener. Comput. Syst.* **2020**, *107*, 49–69. [CrossRef]
78. Yi, J.-H.; Deb, S.; Dong, J.; Alavi, A.H.; Wang, G.-G. An improved NSGA-III algorithm with adaptive mutation operator for Big Data optimization problems. *Futur. Gener. Comput. Syst.* **2018**, *88*, 571–585. [CrossRef]
79. Zhao, F.; Di, S.; Cao, J.; Tang, J. Jonrinaldi A Novel Cooperative Multi-Stage Hyper-Heuristic for Combination Optimization Problems. *Complex Syst. Model. Simul.* **2021**, *1*, 91–108. [CrossRef]
80. Hu, Z.; Wang, J.; Zhang, C.; Luo, Z.; Luo, X.; Xiao, L.; Shi, J. Uncertainty Modeling for Multi center Autism Spectrum Disorder Classification Using Takagi-Sugeno-Kang Fuzzy Systems. *IEEE Trans. Cogn. Dev. Syst.* **2021**, 1. [CrossRef]
81. Chen, C.Z.; Wu, Q.; Li, Z.Y.; Xiao, L.; Hu, Z.Y. Diagnosis of Alzheimer's disease based on Deeply-Fused Nets. *Comb. Chem. High Throughput Screen.* **2020**, *24*, 781–789. [CrossRef]
82. Fei, X.; Wang, J.; Ying, S.; Hu, Z.; Shi, J. Projective parameter transfer based sparse multiple empirical kernel learning Machine for diagnosis of brain disease. *Neurocomputing* **2020**, *413*, 271–283. [CrossRef]
83. Saber, A.; Sakr, M.; Abo-Seida, O.M.; Keshk, A.; Chen, H. A Novel Deep-Learning Model for Automatic Detection and Classification of Breast Cancer Using the Transfer-Learning Technique. *IEEE Access* **2021**, *9*, 71194–71209. [CrossRef]
84. Wu, Z.; Li, G.; Shen, S.; Lian, X.; Chen, E.; Xu, G. Constructing dummy query sequences to protect location privacy and query privacy in location-based services. *World Wide Web* **2021**, *24*, 25–49. [CrossRef]
85. Wu, Z.; Wang, R.; Li, Q.; Lian, X.; Xu, G.; Chen, E.; Liu, X. A Location Privacy-Preserving System Based on Query Range Cover-Up or Location-Based Services. *IEEE Trans. Veh. Technol.* **2020**, *69*, 5244–5254. [CrossRef]
86. Xue, X.; Zhou, D.; Chen, F.; Yu, X.; Feng, Z.; Duan, Y.; Meng, L.; Zhang, M. From SOA to VOA: A Shift in Understanding the Operation and Evolution of Service Ecosystem. *IEEE Trans. Serv. Comput.* **2021**, 1. [CrossRef]
87. Zhang, L.; Zou, Y.; Wang, W.; Jin, Z.; Su, Y.; Chen, H. Resource allocation and trust computing for blockchain-enabled edge computing system. *Comput. Secur.* **2021**, *105*, 102249. [CrossRef]
88. Zhang, L.; Zhang, Z.; Wang, W.; Waqas, R.; Zhao, C.; Kim, S.; Chen, H. A Covert Communication Method Using Special Bitcoin Addresses Generated by Vanitygen. *Comput. Mater. Contin.* **2020**, *65*, 597–616.
89. Zhang, L.; Zhang, Z.; Wang, W.; Jin, Z.; Su, Y.; Chen, H. Research on a Covert Communication Model Realized by Using Smart Contracts in Blockchain Environment. *IEEE Syst. J.* **2021**, 1–12. [CrossRef]
90. Qiu, S.; Hao, Z.; Wang, Z.; Liu, L.; Liu, J.; Zhao, H.; Fortino, G. Sensor Combination Selection Strategy for Kayak Cycle Phase Segmentation Based on Body Sensor Networks. *IEEE Internet Things J.* **2021**, 1. [CrossRef]
91. Zhang, X.; Wang, T.; Wang, J.; Tang, G.; Zhao, L. Pyramid Channel-based Feature Attention Network for image dehazing. *Comput. Vis. Image Underst.* **2020**, *197–198*, 103003. [CrossRef]
92. Liu, H.; Li, X.; Zhang, S.; Tian, Q. Adaptive Hashing With Sparse Matrix Factorization. *IEEE Trans. Neural Networks Learn. Syst.* **2019**, *31*, 4318–4329. [CrossRef]
93. Wu, Z.; Li, R.; Zhou, Z.; Guo, J.; Jiang, J.; Su, X. A user sensitive subject protection approach for book search service. *J. Assoc. Inf. Sci. Technol.* **2020**, *71*, 183–195. [CrossRef]
94. Wu, Z.; Shen, S.; Lian, X.; Su, X.; Chen, E. A dummy-based user privacy protection approach for text information retrieval. *Knowl. Based Syst.* **2020**, *195*, 105679. [CrossRef]
95. Wu, Z.; Shen, S.; Zhou, H.; Li, H.; Lu, C.; Zou, D. An effective approach for the protection of user commodity viewing privacy in e-commerce website. *Knowl. Based Syst.* **2021**, *220*, 106952. [CrossRef]
96. Liu, H.; Liu, L.; Le, T.D.; Lee, I.; Sun, S.; Li, J. Nonparametric Sparse Matrix Decomposition for Cross-View Dimensionality Reduction. *IEEE Trans. Multimedia* **2017**, *19*, 1848–1859. [CrossRef]
97. Qiu, S.; Zhao, H.; Jiang, N.; Wu, D.; Song, G.; Zhao, H.; Wang, Z. Sensor network oriented human motion capture via wearable intelligent system. *Int. J. Intell. Syst.* **2021**, *37*, 1646–1673. [CrossRef]
98. Liu, P.; Gao, H. A novel green supplier selection method based on the interval type-2 fuzzy prioritized choquet bonferroni means. *IEEE/CAA J. Autom. Sin.* **2020**, 1–17. [CrossRef]
99. Han, X.; Han, Y.; Chen, Q.; Li, J.; Sang, H.; Liu, Y.; Pan, Q.; Nojima, Y. Distributed Flow Shop Scheduling with Sequence-Dependent Setup Times Using an Improved Iterated Greedy Algorithm. *Complex Syst. Model. Simul.* **2021**, *1*, 198–217. [CrossRef]
100. Gao, D.; Wang, G.-G.; Pedrycz, W. Solving Fuzzy Job-Shop Scheduling Problem Using DE Algorithm Improved by a Selection Mechanism. *IEEE Trans. Fuzzy Syst.* **2020**, *28*, 3265–3275. [CrossRef]
101. Cao, X.; Cao, T.; Gao, F.; Guan, X. Risk-Averse Storage Planning for Improving RES Hosting Capacity Under Uncertain Siting Choices. *IEEE Trans. Sustain. Energy* **2021**, *12*, 1984–1995. [CrossRef]

102. Cao, X.; Wang, J.; Wang, J.; Zeng, B. A Risk-Averse Conic Model for Networked Microgrids Planning with Reconfiguration and Reorganizations. *IEEE Trans. Smart Grid* **2020**, *11*, 696–709. [CrossRef]
103. Ramadan, A.; Kamel, S.; Taha, I.B.M.; Tostado-Véliz, M. Parameter Estimation of Modified Double-Diode and Triple-Diode Photovoltaic Models Based on Wild Horse Optimizer. *Electronics* **2021**, *10*, 2308. [CrossRef]
104. Liu, Y.; Ran, J.; Hu, H.; Tang, B. Energy-Efficient Virtual Network Function Reconfiguration Strategy Based on Short-Term Resources Requirement Prediction. *Electronics* **2021**, *10*, 2287. [CrossRef]
105. Shafqat, W.; Malik, S.; Lee, K.-T.; Kim, D.-H. PSO Based Optimized Ensemble Learning and Feature Selection Approach for Efficient Energy Forecast. *Electronics* **2021**, *10*, 2188. [CrossRef]
106. Choi, H.-T.; Hong, B.-W. Unsupervised Object Segmentation Based on Bi-Partitioning Image Model Integrated with Classification. *Electronics* **2021**, *10*, 2296. [CrossRef]
107. Saeed, U.; Shah, S.Y.; Shah, S.A.; Ahmad, J.; Alotaibi, A.A.; Althobaiti, T.; Ramzan, N.; Alomainy, A.; Abbasi, Q.H. Discrete Human Activity Recognition and Fall Detection by Combining FMCW RADAR Data of Heterogeneous Environments for Independent Assistive Living. *Electronincs* **2021**, *10*, 2237. [CrossRef]

electronics

MDPI

Article

Random Replacement Crisscross Butterfly Optimization Algorithm for Standard Evaluation of Overseas Chinese Associations

Hanli Bao [1], Guoxi Liang [2,*], Zhennao Cai [3] and Huiling Chen [3,*]

1. College of Overseas Chinese, Wenzhou University, Wenzhou 325035, China; baohl@wzu.edu.cn
2. Department of Information Technology, Wenzhou Polytechnic, Wenzhou 325035, China
3. College of Computer Science and Artificial Intelligence, Wenzhou University, Wenzhou 325035, China; cznao@wzu.edu.cn

* Correspondence: guoxiliang2017@gmail.com (G.L.); chenhuiling.jlu@gmail.com (H.C.)

Abstract: The butterfly optimization algorithm (BOA) is a swarm intelligence optimization algorithm proposed in 2019 that simulates the foraging behavior of butterflies. Similarly, the BOA itself has certain shortcomings, such as a slow convergence speed and low solution accuracy. To cope with these problems, two strategies are introduced to improve the performance of BOA. One is the random replacement strategy, which involves replacing the position of the current solution with that of the optimal solution and is used to increase the convergence speed. The other is the crisscross search strategy, which is utilized to trade off the capability of exploration and exploitation in BOA to remove local dilemmas whenever possible. In this case, we propose a novel optimizer named the random replacement crisscross butterfly optimization algorithm (RCCBOA). In order to evaluate the performance of RCCBOA, comparative experiments are conducted with another nine advanced algorithms on the IEEE CEC2014 function test set. Furthermore, RCCBOA is combined with support vector machine (SVM) and feature selection (FS)—namely, RCCBOA-SVM-FS—to attain a standardized construction model of overseas Chinese associations. It is found that the reasonableness of bylaws; the regularity of general meetings; and the right to elect, be elected, and vote are of importance to the planning and standardization of Chinese associations. Compared with other machine learning methods, the RCCBOA-SVM-FS model has an up to 95% accuracy when dealing with the normative prediction problem of overseas Chinese associations. Therefore, the constructed model is helpful for guiding the orderly and healthy development of overseas Chinese associations.

Keywords: butterfly optimization algorithm; random replacement; crisscross search; overseas Chinese associations; support vector machine

Citation: Bao, H.; Liang, G.; Cai, Z.; Chen, H. Random Replacement Crisscross Butterfly Optimization Algorithm for Standard Evaluation of Overseas Chinese Associations. *Electronics* **2022**, *11*, 1080. https://doi.org/10.3390/electronics11071080

Academic Editor: Maciej Ławryńczuk

Received: 9 February 2022
Accepted: 23 March 2022
Published: 29 March 2022

1. Introduction

As an important organizational form of overseas Chinese society and a direct participant and promoter of the great rejuvenation of China, the Overseas Chinese Association has shown a good development momentum in the new era. In recent years, Zhejiang overseas Chinese groups have gradually increased in number and expanded in scale. However, most overseas Chinese associations are still irregular in terms of their establishment, operation, and management, meaning that they cannot become exemplary and representative featured overseas Chinese associations. The problems of irregularities in overseas Chinese associations are mainly manifested in ten aspects: legality, incomplete constitutions, irregular elections, the phenomenon of the "one-man meeting", prominent concurrent roles, unsound teams, significantly bad records held by the head of the association, lack of innovation awareness caused by aging, the existence of zombie groups, and a lack of professionalism. Both internal contradictions and external pressures lead to the emergence of these problems, which hinder the development of the overseas Chinese associations and undermine their harmonious atmosphere. Carrying out the standardization of overseas Chinese associations is a necessary and urgent task in order to safeguard the rights and interests of overseas Chinese citizens. Therefore, it is necessary to conduct an in-depth analysis of the factors affecting the standardized construction of overseas Chinese associations and establish an evaluation model to help guide their orderly and healthy development.

The study of overseas Chinese associations has been a hot topic in academic circles. Many scholars have studied the development of overseas Chinese associations around the world from multiple perspectives. Li et al. [1] studied the reasons, characteristics, and influence of explosive growth of overseas Chinese associations in Europe from the 1980s to the 1990s. Fei et al. [2] studied how "new" overseas Chinese immigrant associations drive the development of "old" overseas Chinese immigrant associations in the Pacific region from a historical perspective. Maurice et al. [3] focused on how the Chinese immigrants and overseas Chinese associations in Singapore adapted to colonial society in the 19th century. Ma et al. [4] elaborated on the formation of overseas Chinese associations in the United States and their influence on local politics and economy. The above scholars have studied overseas Chinese associations from the perspectives of history, politics, and international relations, but no scholars have automated their analysis by means of computer algorithms yet [5]. Previously, research has only been conducted to classify non-profit organizations by their fields of activity through computer algorithms. In this paper, we will use intelligent algorithms to calculate 1050 valid questionnaires made by overseas Chinese of Zhejiang nationality from all over the world and establish a program that can quickly identify the regularity of overseas Chinese associations so as to provide a reference allowing Chinese embassies and consulates abroad to accurately grasp the latest trends of overseas Chinese associations and examine whether or not they are legitimate. This procedure can provide a standard for the standardization of overseas Chinese associations, so that the rights and interests of every overseas Chinese can be protected and avoid irregular overseas Chinese associations from harming the rights and interests of them.

Based on existing data, this paper first proposes RCCBOA with SVM, the core of which is mainly combined with random replacement and a crisscross search strategy to better predict the standardized construction of overseas Chinese associations, effectively boosting the accuracy of the solution of BOA. BOA is one of many intelligent optimization algorithms. It is a global optimization algorithm inspired by butterfly foraging behavior that was proposed by Sankalap Arora and Satvir Singh in 2019 [6]. BOA has a simple structure, few parameters, and is based on a novel idea; it is suitable for solving high-dimensional optimization problems. Compared with other optimization algorithms proposed recently, the optimization performance is stronger and the influence of dimensional changes is smaller, so it has relevant research potential. However, BOA has the problems of a slow convergence speed and low solution accuracy on some benchmark functions. According to "no free lunch" (NFL) theorems [7], no one algorithm can be applied to all problems. Similarly, the main reason why we conducted this research was to find a type of benchmark problem or practical problem that would be suitable for the algorithm. The relevant work on BOA has attracted a wide range of scholars both at home and abroad. Long et al. [8] designed an enhanced adaptable BOA (EABOA) to improve the parameter estimation of PV models; this was also tested on 12 classical benchmark function sets to verify its superior performance. Sharma et al. [9] proposed a boosted BOA with bidirectional search (BBOA) and tested it on seven unimodal benchmark functions and three practical engineering optimization problems. Mortazavi et al. [10] used a fuzzy BOA (FBOA) and tested it on resolving certain constrained and non-constrained optimization problems. Sundaravadivel et al. [11] proposed a weighted BOA (WBOA) with an intuitionistic fuzzy gaussian function to predict the outcome of infection with COVID-19. Zhou et al. [12] proposed an improved BOA to apply numerical examples of a simply-supported beam and truss structure. Thawkar et al. [13] introduced an ant lion optimizer into BOA (BOAALO), which was used to predict the benign or malignant status of breast tissue. Long et al. [14] designed a hybrid BOA with an adaptive gbest-guided search strategy and pinhole-imaging-based learning (PIL-BOA) to deal with feature selection problems. Sowjanya et al. [15] utilized BOA and gas Brownian motion optimization to obtain the optimal threshold levels for image segmentation. Some other improved algorithms have also been widely used to solve complex problems in various fields. Descriptions of the novel improved algorithms are provided in Table 1.

Table 1. Description of other novel improved algorithms.

Author	Year	Proposed Algorithm	Description
Hu et al. [16]	2021	An improved Harris hawk optimization (HHOSRL)	It was applied to feature selection and gained a high accuracy.
Fan et al. [17]	2021	A new particle swarm optimization (PSOCS)	It was used to estimate the parameters of PV models.
Shi et al. [18]	2021	An enhanced colony predation algorithm (ECPA)	It used a kernel extreme learning machine to classify the severity of COVID-19.
Zhou et al. [19]	2021	A novel enhanced spherical evolution algorithm (DSCSE)	It was applied to evaluate unknown parameters of PV models.
Yu et al. [20]	2021	An improved slime mould algorithm (WQSMA)	It provided effective assistance in solving optimization problems.
Zhou et al. [21]	2021	A new gradient-based optimizer with random learning mechanism (RLGBO)	It was employed to gain the PV model parameters of different conditions.
Xu et al. [22]	2021	An innovative binary version of moth-flame optimizer (ESAMFO)	It was utilized to select optimal feature subsets with the K-Nearest Neighbor Classifier.
Liu et al. [23]	2021	A novel ant colony optimizer (CLACO)	It was verified by comparison with certain excellent peers.
Zhao et al. [24]	2021	An improved salp swarm algorithm (EHSSA)	It was tested on CEC2014 functions and image segmentation problems.
Yu et al. [25]	2020	A dynamic bare-bones fruit fly optimization algorithm (BareFOA)	It was tested on with CEC2017 and seven engineering optimization problems.
Chen et al. [26]	2020	An enhanced DE-driven multi-population Harris hawks optimization (CMDHHO)	It was compared with plenty of other algorithms based on CEC2017 and CEC2011.
Chen et al. [27]	2020	A boosted sine cosine algorithm (OMGSCA)	It was applied in CEC2014 and used to solve some real-world optimization problems.
Tu et al. [28]	2019	An adaptive support vector machine framework (RF-CSCA-SVM)	It was regarded as being successful for predicting student graduation.
Chen et al. [29]	2019	A hybrid whale optimization algorithm (BWOA)	It was treated as an effective means to solve complex problems.

Aiming at addressing the deficiencies of BOA, we propose an improved butterfly algorithm combining random replacement and a crisscross search strategy. The combination of these two strategies effectively boosts the performance of the original one, enabling it to have a better performance on complex problems. In order to evaluate the performance of the proposed algorithm, it is compared with nine recently proposed advanced algorithms on the CEC 2014 benchmark test set, including CDLOBA [30], CBA [31], RCBA J [32], MWOA [33], LWOA [34], IWOA [35], CEFOA [36], CIFOA [37], and AMFOA [38]. In addition, RCCBOA was also combined with SVM to solve the problem of predicting the standardized construction of overseas Chinese associations. Experimental results show that the convergence speed and convergence accuracy of this algorithm are better than those of other advanced algorithms, and the best SVM optimized by RCCBOA had an accuracy rate of 95% on the relevant data set. Therefore, the proposed RCCBOA has very broad application prospects. The main contributions of this paper are as follows:

1. A new type of enhanced BOA which combines a random replacement strategy and cross search strategy is proposed.
2. RCCBOA is compared with nine other advanced algorithms on the CEC 2014 benchmark function test.

3. RCCBOA is combined with SVM to solve the problem of predicting overseas Chinese associations.

The organizational structure of the thesis is as follows. Section 2 describes the SVM and BOA. Sections 3 and 4 introduce the proposed RCCBOA and RCCBOA-SVM. Section 5 describes the data sources and experimental settings used. Section 6 shows the experimental results. Section 7 discusses the experimental results, and the last section summarizes the full paper and related future prospects.

2. Backgrounds

2.1. Overseas Chinese Associations

The full name of "Qiao Tuan" is "Overseas Chinese Association". It is a formal group made up of overseas Chinese nationals due to their certain related attributes and is an important organizational form of overseas Chinese society, whose related attributes include factors such as living area, work industry, academic field, language exchange, ethnic blood relationship, etc. At present, the number of overseas Chinese associations exceeds 25,700. Overseas Chinese associations have the functions of economic construction, safeguarding rights and interests, overseas friendship, political participation, cultural dissemination, and public welfare dedication. Overseas Chinese groups have participated in China's economic construction for a long time to achieve mutual benefits, contributing to the masses and earnestly safeguard the basic rights and interests of overseas Chinese. Moreover, overseas Chinese associations organize networking activities for overseas Chinese nationals to promote communication and interaction among overseas Chinese people; they pay attention to political changes, keep abreast of current trends, strive for resources, and serve overseas Chinese nationals. Overseas Chinese associations are an important part of the overseas dissemination of Chinese culture, inheriting culture vertically and spreading culture horizontally. Moreover, overseas Chinese associations are parts of the country where they are located, and it is the basic responsibility of the overseas Chinese associations to serve local society and participate in public welfare matters.

Overseas Chinese associations are known as one of the three pillars of overseas Chinese society and an important organizational form for maintaining its orderly operation. Overseas Chinese associations have functions such as safeguarding the rights and interests of overseas Chinese, building overseas friendships, promoting cultural dissemination, and contributing to public welfare. Currently, the number of overseas Chinese nationals exceeds 60 million, and the number of overseas Chinese associations around the world has reached 25,700. The total number of overseas Chinese nationals from Zhejiang Province is 3.792 million, ranking fifth in the country. There are also a large number of overseas Chinese associations composed mostly of Zhejiang nationals. According to incomplete statistics, there are 865 overseas Chinese associations, which are mainly distributed in 66 countries including Italy, Spain, the United States, and Australia.

2.2. Butterfly Optimization Algorithm (BOA)

In recent years, many optimization algorithms have been proposed for solving approximate optimal problems, such as hunger games search (HGS) [39], Harris hawks optimization (HHO) [40], the slime mould algorithm (SMA) [41], the Runge–Kutta optimizer (RUN) [42], the colony predation algorithm (CPA) [43], and the weighted mean of vectors (INFO) [44].

These algorithms have a strong search ability and can solve many practical problems, with applications such as medical diagnosis [45,46], economic emission dispatch problems [47], engineering design [48–50], parameter tunning for machine learning models [18,51,52], image segmentation [53–55], plant disease recognition [56], feature selection [57,58], bankruptcy prediction [59,60], prediction problems in the educational field [61,62], PID optimization control [63,64], the detection of foreign fibers in cotton [65,66], expensive optimization problems [67,68], multi-objective or many optimization problems [69–71], the fault diagnosis of rolling bearings [72,73], gate resource allocation [74,75], combination optimization problems [76], big data optimization problems [77], green supplier selection [78], and scheduling problems [79,80].

BOA [6] is a newly proposed optimization algorithm which is based on imitating the foraging behavior of butterflies in nature [6]. Since its introduction, it has been applied to many problems, such as fault diagnosis [81] and disease diagnosis [82]. Each butterfly acts as a search operator and performs an optimization process in the search space. The butterfly can perceive and distinguish different fragrance intensities, and the fragrance emitted by each butterfly has a certain level of intensity. One must assume that the intensity of the fragrance produced by the butterfly is related to its fitness; when the butterfly moves from one place to another, its fitness will also change accordingly. The scent emitted by the butterfly will spread in the air and be felt by other butterflies. This is the process by which individual butterflies share personal information with other individual butterflies, thus forming a collective social knowledge network. When a butterfly detects the scent of other butterflies, it will move to the butterfly with the most scent, which is called a global search. Conversely, when a butterfly cannot perceive the fragrance of other butterflies, it will move randomly, which is called a local search.

If $X_i = (x_{i1}, x_{i2}, \ldots, x_{iD})$ is the i-th ($i = 1, 2, \ldots, N$) butterfly individual, D is the search space dimension, N is the butterfly population size, and the position update of the butterfly individual is as shown in Equation (1).

$$x_i^{t+1} = \begin{cases} x_i^t + \left(r^2 \times g^* - x_i^t\right) f_i \\ x_i^t + \left(r^2 \times x_j^t - x_k^t\right) f_i \end{cases} \quad (1)$$

where x_i^{t+1} is the solution vector of the i-th butterfly in $t+1$ iterations; r is a random number between 0 and 1; g^* represents the global optimal individual in the current iteration; and x_i^t and x_i^t are randomly generated butterfly individuals, representing the solution vector of the j-th butterfly and the k-th butterfly in the solution space. The fragrance emitted by the i-th butterfly is denoted by f_i, and the specific expression of f_i is shown in Equation (2).

$$f = cI^a \quad (2)$$

where f is the level of fragrance perception, c is the form of perception, and a is the power exponent, which depends on the form of perception, reflecting the different degrees of scent absorption.

The BOA is divided into three stages; the pseudo-code is shown in Algorithm 1.

(1) Initial stage. The parameter values used in BOA are assigned, and when these values are set the algorithm proceeds to create an initial butterfly population for optimization. The positions of the butterflies are randomly generated in the search space and their scent and fitness values are calculated and stored.

(2) Iterative stage. In each iteration, all butterflies in the solution space are moved to new positions and their fitness values are re-evaluated. The algorithm m first calculates the fitness values of all butterflies at different positions of the solution space. Then, these butterflies will use Equation (1) to generate fragrance in their place.

(3) End stage. Iteration continues until the maximum number of iterations is reached. When the iteration phase ends, the algorithm outputs the optimal solution with the best fitness.

Algorithm 1: Pseudo-code of BOA.

Initialize population number n, dimensions d, max evaluations Max_{FEs}, objective function $f(x)$;
Initialize sensor modality c, power exponent a, switch probability p, and evaluations t;
Initialize the population of butterflies x_i $(i = 1, 2, \ldots, n)$;
Gain intensity I_i by $f(x_i)$;
While $(t \leq Max_{FEs})$:
Calculate the fragrance bf of each butterfly using Equation (2);
Gain the best bf;
For $i = 1$ *to* n
Update r in [0, 1];
    If $r < p$
        Move to the best solution with Equation (1);
    Else
        Move randomly using Equation (1);
    End if
  End for
  $t = t + 1$;
Update parameter a;
End while
Output best solution.

2.3. Support Vector Machine

The purpose of the support vector machine (SVM) is to find the hyperplane that is the furthest away from various sample points. SVM is a supervised learning method used for classification problems, with the goal of finding the hyperplane that can most accurately separate positive and negative samples. Assuming given sample data $G = (x_i, y_i)$, $i = 1, \ldots, N, x \in R^d, y \in \{\pm 1\}$, the hyperplane is expressed as follows:

$$g(x) = \omega^{T} x + b \tag{3}$$

The SVM model according to the existing standard is as follows:

$$\begin{cases} \min(\omega) = \frac{1}{2}\omega^2 + c\sum\limits_{i=1}^{N} \xi_i{}^2 \\ s.t\ y_i(\omega^{T} x_i + b) \geq 1 - \xi_i, i = 1, 2, \ldots, N \end{cases} \tag{4}$$

where ω is the inertia weight, b is a constant, ξ_i is a slack variable, and c is a disciplinary factor.

The initial low-dimensional sample set is mapped to a high-dimensional space H by introducing a kernel function; then, the optimal classification surface is established using a linear method. The conversion formula is shown below.

$$\begin{cases} Q(\alpha) = \frac{1}{2}\sum\limits_{i=1}^{N} \alpha_i \alpha_j y_i y_j k(x_i, x_j) - \sum\limits_{i=1}^{N} \alpha_i \\ s.t\ \sum\limits_{i=1}^{N} a_i y_i = 0, 0 \leq a_i \leq C, i = 1, 2, \ldots, N \end{cases} \tag{5}$$

where α_i is the Lagrange multiplier and $k(x_i, x_j)$ is the kernel function, which can be expressed in Equation (6).

$$k(x, y) = e^{-\gamma \|x_i - x_j\|} \tag{6}$$

where γ is a kernel parameter, which represents the interaction width of the kernel function.

3. Suggested RCCBOA

3.1. Random Replacement Strategy

Most optimization algorithms will show global exploration behavior in the early stage [83]. When the algorithm exploration is weak, the convergence speed will be slow and it will be easy to fall into the local optimum [84]. Thus, we introduce a random replacement strategy to BOA, which effectively helps the individuals of the population to move closer to the food source, thereby improving the algorithm's convergence speed. The individuals of the population are compatible with the optimal individual in some dimensions, and it is possible that some of the dimensions of the individual will deviate from those of the optimal individual. In this case, the current position is replaced with the position of the optimal solution with some probability. The probability value is mainly determined by comparing the ratio of the remaining time of the algorithm to the total running time and the Cauchy random number. The random replacement strategy can easily be replaced in the early stage of the algorithm, and it is less likely to be replaced in the later stage. In short, the random replacement strategy can effectively improve the convergence speed of the algorithm and prevent the algorithm from falling into the local optimum prematurely.

3.2. Crisscross Search

The crisscross search strategy is derived from the crisscross optimization algorithm (CSO) proposed by Meng et al. in 2014 [85], which includes vertical crossover and horizontal crossover. Its effectiveness has been demonstrated in many optimization algorithms [86]; for example, Zhao et al. [87] used it in ant colony optimization to solve the problem of multi-threshold image segmentation. Liu et al. [88] designed an improved Harris hawks optimizer (HHO) with the crisscross search strategy to estimate the parameters of PV models.

3.2.1. Vertical Crossover Search

The function of vertical crossover is mainly to increase the diversity of the population to avoid it falling into a stagnant state; it mainly uses two dimensions to achieve crossover. Assuming that the $j1$th and $j2$th dimensions of the i-th individual are selected and that a vertical crossover operation is then performed, the new offspring Mvc_i can be obtained by Equations (7) and (8).

$$Mvc_{i,j1} = r \times M_{i,j1} + (1 - r) \times M_{i,j2} \tag{7}$$

$$r = unifrnd(0,1) \tag{8}$$

where $i = 1, 2, \ldots, N$; $j1, j2 = 1, 2, \ldots, D$; $M_{i,j1}$ represent the $j1$ and $j2$ dimensions of the i agent M_i. Additionally, r is a random number uniformly distributed from 0 to 1. Special attention must be paid to the normalization of the lower and upper bounds of each dimension to ensure that the individual remains within the bounds after the operation before implementing the vertical crossover operation. After performing vertical crossing, a reverse normalization operation needs to be performed to ensure that the offspring is still within the given boundary.

3.2.2. Horizontal Crossover Search

Horizontal crossover can further improve the search and development of algorithms. It mainly involves crossover operations being performed on all dimensions of two different agents. Assuming that the agents M_{i1} and M_{i2} are selected to perform level crossing, the new agents Mhc_{i1} and Mhc_{i2} can be obtained using Equations (9) and (10).

$$Mhc_{i1,j} = r_1 \times M_{i1,j} + (1 - r1) \times M_{i2,j} + c1 \times \left(M_{i1,j} - M_{i2,j}\right) \tag{9}$$

$$Mhc_{i2,j} = r_2 \times M_{i2,j} + (1 - r1) \times M_{i1,j} + c2 \times \left(M_{i2,j} - M_{i1,j}\right) \tag{10}$$

where $Mhc_{i1,j}$ and $Mhc_{i2,j}$ are j-th dimension of Mhc_{i1} and Mhc_{i2}; $M_{i1,j}$ and $M_{i2,j}$ is the j-th dimension of M_{i1} and M_{i2}; and r_1 and r_2 are uniformly distributed random numbers from 0 to 1. $c1$ and $c2$ are uniformly distributed random numbers in $[-1, 1]$.

3.3. Proposed RCCBOA

The idea of the RCCBOA is to introduce random replacement and crossover strategies on the basis of the BOA. Regarding the early exploration and later development of the algorithm, it can be determined by including the ratio of the current number of iterations to the total number of iterations. In the exploration phase, the random replacement strategy uses the location of the optimal solution to replace the current solution, which can improve the convergence speed of the BOA. In the exploitation stage, a cross-search strategy is introduced to improve the exploration and exploitation capabilities of the original algorithm, which makes it possible get rid of the local optimal solution as much as possible. The effective combination of the two greatly improves the performance of the original BOA. The pseudo-code of the RCCBOA is shown in Algorithm 2. Based on the BOA, random replacement and crisscross search mechanisms are mainly used in the second stage of the BOA. At the beginning of each iteration, a random replacement mechanism is used to replace the current position with the position of the optimal solution with a certain probability and then evaluate it. At the end of each iteration, the population is updated and evaluated again using the crisscross search mechanism. For a better view, the flowchart of the RCCBOA is offered in Figure 1.

Algorithm 2: Pseudo-code of the RCCBOA.

```
Initialize population number n, dimensions d, max evaluations Max_FEs, objective function f(x);
Initialize sensor modality c, power exponent a, switch probability p and evaluations t;
Initialize the population of butterflies x_i (i = 1,2,...,n);
Gain intensity I_i by f(x_i)
While (t ≤ Max_FEs)
Calculate the fragrance bf of each butterfly using Equation (2);
Gain the best bf;
Gain butterfly individuals by the random replacement strategy;
Update the best solution and position;
For i = 1 to n
Update r in [0,1];
        If r < p
            Move to the best solution using Equation (1);
        Else
            Move randomly using Equation (1);
        End if
    End for
    Update the population of butterflies using Equations (9) and (10);
    t = t + 1;
Update parameter a;
End while
Output best solution;
```

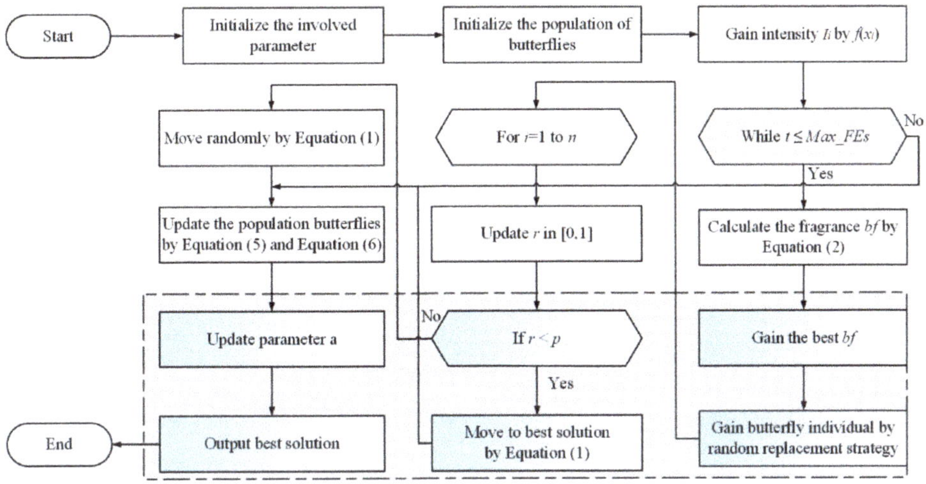

Figure 1. The flowchart of the RCCBOA.

4. Proposed RCCBOA-SVM Method

Firstly, we subject the proposed RCCBOA to data-level feature selection, aiming to obtain effective features in the dataset. Secondly, it is used to optimize the penalty factor C and kernel parameter γ of the SVM. The framework of RCCBOA-SVM is shown in Figure 2. Finally, the model mainly includes two important components in the left half for feature selection and uses the RCCBOA to optimize the two parameters C and γ in the SVM model. In the right half, the best model obtains the classification accuracy (ACC) through 10-fold cross-validation.

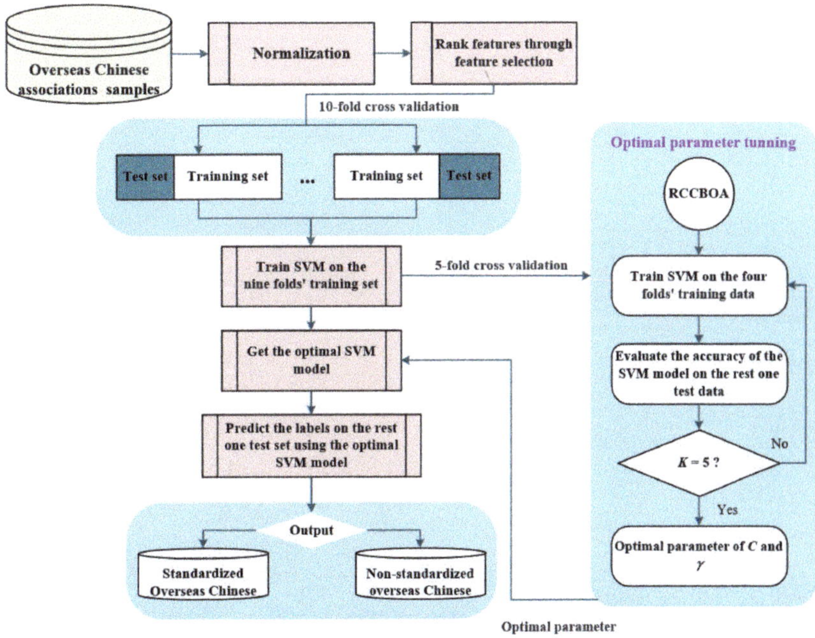

Figure 2. Flowchart of the suggested RCCBOA-SVM model.

For feature selection problems, the focus of the algorithm is to select or not to select a certain feature in the dataset, thus maximizing the classification accuracy of the most effective feature. RCCBOA is inconsistent with the two-dimensionality required by the feature selection problem when solving the problem, and these algorithms cannot be used to directly solve the feature selection problem. Therefore, it is necessary to convert each solution vector in this algorithm to binary form through the sigmoid transfer function, which consists of only '0' and '1'. To achieve this transformation, an S-shaped transformation function is used, which gives the probability of selecting a particular feature in the solution vector.

Through feature selection, the minimum number of key features can be successfully obtained. However, the fitting accuracy of the SVM depends on the values of the parameters (C, γ), and different parameters are suitable for different sample data sets. Therefore, it is necessary to further optimize the SVM parameters using RCCBOA to achieve the optimal effect.

5. Experiments

5.1. Collection of Data

The data involved in this paper were mainly obtained from overseas Chinese citizens; 1050 people were selected as the research objects. The 28 attributes of the test subjects were gender; age range; location of hometown; current identity; place of birth; when they went abroad; reason for going abroad; in which year they became permanent residents in their country of residence; highest level of education (degree); major; type of work currently engaged in; whether they had relatives living together in their native country; their position held in their native country; whether they had joined an overseas Chinese association; whether they were a founder of an overseas Chinese association; their reason for founding an overseas Chinese association; their motivation for joining an overseas Chinese association; their position held in the overseas Chinese association; whether their overseas Chinese association had a clear division of duties; whether their overseas Chinese association is harmonious; whether their overseas Chinese association is a non-profit organization; whether the charter of the overseas Chinese association is reasonable; whether the overseas Chinese association holds regular meetings; whether every member of the association has the right to vote and be elected; whether every member of the association has the right to criticize, make suggestions, and supervise the overseas Chinese association; whether the membership fee of the association is paid according to the regulations; the main source of funding for the overseas Chinese association; and, lastly, their expectations and suggestions for the overseas Chinese association. The importance of these 28 attributes and their internal connections were explored, and based on this a model was built. Table 2 details the 28 attributes.

Table 2. Description of the 28 attributes.

Attributes	Name	Description
A1	Gender	Male—1; Female—0
A2	Age range	0–30 years old—1; 30–60 years old—2; 60–80 years old—3.
A3	Location of hometown	Wenzhou—1; Lishui—2; other places—3.
A4	Current identity	Overseas Chinese with Chinese nationality or international students—1; descendants of overseas Chinese who were born in the country of residence and obtained the nationality of the residence country—2; returned overseas Chinese and family members of returned overseas Chinese—3.

Table 2. *Cont.*

Attributes	Name	Description
A5	Place of birth	China (Hong Kong, China)—1; foreign countries—2; unable to confirm—0.
A6	When they went abroad	Before 1949—1; between 1950 and 1978—2; 1979 till now—3.
A7	Reasons for going abroad	Family visit—1; work, trade, or investment—2; studying abroad or skilled immigrants—3; other reasons or no reasons—4.
A8	In which year they became permanent residents	Before 2000—1; 2000 till now—2; no permanent residence—0
A9	Highest level of education (degree)	High school/technical secondary school/secondary vocational school/technical school or below—1; junior college/higher vocational college—2; Bachelor degree or above—3.
A10	Major	Liberal arts—1; science—2; others (catering, party school cadre training, entertainment, tourism, army academy)—3; none or unknown—0
A11	Type of work currently engaged in	worker, trade, business and investment—1; research and development, knowledge or education—2; catering—3; other or none—4.
A12	Relatives living together in China	Immediate family member—1; non-immediate relatives—2; both 1 and 2–3; none—0.
A13	Their position held in China	National level—1; provincial level—2; city/county level—3; other or none—0.
A14	Whether they had joined an overseas Chinese association	Yes—1; no—0.
A15	Whether they had acted as the founder of an overseas Chinese association	Yes—1; no—0.
A16	Reason for founding the overseas Chinese association	Business relations—1; geographical relations—2; learning relations—3; other reasons or none—4.
A17	Motivation for joining the overseas Chinese association	Active—1; passive (invited by friends)—2; other or skip—3.
A18	Position held in the overseas Chinese association	Positive—1; deputy—2; others—3; none—0.
A19	Whether the overseas Chinese association has a clear division of duties	Yes—1; none or other—0.
A20	Whether the overseas Chinese association is harmonious	Harmony—1; so-so or unknown—2; skip—0.
A21	Whether the overseas Chinese association is a non-profit organization	Yes—1; no—2; unknown or skip—0.
A22	Whether the charter of the overseas Chinese association is reasonable	Yes—1; none—0.
A23	Whether the overseas Chinese association holds regular meetings	Yes—1; no—0.
A24	Whether every member of the association has the right to vote and be elected	Yes—1; no—0.
A25	Whether every member of the association has the right to criticize, make suggestions, and supervise the overseas Chinese association	Yes—1; no—0.
A26	Whether the membership fee of the association is paid according to the regulations	Yes—1; no—0.
A27	Main source of funding for the overseas Chinese association	Other fixed income—1; membership fees (including membership fees and donation)—2; no dues, only donations—3; unknown or other—0.
A28	Expectations and suggestions for the overseas Chinese association	Positive—1; so-so or no (skip)—2; negative—3.

Among these, whether the overseas Chinese group/association is standardized is considered among the 28 attributes (attribute A17 (motivation of joining the overseas Chinese association), A19 (whether the overseas Chinese association has a clear division of duties), A21 (whether the overseas Chinese association is a non-profit organization), A22 (whether the charter of the overseas Chinese association is reasonable), A23 (whether the overseas Chinese association holds regular meetings), A24 (whether every member of the association has the right to vote and be elected)) are the basis of determining whether the overseas Chinese association is standard.

5.2. Experimental Setup

Ensuring the independence of experimental procedures is extremely important, as in computational science and molecular characterization [89,90], location-based services [91,92], drug discovery [93,94], pharmacoinformatic data mining [95,96], and information retrieval services [97–99]. The comparison test described in this section was carried out on a computer with a main central processing unit (CPU) frequency of 3.4 GHz and the win10 operating system (Microsoft, Redmond, WA, USA). Simulation experiments were performed in MATLAB R2016a (MathWorks, Natick, MA, USA). In benchmarking experiments, each comparison algorithm ran 30 experiments simultaneously. When dealing with classification problems, the data were scaled to [−1,1]. k-fold cross-validation (CV) was used to split the data, where k was set to 10.

6. Experimental Results

6.1. Benchmark Function Validation

We mainly conducted test experiments using RCCBOA on the CEC2014 benchmark function set, including mechanism combination experiments and comparative experiments with existing advanced algorithms. Detailed information about the CEC2014 benchmark set can be found in Appendix A (see Table A1), coming from congress on evolutionary computation of the world's highest conference. The experimental results obtained from 30 independent repeated experiments under the same conditions were analyzed, including the average and standard results obtained by the algorithm on each benchmark function. We used the Wilcoxon signed-rank test non-parametric statistical test and the Friedman test, which have used in many other works, to estimate the performance [100–104].

6.1.1. The Component Foundation

To assess the contribution of random replacement and horizontal crossover search mechanisms to the original BOA, a mechanism combination experiment was conducted. By randomly combining the two mechanisms, three additional algorithms were developed—namely, RCCBOA, RBOA, and CCBOA. As shown in Table 3, where "R" and "CC" represent the random replacement strategy and the crossover strategy, respectively, "1" indicates that the BOA incorporates the policy and "0" indicates that the BOA does not incorporate the policy. For example, RCCBOA means that the BOA combines both the random replacement strategy and the horizontal crossover search strategy. Each algorithm was tested on the CEC 2014 benchmark function test set. The experimental results are shown in Table 4. For fair comparison, the parameters commonly used in the experiment were not set to 30. In addition, we utilized the Wilcoxon signed-rank test and average value (ARV) to examine the average ranking values of the algorithms involved to further investigate the difference between the two. It can be seen that the average performance of the RCCBOA combining both strategies was the best.

To further visualize the performance of RCCBOA, Figure 3 shows the convergence curves of RCCBOA, CCBOA, RBOA, and BOA on F3, F7, F11, F13, F16, F20, F23, F27, and F29. Obviously, RCCBOA had a faster convergence speed and smaller convergence value on these benchmark functions than on the other algorithms. In conclusion, the BOA performance achieved by combining both the random replacement strategy and the horizontal crossover search strategy was the best.

Table 3. Four BOA variants with two strategies.

Algorithms	R	CC
RCCBOA	1	1
CCBOA	0	1
RBOA	1	0
BOA	0	0

Table 4. Average ranking of four BOA variants.

Algorithm	Rank	ARV
RCCBOA	1	1.426667
CCBOA	2	1.704444
RBOA	3	2.908889
BOA	4	3.96

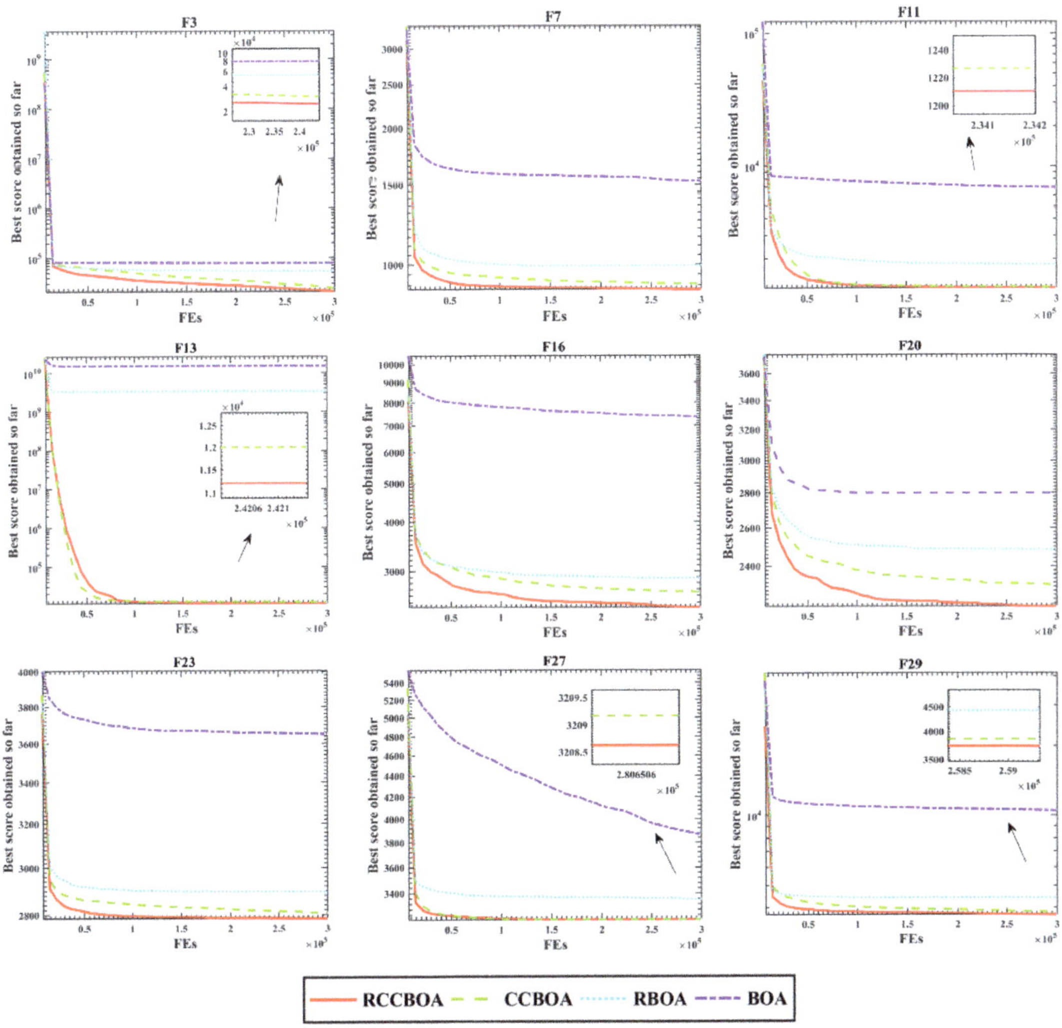

Figure 3. Mechanism combination experiment.

6.1.2. Comparison with Advanced Methods

To evaluate the superiority of the RCCBOA algorithm, this section compares the RCCBOA algorithm with nine improved optimization algorithms, including CDLOBA [30], CBA [31], RCBA J [32], MWOA [33], LWOA [34], IWOA [35], CEFOA [36], CIFOA [37], and AMFOA [38]. These nine advanced algorithms are improved compared to classic algorithms and have strong optimization abilities. We chose to use the CEC 2014 benchmark function as the test set and set the search agent to 30, the dimension to 30, and the maximum number of evaluations to 300,000. In addition, each algorithm was run separately for 30 experiments to obtain the average value; the parameter settings are shown in Table 5.

Table 5. Parameters setting of the RCCBOA and other algorithms.

Method	Parameter
RCCBOA	$p = 0.8$
CDLOBA	$r = [1, 30];\ Q_{min} = 0; Q_{max} = 2$
CBA	$p = [0, 1];\ Q_{min} = 0; Q_{max} = 2$
RCBA	$A = [0, 1]; r = 0.5; Q_{min} = 0; Q_{max} = 2$
MWOA	$a_1 = [0, 2]; a_2 = [-2, -1]$
LWOA	$a_1 = [0, 2]; a_2 = [-2, -1]; b = 1$
IWOA	$a_1 = [0, 2]; a_2 = [-2, -1]; b = 1; C_r = 0.1$
CEFOA	$initiallocation\ \epsilon[-10, 10]$
CIFOA	$mr = 0.8$
AMFOA	$\sigma_1 = 0; \sigma_2 = 0$

Table 6 shows the average fitness value and standard deviation of each algorithm on the 30 benchmark function test sets. It can be seen that the performance of RCCBOA on some test functions is better than that of other algorithms. It is proven that the proposed algorithm has significant advantages compared with other algorithms on the IEEE CEC2014 test set. First, the average result (Avg) and standard deviation (Std) of the optimization values were used to evaluate the potential of the relevant optimizer. Furthermore, we employed the Wilcoxon signed-rank test to evaluate whether the performance of RCCBOA was significantly better than that of other state-of-the-art algorithms in this experiment. It can be seen that the p-values calculated on most test functions were all lower than 0.05, indicating that the RCCBOA had a good performance on most benchmark functions. Furthermore, we screened nine representative convergence plots on the IEEE CEC2014 test benchmark function, as shown in Figure 4. It can be seen that the RCCBOA had an excellent convergence speed and convergence value on nine test functions.

Table 6. Comparison of the results of the RCCBOA and different advanced algorithms.

Fuctions	Indicators	RCCBOA	CDLOBA	CBA	RCBA	MWOA	LWOA	IWOA	CEFOA	CIFOA	AMFOA
F1	Avg	1.30×10^{7}	1.25×10^{6}	1.23×10^{7}	4.81×10^{6}	5.72×10^{9}	1.02×10^{7}	1.85×10^{9}	1.64×10^{10}	1.21×10^{10}	1.31×10^{10}
	Std	3.36×10^{6}	5.59×10^{5}	4.89×10^{6}	1.75×10^{6}	1.90×10^{9}	2.99×10^{6}	5.26×10^{8}	7.19×10^{8}	2.46×10^{9}	8.28×10^{8}
	Rank	5	1	4	2	7	3	6	10	8	9
	p-value	-	1.73×10^{-6}	3.71×10^{-1}	1.73×10^{-6}	1.73×10^{-6}	2.25×10^{-3}	1.73×10^{-6}	1.73×10^{-6}	1.73×10^{-6}	1.73×10^{-6}
F2	Avg	4.41×10^{3}	1.51×10^{4}	1.87×10^{5}	1.38×10^{5}	2.01×11	3.52×06	1.25×10^{11}	1.99×10^{11}	1.91×10^{11}	1.89×10^{11}
	Std	3.71×10^{3}	1.12×10^{4}	9.54×10^{5}	3.89×10^{4}	2.85×10^{10}	8.38×10^{5}	1.33×10^{10}	5.37×10^{8}	5.11×10^{8}	6.09×10^{9}
	Rank	1	2	4	3	10	5	6	9	8	7
	p-value	-	1.74×10^{-4}	4.71×10^{-2}	1.73×10^{-6}	1.73×10^{-6}	1.73×10^{-6}	1.73×10^{-6}	1.73×10^{-6}	1.73×10^{-6}	1.73×10^{-6}
F3	Avg	3.69×10^{4}	1.12×10^{-5}	8.37×10^{3}	3.89×10^{2}	5.33×10^{5}	3.35×10^{3}	1.73×10^{5}	6.81×10^{8}	1.31×10^{7}	5.27×10^{7}
	Std	4.89×10^{3}	2.36×10^{4}	8.78×10^{3}	3.63×10^{1}	1.83×10^{5}	1.50×10^{3}	3.12×10^{4}	1.95×10^{7}	2.26×10^{7}	5.71×10^{7}
	Rank	4	5	3	1	7	2	6	10	8	9
	p-value	-	1.73×10^{-6}	1.92×10^{-6}	1.73×10^{-6}	1.73×10^{-6}	1.73×10^{-6}	1.73×10^{-6}	1.73×10^{-6}	1.73×10^{-6}	1.73×10^{-6}
F4	Avg	5.28×10^{2}	5.19×10^{2}	5.30×10^{2}	5.16×10^{2}	6.41×10^{4}	5.51×10^{2}	3.16×10^{4}	7.18×10^{4}	6.65×10^{4}	6.02×10^{4}
	Std	2.91×10^{1}	3.84×10^{1}	4.41×10^{1}	4.58×10^{1}	1.58×10^{4}	4.31×10^{1}	7.29×10^{3}	7.73×10^{2}	3.85×10^{2}	1.95×10^{3}
	Rank	3	2	4	1	8	5	6	10	9	7
	p-value	-	3.18×10^{-1}	8.77×10^{-1}	1.30×10^{-1}	1.73×10^{-6}	2.18×10^{-2}	1.73×10^{-6}	1.73×10^{-6}	1.73×10^{-6}	1.73×10^{-6}
F5	Avg	5.21×10^{2}	5.21×10^{2}	5.20×10^{2}	5.20×10^{2}	5.21×10^{2}	5.21×10^{2}	5.21×10^{2}	5.22×10^{2}	5.21×10^{2}	5.21×10^{2}
	Std	4.65×10^{-2}	2.38×10^{-1}	2.30×10^{-1}	9.69×10^{-2}	4.02×10^{-2}	9.07×10^{-2}	4.58×10^{-2}	4.33×10^{-2}	2.94×10^{-2}	6.99×10^{-2}
	Rank	5	4	2	1	9	3	7	10	6	8
	p-value	-	2.25×10^{-3}	1.73×10^{-6}	1.73×10^{-6}	1.73×10^{-6}	1.73×10^{-6}	1.73×10^{-6}	1.73×10^{-6}	2.35×10^{-6}	1.73×10^{-6}
F6	Avg	6.35×10^{2}	6.65×10^{2}	6.73×10^{2}	6.70×10^{2}	6.80×10^{2}	6.55×10^{2}	6.71×10^{2}	6.90×10^{2}	6.85×10^{2}	6.85×10^{2}
	Std	1.19×10^{1}	3.32×10^{0}	4.38×10^{0}	4.81×10^{0}	3.63×10^{0}	4.89×10^{0}	3.59×10^{0}	3.78×10^{-1}	1.31×10^{0}	7.42×10^{1}
	Rank	1	3	6	4	7	2	5	10	8	9
	p-value	-	1.73×10^{-6}	1.73×10^{-6}	1.73×10^{-6}	1.73×10^{-6}	1.73×10^{-6}	1.73×10^{-6}	1.73×10^{-6}	1.73×10^{-6}	1.73×10^{-6}
F7	Avg	7.00×10^{2}	7.00×10^{2}	7.00×10^{2}	7.00×10^{2}	2.56×10^{3}	7.01×10^{2}	1.84×10^{3}	2.56×10^{3}	2.49×10^{3}	2.43×10^{3}
	Std	3.53×10^{-3}	1.09×10^{-2}	1.57×10^{-1}	5.11×10^{-2}	3.60×10^{2}	1.58×10^{-2}	1.35×10^{2}	2.14×10^{1}	4.91×10^{0}	1.53×10^{1}
	Rank	1	2	3	4	9	5	6	10	8	7
	p-value	-	1.6×10^{-4}	9.91×10^{-3}	1.73×10^{-6}	1.73×10^{-6}	1.73×10^{-6}	1.73×10^{-6}	1.73×10^{-6}	1.73×10^{-6}	1.73×10^{-6}
F8	Avg	9.55×10^{2}	1.30×10^{3}	1.17×10^{3}	1.19×10^{3}	1.60×10^{3}	9.95×10^{2}	1.38×10^{3}	1.69×10^{3}	1.60×10^{3}	1.65×10^{3}
	Std	1.26×10^{1}	5.35×10^{1}	6.30×10^{1}	5.38×10^{1}	6.30×10^{1}	3.24×10^{1}	4.30×10^{1}	1.59×10^{1}	6.13×10^{0}	2.20×10^{1}
	Rank	1	5	3	4	8	2	6	10	7	9
	p-value	-	1.73×10^{-6}	1.73×10^{-6}	1.73×10^{-6}	1.73×10^{-6}	1.02×10^{-5}	1.73×10^{-6}	1.73×10^{-6}	1.73×10^{-6}	1.73×10^{-6}

Table 6. *Cont.*

Fuctions	Indicators	RCCBOA	CDLOBA	CBA	RCBA	MWOA	LWOA	IWOA	CEFOA	CIFOA	AMFOA
F9	Avg	1.26×10^{3}	1.61×10^{3}	1.41×10^{3}	1.42×10^{3}	1.84×10^{3}	1.32×10^{3}	1.57×10^{3}	1.88×10^{3}	1.78×10^{3}	1.78×10^{3}
	Std	2.84×10^{1}	9.97×10^{1}	8.96×10^{1}	9.41×10^{1}	1.02×10^{2}	7.42×10^{1}	5.94×10^{1}	2.15×10^{1}	7.13×10^{0}	3.09×10^{1}
	Rank	1	6	3	4	9	2	5	10	7	8
	p-value	-	1.73×10^{-6}	2.35×10^{-6}	1.92×10^{-6}	1.73×10^{-6}	1.18×10^{-3}	1.73×10^{-6}	1.73×10^{-6}	1.73×10^{-6}	1.73×10^{-6}
F10	Avg	3.62×10^{3}	8.95×10^{3}	8.81×10^{3}	8.87×10^{3}	1.68×10^{4}	4.81×10^{3}	1.37×10^{4}	1.92×10^{4}	1.69×10^{4}	1.82×10^{4}
	Std	2.63×10^{2}	9.69×10^{2}	9.68×10^{2}	1.08×10^{3}	6.87×10^{2}	7.58×10^{2}	7.53×10^{2}	1.70×10^{2}	3.12×10^{2}	4.85×10^{2}
	Rank	1	5	3	4	7	2	6	10	8	9
	p-value	-	1.73×10^{-6}	1.73×10^{-6}	1.73×10^{-6}	1.73×10^{-6}	3.88×10^{-6}	1.73×10^{-6}	1.73×10^{-6}	1.73×10^{-6}	1.73×10^{-6}
F11	Avg	1.14×10^{4}	8.54×10^{3}	8.80×10^{3}	8.92×10^{3}	1.70×10^{4}	8.57×10^{3}	1.52×10^{4}	1.91×10^{4}	1.67×10^{4}	1.77×10^{4}
	Std	6.78×10^{2}	6.41×10^{2}	9.33×10^{2}	7.13×10^{2}	8.18×10^{2}	9.88×10^{2}	1.04×10^{3}	2.33×10^{2}	1.28×10^{2}	1.36×10^{2}
	Rank	5	1	3	4	8	2	6	10	7	9
	p-value	-	1.73×10^{-6}	1.92×10^{-6}	1.73×10^{-6}	1.73×10^{-6}	1.92×10^{-6}	1.73×10^{-6}	1.73×10^{-6}	1.73×10^{-6}	1.73×10^{-6}
F12	Avg	1.20×10^{3}	1.20×10^{3}	1.20×10^{3}	1.20×10^{3}	1.21×10^{3}	1.20×10^{3}	1.20×10^{3}	1.21×10^{3}	1.20×10^{3}	1.21×10^{3}
	Std	4.15×10^{-1}	4.24×10^{-1}	6.59×10^{-1}	6.04×10^{-1}	9.86×10^{-1}	4.91×10^{-1}	8.92×10^{-1}	9.80×10^{-1}	2.76×10^{-1}	9.65×10^{-1}
	Rank	5	1	4	2	9	3	6	10	7	8
	p-value	-	1.73×10^{-6}	4.90×10^{-4}	2.35×10^{-6}	1.73×10^{-6}	2.13×10^{-6}	1.73×10^{-6}	1.73×10^{-6}	1.73×10^{-6}	1.73×10^{-6}
F13	Avg	1.30×10^{3}	1.30×10^{3}	1.30×10^{3}	1.30×10^{3}	1.31×10^{3}	1.30×10^{3}	1.31×10^{3}	1.31×10^{3}	1.31×10^{3}	1.31×10^{3}
	Std	4.30×10^{-2}	9.22×10^{-2}	8.32×10^{-2}	1.08×10^{-1}	9.09×10^{-1}	1.08×10^{-1}	5.31×10^{-1}	1.25×10^{-2}	1.48×10^{-2}	7.65×10^{-2}
	Rank	3	1	2	4	9	5	6	10	8	7
	p-value	-	2.37×10^{-1}	2.29×10^{-1}	6.00×10^{-1}	1.73×10^{-6}	3.16×10^{-2}	1.73×10^{-6}	1.73×10^{-6}	1.73×10^{-6}	1.73×10^{-6}
F14	Avg	1.40×10^{3}	1.40×10^{3}	1.40×10^{3}	1.40×10^{3}	1.83×10^{3}	1.40×10^{3}	1.65×10^{3}	1.87×10^{3}	1.86×10^{3}	1.82×10^{3}
	Std	3.13×10^{-2}	6.86×10^{-2}	1.13×10^{-1}	1.13×10^{-1}	5.66×10^{1}	1.93×10^{-1}	2.58×10^{1}	5.45×10^{0}	1.46×10^{0}	1.38×10^{1}
	Rank	1	4	2	3	8	5	6	10	9	7
	p-value	-	3.38×10^{-3}	1.78×10^{-1}	3.68×10^{-2}	1.73×10^{-6}	8.94×10^{-4}	1.73×10^{-6}	1.73×10^{-6}	1.73×10^{-6}	1.73×10^{-6}
F15	Avg	1.53×10^{3}	2.04×10^{3}	1.63×10^{3}	1.58×10^{3}	4.03×10^{7}	1.55×10^{3}	3.66×10^{6}	2.54×10^{7}	1.59×10^{7}	7.64×10^{6}
	Std	2.08×10^{0}	1.49×10^{2}	3.91×10^{1}	2.18×10^{1}	3.26×10^{7}	1.09×10^{1}	1.87×10^{6}	1.19×10^{6}	5.14×10^{5}	2.07×10^{6}
	Rank	1	5	4	3	10	2	6	9	8	7
	p-value	-	1.73×10^{-6}	1.73×10^{-6}	1.73×10^{-6}	1.73×10^{-6}	2.35×10^{-6}	1.73×10^{-6}	1.73×10^{-6}	1.73×10^{-6}	1.73×10^{-6}
F16	Avg	1.62×10^{3}	1.62×10^{3}	1.62×10^{3}	1.62×10^{3}	1.62×10^{3}	1.62×10^{3}	1.62×10^{3}	1.62×10^{3}	1.62×10^{3}	1.62×10^{3}
	Std	3.17×10^{-1}	4.08×10^{-1}	4.78×10^{-1}	4.66×10^{-1}	2.09×10^{-1}	6.33×10^{-1}	3.88×10^{-1}	1.80×10^{-1}	6.98×10^{-2}	1.80×10^{-1}
	Rank	1	5	7	4	9	2	3	10	6	8
	p-value	-	1.73×10^{-6}	1.73×10^{-6}	1.73×10^{-6}	1.73×10^{-6}	2.30×10^{-2}	2.35×10^{-6}	1.73×10^{-6}	1.73×10^{-6}	1.73×10^{-6}

Table 6. *Cont.*

Fuctions	Indicators	RCCBOA	CDLOBA	CBA	RCBA	MWOA	LWOA	IWOA	CEFOA	CIFOA	AMFOA
F17	Avg	1.54×10^{6}	7.30×10^{4}	9.46×10^{5}	5.26×10^{5}	9.85×10^{8}	1.67×10^{6}	5.08×10^{8}	3.87×10^{9}	1.98×10^{9}	1.88×10^{9}
	Std	8.74×10^{5}	3.66×10^{4}	4.18×10^{5}	2.53×10^{5}	4.56×10^{8}	7.03×10^{5}	2.87×10^{8}	1.97×10^{7}	7.28×10^{8}	4.50×10^{8}
	Rank	4	1	3	2	7	5	6	10	9	8
	p-value	-	1.73×10^{-6}	1.04×10^{-3}	2.35×10^{-6}	1.73×10^{-6}	3.82×10^{-1}	1.73×10^{-6}	1.73×10^{-6}	1.73×10^{-6}	1.73×10^{-6}
F18	Avg	3.54×10^{3}	5.03×10^{3}	4.57×10^{3}	5.52×10^{3}	2.09×10^{10}	2.84×10^{4}	6.75×10^{9}	3.78×10^{10}	3.45×10^{10}	3.14×10^{10}
	Std	1.03×10^{3}	1.44×10^{3}	2.11×10^{3}	1.87×10^{3}	5.86×10^{9}	1.57×10^{4}	2.90×10^{9}	3.15×10^{8}	2.44×10^{9}	2.13×10^{9}
	Rank	1	3	2	4	7	5	6	10	9	8
	p-value	-	3.59×10^{-4}	2.85×10^{-2}	9.71×10^{-5}	1.73×10^{-6}	1.73×10^{-6}	1.73×10^{-6}	1.73×10^{-6}	1.73×10^{-6}	1.73×10^{-6}
F19	Avg	1.95×10^{3}	1.99×10^{3}	1.96×10^{3}	1.96×10^{3}	5.04×10^{3}	1.95×10^{3}	2.94×10^{3}	1.08×10^{4}	9.08×10^{3}	8.12×10^{3}
	Std	3.14×10^{1}	2.77×10^{1}	2.58×10^{1}	2.85×10^{1}	1.53×10^{3}	2.84×10^{1}	2.73×10^{2}	1.23×10^{2}	7.93×10^{2}	6.17×10^{2}
	Rank	1	5	3	4	7	2	6	10	9	8
	p-value	-	2.84×10^{-5}	7.03×10^{-1}	2.71×10^{-1}	1.73×10^{-6}	7.81×10^{-1}	1.73×10^{-6}	1.73×10^{-6}	1.73×10^{-6}	1.73×10^{-6}
F20	Avg	1.13×10^{4}	4.43×10^{4}	4.24×10^{3}	2.82×10^{3}	9.50×10^{6}	8.83×10^{3}	6.70×10^{5}	3.22×10^{9}	7.50×10^{7}	2.69×10^{9}
	Std	2.99×10^{3}	1.83×10^{4}	3.14×10^{3}	1.52×10^{2}	7.96×10^{6}	5.66×10^{3}	8.59×10^{5}	0.00×10^{0}	4.87×10^{7}	9.58×10^{8}
	Rank	4	5	2	1	7	3	6	10	8	9
	p-value	-	1.73×10^{-6}	8.47×10^{-6}	1.73×10^{-6}	1.73×10^{-6}	3.68×10^{-2}	1.73×10^{-6}	1.73×10^{-6}	1.73×10^{-6}	1.73×10^{-6}
F21	Avg	1.67×10^{6}	8.75×10^{4}	5.42×10^{5}	3.43×10^{5}	2.19×10^{8}	8.61×10^{5}	8.04×10^{7}	1.80×10^{9}	8.18×10^{8}	4.95×10^{8}
	Std	4.61×10^{5}	3.81×10^{4}	2.70×10^{5}	1.79×10^{5}	1.04×10^{8}	4.24×10^{5}	4.69×10^{7}	3.19×10^{7}	3.40×10^{8}	1.83×10^{8}
	Rank	5	1	3	2	7	4	6	10	9	8
	p-value	-	1.73×10^{-6}	1.73×10^{-6}	1.73×10^{-6}	1.73×10^{-6}	1.36×10^{-5}	1.73×10^{-6}	1.73×10^{-6}	1.73×10^{-6}	1.73×10^{-6}
F22	Avg	3.13×10^{3}	4.09×10^{3}	4.26×10^{3}	4.19×10^{3}	3.27×10^{5}	3.93×10^{3}	1.91×10^{4}	6.11×10^{6}	1.89×10^{6}	4.64×10^{6}
	Std	2.26×10^{2}	3.71×10^{2}	3.50×10^{2}	4.44×10^{2}	5.28×10^{5}	3.76×10^{2}	2.20×10^{4}	2.01×10^{4}	1.45×10^{6}	6.69×10^{5}
	Rank	1	3	5	4	7	2	6	10	8	9
	p-value	-	1.73×10^{-6}	1.73×10^{-6}	1.73×10^{-6}	1.73×10^{-6}	1.92×10^{-6}	1.73×10^{-6}	1.73×10^{-6}	1.73×10^{-6}	1.73×10^{-6}
F23	Avg	2.63×10^{3}	2.65×10^{3}	2.65×10^{3}	2.64×10^{3}	4.89×10^{3}	2.64×10^{3}	3.11×10^{3}	2.50×10^{3}	2.50×10^{3}	2.50×10^{3}
	Std	4.39×10^{1}	3.11×10^{0}	2.16×10^{0}	2.12×10^{-1}	6.99×10^{2}	2.96×10^{-1}	5.70×10^{2}	0.00×10^{0}	4.05×10^{-1}	5.09×10^{-4}
	Rank	4	7	8	5	10	6	9	1	3	2
	p-value	-	1.73×10^{-6}	1.73×10^{-6}	1.73×10^{-6}	1.73×10^{-6}	1.73×10^{-6}	3.61×10^{-3}	1.73×10^{-6}	1.73×10^{-6}	3.18×10^{-6}
F24	Avg	2.60×10^{3}	2.79×10^{3}	2.77×10^{3}	2.76×10^{3}	3.07×10^{3}	2.62×10^{3}	2.60×10^{3}	2.60×10^{3}	2.60×10^{3}	2.60×10^{3}
	Std	1.14×10^{-8}	8.48×10^{1}	5.14×10^{1}	5.89×10^{1}	6.62×10^{1}	2.11×10^{1}	7.38×10^{-1}	0.00×10^{0}	1.21×10^{-1}	2.82×10^{-1}
	Rank	2	9	8	7	10	6	5	1	3	4
	p-value	-	1.73×10^{-6}	1.73×10^{-6}	1.73×10^{-6}	1.73×10^{-6}	1.73×10^{-6}	1.73×10^{-6}	1.73×10^{-6}	3.11×10^{-5}	1.73×10^{-6}

Table 6. *Cont.*

Fuctions	Indicators	RCCBOA	CDLOBA	CBA	RCBA	MWOA	LWOA	IWOA	CEFOA	CIFOA	AMFOA
F25	Avg	2.70×10^{3}	2.75×10^{3}	2.76×10^{3}	2.76×10^{3}	3.00×10^{3}	2.72×10^{3}	2.70×10^{3}	2.70×10^{3}	2.70×10^{3}	2.70×10^{3}
	Std	0.00×10^{0}	1.84×10^{1}	2.67×10^{1}	2.14×10^{1}	8.62×10^{1}	2.48×10^{1}	0.00×10^{0}	0.00×10^{0}	0.00×10^{0}	1.41×10^{-5}
	Rank	1	7	9	8	10	6	1	1	1	5
	p-value	-	1.73×10^{-6}	1.73×10^{-6}	1.73×10^{-6}	1.73×10^{-6}	1.73×10^{-6}	1.00×10^{0}	1.00×10^{0}	1.00×10^{0}	1.73×10^{-6}
F26	Avg	2.76×10^{3}	2.75×10^{3}	2.73×03	2.80×10^{3}	2.91×10^{3}	2.70×10^{3}	2.77×10^{3}	2.80×10^{3}	2.80×10^{3}	2.80×10^{3}
	Std	5.00×10^{1}	1.18×10^{2}	1.02×10^{2}	1.43×10^{2}	1.96×10^{2}	1.82×10^{1}	4.16×10^{1}	0.00×10^{0}	0.00×10^{0}	7.13×10^{-8}
	Rank	4	3	2	9	10	1	5	6	6	8
	p-value	-	4.78×10^{-1}	4.95×10^{-2}	1.98×10^{-1}	1.28×10^{-3}	7.16×10^{-4}	2.78×10^{-2}	2.44×10^{-4}	2.44×10^{-4}	1.73×10^{-6}
F27	Avg	3.13×10^{3}	4.89×10^{3}	5.04×10^{3}	4.98×10^{3}	5.55×10^{3}	4.48×10^{3}	5.01×10^{3}	2.90×10^{3}	2.90×10^{3}	2.90×10^{3}
	Std	5.64×10^{1}	1.26×10^{2}	3.82×10^{2}	3.88×10^{2}	2.81×10^{2}	2.98×10^{2}	1.34×10^{2}	1.39×10^{-12}	1.39×10^{-12}	7.74×10^{-5}
	Rank	4	6	9	7	10	5	8	1	1	3
	p-value	-	1.73×10^{-6}	1.92×10^{-6}	1.73×10^{-6}	1.73×10^{-6}	1.73×10^{-6}	1.73×10^{-6}	1.73×10^{-6}	1.73×10^{-6}	1.73×10^{-6}
F28	Avg	3.99×10^{3}	8.63×10^{3}	8.77×10^{3}	9.10×10^{3}	1.56×10^{4}	7.03×10^{3}	8.46×10^{3}	3.00×10^{3}	3.00×10^{3}	3.00×10^{3}
	Std	7.15×10^{1}	1.77×10^{3}	1.75×10^{3}	1.66×10^{3}	2.30×10^{3}	1.06×10^{3}	4.44×10^{3}	1.39×10^{-12}	3.10×10^{-1}	8.84×10^{-4}
	Rank	4	7	8	9	10	5	6	1	3	2
	p-value	-	1.73×10^{-6}	1.73×10^{-6}	1.73×10^{-6}	1.73×10^{-6}	1.73×10^{-6}	6.16×10^{-4}	1.73×10^{-6}	1.73×10^{-6}	1.73×10^{-6}
F29	Avg	5.34×10^{3}	1.49×10^{8}	4.27×10^{8}	1.73×10^{8}	8.94×10^{8}	2.16×10^{7}	2.25×10^{8}	3.10×10^{3}	1.68×10^{5}	7.96×10^{3}
	Std	5.24×10^{2}	1.20×10^{8}	2.17×10^{8}	1.10×10^{8}	3.63×10^{8}	2.08×10^{7}	1.94×10^{8}	0.00×10^{0}	9.01×10^{5}	1.05×10^{3}
	Rank	2	6	9	7	10	5	8	1	4	3
	p-value	-	1.73×10^{-6}	1.73×10^{-6}	1.73×10^{-6}	1.73×10^{-6}	1.73×10^{-6}	5.31×10^{-5}	1.73×10^{-6}	3.11×10^{-5}	1.73×10^{-6}
F30	Avg	1.41×10^{4}	2.32×10^{5}	4.93×10^{5}	3.12×10^{4}	3.35×10^{7}	2.34×10^{4}	7.34×10^{6}	3.20×10^{3}	5.60×10^{3}	3.44×10^{3}
	Std	1.71×10^{3}	5.90×10^{5}	1.55×10^{6}	7.14×10^{3}	1.68×10^{7}	4.07×10^{3}	3.89×10^{6}	0.00×10^{0}	1.32×10^{4}	6.38×10^{1}
	Rank	4	7	8	6	10	5	9	1	3	2
	p-value	-	1.73×10^{-6}	1.73×10^{-6}	1.73×10^{-6}	1.73×10^{-6}	1.73×10^{-6}	1.73×10^{-6}	1.73×10^{-6}	3.11×10^{-5}	1.73×10^{-6}

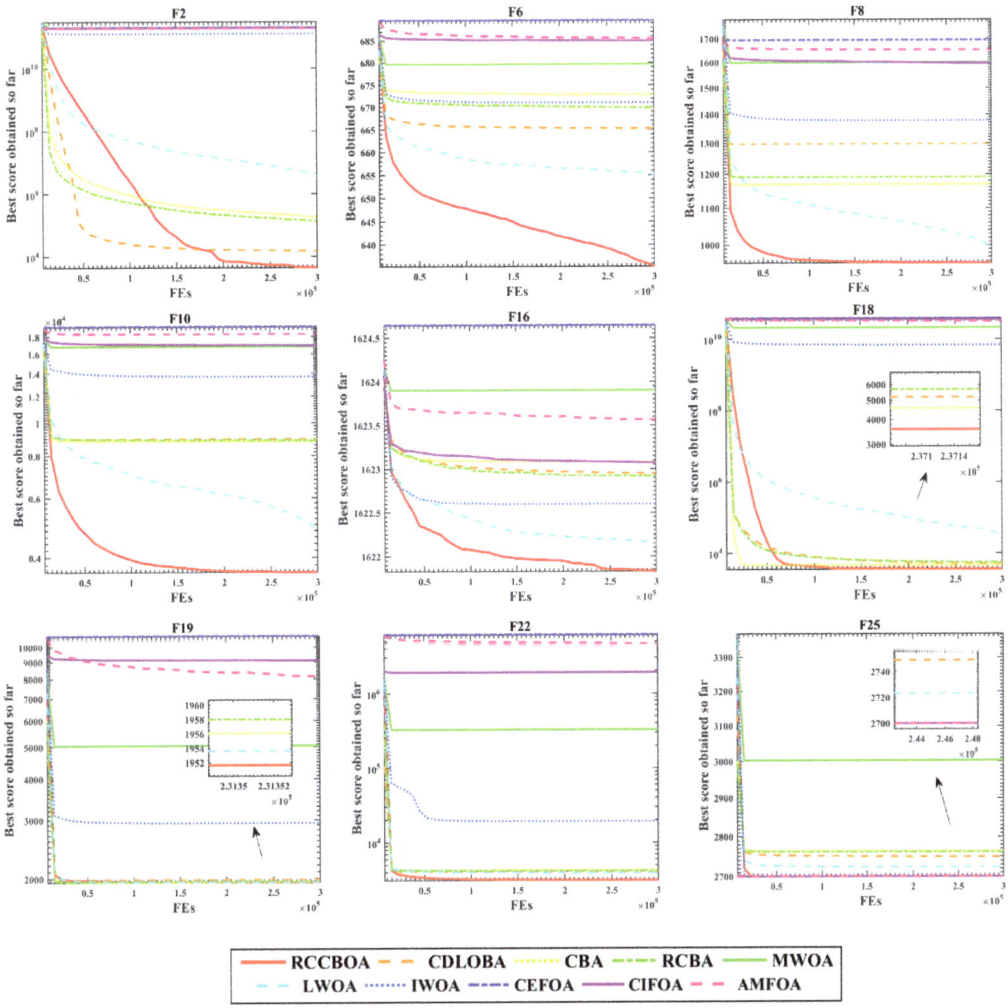

Figure 4. Convergence tendency of the RCCBOA and other advanced algorithms.

In order to further study the effect of random replacement and the crisscross search mechanism on the computation time of BOA, computation time experiments were conducted under the same environment. The experimental results are shown in Figure 5. It can be seen that, compared with the original BOA, the calculation time of RCCBOA was greatly improved in the IEEE CEC2014 benchmark test set. Overall, MWOA was the most time-consuming to calculate, followed by RCCBOA. In addition, the times taken by CDLOBA, CBA, RCBA, LWOA, IWOA, CEFOA, CIFOA, and AMFOA were very close. In conclusion, the introduction of the two mechanisms effectively improved the performance of the BOA as well as improving the execution time. Therefore, when solving practical problems, there is a trade-off between performance and time consumption.

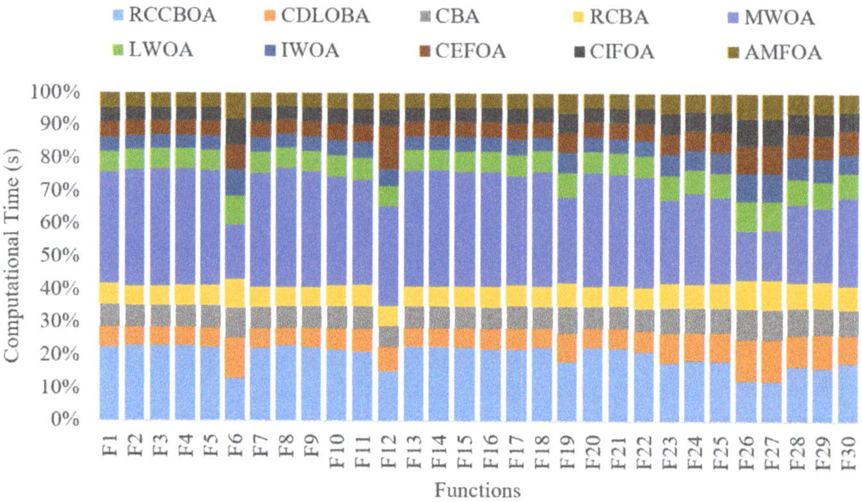

Figure 5. The percentage of computational time taken by RCCBOA and other advanced algorithms.

6.2. Research of Overseas Chinese Associations

In this section, we describe ten independent experimental evaluations of the RCCBOA-SVM (RCCBOA-SVM-FS) model with feature selection, the detailed results of which are shown in Table 7. It can easily be seen that the average accuracy obtained using RCCBOA-SVM-FS was 95%, the sensitivity was 99%, the specificity was 91%, and the MCC index was 90%, with mean standard deviations of 0.02, 0.02, 0.04, and 0.03. Furthermore, the optimal parameters and feature subsets in this experiment were obtained directly through the RCCBOA method, indicating that the constructed model was helpful for guiding the orderly and healthy development of overseas Chinese groups.

Table 7. Classification results obtained for RCCBOA-SVM-FS with four metrics.

Fold	Accuracy	Sensitivity	Specificity	MCC
Num.1	0.94	1.00	0.89	0.89
Num.2	0.91	1.00	0.82	0.84
Num.3	0.99	1.00	0.97	0.97
Num.4	0.96	0.96	0.93	0.91
Num.5	0.96	0.97	0.94	0.91
Num.6	0.94	0.97	0.90	0.88
Num.7	0.93	1.00	0.86	0.86
Num.8	0.96	1.00	0.92	0.91
Num.9	0.96	1.00	0.92	0.91
Num.10	0.94	1.00	0.97	0.88
Avg	0.95	0.99	0.91	0.90
Std	0.02	0.02	0.04	0.03

To further verify the performance of the algorithm, we conducted comparative experiments with another five machine learning models, RCCBOA-SVM, BOA-SVM, ANN, RF, and KELM; the detailed results are shown in Figure 6. The experimental results show that RCCBOA-SVM-FS was better than RCCBOA-SVM, ANN, RF, and KELM in all four evaluation metrics. For the accuracy rate, RCCBOA-SVM-FS had an accuracy rate of about 95%, while the accuracy rates of the other five comparison models were 93%, 91%, 87%, 93%, and 91%, respectively. Regarding the sensitivity index, both RCCBOA-SVM-FS and KELM had values of 99%, 0.06% higher than the that of the lowest ANN. For the specificity index, RCCBOA-SVM-FS, RCCBOA-SVM, and RF surpassed the proportion of 91%. The specificity values of RBOA-SVM, ANN, and KELM were 83%, 80%, and 83%, respectively. In terms of the MCC indicator, RCCBOA-SVM-FS performed the best, with a value of up to 90%. The worst performer was ANN, with a value of 73%. In short, from the above four indicators, it can be seen that the performance of RCCBOA-SVM-FS was better than that of the other five models, and the model accuracy rate was as high as 95%. Therefore, RCCBOA-SVM-FS was effective and reliable for constructing a standardized construction model of overseas Chinese communities.

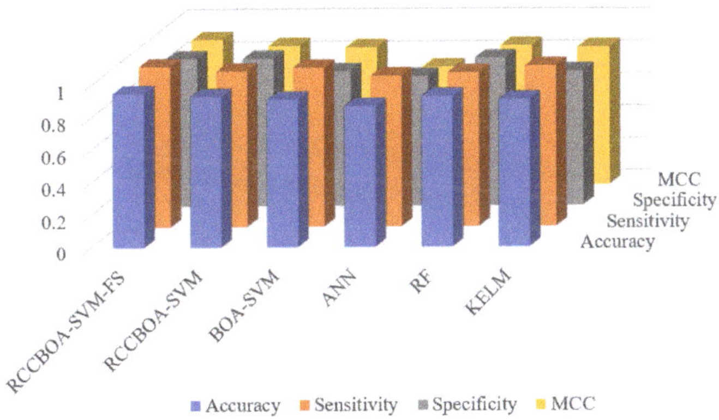

Figure 6. Classification results obtained by the five models in terms of four metrics.

Moreover, the proposed RCCBOA obtained the optimal settings of the SVM hyperparameters as well as the optimal feature set. Here, we used the 10-fold cross-validation technique combined with the RCCBOA algorithm to identify features that have an important impact on the normalization of overseas Chinese groups. Figure 7 illustrates the frequencies of dominant features identified by RCCBOA-SVM-FS via 10-fold cross-validation.

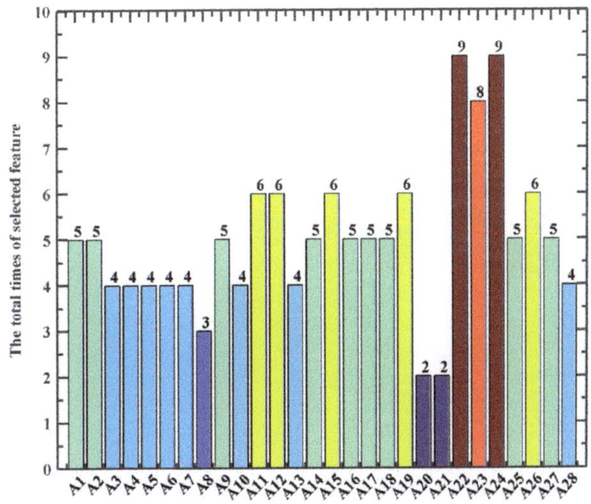

Figure 7. Frequency of the feature selection from RCCBOA-SVM through the 10-fold CV procedure.

As shown in Figure 7, whether the charter of the overseas Chinese association is reasonable (A22), whether the overseas Chinese association holds regular meetings (A23), and whether every member of the association has the right to vote and be elected (A24) were the top three features in terms of frequency, appearing 9, 8, and 9 times, respectively.

Therefore, it can be concluded that these characteristics may play an important role in the standardized construction of overseas Chinese groups.

7. Discussion

The normative nature of overseas Chinese associations is subject to various conditions. Based on the data of overseas Chinese associations, this paper obtained the most important features and models by combining the support vector machine model with RCCBOA. The RCCBOA was introduced and compared with advanced algorithms. It can be seen that when solving related benchmark problems, it had a strong performance. The performance of SVM models can easily be affected by hyperparameters. Therefore, the RCCBOA was combined with SVM and used to extract important features and obtain the best model. From the experimental results found in the study, it can be seen that three attributes—namely, attributes A22, A23, and A24—made up the most important characteristics of overseas Chinese associations, having prominent impacts on the standardized construction of overseas Chinese associations. Generally speaking, an overseas Chinese association which formulates reasonable policies; holds regular meetings; and grants every member of the association the right to vote, stand for election, and vote is standardized. Taking these three features as the main reference attributes, combined with other attributes, a fast judgement of the formality and regularity of an overseas Chinese association can be made using computer algorithm calculation. The advantage of the proposed RCCBOA method is that it can fully mine the key features of data. Based on this advantage, this method also has potential applications in other problems, such as kayak cycle phase segmentation [105], recommender systems [106–109], text clustering [110], human motion capture [111], energy storage planning and scheduling [112], urban road planning [113], microgrid planning [114], active surveillance [115], image super resolution [116,117], anomaly behavior detection [118], and multivariate time series analysis [119].

This study still has several limitations that need to be further discussed. First of all, the samples used in this study were limited; in order to obtain more accurate results, more continuous samples need to be collected to train a more unbiased learning model. Secondly, this study mainly focused on overseas Chinese associations composed of Zhejiang nationals, most of whom were Chinese citizens newly overseas and living mostly in Europe and the United States; therefore, the research data obtained for global overseas Chinese associations were not sufficient and had regional limitations. The determination of the model used in multicenter research made the model more reliable for decision support. In addition, the attributes involved in the study were limited, and future research should seek to use more attributes that have an impact on the standardization construction of overseas Chinese associations.

8. Conclusions and Future Work

In this paper, an improved BOA algorithm combining random replacement and criss-cross search is proposed to study the normalized construction of overseas Chinese groups. The main innovation of the proposed RCCBOA is the introduction of two mechanisms, which effectively improves the convergence speed and convergence accuracy of the original BOA. The comparison experiments performed with other nine advanced algorithms on the CEC2014 benchmark function test set show that the RCCBOA can obtain better solutions and a better stability. Further, the RCCBOA is combined with SVM for better hyperparameter combinations and feature subsets. From the experimental results, it can be seen that the features of A22, A23 and A24 are of great significance to its planning and standardized construction. Compared with other machine learning methods, the proposed method is 95% accurate when dealing with the normative prediction problem of overseas Chinese citizens.

In follow-up studies, the RCCBOA-SVM-FS model will be applied to other problems, such as disease diagnosis and bankruptcy prediction. Of course, it is expected that the proposed RCCBOA can be extended to solve optimization problems in other fields, such as photovoltaic cell parameter identification and image segmentation.

Author Contributions: Funding acquisition, G.L. and H.C.; Writing—original draft, H.B.; Writing—review & editing, Z.C. All authors have read and agreed to the published version of the manuscript.

Funding: This article contains the phased research results of "Research on the Formation and Cultivation Mechanism of Overseas Chinese's Home and Country Feelings from the Perspective of Embodiment Theory (project code: 22JCXK02ZD)", an emerging (intersecting) major project on philosophy and social sciences in Zhejiang Province, and the phased research results of "Research on the Mechanism of Contributions that Overseas Chinese Schools Make to Public Diplomacy", a 2021 Overseas Chinese Characteristic Research Project of Wenzhou University (project code: WDQT21-YB008)".

Data Availability Statement: The data involved in this study are all public data, which can be downloaded through public channels.

Conflicts of Interest: The authors declare no conflict of interest.

Appendix A

Table A1. Summary of the CEC2014 benchmark problem [120].

Type	No.	Function	Optimum Value
Unimodal Functions	1	Rotated High Conditioned Elliptic Function	$f_1\{X_{min}\} = 100$
	2	Rotated Bent Cigar Function	$f_2\{X_{min}\} = 200$
	3	Rotated Discus Function	$f_3\{X_{min}\} = 300$
Multimodal Functions	4	Shifted and Rotated Rosenbrock's Function	$f_4\{X_{min}\} = 400$
	5	Shifted and Rotated Ackley's Function	$f_5\{X_{min}\} = 500$
	6	Shifted and Rotated Weierstrass Function Function	$f_6\{X_{min}\} = 600$
	7	Shifted and Rotated Griewank's Function	$f_7\{X_{min}\} = 700$
	8	Shifted Rastrigin's Function	$f_8\{X_{min}\} = 800$
	9	Shifted and Rotated Rastrigin's Function	$f_9\{X_{min}\} = 900$
	10	Shifted Schwefel's Function	$f_{10}\{X_{min}\} = 1000$
	11	Shifted and Rotated Schwefel's Function	$f_{11}\{X_{min}\} = 1100$
	12	Shifted and Rotated Katsuura Function	$f_{12}\{X_{min}\} = 1200$
	13	Shifted and Rotated Happycat Function	$f_{13}\{X_{min}\} = 1300$
	14	Shifted and Rotated Hgbat Function	$f_{14}\{X_{min}\} = 1400$
	15	Shifted and Rotated Expanded Griewank's Plus Rosenbrock's Function	$f_{15}\{X_{min}\} = 1500$
	16	Shifted and Rotated Expanded Scaffer's Function	$f_{16}\{X_{min}\} = 1600$
Hybrid Functions	17	Hybrid Function 1 (N = 3)	$f_{17}\{X_{min}\} = 1700$
	18	Hybrid Function 2 (N = 3)	$f_{18}\{X_{min}\} = 1800$
	19	Hybrid Function 3 (N = 4)	$f_{19}\{X_{min}\} = 1900$
	20	Hybrid Function 4 (N = 4)	$f_{20}\{X_{min}\} = 2000$
	21	Hybrid Function 5 (N = 5)	$f_{21}\{X_{min}\} = 2100$
	22	Hybrid Function 6 (N = 5)	$f_{22}\{X_{min}\} = 2200$
Composition Functions	23	Composition Function 1 (N = 5)	$f_{23}\{X_{min}\} = 2300$
	24	Composition Function 2 (N = 3)	$f_{24}\{X_{min}\} = 2400$
	25	Composition Function 3 (N = 3)	$f_{25}\{X_{min}\} = 2500$
	26	Composition Function 4 (N = 5)	$f_{26}\{X_{min}\} = 2600$
	27	Composition Function 5 (N = 5)	$f_{27}\{X_{min}\} = 2700$
	28	Composition Function 6 (N = 5)	$f_{28}\{X_{min}\} = 2800$
	29	Composition Function 7 (N = 3)	$f_{29}\{X_{min}\} = 2900$
	30	Composition Function 8 (N = 3)	$f_{30}\{X_{min}\} = 3000$

References

1. Li, M. Transnational Links among the Chinese in Europe: A Study on European-wide Chinese Voluntary Associations. In *The Chinese in Europe*; Palgrave Macmillan: London, UK, 1998.
2. Sheng, F.; Smith, G. The Shifting Fate of China's Pacific Diaspora. 2021: The China Alternative: Changing Regional Order in the Pacific Islands. *China Altern.* **2021**, *1*, 142.

3. Freedman, M. Immigrants and Associations: Chinese in nineteenth-century Singapore. *Comp. Stud. Soc. Hist.* **1960**, *3*, 25–48. [CrossRef]
4. Ma, L.E.A. *Revolutionaries, Monarchists, and Chinatowns Chinese Politics in the Americas and the 1911 Revolution*; University Hawai'i Press: Honolulu, HI, USA, 1990.
5. Litofcenko, J.; Karner, D.; Maier, F. Methods for Classifying Nonprofit Organizations According to their Field of Activity: A Report on Semi-automated Methods Based on Text. *Volunt. Int. J. Volunt. Nonprofit Organ.* **2020**, *31*, 227–237. [CrossRef]
6. Arora, S.; Singh, S. Butterfly optimization algorithm: A novel approach for global optimization. *Soft Comput.* **2019**, *23*, 715–734. [CrossRef]
7. Wolpert, D.H.; Macready, W.G. No free lunch theorems for optimization. *IEEE Trans. Evol. Comput.* **1997**, *1*, 67–82. [CrossRef]
8. Long, W.; Wu, T.; Xu, M.; Tang, M.; Cai, S. Parameters identification of photovoltaic models by using an enhanced adaptive butterfly optimization algorithm. *Energy* **2021**, *229*, 120750. [CrossRef]
9. Sharma, T.K.; Sahoo, A.K.; Goyal, P. Bidirectional butterfly optimization algorithm and engineering applications. *Mater. Today Proc.* **2021**, *34*, 736–741. [CrossRef]
10. Mortazavi, A.; Moloodpoor, M. Enhanced Butterfly Optimization Algorithm with a New fuzzy Regulator Strategy and Virtual Butterfly Concept. *Knowl. Based Syst.* **2021**, *228*, 107291. [CrossRef]
11. Sundaravadivel, T.; Mahalakshmi, V. Weighted butterfly optimization algorithm with intuitionistic fuzzy Gaussian function based adaptive-neuro fuzzy inference system for COVID-19 prediction. *Mater. Today Proc.* **2021**, *42*, 1498–1501. [CrossRef]
12. Zhou, H.; Zhang, G.; Wang, X.; Ni, P.; Zhang, J. Structural identification using improved butterfly optimization algorithm with adaptive sampling test and search space reduction method. *Structures* **2021**, *33*, 2121–2139. [CrossRef]
13. Thawkar, S.; Sharma, S.; Khanna, M.; Singh, L.K. Breast cancer prediction using a hybrid method based on Butterfly Optimization Algorithm and Ant Lion Optimizer. *Comput. Biol. Med.* **2021**, *139*, 104968. [CrossRef]
14. Long, W.; Jiao, J.; Liang, X.; Wu, T.; Xu, M.; Cai, S. Pinhole-imaging-based learning butterfly optimization algorithm for global optimization and feature selection. *Appl. Soft Comput.* **2021**, *103*, 107146. [CrossRef]
15. Sowjanya, K.; Injeti, S.K. Investigation of butterfly optimization and gases Brownian motion optimization algorithms for optimal multilevel image thresholding. *Expert Syst. Appl.* **2021**, *182*, 115286. [CrossRef]
16. Hu, J.; Han, Z.; Heidari, A.A.; Shou, Y.; Ye, H.; Wang, L.; Huang, X.; Chen, H.; Chen, Y.; Wu, P. Detection of COVID-19 severity using blood gas analysis parameters and Harris hawks optimized extreme learning machine. *Comput. Biol. Med.* **2021**, *142*, 105166. [CrossRef] [PubMed]
17. Fan, Y.; Wang, P.; Heidari, A.A.; Chen, H.; Turabieh, H.; Mafarja, M. Random reselection particle swarm optimization for optimal design of solar photovoltaic modules. *Energy* **2022**, *239*, 121865. [CrossRef]
18. Shi, B.; Ye, H.; Zheng, L.; Lyu, J.; Chen, C.; Heidari, A.A.; Hu, Z.; Chen, H.; Wu, P. Evolutionary warning system for COVID-19 severity: Colony predation algorithm enhanced extreme learning machine. *Comput. Biol. Med.* **2021**, *136*, 104698. [CrossRef]
19. Zhou, W.; Wang, P.; Heidari, A.A.; Zhao, X.; Turabieh, H.; Mafarja, M.; Chen, H. Metaphor-free dynamic spherical evolution for parameter estimation of photovoltaic modules. *Energy Rep.* **2021**, *7*, 5175–5202. [CrossRef]
20. Yu, C.; Heidari, A.A.; Xue, X.; Zhang, L.; Chen, H.; Chen, W. Boosting quantum rotation gate embedded slime mould algorithm. *Expert Syst. Appl.* **2021**, *181*, 115082. [CrossRef]
21. Zhou, W.; Wang, P.; Heidari, A.A.; Zhao, X.; Turabieh, H.; Chen, H. Random learning gradient based optimization for efficient design of photovoltaic models. *Energy Convers. Manag.* **2021**, *230*, 113751. [CrossRef]
22. Xu, Y.; Huang, H.; Heidari, A.A.; Gui, W.; Ye, X.; Chen, Y.; Chen, H.; Pan, Z. MFeature: Towards high performance evolutionary tools for feature selection. *Expert Syst. Appl.* **2021**, *186*, 115655. [CrossRef]
23. Liu, L.; Zhao, D.; Yu, F.; Heidari, A.A.; Li, C.; Ouyang, J.; Chen, H.; Mafarja, M.; Turabieh, H.; Pan, J. Ant colony optimization with Cauchy and greedy Levy mutations for multilevel COVID 19 X-ray image segmentation. *Comput. Biol. Med.* **2021**, *136*, 104609. [CrossRef] [PubMed]
24. Zhao, S.; Wang, P.; Heidari, A.A.; Chen, H.; He, W.; Xu, S. Performance optimization of salp swarm algorithm for multi-threshold image segmentation: Comprehensive study of breast cancer microscopy. *Comput. Biol. Med.* **2021**, *139*, 105015. [CrossRef]
25. Yu, H.; Li, W.; Chen, C.; Liang, J.; Gui, W.; Wang, M.; Chen, H. Dynamic Gaussian bare-bones fruit fly optimizers with abandonment mechanism: Method and analysis. *Eng. Comput.* **2020**, 1–29. [CrossRef]
26. Chen, H.; Heidari, A.A.; Chen, H.; Wang, M.; Pan, Z.; Gandomi, A.H. Multi-population differential evolution-assisted Harris hawks optimization: Framework and case studies. *Future Gener. Comput. Syst.* **2020**, *111*, 175–198. [CrossRef]
27. Chen, H.; Heidari, A.A.; Zhao, X.; Zhang, L.; Chen, H. Advanced orthogonal learning-driven multi-swarm sine cosine optimization: Framework and case studies. *Expert Syst. Appl.* **2020**, *144*, 113113. [CrossRef]
28. Tu, J.; Lin, A.; Chen, H.; Li, Y.; Li, C. Predict the Entrepreneurial Intention of Fresh Graduate Students Based on an Adaptive Support Vector Machine Framework. *Math. Probl. Eng.* **2019**, *2019*, 2039872. [CrossRef]
29. Chen, H.; Xu, Y.; Wang, M.; Zhao, X. A balanced whale optimization algorithm for constrained engineering design problems. *Appl. Math. Model.* **2019**, *71*, 45–59. [CrossRef]
30. Yong, J.; He, F.; Li, H.; Zhou, W. A Novel Bat Algorithm based on Collaborative and Dynamic Learning of Opposite Population. In Proceedings of the 2018 IEEE 22nd International Conference on Computer Supported Cooperative Work in Design (CSCWD), Nanjing, China, 9–11 May 2018.
31. Zhou, Y.; Xie, J.; Li, L.; Ma, M. Cloud Model Bat Algorithm. *Sci. World J.* **2014**, *2014*, 237102. [CrossRef]

32. Liang, H.; Liu, Y.; Shen, Y.; Li, F.; Man, Y. A Hybrid Bat Algorithm for Economic Dispatch with Random Wind Power. *IEEE Trans. Power Syst.* **2018**, *33*, 5052–5061. [CrossRef]
33. Sun, Y.; Wang, X.; Chen, Y.; Liu, Z. A modified whale optimization algorithm for large-scale global optimization problems. *Expert Syst. Appl.* **2018**, *114*, 563–577. [CrossRef]
34. Ling, Y.; Zhou, Y.; Luo, Q. Lévy Flight Trajectory-Based Whale Optimization Algorithm for Global Optimization. *IEEE Access* **2017**, *5*, 6168–6186. [CrossRef]
35. Tubishat, M.; Abushariah, M.; Idris, N.; Aljarah, I. Improved whale optimization algorithm for feature selection in Arabic sentiment analysis. *Appl. Intell.* **2019**, *49*, 1688–1707. [CrossRef]
36. Han, X.; Liu, Q.; Wang, H.; Wang, L. Novel fruit fly optimization algorithm with trend search and co-evolution. *Knowl. Based Syst.* **2018**, *141*, 1–17. [CrossRef]
37. Ye, F.; Lou, X.Y.; Sun, L.F. An improved chaotic fruit fly optimization based on a mutation strategy for simultaneous feature selection and parameter optimization for SVM and its applications. *PLoS ONE* **2017**, *12*, e0173516. [CrossRef] [PubMed]
38. Wang, W.; Liu, X. Melt index prediction by least squares support vector machines with an adaptive mutation fruit fly optimization algorithm. *Chemom. Intell. Lab. Syst.* **2015**, *141*, 79–87. [CrossRef]
39. Yang, Y.; Chen, H.; Heidari, A.A.; Gandomi, A.H. Hunger games search: Visions, conception, implementation, deep analysis, perspectives, and towards performance shifts. *Expert Syst. Appl.* **2021**, *177*, 114864. [CrossRef]
40. Heidari, A.A.; Mirjalili, S.; Faris, H.; Aljarah, I.; Mafarja, M.; Chen, H. Harris hawks optimization: Algorithm and applications. *Future Gener. Comput. Syst. Int. J. Escience* **2019**, *97*, 849–872. [CrossRef]
41. Li, S.; Chen, H.; Wang, M.; Heidari, A.A.; Mirjalili, S. Slime mould algorithm: A new method for stochastic optimization. *Future Gener. Comput. Syst.* **2020**, *111*, 300–323. [CrossRef]
42. Ahmadianfar, I.; Heidari, A.A.; Gandomi, A.H.; Chu, X.; Chen, H. RUN beyond the metaphor: An efficient optimization algorithm based on Runge Kutta method. *Expert Syst. Appl.* **2021**, *181*, 115079. [CrossRef]
43. Tu, J.; Chen, H.; Wang, M.; Gandomi, A.H. The Colony Predation Algorithm. *J. Bionic Eng.* **2021**, *18*, 674–710. [CrossRef]
44. Ahmadianfar, I.; Heidari, A.A.; Noshadian, S.; Chen, H.; Gandomi, A.H. INFO: An efficient optimization algorithm based on weighted mean of vectors. *Expert Syst. Appl.* **2022**, *195*, 116516. [CrossRef]
45. Xia, J.; Wang, Z.; Yang, D.; Li, R.; Liang, G.; Chen, H.; Heidari, A.A.; Turabieh, H.; Mafarja, M.; Pan, Z. Performance optimization of support vector machine with oppositional grasshopper optimization for acute appendicitis diagnosis. *Comput. Biol. Med.* **2022**, *143*, 105206. [CrossRef] [PubMed]
46. Xia, J.; Yang, D.; Zhou, H.; Chen, Y.; Zhang, H.; Liu, T.; Heidari, A.A.; Chen, H.; Pan, Z. Evolving kernel extreme learning machine for medical diagnosis via a disperse foraging sine cosine algorithm. *Comput. Biol. Med.* **2021**, *141*, 105137. [CrossRef] [PubMed]
47. Dong, R.; Chen, H.; Heidari, A.A.; Turabieh, H.; Mafarja, M.; Wang, S. Boosted kernel search: Framework, analysis and case studies on the economic emission dispatch problem. *Knowl.-Based Syst.* **2021**, *233*, 107529. [CrossRef]
48. Abbasi, A.; Firouzi, B.; Sendur, P.; Heidari, A.A.; Chen, H.; Tiwari, R. Multi-strategy Gaussian Harris hawks optimization for fatigue life of tapered roller bearings. *Eng. Comput.* **2021**, 1–27. [CrossRef] [PubMed]
49. Nautiyal, B.; Prakash, R.; Vimal, V.; Liang, G.; Chen, H. Improved Salp Swarm Algorithm with mutation schemes for solving global optimization and engineering problems. *Eng. Comput.* **2021**, 1–23. [CrossRef]
50. Zhang, H.; Liu, T.; Ye, X.; Heidari, A.A.; Liang, G.; Chen, H.; Pan, Z. Differential evolution-assisted salp swarm algorithm with chaotic structure for real-world problems. *Eng. Comput.* **2022**, 1–35. [CrossRef]
51. Zhang, Y.; Liu, R.; Heidari, A.A.; Wang, X.; Chen, Y.; Wang, M.; Chen, H. Towards augmented kernel extreme learning models for bankruptcy prediction: Algorithmic behavior and comprehensive analysis. *Neurocomputing* **2021**, *430*, 185–212. [CrossRef]
52. Wu, S.; Mao, P.; Li, R.; Cai, Z.; Heidari, A.A.; Xia, J.; Chen, H.; Mafarja, M.; Turabieh, H.; Chen, X. Evolving fuzzy k-nearest neighbors using an enhanced sine cosine algorithm: Case study of lupus nephritis. *Comput. Biol. Med.* **2021**, *135*, 104582. [CrossRef]
53. Hussien, A.G.; Heidari, A.A.; Ye, X.; Liang, G.; Chen, H.; Pan, Z. Boosting whale optimization with evolution strategy and Gaussian random walks: An image segmentation method. *Eng. Comput.* **2022**, 1–45. [CrossRef]
54. Chen, X.; Huang, H.; Heidari, A.A.; Sun, C.; Lv, Y.; Gui, W.; Liang, G.; Gu, Z.; Chen, H.; Li, C.; et al. An efficient multilevel thresholding image segmentation method based on the slime mould algorithm with bee foraging mechanism: A real case with lupus nephritis images. *Comput. Biol. Med.* **2022**, *142*, 105179. [CrossRef] [PubMed]
55. Yu, H.; Song, J.; Chen, C.; Heidari, A.A.; Liu, J.; Chen, H.; Zaguia, A.; Mafarja, M. Image segmentation of Leaf Spot Diseases on Maize using multi-stage Cauchy-enabled grey wolf algorithm. *Eng. Appl. Artif. Intell.* **2022**, *109*, 104653. [CrossRef]
56. Yu, H.; Cheng, X.; Chen, C.; Heidari, A.A.; Liu, J.; Cai, Z.; Chen, H. Apple leaf disease recognition method with improved residual network. *Multimed. Tools Appl.* **2022**, *81*, 7759–7782. [CrossRef]
57. Hu, J.; Chen, H.; Heidari, A.A.; Wang, M.; Zhang, X.; Chen, Y.; Pan, Z. Orthogonal learning covariance matrix for defects of grey wolf optimizer: Insights, balance, diversity, and feature selection. *Knowl. Based Syst.* **2021**, *213*, 106684. [CrossRef]
58. Hu, J.; Gui, W.; Heidari, A.A.; Cai, Z.; Liang, G.; Chen, H.; Pan, Z. Dispersed foraging slime mould algorithm: Continuous and binary variants for global optimization and wrapper-based feature selection. *Knowl. Based Syst.* **2022**, *237*, 107761. [CrossRef]
59. Cai, Z.; Gu, J.; Luo, J.; Zhang, Q.; Chen, H.; Pan, Z.; Li, Y.; Li, C. Evolving an optimal kernel extreme learning machine by using an enhanced grey wolf optimization strategy. *Expert Syst. Appl.* **2019**, *138*, 112814. [CrossRef]

60. Xu, Y.; Chen, H.; Heidari, A.A.; Luo, J.; Zhang, Q.; Zhao, X.; Li, C. An efficient chaotic mutative moth-flame-inspired optimizer for global optimization tasks. *Expert Syst. Appl.* **2019**, *129*, 135–155. [CrossRef]
61. Wei, Y.; Lv, H.; Chen, M.; Wang, M.; Heidari, A.A.; Chen, H.; Li, C. Predicting Entrepreneurial Intention of Students: An Extreme Learning Machine with Gaussian Barebone Harris Hawks Optimizer. *IEEE Access* **2020**, *8*, 76841–76855. [CrossRef]
62. Wei, Y.; Ni, N.; Liu, D.; Chen, H.; Wang, M.; Li, Q.; Cui, X.; Ye, H. An Improved Grey Wolf Optimization Strategy Enhanced SVM and Its Application in Predicting the Second Major. *Math. Probl. Eng.* **2017**, *2017*, 9316713. [CrossRef]
63. Zeng, G.-Q.; Lu, K.; Dai, Y.-X.; Zhang, Z.; Chen, M.-R.; Zheng, C.-W.; Wu, D.; Peng, W.-W. Binary-coded extremal optimization for the design of PID controllers. *Neurocomputing* **2014**, *138*, 180–188. [CrossRef]
64. Zeng, G.-Q.; Chen, J.; Dai, Y.-X.; Li, L.-M.; Zheng, C.-W.; Chen, M.-R. Design of fractional order PID controller for automatic regulator voltage system based on multi-objective extremal optimization. *Neurocomputing* **2015**, *160*, 173–184. [CrossRef]
65. Zhao, X.; Li, D.; Yang, B.; Ma, C.; Zhu, Y.; Chen, H. Feature selection based on improved ant colony optimization for online detection of foreign fiber in cotton. *Appl. Soft Comput.* **2014**, *24*, 585–596. [CrossRef]
66. Zhao, X.; Li, D.; Yang, B.; Chen, H.; Yang, X.; Yu, C.; Liu, S. A two-stage feature selection method with its application. *Comput. Electr. Eng.* **2015**, *47*, 114–125. [CrossRef]
67. Wu, S.-H.; Zhan, Z.-H.; Zhang, J. SAFE: Scale-Adaptive Fitness Evaluation Method for Expensive Optimization Problems. *IEEE Trans. Evol. Comput.* **2021**, *25*, 478–491. [CrossRef]
68. Li, J.-Y.; Zhan, Z.-H.; Wang, C.; Jin, H.; Zhang, J. Boosting Data-Driven Evolutionary Algorithm with Localized Data Generation. *IEEE Trans. Evol. Comput.* **2020**, *24*, 923–937. [CrossRef]
69. Hua, Y.; Liu, Q.; Hao, K.; Jin, Y. A Survey of Evolutionary Algorithms for Multi-Objective Optimization Problems with Irregular Pareto Fronts. *IEEE/CAA J. Autom. Sin.* **2021**, *8*, 303–318. [CrossRef]
70. Liu, X.-F.; Zhan, Z.-H.; Gao, Y.; Zhang, J.; Kwong, S.; Zhang, J. Coevolutionary Particle Swarm Optimization with Bottleneck Objective Learning Strategy for Many-Objective Optimization. *IEEE Trans. Evol. Comput.* **2018**, *23*, 587–602. [CrossRef]
71. Deng, W.; Zhang, X.; Zhou, Y.; Liu, Y.; Zhou, X.; Chen, H.; Zhao, H. An enhanced fast non-dominated solution sorting genetic algorithm for multi-objective problems. *Inf. Sci.* **2021**, *585*, 441–453. [CrossRef]
72. Deng, W.; Liu, H.; Xu, J.; Zhao, H.; Song, Y. An Improved Quantum-Inspired Differential Evolution Algorithm for Deep Belief Network. *IEEE Trans. Instrum. Meas.* **2020**, *69*, 7319–7327. [CrossRef]
73. Zhao, H.; Liu, H.; Xu, J.; Deng, W. Performance Prediction Using High-Order Differential Mathematical Morphology Gradient Spectrum Entropy and Extreme Learning Machine. *IEEE Trans. Instrum. Meas.* **2019**, *69*, 4165–4172. [CrossRef]
74. Deng, W.; Xu, J.; Zhao, H.; Song, Y. A Novel Gate Resource Allocation Method Using Improved PSO-Based QEA. *IEEE Trans. Intell. Transp. Syst.* **2020**, 1–9. [CrossRef]
75. Deng, W.; Xu, J.; Song, Y.; Zhao, H. An Effective Improved Co-evolution Ant Colony Optimization Algorithm with Multi-Strategies and Its Application. *Int. J. Bio-Inspired Comput.* **2020**, *16*, 158–170. [CrossRef]
76. Zhao, F.; Di, S.; Cao, J.; Tang, J. Jonrinaldi A Novel Cooperative Multi-Stage Hyper-Heuristic for Combination Optimization Problems. *Complex Syst. Model. Simul.* **2021**, *1*, 91–108. [CrossRef]
77. Yi, J.-H.; Deb, S.; Dong, J.; Alavi, A.H.; Wang, G.-G. An improved NSGA-III algorithm with adaptive mutation operator for Big Data optimization problems. *Future Gener. Comput. Syst.* **2018**, *88*, 571–585. [CrossRef]
78. Liu, P.; Gao, H. A Novel Green Supplier Selection Method Based on the Interval Type-2 Fuzzy Prioritized Choquet Bonferroni Means. *IEEE/CAA J. Autom. Sin.* **2020**, *8*, 1549–1566. [CrossRef]
79. Han, X.; Han, Y.; Chen, Q.; Li, J.; Sang, H.; Liu, Y.; Pan, Q.; Nojima, Y. Distributed Flow Shop Scheduling with Sequence-Dependent Setup Times Using an Improved Iterated Greedy Algorithm. *Complex Syst. Model. Simul.* **2021**, *1*, 198–217. [CrossRef]
80. Gao, D.; Wang, G.-G.; Pedrycz, W. Solving Fuzzy Job-Shop Scheduling Problem Using DE Algorithm Improved by a Selection Mechanism. *IEEE Trans. Fuzzy Syst.* **2020**, *28*, 3265–3275. [CrossRef]
81. Yu, H.; Yuan, K.; Li, W.; Zhao, N.; Chen, W.; Huang, C.; Chen, H.; Wang, M. Improved Butterfly Optimizer-Configured Extreme Learning Machine for Fault Diagnosis. *Complexity* **2021**, *2021*, 6315010. [CrossRef]
82. Liu, G.; Jia, W.; Luo, Y.; Wang, M.; Heidari, A.A.; Ouyang, J.; Chen, H.; Chen, M. Prediction Optimization of Cervical Hyperextension Injury: Kernel Extreme Learning Machines with Orthogonal Learning Butterfly Optimizer and Broyden-Fletcher-Goldfarb-Shanno Algorithms. *IEEE Access* **2020**, *8*, 119911–119930. [CrossRef]
83. Ren, H.; Li, J.; Chen, H.; Li, C. Stability of salp swarm algorithm with random replacement and double adaptive weighting. *Appl. Math. Model.* **2021**, *95*, 503–523. [CrossRef]
84. Chen, H.; Yang, C.; Heidari, A.A.; Zhao, X. An efficient double adaptive random spare reinforced whale optimization algorithm. *Expert Syst. Appl.* **2019**, *154*, 113018. [CrossRef]
85. Meng, A.-B.; Chen, Y.-C.; Yin, H.; Chen, S.-Z. Crisscross optimization algorithm and its application. *Knowl.-Based Syst.* **2014**, *67*, 218–229. [CrossRef]
86. Su, H.; Zhao, D.; Yu, F.; Heidari, A.A.; Zhang, Y.; Chen, H.; Li, C.; Pan, J.; Quan, S. Horizontal and vertical search artificial bee colony for image segmentation of COVID-19 X-ray images. *Comput. Biol. Med.* **2022**, *142*, 105181. [CrossRef] [PubMed]
87. Zhao, D.; Liu, L.; Yu, F.; Heidari, A.A.; Wang, M.; Oliva, D.; Muhammad, K.; Chen, H. Ant colony optimization with horizontal and vertical crossover search: Fundamental visions for multi-threshold image segmentation. *Expert Syst. Appl.* **2021**, *167*, 114122. [CrossRef]

88. Liu, Y.; Chong, G.; Heidari, A.A.; Chen, H.; Liang, G.; Ye, X.; Cai, Z.; Wang, M. Horizontal and vertical crossover of Harris hawk optimizer with Nelder-Mead simplex for parameter estimation of photovoltaic models. *Energy Convers. Manag.* **2020**, *223*, 113211. [CrossRef]
89. Fu, J.; Zhang, Y.; Wang, Y.; Zhang, H.; Liu, J.; Tang, J.; Yang, Q.; Sun, H.; Qiu, W.; Ma, Y.; et al. Optimization of metabolomic data processing using NOREVA. *Nat. Protoc.* **2022**, *17*, 129–151. [CrossRef]
90. Li, B.; Tang, J.; Yang, Q.; Li, S.; Cui, X.; Li, Y.H.; Chen, Y.Z.; Xue, W.; Li, X.; Zhu, F. NOREVA: Normalization and evaluation of MS-based metabolomics data. *Nucleic Acids Res.* **2017**, *45*, W162–W170. [CrossRef]
91. Wu, Z.; Li, G.; Shen, S.; Lian, X.; Chen, E.; Xu, G. Constructing dummy query sequences to protect location privacy and query privacy in location-based services. *World Wide Web* **2021**, *24*, 25–49. [CrossRef]
92. Wu, Z.; Wang, R.; Li, Q.; Lian, X.; Xu, G.; Chen, E.; Liu, X. A Location Privacy-Preserving System Based on Query Range Cover-Up or Location-Based Services. *IEEE Trans. Veh. Technol.* **2020**, *69*, 5244–5254. [CrossRef]
93. Li, Y.H.; Li, X.X.; Hong, J.J.; Wang, Y.X.; Fu, J.B.; Yang, H.; Yu, C.Y.; Li, F.C.; Hu, J.; Xue, W.; et al. Clinical trials, progression-speed differentiating features and swiftness rule of the innovative targets of first-in-class drugs. *Brief. Bioinform.* **2020**, *21*, 649–662. [CrossRef]
94. Zhu, F.; Li, X.X.; Yang, S.Y.; Chen, Y.Z. Clinical Success of Drug Targets Prospectively Predicted by In Silico Study. *Trends Pharmacol. Sci.* **2018**, *39*, 229–231. [CrossRef] [PubMed]
95. Yin, J.; Sun, W.; Li, F.; Hong, J.; Li, X.; Zhou, Y.; Lu, Y.; Liu, M.; Zhang, X.; Chen, N.; et al. VARIDT 1.0: Variability of drug transporter database. *Nucleic Acids Res.* **2020**, *48*, D1042–D1050. [CrossRef] [PubMed]
96. Zhu, F.; Shi, Z.; Qin, C.; Tao, L.; Liu, X.; Xu, F.; Zhang, L.; Song, Y.; Zhang, J.; Han, B.; et al. Therapeutic target database update 2012: A resource for facilitating target-oriented drug discovery. *Nucleic Acids Res.* **2012**, *40*, D1128–D1136. [CrossRef]
97. Wu, Z.; Li, R.; Zhou, Z.; Guo, J.; Jiang, J.; Su, X. A user sensitive subject protection approach for book search service. *J. Assoc. Inf. Sci. Technol.* **2020**, *71*, 183–195. [CrossRef]
98. Wu, Z.; Shen, S.; Lian, X.; Su, X.; Chen, E. A dummy-based user privacy protection approach for text information retrieval. *Knowl. Based Syst.* **2020**, *195*, 105679. [CrossRef]
99. Wu, Z.; Shen, S.; Zhou, H.; Li, H.; Lu, C.; Zou, D. An effective approach for the protection of user commodity viewing privacy in e-commerce website. *Knowl. Based Syst.* **2021**, *220*, 106952. [CrossRef]
100. Yu, H.; Zhao, N.; Wang, P.; Chen, H.; Li, C. Chaos-enhanced synchronized bat optimizer. *Appl. Math. Model.* **2020**, *77*, 1201–1215. [CrossRef]
101. Gupta, S.; Deep, K.; Heidari, A.A.; Moayedi, H.; Chen, H. Harmonized salp chain-built optimization. *Eng. Comput.* **2021**, *37*, 1049–1079. [CrossRef]
102. Zhang, H.; Cai, Z.; Ye, X.; Wang, M.; Kuang, F.; Chen, H.; Li, C.; Li, Y. A multi-strategy enhanced salp swarm algorithm for global optimization. *Eng. Comput.* **2020**, 1–27. [CrossRef]
103. Chen, H.; Li, S.; Heidari, A.A.; Wang, P.; Li, J.; Yang, Y.; Wang, M.; Huang, C. Efficient multi-population outpost fruit fly-driven optimizers: Framework and advances in support vector machines. *Expert Syst. Appl.* **2020**, *142*, 112999. [CrossRef]
104. Zhang, Q.; Chen, H.; Heidari, A.A.; Zhao, X.; Xu, Y.; Wang, P.; Li, Y.; Li, C. Chaos-Induced and Mutation-Driven Schemes Boosting Salp Chains-Inspired Optimizers. *IEEE Access* **2019**, *7*, 31243–31261. [CrossRef]
105. Qiu, S.; Hao, Z.; Wang, Z.; Liu, L.; Liu, J.; Zhao, H.; Fortino, G. Sensor Combination Selection Strategy for Kayak Cycle Phase Segmentation Based on Body Sensor Networks. *IEEE Internet Things J.* **2021**, *9*, 4190–4201. [CrossRef]
106. Wang, D.; Liang, Y.; Xu, D.; Feng, X.; Guan, R. A content-based recommender system for computer science publications. *Knowl. Based Syst.* **2018**, *157*, 1–9. [CrossRef]
107. Li, J.; Chen, C.; Chen, H.; Tong, C. Towards Context-aware Social Recommendation via Individual Trust. *Knowl. Based Syst.* **2017**, *127*, 58–66. [CrossRef]
108. Li, J.; Lin, J. A probability distribution detection based hybrid ensemble QoS prediction approach. *Inf. Sci.* **2020**, *519*, 289–305. [CrossRef]
109. Li, J.; Zheng, X.-L.; Chen, S.-T.; Song, W.-W.; Chen, D.-R. An efficient and reliable approach for quality-of-service-aware service composition. *Inf. Sci.* **2014**, *269*, 238–254. [CrossRef]
110. Guan, R.; Zhang, H.; Liang, Y.; Giunchiglia, F.; Huang, L.; Feng, X. Deep Feature-Based Text Clustering and Its Explanation. *IEEE Trans. Knowl. Data Eng.* **2020**, *99*, 1. [CrossRef]
111. Qiu, S.; Zhao, H.; Jiang, N.; Wu, D.; Song, G.; Zhao, H.; Wang, Z. Sensor network oriented human motion capture via wearable intelligent system. *Int. J. Intell. Syst.* **2021**, *37*, 1646–1673. [CrossRef]
112. Cao, X.; Cao, T.; Gao, F.; Guan, X. Risk-Averse Storage Planning for Improving RES Hosting Capacity Under Uncertain Siting Choices. *IEEE Trans. Sustain. Energy* **2021**, *12*, 1984–1995. [CrossRef]
113. Ran, X.; Zhou, X.; Lei, M.; Tepsan, W.; Deng, W. A Novel K-Means Clustering Algorithm with a Noise Algorithm for Capturing Urban Hotspots. *Appl. Sci.* **2021**, *11*, 11202. [CrossRef]
114. Cao, X.; Wang, J.; Wang, J.; Zeng, B. A Risk-Averse Conic Model for Networked Microgrids Planning With Reconfiguration and Reorganizations. *IEEE Trans. Smart Grid* **2020**, *11*, 696–709. [CrossRef]
115. Pei, H.; Yang, B.; Liu, J.; Chang, K. Active Surveillance via Group Sparse Bayesian Learning. *IEEE Trans. Pattern Anal. Mach. Intell.* **2022**, *44*, 1133–1148. [CrossRef] [PubMed]

116. Zhu, X.; Guo, K.; Fang, H.; Chen, L.; Ren, S.; Hu, B. Cross View Capture for Stereo Image Super-Resolution. *IEEE Trans. Multimed.* **2021**, *99*, 1. [CrossRef]
117. Zhu, X.; Guo, K.; Ren, S.; Hu, B.; Hu, M.; Fang, H. Lightweight Image Super-Resolution with Expectation-Maximization Attention Mechanism. *IEEE Trans. Circuits Syst. Video Technol.* **2021**, *32*, 1273–1284. [CrossRef]
118. Guo, K.; Hu, B.; Ma, J.; Ren, S.; Tao, Z.; Zhang, J. Toward Anomaly Behavior Detection as an Edge Network Service Using a Dual-Task Interactive Guided Neural Network. *IEEE Internet Things J.* **2020**, *8*, 12623–12637. [CrossRef]
119. Zhang, Z.-H.; Min, F.; Chen, G.-S.; Shen, S.-P.; Wen, Z.-C.; Zhou, X.-B. Tri-Partition State Alphabet-Based Sequential Pattern for Multivariate Time Series. *Cogn. Comput.* **2021**, 1–19. [CrossRef]
120. Liang, J.J.; Qu, B.Y.; Suganthan, P.N. *Problem Definitions and Evaluation Criteria for the CEC 2014 Special Session and Competition on Single Objective Real-Parameter Numerical Optimization*; Technical Report for Computational Intelligence Laboratory, Zhengzhou University: Zhengzhou, China; Nanyang Technological University: Singapore, 2013.

electronics

MDPI

Article

Improved Multi-Strategy Matrix Particle Swarm Optimization for DNA Sequence Design

Wenyu Zhang [1], Donglin Zhu [2], Zuwei Huang [3] and Changjun Zhou [2,*]

1 Faculty of Information Science and Engineering, Ocean University of China, Qingdao 266100, China
2 College of Mathematics and Computer Science, Zhejiang Normal University, Jinhua 321004, China
3 School of Information Engineering, Jiangxi University of Science and Technology, Ganzhou 341000, China
* Correspondence: zhouchangjun@zjnu.edu.cn

Abstract: The efficiency of DNA computation is closely related to the design of DNA coding sequences. For the purpose of obtaining superior DNA coding sequences, it is necessary to choose suitable DNA constraints to prevent potential conflicting interactions in different DNA sequences and to ensure the reliability of DNA sequences. An improved matrix particle swarm optimization algorithm, referred to as IMPSO, is proposed in this paper to optimize DNA sequence design. In addition, this paper incorporates centroid opposition-based learning to fully preserve population diversity and develops and adapts a dynamic update on the basis of signal-to-noise ratio distance to search for high-quality solutions in a sufficiently intelligent manner. The results show that the proposal of this paper achieves satisfactory results and can obtain higher computational efficiency.

Keywords: DNA computing; DNA sequences design; improved matrix particle swarm optimization algorithm (IMPSO); opposition-based learning; signal-to-noise ratio distance

Citation: Zhang, W.; Zhu, D.; Huang, Z.; Zhou, C. Improved Multi-Strategy Matrix Particle Swarm Optimization for DNA Sequence Design. *Electronics* **2023**, *12*, 547. https://doi.org/10.3390/electronics12030547

Academic Editor: Janos Botzheim

Received: 24 December 2022
Revised: 16 January 2023
Accepted: 18 January 2023
Published: 20 January 2023

1. Introduction

DNA is a macromolecular polymer composed of deoxyribonucleotides, which are composed of deoxyribose, phosphate and bases including adenine (A), guanine (G), thymine (T) and cytosine (C). In 1953, after experimentational analysis, Watson and Crick proposed a molecular model of the double-helix structure of DNA [1] and first proposed the principle of base complementary pairing, in which the bases of the nucleotide residues in a nucleic acid molecule are linked to each other by hydrogen bonds in the correspondence between A and T and G and C. That is to say four possible base pairs for the $A = T$, $T = A$, $G \equiv C$ and $C \equiv G$. A and T form two hydrogen bonds between; G and C constitute the three hydrogen bonds between. In 1994, Turing Award-winner Adleman [2] proposed a simple problem computation using the principle of the base complementary pairing of DNA, thus inaugurating DNA computing. DNA computing then continued to evolve toward generalization. In 2006, Winfree [3] proposed the DNA strand replacement reaction, which was a new way to construct logic circuits. In addition to circuit computing, DNA computing can be combined with a variety of intelligent computing methods, such as neural network chaotic systems, and used in different fields.

According to the biological composition of DNA, DNA can be considered a long string of four symbols, they are A, G, C and T. Through the alphabet of $\sum = \{A, G, C, T\}$, two binary numbers or one quadratic number can be used to encode DNA to store information. In 2012, Church [4] led the first team to store a book of 659 kb in DNA, demonstrating the storage capacity of DNA. In 2016, Extance [5] showed that 1 g of DNA can hold the contents of 100 billion DVDs and that 1 kg of DNA can even hold all the information data in the world. In the same year, Zhirnov et al. [6] found that DNA information storage density is 10 million terabytes per cubic centimeter and that even simple E. coli have a storage density of about 1019 bits per cubic centimeter, further validating the powerful storage capacity of DNA. In addition, due to the inherent parallel mechanism of DNA, i.e., the phenomenon that the leading strand and the trailing strand are replicated simultaneously, DNA computation can be performed simultaneously on many DNA strands, which greatly enhances the speed of DNA computation.

DNA coding sequence design is a key step in DNA computation, which realizes the computation and transformation of data stored in it through specific reactions between DNA molecules. The rationality of DNA coding is directly related to whether the model can be successfully validated by biochemical experimentations and the accuracy of DNA computation. However, DNA encoding needs to satisfy molecular biology constraints, including physical constraints such as GC content constraints and thermodynamic constraints such as melting temperature (Tm).

Efficient DNA computation cannot be carried out without excellent DNA coding. Optimal DNA coding can be obtained by optimal coding algorithms, but the cost required for optimal coding may not be satisfied in a large problem space. Therefore, in order to provide efficient and suitable DNA coding in acceptable computational time and space, heuristic algorithms are widely applied to the design of DNA sequences in recent years as a shortcut algorithm. Zhu et al. [7] proposed an IBPSO algorithm to solve the DNA sequence design problem, as well as further improving the quality of DNA sequences. Chaves-González et al. [8] fused artificial bee colony algorithms to propose a new evolutionary approach to create a DNA sequence on the strength of multi-objective swarm intelligence to automatically generate reliable DNA strands that can be applied to molecular computing. Yang et al. [9] improved the spatial dispersion in the traditional IWO algorithm and used the IWO algorithm and the niche crowding in the algorithm to solve the DNA sequence design problem. Zhang et al. [10] used an improved taboo search algorithm for improving the means for the systematic design of equal-length DNA strands, which conduces the discovery of a range of good DNA sequences that satisfy the required certain combinatorial and thermodynamic constraints. Cervantes-Salido et al. [11] proposed a multi-objective evolutionary algorithm for designing a DNA sequence, taking advantage of a matrix-based GA along with specific genetic operators to improve the performance for DNA sequence optimization compared to previous methods. Chaves-González et al. [12] proposed an adapted multi-objective version of the differential evolution (DE) metaheuristics approach incorporating a multi-objective standard fast non-dominated sorting genetic algorithm to produce high-quality DNA sequences. Vega-Rodríguez et al. [13] made several rectifications in the noted fast non-dominated sorting genetic algorithm in conjunction with a novel multi-objective algorithm in accordance with the behavior of fireflies and proposed a new DNA sequence design method based on multi-objective firefly algorithm for generating reliable DNA sequences for molecular computing. The metaheuristic algorithm as a general heuristic algorithm can greatly reduce the number of attempts in a limited searching space, can achieve the problem solution rapidly and is heavily applied to generate reliable DNA coding sequences by virtue of its high efficiency. However, metaheuristic algorithms, as a product of combining random algorithms with local search algorithms, are susceptible to randomness or fall into a local optimum due to premature search and do not necessarily guarantee the feasibility and reliability of the resulting DNA sequences. In recent years, in order to improve the metaheuristic algorithm, which is prone to being caught in a local optimality, many scholars have done a lot of corresponding research and proposed various improved metaheuristic algorithms, among which the particle swarm algorithm is a theoretically mature and widely used emerging metaheuristic algorithm to find the optimal solution through collaboration and information-sharing among individuals in the population.

Particle swarm optimization [14] (PSO) is a method to seek out the global optimum by following the current searched optimum based on the observation of the regular behavior of the flock activity. This algorithm has appealed to the academics with the strong points of easy implementation, high-accuracy and fast convergence and has shown advantages in solving practical problems. However, if the parameters are not chosen reasonably, the particles may miss the optimal solution and subsequently appear to be non-converging.

Even if all particles move in the direction of convergence, homogenization can occur. Due to the loss of the diversity of the population in the search space, premature convergence, poor local search ability, etc., can occur, leading to a lack of further improvement in accuracy as well as falling into a local optimum. In specific problems, the PSO needs to be analyzed and improved in order to achieve better results. Houssein et al. [15] experimentally demonstrated that the PSO algorithm suffers from premature convergence, being trapped in a local optimum and poor performance in multi-objective optimization. Ghatasheh et al. [16] used innovative optimization paradigms to improve the prediction power of bankruptcy modeling to generate prediction models. Zhang et al. [17] proposed a new vector co-evolutionary particle swarm optimization algorithm (VCPSO) to enhance population diversity and avoid premature convergence, but it suffers from falling into local optima or inefficient execution. The multi-objective particle swarm optimization algorithm (MOPSO) proposed by Coello et al. [18] has good search performance but only focuses on the generation of non-dominated vectors and maintaining population diversity, without considering the constraint functions. The region-based selection algorithm (PESA-II) in evolutionary multi-objective optimization proposed by Corne et al. [19] shows outstanding performance in region-based selection multi-objective algorithms but does not deal with runtime complexity. Eberhart et al. [20] used a dynamic neighborhood particle swarm optimization approach to solve multi-objective optimization problems, which is easy to implement and requires few parameters to be tuned but only deals with unconstrained multi-objective optimization problems. Deb et al. [21] developed a fast and elitist multi-objective genetic algorithm (NSGA-II) based on multi-objective evolutionary algorithm (MOEA), which is able to find better solution diffusion and better convergence for most of the problems but NSGA-II algorithm uses the no-penalty parameter constraint processing method, which has some limitations.

In this study, an improved multi-strategy matrix particle swarm-based optimization algorithm, referred to as IMPSO, is proposed. Compared with the previous matrix particle swarm algorithm, the running time under the same conditions is significantly reduced and the values of the constraints on the DNA sequences are well maintained. In addition, centroid opposition-based learning strategy is incorporated to preserve population diversity and to obtain global and sufficient results; at the same time, this strategy is used to reinitialize the population when the iteration numbers is a multiple of 100 to prevent the algorithm falling into the local optimal solution, while a dynamic update in accordance with signal-to-noise ratio distance is developed and adapted to search for high-quality solutions in a sufficiently intelligent manner and enable every individual to search for the best position within its own near neighborhood. The application of these two strategies puts the global optimal solution into effect. What is more, suitable DNA constraints are chosen to avoid potential conflicting interactions between DNA molecules to prevent the generation of secondary structures, to control non-specific hybridization and to ensure the reliability of DNA sequences. To verify the feasibility of the IMPSO algorithm, the DNA sequences, the values of each constraint and their running times obtained from the optimization of IMPSO with MPSO [22], IWO [23], PSO [24] and HS [25] were compared. MPSO continues the search processes by introducing the speed and position update mechanism of the global best particle, effectively ensuring the convergence. IWO is a simple but effective algorithm employed for finding a solution for an engineering problem. PSO is a typical SI that reproduces the new population by learning from personal and global guidance information. HS is a optimization algorithm to solve TSP and a specific academic optimization problem, etc., by mimicking the improvisation of music players. To show the competitiveness of the IMPSO algorithm in solving the DNA sequence design problem, this paper compares the experimentational DNA sequence design results of IMPSO with those of NCIWO, HSWOA [26], MO-ABC, CPSO [27] and DMEA [28]. NCIWO and MO-ABC are mentioned above when introducing particle swarm optimization. HSWOA [26] is used to design DNA sequences that meet the new combination constraint. CPSO [27] is used to solve precocious phenomena and the local optimum of PSO by chaotic mapping.

DMEA [28] is proposed to solve the DNA sequences design and to mitigate an NP-hard problem. With the same number of iterations, the experimental results show that the scheme is more competitive and has higher computational efficiency in solving the DNA sequence design problem. The main contributions of this study are as follows:

(1) The matrix particle swarm optimization is introduced to improve the efficiency of the traditional PSO.
(2) On the basis of the centroid opposition-based learning strategy, the influence of the optimal and worst position is considered to make the position update more reasonable.
(3) The concept of signal-to-noise ratio distance is led into, and a formula conforming to the internal state of the population is designed.
(4) During DNA sequence optimization design experimentation, the rationality and effectiveness of IMPSO are verified by comparing with the variations of various algorithms.

The rest of the paper is arranged in the following way. Section 2 presents the constraints associated with designing DNA coding sequences. Section 3 describes the strategy along with the algorithm flow of the IMPSO. Section 4 introduces the comparison and analysis of the IMPSO algorithm with other optimization algorithms for DNA sequence design. Section 5 outlines the conclusions of this paper and indicates the next steps.

2. Constraints Formulation for DNA Sequence Design

Reliable DNA sequence design is a two-dimensional discrete optimization problem, and the relevant constraints can be partitioned into two categories, one is the combination of constraints including continuity, hairpin, H-measure and similarity, aiming to improve the specificity of DNA molecule recognition, and the other is thermodynamic constraints, mainly including melting temperature (Tm) and free energy, aiming to ensure the consistency of the physicochemical properties of DNA molecules.

This section describes in detail the constraints associated with designing DNA sequences. In the following constraint equation, S stands for the DNA sequence set; u and v, respectively, represent two certain DNA sequences selected from the DNA sequence set S; α is the DNA sequences number contained in DNA sequence set S, and β is the number of bases contained in a given DNA sequence in S. $T(a,\ T_{value})$ is a threshold function that returns a when the value is $a > T_{value}$, and 0 otherwise. If u and v are complementary, the function $cd(u,\ v)$ returns 1; otherwise, the result of the equation is 0.

2.1. Continuity

Continuity is the amount of contiguous identical bases (A,C,G,T) in a given single strand of DNA. Too large a continuity value in the DNA sequence makes the DNA sequence easily twisted and folded in the hybridization process, thus creating a secondary structure that is not conducive to DNA computation. Assuming the continuity threshold is 3, for the DNA sequence CAATGCGTTAGCCCCGATCTTAC, it reaches the continuity threshold, after which the sequence will use the continuity function to calculate its continuity value, and other sequences that do not trigger the threshold will be considered discontinuous. The formula to calculate the continuity of a certain DNA strand is as shown below [12].

$$f_{continuity}(S) = \sum_{\rho=1}^{\alpha} Continuity(S_\rho) \tag{1}$$

$$Continuity(u) = \sum_{i=1}^{\beta-CT} T(cont_\sigma(u,i), CT) \tag{2}$$

$$cont_\sigma(u,i) = \begin{cases} \theta, & if\ \exists\theta\ s.t.u_i \neq \sigma, u_{i+\theta+1} \neq \sigma, u_{i+j} = \sigma\ for\ 1 < j \leq \theta \\ 0, & \text{otherwise} \end{cases} \tag{3}$$

$\sigma \in \{A, G, C, T\}$; CT is the threshold value; $T(A, CY)$ is a count of the number of contiguous bases in DNA above a threshold; if $A > CY$, then return A; otherwise, return 0. $cont_\sigma(u, i)$ returns the number of consecutive bases of sequence u.

2.2. Hairpin

During the process of DNA sequence self-hybridization, the overlapping part of the sequence will fold and the corresponding bases will complementarily pair, and the pairing forms a secondary structure called a hairpin structure. The hairpin structure consists of a hair stem and a hair loop. If the hairpin structure is present in the DNA sequence, it will undergo self-folding in the biochemical reaction. For avoiding self-hybridization in DNA sequences, making the hairpin structure in DNA sequences as small as possible is of great importance. There are two types of hairpin structures, hair stem and hair loop. L_{min} is the minimum hair loop length required for the hairpin structure; T_{min} is the minimum hairpin stem length required; l is the length of the hair loop; t is the length of the hair stem, and the formula to calculate a DNA hairpin is as shown below [12].

$$f_{hairpin}(S) = \sum_{\rho=1}^{\alpha} Hairpin(S_\rho) \tag{4}$$

$$Hairpin(u) = \sum_{t=T_{min}}^{(\beta-L_{min})} \sum_{l=L_{min}}^{\beta-2t} \sum_{i=1}^{\beta-2t-l} T\left(\sum_{j=1}^{PL_{til}} cb(u_{i+j}, u_{\beta-j}), \frac{PL_{til}}{2}\right) \tag{5}$$

where $PL_{til} = \min(t+i, \beta-l-i-t)$ represents the maximum number of base pairs possible when $t+i+\frac{l}{2}$ is the center of the hairpin structure. $cb(u,\ v)$ determines whether u and v are complementary; if u and v are complementary, the result is 1; otherwise, the result is 0.

2.3. H-Measure

In DNA sequences, *H-Measure* is adapted to count the Hamming distance, which indicates the number of different bases at the same position of two complementary DNA sequences. The likelihood of hybridization between complementary strands of the same DNA molecule is closely linked to the *H-Measure*, showing a positive correlation. With this constraint, non-specific hybridization between a DNA sequence and its complementary sequences can be controlled. *H-Measure* is calculated by the following formula [12].

$$f_{H-measure}(S) = \sum_{\rho=1}^{\alpha} \sum_{\theta=1, \rho\neq\theta}^{\alpha} H-measure(S_\rho, S_\theta) \tag{6}$$

where S_ρ, S_θ respectively represent two reverse parallel DNA sequences. *H-Measure* calculation consists of two parts: continuous and discontinuous calculations.

$$\begin{aligned} H-measure(u,v) = Max_{g,t}(h_{dis}(u, FShift(v(-)^g v, t)) \\ + h_{cont}(u, FShift(v(-)^g v, t)) \end{aligned} \tag{7}$$

$$h_{dis}(u,v) = T\left(\sum_{i=1}^{\beta} cb(u_i, v_i), DH \times \beta\right) \tag{8}$$

$$h_{cont}(u,v) = \sum_{i=1}^{\beta} T(subcb(u,v,i), CH) \tag{9}$$

$h_{dis}(u,\ v)$ calculates the number of complementary bases in the DNA sequence u, v. $h_{cont}(u, v)$ figures the penalty value of the consecutive base pairing of DNA sequences u and v. $v(-)^g v$ is a sequence formed by splicing two fragments of sequence v with a splice gap of g. *H-Measure* is the maximum value after the summation of the above two functions. $subcb(u,\ v,\ i)$ defines the number of consecutive complementary paired bases of the u, v

sequence to begin with position i. DH is a real number in [0, 1], and CH is a positive integer in [1, N].

2.4. Similarity

In DNA calculations, similarity indicates how close two DNA sequences are to each other in terms of bases at the same position. Similarity takes into account the complementary Hamming distance after shifting in addition to the Hamming distance. The similarity value is the maximum value of the totality of the amount of bases with the same displacement and the amount of consecutive identical bases between sequences u and splicing sequence $v(-)_g v$. The similarity is calculated as follows [12].

$$f_{similarity}(S) = \sum_{\varepsilon=1}^{\alpha} \sum_{\delta=1,\varepsilon\neq\delta}^{\alpha} Similarity(S_\varepsilon, S_\delta) \tag{10}$$

where $S\varepsilon$, $S\delta$ denotes two sequences in the DNA sequence set S. The similarity is calculated in two parts: the similarity of discontinuous sequences and the similarity of the largest continuous common subset.

$$\begin{aligned} Similarity(u,v) = Max_{g,t}(&s_{dis}(u, FShift(v(-)^g v, t)) \\ &+ s_{cont}(u, FShift(v(-)^g v, t))) \end{aligned} \tag{11}$$

$$s_{dis}(u,v) = T(\sum_{i=1}^{\beta} eq(u_i, v_i), DS \times \beta) \tag{12}$$

$$s_{cont}(u,v) = \sum_{i=1}^{\beta} T(subeb(u,v,i), CS) \tag{13}$$

$FShift(v(-)^g v, t)$ denotes the shift of $v(-)^g v$ by t positions, $eq(u,\ v)$ is used to determine whether u and v are equal; equal returns 1; otherwise, the result is 0; DS is a real number in [0, 1], and CS is a positive integer in [1, N]. $subeb(u,\ v,\ i)$ shows the amount of consecutive equal bases from DNA sequence u and v starting from position i. $S_{dis}(u,v)$ calculates the Hamming distance of two DNA strands; $S_{cont}(u,v)$ calculates the sum of the consecutive equal numbers of bases starting from positions 1 to β.

2.5. GC Content [29]

GC content stands for the amount of guanines as well as cytosines in the DNA sequence as a percentage of the overall number of bases. GC content is directly related to the biochemical stability of DNA sequences because $G \equiv C$ base pairs contain three hydrogen bonds and release more heat energy when broken than $A = T$ base pairs containing two hydrogen bonds, so GC content also influences the melting temperature of DNA sequences. For the DNA sequence ACGTCGTTCGTACGC, the GC content is 60% (9/15). The GC content (in percentage form) is calculated by the following formula.

$$GC(u) = 100 \sum_{i=1}^{\beta} \frac{GC(u_i)}{\beta} \tag{14}$$

$$GC(\tau) = \begin{cases} 1, & \tau = G \text{ or } \tau = C \\ 0, & \tau = A \text{ or } \tau = T \end{cases} \tag{15}$$

2.6. Melting Temperature (Tm)

Melting temperature is the temperature required for half of the base pairs of a DNA double-stranded structure to be disrupted into a single-stranded structure. Melting temperature is an important thermodynamic constraint of DNA molecules that influences the reaction efficiency of DNA sequences, and a steady Tm allows for the better control of hybridization reactions between DNA molecules. The $G \equiv C$ base pair contains three hydrogen bonds and releases more thermal energy upon breaking than the $A = T$ base pair containing two hydrogen bonds. Tm is usually calculated in accordance with the nearest-neighbor thermodynamic model [30], with the following relevant equation.

$$f_{T_m}(S) = \frac{\Delta H^{\circ}}{\Delta S^{\circ} + R\ln\left(\frac{[C_T]}{4}\right)} - 273.15 \tag{16}$$

where ΔH° represents the enthalpy change from reactants to products, which is the total enthalpy of adjacent bases; ΔS° represents the entropy change from reactants to products, which is the total entropy of adjacent bases. R represents the gas constant (1.987 cal/kmol), and C_T is the concentration of DNA molecules.

2.7. Fitness Function

The optimization problem of this paper belongs to the minimum optimization problem. The fitness function of the DNA sequence is determined by the constraint function described above and is the minimum of the above constraint functions, expressed by the following formula.

$$\begin{gathered} Minimize\ f_i(x), i \in \{Continuity, Hairpin, H-measure, \\ Similarity\}\ subject\ to\ GC = 50\%,\ Tm \end{gathered} \tag{17}$$

3. Improved Multi-Strategy Matrix Particle Swarm Optimization

3.1. Basic Information of Matrix Particle Swarm

In order to describe the IMPSO algorithm more clearly, this section first introduces information about matrix particle swarm, some important formulas used by the algorithm and the operations between matrices.

3.1.1. Representation Information

Assume there exists a N individuals population to solve the D-dimensional problem. This population is represented by a matrix X of size $N \times D$, defined as follows.

$$X = \begin{pmatrix} x_{11} & \cdots & x_{1D} \\ \vdots & \ddots & \vdots \\ x_{N1} & \cdots & x_{ND} \end{pmatrix} \tag{18}$$

where x_{ij} represents the individual i and dimension j.

To accommodate the matrix-based representation, the upper bound of the variables is represented by a matrix XB of size $1 \times D$, the lower bound of the variables is represented by a matrix XM of size $1 \times D$, and the fitness values of every individual are represented by a matrix Fit of size $N \times 1$. The matrix $Ones$ is an all-1 matrix, and the matrix R is a matrix consisting of random numbers of [0, 1].

3.1.2. Common Matrix Operations

Table 1 lists the relevant matrix operations used in this paper and shows their corresponding descriptions. For convenience of description, the size of matrices A and B defaults to $N \times D$ if not specifically mentioned.

3.1.3. Initialization of Particle Swarm Related Variables

Matrix X, also called the population matrix, represents the position of individuals. Matrix V represents the velocity, and $pBest$ represents the personal best positions of all the individuals in the population, respectively. Where X is initialized as follows.

$$X_{N\times D} = Ones_{N\times 1} \times (XB - XM) \circ R_{N\times D} + Ones_{N\times 1} \times XM \tag{19}$$

Table 1. Typical operations in matrix and their notations [22].

Name	Description
Addition operation $(+)$	$A+B=\begin{pmatrix} a_{11}+b_{11} & \cdots & a_{1D}+b_{1D} \\ \vdots & \ddots & \vdots \\ a_{N1}+b_{N1} & \cdots & a_{ND}+b_{ND} \end{pmatrix}$
Subtraction operation $(-)$	$A-B=\begin{pmatrix} a_{11}-b_{11} & \cdots & a_{1D}-b_{1D} \\ \vdots & \ddots & \vdots \\ a_{N1}-b_{N1} & \cdots & a_{ND}-b_{ND} \end{pmatrix}$
Multiplication operation $(\times)$	$A_{N\times D}\times B_{D\times N}=\begin{pmatrix} \sum_{i=1}^{D} a_{1i}\times b_{i1} & \cdots & \sum_{i=1}^{D} a_{1i}\times b_{iN} \\ \vdots & \ddots & \vdots \\ \sum_{i=1}^{D} a_{Ni}\times b_{i1} & \cdots & \sum_{i=1}^{D} a_{Ni}\times b_{iN} \end{pmatrix}$
Scalar multiplication $(\cdot)$	$c\cdot A=\begin{pmatrix} c\times a_{11} & \cdots & c\times a_{1D} \\ \vdots & \ddots & \vdots \\ c\times a_{N1} & \cdots & c\times a_{ND} \end{pmatrix}$
Hadamard product $(\circ)$	$A\circ B=\begin{pmatrix} a_{11}\times b_{11} & \cdots & a_{1D}\times b_{1D} \\ \vdots & \ddots & \vdots \\ a_{N1}\times b_{N1} & \cdots & a_{ND}\times b_{ND} \end{pmatrix}$
Transposition operation (X^T)	$A^T=\begin{pmatrix} a_{11} & \cdots & a_{D1} \\ \vdots & \ddots & \vdots \\ a_{1N} & \cdots & a_{DN} \end{pmatrix}$
Logical operation $(\leq)$	$A\leq B=C,\ c_{i,j}=\begin{cases} 1, & if\ a_{i,j}\leq b_{i,j} \\ 0, & otherwise \end{cases}$
Maximum operation (max)	$a=max(A)$, where a is the maximum element in A
Minimum operation (min)	$a=min(A)$, where a is the minimum element in A
Maximum indexing $(maxind)$	$k=maxind(A_{N\times 1})$, where k is the row index of the maximum element in $A_{N\times 1}$
Minimum indexing $(minind)$	$k=minind(A_{N\times 1})$, where k is the row index of the minimum element in $A_{N\times 1}$
Index operation $(X[I\mid J])$	$X[I\mid J]=\begin{pmatrix} X_{I_1J_1} & \cdots & X_{I_1J_j} \\ \vdots & \ddots & \vdots \\ X_{I_iJ_1} & \cdots & X_{I_iJ_j} \end{pmatrix}$

The initialization process of V is as follows.

$$V_{N\times D}=Ones_{N\times 1}\times(VB-VM)\circ R_{N\times D}+Ones_{N\times 1}\times VM \tag{20}$$

After the initialization of matrices X and V is completed, IMPSO obtains the fitness values of all individuals, represented by a matrix Fit of size $N\times 1$, according to the following equation.

$$Fit_{N\times 1}=f(X) \tag{21}$$

The initialization process of $pBest$ is as follows.

$$pBest_{N\times D}=Ones_{N\times 1}\times(XB-XM)\circ R_{N\times D}+Ones_{N\times 1}\times XM \tag{22}$$

The initialization process of *pBest_Fit* is as follows.

$$pBest\_Fit_{N\times D} = Ones_{N\times 1} \times (VB - VM) \circ R_{N\times D} + Ones_{N\times 1} \times VM \tag{23}$$

After completing the above variable initialization process, the globally best fitness value can be obtained by the following formula, represented by *gBest_Fit*.

$$gBest\_Fit = \begin{cases} \min(Fit), & if\ it\ is\ a\ minimum\ problem \\ \max(Fit), & if\ it\ is\ a\ maximum\ problem \end{cases} \tag{24}$$

Furthermore, the optimization problem considered in this experimentation is the minimum value problem; IMPSO can use *minind()* formula in Table 1 to obtain the corresponding number of rows for individuals with the best *pBest* fitness value, as follows.

$$Index = minind\ (pBest\_Fit) \tag{25}$$

3.1.4. Velocity and Position Update

In the process of IMPSO iterations, the population continuously performs velocity update as well as position updates from generation to generation in order to get as close as possible to the global optimum, and the equations for velocity and position updates are shown below.

$$V = \omega \times V + c_1 \times R_1 \circ (pBest - X) + c_2 \times R_2 \circ (Ones \times gBest - X) \tag{26}$$

$$X = X + V \tag{27}$$

It is worth noting that the matrix *gBest* of size $1 \times D$ is actually the individual with the best fitness value in the matrix *pBest* of $N \times D$, which is the index row corresponding to *pBest*. The $N \times D$ matrix X extended from the $1 \times D$ matrix *gBest* can be obtained by the following matrix multiplication formula, which shows that the value of each row of the matrix X is equal to the value of *gBest*.

$$X_{N\times D} = Ones_{N\times 1} \times gBest_{1\times D} \tag{28}$$

In order to avoid the elements of matrices V and X to exceed the space boundary, the boundary should be detected and processed once the matrix V or X is updated. The specific method can be implemented by logical operations and Hadamard products. For a more visual description, IMPSO is illustrated with the matrix X as an example, where *XB* is the upper boundary, and the detection and processing of the upper boundary can be based on the following equation.

$$LOGIC_{N\times D} = X > (Ones \times XB) \tag{29}$$

where the $1 \times D$ matrix *XB* is first expanded into an $N \times D$ matrix with each row equal to *XB*. Further, it is then compared with the $N \times D$ matrix X. If the elements of the matrix X at the corresponding position are greater than the value of the upper boundary, the corresponding element position of the $N \times D$ matrix *LOGIC* is set to 1, and otherwise 0. With reference to this approach, the processing of the upper boundary can be implemented with the following equation.

$$X = LOGIC \circ XB + (1 - LOGIC) \circ X \tag{30}$$

The result of the operation is the element of matrix X that is greater than the upper bound is set to the value of the upper bound. More specifically, the element of the matrix X that is greater than the upper bound is set to 1 at the corresponding position in the matrix *LOGIC*, and thus the element of the matrix X needs to be set to the value of the upper bound. Conversely, if an element of the matrix *LOGIC* is 0, it means that the element in the corresponding position of the matrix X does not exceed the upper bound, then the element of the matrix X in the corresponding position of that element does not need to be changed either. The elements of the matrix X that are smaller than the lower bound also need to be set to the value of the lower bound by a similar operation, which is not repeated here.

The next subsection describes in detail the two strategies used by the IMPSO algorithm to improve the population best fitness value, wherein the signal-to-noise distance is used to further update population best position on top of the basic update population position, and improved centroid opposition-based learning strategy is used to reinitialize population-related variables when the number of iterations is a multiple of 100 to exclude the influence of extreme values on the best fitness value, making the center of gravity of the population more representative.

3.2. Improved Opposition-Based Learning to Reinitialize the Population-Related Parameters

Opposition-based learning is a computational intelligence scheme proposed by Tizhoosh [31] in 2005, which has been successfully applied to a variety of population-based evolutionary algorithms. Traditional learning strategies are essentially based on randomness, and once the worst-case scenario occurs, the search or optimization becomes unmanageable and the results take a lot of time to converge. The main idea of OBL is to consider both the points in the current space and their opposites and to select them meritedly with a view to obtaining results closer to the global optimum. In order to fully explore the current space and to make full use of the favorable information carried by the population as a merit-seeking whole, the COBL centroid opposition-based learning proposed by Rahnamayan et al. [32] was introduced on the basis of OBL.

Theorem 1. *The opposite point.*

Suppose there exists a number x in $[l, u]$, then the opposite point of x is defined as

$$x' = l + u - x \tag{31}$$

Extending the definition of the opposite point to the D-dimension space, let $p = (x_1, x_2, \ldots, x_D)$ be a point in the D-dimension space, where $x_i \in [l_i, u_i]$, $i = 1, 2, \ldots, D$, then its opposite point is defined as

$$p' = (x'_1,\ x'_2, \ldots, x'_D) \tag{32}$$

where $x'_i = l_i + u_i - x_i$.

Theorem 2. *Center of gravity.*

$(X1, \ldots, Xn)$ is a group of n points with unit mass distributed in D-dimension space, and the center of gravity of the group is defined as

$$M = \frac{(X1 + X2 + \ldots + Xn)}{n} \tag{33}$$

It can also be expressed as.

$$\frac{1}{n}\sum_{i=1}^{n} X_{i,j},\quad j = 1, 2, \ldots, D \tag{34}$$

Theorem 3. *Center of gravity of the opposite point.*

If the location of the center of gravity of a discrete uniform whole is M, then the opposite point of a point X_i in the group is defined as

$$X'_i = 2M - X_i,\ i = 1,\ 2,\ \ldots,\ n \tag{35}$$

The opposite point is located in a search space with dynamic boundary, denoted $X_{i,j} \in [a_j, b_j]$. *The dynamic boundary allows the search space to shrink continuously, which is calculated as*

$$a_j = min(X_{i,j}),\ b_j = max(X_{i,j}) \tag{36}$$

where a_j *is the lower boundary of the search space, and* b_j *is the upper boundary of the search space.*

If the opposite point is outside the search boundary, the opposite point can be recalculated according to the following formula.

$$\begin{cases} a_j + rand(0,1) \times (M_j - a_j), & if\ X_{i,j} < a_j \\ M_j + rand(0,1) \times (b_j - M_j), & if\ X_{i,j} > b_j \end{cases} \tag{37}$$

From the above, it is clear that the center-of-gravity position is chosen from the information of the average position of the population. In real life, people calculate the average value by removing the maximum and minimum values, so as to get rid of the influence of extreme values. In this paper, the center-of-gravity position is also calculated by subtracting the optimal position and the worst position to make the center-of-gravity position more representative. Using it for the initialization of the population will produce individuals that will be spread throughout the space, which is well prepared for the subsequent search for the best.

3.3. Signal-to-Noise Ratio Distance for Further Update the Position

In the field of computer artificial intelligence, distance is a frequent and fundamental concept that has important applications in subfields such as natural language processing and computer vision. The concept of distance originates from the concepts of metrics and measurement in the field of mathematics. Distance is used in the computer field to represent the similarity between data; the greater the distance, the greater the degree of difference between the data. Common distance algorithms are Euclidean distance, Mahalanobis distance, Minkowski distance, etc. Among them, Euclidean distance is the most common representation of the distance between two or more points, but as the number of dimensions increases, the computation of the Euclidean distance increases substantially, which greatly increases the time overhead, and the difference between any two points in the space becomes weaker, leading to a uniform distribution of the data [33]. Hassanat et al. [34] uses the Euclidean norms and greedy algorithm to find the furthest pair of points (diameter) of a set of points in d-dimensional Euclidean feature space. On the other hand, the Euclidean distance treats the differences between the various dimensions of points in a space as equivalent, which sometimes does not satisfy the practical requirements. The Mahalanobis distance is a representation of the covariance distance of the data, and the Minkowski distance is a generalization of the Euclidean distance. In other words, the Minkowski distance can be expressed by a generalized formulation of several distance metric formulas, which can be degraded to Manhattan distance or Euclidean distance depending on the parameters, and the Chebyshev distance is the form in which the Minkowski distance takes its limit. Gueorguieva et al. [35] proposed an optimized fuzzy C-means clustering algorithm to improve the FCM clustering results by combining Mahalanobis distances and Minkowski distance metrics. Yang et al. [36] introduced signal-to-noise distance to measure the degree of difference between data, which can produce more discriminative features than the distance metric based on Euclidean distance [37], and the SNR distances of a pair of data p_i and p_j are defined as

$$d_S(p_i, p_j) = \frac{var(p_j - p_i)}{var(p_i)} = \frac{var(h_{ij})}{var(p_i)} \tag{38}$$

where $var(x) = \frac{\sum_{i=1}^{n}(x_i-\mu)^2}{n}$ denotes the variance of x, $\mu = \frac{\sum_{i=1}^{n} xi}{n}$ denotes the mean of x, and n denotes the dimension of x. The larger the SNR distance, the greater the degree of variance between the anchored and compared data.

Therefore, a new update mechanism that uses signal-to-noise ratio distance to determine the distance information between individuals and the optimal position was proposed in this paper. Through this distance, the worst position can be moved away from. The specific design formula is as follows.

$$d = var(x_i(t) - best(t))/var(best(t)) \tag{39}$$

$$x_i(t+1) = x_i(t) + sigmod(d) \cdot (x_i(t) - worst(t)) \tag{40}$$

In the formula, $x_i(t)$ denotes the position of the i-th individual in the tth generation, $best(t)$ denotes the best position in the tth generation, and $worst(t)$ denotes the worst position in the tth generation. It can be seen that d determines the magnitude of individual search; the smaller d is, the smaller the distance of individual $x_i(t)$ away from the worst position. On the contrary, the larger d is, the larger the distance is. By adjusting individual position in this dynamic update, high-quality solutions can be searched adequately. The intelligence of the search is enhanced.

3.4. IMPSO Algorithm

3.4.1. IMPSO Algorithm Process

Input: The size of population *PopSize*, the dimension of the problem *PerLen*, the parameters ω, $c1$, $c2$, maximal generation *max_iterations*.

Step 1. Initialize the matrices X and V according to Equations (19) and (20), control the elements of the matrix X no greater than *XB* and no less than *XM*; the elements of the matrix V no greater than *VB* and no less than *VM*.

Step 2. The fitness value of each individual of the matrix X, represented by the matrix *Fit*, is obtained from the Equation (21) in terms of individuals within the population.

Step 3. Update the best solution in terms of dimensions and select the individual with the best adaptation value for each dimension, i.e., each column, to form a matrix *gBest* of size $PopSize \times 1$.

Step 4. The best fitness value *gBest_Fit* is updated by the element with the best fitness value from the fitness value matrix *Fit*.

Step 5. Update the best position of an individual, specifically by using the matrix X representing the position of the individual to obtain the personal best position matrix *pBest*.

Step 6. Update the matrix *pBest_Fit*, which represents the fitness values of the personal best positions with the matrix *Fit* representing the fitness values of all the individuals in the population.

Step 7. Perform *max_iterations* iterations for the following operations.

Step 8. Update velocity according to Equation (26).

Step 9. Using matrix V as reference, if the element in matrix V is greater than *VB*, set the element in the corresponding position in matrix *LOGIC* to 1; otherwise, set it to 0.

Step 10. Using the matrix *LOGIC*, the elements of the matrix V greater than *VB* are set to *VB*; otherwise, they remain unchanged.

Step 11. Using matrix V as reference, if the element in matrix V is smaller than *VM*, set the element in the corresponding position in matrix *LOGIC* to 1; otherwise, set it to 0.

Step 12. Using the matrix *LOGIC*, the elements of the matrix V smaller than *VM* are set to *VM*; otherwise, they remain unchanged.

Step 13. The personal position matrix X is updated with the matrix X and the latest obtained matrix V according to Equation (27).

Step 14. Using matrix X as reference, if the element in matrix X is greater than *XB*, set the element in the corresponding position in matrix *LOGIC* to 1; otherwise, set it to 0.

Step 15. Using the matrix *LOGIC*, the elements of the matrix X greater than *XB* are set to *XB*; otherwise, they remain unchanged.

Step 16. Using matrix X as reference, if the element in matrix X is smaller than *XM*, set the element in the corresponding position in matrix *LOGIC* to 1; otherwise, set it to 0.

Step 17. Using the matrix *LOGIC*, the elements of the matrix X smaller than *XM* are set to *XM*; otherwise, they remain unchanged.

Step 18. Update the matrix *Fit* representing the fitness values of all the individuals with the latest obtained matrix X according to Equation (21).

Step 19. Update the matrix *pBest* and the matrix *pBest_Fit*. If the matrix *pBest_Fit* is larger than the corresponding value in the matrix *Fit*, the corresponding element in the matrix *LOGIC* is set to 1; otherwise, it is set to 0.

Step 20. If the matrix *pBest_Fit* is smaller than the corresponding value in the matrix *Fit*, it means that the updated personal position matrix is not as good as the previous personal position matrix, so the matrix *pBest* that represents the personal best positions of all the individuals in the population does not need to be updated. Conversely, it means that the latest personal position matrix is better than the previous individual matrix, because the personal best fitness value is optimized, so it needs to be updated to the latest personal position matrix X.

Step 21. The matrix *Fit* corresponds to the personal best fitness values of the population matrix X. The matrix *pBest_Fit* corresponds to the matrix *pBest*, and the best personal fitness values matrix is updated based on the personal best position matrix by comparing the previous equation.

Step 22. Using Equations (38)–(40) to further update the position of the population particles.

Step 23. Individuals with the best fitness values are selected in terms of dimensions, and the corresponding elements are assigned to the matrix *gBest* according to the obtained individuals and dimensions in the matrix *pBest*.

Step 24. The element with the best fitness value is selected in the personal best fitness value matrix *pBest*, which is the best solution fitness value.

Step 25. When the number of iterations is a multiple of 100, the population-related variables are reinitialized using Equations (31)–(37). Exit the loop at the end of the iteration count; otherwise, go back to step8 to continue the iterations.

Output: The found best solution fitness *gBest_Fit*.

The matrix *pBest* represents the best personal positions of all the individuals in the IMPSO population. *pBest_Fit* is a matrix that selects the element with the best fitness value in all dimensions in terms of individuals, with a matrix size of $PopSize \times 1$. *gBest* is a matrix that finds the corresponding row number of the best personal fitness value matrix *pBest_Fit*, i.e., the individual with the best personal fitness value, in terms of dimensions, to achieve the goal of finding the individual with the best fitness value for each dimension, and the matrix size is $1 \times PerLen$. *gBest_Fit* is the matrix with the best fitness value in the personal best fitness value matrix *pBest_Fit*.

3.4.2. Flowchart Based on IMPSO Algorithm to Optimize DNA Sequence

To solve the problem of excessive time consumption and low quality in DNA sequence design optimization problems, this study proposes a multi-strategy matrix particle swarm and introduces an efficient matrix particle swarm to reduce the time consumption of the algorithm, then introduces novel centroid opposition-based learning to initialize the population during the optimization search to avoid the population falling into local states and finally introduces a signal-to-noise ratio to judge the distance between individuals for updates with high quality. The efficiency and reliability of DNA computing are inseparable from the design of the DNA chain. In order to design more excellent DNA sequences, it can be effective to combine the objective function and the constraints of the DNA chain. Before applying the objective function for calculation, the population particles are coded by dividing them by four, so that the matrix particle swarm can be coded with the four bases (A, C, G, T) of DNA. The specific algorithm flowchart is shown as Figure 1.

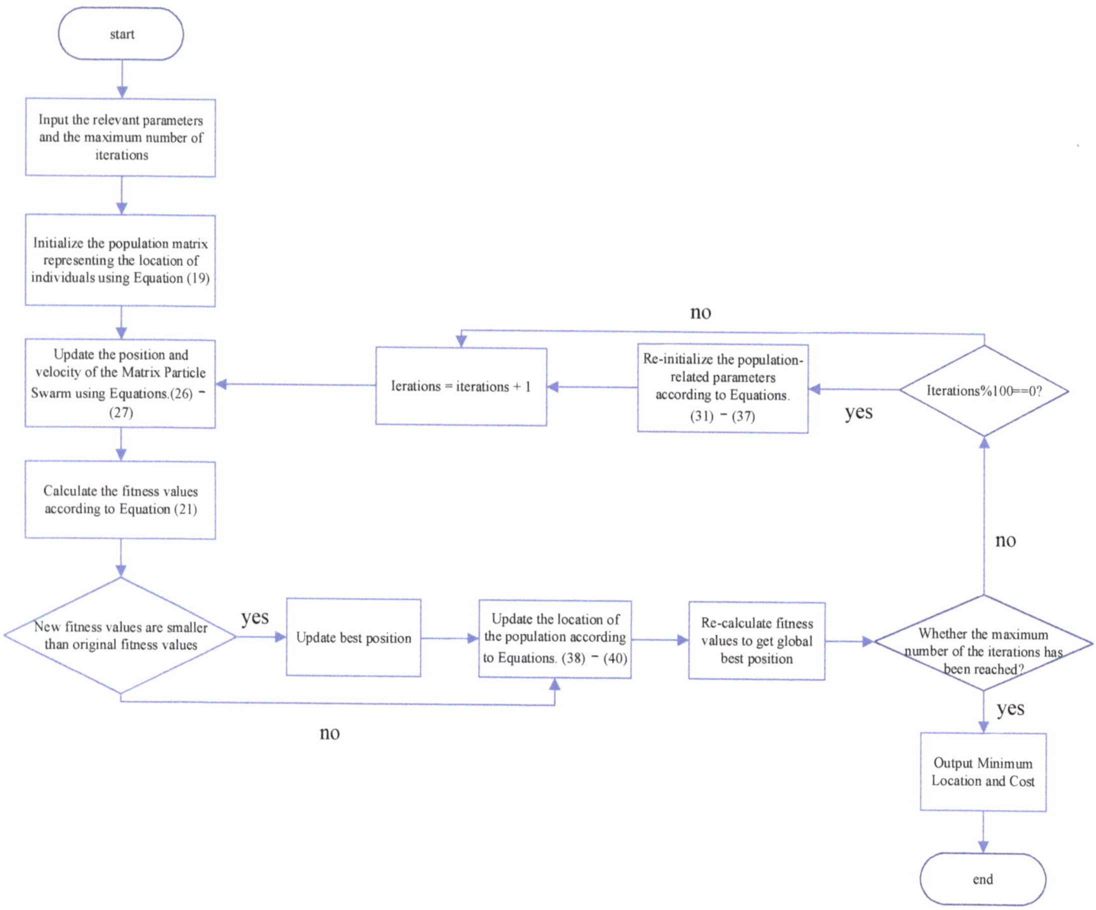

Figure 1. IMPSO algorithm flowchart.

4. Results and Analysis

4.1. Algorithm Parameters

In this section, IMPSO is applied to DNA sequence design experimentation to demonstrate the high efficiency of the IMPSO in solving the DNA coding sequence design problem. All experiments were carried out on a computer with Intel (R) Core (TM) i5-10200H (2.40 Ghz) CPU, 16 GB RAM, 64-bit OS, and MATLAB R2020b simulation platform. In this experiment, the DNA molecule concentration is set to 10 nm, the salt solution concentration in the experimentation is set to 1 mol/L, the minimum values of the hair stem and hair loop were set to 6, and in the experiment on similarity and *H-Measure*, the penalty threshold for base continuity equality is set to 6, and, for discontinuity, it is set to 0.17. The continuity threshold for a single DNA strand is set to 2. The other parameters used in this study are described in Table 2.

4.2. Algorithm Results

4.2.1. Experimentation on the Effectiveness of IMPSO in Solving DNA Coding

To verify the feasibility of the IMPSO algorithm, the DNA sequences, the values of each constraint and their running times obtained from the optimization of IMPSO with MPSO, IWO, PSO and HS were compared. The results in Table 3 show that the IWO, PSO and HS algorithms take a long time to solve the DNA sequence design problem, all above 20,000 s, and IWO even takes more than 35,000 s. The performance of MPSO shows that the running time of the swarm intelligence algorithm based on matrix operations is significantly reduced under the same conditions and that the values of each constraint of the DNA sequence do not become worse. The IMPSO algorithm requires more than two times more time compared to MPSO, which is due to the time required to add the improvement strategy. Although the time consumed increases, all the metrics of the DNA sequences obtained by IMPSO are better than those of MPSO, so the extra time consumption is worthwhile to obtain higher computational efficiency.

Table 2. Related parameters in IMPSO algorithm.

Symbol	Implication	Value
Max_iteration	Maximum number of iterations	3000
PopSize	Size of the population	20
PerLen	Length of the individual	20
XB	Upper bound	3
XM	Lower bound	0
VB	Maximum velocity constraint	3
VM	Minimum velocity constraint	0
ω_{min}	Minimum number of dynamic constant	0.4
ω_{max}	Maximum number of dynamic constant	0.9
$C1_0$	Initial factor for self-learning	2.5
$C1_{min}$	Minimum factor for self-learning	0.5
$C2_0$	Initial factor for social learning	2.5
$C2_{min}$	Minimum factor for social learning	0.5
D	The size of Hamming Distance	11

4.2.2. Experimentations on the Competitiveness of IMPSO in Designing DNA Sequence

For demonstrating the competitiveness of IMPSO to solve DNA sequence design, this paper compares the experimentational DNA sequence design results of IMPSO with those of NCIWO, HSWOA, MO-ABC, CPSO and DMEA by comparing the average values of continuity, hairpin, *H-Measure*, similarity and the variance of Tm to assess sequence quality. Among these metrics, *H-Measure* and similarity are beneficial in preventing DNA strands from mismatching, and hairpin and continuity are beneficial in avoiding secondary structures in DNA strands. To ensure the fairness of the experimentations, parameters in the mentioned algorithm are set in accordance with their relevant references, and population size and iterations numbers were kept consistent.

4.3. *Comparisons and Analysis*

Controlling continuity and hairpin structure in DNA sequences can prevent self-hybridization in DNA molecules to produce secondary structures and to ensure the reliability of DNA calculations. By constraining similarity and *H-Measure*, non-specific hybridization between a DNA sequence and its complementary sequences can be controlled. Melting temperature and free energy are important thermodynamic constraints of DNA molecules, and maintaining their stability is conducive to control the hybridization reaction between DNA molecules and to improve the reaction efficiency of DNA sequences.

4.3.1. Control Secondary Structures

From the results in Table 4 and Figure 2, it can be seen that the continuity and hairpin of IMPSO and HSWOA are 0; however, the continuity or hairpin structures of NCIWO, MO-ABC, CPSO and DMEA exceed 0. This indicates that the DNA sequences created by IMPSO and HSWOA prevent secondary structures with advantage.

Table 3. Comparison of DNA sequences and their constraint values and Cputime.

DNA Sequences (5′-3′)	Continuity	Hairpin	H-Measure	Similarity	Tm	GC%
IWO [23]						
CCAACCTCCGAACCTACATA	0	0	50	57	63.24	50
CAGAACCAGAACAACGCCAA	0	0	52	56	65.76	50
ATTAACCACCTGCCTCTCTG	0	0	54	54	63.85	50
CGATTACACTCCTCACACCA	0	0	51	56	63.78	50
CAGCCAGGTGAAGATAAGAC	0	0	59	53	62.33	50
ACGGTGCTACCTGTTCCTAT	0	0	61	54	65.13	50
AGTATTGCGACGGCCTTCAA	0	0	61	50	66.89	50
Average	0	0	55.43	54.29	64.42	50
Cputime(s)	35,379.59					
PSO [24]						
TACCTCCGTTCTTGCCACTT	0	0	58	49	65.91	50
CGGTGAGAGATGACGATTAG	0	0	60	48	61.85	50
ATAGCGTGACCAGCCAACAA	0	0	63	49	66.88	50
GTTGGATTGCGTACTCTCTG	0	0	61	47	62.92	50
TGTTGGTCAACCTGATGCTG	0	0	64	49	65.25	50
AGTTCTTAGGAGCGTGCAGA	0	0	61	49	65.64	50
CCGCCACACGAATCAATCTA	0	0	63	47	64.81	50
Average	0	0	61.43	48.29	64.75	50
Cputime(s)	20,814.27					
HS [25]						
AGGAGAGACCTGGATTGAGT	0	0	60	51	64.16	50
TGTAGGAAGAGTGTGAACGG	0	0	61	46	63.71	50
GCAACCAACCATTACTCGAC	0	0	57	50	63.78	50
CCTTCCTTCCGCCTTATATC	0	0	64	44	61.9	50
AGGACATGAGAATCACACGG	0	0	60	52	64.15	50
GCAGAGACAATAACAAGCGG	0	0	56	53	63.83	50
GCCAATCAACATCGACACCT	0	0	58	54	65.35	50
Average	0	0	59.43	50	63.84	50
Cputime(s)	21,364.64					
MPSO [22]						
TCCAAGCACACCATACCTCT	0	0	58	50	65.39	50
CGGAGAAGAAGTAGAACTGG	0	0	55	51	61.66	50
GACCACACTCAGGATCCATA	0	0	58	55	62.96	50
GCCAATATAGGCCACAGAGA	0	0	64	50	63.69	50
TCGCGTATCGTTGGTGTCTA	0	0	65	48	65.66	50
TTAACCGAGAATCTCGCAGG	0	0	61	51	64.18	50
ACATGAAGGTGCGGAAGCTT	0	0	61	51	67.18	50
Average	0	0	60.29	50.86	64.39	50
Cputime(s)	6691.05					
IMPSO						
GGAGGTTAGGTTAGTGTTGG	0	0	53	53	61.90	50
CGACAAGAGATGAGAACACC	0	0	54	49	62.57	50
GAGTAGGTGAGATGGTAAGG	0	0	47	55	60.80	50
CAACGAACACGAACCAGTCA	0	0	64	45	65.40	50
GTTGGTGGTTGGTCCTTGTA	0	0	58	47	64.57	50
TATACCTAGAGTGAACGGCG	0	0	61	50	63.04	50
CCGCCATGAGGAAGTGTATA	0	0	59	51	63.66	50
Average	0	0	56.57	50	63.13	50
Cputime(s)	15,008.97					

Table 4. Comparison of DNA sequences and corresponding constraint values.

DNA Sequences (5′-3′)	Continuity	Hairpin	H-Measure	Similarity	Tm	GC%
IMPSO						
GGAGGTTAGGTTAGTGTTGG	0	0	53	53	61.90	50
CGACAAGAGATGAGAACACC	0	0	54	49	62.57	50
GAGTAGGTGAGATGGTAAGG	0	0	47	55	60.80	50
CAACGAACACGAACCAGTCA	0	0	64	45	65.40	50
GTTGGTGGTTGGTCCTTGTA	0	0	58	47	64.57	50
TATACCTAGAGTGAACGGCG	0	0	61	50	63.04	50
CCGCCATGAGGAAGTGTATA	0	0	59	51	63.66	50
Average	0	0	56.57	50	63.13	50
HSWOA [26]						
CTCGTCTAACCTTCTTCAGC	0	0	63	51	62.28	50
CTGTGTGGAATGCAAGGATG	0	0	64	48	63.82	50
CGAGCGTAGTGTAGTCATCA	0	0	63	69	63.56	50
AGTTACAGGACACCACCGAT	0	0	65	51	66.39	50
CAGTAGCAGTCATAACGAGC	0	0	64	56	62.69	50
GCATAGCACATCGTAGCGTA	0	0	59	54	64.60	50
TGGACCTTGAGAGTGGAGAT	0	0	62	50	64.44	50
Average	0	0	62.86	54.14	63.97	50
NCIWO [9]						
ACACCAGCACACAGAAACA	9	0	55	46	66.99	50
GTTCAATCGCCTCTCGGTAT	0	0	57	52	64.26	50
GCTACCTCTTCCACCATTCT	0	0	55	53	63.55	50
GAATCAATGGCGGTCAGAAG	0	0	66	47	63.58	50
TTGGTCCGGTTATTCCTTCG	0	0	65	52	64.44	50
CCATCTTCCGTACTTCACTG	0	0	56	56	62.30	50
TTCGACTCGGTTCCTTGCTA	0	0	58	54	65.61	50
Average	1.29	0	58.86	51.43	64.39	50
MO-ABC [8]						
GTAAGGAAGGCAAGGCAGAA	0	0	42	54	64.70	50
GTTGGTGGTTGTTGGTGGTT	0	0	46	36	66.00	50
GGAGACGGAATGGAAGAGTA	0	0	44	55	62.93	50
CCATTCTTCTCTTCTCTCCC	9	0	67	22	61.39	50
AGGAGAGGAGAGGAGGAAAA	16	0	31	53	63.80	50
ATAAGAGAGAGAGAGAGGGG	16	0	34	51	61.11	50
GAGCCAACAGCCAACCAAAA	16	0	48	45	66.40	50
Average	8.14	0	44.57	45.14	63.76	50
CPSO [27]						
GACCGGTAAGATGAAGAGGA	0	0	60	50	62.94	50
CTATGCTTCTATCGCCTTCC	0	0	61	51	62.23	50
TAGTTGCACGAGAGAAGCAG	0	0	60	51	64.38	50
CGTGTACGAGCCTAATAAGG	0	0	64	54	62.14	50
CTTTGTCCATTGCACATCCG	9	0	61	53	64.42	50
TCCTATCCGAGATGATCCGT	0	3	63	55	64.08	50
TTCAACTTACGCTGTACGGC	0	6	63	54	65.25	50
Average	1.29	1.29	61.71	52.57	63.63	50
DMEA [28]						
TGAGTTGGAACTTGGCGGAA	0	0	70	52	66.76	50
CAGCATGTTAGCCAGTACGA	0	0	60	55	64.65	50
TTGAGTCCGCGTGGTTGGTC	0	0	63	53	69.79	60
AATTGACACTCTGATTCCGC	0	0	73	58	62.89	45
CATACATTGCATCAACGGCG	0	0	67	53	64.84	50
ATACACGCACCTAGCCACAC	0	0	59	50	66.93	55
GTTCCACAACAGGTCTAATG	0	3	61	53	60.65	45
Average	0	0.43	64.71	53.43	65.22	50.71

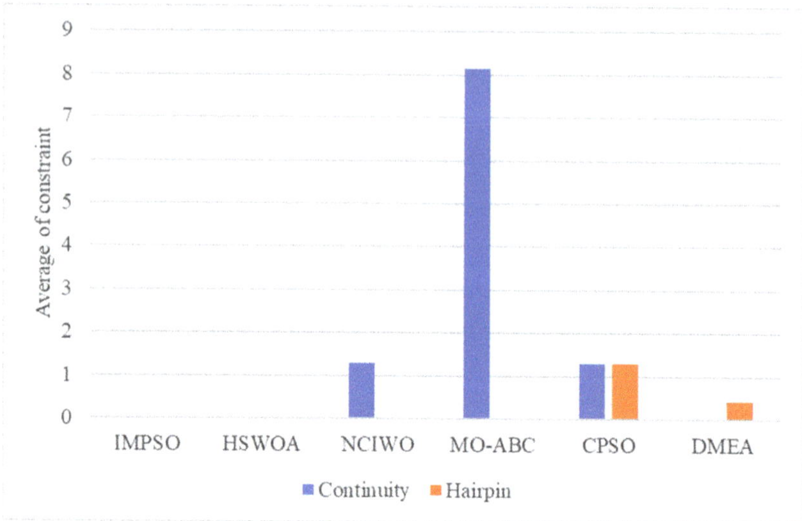

Figure 2. Comparison results among average values of IMPSO, HSWOA, NCIWO, MO-ABC, CPSO, DMEA and IMPSO in continuity and hairpin.

4.3.2. Control Nonspecific Hybridization

From Table 4 and Figure 3, *H-Measure* and similarity values of IMPSO are more desirable than other algorithms, only second to MO-ABC, due to their priority to the constraints set of *H-Measure* and similarity at the expense of continuity and hairpin structure, so the sequences of IMPSO are overall superior to those of MO-ABC.

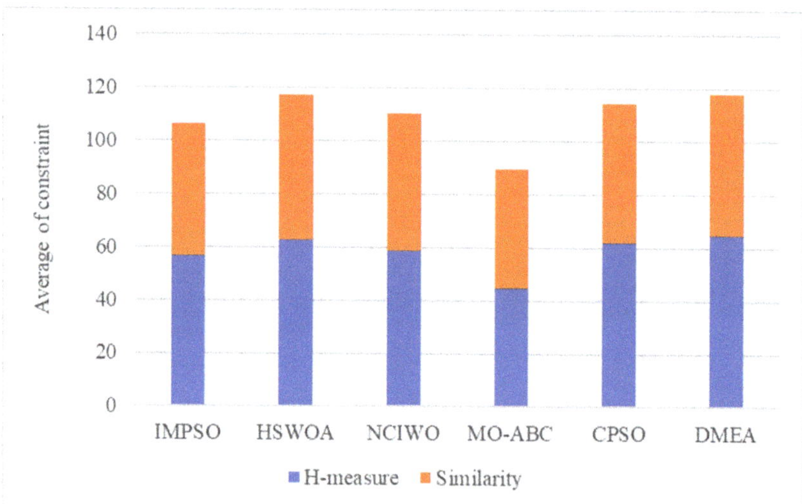

Figure 3. Comparison results among average values of HSWOA, NCIWO, MO-ABC, CPSO, DEMA and IMPSO in *H-Measure* and similarity.

4.3.3. Thermodynamics of Tm

In DNA calculation, DNA sequences need to be as consistent as possible in terms of Tm to dominate biochemical reactions. In this experiment, the variance was used to measure the fluctuation of the Tm of the DNA sequences generated by each algorithm.

From Table 4 and Figure 4, the variance of Tm of IMPSO is superior to MO-ABC and DMEA and slightly inferior to CPSO, HSWOA and NCIWO.

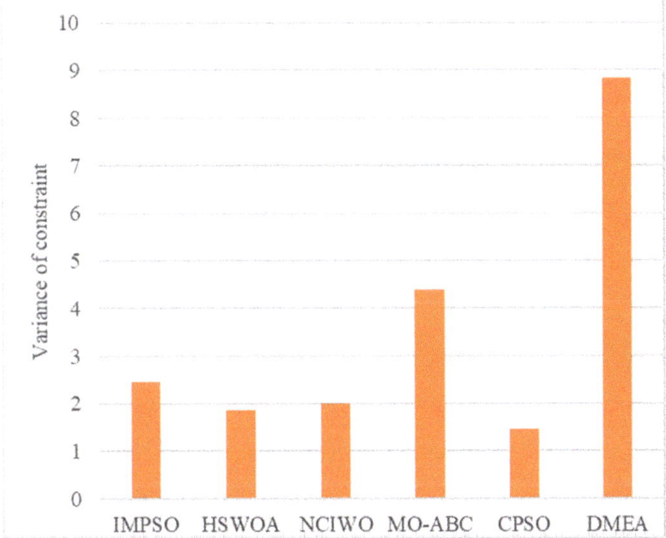

Figure 4. Comparison results among average values of HSWOA, NCIWO, MO-ABC, CPSO, DEMA and IMPSO in Tm variance.

5. Conclusions

To preferably solving the problem of DNA sequence optimization design, an improved multi-strategy matrix particle swarm optimization algorithm is proposed in this paper, which uses an approach in accordance with the signal-to-noise ratio distance to dynamically update the optimal and worst positions of individuals within the population and can adequately search for high-quality solutions. The centroid opposition-based learning strategy is introduced to improve the search range of the algorithm and to exclude the extreme differences brought by the optimal and worst positions when calculating the center-of-gravity positions, so that the center-of-gravity positions are more representative. The individuals generated in the initialization of the population of matrix particles can be spread over the whole space, making full use of the favorable information carried by the population as a whole in the search for the global best, avoiding the premature convergence of the population into a local optimum and fully preparing for the subsequent search for the global optimum. Finally, matrix operations are used to greatly reduce the algorithm running time and to obtain higher computational efficiency without sacrificing the DNA constraint values. Experiments comparing with other particle swarm algorithms confirm that, excluding the MPSO algorithm, the runtime of the swarm intelligence algorithm based on matrix operations is significantly reduced under the same conditions, that various constraint values of DNA sequences do not become worse compared with other algorithms and that the comprehensive capability and reliability of DNA computation are outstanding. The improved multi-strategy matrix particle swarm algorithm (IMPSO) does not underperform in terms of DNA constraint values compared with other DNA sequence design experiments, taking into account the global picture and obtaining optimized sequences of high quality, verifying the effectiveness of the algorithm and meeting the requirements for application to DNA computation. However, the individual capabilities under the combined capability, especially the melting temperature variance, need to be improved. By not sacrificing the DNA constraint values and making full use of the whole population diversity, the CPU running time will also be increased. How to find a breakthrough point to gradually improve the single-item capability without sacrificing any necessary constraint to achieve a more excellent DNA computation capability is also something that needs further consideration in future work.

Author Contributions: Data curation, W.Z.; formal analysis, W.Z. and Z.H.; funding acquisition, C.Z.; software, W.Z. and D.Z.; supervision, D.Z.; validation, C.Z. and Z.H.; writing—review and editing, W.Z. and D.Z. All authors have read and agreed to the published version of the manuscript.

Funding: This work was supported in part by the National Natural Science Foundation of China under Grant numbers 62272418, and 62002046.

Data Availability Statement: Dataset used in this study may be available on demand.

Conflicts of Interest: The authors declare no conflict of interest.

References

1. Watson, J.D.; Crick, F.H. Molecular Structure of Nucleic Acids: A Structure for Deoxyribose Nucleic Acid. *Nature* **1953**, *171*, 737–738. [CrossRef]
2. Adleman, L.M. Molecular Computation of Solutions to Combinatorial Problems. *Science* **1994**, *266*, 1021–1024. [CrossRef] [PubMed]
3. Seelig, G.; Soloveichik, D.; Zhang, D.Y.; Winfree, E. Enzyme-free Nucleic Acid Logic Circuits. *Science* **2006**, *314*, 1585–1588. [CrossRef] [PubMed]
4. Church, G.M.; Gao, Y.; Kosuri, S. Next-generation Digital Information Storage in DNA. *Science* **2012**, *337*, 1628. [CrossRef] [PubMed]
5. Extance, A. How DNA Could Store All the World's Data. *Nature* **2016**, *537*, 22–24. [CrossRef]
6. Zhirnov, V.; Zadegan, R.M.; Sandhu, G.S.; Church, G.M.; Hughes, W.L. Nucleic Acid Memory. *Nat. Mater.* **2016**, *15*, 366–370. [CrossRef]
7. Zhu, D.L.; Huang, Z.W.; Liao, S.G.; Zhou, C.J.; Yan, S.Q.; Chen, G. Improved Bare Bones Particle Swarm Optimization for DNA Sequence Design. *IEEE Trans. NanoBioscience* **2022**. [CrossRef]
8. Chaves-González, J.M.; Vega-Rodríguez, M.A.; Granado-Criado, J.M. Multiobjective Swarm Intelligence Approach Based on Artificial Bee Colony for Reliable DNA Sequence Design. *Eng. Appl. Artif. Intell.* **2013**, *26*, 2045–2057. [CrossRef]
9. Yang, G.J.; Wang, B.; Zheng, X.; Zhou, C.J.; Zhang, Q. IWO Algorithm Based on Niche Crowding for DNA Sequence Design. *Interdiscip. Sci. Comput. Life Sci.* **2017**, *9*, 341–349. [CrossRef]
10. Zhang, K.; Xu, J.; Geng, X.T.; Xiao, J.H.; Pan, L.Q. Improved Taboo Search Algorithm for Designing DNA Sequences. *Prog. Nat. Sci.* **2008**, *18*, 623–627. [CrossRef]
11. Cervantes-Salido, V.M.; Jaime, O.; Brizuela, C.A.; Martínez-Pérez, I.M. Improving the Design of Sequences for DNA Computing: A Multiobjective Evolutionary Approach. *Appl. Soft Comput.* **2013**, *13*, 4594–4607. [CrossRef]
12. Chaves-González, J.M.; Vega-Rodríguez, M.A. DNA Strand Generation for DNA Computing by Using A Multi-objective Differential Evolution Algorithm. *Biosystems* **2014**, *116*, 49–64. [CrossRef]
13. Chaves-González, J.M.; Vega-Rodríguez, M.A. A Multiobjective Approach Based on The Behavior of Fireflies to Generate Reliable DNA Sequences for Molecular Computing. *Appl. Math. Comput.* **2014**, *227*, 291–308. [CrossRef]
14. Eberhart, R.; Kennedy, J. A New Optimizer Using Particle Swarm Theory. In Proceedings of the Sixth International Symposium on Micro Machine and Human Science, Nagoya, Japan, 4–6 October 1995; pp. 39–43. [CrossRef]
15. Houssein, E.H.; Gad, A.G.; Hussain, K.; Suganthan, P.N. Major Advances in Particle Swarm Optimization: Theory, Analysis, and Application. *Swarm Evol. Comput.* **2021**, *63*, 100868. [CrossRef]
16. Ghatasheh, N.; Faris, H.; Abukhurma, R.; Castillo, P.A.; Al-Madi, N.; Mora, A.M.; Al-Zoubi, A.M.; Hassanat, A. Cost-sensitive Ensemble Methods for Bankruptcy Prediction in A Highly Imbalanced Data Distribution: A Real Case from the Spanish Market. *Prog. Artif. Intell.* **2020**, *9*, 361–375. [CrossRef]
17. Zhang, Q.K.; Liu, W.G.; Meng, X.X.; Yang, B.; Vasilakos, A.V. Vector coevolving particle swarm optimization algorithm. *Inf. Sci.* **2017**, *394*, 273–298. [CrossRef]
18. Coello, C.A.C.; Lechuga, M.S. MOPSO: A Proposal for multiple objective particle swarm optimization. In Proceedings of the 2002 Congress on Evolutionary Computation Part of the 2002 IEEE World Congress on Computational Intelligence, Honolulu, HI, USA, 12–17 May 2002; Volume 2, pp. 1051–1056. [CrossRef]
19. Corne, D.W.; Jerram, N.R.; Knowles, J.D.; Oates, M.J. PESA-II: Region-based Selection in Evolutionary Multiobjective Optimization. In Proceedings of the 3rd Annual Conference on Genetic And Evolutionary Computing Conference, San Francisco, CA, USA, 7–11 July 2001; pp. 283–290. [CrossRef]
20. Hu, X.H.; Eberhart, R. Multiobjective Optimization Using Dynamic Neighborhood Particle Swarm Optimization. In Proceedings of the 2002 Congress on Evolutionary Computation, Honolulu, HI, USA, 12–17 May 2002; pp. 1677–1681. [CrossRef]
21. Deb, K.; Pratap, A.; Agarwal, S.; Meyarivan, T.A.M.T. A Fast And Elitist Multiobjective Genetic Algorithm: NSGA-II. *IEEE Trans. Evol. Comput.* **2022**, *6*, 182–197. [CrossRef]

22. Zhan, Z.H.; Zhang, J.; Lin, Y.; Li, J.Y.; Huang, T.; Guo, X.Q.; Wei, F.F.; Kuang, S.X.; Zhang, X.Y.; You, R. Matrix-Based Evolutionary Computation. *IEEE Trans. Emerg. Top. Comput. Intell.* **2021**, *6*, 315–328. [CrossRef]
23. Mehrabian, A.R.; Lucas, C. A Novel Numerical Optimization Algorithm Inspired from Weed Colonization. *Ecol. Inform.* **2006**, *1*, 355–366. [CrossRef]
24. Poli, R.; Kennedy, J.; Blackwell, T. Particle Swarm Optimization. *Swarm Intell.* **2007**, *1*, 33–57. [CrossRef]
25. Geem, Z.W.; Kim, J.H.; Loganathan, G.V. A New Heuristic Optimization Algorithm: Harmony Search. *Simulation* **2001**, *76*, 60–68. [CrossRef]
26. Xue, L.; Wang, B.; Lv, H.; Yin, Q.; Zhang, Q.; Wei, X.P. Constraining DNA Sequences with A Triplet-bases Unpaired. *IEEE Trans. NanoBiosci.* **2020**, *19*, 299–307. [CrossRef]
27. Liu, Y.Y.; Zheng, X.D.; Wang, B.; Zhou, S.H. The Optimization of DNA Encoding Based on Chaotic Optimization Particle Swarm Algorithm. *J. Comput. Theor. Nanosci.* **2016**, *13*, 443–449. [CrossRef]
28. Xiao, J.H.; Jiang, Y.; He, J.J.; Cheng, Z. A Dynamic Membrane Evolutionary Algorithm for Solving DNA Sequences Design with Minimum Free Energy. *MATCH Commun. Math. Comput. Chem.* **2013**, *70*, 971–986.
29. Shin, S.Y.; Lee, I.H.; Kim, D.; Zhang, B.T. Multiobjective Evolutionary Optimization of DNA Sequences for Reliable DNA Computing. *IEEE Trans. Evol. Comput.* **2005**, *9*, 143–158. [CrossRef]
30. Watkins, N.E., Jr.; SantaLucia, J., Jr. Nearest-neighbor Thermodynamics of Deoxyinosine Pairs in DNA Duplexes. *Nucleic Acids Res.* **2005**, *33*, 6258–6267. [CrossRef] [PubMed]
31. Tizhoosh, H.R. Opposition-based Learning: A New Scheme for Machine Intelligence. In Proceedings of the International Conference on Computational Intelligence for Modelling, Control and Automation and International Conference on Intelligent Agents, Web Technologies and Internet Commerce, Vienna, Austria, 28–30 November 2005; Volume 1, pp. 695–701. [CrossRef]
32. Rahnamayan, S.; Jesuthasan, J.; Bourennani, F.; Salehinejad, H.; Naterer, G.F. Computing Opposition by Involving Entire Population. In Proceedings of the IEEE Congress on Evolutionary Computation, Beijing, China, 6–11 July 2014; pp. 1800–1807. [CrossRef]
33. Milman, V.D. New Proof of the Theorem of A. Dvoretzky on Intersections of Convex Bodies. *Funct. Anal. Its Appl.* **1971**, *5*, 288–295. [CrossRef]
34. Hassanat, A.B.A. Furthest-Pair-Based Decision Trees: Experimentational Results on Big Data Classification. *Information* **2018**, *9*, 284. [CrossRef]
35. Gueorguieva, N.; Valova, I.; Georgiev, G. M&MFCM: Fuzzy C-means Clustering with Mahalanobis and Minkowski Distance Metrics. *Procedia Comput. Sci.* **2017**, *114*, 224–233. [CrossRef]
36. Yang, J.H.; Yu, J.H.; Huang, C. Adaptive Multistrategy Ensemble Particle Swarm Optimization with Signal-to-Noise Ratio Distance Metric. *Inf. Sci.* **2022**, *612*, 1066–1094. [CrossRef]
37. Yuan, T.T.; Deng, W.H.; Tang, J.; Tang, Y.N.; Chen, B.H. Signal-To-Noise Ratio: A Robust Distance Metric for Deep Metric Learning. In Proceedings of the Conference on Computer Vision and Pattern Recognition (CVPR), Long Beach, CA, USA, 15–20 June 2019; pp. 4810–4819. [CrossRef]

electronics

MDPI

Article

A Multi-Strategy Adaptive Particle Swarm Optimization Algorithm for Solving Optimization Problem

Yingjie Song [1], Ying Liu [2], Huayue Chen [3,*] and Wu Deng [4,5,*]

[1] School of Computer Science and Technology, Shandong Technology and Business University, Yantai 264005, China
[2] School of Statistics, Shandong Technology and Business University, Yantai 264005, China
[3] School of Computer Science, China West Normal University, Nanchong 637002, China
[4] Traction Power State Key Laboratory, Southwest Jiaotong University, Chengdu 610031, China
[5] College of Electronic Information and Automation, Civil Aviation University of China, Tianjin 300300, China
* Correspondence: sunnyxiaoyue20@cwnu.edu.cn (H.C.); wdeng@cauc.edu.cn (W.D.)

Abstract: In solving the portfolio optimization problem, the mean-semivariance (MSV) model is more complicated and time-consuming, and their relations are unbalanced because they conflict with each other due to return and risk. Therefore, in order to solve these existing problems, multi-strategy adaptive particle swarm optimization, namely APSO/DU, has been developed to solve the portfolio optimization problem. In the present study, a constraint factor is introduced to control velocity weight to reduce blindness in the search process. A dual-update (DU) strategy is based on new speed, and position update strategies are designed. In order to test and prove the effectiveness of the APSO/DU algorithm, test functions and a realistic MSV portfolio optimization problem are selected here. The results demonstrate that the APSO/DU algorithm has better convergence accuracy and speed and finds the least risky stock portfolio for the same level of return. Additionally, the results are closer to the global Pareto front (PF). The algorithm can provide valuable advice to investors and has good practical applications.

Keywords: PSO; multi-strategy; dual-update strategy; mean-semivariance model; portfolio optimization

Citation: Song, Y.; Liu, Y.; Chen, H.; Deng, W. A Multi-Strategy Adaptive Particle Swarm Optimization Algorithm for Solving Optimization Problem. *Electronics* **2023**, *12*, 491. https://doi.org/10.3390/electronics12030491

Academic Editor: Young-Koo Lee

Received: 22 December 2022
Revised: 14 January 2023
Accepted: 16 January 2023
Published: 17 January 2023

1. Introduction

The portfolio optimization problem (POP) aims to improve portfolio returns and reduce portfolio risk in the complex financial market. The mean-variance (MV) model was first proposed by economist Markowitz in 1952 to calculate the POP [1,2] and is a cornerstone of financial theory, providing a theoretical basis for investors to choose the optimal portfolio. However, there are significant limitations in its practical application. The use of variance to assess risk usually requires the calculation of a covariance matrix for all stocks, which is difficult to use in practice due to its computational complexity. Additionally, this risk measurement only considers the extent to which actual returns deviate from expected returns, whereas true losses refer to fluctuations below the mean of returns [3–8]. In order to be more in line with social reality, mean-semivariance portfolio models have been proposed and are widely used [9–12].

Traditional optimization algorithms for solving POPs require the application of many complex statistical methods and reference variables provided by experts, so solving large-scale POPs suffers from slow computational speed and poor solution accuracy, while heuristic algorithms can solve these problems well. In recent years, many scholars have used evolutionary computation algorithms to solve POPs, including the genetic algorithm (GA) [13], particle swarm optimization (PSO) [14,15], artificial bee colony algorithm (ABC) [16], and squirrel search algorithm (SSA) [17]. The particle swarm optimization algorithm (Eberhart & Kennedy, 1995) belongs to a class of swarm intelligence algorithms, which are designed by simulating the predatory behavior of a flock of birds [18–23]. Due to its simple structure, fast convergence, and good robustness, it has been widely used in complex nonlinear portfolio optimization [24–29]. In addition, some new methods have also been proposed in some fields in recent years [30–39].

The improvement directions of the PSO algorithm are mainly divided into parameter improvement, update formula improvement, and integration with other intelligent algorithms. Setting the algorithm's parameters is the key to ensuring the reliability and robustness of the algorithm. With the determined population size and iteration time, the search capability of the algorithm is mainly decided by three core control parameters, namely the inertia weight (w), the self-learning factor (C_1), and the social-learning factor (C_2). To improve the performance of the algorithm, PSO algorithms based on the dual dynamic adaptation mechanism of inertia weights and learning factors have been proposed successively in recent years [40–42], considering that adjusting the core parameters alone weakens the uniformity of the algorithm evolution process and make it difficult to adapt to complex nonlinear optimization problems. Clerc et al. [43] proposed the concept of the shrinkage factor, and this method adds a multiplicative factor to the velocity formulation in order to allow the three core parameters to be tuned simultaneously, ultimately resulting in better algorithm convergence performance. Since then, numerous scholars have explored the full-parameter-tuning strategy to mix the three core parameters for tuning experiments. Zhang et al. [44] used control theory to optimize the core parameters of the standard PSO. Harrison et al. [45] empirically investigated the convergence behavior of 18 adaptive optimization algorithms.

The parameter improvement of PSO only involves improving the velocity update and does not consider the position update. Different position-updating strategies have different exploration and exploitation capabilities. In position updating, because the algorithm's convergence is highly dependent on the position weighting factor, a constraint factor needs to be introduced to control the velocity weight and reduce blindness in the search process. Liu et al. [46] proposed that the position weighting factor facilitates global algorithm exploration. The paper synthesizes the advantages of the two improvement methods and proposes a dual-update (DU) strategy. The method not only adjusts the core parameters of velocity update to make the algorithm more adaptable to nonlinear complex optimization problems, it also considers the position update formula and introduces a constraint factor to control the weight of velocity to reduce blindness in the search process and improve the convergence accuracy and convergence speed of the algorithm.

The main contributions of this paper are described as follows.

(1) This paper makes improvements based on fundamental particle swarm and proposes a multi-strategy adaptive particle swarm optimization algorithm, namely APSO/DU, to solve the portfolio optimization problem. Modern portfolio models are typically complex nonlinear functions, which are more challenging to solve.

(2) A dual-update strategy is designed based on new speed and position update strategies. The approach uses inertia weights to modify the learning factor, which can balance the capacity for learning individual particles and the capacity for learning the population and enhance the algorithm's optimization accuracy.

(3) A position update approach is also considered to lessen search blindness and increase the algorithm's convergence rate.

(4) Experimental findings show that the two strategies work better together than they do separately.

2. Multi-Strategy Adaptive PSO

2.1. Basic PSO Algorithm

The PSO algorithm is a population-based stochastic search algorithm in which the position of each particle represents a feasible solution to the problem to be optimized, and the position of the particle is evaluated in terms of its merit by the fitness value derived from the optimization function. The particle population is initialized randomly as a set of random candidate solutions in the PSO algorithm, and then each particle moves in the search space with a certain speed, which is dynamically adjusted according to its own and its companion's flight experience. The optimal solution is obtained after cyclic iterations until the convergence condition is satisfied.

Suppose a population $X = \{x_1, \ldots, x_i, \ldots, x_n\}$ of n particles without weight and volume in a *D*-dimensional search space, at the *t*th iteration, $x_i(t) = [x_{i1}(t), x_{i2}(t), \ldots, x_{iD}(t)]$ denotes the position of *i*th particle, $V_i(t) = [v_{i1}(t), v_{i2}(t), \ldots, v_{iD}(t)]$ denotes the velocity of *i*th particle. Up to generation t, $p_i(t) = [pbest_{i1}(t), pbest_{i2}(t), \ldots, pbest_{iD}(t)]$ denotes the personal best position particle i has visited since the first-time step. *gbest* denote the best position discovered by all particles so far. In every generation, the evolution process of the *i*th particle is formulated as

$$v_i(t+1) = wv_i(t) + c_1 \times rand() \times (p_i(t) - x_i(t) + c_2 \times rand() \times (gbest(t) - x_i(t)) \quad (1)$$

$$x_i(t+1) = x_i(t) + v_i(t+1) \quad (2)$$

where $i = 1, 2, \ldots, D$. w is the inertia weight. c_1 and c_2 are constants of the PSO algorithm with a value range of [0, 2], while rand () represents the random numbers in [0, 1].

An iteration of PSO-based particle movement is demonstrated in Figure 1.

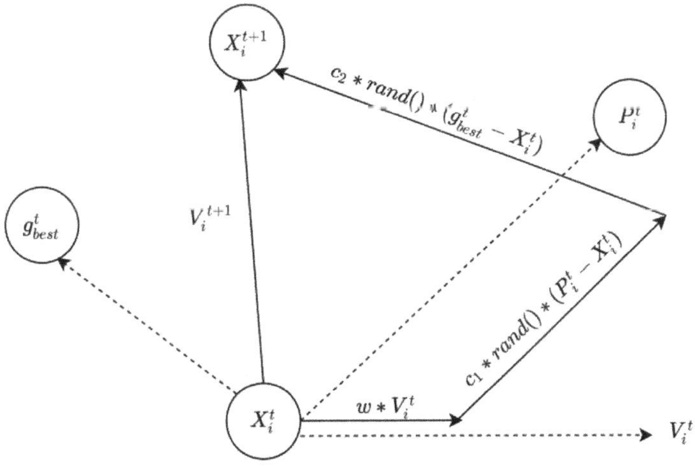

Figure 1. An iterative particle movement in PSO.

2.2. *APSO/DU*

PSO is an intelligent algorithm with global convergence, which requires fewer parameters to be adjusted. However, basic PSO has the problem of easily falling into local optimum and slow convergence. The APSO/DU algorithm can reduce the blindness of the search process and improve the convergence accuracy and speed of the algorithm, making it more adaptable to complex optimization problems. The APSO/DU algorithm can reduce the blindness of the search process and make the algorithm more adaptable to complex optimization problems.

2.2.1. Speed Update Strategy

The improvement strategies for inertia weights (w) and learning factors (c_1, c_2) can be classified as constant or stochastic, linear or nonlinear, and adaptive. The existing research on the dual dynamic adaptation mechanism has experimentally shown that using nonlinear decreasing weights is better than using linear decreasing weights. The functional relationship with nonlinear learning factors can be more adapted to complex optimization objectives. The strategy uses inertia weights to adjust the learning factors, which can balance the learning ability of individual particles and the group's learning ability and improve the algorithm's optimization accuracy. This paper uses a combination of the two with better results.

- Nonlinear Decreasing w

w is the core parameter that affects the performance and efficiency of the PSO algorithm. Smaller weights can strengthen the local search ability and improve convergence accuracy, while larger weights are beneficial to the global search and prevent the particles from falling into the optimal local position, but the convergence speed is slow. Most of the current improvements are related to the adjustment of w. In this paper, we use the nonlinear w exponential function decreasing way, and the formula is as follows.

$$w = w_{min} + (w_{max} + w_{min}) \times \exp\left[-20 \times \left(\frac{t}{T}\right)^6\right] \tag{3}$$

where T is the maximum number of time steps, usually $w_{max} = 0.9$, $w_{min} = 0.4$.

- The learning factor (c_1, c_2) varies according to w

c_1 and c_2 in the velocity update formula determine the size of the amount of learning of the particle in the optimal position. c_1 is used to adjust the amount of self-learning of the particle and c_2 is used to adjust the amount of social learning of the particle, and the change of the learning factor coefficient is used to change the trajectory of the particle. In this paper, referring to the previous summary, the adjustment strategy is better when the learning factor and inertia weights are a nonlinear function. The coefficient combination is A = 0.5, B = 1, C = 0.5, and the formula is described as follows.

$$\begin{aligned} C_1 &= Aw^2 + Bw + C \\ C_2 &= 2.5 - C_1 \end{aligned} \tag{4}$$

2.2.2. Position Update Policy

The convergence and convergence speed of the algorithm are greatly related to the position weighting factor, and the core parameter-tuning strategy only considers improving the velocity update without considering the position update. In order to control the influence of velocity on position, the constraint factor (α) is added to the position update formula, and α is introduced in order to achieve the weight of the control velocity to reduce blindness in the search process and improve the convergence rate.

- The Constraint Factors

In basic PSO, the new position of a particle is equal to its current position plus the current velocity, but the position vector and velocity vector cannot be added directly, so there must be a constraint factor between the two in the position update formula, and the constraint factor in the traditional PSO algorithm is equal to 1. α guides the particle to hover around the best position, and the improvement of α controls the influence of velocity on position so that the convergence of the algorithm is better improved. α based on w change is used in this paper which, in the early stage, is influenced by particle velocity and has strong exploration ability. In the later stage, it is less influenced by particle velocity and has strong local search ability.

$$\begin{aligned} x_{ij}(t+1) &= x_{ij}(t) + \alpha v_{ij}(t+1) \\ \alpha &= 0.1 + w \end{aligned} \tag{5}$$

2.2.3. Model of APSO/DU

The flow of the APSO/DU is shown in Figure 2.

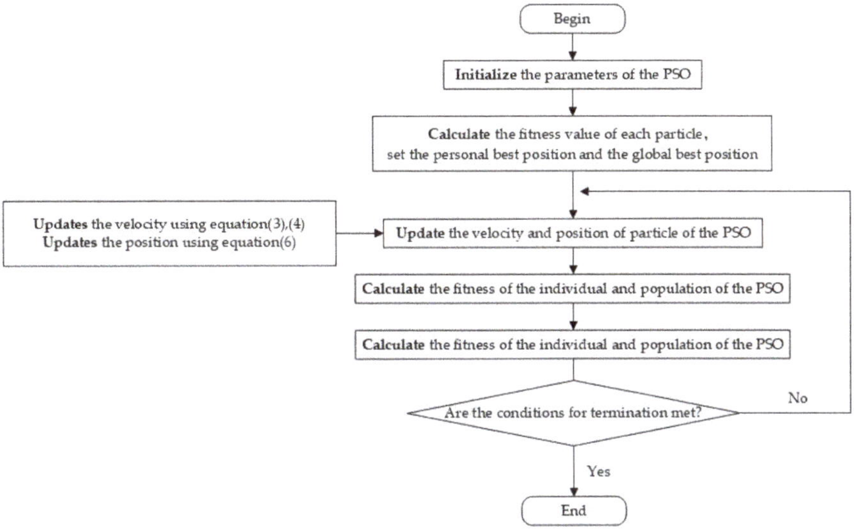

Figure 2. The flow of the APSO/DU.

2.3. Numerical Experiments and Analyses

In order to test the performance of the APSO/DU algorithm, three commonly used test functions were selected for the experiment. The test functions are shown in Table 1.

Table 1. Three test functions.

Selection	Function	Search Range	Global Optimum
Sphere	$f_1 = \sum_{i=1}^{D} x_i^2$	[−100, 100]	$minf_1 = 0$
Schwefel/sp2.22	$f_2 = \sum_{i=1}^{D} \lvert x_i \rvert + \prod_{i=1}^{D} \lvert x_i \rvert$	[−10, 10]	$minf_2 = 0$
Griewank	$f_3 = \frac{1}{4000} \sum_{i=1}^{D} x_i^2 - \prod_{i=1}^{D} cos\left(\frac{x_i}{\sqrt{i}} + 1\right)$	[−600, 600]	$minf_3 = 0$

- Contrast algorithms

The parameters of each PSO algorithm are shown in Table 2. To facilitate the comparison of the effectiveness of the APSO/DU algorithm, this paper chose to compare it with three classical adaptive improved PSO algorithms: PSO-TVIW; PSO-TVAC; and PSOCF. The parameter settings summarized in the literature of Kyle Robert Harrison (2018) [45] were also used, where the time-varying inertia weight values of the PSO-TVIW algorithm are set according to the study in Harrison's paper. The PSO-TVIW algorithm is also known as the standard particle swarm algorithm. The PSO-TVAC algorithm with time-varying acceleration coefficient adjusts the values of the w, c_1, and c_2 parameters and introduces six additional control parameters. Clerc's proposed PSO algorithm with shrinkage factor (PSOCF) has good convergence, but its computational accuracy is not high and its stability is not as good as that of standard PSO, so Eberhart proposed to limit the speed parameter $V_{max} = X_{max}$ of the algorithm so as to improve the convergence speed and search performance of the algorithm, and the PSOCF algorithm used this improved method for comparison experiments.

The new algorithm is based on a combination of two strategies. In order to verify whether the combination of two strategies is superior to one strategy, namely PSO/D (which updates only the core parameters), the formula and parameters are detailed in Section 2.2.1. Additionally, PSO/U, which only updates the velocity update formula, is improved is by adding a constraint factor to the position formula, which needs to be combined with inertia weights. The basic particle swarm does not contain inertia weights, so the standard particle swarm algorithm (PSO-TVIW), by adding a constraint factor, can verify that the combination of update strategies proposed in this paper is superior.

Table 2. Parameter setting of each PSO algorithm.

Algorithm	w	c_1	c_2
PSO-TVAC	[0.4, 0.9]	[0.5, 2.5]	[0.5, 2.5]
PSO-TVIW	[0.4, 0.9]	1.49618	1.49618
PSOCF	0.729	2.8	1.3
PSO/D	[0.4, 0.9]	[0.695, 1.805]	[0.695, 1.805]
PSO/U	[0.4, 0.9]	1.49618	1.49618
APSO/DU	[0.4, 0.9]	[0.695, 1.805]	[0.695, 1.805]

In the experiments, to ensure fairness in the testing of each algorithm, different PSO algorithms were set with the same population size (N = 30), maximum number of iterations ($T_{max} = 500$), and variable dimension (D = 15). Each algorithm was run 30 times, and the test results are shown in Table 3. The bold part of the text indicates the best optimization results.

- Test Results:

Table 3. Optimization results of six algorithms.

Function	Contrast Algorithms	f_{min}	f_{mean}	f_{max}	f_{sd}
F1	PSO − TVAC	2.26×10^{-3}	6.84×10^{-3}	1.37×10^{-2}	2.86×10^{-3}
	PSO − TVIW	1.22×10^{-3}	5.47×10^{-3}	1.16×10^{-2}	2.65×10^{-3}
	PSOCF	1.65×10^{-1}	1.90	6.08	1.69
	PSO/D	9.58×10^{-5}	4.92×10^{-3}	3.83×10^{-2}	8.08×10^{-3}
	PSO/U	3.36×10^{-3}	8.90×10^{-3}	2.10×10^{-2}	3.68×10^{-3}
	APSO/DU	$\mathbf{4.57 \times 10^{-5}}$	$\mathbf{2.43 \times 10^{-3}}$	$\mathbf{1.37 \times 10^{-2}}$	$\mathbf{2.54 \times 10^{-3}}$
F2	PSO − TVAC	2.0110	3.8211	7.2185	0.9937
	PSO − TVIW	1.7082	3.3503	5.0343	0.7761
	PSOCF	0.8047	2.5967	4.6286	0.9490
	PSO/D	**0.4045**	1.3387	3.0566	3.0566
	PSO/U	2.1516	3.6639	5.1408	5.1408
	APSO/DU	0.4246	**1.3033**	**2.5632**	**0.5253**
F3	PSO − TVAC	1.2174	1.8248	2.9463	4.03×10^{-1}
	PSO − TVIW	1.2724	1.9454	2.8534	4.06×10^{-1}
	PSOCF	1.0061	1.1614	1.5375	1.49×10^{-1}
	PSO/D	1.0018	1.0229	1.0959	$\mathbf{2.57 \times 10^{-2}}$
	PSO/U	1.5057	2.1557	3.3043	4.79×10^{-1}
	APSO/DU	**0.8140**	**1.0119**	**1.0913**	4.18×10^{-2}

It can be seen from Table 3 that APSO/DU outperforms the other algorithms overall. (i) The APSO/DU algorithm is compared with the classical adaptive algorithms (PSO-TVAC, PSO-TVIW, and PSOCF). APSO/DU takes the smallest optimal value in the three test functions and is closest to the optimal solution. The standard deviation is also the best among the three algorithms, which indicates that APSO/DU has a stable performance. (ii) To verify whether the combination of two strategies is better than one, the APSO/DU algorithm is compared with a single-strategy algorithm (PSO/D and PSO/U), and the results of PSO/U and APSO/DU are closer to each other. In the Griewank function, APSO/DU takes the smallest optimal value and is closest to the optimal solution with a standard deviation not much different from PSO/D. On balance, the APSO/DU algorithm outperforms the comparison algorithm.

In order to reflect more intuitively on the solution accuracy and convergence speed of each algorithm, the variation curves of the fitness values when each algorithm solves the three test functions are given in Figure 3. The horizontal coordinate indicates the number of iterations, and the vertical coordinate indicates the fitness value.

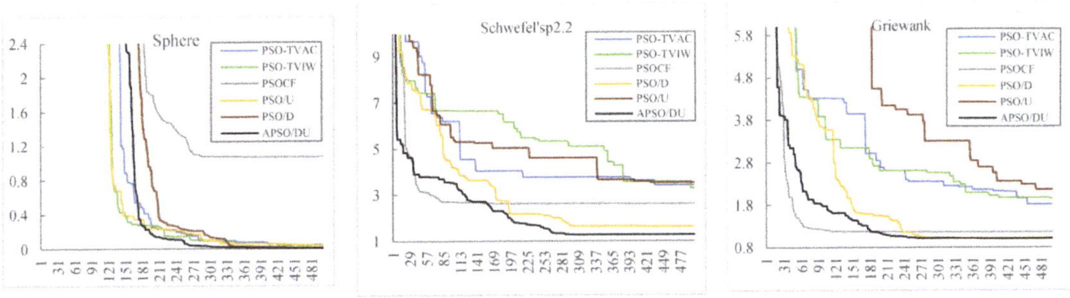

(a) F1 (b) F2 (c) F3

Figure 3. Curves of the convergence process of the benchmark test functions F1–F3.

The average convergence curves of each algorithm for the three tested functions are given in Figure 3. The single-peak test function shows whether the algorithm achieves the target value of the search accuracy On single-peak functions F1 (sphere) and F2 (Schwefel'sp2.22), the relatively high convergence accuracy is achieved by the APSO/DU algorithm and the PSO/D algorithm, with PSOCF easily falling into local optimality.

A multi-peaked test function can test the global searchability of an algorithm. In multi-peak function F3 (Griewank) optimization, the APSO/DU algorithm performs best, followed by the PSO/D algorithm and the PSOCF algorithm, in that order. Among the different functions, APSO/DU has the fastest convergence speed and the highest convergence accuracy and, collectively, the APSO/DU algorithm is the best in terms of finding the best results and showing better stability.

3. Portfolio Optimization Problem

3.1. Related Definitions

The essential parameters in the POP are expected return and risk, and investors usually prefer to maximize return and minimize risk. Assuming a fixed amount of money to buy n stocks, the POP can be described as how to choose the proportion of investments that minimizes ρ the investor's risk (variance or standard deviation) given a minimum rate of return, or how to choose the proportion of investments that maximizes the investor's return given a level of risk.

The investor holds fixed assets invested in n stocks $A_i(i = 1, 2, .., m)$, let R_i be the return rate of A_i, which is a random variable. μ_i is the expected return on stock A_i. Let $E(R_i)$ denote the mathematical expectation of a random variable R. Define

$$\mu_i = E(R_i) \tag{6}$$

In a certain period, the stock return is the relative number of the difference between the opening and closing prices of that stock, where V_{ij} is the return of stock i in period t, as in Equation (7).

$$V_{ij} = \frac{p_{i,t} - p_{i,t-1}}{p_{i,t-1}}, \quad i = 1, 2, \ldots, T \tag{7}$$

where $p_{i,t}$ and $p_{i,t-1}$ are the closing prices of stock i in periods t and $t-1$, respectively. The expected return on the ith stock is given by Equation (8)

$$\mu_i = \frac{1}{T}\sum_{j=1}^{T} V_{ij} \tag{8}$$

3.2. Mean-Semivariance Model

A large number of empirical analysis results show that asset returns are characterized by spikes and thick tails, which contradicts the assumption that asset returns are normally distributed in the standard mean-variance model. Additionally, the variance reflects the degree of deviation between actual returns and expected returns, while actual losses (loss risk) are fluctuations below the mean of returns. Thus, the portfolio optimization model based on the lower half-variance risk function is more realistic. Equations (9)–(12) present the mean-semivariance model. Assume that the short selling of assets is not allowed.

$$minf = \frac{1}{T}\sum_{t=1}^{T}[(\sum_{i=1}^{m}x_i r_{it} - \rho)^{-}]^2 \tag{9}$$

Subject to

$$E(\mu_P) = \sum_{i=1}^{m}\mu_i x_i \geq \rho \tag{10}$$

$$\sum_{i=1}^{m} 0 \leq x_i \leq 1, i = 1, 2, \ldots, m \tag{11}$$

$$\sum_{i=1}^{m} x_i = 1 \tag{12}$$

where:

m is the number of stocks in the portfolio;
ρ is the rate of return required by the investor;
x_i is the proportion $(0 \leq x_i \leq 1)$ of the portfolio held in assets $i(i = 1, 2, \ldots, m)$;
μ_i is the mean return of asset i in the targeted period;
μ_p is the mean return of the portfolio in the targeted period.

Equation (9) is the objective function of the model and represents minimizing the risk of the portfolio (the lower half of the variance); Equation (10) ensures that the return of the portfolio is greater than the investor's expected return ρ; and Equations (11) and (12) indicate that the variables take values in the range [0, 1], and the total investment ratio is 1.

4. Case Analysis

4.1. Experiment Settings

(1) Individual composition

The vector $X = (X_1, X_2, \ldots, X_n)$ represents a portfolio strategy whose ith dimensional component x_i represents the allocation of funds to hold the ith stock in that portfolio, namely the weight of that asset in the portfolio.

(2) Variable constraint processing

Equation (10): the feasibility of the particle is checked after the initial assignment of the algorithm and the update of the position vector and if it does not work, the position vector of the particle is recalculated until it is satisfied before the calculation of the objective function is carried out.

Equation (11): the variables take values in the interval [0, 1] and the iterative process uses the boundary to restrict within the interval.

Equation (12): variables on a non-negative constraint basis, sets $= x_1 + x_2 + \ldots + x_n$ when $s = 0$, so that all variables in the portfolio are $\frac{1}{n}$; when $s \neq 0$, let $x_i = \frac{x_i}{n}$, $i = 1, 2, \ldots, n$.

(3) Parameter values

The particle dimension D is the number of stocks included in the portfolio, and the number of stocks selected in this paper is 15, hence D = 15. The parameters of this experimental algorithm are set as described in Section 2.3. of this paper, and the results show the average of 30 independent runs of each algorithm. All PSO algorithms in this paper were written in Python and run on a Windows system for testing.

4.2. Sample Selection

Regarding the selection of stock data, firstly, recent stock data should be selected for analysis to have a certain practical reference value. Secondly, the number of shares is too small to be credible, and the number of shares is too large for the average investor to be distracted with at the same time. Finally, Markowitz's investment theory states that the risk of a single asset is fixed and cannot be reduced on its own, whereas investing in portfolio form diversifies risk without reducing returns. The lower the correlation between any two assets in a portfolio (preferably negative), the more significant the reduction in overall portfolio unsystematic risk [47]. Some methods can be used to solve this problem [48–52].

Based on the above considerations, 30 stocks from different sectors were selected from Choice Financial Terminal, with a time range of 1 January 2019 to 31 December 2021, for a total of 155 weeks of closing price data. Correlation analysis was conducted on the stock data, and 15 stocks with relatively low correlation coefficients were selected for empirical analysis. The price trend charts and correlation coefficients for the 15 stocks are given in Figures 4 and 5.

Figure 4 shows the weekly closing price trend for the 15 stocks data, which provides a visual indication of the trend in stock data. Stocks vary widely in price from one another, with 600612 being the most expensive. As shown in Figure 5, the fifteen stocks have low correlations, with only two portfolios having correlation coefficients greater than 0.5 for any two stocks. Stock 6 and Stock 8 have strong correlations with Stock 12, with correlation coefficients of 0.6 and 0.5, respectively, while all other correlations are below 0.5. Stock 4 and Stock 14 have the lowest correlation, with a correlation coefficient of −0.045. After calculation, the correlation of the stock data in this paper is low, and the mean correlation coefficient is only 0.198. The lower the correlation between stocks is, the more effective the portfolio choice is in reducing unsystematic risk, thus indicating that investing with a portfolio strategy is effective in reducing risk.

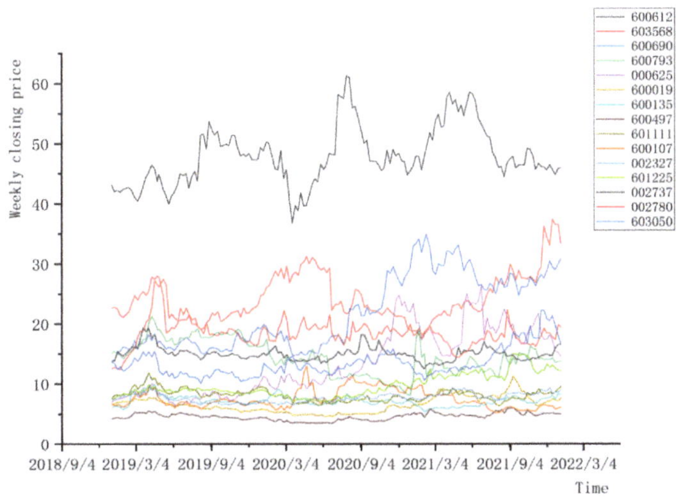

Figure 4. The weekly closing price trend for the 15 stocks data.

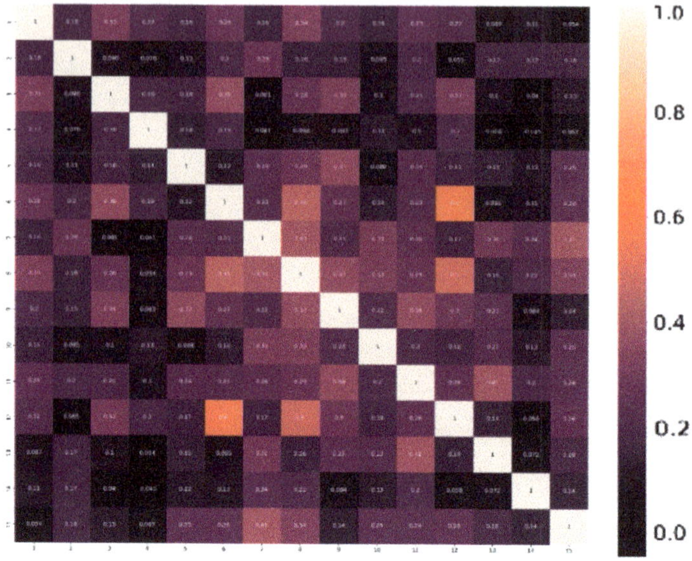

Figure 5. Correlation matrix of 15 stocks.

Table 4 gives the basic statistical characteristics of 15 stocks for 2019–2021, and the returns are the weekly averages of the relative number of closing prices of the stock data. The *p*-values for most of the stock returns in Table 4 are less than 0.05, which should reject the original hypothesis and indicates that the stock returns do not conform to a normal distribution at the 5% significance level. The *p*-values for 600793 and 600135 are greater than 0.05 at a level that does not present significance and cannot reject the original hypothesis, so the data satisfies a normal distribution.

Table 4. Basic characteristics and normality test of 15 stocks from 2019 to 2021.

NO.	Code	Price/(yuan)	Return (%)	Std	Prob	Conclusion at the (5%) Level
1	600612	47.930	0.131	0.043	0.003 ***	Distribution not normally distributed
2	603568	24.329	0.408	0.055	0.000 ***	Distribution not normally distributed
3	600690	21.992	0.662	0.052	0.000 ***	Distribution not normally distributed
4	600793	14.396	0.288	0.087	0.060 *	Normality cannot be ruled out
5	000625	13.432	0.810	0.082	0.000 ***	Distribution not normally distributed
6	600019	6.526	0.207	0.046	0.000 ***	Distribution not normally distributed
7	600135	7.158	0.368	0.060	0.069 *	Normality cannot be ruled out
8	600497	4.558	0.253	0.053	0.030 **	Distribution not normally distributed
9	601111	8.095	0.259	0.049	0.000 ***	Distribution not normally distributed
10	600107	7.522	0.221	0.075	0.000 ***	Distribution not normally distributed
11	002327	7.704	0.208	0.038	0.000 ***	Distribution not normally distributed
12	601225	9.689	0.432	0.049	0.000 ***	Distribution not normally distributed
13	002737	14.959	0.204	0.042	0.000 ***	Distribution not normally distributed
14	002780	18.442	0.474	0.063	0.000 ***	Distribution not normally distributed
15	603050	13.506	0.304	0.060	0.000 ***	Distribution not normally distributed

Note: ***, **, and * represent the significance level of 1%, 5%, and 10%, respectively.

Figure 6 shows the histogram of the normality test for 15 stocks. If the normality plot is roughly bell-shaped (high in the middle and low at the ends), the data are largely accepted as normally distributed. It can be seen from the figure that the normal distribution plots of the 600793 and 600135 stock data roughly show a bell shape, which is consistent with normal distribution. However, the normal distribution of most stocks does not show a bell shape and does not conform to normal distribution.

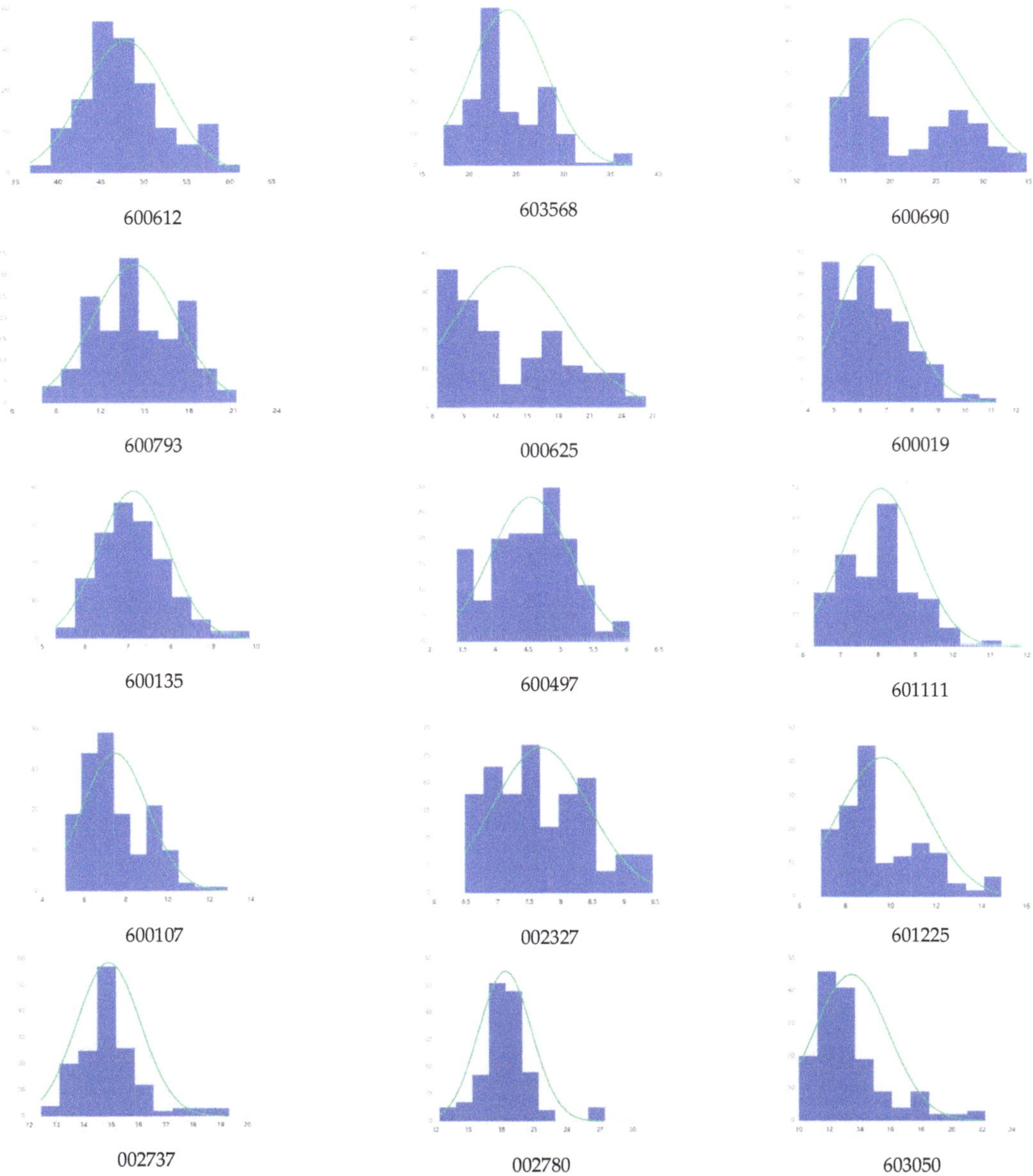

Figure 6. Histogram of normality test.

It is difficult for all the stock data to conform to the assumption that asset returns are normally distributed in MV. Secondly, the real loss refers to the fluctuation below the mean of returns; thus, the portfolio model based on the lower half-variance risk function is more realistic, so the MSV model is used for empirical analysis later in the paper.

4.3. Interpretation of Result

In order to verify the effectiveness of the semi-variance risk measure in practice, six different levels of return (0.005 to 0.0030) are set in this paper. Table 5 gives the risk values obtained by different algorithms at the same return level, and the best results are identified in bold font. A visualization of the Pareto frontier (PF) obtained by solving the four algorithms is given in Figure 7. The optimal investment ratios derived from each algorithm solved at the expected return level of 0.03 are given in Table 6 to visually compare the effectiveness of the APSO/DU algorithm in solving the MSVPOP.

Table 5. Experimental results of five algorithms.

NO.	μ	MSV			
		PSO-TVIW	PSO-TVAC	PSOCF	APSO/DU
1	0.0030	3.70×10^{-4}	3.82×10^{-4}	3.63×10^{-4}	$\mathbf{3.33 \times 10^{-4}}$
2	0.0025	3.52×10^{-4}	3.69×10^{-4}	3.57×10^{-4}	$\mathbf{3.16 \times 10^{-4}}$
3	0.0020	3.34×10^{-4}	3.48×10^{-4}	3.37×10^{-4}	$\mathbf{3.08 \times 10^{-4}}$
4	0.0015	3.20×10^{-4}	3.37×10^{-4}	3.27×10^{-4}	$\mathbf{3.02 \times 10^{-4}}$
5	0.0010	3.16×10^{-4}	3.26×10^{-4}	3.10×10^{-4}	$\mathbf{2.84 \times 10^{-4}}$
6	0.0005	2.99×10^{-4}	3.04×10^{-4}	2.95×10^{-4}	$\mathbf{2.78 \times 10^{-4}}$

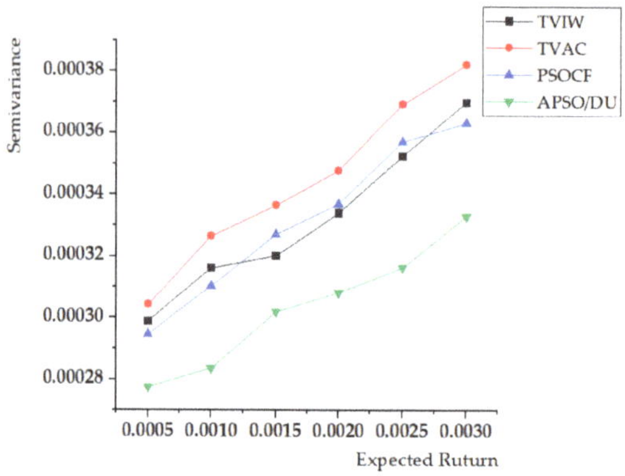

Figure 7. The obtained PF by five algorithms.

Table 6. The optimal investment ratio solved by each algorithm at μ = 0.03.

	Code	PSO-TVIW	PSO-TVAC	PSOCF	APSO/DU
1	600612	0.1166	0.0782	0.0932	0.0729
2	603568	0.0000	0.0655	0.0817	0.1129
3	600690	0.1015	0.0861	0.0055	0.0277
4	600793	0.0000	0.0832	0.0062	0.0168
5	000625	0.0000	0.0186	0.0360	0.0078
6	600019	0.0767	0.0504	0.0091	0.0692
7	600135	0.0692	0.0317	0.1147	0.0038
8	600497	0.0000	0.0812	0.0000	0.0031
9	601111	0.0563	0.0802	0.0573	0.0611
10	600107	0.1311	0.0589	0.0518	0.0081
11	002327	0.1057	0.0760	0.1935	0.1773
12	601225	0.0276	0.0839	0.1324	0.0992
13	002737	0.1338	0.0718	0.0019	0.1273
14	002780	0.1322	0.0746	0.1378	0.0725
15	603050	0.0493	0.0596	0.0790	0.1402
	MSV	3.70×10^{-4}	3.82×10^{-4}	3.63×10^{-4}	$\mathbf{3.33 \times 10^{-4}}$

Table 5 and Figure 7 show that as returns increase, the portfolio's risk also increases, in line with the law of high returns accompanied by high risk in the equity market. Taking the expected return $u = 0.003$ as an example, APSO/DU has the smallest value of risk (2.78×10^{-4}) and the PSO-TVAC algorithm has the largest value of risk (3.82×10^{-4}), so the portfolio solved by the APSO/DU algorithm is chosen at the expected return level of 0.03, corresponding to the smallest value of risk. A sensible person should choose this portfolio. Similar to the other return levels analyzed, the APSO/DU algorithm proposed in this paper is always lower than the results calculated by the other algorithms. The APSO/DU algorithm calculates a lower value of risk than the three classical adaptive improved particle swarm algorithms when the expected returns are the same, indicating that the combination of improved particle swarm solutions obtains relatively better results at the same expected return, and APSO/DU has stronger global search capability and more easily finds the optimal global solution.

The optimal investment ratios derived from each algorithm solved at the expected return level of 0.03 are given in Table 6 to visually compare the effectiveness of the APSO/DU algorithm in solving the MSVPOP.

5. Conclusions

In order to cope with the POPMSV challenge well, a multi-strategy adaptive particle swarm optimization, namely APSO/DU, was developed, which has the following two advantages. Firstly, the variable constraint (1) is set to better represent the stock selection, and asset weights of the solution in the POP help to cope with the MSVPOP challenge efficiently. Secondly, an improved particle swarm optimization algorithm (APSO/DU) with adaptive parameters was proposed by adopting a dual-update strategy. It can adaptively adjust the relevant parameters so that the search behavior of the algorithm can match the current search environment to avoid falling into local optimality and effectively balance global and local search. The sole adjustment of w and c_1 and c_2 would weaken the uniformity of the algorithm's evolutionary process and make it difficult to adapt to complex nonlinear optimization, so a dual dynamic adaptation mechanism is chosen to adjust the core parameters. The APSO/DU algorithm is more adaptable to nonlinear complex optimization problems, improving solution accuracy and approximating the global PF. The results show that APSO/DU exhibits stronger solution accuracy than the comparison algorithm, i.e., the improved algorithm finds the portfolio with the least risk at the same level of return, more closely approximating PF. The above research results can be used for investors to invest in low-risk portfolios with valuable suggestions with good practical applications.

Author Contributions: Conceptualization, Y.S. and Y.L.; methodology, Y.S. and W.D.; software, Y.L.; validation, H.C. and Y.L.; resources, Y.S.; data curation, Y.S.; writing—original draft preparation, Y.S. and Y.L.; writing—review and editing, H.C.; visualization, Y.S.; supervision, H.C.; project administration, H.C.; funding acquisition, H.C. and W.D. All authors have read and agreed to the published version of the manuscript.

Funding: This work was supported by the National Natural Science Foundation of China (61976124, 61976125, U2133205), the Yantai Key Research and Development Program (2020YT06000970), Wealth management characteristic construction project of Shandong Technology and Business University (2022YB10), the Natural Science Foundation of Sichuan Province under Grant 2022NSFSC0536; and the Open Project Program of the Traction Power State Key Laboratory of Southwest Jiaotong University (TPL2203).

Institutional Review Board Statement: Not applicable.

Informed Consent Statement: Not applicable.

Data Availability Statement: Not applicable.

Conflicts of Interest: The authors declare no conflict of interest.

References

1. Markowitz, H. Portfolio selection. *J. Financ.* **1952**, *7*, 77–79.
2. Markowitz, H. *Portfolio Selection: Efficient Diversification of Investments*; Wiley: New York, NY, USA, 1959.
3. Xu, G.; Bai, H.; Xing, J.; Luo, T.; Xiong, N.N.; Cheng, X.; Liu, S.; Zheng, X. SG-PBFT: A secure and highly efficient distributed blockchain PBFT consensus algorithm for intelligent Internet of vehicles. *J. Parallel Distrib. Comput.* **2022**, *164*, 1–11. [CrossRef]

4. Yu, C.; Liu, C.; Yu, H.; Song, M.; Chang, C.-I. Unsupervised Domain Adaptation with Dense-Based Compaction for Hyperspectral Imagery. *IEEE J. Sel. Top. Appl. Earth Obs. Remote Sens.* **2021**, *14*, 12287–12299. [CrossRef]
5. Jin, T.; Yang, X. Monotonicity theorem for the uncertain fractional differential equation and application to uncertain financial market. *Math. Comput. Simul.* **2021**, *190*, 203–221. [CrossRef]
6. Li, N.; Huang, W.; Guo, W.; Gao, G.; Zhu, Z. Multiple Enhanced Sparse Decomposition for Gearbox Compound Fault Diagnosis. *IEEE Trans. Instrum. Meas.* **2019**, *69*, 770–781. [CrossRef]
7. Bi, J.; Zhou, G.; Zhou, Y.; Luo, Q.; Deng, W. Artificial Electric Field Algorithm with Greedy State Transition Strategy for Spherical Multiple Traveling Salesmen Problem. *Int. J. Comput. Intell. Syst.* **2022**, *15*, 5. [CrossRef]
8. Zhong, K.; Zhou, G.; Deng, W.; Zhou, Y.; Luo, Q. MOMPA: Multi-objective marine predator algorithm. *Comput. Methods Appl. Mech. Eng.* **2021**, *385*, 114029.
9. Venkataraman, S.V. A remark on mean: Emivariance behaviour: Downside risk and capital asset pricing. *Int. J. Financ. Econ.* **2021**. [CrossRef]
10. Kumar, R.R.; Stauvermann, P.J.; Samitas, A. An Application of Portfolio Mean-Variance and Semi-Variance Optimization Techniques: A Case of Fiji. *J. Risk Financial Manag.* **2022**, *15*, 190. [CrossRef]
11. Wu, Q.; Gao, Y.; Sun, Y. Research on Probability Mean-Lower Semivariance-Entropy Portfolio Model with Background Risk. *Math. Probl. Eng.* **2020**, *2020*, 2769617. [CrossRef]
12. Wu, X.; Gao, A.; Huang, X. Modified Bacterial Foraging Optimization for Fuzzy Mean-Semivariance-Skewness Portfolio Selection. In Proceedings of the International Conference on Swarm Intelligence, Cham, Switzerland, 13 July 2020; pp. 335–346. [CrossRef]
13. Ivanova, M.; Dospatliev, L. Constructing of an Optimal Portfolio on the Bulgarian Stock Market Using Hybrid Genetic Algorithm for Pre and Post COVID-19 Periods. *Asian-Eur. J. Math.* **2022**, *15*, 2250246. [CrossRef]
14. Sun, Y.; Ren, H. A GD-PSO Algorithm for Smart Transportation Supply Chain ABS Portfolio Optimization. *Discret. Dyn. Nat. Soc.* **2021**, *2021*, 6653051. [CrossRef]
15. Zhao, H.; Chen, Z.G.; Zhan, Z.H.; Kwong, S.; Zhang, J. Multiple populations co-evolutionary particle swarm optimization for multi-objective cardinality constrained portfolio optimization problem. *Neurocomputing* **2021**, *430*, 58–70.
16. Deng, X.; He, X.; Huang, C. A new fuzzy random multi-objective portfolio model with different entropy measures using fuzzy programming based on artificial bee colony algorithm. *Eng. Comput.* **2021**, *39*, 627–649. [CrossRef]
17. Dhaini, M.; Mansour, N. Squirrel search algorithm for portfolio optimization. *Expert Syst. Appl.* **2021**, *178*, 114968. [CrossRef]
18. Shi, Y.; Eberhart, R.C. Empirical study of particle swarm optimization. In Proceedings of the 1999 Congress on Evolutionary Computation-CEC99 (Cat. No. 99TH8406), Washington, DC, USA, 6–9 July 1999; pp. 1945–1950. [CrossRef]
19. Shi, Y.H.; Eberhart, R.C. A modified particle swarm optimizer. In Proceedings of the 1998 IEEE International Conference on Evolutionary Computation, Anchorage, AK, USA, 4–9 May 1998; pp. 69–73.
20. Xiao, Y.; Shao, H.; Han, S.; Huo, Z.; Wan, J. Novel Joint Transfer Network for Unsupervised Bearing Fault Diagnosis from Simulation Domain to Experimental Domain. *IEEE/ASME Trans. Mechatron.* **2022**, *27*, 5254–5263. [CrossRef]
21. Yan, S.; Shao, H.; Xiao, Y.; Liu, B.; Wan, J. Hybrid robust convolutional autoencoder for unsupervised anomaly detection of machine tools under noises. *Robot. Comput. Manuf.* **2023**, *79*, 102441. [CrossRef]
22. Deng, W.; Zhang, L.; Zhou, X.; Zhou, Y.; Sun, Y.; Zhu, W.; Chen, H.; Deng, W.; Chen, H.; Zhao, H. Multi-strategy particle swarm and ant colony hybrid optimization for airport taxiway planning problem. *Inf. Sci.* **2022**, *612*, 576–593. [CrossRef]
23. Wei, Y.; Zhou, Y.; Luo, Q.; Deng, W. Optimal reactive power dispatch using an improved slime mould algorithm. *Energy Rep.* **2021**, *7*, 8742–8759. [CrossRef]
24. Song, Y.; Cai, X.; Zhou, X.; Zhang, B.; Chen, H.; Li, Y.; Deng, W.; Deng, W. Dynamic hybrid mechanism-based differential evolution algorithm and its application. *Expert Syst. Appl.* **2023**, *213*, 118834. [CrossRef]
25. Jin, T.; Zhu, Y.; Shu, Y.; Cao, J.; Yan, H.; Jiang, D. Uncertain optimal control problem with the first hitting time objective and application to a portfolio selection model. *J. Intell. Fuzzy Syst.* **2022**. [CrossRef]
26. Zhang, X.; Wang, H.; Du, C.; Fan, X.; Cui, L.; Chen, H.; Deng, F.; Tong, Q.; He, M.; Yang, M.; et al. Custom-Molded Offloading Footwear Effectively Prevents Recurrence and Amputation, and Lowers Mortality Rates in High-Risk Diabetic Foot Patients: A Multicenter, Prospective Observational Study. *Diabetes Metab. Syndr. Obesity Targets Ther.* **2022**, *15*, 103–109. [CrossRef]
27. Zhao, H.; Zhang, P.; Zhang, R.; Yao, R.; Deng, W. A novel performance trend prediction approach using ENBLS with GWO. *Meas. Sci. Technol.* **2023**, *34*, 025018. [CrossRef]
28. Ren, Z.; Han, X.; Yu, X.; Skjetne, R.; Leira, B.J.; Sævik, S.; Zhu, M. Data-driven simultaneous identification of the 6DOF dynamic model and wave load for a ship in waves. *Mech. Syst. Signal Process.* **2023**, *184*, 109422. [CrossRef]
29. Zhang, Z.; Huang, W.; Liao, Y.; Song, Z.; Shi, J.; Jiang, X.; Shen, C.; Zhu, Z. Bearing fault diagnosis via generalized logarithm sparse regularization. *Mech. Syst. Signal Process.* **2021**, *167*, 108576. [CrossRef]
30. Yu, Y.; Hao, Z.; Li, G.; Liu, Y.; Yang, R.; Liu, H. Optimal search mapping among sensors in heterogeneous smart homes. *Math. Biosci. Eng.* **2022**, *20*, 1960–1980. [CrossRef]
31. Chen, H.Y.; Fang, M.; Xu, S. Hyperspectral remote sensing image classification with CNN based on quantum genetic-optimized sparse representation. *IEEE Access* **2020**, *8*, 99900–99909. [CrossRef]
32. Zhao, H.; Yang, X.; Chen, B.; Chen, H.; Deng, W. Bearing fault diagnosis using transfer learning and optimized deep belief network. *Meas. Sci. Technol.* **2022**, *33*, 065009. [CrossRef]

33. Xu, J.; Zhao, Y.; Chen, H.; Deng, W. ABC-GSPBFT: PBFT with grouping score mechanism and optimized consensus process for flight operation data-sharing. *Inf. Sci.* **2023**, *624*, 110–127. [CrossRef]
34. Duan, Z.; Song, P.; Yang, C.; Deng, L.; Jiang, Y.; Deng, F.; Jiang, X.; Chen, Y.; Yang, G.; Ma, Y.; et al. The impact of hyperglycaemic crisis episodes on long-term outcomes for inpatients presenting with acute organ injury: A prospective, multicentre follow-up study. *Front. Endocrinol.* **2022**, *13*, 1057089. [CrossRef]
35. Chen, H.; Li, C.; Mafarja, M.; Heidari, A.A.; Chen, Y.; Cai, Z. Slime mould algorithm: A comprehensive review of recent variants and applications. *Int. J. Syst. Sci.* **2022**, *54*, 204–235. [CrossRef]
36. Liu, Y.; Heidari, A.A.; Cai, Z.; Liang, G.; Chen, H.; Pan, Z.; Alsufyani, A.; Bourouis, S. Simulated annealing-based dynamic step shuffled frog leaping algorithm: Optimal performance design and feature selection. *Neurocomputing* **2022**, *503*, 325–362. [CrossRef]
37. Dong, R.; Chen, H.; Heidari, A.A.; Turabieh, H.; Mafarja, M.; Wang, S. Boosted kernel search: Framework, analysis and case studies on the economic emission dispatch problem. *Knowl. Based Syst.* **2021**, *233*, 107529. [CrossRef]
38. Chen, M.; Shao, H.; Dou, H.; Li, W.; Liu, B. Data Augmentation and Intelligent Fault Diagnosis of Planetary Gearbox Using ILoFGAN Under Extremely Limited Samples. *IEEE Trans. Reliab.* **2022**. [CrossRef]
39. Tian, C.; Jin, T.; Yang, X.; Liu, Q. Reliability analysis of the uncertain heat conduction model. *Comput. Math. Appl.* **2022**, *119*, 131–140. [CrossRef]
40. Thakkar, A.; Chaudhari, K. A Comprehensive Survey on Portfolio Optimization, Stock Price and Trend Prediction Using Particle Swarm Optimization. *Arch. Comput. Methods Eng.* **2020**, *28*, 2133–2164. [CrossRef]
41. Harrison, K.R.; Engelbrecht, A.P.; Ombuki-Berman, B.M. Self-adaptive particle swarm optimization: A review and analysis of convergence. *Swarm Intell.* **2018**, *12*, 187–226. [CrossRef]
42. Boudt, K.; Wan, C. The effect of velocity sparsity on the performance of cardinality constrained particle swarm optimization. *Optim. Lett.* **2019**, *14*, 747–758. [CrossRef]
43. Clerc, M. *Particle Swarm Optimization*; John Wiley & Sons: Hoboken, NJ, USA, 2010.
44. Zhang, W.; Jin, Y.; Li, X.; Zhang, X. A simple way for parameter selection of standard particle swarm optimization. In Proceedings of the International Conference on Artificial Intelligence and Computational Intelligence, Berlin, Germany, 11–13 November 2011; pp. 436–443.
45. Huang, C.; Zhou, X.B.; Ran, X.J.; Liu, Y.; Deng, W.Q.; Deng, W. Co-evolutionary competitive swarm optimizer with three-phase for large-scale complex optimization problem. *Inf. Sci.* **2023**, *619*, 2–18. [CrossRef]
46. Liu, H.; Zhang, X.W.; Tu, L.P. A modified particle swarm optimization using adaptive strategy. *Expert Syst. Appl.* **2020**, *152*, 113353. [CrossRef]
47. Silva, Y.L.T.; Herthel, A.B.; Subramanian, A. A multi-objective evolutionary algorithm for a class of mean-variance portfolio selection problems. *Expert Syst. Appl.* **2019**, *133*, 225–241. [CrossRef]
48. Heidari, A.A.; Mirjalili, S.; Faris, H.; Aljarah, I.; Mafarja, M.; Chen, H. Harris hawks optimization: Algorithm and applications. *Futur. Gener. Comput. Syst.* **2019**, *97*, 849–872. [CrossRef]
49. Li, S.; Chen, H.; Wang, M.; Heidari, A.A.; Mirjalili, S. Slime mould algorithm: A new method for stochastic optimization. *Future Gener. Comput. Syst.* **2020**, *111*, 300–323. [CrossRef]
50. Yu, C.; Zhou, S.; Song, M.; Chang, C.-I. Semisupervised Hyperspectral Band Selection Based on Dual-Constrained Low-Rank Representation. *IEEE Geosci. Remote Sens. Lett.* **2021**, *19*, 5503005. [CrossRef]
51. Li, W.; Zhong, X.; Shao, H.; Cai, B.; Yang, X. Multi-mode data augmentation and fault diagnosis of rotating machinery using modified ACGAN designed with new framework. *Adv. Eng. Inform.* **2022**, *52*, 101552. [CrossRef]
52. He, Z.Y.; Shao, H.D.; Wang, P.; Janet, L.; Cheng, J.S.; Yang, Y. Deep transfer multi-wavelet auto-encoder for intelligent fault diagnosis of gearbox with few target training samples. *Knowl.-Based Syst.* **2019**, *191*, 105313. [CrossRef]

electronics

MDPI

Article

An Improved Whale Optimizer with Multiple Strategies for Intelligent Prediction of Talent Stability

Hong Li [1], Sicheng Ke [1], Xili Rao [1], Caisi Li [1], Danyan Chen [1], Fangjun Kuang [2,*], Huiling Chen [3,*], Guoxi Liang [4,*] and Lei Liu [5]

[1] Wenzhou Vocational College of Science and Technology, Wenzhou 325006, China
[2] School of Information Engineering, Wenzhou Business College, Wenzhou 325035, China
[3] College of Computer Science and Artificial Intelligence, Wenzhou University, Wenzhou 325035, China
[4] Department of Artificial Intelligence, Wenzhou Polytechnic, Wenzhou 325035, China
[5] College of Computer Science, Sichuan University, Chengdu 610065, China
* Correspondence: kfj@wzbc.edu.cn (F.K.); chenhuiling.jlu@gmail.com (H.C.); guoxiliang2017@gmail.com (G.L.)

Abstract: Talent resources are a primary resource and an important driving force for economic and social development. At present, researchers have conducted studies on talent introduction, but there is a paucity of research work on the stability of talent introduction. This paper presents the first study on talent stability in higher education, aiming to design an intelligent prediction model for talent stability in higher education using a kernel extreme learning machine (KELM) and proposing a differential evolution crisscross whale optimization algorithm (DECCWOA) for optimizing the model parameters. By introducing the crossover operator, the exchange of information regarding individuals is facilitated and the problem of dimensional lag is improved. Differential evolution operation is performed in a certain period of time to perturb the population by using the differences in individuals to ensure the diversity of the population. Furthermore, 35 benchmark functions of 23 baseline functions and CEC2014 were selected for comparison experiments in order to demonstrate the optimization performance of the DECCWOA. It is shown that the DECCWOA can achieve high accuracy and fast convergence in solving both unimodal and multimodal functions. In addition, the DECCWOA is combined with KELM and feature selection (DECCWOA-KELM-FS) to achieve efficient talent stability intelligence prediction for universities or colleges in Wenzhou. The results show that the performance of the proposed model outperforms other comparative algorithms. This study proposes a DECCWOA optimizer and constructs an intelligent prediction of talent stability system. The designed system can be used as a reliable method of predicting talent mobility in higher education.

Keywords: swarm intelligence; whale optimization algorithm; extreme learning machine; talent stability prediction; machine learning

Citation: Li, H.; Ke, S.; Rao, X.; Li, C.; Chen, D.; Kuang, F.; Chen, H.; Liang, G.; Liu, L. An Improved Whale Optimizer with Multiple Strategies for Intelligent Prediction of Talent Stability. *Electronics* **2022**, *11*, 4224. https://doi.org/10.3390/electronics11244224

Academic Editor: Maciej Ławryńczuk

Received: 18 November 2022
Accepted: 15 December 2022
Published: 18 December 2022

1. Introduction

Talent resources are the core resources on which universities rely for survival and development. A reasonable flow of talent can stimulate the vitality of the organization, improve the quality of talent, form a virtuous cycle and promote the complementary advantages of talent resources among universities. However, the "war for talents" against the background of "double tops" has led to the disorderly and utilitarian flow of talents in colleges and universities, an increase in the introduction to talents, a continuous increase in local competition, an accelerated frequency of talent flow and a structural imbalance of talent flow among colleges and universities in the region. This has engendered many negative effects on the development of universities. Therefore, a reasonable forecast in stable trends of university talent is crucial to the survival and development of universities. However, traditional methods have some limitations on predicting the stability of talent.

It is a new trend to use artificial intelligence algorithms to achieve accurate predictions of talent stability. There have been few studies that have used the artificial intelligence tools to solve the prediction issue of talent stability, so we have summarized some related works which used artificial intelligence tools to tackle the prediction problems for students; this is shown in Table 1.

Table 1. The latest research status of prediction issues for students.

Authors	Methods	Overview
Yang et al. [1]	The theory of planned behavior	They suggested that attitudes, subjective norms, perceived behavior, gender and parental experience have a significant impact on students' entrepreneurial intentions.
Gonzalez-Serrano et al. [2]	Questionnaire method	They demonstrated that attitudes and perceived behaviors were statistically significant.
Gorgievski et al. [3]	Values theory and planned behavior theory	They found a strong link between personal values and entrepreneurial career intentions.
Nawaz et al. [4]	Partial least squares structural equation modeling (PLS-SEM)	They found emotional intelligence, entrepreneurial self-efficacy and self-regulation also directly affect college students' entrepreneurial intentions.
Yang et al. [5]	Decision tree	They extracted four key attributes that affect students' intentions to start a career.
Djordjevic et al. [6]	Data analysis approach	They predicted the entrepreneurial intentions of youth in Serbia based on demographic characteristics, social environment, attitudes, awareness of incentives and environmental assessment.
Wei et al. [7]	Kernel extreme learning machine	They provided a reasonable reference for the formulation of talent training programs and guidance for the entrepreneurial intention of students.
Bhagavan et al. [8]	Data mining tools and methods	They predicted the current performance of students through their early performance and awareness, and identified students' expected abilities.
Huang et al. [9]	Artificial intelligence algorithms and fuzzy logic models	They designed a diversified employment recommendation system, combined with students' personal interests, and provided employment plans.
Li et al. [10]	The cluster analysis technology model	They achieved accurate predictions of the employment situation of graduates.

The swarm intelligence algorithm (SIA) is a crucial optimization method by which to predict traditional talent stability. SIA is derived from natural phenomena or group behaviors, etc., such as group predation and physical phenomena. Optimization principles exist within these phenomena. As a kind of SIA, the whale optimization algorithm (WOA) [11] has a clear algorithm structure and good performance, which was proposed in 2016. It was designed by simulating the hunting behavior of whales. During foraging, the whales use bubbles as tools to surround their prey. Furthermore, the algorithm has been used in many natural science fields, such as shop scheduling problems [12,13] and engineering design problems. Navarro et al. [14] proposed a version of the WOA with the K-means mechanism to explore the algorithm's search space. The proposed model was effective against resolving complex optimization issues. Abbas et al. [15] proposed a combination of the technique of an extremely randomized tree with the WOA for the detection and prediction of medical diseases. Abd et al. [16] introduced a novel WOA version application for multilevel threshold image segmentation. Abdel-Basset et al. [17] presented a new WOA version based on local search mechanisms to optimize a scheduling problem with the multimedia data objects field. Qiao et al. [18] presented a novel version of the WOA, which combined the worst individual disturbance and the neighborhood mutation search strategy for solving engineering design problems. Peng et al. [19] introduced an enhanced WOA, which combined the information-sharing search strategy and the Nelder-Mead simplex strategy, to evaluate the parameters of solar cells and photovoltaic modules. Abderazek et al. [20] presented the WOA and a moth-flame optimizer for optimizing spur gear design.

For the high-quality training of talent, in addition to focusing on the employment and entrepreneurship of university students, the stability of talents is also an important foundation for social and economic development. Employment stability reflects psychological satisfaction with practitioners regarding the employment unit, employment environment, remuneration package and career development. In the past five years, the average turnover rate of several colleges and universities in Wenzhou was 28.1%. An appropriate turnover rate is conducive to the "catfish effect" in enterprises and institutions, and stimulates the vitality and competitiveness of the organization; however, an excessive turnover rate has a negative impact on the human resource costs and economic efficiency of universities, as well as their social reputation and the quality development of the economy and society.

Big data has a wide scope of application in the field of talent mobility management. Through the effective mining of big data onto talent flows in a university, the stability of talent employment is analyzed, and the correlation hypothesis is verified by integrating an intelligent optimization algorithm, neural network, support vector machine and other machine learning methods; an intelligent prediction model is then constructed. At the same time, key factors affecting the stability of talent employment are mined, and the key influencing factors are analyzed in depth to explore the main features affecting the stability of talent employment and to provide reference for government decision-making and policy formulation. The main contributions are shown as bellow:

(1) A multi-strategy hybrid modified whale optimization algorithm is proposed.
(2) Introducing the crossover operator to facilitate the exchange of information and improve the problem of dimensional lag.
(3) DECCWOA is verified on the 35 benchmark functions to demonstrate the optimization performance.
(4) DECCWOA is combined with KELM and feature selection to achieve efficient talent stability intelligence prediction.
(5) Results show the proposed methods surpass other reported approaches.

The remainder of this paper is structured as follows. Section 2 reviews the whale optimization algorithm. Section 3 provides a comprehensive description of the proposed method. The proposed method is verified and applied using benchmark function experiments and feature selection experiments in Section 4. The conclusion and future work are outlined in Section 5.

2. Relate Work

In recent years, swarm intelligence optimization algorithms have emerged, such as the Runge Kutta optimizer (RUN) [21], the slime mold algorithm (SMA) [22], the Harris hawks optimization (HHO) [23], the hunger games search (HGS) [24], the weighted mean of vectors (INFO) [25], and the colony predation algorithm (CPA) [26]. Moreover, they have achieved very good results in many fields, such as feature selection [27,28], image segmentation [29,30], bankruptcy prediction [31,32], plant disease recognition [33], medical diagnosis [34,35], the economic emission dispatch problem [36], robust optimization [37,38], expensive optimization problems [39,40], the multi-objective problem [41,42], scheduling problems [43–45], optimization of a machine learning model [46], gate resource allocation [47,48], solar cell parameter identification [49] and fault diagnosis [50]. In addition to the above, the whale optimization algorithm (WOA) [11] is an optimization algorithm simulating the behaviors of whales rounding up their prey. During feeding, whales surround their prey in groups and move in a whirling motion, releasing bubbles in the process, and thus, closing in on their prey. In the WOA, the feeding process of whales can be divided into two behaviors, including encircling prey and forming bubble nets. During each generation of swimming, the whale population will randomly choose between these two behaviors to hunt. In d-dimensional space, suppose that the position of each individual in the whale population is expressed as $X = (x1,\ x2,\ \ldots,\ xD)$.

Agrawal et al. [51] proposed an improved WOA and applied it to the field of feature selection [52]. Bahiraei et al. [53] proposed a novel perceptron neural network, which combined the WOA and other algorithms, and was applied to the field of polymer materials. Qi et al. [54] introduced a new WOA with a directional crossover strategy, directional mutation strategy, and levy initialization strategy. The potential for using the suggested approach to address engineering issues is very high. Bui et al. [55] proposed a neural-network-model-based WOA, which also integrated a dragonfly optimizer and an ant colony optimizer, and was applied to the construction field. Butti et al. [56] presented an effective version of the WOA to optimize the stability of power systems. Cao et al. [57] also proposed a new WOA to improve the efficiency of the proton exchange of membrane fuel cells. Cercevik et al. [58] presented an optimization model, combined with the WOA and others, to improve the parameters of seismic isolated structures. Zhao et al. [59] presented a susceptible-exposed-infected-quarantined (hospital or home)-recovered model based on the WOA and human intervention strategies to simulate and predict recent outbreak transmission trends and peaks in Changchun. A brand-new hybrid optimizer was developed by Fan et al. [60] to solve large-scale, complex practical situations. The proposed hybrid optimization algorithm combined a fruit flew optimizer with the WOA. Raj et al. [61] proposed the application of the WOA as a solution to reactive power planning with flexible transmission systems. Guo et al. [62] proposed an improved WOA with two strategies to improve the exploration and exploitation abilities of the WOA, including the random hopping update mechanism and random control parameter mechanism. To improve the algorithm's convergence rate and accuracy, a new version of the WOA was presented by Jiang et al. [63] to apply constraints to engineering tasks.

Although the WOA has obtained good results in many fields, the algorithm easily falls into the local optimum in the face of complex problems. Therefore, many excellent improvement algorithms have been proposed. For example, Hussien et al. [29] proposed a novel version of the whale optimizer with the gaussian walk mechanism and the virus colony search strategy to improve convergence accuracy. To solve the WOA's susceptibility to falling into the local optimum with slow convergence speeds, an improved WOA with a communication strategy and the biogeography-based model was proposed by Tu et al. [64]. Wang et al. [65] presented a novel-based elite mechanism WOA, with a spiral motion strategy to improve the original algorithm. Ye et al. [49] introduced an enhanced WOA version of the levy flight strategy and search mechanism to improve the algorithm's balance. Abd et al. [66] presented an innovative method to enhance the WOA, including the differential evolution exploration strategy. Abdel-Basset et al. [67] introduced an enhanced whale optimizer, which was combined with a slime mold optimizer to improve the performance of the algorithm. To enhance the WOA's search ability and diversity, a novel version of the WOA with an information exchange mechanism was proposed by Chai et al. [68]. Heidari et al. [69] presented a whale optimizer with two strategies, including an associative learning method and a hill-climbing algorithm. Jin et al. [70] proposed a dual operation mechanism based on the WOA to solve the slow convergence speed problem. Therefore, the WOA is an effective optimizer by which to improve the performance of traditional talent stability prediction.

3. Materials and Methods

This section will improve the problems existing in the traditional whale optimization algorithm, so as to propose a new version of the algorithm. During the process of the whale population continuously approaching the optimal position, the population appears in an aggregation state, which is the main reason for the algorithm falling into the local optimal. Based on this, the DE operation is performed on the whale population during a certain period, and the whale population is disturbed by the differential information of multiple individuals, so as to ensure the diversity of the population. By introducing the idea of a crisscross optimization algorithm, a vertical crossover is performed in dimensions to improve dimensional stagnation as iterations progress, and horizontal crossover is performed between individuals to fully facilitate the exchange of information between individuals, allowing the problem space to be fully searched, effectively improving the search capability of the algorithm. Overall, the proposed algorithm is named as the DE-based crisscross whale algorithm (DECCWOA).

3.1. Whale Optimization Algorithm

3.1.1. Encircling Prey

In the process of encircling the prey, each individual will choose the position closest to the prey in the group, that is, the global optimal solution, or will randomly select a whale and approach it. The equation for updating the position of the whale is shown in Equation (1).

$$X_i^{t+1} = X_{best}^t - A\left| C \times X_q^t - X_{it}^t \right| \tag{1}$$

where X_q^t is X_{best}^t when the whale swims toward the optimal whale position, and X_{rand}^t when the whale swims toward the random whale position. A is a random number with a uniform distribution between $(-a,\ a)$, and the initial value of a is 2, which linearly decreases to 0 with the number of iterations. C is a random number that satisfies the uniform distribution, and its value is between (0, 2). The choice of whether the whale individual swims toward the optimal whale or random position is up to the value of A. When $|A < 1|$, the whale decides to swim toward the optimal individual; otherwise, the whales will select a random location in the population and approach it.

3.1.2. Forming Bubble Nets

Whales release bubbles while hunting, thus forming a spiraling, blistering net to repel the prey. If bubble feeding is chosen, the whale first calculates the distance between itself and the best whale, then swims upwards in a spiral and spits out bubbles of varying sizes to feed on the fish and prawns. At this point, the position of the whale is updated by the equation shown in Equation (2).

$$X_i^{t+1} = |X_{best}^t - X_i^t| \times e^{bl} \times cos(2\pi l) + X_{best}^t \tag{2}$$

where b is a constant, and l is a random number between $[-1, 1]$, meeting a uniform distribution.

3.2. Differential Evolution Algorithm (DE)

The differential evolution algorithm (DE) [71] was proposed in 1997 based on the idea of evolutionary algorithms, such as genetic algorithms, which are essentially multi-objective optimization algorithms that can be used to solve the overall optimal solution in a multi-dimensional space. The DE is the same as other genetic algorithms in that the main process consists of three steps: mutation, crossover and selection. However, the variance vector of the differential DE is generated from the parent differential vector and is crossed with the parent individual vector to generate a new individual vector, which is directly selected with its parent individual. Suppose the position vector of the i-th individual in the population is X_i.

3.2.1. Crossover Operations

The basic variance vector is generated by Equation (3), and $r_1 \neq r_2 \neq r_3$. Therefore, in the DE algorithm, the population must be greater than 3. F is the crossover operator, with a value usually between [0, 2], which controls the amplification of the deviation vector. Commonly, the difference between the two vectors is multiplied by the crossover operator and added to the third vector to generate a new mutation vector.

$$X_i = X_{r_1} + F \times (X_{r_2} - X_{r_3}) \tag{3}$$

In this article, in order to allow for faster convergence of the population algorithm while maintaining population diversity, we attempt to calculate the difference between the position of the current population and the optimal population position (X_{best}), on the basis of which a new variant population is generated. Therefore, Equation (3) is rewritten as shown in Equation (4).

$$V_i = X_i + F \times (X_{best} - X_i) \tag{4}$$

3.2.2. Mutation Operations

To increase the diversity of the interference vectors, crossover operations are introduced. Equation (5) presents the principle of the crossover operation.

$$U_{i,j} = \begin{cases} V_{ji} \ \ if \ randb(j) \ \leq \ CR \ or \ j = rnbr(i) \\ X_{i,j} \ \ if \ randb(j) > CR \ or \ j \ \neq \ rnbr(i) \end{cases}, \ i = 1,\ 2,\ \ldots,\ NP;\ j = 1,2,\ \ldots,\ D \tag{5}$$

$randb(j)$ denotes the generation of the j-th estimate of a random number between [0, 1] and *rnbr* denotes a randomly chosen sequence. CR is the crossover operator. In simple terms, if the randomly generated *randb(j)* is less than CR or $j = r$, then the variant population is placed in the selection population; if not, the original population is placed in the selection population.

3.2.3. Selection Operation

In order to decide whether the vectors in the selection population can become part of the next generation, the newly generated position vectors are compared with the current target vectors, and, if it appears that the objective function is further optimized or the original state is maintained, then, the newly generated individuals will appear in the next generation. The selection operation is defined as shown in Equation (6).

$$X_i = \begin{cases} U_i \ \ if \ f(U_i) < f(X_i) \\ X_i \ \ if \ f(U_i) \ \geq f(X_i) \end{cases} \tag{6}$$

The DE is a simple and easy-to-implement algorithm that mainly performs genetic operations by means of differential variation operators. The algorithm has shown good robustness and efficiency in solving most optimization problems [72–75]. Furthermore, the algorithm is intrinsically parallel and can coordinate searches, so that the DE has a faster convergence rate for the same requirement.

3.3. *Crisscross Optimization Algorithm*

The crisscross optimization algorithm (CSO) [76] is a new population-based stochastic search algorithm that performs both horizontal and vertical crossover in each generation during each iteration, thus allowing certain dimensions of the population that are trapped in a pseudo-optimal a chance to jump out. The new individuals obtained after each crossover need to go through competition, and only the individuals better than the parent generation will be retained for the next iteration.

3.3.1. Horizontal Crossover Operator

A horizontal crossover operation is similar to crossover operations in genetic algorithms, a kind of arithmetic crossover between the same dimension of two different individual particles in a population. Assuming a horizontal crossover in the d-th dimension for the i-th and j-th parent individual particles, the formula for generating offspring is shown in Equations (7) and (8).

$$MS_{hc}(i,d) = r_1 \times X(i,\ d) + (1 - r_1) \times X(j,\ d) + c_1 \times (X(i,\ d) - X(j,\ d)) \tag{7}$$

$$MS_{hc}(j,d) = r_1 \times X(j,\ d) + (1 - r_1) \times X(i,\ d) + c_1 \times (X(j,\ d) - X(i,\ d)) \tag{8}$$

where r_1 and r_2 are random numbers between [0, 1], and, c_1 and c_2 are random numbers between $[-1, 1]$. $X(i,\ d)$ and $X(j,\ d)$ represent the d-th dimension of the i-th and j-th individuals in the population, respectively. $MS_{hc}(i,d)$ and $MS_{hc}(j,d)$ are the d-th dimension of the offspring generated by $X(i,\ d)$ and $X(j,\ d)$ via horizontal crossover, respectively. From a sociological point of view, $r_1 \times X(i,\ d)$ is the memory term of particle $X(i)$. $(1 - r_1) \times X(j,\ d)$ is the group cognitive term of particles $X(i)$ and $X(j)$, representing the interaction between different particles. c_1 is the learning factor, $c_1 \times (X(i,\ d) - X(j,\ d))$ can effectively enlarge the search interval and search for optimization at the edge. The schematic diagram of the horizontal crossover operation is shown in Figure 1.

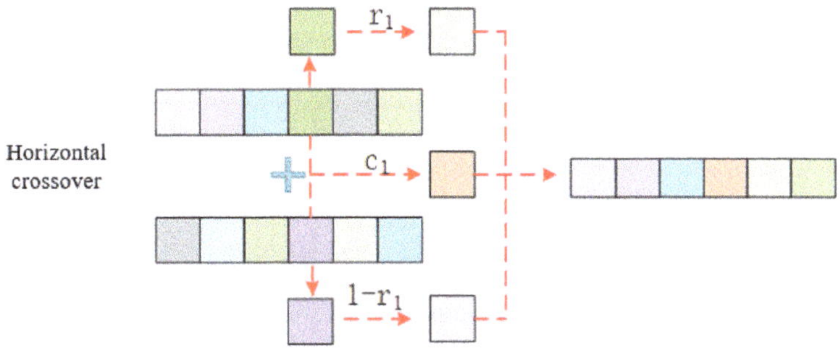

Figure 1. Schematic of horizontal crossover.

3.3.2. Vertical Crossover Operator

A vertical crossover is an arithmetic crossover between two different dimensions of a particle in a population. Since different dimensional elements have different ranges of values, the two dimensions need to be normalized before crossover. Furthermore, in order to allow the dimension that has stalled in the local optimum to jump out of the local optimum without destroying the information of the other dimension, only one child particle is generated for each vertical crossover operation, and only one of the dimensions is updated. The vertical crossover operation is defined by Equation (9).

$$MS_{vc}(i,\ d_1) = r \times X(i,\ d_1) + (1 - r) \times X(i,\ d_2),\ i \in\ N(1,\ M),\ d_1\, d_2\ \in\ N(1,\ D) \quad (9)$$

where r is a random number between [0, 1]. $MS_{vc}(i,\ d_1)$ is the d_1-th dimension of the offspring produced by the d_1-th and d_2-th dimensions of individual $X(i)$ by vertical crossover. The new individual contains not only the information of the d_1-th dimension of the parent particle, but also the information of the d_2-th dimension with a certain probability, and the information of the d_2-th dimension will not be destroyed during the crossover. A schematic diagram of the vertical crossover is shown in Figure 2.

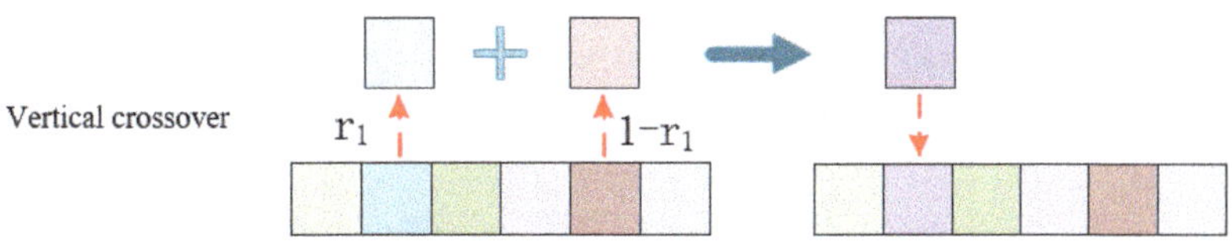

Figure 2. Schematic of vertical crossover.

3.4. Framework of Proposed DECCWOA

The whale algorithm, the crossover and mutation operations in the DE and the criss-cross operators together form the overall framework of the DECCWOA. We consider a positive population renewal to be complete when a location closer to a food source is found in one iteration. When the entire whale population has completed S positive updates, we consider the population to have been concentrated and to be losing population diversity. In one iteration, after the whales have completed one location update, it is determined whether the population has completed S positive updates, and, if so, the crossover and mutation operations of the DE algorithm are performed, resulting in a perturbation of the whale population, further ensuring population quality. Moreover, vertical crossover is performed in dimensions to improve dimensional stagnation as iterations progress, and, when the entire population has completed one location update, horizontal crossover is performed between individuals to fully facilitate the exchange of information between individuals, allowing the problem space to be fully searched, effectively improving the search capability of the algorithm. The pseudo-code of the DECCWOA can be seen in Algorithm 1, and a flow chart of the overall DECCWOA framework is shown in Figure 3.

Algorithm 1: The pseudo-code of the DECCWOA

Input: Number of populations N, maximum number of iterations T, , objective function $fobj$;
Output: Optimum whale position X_{best};

Initialize the whale population positions X;
Calculate fitness values for all individuals in the whale population and sort them;
Set the position of the individual with the smallest fitness value f_{best} to X_{best};
Set s = 0;
while ($t < T$)
for each agent
    Update a, A, C l, S and p;
    if $p < 0.5$
      if $|A| < 1$ && $s < S$
          Update the position of agent using Equation (1), and set X_q as X_{best};
      elseif $|A| \geq 1$ && $s < S$
          Select a random search agent as X_{rand};
          Update the position of agent using Equation (1), and set X_q as X_{rand};
      elseif $s \geq S$
          Performing crossover and mutation operations in DE;
      end if
    else
      Update the position of agent using Equation (2);
    end if
    Perform vertical crossover operator using Equation (9);
  end for
  Perform horizontal crossover operation using Equations (7) and (8);
  Calculate fitness values for all individuals in the whale population and sort them;
  Set the position of the individual with the smallest fitness value g_{best} to Xg_{best};
  if $g_{best} < f_{best}$
   Updates X_{best} and f_{best};
  s = s + 1;
  end if
  t = t + 1;
end while
Return X_{best}

In the basic whale algorithm, only each individual in the population is updated according to the corresponding situation in each iteration, excluding other complex operations. Therefore, the time complexity of the algorithm is only related to the maximum number of iterations T and the population size N; that is, the time complexity of the whale algorithm is $O(T * N)$. When executing the vertical crossover algorithm, the time complexity of the vertical crossover is O(D); a vertical crossover is performed at the end of each individual update as the vertical crossover occurs in dimension D. When the horizontal crossover is executed after the whole population has been updated, the time complexity of the horizontal crossover is $O(N * D)$ depending on the size of the individuals and the dimension of the problem, as the horizontal crossover is performed by communicating between individuals and updating the dimensional information in turn. In DE, a crossover is performed, and the mutation and selection operations are only related to dimensions, so the time complexity of an iteration is $O(D)$. In this work, only when the position of the population is updated every time and a certain period is met, we carry out an operation of crossover, mutation and selection for the population. Therefore, the operation of theoretically introducing the DE does not add a high time cost to the algorithm. In summary, the time complexity of the proposed algorithm DECCWOA is $O(T * (O(N * D) + O(N)))$.

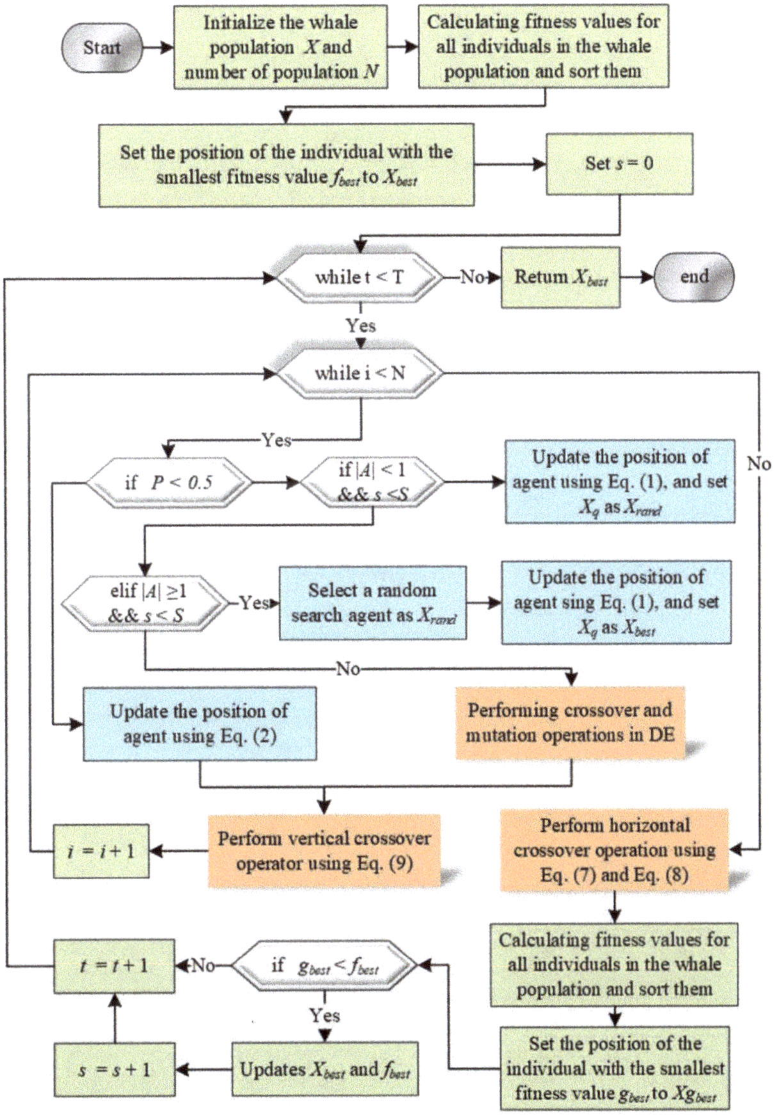

Figure 3. Flow chart of the DECCWOA framework.

4. Experimental Results

This section presents a quantitative analysis of the introduced DE and CSO mechanisms and presents the experimental results comparing the proposed algorithm, DECCWOA, with other improved WOA algorithms and improved swarm intelligence algorithms that have better performance on 35 benchmark functions. Furthermore, to show that the proposed algorithm is still valid for practical applications, the DECCWOA is applied to the intelligent prediction of talent stability in universities. All experiments were carried out on a Windows Server 2012 R2 operating system with Intel(R) Xeon(R) Silver 4110 CPU (2.10 GHz) and 32.GB RAM. All algorithms were coded and run on MATLAB 2014b.

To ensure fairness of the experiment, all algorithms were executed in the same environment. For all algorithms, the population size was set to 30, the maximum number of function evaluations was set to 300,000 and, to avoid the effect of randomness on the results, each algorithm was individually executed 30 times on each benchmark function. *avg* and *std* reflect the average ability and stability of each algorithm after 30 independent experiments. To allow a more visual presentation of the average performance of all the algorithms, the Freidman test is used to evaluate the experimental results of all algorithms on the benchmark function and the final ranking is recorded.

4.1. Experimental Results of the DECCWOA on Benchmark Functions

The DECCWOA and its related comparison algorithm conducted comparison experiments on 35 benchmark functions selected from 23 benchmark functions and CEC2014. In detail, Table A1 of the Appendix A shows a summary of the 35 test functions, which can be divided into three categories, including unimodal functions, multimodal functions and hybrid functions.

4.1.1. Parameter Sensitivity Analysis

Not every dimension of an individual is selected for crossover in a vertical crossover operation. In the vertical crossover operation, there is a key parameter p_2. When the random probability is less than p_2, the crossover operation is performed in the corresponding dimension of the individual, as shown in Equation (9). Otherwise, the operation is considered not to be performed in that dimension. The possible values of p_2 are 0.1, 0.2, 0.3, 0.4, 0.5, 0.6, 0.7, 0.8, 0.9 and 1.0. In order to visually present the impact of p_2 on the optimization capabilities of the DECCWOA, we conducted comparative experiments using different versions. The names corresponding to the different algorithm versions are shown in Table 2.

Table 2. Names of different algorithm versions when p_2 is different.

Algorithm	DEC CWOA1	DEC CWOA2	DEC CWOA3	DEC CWOA4	DEC CWOA5	DEC CWOA6	DEC CWOA7	DEC CWOA8	DEC CWOA9	DEC CWOA10
p_2	0.1	0.2	0.3	0.4	0.5	0.6	0.7	0.8	0.9	1.0

Different values of p_2 have a direct impact on the optimization of the DECCWOA. Table 3 shows the results of the DECCWOA2 and Table A2 in Appendix A shows the detailed results when p_2 is taken to different values. The rankings generated by the Friedman test show that when the value of p_2 is too large, the less effective the average optimization is. The significance of introducing a longitudinal crossover operator is to help the population change dimensional stagnation. That is because when p_2 takes a larger value, it means that each dimension of the individual changes with a high probability. This not only changes the dimension of stagnation, but also the dimension of having good performance along with it. Notably, when p_2 is taken as 0.1 versus 0.2, the performance is similar for 28 of the 35 benchmark functions, but, for overall performance, the DECCWOA2 is slightly better. This is because, when the value of p_2 is too small, only a few dimensions are adjusted after the individual enters the vertical crossover operator, which does not have the problem of falling into local optima due to dimensional stagnation being significantly improved, especially in solving multimodal functions and hybrid functions. Therefore, in the course of the next experiments, p_2 was set to 0.2.

Table 3. Experimental results for analysis of different versions.

Algorithms	Overall	
	+/−/=	Rank
DECCWOA1	~	2
DECCWOA2	4/3/28	1
DECCWOA3	14/3/18	6
DECCWOA4	12/2/21	5
DECCWOA5	14/4/17	3
DECCWOA6	14/5/16	4
DECCWOA7	14/4/17	7
DECCWOA8	15/4/16	8
DECCWOA9	17/2/16	9
DECCWOA10	16/2/17	10

4.1.2. Comparison of Mechanisms

In order to verify the effectiveness of the introduced mechanism in improving the optimization capabilities of the WOA, ablation studies on the integrated DE and CSO were conducted. Table 4 presents the comparison results for the introduced mechanisms. The detailed results can be found in Table A3 of Appendix A. Notably, on most of the benchmark functions, the DECCWOA has the best optimization capability by performing the Friedman test on 30 times randomized trials. Furthermore, the CCWOA with the introduction of CSO outperforms the WOA on more than 90% of the benchmark functions. However, for the DEWOA, which introduces the DE into the WOA, although the overall results are not significantly improved, a comparison of the optimization performance of CCWOA and DECCWOA shows that the combination of CSO with the DE makes the WOA more optimizable.

Table 4. Comparison results for the introduced mechanisms.

Algorithms	Overall	
	+/−/=	Rank
DECCWOA	~	1
DEWOA	31/1/3	4
CCWOA	18/4/13	2
WOA	25/1/9	3

Convergence curves of the comparison results for the introduced mechanisms are shown in Figure 4. Among them, the CCWOA excels in both optimization accuracy and convergence speed on F4, F6 (from 23 benchmark functions) and unimodal functions of F14 (from CEC2014), F12, F13 (from 23 benchmark functions), multimodal functions of F18, F19, F23, F24, F29 (from CEC2014) and hybrid functions of F30 and F32. In particular, CCWOA also has stronger search ability in multimodal functions and hybrid functions. This shows that CSO effectively improves the problem that the basic WOA is prone to falling into the local optimum. It is also worth noting that the introduction of the DE did not give the desired results on most of the benchmark functions. However, when acting together with a CSO on the WOA, the convergence speed and optimization accuracy of the DECCWOA are significantly improved. Especially in F4, F6 and F13, it is obvious that the DECCWOA has better performance than the CCWOA. This is because we perform the DE crossover and mutation operations over a period of time in order to take advantage of differences between individuals to disturb the population, but do not perform the rounding up of prey in the basic WOA at this time, thus slowing the efficiency of the whale population towards the food source. However, when CSO is applied to the whole population, not only is the information between individuals utilized, but also the information in the spatial dimension is considered. Combined with the periodic perturbation of the DE, the whale population can search the whole problem space more efficiently.

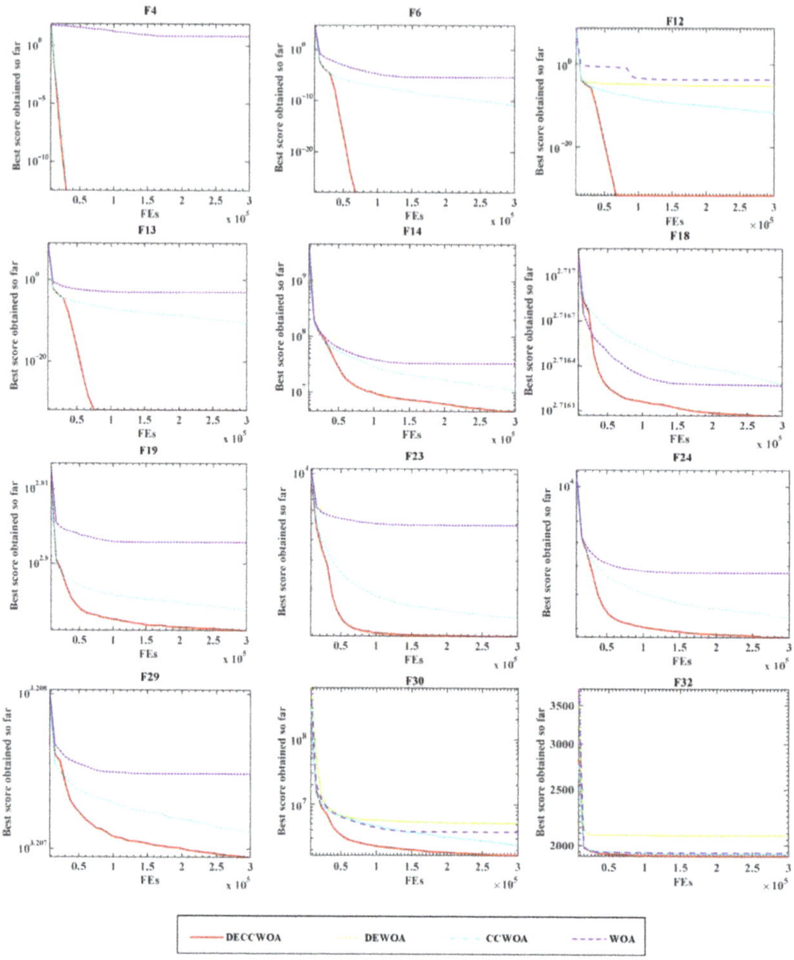

Figure 4. Convergence curves of the comparison results for the introduced mechanisms.

4.1.3. Comparison with Improved WOA Versions

In order to provide a clearer picture of the results of the experiments comparing the DECCWOA with other improved WOA algorithms for 35 benchmark functions, *avg* and *std* of all functions obtained after 30 independent experiments on the corresponding benchmark functions and the average ranking results of the Friedman test on the average results are recorded in Table 5. The detailed results are shown in Table A4 of Appendix A. The composite average ranking of the DECCWOA is the highest, followed by the RDWOA and the CCMWOA with the lowest. Among them, +/−/= respectively records the number of benchmark functions that the DECCWOA is superior to, inferior to and similar to in terms of performance to other competing algorithms among the 35 test functions. For the worst performing CCMWOA, the DECCWOA outperforms it for twenty-eight benchmark functions, has the same performance on five functions, and performs slightly worse on only two functions. Moreover, compared to the RDWOA, which ranks second overall, the DECCWOA has better performance for sixteen benchmark functions, has the same optimization ability for thirteen functions and only has poor performance for six functions. This proves that the DECCWOA has better performance than other improved WOA algorithms for most of the optimization problems, further demonstrating that the introduced CSO and DE have a positive steering effect on improving the basic WOA, such as slow convergence speed and poor accuracy guiding role.

Table 5. Comparison results for DECCWOAs with improved WOA versions.

Algorithms	Overall	
	+/−/=	Rank
DECCWOA	~	1
RDWOA	16/6/13	2
ACWOA	26/3/6	6
CCMWOA	28/2/5	9
CWOA	29/0/6	8
BMWOA	32/1/2	7
BWOA	23/3/9	4
LWOA	26/4/5	5
IWOA	24/1/10	3

In this section, the performance of the DECCWOA is compared with other improved versions of the WOA, including the RDWOA, the ACWOA, the CCMWOA [77], the CWOA [78], the BMWOA, the BWOA, the LWOA [79] and the IWOA [80]. Figure 5 shows the convergence curves of the average results obtained after 30 operations for all algorithms. On unimodal functions such as F6, it can be intuitively observed that the DECCWOA has the strongest search capability, with the RDWOA in second place, but the DECCWOA has a better performance than the RDWOA in terms of both accuracy and convergence speed. For both F12 and F13, the optimal values found by the other improved WOA algorithms are similar and more concentrated; however, the accuracy of the optimization obtained by the DECCWOA calculation is substantially improved. On F18, F19, F21, F23, F25 and F29, the DECCOWA can still search for more satisfactory optimal values compared to the other improved WOA algorithms. This demonstrates that the improvements to the WOA in this experiment are relatively more effective, and that, even when solving for multimodal functions, the DECCWOA can still jump out of the local optimum in time to obtain a high-quality optimal solution.

4.1.4. Comparison with Advanced Algorithms

Table 6 presents the comparison results for the DECCWOA with advanced algorithms. The detailed results can be found in Table A5 of Appendix A. *avg* reflects the average optimization ability of the algorithm after independently running on the benchmark function for 30 times, and *std* represents the influence of randomness on the optimization ability of the algorithm, which further reflects the stability of the algorithm to solve problems. From Table 6, the DECCWOA is superior to the IGWO on twenty functions and is inferior to the IGWO on eight functions (F3, F5, F22, F28, F30, F33, F34, F35). The DECCWOA beats the OBLGWOA on nineteen functions and loses to the OBLGWO on five functions (F3, F7, F28, F30, F33). For the CGPSO, ALCPSO and RCBA, the DECCWOA is inferior to them on nine functions, and outperforms most of the others. In detail, the DECCWOA is worse than the CGPSO at F5, F7, F8, F16, F17, F26, F30, F33 and F34, worse than the ALCPSO at F15, F16, F20, F21, F22, F28, F30, F33 and F34 and is worse than the RCBA at F14, F15, F16, F17, F20, F26, F30, F33 and F34. The DECCWOA beats the CBA on twenty-four functions, and loses to the CBA in six functions (F15, F16, F20, F30, F33, F34). The DECCWOA outperforms the OBSCA on 32 functions and only performs worse than the OBSCA on one function of F3. The DECCWOA is worse than the SCADE on F3 and F6. Based on the analysis above, the DECCWOA did not perform as well as the ALPSO, RCBA and CBA on the three unimodal functions (F14~F16) selected in CEC2014, but demonstrated competitive performance on the seven unimodal functions (F1~F7) selected from the twenty-three benchmark functions. The DECCWOA does not perform more competitively than the other comparison algorithms in terms of hybrid functions, but the DECCWOA performs well on most of the multimodal functions.

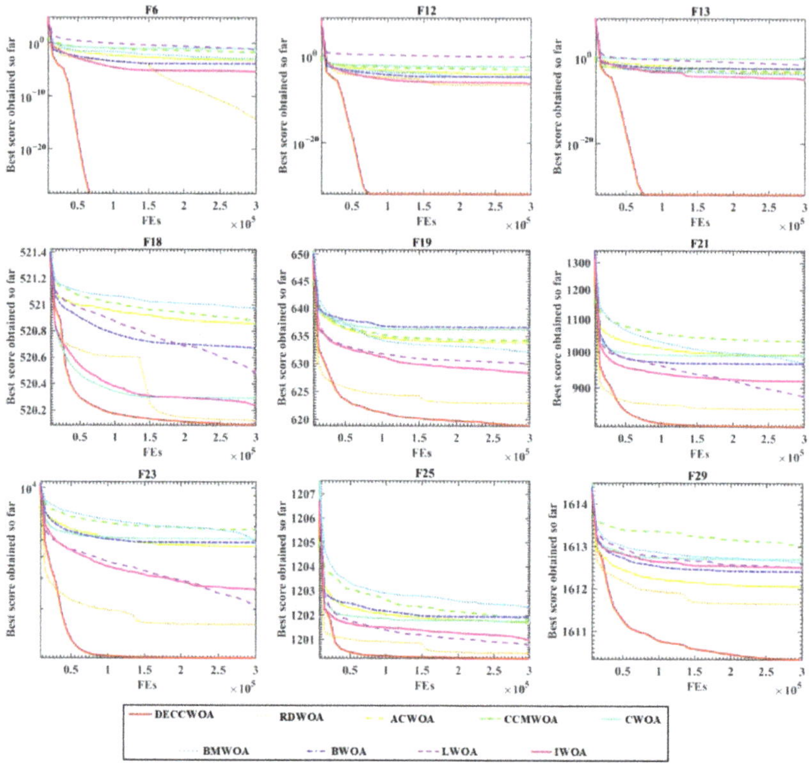

Figure 5. Convergence curves of comparison with improved WOA versions.

Table 6. Comparison results for the DECCWOA and advanced algorithms.

Algorithms	Overall	
	+/−/=	Rank
DECCWOA	~	1
IGWO	20/8/7	2
OBLGWO	19/5/11	5
CGPSO	23/9/3	4
ALPSO	19/9/7	3
RCBA	23/9/3	6
CBA	24/6/5	7
OBSCA	32/1/2	9
SCADE	28/2/5	7

In order to verify the effectiveness of the proposed DECCWOA compared to other advanced algorithms, comparison experiments were carried out. Among them, an enhanced GWO with a new hierarchical structure (IGWO) [81], boosted GWO (OBLGWO) [82], cluster guide PSO (CGPSO) [83], hybridizing sine cosine algorithm with differential evolution (SCADE) [84], particle swarm optimization with an aging leader and challengers (ALPSO) [85], hybrid bat algorithm (RCBA) [86], chaotic BA (CBA) [87] and opposition-based SCA (OBSCA) [88] were selected as the comparison algorithms. Convergence curves for comparison with the advanced algorithms are displayed in Figure 6. In particular, for unimodal functions, the DECCWOA has the same search capability as the IGWO, OBLGWO, CGPSO and SCADE in F1. For F6, the DECCWOA has the strongest optimization capability and, as can be seen in Figure 6, the DECCWOA maintains a satisfactory convergence rate for F6. On the multimodal functions, such as F12, F13, F21, F23, F24 and F29, the DECCWOA also shows strong optimization ability. Compared with the classic ALPSO, the optimization performance of the DECCWOA is not inferior, and it can even converge to a better solution at a faster convergence rate. When solving a hybrid optimization problem, such as F31, although the IGWO can still obtain better solutions in the late iteration, its convergence speed is slow and the search ability is poor in the early iteration. The OBLGWOA, CGPAO, SCADE and OBSCA are unsatisfactory in terms of their optimization ability and convergence speeds during the entire iterative process, while the ALPSO, RCBA and CBA are relatively better; however, the DECCWOA showed better optimization than them.

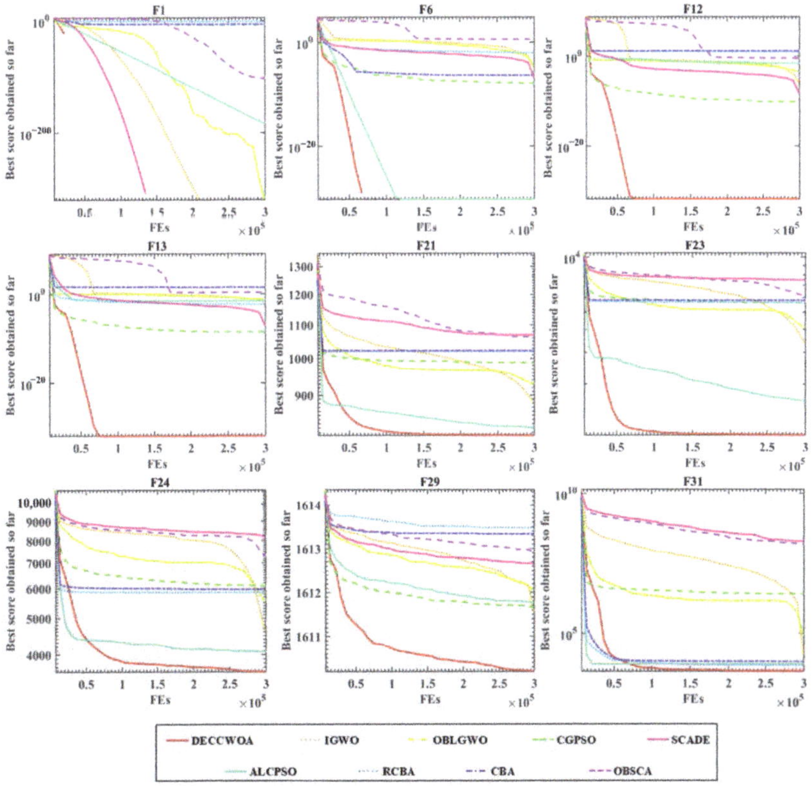

Figure 6. Convergence curves of the DECCWOA and advanced algorithms.

4.2. *Experiments on Application of the DECCWOA in Predicting Talent Stability in Higher Education*

4.2.1. Description of the Selected Data

The subjects studied in this paper were 69 talented individuals who left several colleges and universities in Wenzhou from 1 January 2015, accounting for 11.5% of the official staff. The following characteristics were examined: subject gender, political status, professional attributes, age, type of place of origin, category of talents above the municipal level, nature of the previous unit, type of location of college and university, year of employment at college and university, type of position at college and university, professional relevance of employment at college and university, annual salary level at college and university, current employment unit, time of introduction of current employment unit, nature of current employment unit and type of location of current employment unit. The indicators, as presented in Table A6 of the Appendix A, were mined and analyzed to explore the importance and interconnectedness of each indicator, and to build an intelligent prediction model based on these indicators. Moreover, the following indicators are bolded as important indicators.

4.2.2. Experimental Results

The proposed DECCWOA was combined with the KELM and the feature selection (DECCWOA-KELM-FS) method to solve the classification problem of employment intention of talent. The experimental results are shown in Tables 7 and 8. The DECCWOA-KELM-FS's results on the ACC, Sensitivity, Specificity and MCC indicators are 95.87%, 94.96%, 96.59% and 91.64%, respectively. The classification results are all superior to other comparison algorithms, including the DECCWOA-KELM, DECCWOA-KELM, WOA-KELM, ANN, RF and SVM. Furthermore, the stability results of the ten experimental results of the proposed model are also superior. The std metrics results of the ACC, Sensitivity, Specificity and MCC indicators are 3.19×10^{-2}, 6.85×10^{-2}, 4.25×10^{-2} and 6.66×10^{-2}. Obviously, the stability of the proposed algorithm is better than that of most comparison algorithms. Therefore, by combining the DECCWOA with the KELM and FS, the talent stability prediction of Wenzhou Vocational College is effectively realized. To further visualize the results, Figure 7 shows a comparison of results between the proposed algorithm and the other five methods, including the average results and standard deviations of the five indicators. Similarly, the average performance and stability of the DECCWOA-KELM-FS in each index are better than most reported algorithms.

Table 7. Four avg metrics results of the proposed model and other models.

Models	ACC	Sensitivity	Specificity	MCC
DECCWOA-KELM-FS	95.87%	94.64%	96.59%	91.64%
DECCWOA-KELM	92.57%	94.05%	92.50%	86.27%
WOA-KELM	90.32%	89.39%	91.05%	81.10%
ANN	88.96%	87.58%	90.51%	77.16%
RF	92.67%	92.85%	91.01%	85.33%
SVM	89.30%	91.92%	86.96%	79.75%

Table 8. Four std metrics results of the proposed model and other models.

Models	ACC	Sensitivity	Specificity	MCC
DECCWOA-KELM-FS	3.19×10^{-2}	6.85×10^{-2}	4.25×10^{-2}	6.66×10^{-2}
DECCWOA-KELM	5.60×10^{-2}	8.31×10^{-2}	8.93×10^{-2}	1.01×10^{-1}
WOA-KELM	4.33×10^{-2}	9.06×10^{-2}	1.02×10^{-1}	8.80×10^{-2}
ANN	4.16×10^{-2}	5.61×10^{-2}	5.91×10^{-2}	9.50×10^{-2}
RF	4.17×10^{-2}	1.08×10^{-1}	6.74×10^{-2}	8.59×10^{-2}
SVM	6.72×10^{-2}	1.12×10^{-1}	1.07×10^{-1}	1.18×10^{-1}

Figure 8 shows the feature selection results of the proposed model. As can be seen, F7 (city-level and above talent categories) and F22 (professional and technical position at the time of leaving) are both screened the most, eight times. It shows that the two key factors affecting the stability of university talents are F7 and F22, which provides some guiding significance of the flow of highly educated talents. Based on the fact that the proposed method has such excellent performance, it can also be applied in many other fields in the future, such as information retrieval services [89,90], named entity recognition [91], road network planning [92], colorectal polyp region extraction [93], image denoising [94], image segmentation [95–97] and power flow optimization [98].

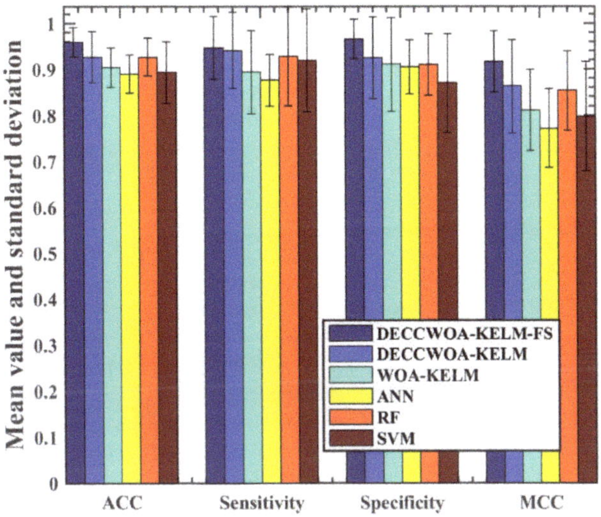

Figure 7. Mean value and standard deviation of four metrics for the DECCWOA and others methods.

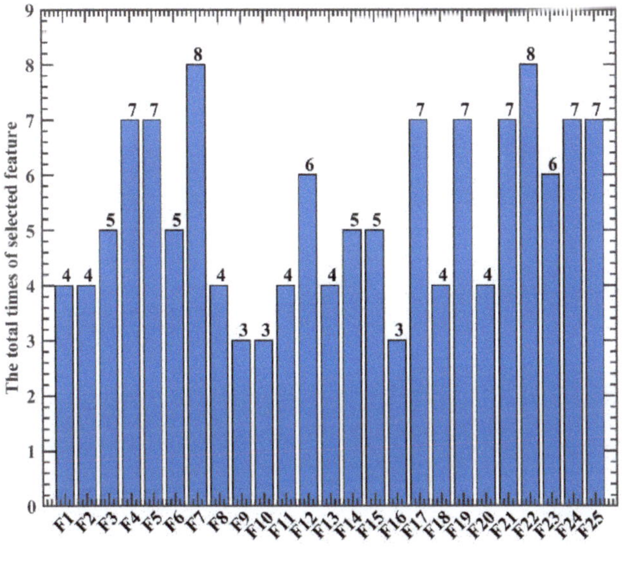

Figure 8. Feature selection results of the proposed model.

5. Conclusions

This paper studied the stability of higher education talent for the first time, and proposed a DECCWOA-KELM-FS model to intelligently predict the stability of higher education talent. By introducing a crossover algorithm, the information exchange between individuals was promoted and the problem of dimension stagnation was improved. The DE operation was carried out in a certain time, and the difference between individuals was used to disturb the population and ensure the diversity of the population. In order to verify the optimization performance of the DECCWOA, 35 benchmark functions were selected from 32 benchmark functions and CEC214 for comparative experiments. Experimental results showed that the DECCWOA algorithm had higher accuracy and faster convergence rates when solving unimodal and multimodal functions; although the mixture function also had very good performance. By combining the DECCWOA with the KELM and feature selection, the stable intelligence of talent in Wenzhou colleges and universities was efficiently predicted. This method can be used as a reliable and high precision method to predict the flow of talent in colleges and universities.

Subsequent studies will further improve the generality of the proposed GLLCSA-KELM-FS and solve more complex classification problems, such as disease diagnosis and financial risk prediction.

Author Contributions: Conceptualization, G.L. and H.C.; methodology, F.K. and H.C.; software, G.L. and F.K.; validation, H.L., S.K., X.R., C.L., G.L., H.C., F.K. and L.L.; formal analysis, F.K., G.L. and L.L.; investigation, H.L., S.K., D.C. and C.L.; resources, F.K., G.L. and L.L.; data curation, F.K., G.L. and C.L.; writing—original draft preparation, H.L., S.K., X.R. and C.L.; writing—review and editing, G.L. and H.C.; visualization, G.L. and H.C.; supervision, F.K., G.L. and L.L.; project administration, F.K., G.L. and L.L.; funding acquisition, F.K., G.L. and H.C. All authors have read and agreed to the published version of the manuscript.

Funding: Zhejiang Provincial universities Major Humanities and social Science project: Innovation and Practice of Cultivating Paths for Leaders in Rural Industry Revitalization under the Background of Common Prosperity (Moderator: Li Hong), Humanities and Social Science Research Planning Fund Project of the Ministry of Education (research on risk measurement and early warning mechanism of science and technology finance based on big data analysis, 20YJA790090), Zhejiang Provincial Philosophy and Social Sciences Planning Project (Research on rumor recognition and dissemination intervention based on automated essay scoring, 23NDJC393YBM).

Data Availability Statement: The data involved in this study are all public data, which can be downloaded through public channels.

Conflicts of Interest: The authors declare no conflict of interest.

Appendix A

Table A1. Details of the selected 35 benchmark functions.

Types	No.	Functions	Rang	f_{min}
Unimodal Functions	F1	$f_1(x) = \sum_{i=1}^{n} x_i^2$	[−100, 100]	0
	F2	$f_2(x) = \sum_{i=1}^{n} \lvert x_i \rvert + \prod_{i=1}^{n} \lvert x_i \rvert$	[−10, 10]	0
	F3	$f_3(x) = \sum_{i=1}^{n} \left(\sum_{j-1}^{i} x_j \right)^2$	[−100, 100]	0
	F4	$f_4(x) = max_i\{\lvert x_i \rvert, 1 \le i \le n\}$	[−100, 100]	0
	F5	$f_5(x) = \sum_{i=1}^{n-1} \left[100\left(x_{i+1} - x_i^2\right)^2 + (x_i - 1)^2 \right]$	[−30, 30]	0
	F6	$f_6(x) = \sum_{i=1}^{n} ([x_i + 0.5])^2$	[−100, 100]	0
	F7	$f_7(x) = \sum_{i=1}^{n} i x_i^4 + random[0, 1]$	[−1.28, 1.28]	0
Multimodal Functions	F8	$f_8(x) = \sum_{i=1}^{n} -x_i \sin\left(\sqrt{\lvert x_i \rvert}\right)$	[−500, 500]	−418.9829 × 5
	F9	$f_9(x) = \sum_{i=1}^{n} \left[x_i^2 - 10\cos(2\pi x_i) + 10\right]$	[−5.12, 5.12]	0
	F10	$f_{10}(x) = -20\exp\left(-0.2\sqrt{\frac{1}{n}\sum_{i=1}^{n} x_i^2}\right) - \exp\left(\frac{1}{n}\sum_{i=1}^{n}\cos(2\pi x_i)\right) + 20 + e$	[−32, 32]	0
	F11	$f_{11}(x) = \frac{1}{4000}\sum_{i=1}^{n} x_i^2 - \prod_{i=1}^{n}\cos\left(\frac{x_i}{\sqrt{i}}\right) + 1$	[−600, 600]	0
	F12	$f_{12}(x) = \frac{\pi}{n}\left\{10\sin(\pi y_1) + \sum_{i=1}^{n-1}(y_i - 1)^2\left[1 + 10\sin^2\left(\pi y_{i+1}\right)\right] + (y_n - 1)^2\right\} + \sum_{i=1}^{n} u(x_i, 10, 100, 4)$ $y_i = 1 + \frac{x_i + 1}{4}, u(x_i, a, k, m) = \begin{cases} k(x_i - a)^m & x_i > a \\ 0 & -a < x_i < a \\ k(-x_i - a)^m & x_i < -a \end{cases}$	[−50, 50]	0
	F13	$f_{13}(x) = 0.1\left\{\sin^2(3\pi x_1) + \sum_{i=1}^{n}(x_i - 1)^2\left[1 + \sin^2(3\pi x_i + 1)\right] + (x_n - 1)^2\left[1 + \sin^2(2\pi x_n)\right]\right\} + \sum_{i=1}^{n} u(x_i, 5, 100, 4)$	[−50, 50]	0
Unimodal Functions	F14	Rotated High Conditioned Elliptic Function	[−100, 100]	100
	F15	Rotated Bent Cigar Function	[−100, 100]	200
	F16	Rotated Discus Function	[−100, 100]	300

Table A1. *Cont.*

Types	No.	Functions	Rang	f_{min}
Simple Multimodal Functions	F17	Shifted and Rotated Rosenbrock's Function	[−100, 100]	400
	F18	Shifted and Rotated Ackley's Function	[−100, 100]	500
	F19	Shifted and Rotated Weierstrass Function	[−100, 100]	600
	F20	Shifted and Rotated Griewank's Function	[−100, 100]	700
	F21	Shifted Rastrigin's Function	[−100, 100]	800
	F22	Shifted and Rotated Rastrigin's Function	[−100, 100]	900
	F23	Shifted Schwefel's Function	[−100, 100]	1000
	F24	Shifted and Rotated Schwefel's Function	[−100, 100]	1100
	F25	Shifted and Rotated Katsuura Function	[−100, 100]	1200
	F26	Shifted and Rotated HappyCat Function	[−100, 100]	1300
	F27	Shifted and Rotated HGBat Function	[−100, 100]	1400
	F28	Shifted and Rotated Expanded Griewank's plus Rosenbrock's Function	[−100, 100]	1500
	F29	Shifted and Rotated Expanded Scaffer's F6 Function	[−100, 100]	1600
Hybrid Function1	F30	Hybrid Function 1 (N = 3)	[−100, 100]	1700
	F31	Hybrid Function 2 (N = 3)	[−100, 100]	1800
	F32	Hybrid Function 3 (N = 4)	[−100, 100]	1900
	F33	Hybrid Function 4 (N = 4)	[−100, 100]	2000
	F34	Hybrid Function 5 (N = 5)	[−100, 100]	2100
	F35	Hybrid Function 6 (N = 5)	[−100, 100]	2200

Table A2. Experimental results for analysis of key parameter p_2.

	F1		F2		F3	
	avg	*std*	*avg*	*std*	*avg*	*std*
DECCWOA1	0.0000×10^0	0.0000×10^0	0.0000×10^0	0.0000×10^0	6.0453×10^{-8}	3.3108×10^{-7}
DECCWOA2	0.0000×10^0	0.0000×10^0	0.0000×10^0	0.0000×10^0	4.2361×10^{-17}	2.3074×10^{-16}
DECCWOA3	0.0000×10^0	0.0000×10^0	0.0000×10^0	0.0000×10^0	6.1871×10^{-28}	1.4243×10^{-27}
DECCWOA4	0.0000×10^0	0.0000×10^0	0.0000×10^0	0.0000×10^0	1.8338×10^{-27}	2.4802×10^{-27}
DECCWOA5	0.0000×10^0	0.0000×10^0	0.0000×10^0	0.0000×10^0	5.5498×10^{-28}	1.4428×10^{-27}
DECCWOA6	0.0000×10^0	0.0000×10^0	0.0000×10^0	0.0000×10^0	5.9383×10^{-28}	1.6120×10^{-27}
DECCWOA7	0.0000×10^0	0.0000×10^0	0.0000×10^0	0.0000×10^0	9.3229×10^{-28}	1.9913×10^{-27}
DECCWOA8	0.0000×10^0	0.0000×10^0	0.0000×10^0	0.0000×10^0	3.4963×10^{-28}	1.1052×10^{-27}
DECCWOA9	0.0000×10^0	0.0000×10^0	0.0000×10^0	0.0000×10^0	5.3287×10^{-28}	1.2818×10^{-27}
DECCWOA10	0.0000×10^0	0.0000×10^0	0.0000×10^0	0.0000×10^0	4.2382×10^{-28}	1.3085×10^{-27}
	F4		F5		F6	
	avg	*std*	*avg*	*std*	*avg*	*std*
DECCWOA1	0.0000×10^0	0.0000×10^0	2.4439×10^1	6.6504×10^0	0.0000×10^0	0.0000×10^0
DECCWOA2	0.0000×10^0	0.0000×10^0	2.4123×10^1	6.5641×10^0	0.0000×10^0	0.0000×10^0
DECCWOA3	0.0000×10^0	0.0000×10^0	2.6257×10^1	2.8910×10^{-1}	0.0000×10^0	0.0000×10^0
DECCWOA4	0.0000×10^0	0.0000×10^0	2.5987×10^1	4.1389×10^{-1}	0.0000×10^0	0.0000×10^0
DECCWOA5	0.0000×10^0	0.0000×10^0	2.6107×10^1	3.0119×10^{-1}	0.0000×10^0	0.0000×10^0
DECCWOA6	0.0000×10^0	0.0000×10^0	2.5412×10^1	4.8062×10^0	0.0000×10^0	0.0000×10^0
DECCWOA7	0.0000×10^0	0.0000×10^0	2.5432×10^1	4.8097×10^0	0.0000×10^0	0.0000×10^0
DECCWOA8	0.0000×10^0	0.0000×10^0	2.4619×10^1	6.6972×10^0	0.0000×10^0	0.0000×10^0
DECCWOA9	0.0000×10^0	0.0000×10^0	2.5539×10^1	4.8318×10^0	0.0000×10^0	0.0000×10^0
DECCWOA10	0.0000×10^0	0.0000×10^0	2.6489×10^1	3.4054×10^{-1}	0.0000×10^0	0.0000×10^0

Table A2. *Cont.*

	F7		F8		F9	
	avg	*std*	*avg*	*std*	*avg*	*std*
DECCWOA1	1.6842×10^{-4}	2.7932×10^{-4}	-1.3963×10^{4}	5.1858×10^{3}	0.0000×10^{0}	0.0000×10^{0}
DECCWOA2	1.5645×10^{-4}	1.7877×10^{-4}	-1.2595×10^{4}	1.3910×10^{2}	0.0000×10^{0}	0.0000×10^{0}
DECCWOA3	6.6910×10^{-5}	9.9764×10^{-5}	-1.2619×10^{4}	2.7031×10^{2}	0.0000×10^{0}	0.0000×10^{0}
DECCWOA4	1.0031×10^{-4}	1.4658×10^{-4}	-1.2530×10^{4}	5.3042×10^{2}	0.0000×10^{0}	0.0000×10^{0}
DECCWOA5	8.9743×10^{-5}	1.0676×10^{-4}	-1.3512×10^{4}	5.1597×10^{3}	0.0000×10^{0}	0.0000×10^{0}
DECCWOA6	5.2519×10^{-5}	5.3764×10^{-5}	-1.2569×10^{4}	1.9404×10^{-12}	0.0000×10^{0}	0.0000×10^{0}
DECCWOA7	6.4531×10^{-5}	6.4938×10^{-5}	-1.3485×10^{4}	2.8423×10^{3}	0.0000×10^{0}	0.0000×10^{0}
DECCWOA8	4.0419×10^{-5}	5.3492×10^{-5}	-1.2805×10^{4}	9.0308×10^{2}	0.0000×10^{0}	0.0000×10^{0}
DECCWOA9	6.3101×10^{-5}	7.7305×10^{-5}	-1.2569×10^{4}	2.0267×10^{-12}	0.0000×10^{0}	0.0000×10^{0}
DECCWOA10	3.8169×10^{-5}	6.1534×10^{-5}	-1.2673×10^{4}	2.7066×10^{2}	0.0000×10^{0}	0.0000×10^{0}
	F10		F11		F12	
	avg	*std*	*avg*	*std*	*avg*	*std*
DECCWOA1	8.8818×10^{-16}	0.0000×10^{0}	0.0000×10^{0}	0.0000×10^{0}	1.5705×10^{-32}	5.5674×10^{-48}
DECCWOA2	8.8818×10^{-16}	0.0000×10^{0}	0.0000×10^{0}	0.0000×10^{0}	1.5705×10^{-32}	5.5674×10^{-48}
DECCWOA3	8.8818×10^{-16}	0.0000×10^{0}	0.0000×10^{0}	0.0000×10^{0}	1.5705×10^{-32}	5.5674×10^{-48}
DECCWOA4	8.8818×10^{-16}	0.0000×10^{0}	0.0000×10^{0}	0.0000×10^{0}	1.5705×10^{-32}	5.5674×10^{-48}
DECCWOA5	8.8818×10^{-16}	0.0000×10^{0}	0.0000×10^{0}	0.0000×10^{0}	1.5705×10^{-32}	5.5674×10^{-48}
DECCWOA6	8.8818×10^{-16}	0.0000×10^{0}	0.0000×10^{0}	0.0000×10^{0}	1.5705×10^{-32}	5.5674×10^{-48}
DECCWOA7	8.8818×10^{-16}	0.0000×10^{0}	0.0000×10^{0}	0.0000×10^{0}	1.5705×10^{-32}	5.5674×10^{-48}
DECCWOA8	8.8818×10^{-16}	0.0000×10^{0}	0.0000×10^{0}	0.0000×10^{0}	1.5705×10^{-32}	5.5674×10^{-48}
DECCWOA9	8.8818×10^{-16}	0.0000×10^{0}	0.0000×10^{0}	0.0000×10^{0}	1.5705×10^{-32}	5.5674×10^{-48}
DECCWOA10	8.8818×10^{-16}	0.0000×10^{0}	0.0000×10^{0}	0.0000×10^{0}	1.5705×10^{-32}	5.5674×10^{-48}
	F13		F14		F15	
	avg	*std*	*avg*	*std*	*avg*	*std*
DECCWOA1	1.3498×10^{-32}	5.5674×10^{-48}	3.6810×10^{6}	2.7334×10^{6}	1.3342×10^{5}	1.9982×10^{5}
DECCWOA2	1.3498×10^{-32}	5.5674×10^{-48}	5.1469×10^{6}	3.7469×10^{6}	2.2654×10^{5}	3.0564×10^{5}
DECCWOA3	1.3498×10^{-32}	5.5674×10^{-48}	1.6507×10^{7}	1.0130×10^{7}	9.4042×10^{7}	8.6776×10^{7}
DECCWOA4	1.3498×10^{-32}	5.5674×10^{-48}	7.2801×10^{6}	5.0765×10^{6}	8.7941×10^{6}	7.2596×10^{6}
DECCWOA5	1.3498×10^{-32}	5.5674×10^{-48}	1.1619×10^{7}	7.8606×10^{6}	3.3924×10^{7}	2.5699×10^{7}
DECCWOA6	1.3498×10^{-32}	5.5674×10^{-48}	1.2238×10^{7}	8.2473×10^{6}	7.4207×10^{7}	5.2159×10^{7}
DECCWOA7	1.3498×10^{-32}	5.5674×10^{-48}	2.5408×10^{7}	1.2932×10^{7}	1.5504×10^{8}	1.0836×10^{8}
DECCWOA8	1.3498×10^{-32}	5.5674×10^{-48}	2.9064×10^{7}	1.2592×10^{7}	3.1493×10^{8}	3.1857×10^{8}
DECCWOA9	1.3498×10^{-32}	5.5674×10^{-48}	2.7983×10^{7}	1.7994×10^{7}	4.3963×10^{8}	5.9410×10^{8}
DECCWOA10	1.3498×10^{-32}	5.5674×10^{-48}	4.0140×10^{7}	2.8456×10^{7}	6.8723×10^{8}	7.6542×10^{8}
	F16		F17		F18	
	avg	*std*	*avg*	*std*	*avg*	*std*
DECCWOA1	7.4819×10^{3}	4.6257×10^{3}	4.9577×10^{2}	4.4816×10^{1}	5.2004×10^{2}	4.4169×10^{-2}
DECCWOA2	5.3970×10^{3}	5.6834×10^{3}	5.2284×10^{2}	4.4110×10^{1}	5.2009×10^{2}	4.7541×10^{-2}
DECCWOA3	5.5074×10^{3}	3.3963×10^{3}	5.9759×10^{2}	5.0184×10^{1}	5.2036×10^{2}	1.4114×10^{-1}
DECCWOA4	4.7987×10^{3}	4.0122×10^{3}	5.5522×10^{2}	5.6453×10^{1}	5.2019×10^{2}	9.2319×10^{-2}
DECCWOA5	3.9879×10^{3}	2.8027×10^{3}	5.6621×10^{2}	4.6593×10^{1}	5.2029×10^{2}	1.0546×10^{-1}
DECCWOA6	4.7947×10^{3}	3.8454×10^{3}	6.0512×10^{2}	3.7665×10^{1}	5.2031×10^{2}	1.4175×10^{-1}
DECCWOA7	4.7025×10^{3}	3.4303×10^{3}	6.4612×10^{2}	5.0034×10^{1}	5.2039×10^{2}	1.4396×10^{-1}
DECCWOA8	5.7773×10^{3}	3.8347×10^{3}	6.9438×10^{2}	9.0009×10^{1}	5.2033×10^{2}	1.7149×10^{-1}
DECCWOA9	6.8194×10^{3}	3.9312×10^{3}	6.7202×10^{2}	7.1887×10^{1}	5.2035×10^{2}	1.6351×10^{-1}
DECCWOA10	7.6046×10^{3}	2.9400×10^{3}	6.9534×10^{2}	7.7239×10^{1}	5.2038×10^{2}	1.7225×10^{-1}

Table A2. *Cont.*

	F19		F20		F21	
	avg	*std*	*avg*	*std*	*avg*	*std*
DECCWOA1	6.2052×10^{2}	3.2852×10^{0}	7.0040×10^{2}	2.2886×10^{-1}	8.0287×10^{2}	1.0273×10^{1}
DECCWOA2	6.1946×10^{2}	2.9290×10^{0}	7.0052×10^{2}	2.0981×10^{-1}	8.1110×10^{2}	2.0103×10^{1}
DECCWOA3	6.2495×10^{2}	2.9171×10^{0}	7.0250×10^{2}	8.3283×10^{-1}	8.7716×10^{2}	1.8311×10^{1}
DECCWOA4	6.2184×10^{2}	2.6694×10^{0}	7.0111×10^{2}	7.1368×10^{-2}	8.2441×10^{2}	7.7346×10^{0}
DECCWOA5	6.2371×10^{2}	2.6632×10^{0}	7.0164×10^{2}	4.2137×10^{-1}	8.4642×10^{2}	1.7780×10^{1}
DECCWOA6	6.2456×10^{2}	3.2157×10^{0}	7.0230×10^{2}	8.4958×10^{-1}	8.7407×10^{2}	2.4081×10^{1}
DECCWOA7	6.2621×10^{2}	3.2713×10^{0}	7.0300×10^{2}	8.7250×10^{-1}	8.9713×10^{2}	1.9802×10^{1}
DECCWOA8	6.2840×10^{2}	3.4007×10^{0}	7.0506×10^{2}	2.9144×10^{0}	9.1581×10^{2}	1.9967×10^{1}
DECCWOA9	6.2897×10^{2}	3.4733×10^{0}	7.0619×10^{2}	2.5650×10^{0}	9.2703×10^{2}	2.5100×10^{1}
DECCWOA10	6.2933×10^{2}	3.7025×10^{0}	7.0753×10^{2}	3.4438×10^{0}	9.4279×10^{2}	2.4986×10^{1}
	F22		F23		F24	
	avg	*std*	*avg*	*std*	*avg*	*std*
DECCWOA1	1.0312×10^{3}	2.7380×10^{1}	1.0420×10^{3}	1.1886×10^{2}	3.9026×10^{3}	6.5371×10^{2}
DECCWOA2	1.0377×10^{3}	3.4427×10^{1}	1.0510×10^{3}	7.9342×10^{1}	4.1202×10^{3}	7.1968×10^{2}
DECCWOA3	1.0545×10^{3}	3.3397×10^{1}	1.7059×10^{3}	3.0416×10^{2}	5.3301×10^{3}	4.9187×10^{2}
DECCWOA4	1.0420×10^{3}	3.5382×10^{1}	1.1956×10^{3}	2.0676×10^{2}	4.2088×10^{3}	6.4091×10^{2}
DECCWOA5	1.0502×10^{3}	2.9746×10^{1}	1.4485×10^{3}	4.7609×10^{2}	4.9060×10^{3}	7.5372×10^{2}
DECCWOA6	1.0708×10^{3}	2.3208×10^{1}	1.6654×10^{3}	3.5966×10^{2}	5.0839×10^{3}	7.4254×10^{2}
DECCWOA7	1.0725×10^{3}	2.9966×10^{1}	2.4340×10^{3}	3.6991×10^{2}	5.5460×10^{3}	6.9622×10^{2}
DECCWOA8	1.0867×10^{3}	3.0093×10^{1}	2.8661×10^{3}	5.7401×10^{2}	5.6002×10^{3}	5.3783×10^{2}
DECCWOA9	1.0862×10^{3}	2.9502×10^{1}	3.4724×10^{3}	6.7538×10^{2}	5.6526×10^{3}	7.8496×10^{2}
DECCWOA10	1.0860×10^{3}	3.5118×10^{1}	3.7201×10^{3}	5.3438×10^{2}	5.6708×10^{3}	5.7142×10^{2}
	F25		F26		F27	
	avg	*std*	*avg*	*std*	*avg*	*std*
DECCWOA1	1.2002×10^{3}	6.7099×10^{-2}	1.3005×10^{3}	1.5439×10^{-1}	1.4003×10^{3}	4.8887×10^{-2}
DECCWOA2	1.2002×10^{3}	6.0385×10^{-2}	1.3005×10^{3}	1.1841×10^{-1}	1.4003×10^{3}	5.5441×10^{-2}
DECCWOA3	1.2008×10^{3}	2.9906×10^{-1}	1.3005×10^{3}	1.1364×10^{-1}	1.4003×10^{3}	4.0852×10^{-2}
DECCWOA4	1.2004×10^{3}	1.2088×10^{-1}	1.3005×10^{3}	1.3836×10^{-1}	1.4003×10^{3}	1.9112×10^{-1}
DECCWOA5	1.2005×10^{3}	1.9004×10^{-1}	1.3005×10^{3}	1.0714×10^{-1}	1.4003×10^{3}	5.7154×10^{-2}
DECCWOA6	1.2008×10^{3}	2.5696×10^{-1}	1.3005×10^{3}	9.5896×10^{-2}	1.4003×10^{3}	1.0255×10^{-1}
DECCWOA7	1.2010×10^{3}	3.2942×10^{-1}	1.3005×10^{3}	1.1984×10^{-1}	1.4003×10^{3}	1.7872×10^{-1}
DECCWOA8	1.2011×10^{3}	3.7909×10^{-1}	1.3005×10^{3}	1.3301×10^{-1}	1.4004×10^{3}	6.0314×10^{-1}
DECCWOA9	1.2012×10^{3}	3.9281×10^{-1}	1.3006×10^{3}	1.4510×10^{-1}	1.4003×10^{3}	5.1310×10^{-2}
DECCWOA10	1.2014×10^{3}	4.0093×10^{-1}	1.3005×10^{3}	1.1885×10^{-1}	1.4004×10^{3}	1.7136×10^{-1}
	F28		F29		F30	
	avg	*std*	*avg*	*std*	*avg*	*std*
DECCWOA1	1.5176×10^{3}	6.4897×10^{0}	1.6104×10^{3}	6.8683×10^{-1}	1.9247×10^{6}	1.1101×10^{6}
DECCWOA2	1.5189×10^{3}	7.4912×10^{0}	1.6104×10^{3}	7.6782×10^{-1}	1.5103×10^{6}	1.1011×10^{6}
DECCWOA3	1.5422×10^{3}	1.3549×10^{1}	1.6115×10^{3}	6.7294×10^{-1}	1.9418×10^{6}	1.2814×10^{6}
DECCWOA4	1.5266×10^{3}	6.5567×10^{0}	1.6110×10^{3}	6.9374×10^{-1}	1.7005×10^{6}	9.1883×10^{5}
DECCWOA5	1.5304×10^{3}	9.0407×10^{0}	1.6113×10^{3}	6.2976×10^{-1}	1.8482×10^{6}	1.1135×10^{6}
DECCWOA6	1.5424×10^{3}	1.2171×10^{1}	1.6116×10^{3}	5.9848×10^{-1}	1.9396×10^{6}	1.2724×10^{6}
DECCWOA7	1.6047×10^{3}	2.9073×10^{2}	1.6119×10^{3}	6.5689×10^{-1}	2.0687×10^{6}	1.2220×10^{6}
DECCWOA8	1.5684×10^{3}	3.2783×10^{1}	1.6120×10^{3}	4.3100×10^{-1}	2.2597×10^{6}	1.5545×10^{6}
DECCWOA9	1.5970×10^{3}	4.5170×10^{1}	1.6121×10^{3}	6.1504×10^{-1}	2.2349×10^{6}	1.2776×10^{6}
DECCWOA10	1.6376×10^{3}	9.5739×10^{1}	1.6119×10^{3}	7.4369×10^{-1}	2.4610×10^{6}	1.6436×10^{6}

Table A2. *Cont.*

	F31		F32		F33	
	avg	*std*	*avg*	*std*	*avg*	*std*
DECCWOA1	5.8087×10^3	6.2358×10^3	1.9179×10^3	2.6069×10^1	1.4137×10^4	7.6746×10^3
DECCWOA2	4.8867×10^3	3.6023×10^3	1.9189×10^3	2.3237×10^1	8.3832×10^3	5.6464×10^3
DECCWOA3	4.0166×10^3	2.5970×10^3	1.9215×10^3	2.4809×10^1	5.5742×10^3	2.6752×10^3
DECCWOA4	5.2173×10^3	3.9350×10^3	1.9267×10^3	3.4926×10^1	4.9997×10^3	2.4000×10^3
DECCWOA5	4.9446×10^3	4.1973×10^3	1.9262×10^3	2.9792×10^1	4.5639×10^3	2.0732×10^3
DECCWOA6	4.5409×10^3	2.8935×10^3	1.9239×10^3	2.2044×10^1	5.0427×10^3	2.4608×10^3
DECCWOA7	4.7751×10^3	2.8190×10^3	1.9279×10^3	2.2825×10^1	4.5396×10^3	1.8457×10^3
DECCWOA8	4.6894×10^4	2.3224×10^5	1.9358×10^3	3.7797×10^1	4.8777×10^3	1.6024×10^3
DECCWOA9	7.0415×10^3	9.3191×10^3	1.9284×10^3	2.5490×10^1	4.5525×10^3	2.1545×10^3
DECCWOA10	6.2462×10^3	9.1482×10^3	1.9295×10^3	2.4829×10^1	3.7338×10^3	1.6768×10^3
Overall rank	F34		F35		overall	
	avg	*std*	*avg*	*std*	+/−/=	rank
DECCWOA1	8.6335×10^5	6.2796×10^5	2.8860×10^3	2.2542×10^2	~	2
DECCWOA2	7.0965×10^5	6.4048×10^5	2.7590×10^3	1.9713×10^2	4/3/28	1
DECCWOA3	8.1850×10^5	6.4237×10^5	2.6856×10^3	1.9951×10^2	14/3/18	6
DECCWOA4	8.3727×10^5	6.0484×10^5	2.8021×10^3	1.8728×10^2	12/2/21	5
DECCWOA5	7.2517×10^5	6.6623×10^5	2.7406×10^3	2.2485×10^2	14/4/17	3
DECCWOA6	5.9730×10^5	4.4486×10^5	2.7219×10^3	1.8356×10^2	14/5/16	4
DECCWOA7	6.4163×10^5	4.3900×10^5	2.6995×10^3	2.0571×10^2	14/4/17	7
DECCWOA8	7.0675×10^5	6.1652×10^5	2.7178×10^3	1.9634×10^2	15/4/16	8
DECCWOA9	6.4914×10^5	5.4887×10^5	2.7793×10^3	1.9480×10^2	17/2/16	9
DECCWOA10	5.9732×10^5	4.7600×10^5	2.7603×10^3	2.1609×10^2	16/2/17	10

Table A3. Comparison results for the introduced mechanisms.

	F1		F2		F3	
	avg	*std*	*avg*	*std*	*avg*	*std*
DECCWOA	0.0000×10^0	0.0000×10^0	0.0000×10^0	0.0000×10^0	4.1255×10^{-18}	2.1806×10^{-17}
DEWOA	1.2420×10^{-10}	3.9129×10^{-10}	8.5083×10^{-6}	1.4067×10^{-5}	7.8264×10^3	1.2145×10^4
CCWOA	0.0000×10^0	0.0000×10^0	0.0000×10^0	0.0000×10^0	0.0000×10^0	0.0000×10^0
WOA	0.0000×10^0	0.0000×10^0	0.0000×10^0	0.0000×10^0	3.2071×10^1	6.1783×10^1
	F4		F5		F6	
	avg	*std*	*avg*	*std*	*avg*	*std*
DECCWOA	0.0000×10^0	0.0000×10^0	2.5968×10^1	3.5258×10^{-1}	0.0000×10^0	0.0000×10^0
DEWOA	6.3178×10^{-3}	1.5936×10^{-2}	4.4319×10^{-3}	4.7769×10^{-3}	9.6180×10^{-5}	1.2811×10^{-4}
CCWOA	0.0000×10^0	0.0000×10^0	2.2483×10^1	6.1153×10^0	1.1597×10^{-11}	1.3565×10^{-11}
WOA	7.5414×10^0	1.7526×10^1	2.3562×10^1	4.4675×10^0	4.7799×10^{-6}	1.8846×10^{-6}
	F7		F8		F9	
	avg	*std*	*avg*	*std*	*avg*	*std*
DECCWOA	1.0328×10^{-4}	1.7676×10^{-4}	-1.2783×10^4	8.3042×10^2	0.0000×10^0	0.0000×10^0
DEWOA	3.4741×10^{-3}	5.8983×10^{-3}	-1.4406×10^4	4.9552×10^3	6.2111×10^{-10}	8.6723×10^{-10}
CCWOA	1.7804×10^{-5}	3.0937×10^{-5}	-1.2569×10^4	5.6938×10^{-7}	0.0000×10^0	0.0000×10^0
WOA	1.5818×10^{-4}	1.8724×10^{-4}	-1.2236×10^4	8.6401×10^2	0.0000×10^0	0.0000×10^0

Table A3. *Cont.*

	F10		F11		F12	
	avg	*std*	*avg*	*std*	*avg*	*std*
DECCWOA	8.8818×10^{-16}	0.0000×10^{0}	0.0000×10^{0}	0.0000×10^{0}	1.5705×10^{-32}	5.5674×10^{-48}
DEWOA	4.8353×10^{-6}	1.1226×10^{-5}	6.9380×10^{-7}	3.7853×10^{-6}	7.0591×10^{-6}	9.9086×10^{-6}
CCWOA	8.8818×10^{-16}	0.0000×10^{0}	0.0000×10^{0}	0.0000×10^{0}	1.6821×10^{-12}	2.1310×10^{-12}
WOA	3.6119×10^{-15}	1.7906×10^{-15}	2.7668×10^{-4}	1.5155×10^{-3}	2.1111×10^{-4}	1.1507×10^{-3}
	F13		F14		F15	
	avg	*std*	*avg*	*std*	*avg*	*std*
DECCWOA	1.3498×10^{-32}	5.5674×10^{-48}	4.6009×10^{6}	3.5694×10^{6}	1.5068×10^{5}	1.7046×10^{5}
DEWOA	1.3742×10^{-4}	2.1617×10^{-4}	4.5118×10^{7}	3.3352×10^{7}	2.6842×10^{9}	2.2525×10^{9}
CCWOA	2.2806×10^{-11}	2.1476×10^{-11}	1.0815×10^{7}	7.8011×10^{6}	6.3006×10^{6}	3.9908×10^{6}
WOA	1.1255×10^{-3}	3.3493×10^{-3}	3.3488×10^{7}	1.7699×10^{7}	3.8002×10^{6}	7.5915×10^{6}
	F16		F17		F18	
	avg	*std*	*avg*	*std*	*avg*	*std*
DECCWOA	6.2884×10^{3}	4.0514×10^{3}	4.9673×10^{2}	3.4575×10^{1}	5.2008×10^{2}	5.6636×10^{-2}
DEWOA	3.6134×10^{4}	3.3572×10^{4}	8.0754×10^{2}	1.5372×10^{2}	5.2033×10^{2}	2.2602×10^{-1}
CCWOA	5.8371×10^{3}	3.4122×10^{3}	5.3911×10^{2}	5.8876×10^{1}	5.2032×10^{2}	8.7776×10^{-2}
WOA	3.6143×10^{4}	2.3682×10^{4}	5.8016×10^{2}	5.5983×10^{1}	5.2032×10^{2}	1.6884×10^{-1}
	F19		F20		F21	
	avg	*std*	*avg*	*std*	*avg*	*std*
DECCWOA	6.1813×10^{2}	3.3345×10^{0}	7.0050×10^{2}	1.8285×10^{-1}	8.0336×10^{2}	5.5434×10^{0}
DEWOA	6.4146×10^{2}	2.3442×10^{0}	7.1656×10^{2}	1.1726×10^{1}	9.7436×10^{2}	2.7480×10^{1}
CCWOA	6.2191×10^{2}	3.1893×10^{0}	7.0106×10^{2}	1.1101×10^{-1}	8.2734×10^{2}	1.6890×10^{1}
WOA	6.3521×10^{2}	3.8342×10^{0}	7.0102×10^{2}	7.0807×10^{-2}	9.8601×10^{2}	3.9020×10^{1}
	F22		F23		F24	
	avg	*std*	*avg*	*std*	*avg*	*std*
DECCWOA	1.0398×10^{3}	4.1688×10^{1}	1.0459×10^{3}	8.8217×10^{1}	3.7945×10^{3}	5.1310×10^{2}
DEWOA	1.1215×10^{3}	3.4240×10^{1}	5.3843×10^{3}	9.7624×10^{2}	7.0285×10^{3}	1.3163×10^{3}
CCWOA	1.0486×10^{3}	3.6153×10^{1}	1.3579×10^{3}	7.6567×10^{2}	4.2859×10^{3}	5.9815×10^{2}
WOA	1.1285×10^{3}	5.7638×10^{1}	4.9219×10^{3}	6.8542×10^{2}	5.7481×10^{3}	9.5752×10^{2}
DECCWOA	1.0398×10^{3}	4.1688×10^{1}	1.0459×10^{3}	8.8217×10^{1}	3.7945×10^{3}	5.1310×10^{2}
	F25		F26		F27	
	avg	*std*	*avg*	*std*	*avg*	*std*
DECCWOA	1.2002×10^{3}	7.7530×10^{-2}	1.3005×10^{3}	1.0640×10^{-1}	1.4003×10^{3}	5.7690×10^{-2}
DEWOA	1.2026×10^{3}	6.7176×10^{-1}	1.3011×10^{3}	8.8278×10^{-1}	1.4032×10^{3}	6.9794×10^{0}
CCWOA	1.2007×10^{3}	1.9439×10^{-1}	1.3005×10^{3}	1.0762×10^{-1}	1.4003×10^{3}	5.6885×10^{-2}
WOA	1.2017×10^{3}	5.8032×10^{-1}	1.3005×10^{3}	1.4025×10^{-1}	1.4003×10^{3}	4.6882×10^{-2}
	F28		F29		F30	
	avg	*std*	*avg*	*std*	*avg*	*std*
DECCWOA	1.5215×10^{3}	7.2851×10^{0}	1.6105×10^{3}	5.6166×10^{-1}	1.6567×10^{6}	1.2704×10^{6}
DEWOA	3.9455×10^{3}	1.8290×10^{3}	1.6129×10^{3}	5.3817×10^{-1}	5.2403×10^{6}	3.8043×10^{6}
CCWOA	1.5263×10^{3}	7.8018×10^{0}	1.6111×10^{3}	5.3599×10^{-1}	2.3769×10^{6}	1.7422×10^{6}
WOA	1.5710×10^{3}	2.7076×10^{1}	1.6124×10^{3}	6.2114×10^{-1}	3.8096×10^{6}	3.0597×10^{6}

Table A3. *Cont.*

	F31		F32		F33	
	avg	*std*	*avg*	*std*	*avg*	*std*
DECCWOA	6.8401×10^{3}	6.1595×10^{3}	1.9156×10^{3}	1.7561×10^{1}	8.1086×10^{3}	4.1891×10^{3}
DEWOA	1.1805×10^{4}	1.9283×10^{4}	2.0808×10^{3}	7.9264×10^{1}	2.3551×10^{4}	1.9203×10^{4}
CCWOA	1.5061×10^{4}	2.0280×10^{4}	1.9305×10^{3}	3.7604×10^{1}	5.3268×10^{3}	3.0310×10^{3}
WOA	6.2265×10^{3}	4.2908×10^{3}	1.9415×10^{3}	3.6716×10^{1}	2.3634×10^{4}	1.4518×10^{4}
Overall rank	F34		F35		overall	
	avg	*std*	*avg*	*std*	+/−/=	rank
DECCWOA	7.8180×10^{5}	6.8802×10^{5}	2.8140×10^{3}	2.4485×10^{2}	~	1
DEWOA	1.1914×10^{6}	1.0362×10^{6}	3.3508×10^{3}	3.5303×10^{2}	31/1/3	4
CCWOA	5.8255×10^{5}	4.9559×10^{5}	2.7967×10^{3}	1.7716×10^{2}	18/4/13	2
WOA	1.3452×10^{6}	1.6988×10^{6}	3.0734×10^{3}	2.6034×10^{2}	25/1/9	3

Table A4. Comparison results for the DECCWOA with improved WOA versions.

	F1		F2		F3	
	avg	*std*	*avg*	*std*	*avg*	*std*
DECCWOA	0.0000×10^{0}	0.0000×10^{0}	0.0000×10^{0}	0.0000×10^{0}	2.7777×10^{-18}	1.2942×10^{-17}
RDWOA	0.0000×10^{0}	0.0000×10^{0}	0.0000×10^{0}	0.0000×10^{0}	0.0000×10^{0}	0.0000×10^{0}
ACWOA	0.0000×10^{0}	0.0000×10^{0}	0.0000×10^{0}	0.0000×10^{0}	0.0000×10^{0}	0.0000×10^{0}
CCMWOA	0.0000×10^{0}	0.0000×10^{0}	4.7501×10^{-286}	0.0000×10^{0}	0.0000×10^{0}	0.0000×10^{0}
CWOA	0.0000×10^{0}	0.0000×10^{0}	0.0000×10^{0}	0.0000×10^{0}	6.9438×10^{0}	1.0312×10^{1}
BMWOA	9.0723×10^{-4}	1.2467×10^{-3}	7.9729×10^{-3}	7.3643×10^{-3}	2.4579×10^{-1}	7.2733×10^{-1}
BWOA	0.0000×10^{0}	0.0000×10^{0}	0.0000×10^{0}	0.0000×10^{0}	0.0000×10^{0}	0.0000×10^{0}
LWOA	4.9293×10^{-2}	1.1696×10^{-2}	1.0756×10^{0}	1.8737×10^{-1}	1.8394×10^{1}	4.5449×10^{0}
IWOA	0.0000×10^{0}	0.0000×10^{0}	0.0000×10^{0}	0.0000×10^{0}	8.8944×10^{1}	1.3312×10^{2}
	F4		F5		F6	
	avg	*std*	*avg*	*std*	*avg*	*std*
DECCWOA	0.0000×10^{0}	0.0000×10^{0}	2.4313×10^{1}	6.6194×10^{0}	0.0000×10^{0}	0.0000×10^{0}
RDWOA	0.0000×10^{0}	0.0000×10^{0}	1.8882×10^{1}	5.1359×10^{0}	5.1469×10^{-15}	3.2926×10^{-15}
ACWOA	0.0000×10^{0}	0.0000×10^{0}	2.4274×10^{1}	4.5690×10^{0}	6.3093×10^{-4}	2.1127×10^{-4}
CCMWOA	4.3891×10^{-289}	0.0000×10^{0}	2.7607×10^{0}	7.6225×10^{0}	2.0854×10^{-2}	8.2250×10^{-3}
CWOA	8.4827×10^{0}	1.6542×10^{1}	2.5501×10^{1}	1.5480×10^{0}	1.0737×10^{-1}	1.6796×10^{-1}
BMWOA	4.4563×10^{-3}	6.7037×10^{-3}	1.2474×10^{-2}	3.0382×10^{-2}	1.2974×10^{-3}	1.8541×10^{-3}
BWOA	0.0000×10^{0}	0.0000×10^{0}	2.3788×10^{1}	6.4677×10^{0}	1.3716×10^{-4}	5.6219×10^{-5}
LWOA	3.5964×10^{-1}	9.6483×10^{-2}	4.8931×10^{1}	4.4527×10^{1}	5.8005×10^{-2}	1.4408×10^{-2}
IWOA	3.0373×10^{-4}	1.4320×10^{-3}	2.3521×10^{1}	7.0061×10^{-1}	3.5922×10^{-6}	1.7322×10^{-6}
	F7		F8		F9	
	avg	*std*	*avg*	*std*	*avg*	*std*
DECCWOA	1.7008×10^{-4}	2.2308×10^{-4}	-1.2569×10^{4}	2.8058×10^{-12}	0.0000×10^{0}	0.0000×10^{0}
RDWOA	2.8442×10^{-5}	3.6777×10^{-5}	-1.2521×10^{4}	1.6733×10^{2}	0.0000×10^{0}	0.0000×10^{0}
ACWOA	5.6623×10^{-6}	5.7698×10^{-6}	-1.2569×10^{4}	2.1881×10^{-3}	0.0000×10^{0}	0.0000×10^{0}
CCMWOA	1.9668×10^{-4}	1.6220×10^{-4}	-1.0928×10^{4}	9.5870×10^{2}	0.0000×10^{0}	0.0000×10^{0}
CWOA	3.1139×10^{-4}	3.9744×10^{-4}	-1.1583×10^{4}	1.6942×10^{3}	0.0000×10^{0}	0.0000×10^{0}
BMWOA	1.0619×10^{-3}	8.6629×10^{-4}	-1.2569×10^{4}	2.9396×10^{-3}	6.3549×10^{-4}	1.1849×10^{-3}
BWOA	2.5018×10^{-5}	3.0399×10^{-5}	-1.2357×10^{4}	4.2512×10^{2}	0.0000×10^{0}	0.0000×10^{0}
LWOA	1.2178×10^{-1}	4.6795×10^{-2}	-1.2382×10^{4}	4.4862×10^{2}	1.0110×10^{2}	2.6929×10^{1}
IWOA	2.6929×10^{-4}	3.2479×10^{-4}	-1.2298×10^{4}	7.5775×10^{2}	0.0000×10^{0}	0.0000×10^{0}

Table A4. *Cont.*

	F10		F11		F12	
	avg	*std*	*avg*	*std*	*avg*	*std*
DECCWOA	8.8818×10^{-16}	0.0000×10^{0}	0.0000×10^{0}	0.0000×10^{0}	1.5705×10^{-32}	5.5674×10^{-48}
RDWOA	8.8818×10^{-16}	0.0000×10^{0}	0.0000×10^{0}	0.0000×10^{0}	2.1668×10^{-7}	1.1868×10^{-6}
ACWOA	1.0066×10^{-15}	6.4863×10^{-16}	0.0000×10^{0}	0.0000×10^{0}	6.7416×10^{-5}	1.9647×10^{-5}
CCMWOA	8.8818×10^{-16}	0.0000×10^{0}	0.0000×10^{0}	0.0000×10^{0}	7.8157×10^{-4}	3.6365×10^{-4}
CWOA	3.0198×10^{-15}	2.0010×10^{-15}	0.0000×10^{0}	0.0000×10^{0}	4.4206×10^{-3}	6.6548×10^{-3}
BMWOA	5.1495×10^{-3}	4.7013×10^{-3}	2.1417×10^{-3}	4.0514×10^{-3}	1.5181×10^{-5}	2.4654×10^{-5}
BWOA	8.8818×10^{-16}	0.0000×10^{0}	0.0000×10^{0}	0.0000×10^{0}	1.8415×10^{-5}	6.4484×10^{-6}
LWOA	6.6177×10^{-1}	6.9724×10^{-1}	1.4973×10^{-2}	1.3048×10^{-2}	5.5931×10^{-1}	1.1387×10^{0}
IWOA	2.5461×10^{-15}	2.0298×10^{-15}	1.8892×10^{-3}	1.0348×10^{-2}	5.1354×10^{-7}	1.4075×10^{-7}
	F13		F14		F15	
	avg	*std*	*avg*	*std*	*avg*	*std*
DECCWOA	1.3498×10^{-32}	5.5674×10^{-48}	5.3971×10^{6}	4.3864×10^{6}	1.1568×10^{5}	1.0038×10^{5}
RDWOA	3.6625×10^{-4}	2.0060×10^{-3}	1.0435×10^{7}	6.2580×10^{6}	2.2927×10^{7}	2.8097×10^{7}
ACWOA	2.5808×10^{-3}	4.6664×10^{-3}	1.4456×10^{8}	5.9824×10^{7}	7.4176×10^{9}	4.5829×10^{9}
CCMWOA	6.6551×10^{-4}	7.9798×10^{-4}	3.1814×10^{8}	1.2548×10^{8}	3.0720×10^{10}	7.7697×10^{9}
CWOA	5.3049×10^{-1}	4.3677×10^{-1}	6.6829×10^{7}	4.6037×10^{7}	2.0499×10^{9}	2.4020×10^{9}
BMWOA	1.2733×10^{-4}	2.2038×10^{-4}	1.0438×10^{8}	3.7729×10^{7}	2.8151×10^{8}	1.3106×10^{8}
BWOA	3.3556×10^{-3}	5.0182×10^{-3}	6.4289×10^{7}	2.9224×10^{7}	2.4303×10^{8}	1.3580×10^{8}
LWOA	2.1065×10^{-2}	7.2157×10^{-3}	3.7558×10^{6}	1.3953×10^{6}	5.2308×10^{5}	1.5679×10^{5}
IWOA	9.8357×10^{-6}	9.6866×10^{-6}	2.4173×10^{7}	1.2055×10^{7}	2.1211×10^{6}	3.5049×10^{6}
	F16		F17		F18	
	avg	*std*	*avg*	*std*	*avg*	*std*
DECCWOA	4.0353×10^{3}	2.8694×10^{3}	4.9768×10^{2}	4.3913×10^{1}	5.2008×10^{2}	6.1713×10^{-2}
RDWOA	6.3763×10^{3}	3.2741×10^{3}	5.3428×10^{2}	3.9426×10^{1}	5.2012×10^{2}	1.1564×10^{-1}
ACWOA	5.0442×10^{4}	6.7503×10^{3}	1.2586×10^{3}	2.9833×10^{2}	5.2085×10^{2}	1.1588×10^{-1}
CCMWOA	5.9083×10^{4}	9.2658×10^{3}	2.7400×10^{3}	1.0666×10^{3}	5.2088×10^{2}	1.7053×10^{-1}
CWOA	5.7234×10^{4}	3.7392×10^{4}	8.0857×10^{2}	2.2460×10^{2}	5.2029×10^{2}	1.3417×10^{-1}
BMWOA	5.7008×10^{4}	9.2535×10^{3}	6.7153×10^{2}	6.3683×10^{1}	5.2097×10^{2}	9.6222×10^{-2}
BWOA	3.2197×10^{4}	1.0637×10^{4}	6.9451×10^{2}	7.2749×10^{1}	5.2067×10^{2}	1.7262×10^{-1}
LWOA	1.0354×10^{3}	4.9837×10^{2}	5.0374×10^{2}	4.8236×10^{1}	5.2048×10^{2}	9.4869×10^{-2}
IWOA	1.6513×10^{4}	9.7529×10^{3}	5.7401×10^{2}	6.2791×10^{1}	5.2023×10^{2}	1.3567×10^{-1}
	F19		F20		F21	
	avg	*std*	*avg*	*std*	*avg*	*std*
DECCWOA	6.1886×10^{2}	2.8176×10^{0}	7.0051×10^{2}	2.4043×10^{-1}	8.0268×10^{2}	2.9674×10^{0}
RDWOA	6.2296×10^{2}	3.2376×10^{0}	7.0097×10^{2}	2.3043×10^{-1}	8.4662×10^{2}	1.2443×10^{1}
ACWOA	6.3380×10^{2}	2.8192×10^{0}	7.4705×10^{2}	2.6949×10^{1}	9.9173×10^{2}	2.8423×10^{1}
CCMWOA	6.3421×10^{2}	3.1895×10^{0}	9.1058×10^{2}	6.8146×10^{1}	1.0329×10^{3}	2.5834×10^{1}
CWOA	6.3619×10^{2}	2.3328×10^{0}	7.1987×10^{2}	1.9833×10^{1}	9.8919×10^{2}	3.5017×10^{1}
BMWOA	6.3209×10^{2}	3.4550×10^{0}	7.0298×10^{2}	7.9463×10^{-1}	9.6429×10^{2}	2.2995×10^{1}
BWOA	6.3659×10^{2}	2.7733×10^{0}	7.0210×10^{2}	4.7824×10^{-1}	9.6653×10^{2}	2.0133×10^{1}
LWOA	6.2990×10^{2}	3.7057×10^{0}	7.0071×10^{2}	1.0660×10^{-1}	8.7689×10^{2}	1.4956×10^{1}
IWOA	6.2837×10^{2}	3.5182×10^{0}	7.0086×10^{2}	1.8091×10^{-1}	9.1853×10^{2}	2.3903×10^{1}

Table A4. *Cont.*

	F22		F23		F24	
	avg	*std*	*avg*	*std*	*avg*	*std*
DECCWOA	1.0349×10^3	3.5053×10^1	1.0502×10^3	7.2638×10^1	3.9773×10^3	5.2239×10^2
RDWOA	1.0862×10^3	3.9406×10^1	1.6401×10^3	2.1508×10^2	4.8765×10^3	4.8011×10^2
ACWOA	1.1353×10^3	2.7084×10^1	4.6113×10^3	7.8646×10^2	6.0951×10^3	8.4619×10^2
CCMWOA	1.1585×10^3	2.0571×10^1	5.7676×10^3	4.6957×10^2	7.0634×10^3	8.3597×10^2
CWOA	1.1502×10^3	5.9550×10^1	5.0935×10^3	8.0866×10^2	6.4607×10^3	8.0167×10^2
BMWOA	1.1247×10^3	3.0558×10^1	4.7543×10^3	7.0975×10^2	7.1260×10^3	9.0188×10^2
BWOA	1.1033×10^3	2.3404×10^1	4.8599×10^3	7.9435×10^2	6.5557×10^3	1.0588×10^3
LWOA	1.1231×10^3	4.1332×10^1	2.1055×10^3	5.0404×10^2	5.3203×10^3	5.2375×10^2
IWOA	1.1290×10^3	5.0219×10^1	2.6021×10^3	4.6117×10^2	5.5791×10^3	7.2655×10^2
	F25		F26		F27	
	avg	*std*	*avg*	*std*	*avg*	*std*
DECCWOA	1.2002×10^3	5.1670×10^{-2}	1.3005×10^3	1.1585×10^{-1}	1.4003×10^3	4.7444×10^{-2}
RDWOA	1.2005×10^3	1.8157×10^{-1}	1.3005×10^3	1.0242×10^{-1}	1.4002×10^3	3.7231×10^{-2}
ACWOA	1.2017×10^3	5.3455×10^{-1}	1.3011×10^3	8.3617×10^{-1}	1.4239×10^3	1.2976×10^1
CCMWOA	1.2018×10^3	4.5129×10^{-1}	1.3041×10^3	8.3728×10^{-1}	1.4661×10^3	1.6380×10^1
CWOA	1.2018×10^3	5.2022×10^{-1}	1.3006×10^3	1.2049×10^{-1}	1.4102×10^3	1.2849×10^1
BMWOA	1.2023×10^3	4.1490×10^{-1}	1.3005×10^3	1.1904×10^{-1}	1.4003×10^3	1.0699×10^{-1}
BWOA	1.2019×10^3	4.8775×10^{-1}	1.3005×10^3	1.3303×10^{-1}	1.4003×10^3	4.2027×10^{-2}
LWOA	1.2008×10^3	3.0020×10^{-1}	1.3005×10^3	1.1102×10^{-1}	1.4003×10^3	9.9959×10^{-2}
IWOA	1.2010×10^3	2.9781×10^{-1}	1.3005×10^3	9.8276×10^{-2}	1.4003×10^3	5.2298×10^{-2}
	F28		F29		F30	
	avg	*std*	*avg*	*std*	*avg*	*std*
DECCWOA	1.5195×10^3	5.8976×10^0	1.6103×10^3	7.8355×10^{-1}	1.4951×10^6	1.0794×10^6
RDWOA	1.5215×10^3	7.2912×10^0	1.6116×10^3	5.9134×10^{-1}	1.1935×10^6	1.0797×10^6
ACWOA	1.8396×10^3	4.0931×10^2	1.6121×10^3	4.8464×10^{-1}	1.3124×10^7	8.0308×10^6
CCMWOA	7.0062×10^3	4.0693×10^3	1.6130×10^3	3.2081×10^{-1}	2.6816×10^7	1.9260×10^7
CWOA	1.9731×10^3	8.2350×10^2	1.6127×10^3	5.6484×10^{-1}	9.8463×10^6	8.8774×10^6
BMWOA	1.5828×10^3	3.6612×10^1	1.6126×10^3	2.1626×10^{-1}	6.9174×10^6	4.9113×10^6
BWOA	1.6258×10^3	4.8685×10^1	1.6124×10^3	4.8342×10^{-1}	7.5640×10^6	5.4938×10^6
LWOA	1.5213×10^3	4.9669×10^0	1.6125×10^3	5.6375×10^{-1}	4.9220×10^5	2.9566×10^5
IWOA	1.5506×10^3	1.6210×10^1	1.6125×10^3	5.1867×10^{-1}	2.7873×10^6	2.0454×10^6
	F31		F32		F33	
	avg	*std*	*avg*	*std*	*avg*	*std*
DECCWOA	9.2541×10^3	2.1036×10^4	1.9198×10^3	2.4744×10^1	8.8237×10^3	5.3819×10^3
RDWOA	4.8557×10^3	3.4855×10^3	1.9194×10^3	2.6942×10^1	6.7743×10^3	3.5180×10^3
ACWOA	4.4984×10^7	4.3109×10^7	2.0047×10^3	3.3162×10^1	3.7129×10^4	1.9130×10^4
CCMWOA	9.8440×10^7	1.2615×10^8	2.0824×10^3	5.0281×10^1	5.7205×10^4	2.2855×10^4
CWOA	3.9424×10^6	1.2057×10^7	2.0018×10^3	6.3131×10^1	5.7907×10^4	5.9437×10^4
BMWOA	1.1037×10^5	1.2099×10^5	1.9467×10^3	4.0178×10^1	3.3436×10^4	1.7890×10^4
BWOA	1.1889×10^5	3.5142×10^5	1.9593×10^3	3.8145×10^1	3.2143×10^4	1.6907×10^4
LWOA	1.0695×10^4	5.9845×10^3	1.9230×10^3	2.4187×10^1	3.0576×10^3	7.5160×10^2
IWOA	5.4855×10^3	4.2910×10^3	1.9348×10^3	3.5960×10^1	1.6411×10^4	1.0072×10^4

Table A4. *Cont.*

Overall rank	F34		F35		overall	
	avg	*std*	*avg*	*std*	+/−/=	rank
DECCWOA	1.0426×10^{6}	8.4796×10^{5}	2.8721×10^{3}	2.1610×10^{2}	~	1
RDWOA	4.2175×10^{5}	3.3446×10^{5}	2.7874×10^{3}	2.1665×10^{2}	16/6/13	2
ACWOA	4.2559×10^{6}	3.6243×10^{6}	3.0278×10^{3}	2.2024×10^{2}	26/3/6	6
CCMWOA	8.7004×10^{6}	5.9791×10^{6}	3.2984×10^{3}	4.4928×10^{2}	28/2/5	9
CWOA	3.1633×10^{6}	2.9730×10^{6}	3.1092×10^{3}	2.3259×10^{2}	29/0/6	8
BMWOA	1.0736×10^{6}	9.0915×10^{5}	3.0014×10^{3}	2.7107×10^{2}	32/1/2	7
BWOA	1.9551×10^{6}	1.5899×10^{6}	2.9774×10^{3}	2.8735×10^{2}	23/3/9	4
LWOA	2.0517×10^{5}	1.6837×10^{5}	2.9007×10^{3}	2.4887×10^{2}	26/4/5	5
IWOA	9.3081×10^{5}	7.7249×10^{5}	2.9329×10^{3}	1.7696×10^{2}	24/1/10	3

Table A5. Comparison results for the DECCWOA with advanced algorithms.

	F1		F2		F3	
	avg	*std*	*avg*	*std*	*avg*	*std*
DECCWOA	0.0000×10^{0}	0.0000×10^{0}	0.0000×10^{0}	0.0000×10^{0}	1.0221×10^{-18}	4.2975×10^{-18}
IGWO	0.0000×10^{0}	0.0000×10^{0}	3.6328×10^{-261}	0.0000×10^{0}	1.3124×10^{-86}	7.1881×10^{-86}
OBLGWO	0.0000×10^{0}	0.0000×10^{0}	3.6589×10^{-142}	2.004×10^{-141}	6.2014×10^{-293}	0.0000×10^{0}
CGPSO	2.3583×10^{-8}	7.7088×10^{-8}	3.9726×10^{-5}	2.8781×10^{-5}	6.3491×10^{-2}	5.1833×10^{-2}
ALPSO	1.1539×10^{-184}	0.0000×10^{0}	2.5959×10^{-8}	7.1555×10^{-8}	2.2102×10^{-11}	2.9723×10^{-11}
RCBA	8.9446×10^{-3}	2.9769×10^{-3}	5.8765×10^{-1}	8.4909×10^{-2}	2.1948×10^{0}	5.2552×10^{-1}
CBA	7.2954×10^{-8}	3.8213×10^{-7}	4.1161×10^{1}	1.3912×10^{2}	1.3118×10^{1}	6.5496×10^{0}
OBSCA	1.0911×10^{-103}	5.5402×10^{-103}	4.3833×10^{-91}	1.1161×10^{-90}	3.1617×10^{-24}	1.1702×10^{-23}
SCADE	0.0000×10^{0}	0.0000×10^{0}	0.0000×10^{0}	0.0000×10^{0}	0.0000×10^{0}	0.0000×10^{0}
	F4		F5		F6	
	avg	*std*	*avg*	*std*	*avg*	*std*
DECCWOA	1.0221×10^{-18}	4.2975×10^{-18}	2.6083×10^{1}	3.1418×10^{-1}	0.0000×10^{0}	0.0000×10^{0}
IGWO	1.3124×10^{-86}	7.1881×10^{-86}	2.3216×10^{1}	1.8144×10^{-1}	1.2448×10^{-5}	3.5159×10^{-6}
OBLGWO	6.2014×10^{-293}	0.0000×10^{0}	2.6052×10^{1}	3.8656×10^{-1}	3.9085×10^{-5}	1.4498×10^{-5}
CGPSO	6.3491×10^{-2}	5.1833×10^{-2}	1.0747×10^{-7}	1.4040×10^{-7}	1.5149×10^{-8}	2.7356×10^{-8}
ALPSO	2.2102×10^{-11}	2.9723×10^{-11}	3.5496×10^{1}	3.2473×10^{1}	5.9288×10^{-31}	2.2626×10^{-30}
RCBA	2.1948×10^{0}	5.2552×10^{-1}	3.6041×10^{1}	4.0444×10^{1}	8.7533×10^{-3}	2.4284×10^{-3}
CBA	1.3118×10^{1}	6.5496×10^{0}	7.3423×10^{1}	1.2319×10^{2}	4.4526×10^{-7}	2.4194×10^{-6}
OBSCA	3.1617×10^{-24}	1.1702×10^{-23}	2.7647×10^{1}	3.8007×10^{-1}	3.8321×10^{0}	2.7513×10^{-1}
SCADE	0.0000×10^{0}	0.0000×10^{0}	1.5398×10^{1}	1.3017×10^{1}	1.7996×10^{-7}	1.6508×10^{-7}
	F7		F8		F9	
	avg	*std*	*avg*	*std*	*avg*	*std*
DECCWOA	1.1896×10^{-4}	1.4262×10^{-4}	-1.3066×10^{4}	2.6313×10^{3}	0.0000×10^{0}	0.0000×10^{0}
IGWO	2.9290×10^{-4}	2.6976×10^{-4}	-7.4319×10^{3}	6.6317×10^{2}	0.0000×10^{0}	0.0000×10^{0}
OBLGWO	2.4381×10^{-5}	2.9727×10^{-5}	-1.2561×10^{4}	4.4545×10^{1}	0.0000×10^{0}	0.0000×10^{0}
CGPSO	1.4906×10^{-5}	1.4183×10^{-5}	-3.7698×10^{4}	6.6756×10^{3}	3.0143×10^{-9}	6.2053×10^{-9}
ALPSO	7.8389×10^{-2}	3.1754×10^{-2}	-1.1531×10^{4}	2.8700×10^{2}	1.9471×10^{1}	7.9710×10^{0}
RCBA	1.1712×10^{-1}	5.5739×10^{-2}	-7.3244×10^{3}	5.4651×10^{2}	2.0111×10^{1}	4.6024×10^{0}
CBA	1.5885×10^{-1}	3.4560×10^{-1}	-7.3445×10^{3}	6.5505×10^{2}	1.2498×10^{2}	4.7753×10^{1}
OBSCA	8.1175×10^{-4}	5.3137×10^{-4}	-4.1274×10^{3}	2.4305×10^{2}	0.0000×10^{0}	0.0000×10^{0}
SCADE	2.9509×10^{-4}	2.0997×10^{-4}	-1.2569×10^{4}	1.1550×10^{-2}	0.0000×10^{0}	0.0000×10^{0}

Table A5. *Cont.*

	F10		F11		F12	
	avg	*std*	*avg*	*std*	*avg*	*std*
DECCWOA	8.8818×10^{-16}	0.0000×10^{0}	0.0000×10^{0}	0.0000×10^{0}	1.5705×10^{-32}	5.5674×10^{-48}
IGWO	4.9146×10^{-15}	1.2283×10^{-15}	0.0000×10^{0}	0.0000×10^{0}	1.1169×10^{-6}	3.8305×10^{-7}
OBLGWO	8.8818×10^{-16}	0.0000×10^{0}	0.0000×10^{0}	0.0000×10^{0}	3.8858×10^{-4}	1.1452×10^{-3}
CGPSO	1.8069×10^{-5}	1.2953×10^{-5}	4.3701×10^{-8}	7.1267×10^{-8}	6.2743×10^{-11}	1.4249×10^{-10}
ALPSO	8.0156×10^{-1}	8.3429×10^{-1}	1.6465×10^{-2}	1.5344×10^{-2}	3.0222×10^{-2}	8.1300×10^{-2}
RCBA	1.0853×10^{-1}	2.7602×10^{-2}	1.0473×10^{-2}	1.0208×10^{-2}	9.1885×10^{0}	2.8806×10^{0}
CBA	1.5880×10^{1}	2.1141×10^{0}	1.3514×10^{-2}	1.8331×10^{-2}	1.4396×10^{1}	4.7131×10^{0}
OBSCA	4.3225×10^{-15}	6.4863×10^{-16}	0.0000×10^{0}	0.0000×10^{0}	3.8964×10^{-1}	4.5185×10^{-2}
SCADE	8.8818×10^{-16}	0.0000×10^{0}	0.0000×10^{0}	0.0000×10^{0}	5.6904×10^{-9}	4.8080×10^{-9}
	F13		F14		F15	
	avg	*std*	*avg*	*std*	*avg*	*std*
DECCWOA	1.3498×10^{-32}	5.5674×10^{-48}	4.2212×10^{6}	3.5710×10^{6}	1.6057×10^{5}	1.7381×10^{5}
IGWO	1.4838×10^{-2}	2.6985×10^{-2}	1.7924×10^{7}	6.8885×10^{6}	2.3989×10^{6}	1.4218×10^{6}
OBLGWO	5.7997×10^{-2}	7.7859×10^{-2}	1.7531×10^{7}	1.0402×10^{7}	1.3307×10^{7}	1.0251×10^{7}
CGPSO	3.1353×10^{-9}	9.6561×10^{-9}	9.8321×10^{6}	2.5274×10^{6}	1.5880×10^{8}	1.6828×10^{7}
ALPSO	2.8346×10^{-2}	9.8785×10^{-2}	5.7145×10^{6}	5.4783×10^{6}	2.9466×10^{3}	3.6264×10^{3}
RCBA	5.1201×10^{-3}	4.6593×10^{-3}	1.0422×10^{6}	3.8825×10^{5}	2.5206×10^{4}	1.0389×10^{4}
CBA	3.1616×10^{1}	2.6961×10^{1}	4.7886×10^{6}	1.8780×10^{6}	1.2467×10^{4}	1.0840×10^{4}
OBSCA	2.1646×10^{0}	1.1588×10^{-1}	3.9908×10^{8}	1.1211×10^{8}	2.4969×10^{10}	3.6691×10^{9}
SCADE	8.3767×10^{-8}	6.6478×10^{-8}	4.7479×10^{8}	9.8759×10^{7}	2.9269×10^{10}	4.2012×10^{9}
	F16		F17		F18	
	avg	*std*	*avg*	*std*	*avg*	*std*
DECCWOA	7.8122×10^{3}	5.4762×10^{3}	5.1647×10^{2}	4.8965×10^{1}	5.2007×10^{2}	5.1181×10^{-2}
IGWO	7.2399×10^{3}	2.0340×10^{3}	5.2142×10^{2}	3.0682×10^{1}	5.2050×10^{2}	1.4090×10^{-1}
OBLGWO	1.0074×10^{4}	3.8144×10^{3}	5.4284×10^{2}	3.5930×10^{1}	5.2095×10^{2}	5.4001×10^{-2}
CGPSO	2.3365×10^{3}	5.0383×10^{2}	4.6884×10^{2}	3.1903×10^{1}	5.2098×10^{2}	4.2868×10^{-2}
ALPSO	3.7450×10^{2}	1.3166×10^{2}	5.4221×10^{2}	5.6842×10^{1}	5.2080×10^{2}	5.8119×10^{-2}
RCBA	3.2947×10^{2}	1.2959×10^{1}	4.7036×10^{2}	3.7390×10^{1}	5.2010×10^{2}	9.5615×10^{-2}
CBA	4.7703×10^{3}	9.2720×10^{3}	4.9781×10^{2}	4.4714×10^{1}	5.2009×10^{2}	1.3697×10^{-1}
OBSCA	5.3939×10^{4}	6.9272×10^{3}	2.2087×10^{3}	5.3362×10^{2}	5.2096×10^{2}	5.8056×10^{-2}
SCADE	5.4785×10^{4}	6.7546×10^{3}	2.2800×10^{3}	4.7955×10^{2}	5.2094×10^{2}	5.3187×10^{-2}
	F19		F20		F21	
	avg	*std*	*avg*	*std*	*avg*	*std*
DECCWOA	6.1885×10^{2}	2.4155×10^{0}	7.0061×10^{2}	3.1716×10^{-1}	8.0279×10^{2}	8.0509×10^{0}
IGWO	6.1887×10^{2}	2.6487×10^{0}	7.0099×10^{2}	5.1420×10^{-2}	8.8181×10^{2}	1.6752×10^{1}
OBLGWO	6.1968×10^{2}	4.3347×10^{0}	7.0117×10^{2}	9.6122×10^{-2}	9.2963×10^{2}	3.9241×10^{1}
CGPSO	6.2402×10^{2}	2.9481×10^{0}	7.0241×10^{2}	2.0187×10^{-1}	9.8743×10^{2}	2.5413×10^{1}
ALPSO	6.1705×10^{2}	2.4439×10^{0}	7.0001×10^{2}	8.9890×10^{-3}	8.2142×10^{2}	9.5753×10^{0}
RCBA	6.3882×10^{2}	3.5284×10^{0}	7.0007×10^{2}	1.8671×10^{-2}	1.0209×10^{3}	3.9995×10^{1}
CBA	6.4033×10^{2}	2.8154×10^{0}	7.0003×10^{2}	3.4856×10^{-2}	1.0228×10^{3}	6.1207×10^{1}
OBSCA	6.3161×10^{2}	1.3106×10^{0}	9.0597×10^{2}	3.6858×10^{1}	1.0643×10^{3}	1.7737×10^{1}
SCADE	6.3355×10^{2}	2.2424×10^{0}	8.9018×10^{2}	3.3322×10^{1}	1.0695×10^{3}	1.2949×10^{1}

Table A5. *Cont.*

	F22		F23		F24	
	avg	*std*	*avg*	*std*	*avg*	*std*
DECCWOA	1.0366×10^3	3.6638×10^1	1.0472×10^3	5.2652×10^1	3.6010×10^3	4.5441×10^2
IGWO	1.0087×10^3	1.9673×10^1	3.3271×10^3	4.7812×10^2	4.6073×10^3	7.5057×10^2
OBLGWO	1.0661×10^3	4.0353×10^1	4.0667×10^3	1.0107×10^3	5.5480×10^3	1.0343×10^3
CGPSO	1.1225×10^3	2.5722×10^1	5.5982×10^3	5.0945×10^2	6.0825×10^3	5.5185×10^2
ALPSO	1.0021×10^3	2.9422×10^1	1.6121×10^3	3.2630×10^2	4.0735×10^3	5.3691×10^2
RCBA	1.1638×10^3	6.3597×10^1	5.6153×10^3	6.4055×10^2	5.8506×10^3	8.1461×10^2
CBA	1.1526×10^3	7.2911×10^1	5.7931×10^3	7.2017×10^2	5.9590×10^3	7.1736×10^2
OBSCA	1.1929×10^3	1.5671×10^1	6.1341×10^3	3.2876×10^2	7.2777×10^3	4.0271×10^2
SCADE	1.2049×10^3	1.8580×10^1	7.4883×10^3	2.3162×10^2	8.2006×10^3	2.9486×10^2
	F25		F26		F27	
	avg	*std*	*avg*	*std*	*avg*	*std*
DECCWOA	1.2002×10^3	7.1972×10^{-2}	1.3005×10^3	1.4836×10^{-1}	1.4003×10^3	4.2207×10^{-2}
IGWO	1.2008×10^3	3.5285×10^{-1}	1.3006×10^3	1.2395×10^{-1}	1.4004×10^3	2.5832×10^{-1}
OBLGWO	1.2023×10^3	6.8209×10^{-1}	1.3005×10^3	1.1878×10^{-1}	1.4005×10^3	2.3451×10^{-1}
CGPSO	1.2025×10^3	2.0362×10^{-1}	1.3004×10^3	1.0032×10^{-1}	1.4003×10^3	1.2353×10^{-1}
ALPSO	1.2013×10^3	5.4330×10^{-1}	1.3005×10^3	7.9368×10^{-2}	1.4006×10^3	2.8021×10^{-1}
RCBA	1.2006×10^3	3.7833×10^{-1}	1.3005×10^3	1.3700×10^{-1}	1.4003×10^3	9.7996×10^{-2}
CBA	1.2011×10^3	7.3537×10^{-1}	1.3005×10^3	1.4823×10^{-1}	1.4003×10^3	1.5708×10^{-1}
OBSCA	1.2023×10^3	4.4892×10^{-1}	1.3036×10^3	2.6704×10^{-1}	1.4695×10^3	1.1814×10^1
SCADE	1.2026×10^3	2.4794×10^{-1}	1.3039×10^3	2.9836×10^{-1}	1.4881×10^3	1.3091×10^1
	F28		F29		F30	
	avg	*std*	*avg*	*std*	*avg*	*std*
DECCWOA	1.5202×10^3	6.6341×10^0	1.6102×10^3	7.9146×10^{-1}	1.8690×10^6	1.1246×10^6
IGWO	1.5176×10^3	3.7863×10^0	1.6116×10^3	5.8466×10^{-1}	9.2251×10^5	5.5675×10^5
OBLGWO	1.5150×10^3	5.7646×10^0	1.6120×10^3	6.3233×10^{-1}	1.3955×10^6	1.0330×10^6
CGPSO	1.5176×10^3	1.3084×10^0	1.6117×10^3	3.2922×10^{-1}	3.3985×10^5	1.7015×10^5
ALPSO	1.5115×10^3	4.1448×10^0	1.6118×10^3	3.2864×10^{-1}	5.5179×10^5	4.6710×10^5
RCBA	1.5371×10^3	9.0984×10^0	1.6135×10^3	3.4738×10^{-1}	1.2165×10^5	7.5279×10^4
CBA	1.5589×10^3	1.7899×10^1	1.6133×10^3	5.2338×10^{-1}	1.8619×10^5	1.2340×10^5
OBSCA	1.4085×10^4	8.8848×10^3	1.6129×10^3	2.5975×10^{-1}	1.1278×10^7	5.7586×10^6
SCADE	1.8666×10^4	6.8024×10^3	1.6127×10^3	1.7040×10^{-1}	1.5787×10^7	8.3755×10^6
	F31		F32		F33	
	avg	*std*	*avg*	*std*	*avg*	*std*
DECCWOA	4.3650×10^3	3.0442×10^3	1.9120×10^3	1.2480×10^1	7.8097×10^3	4.1959×10^3
IGWO	1.7080×10^4	2.1685×10^4	1.9180×10^3	1.4497×10^1	3.0438×10^3	6.5561×10^2
OBLGWO	8.3234×10^4	1.7925×10^5	1.9215×10^3	2.3356×10^1	5.5656×10^3	2.6114×10^3
CGPSO	2.4871×10^6	7.6618×10^5	1.9170×10^3	2.9535×10^0	2.4762×10^3	1.6062×10^2
ALPSO	7.8595×10^3	7.3812×10^3	1.9170×10^3	2.0884×10^1	3.0087×10^3	4.2177×10^2
RCBA	6.9919×10^3	7.0401×10^3	1.9292×10^3	2.9262×10^1	2.4379×10^3	1.3813×10^2
CBA	9.7529×10^3	9.7686×10^3	1.9246×10^3	2.6288×10^1	2.9663×10^3	1.2059×10^3
OBSCA	1.5585×10^8	1.0228×10^8	2.0091×10^3	1.5224×10^1	3.1925×10^4	1.3827×10^4
SCADE	1.9277×10^8	9.7379×10^7	2.0133×10^3	1.3639×10^1	2.7694×10^4	1.1641×10^4

Table A5. *Cont.*

Overall rank	F34		F35		overall	
	avg	*std*	*avg*	*std*	+/−/=	rank
DECCWOA	5.6266×10^5	5.3759×10^5	2.8005×10^3	2.0373×10^2	~	1
IGWO	2.5893×10^5	2.2750×10^5	2.5846×10^3	1.4319×10^2	20/8/7	2
OBLGWO	5.1886×10^5	3.8300×10^5	2.6930×10^3	1.9962×10^2	19/5/11	5
CGPSO	1.2464×10^5	6.8764×10^4	2.9020×10^3	2.0995×10^2	23/9/3	4
ALPSO	1.1929×10^5	2.9474×10^5	2.7316×10^3	2.0662×10^2	19/9/7	3
RCBA	8.3440×10^4	3.8322×10^4	3.3862×10^3	3.5129×10^2	23/9/3	6
CBA	1.2013×10^5	7.8959×10^4	3.4067×10^3	2.6854×10^2	24/6/5	7
OBSCA	1.9523×10^6	9.0846×10^5	3.1622×10^3	1.4760×10^2	32/1/2	9
SCADE	2.4532×10^6	1.2033×10^6	3.1167×10^3	1.2839×10^2	28/2/5	7

Table A6. Description of each attribute for the talent stability data.

Attributes	Name	Description
F1	Sex	1 for male and 2 for female.
F2	Political affiliation	There are five categories: Communist Party members, reserve party members, democratic party members, Communist Youth League members and the masses, denoted by 1, 2, 3, 4 and 13, respectively.
F3	Professional attributes	1 indicates arts, 2 indicates science and 3 indicates less than junior college (junior college not divided into arts and science subjects)
F4	Age	Ages 25–30, 31–35, 36–40, 41–45, 46–50, 51–55 and 56–60 are indicated by 1, 2, 3, 4, 5, 6 and 7, respectively. Young and middle-aged people have a strong level of competence and a strong tendency to move because of upward mobility, life pressures, etc.
F5	Household Registration	There are three categories: in-city, in-province and out-of-province, indicated by 0, 1 and 2, respectively.
F6	Type of place of origin	There are three categories: urban, township and rural, denoted by 1, 2 and 3, respectively.
F7	City-level and above talent categories	There are categories A, B, C, D and E, denoted by 1, 2, 3, 4 and 5, respectively. 6 is for talent category F and no talent category is denoted by 10.
F8	Nature of previous unit	0 indicating pending employment, 10 indicating state institutions, 20 indicating scientific research institutions, 21 indicating higher education institutions, 22 indicating secondary and junior education institutions, 23 indicating health and medical institutions, 29 indicating other institutions, 31 indicating state-owned enterprises, 32 indicating foreign-funded enterprises, 39 indicating private enterprises, 40 indicating the army, 55 indicating rural organizations, and 99 indicating self-employment. No previous unit is denoted by 100.
F9	Wenzhou colleges and university's location type	Prefectural level cities, denoted by 2.
F10	Year of employment at Wenzhou colleges and universities	This is a measure of stability in the unit of employment. 1 is used for entry before 2000 (merger), 2 for entry from 2001–2006 (preparation), 3 for entry from 2007–2008 (de-preparation), 4 for entry from 2009–2014 (school introduction policy), 5 for entry from 2015–2017 (city introduction policy) and 6 for entry from 2018. To date (increased introduction by the school) entry is indicated by 6. (It can also be described in terms of stable years 3, 4–6, 7–10, 11+ years).
F11	Types of positions at Wenzhou colleges and universities	Teaching staff are represented by 24, PhD students and research staff by 11, professional and technical staff by 29, administrative staff by 101 and counsellors by 102.

Table A6. *Cont.*

Attributes	Name	Description
F12	Professional relevance of employment at Wenzhou colleges and universities	It is used to measure the relevance of the major studied to the job, with higher percentages indicating higher relevance.
F13	Monthly salary level for employment at Wenzhou colleges and universities: RMB	It is used to measure the average monthly salary received, with higher values indicating higher salary levels.
F14	Current employment	Current employment is indicated by 1 for Wenzhou undergraduate institutions; 2 for civil servants or institutions; 3 for undergraduate institutions (including doctoral studies); 4 for vocational institutions in other cities; 5 for vocational institutions in the city, 6 for enterprises, 7 for going abroad and 8 for pending employment.
F15	Time of introduction at current employment	Indicated by 1 for entry before 2000 (merger), 2 for entry from 2001–2006 (preparation), 3 for entry from 2007–2008 (de-preparation), 4 for entry from 2009–2014 (school introduction policy), 5 for entry from 2015–2017 (city introduction policy) and 6 for entry from 2018-present (increased school introduction).
F16	Nature of current employment	0 indicating pending employment, 10 indicating state institutions, 20 indicating scientific research institutions, 21 indicating higher education institutions, 22 indicating secondary and junior education institutions, 23 indicating health and medical institutions, 29 indicating other institutions, 31 indicating state-owned enterprises, 32 indicating foreign funded enterprises, 39 indicating private enterprises, 40 indicating the army, 55 indicating rural organizations and 99 indicating self-employment.
F17	Type of location of current employment unit	Pending employment is represented by 0, sub-provincial and large cities by 1, prefecture-level cities by 2 and counties and villages by 3.
F18	Type of current employment	The type of position currently employed is expressed in the same way as the type of position in the previous employment unit indicated in F11. Pending employment is indicated by 0, civil servants by 10, doctoral students and researchers by 11, engineers and technicians by 13, teaching staff by 24, professional and technical staff by 29, commercial service staff and clerks by 30, military personnel by 80, administrative staff by 101 and counsellors by 102.
F19	Relevance of current employment profession	The professional relevance of current employment is expressed in the same way as the type of position in the previous employment unit indicated by F11.
F20	Monthly salary level in current employment unit: RMB	The current employment monthly salary level is expressed in the same way as the previous employment monthly salary level in F13.
F21	Salary differential	It is used to measure the change in the monthly salary of the current employment unit from that of the previous employment unit, that is, the difference between the monthly salary level of the current employment unit expressed in F21 and the monthly salary level of the previous employment unit expressed in F13, with a larger value indicating a larger increase in monthly salary.
F22	Professional and technical position at the time of leaving	Positive senior, deputy senior, intermediate, primary and none are represented by 1, 2, 3, 4, and 5, respectively.
F23	Double first-rate	1 means double first-rate, 2 means not.
F24	Highest Education	College, university and postgraduate are denoted by 0, 1 and 2, respectively. Below junior college, it is denoted by 5.
F25	Highest degrees	Tertiary, bachelor, master and doctoral degrees are denoted by 0, 1, 2 and 3, respectively. Below the tertiary level, they are denoted by 5.
F26	Change in place of employment	A variation is indicated by 1 and no variation is indicated by 0.

References

1. Yang, J. The Theory of Planned Behavior and Prediction of Entrepreneurial Intention Among Chinese Undergraduates. *Soc. Behav. Pers. Int. J.* **2013**, *41*, 367–376. [CrossRef]
2. González-Serrano, M.H.; Moreno, F.C.; Hervás, J.C. Prediction model of the entrepreneurial intentions in pre-graduated and post-graduated Sport Sciences students. *Cult. Cienc. Y Deporte* **2018**, *13*, 219–230. [CrossRef]
3. Gorgievski, M.J.; Stephan, U.; Laguna, M.; Moriano, J.A. Predicting Entrepreneurial Career Intentions: Values and the theory of planned behavior. *J. Career Assess.* **2018**, *26*, 457–475. [CrossRef] [PubMed]
4. Nawaz, T.; Khattak, B.K.; Rehman, K. New look of predicting entrepreneurial intention: A serial mediation analysis. *Dilemas Contemp. Educ. Polit. Y Valor.* **2019**, *6*, 126.
5. Yang, F. Decision Tree Algorithm Based University Graduate Employment Trend Prediction. *Informatica* **2019**, *43*. [CrossRef]
6. Djordjevic, D.; Cockalo, D.; Bogetic, S.; Bakator, M. Predicting Entrepreneurial Intentions among the Youth in Serbia with a Classification Decision Tree Model with the QUEST Algorithm. *Mathematics* **2021**, *9*, 1487. [CrossRef]
7. Wei, Y.; Lv, H.; Chen, M.; Wang, M.; Heidari, A.A.; Chen, H.; Li, C. Predicting Entrepreneurial Intention of Students: An Extreme Learning Machine with Gaussian Barebone Harris Hawks Optimizer. *IEEE Access* **2020**, *8*, 76841–76855. [CrossRef]
8. Bhagavan, K.S.; Thangakumar, J.; Subramanian, D.V. RETRACTED ARTICLE: Predictive analysis of student academic performance and employability chances using HLVQ algorithm. *J. Ambient. Intell. Humaniz. Comput.* **2020**, *12*, 3789–3797. [CrossRef]
9. Huang, Z.; Liu, G. Prediction model of college students entrepreneurship ability based on artificial intelligence and fuzzy logic model. *J. Intell. Fuzzy Syst.* **2021**, *40*, 2541–2552. [CrossRef]
10. Li, X.; Yang, T. Forecast of the Employment Situation of College Graduates Based on the LSTM Neural Network. *Comput. Intell. Neurosci.* **2021**, *2021*, 5787355. [CrossRef]
11. Mirjalili, S.; Lewis, A. The Whale Optimization Algorithm. *Adv. Eng. Softw.* **2016**, *95*, 51–67. [CrossRef]
12. Li, J.; Guo, L.; Li, Y.; Liu, C.; Wang, L.; Hu, H. Enhancing Whale Optimization Algorithm with Chaotic Theory for Permutation Flow Shop Scheduling Problem. *Int. J. Comput. Intell. Syst.* **2021**, *14*, 651–675. [CrossRef]
13. Luan, F.; Cai, Z.; Wu, S.; Jiang, T.; Li, F.; Yang, J. Improved Whale Algorithm for Solving the Flexible Job Shop Scheduling Problem. *Mathematics* **2019**, *7*, 384. [CrossRef]
14. Navarro, M.A.; Oliva, D.; Ramos-Michel, A.; Zaldívar, D.; Morales-Castañeda, B.; Pérez-Cisneros, M.; Valdivia, A.; Chen, H. An improved multi-population whale optimization algorithm. *Int. J. Mach. Learn. Cybern.* **2022**, *13*, 2447–2478. [CrossRef]
15. Abbas, S.; Jalil, Z.; Javed, A.R.; Batool, I.; Khan, M.Z.; Noorwali, A.; Gadekallu, T.R.; Akbar, A. BCD-WERT: A novel approach for breast cancer detection using whale optimization based efficient features and extremely randomized tree algorithm. *PeerJ Comput. Sci.* **2021**, *7*, e390. [CrossRef]
16. Elaziz, M.A.; Nabil, N.; Moghdani, R.; Ewees, A.A.; Cuevas, E.; Lu, S. Multilevel thresholding image segmentation based on improved volleyball premier league algorithm using whale optimization algorithm. *Multimed. Tools Appl.* **2021**, *80*, 12435–12468. [CrossRef]
17. Abdel-Basset, M.; El-Shahat, D.; El-Henawy, I. A modified hybrid whale optimization algorithm for the scheduling problem in multimedia data objects. *Concurr. Comput. Pr. Exp.* **2020**, *32*, e5137. [CrossRef]
18. Qiao, S.; Yu, H.; Heidari, A.A.; El-Saleh, A.A.; Cai, Z.; Xu, X.; Mafarja, M.; Chen, H. Individual disturbance and neighborhood mutation search enhanced whale optimization: Performance design for engineering problems. *J. Comput. Des. Eng.* **2022**, *9*, 1817–1851. [CrossRef]
19. Peng, L.; He, C.; Heidari, A.A.; Zhang, Q.; Chen, H.; Liang, G.; Aljehane, N.O.; Mansour, R.F. Information sharing search boosted whale optimizer with Nelder-Mead simplex for parameter estimation of photovoltaic models. *Energy Convers. Manag.* **2022**, *270*, 116246. [CrossRef]
20. Abderazek, H.; Hamza, F.; Yildiz, A.R.; Sait, S.M. Comparative investigation of the moth-flame algorithm and whale optimization algorithm for optimal spur gear design. *Mater. Test.* **2021**, *63*, 266–271. [CrossRef]
21. Ahmadianfar, I.; Asghar Heidari, A.; Gandomi, A.H.; Chu, X.; Chen, H. RUN Beyond the Metaphor: An Efficient Optimization Algorithm Based on Runge Kutta Method. *Expert Syst. Appl.* **2021**, *181*, 115079. [CrossRef]
22. Zhang, H.; Li, R.; Cai, Z.; Gu, Z.; Heidari, A.A.; Wang, M.; Chen, H.; Chen, M. Advanced orthogonal moth flame optimization with Broyden–Fletcher–Goldfarb–Shanno algorithm: Framework and real-world problems. *Expert Syst. Appl.* **2020**, *159*, 113617. [CrossRef]
23. Heidari, A.A.; Mirjalili, S.; Faris, H.; Aljarah, I.; Mafarja, M.; Chen, H. Harris hawks optimization: Algorithm and applications. *Futur. Gener. Comput. Syst.* **2019**, *97*, 849–872. [CrossRef]
24. Yang, Y.; Chen, H.; Heidari, A.A.; Gandomi, A.H. Hunger games search: Visions, conception, implementation, deep analysis, perspectives, and towards performance shifts. *Expert Syst. Appl.* **2021**, *177*, 114864. [CrossRef]
25. Ahmadianfar, I.; Asghar Heidari, A.; Noshadian, S.; Chen, H.; Gandomi, A.H. INFO: An Efficient Optimization Algorithm based on Weighted Mean of Vectors. *Expert Syst. Appl.* **2022**, *195*, 116516. [CrossRef]
26. Tu, J.; Chen, H.; Wang, M.; Gandomi, A.H. The Colony Predation Algorithm. *J. Bionic Eng.* **2021**, *18*, 674–710. [CrossRef]
27. Hu, J.; Gui, W.; Heidari, A.A.; Cai, Z.; Liang, G.; Chen, H.; Pan, Z. Dispersed foraging slime mould algorithm: Continuous and binary variants for global optimization and wrapper-based feature selection. *Knowl. Based Syst.* **2022**, *237*, 107761. [CrossRef]

28. Liu, Y.; Heidari, A.A.; Cai, Z.; Liang, G.; Chen, H.; Pan, Z.; Alsufyani, A.; Bourouis, S. Simulated annealing-based dynamic step shuffled frog leaping algorithm: Optimal performance design and feature selection. *Neurocomputing* **2022**, *503*, 325–362. [CrossRef]
29. Hussien, A.G.; Heidari, A.A.; Ye, X.; Liang, G.; Chen, H.; Pan, Z. Boosting whale optimization with evolution strategy and Gaussian random walks: An image segmentation method. *Eng. Comput.* **2022**, 1–45. [CrossRef]
30. Yu, H.; Song, J.; Chen, C.; Heidari, A.A.; Liu, J.; Chen, H.; Zaguia, A.; Mafarja, M. Image segmentation of Leaf Spot Diseases on Maize using multi-stage Cauchy-Enabled grey wolf algorithm. *Eng. Appl. Artif. Intell.* **2022**, *109*, 104653. [CrossRef]
31. Xu, Y.; Chen, H.; Heidari, A.A.; Luo, J.; Zhang, Q.; Zhao, X.; Li, C. An efficient chaotic mutative moth-flame-inspired optimizer for global optimization tasks. *Expert Syst. Appl.* **2019**, *129*, 135–155. [CrossRef]
32. Zhang, Y.; Liu, R.; Heidari, A.A.; Wang, X.; Chen, Y.; Wang, M.; Chen, H. Towards augmented kernel extreme learning models for bankruptcy prediction: Algorithmic behavior and comprehensive analysis. *Neurocomputing* **2021**, *430*, 185–212. [CrossRef]
33. Yu, H.; Cheng, X.; Chen, C.; Heidari, A.A.; Liu, J.; Cai, Z.; Chen, H. Apple leaf disease recognition method with improved residual network. *Multimed. Tools Appl.* **2022**, *81*, 7759–7782. [CrossRef]
34. Wang, M.; Chen, H.; Yang, B.; Zhao, X.; Hu, L.; Cai, Z.; Huang, H.; Tong, C. Toward an optimal kernel extreme learning machine using a chaotic moth-flame optimization strategy with applications in medical diagnoses. *Neurocomputing* **2017**, *267*, 69–84. [CrossRef]
35. Chen, H.L.; Wang, G.; Ma, C.; Cai, Z.N.; Liu, W.B.; Wang, S.J. An efficient hybrid kernel extreme learning machine approach for early diagnosis of Parkinson's disease. *Neurocomputing* **2016**, *184*, 131–144. [CrossRef]
36. Dong, R.; Chen, H.; Heidari, A.A.; Turabieh, H.; Mafarja, M.; Wang, S. Boosted kernel search: Framework, analysis and case studies on the economic emission dispatch problem. *Knowl. Based Syst.* **2021**, *233*, 107529. [CrossRef]
37. He, Z.; Yen, G.G.; Ding, J. Knee-Based Decision Making and Visualization in Many-Objective Optimization. *IEEE Trans. Evol. Comput.* **2020**, *25*, 292–306. [CrossRef]
38. He, Z.; Yen, G.G.; Lv, J. Evolutionary Multiobjective Optimization with Robustness Enhancement. *IEEE Trans. Evol. Comput.* **2019**, *24*, 494–507. [CrossRef]
39. Wu, S.-H.; Zhan, Z.-H.; Zhang, J. SAFE: Scale-Adaptive Fitness Evaluation Method for Expensive Optimization Problems. *IEEE Trans. Evol. Comput.* **2021**, *25*, 478–491. [CrossRef]
40. Li, J.Y.; Zhan, Z.H.; Wang, C.; Jin, H.; Zhang, J. Boosting data-driven evolutionary algorithm with localized data generation. *IEEE Trans. Evol. Comput.* **2020**, *24*, 923–937. [CrossRef]
41. Deng, W.; Zhang, X.; Zhou, Y.; Liu, Y.; Zhou, X.; Chen, H.; Zhao, H. An enhanced fast non-dominated solution sorting genetic algorithm for multi-objective problems. *Inf. Sci.* **2022**, *585*, 441–453. [CrossRef]
42. Hua, Y.; Liu, Q.; Hao, K.; Jin, Y. A Survey of Evolutionary Algorithms for Multi-Objective Optimization Problems with Irregular Pareto Fronts. *IEEE/CAA J. Autom. Sin.* **2021**, *8*, 303–318. [CrossRef]
43. Han, X.; Han, Y.; Chen, Q.; Li, J.; Sang, H.; Liu, Y.; Pan, Q.; Nojima, Y. Distributed Flow Shop Scheduling with Se-quence-Dependent Setup Times Using an Improved Iterated Greedy Algorithm. *Complex Syst. Model. Simul.* **2021**, *1*, 198–217. [CrossRef]
44. Gao, D.; Wang, G.-G.; Pedrycz, W. Solving Fuzzy Job-Shop Scheduling Problem Using DE Algorithm Improved by a Selection Mechanism. *IEEE Trans. Fuzzy Syst.* **2020**, *28*, 3265–3275. [CrossRef]
45. Wang, G.-G.; Gao, D.; Pedrycz, W. Solving Multiobjective Fuzzy Job-Shop Scheduling Problem by a Hybrid Adaptive Differential Evolution Algorithm. *IEEE Trans. Ind. Informatics* **2022**, *18*, 8519–8528. [CrossRef]
46. Chen, H.L.; Yang, B.; Wang, S.J.; Wang, G.; Liu, D.Y.; Li, H.Z.; Liu, W. Towards an optimal support vector machine classifier using a parallel particle swarm optimization strategy. *Appl. Math. Comput.* **2014**, *239*, 180–197. [CrossRef]
47. Deng, W.; Xu, J.; Zhao, H.; Song, Y. A Novel Gate Resource Allocation Method Using Improved PSO-Based QEA. *IEEE Trans. Intell. Transp. Syst.* **2020**, *23*, 1737–1745. [CrossRef]
48. Deng, W.; Xu, J.; Song, Y.; Zhao, H. An effective improved co-evolution ant colony optimisation algorithm with multi-strategies and its application. *Int. J. Bio Inspir. Comput.* **2020**, *16*, 158–170. [CrossRef]
49. Ye, X.; Liu, W.; Li, H.; Wang, M.; Chi, C.; Liang, G.; Chen, H.; Huang, H. Modified Whale Optimization Algorithm for Solar Cell and PV Module Parameter Identification. *Complexity* **2021**, *2021*, 8878686. [CrossRef]
50. Yu, H.; Yuan, K.; Li, W.; Zhao, N.; Chen, W.; Huang, C.; Chen, H.; Wang, M. Improved Butterfly Optimizer-Configured Extreme Learning Machine for Fault Diagnosis. *Complexity* **2021**, *2021*, 6315010. [CrossRef]
51. Agrawal, R.; Kaur, B.; Sharma, S. Quantum based Whale Optimization Algorithm for wrapper feature selection. *Appl. Soft Comput.* **2020**, *89*, 106092. [CrossRef]
52. Bai, L.; Han, Z.; Ren, J.; Qin, X. Research on feature selection for rotating machinery based on Supervision Kernel Entropy Component Analysis with Whale Optimization Algorithm. *Appl. Soft Comput.* **2020**, *92*, 106245. [CrossRef]
53. Bahiraei, M.; Foong, L.K.; Hosseini, S.; Mazaheri, N. Predicting heat transfer rate of a ribbed triple-tube heat exchanger working with nanofluid using neural network enhanced by advanced optimization algorithms. *Powder Technol.* **2021**, *381*, 459–476. [CrossRef]
54. Qi, A.; Zhao, D.; Yu, F.; Heidari, A.A.; Chen, H.; Xiao, L. Directional mutation and crossover for immature performance of whale algorithm with application to engineering optimization. *J. Comput. Des. Eng.* **2022**, *9*, 519–563. [CrossRef]
55. Bui, D.T.; Abdullahi, M.M.; Ghareh, S.; Moayedi, H.; Nguyen, H. Fine-tuning of neural computing using whale optimization algorithm for predicting compressive strength of concrete. *Eng. Comput.* **2021**, *37*, 701–712. [CrossRef]

56. Butti, D.; Mangipudi, S.K.; Rayapudi, S. Model Order Reduction Based Power System Stabilizer Design Using Improved Whale Optimization Algorithm. *IETE J. Res.* **2021**, 1–20. [CrossRef]
57. Cao, Y.; Li, Y.; Zhang, G.; Jermsittiparsert, K.; Nasseri, M. An efficient terminal voltage control for PEMFC based on an improved version of whale optimization algorithm. *Energy Rep.* **2020**, *6*, 530–542. [CrossRef]
58. Çerçevik, A.E.; Avşar, Ö.; Hasançebi, O. Optimum design of seismic isolation systems using metaheuristic search methods. *Soil Dyn. Earthq. Eng.* **2019**, *131*, 106012. [CrossRef]
59. Zhao, S.; Song, J.; Du, X.; Liu, T.; Chen, H.; Chen, H. Intervention-Aware Epidemic Prediction by Enhanced Whale Optimization. In *International Conference on Knowledge Science, Engineering and Management*; Memmi, G., Yang, B., Kong, L., Zhang, T., Qiu, M., Eds.; Springer: Berlin/Heidelberg, Germany, 2022; pp. 457–468.
60. Fan, Y.; Wang, P.; Heidari, A.A.; Wang, M.; Zhao, X.; Chen, H.; Li, C. Boosted hunting-based fruit fly optimization and advances in real-world problems. *Expert Syst. Appl.* **2020**, *159*, 113502. [CrossRef]
61. Raj, S.; Bhattacharyya, B. Optimal placement of TCSC and SVC for reactive power planning using Whale optimization algorithm. *Swarm Evol. Comput.* **2018**, *40*, 131–143. [CrossRef]
62. Guo, Y.; Shen, H.; Chen, L.; Liu, Y.; Kang, Z. Improved whale optimization algorithm based on random hopping update and random control parameter. *J. Intell. Fuzzy Syst.* **2021**, *40*, 363–379. [CrossRef]
63. Jiang, R.; Yang, M.; Wang, S.; Chao, T. An improved whale optimization algorithm with armed force program and strategic adjustment. *Appl. Math. Model.* **2020**, *81*, 603–623. [CrossRef]
64. Tu, J.; Chen, H.; Liu, J.; Heidari, A.A.; Zhang, X.; Wang, M.; Ruby, R.; Pham, Q.V. Evolutionary biogeography-based whale optimization methods with communication structure: Towards measuring the balance. *Knowl. Based Syst.* **2021**, *212*, 106642. [CrossRef]
65. Wang, G.; Gui, W.; Liang, G.; Zhao, X.; Wang, M.; Mafarja, M.; Turabieh, H.; Xin, J.; Chen, H.; Ma, X. Spiral Motion Enhanced Elite Whale Optimizer for Global Tasks. *Complexity* **2021**, *2021*, 1–33. [CrossRef]
66. Abd Elazim, S.M.; Ali, E.S. Optimal network restructure via improved whale optimization approach. *Int. J. Commun. Syst.* **2021**, *34*, e4617. [CrossRef]
67. Abdel-Basset, M.; Chang, V.; Mohamed, R. HSMA_WOA: A hybrid novel Slime mould algorithm with whale optimization algorithm for tackling the image segmentation problem of chest X-ray images. *Appl. Soft Comput.* **2020**, *95*, 106642. [CrossRef]
68. Chai, Q.-W.; Chu, S.-C.; Pan, J.-S.; Hu, P.; Zheng, W.-M. A parallel WOA with two communication strategies applied in DV-Hop localization method. *EURASIP J. Wirel. Commun. Netw.* **2020**, *2020*, 1–10. [CrossRef]
69. Heidari, A.A.; Aljarah, I.; Faris, H.; Chen, H.; Luo, J.; Mirjalili, S. An enhanced associative learning-based exploratory whale optimizer for global optimization. *Neural Comput. Appl.* **2020**, *32*, 5185–5211. [CrossRef]
70. Jin, Q.; Xu, Z.; Cai, W. An Improved Whale Optimization Algorithm with Random Evolution and Special Reinforcement Dual-Operation Strategy Collaboration. *Symmetry* **2021**, *13*, 238. [CrossRef]
71. Qin, A.K.; Huang, V.L.; Suganthan, P.N. Differential Evolution Algorithm with Strategy Adaptation for Global Numerical Optimization. *IEEE Trans. Evol. Comput.* **2008**, *13*, 398–417. [CrossRef]
72. Wan, X.; Zuo, X.; Zhao, X. A differential evolution algorithm combined with linear programming for solving a closed loop facility layout problem. *Appl. Soft Comput.* **2022**, 108725. [CrossRef]
73. Yuan, Y.; Cao, J.; Wang, X.; Zhang, Z.; Liu, Y. Economic-effectiveness analysis of micro-fins helically coiled tube heat exchanger and optimization based on multi-objective differential evolution algorithm. *Appl. Therm. Eng.* **2022**, *201*, 117764. [CrossRef]
74. Liu, D.; Hu, Z.; Su, Q.; Liu, M. A niching differential evolution algorithm for the large-scale combined heat and power economic dispatch problem. *Appl. Soft Comput.* **2021**, *113*, 108017. [CrossRef]
75. He, Z.; Ning, D.; Gou, Y.; Zhou, Z. Wave energy converter optimization based on differential evolution algorithm. *Energy* **2022**, *246*, 123433. [CrossRef]
76. Meng, A.-B.; Chen, Y.-C.; Yin, H.; Chen, S.-Z. Crisscross optimization algorithm and its application. *Knowl. Based Syst.* **2014**, *67*, 218–229. [CrossRef]
77. Luo, J.; Chen, H.; Heidari, A.A.; Xu, Y.; Zhang, Q.; Li, C. Multi-strategy boosted mutative whale-inspired optimization approaches. *Appl. Math. Model.* **2019**, *73*, 109–123. [CrossRef]
78. Yousri, D.; Allam, D.; Eteiba, M. Chaotic whale optimizer variants for parameters estimation of the chaotic behavior in Permanent Magnet Synchronous Motor. *Appl. Soft Comput.* **2019**, *74*, 479–503. [CrossRef]
79. Ling, Y.; Zhou, Y.; Luo, Q. Lévy Flight Trajectory-Based Whale Optimization Algorithm for Global Optimization. *IEEE Access* **2017**, *5*, 6168–6186. [CrossRef]
80. Tubishat, M.; Abushariah, M.A.M.; Idris, N.; Aljarah, I. Improved whale optimization algorithm for feature selection in Arabic sentiment analysis. *Appl. Intell.* **2018**, *49*, 1688–1707. [CrossRef]
81. Cai, Z.; Gu, J.; Luo, J.; Zhang, Q.; Chen, H.; Pan, Z.; Li, Y.; Li, C. Evolving an optimal kernel extreme learning machine by using an enhanced grey wolf optimization strategy. *Expert Syst. Appl.* **2019**, *138*, 112814. [CrossRef]
82. Heidari, A.A.; Ali Abbaspour, R.; Chen, H. Efficient boosted grey wolf optimizers for global search and kernel extreme learning machine training. *Appl. Soft Comput.* **2019**, *81*, 105521. [CrossRef]
83. Sun, T.-Y.; Liu, C.-C.; Tsai, S.-J.; Hsieh, S.-T.; Li, K.-Y. Cluster Guide Particle Swarm Optimization (CGPSO) for Underdetermined Blind Source Separation with Advanced Conditions. *IEEE Trans. Evol. Comput.* **2010**, *15*, 798–811. [CrossRef]

84. Nenavath, H.; Jatoth, R.K. Hybridizing sine cosine algorithm with differential evolution for global optimization and object tracking. *Appl. Soft Comput.* **2018**, *62*, 1019–1043. [CrossRef]
85. Singh, R.P.; Mukherjee, V.; Ghoshal, S.P. Optimal power flow by particle swarm optimization with an aging leader and challengers. *Int. J. Eng.* **2015**, *7*, 123–132. [CrossRef]
86. Liang, H.; Liu, Y.; Shen, Y.; Li, F.; Man, Y. A Hybrid Bat Algorithm for Economic Dispatch with Random Wind Power. *IEEE Trans. Power Syst.* **2018**, *33*, 5052–5061. [CrossRef]
87. Adarsh, B.R.; Raghunathan, T.; Jayabarathi, T.; Yang, X.-S. Economic dispatch using chaotic bat algorithm. *Energy* **2016**, *96*, 666–675. [CrossRef]
88. Abd Elaziz, M.; Oliva, D.; Xiong, S. An improved Opposition-Based Sine Cosine Algorithm for global optimization. *Expert Syst. Appl.* **2017**, *90*, 484–500. [CrossRef]
89. Wu, Z.; Li, R.; Xie, J.; Zhou, Z.; Guo, J.; Jiang, J.; Su, X. A user sensitive subject protection approach for book search service. *J. Assoc. Inf. Sci. Technol.* **2020**, *71*, 183–195. [CrossRef]
90. Wu, Z.; Shen, S.; Zhou, H.; Li, H.; Lu, C.; Zou, D. An effective approach for the protection of user commodity viewing privacy in e-commerce website. *Knowl. Based Syst.* **2021**, *220*, 106952. [CrossRef]
91. Yang, Z.; Chen, H.; Zhang, J.; Chang, Y. Context-aware Attentive Multilevel Feature Fusion for Named Entity Recognition. *IEEE Trans. Neural Netw. Learn. Syst.* **2022**. [CrossRef]
92. Huang, L.; Yang, Y.; Chen, H.; Zhang, Y.; Wang, Z.; He, L. Context-aware road travel time estimation by coupled tensor decomposition based on trajectory data. *Knowl. Based Syst.* **2022**, *245*. [CrossRef]
93. Hu, K.; Zhao, L.; Feng, S.; Zhang, S.; Zhou, Q.; Gao, X.; Guo, Y. Colorectal polyp region extraction using saliency detection network with neutrosophic enhancement. *Comput. Biol. Med.* **2022**, *147*, 105760. [CrossRef] [PubMed]
94. Zhang, X.; Zheng, J.; Wang, D.; Zhao, L. Exemplar-Based Denoising: A Unified Low-Rank Recovery Framework. *IEEE Trans. Circuits Syst. Video Technol.* **2020**, *30*, 2538–2549. [CrossRef]
95. Qi, A.; Zhao, D.; Yu, F.; Heidari, A.A.; Wu, Z.; Cai, Z.; Alenezi, F.; Mansour, R.F.; Chen, H.; Chen, M. Directional mutation and crossover boosted ant colony optimization with application to COVID-19 X-ray image segmentation. *Comput. Biol. Med.* **2022**, *148*, 105810. [CrossRef]
96. Ren, L.; Zhao, D.; Zhao, X.; Chen, W.; Li, L.; Wu, T.; Liang, G.; Cai, Z.; Xu, S. Multi-level thresholding segmentation for pathological images: Optimal performance design of a new modified differential evolution. *Comput. Bio. Med.* **2022**, *148*, 105910. [CrossRef]
97. Su, H.; Zhao, D.; Elmannai, H.; Heidari, A.A.; Bourouis, S.; Wu, Z.; Cai, Z.; Gui, W.; Chen, M. Multilevel threshold image segmentation for COVID-19 chest radiography: A framework using horizontal and vertical multiverse optimization. *Comput. Biol. Med.* **2022**, *146*, 105618. [CrossRef]
98. Cao, X.; Wang, J.; Zeng, B. A Study on the Strong Duality of Second-Order Conic Relaxation of AC Optimal Power Flow in Radial Networks. *IEEE Trans. Power Syst.* **2021**, *37*, 443–455. [CrossRef]

electronics

MDPI

Article

A Novel Multistrategy-Based Differential Evolution Algorithm and Its Application

Jinyin Wang [1], Shifan Shang [2,3], Huanyu Jing [2], Jiahui Zhu [2], Yingjie Song [4], Yuangang Li [5,*] and Wu Deng [2,6,*]

1 UNI-FI Credit Solutions Co., Ltd., Beijing 100083, China
2 School of Electronic Information and Automation, Civil Aviation University of China, Tianjin 300300, China
3 HAI ROBOTICS Co., Ltd., Shenzhen 518000, China
4 College of Computer Science and Technology, Shandong Technology and Business University, Yantai 264005, China
5 Faculty of Business Information, Shanghai Business School, Shanghai 200235, China
6 Traction Power State Key Laboratory, Southwest Jiaotong University, Chengdu 610031, China
* Correspondence: yuangang.li@sbs.edu.cn (Y.L.); wdeng@cauc.edu.cn (W.D.)

Abstract: To address the poor searchability, population diversity, and slow convergence speed of the differential evolution (DE) algorithm in solving capacitated vehicle routing problems (CVRP), a new multistrategy-based differential evolution algorithm with the saving mileage algorithm, sequential encoding, and gravitational search algorithm, namely SEGDE, is proposed to solve CVRP in this paper. Firstly, an optimization model of CVRP with the shortest total vehicle routing is established. Then, the saving mileage algorithm is employed to initialize the population of the DE to improve the initial solution quality and the search efficiency. The sequential encoding approach is used to adjust the differential mutation strategy to legalize the current solution and ensure its effectiveness. Finally, the gravitational search algorithm is applied to calculate the gravitational relationship between points to effectively adjust the evolutionary search direction and further improve the search efficiency. Four CVRPs are selected to verify the effectiveness of the proposed SEGDE algorithm. The experimental results show that the proposed SEGDE algorithm can effectively solve the CVRPs and obtain the ideal vehicle routing. It adopts better search speed, global optimization ability, routing length, and stability.

Keywords: differential evolution; capacitated vehicle routing planning; saving mileage; gravity search

Citation: Wang, J.; Shang, S.; Jing, H.; Zhu, J.; Song, Y.; Li, Y.; Deng, W. A Novel Multistrategy-Based Differential Evolution Algorithm and Its Application. *Electronics* **2022**, *11*, 3476. https://doi.org/10.3390/electronics11213476

Academic Editor: João Soares

Received: 5 October 2022
Accepted: 25 October 2022
Published: 26 October 2022

1. Introduction

The vehicle routing problem (VRP) was formally presented in 1959 by Dantzig [1]. The problem is defined as finding the optimal route of a vehicle under certain constraint conditions (such as vehicle capacity, customer demand, transportation process, etc.), so as to minimize the transportation cost or find the shortest transportation distance [2–4]. VRP is a NP-hard problem and is one of the hotspots in operations research and combinatorial optimization. In recent years, heuristic algorithms have been widely explored in solving large-scale VRPs [5–8]. Therefore, a new algorithm for VRP has a certain theoretical significance and practical value.

The algorithms for solving VRP can be broadly divided into exact algorithms and heuristic algorithms (including metaheuristics). The exact algorithm can obtain the optimal solution, but its high computational complexity makes it unsuitable for solving large-scale VRPs [9–11]. Heuristic algorithms can be further divided into neighborhood-based algorithms and population-based algorithms [12–14]. The neighborhood-based algorithms maintain a single solution during the search process and seek a more optimal solution by iterating between neighborhood solutions according to the strategy. The algorithms include iterative local search, Tabu search, and so on.

The differential evolution (DE) algorithm is a heuristic search algorithm based on population, and each individual in the population corresponds to a solution vector [15]. The evolution process of DE favors that of GA, which includes mutation, crossover, and selection, but its specific definition is different from that of GA. Since the DE has a simple structure, fast convergence, and so on, it is applied in data mining, pattern recognition, electromagnetics, and so on. However, the DE algorithm also has some defects in solving large-scale VRPs, such as poor searchability and population diversity, slow convergence speed, and so on. Therefore, some variants of DE algorithms are proposed from the different aspects of algorithm, such as parameter adaption, new mutation strategies, crossover strategy strategies, population initialization, hybrid DE with the other algorithms, and so on.

To some extent, these improved DE algorithms have improved the searchability, accelerated the convergence, strengthened avoidance of falling into local optimum, and so on, which can help better obtain optimization results in solving the complex optimization problems and the different VRPs. However, there still exists some defects in solving the complex optimization problems, such as poor population diversity, low search accuracy, easily falling into local optimum, and so on. To solve these problems, a new multistrategy-based differential evolution algorithm with the saving mileage algorithm, sequential encoding, and gravitational search algorithm, namely SEGDE, is proposed to solve the CVRP. A planning method of the CVRP based on SEGDE is implemented to solve the actual CVRP for obtaining the ideal results of the vehicle routing problems.

The main contributions of this study are described as follows:

(1) A new multistrategy DE algorithm, namely SEGDE, is developed to improve the solution quality and the search efficiency in solving the CVRPs.
(2) The saving mileage algorithm is used to initialize the population of the DE to ensure the initial solution quality and improve the search efficiency.
(3) The sorting and coding strategy is used to adjust the differential mutation strategy, and the vectors are added and subtracted.

The structure of this paper is as follows: In Section 3, the related works are reviewed, and the basic DE is introduced. In Section 4, the capacitated vehicle routing model is constructed. Section 5 develops a new multistrategy DE algorithm, and the idea, model, and steps are described in detail. The experimental calculation and analysis are executed in the Section 6. Finally, the conclusions are summarized in Section 7.

2. Related Works

Since the VRP was proposed, many researchers have made in-depth explorations and solved VRPs. When the traditional methods, the exact algorithm, heuristic algorithms, and so on are used to solve the VRPs, a slow solving speed and excessive calculation will occur. In recent years, the focus for solving VRPs has been on combining heuristic algorithms with artificial intelligence technology, such as simulated annealing (SA), tabu search (TS), genetic algorithm (GA), ant colony optimization (ACO), different improvements, and so on. Yusuf et al. [16] studied the GA to solve a combinatorial problem of VRP. Akpinar [17] presented a hybrid algorithm with a large neighborhood search and ACO for CVRP. Zhang et al. [18] presented a hybrid approach with Tabu search and ABC to solve VRP. Dechampai et al. [19] presented a MESOMDE_G-Q-DVRP-FD for solving GQDVRP. Gutierrez et al. [20] presented a new memetic algorithm with multipopulation to solve VRP. Fallah et al. [21] presented a robust algorithm to solve the competitive VRP. Altabeeb et al. [22] presented a new CVRP-firefly algorithm. Altabeeb et al. [23] presented a cooperative hybrid FA with multipopulation to solve VRP. Xiao et al. [24] presented a heuristic EMRG-HA to solve CVRP with a large scale. Jia et al. [25] presented a novel bilevel ACO to solve the CEVRP. Jiang et al. [26] presented a fast evolutionary algorithm called RMEA to accelerate convergence for CVRP. Deng et al. [27] presented an ACDE/F for the gate allocation problem. Zhang et al. [28] presented a branch-and-cut algorithm to solve the two-dimensional loading constraint VRP. Song et al. [29] presented a dynamic hybrid mechanism CDE to solve the complex optimization problem. Niu et al. [30] presented a multiobjective EA to tackle the MO-VRPSD. Deng et al. [31] presented a new MPSACO with CWBPSO and ACO for solving the taxiway planning problem. Gu et al. [32] presented a hierarchical solution evaluation approach for a general VRPD. Azad et al. [33] presented a QAOA to solve VRP. Lai et al. [34] presented a data-driven flexible transit method with the origin-destination insertion and mixed-integer linear programming for scheduling vehicles. Voigt et al. [35] presented a hybrid adaptive large neighborhood search method to solve three variants of VRP. Seyfi et al. [36] presented a matheuristic method with a variable neighborhood search with mathematical programming to solve multimode HEVRP. Cai et al. [37] presented a hybrid evolutionary multitask algorithm to solve multiobjective VRPTWs. Wen et al. [38] presented an improved adaptive large neighborhood search algorithm to efficiently solve large-scale instances of the multidepot green VRP with time windows. Ma et al. [39] presented an adaptive large neighborhood search algorithm to find near-optimal solutions for larger-size time-dependent VRPs. In addition, some other algorithms are also presented for solving VRPs and the other optimization problems [40–51].

The DE algorithm is widely applied in solving different VRPs. For solving large-scale VRPs, there exist poor searchability, worsened population diversity, a slow convergence speed, and so on. Many researchers have deeply studied and proposed some improvements to the DE algorithm. Zhang et al. [52] presented a new constrained DE to obtain an optimal feasible routing. Teoh et al. [53] presented a local search-based DE to solve CVRP. Pitakaso et al. [54] presented five modified DEs for solving three subproblems. Xing et al. [55] presented a hybrid discrete DE for solving the split delivery VRP in the logistic distribution. Sethanan et al. [56] presented a novel hybrid DE with a genetic operator to solve the multitrip VRP with backhauls. Hameed et al. [57] presented a hybrid algorithm based on discrete DE and TS for solving many instances of QAP. Liu et al. [58] presented a mixed-variable DE for solving the hierarchical mixed-variable optimization problem. Moonsri et al. [59] presented a hybrid and self-adaptive DE for solving an EGG distribution problem. Chai et al. [60] presented a multi-strategy fusion DE with multipopulation, self-adaption and interactive mutation to solve the path planning of UAV. Wu et al. [61] presented a fast and effective improved DE to solve the integer linear programming model. Hou et al. [62] presented a multistate-constrained MODE with a variable neighborhood to solve the real-world-constrained multiobjective problem. Chen et al. [63] presented a fast-neighborhood algorithm based on crowding DE. In addition, some other DE algorithms are also improved for solving the complex optimization problems [64–66]. A summary of the main works is shown Table 1.

Table 1. A summary of the main works.

Name	Key Points	Advantages	Disadvantages
Zhang et al. [52]	Constrained DE	Improve optimization performance	Lack of population diversity
Teoh et al. [53]	Local search-based DE	Explore new search areas	Lack of global searchability
Pitakaso et al. [54]	Five modified DE	Improve population diversity	Fall into local optimal value
Xing et al. [55]	Hybrid discrete DE	Avoid the prematurity and ensure the solution quality	Slow convergence to some extent
Sethanan et al. [56]	Hybrid DE with a genetic operator	Balance the exploration ability	Fall into local optimal value
Hameed et al. [57]	Hybrid algorithm	Enhance solutions, to reduce the distances between the locations	Increase the time complexity
Liu et al. [58]	Mixed-variable DE	Hierarchical mixed-variable mutation operator	Lack of population diversity

Table 1. *Cont.*

Name	Key Points	Advantages	Disadvantages
Moonsri et al. [59]	Hybrid and self-adaptive DE	Self-adaptive mutation strategy	Fall into local optimal value
Chai et al. [60]	Multistrategy fusion DE	Enhance population diversity	Slow convergence to some extent
Hou et al. [62]	Multistate-constrained MODE	Enhance the optimization effectiveness	Increase the time complexity
Chen et al. [63]	Fast-neighborhood DE	Faster convergence	Lack of population diversity

Through these variants of DE, algorithms from various aspects have improved its performance by parameter adaption, designing new mutation/crossover strategy, and hybridity with the other algorithms, and so on. However, some defects, such as poor population diversity and low search accuracy, still exist in solving the complex optimization. Therefore, the DE algorithm needs to be further and more deeply studied in order to solve the large-scale complex optimization problem.

3. Differential Evolution Algorithm

DE is an efficient evolutionary algorithm with a simple and clear structure and idea. It combines parent individuals with other individuals in a population to produce new offspring, which will continue to evolve in place of the parent if they possess better fitness values. In brief, DE consists of the following parts:

3.1. Initialization

The parameters of DE are initialized and generally include: population (Np), dimension (D), mutation factor (F), crossover factor (CR), and the maximum number of iteration (Gm). In addition, the individuals are initialized randomly within the specified range:

$$\left\{x_{i,1}^{(G)}, x_{i,2}^{(G)}, \ldots, x_{i,D}^{(G)}\right\}, x_{i,D} \in R^D, i = 1,2,\ldots,NP.$$

3.2. Mutation

In each iteration of evolution, the parent generation generates Np mutation vectors through certain mutation strategies. The mutation strategy is usually expressed as DE/x/y, where x represents the vector to be mutated and Y represents the number of vectors to be mutated during the mutation process. There are five variation strategies that are commonly used in DE:

(1) DE/rand/1

(2) DE/Rand/1

$$V_i^g = X_{r_1}^g + F \times (X_{r_2}^g - X_{r_3}^g) \tag{1}$$

(3) DE/best/1

(4) DE/Best/1

$$V_i^g = X_{best}^g + F \times (X_{r_1}^g - X_{r_2}^g) \tag{2}$$

(5) DE/rand-to-best/1

(6) DE/Rand-to-best/1

$$V_i^g = X_i^g + F \times (X_{best}^g - X_i^g) + F \times (X_{r_1}^g - X_{r_2}^g) \tag{3}$$

(7) DE/current-to-rand/1

(8) DE/Current-to-rand/1

$$V_i^g = X_i^g + K \times (X_{r_1}^g - X_i^g) + F \times (X_{r_2}^g - X_{r_3}^g) \tag{4}$$

(9) DE/current-to-best/1
(10) DE/Current-to-best/1

$$V_i^g = X_i^g + F_1 \times (X_{best}^g - X_i^g) + F_2 \times (X_{r_1}^g - X_{r_2}^g) \quad (5)$$

where r_1, r_2 and r_3 are individuals selected randomly from 1 to Np individuals, and X is the individual with the best adaptation in the gth iteration.

3.3. Crossover

After the mutation is executed, a crossover operation is performed to generate the final experimental vector U by crossing the parent vector X with the mutation vector V with a certain probability:

$$U_{i,j}^g = \begin{cases} V_{i,j}^g, if \text{ rand}(0,1) \leq \text{CR or j} = \text{j}_{\text{rand}} \\ X_{i,j}^g, otherwise \end{cases} \quad (6)$$

where $j \in [1, D]$.

3.4. Selection

If the experimental vector U performs better in fitness than the parent individual X, then the parent individual is replaced with it:

$$X_i^{g+1} = \begin{cases} U_i^g, if\ f\left(U_i^g\right) \leq f(X_i^g) \\ X_i^g, otherwise \end{cases} \quad (7)$$

where X will be the parent individual of the next generation evolution, and $f(U)$ and $f(X)$ represent the adaptation values of the current generation experiment vector and the parent individual, respectively.

4. Modeling Capacitated Vehicle Routing

VRP generally refers to organizing and calling a certain number of vehicles to a series of shipping and receiving points, arranging appropriate travel routes so that the vehicles pass through them in an orderly manner [67]. Under specified constraints (e.g., demand and delivery of goods, delivery time, vehicle capacity limits, mileage limits, travel time limits, etc.), we strive to achieve certain goals (e.g., shortest total vehicle miles driven, lowest total transportation costs, vehicles arriving at a certain time, minimum number of vehicles used, and so on.) [68–71].

4.1. Model Assumptions

The following assumptions are made for the model based on the actual problem:

(1) The distribution center is assigned to complete a series of demand point distribution services.
(2) The relative geographical location and the corresponding demand quantity of the distribution center and each demand point are given clearly.
(3) Vehicle distribution is completed and returned to the designated distribution center.
(4) The vehicles have the same specifications, and there are no errors.
(5) There is no consideration of urban traffic congestion.
(6) The distribution vehicles always travel at a constant speed, and the distribution cost is equal within the unit distance, so the travel distance can represent the distribution cost.
(7) Each demand point shall be served by only one delivery vehicle, and the sum of the requirements of all the demand points of the vehicle service shall be less than or equal to the rated load limit of the vehicle.

4.2. Symbolic Description

The relevant symbols are described in Table 2.

Table 2. List of symbols involved in the CVRP model.

Symbols	Meaning
m	Number of vehicles in distribution center
n	Number of customer points
Q	Vehicle capacity
di	The requirement for customer points I, di > 0 (i > 0), and D0 = 0
cij	The distance from point i to point j
Xijk	The degree of delivery requirements from the k vehicle distribution Point i to point j
V	A collection of distribution centers and customer points

4.3. Objective Optimization Function

The CVRP model can be constructed based on the mentioned distribution objectives and distribution requirements as follows:

Distribution objective:

$$Min\ Z = \sum_{i=0}^{n}\sum_{j=0}^{n}\sum_{k=1}^{m} c_{ij}x_{ijk} \tag{8}$$

Constraints:

$$\sum_{i(j)=0}^{n}\sum_{k=1}^{m} x_{ijk} = 1\ ,\ i,j = 0,1,2,\ldots,n \tag{9}$$

$$\sum_{i=0}^{n} x_{ipk} - \sum_{j=0}^{n} x_{ijk} = 0\ ,\ k = 1,2,\ldots,m\ ,\ p = 0,1,\ldots,n \tag{10}$$

$$\sum_{i=0}^{n}\sum_{j=0}^{n} d_i x_{ijk} \leq Q, k = 1,2,\ldots,m \tag{11}$$

$$\sum_{i=1}^{n}\sum_{j=1}^{n} x_{ijk} \leq |V| - 1\ , k = 1,2,\ldots,m \tag{12}$$

$$x_{ijk} \in \{0,1\}\ ,\ i,j = 0,1,2,\ldots,n\ ,\ k = 1,2,\ldots,m \tag{13}$$

The optimization goal is represented by an Equation (8) to minimize the total distance traveled. The constraint (9) represents the availability of one and only one vehicle per customer point to provide service. The constraint (10) ensures that a customer point is visited the same number of times as it is left. The constraint (11) ensures that the vehicle works within its maximum load. The constraint (12) means that the subtour is eliminated. The constraint (13) provides a mutable limit.

5. A Multistrategy-Based Differential Evolution Algorithm

The DE is a population-based adaptive global optimization algorithm with a simple structure and high robustness. However, there are some problems in solving optimization problems, such as poor searchability, slow convergence, and a tendency to fall into local optimality. Therefore, a multistrategy DE algorithm, namely SEGDE, is proposed by introducing the population initialization strategy, the differential mutation strategy, and the gravity search algorithm. The mileage saving method is used to initialize the population of the DE to improve the initial solution quality and the search efficiency. The differential mutation strategy is adjusted by using a sequential encoding approach to perform a legalization operation on the current solution to ensure that the solution is valid. Finally, the gravity search algorithm (GSA) is introduced to calculate the gravitational relationship between points, which can be used to legitimize the solution, reinsert the points, effectively adjust the search direction of evolution, optimize the search efficiency, and prevent the algorithm from falling into local optimum, to obtain better optimization ability of complex optimization problems.

These strategies in the SEGDE are described in detail as follows.

5.1. Population Initialization Strategy

Traditional DE algorithms usually use population random initialization to randomly distribute the initial population in the feasible domain. In this way, the algorithm does not depend on the initial population solution, but the quality of the initial population often affects the efficiency and accuracy of the global search algorithm. The saving mileage method is a heuristic algorithm for solving transportation problems [72]. The key idea of the heuristic method is to combine the two circuits of the transportation problem according to the distance table, which can reduce the total transportation distance and make the distribution more efficient. Therefore, the initial population is a combination of the solution of the mileage-saving method and the random individuals, which ensures the initial population solution quality and allows the algorithm to carry out the follow-up search around the individuals with better quality, to improve search efficiency.

5.2. Differential Mutation Strategy

Since the CVRP is discrete, a ranking encoding approach is used to adjust the operation of the differential variation strategy DE/neighbor to neighbor/1 by using ranking numbers instead of vectors for addition and subtraction. In addition, the solution after mutation operation is not necessarily the legal solution to meet the requirements; after the mutation operation, the current solution should be legal operation to ensure the effectiveness of the solution. The solutions are searched from right to left, the repeated points are set to zero, and the zero positions are re-inserted by using contemporary evolutionary individuals. The individual variation was calculated using Equation (14), and the adjusted variation process is shown in Table 3.

$$V_{i,j}^{g} = \begin{cases} \mathrm{mod}\left(X_{r_3,j}^{g} + \left(X_{best,j}^{g} - X_{r_3,j}^{g}\right) + \left(X_{r_1,j}^{g} - X_{r_2,j}^{g}\right) + j - 1, j\right), if\ rand < F \\ X_{best,j}^{g}, if\ rand \geq F \end{cases} \tag{14}$$

Table 3. Examples of variant operations (F = 0.5).

				$X_{r_1}^{g} - X_{r_2}^{g}$			
$X_{r_1}^{g}$	7	4	3	5	2	1	6
$X_{r_2}^{g}$	5	2	1	3	7	4	6
$X_{r_1}^{g} - X_{r_2}^{g}$	2	2	2	2	−5 −5	−3 −3	0
				$X_{best}^{g} - X_{r_3}^{g}$			
X_{best}^{g}	5	1	3	4	2	7	6
$X_{r_3}^{g}$	2	3	5	1	7	6	4
$X_{best}^{g} - X_{r_3}^{g}$	3	−2 −2	−2 −2	3	−5 −5	1	2
				V_{i}^{g}			
rand Rand	0.18	0.22	0.53	0.78	0.61	0.39	0.42
U_{i}^{g}	1	1	3	1	1	5	1
V_{i}^{g}	5	1	3	1	1	7	6

5.3. Variable Correlation Using GSA

VRP is an optimization problem with point-line network topology. The key to solving this problem is discovering the correlation between the points and connecting them. The gravitational search algorithm (GSA) is used to calculate the gravitational relationship between points, and the point-point relationship table is used for the legitimization of the solution and the reinsertion link of points, which can effectively adjust the evolutionary search direction and optimize the search efficiency. GSA is a bionic algorithm based on the laws of Newton's law of gravity and the laws of kinematics [73]. The core idea of the algorithm is to calculate the value of the gravitational force between points according to Newton's universal gravity formula, update the gravitational table, adjust the mass of the points according to the gravitational table, and use the mass table updated in the current generation to guide the next generation solution.

Define the attraction between individual i and individual j as follows:

$$F_{ij}^{d}(t)= G(t)\frac{M_{pi}(t) \times M_{aj}(t)}{R_{ij}(t)+\varepsilon}\left(x_j^d(t) - x_i^d(t)\right) \tag{15}$$

where M_{aj} is the related active gravitational mass of individual j, and M_{pj} is the related passive gravitational mass of individual i. ε is a variable to prevent variables with denominators. $R_{ij}(t)$ is the Euclidean distance between individuals i and j.

$$R_{ij}(t) = \|X_i(t) \cdot X_j(t)\|_2 \tag{16}$$

In the d-dimension space, the exerted force on any particle is the exerted resultant force on it by other particles, and the random weighted sum of the gravitational forces of each particle is expressed as follows:

$$F_i^d(t) = \sum_{j=1, j\neq i}^{N} rand_j F_{ij}^d(t) \tag{17}$$

where $rand_j$ is a random value in [0,1].

Therefore, the acceleration of an individual i in the d-dimension is described as follows:

$$a_i^d(t) = \frac{F_i^d(t)}{M_{ii}(t)} \tag{18}$$

where M_{ii} is the inertial gravity of individual i at iteration t.

Based on the above model, the position update of individuals can be obtained as follows:

$$v_i^d(t+1) = rand_i \times v_i^d(t) + a_i^d(t) \tag{19}$$

$$x_i^d(t+1) = x_i^d(t) + v_i^d(t+1) \tag{20}$$

where $rand_i$ is a random value in [0,1].

The GSA algorithm framework is shown in Figure 1.

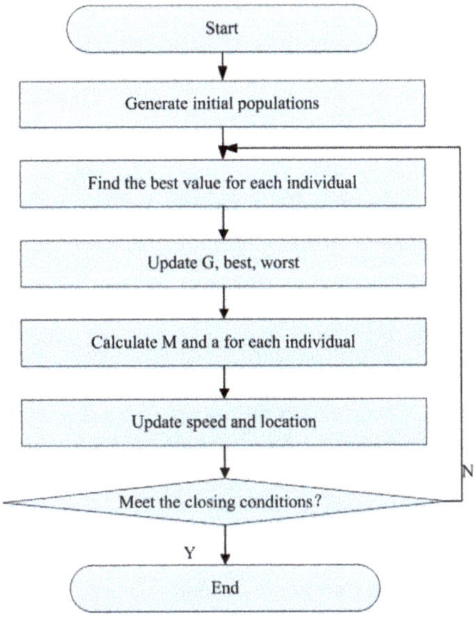

Figure 1. The framework of the GSA.

5.4. Model of the SEGDE

The flow of the SEGDE algorithm is shown in Figure 2.

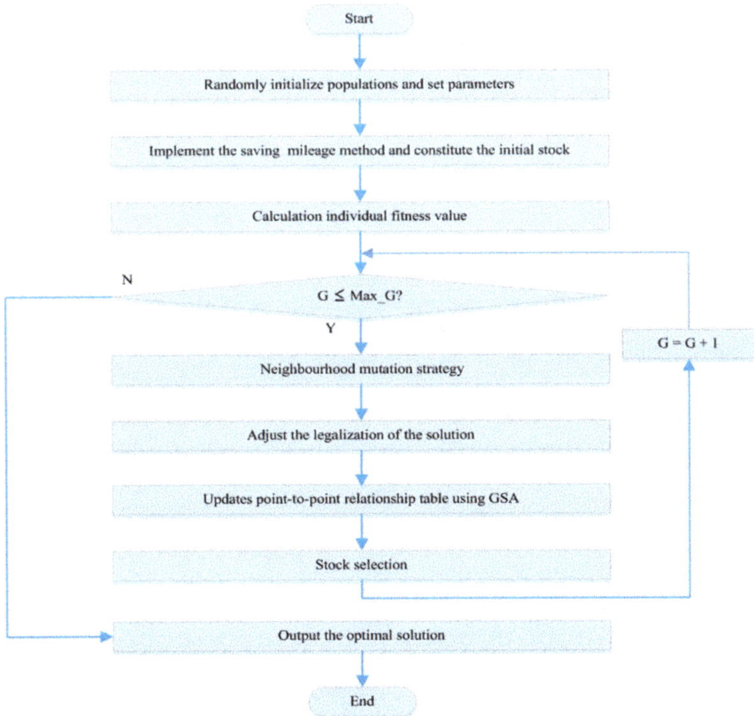

Figure 2. The flow of the SEGDE algorithm.

The implementation steps of the SEGDE are described as follows:

Step 1. The initial population is randomly generated by sequence coding, and the size of the initial population is NP, the dimension D, the maximum evolutionary iteration number Max, and the iteration number G = 1.

Step 2. The initial population is composed of the solution of the mileage saving method and the random solution of the mileage saving method.

Step 3. Calculate the initial fitness values of the individuals.

Step 4. If the number of iterations G is less than the maximum number of evolutionary iterations Max, enter Step 5; otherwise, proceed to Step 10.

Step 5. The strategy of neighborhood mutation is implemented to legalize the solution of the mutated population.

Step 6. The neighborhood search is carried out for the individual population, and the optimal solution in the local search is preserved.

Step 7. The gravity search algorithm is used to explore the relationship between variables and update the table of point-point relations, preserving the optimal solution.

Step 8. A population selection operation is performed.

Step 9. If the number of iterations G = G + 1, return to Step 4.

Step 10. The output evolutionary optimal solution is obtained.

6. Experimental Calculation and Analysis

6.1. Experimental Data

In order to verify the effectiveness of the SEGDE algorithm in solving the CVRP, data sets were selected from the operational research database OR-LIBRARY and the VRP database I NEO Research Group (uma.es). A total of 41 data instances with fewer than 50 dimensions were selected from among four test data sets.

6.2. Experimental Environment and Parameter Settings

The experimental environment included CPU-intel Core I5-4200H, Windows-Win8, RAM-4GB, and MATLAB R2018B. In the experiment, many alternative values are tested, and some classical values were selected from the literature; these parameter values were experimentally modified until the most reasonable parameter values were determined. These selected parameter values obtained the optimal solution, so that they could accurately and efficiently verify the effectiveness of the proposed SEGDE algorithm. Each experiment was carried out 25 times independently, and the optimal solution of 25 experiments was selected to compare with the other five algorithms. The five comparison algorithms were standard DE, GA, SA, the mileage-saving method (MS), and the improved MS(IMS) method. The settings of the parameters are shown in Table 4.

Table 4. The initial parameters of all algorithms.

Algorithms	Parameter Settings
SA	delta = 0.85, T = 150, Np = 100
GA	CR = 0.7, F = 0.5, Np = 100
DE	CR = 0.9, F = 0.5, Np = 100
SEGDE	Fmin = 0.5, Fmax = 0.9, CR = 0.9, Np = 100

6.3. Experimental Results and Analysis

The obtained experimental results are shown in Tables 5–8.

Table 5. The experimental results of six algorithms in solving set A.

Test Data	Opt.	SA	GA	MS	IMS	DE	SEGDE
A32_5	784	**739**	850	842	827	1426	813
A33_5	661	740	700	713	700	1194	**680**
A33_6	742	924	798	775	**743**	1233	746
A34_5	778	895	856	810	793	1347	**789**
A36_5	799	814	897	826	806	1367	**805**
A37_5	669	806	752	705	708	1366	**685**
A37_6	949	**949**	1047	975	974	1595	954
A38_5	730	908	789	765	751	1497	**734**
A39_5	822	1009	954	898	894	1575	**871**
A39_6	831	1011	940	861	**848**	1618	852
A44_6	937	1021	974	985	1785	1534	943
A45_6	944	1231	1111	1005	**955**	2093	963
A45_7	1146	1431	1282	1200	**1178**	1968	1203
A46_7	914	1431	1068	940	**934**	1862	935
A48_7	1073	1343	1280	1110	**1102**	2180	1129

Table 6. The experimental results of six algorithms in solving set E.

Test Data	Opt.	SA	GA	MS	IMS	DE	SEGDE
E22_K4	375	394	**375**	388	**375**	441	**375**
E23_K3	569	575	575	621	574	888	**569**
E30_K3	508	564	557	532	-	976	**508**
E33_K4	835	929	904	**841**	**841**	1180	**841**
E51_K5	521	697	685	582	-	1315	**575**

Table 7. The experimental results of six algorithms in solving set P.

Test Data	Opt.	SA	GA	MS	IMS	DE	SEGDE
P16_K8	450	889	**451**	478	472	452	**451**
P19_K2	212	**213**	**213**	237	219	276	**213**
P20_K2	216	**217**	218	234	247	452	**217**
P21_K2	211	**213**	**213**	236	233	318	**213**
P22_K2	216	222	219	240	234	317	**218**
P22_K8	589	**589**	**589**	591	590	624	**589**
P23_K8	529	541	532	537	537	633	**531**
P40_K5	458	561	526	516	**484**	629	508
P45_K5	510	616	614	569	**519**	1142	563

Table 8. The experimental results of six algorithms in solving set B.

Test Data	Opt.	SA	GA	DE	SEGDE
B31_K5	672	697	706	886	**679**
B34_K5	788	839	799	1186	**790**
B35_K5	955	1021	991	1665	**970**
B38_K6	805	887	845	1343	**825**
B39_K5	549	649	577	1314	**563**
B41_K6	829	989	880	1565	**838**
B43_K6	742	907	833	1387	**775**
B44_K7	909	1139	1058	1725	**931**
B45_K5	751	918	880	1631	**755**
B45_K6	678	888	791	1317	**698**
B50_K7	741	1006	879	1875	**766**
B50_K8	1312	1462	1401	2132	**1352**
B31_K5	672	697	706	886	**679**
B34_K5	788	839	799	1186	**790**
B35_K5	955	1021	991	1665	**970**

As can be observed from Tables 5–8, for set A, the proposed SEGDE algorithm has the best solutions of A33_5, A34_5, A36_5, A37_5, A38_5, and A39_5, and the IMS has the best solutions of A33_6, A39_6, A45_6, A45_7, A46_7, and A48_7. SA has the best solutions of A32_5 and A37_6. The IMS and SEGDE algorithm have obtained the best solutions of six cases. The obtained best solutions of A33_6, A34_5, A37_6, A38_5, and A44_6 are close to the optimal values by using the proposed SEGDE algorithm. For set E, the proposed SEGDE algorithm has obtained the best solutions of all cases. In particular, the optimal solutions of E22_K4, E23_K3, and E30_K3 are obtained using the proposed SEGDE algorithm. The best solutions of the other cases are also close to the optimal values using the proposed SEGDE algorithm. For set P, the proposed SEGDE algorithm has obtained the best solutions, except those of P40_K5 and P45_K5. The optimal solution of P22_K8 is obtained, and the obtained other solutions are also infinitely close to the optimal values using the proposed SEGDE algorithm. The IMS has obtained the best solutions of P40_K5 and P45_K5. For set B, the proposed SEGDE algorithm has obtained all best solutions of all cases. The obtained best solutions of B31_K5, B34_K5, B45_K5, and B34_K5 are infinitely close to the optimal values using the proposed SEGDE algorithm. The experimental results demonstrate that the proposed SEGDE algorithm can better solve these CVRPs from the operational research database OR-LIBRARY and the VRP database, and the optimized solutions are the optimal values, or are (infinitely) close to the optimal values. Therefore, the proposed SEGDE algorithm takes on a better global optimization ability in solving these different CVRPs. The reason for this is that the proposed SEGDE algorithm optimizes the abilities of the saving mileage algorithm, the sequential encoding approach, and the differential mutation strategy.

The routing comparison curves for generations 1 and 200 in the A33-K6 and B34-K5 optimization iterations are shown in Figures 3 and 4.

As can be observed from the optimization curves of the A33-K6 and B34-K5 cases in Figures 3 and 4, the obtained optimization paths by using the proposed SEGDE algorithm overlap to lessen, eliminate the path knot phenomenon, and effectively connect the adjacent points. In addition, the paths gradually become localized, which achieves the total path reduction. Through the experimental results of the test data, it can be observed that the proposed SEGDE algorithm possesses an advantage in addressing the vehicle path planning problem, and can approach the optimal solution to a great extent when the problem of fewer than 30 dimensions are processed. It also performs well on most of the problems with fewer than 50 dimensions, which proves the effectiveness of the proposed SEGDE algorithm in solving the different CVRPs. Therefore, the proposed SEGDE algorithm can effectively solve the CVRPs and obtain the optimized vehicle routing, as well as eliminate the path knotting, thus avoiding overlap. It is an effective algorithm for solving the CVRPs and the complex optimization problems.

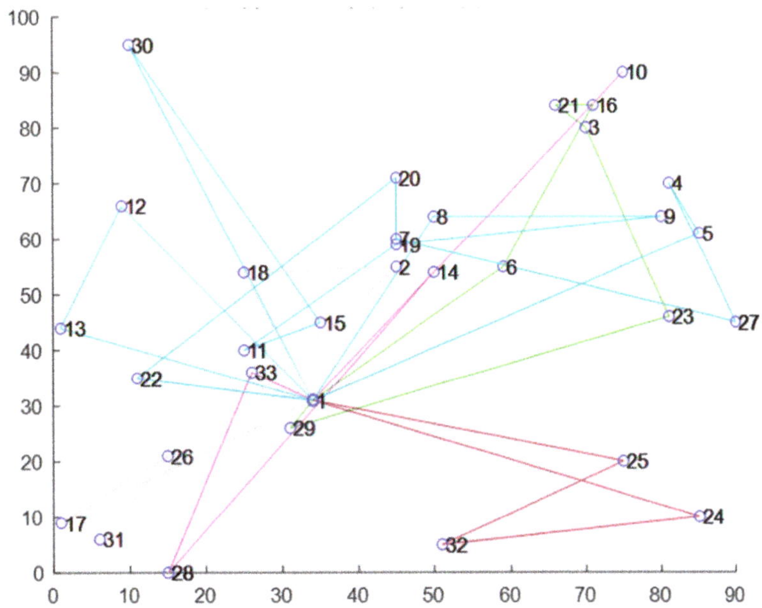

(a)

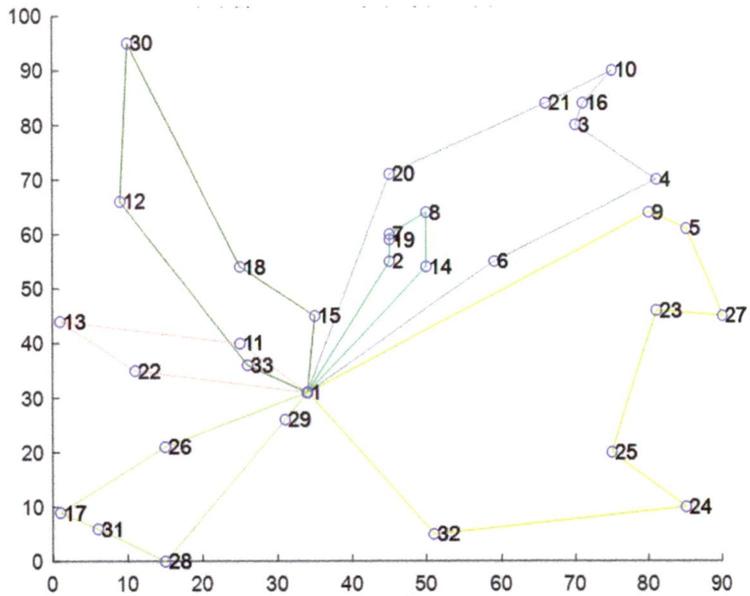

(b)

Figure 3. The optimization effect of A33-K6. (**a**) Optimization curve at Generation 1(1336.2577). (**b**) Optimization curve at Generation 200(745.6772).

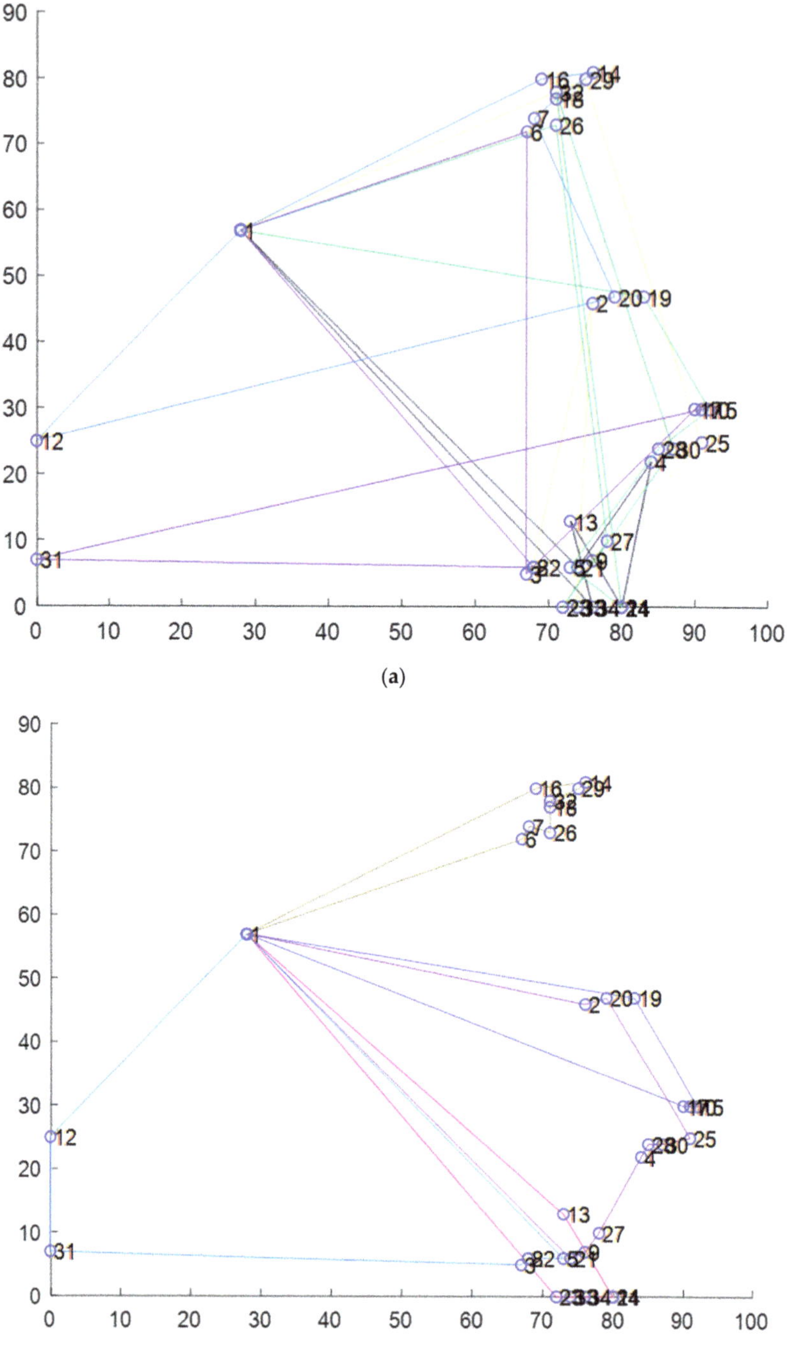

(**a**)

(**b**)

Figure 4. The optimization effect of B34-K5. (**a**) Optimization curve at Generation 1(1492.6296). (**b**) Optimization curve at Generation 200(790.3643).

6.4. Discussion

As can be observed from Tables 5–8 and Figures 3 and 4, the proposed SEGDE algorithm is used to solve CVRPs of set A, set B, set E, and set P; the obtained best solutions of E22_K4, E23_K3, E30_K3, and P22_K8 are the optimal values, and the obtained best solutions of A36_5, A38_5, E33_K4, P16_K8, P19_K2, P20_K2, P21_K2, P22_K2, and P23_K8 are (infinitely) close to the optimal values. Compared with the SA, GA, MS, IMS, and DE, the proposed SEGDE algorithm can effectively solve these various CVRPs and obtain the ideal vehicle routing, as well as eliminate the path knotting, avoiding overlap. Therefore, the proposed SEGDE algorithm adopts a better global optimization ability. The reason is that the proposed SEGDE algorithm is based on the saving mileage algorithm, the sequential encoding approach, and the differential mutation strategy. It optimizes the abilities of the saving mileage algorithm, the sequential encoding approach, and the differential mutation strategy. The saving mileage algorithm can improve the initial solution quality and the search efficiency by initializing the population of the DE. The sequential encoding approach can legalize the current solution and ensure its effectiveness by adjusting the differential mutation strategy. The gravitational search algorithm can effectively adjust the evolutionary search direction and further improve the search efficiency by calculating the gravitational relationship between points.

7. Conclusions

In this paper, a new multistrategy DE, namely SEGDE, is proposed to solve various CVRPs. In order to improve the search efficiency, the saving mileage algorithm is employed to initialize the population of DE. The sequential encoding method is used to adjust the differential mutation strategy to legalize the current solution and ensure its effectiveness. The GSA is applied to calculate the gravitational relationship between points for solution legalization and point reinsertion, which can effectively adjust the evolutionary search direction and optimize the search efficiency. Finally, the CVRP example from the operational research database is selected to verify the effectiveness of the proposed SEGDE algorithm. The obtained best solutions of E22_K4, E23_K3, E30_K3, and P22_K8 are the optimal values, and the obtained best solutions of A36_5, A38_5, E33_K4, P16_K8, P19_K2, P20_K2, P21_K2, P22_K2, and P23_K8 are (infinitely) close to the optimal values. Compared with the SA, GA, MS, IMS, and DE, the proposed SEGDE algorithm can effectively solve these different CVRPs and obtain the ideal vehicle routing, as well as eliminate the path knotting, avoiding overlap. Therefore, the experimental results demonstrate that the proposed SEGDE algorithm has a good optimization ability, search speed, and routing length. In addition, the stability of the SEGDE also possesses a good advantage.

Author Contributions: Conceptualization, J.W. and S.S.; methodology, S.S.; software, H.J.; validation, J.Z., H.J. and Y.S.; formal analysis, H.J.; resources, Y.L.; data curation, Y.L.; writing—original draft preparation, J.W. and S.S.; writing—review and editing, Y.L. and W.D.; visualization, J.Z.; supervision, H.J.; project administration, J.W.; funding acquisition, W.D. All authors have read and agreed to the published version of the manuscript.

Funding: This research was funded by the National Natural Science Foundation of China under grant numbers U2133205 and 61771087, the Innovation and Entrepreneurship Training Program of Civil Aviation University of China under grant number IECAUC2022126, the Traction Power State Key Laboratory of Southwest Jiaotong University under Grant TPL2203, and the Research Foundation for Civil Aviation University of China under grant numbers 3122022PT02 and 2020KYQD123.

Institutional Review Board Statement: Not applicable.

Informed Consent Statement: Not applicable.

Data Availability Statement: Not applicable.

Conflicts of Interest: The authors declare no conflict of interest.

References

1. Hulagu, S.; Celikoglu, H.B. An electric vehicle routing problem with intermediate nodes for shuttle fleets. *IEEE Trans. Intell. Transp. Syst.* **2022**, *23*, 1223–1235. [CrossRef]
2. Yang, Y.; Chen, H.; Heidari, A.A.; Gandomi, A.H. Hunger games search: Visions, conception, implementation, deep analysis, perspectives, and towards performance shifts. *Expert Syst. Appl.* **2021**, *177*, 114864. [CrossRef]
3. Felipe, A.; Ortuno, M.T.; Righini, G.; Tirado, G. A heuristic approach for the green vehicle routing problem with multiple technologies and partial recharges. *Transp. Res. Part E-Logist. Transp. Rev.* **2014**, *71*, 111–128. [CrossRef]
4. Ahmadianfar, I.; Heidari, A.A.; Noshadian, S.; Chen, H.; Gandomi, A.H. INFO: An efficient optimization algorithm based on weighted mean of vectors. *Expert Syst. Appl.* **2022**, *195*, 116516. [CrossRef]
5. Wang, Z.; Sheu, J.B. Vehicle routing problem with drones. *Transp. Res. Part B-Methodol.* **2019**, *122*, 350–364. [CrossRef]
6. Dorling, K.; Heinrichs, J.; Messier, G.G.; Magierowski, S. Vehicle routing problems for drone delivery. *IEEE Trans. Syst. Man Cybern.-Syst.* **2017**, *47*, 70–85. [CrossRef]
7. Wang, X.Y.; Shao, S.; Tang, J.F. Iterative local-search heuristic for weighted vehicle routing problem. *IEEE Trans. Intell. Transp. Syst.* **2021**, *22*, 3444–3454. [CrossRef]
8. Wang, H.; Li, M.H.; Wang, Z.Y.; Li, W.; Hou, T.J.; Yang, X.Y.; Zhao, Z.Z.; Wang, Z.F.; Sun, T. Heterogeneous fleets for green vehicle routing problem with traffic restrictions. In *IEEE Transactions on Intelligent Transportation Systems*; IEEE: Piscataway, NJ, USA, 2022. [CrossRef]
9. Khaitan, A.; Mehlawat, M.K.; Gupta, P.; Pedrycz, W. Socially aware fuzzy vehicle routing problem: A topic modeling based approach for driver well-being. *Expert Syst. Appl.* **2022**, *205*, 117655. [CrossRef]
10. Ahmadianfar, I.; Heidari, A.A.; Gandomi, A.H.; Chu, X.; Chen, H. Run beyond the metaphor: An efficient optimization algorithm based on Runge Kutta method. *Expert Syst. Appl.* **2021**, *181*, 115079. [CrossRef]
11. Oztas, T.; Tus, A. A hybrid metaheuristic algorithm based on iterated local search for vehicle routing problem with simultaneous pickup and delivery. *Expert Syst. Appl.* **2022**, *202*, 117401. [CrossRef]
12. Feng, B.; Wei, L.X. An improved multi-directional local search algorithm for vehicle routing problem with time windows and route balance. *Appl. Intell.* **2022**, 1–13. [CrossRef]
13. Thiebaut, K.; Pessoa, A. Approximating the chance-constrained capacitated vehicle routing problem with robust optimization. *4OR-A Q. J. Oper. Res.* **2022**, 1–19. [CrossRef]
14. Li, S.; Chen, H.; Wang, M.; Heidari, A.A.; Mirjalili, S. Slime mould algorithm: A new method for stochastic optimization. *Futur. Gener. Comput. Syst.* **2020**, *111*, 300–323. [CrossRef]
15. Storn, R.; Price, K. *Differential Evolution: A Simple and Efficient Adaptive Scheme for Global Optimization over Continuous Spaces; Technical Report; TR-95-012*; International Computer Science Institute: California, CA, USA, 1995.
16. Yusuf, I.; Baba, M.S.; Iksan, N. Applied genetic algorithm for solving rich VRP. *Appl. Artif. Intell.* **2014**, *28*, 957–991. [CrossRef]
17. Akpinar, S. Hybrid large neighbourhood search algorithm for capacitated vehicle routing problem. *Expert Syst. Appl.* **2016**, *61*, 28–38. [CrossRef]
18. Zhang, D.F.; Cai, S.F.; Ye, F.R.; Si, Y.W.; Nguyen, T.T. A hybrid algorithm for a vehicle routing problem with realistic constraints. *Inf. Sci.* **2017**, *394*, 167–182. [CrossRef]
19. Dechampai, D.; Tanwanichkul, L.; Sethanan, K.; Pitakaso, R. A differential evolution algorithm for the capacitated VRP with flexibility of mixing pickup and delivery services and the maximum duration of a route in poultry industry. *J. Intell. Manuf.* **2017**, *28*, 1357–1376. [CrossRef]
20. Gutierrez, A.; Dieulle, L.; Labadie, N.; Velasco, N. A multi-population algorithm to solve the VRP with stochastic service and travel times. *Comput. Ind. Eng.* **2018**, *125*, 144–156. [CrossRef]
21. Fallah, M.; Tavakkoli-Moghaddam, R.; Alinaghian, M.; Salamatbakhsh-Varjovi, A. A robust approach for a green periodic competitive VRP under uncertainty: DE and PSO algorithms. *J. Intell. Fuzzy Syst.* **2019**, *36*, 5213–5225. [CrossRef]
22. Altabeeb, A.M.; Mohsen, A.M.; Ghallab, A. An improved hybrid firefly algorithm for capacitated vehicle routing problem. *Appl. Soft Comput.* **2019**, *84*, 105728. [CrossRef]
23. Altabeeb, A.M.; Mohsen, A.M.; Abualigah, L.; Ghallab, A. Solving capacitated vehicle routing problem using cooperative firefly algorithm. *Appl. Soft Comput.* **2021**, *108*, 107403. [CrossRef]
24. Xiao, J.H.; Zhang, T.; Du, J.G.; Zhang, X.Y. An evolutionary multiobjective route grouping-based heuristic algorithm for large-scale capacitated vehicle routing problems. *IEEE Trans. Cybern.* **2021**, *51*, 4173–4186. [CrossRef] [PubMed]
25. Jia, Y.H.; Mei, Y.; Zhang, M.J. A bilevel ant colony optimization algorithm for capacitated electric vehicle routing problem. *IEEE Trans. Cybern.* **2022**, *52*, 10855–10868. [CrossRef] [PubMed]
26. Jiang, H.; Lu, M.X.; Tian, Y.; Qiu, J.F.; Zhang, X.Y. An evolutionary algorithm for solving Capacitated Vehicle Routing Problems by using local information. *Appl. Soft Comput.* **2022**, *117*, 108431. [CrossRef]
27. Deng, W.; Ni, H.C.; Liu, Y.; Chen, H.L.; Zhao, H.M. An adaptive differential evolution algorithm based on belief space and generalized opposition-based learning for resource allocation. *Appl. Soft Comput.* **2022**, *127*, 109419. [CrossRef]
28. Zhang, X.Y.; Chen, L.; Gendreau, M.; Langevin, A. A branch-and-cut algorithm for the vehicle routing problem with two-dimensional loading constraints. *Eur. J. Oper. Res.* **2022**, *302*, 259–269. [CrossRef]
29. Song, Y.J.; Cai, X.; Zhou, X.; Zhang, B.; Chen, H.; Li, Y.G.; Deng, W.; Deng, W. Dynamic hybrid mechanism-based differential evolution algorithm and its application. *Expert Syst. Appl.* **2023**, *213*, 118834. [CrossRef]

30. Niu, Y.Y.; Shao, J.; Xiao, J.H.; Song, W.; Cao, Z.G. Multi-objective evolutionary algorithm based on RBF network for solving the stochastic vehicle routing problem. *Inf. Sci.* **2022**, *609*, 387–410. [CrossRef]
31. Deng, W.; Zhang, L.R.; Zhou, X.B.; Zhou, Y.Q.; Sun, Y.Z.; Zhu, W.H.; Chen, H.Y.; Deng, W.Q.; Chen, H.L.; Zhao, H.M. Multi-strategy particle swarm and ant colony hybrid optimization for airport taxiway planning problem. *Inf. Sci.* **2022**, *612*, 576–593. [CrossRef]
32. Gu, R.X.; Poon, M.; Luo, Z.H.; Liu, Y.; Liu, Z. A hierarchical solution evaluation method and a hybrid algorithm for the vehicle routing problem with drones and multiple visits. *Transp. Res. Part C Emerg. Technol.* **2022**, *141*, 103733. [CrossRef]
33. Azad, U.; Behera, B.K.; Ahmed, E.A.; Panigrahi, P.K.; Farouk, A. Solving vehicle routing problem using quantum approximate optimization algorithm. In *IEEE Transactions on Intelligent Transportation Systems*; IEEE: Piscataway, NJ, USA, 2022. [CrossRef]
34. Lai, Y.X.; Yang, F.; Meng, G.; Lu, W. Data-driven flexible vehicle scheduling and route optimization. In *IEEE Transactions on Intelligent Transportation Systems*; IEEE: Piscataway, NJ, USA, 2022. [CrossRef]
35. Voigt, S.; Frank, M.; Fontaine, P.; Kuhn, H. Hybrid adaptive large neighborhood search for vehicle routing problems with depot location decisions. *Comput. Oper. Res.* **2022**, *146*, 105856. [CrossRef]
36. Seyfi, M.; Alinaghian, M.; Ghorbani, E.; Catay, B.; Sabbagh, M.S. Multi-mode hybrid electric vehicle routing problem. *Transp. Res. Part E-Logist. Transp. Rev.* **2022**, *166*, 102882. [CrossRef]
37. Cai, Y.Q.; Cheng, M.Q.; Zhou, Y.; Liu, P.Z.; Guo, J.M. A hybrid evolutionary multitask algorithm for the multiobjective vehicle routing problem with time windows. *Inf. Sci.* **2022**, *612*, 168–187. [CrossRef]
38. Wen, M.Y.; Sun, W.; Yu, Y.; Tang, J.F.; Ikou, K. An adaptive large neighborhood search for the larger-scale multi depot green vehicle routing problem with time windows. *J. Clean. Prod.* **2022**, *374*, 133916. [CrossRef]
39. Ma, B.S.; Hu, D.W.; Wang, Y.; Sun, Q.; He, L.W.; Chen, X.Q. Time-dependent vehicle routing problem with departure time and speed optimization for shared autonomous electric vehicle service. *Appl. Math. Model.* **2023**, *113*, 333–357. [CrossRef]
40. Zhou, X.B.; Ma, H.J.; Gu, J.G.; Chen, H.L.; Deng, W. Parameter adaptation-based ant colony optimization with dynamic hybrid mechanism. *Eng. Appl. Artif. Intell.* **2022**, *114*, 105139. [CrossRef]
41. Ren, Z.; Han, X.; Yu, X.; Skjetne, R.; Leira, B.J.; Sævik, S.; Zhu, M. Data-driven simultaneous identification of the 6DOF dynamic model and wave load for a ship in waves. *Mech. Syst. Signal Process.* **2023**, *184*, 109422. [CrossRef]
42. Chen, H.Y.; Miao, F.; Chen, Y.J.; Xiong, Y.J.; Chen, T. A hyperspectral image classification method using multifeature vectors and optimized KELM. *IEEE J. Sel. Top. Appl. Earth Obs. Remote Sens.* **2021**, *14*, 2781–2795. [CrossRef]
43. Xu, G.; Dong, W.; Xing, J.; Lei, W.; Liu, J. Delay-CJ: A novel cryptojacking covert attack method based on delayed strategy and its detection. *Digit. Commun. Netw.* 2022, in press. [CrossRef]
44. Yao, R.; Guo, C.; Deng, W.; Zhao, H.M. A novel mathematical morphology spectrum entropy based on scale-adaptive techniques. *ISA Trans.* **2022**, *126*, 691–702. [CrossRef]
45. Li, T.Y.; Shi, J.Y.; Deng, W.; Hu, Z.D. Pyramid particle swarm optimization with novel strategies of competition and cooperation. *Appl. Soft Comput.* **2022**, *121*, 108731. [CrossRef]
46. Zhao, H.M.; Liu, J.; Chen, H.Y.; Chen, J.; Li, Y.; Xu, J.J.; Deng, W. Intelligent diagnosis using continuous wavelet transform and gauss convolutional deep belief network. In *IEEE Transactions on Reliability*; IEEE: Piscataway, NJ, USA, 2022. [CrossRef]
47. Wu, D.; Wu, C. Research on the time-dependent split delivery green vehicle routing problem for fresh agricultural products with multiple time windows. *Agriculture* **2022**, *12*, 793. [CrossRef]
48. Xu, G.; Bai, H.; Xing, J.; Luo, T.; Xiong, N.N. SG-PBFT: A secure and highly efficient distributed blockchain PBFT consensus algorithm for intelligent Internet of vehicles. *J. Parallel Distrib. Comput.* **2022**, *164*, 1–11. [CrossRef]
49. Wei, Y.Y.; Zhou, Y.Q.; Luo, Q.F.; Deng, W. Optimal reactive power dispatch using an improved slime Mould algorithm. *Energy Rep.* **2021**, *7*, 8742–8759. [CrossRef]
50. Zhang, Z.; Huang, W.G.; Liao, Y.; Song, Z.; Shi, J.; Jiang, X.; Shen, C.; Zhu, Z. Bearing fault diagnosis via generalized logarithm sparse regularization. *Mech. Syst. Signal Process.* **2022**, *167*, 108576. [CrossRef]
51. Chen, H.Y.; Fang, M.; Xu, S. Hyperspectral remote sensing image classification with CNN based on quantum genetic-optimized sparse representation. *IEEE Access* **2020**, *8*, 99900–99909. [CrossRef]
52. Zhang, X.Y.; Duan, H.B. An improved constrained differential evolution algorithm for unmanned aerial vehicle global route planning. *Appl. Soft Comput.* **2014**, *26*, 270–284. [CrossRef]
53. Teoh, B.E.; Ponnambalam, S.G.; Kanagaraj, G. Differential evolution algorithm with local search for capacitated vehicle routing problem. *Int. J. Bio-Inspired Comput.* **2015**, *7*, 321–342. [CrossRef]
54. Pitakaso, R.; Sethanan, K.; Srijaroon, N. Modified differential evolution algorithms for multi-vehicle allocation and route optimization for employee transportation. *Eng. Optim.* **2019**, *52*, 1225–1243. [CrossRef]
55. Xing, L.N.; Liu, Y.Y.; Li, H.Y.; Wu, C.C.; Lin, W.C.; Song, W. A hybrid discrete differential evolution algorithm to solve the split delivery vehicle routing problem. *IEEE Access* **2020**, *8*, 207962–207972. [CrossRef]
56. Sethanan, K.; Jamrus, T. Hybrid differential evolution algorithm and genetic operator for multi-trip vehicle routing problem with backhauls and heterogeneous fleet in the beverage logistics industry. *Comput. Ind. Eng.* **2020**, *146*, 106571. [CrossRef]
57. Hameed, A.S.; Aboobaider, B.M.; Mutar, M.L.; Choon, N.H. A new hybrid approach based on discrete differential evolution algorithm to enhancement solutions of quadratic assignment problem. *Int. J. Ind. Eng. Comput.* **2020**, *11*, 51–72. [CrossRef]
58. Liu, W.L.; Gong, Y.J.; Chen, W.N.; Liu, Z.Q.; Wang, H.; Zhang, J. Coordinated charging scheduling of electric vehicles: A mixed-variable differential evolution approach. *IEEE Trans. Intell. Transp. Syst.* **2020**, *21*, 5094–5109. [CrossRef]

59. Moonsri, K.; Sethanan, K.; Worasan, K.; Nitisiri, K. A hybrid and self-adaptive differential evolution algorithm for the multi-depot vehicle routing problem in EGG distribution. *Appl. Sci.* **2022**, *12*, 35. [CrossRef]
60. Chai, X.Z.; Zheng, Z.S.; Xiao, J.M.; Yan, L.; Qu, B.Y.; Wen, P.W.; Wang, H.Y.; Zhou, Y.; Sun, H. Multi-strategy fusion differential evolution algorithm for UAV path planning in complex environment. *Aerosp. Sci. Technol.* **2022**, *121*, 107287. [CrossRef]
61. Wu, P.; Xu, L.; D'Ariano, A.; Zhao, Y.X.; Chu, C.B. Novel formulations and improved differential evolution algorithm for optimal lane reservation with task merging. In *IEEE Transactions on Intelligent Transportation Systems*; IEEE: Piscataway, NJ, USA, 2022. [CrossRef]
62. Hou, Y.; Wu, Y.L.; Han, H.G. Multistate-constrained multiobjective differential evolution algorithm with variable neighborhood strategy. In *IEEE Transactions on Cybernetics*; IEEE: Piscataway, NJ, USA, 2022. [CrossRef]
63. Chen, M.C.; Yerasani, S.; Tiwari, M.K. Solving a 3-dimensional vehicle routing problem with delivery options in city logistics using fast-neighborhood based crowding differential evolution algorithm. *J. Ambient. Intell. Humaniz. Comput.* **2022**, 1–14. [CrossRef]
64. Deng, W.; Xu, J.; Song, Y.; Zhao, H.M. Differential evolution algorithm with wavelet basis function and optimal mutation strategy for complex optimization problem. *Appl. Soft Comput.* **2021**, *100*, 106724. [CrossRef]
65. Li, X.; Zhao, H.; Yu, L.; Chen, H.; Deng, W.Q.; Deng, W. Feature extraction using parameterized multisynchrosqueezing transform. *IEEE Sens. J.* **2022**, *22*, 14263–14272. [CrossRef]
66. Wu, X.; Wang, Z.C.; Wu, T.H.; Bao, X.G. Solving the family traveling salesperson problem in the adleman–lipton model based on DNA computing. *IEEE Trans. NanoBioscience* **2021**, *21*, 75–85. [CrossRef]
67. Abu-Monshar, A.; Al-Bazi, A. A multi-objective centralised agent-based optimisation approach for vehicle routing problem with unique vehicles. *Appl. Soft Comput.* **2022**, *125*, 109187. [CrossRef]
68. Torres, F.; Gendreau, M.; Rei, W. Vehicle routing with stochastic supply of crowd vehicles and time windows. *Transp. Sci.* **2021**, *56*, 631–653. [CrossRef]
69. Kuo, R.J.; Lu, S.H.; Mara, S.T.W. Vehicle routing problem with drones considering time windows. *Expert Syst. Appl.* **2022**, *191*, 116264. [CrossRef]
70. Ochelska-Mierzejewska, J.; Poniszewska-Maranda, A.; Maranda, W. Selected genetic algorithms for vehicle routing problem solving. *Electronics* **2022**, *10*, 3147. [CrossRef]
71. Lei, D.M.; Cui, Z.Z.; Li, M. A dynamical artificial bee colony for vehicle routing problem with drones. *Eng. Appl. Artif. Intell.* **2022**, *107*, 104510. [CrossRef]
72. Sheng, Y.K.; Lan, W.L. *Application of Clarke-Wright Saving Mileage Heuristic Algorithm in Logistics Distribution Route Optimization*; Trans Tech Publications Ltd.: Baech, Switzerland, 2011.
73. Hosseinabadi, A.A.R.; Vahidi, J.; Balas, V.E.; Mirkamali, S.S. OVRP_GELS: Solving open vehicle routing problem using the gravitational emulation local search algorithm. *Neural Comput. Appl.* **2017**, *29*, 955–968. [CrossRef]

electronics

MDPI

Article

Fine-Grained Classification of Announcement News Events in the Chinese Stock Market

Feng Miu [1,*], Ping Wang [2], Yuning Xiong [3], Huading Jia [2] and Wei Liu [1]

1 School of Artificial Intelligence and Law, Southwest University of Political Science & Law, Chongqing 401120, China; liuwei@swupl.edu.cn
2 School of Economic Information Engineering, Southwestern University of Finance and Economics, Chengdu 611130, China; wangping8139@swufe.edu.cn (P.W.); jiahd_t@swufe.edu.cn (H.J.)
3 School of Economics, Xihua University, Chengdu 610039, China; xiongyn@mail.xhu.edu.cn
* Correspondence: miu_feng@swupl.edu.cn; Tel.: +86-189-8371-5062

Abstract: Determining the event type is one of the main tasks of event extraction (EE). The announcement news released by listed companies contains a wide range of information, and it is a challenge to determine the event types. Some fine-grained event type frameworks have been built from financial news or stock announcement news by domain experts manually or by clustering, ontology or other methods. However, we think there are still some improvements to be made based on the existing results. For example, a legal category has been created in previous studies, which considers violations of company rules and violations of the law the same thing. However, the penalties they face and the expectations they bring to investors are different, so it is more reasonable to consider them different types. In order to more finely classify the event type of stock announcement news, this paper proposes a two-step method. First, the candidate event trigger words and co-occurrence words satisfying the support value are extracted, and they are arranged in the order of common expressions through the algorithm. Then, the final event types are determined using three proposed criteria. Based on the real data of the Chinese stock market, this paper constructs 54 event types ($p = 0.927, f = 0.946$), and some reasonable and valuable types have not been discussed in previous studies. Finally, based on the unilateral trading policy of the Chinese stock market, we screened out some event types that may not be valuable to investors.

Keywords: event extraction; event type; event trigger words; stock announcement news; stock return

Citation: Miu, F.; Wang, P.; Xiong, Y.; Jia, H.; Liu, W. Fine-Grained Classification of Announcement News Events in the Chinese Stock Market. *Electronics* **2022**, *11*, 2058. https://doi.org/10.3390/electronics11132058

Academic Editor: Arkaitz Zubiaga

Received: 4 June 2022
Accepted: 28 June 2022
Published: 30 June 2022

1. Introduction

Much empirical research has shown that news events have important impacts on the stock market. According to Yin's classification standard [1], news can be divided into specific news and general news. Specific news refers to news stories where the affected stock entities are clearly specified in the news text, while general news refers to stories where they are not. Specific news usually involves announcements about stocks. General news includes industry news, policy news, microeconomic news and so on. When analyzing the impact of general news on the stock market, we first need to determine the stock entities that may be affected by such news. Due to the different processing methods for the two kinds of news, this paper only focuses on stock announcement news, which can reflect the recent development of a listed company. It can assist investors in making decisions and can be used in stock return predictions. Determining the event type is one of the main tasks of event extraction. The announcement news covers all aspects of information about listed companies, so it is a challenge to build a maturity event type framework from stock announcement news.

To date, there is no unified classification framework or standard for the news announcements regarding the Chinese stock market. Therefore, the existing research has constructed various event type frameworks using expert domain knowledge and experience [2], clustering [3], ontology [4], and other methods [5]. Some studies have implemented a fine-grained event type framework. Following an analysis of the existing studies, we believe that there is still room for improvement. The existing methods usually focus on the event types that occur more frequently or are generally considered important. Some low-frequency event types are usually neglected, and some event types can be further subdivided. For example, an event type called legal has been constructed in many studies, which regards violations of company policy and violations of the law as being the same. However, the two will receive different penalties, resulting in different expected impacts on investors. Therefore, we think it is more reasonable to regard the two as different event types.

Inspired by the "IDA-CLUSTERING + HUMAN-IDENTIFICATION" strategy [6], we propose a two-step method to divide stock announcement news into more detailed types. Event trigger words play an important role in the event type, which are usually verbs. By combining these with the conventional expressions of a certain kind of announcement news, we can extract expressions from an announcement news set containing event trigger words. In order to take into account the industry characteristics and event emotional tendency, we propose three event type judgment criteria to determine the final event type. The experimental results of real data on the Chinese stock market show that the event type framework constructed in this paper is reasonable and consistent with people's cognition. Compared with the existing related research results, our method finds some reasonable and valuable event types that have not been discussed yet. Our work enriches the existing research, and the results will help investors.

After extracting all kinds of announcement news events, we did not choose to conduct the stock prediction work in the traditional way. This is due to the fact that we think it is inappropriate to rely solely on announcement news for prediction without considering other types of news such as industry news, financial news and so on. Instead, considering the unilateral trading policy in the Chinese stock market, we screened out some event types that are not valuable to investors.

2. Related Research

Event extraction is a typical task in the field of NLP that has been widely studied in the past. Due to the subject of this paper, we focus on event extraction methods in the economic domain. The ACE event typology "business", which has four subtypes (Start-org, Merge-org, Declare-bankruptcy, and End-org), is relevant to the economic domain. The ACE event type definition does not meet our requirements. Therefore, researchers have proposed various methods to categorize types of financial events according to actual situations.

Fung et al. [7] classified financial news into two simple types of events: stimulating stock rise and stimulating stock fall. Wong et al. [8] carried out similar work, which used a method based on feature words and template rules to identify three types of stock opinion events (rising, stable and falling). Du et al. [9] proposed a PULS business intelligence system, which detected 15 event types pre-categorized as "positive" or "negative". Chen et al. [10] proposed a fine-grained event extraction method and applied it to the stock price prediction model. Firstly, a professional financial event dictionary (TFED) was constructed manually by experts. The event type, event trigger word and event role were determined by the dictionary, and the event was extracted using the template rules. The abovementioned research did not separate stock announcement news from financial news, opting instead to combine the two. Events are classified and used as inputs for the prediction model, so the classification is usually rough. Liu [11] proposed a method for discovering financial events that affect stock movements. Firstly, 13 types of financial events were manually determined according to the industry characters; then, the keywords in the constructed financial ontology were used to annotate the text. Liu's work classified financial news according to industry characteristics, and the types of construction are biased towards industry news.

The Stock Sonar project expert-created event typology identified eight event types: "Legal", "Analyst Recommendation", "Financial", "Stock Price Change", "Deals", "Mergers and Acquisitions", "Partnerships", "Product", and "Employment" [12]. The author focused on the event types in stock announcement news, but the number of designed event types are small and the coverage is not wide. He [13] constructed a stock market theme event case base through ontology. The theme event types included financial policy events, monetary policy events and market rule adjustment with multiple subtypes. The subject event was defined in triple (event description, market description, event result). It can be seen from the construction types that the author focuses on three types of macroeconomic events and did not focus on the stock announcement news. Wang [14] constructed a corpus of 2500 news texts that were manually divided into two categories and six sub-categories. Then, based on semantic, grammatical and syntactic features, the SVM method was used to identify event types. Chen [15] implemented an event extraction system in the financial field. Firstly, the system manually determined eight event types and selected seed event sentences for each event type. Then, the seed event trigger words were extracted using verb object relationship and subject predicate relationship and were extended by word2vec to obtain the event trigger word dictionary. Han et al. [16] proposed a method for event extraction in the business field by combining machine learning and template rules. Firstly, a business event type framework was defined manually, in which business events were divided into 8 categories and 16 sub-categories, and a small number of event triggers were constructed. Then, the trigger word dictionary was extended via word embedding to identify event types through multiple classification models combined with the trigger word dictionary. References [14–16] manually classified the event type from the financial news while paying close attention to the design of the event recognition model. Boudoukh et al. [17] identified 18 event categories based on Capital-IQ types and a cross-section of academics.

Arenarenarenko et al. [18] proposed an event extraction system named BEECON (Business Events Extractor Component based on the ONtology) for business intelligence. The system can identify 11 types and 41 sub-types of business events from news texts using template rules. The experimental results verified that the system had high accuracy (95%). Although the author built a rich and fine-grained event type framework, which includes some news on the stock market, he focused on the events in the business domain, and the coverage of stock announcement news event types was not comprehensive enough. Zhang [19] proposed an event-driven stock recommendation model. The financial events are manually classified into 12 categories and 30 sub-categories. The fine-grained event type framework constructed by the author was all centered around stock announcement news. It covered most of the events in the announcement news, but also ignored some low-frequency event types, such as winning bid events. In addition, as we mentioned earlier, some event types can be further subdivided. In terms of event recognition, the author's accuracy on the domain data set (67.3%) was much lower than the method that used template rules in [18] (95%). The template rules method can usually achieve high precision but requires much energy and expert experience. Some researchers consider automatic template rule generation and use a small amount of training corpus and seed templates through weak supervision, bootstrapping or other methods to automatically generate more templates [20].

Zhou [21] implemented a financial event extraction system based on deep learning. In the system, experts manually divided types of financial events (4 categories and 34 sub-categories) and built two kinds of relationships tables between financial entities (personnel to enterprise, enterprise to enterprise). The author constructed a detailed event type framework around stock announcement news. From the classification of the first layer types, the coverage was not wide (far less so than that of [19]). However, the author divided the sub-types in a very detailed way, which is better than the divisions used in [19]. Wang et al. [22] proposed a bond event element extraction method based on CRF. The event element framework was manually predefined and included bond event type and an event element list. Ding et al. [23] proposed a method to extract events from financial reports. Due to the standard writing of the financial report text, it takes the titles at all levels as the event category and the paragraphs under the title as the extraction unit. The author constructs event types according to the characteristics of financial reports, and the method is not suitable for stock announcement news. Wu [20] used the improved TFIDF algorithm to calculate the weight of text eigenvalues, then clustered the text using the K-means method. Finally, the most appropriate K = 13 value was selected by listing. The 13 event types included: issuance, dividend, event prompt, pledge, performance notice, suspension and resumption of trading, fund-raising, increase or decrease of holdings, financial report, investment in subsidiaries, abnormal fluctuation, asset reorganization and change registration. The author used the clustering method to construct event types from stock news. Although some event types could be found, some event types, especially those with low frequency, are easily ignored.

The event study method is also widely used by researchers to analyze the impact of news events on the Chinese stock market, which was initiated by Ball and Brown (1968) and Fama et al. (1969). It is essentially a statistical analysis method. The basic idea of the event study method is to select a certain type of specific event according to the research purpose, calculate the abnormal return index in the event window period, and then explain the impact of specific events on the change in sample stock price and return. There have been many achievements in the research on the Chinese stock market involving many types of stock announcement news events, such as monetary policy, industry related policies, epidemic situations, explosion accidents, earthquakes, avian influenza, the Shenzhou spacecraft launch, negative reputations, food safety accidents, environmental pollution, performance forecast events, corporate mergers and acquisitions, the lifting of stock bans and so on [24–26]. Besides stock markets, news event study also plays an important role in commodity markets [27,28].

3. Proposed Method

3.1. Extracting Event Trigger Words

Event trigger words are key words that help us to identify event types, which are usually verbs. Firstly, this paper proposes an algorithm and a support calculation formula, which takes all stock announcement news texts as the input, extracts all verbs from the text, marks the emotional polarity according to the emotional dictionary, takes the verb as a candidate event trigger word and takes the announcement news containing the verb as a class. Then it calculates the support between the other words and the verb, takes the other words that meet the threshold as collocations and judges the word order between collocations. Finally, it extracts candidate event trigger words and co-occurrence words and arranges them in the order of common expressions. It can be described as Algorithm 1:

The function of Formula (1) involves calculating the support between words and the verb, where *CountB()* represents the number of times the word appears before the verb, and *CountA()* stands for the opposite. If the absolute value of Formula (1) exceeds the threshold, this means that it is a conventional expression with a verb in the announcement news. If the result is positive, it means the word is usually in front of the verb. If the probability of a word appearing before and after the verb is close, it means that the word has no value in the representation of the event type.

$$Support(w_i, w_t) = \frac{CountB(w_i, w_t)}{Count(w_t)} - \frac{CountA(w_i, w_t)}{Count(w_t)} \tag{1}$$

Algorithm 1: Extract Candidate Trigger Words and Collocations from Announcement News

```
Input: Announcement news text set C
Output: event trigger words and collocation sequence set E
#Text preprocessing, put all verbs into set V, and judge the emotional polarity of verbs
For text in C
wlist1 = Segment(text)
wlist2 = NotStopWord(word1)
Postagger(wlist2)
if Postagger[word in wlist2] = Verb
add word in V
SentimentTag(V)
#From the announcement news set Ci containing verb Vi, judge the best position of other words,
trigger words and the best order between other words according to word co-occurrence, and calculate
the support
For word in Ci
beforeValueocc = Count(word, vi)/Count(Ci)
afterValueocc = Count(vi, word)/Count(Ci)
if beforeValueocc > afterValueocc and beforeValueocc > thred1
add word in listbfi
else if afterValueocc > thred1
add word in listafi
Add first word from listbfi, listafi in Ebefore and Eafter
For every word as w1 in listbfi
For every word as w2 in Ebefore
if Count(w1, w2) > Count(w2,w1)
add w1 in Ebefore and put w1 before w2
else
for next w2
if w1 not in Ebefore
// If a word is not placed before the sorted word, it is placed in the last position
add w1 in Ebefore and put in the last location.
Judge the position of words in listafi as shown above
Connect the words in ebefore and eafter in order to form vlist
if exist (beforeValueocc or afterValueocc)>thred2
put vList in E
else if Sentiment(vi) > 0
put vList in E
END
```

3.2. Three Classification Criteria

Based on the results of the above Algorithm 1, this paper puts forward three criteria to judge whether it constitutes the final event type from the perspective of data mining. The purpose of the three criteria is to select regular announcement news from the stock announcement news set containing verbs and construct it into a type. When constructing the event type of stock announcement news, the event extraction template is determined according to the criteria. The three criteria are as follows:

(1) For verbs without emotional polarity, if there is a collocation with more than 0.95 support around the verb, combine the collocation with the verb as the event trigger words. For example, there is a collocation of "扩股/capital increase (0.99)" to the left of the verb "增资/share expansion", so "增资扩股/capital increase and share expansion" is used as the event trigger words. If the event trigger word itself contains independent type information, it is determined as a type of event, and the event trigger word is directly used as the extraction template. If the event trigger word does not constitute independent type information, the collocation words whose support exceeds the threshold and the event trigger words are constructed as a type of event. The form of word combination is used as the event extraction template.

We take the "垃圾焚烧/garbage burn" event as an example to illustrate the advantages of the classification method. Firstly, through the algorithm, the output results about the verb "焚烧/burn" are as follows:

[中标/winning the bid (0.38)-环保/environmental protection (0.54)-生活/domestic (0.61)-垃圾/garbage (0.93)-**焚烧/burn**-发电/power generation (0.90)-项目/project (0.93)-投资/investment (0.45)-建设/construction (0.24)-处理/treatment (0.32)]

Since there is no combinatorial collocation around the word "焚烧/burn", the word "burn" itself is used as an event trigger word, and the word itself does not constitute independent type information. Therefore, the type is constructed together with the collocation whose support exceeds the threshold, which forms the "garbage burn" event. The extraction template is:

[. . .]垃圾/garbage [. . .]焚烧/burn [. . .]发电/power generation [. . .]项目/project [. . .]

Through the event extraction template, the announcement news events related to environmental protection can be screened out from the meaningless stock announcement news containing "burn". For example, announcement news 3 (shown below) can be excluded.

公告新闻3：辉丰股份(002496)公告，子公司华通化学收到环保局出具行政处罚决定书，要求对吡氟酰草胺项目责令限期改正，对RTO废气焚烧装置责令立即停止建设。

NEWS 3: (SZ002496) announced that Huatong Chemical, a subsidiary company, received the decision of administrative punishment issued by the Environmental Protection Bureau, demanding it to order the correction within a time limit for the pyruvic oxalamide project and order the construction of RTO waste gas burn unit to stop immediately.

At the same time, the advantages of the event extraction template compared with the word similarity calculation method and clustering method can be seen in announcements 4 and 5.

公告新闻4：绿色动力(601330)11月28日晚间公告，公司成为葫芦岛东部垃圾焚烧发电综合处理厂生活垃圾焚烧发电项目的社会资本合作方，项目估算总投资不超过6.3亿元。

NEWS 4: (SH601330) announced on the evening of November 28 that the company has become a social capital partner of the domestic garbage burn power generation project of the waste incineration power generation comprehensive treatment plant in the east of Huludao, with an estimated total investment of no more than 630 million yuan.

公告新闻5：城发环境(000885)9月26日晚间公告，公司为宜阳县生活垃圾焚烧发电项目的中标人，项目总投资3.60亿元，项目合作期30年，含2年建设期。

News 5: (SZ000885) announced on the evening of September 26 that the company is the bid winner of Yiyang domestic garbage burn power generation project, with a total investment of 360 million yuan and a cooperation period of 30 years, including a two-year construction period.

(2) If the verb or the collocation around the verb is combined to form an industry characteristic word, the verb or the combined words are used as the event trigger words, and the event trigger words are used as the event extraction template. Taking the "评价/evaluate" verb as an example, the output of the algorithm is:

仿制/Imitation (0.50)-药/medical (0.76)-药品/drug (0.41)-制药/Pharmaceutical (0.57)-通过/pass (0.67)-收到/receive (0.33)-一致性/consistency (0.78)-评价/Evaluate

The word "evaluate" itself does not constitute event type information, but it becomes a word with the characteristics of the pharmaceutical industry after being combined with the collocation word "consistency". The introduction of "consistency evaluate" in Baidu Encyclopedia is as follows:

> "Drug consistency evaluation" is a drug quality requirement in the 12th Five Year Plan for national drug safety, that is, the state requires that the imitated drugs should be consistent with the quality and efficacy of the original drugs. Specifically, it is required that the impurity spectrum is consistent, the stability is consistent, and the dissolution law in vivo and in vitro is consistent.

(3) The verbs with emotional polarity are screened, the words with clear semantics are retained as event trigger words and the event recognition template is constructed according to the trigger words. For example, emotional words such as "支持/support, 通过/pass and 指导/guide" are filtered out, and words such as "犯罪/crime" and "违纪/violation of discipline" are retained.

3.3. *Event Types of Chinese Stock Announcement News*

We used data on Chinese stock announcements collected from the EASTMONEY website (https://www.eastmoney.com/ (accessed on 21 April 2020)) from March 2015 to December 2019. A total of 59 types of events were constructed using the proposed method. After sorting and merging, 54 types of events were finally obtained. The types of events can also be optimized by evolutionary algorithms [29,30]. Table 1 shows the "put into production" event type as an example.

Table 1. "投产/Put into production" event.

Event Type:	"投产/Put into Production"
Event trigger word:	投产/Put into production
Words matching list extracted by the algorithm	期/Phase (0.33)-项目/Project (0.75)-建成/completed (0.23)-投产/put into operation
Event identification template:	[. . .]投产/Put into production [. . .]
Event example:	(SZ000952): The VB2 production line of the industrial park will be officially put into production, and the performance is expected to achieve restorative growth.

The processing flow of our model is shown in Figure 1.

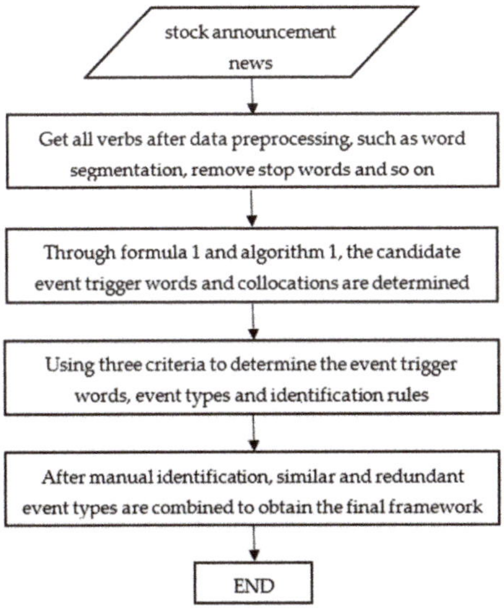

Figure 1. Processing flow of the proposed method.

4. Experimental Verification

4.1. *Data Description*

The main problem faced by experiments on event extraction methods in a specific domain is a lack of a unified corpus and type division standards. Existing studies generally label the experimental data manually and then verify the event extraction method on the labeled data set. The purpose of this section is to verify whether the classification of event types in the stock announcement news proposed by this paper is reasonable. Therefore, we first build the evaluation dataset and randomly select 60 stock announcements for each type of event, of which 30 meet the event identification template (the actual number shall prevail if less than 30). If the event identification template is in the form of a word combination, then the remaining announcement news is extracted from the announcement news that contains event trigger words but does not meet the identification template. For example, the announcement news that contains the word "焚烧/burn" is selected from the garbage burn event. If the recognition template is in the form of non-compound words, it will be randomly selected from other announcements. Finally, each evaluation sample contains $54 \times 60 = 3240$ announcements.

We select five teachers from Southwest University of Political Science and Law who hold doctoral degrees or a vice senior academic title and have more than three years of practical experience in the stock market as the evaluators. A random evaluation sample is generated for each evaluator. In the evaluation sample, an example announcement is provided for each type of event. The evaluator marks the announcement news similar to the example as 1 and those not similar as 0.

4.2. Evaluation Results

In this paper, the precision p, recall R and F values are used to calculate the results of five evaluation samples, and then the average value is taken as the final experimental result. The formal definitions of p, R, and F are as follows:

$$p = \frac{\begin{matrix} The\ number\ of\ announcements\ that\ the\ evaluators\ consider \\ consistent\ with\ the\ method\ in\ this\ paper \end{matrix}}{\begin{matrix} The\ number\ of\ announcements\ of\ this\ event\ type \\ identified\ in\ this\ paper \end{matrix}} \tag{2}$$

$$R = \frac{\begin{matrix} The\ number\ of\ announcements\ that\ the\ evaluators\ consider \\ consistent\ with\ the\ method\ in\ this\ paper \end{matrix}}{\begin{matrix} The\ number\ of\ announcements\ of\ this\ event\ type \\ maked\ as\ 1\ by\ the\ evalutor \end{matrix}} \tag{3}$$

$$F = \frac{2PR}{(P+R)} \tag{4}$$

The final experimental results are shown in Table 2.

From the overall results, all event types identified have an average p value of 0.927, R value of 0.969 and F value of 0.946, which shows that the type of announcements constructed in this paper is reasonable. From the individual point of view, the p value of some event types is poor, far lower than the average value. Through discussions with the evaluators, we found that the reasons are as follows:

1. The average p value of "signing" is 0.796, which is far lower than the average since one of the legal professional evaluators believes that there is a semantic difference between the word "签署" and the word "签订", although the two meanings are very similar.
2. The average p value of the "profit" event is 0.743, which is much lower than the average since the specific information about the "profit" event is described in detail in the sample text. Announcements that do not describe the specific information of profit in detail shifted the focus of the evaluators.
3. The average p value of "impairment" is 0.752, which is much lower than the average value. The reason is that the sample text describes the event of "asset impairment". The evaluators exclude the remaining "goodwill impairment" and "accrued (excluding asset) impairment" from the type, so the p value is low.
4. The average p value of "planning" is 0.759, which is much lower than the average since the example text contains the word "major event" in addition to the planning

trigger word. The evaluators believe that "major events" play an important role in representing the planned events, so they marked the evaluation text without "major events" as different.

Table 2. Experimental results of event types.

ID	Event Type	*p*	*R*	*F*
1	"垃圾焚烧"事件/garbage burn	0.977	0.95	0.963
2	"增资扩股"事件/Capital increase and share expansion	0.903	0.920	0.912
3	"业绩预告"事件/Performance forecast	0.910	0.947	0.928
4	"责令改正"事件/Order to correct	0.936	1.000	0.967
5	"权益分派"事件/Equity distribution	1.000	0.952	0.975
6	"股票解禁"事件/lifting the ban on stocks	0.901	1.000	0.948
7	"到期失效"事件/Expiration	1.000	0.988	0.994
8	"不确定性"事件/Uncertain	0.957	0.989	0.972
9	"届满"事件/Expiration	0.879	0.944	0.911
10	"可转换债券"事件/Convertible bond	0.925	0.966	0.945
11	"补助"事件/Subsidy	0.935	0.974	0.955
12	"犯罪"事件/Crime	0.917	0.927	0.922
13	"辞职"事件/Resignation	0.962	0.989	0.975
14	"一致性评价"事件/Consistency evaluation	0.871	1.000	0.931
15	"侦查"事件/Investigation incident	1.000	0.933	0.966
16	"违纪"事件/Violation of discipline	0.897	0.977	0.935
17	"行政处罚"事件/Administrative punishment	0.946	0.891	0.918
18	"拨付款"事件/Payment allocation	0.879	1.000	0.935
19	"投产"事件/Put into production	0.871	0.989	0.926
20	"拘留"事件/Detention	1.000	1.000	1.000
21	"盈利"事件/Profit	0.743	0.909	0.818
22	"预增"事件/Pre increase	0.978	0.940	0.959
23	"改制"事件/Restructuring	1.000	1.000	1.000
24	"减值"事件/Devaluation	0.752	0.989	0.854
25	"减持"事件/Reduction	0.968	0.968	0.968
26	"建成"事件/Completion	0.853	1.000	0.921
27	"清仓"事件/Clearance	0.849	0.939	0.892
28	"吞吐量"事件/Throughput	1.000	1.000	1.000
29	"预中标"事件/Pre bid winning	1.000	1.000	1.000
30	"转增股"事件/Conversion to share capital	0.827	0.990	0.901
31	"中标"事件/Winning the bid	1.000	1.000	1.000
32	"吸收合并"事件/Absorb merge	0.957	0.937	0.947
33	"扩建"事件/Expansion	0.882	0.978	0.927

Table 2. *Cont.*

ID	Event Type	*p*	*R*	*F*
34	"诉讼"事件/Litigation	0.957	1.000	0.978
35	"发起设立"事件/Initiate establishment	0.875	0.893	0.884
36	"投建"事件/Investment and construction	0.978	0.989	0.984
37	"罢免"事件/Recall	0.967	1.000	0.983
38	"药品临床"事件/Drug clinical	0.817	1.000	0.899
39	"筹划"事件/Planning	0.759	0.908	0.827
40	"并购"事件/Merger and acquisition	0.925	0.976	0.950
41	"转让"事件/Transfer	0.829	0.823	0.826
42	"净利"事件/Net profit	1.000	0.979	0.989
43	"补贴"事件/Subsidy	0.913	1.000	0.955
44	"收购"事件/Acquisition	0.968	0.958	0.963
45	"增持"事件/Overweight	0.989	0.924	0.956
46	"质押"事件/Pledge	0.989	0.969	0.979
47	"罚款"事件/Fine	0.975	1.000	0.988
48	"违法"事件/Illegal	0.914	1.000	0.955
49	"冻结"事件/Freeze	1.000	1.000	1.000
50	"签署签订"事件/Signing	0.796	1.000	0.886
51	"回购"事件/Repurchase	0.978	0.989	0.984
52	"出售"事件/Sale	1.000	0.990	0.995
53	"设立公司"事件/Establishment of company	0.925	0.943	0.934
54	"股票激励"事件/Stock incentive	0.968	0.949	0.959
	Total	*0.927*	*0.969*	*0.946*

4.3. Comparison to Existing Results

At present, there have been few studies focusing on event extraction from stock announcements [12,19,21], and more studies are focusing on event extraction from financial news. We select some representative related studies and list them in Table 3 for comparison. We roughly divide the methods of generating the event type framework into two categories: full-manual and semi-manual. Full-manual means that the event type framework is completely determined by domain experts; semi-manual means a combination of some model or algorithm and manual identification.

Due to a lack of a unified event type standard and framework for stock announcements, we cannot compare our work with the existing related studies with numerical indicators. Based on the fact that we have built a fine-grained event type framework, we mainly compare our work with [18,19,21].

The model proposed in [18] identifies 11 types and 41 sub-types of events from the business domain. The p value is 0.95, and the F value is 0.79. References [19,21] both use full-manual methods to determine the event types, and both focus on the stock announcement news. The event type framework in [19] includes 12 types and 30 sub-types of events. Reference [21] includes 4 types and 34 sub-types of events. The p value in [19] is 0.673, and the F1 value is 0.60. The p value in [21] is 0.967, and the F value is missing in the paper. The p value of our work is 0.927, which is lower than those of [18,21] but higher than that of [19]. The F value of our work is 0.946, higher than [18,19].

Table 3. The results of related studies.

Source	Method	Event Type Framework
ACE event typology	full-manual	1 type: Business 4 subtypes: Start-org, Merge-org, Declare bankruptcy and End-org
The Stock Sonar project [12]	semi-manual	8 types: Legal, Analyst Recommendation, Financial, Stock Price Change, Deals, Mergers and Acquisitions, Partnerships, Product and Employment
BEECON [18]	semi-manual	11 types: Analyst Event, Bankruptcy, Company Basic Information Change, Company Collaboration, Company Growth, Product Event, etc. 41 subtypes: reorganizations and changes in employment, company changing its stock listing, name or accounting procedures, debt financing, etc.
He [13]	semi-manual	3 types: Financial Policy Events, Monetary Policy Events and Market Rule Adjustment multiple subtypes: tax rate adjustment, deposit and loan interest rate adjustment, national debt adjustment, etc.
Wang [14]	semi-manual	2 types: Macro-Events, Individual Stock Event 6 subtypes: policy events, social emergencies, mergers and acquisitions, profitability, personnel changes and refinancing
Chen [15]	full-manual	8 types: Major contracts, Raw Materials, Major Conferences, Company Financial Statements, Major Policies, Mergers and Acquisitions, Personnel Changes and Additional Allotments
Han et al. [16]	full-manual	8 types: Product Transformation, Equity Change, Share Price Movement, Personnel Changes, Financial Status, etc. 16 subtypes: win bidding, shareholding increase, stock suspension, profit, debt, etc.
Zhang [19]	full-manual	12 types: Major Events, Major Risks, Shareholding Changes, Capital Changes, Emergencies, Special Treatment, etc. 30 subtypes: enterprise cooperation, product release, senior management change, reorganization and merger, government support, etc.
Boudoukh et al. [17]	full-manual	18 types: Business Trend, Deal, Employment, Financial, Mergers and Acquisitions, Earnings Factors, Ratings, Legal, Product, Investment, etc.
Wu [20]	semi-manual	13 types: Issuance, Dividend, Event Prompt, Pledge, Performance Notice, Suspension and Resumption of Trading, Fund-Raising, etc.
Zhou [21]	full-manual	4 types: Share Change, Debt, Market Transaction, Enterprise Change 34 subtypes: senior management change, performance change, product release, related party transactions, equity auction, debt overdue, etc.

From the classification results, the fine-grained event type framework built in this paper finds some reasonable and valuable event types that have not been discussed yet. An example is the violation of company policies and violation of the law discussed in the previous section. Another interesting example is that we built an event called throughput, which is usually issued by listed companies in the airline or port sector. According to our knowledge, this event type has not been discussed in existing studies, which only list a related type called "performance change". Technically, the throughput event is actually a sub-type of the performance change. Performance change news is usually announced on a quarterly, semi-annual and annual basis, and thus cannot reflect short-term changes. Unlike listed companies in other sectors, throughput events usually involve the company's main business. For example, Air China's (SH601111) passenger transport business achieved a revenue of 58.317 billion yuan in 2021, accounting for 78.24% of the operation revenue; CMB Shekou's (SZ001872) port business accounted for 95.76% of its revenue in 2021.

Due to various limitations, the event type framework constructed in this paper cannot be directly compared with those of other studies. However, through the analysis of the results we did find some event types that have not been discussed in the existing literature, and these types are effective and reasonable. Therefore, we can say that the event type framework constructed by the method proposed in this paper enriches the existing research, and thus it has certain value and significance.

5. Filtering of Event Types

In this section, we did not choose to conduct stock return prediction in the traditional way since we think it is inappropriate to rely solely on stock announcement news. Instead, based on the unilateral trading policy of the Chinese stock market, we filtered out some event types that are not valuable to investors. Our approach is that after the announcement news is released, we enter the market and sell at the highest price in the short term. Although in real life, the possibility of selling at the highest price is very low, here we are describing the best-case scenario. We believe that if in the ideal situation the return obtained based on a certain event type is small or the probability is low, combined with the unilateral trading policy, we think such an event type is not valuable to investors.

5.1. Return Calculation Method

If the official announcement time of the event is between the opening in the morning of the day and the closing in the afternoon of the day, mark the day as $t = 1$. Otherwise, mark the first trading day after the event as $t = 1$. We consider the best return and probability of entering the market at $t = 1$ and selling stocks at $t = 2$ or $t = 3$. This paper selects three kinds of entry prices: opening price, closing price and the highest price in the worst case, and sells the stock at the highest price of the day at $t = 2$ or $t = 3$ to obtain the return. Although the probability of selling at the highest price is small in reality, the purpose of this paper is to provide a reference for investors according to historical data based on the probability of obtaining the best return value. If the return value or proportion is low under the best-case scenario, this indicates that making decisions based on this kind of news is risky and has no investment value. The calculation of the best return obtained at three entry prices is as follows:

$$DRET_{OH2} = (Hipr_{T=2} - Oppr_{T=1})/Oppr_{T=1} \tag{5}$$

$$DRET_{HH2} = (Hipr_{T=2} - Hipr_{T=1})/Hipr_{T=1} \tag{6}$$

$$DRET_{CH2} = (Hipr_{T=2} - Clpr_{T=1})/Clpr_{T=1} \tag{7}$$

$DRET_{OH2}$ refers to the return from selling stocks at the highest price at $t = 2$ after entering the market at $t = 1$; similarly, $DRET_{OH3}$ is expressed as the return of entering the market at $t = 1$ and selling stocks at the highest price at $t = 3$. $DRET_{HH2}$ and $DRET_{CH2}$ represent the best return when entering the market at the highest price and closing price of the day at $t = 1$ and when selling stocks at the highest price at $t = 2$, respectively.

5.2. Investment Results

According to the 54 types of announcement events constructed in this paper, the transaction data within the time span from March 2015 to December 2019 are selected. In this period, by using the above return calculation method, the results of some types of investment return are shown in Table 4.

Table 4. Investment return for some event types.

Event Type	Selling Time t	Purchase Price: Opening Price			Purchase Price: Highest Price			Purchase Price: Closing Price			Sample Sizes
		Probability of Positive Return	Average Return	Variance	Probability of Positive Return	Average Return	Variance	Probability of Positive Return	Average Return	Variance	
Capital increase and share expansion	2	69.4%	3.3%	0.4%	42.2%	0.3%	0.3%	78.2%	2.6%	0.2%	147
	3	61.2%	3.2%	1.1%	42.2%	0.2%	0.9%	63.3%	2.5%	0.7%	147
	The experimental results show that the best investment scheme for such events is to buy at the closing price and sell on the second day. It has a probability of 78.2% and can obtain a positive return with an average of 2.6%.										
Expiration	2	77.1%	3.1%	0.1%	62.9%	0.7%	0.1%	91.4%	2.5%	0.1%	35
	3	91.4%	3.4%	0.1%	77.1%	1.0%	0.1%	91.4%	2.8%	0.1%	35
	The experimental results show that the best investment scheme for such events is to buy at the opening price and sell on the third day, with a probability of 91.4% and a positive return with an average of 3.4%; the second best investment scheme is to buy at the closing price and sell on the third day, with a probability of 91.4% and a positive return with an average of 2.8% or to sell on the second day with a probability of 91.4% and a positive return with an average of 2.5%.										
Restructuring	2	71.6%	1.0%	0.3%	33.8%	−1.4%	0.2%	79.7%	0.8%	0.1%	74
	3	58.1%	−0.1%	0.6%	35.1%	−2.5%	0.4%	51.4%	−0.4%	0.3%	74
	It can be seen from the experimental results that the positive average return of this kind of event sample is low. Therefore, such events have no investment value.										
Throughput	2	68.8%	1.5%	0.1%	42.5%	0.1%	0.1%	83.8%	1.4%	0.0%	80
	3	60.0%	1.5%	0.2%	42.5%	0.2%	0.2%	70.0%	1.4%	0.1%	80
	The experimental results show that the best investment scheme for such events is to buy at the closing price and sell on the second day, with a probability of 83.8% and a positive return with an average of 1.4%.										
Conversion to share capital	2	64.4%	2.3%	0.5%	42.5%	0.1%	0.4%	78.3%	2.9%	0.3%	811
	3	59.4%	2.5%	1.0%	45.4%	0.3%	1.0%	67.2%	3.0%	0.8%	811
	The experimental results show that the best investment scheme for such events is to buy at the closing price and sell on the second day. It has a probability of 78.3% and can obtain a positive return with an average of 2.9%.										
Winning the bid	2	66.7%	1.5%	0.2%	36.3%	−0.5%	0.1%	79.6%	1.5%	0.1%	1990
	3	59.6%	1.3%	0.4%	37.3%	−0.7%	0.3%	64.2%	1.3%	0.3%	1990
	The experimental results show that the best investment scheme for such events is to buy at the closing price and sell on the second day. It has a probability of 79.6% and can obtain a positive return with an average of 1.5%.										
Subsidy	2	73.8%	2.3%	0.2%	42.1%	0.2%	0.1%	82.2%	1.9%	0.1%	107
	3	67.3%	2.3%	0.3%	40.2%	0.2%	0.2%	70.1%	2.0%	0.2%	107
	The experimental results show that the best investment scheme for such events is to buy at the closing price and sell on the second day, with a probability of 82.2% and a positive return with an average of 1.9%.										

Table 4. *Cont.*

Event Type	Selling Time t	Purchase Price: Opening Price			Purchase Price: Highest Price			Purchase Price: Closing Price			Sample Sizes
		Probability of Positive Return	Average Return	Variance	Probability of Positive Return	Average Return	Variance	Probability of Positive Return	Average Return	Variance	
Acquisition	2	64.5%	2.3%	0.5%	42.0%	0.0%	0.4%	73.8%	2.3%	0.3%	3555
	3	58.6%	2.3%	1.1%	42.0%	0.0%	1.0%	60.0%	2.3%	0.9%	3555
	The experimental results show that the best investment scheme for such events is to buy at the closing price and sell on the second day, with a probability of 73.8% and a positive return with an average of 2.3%.										
Overweight	2	72.8%	3.2%	0.4%	43.1%	0.2%	0.2%	81.7%	2.6%	0.2%	3268
	3	66.7%	3.3%	0.7%	43.5%	0.3%	0.5%	67.1%	2.7%	0.5%	3268
	The experimental results show that the best investment scheme for such events is to buy at the closing price and sell on the second day, with a probability of 81.7% and a positive return with an average of 2.6%.										
Illegal	2	65.2%	1.5%	0.4%	35.0%	−1.2%	0.2%	70.5%	0.8%	0.2%	397
	3	60.5%	1.5%	0.8%	39.3%	−1.2%	0.5%	63.2%	0.8%	0.5%	397
	It can be seen from the experimental results that although the probability of obtaining a positive return is 70.5%, the average positive return is small. Therefore, on the whole, such events are not good for investment.										
Signing	2	65.7%	2.1%	0.3%	39.2%	−0.2%	0.2%	77.4%	2.1%	0.1%	1809
	3	58.5%	1.9%	0.6%	40.3%	−0.5%	0.6%	64.0%	1.8%	0.5%	1809
	The experimental results show that the best investment scheme for such events is to buy at the closing price and sell on the second day, with a probability of 77.4% and a positive return with an average of 2.1%.										
Stock incentive	2	72.1%	2.9%	0.3%	40.4%	0.0%	0.2%	81.9%	2.3%	0.1%	408
	3	66.4%	2.8%	0.7%	42.9%	−0.2%	0.5%	67.4%	2.1%	0.5%	408
	The experimental results show that the best investment scheme for such events is to buy at the closing price and sell on the second day, with a probability of 81.9% and a positive return with an average of 2.3% or buy at the opening price and sell on the second day, with a probability of 72.1% and a positive return with an average of 2.9%.										

Due to space limitations, we only list the results of several event types in Table 4. Eight event types have low returns or probabilities, even under the best conditions. Among these eight types of events, the event types with the smallest benefit value are "Illegal" and "Restructuring" which only come with an average positive return of 0.8%. The remaining events in the order of return from small to large are: "Expiration" (0.9%), "order to correct" (1.0%), "Resignation" (1.1%), "Recall" (1.1%), "Freeze" (1.3%) and "Planning" (2.7%).

The event type with the smallest probability value is "Freeze" (68.8%). The remaining events in the order of probability from small to large are: "Planning" (69%), "Illegal" (70.5%), "Recall" (75%), "Resignation" (75.9%), "Expiration" (78%), "order to correct" (78.9%), and "Restructuring" (79.7%).

6. Conclusions

Stock announcements contains much information about all aspects of a company, which are important for investors and stock forecasting. It is difficult to determine the event types from stock announcements. As there is no unified classification standard, existing studies have constructed various event type frameworks based on domain experts' experience, Clustering, ontology and other methods. Some studies have resulted in a fine-grained classification framework. However, we believe that there is still room for improvement on the basis of the existing research (e.g., the abovementioned violations of laws and violations of company policy events). Based on different punishments and expectations for investors, we think that it is more reasonable to classify events into different types rather than into one type in the manner of the extant literature.

In order to obtain more detailed event types of stock announcement news, we proposed a two-step method. First, all verbs extracted from the announcement news are used as candidate event triggers. Due to the common expressions in Chinese announcement news, if there is an event type, it usually has a conventional expression form. On the contrary, if a candidate event trigger word (verb) does not suggest an event type, the expression of the news containing the verb is chaotic. Therefore, we combine co-occurrence words with the candidate event trigger words and express them in an ordered sequence of words. Then, we use three proposed criteria to determine the final event types.

Based on real data on the Chinese stock market, we finally constructed 54 event types from the announcement news. The verification results of the constructed event types ($p = 0.927, f = 0.946$) show that it is reasonable and consistent with people's cognition. Further, we compare our work with other similar studies (summarized in Table 3). First of all, most of the existing studies focus on the event types in the financial news, and only regard the stock announcement news as part of this greater whole. Therefore, the event type frameworks built are usually rough. Then, we compared our results with those of [18,19,21], which also constructed fine-grained event types from stock announcement news. The p value of our work is lower than those of [18] (0.95) and [21] (0.967) but higher than that of [19] (0.673). The F value of our work is higher than those of [18] (0.79) and [19] (0.6). From the results of the constructed event types, our method has found some reasonable and valuable event types that have not been discussed yet. For example, an event type named "throughput" is constructed in this paper. To the best of our knowledge, this is the first of its kind, and only one similar event type called "performance change" can be found in the existing research. In the Chinese stock market, companies usually release quarterly, semi-annual or annual performance change news, so this method cannot reflect short-term changes. "Throughput" events are released by airline or port sector stocks. Unlike in other stock sectors, a "throughput" event is usually the main business. For example, CMB Shekou's (SZ001872) port business accounted for 95.76% of its revenue in 2021. "Throughput" events can reflect the short-term performance changes of these companies and are valuable for investors.

In conclusion, our research on event extraction from stock announcements has enriched the existing literature, so it is of value and significance.

After constructing a fine-grained announcement news event type framework, we did not choose to conduct stock prediction work in a traditional way since we believe that it is inappropriate to consider announcements without other types of news, such as industry news and macroeconomic news (as has been proven in the literature). Instead, based on the unilateral trading policy of the Chinese stock market, we screen out some event types that are not valuable to investors according to their performance under the best-case scenario. We did not carry out a precise calculation here but consider the event performance under special circumstances.

Author Contributions: Formal analysis, F.M.; investigation, P.W.; methodology, F.M., Y.X. and W.L.; software, F.M. and H.J.; supervision, F.M.; writing—original draft, F.M., P.W., Y.X., H.J. and W.L.; writing—review and editing, F.M., Y.X., P.W., H.J. and W.L. All authors have read and agreed to the published version of the manuscript.

Funding: This research received no external funding.

Data Availability Statement: The data presented in this study are available on request from the corresponding author.

Conflicts of Interest: The authors declare no conflict of interest.

References

1. Yi, Z. Research on Deep Learning Based Event-Driven Stock Prediction. Ph.D. Thesis, Harbin Institute of Technology, Harbin, China, 2019.
2. Yang, H.; Chen, Y.; Liu, K.; Xiao, Y.; Zhao, J. DCFEE: A document-level Chinese financial event extraction system based on automatically labeled training data. In Proceedings of the ACL 2018, System Demonstrations, Melbourne, Australia, 15–20 July 2018; pp. 50–55.
3. Linghao, W.U. Empirical Analysis of the Impact of Stock Events on Abnormal Volatility of Stock Prices. Master's Thesis, Huazhong University of Science and Technology, Wuhan, China, 2019.
4. Balali, A.; Asadpour, M.; Jafari, S.H. COfEE: A Comprehensive Ontology for Event Extraction from text. *arXiv* **2021**, arXiv:2107.10326. [CrossRef]
5. Guda, V.; Sanampudi, S.K. Rules based event extraction from natural language text. In Proceedings of the 2016 IEEE International Conference on Recent Trends in Electronics, Information & Communication Technology (RTEICT), Bangalore, India, 20–21 May 2016; pp. 9–13.
6. Ritter, A.; Etzioni, O.; Clark, S. Open domain event extraction from twitter. In Proceedings of the 18th ACM SIGKDD International Conference on Knowledge Discovery and Data Mining, Beijing, China, 12–16 August 2012; pp. 1104–1112.
7. Fung GP, C.; Yu, J.X.; Lam, W. Stock prediction: Integrating text mining approach using real-time news. In Proceedings of the 2003 IEEE International Conference on Computational Intelligence for Financial Engineering, Hong Kong, China, 20–23 March 2003; pp. 395–402.
8. Wong, K.F.; Xia, Y.; Xu, R.; Wu, M.; Li, W. Pattern-based opinion mining for stock market trend prediction. *Int. J. Comput. Processing Lang.* **2008**, *21*, 347–361. [CrossRef]
9. Du, M.; Pivovarova, L.; Yangarber, R. PULS: Natural language processing for business intelligence. In Proceedings of the 2016 Workshop on Human Language Technology, New York, NY, USA, 9–15 July 2016; pp. 1–8.
10. Chen, C.; Ng, V. Joint modeling for Chinese event extraction with rich linguistic features. In Proceedings of the COLING 2012, Mumbai, India, 8–15 December 2012; pp. 529–544.
11. Liu, L. Heterogeneous Information Based Financial Event Detection. Ph.D. Thesis, Harbin Institute of Technology, Harbin, China, 2010.
12. Feldman, R.; Rosenfeld, B.; Bar-Haim, R.; Fresko, M. The stock sonar—sentiment analysis of stocks based on a hybrid approach. In Proceedings of the AAAI Conference on Artificial Intelligence, San Francisco, CA, USA, 7–11 August 2011; pp. 1642–1647.
13. He, Y. Research on Ontology-Based Case Base Building and Reasoning for Theme Events in the Stock Markets. Master's Thesis, Hefei University of Technology, Hefei, China, 2017.
14. Wang, Y. Research on Financial Events Detection by Incorporating Text and Time-Series Data. Master's Thesis, Harbin Institute of Technology, Harbin, China, 2015.
15. Chen, H. Research and Application of Event Extraction Technology in Financial Field. Ph.D. Thesis, Beijing Institute of Technology, Beijing, China, 2017.
16. Han, S.; Hao, X.; Huang, H. An event-extraction approach for business analysis from online Chinese news. *Electron. Commer. Res. Appl.* **2018**, *28*, 244–260. [CrossRef]
17. Boudoukh, J.; Feldman, R.; Kogan, S.; Richardson, M. Information, trading, and volatility: Evidence from firm-specific news. *Rev. Financ. Stud.* **2019**, *32*, 992–1033. [CrossRef]

18. Arendarenko, E.; Kakkonen, T. Ontology-based information and event extraction for business intelligence. In Proceedings of the 2012 International Conference on Artificial Intelligence: Methodology, Systems, and Applications, Varna, Bulgaria, 13–15 September 2012; pp. 89–102.
19. Zhang, W. Research on key technologies of event-driven stock market prediction. Ph.D. Thesis, Harbin Institute of Technology, Harbin, China, 2018.
20. Turchi, M.; Zavarella, V.; Tanev, H. Pattern learning for event extraction using monolingual statistical machine translation. In Proceedings of the International Conference Recent Advances in Natural Language Processing 2011, Hissar, Bulgaria, 10–16 September 2011; pp. 371–377.
21. Zhou, X. Research on Financial Event Extraction Technology Based on Deep Learning. Ph.D. Thesis, University of Electronic Science and Technology of China, Chengdu, China, 2020.
22. Wang, Y.; Luo, S.; Hu, Z.; Han, M. A Study of event elements extraction on Chinese bond news texts. In Proceedings of the 2018 IEEE International Conference on Progress in Informatics and Computing (PIC), Suzhou, China, 14–16 December 2018; pp. 420–424.
23. Ding, P.; Zhuoqian, L.; Yuan, D. Textual information extraction model of financial reports. In Proceedings of the 2019 7th International Conference on Information Technology: IoT and Smart City, Shanghai, China, 20–23 December 2019; pp. 404–408.
24. Wang, A. Study on the Impact of Change on Interest Rate on Real Estate Listed Companies Stock Price. Master's thesis, Southwestern University of Finance and Economics, Chengdu, China, 2012.
25. Jinmei Zhao Yu Shen Fengyun, W.U. Natural disasters, man-made disasters and stock prices: A study based on earthquakes and mass riots. *J. Manag. Sci. China* **2014**, *17*, 19–33.
26. Yi, Z.; Lu, H.; Pan, B. The impact of Sino US trade war on China's stock market—An analysis based on event study method. *J. Manag.* **2020**, *33*, 18–28.
27. Li, T.; Qian, Z.; Deng, W.; Zhang, D.; Lu, H.; Wang, S. Forecasting crude oil prices based on variational mode decomposition and random sparse Bayesian learning. *Appl. Soft Comput.* **2021**, *113*, 108032. [CrossRef]
28. Brandt, M.; Gao, L. Macro fundamentals or geopolitical events? A textual analysis of news events for crude oil. *J. Empir. Financ.* **2019**, *51*, 64–94. [CrossRef]
29. Li, T.; Shi, J.; Deng, W.; Hu, Z. Pyramid particle swarm optimization with novel strategies of competition and cooperation. *Appl Soft Comput.* **2022**, *121*, 108731. [CrossRef]
30. Deng, W.; Shang, S.; Cai, X.; Zhao, H.; Zhou, Y.; Chen, H.; Deng, W. Quantum differential evolution with cooperative coevolution framework and hybrid mutation strategy for large scale optimization. *Knowl.-Based Syst.* **2021**, *224*, 107080. [CrossRef]

electronics

Article

Hybrid Graph Neural Network Recommendation Based on Multi-Behavior Interaction and Time Sequence Awareness

Mingyu Jia, Fang'ai Liu *, Xinmeng Li and Xuqiang Zhuang

School of Information Science and Engineering, Shandong Normal University, Jinan 250358, China
* Correspondence: lfa@sdnu.edu.cn

Abstract: In recent years, mining user multi-behavior information for prediction has become a hot topic in recommendation systems. Usually, researchers only use graph networks to capture the relationship between multiple types of user-interaction information and target items, while ignoring the order of interactions. This makes multi-behavior information underutilized. In response to the above problem, we propose a new hybrid graph network recommendation model called the User Multi-Behavior Graph Network (UMBGN). The model uses a joint learning mechanism to integrate user–item multi-behavior interaction sequences. We designed a user multi-behavior information-aware layer to focus on the long-term multi-behavior features of users and learn temporally ordered user–item interaction information through BiGRU units and AUGRU units. Furthermore, we also defined the propagation weights between the user–item interaction graph and the item–item relationship graph according to user behavior preferences to capture more valuable dependencies. Extensive experiments on three public datasets, namely MovieLens, Yelp2018, and Online Mall, show that our model outperforms the best baselines by 2.04%, 3.82%, and 3.23%.

Keywords: multi-behavior recommendation; sequential recommendation; graph neural network; embedding propagation

Citation: Jia, M.; Liu, F.; Li, X.; Zhuang, X. Hybrid Graph Neural Network Recommendation Based on Multi-Behavior Interaction and Time Sequence Awareness. *Electronics* **2023**, *12*, 1223. https://doi.org/10.3390/electronics12051223

Academic Editor: Alberto Fernandez Hilario

Received: 25 December 2022
Revised: 27 February 2023
Accepted: 2 March 2023
Published: 3 March 2023

1. Introduction

Recommendation systems have been widely used in various Internet business services in the era of big data. The recommendation model can recommend products that match its users for various businesses and find suitable user groups for enterprises [1]. In order to better personalize recommendations for each user, it is crucial to fully understand the interests and behavioral preferences of users. For sales platforms, understanding users' purchasing interests and behavioral preferences can increase sales and profit margins. For the user him/herself, identifying the user's shopping interests and behavioral preferences on the client side can improve the user's experience and save unnecessary browsing time. The early popular collaborative filtering algorithm (CF) decomposes a single user–item interaction into latent representations for finding similar users and related items and then predicting the next user behavior [2,3]. However, since traditional CF cannot model user attributes and item auxiliary information, there are data-sparsity and cold-start problems in practical application scenarios [4]. To address these issues, supervised learning (SL) models such as Factorization Machine (FM) [5] and NFM (Neural FM) [6] have emerged one after another. With the development of neural network techniques, collaborative filtering architectures for enhancing nonlinear feature interactions utilize multilayer perceptrons to handle advanced nonlinear relationships, such as NCF [7] and DMF [8].

In recent years, deep neural networks based on graph data have received extensive attention, showing good results in processing high-dimensional sparse user interaction data. These neural network structures, called graph neural networks [9,10], are used to learn meaningful representations in graph data structures. Since user–item interactions are often sparse non-Euclidean data, graph data structures can be used to store their interactions. In addition, the introduction of external Knowledge Graph (KG) data can expand the additional information about users and items [11]. This provides a feasible solution for improving the accuracy and interpretability of recommendation systems. Given the strong performance of graph neural networks in aggregating and propagating graph-structured data, it provides an unprecedented opportunity to improve the performance of recommendation systems.

However, recommendation systems based on graph neural networks also face many problems: (1) Different graph data provide user and item information from different perspectives. How to aggregate and learn more accurate node representations from different types of graphs is crucial for recommendation models [12]. (2) The connections between nodes are diverse rather than single [13]. The assignment of weights to different connection methods requires more consideration. (3) Graph neural networks show good performance in learning the relationships between nodes. However, it is difficult for them to process sequence information [14]. Therefore, it is worth considering how to incorporate temporal information into the model. In this paper, our research question is how to utilize multi-behavior interaction time-series information for an accurate recommendation.

Because of the limitations of existing graph network methods, it is crucial to develop a hybrid graph neural network model that focuses on user behavioral characteristics and user–item interaction habits. Therefore, we designed a user multi-behavior awareness module and an item-information-relation module based on the graph neural network. Specifically, we propose a new method called the User Multi-Behavior Graph Neural Network (UMBGN) Hybrid Model, which has four sections. (1) User–item connection weight calculation: It provides unique weight information for each edge to describe the connection relationship between nodes according to the multi-behavior interaction information between users and items. (2) User–item graph network information transfer: It aggregates the feature information of the node's neighbors according to the edge weights to obtain the final feature representation. (3) Information perception based on user behavior sequence: It uses a behavior-aware network module with bidirectional GRU and AUGRU to enrich the user's behavioral information representation, fully considering the user's behavioral characteristics. (4) Information aggregation between items. It aggregates user–item interaction information by using an attention mechanism and considers the order of interactions between items. Compared with traditional graph network models, our model computes weight information between nodes according to different behavioral interactions. This allows for a more accurate dissemination of information between neighboring nodes. Furthermore, compared with the existing state-of-the-art graph neural network recommendation models, our proposed method introduces user multi-behavior sequential information perception, achieving more accurate recommendation performance. This benefits from the fact that our model considers not only the global nature of multi-behavior interactions but also each user's personality. Therefore, the contributions of the paper can be summarized as follows:

(1) We constructed a user multi-behavior awareness module with bidirectional GRU and AUGRU to enrich user-behavior-information representation. We input the user's interaction with items into the network in chronological order to obtain the user's interaction behavior feature vector, which helps us understand the user's behavioral preferences. Then we integrate the interaction behavior feature vector with the user's feature vector to more accurately locate the user's interest.

(2) We propose the connection weights between user–item nodes by focusing on user–item multi-behavior interaction information to make information aggregation and dissemination more accurate. In addition, we design an item-information relation module based on the user's dependencies on items. Then we use the attention mechanism to aggregate the item–item connections information to further enrich the embedding representation of items.

(3) The experiments performed on three real datasets indicate that our UMBGN model achieves significant improvements over existing models. In addition, we also

extensively studied the overall impact of different modules on the experiments to prove the effectiveness of our method.

2. Methods

In this section, we elaborate on our method; the basic architecture is shown in Figure 1. Our model consists of four modules: (1) a user–item interaction information module, which mines user–item multi-behavior interaction information; (2) a user multi-behavior awareness module, which further learns the strength of each user interaction behavior and extracts long-term user behavior preference; (3) an item-information-relation module, which, according to the user–item interaction information, calculates the information of other items related to an item; and (4) a joint prediction module, which combines the information of each module to obtain the final output result.

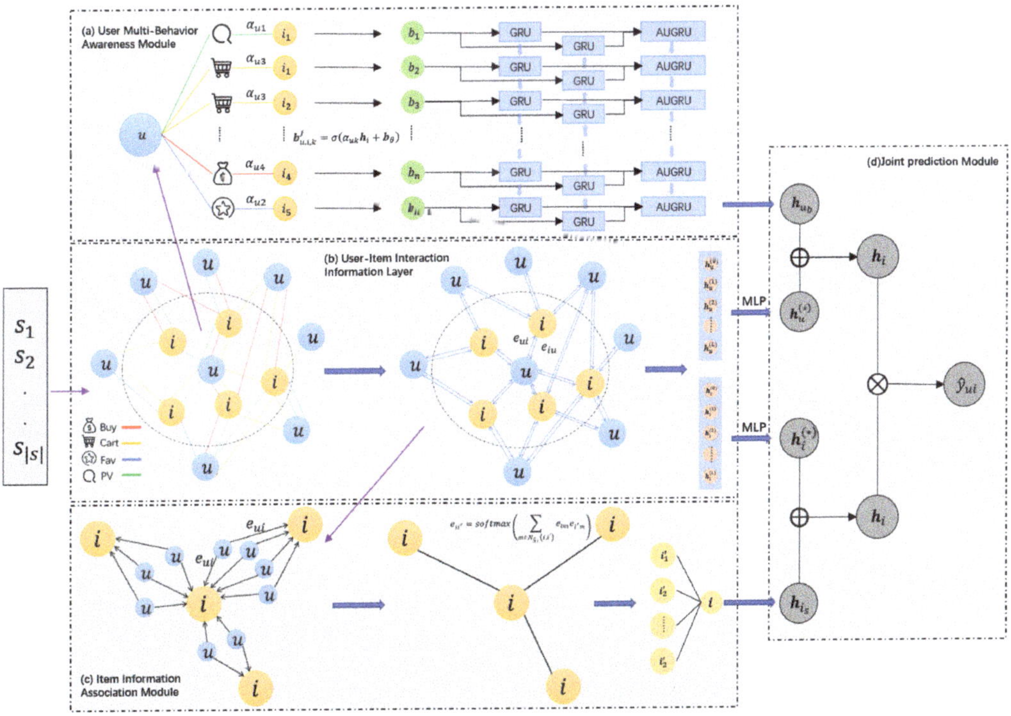

Figure 1. The framework of the UMBGN model. It contains four modules: module (**a**) is used to extract user multi-behavior interaction information, module (**b**) is used to extract user long-term multi-behavior preferences, module (**c**) is used to extract association information between items, and module (**d**) is used to output the result.

2.1. Symbol Description

We use the set $U = \{u_1, u_2, \ldots, u_m\}$ to represent user information, where m is the total number of users. Similarly, we use set $I = \{i_1, i_2, \ldots, i_n\}$ to represent item information, where n is the total number of items. The set $K = \{k_1, k_2, \ldots, k_L\}$ is used to represent user interactions with items (e.g., favorites, purchases, and clicks).

User–item interaction sequence: In the recommendation scenario, we usually obtain the historical sequence of user–item interactions and the time information of their interactions, defined as $S = \left\{s_1, s_2, \ldots, s_{|S|}\right\}$. Moreover, $s_i = (u, i, k, t)$ indicates that user u interacts with item i through behavior k at time t.

Input: User–item multi-behavior interaction sequence $S = \left\{s_1, s_2, \ldots, s_{|S|}\right\}$.

Output: The probability, $\hat{y}_{(p,q)}$, that user u_p interacts with item i_q, with which he/she has no interaction.

2.2. Preliminary Preparation

2.2.1. Generation of the Bipartite Graph of User–Item Interactions

Our task is to use various interaction information to make recommendations for target users. According to Zhang et al. [15,16], the interaction information between users and items is sparse with non-Euclidean data, and building a knowledge graph can better represent the relationship between them. Therefore, we generate an extended user–item interaction graph $\mathcal{G}_0 = (\mathcal{V}_0, \mathcal{E}_0)$, using the user–item interaction data $S = \left\{s_1, s_2, \ldots, s_{|S|}\right\}$, where node $\mathcal{V}_0$ consists of user node $u \in U$ and item node $i \in I$. Similar to existing graph models in [17], the d-dimensional vectors $\mathrm{p_u}$ and $\mathrm{p_i}$ are used to represent the user and item embeddings. The edge, set $\mathcal{E}_0$, is a two-tuple composed of interaction type, $k \in K$, and timestamp information, t, denoted as $\mathcal{E}_0 = (k, t)$. Different edges represent different behaviors. This can help extract behavior-based information between users and items.

2.2.2. User's Behavior Interaction Information Extraction

Traditional graph network recommendation often only focuses on the users' single-behavior interaction information, ignoring the influence of edge sets on information dissemination in the graph network. In this paper, we add a certain weight to the edge set of the graph network according to the user–item interaction behavior. This optimizes the process of information transfer in the graph network. We consider two factors that affect user–item interaction preferences: the relative importance of interactions and the temporal order of interactions. On the one hand, users have their own unique interactive behavior habits. For example, user u_1 likes to favorite items, but user u_2 prefers to put the favorite items in the shopping cart. Then their unique behaviors have different relative importance. On the other hand, items also have unique interactions with users. For example, item i_1 is usually favorited by users, but item i_2 is usually added to the shopping cart by users. Therefore, we design different interactive behavior weights, α_{uk} and α_{ik}, between users and items:

$$\alpha_{uk} = \frac{w_k^u \cdot n_{uk}}{\sum_{m \in N(u)} w_m^u \cdot n_{um}}; \; \alpha_{ik} = \frac{w_k^i \cdot n_{ik}}{\sum_{m \in N(i)} w_m^i \cdot n_{um}}, \tag{1}$$

where w_k^u and w_k^i are learnable parameters, representing the degree of influence of users and items on behavior, k; n_{uk} represents the number of items that user u interacts with through type k; n_{ik} represents the number of users that user i interacts with through type k; $N(u)$ represents all items interacting with user u; and $N(i)$ represents all users interacting with the item i.

2.3. User–Item Multi-Behavior Interaction Information Transfer and Aggregation

2.3.1. Construction of User–Item Relationship Graph

We not only pay attention to local relations but also global interaction relations to learn user–item interaction multi-behavior information. According to the user–item interaction graph, $\mathcal{G}_0$, we calculate the weight of the edge set through normalization and obtain the connection strength information, e_{ui} and e_{iu}, between every two points:

$$e_{ui} = \sigma\left(\sum_{k \in N(u,i)} \alpha_{uk} + b\right); \; e_{iu} = \sigma\left(\sum_{k \in N(i,u)} \alpha_{ik} + b\right), \tag{2}$$

where σ is the sigmoid function, b is the bias, and $N(i,u)$ is the sum of the interaction types between i and u. Then $e_{ui}, e_{iu} \in \mathcal{E}_1$, and point set $\mathcal{V}_0$ are combined to obtain the user–item bidirectional relationship graph, $\mathcal{G}_1 = (\mathcal{V}_0, \mathcal{E}_1)$. Compared with the traditional undirected graph network, the bidirectional graph network with weight information has better performance in information transmission.

2.3.2. Information Dissemination of User–Item Relationship Graph

It is an effective method to use graph neural networks to analyze graph data structures [9,10]. These networks used an iterative message aggregation method to mine structural information within node neighborhoods. According to the method of Xiang et al. [18], our graph network has a total of L layers and follows their aggregation and propagation method. Firstly, the nodes in the graph network aggregate the information of their neighbor nodes in the previous layer. Then they update themselves by combining the aggregated information with their original information. Different from Xiang et al., we designed a propagation weight according to the connection strength of nodes in order to achieve a better information transmission effect. Specifically,

$$h_u^{(l)} = \varphi\left(W_1 h_u^{(l-1)} + \sum_{i \in N(u)} \lambda_u^i W_2 h_i^{(l-1)}\right), \tag{3}$$

where $h_u^{(l)} \in R^d$ is the user's embedding in the l-th layer, $h_i^{(l)} \in R^d$ is the item's embedding in the l-th layer; $h_u^{(0)} = p_u, h_i^{(0)} = p_i$; φ represents the LeakyReLU function for information transformation; and W_1 and $W_2 \in R^{d \times d}$ are learnable weight matrices. Moreover, λ_u^i is the attention coefficient of user u to item i, and its calculation formula is as follows:

$$\lambda_u^i = \frac{exp(e_{ui})}{exp\left(\sum_{j \in N(u)} e_{uj}\right)}. \tag{4}$$

Similarly, we can obtain the l-th embedding information, $h_i^{(l)}$, of item node i. After embedding propagation, neighborhood information is fused into each node's embedding information. To obtain a better representation of the nodes' information, we use a standard multilayer perceptron (*MLP*) to combine the L layers embedding representations of nodes into the final embedding representation. Among them, all the embedding information of the L layers is concatenated together before being input into the *MLP*. The specific form is as follows:

$$h_u^{(*)} = MLP\left(h_u^{(0)} \| h_u^{(1)} \ldots \| h_u^{(L)}\right);\ h_i^{(*)} = MLP\left(h_i^{(0)} \| h_i^{(1)} \ldots \| h_i^{(L)}\right), \tag{5}$$

where *MLP* is a multilayer perceptron; $\|$ represents the concatenation operation of vectors; and $h_u^{(*)}$ and $h_i^{(*)} \in R^d$ are the final embedding representations of user u and item i, respectively.

2.4. *Perceptron Module Based on User–Item Multi-Behavior Interaction Sequence*

2.4.1. User Multi-Behavior Feature Extraction

The purpose of this module is to aggregate heterogeneous information generated by multi-behavior patterns between users and their interacting items. Different from the extraction of neighbor information, we also mine user multi-behavior embedding features based on user historical interaction behavior sequences. To obtain the preference information of users interacting with items, we designed a user multi-behavior awareness module. This module extracts the target user, u, and the neighbor nodes, $i \in N(u)$, interacting with it, and it arranges them into $S_u = \left\{s_u^1, s_u^2, \ldots, s_u^{|S_u|}\right\}$ according to the time sequence.

According to the embedding information of item nodes and edge nodes, we can obtain the behavior characteristics of user u:

$$b^{j}_{u,i,k} = \sigma(\alpha_{uk}h_i + b_\theta), \tag{6}$$

where $h_v \in R^d$ is the embedding representation information of item v, σ is the sigmoid function, and b_θ is the bias. By using Formula (6), we can obtain an embedding interaction sequence $\left(b^1, b^2, \ldots, b^{|N(u)|}\right)$ of user u.

2.4.2. Bi-GRU-Based Behavior Feature Extraction

To mine the overall features of user-embedded behavioral feature sequences, we use an RNN model to explore their temporal information and generate a single representation to encode their overall semantics. Different from basic RNN units, *GRU* units can memorize long-term dependencies sequentially [19]. Therefore, in this module, we use *GRU* to capture the user's multi-behavior preferences. Guo et al. [20] demonstrate that Bi-LSTM and *Bi-GRU* can achieve better results in sequential problems than LSTM and *GRU*. Therefore, we input the embedding interaction sequence $\left(b^1, b^2, \ldots, b^{|N(u)|}\right)$ into a *Bi-GRU* network:

$$\left(h_b^1, h_b^2, \ldots, h_b^{|N(u)|}\right) = Bi - GRU\left(b^1, b^2, \ldots, b^{|N(u)|}\right). \tag{7}$$

where $h_b^i \in R^d$.

We obtain the user's multi-behavior preference sequence based on the user's behavior information and interactive item information. As we all know, users' way of thinking and external market conditions change over time. If the model does not pay attention to changes in the user's core behavior, it will cause errors in subsequent recommendations. Inspired by Chang et al. [21], we input the user multi-behavior preference sequence into a *GRU* network with an attention update gate (*AUGRU*) to obtain the user's final multi-behavior preference representation:

$$h_{u_b} = AUGRU\left(h_b^1, h_b^2, \ldots, h_b^{|N(u)|}\right) \tag{8}$$

The AUGRU model uses an attention mechanism to process differentiated multi-behavior information. It scales the individual multi-behavior features of the update gates by using attention scores. Therefore, behavior features with less correlation have less influence on the hidden state. This makes the acquired multi-behavior information changes more accurate.

2.5. Item–Item Multi-Behavior Interaction Information Aggregation

2.5.1. Construction of Item–Item Relationship Graph

Even for the same item, different users may show different meanings when interacting with the item. We can mine the connection between items from the perspective of users, and then obtain the potential representation of items. Therefore, we extract the item set with the same interaction type as the target item, i_j, in the user–item interaction graph, $\mathcal{G}_0$ (Figure 2). Then we construct the item–item multi-behavior interaction graph, $\mathcal{G}_i$, for further learning the latent factors of items. The weight of each interactive edge is expressed as follows:

$$e^*_{jj'} = \sum_{m \in N_{\mathcal{G}_1}(j,j')} e_{jm} e_{j'm}, e_{jj'} = \frac{\exp\left(e^*_{jj'}\right)}{\exp\left(\sum_{r \in N_{\mathcal{G}_1}(j)} e^*_{jr}\right)}, \tag{9}$$

where α_{jm} and $\alpha_{j'm}$ are the weight information calculated by Formula (1). $N_{\mathcal{G}_1}(j, j')$ represents the users adjacent to j and j' in graph $\mathcal{G}_1$. $N_{\mathcal{G}_1}(j)$ represents the items that are second-order adjacent to $\mathcal{G}_1$ and j in the graph. The final attention weight $e_{jj'}$ is obtained by normalizing $e^*_{jj'}$ using the Softmax function.

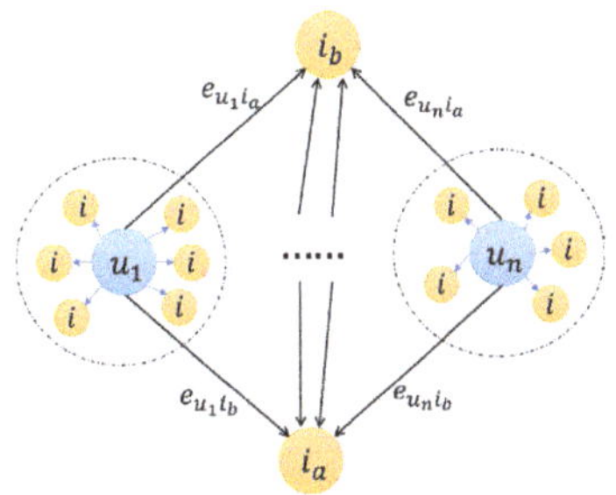

Figure 2. The principal figure of item–item relation information aggregation.

2.5.2. Information Propagation of Item–Item Relationship Graph

Through the weight information, we can define the information propagation method of neighbor item j to item i. Based on the weight information of the item-relation graph and the feature information of neighbor items, we obtain the extended representation of item i:

$$h_{i_s} = f\Big(\sum_{j \in N^i(i)} e_{ij} h_i^{(*)} + h_i\Big), \tag{10}$$

where $N^i(i)$ represents the neighborhood of i in the item–item interaction graph $\mathcal{G}_i$, and h_{i_s} is the aggregated information of i. Moreover, $f(\,)$ is an activation function similar to LeakyReLU.

2.6. Joint Prediction Module

After the above three modules, we obtain the user's preference behavior, h_{u_b}; the user interest feature, $h_u^{(*)}$; the clustering feature, h_{i_s}; and the feature, $h_i^{(*)}$, of the item. We combine the above feature information to obtain the final embeddings of users and items for the final prediction:

$$h_u = h_u^{(*)} \oplus h_{u_b}, h_i = h_i^{(*)} \oplus h_{u_s}, \tag{11}$$

where $\oplus$ represents addition between vector elements. Finally, we inner-product the final representations of users and items to predict their match scores:

$$\hat{y}_{(u,i)} = {h_u}^{\mathrm{T}} \cdot h_i, \tag{12}$$

2.7. Model Learning

Given a user–item interaction sequence, $S = \left\{s_1, s_2, \ldots, s_{|S|}\right\}$, we extract its top $|S| - x$ items to predict the x items of its last interaction. To optimize our UMBGN model, we choose BPRloss [22], which is widely used in recommendation systems [9,17]. Specifically, the final loss function is denoted as follows:

$$\mathcal{L} = \sum_{(u,i^+,i^-) \in O} -ln\sigma\left(\hat{y}_{(u,i^+)} - \hat{y}_{(u,i^-)}\right) + \lambda \|\Theta\|_2^2, \tag{13}$$

where $O = \{(u,i^+,i^-) | (u,i^+) \in R^+, (u,i^-) \in R^-\}$ represents the paired target behavior training dataset; R^+ and R^- refer to target behaviors that have occurred and target behaviors that have not occurred, respectively; σ refers to the sigmoid function; Θ is a parameter that can be trained in the network; and λ is the L2 normalization coefficient.

3. Experiment

In this section, we recount the experiments we conducted on three real datasets, namely MovieLens, Yelp2018, and Online Mall, to evaluate our UMBGN model. We explore the following four questions:

RQ1: In this paper, we consider user multi-behavior information. Does this improve recommendation performance? How does UMBGN perform compared to existing models?

RQ2: We also set the propagation weight among network nodes according to the behavior information. Does this improve the performance of the model? If the weight information is not considered, what will be the effect on the experimental results?

RQ3: How does each module of the model contribute to the improvement of the accuracy of the prediction results?

RQ4: What are the effects of various parameters of the model on the final performance of our proposed method?

3.1. Experimental Environment

3.1.1. Datasets

To evaluate the performance of UMBGN, we conduct experiments on MovieLens, Yelp2018, and the real e-commerce dataset Online Mall, respectively.

MovieLens is a widely used benchmark dataset in recommendation systems containing 20 million movie ratings (accessed at https://grouplens.org/datasets/movielens/20m/, accessed on 15 April 2022). In the experiment, we divided user ratings into multiple behavior types: (1) dislike behavior, (2) neutral behavior, and (3) like behavior.

Yelp2018 is a famous merchant-review website in the US (accessed at https://www.yelp.com/dataset/download, accessed on 16 April 2022). Users can rate merchants, submit reviews, and give tips on the Yelp website. We divided the Yelp dataset into four behaviors (like, dislike, neutral, and tip), using the same criteria as we did for MovieLens.

Online Mall is provided by JD.com, a commerce company with a huge number of users and a full range of goods (accessed at https://jdata.jd.com/html/detail.html?id=8, accessed on 16 April 2022). User-behavior types include click, favorite, add to cart, and purchase.

To ensure the accuracy of the experiments, we performed basic preprocessing on the dataset. We removed users and items with fewer than 10 interactions. Then we divide the dataset into the training set, validation set, and test set according to 80%, 10%, and 10%. The dataset information after data preprocessing is shown in Table 1.

Table 1. Experimental dataset statistics.

Dataset	User	Item	Interaction	Behavior Type
Yelp2018	31,668	38,048	1,561,406	{Tip, Dislike, Neutral, Like}
ML-20 M	138,493	26,744	19,989,593	{Dislike, Neutral, Like}
Online Mall	102,703	24,677	37,059,872	{Buy, Cart, Fav, PV}

3.1.2. Comparison Methods

To evaluate our method, we adopted two evaluation metrics that were widely used in previous work: recall@K and NDCG@K [18]. They are defined as follows:

Recall@K: It is used to measure the probability that the actual interaction item appears in the top-K leaderboard recommendation task. Recall@K does not pay attention to the order in which the user actually clicks an item in the recommended task list; it only considers whether the item appears in the top N positions of the recommended task list.

NDCG@K: In the top-K ranking list, NDCG@K evaluates the quality of the recommendation list according to the rank order of correct items. It assigns higher scores to higher-ranked positions, which means that test items should be ranked as high as possible.

For each user in the test set, we adopt the next-item-recommendation task [23]. For each user, we pair the ground-truth items in the test set with other negative items that are interactions, obtain the user's preference score for all items, and then rank them. In this paper, we set K=10. It is known that higher HR and NDCG scores indicate a better model performance.

3.1.3. Parameter Setting

In this paper, we use TensorFlow to implement the UMBGN model and use the Adam optimizer to infer the model parameters. We performed experiments on two NVIDIA GeForce GTX2080 Ti GPUs. Firstly, we initialized the user–item embedding matrix and the weights of each item in the mixture model. The embedding dimension of users and items is set to 32. The initial learning rate and the batch size are set to 0.01 and 64, respectively. Secondly, a regularization strategy with weight decay selected from the set of {0.1, 0.05, 0.01, 0.005, 0.001} was used to alleviate the overfitting problem during the training phase. In our evaluation, we employed early stopping to terminate training when the performance on the validation data degraded for 5 consecutive epochs.

3.1.4. Baseline

To verify the effectiveness of the UMBGN model, we compare it with six baseline models: two traditional recommendation methods, two RNN-based methods, and two graph network recommendation methods. We briefly describe the six baseline models as follows:

BPR-MF [24]: It optimizes the latent factor of implicit feedback, using pairwise ranking loss in Bayesian methods to maximize the gap between positive and negative terms.

FPMC [25]: This is a classic mixed model that captures sequential effects and the general interest of users. FPMC fuses sequence and personalized information for recommendation by constructing a Markov transition matrix.

GRURec [19]: It is a GRU model trained based on a parallel mini-batch top1 loss function. GRURec uses parallel computation, as well as mini-batch computation, to learn model parameters.

GRU4Rec+ [26]: This is an improved version of GRURec, which concatenates the hot term vector and the feature vector as the input GRU network and has a new loss function and sampling strategy.

GraphRec [27]: It is a deep graph neural network model that enriches the information representation of nodes through embedding propagation and aggregation. GraphRec also aggregates social relations among users through a graph neural network structure.

NGCF [18]: It is an advanced graph neural network model. NGCF has some special designs that can combine traditional collaborative filtering with graph neural networks for application in recommendation systems.

Among all of these methods, BPR-MF and FPMC are traditional recommendation methods, GRURec and GRU4Rec+ are RNN-based methods, and GrahRec and NGCF are graph-network-based methods.

3.2. Performance Comparison

We demonstrate the performance of the above methods in predicting target types for user–item interactions on three real datasets. As shown in Table 2, UMBGN achieves significant performance improvement on different types of datasets. This improvement benefits from our consideration of the user's multi-behavior interaction sequence and the relationship between items.

Table 2. Performance comparison of all methods in terms of Recall@10 and NDCG@10 on all datasets.

Dataset	Metric	BPR-MF	FPMC	GRURec	GRU4Rec+	NGCF	GraphRec	UMBGN
Yelp2018	Recall	0.2136	0.3811	0.6689	0.7051	0.7813	0.7594	0.7938
	NDCG	0.1208	0.2439	0.3607	0.4518	0.5232	0.4943	0.5362
ML-20 M	Recall	0.2608	0.3404	0.3748	0.4976	0.6020	0.5891	0.6351
	NDCG	0.1365	0.3163	0.2942	0.4129	0.5029	0.4982	0.5137
Online Mall	Recall	0.2679	0.3637	0.4474	0.6786	0.7801	0.7546	0.8293
	NDCG	0.1347	0.2692	0.3815	0.4641	0.5832	0.5317	0.5841

The experimental result shows that BPR-MF performed poorly overall. This may be because it cannot consider the user's long-term preference information. It proves that some traditional matrix factorization methods are not suitable for multi-behavior recommendation tasks. Although FPMC has an improved performance compared with BPR-MF, it still has not achieved satisfactory results. RNN-based models (GRURec, GRU4Rec+) have been greatly improved compared to traditional methods because RNN-based models can capture users' long-term preferences more effectively. In addition, GRU4Rec+ performs better than GRURec. This may be attributed to GRU4Rec+ considering personalized information.

Graph-network-based models (GraphRec, NGCF, and UMBGN) significantly outperform traditional methods and RNN-based methods. This shows that using the graph network method can better mine user–item connections and have a better ability to recommend the next item. Furthermore, we observe that UMBGN outperforms other datasets in the Online Mall dataset. One possible explanation is that Online Mall has a large amount of data and rich types of user–item interactions. In addition, the number of users in the Online Mall dataset is relatively large, thus enabling the model to better model user preference information. Therefore, UMBGN is more practical in the real world with massive user data, such as online shopping platforms and social platforms. This shows that considering the multi-behavior information of users improves the recommendation performance.

3.3. Ablation Experiments

3.3.1. The Influence of Different Behavioral Weights on the Experimental Results

To evaluate the impact of different behavioral information on user purchase intention, we compared the performance of our method on the Online Mall dataset. We designed the following controlled experiments: (1) setting the behavior weight of each user to the same weight, α_{uk}; and (2) setting each interaction behavior to the same weight, w_k^u. We present the results of the ablation experiments in Table 3. It shows that our UMBGN model with learnable behavior weight information is 68.45% higher than the model with the same α_{uk} and 34.10% higher than the model with the same w_k^u on recall@10. It is 50.27% higher than the model with the same α_{uk} and 25.21% higher than the model with the same w_k^u on NDCG@10. This indicates that focusing on multi-behavior weights is necessary and should be learned by the model itself. Therefore, setting the propagation weights between graph network nodes according to multi-behavior information improves the performance of the model.

Table 3. Results of ablation experiments with multi-behavior weights on Online Mall.

Model	UMBGN	Same α_{uk}	Same w_k^u
Recall@10	0.8293	0.4923	0.6182
NDCG@10	0.5841	0.3887	0.4665

3.3.2. The Influence of Each Module in UMBGN on the Experimental Results

The user multi-behavior awareness module aims to obtain the user's behavior preference information, and the item-information-relation module aims to obtain the relevant information between items. They are both complementary to the user–item interaction information module. We conducted an ablation study to test the effectiveness of the user multi-behavior awareness module and the item information relation module in our UMBGN. The results are shown in Table 4.

Table 4. Performance of user multi-behavior awareness module and item information relation module.

Dataset	Yelp2018		ML-20 M		Online Mall	
	Recall	NDCG	Recall	NDCG	Recall	NDCG
UMBGN	0.7938	0.5362	0.6351	0.5137	0.8293	0.5841
w/o UBAM	0.6905	0.4803	0.5960	0.4821	0.7294	0.5267
w/o IIRM	0.7513	0.5189	0.6127	0.5088	0.7687	0.5502

The results of the ablation experiments (Figure 3) show that the UMBGN model has a higher recall rate and NDCG than the model without the user multi-behavior awareness model and the item-information-relation model. Especially on the Online Mall dataset, it improves the recall rate by 13.70% and 7.88%, respectively. Moreover, it improves the NDCG by 9.83% and 5.80%, respectively. This shows that taking into account the user's multi-behavior interaction sequence and the relationship between items can make more accurate recommendations to users. This shows that each module is necessary to improve the accuracy of the prediction results.

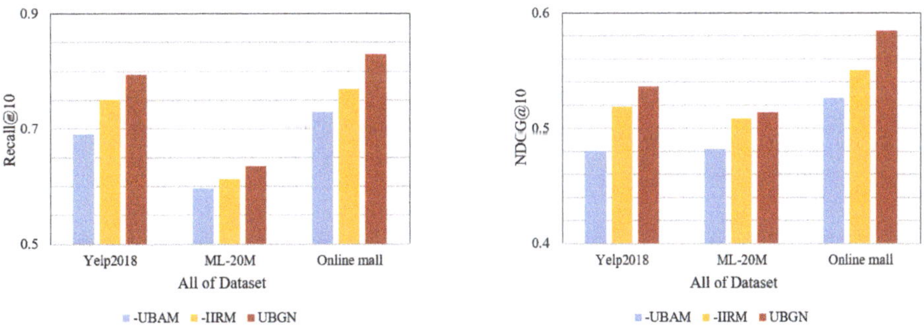

Figure 3. Results of ablation experiments of sub-modules in UMBGN.

3.4. *Parametric Analysis*

3.4.1. The Effect of Sequence Length on Prediction Results

We also explored the effect of the maximum length, N, of user–item interaction sequences on the model recommendation performance. Figure 4 shows the impact of the maximum length, N, on the recommendation performance on the ML-20 m dataset and the Online Mall dataset, respectively. We observe that the recommendation performance improves as the N increases until the N is less than 40. This indicates that the length of the user's behavior sequence has an impact on the recommendation performance. However, when N exceeds 40, the recommendation performance on the ML-20 m dataset no longer increases significantly. Moreover, the recommendation performance of the Online Mall dataset has also declined. This suggests that the model does not always benefit from larger N, as larger N tends to introduce more noise. However, our model remains stable when the length N becomes larger. This also proves that our model can handle noisy behavioral sequence information well.

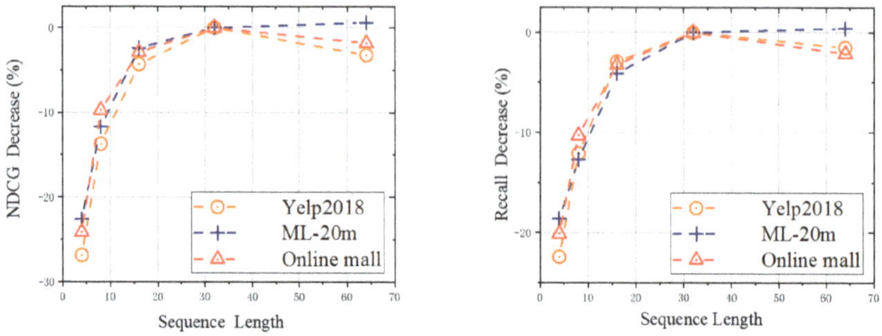

Figure 4. Performance comparison of methods with different behavior sequence lengths, N, on three datasets.

3.4.2. The Influence of the Number of Layers of the Graph Neural Network on the Prediction Results

We wish to test the effect of the number of layers of the GNN on the UMBGN model. In the user–item interaction information module, UMBGN, with two recursive message propagation layers, achieves the best results. This shows that it is essential to model higher-order relationships between items and features via GNNs. However, as shown in Figure 5, the performance starts to degrade as the depth of the graph model increases. This is because multiple embedded propagation layers may contain some noisy signals, resulting in over-smoothing [28]. This shows that determining the optimal parameters of the model through a large number of experiments is conducive to improving the performance of the model.

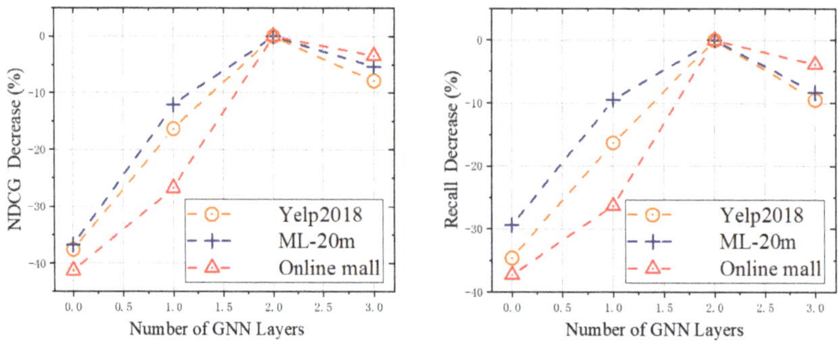

Figure 5. Performance comparison of methods with different numbers of GNN layers on three datasets.

4. Related Work

4.1. Recommendation Based on Graph Neural Network

In recent years, graph networks that can naturally aggregate node information and topology have attracted extensive attention. Especially in recommendation systems, the use of graph networks to mine user–item interaction data has achieved remarkable results [29–31]. Yang et al. [32] constructed a Hierarchical Attention Convolutional Network (HAGERec) combined with a knowledge graph. They exploited the high-order connectivity relationship of heterogeneous knowledge graphs to mine users' latent preferences. In addition, information aggregation was performed on user and item entities through local proximity and attention mechanisms. Gwadabe et al. [33] proposed a GNN-based recommendation model, GRASER, for the session-based recommendation. It used GNN to learn the sequential and non-sequential complex transformation relationship between items in each session, which improved the performance of the recommendation. Zhang et al. [34]

proposed a dynamic graph neural network (DGSR) for the sequential recommendation. It explicitly modeled the dynamic collaboration information between different user sequences in sequential recommendations. Therefore, it could transform the task of the next prediction in sequential recommendation into a link prediction between user nodes and item nodes in a dynamic graph. Fan et al. [27] designed a graph network framework (GraphRec) for the social recommendation. The method jointly captured users' purchase preferences from the user's social graph and the user–item interaction graph. The SURGE graph neural network frame proposed by Chang et al. [21] combined the sequential recommendation model and the graph neural network model. This method first integrated the different preferences in the user's long-term behavior sequence into the graph structure, and then it performed operations such as perception, propagation, and pooling of the graph network. It could dynamically extract the core interests of the current user from noisy user behavior sequences. Different from their work, our work defines new multi-behavior information weights for information propagation in graph neural networks.

4.2. Multi-Behavior Recommendation

Traditional recommendation systems usually rely only on a single type of user–item interaction, which limits the performance of the methods. Recommendation methods utilizing multiple behaviors can more accurately capture user preference information. Guo et al. [20] designed a Deep Intent Prediction Network (DIPN) to predict users' purchase intentions from multiple perspectives. They combined touch interaction behavior with traditional browsing behavior and introduced multi-task learning to differentiate user behavior. Experiments on large-scale datasets showed that the network significantly outperforms traditional methods that used only browsing interaction behavior. Rosaci [35,36] proposed a CILIOS method to determine inter-ontology similarities between agents. It monitored user behavior and interests to extend the recommendation dataset generated by traditional methods. In addition, this method extracted logical knowledge in recommendation scenarios to support web recommendations. Wu et al. [37] constructed a new multi-behavior multi-view contrastive learning recommendation model (MMCLR) to solve the data sparsity and cold-start problems in traditional recommender models. They considered the similarities and differences between different user behaviors and views through three tasks. Experiments on real datasets indicate that MMCLR significantly improved the performance of recommendations. Pan et al. [38] designed a Spatiotemporal Interaction Augmented Graph Neural Network (SIGMA). It encoded a mobile graph to represent individual mobile behavior and used a stacked scoring approach to generate recommendation scores. This showed that the mobile behavior of individuals and groups played an important role in location recommender systems. Xia et al. [39] developed a Multi-Behavior Graph Meta Network (MB-GMN) to extract the interaction information of multiple behavior types between users and items. The proposed method jointly models behavioral heterogeneity and interaction behavioral diversity, combined with the meta-learning paradigm. A large number of comparative experiments on three datasets demonstrated the effectiveness of their method. Inspired by the above research work, we propose a new multi-behavior awareness module to further mine time-series based user multi-behavior information.

5. Conclusions

In this paper, we explored the problem of graph network recommendation, focusing on user multi-behavior interaction sequences, and proposed a UMBGN model. Compared with the traditional GNN model, our model updates the node connection weights of the user–item interaction graph according to the multi-behavior interaction information, so that it can capture the user's interest in specific items under different behavioral information. In this study, we designed two modules to further mine the user's multi-behavior preference information. Firstly, we put the multi-behavior sequence information of the target user into an improved Bi-GRU model, the AUGRU model, to enrich the user's embedding representation. Secondly, we built an item–item graph based on the user's dependencies on items to further enrich the embedding representation of items. The comparative experiments that we performed on three real datasets demonstrate the effectiveness of the UMBGN model. Further ablation experiments prove the necessity of the user multi-behavior awareness module and item information awareness module in our UMBGN model. In addition, we also evaluated the impact of different parameters on recommendation performance, confirming the applicability of UMBGN in practical applications. However, our approach does not consider potential connections among users. In the future, we plan to introduce users' social relations into our method to improve the accuracy of the next-item recommendation.

Author Contributions: Conceptualization, M.J. and F.L.; methodology, M.J. and X.Z.; software, M.J. and X.L.; validation, X.Z.; investigation, M.J.; resources, F.L.; data curation, M.J.; writing—original draft preparation, M.J. and X.L.; writing—review and editing, F.L. and X.Z.; visualization, M.J.; supervision, F.L.; funding acquisition, F.L.; All authors have read and agreed to the published version of the manuscript.

Funding: This work was funded by the National Natural Science Foundation of Shandong (ZR202011 020044) and the National Natural Science Foundation of China (61772321).

Data Availability Statement: Publicly available datasets were analyzed in this study. The data of MovieLens can be found here: https://grouplens.org/datasets/movielens/20m/ (accessed on 15 April 2022). The data of Yelp2018 can be found here: https://www.yelp.com/dataset/download (accessed on 16 April 2022). The data of Online Mall can be found here: https://jdata.jd.com/html/detail.html?id=8 (accessed on 16 April 2022).

Conflicts of Interest: The authors declare no conflict of interest.

References

1. Gu, Y.; Ding, Z.; Wang, S.; Yin, D. Hierarchical user profiling for e-commerce recommender systems. In Proceedings of the 13th International Conference on Web Search and Data Mining, Houston, TX, USA, 3–7 February 2020; pp. 223–231.
2. Schafer, J.B.; Frankowski, D.; Herlocker, J.; Sen, S. Collaborative filtering recommender systems. In *The Adaptive Web: Methods and Strategies of Web Personalization*; Springer: Berlin/Heidelberg, Germany, 2007; pp. 291–324.
3. Koren, Y. Factorization meets the neighborhood: A multifaceted collaborative filtering model. In Proceedings of the 14th ACM SIGKDD International Conference on Knowledge Discovery and Data Mining, Las Vegas, NV, USA, 24–27 August 2008; pp. 426–434.
4. Ning, X.; Karypis, G. Slim: Sparse linear methods for top-n recommender systems. In Proceedings of the 2011 IEEE 11th international conference on data mining, Vancouver, BC, Canada, 11–14 December 2011; IEEE: Piscataway, NJ, USA, 2011; pp. 497–506.
5. Rendle, S.; Gantner, Z.; Freudenthaler, C.; Schmidt-Thieme, L. Fast context-aware recommendations with factorization machines. In Proceedings of the 34th international ACM SIGIR conference on Research and development in Information Retrieval, Beijing, China, 25–29 July 2011; pp. 635–644.
6. He, X.; Chua, T.S. Neural factorization machines for sparse predictive analytics. In Proceedings of the 40th International ACM SIGIR conference on Research and Development in Information Retrieval, Tokyo, Japan, 7–11 August 2017; pp. 355–364.
7. He, X.; Liao, L.; Zhang, H.; Nie, L.; Hu, X.; Chua, T.S. Neural collaborative filtering. In Proceedings of the 26th International Conference on World Wide Web, Perth, Australia, 3–7 April 2017; pp. 173–182.
8. Xue, H.J.; Dai, X.; Zhang, J.; Huang, S.; Chen, J. Deep matrix factorization models for recommender systems. In Proceedings of the IJCAI International Joint Conference on Artificial Intelligence, Melbourne, Australia, 19–25 August 2017; pp. 3203–3209.
9. Fan, S.; Zhu, J.; Han, X.; Shi, C.; Hu, L.; Ma, B.; Li, Y. Metapath-guided heterogeneous graph neural network for intent recommendation. In Proceedings of the 25th ACM SIGKDD International Conference on Knowledge Discovery & Data Mining, Anchorage, AK, USA, 4–8 August 2019; pp. 2478–2486.
10. Wu, S.; Tang, Y.; Zhu, Y.; Wang, L.; Xie, X.; Tan, T. Session-based recommendation with graph neural networks. In Proceedings of the AAAI Conference on Artificial Intelligence, Honolulu, HI, USA, 27 January–1 February 2019; Volume 33, pp. 346–353.
11. Wang, X.; He, X.; Cao, Y.; Liu, M.; Chua, T.S. Kgat: Knowledge graph attention network for recommendation. In Proceedings of the 25th ACM SIGKDD International Conference on Knowledge Discovery & Data Mining, Anchorage, AK, USA, 4–8 August 2019; pp. 950–958.
12. Yang, L.; Liu, Z.; Dou, Y.; Ma, J.; Yu, P.S. Consisrec: Enhancing gnn for social recommendation via consistent neighbor aggregation. In Proceedings of the 44th International ACM SIGIR Conference on Research and Development in Information Retrieval, Montreal, QC, Canada, 11–15 July 2021; pp. 2141–2145.

13. Gao, C.; Zheng, Y.; Li, N.; Li, Y.; Qin, Y.; Piao, J.; Quan, Y.; Chang, J.; Jin, D.; He, X.; et al. Graph neural networks for recommender systems: Challenges, methods, and directions. *arXiv* **2021**, arXiv:2109.12843, 2021. [CrossRef]
14. Fan, Z.; Liu, Z.; Zhang, J.; Xiong, Y.; Zheng, L.; Yu, P.S. Continuous-time sequential recommendation with temporal graph collaborative transformer. In Proceedings of the 30th ACM International Conference on Information & Knowledge Management, Gold Coast, Australia, 1–5 November 2021; pp. 433–442.
15. Zhao, Y.; Ou, M.; Zhang, R.; Li, M. Attributed Graph Neural Networks for Recommendation Systems on Large-Scale and Sparse Graph. *arXiv* **2021**, arXiv:2112.13389.
16. Tan, Q.; Zhang, J.; Yao, J.; Liu, N.; Zhou, J.; Yang, H.; Hu, X. Sparse-interest network for sequential recommendation. In Proceedings of the 14th ACM International Conference on Web Search and Data Mining, Jerusalem, Israel, 8–12 March 2021; pp. 598–606.
17. Zheng, Y.; Gao, C.; He, X.; Li, Y.; Jin, D. Price-aware recommendation with graph convolutional networks. In Proceedings of the 2020 IEEE 36th International Conference on Data Engineering (ICDE), Dallas, TX, USA, 20–24 April 2020; IEEE: Piscataway, NJ, USA, 2020; pp. 133–144.
18. Wang, X.; He, X.; Wang, M.; Feng, F.; Chua, T.-S. Neural graph collaborative filtering. In Proceedings of the SIGIR, Paris, France, 21–25 July 2019; pp. 165–174.
19. Hidasi, B.; Karatzoglou, A.; Baltrunas, L.; Tikk, D. Session-based recommendations with recurrent neural networks. *arXiv* **2015**, arXiv:1511.06939.
20. Guo, L.; Hua, L.; Jia, R.; Zhao, B.; Wang, X.; Cui, B. Buying or browsing? Predicting real-time purchasing intent using attention-based deep network with multiple behavior. In Proceedings of the 25th ACM SIGKDD International Conference on Knowledge Discovery & Data Mining, Anchorage, AK, USA, 4–8 August 2019; pp. 1984–1992.
21. Chang, J.; Gao, C.; Zheng, Y.; Hui, Y.; Niu, Y.; Song, Y.; Jin, D.; Li, Y. Sequential recommendation with graph neural networks. In Proceedings of the 44th International ACM SIGIR Conference on Research and Development in Information Retrieval, Montreal, QC, Canada, 11–15 July 2021; pp. 378–387.
22. Rendle, S.; Freudenthaler, C.; Gantner, Z.; Schmidt-Thieme, L. BPR: Bayesian personalized ranking from implicit feedback. *arXiv* **2012**, arXiv:1205.2618.
23. Sun, F.; Liu, J.; Wu, J.; Pei, C.; Lin, X.; Ou, W.; Jiang, P. BERT4Rec: Sequential recommendation with bidirectional encoder representations from transformer. In Proceedings of the 28th ACM International Conference on Information and Knowledge Management, Beijing, China, 3–7 November 2019; pp. 1441–1450.
24. Guo, H.; Tang, R.; Ye, Y.; Li, Z.; He, X. DeepFM: A factorization-machine based neural network for CTR prediction. *arXiv* **2017**, arXiv:1703.04247.
25. Rendle, S.; Freudenthaler, C.; Schmidt-Tieme, L. Fac-torizing personalized Markov chains for next-basket recommendation. In Proceedings of the 19th International Conference on World Wide Web-WWW'10, Raleigh, NC, USA, 26–30 April 2010.
26. Hidasi, B.; Karatzoglou, A. Recurrent neural networks with top-k gains for session-based recommendations. In Proceedings of the 27th ACM International Conference on Information and Knowledge Management, Torino, Italy, 22–26 October 2018; pp. 843–852.
27. Fan, W.; Ma, Y.; Li, Q.; He, Y.; Zhao, E.; Tang, J.; Yin, D. Graph neural networks for social recommendation. In Proceedings of the World Wide Web Conference, San Francisco, CA, USA, 13–17 May 2019; ACM: New York, NY, USA, 2019; pp. 417–426.
28. Chen, D.; Lin, Y.; Li, W.; Li, P.; Zhou, J.; Sun, X. Measuring and relieving the over-smoothing problem for graph neural networks from the topological view. In Proceedings of the AAAI Conference on Artificial Intelligence, New York, NY, USA, 7–12 February 2020; Volume 34, pp. 3438–3445.
29. Kipf, T.N.; Welling, M. Semi-supervised classification with graph convolutional networks. *arXiv* **2016**, arXiv:1609.02907.
30. Wu, Z.; Pan, S.; Chen, F.; Long, G.; Zhang, C.; Philip, S.Y. A comprehensive survey on graph neural networks. *IEEE Trans. Neural Netw. Learn. Syst.* **2020**, 32, 4–24. [CrossRef] [PubMed]
31. Yin, R.; Li, K.; Zhang, G.; Lu, J. A deeper graph neural network for recommender systems. *Knowl. Based Syst.* **2019**, *185*, 105020. [CrossRef]
32. Yang, Z.; Dong, S. HAGERec: Hierarchical attention graph convolutional network incorporating knowledge graph for explainable recommendation. *Knowl. Based Syst.* **2020**, *204*, 106194. [CrossRef]
33. Gwadabe, T.R.; Liu, Y. Improving graph neural network for session-based recommendation system via non-sequential interactions. *Neurocomputing* **2022**, *468*, 111–122. [CrossRef]
34. Zhang, M.; Wu, S.; Yu, X.; Liu, Q.; Wang, L. Dynamic graph neural networks for sequential recommendation. *IEEE Trans. Knowl. Data Eng.* **2022**. [CrossRef]
35. Rosaci, D. Web Recommender Agents with Inductive Learning Capabilities. In *Emergent Web Intelligence: Advanced Information Retrieval*; Springer: London, UK, 2010; pp. 233–267. [CrossRef]
36. Rosaci, D. CILIOS: Connectionist inductive learning and inter-ontology similarities for recommending information agents. *Inf. Syst.* **2007**, *32*, 793–825. [CrossRef]
37. Wu, Y.; Xie, R.; Zhu, Y.; Ao, X.; Chen, X.; Zhang, X.; Zhuang, F.; Lin, L.; He, Q. Multi-view Multi-behavior Contrastive Learning in Recommendation. In Proceedings of the International Conference on Database Systems for Advanced Applications, Virtual Event, 11–14 April 2022; Springer: Cham, Switzerland, 2022; pp. 166–182.

38. Pan, X.; Cai, X.; Song, K.; Baker, T.; Gadekallu, T.R.; Yuan, X. Location recommendation based on mobility graph with individual and group influences. *IEEE Trans. Intell. Transp. Syst.* **2022**. [CrossRef]
39. Xia, L.; Xu, Y.; Huang, C.; Dai, P.; Bo, L. Graph meta network for multi-behavior recommendation. In Proceedings of the 44th International ACM SIGIR Conference on Research and Development in Information Retrieval, Montreal, QC, Canada, 11–15 July 2021; pp. 757–766.

electronics

MDPI

Article

Towards Adversarial Attacks for Clinical Document Classification

Nina Fatehi [1], Qutaiba Alasad [2] and Mohammed Alawad [1,*]

[1] Electrical and Computer Engineering Department, Wayne State University, Detroit, MI 48202, USA
[2] Department of Petroleum Processing Engineering, Tikrit University, Al Qadisiyah P.O. Box 42, Iraq
* Correspondence: alawad@wayne.edu

Abstract: Regardless of revolutionizing improvements in various domains thanks to recent advancements in the field of Deep Learning (DL), recent studies have demonstrated that DL networks are susceptible to adversarial attacks. Such attacks are crucial in sensitive environments to make critical and life-changing decisions, such as health decision-making. Research efforts on using textual adversaries to attack DL for natural language processing (NLP) have received increasing attention in recent years. Among the available textual adversarial studies, Electronic Health Records (EHR) have gained the least attention. This paper investigates the effectiveness of adversarial attacks on clinical document classification and proposes a defense mechanism to develop a robust convolutional neural network (CNN) model and counteract these attacks. Specifically, we apply various black-box attacks based on concatenation and editing adversaries on unstructured clinical text. Then, we propose a defense technique based on feature selection and filtering to improve the robustness of the models. Experimental results show that a small perturbation to the unstructured text in clinical documents causes a significant drop in performance. Performing the proposed defense mechanism under the same adversarial attacks, on the other hand, avoids such a drop in performance. Therefore, it enhances the robustness of the CNN model for clinical document classification.

Keywords: adversarial attacks; document classification; CNN; NLP

Citation: Fatehi, N.; Alasad, Q.; Alawad, M. Towards Adversarial Attacks for Clinical Document Classification. *Electronics* **2023**, *12*, 129. https://doi.org/10.3390/electronics12010129

Academic Editors: Taiyong Li, Wu Deng and Jiang Wu

Received: 15 November 2022
Revised: 21 December 2022
Accepted: 22 December 2022
Published: 28 December 2022

1. Introduction

Although DL models for NLP have achieved remarkable success in various domains, such as text classification [1], sentiment analysis [2] and Named Entity Recognition (NER) [3], recent studies have demonstrated that DL models are susceptible to adversarial attacks, small perturbations and named adversarial examples (AEs), crafted to fool the DL model to make false predictions [4]. Such attacks are crucial in sensitive environments like healthcare where such vulnerabilities can directly threaten human life. Similar to other domains, DL in healthcare has obtained diagnostic parity with human physicians on various health information tasks such as pathology [5] and radiology [6]. The issue of AEs has emerged as a pervasive challenge in even state-of-the-art learning systems for health and has raised concerns about the practical deployment of DL models in such a domain. However, in comparison to non-clinical NLP tasks, adversarial attacks on Electronic Health Records (EHR) and tasks such as clinical document classification have gained the least attention.

Various approaches based on concatenation [7] or editing [8] perturbations have been proposed to attack NLP models. Attacking these models by manipulating characters in a word to generate AEs seems unnatural for some applications due to grammatical disfluency. Also, generating AEs is challenging in the text compared to images, due to the discrete space of input data as well as the fact that generating perturbations which can fool the DL model and at the same time be unperceivable for humans is not easy in text [4]. However, these approaches apply very well to the target application of this paper, i.e., pathology report classification based on the cancer type. The unstructured text in pathology reports is ungrammatical, fragmented, and marred with typos and abbreviations. Also, the document text is usually long and results from the concatenation of several fields, such as microscopic description, diagnosis, and summary. Whenever they are combined, the human cannot easily differentiate between the beginning and end of each field. Moreover, the text in pathology reports exhibits linguistic variability across pathologists even when describing the same cancer characteristics [9,10].

Perturbations can occur at all stages of the DL pipeline from data collection to model training to the post-processing stage. In this paper, we will focus on two aspects of it. The first part will be the robustness during the training time. The case when the training set is unvetted, e.g., when the training set has arbitrarily chosen an outlier that the model is biased towards. The second aspect is the robustness during the test time. The case when the adversary is trying to fool the model.

The AEs that we will use in this paper are the class label names. These words are known to the attacker without accessing the target DL model. Also, due to the unbalanced nature of this dataset, the model is biased to majority classes or to specific keywords, which are mostly the class label names, that appear in their corresponding samples. Then, we propose a novel defense method against adversarial attacks. Specifically, we select and filter specific features during the training phase. Two criteria are followed when determining these features: (1) the DL model has to be biased to them, and (2) filtering them does not impact the overall model accuracy. We focus on the CNN model to carry out the adversarial evaluation on the clinical document classification, i.e., classifying cancer pathology reports based on their associated cancer type. This model performs equally or better than state-of-the-art natural language models, i.e., BERT [11]. This is mainly because in clinical text classification tasks on documents in which only very few words contribute toward a specific label, most of these subtle word relationships may not be necessary or even relevant to the task at hand [12].

The main contributions of the paper include:

- We compare the effectiveness of different black-box adversarial attacks on the robustness of the CNN model for document classification on long clinical texts.
- We evaluate the effectiveness of using class label names as AEs by either concatenating these examples to the unstructured text or editing them whenever they appear in the text.
- We propose a novel defense technique based on feature selection and filtering to enhance the robustness of the CNN model.
- We evaluate the robustness of the proposed approach on clinical document classification.

The rest of the paper is organized as follows: related works are briefly outlined in Section 2. Sections 3 and 4 present the method and experimental setup, respectively. In Section 5, the results are discussed. Finally, we conclude our paper in Section 6.

2. Related Works

Numerous methods have been proposed in the area of computer vision and NLP for adversarial attacks [4,13,14]. Since our case study focuses on adversarial attacks and defense for clinical document classification, we mainly review state-of-the-art approaches in the NLP domain. Zhang et al. present a comprehensive survey of the latest progress and existing adversarial attacks in various NLP tasks and textual DL models [4]. They categorize adversarial attacks on textual DL as follows:

- Model knowledge determines if the adversary has access to the model information (white-box attack) or if the model is unknown and inaccessible to the adversary (black-box attack).
- Target type determines the aim of the adversary. If the attack can alter the output prediction to a specific class, it is called a targeted attack, whereas an untargeted attack tries to fool the DL model into making any incorrect prediction.

- Semantic granularity refers to the level to which the perturbations are applied. In other words, AEs are generated by perturbing sentences (sentence-level), words (word-level) or characters (character-level).

The work investigated in this paper relates to the adversarial attack on document classification tasks in the healthcare domain and focuses on the targeted/untargeted black-box attack using word/character-level perturbations. We choose black-box attacks as they are more natural than white-box attacks.

In the following subsections, we first present the popular attack strategies with respect to the three above-mentioned categories. Then, we discuss the adversarial defense techniques.

2.1. Adversarial Attack Strategies

Adversarial attacks have been widely investigated for various NLP tasks, including NER [15,16], semantic textual similarity [16], and text classification [17]. These attacks generate perturbations by modifying characters within a word [18], adding or removing words [16], replacing words with semantically similar and grammatically correct synonyms using a word embedding optimized for synonyms replacement [17], or by synonym substitution using WordNet where the replaced word has the same Part of Speech (POS) as the original one [19]. The drawback of character-level methods is that generated AEs can easily be perceived by human and impact the readability of the text [20], and the drawback of AEs generated in word-level is their dependency on the original sentences [20].

Clinical text comes with its unique challenges, where the findings in the non-clinical text might not be applied to clinical text. For instance, in non-clinical text, character-level or word-level perturbations that change the syntax can be easily detected and defended against by the spelling or syntax check. However, this does not apply to clinical text, which often contains incomplete sentences, typographical errors, and inconsistent formatting. Thus, domain-specific strategies for adversarial attacks and defense are required [21].

There are relatively few works that have examined the area of clinical NLP. Mondal et al. propose BBAEG, which is a black-box attack on biomedical text classification tasks using both character-level and word-level perturbations [22]. BBAEG is benchmarked on a simple binary classification and the text is relatively short and clean when compared to real-world clinical text. There are also some works that investigate adversarial attack on EHR including [23,24]. However, they are different from this paper's work as they use the temporal property of EHR to generate AEs and none of them investigates the adversarial attack on unstructured clinical text.

2.2. Adversarial Defense Strategies

As explained in the previous section, detection-based approaches, such as spelling check, have been used as a defense strategy. Gao et al. use python's spelling check to detect the adversarial perturbations in character level; however, this detection method can be performed only on character-level AEs [18]. Another approach in detection to evaluate the model's robustness under adversarial attacks is discriminator training. Xu et al. train another discriminator using a portion of original samples plus AEs to discriminate AEs from original examples [25]. Adversarial training has also been used to enhance the robustness of DL model [4,26]. In adversarial training, adversarial perturbations are involved in the training process [27]. The authors of [15–17] utilize an augmentation strategy to evaluate or enhance the robustness of DL models. In this approach, the model is trained on the augmented dataset that includes original samples plus AEs. The drawback of adversarial training, which makes it an ineffective defense strategy against some adversarial attacks, is overfitting [27]. If the model is biased to the AEs, as in the case of our paper, augmentation will make the bias issue worse.

3. Method

In this section, we first formalize the adversarial attack in a textual CNN context and then describe two methods, namely concatenation adversaries and edit adversaries to generate AEs.

3.1. Problem Formulation

Let us assume that a dataset consists of N documents $X = \{X_1, X_2, ..., X_N\}$ and a corresponding set of N labels $y = \{Y_1, Y_2, ..., Y_N\}$. On such a dataset, $F : X \rightarrow y$ is the CNN model which maps input space X to the output space y. Adversarial attack and adversarial example can be formalized as follows:

$$X_{adv} = X + \Delta$$

$$F(X_{adv}) \neq y \quad (Untargeted \quad attack)$$

$$F(X_{adv}) = y\prime, \quad y \neq y\prime \quad (Targeted \quad attack)$$

3.2. Concatenation Adversaries

Given an input document X_1 of n words $X_1 = \{w_1, w_2, ..., w_n\}$, in concatenation adversaries, there is a list of selected perturbation words that are supposed to be added (one at a time) to different locations of documents. In this paper, we consider three locations: "random", "end", and "beginning".

- Adding perturbation words at random locations: In this attack, we attempt to add a various number of a specific perturbation word in random locations of input documents. If w_{adv} denotes the added word, the adversarial input would be as $X = \{w_1,\ w_2,\ w_{adv},\ \ldots,\ w_{adv},\ w_{n-4},\ w_{adv},\ w_n\}$. The location of each w_{adv} is determined randomly.
- Adding perturbation words in the beginning: In this attack, the aim is to append a various number of a specific perturbation word in the beginning of each input document. In this way, the adversarial input would be as $X = \{w_{adv},\ w_{adv},\ w_{adv},\ \ldots,\ w_1,\ w_2,\ \ldots,\ w_n\}$.
- Adding perturbation words at the end: This attack is carried out to add a various number of a specific perturbation word at the end of each document. The adversarial inputs would be as $X = \{w_1,\ w_2,\ \ldots,\ w_n,\ w_{adv},\ w_{adv},\ w_{adv}\}$.

3.3. Edit Adversaries

Instead of adding perturbation words to the input document text, edit adversaries manipulate specific words in the input document text. In this paper, we apply two edit adversaries forms: synthetic perturbation, which is an untargeted attack; and replacing strategy, which in contrast is a targeted attack [28].

- Synthetic perturbation: In this attack, AEs are generated by perturbing characters including swapping all two neighboring characters $(breast \rightarrow rbaets)$, randomly deleting one character $(breast \rightarrow breat)$, randomly changing orders of all characters $(breast \rightarrow reastb)$ and randomly changing orders of characters except the first and last ones $(breast \rightarrow beasrt)$ of a specific list words. Each of these perturbations are performed on all selected words at the same time.
- Replacing strategy: This attack is a targeted attack in which the selected words in all input documents are replaced with a specific word that leads to the targeted prediction (for instance $F(X_{tadv}) = Y_d$ and w_{adv} is the perturbation word that makes the prediction Y_d instead of Y_t).

3.4. Defense Strategy

For the defense mechanism, we propose a novel method called feature selection and filtering, in which features are selected and filtered from input documents during the model training. These features are selected based on two criteria: (1) the CNN model has to be biased to them, and (2) filtering them does not impact the overall model accuracy. In this paper, we select the class label names as the target features. Other techniques can also be used to determine which features should be selected, such as model interpretability tools, attention weights, scoring functions, etc.

3.5. Evaluation Metrics

The focus of this study is to evaluate the performance of the CNN model for document classification against adversarial examples. The following common performance metrics for classification tasks are used for model evaluation:

F1 Score: The overall accuracy is calculated using the standard micro- and macro- F1 scores as follows:

$$Micro\ F1 = 2(\frac{Precision * Recall}{Precision + Recall})$$

$$Macro\ F1 = \frac{1}{|c|}\Sigma_{c_i}^{c} Micro\ F1(c_i)$$

where $|C|$ is the total number of classes and c_i represents the number of samples belonging to class i.

Accuracy per class: To evaluate the vulnerability of the model per class, we use the accuracy per class metric, which is the percentage of correctly predicted classes after an attack to the number of all samples of the class.

$$Accuracy = \frac{TP_i}{c_i}$$

Number of Perturbed Words: For the attack itself, we include a metric to measure the amount of required perturbations to fool the CNN model. We call this metric "number of perturbed words". In this way, we can determine the minimum number of perturbation words, in concatenation adversaries, that leads to a significant degradation in accuracy.

4. Experimental Setup

4.1. Data

In this paper, we benchmark the proposed adversarial attack and defense on a clinical dataset, specifically The Cancer Genome Atlas Program pathology reports dataset (TCGA) (https://www.cancer.gov/tcga, accessed on 1 October 2021).

The original TCGA dataset consists of 6365 cancer pathology reports; five of which are excluded because they are unlabeled. Therefore, the final dataset consists of 6360 documents. Each document is assigned a ground truth label for the site of the cancer, the body organ where the cancer is detected. In the TCGA dataset, there is a total of 25 classes for the site label. Figure A1 in Appendix A shows the histograms of the number of occurrences per class. Standard text cleaning, such as lowercasing and tokenization, is applied to the unstructured text in the documents. Then, a word vector of size 300 is chosen. The maximum length of 1500 is chosen to limit the length of documents in pathology reports. In this way, reports containing more than 1500 tokens are truncated and those with less than 1500 tokens are zero-padded. Also, we choose 80%/20% data splitting strategy.

4.2. Target Model

In this paper, we use a CNN network as the DL model. ADAM adaptive optimization is used to train the network weights. For all the experiments, the embedding layer is followed by three parallel 1-D convolutional layers. The number of filters in each convolution layer is 100, and the kernel sizes are 3, 4, and 5. ReLU is employed as the activation function and a dropout of 50% is applied to the global max pooling at the output layer. Finally, a fully connected softmax layer is used for the classification task. These parameters are optimized following previous studies [29,30]. We use NVIDIA V100 GPU for all the experiments.

4.3. Adversarial Attack

In this subsection we go through the details of each adversarial attack. For these attacks, we use the dataset class label names, which are the cancer types, as the selected perturbation words to perform concatenation and edit adversaries. The reasons for selecting the label names as the AEs are as follows:

1. Using these names is considered a black-box attack as the adversary does not need to know the CNN model details.
2. The presence of a specific class label name frequently in the unstructured text of pathology reports biases the CNN model to a specific prediction [29,30].
3. As we will see later, filtering these names during the model training does not impact the overall model accuracy.

Therefore, we note that the practical dataset is the one whose class label names exist in the document text.

Since there are 25 different labels' class names in the dataset, we select three of them as AEs to report the result in this paper. Specifically, we select one of the majority classes (breast), one of the minority classes (leukemia/lymphoma- in short lymphoma) and one of the moderate classes (sarcoma). From that selection, we can see how the classes with different distributions can impact the performance of the CNN model under adversarial attacks. In this way, the impact of class distribution on the CNN model's performance can be evaluated as well.

4.3.1. Concatenation Adversaries

In this attack, we investigate the impact of adding selected class names (breast, leukemia/lymphoma, and sarcoma) to the input documents as perturbation words. Three different concatenation adversaries are used:

- Concat-Random: For each selected perturbation word, 1, 2, 3, 5, 10, 20 or 40 words are randomly added to all input documents. For instance, Concat-Random-Breast-1 means randomly adding one "breast" perturbation word to the documents.
- Concat-Begin: For each selected perturbation word, 1, 2, 3, 5, 10, 20 or 40 words are added at the beginning of all input documents. For instance, Concat-Begin-Breast-1 denotes appending one "breast" perturbation word at the beginning of all documents.
- Concat-End: For each selected perturbation word, 1, 2, 3, 5, 10, 20 or 40 words are appended at the end of all input documents. For instance, Concat-End-Breast-1 means adding one "breast" perturbation word at the end of the documents.

4.3.2. Edit Adversaries

We apply the following edit adversaries to the text dataset:

- Edit-Synthetic: In this attack, we perturb the letters of all the selected words (breast, leukemia/lymphoma, and sarcoma) whenever they appear in the document text. Different approaches are applied to edit the targeted tokens, such as swapping all two neighboring characters (Swap), randomly deleting one character (Delete), randomly changing orders of all character (Fully Random), or randomly changing orders of characters except the first and last ones (Middle Random).
- Edit-Replacing: In this attack, all class label names (the 25 different labels) are replaced with one of the target words (breast, leukemia/lymphoma, or sarcoma) whenever they appear in the unstructured text. For instance, Edit-Replacing-Breast means all class label names that appear in the input document text are replaced with the word "breast" as the perturbed word.

4.4. Defense

In defense, all class label names are filtered from the input documents during the new model training. Then, we attack the model using the same AEs as before to investigate the word-level and character-level adversarial training impacts on enhancing the CNN model's robustness.

5. Results

In this section, we present the results related to each experiment.

5.1. Concatenation Adversaries

Figure 1 illustrates the impact of increasing the number of perturbed words on the overall accuracy. We can see, as expected, that the drop in accuracy increases when adding more perturbation words to the document text.

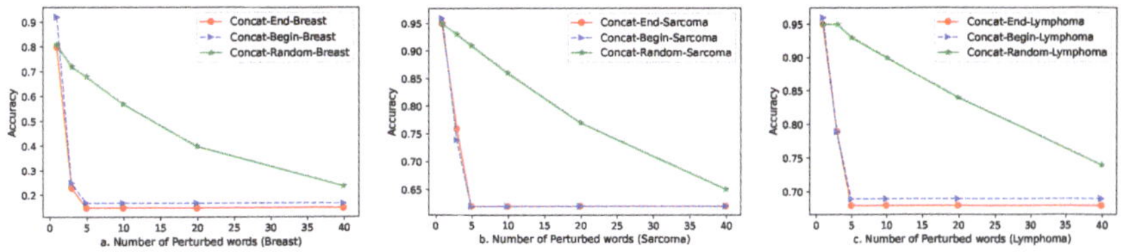

Figure 1. Impact of increasing number of words in Concatenation adversaries; for (**a**). breast, (**b**). sarcoma and (**c**). lymphoma.

The figure also shows that Concat-Random's accuracy degrades slowly with an increasing number of perturbation words; however, in Concat-Begin and Concat-End, there is a sharp drop in accuracy by adding only 3 perturbation words and this decrease continues until adding 5 words. Adding more than 5 words does not change the accuracy. This indicates that if the perturbation words are adjacent in the input text, they have higher impact on the model predictions.

Another observation is the different impact of the selected perturbed words (breast, sarcoma and lymphoma) on the overall model accuracy. From the accuracy values for each class, we see that accuracy drop in breast as a majority class is significant, as adding 3 words causes accuracy to become less than 30%. However, in lymphoma and sarcoma as minority and moderate classes, accuracy drops to 79% and 74%, respectively.

In Table 1, a comparison between different concatenation adversaries is provided. In this table, we consider 3 perturbed words. Compared with the baseline model, we can see that adding only 3 words can reduce the accuracy significantly, which is an indication of the effectiveness of the attack. From the results of Table 1, we came to conclude that in an imbalanced dataset and under an adversarial attack, majority classes contribute at least 3 times more than the minority classes. This conclusion is drawn from the fact that the CNN model is biased towards the majority classes in an imbalanced dataset; therefore, minority classes contribute less to the overall accuracy than majority classes.

Table 1. Comparison between different concatenation adversaries attack strategies.

Model	Micro F1			Macro F1		
	Beginning	End	Random	Beginning	End	Random
Baseline		0.9623			0.9400	
Concat-Breast-3	0.2461	0.2335	0.7193	0.2337	0.2029	0.7501
Concat-Sarcoma-3	0.7429	0.7594	0.9261	0.6666	0.6818	0.8794
Concat-lymphoma-3	0.7862	0.7932	0.9465	0.7262	0.7367	0.9028

To gain more insight on the impact of concatenation adversaries, we investigate the accuracy per class. Figure 2 illustrates the accuracy of each class when the perturbed word is "breast" for Concat-End attack. The figures for the other two perturbation words and the other concatenation strategies are included in Appendix B. The interesting observation is that adding the perturbed word contributes in an accuracy drop of all classes except the "breast" class. In other words, the adversarial attack was able to fool the CNN model to a target attack and give 100% accuracy for the perturbed class word.

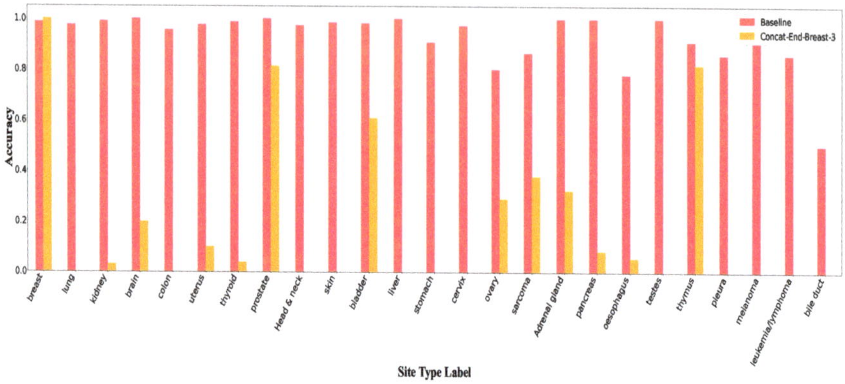

Figure 2. Accuracy per class in Concat-End for breast.

With further analysis, we also realize that adding the perturbed word causes an increase in number of false predictions such that the CNN model is most likely to classify the documents of other classes as the class equal to the perturbed word. Table 2 shows the number of documents classified as the perturbed word after an adversarial attack.

While analysing the two-term word class names, such as "leukemia/lymphoma", "bile duct" and "head and neck", we noticed that such classes seem to have one term neutral which does not cause any changes in the accuracy; however, the other term follows almost the same pattern as the other single-term word class names in the dataset. To find the reason, we looked into the input dataset to see the occurrence of each word in the whole dataset (Table A1 in Appendix B). We found that the term that occurred more often is likely to impact the performance more under adversarial attacks.

Table 2. Number of documents classified as the perturbed word before and after adversarial attack.

	Number of Documents
Baseline-breast	134 out of 1272
Concat-Random-Breast-1	359
Concat-Random-Breast-10	671
Concat-Random-Breast-20	878
Baseline-sarcoma	31 out of 1272
Concat-Random-sarcoma-1	61
Concat-Random-sarcoma-10	196
Concat-Random-sarcoma-20	312
Baseline-lymphoma	6 out of 1272
Concat-Random-lymphoma-1	22
Concat-Random-sarcoma-10	90
Concat-Random-lymphoma-20	179

5.2. Edit Adversaries

Table 3 depicts the comparison of accuracy on different edit adversaries attacks. As we can see from the results, compared to the baseline model, all edit adversaries attack strategies degrade the accuracy. We also see that all character-level perturbations cause the same amount of drop in accuracy (4% in micro F1 and 6% in macro F1). The reason is that, only class names have been targeted in this set of experiments and no matter how they are edited, the CNN model interprets them all as unknown words; therefore, they all contribute in the same amount of accuracy drop. This also confirms that there are keywords other than the class names that are critical to the class prediction. On the contrary, Edit-Replacing strategies result in a significant decrease in accuracy (12% in micro F1 and 17% in macro F1) and (58% in micro F1 and 44% in macro F1) when all 25 class names in the text are replaced with "lymphoma" and "breast" perturbation words, respectively. It shows that although the CNN model is biased towards all class names, majority classes seem to have a more significant impact than the minority. Figure 3 shows accuracy per class under Edit-Synthetic adversarial attack. From the figure, we see that minority classes are impacted more than majority classes. Figures of accuracy per class in Edit-Replacing attacks for breast, sarcoma and lymphoma are included in Appendix B.

Table 3. Comparison between different edit adversaries attack strategy.

	Micro F1	Macro F1
Baseline	0.9623	0.9400
Swap	0.9230	0.8815
Delete	0.9230	0.8815
Fully Random	0.9230	0.8815
Middle Random	0.9230	0.8815
Edit-Replacing-Breast	0.3774	0.4209
Edit-Replacing-Sarcoma	0.7987	0.7366
Edit-Replacing-Lymphoma	0.8373	0.7648

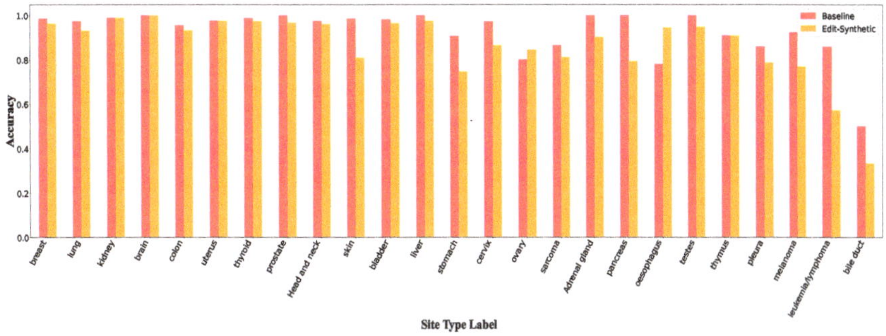

Figure 3. Accuracy per class in Edit-Synthetic.

5.3. Defense

Tables 4 and 5 demonstrate the performance results of the CNN model after filtering the class names from the text during the training, as well as the model performance under adversarial attacks using the concatenation and edit adversaries. From the result, we can easily see that the defense strategy was able to successfully defend against adversarial attacks with little to no degradation of the performance of the baseline CNN model under the same adversarial attack. From the macro-F1 score, we see that after performing the defense strategy, the accuracy of minority classes increases while the accuracy of majority classes remains unchanged; so, we came to conclude that the defense strategy is able to enhance the CNN model's robustness not only by immunizing the model against adversarial attack but also by tackling the class imbalance problem as well.

Table 4. Comparison between different concatenation adversaries attack strategies while defense strategy is imposed.

Model	Micro F1			Macro F1		
	Beginning	End	Random	Beginning	End	Random
Baseline		0.9544			0.9240	
Concat-Breast-3	0.9544	0.9544	0.9544	0.9240	0.9240	0.9243
Concat-Sarcoma-3	0.9544	0.9544	0.9544	0.9240	0.9243	0.9243
Concat-lymphoma-3	0.9544	0.9544	0.9544	0.9240	0.9240	0.9243

Table 5. Overall micro/macro F1 by performing defense.

	Micro F1	Macro F1
Baseline	0.9544	0.9240
Swap	0.9583	0.9369
Delete	0.9583	0.9369
Fully Random	0.9583	0.9369
Middle Random	0.9583	0.9369
Edit-Replacing-Breast	0.9583	0.9369
Edit-Replacing-Sarcoma	0.9583	0.9369
Edit-Replacing-Lymphoma	0.9583	0.9369

6. Conclusions

In this paper, we investigate the problem of adversarial attacks on unstructured clinical datasets. Our work demonstrates the vulnerability of the CNN model in clinical document classification tasks, specifically cancer pathology reports. We apply various black-box attacks based on concatenation and edit adversaries; then, using the proposed defense technique, we are able to enhance the robustness of the CNN model under adversarial attacks. Experimental results show that adding a few perturbation words as AEs to the input data will drastically decrease the model accuracy. We also indicate that by filtering the class names in the input data, the CNN model will be robust to such adversarial attacks. Furthermore, this defense technique is able to mitigate the bias of the CNN model towards the majority classes in the imbalanced clinical dataset.

Author Contributions: Conceptualization, M.A. and Q.A.; methodology, M.A. and N.F.; software, M.A. and N.F.; validation, M.A., N.F. and Q.A.; formal analysis, M.A.; investigation, M.A. and N.F.; resources, M.A.; writing, review, and editing, M.A., N.F., Q.A.; visualization, M.A. and N.F.; supervision, M.A.; project administration, M.A. All authors have read and agreed to the published version of the manuscript.

Funding: This research received no external funding.

Acknowledgments: The results published here are in whole or part based upon data generated by the TCGA Research Network: https://www.cancer.gov/tcga, accessed on 1 October 2021.

Conflicts of Interest: The authors declare no conflict of interest.

Appendix A. TCGA Dataset

Figure A1 shows the histograms of the number of occurrences per class for the cancer site.

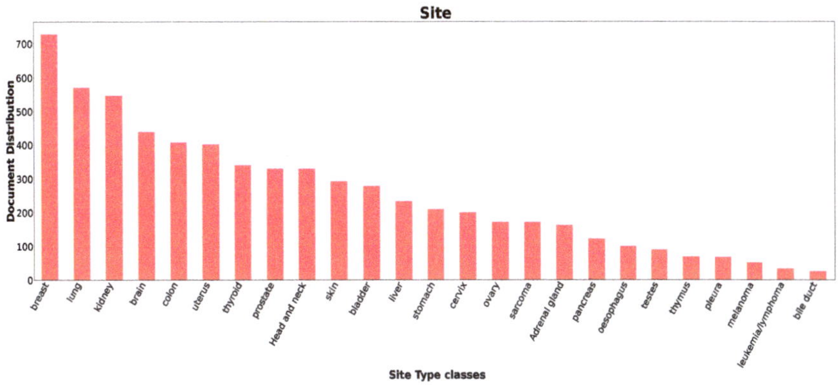

Figure A1. Classes Distribution in TCGA Dataset for Site.

Appendix B. Adversarial Attack

Table A1 shows the frequency of each term of two-term Label's classes word in the whole dataset.

Table A1. Two-term word Labels' occurrence in whole dataset.

	Occurrence
duct	1542
bile	1012
gland	2589
adrenal	1786
lymphoma	90
leukemia	3
neck	2817
head	356

Appendix B.1. Concatenation Adversaries

The overall micro- and macro- F1 scores for various number of perturbed words in Concat-End-Breast, Concat-End-sarcoma and Concat-End-lymphoma adversarial attacks are depicted in Tables A2–A4.

Table A2. Overall micro/macro F1 in Concat-End-Breast adversarial attack for various number of perturbed word.

	Micro F1	Macro F1
Baseline	0.9623	0.9400
Concat-End-Breast-1	0.8003	0.8156
Concat-End-Breast-3	0.2335	0.2029
Concat-End-Breast-5	0.1486	0.0915
Concat-End-Breast-20	0.1486	0.0915

Table A3. Overall micro/macro F1 in Concat-End-sarcoma adversarial attack for various number of perturbed word.

	Micro F1	Macro F1
Baseline	0.9623	0.9400
Concat-End-sarcoma-1	0.9520	0.9172
Concat-End-sarcoma-3	0.7594	0.6818
Concat-End-sarcoma-5	0.6156	0.5506
Concat-End-sarcoma-20	0.6156	0.5506

Table A4. Overall micro/macro F1 in Concat-End-lymphoma adversarial attack for various number of perturbed word.

	Micro F1	Macro F1
Baseline	0.9623	0.9400
Concat-End-lymphoma-1	0.9520	0.9091
Concat-End-lymphoma-3	0.7932	0.7367
Concat-End-lymphoma-5	0.6824	0.6203
Concat-End-lymphoma-20	0.6824	0.6203

The overall micro- and macro- F1 scores for various number of perturbed words in Concat-Begin-Breast, Concat-Begin-sarcoma and Concat-Begin-lymphoma adversarial attacks are depicted in Tables A5–A7.

Table A5. Overall micro/macro F1 in Concat-Begin-Breast adversarial attack for various number of perturbed word.

	Micro F1	Macro F1
Baseline	0.9623	0.9400
Concat-Begin-Breast-1	0.9198	0.9157
Concat-Begin-Breast-3	0.2461	0.2337
Concat-Begin-Breast-5	0.1682	0.1332
Concat-Begin-Breast-20	0.1682	0.1332

Table A6. Overall micro/macro F1 in Concat-Begin-sarcoma adversarial attack for various number of perturbed word.

	Micro F1	Macro F1
Baseline	0.9623	0.9400
Concat-Begin-sarcoma-1	0.9615	0.9157
Concat-Begin-sarcoma-3	0.7429	0.6666
Concat-Begin-sarcoma-5	0.6211	0.5684
Concat-Begin-sarcoma-20	0.6211	0.5684

Table A7. Overall micro/macro F1 in Concat-Begin-lymphoma adversarial attack for various number of perturbed word.

	Micro F1	Macro F1
Baseline	0.9623	0.9400
Concat-Begin-lymphoma-1	0.9638	0.9289
Concat-Begin-lymphoma-3	0.7862	0.7262
Concat-Begin-lymphoma-5	0.6863	0.6209
Concat-Begin-lymphoma-20	0.6863	0.6209

The overall micro- and macro- F1 scores for various number of perturbed words in Concat-Random-Breast, Concat-Random-lymphoma and Concat-Random-sarcoma adversarial attacks are depicted in Tables A8–A10.

Table A8. Overall micro/macro F1 in Concat-Random-Breast adversarial atttack for various number of perturbed word.

	Micro F1	Macro F1
Baseline	0.9623	0.9400
Concat-Random-Breast-1	0.8066	0.8240
Concat-Random-Breast-10	0.5660	0.6006
Concat-Random-Breast-20	0.4049	0.3992

Table A9. Overal micro/macro F1 in Concat-Random-lymphoma adversarial atttack for various number of perturbed word.

	Micro F1	Macro F1
Baseline	0.9623	0.9400
Concat-Random-lymphoma-1	0. 9520	0.9105
Concat-Random-lymphoma-10	0.9033	0.8567
Concat-Random-lymphoma-20	0.8381	0.7924

Table A10. Overall micro/macro F1 in Concat-Random-sarcoma adversarial attack for various number of perturbed word.

	Micro F1	Macro F1
Baseline	0.9623	0.9400
Concat-Random-sarcoma-1	0. 4049	0.3992
Concat-Random-sarcoma-10	0.8585	0.8051
Concat-Random-sarcoma-20	0.7720	0.7148

Figures A2–A9 illustrates the accuracy per class for each perturbed word (breast, sarcoma and lymphoma) in concatenation adversaries.

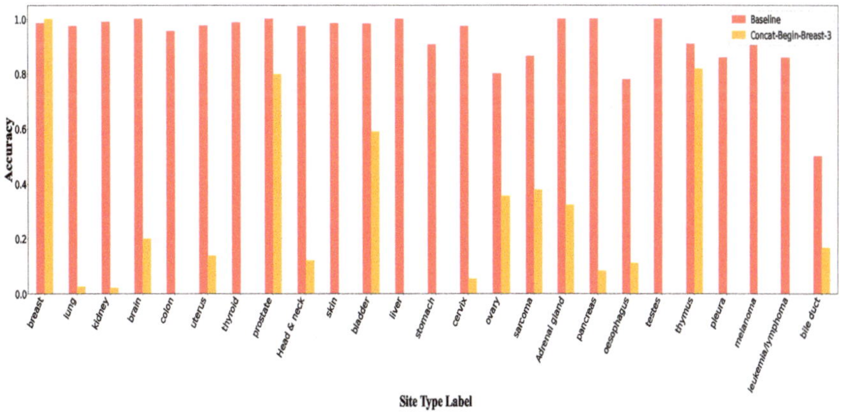

Figure A2. Accuracy per class in Concat-Begin for breast.

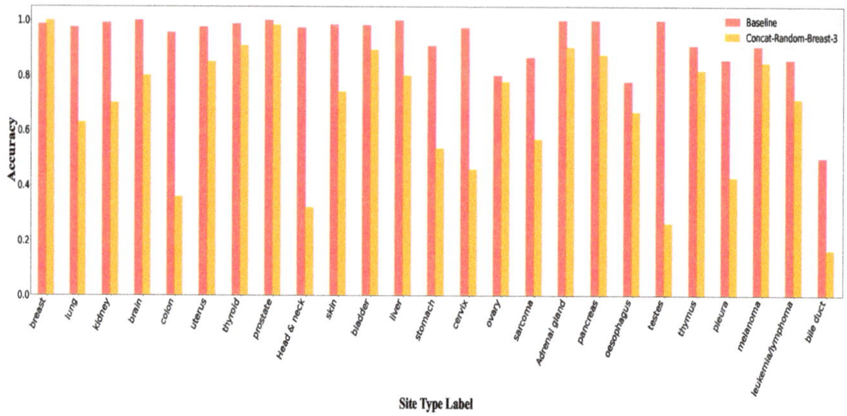

Figure A3. Accuracy per class in Concat-Random for breast.

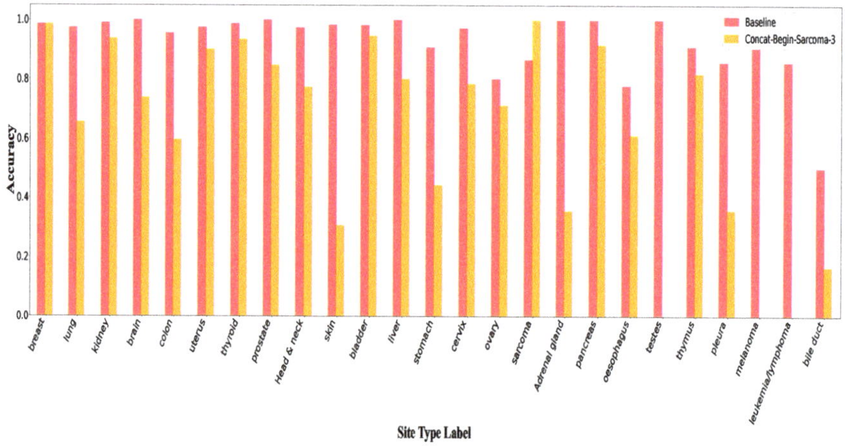

Figure A4. Accuracy per class in Concat-Begin for sarcoma.

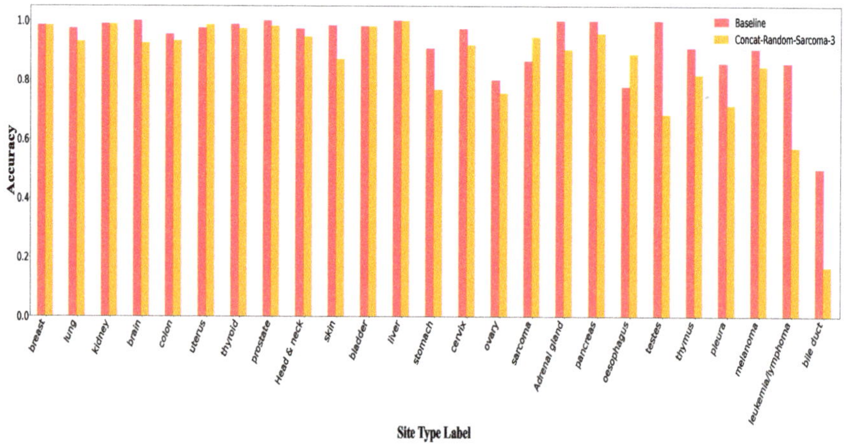

Figure A5. Accuracy per class in Concat-Random for sarcoma.

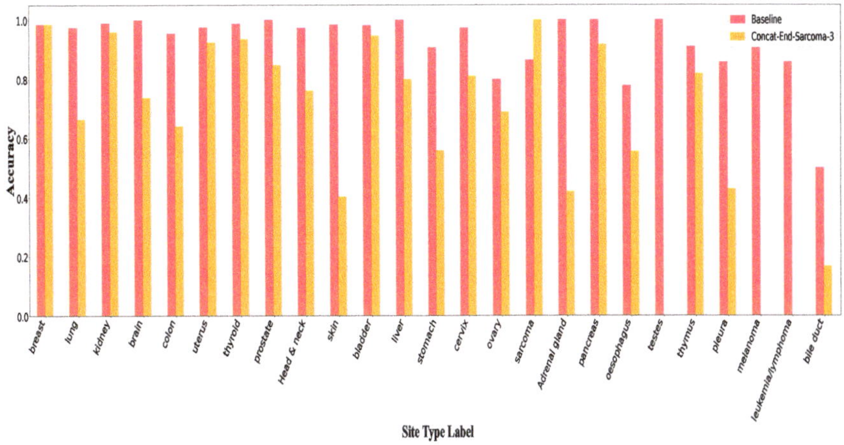

Figure A6. Accuracy per class in Concat-End for sarcoma.

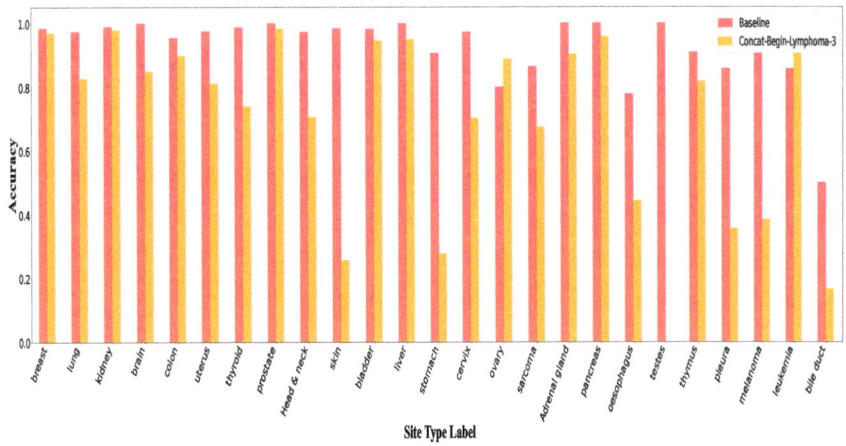

Figure A7. Accuracy per class in Concat-Begin for lymphoma.

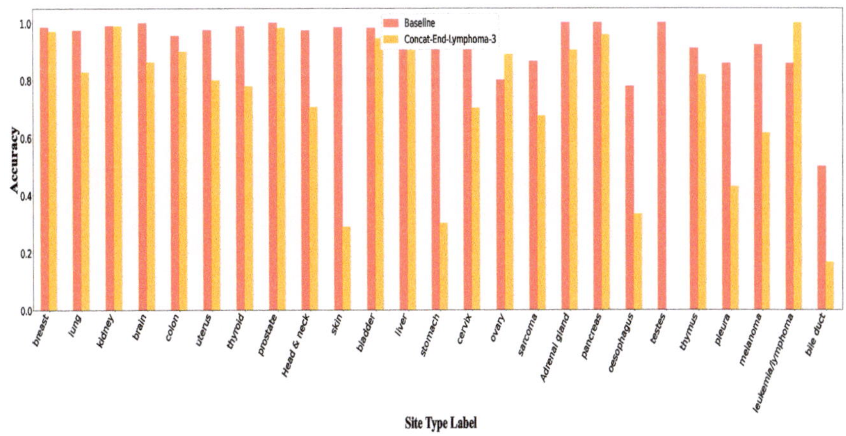

Figure A8. Accuracy per class in Concat-End for lymphoma.

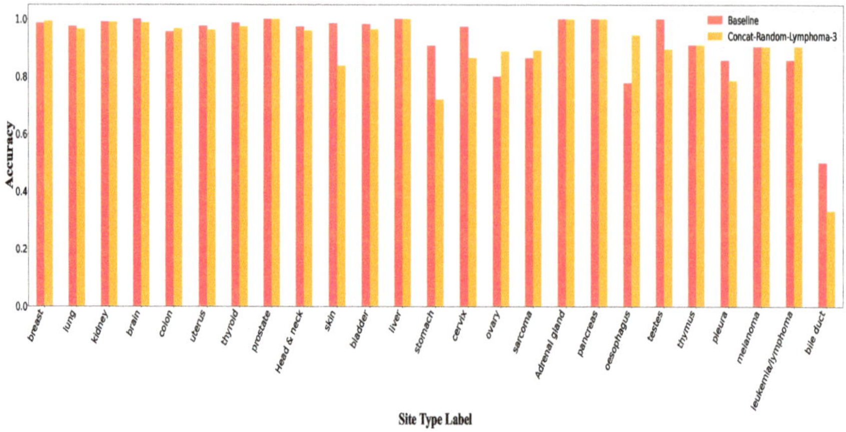

Figure A9. Accuracy per class in Concat-Random for lymphoma.

Appendix B.2. Edit Adversaries

Figures A10–A12 show accuracy per class in Edit-Replacing-Breast, Edit-Replacing-Sarcoma and Edit-Replacing-Lymphoma attacks, respectively.

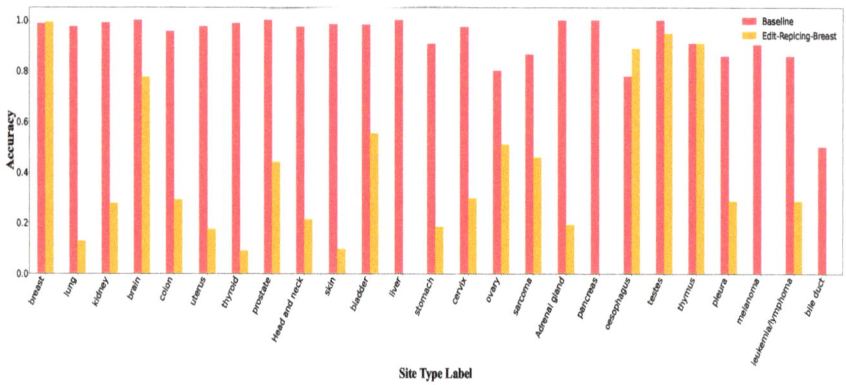

Figure A10. Accuracy per class in Edit-Replacing-breast.

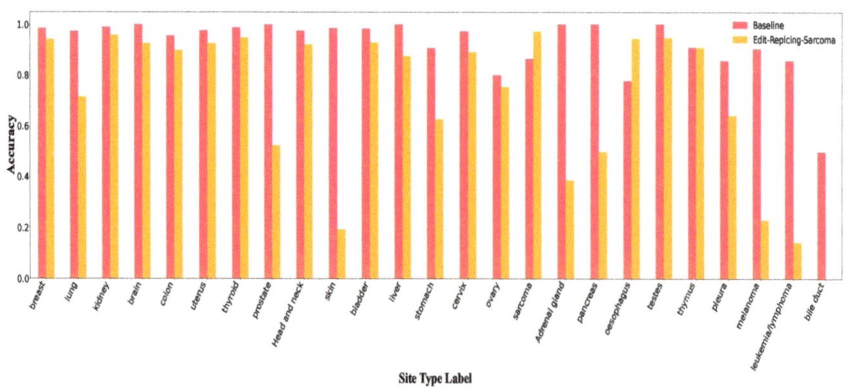

Figure A11. Accuracy per class in Edit-Replacing-Sarcoma.

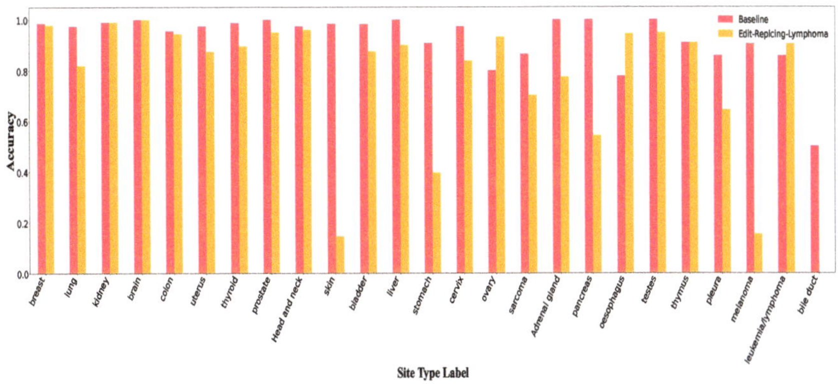

Figure A12. Accuracy per class in Edit-Replacing-lymphoma.

Appendix C. Defense

In this section, we provide figures and tables that are related to the defense under different adversarial attacks. Figures A13–A15 illustrate accuracy per class under concatenation and edit adversaries attacks when defense strategy is applied.

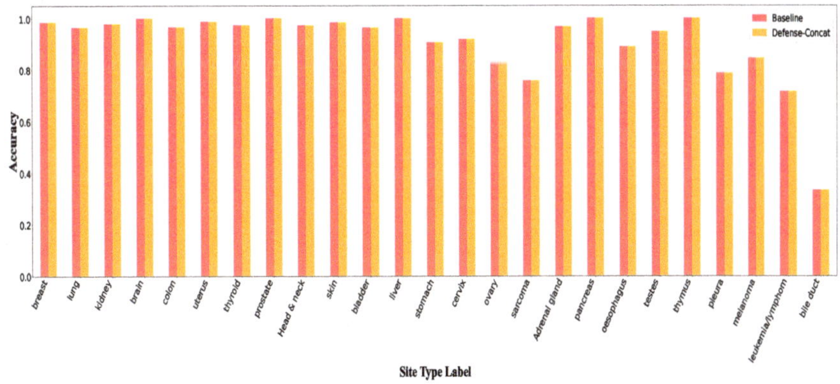

Figure A13. Accuracy per class in Defense-Concat.

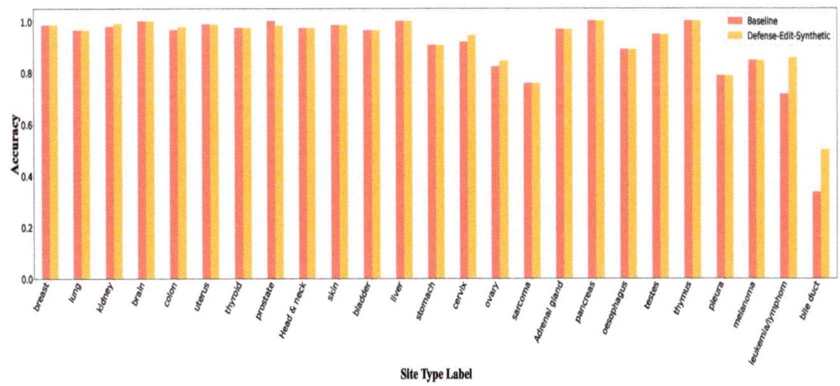

Figure A14. Accuracy per class in Defense-Edit-Synthetic.

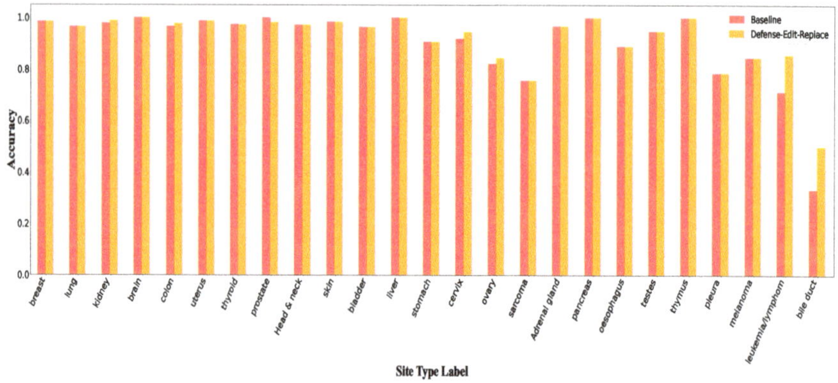

Figure A15. Accuracy per class in Defense-Edit-Replace.

Table A11 lists the results of overall micro/macro F1 by performing defense on Edit-Replacing for all classes names. From the result, we can easily see that defense strategy enhance the robustness of the CNN model.

Table A11. Overall micro/macro F1 by performing defense.

	Micro F1	**Macro F1**
Baseline	0.9544	0.9240
Edit-Replacing-Breast	0.9583	0.9369
Edit-Replacing-Lung	0.9583	0.9369
Edit-Replacing-Kidney	0.9583	0.9369
Edit-Replacing-Brain	0.9583	0.9369
Edit-Replacing-colon	0.9583	0.9369
Edit-Replacing-uterus	0.9583	0.9369
Edit-Replacing-thyroid	0.9583	0.9369
Edit-Replacing-prostate	0.9583	0.9369
Edit-Replacing-head and neck	0.9583	0.9369
Edit-Replacing-skin	0.9583	0.9369
Edit-Replacing-bladder	0.9583	0.9369
Edit-Replacing-liver	0.9583	0.9369
Edit-Replacing-stomach	0.9583	0.9369
Edit-Replacing-cervix	0.9583	0.9369
Edit-Replacing-ovary	0.9583	0.9369
Edit-Replacing-sarcoma	0.9583	0.9369
Edit-Replacing-adrenal gland	0.9583	0.9369
Edit-Replacing-pancreas	0.9583	0.9369
Edit-Replacing-oesophagus	0.9583	0.9369
Edit-Replacing-testes	0.9583	0.9369
Edit-Replacing-thymus	0.9583	0.9369
Edit-Replacing-melanoma	0.9583	0.9369
Edit-Replacing-leukemia/lymphoma	0.9583	0.9369
Edit-Replacing-bile duct	0.9583	0.9369

References

1. Köksal, Ö.; Akgül, Ö. A Comparative Text Classification Study with Deep Learning-Based Algorithms. In Proceedings of the 2022 9th International Conference on Electrical and Electronics Engineering (ICEEE), Alanya, Turkey, 29–31 March 2022; IEEE: New York, NY, USA, 2022; pp. 387–391.
2. Varghese, M.; Anoop, V. Deep Learning-Based Sentiment Analysis on COVID-19 News Videos. In Proceedings of the International Conference on Information Technology and Applications, Lisbon, Portugal, 20–22 October 2022; Spinger: Berlin/Heidelberg, Germany, 2022; pp. 229–238.

3. Affi, M.; Latiri, C. BE-BLC: BERT-ELMO-Based deep neural network architecture for English named entity recognition task. *Procedia Comput. Sci.* **2021**, *192*, 168–181. [CrossRef]
4. Zhang, W.E.; Sheng, Q.Z.; Alhazmi, A.; Li, C. Adversarial attacks on deep-learning models in natural language processing: A survey. *ACM Trans. Intell. Syst. Technol. (TIST)* **2020**, *11*, 1–41. [CrossRef]
5. Alawad, M.; Yoon, H.J.; Tourassi, G.D. Coarse-to-fine multi-task training of convolutional neural networks for automated information extraction from cancer pathology reports. In Proceedings of the 2018 IEEE EMBS International Conference on Biomedical & Health Informatics (BHI), Las Vegas, NV, USA, 4–7 March 2018; pp. 218–221. [CrossRef]
6. Olthof, A.W.; van Ooijen, P.M.A.; Cornelissen, L.J. Deep Learning-Based Natural Language Processing in Radiology: The Impact of Report Complexity, Disease Prevalence, Dataset Size, and Algorithm Type on Model Performance. *J. Med. Syst.* **2021**, *45*. [CrossRef] [PubMed]
7. Wang, Y.; Bansal, M. Robust machine comprehension models via adversarial training. *arXiv* **2018**, arXiv:1804.06473.
8. Suya, F.; Chi, J.; Evans, D.; Tian, Y. Hybrid batch attacks: Finding black-box adversarial examples with limited queries. In Proceedings of the 29th USENIX Security Symposium (USENIX Security 20), Boston, MA, USA, 12–14 August 2020; pp. 1327–1344.
9. Yala, A.; Barzilay, R.; Salama, L.; Griffin, M.; Sollender, G.; Bardia, A.; Lehman, C.; Buckley, J.M.; Coopey, S.B.; Polubriaginof, F.; et al. Using Machine Learning to Parse Breast Pathology Reports. *bioRxiv* **2016**. [CrossRef] [PubMed]
10. Buckley, J.M.; Coopey, S.B.; Sharko, J.; Polubriaginof, F.C.G.; Drohan, B.; Belli, A.K.; Kim, E.M.H.; Garber, J.E.; Smith, B.L.; Gadd, M.A.; et al. The feasibility of using natural language processing to extract clinical information from breast pathology reports. *J. Pathol. Inform.* **2012**, 3, 23. [CrossRef] [PubMed]
11. Devlin, J.; Chang, M.W.; Lee, K.; Toutanova, K. BERT: Pre-training of Deep Bidirectional Transformers for Language Understanding. In Proceedings of the NAACL-HLT, Minneapolis, MN, USA, 2–7 June 2019; pp. 4171–4186.
12. Gao, S.; Alawad, M.; Young, M.T.; Gounley, J.; Schaefferkoetter, N.; Yoon, H.J.; Wu, X.C.; Durbin, E.B.; Doherty, J.; Stroup, A.; et al. Limitations of Transformers on Clinical Text Classification. *IEEE J. Biomed. Health Inform.* **2021**, *25*, 3596–3607. [CrossRef] [PubMed]
13. Chakraborty, A.; Alam, M.; Dey, V.; Chattopadhyay, A.; Mukhopadhyay, D. Adversarial Attacks and Defences: A Survey, 2018. *arXiv* **2018**, arXiv:1810.00069.
14. Long, T.; Gao, Q.; Xu, L.; Zhou, Z. A survey on adversarial attacks in computer vision: Taxonomy, visualization and future directions. *Comput. Secur.* **2022**, *121*, 102847. [CrossRef]
15. Simoncini, W.; Spanakis, G. SeqAttack: On adversarial attacks for named entity recognition. In Proceedings of the 2021 Conference on Empirical Methods in Natural Language Processing: System Demonstrations, Punta Cana, Dominican Republic, 7–11 November 2021; pp. 308–318.
16. Araujo, V.; Carvallo, A.; Aspillaga, C.; Parra, D. On adversarial examples for biomedical nlp tasks. *arXiv* **2020**, arXiv:2004.11157.
17. Jin, D.; Jin, Z.; Zhou, J.T.; Szolovits, P. Is bert really robust? a strong baseline for natural language attack on text classification and entailment. In Proceedings of the AAAI Conference on Artificial Intelligence, New York, NY, USA, 7–12 February 2020; Volume 34, pp. 8018–8025.
18. Gao, J.; Lanchantin, J.; Soffa, M.L.; Qi, Y. Black-box generation of adversarial text sequences to evade deep learning classifiers. In Proceedings of the 2018 IEEE Security and Privacy Workshops (SPW), San Francisco, CA, USA, 24–24 May 2018; IEEE: New York, NY, USA, 2018; pp. 50–56.
19. Yuan, L.; Zheng, X.; Zhou, Y.; Hsieh, C.J.; Chang, K.W. On the Transferability of Adversarial Attacksagainst Neural Text Classifier. *arXiv* **2020**, arXiv:2011.08558.
20. Pei, W.; Yue, C. Generating Content-Preserving and Semantics-Flipping Adversarial Text. In Proceedings of the 2022 ACM on Asia Conference on Computer and Communications Security, Nagasaki, Japan, 30 May–3 June 2022; pp. 975–989.
21. Finlayson, S.G.; Kohane, I.S.; Beam, A.L. Adversarial Attacks Against Medical Deep Learning Systems. *CoRR* **2018**, *abs/1804.05296*. Available online: http://xxx.lanl.gov/abs/1804.05296 (accessed on 1 December 2022).
22. Mondal, I. BBAEG: Towards BERT-based biomedical adversarial example generation for text classification. *arXiv* **2021**, arXiv:2104.01782.
23. Zhang, R.; Zhang, W.; Liu, N.; Wang, J. Susceptible Temporal Patterns Discovery for Electronic Health Records via Adversarial Attack. In Proceedings of the International Conference on Database Systems for Advanced Applications, Taipei, Taiwan, 11–14 April; Springer: Berlin/Heidelberg, Germany, 2021; pp. 429–444.
24. Sun, M.; Tang, F.; Yi, J.; Wang, F.; Zhou, J. Identify susceptible locations in medical records via adversarial attacks on deep predictive models. In Proceedings of the 24th ACM SIGKDD International Conference on Knowledge Discovery & Data Mining, London, UK, 19–23 August 2018; pp. 793–801.
25. Xu, H.; Ma, Y.; Liu, H.C.; Deb, D.; Liu, H.; Tang, J.L.; Jain, A.K. Adversarial attacks and defenses in images, graphs and text: A review. *Int. J. Autom. Comput.* **2020**, *17*, 151–178. [CrossRef]
26. Szegedy, C.; Zaremba, W.; Sutskever, I.; Bruna, J.; Erhan, D.; Goodfellow, I.; Fergus, R. Intriguing properties of neural networks. *arXiv* **2013**, arXiv:1312.6199.
27. Wang, W.; Park, Y.; Lee, T.; Molloy, I.; Tang, P.; Xiong, L. Utilizing Multimodal Feature Consistency to Detect Adversarial Examples on Clinical Summaries. In Proceedings of the 3rd Clinical Natural Language Processing Workshop, Online, 19 November 2020; pp. 259–268.
28. Belinkov, Y.; Bisk, Y. Synthetic and natural noise both break neural machine translation. *arXiv* **2017**, arXiv:1711.02173.

29. Alawad, M.; Gao, S.; Qiu, J.; Schaefferkoetter, N.; Hinkle, J.D.; Yoon, H.J.; Christian, J.B.; Wu, X.C.; Durbin, E.B.; Jeong, J.C.; et al. Deep transfer learning across cancer registries for information extraction from pathology reports. In Proceedings of the 2019 IEEE EMBS International Conference on Biomedical & Health Informatics (BHI), Chicago, IL, USA, 19–22 May 2019; IEEE: New York, NY, USA, 2019; pp. 1–4. [CrossRef]
30. Gao, S.; Alawad, M.; Schaefferkoetter, N.; Penberthy, L.; Wu, X.C.; Durbin, E.B.; Coyle, L.; Ramanathan, A.; Tourassi, G. Using case-level context to classify cancer pathology reports. *PLoS ONE* **2020**, *15*, e0232840. [CrossRef] [PubMed]

electronics

MDPI

Article

An Improved Hierarchical Clustering Algorithm Based on the Idea of Population Reproduction and Fusion

Lifeng Yin [1], Menglin Li [2], Huayue Chen [3,*] and Wu Deng [4,5,*]

1 School of Software, Dalian Jiaotong University, Dalian 116000, China
2 School of Computer and Communication Engineering, Dalian Jiaotong University, Dalian 116028, China
3 School of Computer Science, China West Normal University, Nanchong 637002, China
4 School of Electronic Information and Automation, Civil Aviation University of China, Tianjin 300300, China
5 Traction Power State Key Laboratory, Southwest Jiaotong University, Chengdu 610031, China
* Correspondence: sunnyxiaoyue20@cwnu.edu.cn (H.C.); wdeng@cauc.edu.cn (W.D.)

Abstract: Aiming to resolve the problems of the traditional hierarchical clustering algorithm that cannot find clusters with uneven density, requires a large amount of calculation, and has low efficiency, this paper proposes an improved hierarchical clustering algorithm (referred to as PRI-MFC) based on the idea of population reproduction and fusion. It is divided into two stages: fuzzy pre-clustering and Jaccard fusion clustering. In the fuzzy pre-clustering stage, it determines the center point, uses the product of the neighborhood radius *eps* and the dispersion degree *fog* as the benchmark to divide the data, uses the Euclidean distance to determine the similarity of the two data points, and uses the membership grade to record the information of the common points in each cluster. In the Jaccard fusion clustering stage, the clusters with common points are the clusters to be fused, and the clusters whose Jaccard similarity coefficient between the clusters to be fused is greater than the fusion parameter *jac* are fused. The common points of the clusters whose Jaccard similarity coefficient between clusters is less than the fusion parameter *jac* are divided into the cluster with the largest membership grade. A variety of experiments are designed from multiple perspectives on artificial datasets and real datasets to demonstrate the superiority of the PRI-MFC algorithm in terms of clustering effect, clustering quality, and time consumption. Experiments are carried out on Chinese household financial survey data, and the clustering results that conform to the actual situation of Chinese households are obtained, which shows the practicability of this algorithm.

Keywords: hierarchical clustering; Jaccard distance; membership grade; community clustering

Citation: Yin, L.; Li, M.; Chen, H.; Deng, W. An Improved Hierarchical Clustering Algorithm Based on the Idea of Population Reproduction and Fusion. *Electronics* **2022**, *11*, 2735. https://doi.org/10.3390/electronics11172735

Academic Editor: Yu-Chen Hu

Received: 29 July 2022
Accepted: 26 August 2022
Published: 30 August 2022

1. Introduction

Clustering [1] is a process of dividing a set of data objects into multiple groups or clusters, so that objects in a cluster have high similarity, but it is very dissimilar to objects in other clusters [2–5]. It is also an unsupervised machine learning technique that does not require labels associated with data points [6–10]. As a data mining and machine learning tool, clustering has been rooted in many application fields, such as pattern recognition, image analysis, statistical analysis, business intelligence, and other fields [11–15]. In addition, the feature selection methods are also proposed to deal with data [16].

The basic idea of the hierarchical clustering algorithm [17] is to construct the hierarchical relationship between data for clustering. The obtained clustering result has the characteristics of a tree structure, which is called a clustering tree. It is mainly performed using two methods, agglomeration techniques such as AGNE (agglomeration analysis) and divisive techniques such as DIANA (division analysis) [18]. Regardless of agglomeration technology or splitting technology, a core problem is measuring the distance between two clusters, and time is basically spent on distance calculation. Therefore, a large number of improved algorithms that use different means to reduce the number of distance calculations have been proposed one after another to improve algorithmic efficiency [19–27].

Guha et al. [28] proposed the CURE algorithm, which considers sampling the data in the cluster and uses the sampled data as representative of the cluster to reduce the amount of calculation of pairwise distances. The Guha team [29] improved CURE and proposed the ROCK algorithm, which can handle non-standard metric data (non-Euclidean space, graph structure, etc.). Karypis et al. [30] proposed the Chameleon algorithm, which uses the K-nearest-neighbor method to divide the data points into many small cluster sub-clusters in a two-step clustering manner before hierarchical aggregation in order to reduce the number of iterations for hierarchical aggregation. Gagolewski et al. [31] proposed the Genie algorithm which calculates the Gini index of the current cluster division before calculating the distance between clusters. If the Gini index exceeds the threshold, the merging of the smallest clusters is given priority to reduce pairwise distance calculation. Another hierarchical clustering idea is to incrementally calculate and update the data nodes and clustering features (abbreviated CF) of clusters to construct a CF clustering tree. The earliest proposed CF tree algorithm BIRCH [32] is a linear complexity algorithm. When a node is added, the number of CF nodes compared does not exceed the height of the clustering tree. While having excellent algorithm complexity, the BIRCH algorithm cannot ensure the accuracy and robustness of the clustering results, and it is extremely sensitive to the input order of the data. Kobren et al. [33] improved this and proposed the PERCH algorithm. This algorithm adds two optimization operations which are the rotation of the binary tree branch and the balance of the tree height. This greatly reduces the sensitivity of the data input order. Based on the PERCH algorithm, the PKobren team proposed the GRINCH algorithm [34] to build a single binary clustering tree. The GRINCH algorithm adds the grafting operation of two branches, allowing the ability to reconstruct, which further reduces the algorithm sensitivity to the order of data input, but, at the same time, it greatly reduces the scalability of the algorithm. Although most CF tree-like algorithms have excellent scalability, their clustering accuracy on real-world datasets is generally lower than that of classical hierarchical aggregation clustering algorithms.

To discover clusters of arbitrary shapes, density-based clustering algorithms are born. Ester et al. [35] proposed a DBSCAN algorithm based on high-density connected regions. This algorithm has two key parameters, *Eps* and *Minpts*. Many scholars at home and abroad have studied and improved the DBSCAN algorithm for the selection of *Eps* and *Minpts*. The VDBSCAN algorithm [36] selects the parameter values under different densities through the K-dist graph and uses these parameter values to cluster clusters of different densities to finally find clusters of different densities. The AF-DBSCAN algorithm [37] is an algorithm for adaptive parameter selection, which adaptively calculates the optimal global parameters *Eps* and *MinPts* according to the KNN distribution and mathematical statistical analysis. The KANN-DBSCAN algorithm [38] is based on the parameter optimization strategy and automatically determines the *Eps* and *Minpts* parameters by automatically finding the change and stable interval of the cluster number of the clustering results to achieve a high-accuracy clustering process. The KLS-DBSCAN algorithm [39] uses kernel density estimation and the mathematical expectation method to determine the parameter range according to the data distribution characteristics. The reasonable number of clusters in the data set is calculated by analyzing the local density characteristics, and it uses the silhouette coefficient to determine the optimal *Eps* and *MinPts* parameters. The MAD-DBSCAN algorithm [40] uses the self-distribution characteristics of the denoised attenuated datasets to generate a list of candidate *Eps* and *MinPts* parameters. It selects the corresponding *Eps* and *MinPts* as the initial density threshold according to the denoising level in the interval where the number of clusters tends to be stable.

To represent the uncertainty present in the data, Zadeh [41] proposed the concept of fuzzy sets, which allow elements to contain rank membership values from the interval [0, 1]. Correspondingly, the widely used fuzzy C-means clustering algorithm [42] is proposed, and many variants have appeared since then. However, membership levels alone are not sufficient to deal with the uncertainty that exists in the data. With the introduction of the hesitation class by Atanassov, Intuitive Fuzzy Sets (IFS) [43] emerge, in which a pair of membership and non-membership values for an element is used to represent the uncertainty present in the data. Due to its better uncertainty management capability, IFS is used in various clustering techniques, such as Intuitionistic Fuzzy C-means (IFCM) [44], improved IFCM [45], probabilistic intuitionistic fuzzy C-means [46,47], Intuitive Fuzzy Hierarchical Clustering (IFHC) [48], and Generalized Fuzzy Hierarchical Clustering (GHFHC) [49].

Most clustering algorithms assign each data object to one of several clusters, and such cluster assignment rules are necessary for some applications. However, in many applications, this rigid requirement may not be what we expect. It is important to study the vague or flexible assignment of which cluster each data object is in. At present, the integration of the DBSCAN algorithm and the fuzzy idea is rarely used in hierarchical clustering research. The traditional hierarchical clustering algorithm cannot find clusters with uneven density, requires a large amount of calculation, and has low efficiency. Using the advantages of the high accuracy of classical hierarchical aggregation clustering and the advantages of the DBSCAN algorithm for clustering data with uneven density, a new hierarchical clustering algorithm is proposed based on the idea of population reproduction and fusion, which we call the hierarchical clustering algorithm of population reproduction and fusion (denoted as PRI-MFC). The PRI-MFC algorithm is divided into the fuzzy pre-clustering stage and the Jaccard fusion clustering stage.

The main contributions of this work are as follows:

1. In the fuzzy pre-clustering stage, the center point is first determined to divide the data. The benchmark of data division is the product of the neighborhood radius *eps* and the dispersion grade *fog*. The overlapping degree of the initial clusters in the algorithm can be adjusted by setting the dispersion grade *fog* so as to avoid misjudging outliers;
2. The Euclidean distance is used to determine the similarity of two data points, and the membership grade is used to record the information of the common points in each cluster. The introduction of the membership grade solves the problem that the data points can flexibly belong to a certain cluster;
3. Comparative experiments are carried out on five artificial data sets to verify that the clustering effect of PRI-MFC is superior to that of the K-means algorithm;
4. Extensive simulation experiments are carried out on six real data sets. From the comprehensive point of view of the measurement indicators of clustering quality, the PRI-MFC algorithm has better clustering quality;
5. Experiments on six real data sets show that the time consumption of the PRI-MFC algorithm is negatively correlated with the parameter *eps* and positively correlated with the parameter *fog*, and the time consumption of the algorithm is also better than that of most algorithms;
6. In order to prove the practicability of this algorithm, a cluster analysis of household financial groups is carried out using the data of China's household financial survey.

The rest of this paper is organized as follows: Section 2 briefly introduces the relevant concepts required in this paper. Section 3 introduces the principle of the PRI-MFC algorithm. Section 4 introduces the implementation steps and flow chart of the PRI-MFC algorithm. Section 5 presents experiments on the artificial datasets, various UCI datasets, and the Chinese Household Finance Survey datasets. Finally, Section 6 contains the conclusion of the work.

2. Related Concepts

This section introduces the related concepts involved in the PRI-MFC algorithm.

2.1. Data Normalization

The multi-index evaluation system, due to the different nature of each evaluation index, usually has different dimensions and orders of magnitude. When the level of each index differs greatly if the original index value is directly used for analysis, the role of the index with a higher numerical value in the comprehensive analysis will be highlighted, and the effect of the index with a lower numerical level will be relatively weakened. Therefore, in order to ensure the reliability of the results, it is necessary to standardize the original indicator data. The normalization of data is performed to scale the data so that it falls into a small specific interval. It removes the unit limitation of the data and converts it into a pure, dimensionless value so that the indicators of different units or magnitudes can be compared and weighted.

Data standardization methods can be roughly divided into three categories; linear methods, such as the extreme value method and the standard deviation method; broken line methods, such as the three-fold line method; and curve methods, such as the half-normal distribution. This paper adopts the most commonly used z-score normalization (zero-mean normalization) method [50], which is defined as Formula (1).

$$x^* = \frac{x - \mu}{\sigma} \tag{1}$$

Among them, x^* are the transformed data, x are the original data, μ is the mean of all sample data, and σ is the standard deviation of all sample data. Normalized data are normally distributed with mean 0 and variance 1.

2.2. Membership Grade

In many clustering cases, the objects in the datasets cannot be divided into clearly separated clusters, and absolutely assigning an object to a specific cluster can go wrong. By assigning a weight to each object and each cluster and using the weight to indicate the degree to which an object belongs to a certain cluster, the accuracy of clustering can be improved.

Fuzzy C-means (FCM) incorporates the essence of fuzzy theory and is a clustering algorithm that uses membership grade to determine the degree to which each data point belongs to a certain cluster. The term ambiguity refers to something that is not clear or ambiguous. Any changing event, process, or function cannot always be defined as true or false. These activities need to be defined in an ambiguous way. Fuzzy logic is similar to human decision-making methods. It is able to deal with vague and imprecise information. Problems in the real world are often oversimplified to represent the existence of things in terms of true or false or Boolean logic. In fuzzy systems, the existence of things is represented by a number between 0 and 1. Fuzzy sets contain elements that satisfy imprecise membership properties, and membership grade [51] is used to determine the degree to which each element belongs to a certain cluster.

Assuming that any mapping from the universe X to the closed interval [0, 1] determines a fuzzy set A on X, then the fuzzy set A can be written as Formula (2).

$$A = \{(x, \mu_A(x)) | x \in X\} \tag{2}$$

Among them, $\mu_A(x)$ is the membership grade of x to fuzzy set A. When a certain point in X makes $\mu_A(x) = 0.5$, the point is called the transition point of fuzzy set A, which has the strongest ambiguity.

2.3. Similarity

In a cluster analysis, the measurement of similarity between different samples is its core. The similarity measurement methods involved in the PRI-MFC algorithm are the Euclidean distance [52] and the Jaccard similarity coefficient [53]. Euclidean distance is a commonly used definition of distance, which refers to the true distance between two points in n-dimensional space. Assuming that there are two points x and y in the n-dimensional space, the Euclidean distance formula is shown in (3). The featured parameters in the Euclidean distance are equally weighted, and different dimensions are treated equally.

$$D(x, y) = \left(\sum_{m=1}^{n} |x_m - y_m|^2\right)^{\frac{1}{2}} \tag{3}$$

The Jaccard similarity coefficient can also be used to measure the similarity of samples. Suppose there are two n-dimensional binary vectors X_1 and X_2, and each dimension of X_1 and X_2 can only be 0 or 1. M_{00} represents the number of dimensions in which both vector X_1 and vector X_2 are 0, M_{01} represents the number of dimensions in which vector X_1 is 0 and vector X_2 is 1, M_{10} represents the number of dimensions in which vector X_1 is 1 and vector X_2 is 0, and M_{11} represents the number of dimensions in which vector X_1 is 1 and vector X_2 are 1. Then each dimension of the n-dimensional vector falls into one of these four classes, so Formula (4) is established.

$$M_{00} + M_{01} + M_{10} + M_{11} = n \tag{4}$$

The Jaccard similarity index is shown in Formula (5). The larger the Jaccard value, the higher the similarity, and the smaller the Jaccard value, the lower the similarity.

$$J(A,B) = \frac{M_{11}}{M_{01} + M_{10} + M_{11}} \tag{5}$$

3. Principles of the PRI-MFC Algorithm

In the behavior of population reproduction and population fusion in nature, it is assumed that there are initially n non-adjacent population origin points. Then new individuals are born near the origin point, and the points close to the origin point are divided into points where races multiply. This cycle continues until all data points have been divided. At this point, the reproduction process ends, and the population fusion process begins. Since data points can belong to multiple populations in the process of dividing, there are common data points between different populations. When the common points between the populations reach a certain number, the populations merge. On the basis of this idea, this section designs and implements an improved hierarchical clustering algorithm with two clustering stages denoted as the PRI-MFC algorithm. The general process of the clustering division of the PRI-MFC is shown in Figure 1.

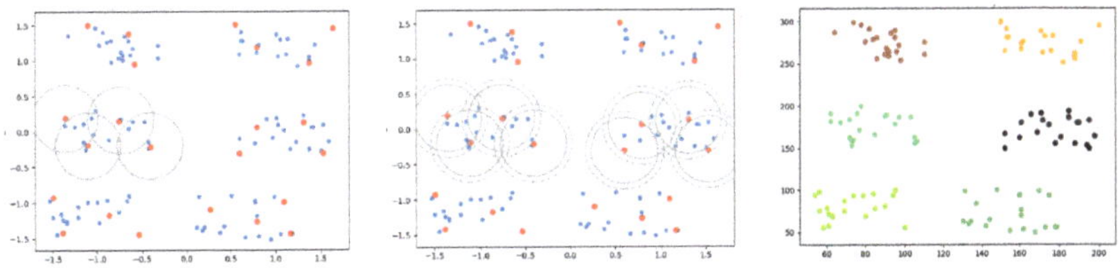

(**a**). Clustering center division (**b**). Fuzzy pre-clustering (**c**). Clustering results partition

Figure 1. Data sample division process.

In the fuzzy pre-clustering stage, based on the neighborhood knowledge of DBSCAN clustering, starting from any point in the overall data, through the neighborhood radius *eps*, multiple initial cluster center points (Suppose there are k) are divided in turn, and the non-center points are divided into the corresponding cluster centers with *eps* as the neighborhood radius, as shown in the Figure 1a. The red point in the figure is the cluster center point, and the solid line is the initial clustering. Once again, the non-central data points are divided into k cluster centers according to the neighborhood dispersion radius *eps*fog*. The same data point can be divided into multiple clusters, and finally, k clusters are formed to complete the fuzzy pre-clustering process. This process is shown in Figure 1b. The radius of the circle drawn by the dotted line in the figure is *eps*fog*, and the point of the overlapping part between the dotted circles is the common point to be divided. The Euclidean distance is used to determine the similarity of two data points, and the membership grade of a cluster to which the common point belongs is recorded. The Euclidean distance between the common point d_i and the center point c_i divided by *eps*fog* is the membership grade of ci to which d_i belongs. The neighborhood radius *eps* is taken from the definition of ε-neighborhood proposed by Stevens. The algorithm parameter *fog* is the dispersion grade. By setting *fog*, the overlapping degree of the initial clusters in the algorithm can be adjusted to avoid the misjudgment of outliers. The value range of the parameter *fog* is [1, 2.5].

In the Jaccard fusion clustering stage, the information of the common points of the clusters is counted and sorted, and the cluster groups to be fused without repeated fusion are found. Then, it sets the parameter *jac* according to the similarity coefficient of Jaccard to perform the fusion operation on the clusters obtained in the clustering fuzzy pre-clustering stage and obtains several clusters formed by the fusion of m pre-clustering small clusters. The sparse clusters with a data amount of less than three in these clusters are individually marked as outliers to form the final clustering result, as shown in Figure 1c.

The fuzzy pre-clustering of the PRI-MFC algorithm can input data in batches to prevent the situation from running out of memory caused by reading all the data into the memory at one time. The samples in the cluster are divided and stored in the records with unique labels. The pre-clustering process coarsens the original data. In the Jaccard fusion clustering stage, only the number of labels needs to be read to complete the statistics, which reduces the computational complexity of the hierarchical clustering process.

4. Implementation of PRI-MFC Algorithm

This section mainly introduces the steps, flowcharts, and pseudocode of the PRI-MFC algorithm.

4.1. Algorithm Steps and Flow Chart

Combined with optimization strategies [54,55] such as the fuzzy cluster membership grade, coarse-grained data, and staged clustering, the PRI-MFC algorithm reduces the computational complexity of the hierarchical clustering process and improves the execution efficiency of the algorithm. The implementation steps are as follows:

Step 1. Assuming that there are n data points in the data set D, it randomly selects one data point x_i, adds it to the cluster center set *centroids*, and synchronously builds the cluster dictionary *clusters* corresponding to the data center *centroids* set.

Step 2. The remaining $n - 1$ data points are compared with the points in the *centroids*, the data points whose distance is greater than the neighborhood radius *eps* are added to the *centroids*, and the *clusters* are updated to obtain all the initial cluster center points in a loop.

Step 3. It performs clustering based on *centroids* and divides the data points x_i in the data set D whose distance from the cluster center point c_i is less than *eps*fog* to the clusters with c_i as the cluster center. In the process, if x_i belongs to multiple clusters, it marks it as the point to be fused and records its belonging cluster k and membership grade in the fusion information statistical dictionary *match_dic*.

Step 4. It counts the number of common points between the clusters, merges the clusters to be fused with repeated clusters to be fused, and calculates the Jaccard similarity coefficient between the clusters to be fused.

Step 5. It fuses the clusters whose similarity between clusters is greater than the fusion parameter *jac* and divides the common points of the clusters whose similarity between clusters is less than the fusion parameter *jac* into the cluster with the largest membership grade.

Step 6. In the clustering result obtained in step 5, the clusters with less than three data in the cluster are classified as outliers.

Step 7. The clustering is completed, and the clustering result is output.

Through the description of the above algorithm steps, the obtained PRI-MFC algorithm flowchart is shown in Figure 2.

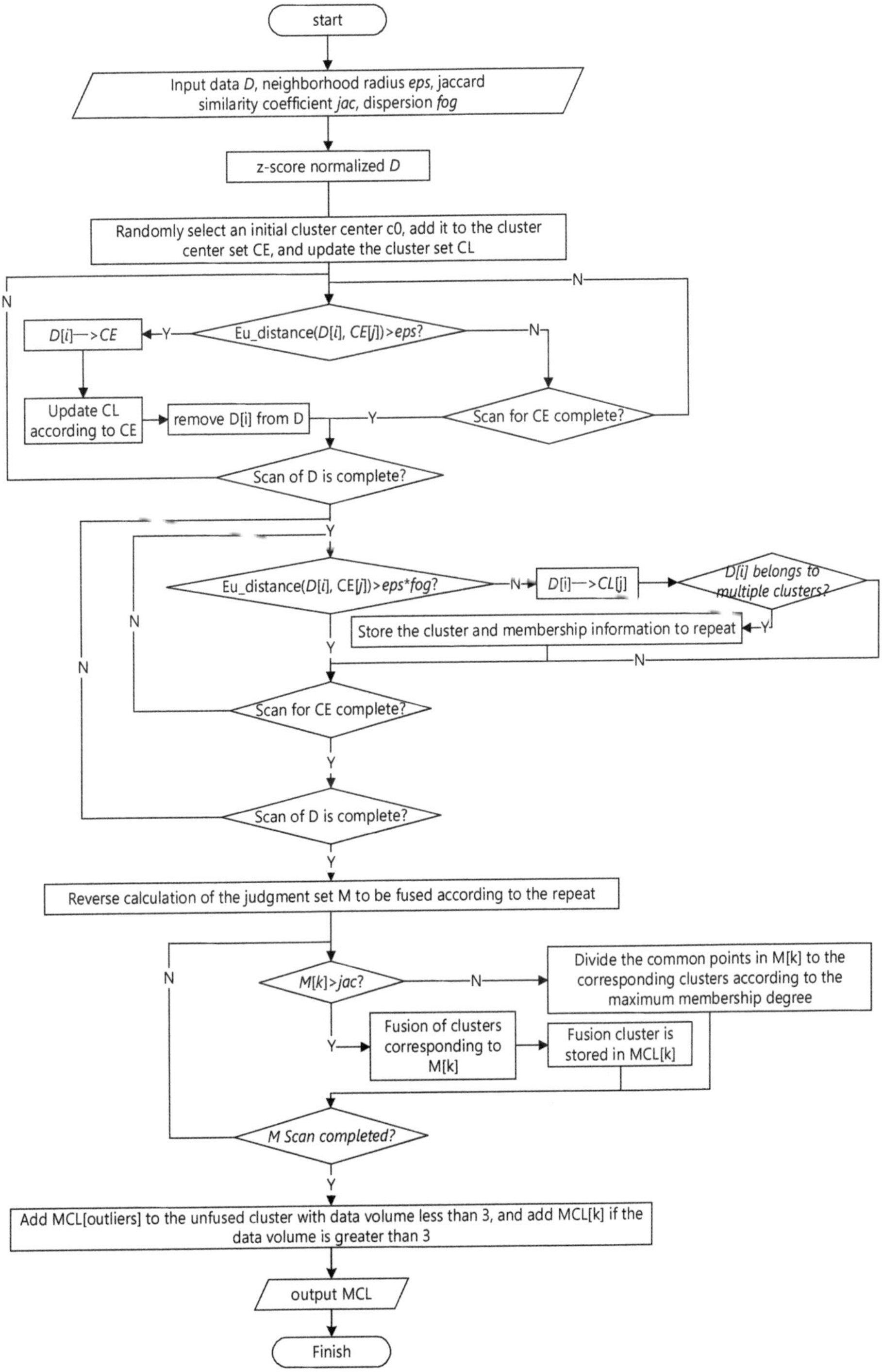

Figure 2. PRI-MFC algorithm flow chart.

4.2. Pseudocode of the Improved Algorithm

The pseudo-code of the PRI-MFC Algorithm 1 is as follows:

Algorithm 1 PRI-MFC

```
Input: Data D, Neighborhood radius eps, Jaccard similarity coefficient jac, Dispersion fog
Output: Clustering results
X = read(D) // read data into X
Zscore(X) // data normalization
for x0 to xn
    if x0 then
        x0 divided into the cluster center set as the first cluster center centers
        Delete x0 from X and added it into cluster[0]
    else
        if Eu_distance(xi, cj) > eps then
            xi as the j_th clustering center divided into centers
            Delete xi from X, and added it into cluster
        end if
end for
for x0 to xn
    if Eu_distance(xi, cj) < eps*fog then
        xi divided into cluster
        if xi ∈ multi clustering centers then
            Recode the Membership information of xi to public point collection repeat
        end if
    end if
end for
According to the information in repeat, reversely count the number of common points between each cluster, save to merge
for m0 to mk // scan the cluster group to be fused in merge
    if the public points of group mi > jac then
        Merge the clusters in group mi, and save it into new clusters
    else
        Divide them into corresponding clusters according to the maximum membership grade
Mark clusters with less than 3 data within clusters as outliers, save in outliers
end for
return clusters
```

5. Experimental Comparative Analysis of PRI-MFC Algorithm

This section introduces the evaluation metrics to measure the quality of the clustering algorithm, designs a variety of experimental methods for different data sets, and illustrates the superiority of the PRI-MFC algorithm by analyzing the experimental results from multiple perspectives.

5.1. Cluster Evaluation Metrics

The experiments in this paper use Accuracy (ACC) [56], Normalized Mutual Information (NMI) [57], and the Adjusted Rand Index (ARI) [58] to evaluate the performance of the clustering algorithm.

The accuracy of the clustering is also often referred to as the clustering purity (purity). The general idea is to divide the number of correctly clustered samples by the total number of samples. However, for the results after clustering, the true category corresponding to each cluster is unknown, so it is necessary to take the maximum value in each case, and the calculation method is shown in Formula (6).

$$\mathrm{ACC}(\Omega, C) = \frac{1}{N}\sum_{k} \max_j |w_k \cap c_j| \tag{6}$$

Among them, N is the total number of samples, $\Omega = \{w_1, w_2, \ldots, w_k\}$ represents the classification of the samples in the cluster, $C = \{c_1, c_2, \ldots, c_j\}$ represents the real class of the samples, w_k denotes all samples in the k-th cluster after clustering, and c_j denotes the real samples in the j-th class. The value range of ACC is [0, 1], and the larger the value, the better the clustering result.

Normalized Mutual Information (NMI), that is, the normalization of the mutual information score, can adjust the result between 0 and 1 using the entropy as the denominator. For the true label, A, of the class in the data sets and a certain clustering result, B, the unique value in A is extracted to form a vector, C, and the unique value in B is extracted to form a vector, S. The calculation of NMI is shown in Formula (7).

$$NMI(A, B) = \frac{I(C, S)}{\sqrt{H(C) \times H(S)}} \tag{7}$$

Among them, I(C, S) is the mutual information of the two vectors, C and S, and H(C) is the information entropy of the C vector. The calculation formulas are shown in Formulas (8) and (9). NMI is often used in clustering to measure the similarity of two clustering results. The closer the value is to 1, the better the clustering results.

$$I(C, S) = \sum_{y \in S} \sum_{x \in C} \log\left(\frac{p(c,s)}{p(c)p(s)}\right) \tag{8}$$

$$H(C) = -\sum_{1}^{n} p(c_i) \log_2 p(c_i) \tag{9}$$

Adjusted Rand Index (ARI) assumes that the super-distribution of the model is a random model, that is, the division of X and Y is random, and the number of data points for each category and each cluster is fixed. To calculate this value, first calculate the contingency table, as shown in Table 1.

Table 1. Contingency table.

	X_1	X_2	...	X_s	Sum
Y_1	n_{11}	n_{12}	...	n_{1s}	a_1
Y_2	n_{21}	n_{22}	...	n_{2s}	a_2
...	...	...	...	...	...
Y_r	n_{r1}	n_{r2}	...	n_{rs}	a_r
sum	b_1	b_2	...	b_s	

The rows in the table represent the actual divided categories, the columns of the table represent the cluster labels of the clustering division, and each value n_{ij} represents the number of files in both class(Y) and class(X) at the same time. Calculate the value of ARI through this table. The calculation formula of ARI is shown in Formula (10).

$$\mathrm{ARI}(X, Y) = \frac{\sum_{ij}\binom{n_{ij}}{2} - \left[\sum_i \binom{a_i}{2} \sum_j \binom{b_j}{2}\right] / (n_2)}{\frac{1}{2}\left[\sum_i \binom{a_i}{2} + \sum_j \binom{b_j}{2}\right] - \left[\sum_i \binom{a_i}{2} \sum_j \binom{b_j}{2}\right] / \binom{n}{2}} \tag{10}$$

The value range of ARI is [−1, 1], and the larger the value, the more consistent the clustering results are with the real situation.

5.2. Experimental Data

For algorithm performance testing, the experiments use five simulated datasets, as shown in Table 2. The tricyclic datasets, bimonthly datasets, and spiral datasets are used to test the clustering effect of the algorithm on irregular clusters, and the C5 datasets and C9 datasets are used to test the clustering effect of the algorithm on common clusters.

Table 2. Artificial datasets.

Serial Number	Datasets	Sample	Feature	Number of Clusters
D1	Three-ring	3600	2	3
D2	Bimonthly	1500	2	2
D3	Spiral	941	2	2
D4	C5	2000	2	5
D5	C9	1009	2	9

In addition, the algorithm performance comparison experiment also uses six UCI real datasets, including Seeds datasets. The details of the data are shown in Table 3.

Table 3. UCI datasets.

Serial Number	Datasets	Sample	Feature	Number of Clusters
D1	Seeds	210	7	3
D2	Iris	150	4	3
D3	Breast	699	10	2
D4	Glass	214	9	6
D5	Ecoli	336	7	7
D6	Pima	768	9	2

5.3. Analysis of Experimental Results

This section contains experiments on the PRI-MFC algorithm on artificial datasets, various UCI datasets, and the China Financial Household Survey datasets.

5.3.1. Experiments on Artificial Datasets

The K-means algorithm [53] and the PRI-MFC algorithm are used for experiments on datasets shown in Table 1, and the experimental clustering results are visualized as shown in Figures 3 and 4, respectively.

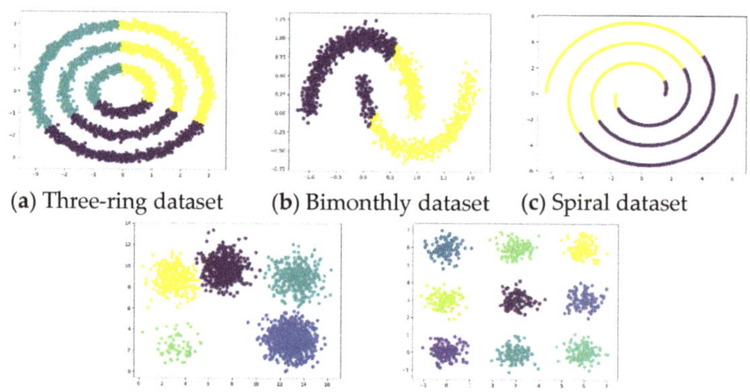

(**a**) Three-ring dataset (**b**) Bimonthly dataset (**c**) Spiral dataset

(**d**) C5 dataset (**e**) C9 dataset

Figure 3. Clustering results of the k-means algorithm on artificial datasets.

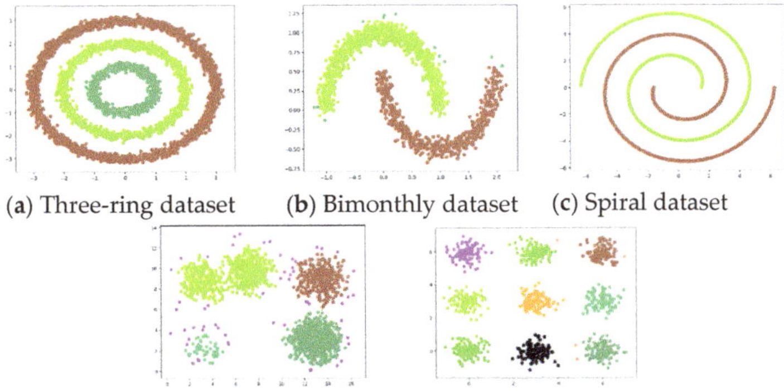

(**a**) Three-ring dataset (**b**) Bimonthly dataset (**c**) Spiral dataset

(**d**) C5 dataset (**e**) C9 dataset

Figure 4. Clustering results of PRI-MFC algorithm on artificial datasets.

It can be seen from the figure that the clustering effect of K-means on the tricyclic datasets, bimonthly datasets, and spiral datasets with uniform density distribution is not ideal. However, K-means has a good clustering effect on both C5 datasets and C9 datasets with uneven density distribution. The PRI-MFC algorithm has a good clustering effect on the three-ring datasets, bimonthly datasets, spiral datasets, and C9 datasets. While accurately clustering the data, it more accurately marks the outliers in the data. However, it fails to distinguish adjacent clusters on the C5 datasets, and the clustering effect is poor for clusters with insignificant clusters in the data.

Comparing the clustering results of the two algorithms, it can be seen that the clustering effect of the PRI-MFC algorithm is better than that of the K-means algorithm on most of the experimental datasets. The PRI-MFC algorithm is not only effective on datasets with uniform density distributions but also has better clustering effects on datasets with large differences in density distributions.

5.3.2. Experiments on UCI Datasets

In this section, experiments on PRI-MFC, K-means [1], ISODATA [59], DBSCAN, and KMM [1] are carried out on various UCI datasets to verify the superiority of the PRI-MFC from the perspective of clustering quality, time, and algorithm parameter influence.

Clustering Quality Perspective

On the UCI data set, PRI-MFC is compared with K-means, ISODATA, DBSCAN, and KMM, and the evaluation index values of the clustering results on various UCI data sets are obtained, which are the accuracy rate (ACC), the standardized mutual information (NMI), and the adjusted Rand coefficient (ARI). The specific experimental results are shown in Table 4.

In order to better observe the clustering quality, the evaluation index data in Table 4 are assigned weight values 5, 4, 3, 2, and 1 in descending order. The ACC index values of the five algorithms on the UCI datasets are shown in Table 5, and the weight values assigned to the ACC index values are shown in Table 6. Taking the ACC of K-means as an example, the weighted average of the ACC of K-means is $(90.95 \times 5 + 30.67 \times 1 + 96.05 \times 5 + 51.87 \times 5 + 56.55 \times 3 + 67.19 \times 5)/24 = 72.11$. Calculated in this way, the weighted average of each algorithm evaluation index is obtained as shown in Table 7.

Table 4. Clustering evaluation index values of five algorithms on the UCI datasets(%).

Datasets	Evaluation Metrics	K-Means	DBSCAN	ISODATA	KMM	PRI-MFC
	ACC	90.95	59.04	35.71	62.86	66.66
Seeds	NMI	70.88	52	76.81	58.16	64.68
	ARI	75.05	39.49	73.98	49.69	51.44
	ACC	30.67	33.33	56.81	36.67	66.66
Iris	NMI	0.8	0.33	73.37	0.44	73.36
	ARI	0.42	0.64	56.81	0.33	76.81
	ACC	96.05	64.57	41.29	1.17	95.16
Breast	NMI	74.68	10.54	2.22	79.21	76.64
	ARI	84.65	9.60	−2.74	87.98	85.43
	ACC	51.87	42.52	50	32.71	34.11
Glass	NMI	42.37	36.07	66.17	28.55	42.19
	ARI	27.66	22.26	48.76	14.10	27.52
	ACC	56.55	42.56	73.51	0.89	57.35
Ecoli	NMI	58.38	11.31	73.63	44.17	61.97
	ARI	46.50	3.80	78.45	36.99	75.34
	ACC	67.19	63.80	32.94	34.51	35.28
Pima	NMI	6.07	0.0	0.31	0.25	6.03
	ARI	11.07	0.13	−0.54	0.43	9.08

Table 5. ACC index values of five algorithms on the UCI datasets (%).

Datasets	K-Means	DBSCAN	ISODATA	KMM	PRI-MFC
Seeds	90.95	59.04	35.71	63.31	74.28
Iris	30.67	34.62	56.81	37.33	69.74
Breast	96.05	64.57	41.29	56.95	95.16
Glass	51.87	42.52	50	35.04	34.11
Ecoli	56.55	42.56	73.51	47.32	57.35
Pima	67.19	63.8	32.94	60.15	35.28

Table 6. The weight values of the ACC of five algorithms on the UCI datasets.

Datasets	K-Means	DBSCAN	ISODATA	KMM	PRI-MFC
Seeds	5	2	1	3	4
Iris	1	2	4	3	5
Breast	5	3	1	2	4
Glass	5	3	4	2	1
Ecoli	3	1	5	2	4
Pima	5	4	1	3	2

Table 7. The weighted averages of the evaluation index of five algorithms (%).

Evaluation Metrics	K-Means	DBSCAN	ISODATA	KMM	PRI-MFC
ACC	72.11	53.76	56.55	50.73	68.03
NMI	40.22	19.61	60.54	45.05	54.21
ARI	42.52	9.93	57.8	44.93	56.54

From Table 7, the weighted average of ACC of K-means is 0.7211 and the weighted average of ACC of PRI-MFC is 0.6803. From the perspective of ACC, the K-means algorithm is the best, and the PRI-MFC algorithm is better. The weighted average of NMI of ISODATA is 0.6054, and the weighted average of NMI of the PRI-MFC algorithm is 0.5424. From the perspective of NMI, the PRI-MFC algorithm is better. Similarly, it can also be seen that the PRI-MFC algorithm has a better effect from the perspective of ARI.

In order to comprehensively consider the quality of the five clustering algorithms, weights 5, 4, 3, 2, and 1 are assigned to each evaluation index data in Table 7 in descending order, and the result is shown in Table 8.

Table 8. The weight values of the weighted averages of the evaluation index of five algorithms.

Evaluation Metrics	K-Means	DBSCAN	ISODATA	KMM	PRI-MFC
ACC	5	2	3	1	4
NMI	2	1	5	3	4
ARI	2	1	5	3	4

The weighted average of the comprehensive evaluation index of each algorithm is calculated according to the above method, and the result is shown as Table 9. It can be seen that the PRI-MFC algorithm proposed in this paper is the best in terms of clustering quality.

Table 9. The weighted averages of comprehensive evaluation index of five algorithms (%).

Evaluation Metrics	K-Means	DBSCAN	ISODATA	KMM	PRI-MFC
comprehensive evaluation index	56.91	34.27	58.57	44.93	58.76

Time Perspective

In order to illustrate the superiority of the algorithm proposed in this paper, the PRI-MFC algorithm, the classical partition-based clustering algorithm, K-means, the commonly used hierarchical clustering algorithm, BIRCH, and Agglomerative are tested on six real data sets, respectively, as shown in Figure 5.

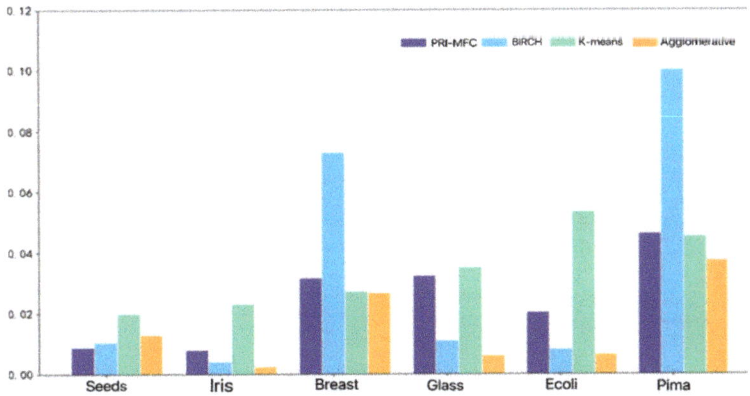

Figure 5. Comparison of running time of clustering algorithm on UCI datasets.

The BIRCH algorithm takes the longest time, with an average time of 34.5 ms. The K-means algorithm takes second place, with an average time of 34.07 ms. The PRI-MFC algorithm takes a shorter time, with an average time of 24.59 ms, and Agglomerative is the shortest, with an average time-consuming of 15.35 ms. The PRI-MFC clustering algorithm wastes time in fuzzy clustering processing so it takes a little longer than Agglomerative. However, the PRI-MFC algorithm only needs to read the number of labels in the Jaccard fusion clustering stage to complete the statistics which saves time. The overall time consumption is shorter than the other algorithms.

Algorithm Parameter Influence Angle

In this section, the PRI-MFC algorithm is tested on UCI, and the *eps* parameter value is modified. The time consumption of the PRI-MFC is shown in Figure 6. It can be seen that with an increase in the *eps* parameter value, the time consumption of the algorithm decreases again. It can be seen that the time of the algorithm is negatively correlated with the *eps* parameter. In the fuzzy pre-clustering stage of the PRI-MFC algorithm, the influence of the *eps* parameter on the time consumption of the algorithm is more obvious.

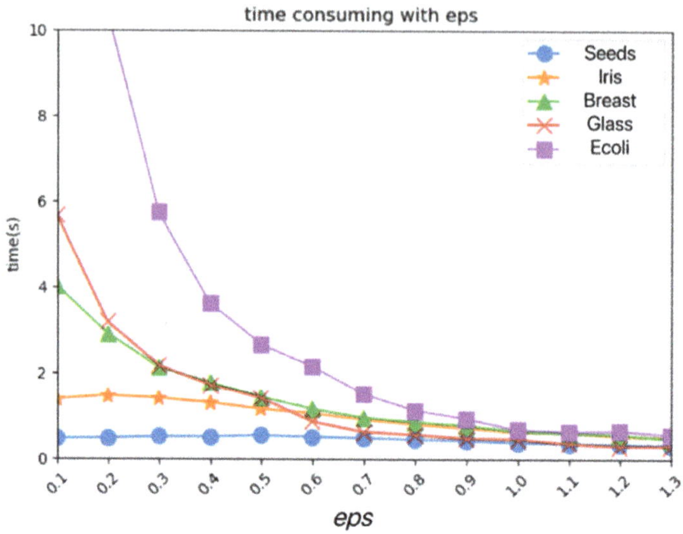

Figure 6. Parameter *eps* and time consumption of PRI-MFC algorithm.

After modifying the *fog* parameter value, the time consumption of the PRI-MFC algorithm is shown in Figure 7. It can be seen that, with the increase of the *fog* parameter value, the time consumption of the algorithm increases again. It can be seen that the time of the algorithm is positively correlated with the *fog* parameter.

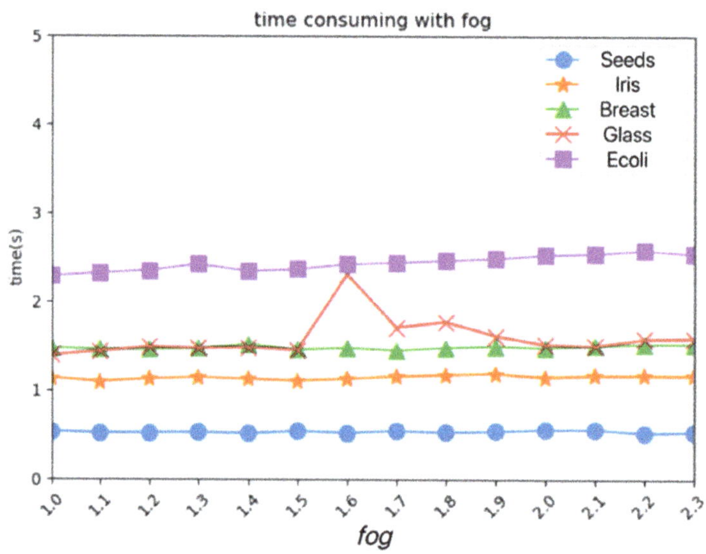

Figure 7. Parameter *fog* and time consumption of PRI-MFC algorithm.

5.3.3. Experiments on China's Financial Household Survey Data

The similarity of the hierarchical clustering algorithm is easy to define. It does not need to pre-determine the number of clusters. It can discover the hierarchical relationship of the classes and cluster them into various shapes, which is suitable for community analysis and market analysis [60]. In this section, the PRI-MFC algorithm conducts experiments on real

Chinese financial household survey data, displays the clustering results, and then analyzes the household financial community to demonstrate the practicability of this algorithm.

Datasets

This section uses the 2019 China Household Finance Survey data, which covers 29 provinces (autonomous regions and municipalities), 343 districts and counties, and 1360 village (neighborhood) committees. Finally, the information of 34,643 households and 107,008 family members is collected. The data are nationally and provincially representative, including three datasets: family datasets, personal datasets, and master datasets. The data details are shown in Table 10.

Table 10. China household finance survey data details from 2019.

Data Name	The Amount of Data	Attributes
family data	34,643	2656
Personal data	107,008	423
master data	107,008	54

The attributes that have high values for the family financial group clustering experiment in the three data sets are selected, redundant irrelevant attributes are deleted, and then duplicate data are removed, and the family data set and master data set are combined into a family data set. The preprocessed data are shown in Table 11.

Table 11. Preprocessed China household finance survey data.

Data Name	The Amount of Data	Attributes
Family	34,643	53
Personal	107,008	13

Experiment

The experiments of the PRI-MFC algorithm are carried out on the two data sets in Table 11. The family data table has a total of 34,643 pieces of data and 53 features, of which there are 16,477 pieces of household data without debt. First, the household data of debt-free urban residents are selected to conduct the PRI-MFC algorithm experiment. The data features are selected as total assets, total household consumption, and total household income. Since there are 28 missing values in each feature of the data, there are 9373 actual experimental data. Secondly, the household data of non-debt rural residents are selected. The selected data features are the same as above. There are 10 missing values for each feature of these data, and the actual experimental data have a total of 7066 items. The clustering results obtained from the two experiments are shown in Table 12.

Table 12. Financial micro-data clustering of Chinese debt-free households.

Area	Cluster	Sample	Proportion (%)	Mean 1	Mean 2	Mean 3	Tag
Town	1	8179	87.26	672,405.82	67,649.73	73,863.01	Well-off
	2	1047	11.17	4,454,086.8	130,084.35	185,686.23	Middle
	3	144	1.53	10,647,847.47	239,064.97	390,016.64	Rich
Rural	1	6956	98.44	307,014.13	52,879.74	41,349.55	Well-off
	2	105	1.48	4,647,506.43	113,648.82	228,253.08	Middle
	3	5	0.07	20,590,783.60	66,631.20	69,243.30	Rich

It can be seen from Table 12 that regardless of urban or rural areas, the population in my country can be roughly divided into three categories: well-off, middle-class, and affluent. The clustering results are basically consistent with the distribution of population income in my country. The total income of middle-class households in urban areas is lower than that of middle-class households in rural areas, but their expenditures are lower and their total assets are higher. It can be seen that the fixed asset value of the urban population is higher, the fixed asset value of the rural population is lower, and the well-off households account for the highest proportion of the total rural households, accounting for 98.44%. Obviously, urban people and a small number of rural people have investment needs, but only a few wealthy families can have professional financial advisors. Most families have minimal financial knowledge and do not know much about asset appreciation and maintaining capital value stability. This clustering result is beneficial for financial managers to make decisions and bring them more benefits.

5.4. Discussion

The experiment on artificial datasets shows that the clustering effect of the PRI-MFC algorithm is better than that of the classical partitioned K-means algorithm regardless of whether the data density is uniform or not. Because the first stage of PRI-MFC algorithm clustering relies on the idea of density clustering, it can cluster uneven density data. Experiments were carried out on the real data set from three aspects: clustering quality, time consumption, and parameter influence. The evaluation metrics of ACC, NMI, and ARI of the five algorithms obtained in the experiment were further analyzed. Calculating the weighted average of each evaluation index of each algorithm, the experiment concludes that the clustering quality of the PRI-MFC algorithm is better. The weighted average of the comprehensive evaluation index of each algorithm was further calculated, and it was concluded that the PRI-MFC algorithm is optimal in terms of clustering quality. The time consumption of each algorithm is displayed through the histogram. The PRI-MFC clustering algorithm wastes time in fuzzy clustering processing, and its time consumption is slightly longer than that of Agglomerative. However, in the Jaccard fusion clustering stage, the PRI-MFC algorithm only needs to read the number of labels to complete the statistics, which saves time, and the overall time consumption is less than other algorithms. Experiments from the perspective of parameters show that the time of this algorithm has a negative correlation with the parameter *eps* and a positive correlation with the parameter *fog*. When the parameter *eps* changes from large to small in the interval [0, 0.4], the time consumption of the algorithm increases rapidly. When the *eps* parameter changes from large to small in the interval [0.4, 0.8], the time consumption of the algorithm increases slowly. When the *eps* parameter in the interval between [0.8, 1.3] changes from large to small, the time consumption of the algorithm tends to be stable. In conclusion, from the perspective of the clustering effect and time consumption, the algorithm is better when the *eps* is 0.8. When the *fog* parameter is set to 1, the time consumption is the lowest, because the neighborhood radius and the dispersion radius are the same at this time. With the increase of the *fog* value, the time consumption of the algorithm gradually increases. In conclusion, from the perspective of the clustering effect and time consumption, the algorithm is better when *fog* is set to 1.8. Experiments conducted on Chinese household finance survey data show that the PRI-MFC algorithm is practical and can be applied in market analysis, community analysis, etc.

6. Conclusions

In view of the problems that the traditional hierarchical clustering algorithm cannot find clusters with uneven density, requires a large amount of calculation and has low efficiency, this paper takes advantage of the benefits of the classical hierarchical clustering algorithm and the advantages of the DBSCAN algorithm for clustering data with uneven density. Based on population reproduction and fusion, a new hierarchical clustering algorithm PRI-MFC is proposed. This algorithm can effectively identify clusters of any shape, and preferentially identify cluster-dense centers. It can effectively remove noise in samples and reduce outlier pairs by clustering and re-integrating multiple cluster centers. By setting different parameters for *eps* and *fog*, the granularity of clustering can be adjusted. Secondly, various experiments are designed on artificial datasets and real datasets, and the results show that this algorithm is better in terms of clustering effect, clustering quality, and time consumption. Due to the uncertainty of objective world data, the next step is to study the fuzzy hierarchical clustering algorithm further. With the advent of the era of big data, running the algorithm on a single computer is prone to bottleneck problems. The next step is to study the improvement of clustering algorithms under the big data platform.

Author Contributions: Conceptualization, L.Y. and M.L.; methodology, L.Y.; software, M.L.; validation, LY., H.C. and M.L.; formal analysis, M.L.; investigation, M.L.; resources, H.C.; data curation, M.L.; writing—original draft preparation, M.L.; writing—review and editing, L.Y.; visualization, M.L.; supervision, L.Y.; project administration, W.D.; funding acquisition, L.Y., H.C. and W.D. All authors have read and agreed to the published version of the manuscript.

Funding: This research was funded by the National Natural Science Foundation of China under grant number U2133205 and 61771087, the Natural Science Foundation of Sichuan Province under Grant 2022NSFSC0536, the Research Foundation for Civil Aviation University of China under Grant 3122022PT02 and 2020KYQD123.

Institutional Review Board Statement: Not applicable.

Informed Consent Statement: Not applicable.

Data Availability Statement: Not applicable.

Conflicts of Interest: The authors declare no conflict of interest.

References

1. Han, J.; Pei, J.; Tong, H. *Data Mining Concepts and Techniques*, 3rd ed.; China Machine Press: Beijing, China, 2016.
2. Li, X.; Zhao, H.; Yu, L.; Chen, H.; Deng, W.; Deng, W. Feature Extraction Using Parameterized Multisynchrosqueezing Transform. *IEEE Sens. J.* **2022**, *22*, 14263–14272. [CrossRef]
3. Wu, D.; Wu, C. Research on the Time-Dependent Split Delivery Green Vehicle Routing Problem for Fresh Agricultural Products with Multiple Time Windows. *Agriculture* **2022**, *12*, 793. [CrossRef]
4. Zhou, X.; Ma, H.; Gu, J.; Chen, H.; Deng, W. Parameter adaptation-based ant colony optimization with dynamic hybrid mechanism. *Eng. Appl. Artif. Intell.* **2022**, *114*, 105139. [CrossRef]
5. Li, T.; Shi, J.; Deng, W.; Hu, Z. Pyramid particle swarm optimization with novel strategies of competition and cooperation. *Appl. Soft Comput.* **2022**, *121*, 108731. [CrossRef]
6. Deng, W.; Xu, J.; Gao, X.-Z.; Zhao, H. An Enhanced MSIQDE Algorithm With Novel Multiple Strategies for Global Optimization Problems. *IEEE Trans. Syst. Man Cybern. Syst.* **2020**, *52*, 1578–1587. [CrossRef]
7. Chen, H.; Miao, F.; Chen, Y.; Xiong, Y.; Chen, T. A Hyperspectral Image Classification Method Using Multifeature Vectors and Optimized KELM. *IEEE J. Sel. Top. Appl. Earth Obs. Remote Sens.* **2021**, *14*, 2781–2795. [CrossRef]
8. Yao, R.; Guo, C.; Deng, W.; Zhao, H. A novel mathematical morphology spectrum entropy based on scale-adaptive techniques. *ISA Trans.* **2021**, *126*, 691–702. [CrossRef]
9. Deng, W.; Li, Z.; Li, X.; Chen, H.; Zhao, H. Compound Fault Diagnosis Using Optimized MCKD and Sparse Representation for Rolling Bearings. *IEEE Trans. Instrum. Meas.* **2022**, *71*, 1–9. [CrossRef]
10. Tian, C.; Jin, T.; Yang, X.; Liu, Q. Reliability analysis of the uncertain heat conduction model. *Comput. Math. Appl.* **2022**, *119*, 131–140. [CrossRef]
11. Zhao, H.; Liu, J.; Chen, H.; Chen, J.; Li, Y.; Xu, J.; Deng, W. Intelligent Diagnosis Using Continuous Wavelet Transform and Gauss Convolutional Deep Belief Network. *IEEE Trans. Reliab.* **2022**, 1–11. [CrossRef]
12. Wei, Y.; Zhou, Y.; Luo, Q.; Deng, W. Optimal reactive power dispatch using an improved slime mould algorithm. *Energy Rep.* **2021**, *7*, 8742–8759. [CrossRef]
13. Jin, T.; Xia, H.; Deng, W.; Li, Y.; Chen, H. Uncertain Fractional-Order Multi-Objective Optimization Based on Reliability Analysis and Application to Fractional-Order Circuit with Caputo Type. *Circuits Syst. Signal Process.* **2021**, *40*, 5955–5982. [CrossRef]
14. He, Z.Y.; Shao, H.D.; Wang, P.; Janet, L.; Cheng, J.S.; Yang, Y. Deep transfer multi-wavelet auto-encoder for intelligent fault diagnosis of gearbox with few target training samples. *Knowl.-Based Syst.* **2019**. [CrossRef]
15. Li, X.; Shao, H.; Lu, S.; Xiang, J.; Cai, B. Highly Efficient Fault Diagnosis of Rotating Machinery Under Time-Varying Speeds Using LSISMM and Small Infrared Thermal Images. *IEEE Trans. Syst. Man Cybern. Syst.* **2022**, 1–13. [CrossRef]
16. An, Z.; Wang, X.; Li, B.; Xiang, Z.; Zhang, B. Robust visual tracking for UAVs with dynamic feature weight selection. *Appl. Intell.* **2022**, 1–14. [CrossRef]
17. Johnson, S.C. Hierarchical clustering schemes. *Psychometrika* **1967**, *32*, 241–254. [CrossRef]
18. Kaufman, L.; Rousseeuw, P.J. *Finding Groups in Data: An Introduction to Cluster Analysis*; John Wiley & Sons: Hoboken, NJ, USA, 2009; Volume 344.

19. Koga, H.; Ishibashi, T.; Watanabe, T. Fast agglomerative hierarchical clustering algorithm using Locality-Sensitive Hashing. *Knowl. Inf. Syst.* **2006**, *12*, 25–53. [CrossRef]
20. Cao, H.; Shao, H.; Zhong, X.; Deng, Q.; Yang, X.; Xuan, J. Unsupervised domain-share CNN for machine fault transfer diagnosis from steady speeds to time-varying speeds. *J. Manuf. Syst.* **2021**, *62*, 186–198. [CrossRef]
21. Deng, W.; Ni, H.; Liu, Y.; Chen, H.; Zhao, H. An adaptive differential evolution algorithm based on belief space and generalized opposition-based learning for resource allocation. *Appl. Soft Comput.* **2022**, *127*, 109419. [CrossRef]
22. Rodrigues, J.; Von Mering, C. HPC-CLUST: Distributed hierarchical clustering for large sets of nucleotide sequences. *Bioinformatics* **2013**, *30*, 287–288. [CrossRef]
23. Li, T.; Qian, Z.; Deng, W.; Zhang, D.; Lu, H.; Wang, S. Forecasting crude oil prices based on variational mode decomposition and random sparse Bayesian learning. *Appl. Soft Comput.* **2021**, *113*, 108032. [CrossRef]
24. Cui, H.; Guan, Y.; Chen, H. Rolling Element Fault Diagnosis Based on VMD and Sensitivity MCKD. *IEEE Access* **2021**, *9*, 120297–120308. [CrossRef]
25. Bouguettaya, A.; Yu, Q.; Liu, X.; Zhou, X.; Song, A. Efficient agglomerative hierarchical clustering. *Expert Syst. Appl.* **2014**, *42*, 2785–2797. [CrossRef]
26. Liu, Q.; Jin, T.; Zhu, M.; Tian, C.; Li, F.; Jiang, D. Uncertain Currency Option Pricing Based on the Fractional Differential Equation in the Caputo Sense. *Fractal Fract.* **2022**, *6*, 407. [CrossRef]
27. Li, G.; Li, Y.; Chen, H.; Deng, W. Fractional-Order Controller for Course-Keeping of Underactuated Surface Vessels Based on Frequency Domain Specification and Improved Particle Swarm Optimization Algorithm. *Appl. Sci.* **2022**, *12*, 3139. [CrossRef]
28. Guha, S.; Rastogi, R.; Shim, K. Cure: An Efficient Clustering Algorithm for Large Databases. *Inf. Syst.* **1998**, *26*, 35–58. [CrossRef]
29. Guha, S.; Rastogi, R.; Shim, K. Rock: A robust clustering algorithm for categorical attributes. *Inf. Syst.* **2000**, *25*, 345–366. [CrossRef]
30. Karypis, G.; Han, E.-H.; Kumar, V. Chameleon: Hierarchical clustering using dynamic modeling. *Computer* **1999**, *32*, 68–75. [CrossRef]
31. Gagolewski, M.; Bartoszuk, M.; Cena, A. Genie: A new, fast, and outlier resistant hierarchical clustering algorithm. *Inf. Sci.* **2017**, *363*, 8–23. [CrossRef]
32. Zhang, T.; Ramakrishnan, R.; Livny, M. BIRCH: A New Data Clustering Algorithm and Its Applications. *Data Min. Knowl. Discov.* **1997**, *1*, 141–182. [CrossRef]
33. Kobren, A.; Monath, N.; Krishnamurthy, A.; McCallum, A. A hierarchical algorithm for extreme clustering. In Proceedings of the 23rd ACM SIGKDD International Conference on Knowledge Discovery and Data Mining, New York, NY, USA, 13–17 August 2017; pp. 255–264.
34. Monath, N.; Kobren, A.; Krishnamurthy, A.; Glass, M.R.; McCallum, A. Scalable hierarchical clustering with tree grafting. In Proceedings of the 25th ACM SIGKDD International Conference on Knowledge Discovery and Data Mining, New York, NY, USA, 4–8 August 2019; pp. 438–1448.
35. Ester, M.; Kriegel, H.; Sander, J.; Xu, X. A density-based algorithm for discovering clusters in large spatial databases with noise. In Proceedings of the 2nd International Conference on Knowledge Discovery and Data Mining, Portland, OR, USA, 2–4 August 1996; Volume 34, pp. 226–231.
36. Zhou, D.; Liu, P. VDBSCAN: Variable Density Clustering Algorithm. *Comput. Eng. Appl.* **2009**, *45*, 137–141.
37. Zhou, Z.P.; Wang, J.F.; Zhu, S.W.; Sun, Z.W. An Improved Adaptive Fast AF-DBSCAN Clustering Algorithm. *J. Intell. Syst.* **2016**, *11*, 93–98.
38. Li, W.; Yan, S.; Jiang, Y.; Zhang, S.; Wang, C. Algorithm research on adaptively determining DBSCAN algorithm parameters. *Comput. Eng. Appl.* **2019**, *55*, 1–7.
39. Wang, G.; Lin, G.Y. Improved adaptive parameter DBSCAN clustering algorithm. *Comput. Eng. Appl.* **2020**, *56*, 45–51.
40. Wan, J.; Hu, D.Z.; Jiang, Y. Algorithm research on multi-density adaptive determination of DBSCAN algorithm parameters. *Comput. Eng. Appl.* **2022**, *58*, 78–85.
41. Zadeh, L.A. Fuzzy sets. *Inf. Control* **1965**, *8*, 338–353. [CrossRef]
42. Bezdek, J.C.; Ehrlich, R.; Full, W. FCM: The fuzzy c-means clustering algorithm. *Comput. Geosci.* **1984**, *10*, 191–203. [CrossRef]
43. Atanassov, K.T. Intuitionistic fuzzy sets. *Fuzzy Sets Syst.* **1986**, *20*, 87–96. [CrossRef]
44. Xu, Z.; Wu, J. Intuitionistic fuzzy C-means clustering algorithms. *J. Syst. Eng. Electron.* **2010**, *21*, 580–590. [CrossRef]
45. Kumar, D.; Verma, H.; Mehra, A.; Agrawal, R.K. A modified intuitionistic fuzzy c-means clustering approach to segment human brain MRI image. *Multimed. Tools Appl.* **2018**, *78*, 12663–12687. [CrossRef]
46. Danish, Q.M.; Solanki, R.; Pranab, K. Novel adaptive clustering algorithms based on a probabilistic similarity measure over atanassov intuitionistic fuzzy set. *IEEE Trans. Fuzzy Syst.* **2018**, *26*, 3715–3729.
47. Varshney, A.K.; Lohani, Q.D.; Muhuri, P.K. Improved probabilistic intuitionistic fuzzy c-means clustering algorithm: Improved PIFCM. In Proceedings of the 2020 IEEE International Conference on Fuzzy Systems, Glasgow, UK, 19–24 July 2020; pp. 1–6.
48. Zeshui, X. Intuitionistic fuzzy hierarchical clustering algorithms. *J. Syst. Eng. Electron.* **2009**, *20*, 90–97.
49. Aliahmadipour, L.; Eslami, E. GHFHC: Generalized Hesitant Fuzzy Hierarchical Clustering Algorithm. *Int. J. Intell. Syst.* **2016**, *31*, 855–871. [CrossRef]
50. Gao, S.H.; Han, Q.; Li, D.; Cheng, M.M.; Peng, P. Representative batch normalization with feature calibration. In Proceedings of the IEEE/CVF Conference on Computer Vision and Pattern Recognition, Virtual, 19–25 June 2021; pp. 8669–8679.

51. Babanezhad, M.; Masoumian, A.; Nakhjiri, A.T.; Marjani, A.; Shirazian, S. Influence of number of membership functions on prediction of membrane systems using adaptive network based fuzzy inference system (ANFIS). *Sci. Rep.* **2020**, *10*, 1–20. [CrossRef]
52. Kumbure, M.M.; Luukka, P. A generalized fuzzy k-nearest neighbor regression model based on Minkowski distance. *Granul. Comput.* **2021**, *7*, 657–671. [CrossRef]
53. Kongsin, T.; Klongboonjit, S. Machine component clustering with mixing technique of DSM, jaccard distance coefficient and k-means algorithm. In Proceedings of the 2020 IEEE 7th International Conference on Industrial Engineering and Applications (ICIEA), Bangkok, Thailand, 16–21 April 2020; pp. 251–255.
54. Karasu, S.; Altan, A. Crude oil time series prediction model based on LSTM network with chaotic Henry gas solubility optimization. *Energy* **2021**, *242*, 122964. [CrossRef]
55. Karasu, S.; Altan, A.; Bekiros, S.; Ahmad, W. A new forecasting model with wrapper-based feature selection approach using multi-objective optimization technique for chaotic crude oil time series. *Energy* **2020**, *212*, 118750. [CrossRef]
56. Cai, D.; He, X.; Han, J. Document clustering using locality preserving indexing. *IEEE Trans. Knowl. Data Eng.* **2005**, *17*, 1624–1637. [CrossRef]
57. Strehl, A.; Ghosh, J. Cluster Ensembles—A Knowledge Reuse Framework for Combining Multiple Partitions. *J. Mach. Learn. Res.* **2003**, *3*, 583–617.
58. Hubert, L.; Arabie, P. Comparing partitions. *J. Classif.* **1985**, *2*, 193–218. [CrossRef]
59. Rajab, M.A.; George, L.E. Stamps extraction using local adaptive k-means and ISODATA algorithms. *Indones. J. Electr. Eng. Comput. Sci.* **2021**, *21*, 137–145. [CrossRef]
60. Renigier-Biłozor, M.; Janowski, A.; Walacik, M.; Chmielewska, A. Modern challenges of property market analysis- homogeneous areas determination. *Land Use Policy* **2022**, *119*, 106209. [CrossRef]

electronics

MDPI

Article

An Intelligent Identification Approach Using VMD-CMDE and PSO-DBN for Bearing Faults

Erbin Yang [1], Yingchao Wang [2,*], Peng Wang [3], Zheming Guan [4] and Wu Deng [5,*]

1 Guoneng Railway Equipment Co., Ltd., Beijing 100120, China
2 Dalian Locomotive & Rolling Stock Co., Ltd., Dalian 116300, China
3 College of Mechanical Engineering, Dalian Jiaotong University, Dalian 116028, China
4 College of Rolling Stock Engineering, Dalian Jiaotong University, Dalian 116028, China
5 School of Electronic Information and Automation, Civil Aviation University of China, Tianjin 300300, China
* Correspondence: yy177@live.cn (Y.W.); wdeng@cauc.edu.cn (W.D.)

Abstract: In order to improve the fault diagnosis accuracy of bearings, an intelligent fault diagnosis method based on Variational Mode Decomposition (VMD), Composite Multi-scale Dispersion Entropy (CMDE), and Deep Belief Network (DBN) with Particle Swarm Optimization (PSO) algorithm—namely VMD-CMDE-PSO-DBN—is proposed in this paper. The number of modal components decomposed by VMD is determined by the observation center frequency, reconstructed according to the kurtosis, and the composite multi-scale dispersion entropy of the reconstructed signal is calculated to form the training samples and test samples of pattern recognition. Considering that the artificial setting of DBN node parameters cannot achieve the best recognition rate, PSO is used to optimize the parameters of DBN model, and the optimized DBN model is used to identify faults. Through experimental comparison and analysis, we propose that the VMD-CMDE-PSO-DBN method has certain application value in intelligent fault diagnosis.

Keywords: fault diagnosis; variational mode decomposition; composite multi-scale dispersion entropy; particle swarm optimization; deep belief network

Citation: Yang, E.; Wang, Y.; Wang, P.; Guan, Z.; Deng, W. An Intelligent Identification Approach Using VMD-CMDE and PSO-DBN for Bearing Faults. *Electronics* **2022**, *11*, 2582. https://doi.org/10.3390/electronics11162582

Academic Editor: George A. Papakostas

Received: 18 July 2022
Accepted: 9 August 2022
Published: 18 August 2022

1. Introduction

Rolling bearing is one of the most commonly used components in rotating machinery. Its working state directly affects the performance of the whole equipment and even the safety of the whole production line [1–4]. Therefore, research on intelligent fault diagnosis technology of rolling bearing has important theoretical value and practical significance in avoiding accidents. The operating conditions of rolling bearing in engineering applications are complex and changeable [5–9]. The collected fault vibration signal is easily disturbed by uncontrollable factors, and the subsequent diagnosis and prediction accuracy will also be reduced [10–14].

The complex problem of signal noise reduction in practical engineering was studied and analyzed by combining with the characteristics of wavelet packet decomposition, leading to a new signal noise reduction method; experimental results show that the method has good noise reduction ability [15–18]. A series of analyses on the problem were carried out, revealing that the initial fault feature information of mechanical equipment is affected by strong background noise, and verifying the effectiveness of the new denoising method of the airspace and neighborhood of wavelet packet transforms [19–22]. In order to solve the problem that the measured vibration signal of the discharge structure is interfered with by noise, the wavelet packet threshold with the optimized empirical mode decomposition was combined, and a new method to eliminate noise interference was proposed [23–28]. On the basis of EMD algorithm, many optimization algorithms with good effects have been derived, which also have good performance in engineering applications. However, they are all based on EMD in essence, so the mode aliasing problem is difficult to solve.

Konstantin [29] proposed Variational Mode Decomposition (VMD) in 2014; the VMD method not only has good a signal-to-noise separation effect for non-stationary vibration signals, but also the decomposition scale can be preset according to the vibration signal itself. If an appropriate scale can be selected, the occurrence of mode aliasing will be effectively suppressed. Mostafa et al. [30] proposed a new complexity theory, namely Dispersion Entropy (DE), for the defects of slow calculation speed and unreasonable measurement methods of general complexity theory. The entropy of a single scale often cannot show more complete information in feature extraction, which leads to the final classification not having ideal results. More signals are analyzed by multi-scale analysis of complexity theory. For example, Zhang et al. [31] extracted fault features by LMD multi-scale approximate entropy. Wang et al. [32] calculated the gear signal with the Variational Mode Decomposition (VMD) method, and selected four modal components after decomposition to calculate permutation entropy to extract features. Li et al. [33] have significantly improved the fault identification by combining Empirical Wavelet Transform (EWT) with various algorithms of dispersion entropy (DE). In 2006, Hinton et al. [34] published a significant paper. In Science, they told many scholars about the concept of deep learning, and specifically expounded the Deep Belief Network (DBN), which stimulated people's enthusiasm for deep learning theory research and learning. Lei et al. [35] have found that training mechanical vibration signals of relevant faults through deep learning neural network is more conducive to fault identification and classification. This paper also points out the advantages of using deep learning theory for fault diagnosis, which is mainly reflected in breaking the researchers' dependence on many types of signal processing technology and fault diagnosis experience. Starting with the statistical characteristics of vibration signals, Shan et al. [36] achieved the simultaneous identification of different types and degrees of bearing faults, and finally obtained a high classification accuracy. It was confirmed that the application of DBN in fault diagnosis has a good effect compared with traditional fault diagnosis. Shi et al. [37], through experimental verification, found that when pattern recognition is carried out on gears, the recognition rate of fault features using Particle Swarm Optimization support vector machine is considerable. Other fault diagnosis methods have also been proposed in recent years [38–47].

In this paper, the data of the Electrical laboratory of Case Western Reserve University have been used for experiments. Through the noise reduction method of variational mode decomposition, the signals of the four states of normal bearing condition, bearing inner ring fault, rolling body fault, and bearing outer ring fault are decomposed into multiple modal components. The reconstructed signals preprocessed by variational mode decomposition were combined with multi-scale permutation entropy, multi-scale dispersion entropy, and composite multi-scale dispersion entropy, and their method principles were analyzed. The rolling bearing data were used for simulation, and the eigenvalues of the three methods were calculated as the input of the classification model. Three kinds of multi-scale entropy values were used as feature vectors and input into the Deep Belief Network (DBN) model for fault pattern recognition. In order to solve the problem that it is time consuming to debug the network layer structure in a deep belief network (DBN) when it is used for bearing fault diagnosis, a fault identification model of DBN bearing based on Particle Swarm Optimization (PSO) was proposed. The model uses particle swarm optimization (PSO) algorithm to find the optimal solution of hidden layer node parameters, and then compares the function between DBN model and PSO-DBN model and draws a conclusion.

2. Composite Multi-Scale Dispersion Entropy Based on VMD

2.1. Variational Mode Decomposition Algorithm

The essence of the VMD decomposition method is related to selecting the number of components (parameter K) to decompose the original signal $f(t)$ into a corresponding number of sub-signal components u_k; these decomposed modal components can ensure the sparsity and reproduce the input signal. In short, the Gaussian smoothing of demodulated signal is used to estimate the bandwidth, and then the constraints are divided into the following:

$$\min_{\{u_k\},\{\omega_k\}}\left\{\sum_k\left\|\partial_t\left[\left(\delta(t)+\frac{j}{\pi t}\right)*u_k(t)\right]e^{-j\omega_k t}\right\|_2^2\right\},\ \text{s.t.}\sum_k u_k = f \tag{1}$$

Most of the optimal solutions of constrained models are solved by alternative direction method of multipliers (ADMM), alternately updating u_k^{n+1}, ω_k^{n+1}, and λ^{n+1} to look for a Lagrangian augmented "saddle point"; the specific steps are as follows:

Initialize $\left\{u_k^l\right\}, \left\{\omega_k^l\right\}, \lambda'$; $n \leftarrow O$; make $n \leftarrow n+1$ for $k = 1:K$ to update U_k:

$$\hat{u}_k^{n+1}(\omega) = \frac{f(\omega) - \sum_{i\neq k} u(\omega) + \frac{\lambda(\hat{\omega})}{2}}{1+2\alpha(\omega-\omega_k)^2} \tag{2}$$

For all $\omega \geq 0$, update $\hat{u}_k$; the formula is as follows:

$$\hat{u}_k^{n+1}(\omega) \leftarrow \frac{\hat{f}(\omega) - \sum_{i<k}\hat{u}_i^{n+1}(\omega) - \sum_{i>k}\hat{u}_i^{n}(\omega) + \frac{\hat{\lambda}^n(\omega)}{2}}{1+2\alpha\left(\omega-\omega_k^n\right)^2}, k \in \{1, K\} \tag{3}$$

Update ω_k:

$$\hat{\lambda}^{n+1}(\omega) \leftarrow \hat{\lambda}^n(\omega) + \tau\left(\hat{f}(\omega) - \sum_k \hat{u}_k^{n+1}(\omega)\right) \tag{4}$$

Repeat (3)~(4) until the following iterative conditions are met:

$$\sum_k \left\|\hat{u}_k^{n+1} - \hat{u}_k^{n}\right\|_2^2 / \|\hat{u}_k^n\|_2^2 < \varepsilon \tag{5}$$

Usually, the u_k^{n+1} problem is transformed into the minimum problem; the same is true for the solution of center frequency ω_k^{n+1}:

$$\omega_k^{n+1} = \underset{\omega_k}{argmin}\left\{\left\|\partial_t\left[\left(\delta(t)+\frac{j}{\pi t}\right)*u_k(t)\right]e^{-j\omega_k t}\right\|_2^2\right\} \tag{6}$$

2.2. *Composite Multi-Scale Dispersion Entropy*

2.2.1. Dispersion Entropy Algorithm

Dispersion Entropy (DE) is an index to measure the complexity of a time series. When it was first proposed, it was mostly applied in the field of biology. The main construction steps and descriptions of DE are described as follows [30]:

Supposing a time series $x = \{x_i, i = 1, 2, \cdots, N\}$ of length N, the normal distribution Function (7) is used to map time series x to $y = \{y_j, j = 1, 2, \cdots, N\}$, $y_j \in (0, 1)$.

$$y_j = \frac{1}{\sigma\sqrt{2\pi}}\int_{-\infty}^{x_i} e^{\frac{-(t-\mu)^2}{2\sigma^2}}dt \tag{7}$$

where μ is mathematical expectation and σ^2 is variance.

The linear transformation is performed using Formula (8), mapping y to the range of $[1, 2, \ldots, c]$:

$$z_j^c = R(c \cdot y_j + 0.5) \tag{8}$$

where R is an integral function and c is the number of categories.

Calculating the embedded vector $z_i^{m,c}$ is as follows:

$$z_i^{m,c} = \left\{z_i^c, z_{i+d}^c, \cdots, z_{i+(m-1)d}^c\right\}, i = 1,2,\ldots,N-(m-1)d \tag{9}$$

where m is an embedded dimension and d is time delay.

Calculating the probability $p\left(\pi_{v_v v_1 \cdots v_{m-1}}\right)$ of $\pi_{v_0 v_1 \cdots v_{m-1}}$ for each dispersion mode is as follows:

$$p\left(\pi_{v_0 v_1 \cdots v_{m-1}}\right) = \frac{\text{Number}\left(\pi_{v_0 \eta_1 \cdots v_{m-1}}\right)}{N-(m-1)d} \tag{10}$$

where Number $\left(\pi_{v_0 v_1 \cdots v_{m-1}}\right)$ represents the number of maps $z_i^{m,c}$ to $\pi_{v_0 v_1 \cdots v_{m-1}}$.

The DE value of the original signal x is

$$DE(x,m,c,d) = -\sum_{\pi=1}^{c^m} p\left(\pi_{v_0 v_1 \cdots v_{m-1}}\right) \ln\left(p\left(\pi_{v_0 v_1 \cdots v_{m-1}}\right)\right) \tag{11}$$

2.2.2. Composite Multi-Scale Dispersion Entropy

The calculation method of composite multi-scale dispersion entropy involves optimizing the multi-scale process on the basis of multi-scale dispersion entropy; the steps and instructions are as follows:

For an initial time series $\{u(i), i = 1,2,\cdots,L\}$, when the time is in the k-th coarsening sequence and the scale factor is τ, $x_k^\tau = \left\{x_{k,1}^{(\tau)}, x_{k,2}^{(\tau)}, \ldots\right\}$ can be given by Formula (12):

$$x_{k,j}^\tau = \frac{1}{\tau}\sum_{i=k+\tau(j-1)}^{k+j\tau-1} u_i, 1 \le j \le L/\tau \tag{12}$$

where $1 \le k \le \tau$.

The CMDE under each scale factor is defined as

$$\text{CMDE}(X,m,c,d,\tau) = \frac{1}{\tau}\sum_{k=1}^{\tau} DE(x_k^\tau, m, c, d) \tag{13}$$

2.3. *Fault Eigenvalue Based on VMD Composite Multi-Scale Entropy*

In this paper, the experimental data of the bearing data center of Case Western Reserve University are selected for the simulation test, and the selection of important parameters is compared and analyzed. The specific data of bearing are as follows: the acquisition frequency is 12,000 Hz; the motor speed is 1797 r/min; and four vibration signals are included, namely an inner ring (IR) fault, outer ring (OR) fault, rolling element (BE) fault, and a local single-point pitting normal state (Norm).

2.3.1. The Process of Fault Eigenvalue Calculation

The specific steps of VMD composite multi-scale dispersion entropy are as follows:

Step 1: Firstly, the original vibration signals (inner and outer ring fault signals, roller fault signals, and normal signals) in the four bearing databases are decomposed and preprocessed by VMD.

Step 2: The kurtosis of the decomposed modal components is calculated and sorted.

Step 3: The first three modal components are selected for signal reconstruction.

Step 4: The composite multi-scale dispersion entropy of the reconstructed four signals is calculated.

2.3.2. Simulation Signal Analysis

In this paper, a vibration signal with a motor speed of 1797r/min in the bearing experiment database of Western Reserve University is decomposed by VMD, where determining the value of modal component K is the primary task. For example, the center frequency of the modal component of the outer ring fault signal is simulated. The value of K in the simulation diagram is reflected in the number of curves in the center frequency diagram of the modal component, and its value is determined by observing the convergence trend of the curve. Selecting K = 4, K = 5, and K = 6, the corresponding center frequency curves are described as follows.

The abscissa in the figure represents the number of iterations, and the ordinate represents the center frequency. The four curves represent the central frequency convergence process of the four modal components, respectively. When K = 4, as shown in Figure 1, the four curves do not overlap, which proves that there is no mode mixing. There are occasional fluctuations in the previous iteration, and the convergence is fast. When the number of components K is 5, the relationship between the center frequency of the modal component and the iteration parameters is as shown in Figure 2. With the increase in the number of abscissa iterations, the center frequencies corresponding to the five modal components converge smoothly and fluctuate less, and there is no curve intersection. When the number of decomposition K = 6 is selected and the same vibration signal is decomposed, the central frequency convergence process of the modal component is as shown in Figures 3 and 4. The abscissa in the figure represents the number of iterations, and the ordinate represents the center frequency. From the curves corresponding to the six modal components, it is obvious that the third, fourth, and fifth curves also correspond to the intersection of the third, fourth, and fifth order modal components, respectively. This proves that there is modal mixing between modal components, and the convergence speed is slow. In summary, in the VMD decomposition preprocessing of this kind of bearing vibration signal, the preset value of the modal component is 5, which is more effective for the signal decomposition effect and helpful for the next feature extraction.

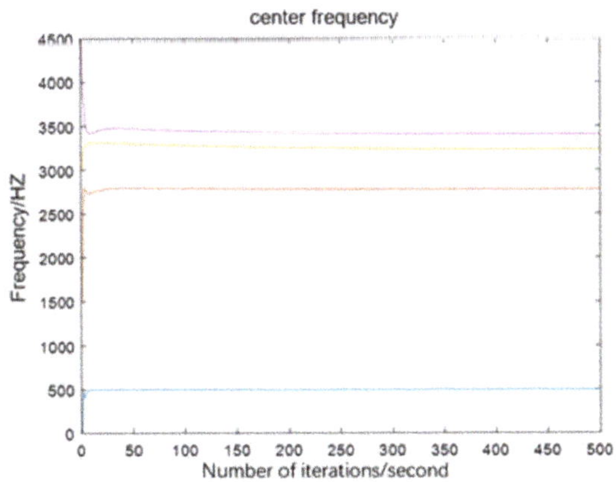

Figure 1. Center frequency of the modal component K = 4.

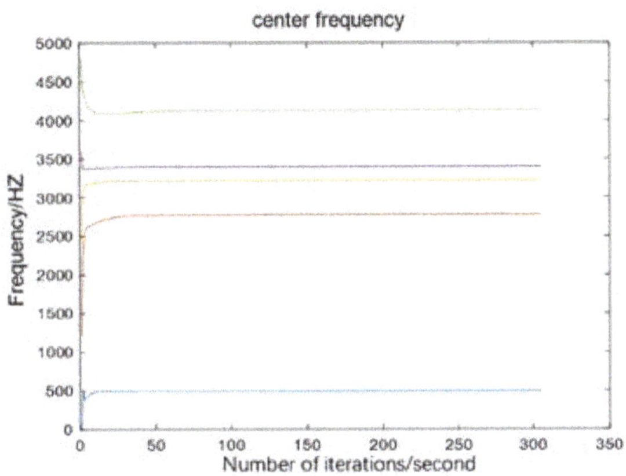

Figure 2. Center frequency of the modal component K = 5.

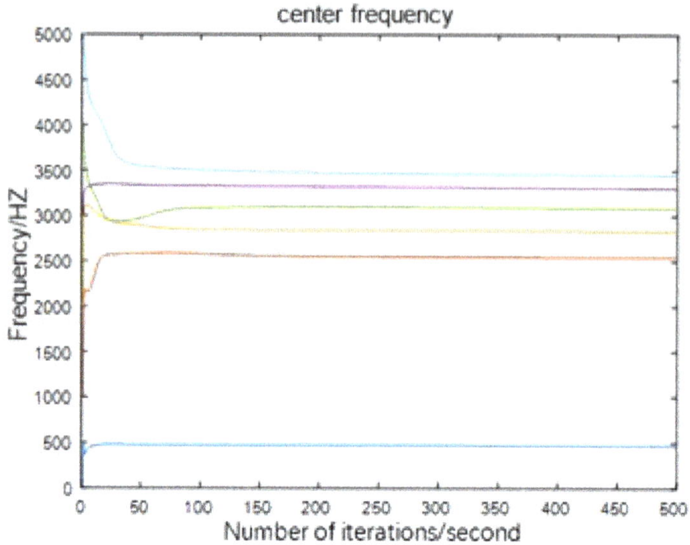

Figure 3. Center frequency of the modal component K = 6.

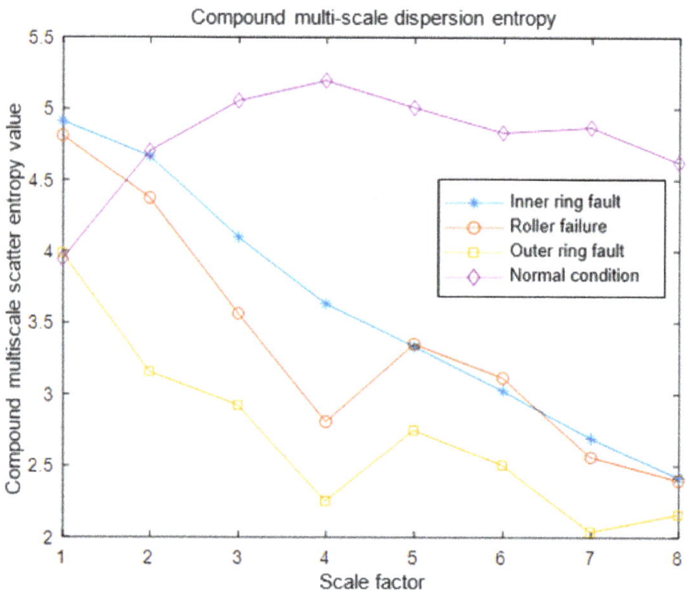

Figure 4. VMD-CMDE at $\tau = 8$.

From the calculation formulas of multi-scale dispersion entropy and composite multi-scale dispersion entropy, it can be seen that five parameters need to be selected. They are the length N of the sequence, the embedding dimension m, the number of categories c, the time delay d and the scale factor τ. In this paper, the length $n = 1024$, the embedding dimension m = 3, the number of categories c = 6, the time delay d = 1, and the scale factor are selected through simulation analysis. Figures 4–6 show a random point entropy curve corresponding to scale factors 8, 10, and 12, respectively.

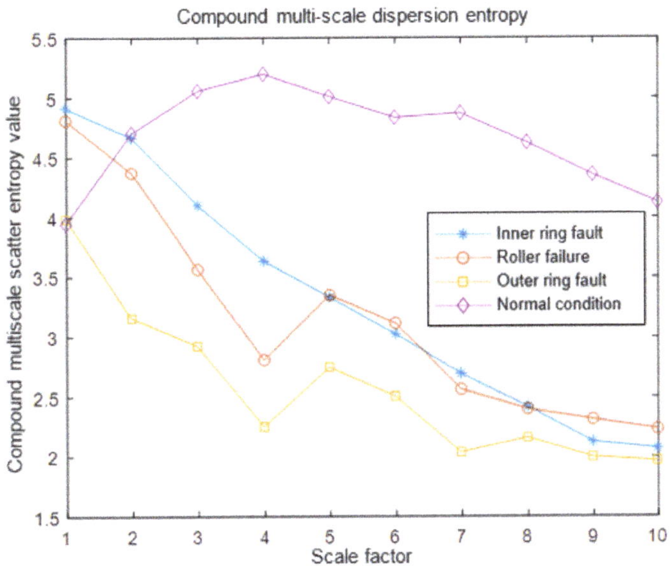

Figure 5. VMD-CMDE at τ = 10.

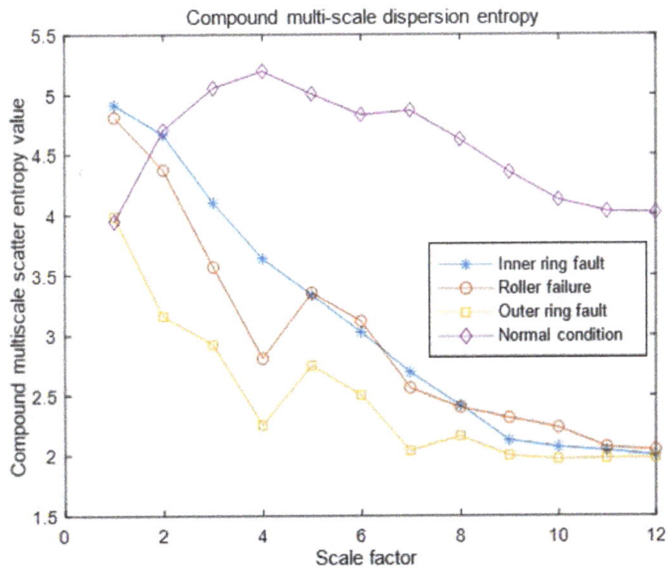

Figure 6. VMD-CMDE at τ = 12.

The abscissa in the figure is the number of scale factors, and the ordinate is the composite multi-scale dispersion entropy. Since the selection of basic theory and parameters and the multi-scale dispersion entropy are roughly the same, the curves are roughly the same as a whole. Except for the normal signals, the overall trend of the vibration signals of the other three faults is to decline first and then flatten. During the change of scale factors from 1 to 4, except when they are in the upward trend under normal conditions, the other three fault signals are in the downward trend, and the downward trend is obvious from the instantaneous change rate. When the scale factor ranges from 4 to 8, the overall decline is relatively gentle, with occasional fluctuations, and the decline of the inner ring fault is more obvious. When the scale factor ranges from 8 to 10, the decline is gentle, and the entropy of the fault signal is slowly approaching. The reason why the normal situation is different from the three fault signals is that there is no periodic vibration similar to the fault signal. When the scale factor ranges from 10 to 12, the entropy of the fault signal has a tendency to coincide, and the CMDE value does not change much, but the simulation time is longer with the increase of parameters.

Combined with the above simulation and analysis of the CMDE of the four pre-processed vibration signals, when the scale factor is 10, it can not only ensure that the deep-seated information of the vibration signal is extracted, but also ensure that the time will not be consumed too much. Therefore, the composite multi-scale dispersion entropy scale factor in this paper is 10.

3. Fault Identification Model Based on PSO-DBN

3.1. DBN Network Structure

As one of the typical deep learning algorithms, the Deep Belief Network (DBN) has good development prospects in the field of fault identification. The Deep Belief Network (DBN) is a probabilistic artificial neural network with multiple hidden layers, constructed by stacking multiple Restricted Boltzmann Machines (RBMs). By looking at the Restricted Boltzmann Machine architecture, we can obtain the associated functions as follows:

$$E(v,h \mid \theta) = -\sum_{i=1}^{n} a_i v_i - \sum_{j=1}^{m} b_i h_i - \sum_{i=1}^{n} \sum_{j=1}^{m} v_i W_{ij} h_j \tag{14}$$

where

θ—node parameters of Restricted Boltzmann Machine and $\theta = \{W_{ij}, a_i, b_j\}$ are all real numbers;

a_i—offset coefficient of visible unit i;

W_{ij}—weight values of hidden unit j and visible unit i;

b_j—offset coefficient of hidden unit j.

When these parameters are constant, based on this function, the joint probability distribution can be obtained, as shown in Formula (15):

$$P(v,h \mid \theta) = \frac{e^{-E(v,h|\theta)}}{Z(\theta)}, Z(\theta) = \sum_{v,h} e^{-E(v,h|\theta)} \tag{15}$$

where

$Z(\theta)$—partition function (Normalization factor);

a_i, b_i—offset coefficient;

h_j, v_i—state variables for hidden and visible units;

W_{ij}—hidden and visible unit weights.

In this energy function, it can be seen from the special structure that there is a connection between the layers of RBM and there is no connection between nodes in layers and star lakes. When the state of the hidden layer is known, the activation states for different visible units are conditionally independent. The probability of visible node activation is shown in Formula (16):

$$P(v_i = 1 \mid h, \theta) = \sigma(a_i + \sum_j W_{ji} h_j) \tag{16}$$

Similarly, the activation probability of the hidden unit is

$$P(h_j = 1 \mid v, \theta) = \sigma(b_j + \sum_i v_i W_{ij}) \tag{17}$$

where $\sigma(x) = \frac{1}{1+\exp(-x)}$ is the Sigmoid activation function. The complete Deep Belief Network structure is shown in Figure 7.

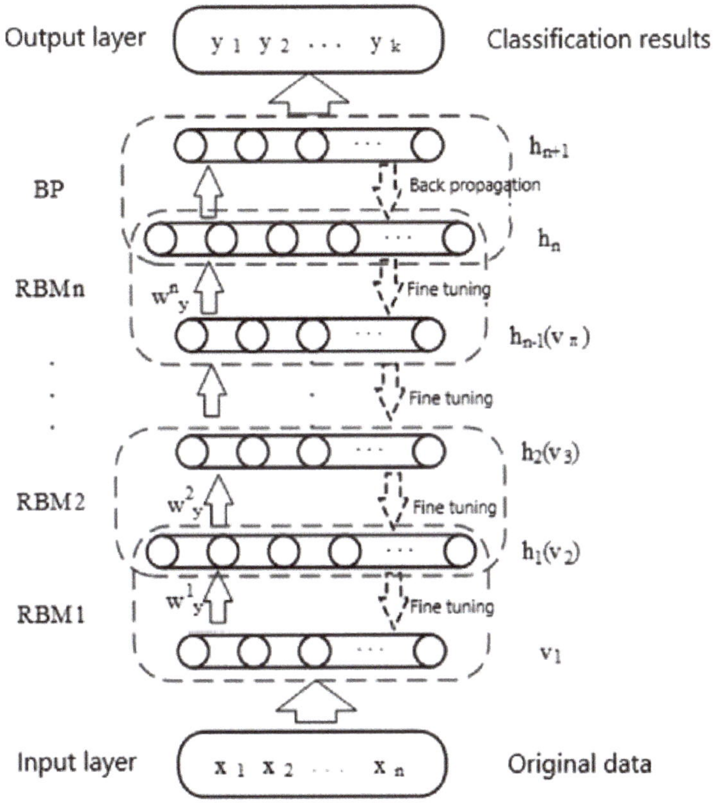

Figure 7. DBN model.

3.2. PSO-Optimized DBN Model

The particle swarm optimization algorithm is the same as many algorithms; that is, after the system initialization, it starts to iterate through a group of solutions, and constantly looks for the optimal solution in the iterative process. Particles (potential solutions) will follow the best particles in space to explore, so the number of iterations required to reach the best solution is relatively small. In the engineering application in the field of bearing diagnosis, particle swarm optimization can be easily employed because of its simple principle, strong universality, and strong anti-interference. Moreover, the algorithm supports group search and takes a short time. Combined with the above advantages, this paper selects the PSO optimization algorithm to improve the DBN model.

Bengio [48] has performed many experiments to illustrate a problem: the application effect of a multi-layer deep confidence network is often higher than that of a single layer. Larochelle and others [49] have proven through many tests that when the hidden layer of the deep confidence network model is about three layers, the classification accuracy reaches the highest value. Before the number of layers reaches four, the recognition rate is directly proportional to the increase in the number of hidden layers. When the number of hidden layers reaches four or more, the classification accuracy of the model will decline. This paper selects three hidden layers, corresponding to m_1, m_2, m_3 neurons. N represents the number of particles, which generally ranges from 10 to 20. In this paper, the number of particles is 10. The maximum iteration number of particle swarm optimization is M. This paper takes 20. The process of the PSO-optimized DBN model is shown in Figure 8.

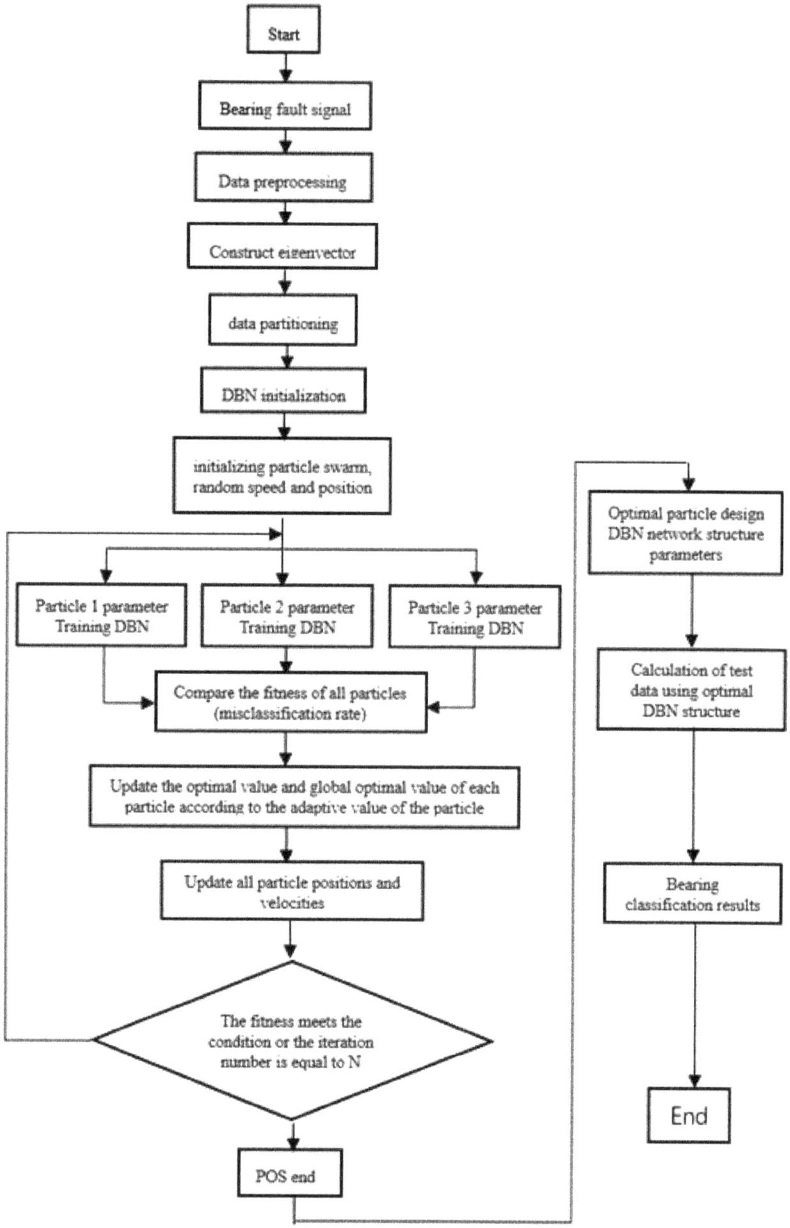

Figure 8. General flow chart of PSO-optimized DBN model.

The specific steps are as follows:

Step 1: Preprocess the original vibration signal of the bearing of Western Reserve University. Because the time and accuracy of training the original vibration signal are directly greatly affected, VMD decomposition is needed to reconstruct the signal according to kurtosis.

Step 2: In order to improve the accuracy of fault identification, the decomposed and reconstructed signals are combined with multi-scale arrangement entropy, multi-scale dispersion entropy, and composite multi-scale dispersion entropy to construct feature vectors.

Step 3: For the test data of four states, 100 samples are taken for each state, and a total of 400 samples are obtained. The fault feature set is P; the 100 samples of each signal in the obtained feature set are randomly divided into 70 training sets, recorded as P1, and 30 test sample sets, recorded as P.

Step 4: Initialize particle swarm velocity $V_i^{k=0}$; initialize the position of the particle swarm $X_i^{k=0}$.

Step 5: Calculate the classification error rate of all particles, and find the optimal particles of this round of particle swarm, including the optimal particles that have completed the search before.

Step 6: The velocity and position of each particle are updated by Formulas (18) and (19).

$$X_I^{k+1} = X_i^k + V_i^{k+1} \tag{18}$$

$$V_i^{k+1} = \omega V_i^k + c_1 r_1 \left(X_{iphest}^k - X_i^k\right) + c_2 r_2 \left(X_{ighest}^k - X_i^k\right) \tag{19}$$

where

ω—inertia weight;

c_1, c_2—acceleration parameters;

r_1, r_2—random value.

Among them, the value range of inertia weight is generally between 0 and 1, and ω = 0.7 is taken in this paper. The acceleration parameters generally range from 0 to 4. Shi et al. have done many tests; it was found that the selection of this parameter will affect the optimization results. In order to make the results not too disturbed by external factors and make the two acceleration parameters equal and have the best effect, parameter $c_1 = c_2 = 2$ is selected in this paper. Random values generally range from 0 to 1.

Step 7: One of two conditions needs to be met when PSO ends optimization. One is that the classification error rate of experimental data is lower than the pre-set value, or the number of iterations reaches the preset value. If one of the two meets, it can be stopped. Otherwise, go to step 5, increase the number of iterations, and repeat step 6 and step 7 until the discrimination conditions are met.

Step 8: The optimized parameters are substituted into the original DBN model, and the rolling bearing fault classification results are obtained by retraining and retesting the data samples.

4. Experimental Verification

The optimized DBN is applied to the experiment to analyze the data and construct the classifier. Aiming at the problem of rolling bearing fault pattern recognition proposed in this paper, the specific experimental steps and instructions are as follows:

Step 1: For the experimental data of four states, take 100 samples at random, with a total of 400 samples. Calculate the eigenvalues according to the VMD-CMDE composition method, combine them into the eigenvector set, and record them as the fault feature P. A total of 70 groups of eigenvalues are randomly selected from P as the training set and recorded as P1. The remaining 30 sets of eigenvalues are divided into test sets, namely P2.

Step2: Input P1 into DBN for training. In order to more comprehensively verify the reliability of the rolling bearing fault identification model, this paper selects the rolling bearing data of 1797r/min speed for research. Different bearing fault types are replaced by different numbers, as shown in Table 1. Here, 1 represents inner ring fault, 2 represents roller ring fault, 3 represents outer ring fault, and 4 represents normal condition.

Table 1. Description of bearing pattern recognition dataset.

Bearing Status	Training Sample Numbers	Test Sample Numbers	Categorization Label
Inner ring fault	70	30	1
Roller ring fault	70	30	2
Outer ring fault	70	30	3
Normal signal	70	30	4

Here, the experimental results of the DBN model input by the composite multi-scale scattered entropy eigenvector obtained after the decomposition of the original signal are analyzed. As shown in Figure 9, the recognition rate of each fault type of rolling bearing can be seen. According to the different numbers marked in this paper, they represent different fault types. Number 1 corresponds to the inner ring fault signal, and the recognition rate is 90%. Number 3 represents the outer ring fault signal, and the recognition rate is 100%. Number 2 corresponds to the roller fault signal, and the recognition rate is 73.33%. Number 4 corresponds to the normal bearing signal, and the recognition rate is 100%. After calculation, the overall recognition accuracy reaches 90.33%.

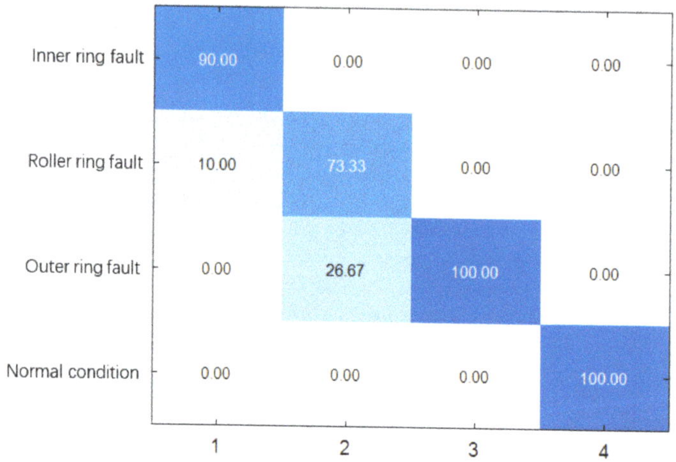

Figure 9. VMD-CMDE-DBN fault recognition rate.

Among them, 27 groups were correctly identified by 30 groups of bearing with inner ring fault, 22 groups were correctly identified by 30 groups of roller fault, 30 groups were correctly identified by 30 groups of bearing with outer ring fault, and 30 groups were correctly identified by 30 groups of bearing under normal conditions. Compared with the previous two models, the overall recognition rate of this group can reach 90.33%, and the roller fault recognition rate has also been greatly improved, but there is still room for improvement. Based on this data, Table 2 is established.

Table 2. Accuracy rate of DBN model with VMD-CMDE as input.

Bearing Status	Total Number of Test Set Samples	Correct Number	Accuracy
Inner ring fault	30	27	90%
Roller ring fault	30	22	73.3%
Outer ring fault	30	30	100%
Normal condition	30	30	100%
Whole bearing	120	109	90.33%

The key parameters of the VMD-CMDE-DBN model are optimized by the particle swarm optimization algorithm to obtain the VMD-CMDE-PSO-DBN model. Through the analysis of the experimental results of the optimized DBN model input by the composite multi-scale dispersion entropy eigenvector obtained after the decomposition of the original signal, as shown in Figure 10, we can see the recognition rate of each fault type of rolling bearing. According to the different numbers marked in this paper, they represent different fault types. Numbers 1, 2, 3, and 4 correspond to inner ring fault signal, roller fault signal, outer ring fault signal, and normal bearing signal, respectively.

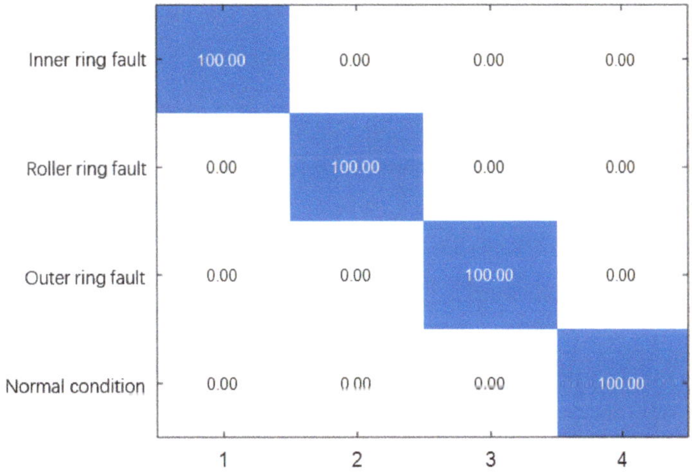

Figure 10. VMD-CMDE-PSO-DBN fault recognition rate.

According to this data, Table 3 is established. From Table 3, we can clearly see the identification number of each fault type; among them, 30 groups of bearing with inner ring fault are correctly identified, 30 groups of roller fault are correctly identified, 30 groups of bearing with outer ring fault are correctly identified, and 30 groups of bearing under normal conditions are correctly identified.

Table 3. PSO-DBN model accuracy with VMD-CMDE as input.

Bearing Status	Total Number of Test Set Samples	Correct Number	Accuracy
Inner ring fault	30	30	100%
Roller ring fault	30	30	100%
Outer ring fault	30	30	100%
Normal condition	30	30	100%
Whole bearing	120	120	100%

In order to fully prove the effectiveness of VMD-CMDE-PSO-DBN fault identification model, Multi-scale Permutation Entropy (MPE) and Multi-scale Dispersion Entropy (MDE) are substituted into the DBN model and optimized model in this paper. Observing and compare the recognition rate, the number of samples in the training set and the test set is the same as above; the recognition rate data input into the DBN model is shown in Table 4.

The number of nodes after particle swarm optimization is substituted into the three models, and the same eigenvalues of the three entropy are used as the input of particle swarm optimization DBN model. The recognition rate data are shown in Table 5.

Table 4. DBN model accuracy.

Bearing Status	VMD-MPE	VMD-MDE	VMD-CMDE
Inner ring fault	100%	100%	90%
Roller ring fault	43.33%	33.33%	73.33%
Outer ring fault	70%	100%	100%
Normal condition	100%	100%	100%
Whole bearing	78.33%	88.33%	90.33%

Table 5. PSO-DBN model accuracy.

Bearing Status	VMD-MPE	VMD-MDE	VMD-CMDE
Inner ring fault	96.67%	100%	100%
Roller ring fault	96.67%	93.33%	100%
Outer ring fault	100%	100%	100%
Normal condition	100%	100%	100%
Whole bearing	98.33%	98.33%	100%

5. The Result Discussion

A total of 70 sets of multi-scale entropy eigenvalues of rolling bearing fault signals were substituted into the DBN model for recognition training. The DBN model was tested with 30 groups of test set data. Through the test, the experimental results show that the recognition accuracy of multi-scale arrangement entropy and DBN is 78.33%, the recognition accuracy of multi-scale dispersion entropy and DBN is 83.33%, and the recognition accuracy of composite multi-scale dispersion entropy and DBN is 90.33%. Each model is not particularly ideal in roller fault recognition. The experimental results show that the recognition accuracy of multi-scale arrangement entropy and optimized DBN is 98.33%, the recognition accuracy of multi-scale dispersion entropy and optimized DBN is 98.33%, and the recognition accuracy of composite multi-scale dispersion entropy and optimized DBN is 100%. Compared vertically, the PSO-DBN classification effect of the DBN model after optimizing parameters by the particle swarm optimization algorithm has been improved in different multi-scale entropy. Compared horizontally, the classification effect of the PSO-DBN model with different multi-scale entropy eigenvectors as input has also been significantly improved. Especially in the identification of roller fault, the three models have been greatly improved.

Through theoretical proof and experimental verification, the combination of VMD, CMDE, DBN, and PSO algorithm is very effective in rolling bearing fault diagnosis and identification. The main conclusions are as follows:

The rolling bearing fault recognition model is established; the eigenvectors are substituted into the DBN and PSO-DBN models, trained and tested; and the final experimental results are obtained. By comparing the recognition accuracy of DBN and PSO-DBN, it can be concluded that the PSO-DBN model has a higher recognition rate than the DBN model. Overall, the recognition rate based on VMD-CMDE-PSO-DBN is the best, which provides new insight for signal pattern recognition.

6. Conclusions

In this paper, an intelligent fault diagnosis method based on Variational Mode Decomposition (VMD), Composite Multi-scale Dispersion Entropy (CMDE), and Deep Belief Network (DBN) with Particle Swarm Optimization (PSO) algorithm—namely VMD-CMDE-PSO-DBN—is proposed. The decomposed number of modal components of VMD is determined by the observation center frequency, reconstructed according to the kurtosis, and the composite multi-scale dispersion entropy of the reconstructed signal is calculated to form the training samples and test samples of pattern recognition.

- The experimental data used in this paper are manually added faults, which may not fully reflect the diversified faults of rolling bearings, single fault forms, and low bearing speed. Under actual working conditions, bearings are mostly in high-speed operation and the fault forms are complex, so the next step should be to focus on the high-speed operation of rolling bearings and the composite fault state.

- VMD multi-scale permutation entropy eigenvector, VMD multi-scale dispersion entropy eigenvector, and VMD composite multi-scale dispersion entropy eigenvector is used as the inputs of the Deep Belief Network classification model. The accuracy of VMD decomposition composite multi-scale dispersion entropy is the best.

Author Contributions: Conceptualization, Z.G.; methodology, E.Y. and Y.W.; writing—original draft preparation, E.Y.; writing—review and editing, P.W. and W.D.; Resources, W.D.; software, P.W. and Z.G.; validation, E.Y., Y.W., P.W. and Z.G. All authors have read and agreed to the published version of the manuscript.

Funding: This research was funded by the Department of Education Foundation of Liaoning Province under grant JDL2020013, the Natural Science Foundation of Liaoning Province under grant 2019ZD0112, and the National Natural Science Foundation of China under grant 62001079, the Research Foundation for Civil Aviation University of China under Grant 3122022PT02 and 2020KYQD123.

Institutional Review Board Statement: Not applicable.

Informed Consent Statement: Not applicable.

Data Availability Statement: All data, models, and code generated or used during the study appear in the submitted article.

Conflicts of Interest: The authors declare no conflict of interest.

References

1. He, Z.; Shao, H.; Wang, P.; Lin, J.; Cheng, J. Deep transfer multi-wavelet auto-encoder for intelligent fault diagnosis of gearbox with few target training samples. *Knowl. Based Syst.* **2020**, *191*, 105313. [CrossRef]
2. Li, X.; Zhao, H.; Yu, L.; Chen, H.; Deng, W.; Deng, W. Feature extraction using parameterized multisynchrosqueezing transform. *IEEE Sens. J.* **2022**, *2*, 14263–14272. [CrossRef]
3. Xu, G.; Bai, H.; Xing, J.; Luo, T.; Xiong, N.N. SG-PBFT: A secure and highly efficient distributed blockchain PBFT consensus algorithm for intelligent Internet of vehicles. *J. Parallel Distrib. Comput.* **2022**, *164*, 1–11. [CrossRef]
4. Zheng, J.J.; Yuan, Y.; Zou, L.; Deng, W.; Guo, C.; Zhao, H. Study on a novel fault diagnosis method based on VMD and BLM. *Symmetry* **2019**, *11*, 747. [CrossRef]
5. Wu, D.; Wu, C. Research on the Time-Dependent Split Delivery Green Vehicle Routing Problem for Fresh Agricultural Products with Multiple Time Windows. *Agriculture* **2022**, *12*, 793. [CrossRef]
6. Zhou, X.B.; Ma, H.J.; Gu, J.G.; Chen, H.L.; Deng, W. Parameter adaptation-based ant colony optimization with dynamic hybrid mechanism. *Eng. Appl. Artif. Intell.* **2022**, *114*, 105139. [CrossRef]
7. Wu, X.; Wang, Z.C.; Wu, T.H.; Bao, X.G. Solving the family traveling salesperson problem in the adleman–lipton model based on DNA computing. *IEEE Trans. NanoBioscience* **2021**, *21*, 75–85. [CrossRef]
8. Li, X.; Shao, H.; Lu, S.; Xiang, J.; Cai, B. Highly-efficient fault diagnosis of rotating machinery under time-varying speeds using LSISMM and small infrared thermal images. *IEEE Trans. Syst. Man Cybern. Syst.* **2022**, *50*, 1–13. [CrossRef]
9. An, Z.; Wang, X.; Li, B.; Xiang, Z.L.; Zhang, B. Robust visual tracking for UAVs with dynamic feature weight selection. *Appl. Intell.* **2022**. [CrossRef]
10. Cao, H.; Shao, H.; Zhong, X.; Deng, Q.; Yang, X.; Xuan, J. Unsupervised domain-share CNN for machine fault transfer diagnosis from steady speeds to time-varying speeds. *J. Manuf. Syst.* **2022**, *62*, 186–198. [CrossRef]
11. Li, T.Y.; Shi, J.Y.; Deng, W.; Hu, Z.D. Pyramid particle swarm optimization with novel strategies of competition and cooperation. *Appl. Soft Comput.* **2022**, *121*, 108731. [CrossRef]
12. Deng, W.; Xu, J.; Gao, X.; Zhao, H. An enhanced MSIQDE algorithm with novel multiple strategies for global optimization problems. *IEEE Trans. Syst. Man Cybern. Syst.* **2022**, *52*, 1578–1587. [CrossRef]
13. Chen, H.Y.; Miao, F.; Chen, Y.J.; Xiong, Y.J.; Chen, T. A hyperspectral image classification method using multifeature vectors and optimized KELM. *IEEE J. Sel. Top. Appl. Earth Obs. Remote Sens.* **2021**, *14*, 2781–2795. [CrossRef]
14. Yao, R.; Guo, C.; Deng, W.; Zhao, H.M. A novel mathematical morphology spectrum entropy based on scale-adaptive techniques. *ISA Trans.* **2022**, *126*, 691–702. [CrossRef]
15. Deng, W.; Li, Z.; Li, X.; Chen, H.; Zhao, H. Compound fault diagnosis using optimized MCKD and sparse representation for rolling bearings. *IEEE Trans. Instrum. Meas.* **2022**, *71*, 3508509. [CrossRef]

16. Tian, C.; Jin, T.; Yang, X.; Liu, Q. Reliability analysis of the uncertain heat conduction model. *Comput. Math. Appl.* **2022**, *119*, 131–140. [CrossRef]
17. Zhao, H.M.; Liu, J.; Chen, H.Y.; Chen, J.; Li, Y.; Xu, J.J.; Deng, W. Intelligent diagnosis using continuous wavelet transform and gauss convolutional deep belief network. *IEEE Trans. Reliab.* **2022**, 1–11. [CrossRef]
18. Zhou, Y.; Zhang, J.; Yang, X.; Ling, Y. Optimal reactive power dispatch using water wave optimization algorithm. *Oper. Res.* **2020**, *20*, 2537–2553. [CrossRef]
19. Cui, H.; Guan, Y.; Chen, H. Rolling element fault diagnosis based on VMD and sensitivity MCKD. *IEEE Access* **2021**, *9*, 120297–120308. [CrossRef]
20. Xu, Y.; Chen, H.; Luo, J.; Zhang, Q.; Jiao, S.; Zhang, X. Enhanced Moth-flame optimizer with mutation strategy for global optimization. *Inf. Sci.* **2019**, *492*, 181–203. [CrossRef]
21. Liu, Q.; Jin, T.; Zhu, M.; Tian, C.; Li, F.; Jiang, D. Uncertain currency option pricing based on the fractional differential equation in the Caputo sense. *Fractal Fract.* **2022**, *6*, 407. [CrossRef]
22. Li, G.; Li, Y.; Chen, H.; Deng, W. Fractional-order controller for course-keeping of underactuated surface vessels based on frequency domain specification and improved particle swarm optimization algorithm. *Appl. Sci.* **2022**, *12*, 3139. [CrossRef]
23. Wei, Y.Y.; Zhou, Y.Q.; Luo, Q.F.; Deng, W. Optimal reactive power dispatch using an improved slime Mould algorithm. *Energy Rep.* **2021**, *7*, 8742–8759. [CrossRef]
24. Jin, T.; Xia, H.; Deng, W.; Li, Y.; Chen, H. Uncertain fractional-order multi-objective optimization based on reliability analysis and application to fractional-order circuit with Caputo type. *Circ. Syst. Signal Pract.* **2021**, *40*, 5955–5982. [CrossRef]
25. Xu, G.; Dong, W.; Xing, J.; Lei, W.; Liu, J. Delay-CJ: A novel cryptojacking covert attack method based on delayed strategy and its detection. *Digit. Commun. Netw.* **2022**. [CrossRef]
26. Wu, E.Q.; Zhou, M.; Hu, D.; Zhu, L.; Tang, Z. Self-paced dynamic infinite mixture model for fatigue evaluation of pilots' brains. *IEEE Trans. Cybern.* **2022**, *52*, 5623–5638. [CrossRef]
27. Jiang, M.; Yang, H. Secure outsourcing algorithm of BTC feature extraction in cloud computing. *IEEE Access* **2020**, *8*, 106958–106967. [CrossRef]
28. Deng, W.; Ni, H.C.; Liu, Y.; Chen, H.L.; Zhao, H.M. An adaptive differential evolution algorithm based on belief space and generalized opposition-based learning for resource allocation. *Appl. Soft Comput.* **2022**. [CrossRef]
29. Zosso, D.; Dragomiretskiy, K. Variational mode decomposition. In *IEEE Transactions on Signal Processing: A Publication of the IEEE Signal Processing Society*; IEEE: Piscataway, NJ, USA, 2014.
30. Rostaghi, M.; Azami, H. Dispersion Entropy: A measure for time series analysis. *IEEE Signal Process. Lett.* **2016**, *23*, 610–614. [CrossRef]
31. Zhang, S.; Sun, G.; Li, L.; Li, X.; Jian, X. Study on mechanical fault diagnosis method based on LMD approximate entropy and fuzzy C-means clustering. *Chin. J. Sci. Instrum.* **2013**, *34*, 714–720.
32. Wang, J.; Shuai, C.; Chao, Z. Fault diagnosis method of gear based on VMD and multi-feature fusion. *J. Mech. Transm.* **2017**, *3*, 032.
33. Li, C. *Research on Rolling Bearing Fault Diagnosis Method Based on Empirical Wavelet Transform and Scattered Entropy*; Anhui University of Technology: Anhui, China, 2019.
34. Hinton, G.; Osindero, S.; Teh, Y. A fast learning algorithm for deep belief nets. *Neura I Comput.* **2006**, *18*, 1527–1554. [CrossRef] [PubMed]
35. Lei, Y.; Jia, F.; Zhou, X.; Lin, J. A Deep learning-based method for machinery health monitoring with big data. *J. Mech. Eng.* **2015**, *51*, 49–56. [CrossRef]
36. Li, W.; Shan, W.; Xu, Z.; Zeng, X. Bearing fault classification and recognition based on deep belief network. *J. Vib. Eng.* **2015**, *29*, 152–159.
37. Shi, P.; Liang, K.; Zhao, N. Gear intelligent fault diagnosis based on deep learning feature extraction and particle swarm support vector machine state recognition. *China Mech. Eng.* **2017**, *28*, 1056–1061.
38. Wang, W.; Carr, M.; Xu, W.; Kobbacy, K. A model for residual life prediction based on Brownian motion with an adaptive drift. *Microelectron. Reliab.* **2010**, *51*, 285–293. [CrossRef]
39. Sun, L.; Tang, X.G.; Zhang, X.H. Study of gearbox residual life prediction based on stochastic filtering model. *Mech. Transm.* **2011**, *35*, 56–60.
40. Deng, W.; Liu, H.; Xu, J.; Zhao, H.; Song, Y. An improved quantum-inspired differential evolution algorithm for deep belief network. *IEEE Trans. Instrum. Meas.* **2020**, *69*, 7319–7327. [CrossRef]
41. Zhao, H.; Liu, H.; Xu, J.; Guo, C.; Deng, W. Research on a fault diagnosis method of rolling bearings using variation mode decomposition and deep belief network. *J. Mech. Sci. Technol.* **2019**, *33*, 4165–4172. [CrossRef]
42. Li, T.; Qian, Z.; Deng, W.; Zhang, D.; Lu, H.; Wang, S. Forecasting crude oil prices based on variational mode decomposition and random sparse Bayesian learning. *Appl. Soft Comput.* **2021**, *113PB*, 108032. [CrossRef]
43. Pan, Y.; Chen, J.; Guo, L. Robust bearing performance degradation assessment method based on improved wavelet packet–support vector data description. *Mech. Syst. Signal. Process.* **2009**, *23*, 669–681. [CrossRef]
44. Dong, S.; Luo, T. Bearing degradation process prediction based on the PCA and optimized LS-SVM model. *Measurement* **2013**, *46*, 3143–3152. [CrossRef]
45. Xue, X.H. Evaluation of concrete compressive strength based on an improved PSO-LSSVM model. *Comput. Concr. Int. J.* **2018**, *21*, 505–511.
46. He, Q. Vibration signal classification by wavelet packet energy flow manifold learning. *J. Sound Vib.* **2013**, *332*, 1881–1894. [CrossRef]
47. Ishaque, K.; Salam, Z.; Amjad, M.; Mekhilef, S. An improved particle swarm optimization (PSO)–Based MPPT for PV with reduced steady-state oscillation. *IEEE Trans. Power Electron.* **2012**, *27*, 3627–3638. [CrossRef]

48. Roux, N.L.; Bengio, Y. Representational Power of Restricted Boltzmann Machines and Deep Belief Networks. *Neural Comput.* **2008**, *20*, 1631–1649. [CrossRef]
49. Larochelle, H.; Bengio, Y.; Louradour, J.; Lamblin, P. Exploring Strategies for Training Deep Neural Networks. *J. Mach. Learn. Res.* **2009**, *1*, 1–40.

electronics

Article

Improved LS-SVM Method for Flight Data Fitting of Civil Aircraft Flying at High Plateau

Nongtian Chen [1,2], Youchao Sun [1,*], Zongpeng Wang [1] and Chong Peng [1]

1 College of Civil Aviation, Nanjing University of Aeronautics and Astronautics, Nanjing 211100, China; chennongtian@nuaa.edu.cn (N.C.); wangzp@nuaa.edu.cn (Z.W.); chong_peng@nuaa.edu.cn (C.P.)
2 College of Aviation Engineering, Civil Aviation Flight University of China, Guanghan 618307, China
* Correspondence: sunyc@nuaa.edu.cn

Abstract: High-plateau flight safety is an important research hotspot in the field of civil aviation transportation safety science. Complete and accurate high-plateau flight data are beneficial for effectively assessing and improving the flight status of civil aviation aircrafts, and can play an important role in carrying out high-plateau operation safety risk analysis. Due to various reasons, such as low temperature and low pressure in the harsh environment of high-plateau flights, the abnormality or loss of the quick access recorder (QAR) data affects the flight data processing and analysis results to a certain extent. In order to effectively solve this problem, an improved least squares support vector machines method is proposed. Firstly, the entropy weight method is used to obtain the index weights. Secondly, the principal component analysis method is used for dimensionality reduction. Finally, the data are fitted and repaired by selecting appropriate eigenvalues through multiple tests based on the LS-SVM. In order to verify the effectiveness of this method, the QAR data related to multiple real plateau flights are used for testing and comparing with the improved method for verification. The fitting results show that the error measurement index mean absolute error of the average error accuracy is more than 90%, and the error index value equal coefficient reaches a high fit degree of 0.99, which proves that the improved least squares support vector machines machine learning model can fit and supplement the missing QAR data in the plateau area through historical flight data to effectively meet application needs.

Keywords: least squares method; support vector machines; principal component analysis; quick access recorder; mean absolute error; high-plateau flight

Citation: Chen, N.; Sun, Y.; Wang, Z.; Peng, C. Improved LS-SVM Method for Flight Data Fitting of Civil Aircraft Flying at High Plateau. *Electronics* **2022**, *11*, 1558. https://doi.org/10.3390/electronics11101558

Academic Editor: Gyu Myoung Lee

Received: 24 March 2022
Accepted: 10 May 2022
Published: 13 May 2022

1. Introduction

High-plateau flights represent an important safety issue for civil aviation, especially for China's civil aviation transportation. High-plateau airports are mainly distributed in China, Nepal, Peru, Bolivia, Ecuador, and other countries. Among the 42 high-plateau airports in the world, 16 are located in China, so their operation safety problems have a profound impact on China's civil aviation [1]. On 14 May 2018, the flight mission of Chinese Sichuan Airlines flight 3U8633 from Chongqing to Lhasa plateau was an example of the typical unsafe event; the front windshield of the cockpit burst and fell off during the flight in high-plateau airspace, and the crew made an emergency descent. Compared with ordinary flight, high-plateau flight has low air density and atmospheric pressure, complex terrain, solar radiation, uneven heating of the terrain facing the sun, and many other environmental characteristics which result in stricter takeoff and landing conditions for aircrafts on high plateaus. The technical requirements of the personnel are more stringent and certain factors such as modification on the basis of ordinary civil aircrafts will cause the flight parameters of high-plateau civil airliners to change from those of civil airliners on general routes. During the entire flight phase, the quick access recorder (QAR) data may be abnormal or lost due to the influence of the high plateau's harsh environment, detection equipment, transmission equipment, or other unknown conditions. QAR is an important data warehouse for post-flight flight technical analysis, engine health analysis, flight safety incident investigation, flight quality analysis, operational quality analysis, and aircraft health management. The abnormality of these data will bring inconvenience and hidden hazards for monitoring and analyzing the safety status of high-plateau flights for theoretical research.

Many scholars have carried out fruitful research on flight data analysis and application, mainly focusing on flight data processing, flight data application, and other application research. Flight data have many applications in aviation operation safety research [2–6]. Some scholars have applied flight data to turbine fault diagnosis, general aviation anomaly detection, aviation safety key landing index prediction [7–11], tower flight data manager man–machine system integration design processes, and new methods for nonlinear aerodynamic modeling of flight data [12–14]. Some scholars also analyze the flight characteristics of QAR data for landing at high-altitude airports, and use it for airline flight data monitoring machine learning methods, generating new operational safety knowledge from existing data, safety science insights gained from black-box-to-flight data monitoring, composite fault diagnosis using optimized MCKD and sparse representation of rolling bearings, rolling elements based on VMD, and sensitivity MCKD fault diagnosis, etc. [15–19]. Some scholars have carried out research on the impact of leveling operation on landing safety based on variance analysis of real flight data, civil aircraft hazard identification and prediction based on deep learning [20,21], unsteady aerodynamic modeling of unstable dynamic processes [22], and small-sample inspection data-driven diagnosis of critical deviation sources in aircraft structural assembly [23].

In the research of flight data processing methods and technologies, many scholars have also carried out a series of studies [24–26]. Some scholars have proposed improved binary gray wolf optimizer and support vector machine methods, arithmetic optimization algorithms, particle swarm optimization, average impact value-support vector machine algorithms, etc., for in-flight data processing and optimization [27–29]. Some scholars combined multiple classifiers to quantitatively sort the impact of anomalies in flight data based on frequency domain specification and improved particle swarm optimization algorithms, as well as enhanced fast non-dominated solution sorting genetic algorithms for multi-objective problems research [30–32].

In short, many scholars have carried out a series of researches on flight data collection and analysis, as well as application methods and technologies, and have also achieved many valuable results. However, research on high-altitude flight data is rare, especially research on the filling and simulation of flight data loss due to high altitude, low temperature, low pressure, and other elements of the special operating environments. To effectively solve the problem of high-plateau QAR flight data padding, an improved least squares support vector machines method is proposed. The entropy weight method is used to obtain the index weights, and the principal component analysis method is used for dimensionality reduction. The flight data are fitted and repaired by selecting appropriate eigenvalues through multiple tests based on LS-SVM. The data are fitted and repaired by selecting appropriate eigenvalues through multiple tests based on LS-SVM. In order to verify the effectiveness of this method, the QAR data related to multiple real plateau flights are used for testing and are compared with the improved method for verification.

2. Principle of Data Restoration Method

2.1. LS-SVM Principle

The support vector machine is a generalized linear classifier proposed to perform binary classification of data in a supervised learning manner. Its decision boundary is the maximum margin hyperplane for the learning sample solution. The basic principle is shown in Figure 1.

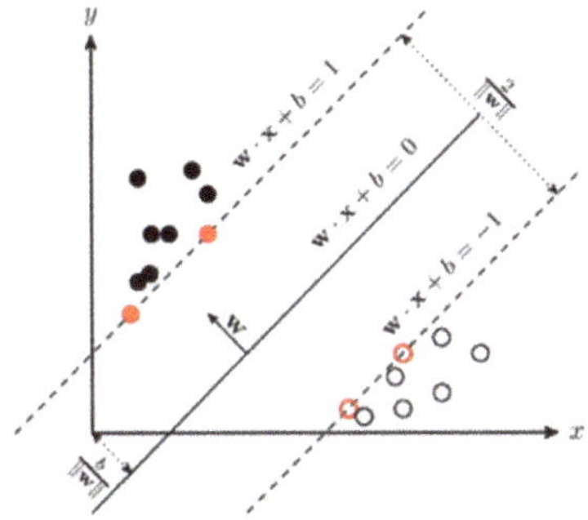

Figure 1. Support vector machine hyperplane conceptual model.

It is a machine learning method that is based on a complete statistical learning theory and has excellent learning capabilities. It has strict mathematical theory support, strong interpretability, and does not rely on statistical methods, thus simplifying the usual problems of classification and regression. It can also find key samples (support vectors) that are critical to the task. After adopting nuclear techniques, it can handle non-linear classification–regression tasks. The final decision function is determined by only a small number of support vectors and the complexity of the calculation depends on the number of support vectors, not the dimensionality of the sample space.

The LS-SVM demonstrates an improvement in the standard support vector machine, a new type of support vector machine method proposed by Suykens and Vandewalb. Compared with the standard SVM, it replaces the inequality constraints in SVM with equality constraints, which increases the convergence speed, improves classification progress in problems with desired goals, and achieves good results [33].

Supposing the data training set of a given LS-SVM is expressed as (1)

$$(x_1, y_1), \ldots, (x_1, y_1), x \in R_n, y \in \{-1, +1\} \tag{1}$$

$x_i \in R_n$ is the n-dimensional system input vector, $y_i \in R_n$ is the system output and $f(x) = \omega^T \varphi(x) + b$ is the unknown function to be estimated. Making a nonlinear mapping γ: $R_n \to H$, where Φ is called the feature map and H is the feature space, the unknown function is estimated to use the function of the form (2).

$$f(x) = \omega^T \varphi(x) + b \tag{2}$$

Among them, ω is the weight vector in R_n space, and $b \in R$ is the bias. The SVM algorithm uses the kernel function of the original space to replace the dot product operation in the high-dimensional feature space, avoids complex operations, and uses structural risk to minimize as a learning rule, which is mathematically described as $\omega T \omega \leq$ constant. The standard SVM algorithm takes the insensitive loss function as the structural risk minimization estimation problem. The meaning of the ε-insensitive loss function is as follows: when the difference between the observed value y of the x point and the predicted value $f(x)$ does not exceed the predetermined ε, it is considered that the predicted value $f(x)$ at this point is lossless, although the predicted value $f(x)$ and the observed value y may not be equal. On the other hand, LS-VSM chooses the second norm e^i of ξ_i as the loss function to make the equation true. Therefore, the optimization equation is established as (3) and (4).

$$\min_{\omega,b,e}(J\omega e) = \frac{1}{2}\omega^T \omega + \frac{1}{2}\gamma \sum_{i=1}^{N} e^2, \gamma > \tag{3}$$

$$y_i = \omega^T \varphi(x_i) + b + e_i{}^2, i = 1, 2, \ldots, N \tag{4}$$

Here, γ is a real constant which determines the relative size of $\frac{1}{2}\omega T\omega$ and $\frac{1}{2}\sum_{i=1}^{N} e^2$, which can be between the training error and the compromised model complexity so that the function can seek better generalization ability. The LS-SVM algorithm defines a loss function that is different from the standard SVM algorithm and changes its inequality constraints to equality constraints, which can obtain ω in the dual space. The Lagrange Function (5) is as follows:

$$L(\omega, b, e, a) = \frac{1}{2}\omega^T\omega + \frac{1}{2}\gamma\sum_{i=1}^{N} e_i{}^2 - \sum_{i=1}^{N} a_i\omega^T\varphi(x_i) + b + e_i - y_i \tag{5}$$

where $\alpha^i \in R$, $\alpha^i > 0$ is the Lagrange multiplier so the optimal solution condition is as follows (6):

$$\begin{aligned} &\tfrac{\delta L}{\delta \omega} = 0, \omega = \sum_{i=1}^{N} a_i\varphi(x_i) \\ &\tfrac{\delta L}{\delta b} = 0, \sum_{i=1}^{N} a_i = 0 \\ &\tfrac{\delta L}{\delta e_i} = 0, a_i = \gamma e_i \\ &\tfrac{\delta L}{\delta a_i} = 0, y_i = \omega^T\varphi(x_i) + b + e_i, i = 1, \ldots, N \end{aligned} \tag{6}$$

After eliminating ω and e_i from Equation (6), this optimization problem is transformed into solving the following equation:

$$\begin{bmatrix} b \\ 0 \end{bmatrix} = \begin{bmatrix} 0 & 1 \\ 1 & B + \gamma^{-1} \end{bmatrix}^{-1} \begin{bmatrix} 0 \\ \gamma \end{bmatrix} \tag{7}$$

Among them, $y = [y_1, y_2, \ldots, y_N]^T$, $a = [a_1, a_2, \ldots, a_N]^T$, $1 = [1, \ldots, 1]^T$, and B represent a square matrix; the element in the i-th column and row j is $B_{ij} = \varphi(x_i)^T\varphi(x_i) = K(x_i, x_j)$, $i, j = 1, \ldots, N$; and $K(x_i, x_j)$ is the kernel function. On the basis of Formula (3), ω can be further obtained, so as to obtain the nonlinear approximation of the training data set

$$f(x) = \sum_{i=1}^{N} a_i K(x_i, x_j) + b \tag{8}$$

2.2. The Choice of Kernel Function

The kernel function is used to prevent the non-linear transformation from mapping its input space to the high-latitude space, causing particularly high-dimensional complex operations. When the support vector machine only needs the inner product operation and looks for a function that represents a low-dimensional input space that is exactly equal to the inner product in the high-dimensional space, the result can be obtained directly to avoid complicated operations. The choice of the kernel function requires Mercer's theorem to be satisfied, that is, any Gram matrix of the kernel function in the sample space is a semi-positive definite matrix (semi-positive definite) [34]. Currently, the commonly used kernel functions in research and practice are as follows:

(1) Linear kernel function:

$$K(x, x_i) = x \cdot x_i \tag{9}$$

(2) Polynomial kernel function:

$$K(x, x_i) = (x \cdot x_i + 1)^d \tag{10}$$

(d value is the order of the polynomial)

(3) Radial basis kernel function:

$$K(x, x_i) = \exp(-\frac{(x - x_i)^2}{2\sigma^2}) \tag{11}$$

(4) B-spline kernel function:

$$K(x, x_i) = B_{2n+1}(x - x_i) \quad (12)$$

(5) Perceptual kernel function:

$$K(x, x_i) = \tan h(\beta x_i + b) \quad (13)$$

2.3. LS-SVM Principle

Entropy comes from physical thermodynamics and is one of the parameters that can characterize matter. It was first introduced into information theory by C.E. Shannony and called information entropy. The entropy weight method (EWM) abstracts information and tests its degree of variation through various eigenvalues. In this way, the weight of each feature is calculated and modified to achieve a more reasonable weight index [35]. The specific process is as follows:

(1) Perform data standardization processing on each feature value. Suppose that k feature quantities $Y_{ij} = \frac{x_{ij} - \min(x_i)}{max(x_i) - \min(x_i)}$ are given, where $X_i = x_1, x_2, \ldots, x_n$, assuming that the standardized value of each feature value is $Y_1, Y_2, \ldots, Y_K$

$$Y_{ij} = \frac{x_{ij} - \min(x_i)}{max(x_i) - \min(x_i)} \quad (14)$$

(2) Find the information entropy of each eigenvalue. According to the definition of information entropy in information theory, the information entropy of a set of data can be written as

$$P_{ij} = \frac{Y_{ij}}{\sum_{i=1}^{n} Y_{ij}} \quad (15)$$

where $p_{ij} = \frac{Y_{ij}}{\sum_{i=1}^{n} Y_{ij}}$, if $\lim_{p_{ij}=0} \sum_{i=1}^{n} P_{ij} ln P_{ij} = 0$, then define $\lim_{p_{ij}=0} \sum_{i=1}^{n} P_{ij} ln P_{ij} = 0$, determine the weight w of each feature quantity:

$$w_i = \frac{1 - E_i}{k - \sum E_i} (i = 1, 2, \ldots, k) \quad (16)$$

2.4. Principles of Principal Component Analysis (PCA)

The principal component analysis (PCA) method is currently the most widely used data dimensionality reduction algorithm. It aims to sequentially find a set of mutually orthogonal coordinate axes from the original high-dimensional space to determine its correlation by comparing the variance of the original data under the new coordinate axis; the degree is used to exclude zero-correlation or low-correlation feature quantities to achieve a dimensionality reduction of data features. Because of the efficiency and simplicity of PCA processing high-dimensional data sets, it is widely used in various fields in practice, especially in the field of compressed data [36].

2.5. Verification Method

In order to judge the conformity of the selected number of feature quantities, the coefficient of determination (R^2) is introduced. The coefficient of determination indicates how much the fluctuation of the dependent variable can be described by the fluctuation of the independent variable. Its expression is as follows:

$$R^2 = \left(\frac{\sum_{i=1}^{n}\left(y_i - \hat{y}\right) * \left(\hat{y}_i - \ddot{y}\right)}{\sqrt{\sum_{i=1}^{n}\left(y_i - \hat{y}\right)^2} * \sqrt{\sum_{i=1}^{n}\left(\hat{y}_i - \ddot{y}\right)^2}}\right)^2 \quad (17)$$

y and $\hat{y}$ represent the actual value and the predicted value of the simulation result. The closer the R^2 value is to 1, the better the correlation between the two.

For the evaluation of the complementation results, four commonly used indicators for data repair are introduced for analysis purposes: mean square error (MSE), root mean square error (RMSE), mean absolute error (MAE), and equal coefficient (EC). The calculation is as follows:

$$\begin{aligned} \text{MSE} &= \frac{1}{N}\sum_{i=1}^{N}\left(y_i - \hat{y}_i\right)^2 \\ \text{RMSE} &= \sqrt{\frac{1}{N}\sum_{i=1}^{N}\left(y_i - \hat{y}_i\right)^2} \\ \text{MAE} &= \frac{1}{N}\left|y_i - \hat{y}_i\right| \\ \text{EC} &= 1 - \frac{\sqrt{\sum_{i=1}^{N}\left(y_i - \hat{y}_i\right)^2}}{\sqrt{\sum_{i=1}^{N} y_i^2} - \sqrt{\sum_{i=1}^{N} \hat{y}_i^2}} \end{aligned} \tag{18}$$

y and $\hat{y}$ still represent the actual value and predict the value of the simulation result, and N represents the number of samples in the training set. The smaller value of MSE, the higher the accuracy of the machine learning simulation results describing the experimental data. EC indicates the degree of fit between the output value and the true value. Generally, any value above 0.9 indicates a good fit.

3. Compensation Model and Simulation of High-Plateau Missing Data

The pseudo-compensation of missing data by other QAR data is essentially based on the existence of a certain functional relationship between QAR parameters. The value of the parameter can be derived from other parameter values. Therefore, the purpose of simulation is to determine this functional relationship. To be more specific, high-plateau flight data padding is essentially a function approximation problem.

This paper takes some flight parameters of QAR flight data as the assumed missing data in order to show the feasibility of this method. According to the actual meaning of the QAR data, the loss parameter $N(\tau) = N_{real}$ is set as the missing QAR parameter, where τ is the current moment of the missing data and other intact QAR parameters are used as the known vector set ${\omega_\tau}^T$ according to the previous setting. Finding a functional relationship between the two or its first approximation such that $N(\tau) = N_{real}$, the relationship model can be written as $N(\tau) = {\omega_\tau}^T\varphi(x,t) + b$, where the parameter requirements are (8) the same, so LS-SVM can be used to complement the QAR loss parameters.

3.1. Data Selection

In order to verify the feasibility of the high-plateau QAR data patching, this paper collects ten flight data of a certain airline's civil transport aircraft in the same time period and the same origin and destination for simulation analysis. In order to reduce irrelevant external factors, interference data selection controls possible related variables, such as changing in crew members, and determines whether it is pre-flight or post-flight to ensure that the accuracy of the simulation is improved. After selection, nine groups were randomly selected as the model training group and the last group was used as the comparison group to test the accuracy of the experimental results.

3.2. Algorithm Improvement

Based on the support vector machine algorithm, an improved method is proposed for the shortcomings of difficulty in training and analyzing large-scale samples. The eigenvalue range definition plays a very important role in training. The input and output are put into a small range and then predicted by the support vector machine model. On the one hand, it can avoid overfitting caused by large-value data dominating small-value data. On the other hand, scaling the data to a small range can avoid the "dimension disaster" and reduce the computational load. The principal component analysis method, as a commonly used dimensionality reduction algorithm, can easily simplify and refine complex data, process the data through the entropy method, and complete the algorithm optimization to achieve concise and accurate data under the premise of ensuring the robustness of the data.

3.3. Algorithm Flow

Before the simulation starts, it is necessary to determine the key parameters γ and the core width σ^2 in advance and then use the above algorithm to perform simulation training to fill in the missing data; the specific details and steps are shown in Figure 2.

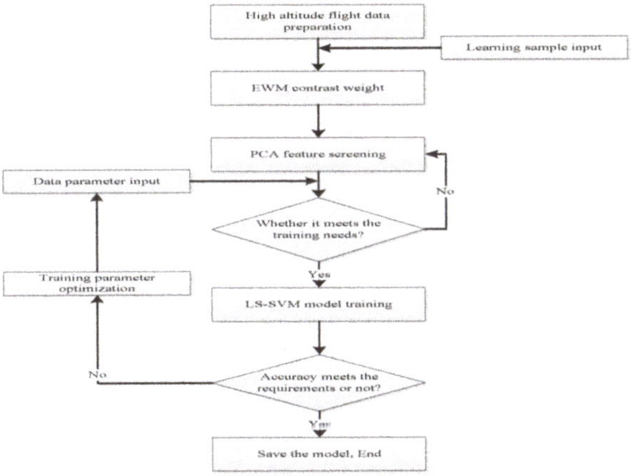

Figure 2. Flow chart of flight data fitting based on improved LS-SVM.

3.4. Simulation Application

QAR's overall data cannot be analyzed due to the existence of text items and 78 data items remain after all text items are excluded. Python is taken as the expected environment, which measures the weight of each item through the EWM method and divides the interval to select the data items for simulation training. After multiple rounds of testing, the coefficient of determination is compared. It is found that when the number of feature quantities is smaller, the coefficient of fit is larger and the change tends to stably increase; thus, few features are prone to overfitting. After weighing and selecting the 17 feature items with the largest weight, they have good accuracy and credibility. The relationship between the number of specific features and the accuracy rate, as well as the weight ratio of the feature quantity, are shown in Figure 3.

Compared to the algorithm without the improved method, the improved algorithm not only improves the fitting effect but also greatly reduces the amount of data in the simulation. The fitting coefficient is increased by 0.64% but the amount of data calculation is reduced by 78.21%. The details are shown in Table 1.

Table 1. Performance table of improved method.

Characteristics of Several	No Improve	Improve	Promotion
R^2	0.991	0.9973	0.64%
The amount of data	778284	169626	78.21%

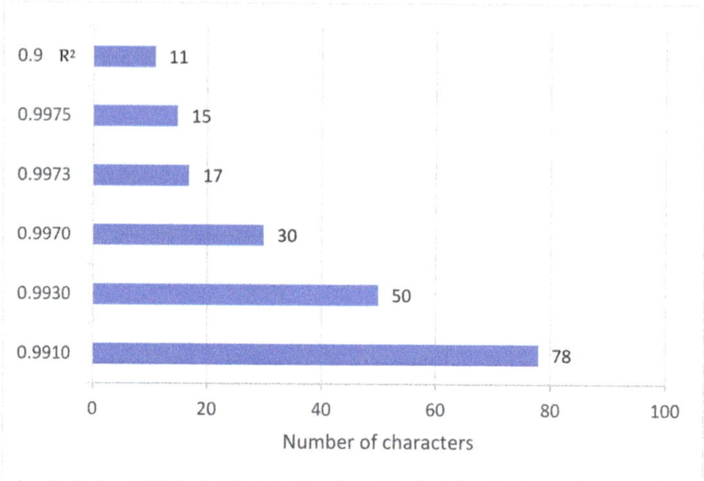

Figure 3. Relationship between the number of features and the correlation coefficient.

Among them, the selected feature quantities and the corresponding weights are shown in Table 2 and Figure 4.

Table 2. Weight-characteristic quantity correspondents.

Serial Number	Abbreviation	Name Connotation	Weight
1	N1_1	Left engine speed	0.041
2	N1_2	Right engine speed	0.042
3	N2_1	Left engine power	0.008
4	N2_2	Right engine power	0.011
5	FLIGHT_PHASE	Flight phase	0.054
6	GS1	True ground speed	0.082
7	GS2	Captain's instrument displays ground speed	0.083
8	GS_FO	The co-pilot's gauge shows ground speed	0.082
9	CAS	Calibrated air speed	0.079
10	DRIFT	Drift angle	0.041
11	TAS	True airspeed	0.099
12	PITCH11	The captain's instrument displays the pitch angle on the left side	0.064
13	PITCH12	The captain's instrument displays the pitch angle on the inner left side	0.064
14	PITCH21	The captain's instrument displays the pitch angle to the outer right	0.064
15	PITCH22	The captain's instrument displays the pitch angle on the inner right side	0.064
16	PITCH_DISP_FO1	The assistant captain's gauge shows the outer left side of the pitch angle	0.061
17	PITCH_DISP_FO2	The assistant captain's instrument displays the pitch angle on the inner left side	0.061

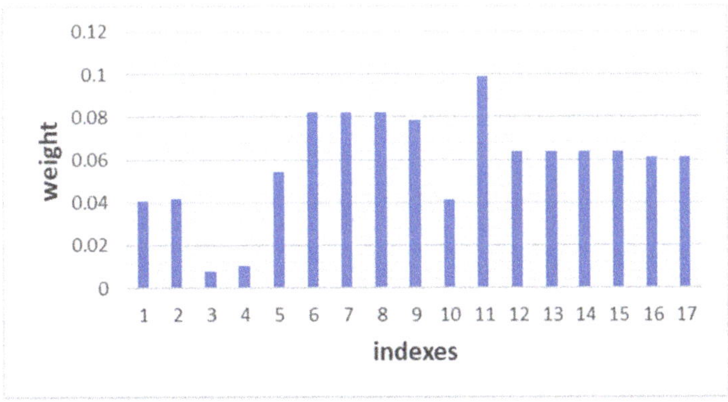

Figure 4. Weight-characteristic quantity correspondents.

Among them, the feature that has the greatest impact on the prediction is the true flight speed (TAS), and the feature that has the least impact is the right engine speed (N2_1). After determining the selection of the feature quantity, due to the large amplitude of the QAR data, in order to reduce the modeling error, the input data and the expected data were normalized on $[-1, 0]$ and $[0, 1]$, respectively. The original interval should be returned to after analysis. In this paper, the kernel function selects the most commonly used radial basis function for data repair:

$$K(x, x_i) = \exp(-\frac{(x - x_i)^2}{2\sigma^2}) \tag{19}$$

The simulation found that the parameters γ and the kernel width σ^2 have a significant impact on the complementation effect, which needs to be determined according to the specific characteristics of the training data. Generally speaking, a reduction in the kernel width σ^2 can improve the training accuracy but can reduce the generalization ability, and an increase in the parameter γ can also improve the training accuracy. The training shows that when the parameter $\gamma = 3$ and the training model is filled with missing data, the data with core width $\sigma^2 = 0.6$ have the best complementation effect. With the left engine speed (N1, unit: RPM), the aircraft pitch angle (pitch, unit: °) and the flap angle (flap angle, unit: °), as examples, intercept the data simulation results of the climb, approach, and landing stages to show the degree of flight data padding. In order to facilitate the analysis and observation, the predicted and actual values of the aircraft inclination angle are placed in (−1,1) interval, the predicted value and actual value of the left engine speed are put in the (1,3) interval, and the predicted value and actual value of the flap angle are put in the (3,5) interval, as shown in Figures 5–7.

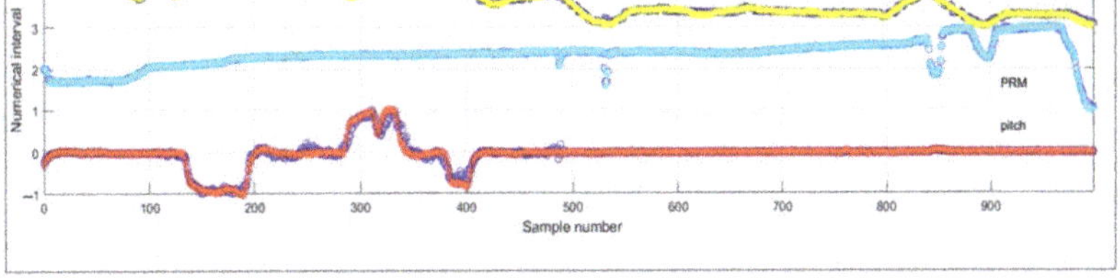

Figure 5. Climbing phase simulation diagram.

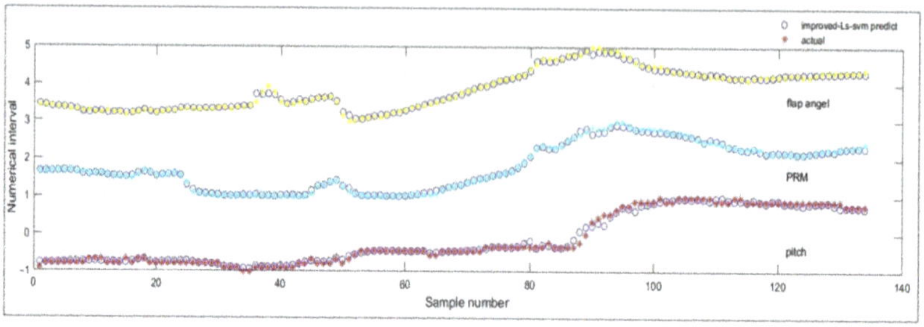

Figure 6. Approach phase simulation diagram.

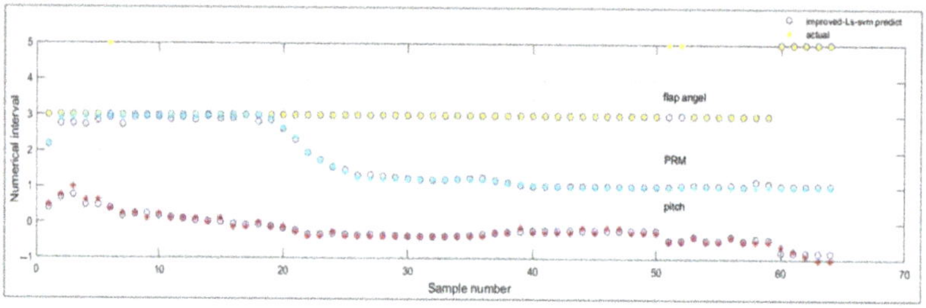

Figure 7. Landing phase simulation diagram.

By observing the image, it is found that the data fitting degree of each factor and each stage is relatively good, so further simulation result analysis can be carried out.

4. Simulation and Discussion

The experimental results are analyzed through simulation methods, and the error indicators of the complement results are shown in Table 3.

Table 3. Error index of missing data completion.

Pitch	**MSE**	**MAE(%)**	**RMSE**	**EC**
climb	-4.81×10^{-17}	3.59%	5.56×10^{-16}	0.99
approach	-5.15×10^{-16}	7.70%	5.95×10^{-15}	0.99
landing	-1.78×10^{-17}	2.64%	2.06×10^{-16}	0.99
N1	MSE	MAE(%)	RMSE	EC
climb	-5.20×10^{-17}	2.93%	6.02×10^{-16}	0.99
approach	3.89×10^{-17}	7.43%	4.51×10^{-16}	0.99
landing	2.40×10^{-17}	2.61%	2.78×10^{-16}	0.99
Flap angle	MSE	MAE(%)	RMSE	EC
climb	-9.53×10^{-17}	4.07%	1.10×10^{-15}	0.99
approach	-5.15×10^{-16}	9.00%	0.99	0.99
landing	7.66×10^{-17}	2.41%	8.87×10^{-16}	0.99

The error measurement index MAE in the table shows that the lower average error accuracy is more than 90% and the error index value EC in the table has reached a high degree of fit of 0.99. It can be seen that the QAR data item is used as the feature value to assign weights through EWM, and the PCA dimensionality reduction method finally uses the LS-SVM algorithm to fill in the missing data of the QAR to great effect. However, since most of the routes sailed by the aircraft are repeated flights of the same route, when faced with multiple losses or overall losses, the same method can be used to simulate the historical data to restore the lost flight data.

5. Conclusions

The previous data processing experience is based on the QAR itself to detect changes in the body or environment and other actual conditions. Few studies have been conducted on the preservation and restoration of the QAR data itself. This work provides some ideas in this regard. In this paper, the improved LS-SVM method based on the entropy weight method (EWM) and principal component analysis (PCA) is shown to effectively fit the missing QAR data. The parameters are gradually stable during the training process, which ensures that the model can be directly applied for data fitting without retraining, achieving the purpose of fast and simple applicability. This article only considers the case of single item loss, since most of the aircraft sailing on the same route repeats the flight; when faced with multiple losses or overall loss, the same method can be used to simulate historical data to restore this loss of flight data.

Due to the uniqueness of flying at high plateaus, there may be differences when flying on normal routes and the same conclusion may not be applicable for the normal flight. Its practical applicability remains to be further studied.

Author Contributions: Conceptualization, N.C. and Y.S.; data curation, N.C. and Z.W.; methodology, N.C. and C.P.; formal analysis, N.C. and Z.W.; writing—original draft preparation, N.C. All authors have read and agreed to the published version of the manuscript.

Funding: This research was funded by the National Natural Science Foundation of China (grant number: U2033202); the Key R&D Program of the Sichuan Provincial Department of Science and Technology (2022YFG0213); and the Safety Capability Fund Project of the Civil Aviation Administration of China (2022J026).

Data Availability Statement: The data used to support the findings of this study are included within the article.

Acknowledgments: The authors would like to thank the National Natural Science Foundation of China (U2033202), the Key R&D Program of the Sichuan Science and Technology Department (2022YFG0213), the Safety Capability Fund Project of the Civil Aviation Administration of China (2022J026), and the Flight Technology and Flight Safety Research Base Open Fund Project (F2019KF08).

Conflicts of Interest: The authors declare no conflict of interest.

References

1. Xu, J.C.; Sun, Y.C. Airworthiness requirement of transportation category aircraft operation on high plateau airports. *Aeronaut. Comput. Tech.* **2018**, *48*, 133–138.
2. Feng, Y.W.; Pan, W.H.; Lu, C. Research on Operation Reliability of Aircraft Power Plant Based on Ma-chine Learning. *Acta Aeronaut. Astronaut. Sin.* **2021**, *42*, 524732. [CrossRef]
3. Ye, B.J.; Bao, X.; Liu, B. Machine learning for aircraft approach time prediction. *Acta Aeronaut. Astronaut. Sin.* **2020**, *41*, 359–370.
4. Fang, G.C.; Jia, D.P.; Liu, Y.F. Military airplane health assessment technique based on data mining of flight parameters. *Acta Aeronaut. Astronaut. Sin.* **2020**, *41*, 296–306.
5. Liu, J.Y.; Wang, D.Q.; Cui, J.W. Research on classification of screw locking results based on improved kernel LS-SVM algorithm. *Ind. Instrum. Autom.* **2020**, *4*, 12–15.
6. Li, S.; Wang, Y.; Xue, Z.L. Grounding resistance monitoring data regression prediction method based on LS-SVM. *Foreign Electron. Meas. Technol.* **2019**, *8*, 19–22.
7. Wu, H.; Li, B.W.; Zhao, S.F.; Yang, X.; Song, H. Research on initial installed power loss of a certain type of turbo-shaft engine using data mining and statistical approach. *Math. Probl. Eng.* **2018**, *2018*, 9412350. [CrossRef]
8. Puranik, T.G.; Mavris, D.N. Anomaly detection in general-aviation operations using energy metrics and flight-data records. *J. Aeros. Comp. Inf. Com.* **2018**, *15*, 22–253. [CrossRef]
9. Puranik, T.G.; Rodriguez, N.; Mavris, D.N. Towards online prediction of safety-critical landing metrics in aviation using supervised machine learning. *Transp. Res. Part C Emerg. Technol.* **2020**, *120*, 102819. [CrossRef]
10. Yildirim, M.T.; Kurt, B. Aircraft gas turbine engine health monitoring system by real flight data. *Int. J. Aerospace Eng.* **2018**, *2018*, 9570873. [CrossRef]
11. Yildirim, M.T.; Kurt, B. Confidence interval prediction of ANN estimated LPT parameters. *Aircr. Eng. Aerosp. Technol.* **2019**, *9*, 101–106. [CrossRef]
12. Martín, F.J.V.; Sequera, J.L.C.; Huerga, M.A.N. Using data mining techniques to discover patterns in an airline's flight hours assignments. *Int. J. Data. Warehous.* **2017**, *13*, 45–62. [CrossRef]

13. Davison Reynolds, H.J.; Lokhande, K.; Kuffner, M.; Yenson, S. Human–Systems integration design process of the air traffic control tower flight data manager. *J. Cogn. Eng. Decis. Mak.* **2013**, *7*, 273–292. [CrossRef]
14. Kumar, A.; Ghosh, K. GPR-based novel approach for non-linear aerodynamic modeling from flight data. *Aeronaut. J.* **2019**, *123*, 79–92. [CrossRef]
15. Lan, C.E.; Wu, K.Y.; Yu, J. Flight characteristics analysis based on QAR data of a jet transport during landing at a high-altitude airport. *Chin. J. Aeronaut.* **2012**, *25*, 13–24. [CrossRef]
16. Oehling, J.; Barry, D.J. Using machine learning methods in airline flight data monitoring to generate new operational safety knowledge from existing data. *Saf. Sci.* **2019**, *114*, 89–104. [CrossRef]
17. Walker, G. Redefining the incidents to learn from: Safety science insights acquired on the journey from black boxes to flight data monitoring. *Saf. Sci.* **2017**, *99*, 14–22. [CrossRef]
18. Deng, W.; Li, Z.; Li, X.; Chen, H.; Zhao, H. Compound fault diagnosis using optimized MCKD and sparse representation for rolling bearings. *IEEE Trans. Instrum. Meas.* **2022**, *71*, 1–9. [CrossRef]
19. Cui, H.; Guan, Y.; Chen, H. Rolling element fault diagnosis based on VMD and sensitivity MCKD. *IEEE Access* **2021**, *9*, 120297–120308. [CrossRef]
20. Wang, L.; Ren, Y.; Wu, C.X. Effects of flare operation on landing safety: A study based on ANOVA of real flight data. *Saf. Sci.* **2018**, *102*, 14–25. [CrossRef]
21. Zhou, D.; Zhuang, X.; Zuo, H.; Wang, H.; Yan, H. Deep learning-based approach for civil aircraft for civil aircraft hazard identification and prediction. *IEEE Access* **2020**, *8*, 103665–103683. [CrossRef]
22. Cheng, S.L.; Gao, Z.H.; Zhu, X.Q. Unsteady aerodynamic modelling of unstable dynamic process. *Acta Aeronaut. Astronaut. Sin.* **2020**, *41*, 238–249.
23. Li, M.; Wu, C. A distance model of intuitionistic fuzzy cross entropy to solve preference problem on alternatives. *Math. Probl. Eng.* **2016**, *2016*, 8324124. [CrossRef]
24. Zhang, X.; Wang, H.; Du, C.; Fan, X.; Cui, L.; Chen, H.; Deng, F.; Tong, Q.; He, M.; Yang, M.; et al. Custom-molded offloading footwear effectively prevents recurrence and amputation, and lowers mortality rates in high-risk diabetic foot patients: A multicenter, prospective observational study. *Diabetes Metab. Syndr. Obes. Targets Ther.* **2022**, *15*, 103–109. [CrossRef] [PubMed]
25. Zhu, Y.; Deng, B.; Huo, Z. Key deviation source diagnosis for aircraft structural component assembly driven by small sample inspection data. *China Mech. Eng.* **2019**, *30*, 2725–2733.
26. Gao, X.; Hou, J. An improved SVM integrated GS-PCA fault diagnosis approach of Tennessee Eastman process. *Neurocomputing* **2016**, *174*, 906–911. [CrossRef]
27. Safaldin, M.; Otair, M.; Abualigah, L. Improved binary gray wolf optimizer and SVM for intrusion detection system in wireless sensor networks. *J. Amb. Intel. Hum. Comp.* **2021**, *12*, 1559–1576. [CrossRef]
28. Abualigah, L.; Diabat, A.; Mirjalili, S.; Elaziz, M.A.; Gandomi, A.H. The arithmetic optimization algorithm. *Comput. Methods Appl. Mech. Eng.* **2021**, *376*, 113609. [CrossRef]
29. Cai, J.; Bao, H.; Huang, Y.; Zhou, D. Risk identification of civil aviation engine control system based on particle swarm optimization-mean impact value-support vector machine. *Proc. Inst. Mech. Eng. Part G J. Aerosp. Eng.* 2022, *in press*. [CrossRef]
30. Smart, E.; Brown, D.; Denman, J. Combining multiple classifiers to quantitatively rank the impact of abnormalities in flight data. *Appl. Soft Comput.* **2012**, *12*, 2583–2592. [CrossRef]
31. Li, G.; Li, Y.; Chen, H.; Deng, W. Fractional-Order Controller for Course-Keeping of Underactuated Surface Vessels Based on Frequency Domain Specification and Improved Particle Swarm Optimization Algorithm. *Appl. Sci.* **2022**, *12*, 3139. [CrossRef]
32. Deng, W.; Zhang, X.X.; Zhou, Y.Q.; Liu, Y.; Zhou, X.B.; Chen, H.L.; Zhao, H.M. An enhanced fast non-dominated solution sorting genetic algorithm for multi-objective problems. *Inf. Sci.* **2022**, *585*, 441–453. [CrossRef]
33. Elisa, Q.M.; Lu, S.; Blazquez, C. Use of data imputation tools to reconstruct incomplete air quality datasets: A case-study in Temuco, Chile. *Atmos. Environ.* **2019**, *200*, 40–49.
34. Hadeed, S.J.; O'Rourke, M.K.; Burgess, J.L.; Harris, R.B.; Canales, R.A. Imputation methods for addressing missing data in short-term monitoring of air pollutants. *Sci. Total Environ.* **2020**, *730*, 139140. [CrossRef]
35. Liu, Z.J.; Wan, J.Q.; Ma, Y.W. Online prediction of effluent COD in the anaerobic wastewater treatment system based on PCA-LS-SVM algorithm. *Environ. Sci. Pollut. Res.* **2019**, *26*, 12828–12841. [CrossRef]
36. Cheolmin, K.; Klabjan, D. A simple and fast algorithm for L1-norm Kernel PCA. *IEEE Trans. Pattern Anal. Mach. Intell.* **2019**, *42*, 1842–1855.

electronics

MDPI

Article

A Hierarchical Heterogeneous Graph Attention Network for Emotion-Cause Pair Extraction

Jiaxin Yu [1], Wenyuan Liu [1,2,*], Yongjun He [3,*] and Bineng Zhong [4]

1 School of Information Science and Engineering, Yanshan University, Qinhuangdao 066004, China
2 The Engineering Research Center for Network Perception & Big Data of Hebei Province, Qinhuangdao 066004, China
3 School of Computer Science and Technology, Harbin Institute of Technology, Harbin 150001, China
4 The Guangxi Key Lab of Multi-Source Information Mining & Security, Guangxi Normal University, Guilin 541004, China
* Correspondence: wyl@ysu.edu.cn (W.L.); heyongjun@hit.edu.cn (Y.H.)

Abstract: Recently, graph neural networks (GNN), due to their compelling representation learning ability, have been exploited to deal with emotion-cause pair extraction (ECPE). However, current GNN-based ECPE methods mostly concentrate on modeling the local dependency relation between homogeneous nodes at the semantic granularity of clauses or clause pairs, while they fail to take full advantage of the rich semantic information in the document. To solve this problem, we propose a novel hierarchical heterogeneous graph attention network to model global semantic relations among nodes. Especially, our method introduces all types of semantic elements involved in the ECPE, not just clauses or clause pairs. Specifically, we first model the dependency between clauses and words, in which word nodes are also exploited as an intermediary for the association between clause nodes. Secondly, a pair-level subgraph is constructed to explore the correlation between the pair nodes and their different neighboring nodes. Representation learning of clauses and clause pairs is achieved by two-level heterogeneous graph attention networks. Experiments on the benchmark datasets show that our proposed model achieves a significant improvement over 13 compared methods.

Keywords: emotion-cause pair extraction; heterogeneous graph; graph attention network; hierarchical model

Citation: Yu, J.; Liu, W.; He, Y.; Zhong, B. A Hierarchical Heterogeneous Graph Attention Network for Emotion-Cause Pair Extraction. *Electronics* **2022**, *11*, 2884. https://doi.org/10.3390/electronics11182884

Academic Editor: George Angelos Papadopoulos

Received: 15 August 2022
Accepted: 7 September 2022
Published: 12 September 2022

1. Introduction

As a research hotspot in natural language processing (NLP), emotion-cause extraction (ECE), aimed at extracting the causes corresponding to the emotions specified in a given document, has been widely utilized in public opinion analysis, human–machine dialogue systems, and so on. Originally, taking events as the causes, Lee et al. [1] regarded ECE as a word-level sequence annotating task. Afterwards, some studies redefined the granularity of annotation in ECE to the clause level to make full use of context information [2,3]. Although annotating emotions in advance contributes to cause extraction, it is very labor-consuming, which limits the real application of the ECE approach. To solve this problem, Xia and Ding [4] put forward a new emotion analysis task called emotion-cause pair extraction (ECPE), which extracts emotion clauses and their corresponding cause clauses in pairs. ECPE does not rely on labeling emotions, so it is preferable, but more challenging, than ECE. Furthermore, they also proposed a two-stage pipelined framework to handle this new task, in which the emotions and causes are first extracted and then paired. Since this two-stage approach may result in cross-stage propagation of errors, a lot of end-to-end approaches are presented and achieve improvements over two-stage approaches. In the end-to-end ECPE approaches, the crucial issue is to learn good representations of semantic elements. GNN [5,6] can learn node representations based on node features and the graph structure; therefore, it is a powerful deep representation learning method and has been widely utilized in many application fields. Inspired by this, a few researchers attempted to apply GNN to the ECPE task. They mostly construct a homogeneous graph with the semantic information of a document and employed GNNs to learn these semantic representations. For example, Wei et al. [7] and Chen et al. [8] model the inter-clause and inter-pair relations, respectively.

Nevertheless, existing GNN-based ECPE approaches only concentrate on one semantic level, ignoring the rich semantic relations between different kinds of semantic elements. Hence, the captured semantic information is local, rather than global. In fact, in the ECPE task, a document involves different semantic granularity of words, clauses, clause pairs, and so on; hence, the constructed text graph should come with multiple types of nodes, also well-known as a heterogeneous graph. Furthermore, all the associations between these nodes can provide clues for extracting causality. Therefore, it is conductive for the joint extraction of emotion clauses, cause clauses, and emotion-cause pairs to take all semantic elements into account and model the global semantic relations between them.

In this study, we propose an end-to-end hierarchical heterogeneous graph attention model (HHGAT). Different from the existing methods that only consider clause or pair nodes, we introduce word nodes into our heterogeneous graph, together with clause and pair nodes, to cover all semantic elements. In particular, the introduced word nodes can not only extract fine-grained clause features by modeling the dependency between clauses and words, but also act as an intermediate node connecting clause nodes to enrich the correlation between clause nodes. Moreover, a fully connected pair-level subgraph is established to capture the relations between a pair node and its neighboring nodes on different semantic paths. Depending on such a hierarchy of "word-clause-pair", we realize a model of the global semantics in a document.

2. Related Work

Emotion analysis is active in the field of NLP. In many application scenarios, it is more important to understand the emotional cause than the emotion itself. Here, we focus on two challenging tasks, namely ECE and ECPE.

2.1. ECE

Different from traditional emotion classification, the purpose of ECE is to extract the causes of specific emotions. Lee et al. [1] first defined the ECE task and introduced a method based on linguistic rules (RB). Subsequently, for different linguistic patterns, a variety of RB methods are proposed [9–11]. In addition, Russo et al. [12] designed a novel method combining RB and common-sense knowledge. However, the performance of these RB methods is usually unsatisfactory. Considering that it is impossible for rules to cover all language phenomena, some machine learning (ML)-based ECE methods are proposed. Gui et al. [13] designed two ML-based methods, combined with 25 rules. Ghazi et al. [14] employed conditional random field (CRF) to tag emotional causes. Moreover, Gui et al. [2] constructed a new clause-level corpus and utilized support vector machine (SVM) to deal with the ECE task. To benefit from the representation learning ability of deep learning (DL), some DL-based methods achieved excellent performance on ECE. Gui et al. [15] presented a new method based on convolutional neural network (CNN). Cheng et al. [16] used long short-term memory networks (LSTM) to model the clauses. To obtain better context representations, a series of hierarchical models [17–24] were explored. Inspired by multitask learning, Chen et al. [25] and Hu et al. [26] focused on the joint extraction of emotion and cause. In addition, Ding et al. [27] and Xu et al. [28] reformulated ECE into a ranking problem. Considering the importance of emotion-independent features, Xiao et al. [29] presented a multi-view attention network. Recently, Hu et al. [30] proposed a graph convolution network (GCN) integrating semantics and structure information, which is the state-of-the-art ECE method.

2.2. ECPE

2.2.1. Pipelined ECPE

ECE requires the manual annotation of emotion clauses before cause extraction, which is labor-consuming. To solve this problem, Xia and Ding [4] proposed a new task called ECPE, and they introduced three two-stage pipelined models, namely Indep, Inter-CE, and Inter-EC. For Inter-EC [4], Shan and Zhu [31] designed a new cause extraction component based on transformer [32] to improve this model. Yu et al. [33] applied the self-distillation method to train a mutually auxiliary multitask model. Jia et al. [34] realized mutual promotion of emotion extraction and cause extraction by recursively modeling clauses. To improve the pairing stage of two-stage pipelined methods, Sun et al. [35] presented a dual-questioning attention network. Moreover, Shi et al. [36] simultaneously enhanced both stages of the pipelined method.

2.2.2. End-to-End ECPE

Although the pipelined approach has been proved to be effective for ECPE, it leads to cross-stage error propagation. To solve this problem, a series of end-to-end ECPE approaches are proposed.

Wu et al. [37] jointly trained the three subtasks in ECPE via a unified framework and had clause features shared to exploit the interaction between subtasks. To make full use of the implicit connection between emotion detection and emotion-cause pair extraction, Tang et al. [38] tackled these two tasks in a joint framework. Concentrating on the interaction between emotion-cause pairs, Ding et al. [39] presented a 2D transformer and its two variants. Fan et al. [40] introduced a scope controller to concentrate the predicted distribution of emotion-cause pair. Ding et al. [41] restricted ECPE to the emotion-centered cause extraction in the sliding window and proposed a multi-label learning method. Cheng et al. [42] took advantage of two symmetrical subnetworks to conduct a local search [43,44] around emotion or cause, respectively. Singh et al. [45] adopted the prediction results of emotion extraction to promote the cause extraction. Considering the importance of order information, Fan et al. [46] captured the sequential features of clauses through three LSTMs: forward LSTM, backward LSTM, and BiLSTM. Yang et al. [47] utilized the consistency of emotion type between the emotion clause and clause pair. Chen et al. [48] achieved the mutual promotion of emotion extraction and cause extraction through iterative learning. Furthermore, some studies [49–52] coincidentally reformulated ECPE as a sequence labeling problem.

Recently, some graph structure-based approaches are proposed. Song et al. [53] treated ECPE as a link prediction task of directed graph; however, they did not adopt a GNN that is more suitable for graph structure modeling. Despite Fan et al. [54] introduced a novel approach that regards ECPE as an action prediction task in directed graph construction; their model is not based on GNN, either. In addition, Wei et al. [7] exploited a graph attention network (GAT) to enhance inter-clause relation modeling and deal with the ECPE task from a ranking perspective. Chen et al. [8] developed an approach based on a graph convolutional network to capture the relevance among local neighboring candidate pairs. However, the above graph-based approaches ignored the relationship between heterogeneous nodes, so they failed to model global semantics.

3. Methodology

3.1. Task Definition

In this section, the ECPE task is formalized as follows. Let $d = [c_1, \cdots c_i \cdots, c_m]$ be a document that contains m clauses, where $c_i = [w_{i,1}, \cdots w_{i,j} \cdots, w_{i,n}]$ is the i-th clause and further decomposed into a sequence of n words. The aim of ECPE is to extract the emotion-cause pairs from d:

$$P = \{p_k\}_{k=1}^{|P|} = \{(c_k^e, c_k^c)\}_{k=1}^{|P|}, \tag{1}$$

where c_i^e is the emotion clause in the k-th emotion-cause pair, c_j^c corresponds to the cause clause, and P represents the candidate pair set.

3.2. Overview

In this work, we first represent a document with a "word-clause-pair" heterogeneous graph, as illustrated in Figure 1. Then, we present a hierarchical heterogeneous graph attention network to model the "word-clause-pair" hierarchical structure and identify the emotion-cause pairs according to the learned node representation. As shown in Figure 2, our proposed model mainly includes three components: (1) the node initialization layer, which utilizes word-level BiLSTM, followed by a self-attention module or pre-trained BERT to obtain the initial semantic representations of word and clause nodes; (2) the clause node encoding layer employs a node-level heterogeneous graph attention network to integrate the inner-clause contextual features into the clause representations by capturing the dependencies between clause nodes and word nodes they contains; (3) the pair node encoding layer is a heterogeneous graph attention network based on meta-path, which first applies a node-level attention and then a meta-path level attention. Finally, three multilayer perceptrons (MLP) are adopted to predict the emotion clauses, cause clauses, and emotion-cause pairs, respectively.

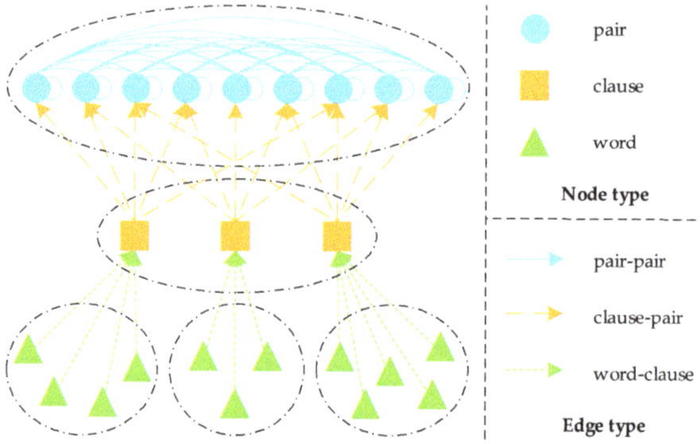

Figure 1. A toy example of heterogeneous graph composed of word, clause, and pair nodes.

3.3. Heterogeneous Graph Construction

We denote our hierarchical heterogeneous graph as $\mathcal{G} = (\mathcal{V}, \mathcal{E})$, where $\mathcal{V} = \mathcal{V}^w \cup \mathcal{V}^c \cup \mathcal{V}^p$ represents a node set that consists of three types of nodes, and $\mathcal{E}$ stands for the edges between all nodes. $\mathcal{V}^w = \cup_{i=1}^m \{w_{i,j}\}_{j=1}^n$, $\mathcal{V}^c = \{c_i\}_{i=1}^m$, and $\mathcal{V}^p = \cup_{i=1}^m \{p_{i,j}\}_{j=1}^m$ indicate the sets of words, clauses, and pair nodes, respectively. As shown in Figure 2, a word-to-clause edge distinctly indicates which clause a word is contained in. The two clause nodes connected with the same pair node together form a candidate emotion-cause pair. Moreover, the association between two pair nodes is represented by a pair-to-pair edge.

On the one hand, most current methods employ two clause-level subtasks (i.e., emotion extraction and cause extraction) in a unified framework to facilitate the detection of emotion-cause pairs. On the other hand, good clause representation is conducive to the feature construction of clause pairs. Hence, in order to learn the semantic representations of clause and pair nodes in detail, we divide our heterogeneous graph into two subgraphs, i.e., word-clause $\mathcal{G}^{wc} = (\mathcal{V}^w \cup \mathcal{V}^c, \mathcal{E}^{wc})$ and pair-level $\mathcal{G}^p = (\mathcal{V}^p, \mathcal{E}^p)$ subgraphs. Here, $\mathcal{E}^{wc}$ denotes the word-to-clause edge set, and $\mathcal{E}^p$ represents the pair-to-pair edge set. Furthermore, $\mathcal{G}^{wc}$ and $\mathcal{G}^p$ are further divided into a series of more fine-grained subgraphs, i.e., $\cup_{i=1}^m \mathcal{G}_i^{wc}$ and $\cup_{i=1}^m \mathcal{G}_i^p$, respectively, to facilitate the formalized description of our algorithm.

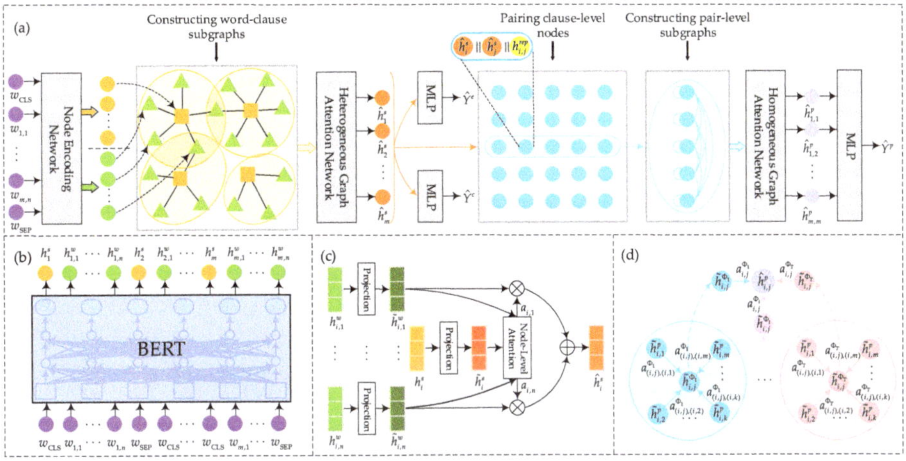

Figure 2. (**a**) An overview of HHGAT; (**b**) node initialization layer; (**c**) clause node encoding layer; (**d**) pair node encoding layer.

3.4. Hierarchical Heterogeneous Graph Attention Network

3.4.1. Node Initialization Layer

In this layer, a word embedding matrix $E_w \in \mathbb{R}^{d_w \times d_v}$ is first applied to transform each word $w_{i,j}$ into a vector $v_{i,j}$. Here, d_w and d_v are the vocabulary size and embedding dimension, respectively. Next, the contextual information for each word is captured through a BiLSTM module:

$$\left[h_{i,1}^w, \cdots h_{i,j}^w \cdots, h_{i,n}^w\right] = \text{BiLSTM}([v_{i,1}, \cdots v_{i,j} \cdots, v_{i,n}]), \tag{2}$$

where, $h_{i,j}^w$ represents the hidden state of the j-th word in the i-th clause. Then, an attention module is adopted to aggregate the word representations in the clause c_i:

$$h_i^s = \text{Attention}([h_{i,1}^w, \cdots h_{i,j}^w \cdots, h_{i,n}^w]), \tag{3}$$

where h_i^s is the vectorization representation of the i-th clause.

Furthermore, inspired by the BERT [55], we implement another version of node initialization layer, which utilizes the pre-trained BERT model to replace above BiLSTM and attention modules. The tokens [CLS] and [SEP] are inserted at the beginning and end of a given clause c_i, respectively, to obtain a sequence $c_i = [w_{\text{CLS}}, w_{i,1}, \cdots w_{i,j} \cdots, w_{i,n}, w_{\text{SEP}}]$. It is worth noting that $w_{i,j}$ represents the j-th token, rather than j-th word of the clause c_i, in the BERT version. Afterwards, the sequences corresponding to all clauses in the document are concatenated to form a whole sequence, and then input it to BERT. Through stacked transformer modules, we can obtain the output vectors $\cup_{i=1}^m \left\{h_{i,1}^w, \cdots h_{i,j}^w \cdots, h_{i,n}^w\right\}$ and $\{h_i^s\}_{i=1}^m$, which are the initialization representations of word and clause nodes, respectively. Here, h_i^s is the output of w_{CLS} corresponding to the clause c_i.

3.4.2. Clause Node Encoding Layer

Inner-clause relationships plays an important role in semantic understanding. In addition, a word can be also treated as a specific relation between the clauses containing it. Therefore, to further learn the semantic representation of a clause node, we extract each clause node and its connected word nodes from the hierarchical graph to build a fine-grained word-clause subgraph. Given a constructed subgraph $\mathcal{G}_i^{wc}$, with the clause node c_i and word nodes $\{w_{i,j}\}_{j=1}^n$, we apply a heterogeneous graph attention network to update the representation of the clause node.

Since two types of nodes exist in the heterogeneous subgraph, different types of nodes may belong to different feature spaces. Consequently, type-specific transformation matrices W_s and W_w are adopted to respectively project the features of clause and word nodes, with possibly different dimensions into the same feature space. The projection process can be shown in the following:

$$\widetilde{h}_i^s = W_s \cdot h_i^s, \widetilde{h}_{i,j}^w = W_w \cdot h_{i,j}^w, \tag{4}$$

where h_i^s is the initialization representation of clause node c_i, and $h_{i,j}^w$ denotes the initialization representation of word node $w_{i,j}$.

The node-level attention mechanism is then applied to learn the importance of different neighboring nodes to each target node. For a word-clause subgraph $\mathcal{G}_i^{wc}$, the clause node $c_i \in \mathcal{V}^c$ is the target node, while the corresponding neighboring nodes come from the word node set $\{w_{i,j}\}_{j=1}^n$. Specifically, importance scores are computed through a linear layer parameterized by $w_1{}^\top$, and then they are normalized to obtain weight coefficients via the softmax function. Next, according to these weight coefficients, the node aggregation over the subgraph is conducted by a weighted summation. In addition, we also apply a residual connection when updating the semantic representation of the clause node c_i. The specific process is as follows:

$$e_{i,j} = \text{LeakyReLU}(w_1{}^\top \cdot \tanh(\widetilde{h}_i^s \parallel \widetilde{h}_{i,j}^w)), \tag{5}$$

$$a_{i,j} = \frac{\exp(e_{i,j})}{\sum_{k=1}^n \exp(e_{i,k})}, \tag{6}$$

$$\hat{h}_i^s = \text{ReLU}(\textstyle\sum_{j=1}^n a_{i,j} \cdot \widetilde{h}_{i,j}^w + b_w) + h_i^s, \tag{7}$$

where w_1 is trainable weight matrix, b_w is the bias parameter, $\parallel$ denotes the concatenation operation, and $\top$ represents the transpose of matrix. As a result, the clause representation $\hat{h}_i^s$ integrating word semantics is generated.

Once obtaining updated node representation $\hat{h}_i^s$, it is fed into the emotion clause classifier to determine whether the clause corresponding to c_i is an emotion clause or not, and the classifier is implemented by a linear layer (parameterized by w_e and b_e) with the sigmoid function:

$$\hat{y}_i^e = \text{sigmoid}(w_e^\top \cdot \hat{h}_i^s + b_e), \tag{8}$$

where $\hat{y}_i^e$ is the predicted probability that the clause node c_i is an emotion clause. The calculation process of obtaining the cause probability $\hat{y}_i^c$ is similar to that of $\hat{y}_i^e$, except that the parameters are replaced by w_c and b_c.

3.4.3. Pair Node Encoding Layer

It can be observed that there are only simple subordinate relationships between the clause and pair nodes, rather than complex semantic relationships. Hence, we just need to consider pair nodes and the correlation between them when performing subgraph segmentation in this section. Furthermore, in a fine-grained pair-level subgraph $\mathcal{G}_i^p$, the neighboring nodes of a node $p_{i,j}$ are restricted to those nodes with the same emotion candidate as this one. Therefore, a pair-level, fully connected subgraph is formalized as $\mathcal{G}_i^p = (\{p_{i,j}\}_{j=1}^m, \mathcal{E}_i^p)$. Moreover, a meta-path Φ_t is described as a kind of path in the forms of $p_{i,k} \rightarrow \cdots p_{i,j-1} \rightarrow p_{i,j}$ and $p_{i,j} \leftarrow p_{i,j+1} \cdots \leftarrow p_{i,k}$, where $t = |k-j|$ represents the number of hops from a source node $p_{i,k}$ to the target node $p_{i,j}$. According to the statistical results of [8], the proportion that the distance between an emotion clause and the corresponding cause clause less than or equal to 2 is 95.8%. Taking into account this, we introduce four kinds of meta-paths: Φ_0, Φ_1, Φ_2, and Φ_3. Different from the other three types of paths, Φ_3 indicates the length of the path from the source node to the target node is ≥ 3.

Given a pair-level subgraph $\mathcal{G}_i^p$, the initial representation $h_{i,j}^p$ of a node $p_{i,j} = (c_i^e, c_j^c)$ in $\mathcal{G}_i^p$ is obtained by concatenating three vectors:

$$h^p_{i,j} = \hat{h}^s_i \parallel \hat{h}^s_j \parallel h^{rep}_{i,j}, \tag{9}$$

where $\hat{h}^s_i$ and $\hat{h}^s_j$ represent the semantic representations of candidate emotion clause c^e_i and candidate cause clause c^c_j, respectively. $h^{rep}_{i,j}$ indicates the relative position embedding, which is randomly initialized by the sampling of a uniform distribution. Considering that the meta-path-based neighbors play different roles in the representation of each node, we apply a meta-path-based graph attention network, which aggregates the features of neighboring nodes from different-typed paths to update the representation of this node. Specifically, two aggregation operations need to be performed.

Firstly, node-level attention is leveraged to aggregate the path-specific node representations. Specifically, for all pair nodes in the subgraph $\mathcal{G}^p_i$, a shared linear transformation, followed by the tanh function, is employed. Given a target node $p_{i,j}$ and meta-path Φ_t, the weight coefficient $e^{\Phi_t}_{(i,j),(i,k)}$ of a neighboring node $p_{i,k}$ that is connected to node $p_{i,j}$ through meta-path Φ_t is calculated. $e^{\Phi_t}_{(i,j),(i,k)}$ reflects the importance of node $p_{i,k}$ to node $p_{i,j}$. The weight coefficients of all Φ_t-based neighboring nodes are then normalized via the softmax function. By weighted summation, Φ_t-specific aggregate representation $\widetilde{h}^{\Phi_t}_{i,j}$ of the node $p_{i,j}$ is generated:

$$\widetilde{h}^p_{i,j} = W_p \cdot h^p_{i,j}, \widetilde{h}^p_{i,k} = W_p \cdot h^p_{i,k}, \tag{10}$$

$$\widetilde{e}^{\Phi_t}_{(i,j),(i,k)} = \text{LeakyReLU}({w_{\Phi_t}}^\top \cdot \tanh(\widetilde{h}^p_{i,j} \parallel \widetilde{h}^p_{i,k})), \tag{11}$$

$$e^{\Phi_t}_{(i,j),(i,k)} = I^{\Phi_t}_{(i,j),(i,k)} \cdot \widetilde{e}^{\Phi_t}_{(i,j),(i,k)}, \quad I^{\Phi_t}_{(i,j),(i,k)} = \begin{cases} 1, & p_{i,k} \in P^{\Phi_t}_{i,j} \\ 0, & p_{i,k} \notin P^{\Phi_t}_{i,j} \end{cases}, \tag{12}$$

$$a^{\Phi_t}_{(i,j),(i,k)} = \frac{\exp(e^{\Phi_t}_{(i,j),(i,k)})}{\sum_{k'=1}^{m} \exp(e^{\Phi_t}_{(i,j),(i,k')})}, \tag{13}$$

$$\widetilde{h}^{\Phi_t}_{i,j} = \text{ReLU}(\sum_{k=1}^{m} a^{\Phi_t}_{(i,j),(i,k)} \cdot \widetilde{h}^p_{i,k} + b_{\Phi_t}), \tag{14}$$

where W_p and w_{Φ_t} are trainable weight matrices, b_{Φ_t} denotes the bias, and $h^p_{i,j}$ represents the initial feature of node $p_{i,j}$. In addition, $I^{\Phi_t}_{(i,j),(i,k)}$ is the node mask, which injects structural information into the model. Additionally, $I^{\Phi_t}_{(i,j),(i,k)} = 1$ means that $p_{i,k}$ belongs to the Φ_t-based neighboring node set $P^{\Phi_t}_{i,j}$ of $p_{i,j}$.

Secondly, path-level attention is applied to measure the importance of different meta-paths to the target node. For this purpose, the path-specific aggregate representations obtained by previous node-level attention are transformed into the weight values through a linear transformation matrix. After that, the softmax function is employed to normalize these weight values, so as to obtain the weight coefficients of different paths. Using the learned weight coefficients, the aggregate representations from different meta-paths are fused with the initial node representation $h^p_{i,j}$. The final semantic representation $\hat{h}^p_{i,j}$ of node $p_{i,j}$ is obtained by:

$$a^{\Phi_t}_{i,j} = \frac{\exp({w_2}^\top \cdot \widetilde{h}^{\Phi_t}_{i,j})}{\sum_{t'=0}^{T} \exp({w_2}^\top \cdot \widetilde{h}^{\Phi_{t'}}_{i,j})}, \tag{15}$$

$$\hat{h}^p_{i,j} = \sum_{t=0}^{T} a^{\Phi_t}_{i,j} \cdot \widetilde{h}^{\Phi_t}_{i,j} + h^p_{i,j}, \tag{16}$$

where ${w_2}^\top$ is a trainable transformation matrix, the meta-path Φ_t belongs to the path set $\Phi = \{\Phi_t\}_{t=0}^{T}$, and $T = |\Phi| - 1$. $a^{\Phi_t}_{i,j}$ represents the weight coefficient of meta-path Φ_t to node $p_{i,j}$. Here, it is worth noting that, if the target nodes are different, the weight distribution of the meta-paths is also different.

Then, a logistic regression layer (parameterized by $w_p^\top$ and b_p) is utilized to identify whether each pair node is a true emotion-cause pair node:

$$\hat{y}_{i,j}^p = \text{sigmoid}(w_p^\top \cdot \hat{h}_{i,j}^p + b_p) . \tag{17}$$

3.5. Model Training and Optimization

The loss function of extracting emotion-cause pairs from a given document d is formulated as follows:

$$L_p = -\frac{1}{m^2} \cdot \sum_{i=1}^{m} \sum_{j=1}^{m} (y_{i,j}^p \cdot \log(\hat{y}_{i,j}^p) + (1 - y_{i,j}^p) \cdot \log(1 - \hat{y}_{i,j}^p)), \tag{18}$$

where $y_{i,j}^p$ is the ground-truth of node $p_{i,j}$. To benefit from the other two subtasks, the loss terms of the emotion extraction and cause extraction are introduced. For simplicity, only the calculation process of loss term for the emotion extraction is provided in the following:

$$L_e = -\frac{1}{m} \cdot \sum_{i=1}^{m} (y_i^e \cdot \log(\hat{y}_i^e) + (1 - y_i^e) \cdot \log(1 - \hat{y}_i^e)), \tag{19}$$

where y_i^e is the emotion annotation of clause c_i. Therefore, the total loss of our model is

$$L_{total} = L_p + L_e + L_c. \tag{20}$$

Finally, the purpose of the model training is to minimize the total loss. The overall process is shown in Algorithm 1.

Algorithm 1: The overall process of HHGAT.

Input : The heterogeneous graph $\mathcal{G} = (\mathcal{V}, \mathcal{E})$, $\mathcal{V} = \mathcal{V}^w \cup \mathcal{V}^c \cup \mathcal{V}^p$,
The initial feature h_i^s of clause node $\forall c_i \in \mathcal{V}^c = \{c_i\}_{i=1}^m$,
The initial feature $h_{i,j}^w$ of word node $\forall w_{i,j} \in \mathcal{V}^w = \cup_{i=1}^m \left\{w_{i,j}\right\}_{j=1}^n$.
Output: The clause node representations $\left\{\hat{h}_i^s\right\}_{i=1}^m$,
The pair node representations $\cup_{i=1}^m \left\{\hat{h}_{i,j}^p\right\}_{j=1}^m$.
for word-clause subgraph $\mathcal{G}_i^{wc} \subset \mathcal{G}^{wc}$ **do**
  Project feature space $\widetilde{h}_i^s = W_s \cdot h_i^s$;
  for word node $w_{i,j} \in \left\{w_{i,j}\right\}_{j=1}^n$ **do**
    Project feature space $\widetilde{h}_{i,j}^w = W_w \cdot h_{i,j}^w$;
    Calculate the node-level weight coefficient $a_{i,j}$;
  Update clause node feature $\hat{h}_i^s = \text{ReLU}(\sum_{j=1}^n a_{i,j} \cdot \widetilde{h}_{i,j}^w + b_w) + h_i^s$;
for pair-level subgraph $\mathcal{G}_i^p \subset \mathcal{G}_i^p$ **do**
  for pair node $p_{i,j} \in \mathcal{G}_i^p$ **do**
    Initialize the node representation $h_{i,j}^p = \hat{h}_i^s \parallel \hat{h}_j^s \parallel h_{i,j}^{rep}$;
    Project feature space $\widetilde{h}_{i,j}^p = W_p \cdot h_{i,j}^p$;
    for meta-path $\Phi_t \in \Phi$ **do**
      for Φ_t−based neighboring node $p_{i,k} \in P_{i,j}^{\Phi_t}$ **do**
        Calculate the node-level weight coefficient $a_{(i,j),(i,k)}^{\Phi_t}$;
      Aggregate node feature $\widetilde{h}_{i,j}^{\Phi_t} = \text{ReLU}(\sum_{k=1}^m a_{(i,j),(i,k)}^{\Phi_t} \cdot \widetilde{h}_{i,k}^p + b_{\Phi_t})$;
      Calculate the weight coefficient $a_{i,j}^{\Phi_t}$ of meta-path Φ_t;
    Update pair node feature $\hat{h}_{i,j}^p = \sum_{t=0}^T a_{i,j}^{\Phi_t} \cdot \widetilde{h}_{i,j}^{\Phi_t} + h_{i,j}^p$;
Calculate the total loss $L_{total} = L_p + L_e + L_c$;
Back propagation and update parameters;
return $\left\{\hat{h}_i^s\right\}_{i=1}^m, \cup_{i=1}^m \left\{\hat{h}_{i,j}^p\right\}_{j=1}^m$.

4. Experiments

4.1. Dataset and Evaluation Metrics

To evaluate our method, we utilized the benchmark ECPE dataset released by Xia and Ding [4], which consists of 1945 Chinese news documents. In these documents, there are a total of 490,367 candidate pairs, of which, the real emotion-cause pairs account for less than 1%, and each document possibly contains more than one emotion corresponding to multiple causes. According to the data-split setting of previous work, the dataset was segmented into 10 equal parts, and they were chosen as the train and test sets in the proportion of 9 to 1. In order to achieve statistically credible verification, we applied 10-fold cross-validation and repeated the experiments 20 times to average the results. Furthermore, precision (P), recall (R), and F1-score (F1) were selected as the evaluation metrics for emotion, cause, and emotion-cause pair extraction.

4.2. Experimental Settings

In our experiments, to make a fair comparison, the word embedding trained in [4] is utilized in our method. The dimensions of word embedding, BiLSTM's hidden state, and relative position embedding were set to 200, 100, and 50, respectively. In addition, for our BERT version model, the output dimension of pre-trained BERT is 768. The weight matrices and bias vectors involved in the two versions of our model were all randomly initialized by a continuous uniform distribution, $U(-0.01, 0.01)$. To avoid overfitting, we applied dropout, and the dropout rate was set to 0.1. Compared to some excellent global optimization algorithms [56–58], Adam [59] is more effective in deep learning. Therefore, in the training process of our model, we utilized the Adam optimizer to update all parameters with the learning rate of 0.005, mini-batch size of 32, and L_2 regularization coefficient of 1×10^{-5}. Our models were performed on the NVIDIA GeForce RTX 2080 Ti GPUs.

4.3. Compared Methods

We compared our method with the following state-of-the-art methods. It is worth noting that the models above the dotted line in Table 1 did not adopt BERT.

Table 1. Comparison of experimental results on the emotion extraction, cause extraction, and ECPE.

Category	Method	Emotion Extraction			Cause Extraction			Emotion-Cause Pair Extraction		
		P	R	F1	P	R	F1	P	R	F1
Pipelined	Inter-EC [4]	0.8364	0.8107	0.8230	0.7041	0.6083	0.6507	0.6721	0.5705	0.6128
	Inter-ECNC [31]	-	-	-	0.6863	0.6254	0.6544	0.6601	0.5734	0.6138
	DQAN [35]	-	-	-	0.7732	0.6370	0.6979	0.6733	0.6040	0.6362
End-to-end	E2EECPE [53]	0.8552	0.8024	0.8275	0.7048	0.6159	0.6571	0.6491	0.6195	0.6315
	MTNECP [37]	0.8662	0.8393	0.8520	0.7400	0.6378	0.6844	0.6828	0.5894	0.6321
	SLSN [42]	0.8406	0.7980	0.8181	0.6992	0.6588	0.6778	0.6836	0.6291	0.6545
	LAE-MANN [38]	0.8990	0.8000	0.8470	-	-	-	0.7110	0.6070	0.6550
	TDGC [54]	0.8716	0.8244	0.8474	0.7562	0.6471	0.6974	0.7374	0.6307	0.6799
	ECPE-2D [39]	0.8627	0.9221	0.8910	0.7336	0.6934	0.7123	0.7292	0.6544	0.6889
	PairGCN [8]	0.8857	0.7958	0.8375	0.7907	0.6928	0.7375	0.7692	0.6791	0.7202
	UTOS [52]	0.8815	0.8321	0.8556	0.7671	0.7320	0.7471	0.7389	0.7062	0.7203
	RANKCP [7]	0.9123	0.8999	0.9057	0.7461	0.7788	0.7615	0.7119	0.7630	0.7360
	RSN [48]	0.8614	0.8922	0.8755	0.7727	0.7398	0.7545	0.7601	0.7219	0.7393
Ours	***w/o* BERT**	**0.8361**	**0.8327**	**0.8337**	**0.7157**	**0.6519**	**0.6811**	**0.7143**	**0.6238**	**0.6654**
	HHGAT	**0.8655**	**0.9181**	**0.8906**	**0.7427**	**0.7988**	**0.7686**	**0.7458**	**0.7631**	**0.7525**

- Inter-EC, which uses emotion extraction to facilitate cause extraction, archives the best performance among the three pipelined methods proposed in [4].
- Inter-ECNC [31], as a variant of Inter-EC, employs transformer to optimize the extraction of cause clauses.
- DQAN [35] is a dual-questioning attention network, separately questioning candidate emotions and causes.

- E2EECPE [53] is an end-to-end link prediction model of directed graph, which establishes the directional links from emotions to causes by a biaffine attention.
- MTNECP [37] is a feature-shared, multi-task model and improves cause extraction with the help of position-aware emotion information.
- SLSN [42] is a symmetrical network composed of two subnetworks. Each subnetwork also performs a local pairing search, while extracting each target clause.
- LAE-MANN [38] explores a hierarchical attention to model the correlation between each pair of clauses.
- TDGC [54] is a transition-based, end-to-end model that regards ECPE as the construction process of directed graph.
- ECPE-2D [39] designs a 2D transformer to model the interaction between candidate pairs.
- PairGCN [8] employs a GCN to learn the dependency relations between candidate pairs.
- UTOS [52] redefines the ECPE task as a unified sequence labeling task, in which each label indicates not only the clause type, but also pairing index.
- RANKCP [7] is a ranking model that introduces a GAT to learn the representations of clauses.
- RSN [48] explicitly realizes the pairwise interaction between the three subtasks through multiple rounds of inference.

4.4. Main Results

The comparative results are shown in Table 1. We can observe that HHGAT achieves the best performance. In general, the end-to-end models obviously perform better than the pipelined models (e.g., Inter-EC, Inter-ECNC, and DQAN) because the end-to-end manner can avoid the cross-stage propagation of errors. In addition, better performance is usually achieved by the models with pre-trained BERT than those without it. Significantly, in terms of the F1-score, the non-BERT version of HHGAT outperforms SLSN (i.e., it is the best-performing model, without employing pre-trained BERT, and is based on LSTM) by 1.09% on emotion-cause pair extraction, which verifies the effectiveness of HHGAT for emotion-cause pair extraction.

By adopting BERT to encode the initial representation of nodes, the performance of HHGAT is further improved. Although LAE-MANN also designs a hierarchical attention network, it is not graph structure oriented, so it is inferior to our graph attention network in modeling the structural features of text. As shown in Table 1, LAE-MANN underperforms HHGAT by 9.75% in the F1-score of emotion-cause pair extraction. Inspired by Inter-EC, which utilizes the prediction results of emotions to promote cause extraction, ECPE-2D, UTOS, and RSN explicitly establish the interaction between emotion and cause, in their respective ways, to improve their performance. However, even without using the measures used in the above three methods, our model still outperforms them. Compared to the best-performing model RSN, the F1-scores of our HHGAT are increased by 1.51%, 1.41%, and 1.32% on emotion, cause, and emotion-cause pair extraction, respectively. This demonstrates that, even if the interaction between emotion and cause is not explicitly constructed, HHGAT can achieve excellent performance because of powerful modeling ability of the graph neural network.

Furthermore, TDGC, PairGCN, and RANKCP all employ graph structures to represent documents. However, TDGC is not realized by the graph neural network, but by LSTM, so its performance is the worst among these graph structure-based methods. Despite PairGCN and RANKCP employ GCN and GAT to learn node representations, respectively, they are all homogeneous graph oriented. This leads them to only focus on learning the correlations between the same kind of semantic elements. Different from them, our heterogeneous graph contains more kinds of nodes and richer semantic information. Compared to these three-graph, structure-based methods, our method improves the F1 score of emotion-cause pair extraction by 7.26%, 3.23%, and 1.65%, respectively. In summary, experimental results indicate that our method, based on the heterogeneous graph, is effective.

4.5. Ablation Study

To further validate the components of our model, we conduct an ablation experiment, where G1 denotes the clause node encoding layer, G2 represents the pair node encoding layer, and H1 and H2 correspond to the heterogeneous design of G1 and G2, respectively. The ablation results are shown in Table 2.

Table 2. Experimental results of structural ablation.

Method	Emotion Extraction			Cause Extraction			Emotion-Cause Pair Extraction		
	P	R	F1	P	R	F1	P	R	F1
HHGAT	0.8655	0.9181	0.8906	0.7427	0.7988	0.7686	0.7458	0.7631	0.7525
w/o G2	0.8553	0.9164	0.8839	0.7365	0.7970	0.7644	0.7093	0.7692	0.7361
w/o G2&H1	0.8625	0.9116	0.8860	0.7300	0.7654	0.7464	0.6895	0.7296	0.7075
w/o G1	0.8618	0.9021	0.8808	0.7381	0.7583	0.7477	0.7113	0.7224	0.7164
w/o G1&H2	0.8375	0.9170	0.8748	0.7308	0.7615	0.7449	0.6884	0.7379	0.7111
w/o G1&G2	0.8596	0.9148	0.8858	0.7296	0.7456	0.7353	0.6831	0.7169	0.6974

Firstly, HHGAT removes G2, resulting in the absence of dependency relations between local neighboring candidate pairs. As a result, the F1-score of emotion-cause pair extraction is decreased by 1.64%. This demonstrates that it is not enough to rely solely on modeling the word-clause connections. Specially, without an explicit interaction between emotion and cause, local context from neighboring pair nodes plays an important role in pairing the emotions and their corresponding causes.

Secondly, HHGAT *w/o* G2&H1 means that it only applies a graph attention network to learn the inter-clause relationships. Compared with HHGAT, the F1-score on emotion-cause pair extraction drops by 4.5%. The significant degradation of performance is mainly caused by the following two aspects. On the one hand, as the basic elements in clauses, words can provide more fine-grained semantic information. On the other hand, word nodes can enrich the correlations among clause nodes.

Then, HHGAT *w/o* G1 underperforms HHGAT by 0.98%, 2.09%, and 3.61% in the F1 scores of the three subtasks, respectively, which shows that our hierarchical design is beneficial to the ECPE task. This is because there is a natural hierarchical relationship between different semantic elements in human language. In addition, in the joint learning of three subtasks, good clause representation is helpful for the extraction of emotion-cause pairs.

Next, we can observe that the performance of HHGAT *w/o* G1&H2 is further dropped, compared with HHGAT *w/o* G1, because HHGAT *w/o* G1&H2 does not consider that the semantic information aggregated from neighboring nodes on different meta-paths is different. Hence, to learn more comprehensive pair node representations, it is necessary to employ a graph attention network based on meta-path on the pair-level subgraphs.

Finally, HHGAT *w/o* G1&G2 uses a clause-level BiLSTM to replace our two-layer graph attention network, which means that it is not a GNN-based method. Consequently, HHGAT *w/o* G1&G2 achieves the worst performance in all ablation models (F1-score dropped by 5.51%). The above results further show that each module of our method is helpful for the ECPE task.

4.6. Evaluation on Emotion-Cause Extraction

To provide a wider comparison, we also evaluate our model on the benchmark ECE corpus [2], and the compared models are as follows:

- Multi-Kernel [2] proposes a convolution kernel-based learning method to train a multi-kernel SVM.
- Memnet [15] is a convolutional deep memory network, which regards ECE as an answer retrieval task.
- PAE-DGL [27], as a reordering model, integrates the relative position and global label, with text content.

- CANN [18] presents a co-attention network based on emotional context awareness.
- MBiAS [24] designs a multi-granularity bidirectional attention network in a machine comprehension frame.
- RTHN [21] introduces a hierarchical neural network composed of RNN and transformer.
- FSS-GCN [30] adopts a graph convolutional network to model the dependency information between clauses.

The comparative results are shown in Figure 3. It can be observed that our model achieves slightly higher F1 than RTHN (i.e., the best-performing one in the models that are not based on graph neural networks). This further verifies the effectiveness of our approach on emotion-cause extraction. Furthermore, the performance of our model and FSS-GCN (i.e., a graph structure-based model) is nearly matched, in terms of the F1-score. Different from FSS-GCN, in which only clause nodes are considered, the heterogeneous graph built by us contains more kinds of nodes, and the structure of our model is more complicated. However, it is worth noting that the compared methods listed in Figure 3 all need to annotate emotions before extracting causes. This is very labor-consuming. Therefore, when the performance is equivalent, our method is more suitable for real applications.

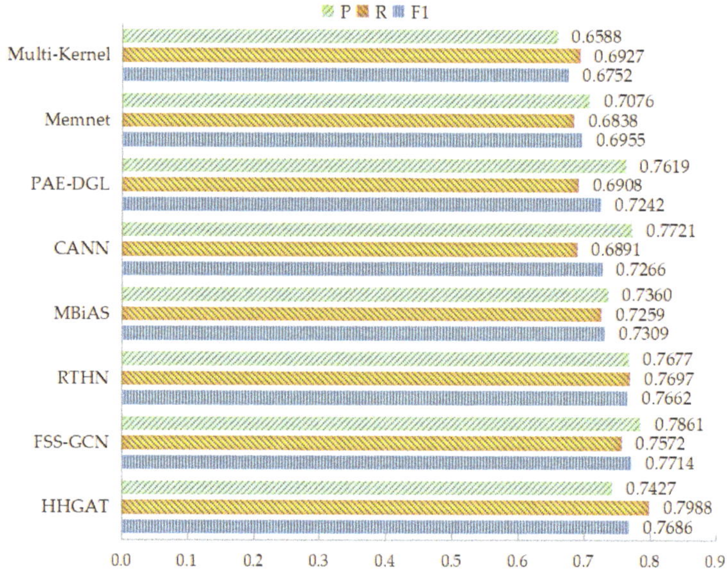

Figure 3. Comparison of experimental results on ECE.

4.7. Case Study

4.7.1. Effect of Word-Clause Graph Attention

As shown in Figure 4, the information regarding the three clauses in one representative case (i.e., Document 41) is introduced, including the word identifier, clause identifier, and details of the clause. This document consists of eight clauses and contains one emotion-cause pair (c_4, c_3), where c_4 and c_3 are the emotion and cause clauses, respectively. To examine the effect of word-clause graph attention, we visualize the weight vector $a_i = [a_{i,1}, \ldots, a_{i,n}]$. The visualization results are shown in Figure 4—where the darker the color is, the higher the relevance is.

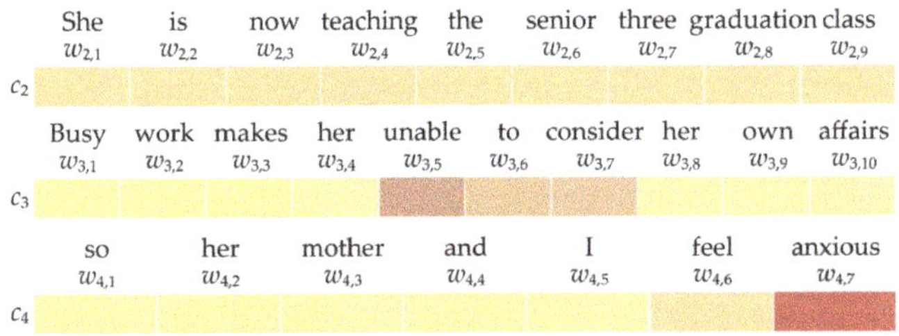

Figure 4. Visualization of word-clause attention.

We can find that the dark color is mainly concentrated around the word "anxious" in the emotion clause c_4, which indicates that HHGAT can effectively capture the emotion keywords and ignore other non-emotion words. Moreover, in the cause clause c_3, the words "unable", "to", and "consider" are significantly darker, which semantically constitutes the cause for triggering the emotion "anxious". This shows that our HHGAT is also able to focus on the cause keywords. In sharp contrast, the color of all words in clause c_2 is very similar, which causes attention to be dispersed because c_2 is neither an emotion clause nor a cause clause. Consequently, HHGAT is effective in learning the features of emotion and cause clauses.

4.7.2. Effect of Meta-Path-Based Attention

In this section, Document 41 is analyzed again to verify the effect of meta-path-based attention. To this end, we visualize the weight coefficients of different-typed meta-paths to each pair node, as shown in Figure 5. Since the document consists of eight clauses, we divide the visualization results into eight subgraphs, and each subgraph shows the attention visualization results of those pair nodes with the same candidate emotion clause. The color instructions are the same as that in the previous section.

	Φ_0	Φ_1	Φ_2	Φ_3
$p_{1,1}$	0.100	0.102	0.684	0.112
$p_{1,2}$	0.181	0.349	0.289	0.180
$p_{1,3}$	0.678	0.124	0.099	0.097
$p_{1,4}$	0.287	0.382	0.166	0.163
$p_{1,5}$	0.194	0.219	0.423	0.162
$p_{1,6}$	0.208	0.213	0.249	0.328
$p_{1,7}$	0.219	0.209	0.229	0.341
$p_{1,8}$	0.227	0.224	0.219	0.328

	Φ_0	Φ_1	Φ_2	Φ_3
$p_{2,1}$	0.118	0.108	0.660	0.112
$p_{2,2}$	0.182	0.382	0.259	0.174
$p_{2,3}$	0.662	0.125	0.109	0.102
$p_{2,4}$	0.271	0.379	0.173	0.175
$p_{2,5}$	0.192	0.214	0.418	0.174
$p_{2,6}$	0.208	0.212	0.240	0.339
$p_{2,7}$	0.221	0.209	0.223	0.344
$p_{2,8}$	0.226	0.227	0.218	0.328

	Φ_0	Φ_1	Φ_2	Φ_3
$p_{3,1}$	0.109	0.118	0.669	0.103
$p_{3,2}$	0.207	0.383	0.242	0.167
$p_{3,3}$	0.678	0.124	0.099	0.097
$p_{3,4}$	0.262	0.372	0.186	0.177
$p_{3,5}$	0.179	0.207	0.430	0.183
$p_{3,6}$	0.204	0.205	0.236	0.354
$p_{3,7}$	0.224	0.210	0.209	0.355
$p_{3,8}$	0.229	0.227	0.214	0.328

	Φ_0	Φ_1	Φ_2	Φ_3
$p_{4,1}$	0.086	0.103	0.721	0.088
$p_{4,2}$	0.208	0.362	0.266	0.163
$p_{4,3}$	0.724	0.112	0.080	0.082
$p_{4,4}$	0.294	0.344	0.185	0.176
$p_{4,5}$	0.165	0.204	0.459	0.171
$p_{4,6}$	0.192	0.199	0.247	0.360
$p_{4,7}$	0.224	0.206	0.201	0.368
$p_{4,8}$	0.234	0.226	0.201	0.337

	Φ_0	Φ_1	Φ_2	Φ_3
$p_{5,1}$	0.092	0.100	0.709	0.097
$p_{5,2}$	0.185	0.350	0.302	0.161
$p_{5,3}$	0.704	0.122	0.087	0.085
$p_{5,4}$	0.316	0.356	0.163	0.163
$p_{5,5}$	0.178	0.214	0.437	0.169
$p_{5,6}$	0.186	0.202	0.266	0.344
$p_{5,7}$	0.214	0.204	0.215	0.365
$p_{5,8}$	0.238	0.219	0.198	0.344

	Φ_0	Φ_1	Φ_2	Φ_3
$p_{6,1}$	0.099	0.106	0.686	0.107
$p_{6,2}$	0.184	0.344	0.303	0.167
$p_{6,3}$	0.679	0.129	0.101	0.089
$p_{6,4}$	0.300	0.388	0.159	0.151
$p_{6,5}$	0.209	0.217	0.407	0.165
$p_{6,6}$	0.196	0.212	0.257	0.333
$p_{6,7}$	0.201	0.200	0.249	0.349
$p_{6,8}$	0.226	0.211	0.210	0.351

	Φ_0	Φ_1	Φ_2	Φ_3
$p_{7,1}$	0.102	0.107	0.680	0.110
$p_{7,2}$	0.188	0.356	0.276	0.178
$p_{7,3}$	0.676	0.124	0.103	0.095
$p_{7,4}$	0.273	0.395	0.175	0.155
$p_{7,5}$	0.205	0.223	0.410	0.161
$p_{7,6}$	0.225	0.212	0.233	0.327
$p_{7,7}$	0.209	0.208	0.243	0.338
$p_{7,8}$	0.209	0.215	0.240	0.334

	Φ_0	Φ_1	Φ_2	Φ_3
$p_{8,1}$	0.102	0.109	0.676	0.111
$p_{8,2}$	0.191	0.357	0.269	0.181
$p_{8,3}$	0.674	0.124	0.098	0.101
$p_{8,4}$	0.274	0.372	0.181	0.171
$p_{8,5}$	0.181	0.216	0.439	0.162
$p_{8,6}$	0.220	0.217	0.234	0.327
$p_{8,7}$	0.241	0.209	0.214	0.334
$p_{8,8}$	0.212	0.243	0.229	0.314

Figure 5. Visualization of meta-path-based attention.

From the visualization results in Figure 5, we can observe that the color distribution on these subgraphs is very similar. In each subgraph, the color of Φ_0 corresponding to the pair node containing the ground-truth cause is the darkest. Additionally, in each row, the path with the largest weight coefficient to the target node is mostly the one where the real cause lies. In addition, as the offset from the central node or path increases, the correlation usually becomes lower. This shows that our method can find pair nodes containing ground-truth causes, according to the meta-paths.

Next, we conduct an inter-graph analysis, comparing the maximum attention coefficients in those rows corresponding to the ground-truth causes. In addition to Document 41, we also select the documents numbered 43, 167, and 151 as representative cases, where their emotion-cause pairs are $p_{5,5}$, $p_{6,4}$, and $p_{5,4}$, respectively. The comparison results are shown in Figure 6. We can notice that the highest point on each fold line is consistent with the ground-truth emotion-cause pair, which indicates that our meta-path-based graph attention network can effectively identify the emotion-cause pairs. It is worth noting that the values of all points on the fold line denoting Document 43 are relatively close. This is because the clause c_5 in Document 43 is both an emotion and cause clause, and each pair node on the fold line includes the clause c_5. The above results further verify that our method is effective for ECPE.

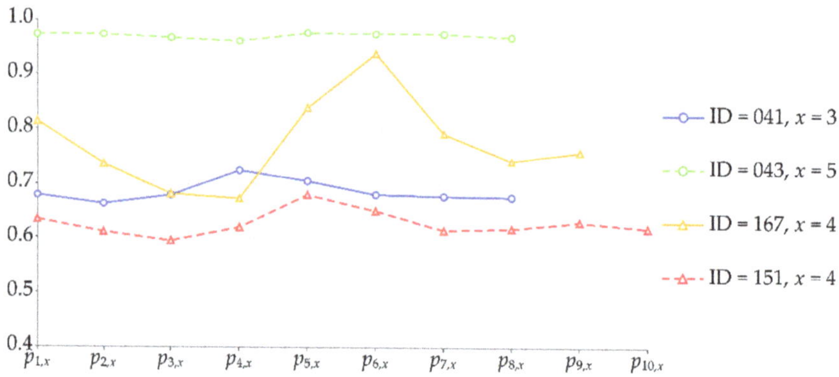

Figure 6. The inter-graph analysis of meta-path-based attention.

4.7.3. Error Analysis

In this section, we collect all emotion-cause pairs that were erroneously predicted on the test set. Inspired by [52], we also classify these errors into four categories, i.e., emotion, cause, both, and missing errors. Depending on the statistical results in Table 3, we can notice that the proportion of cause errors is the largest, followed by both errors. However, we can find that most of both errors are due to unlabeled emotions, which are usually irrelevant to the topic of the document. Furthermore, the proportion of missing errors is also relatively large. Therefore, we select two cases to analyze the cause and missing errors, respectively.

Table 3. The statistics of error emotion-cause pairs.

Category	Emotion Error	Cause Error	Both Error	Missing Error
Proportion	3.3%	46.2%	30.8%	19.7%

For the first case in Table 4, our model correctly predicts the emotion-cause pair $p_{8,8}$, while it identifies Clause 8 as the cause clause in the emotion-cause pair $p_{10,9}$ by mistake. It may be the cause of the prediction error that Clause 8 triggers the occurrence of the event described in Clause 9. Therefore, the ability of our model in distinguishing the indirect causes from direct causes needs to be further strengthened. Furthermore, in the prediction result of Case 2, the ground-truth emotion-cause pair $p_{3,5}$ is missing. We observe that the clause "it feels like the sky is falling down" is a metaphor, so it expresses an implicit emotion. Obviously, there are no emotion keywords in implicit emotional expression, and the identification of such emotions needs to comprehensively consider language style, rhetoric, metaphor, and so on, so it is more difficult to identify implicit emotions.

Table 4. Two error cases.

Case	Truth	Prediction
[...]. [Xiao was holding Long's 2-year-old son][7]. [Fearing that Long would hurt the child][8], [he knocked Long on the head with a lid][9], [and then Long became angry][10]. [...].	[8, 8] [10, 9]	[8, 8] [10, 8]
[...]. ["It feels like the sky is falling down"][3]. [Xu Ping described how she felt when she learned that her husband was ill][4], [he knocked Long on the head with a lid][5]. [...].	[3, 5]	[ ]

The superscript number at the end of a clause indicates the clause number.

5. Conclusions and Future Work

In this paper, we propose HHGAT to capture the global semantic information contained in the documents. Specifically, we first constructed a heterogeneous graph that considers all types of semantic elements involved in the ECPE and models the global semantic relations between these elements. Secondly, we proposed a hierarchical heterogeneous graph attention network to learn the representations of clauses and clause pairs with global semantic information. Thirdly, we conducted extensive experiments on the benchmark ECPE dataset. The experimental results show that our proposed method achieves a better performance than the 13 compared methods and out performs the best competitor, RSN, by a 1.32% F1-score

In addition, the essence of pairing emotions and causes is to calculate the similarity between them. Nevertheless, similarity is a fuzzy, and not clearly defined, concept. It is difficult for traditional graph neural networks to handle the fuzzy relationship. Therefore, we will introduce fuzzy graph theory [60–62] into graph neural networks in our future work, so as to effectively learn the fuzzy relation between clauses.

Author Contributions: J.Y.: conceptualization, methodology, formal analysis, software, validation, visualization, and writing original draft; W.L.: resources, supervision, project administration, and writing review; Y.H.: conceptualization, formal analysis, writing review, and editing; B.Z.: funding acquisition, data curation, and visualization. All authors have read and agreed to the published version of the manuscript.

Funding: This work was funded in part by the National Natural Science Foundation of China, under grant No. 61672448, grant No. 61673142, and grant No. 61972167, as well as, in part, by the Key R&D project of Hebei Province, under grant No. 18270307D, and Natural Science Foundation of Heilongjiang Province of China, under grant No. JJ2019JQ0013.

Data Availability Statement: Not applicable.

Conflicts of Interest: The authors declare no conflict of interest.

References

1. Lee, S.Y.M.; Chen, Y.; Huang, C.-R. A Text-Driven Rule-Based System for Emotion Cause Detection. In Proceedings of the 2010 North American Chapter of the Association for Computational Linguistics (NAACL), Los Angeles, CA, USA, 5 June 2010; pp. 45–53.
2. Gui, L.; Wu, D.; Xu, R.; Lu, Q.; Zhou, Y. Event-Driven Emotion Cause Extraction with Corpus Construction. In Proceedings of the 2016 Empirical Methods in Natural Language Processing, Austin, TX, USA, 1–5 November 2016; pp. 1639–1649.
3. Xu, R.; Hu, J.; Lu, Q.; Wu, D.; Gui, L. An Ensemble Approach for Emotion Cause Detection with Event Extraction and Multi-Kernel SVMs. *Tsinghua Sci. Technol.* **2017**, *22*, 646–659. [CrossRef]
4. Xia, R.; Ding, Z. Emotion-Cause Pair Extraction: A New Task to Emotion Analysis in Texts. In Proceedings of the 57th Association for Computational Linguistics, Florence, Italy, 28 July 2019; pp. 1003–1012.
5. Gori, M.; Monfardini, G.; Scarselli, F. A New Model for Learning in Graph Domains. In Proceedings of the 2005 Neural Networks, Montreal, QC, Canada, 31 July 2005–4 August 2005; Volume 2, pp. 729–734.

6. Scarselli, F.; Gori, M.; Tsoi, A.C.; Hagenbuchner, M.; Monfardini, G. The Graph Neural Network Model. *IEEE Trans. Neural Netw.* **2009**, *20*, 61–80. [CrossRef] [PubMed]
7. Wei, P.; Zhao, J.; Mao, W. Effective Inter-Clause Modeling for End-to-End Emotion-Cause Pair Extraction. In Proceedings of the 58th Association for Computational Linguistics, Online, 5–10 July 2020; pp. 3171–3181.
8. Chen, Y.; Hou, W.; Li, S.; Wu, C.; Zhang, X. End-to-End Emotion-Cause Pair Extraction with Graph Convolutional Network. In Proceedings of the 28th Computational Linguistics, Barcelona, Spain, 8–13 December 2020; pp. 198–207.
9. Chen, Y.; Lee, S.Y.M.; Li, S.; Huang, C.-R. Emotion Cause Detection with Linguistic Constructions. In Proceedings of the 23rd Computational Linguistics (Coling 2010), Beijing, China, 23–27 August 2010; pp. 179–187.
10. Gao, K.; Xu, H.; Wang, J. *Emotion Cause Detection for Chinese Micro-Blogs Based on ECOCC Model. Advances in Knowledge Discovery and Data Mining*; Springer International Publishing: Cham, Switzerland, 2015; pp. 3–14.
11. Gao, K.; Xu, H.; Wang, J. A Rule-Based Approach to Emotion Cause Detection for Chinese Micro-Blogs. *Expert Syst. Appl.* **2015**, *42*, 4517–4528. [CrossRef]
12. Russo, I.; Caselli, T.; Rubino, F.; Boldrini, E.; Martínez-Barco, P. EMOCause: An Easy-Adaptable Approach to Emotion Cause Contexts. In Proceedings of the 2nd Computational Approaches to Subjectivity and Sentiment Analysis, Portland, OR, USA, 24 June 2011; pp. 153–160.
13. Gui, L.; Yuan, L.; Xu, R.; Liu, B.; Lu, Q.; Zhou, Y. Emotion Cause Detection with Linguistic Construction in Chinese Weibo Text. In Proceedings of the Natural Language Processing and Chinese Computing, Shenzhen, China, 5–9 December 2014; pp. 457–464.
14. Ghazi, D.; Inkpen, D.; Szpakowicz, S. Detecting Emotion Stimuli in Emotion-Bearing Sentences. In *Computational Linguistics and Intelligent Text Processing*; Springer International Publishing: Cham, Switzerland, 2015; pp. 152–165.
15. Gui, L.; Hu, J.; He, Y.; Xu, R.; Lu, Q.; Du, J. A Question Answering Approach for Emotion Cause Extraction. In Proceedings of the 2017 Empirical Methods in Natural Language Processing, Copenhagen, Denmark, 7–11 September 2017; pp. 1593–1602.
16. Cheng, X.; Chen, Y.; Cheng, B.; Li, S.; Zhou, G. An Emotion Cause Corpus for Chinese Microblogs with Multiple-User Structures. *ACM Trans. Asian Low-Resour. Lang. Inf. Process.* **2017**, *17*, 1–19. [CrossRef]
17. Chen, Y.; Hou, W.; Cheng, X. Hierarchical Convolution Neural Network for Emotion Cause Detection on Microblogs. In Proceedings of the International Conference on Artificial Neural Networks and Machine Learning (ICANN 2018), Cham, Switzerland, 2018; pp. 115–122.
18. Li, X.; Song, K.; Feng, S.; Wang, D.; Zhang, Y. A Co-Attention Neural Network Model for Emotion Cause Analysis with Emotional Context Awareness. In Proceedings of the 2018 Empirical Methods in Natural Language Processing, Brussels, Belgium, 31 October–4 November 2018; pp. 4752–4757.
19. Li, X.; Feng, S.; Wang, D.; Zhang, Y. Context-Aware Emotion Cause Analysis with Multi-Attention-Based Neural Network. *Knowl. Based Syst.* **2019**, *174*, 205–218. [CrossRef]
20. Yu, X.; Rong, W.; Zhang, Z.; Ouyang, Y.; Xiong, Z. Multiple Level Hierarchical Network-Based Clause Selection for Emotion Cause Extraction. *IEEE Access* **2019**, *7*, 9071–9079. [CrossRef]
21. Xia, R.; Zhang, M.; Ding, Z. RTHN: A RNN-Transformer Hierarchical Network for Emotion Cause Extraction. In Proceedings of the 28th International Joint Conference on Artificial Intelligence, IJCAI-19, Macao, China, 10–16 August 2019; pp. 5285–5291.
22. Fan, C.; Yan, H.; Du, J.; Gui, L.; Bing, L.; Yang, M.; Xu, R.; Mao, R. A Knowledge Regularized Hierarchical Approach for Emotion Cause Analysis. In Proceedings of the 2019 Empirical Methods in Natural Language Processing and the 9th International Joint Conference on Natural Language Processing (EMNLP-IJCNLP), Hong Kong, China, 3–7 November 2019; pp. 5614–5624.
23. Hu, J.; Shi, S.; Huang, H. Combining External Sentiment Knowledge for Emotion Cause Detection. In Proceedings of the Natural Language Processing and Chinese Computing, Dunhuang, China, 9–14 October 2019; pp. 711–722.
24. Diao, Y.; Lin, H.; Yang, L.; Fan, X.; Chu, Y.; Wu, D.; Xu, K.; Xu, B. Multi-Granularity Bidirectional Attention Stream Machine Comprehension Method for Emotion Cause Extraction. *Neural. Comput. Applic.* **2020**, *32*, 8401–8413. [CrossRef]
25. Chen, Y.; Hou, W.; Cheng, X.; Li, S. Joint Learning for Emotion Classification and Emotion Cause Detection. In Proceedings of the 2018 Empirical Methods in Natural Language Processing, Brussels, Belgium, 3 October–4 November 2018; pp. 646–651.
26. Hu, G.; Lu, G.; Zhao, Y. Emotion-Cause Joint Detection: A Unified Network with Dual Interaction for Emotion Cause Analysis. In Proceedings of the Natural Language Processing and Chinese Computing, Zhengzhou, China, 14–18 October 2020; pp. 568–579.
27. Ding, Z.; He, H.; Zhang, M.; Xia, R. From Independent Prediction to Reordered Prediction: Integrating Relative Position and Global Label Information to Emotion Cause Identification. *Proc. AAAI Conf. Artif. Intell.* **2019**, *33*, 6343–6350. [CrossRef]
28. Xu, B.; Lin, H.; Lin, Y.; Diao, Y.; Yang, L.; Xu, K. Extracting Emotion Causes Using Learning to Rank Methods from an Information Retrieval Perspective. *IEEE Access* **2019**, *7*, 15573–15583. [CrossRef]
29. Xiao, X.; Wei, P.; Mao, W.; Wang, L. Context-Aware Multi-View Attention Networks for Emotion Cause Extraction. In Proceedings of the 2019 Intelligence and Security Informatics (ISI), Shenzhen, China, 1–3 July 2019; pp. 128–133.
30. Hu, G.; Lu, G.; Zhao, Y. FSS-GCN: A Graph Convolutional Networks with Fusion of Semantic and Structure for Emotion Cause Analysis. *Knowl. Based Syst.* **2021**, *212*, 106584. [CrossRef]
31. Shan, J.; Zhu, M. A New Component of Interactive Multi-Task Network Model for Emotion-Cause Pair Extraction. In Proceedings of the 3rd Computer Information Science and Artificial Intelligence (CISAI), Inner Mongolia, China, 25–27 September 2020; pp. 12–22.

32. Vaswani, A.; Shazeer, N.; Parmar, N.; Uszkoreit, J.; Jones, L.; Gomez, A.N.; Kaiser, Ł.; Polosukhin, I. Attention Is All You Need. In Proceedings of the Advances in Neural Information Processing Systems (NIPS), Long Beach, CA, USA, 4–9 December 2017; pp. 5998–6008.
33. Yu, J.; Liu, W.; He, Y.; Zhang, C. A Mutually Auxiliary Multitask Model with Self-Distillation for Emotion-Cause Pair Extraction. *IEEE Access* **2021**, *9*, 26811–26821. [CrossRef]
34. Jia, X.; Chen, X.; Wan, Q.; Liu, J. A Novel Interactive Recurrent Attention Network for Emotion-Cause Pair Extraction. In Proceedings of the 3rd Algorithms, Computing and Artificial Intelligence, New York, NY, USA, 24 December 2020; pp. 1–9.
35. Sun, Q.; Yin, Y.; Yu, H. A Dual-Questioning Attention Network for Emotion-Cause Pair Extraction with Context Awareness. In Proceedings of the 2021 International Joint Conference on Neural Networks (IJCNN), Online, 18–22 July 2021; pp. 1–8.
36. Shi, J.; Li, H.; Zhou, J.; Pang, Z.; Wang, C. Optimizing Emotion–Cause Pair Extraction Task by Using Mutual Assistance Single-Task Model, Clause Position Information and Semantic Features. *J. Supercomput.* **2021**, *78*, 4759–4778. [CrossRef]
37. Wu, S.; Chen, F.; Wu, F.; Huang, Y.; Li, X. A Multi-Task Learning Neural Network for Emotion-Cause Pair Extraction. In Proceedings of the 24th European Conference on Artificial Intelligence—ECAI 2020, Santiago de Compostela, Spain, 29 August–8 September 2020.
38. Tang, H.; Ji, D.; Zhou, Q. Joint Multi-Level Attentional Model for Emotion Detection and Emotion-Cause Pair Extraction. *Neurocomputing* **2020**, *409*, 329–340. [CrossRef]
39. Ding, Z.; Xia, R.; Yu, J. ECPE-2D: Emotion-Cause Pair Extraction Based on Joint Two-Dimensional Representation, Interaction and Prediction. In Proceedings of the 58th Annual Meeting of the Association for Computational Linguistics, Online, 6–8 July 2020; pp. 3161–3170.
40. Fan, R.; Wang, Y.; He, T. An End-to-End Multi-Task Learning Network with Scope Controller for Emotion-Cause Pair Extraction. In Proceedings of the Natural Language Processing and Chinese Computing, Zhengzhou, China, 14–18 October 2020; pp. 764–776.
41. Ding, Z.; Xia, R.; Yu, J. End-to-End Emotion-Cause Pair Extraction Based on Sliding Window Multi-Label Learning. In Proceedings of the 2020 Empirical Methods in Natural Language Processing (EMNLP), Online, 16–20 November 2020; pp. 3574–3583.
42. Cheng, Z.; Jiang, Z.; Yin, Y.; Yu, H.; Gu, Q. A Symmetric Local Search Network for Emotion-Cause Pair Extraction. In Proceedings of the 28th International Conference on Computational Linguistics, Barcelona, Spain, 8–13 December 2020; pp. 139–149.
43. Wu, D.; Wu, C. Research on the Time-Dependent Split Delivery Green Vehicle Routing Problem for Fresh Agricultural Products with Multiple Time Windows. *Agriculture* **2022**, *12*, 793. [CrossRef]
44. Deng, W.; Ni, H.; Liu, Y.; Chen, H.; Zhao, H. An Adaptive Differential Evolution Algorithm Based on Belief Space and Generalized Opposition-Based Learning for Resource Allocation. *Appl. Soft Comput.* **2022**, *127*, 109419. [CrossRef]
45. Singh, A.; Hingane, S.; Wani, S.; Modi, A. An End-to-End Network for Emotion-Cause Pair Extraction. In Proceedings of the Eleventh Workshop on Computational Approaches to Subjectivity, Sentiment and Social Media Analysis, Online, 19 April 2021; pp. 84–91.
46. Fan, W.; Zhu, Y.; Wei, Z.; Yang, T.; Ip, W.H.; Zhang, Y. Order-Guided Deep Neural Network for Emotion-Cause Pair Prediction. *Appl. Soft Comput.* **2021**, *112*, 107818. [CrossRef]
47. Yang, X.; Yang, Y. Emotion-Type-Based Global Attention Neural Network for Emotion-Cause Pair Extraction. In Proceedings of the International Conference on Natural Computation, Fuzzy Systems and Knowledge Discovery (ICNC-FSKD), Fuzhou, China, July 30–August 1 2022; pp. 546–557.
48. Chen, F.; Shi, Z.; Yang, Z.; Huang, Y. Recurrent Synchronization Network for Emotion-Cause Pair Extraction. *Knowl. Based Syst.* **2022**, *238*, 107965. [CrossRef]
49. Yuan, C.; Fan, C.; Bao, J.; Xu, R. Emotion-Cause Pair Extraction as Sequence Labeling Based on A Novel Tagging Scheme. In Proceedings of the Empirical Methods in Natural Language Processing (EMNLP), Online, 16–20 November 2020; pp. 3568–3573.
50. Fan, C.; Yuan, C.; Gui, L.; Zhang, Y.; Xu, R. Multi-Task Sequence Tagging for Emotion-Cause Pair Extraction Via Tag Distribution Refinement. *IEEE/ACM Trans. Audio Speech Lang. Process.* **2021**, *29*, 2339–2350. [CrossRef]
51. Chen, X.; Li, Q.; Wang, J. A Unified Sequence Labeling Model for Emotion Cause Pair Extraction. In Proceedings of the 28th International Conference on Computational Linguistics, Barcelona, Spain, 8–13 December 2020; pp. 208–218.
52. Cheng, Z.; Jiang, Z.; Yin, Y.; Li, N.; Gu, Q. A Unified Target-Oriented Sequence-to-Sequence Model for Emotion-Cause Pair Extraction. *IEEE/ACM Trans. Audio Speech Lang. Process.* **2021**, *29*, 2779–2791. [CrossRef]
53. Song, H.; Zhang, C.; Li, Q.; Song, D. An End-to-End Multi-Task Learning to Link Framework for Emotion-Cause Pair Extraction. *arXiv* **2020**, arXiv:2002.10710.
54. Fan, C.; Yuan, C.; Du, J.; Gui, L.; Yang, M.; Xu, R. Transition-Based Directed Graph Construction for Emotion-Cause Pair Extraction. In Proceedings of the 58th Annual Meeting of the Association for Computational Linguistics, Online, 5–10 July 2020; pp. 3707–3717.
55. Devlin, J.; Chang, M.-W.; Lee, K.; Toutanova, K. BERT: Pre-Training of Deep Bidirectional Transformers for Language Understanding. In Proceedings of the 2019 Conference of the North American Chapter of the Association for Computational Linguistics: Human Language Technologies, Volume 1 (Long and Short Papers), Minneapolis, MN, USA, 2–7 June 2019; pp. 4171–4186.
56. Yao, R.; Guo, C.; Deng, W.; Zhao, H. A Novel Mathematical Morphology Spectrum Entropy Based on Scale-Adaptive Techniques. *ISA Trans.* **2022**, *126*, 691–702. [CrossRef] [PubMed]
57. Zhou, X.; Ma, H.; Gu, J.; Chen, H.; Deng, W. Parameter Adaptation-Based Ant Colony Optimization with Dynamic Hybrid Mechanism. *Eng. Appl. Artif. Intell.* **2022**, *114*, 105139. [CrossRef]

58. Deng, W.; Xu, J.; Gao, X.-Z.; Zhao, H. An Enhanced MSIQDE Algorithm with Novel Multiple Strategies for Global Optimization Problems. *IEEE Trans. Syst. Man Cybern. Syst.* **2022**, *52*, 1578–1587. [CrossRef]
59. Kingma, D.P.; Ba, J. Adam: A Method for Stochastic Optimization. In Proceedings of the 3rd International Conference for Learning Representations (ICLR), San Diego, CA, USA, 7–9 May 2015.
60. Akram, M. *M−Polar Fuzzy Graphs: Theory, Methods & Applications*; Studies in Fuzziness and Soft Computing; Springer International Publishing: Cham, Switzerland, 2019; pp. 1–3, ISBN 978-3-030-03750-5.
61. Akram, M.; Zafar, F. *Hybrid Soft Computing Models Applied to Graph Theory*; Studies in Fuzziness and Soft Computing; Springer International Publishing: Cham, Switzerland, 2020; Volume 380, ISBN 978-3-030-16019-7.
62. Akram, M.; Luqman, A. *Fuzzy Hypergraphs and Related Extensions*; Studies in Fuzziness and Soft Computing; Springer: Singapore, 2020; Volume 390, ISBN 9789811524028.

electronics

MDPI

Article

Application of Improved YOLOv5 in Aerial Photographing Infrared Vehicle Detection

Youchen Fan [1,†], Qianlong Qiu [1,†], Shunhu Hou [1], Yuhai Li [1], Jiaxuan Xie [1], Mingyu Qin [2] and Feihuang Chu [1,*]

[1] School of Space Information, Space Engineering University, Beijing 101416, China; love193777@sina.com (Y.F.); overwindstarlimpid@163.com (Q.Q.); hshhsc2022@163.com (S.H.); lyth17739854226@163.com (Y.L.); 18543268891@139.com (J.X.)

[2] Graduate School, Department of Electronic and Optical Engineering, Space Engineering University, Beijing 101416, China; 15544029418m@sina.cn

* Correspondence: fhchu@126.com; Tel.: +86-189-5518-5670

† These authors contributed equally to this work.

Abstract: Aiming to solve the problems of false detection, missed detection, and insufficient detection ability of infrared vehicle images, an infrared vehicle target detection algorithm based on the improved YOLOv5 is proposed. The article analyzes the image characteristics of infrared vehicle detection, and then discusses the improved YOLOv5 algorithm in detail. The algorithm uses the DenseBlock module to increase the ability of shallow feature extraction. The Ghost convolution layer is used to replace the ordinary convolution layer, which increases the redundant feature graph based on linear calculation, improves the network feature extraction ability, and increases the amount of information from the original image. The detection accuracy of the whole network is enhanced by adding a channel attention mechanism and modifying loss function. Finally, the improved performance and comprehensive improved performance of each module are compared with common algorithms. Experimental results show that the detection accuracy of the DenseBlock and EIOU module added alone are improved by 2.5% and 3% compared with the original YOLOv5 algorithm, respectively, and the addition of the Ghost convolution module and SE module alone does not increase significantly. By using the EIOU module as the loss function, the three modules of DenseBlock, Ghost convolution and SE Layer are added to the YOLOv5 algorithm for comparative analysis, of which the combination of DenseBlock and Ghost convolution has the best effect. When adding three modules at the same time, the mAP fluctuation is smaller, which can reach 73.1%, which is 4.6% higher than the original YOLOv5 algorithm.

Keywords: target detection; infrared; deep learning; YOLOv5 algorithm

Citation: Fan, Y.; Qiu, Q.; Hou, S.; Li, Y.; Xie, J.; Qin, M.; Chu, F. Application of Improved YOLOv5 in Aerial Photographing Infrared Vehicle Detection. *Electronics* **2022**, *11*, 2344. https://doi.org/10.3390/electronics11152344

Academic Editor: José L. Abellán

Received: 14 June 2022
Accepted: 13 July 2022
Published: 27 July 2022

1. Introduction

With the gradual development of deep learning research, in-depth research in the field of computer vision constitutes not only a new change and development for people's daily lives, but also gives prospects for development in war and military training [1]. Among these prospects, the infrared imaging detection system is often used to detect and track local targets in military reconnaissance, to collect enemy military intelligence, and to provide guidance information for individual soldiers or conventional weapons to quickly obtain battlefield intelligence. In recent years, land-vehicle reconnaissance technology is the key research direction of battlefield control and surveillance capacity building, because in the actual combat environment [2,3], the ground environment is very complex. Vehicle targets may have the characteristics of occlusion, overlap, blur, etc., and so through infrared vehicle detection technology, ground vehicle targets and deployment can be more effectively found, which is conducive to the control of the battlefield and the overall situation.

In terms of infrared vehicle detection, in 2013, Iwasaki et al. proposed an algorithm to detect vehicle position and motion by using thermal imaging obtained with an infrared imaging sensor [4]. The algorithm specifies the vehicle position by applying a pattern-recognition algorithm according to the change of pixel values. The algorithm uses Haar-like features in each frame of the image, adopts a correction program for vehicle misidentification. The two detections can be combined to obtain vehicle position and motion information, and the vehicle detection accuracy is 96.3%. In 2017, Tang Tianyu proposed an improved aerial vehicle detection method based on Faster R-CNN, which was evaluated on the Munich vehicle dataset and the collected vehicle dataset, which improved accuracy and robustness compared with existing methods [5]. In 2018, Liu Xiaofei proposed a new method for ground-vehicle detection in aerial infrared images based on convolutional neural network [6], and experiments on four different scenarios on the NPU_CS_UAV_IR_DATA dataset showed that the proposed method was effective and efficient for the identification of ground vehicles. The overall recognition accuracy rate could reach 91.34%. In 2019, Lecheng Ouyang et al. [7] aimed at solving the problem of the low accuracy of traditional vehicle target-detection methods in complex scenarios, by combining them with the current hot development of deep learning. The YOLOv3 algorithm framework is used to achieve vehicle target detection, and by using the PASCAL VOC2007 and VOC2012 datasets, the images containing vehicle targets are screened out to form the VOC car dataset, and the target detection problem is transformed into a binary classification problem. Compared with the traditional target detection algorithm, the recognition accuracy of this method can reach 89.16%, and the average operating speed is 21FPS. In 2020, H. Li et al. proposed an incremental learning infrared vehicle-detection method based on (single-hot multiBox detector (SSD) for problems related to the lack of details in infrared vehicle images [8], the difficulty in extracting feature information, and low detection accuracy. This detection method can effectively identify and locate infrared vehicles, compared with the results of infrared vehicle detection using incremental datasets and non-incremental datasets. Experimental results show that the use of incremental datasets has significantly improved the error detection and missed detection of infrared vehicles, and the mAP has increased by 10.61%. In the same year, Mohammed Thakir Mahmood et al. proposed an infrared image vehicle-detection system by using YOLO's computer, combined with YOLO to propose an infrared-based technology [9]. Compared with the machine learning technique of K-means++ clustering algorithm, multi-object detection using convolutional neural networks, and the deep learning mechanism of infrared images, the method can run at a speed of 18.1 frames per second, with good performance. In 2022, Zhu Zijian et al. proposed a small target detection method for aerial infrared vehicles based on parallel fusion network [10]. An improved YOLOv3 algorithm based on cross-layer connection is proposed, which can accurately detect small targets of infrared vehicles in the background of complex motion, and achieve higher detection accuracy in the case of low false alarm rate, of which the false alarm rate is only 0.01% and the missed detection rate is only 1.36%.

Existing technologies have proven that the YOLOv3 algorithm has a good recognition performance for infrared vehicles [11–17]; however, on the basis of the YOLOv3 algorithm, in order to further improve the extraction ability of small targets, the YOLOv5 algorithm is generated [18–20]. In 2021, Kasper–Eulaers used the YOLOv5 algorithm to detect heavy trucks in winter rest areas, and the results showed that the trained algorithm could detect the front cabin of heavy trucks with high confidence. This article will also use the vehicle as an identification object for experiments under the improved YOLOv5 model. In the same year, Wu et al. combined local FCN and YOLOv5 to the detection of small targets in remote sensing images [20]. The application effects of R-CNN, FRCN, and R-FCN in image feature extraction are analyzed, and the high adaptability of the YOLOv5 algorithm to different scenarios is realized, and the proposed YOLOv5 algorithm + R-FCN detection method is compared with other algorithms. Experimental results show that the YOLOv5+R-FCN detection method has better detection ability among many algorithms.

Although the above literature has proven the applicability and advanced nature of the existing YOLOv3 and YOLOv5 infrared vehicle-detection algorithms, there is no unified and efficient detection method for the problems of false detection, missed detection, and detection accuracy in the multi-target and small target scenarios in the infrared vehicle images, so this paper proposes an infrared vehicle target detection algorithm based on improved YOLoOv5. The algorithm uses the EnseBlock module to improve the missed detection rate and detection accuracy through the dense characteristics between the feature layers. The use of Ghost convolutional layers to replace ordinary convolutional layers reduces the amount of parameters under the same characteristics, reduces the size of the model, and increases the amount of information in the original image. By adding channel attention mechanisms and changing the loss function, the inter-channel features are interrelated, and the anchor frame description is more accurate, which enhances the detection accuracy of the overall network, reduces the rate of missed detection, and is experimented and verified on the public infrared vehicle dataset.

2. Infrared Vehicle Image Data and Characteristic Analysis

2.1. Dataset Introduction

The dataset is derived from the public dataset used in the Space Cup competition [21], consisting of 16,000 images of infrared vehicles captured by drones equipped with infrared cameras. The dataset contains images of a single infrared vehicle target, as well as multi-target images. Some of the images contain false targets similar to vehicle targets, whereas others have the phenomenon of vehicles obscured by complex environments. Therefore, this dataset can be used for multi-target detection, as well as detection under complex ambient occlusion. At the same time, the pixel ratio of the ground truth of the detection target is between 0.04 and 0.1 in the training set, and most of them are small targets, due to the blurry edge characteristics of infrared images. Most target recognition is difficult, so it is a relatively complete dataset in general. Part of the dataset image is shown in Figure 1.

Figure 1. Dataset partial image example. (**a**) Single target. (**b**) Multi-target. (**c**) Single target in complex environment. (**d**) Multi-target in complex environment.

2.2. Image Characteristic Analysis

The images in the dataset are infrared vehicle images, which are single-channel grayscale images from 0 to 255. For this kind of image, a three-dimensional coordinate system is used to visualize the gray value information of the entire image. The xoy plane is used as the image plane, and the value of the z axis represents the gray value of the corresponding coordinate pixel. Secondly, the grayscale histogram is used for data analysis, reflecting the frequency of each gray level in the image. In the histogram, the abscissa is the gray level and the ordinate is the frequency of the gray level in the image, as shown in Figure 2.

As can be seen from Figure 2a, when the drone is closer to the target, its characteristics are apparent. The target image can be seen in the original image, and the target three-dimensional grayscale plot in Figure 2b is significantly higher than that of the background image, and the frequency of pixels is close to the actual target gray value in the grayscale histogram. Figure 2c is less high, making it easier to detect such a target. In Figure 2d, when the target shooting distance is far away, and the target is in a complex environment, the gray value of the three-dimensional grayscale plot Figure 2e is relatively more chaotic. The pixel frequency is similar to the actual target gray value in the grayscale histogram.

Figure 2f is higher, so that the target is easily submerged in the background of the similar gray value, and the detection is more difficult.

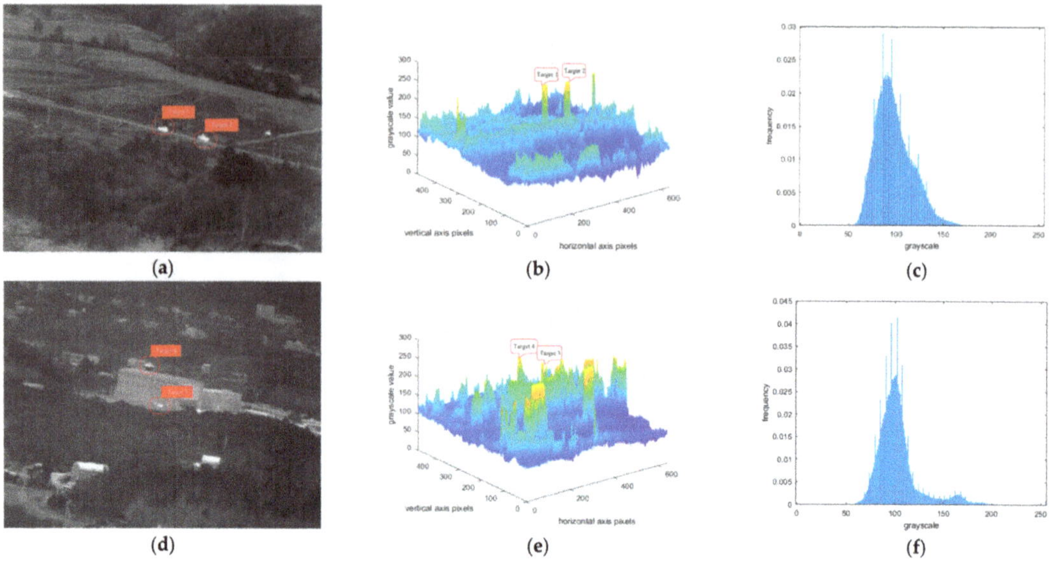

Figure 2. Image Characteristic analysis. (**a**) Original image. (**b**) 3D grayscale plot. (**c**) Grayscale histogram. (**d**) Original image in complex environment. (**e**) 3D grayscale plot. (**f**) Grayscale histogram.

Because the drone shoots at a distance, the infrared vehicle pixels in the figure account for a relatively small proportion of the entire image, as shown in Figure 2d, where the ground truth of a single target vehicle occupies 0.04% of the entire image in the training set. Therefore, the image has the characteristics of both infrared grayscale images and small targets, and is accompanied by the influence of multi-target and false targets. As shown in Figure 2d, target 4 is a false target, which increases the difficulty of infrared vehicle detection and not only reduces the accuracy of the detection algorithm, but also the feature extraction quality of the target detection network will be affected by different data content, resulting in a certain randomness of the training model. That is, for the training sets and verification sets for different images, the detection probability of the infrared vehicle target will fluctuate randomly within a certain range.

3. Improved Algorithm for YOLOv5

3.1. Model Improvement Ideas

The improvement of neural networks is an important field in neural networks [22,23], based on a baseline, adding, replacing, and deleting the middle layer on the original network, improving the loss function, optimizer, and related parameters, or combining other target processing techniques. Its purpose is to fuse and optimize various neural networks to improve the positioning accuracy, classification accuracy, classification speed and model size of the data.

The improved algorithm uses the main module of DenseNet to increase the extraction ability of shallow features by linking the dense superposition between the feature layers; it replaces the ordinary convolution layer with the Ghost convolution layer, to improve the network redundant feature-extraction ability and increase the amount of information in the original image by extracting the redundant feature map obtained by linear calculation of input images based on different parameters. By adding the channel attention mechanism, the features between the channels can be correlated with each other to improve the detection accuracy of the network layer and change the loss function to more accurately describe the relationship between the prediction box and the real box, and enhance the detection accuracy of the overall network anchor frame.

3.2. Dense Convolutional Network (DenseNet)

The Dense Convolutional Network (DenseNet) has four main advantages, namely alleviating the gradient disappearance problem, enhancing feature propagation (retain low-latitude features), promoting feature reuse, and greatly reducing the number of parameters. When the CNN layers get deeper, the path from output to input will become longer, which will cause a problem: the gradient will probably disappear when it is backpropagated to the input through such a long path, DenseNet proposes a very simple way to make the network deep and the gradient does not disappear by establishing dense connections to reuse features. To solve this problem, the following is the schematic diagram of DenseBlock, as shown in Figure 3.

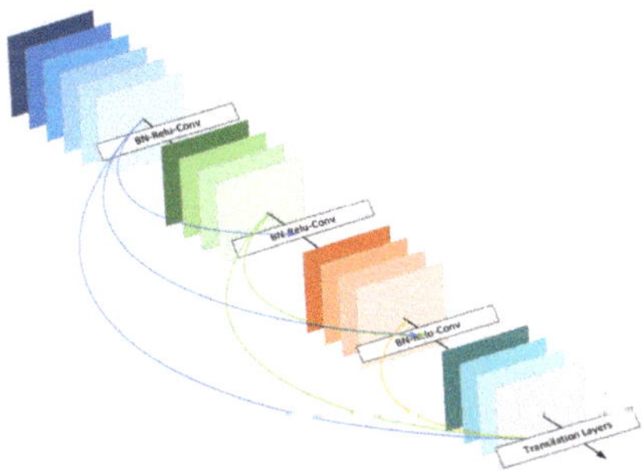

Figure 3. Schematic diagram of the DenseBlock structure.

As can be seen from Figure 3, the output of each layer is connected to the input of the latter layer, for an L layer network, there will be connections. For each layer, all the previous feature layers are the inputs of the current layer, and the feature layers are the subsequent inputs, forming a full interlink, and the feature maps extracted by each layer can be used by subsequent layers.

DenseNet consists of four DenseBlocks and the connected translation layers. The text additionally extracts DenseBlock as a pluggable module for acquiring and connecting denser image features at the beginning of the network structure, but due to its own characteristics, the number of output channels is determined by the number of input channels, module layers, and the learning multiple, which cannot be freely defined. The robustness is poor, and specific parameters need to be adjusted to join the network as a module.

3.3. End-Side Neural Networks (GhostNet)

In CNN models, redundancy in feature maps is very important, but few people consider the problem of redundancy in feature maps in the model structure design. In 2021, He Kaiming et al. proposed a novel Ghost module that can use fewer parameters to generate more feature maps. In the Ghost module, the feature map generated by the linear operation is called the Ghost feature maps, and the feature map manipulated is called the intrinsic feature maps. Obviously, the Ghost module's computation is significantly reduced compared to using conventional convolution directly. From another point of view, it can be considered that the feature map obtained by convolution has been enhanced, similar to the data augmentation. The Ghost convolutional structure is shown in Figure 4 below.

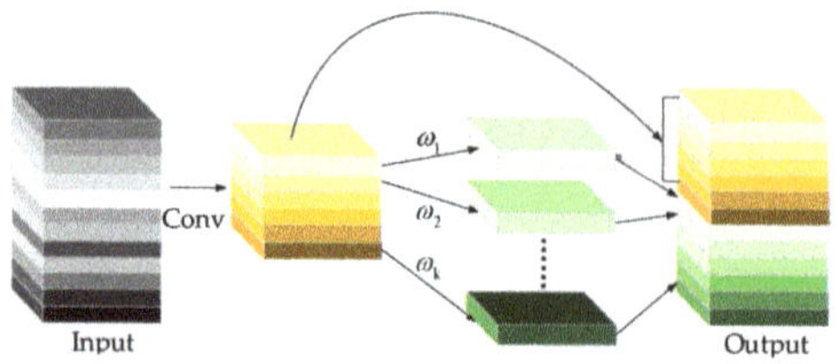

Figure 4. Schematic diagram of Ghost convolutional structure.

3.4. Squeeze-and-Excitation Networks (SENet)

Squeeze-and-Excitation Networks (SENet) constitute a new image recognition structure announced by the autonomous driving company Momenta in 2017, which improves accuracy by modeling correlations between feature channels and enhancing important features. This structure is the winner of the 2017 ILSVR competition, with a top 5 error rate of 2.251%, 25% lower than the first place in 2016. SENet strengthens the characteristics of important channels and weakens the characteristics of non-important channels, which has obtained good results. The SE layer structure is shown in Figure 5 below.

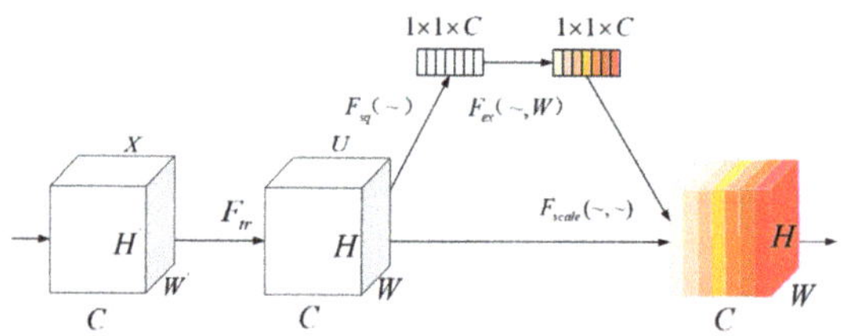

Figure 5. Schematic diagram of the structure of the SE layer.

3.5. EIOU Loss

YOLOv5 uses a combination of IOU Loss, GIOU Loss, and CIOU Loss, although CIOU considers the overlapping area, center point distance, and aspect ratio of bounding box regression. However, the difference in aspect ratio reflected by v in the formula is not the true difference between the width and height and its confidence, so it sometimes hinders the effective optimization similarity of the model. In response to this problem, in 2021, Yi-Fan Zhang, Weiqiang Ren, Zhang Zhang, etc. took apart the aspect ratio on the basis of CIOU, proposed EIOU Leoss, and added Focal and Efficient IOU Loss for Accurate Bounding Box Regression.

The formula for the loss function EIOU Loss is as follows:

$$\begin{aligned} L_{EIOU} &= L_{IOU} + L_{dis} + L_{asp} \\ &= 1 - IOU + \frac{\rho^2(b,b^{gt})}{c^2} + \frac{\rho^2(w,w^{gt})}{C_w{}^2} + \frac{\rho^2(h,h^{gt})}{C_h{}^2} \end{aligned} \tag{1}$$

The EIOU formula consists of three parts, namely the overlap loss, the center point distance loss, and the width and height loss. The first part of the overlapping area loss is the definition of the IOU itself: the area where the prediction box and the real box are combined with the area ratio intersection, and the second part continues the center distance loss in CIOU, that is, the Euclidean distance ratio between the prediction box and the real box contains the square of the diagonal distance of the minimum external box of the prediction box and the real box. The third part innovatively uses the Euclidean distance of the width and height difference between the target box and the real box divided by the square of the width and height of the minimum external box.

In summary, EIOU Loss describes the image overlapping area, the center point distance, the true difference between the length and width of the sides, solves the blurry definition of aspect ratio based on CIOU, and adds Focal Loss to solve the sample imbalance problem in BBox regression.

3.6. Improved YOLOv5 Network

To describe improvement ideas, the improvement of the YOLOv5 network in this paper is mainly divided into four parts:

1. For the image input network layer, the DenseBlock module is used to strengthen the extraction of strong correlation features for shallow images, and reduce the image correlation features lost in the initial stage of the network through multi-layer dense networks.
2. For the backbone network, the Ghost convolution layer is used to replace the first two general convolution layers, which increases the feature redundancy, reduces the computation amount of the overall network, and increases the detection speed.
3. For the feature extraction network, the channel attention mechanism is introduced by using the SE network layer, which strengthens the network detection capability on the basis of the integration of image channel features.
4. For the loss function, the latest EIOU is used to replace the original CIOU of YOLOv5, which improves the accuracy of the description relationship between the prediction box and the GT box, and improves the network binding ability.

The four improved modules in this article are pluggable modules as shown in Figure 6. The corresponding modules can be selected and added to the target detection network according to the needs.

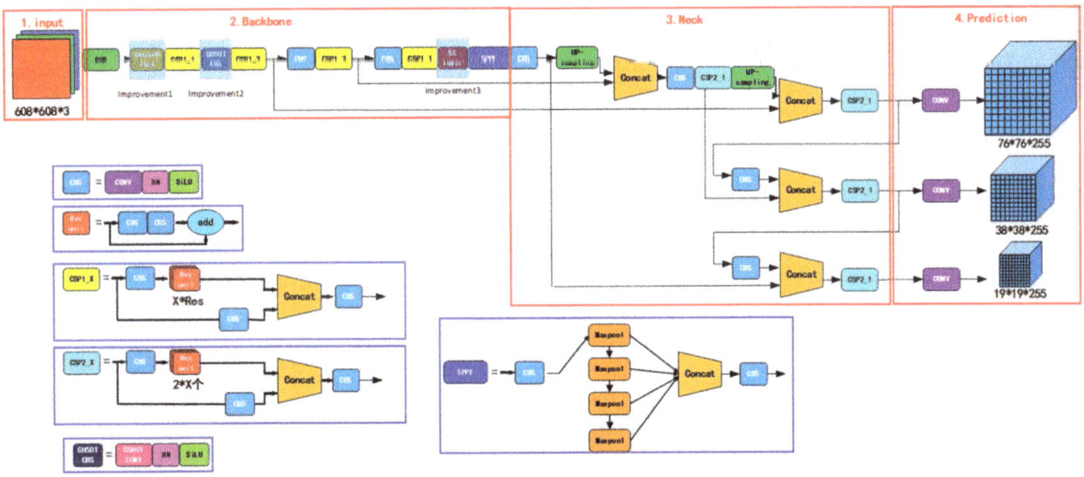

Figure 6. Improved YOLOv5 network.

4. Experiments on Improved Algorithms for Each Module

4.1. Training Environment Configuration

The specific experimental parameters are configured as shown in Table 1.

Table 1. Experimental parameter configuration.

Parameter	Disposition
Operating system	Linux
Redaction language	Python 3.8
CUDA version	10.2
Pytorch	1.8.1
YOLOv5	6.0
GPU	TITAN RTX
CPU	Intel i9-10900K
Internal storage	125.8GB

4.2. *Experiments with Dense Convolutional Networks (DenseBlock)*

In the experiment, first, the parameter adjustment experiment is carried out for each improved module in the text, and then a single improved network is compared with YOLOv5s. Finally the improved modules are synthesized and compared with the original network and the current mainstream target detection network.

4.2.1. Experimental Parameters

Under the dataset, optimize the parameter settings of the DenseBlock module, i.e., Grow_rate and layers. Grow_rate represents how many feature layers are connected to the previous feature layer and how many are connected to the back. The layers represent how many DenseBlock dense link layers are used.

The DenseBlock module has the characteristics of the number of input channels and parameter settings that determine the number of output channels, so there are two sets of parameter settings for matching the number of channels before and after the experiment, as shown in Table 2.

Table 2. DenseBlock module experimental parameter table.

Parameter Settings	8-3	16-1
Training times	100	100
Recognition rate(mAP)	0.616	0.602
Model size(mb)	14.43	14.43
Inference time(ms)	4.8	4.5

4.2.2. Training Results

The training results for different parameter selections are shown in Figure 7.

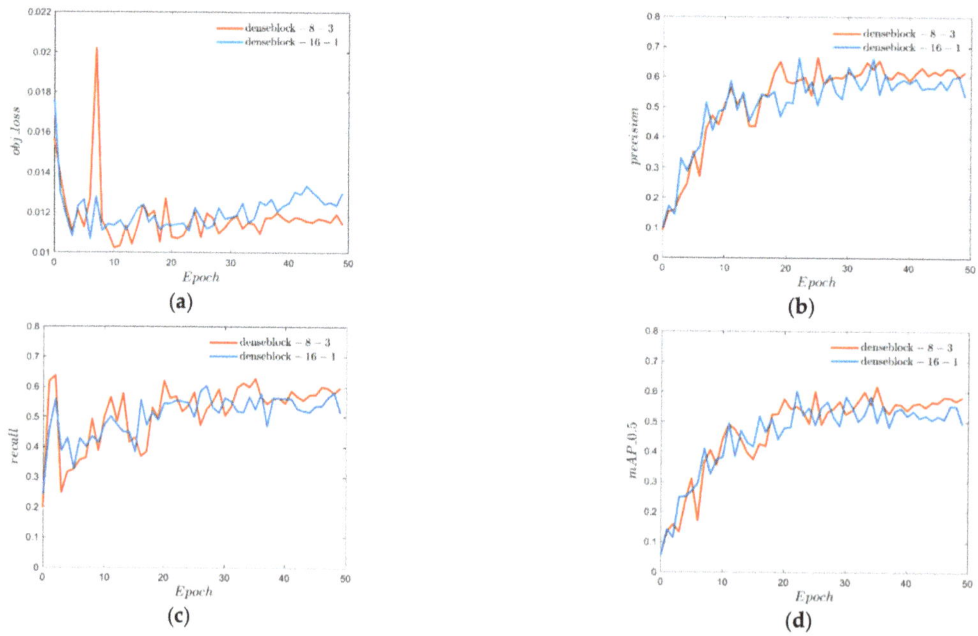

Figure 7. Comparison results of DenseBlock parameters. (**a**) Target loss. (**b**) Accuracy rate. (**c**) Recalling rate. (**d**) mAP value.

From Figure 7a, it can be seen that the target loss value of the 8-3 experimental group is lower than that of the 16-1 experimental group. That is, the target anchor frame classification is more accurate, and from Figure 7b,d, it can be seen that the detection accuracy of the 8-3 experimental group in the first 20 epochs is lower than that of the 16-1 experimental group, but with the increase of the number of trainings. When the epoch reaches more than 40 times and the experimental result tends to stabilize, the detection accuracy of the 8-3 experimental group is higher. As can be seen from Figure 7c, there is no significant difference in recall rates.

For the parameter growth_rate and num_layers used in the DenseBlock module, due to the limitation of input and output channels, a total of 2 parameter combinations were used for comparative experiments. It can be seen that under the premise of the same model size, the DenseBlock module with more dense layers and lower learning rate has an obvious performance advantage, but it is worth mentioning that the training time of adding the DenseBlock module is longer, the training configuration requirements are higher, and the amount of computation is greater.

4.2.3. Testing Results

The detection results before and after adding the DenseBlock module are shown in Figures 8–10.

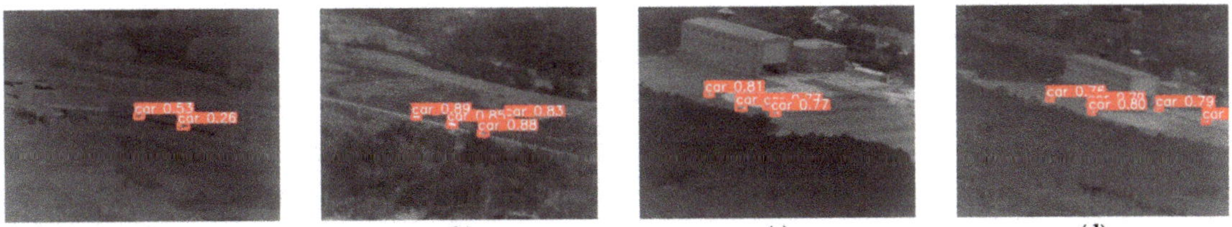

(a) (b) (c) (d)

Figure 8. YOLOv5s detection map. (**a**) Scene 1. (**b**) Scene 2. (**c**) Scene 3. (**d**) Scene 4.

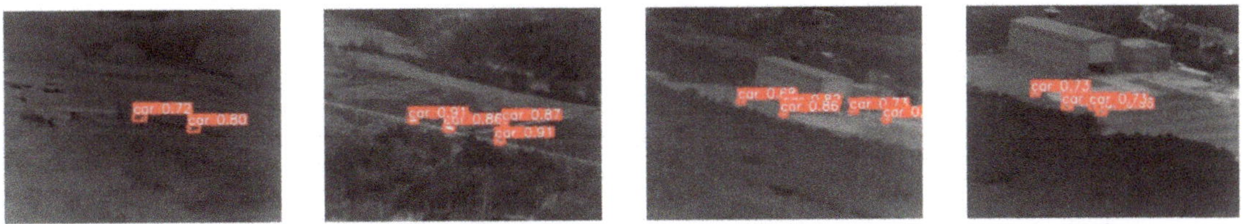

(a) (b) (c) (d)

Figure 9. 8-3 DenseBlock detection results. (**a**) Scene 1. (**b**) Scene 2. (**c**) Scene 3. (**d**) Scene 4.

(a) (b) (c) (d)

Figure 10. 16-1 DenseBlock detection results. (**a**) Scene 1. (**b**) Scene 2. (**c**) Scene 3. (**d**) Scene 4.

As can be seen from Figures 8–10, whether the 8-3 experimental group or the 16-1 experimental group, the average confidence in detecting infrared small target vehicles is higher than that of the original algorithm, and the experimental group of 8-3 performed better than the experimental group of 16-1. This shows that the DenseBlock module with

8-3 parameters is more suitable for the detection of this dataset, and this parameter group is used in the comprehensive module of subsequent experiments.

4.3. Experiments with End-Side Neural Networks (GhostNet)

4.3.1. Experimental Parameters

According to the feature map redundancy of the Ghost convolutional layer, it can be inferred that the deep feature map is not suitable for feature redundancy inference by using linear calculation. Therefore, the replaced convolutional layers are close to the input layer, which are the backbone network convolutional layers. The parameter settings such as the number of Ghost convolutional layers replaced, training time, and recognition rate in the experiment are shown in Table 3.

Table 3. Ghost module experimental parameter table.

Ghost Convolutional Replacement Quantity	**1**	**2**	**3**	**4**
Training times	100	100	100	100
Recognition rate(mAP)	0.64	0.655	0.613	0.599
Model size(mb)	14.05	13.99	13.71	12.57
Inference time(ms)	4.2	4.3	4.4	4.4

4.3.2. Training Results

The training results for different parameter selections are shown in Figure 11.

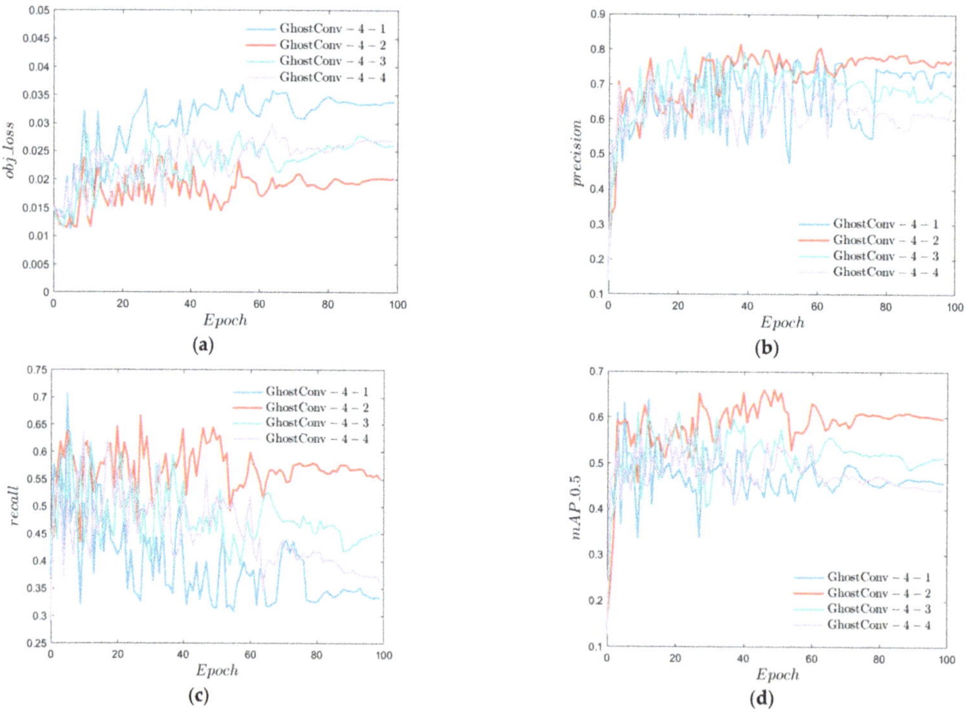

Figure 11. Comparison of the number of GhostConv replacements. (**a**) Confidence loss. (**b**) Accuracy rate. (**c**) Recalling rate. (**d**) mAP value.

From Figure 11a, it can be seen that the Ghost experimental group replacing the two convolution layers had lower target loss values during training, and it can be seen from Figure 11b that the detection accuracy of the 4-2 experimental group was higher in the 30 epochs after the training results tended to stabilize. From Figure 11c,d, it can be seen that the recall rate and detection accuracy of the 4-2 experimental group in a total of 100 epoch training are always higher than that of other experimental groups, and the gap is noticeable.

For a single Ghost module, although the model size is effectively reduced with the increase of the number of substitutions, after replacing three ordinary convolution layers, the recognition rate shows a downward trend. That is, too much feature map redundancy harms the detection accuracy, and in terms of model size and inference time, the more Ghost convolutional replacements, the smaller the model, and the slower the inference time.

When replacing two convolution layers, the network recognition rate shows a peak due to the increase of the redundancy feature map, which proves that the redundancy of the feature map is not always positive for the recognition rate, at the same time, the inference time is faster, and the model size increases less. It is the best choice to replace the two convolution layers, so the subsequent Ghost modules use a replacement number of two Ghost convolutional modules by default.

4.3.3. Testing Results

After adding the corresponding Ghost module, the test result is shown in Figure 12.

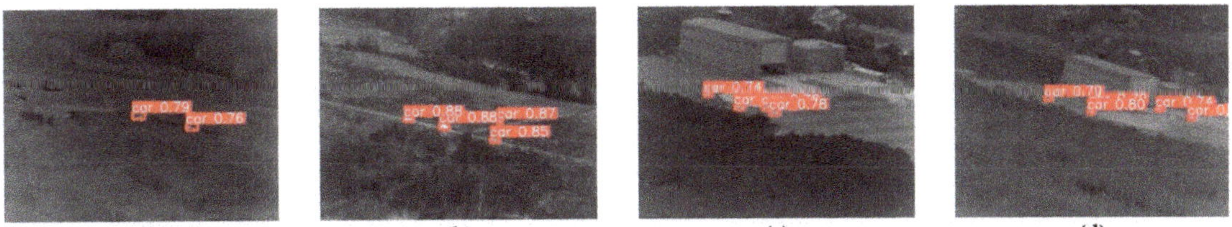

(a) (b) (c) (d)

Figure 12. Ghost convolutional test results. (**a**) Scene 1. (**b**) Scene 2. (**c**) Scene 3. (**d**) Scene 4.

From the comparison of Figures 8 and 12, it can be seen that the network that joins the Ghost convolution can accurately detect the vehicle target, and the detection accuracy has been improved in each scene. Among the two targets in scene 1, the detection accuracy was the highest, increasing by 26% and 50% respectively.

4.4. *Experiments with the Squeeze-and-Excitation Layer (SE Layer)*

In the SE layer, the module position of the SE layer is optimized by parameter reduction, and the more suitable module position and parameters have been pre-selected according to the previous experiments. See Table 4 for experimental parameters.

Table 4. SE module experimental parameter table.

Module Position and Parameters	Before SPPF	After SPPF	Reduction = 16	Reduction = 4
Training times	50	50	50	50
Recognition rate (mAP)	0.661	0.655	0.612	0.667
Model size (mb)	14.67	14.67	14.67	14.47
Extrapolation time (ms)	4.5	4.4	4.4	4.4

4.4.1. Training Results

The training results for different parameter selections are shown in Figures 13 and 14.

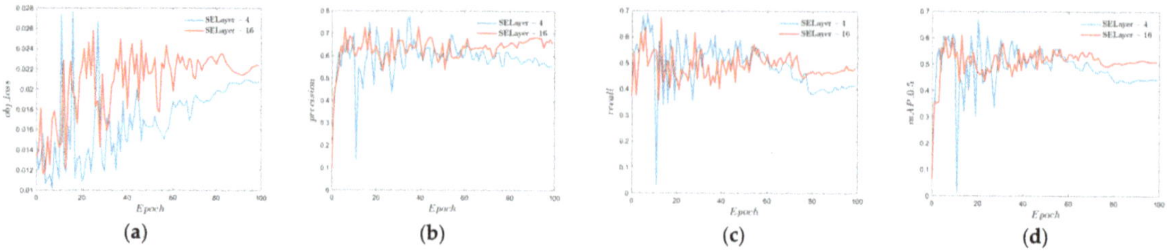

Figure 13. SE reduction parameter comparison results. (**a**) Target loss. (**b**) Accuracy rate. (**c**) Recalling rate. (**d**) mAP value.

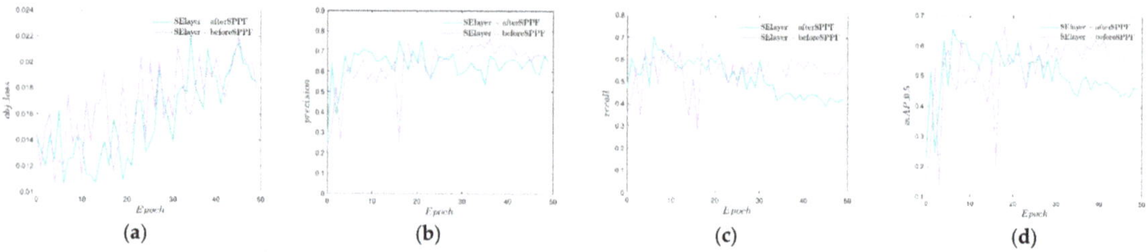

Figure 14. SE module position comparison result. (**a**) Target loss. (**b**) Accuracy rate. (**c**) Recalling rate. (**d**) mAP value.

From Figure 13a, it can be seen that the target loss value is higher when the reduction parameter is taken with reduction = 16. From Figure 13b–d, it can be seen that the totality is relatively stable after 40 epochs, and the experimental group with a parameter of 16 has a higher detection accuracy. As can be seen from Figure 14a,b, the target loss values of the two experimental control groups are similar. The detection accuracy is generally similar. As can be seen from Figure 14c,d, the overall mAP value of the target detection in the pre-SPPF experimental group was higher due to the higher recall rate in the pre-SPPF experimental group.

In terms of attention parameters, try where different SE layers are added, and finally select SPPF before and after doing the comparison experiment. It can be seen that the SE module is more suitable before the SPPF, according to the analysis of the role of SPPF can be obtained. The SE module for the high-level features of the channel attention mechanism is more biased toward the image features before the pooling layer rather than the semantic features after the pooling layer. At the same time, according to the comparison of reduction parameters, the SE model with a reduction of 4 performs prominently in a single epoch but is not stable overall, whereas the overall trend results with a parameter of 16 perform better. That is to say, increasing the decline rate of the hidden layer channel can improve the detection rate of the image attention mechanism. Finally, the parameter reduction of 16 is selected according to the image.

4.4.2. Testing Results

When the SE module is added to the SPPF and the reduction parameter is selected 16, the detection results are shown in Figure 15.

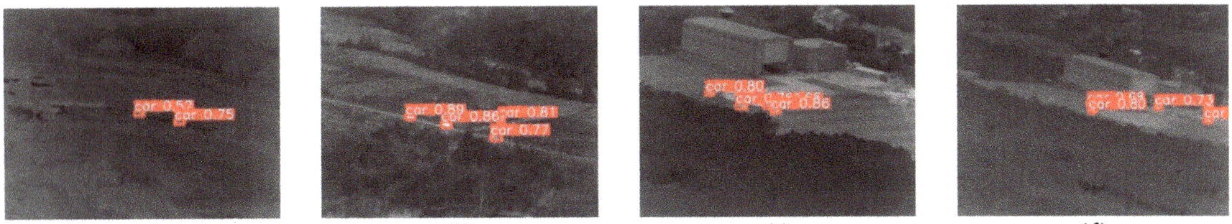

(a) (b) (c) (d)

Figure 15. SE layer detection results. (**a**) Scene 1. (**b**) Scene 2. (**c**) Scene 3. (**d**) Scene 4.

Compared with Figures 8 and 15, the average detection accuracy of the network with the addition of an attention mechanism is significantly improved in each scene.

4.5. Experiments with EIOU

For the replacement loss function, because YOLOv5 used a total of GIOU, DIOU, and CIOU, three kinds of loss functions, along with the development of the loss function research, now YOLOv5 mainly uses CIOU. This article uses EIOU to replace CIOU. For improved models, replacement loss function increases the detection recognition rate, so the subsequent experiments are all replaced with EIOU loss functions. Training results are shown in Figure 16 below.

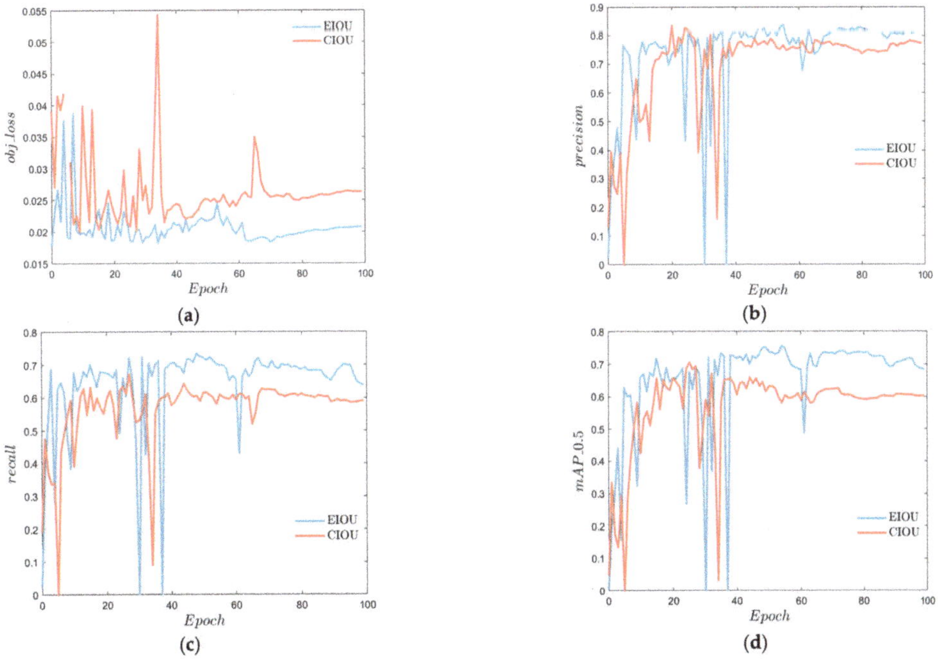

Figure 16. EIOU detection results. (**a**) Target loss. (**b**) Accuracy rate. (**c**) Recalling rate. (**d**) mAP value.

As can be seen from Figure 16, compared with CIOU, the recall value increases and the object loss value decreases in the detection results by using EIOU, and the mAP value of the EIOU group in the overall model detection is significantly improved.

5. Modular Combination Improved Algorithm Experiment

5.1. Improved YOLOv5 Network Experiment

In order to improve the detection effect of the comprehensive improved model, the single module is compared, and they are added to the original YOLOv5 algorithm in pairs. The results are shown in Tables 5 and 6. Refer to [24,25] for a graphical representation of the optimization results. Convert the mAP column in Table 5 to a histogram as Figure 17 shows and convert the mAP column in Table 6 to a histogram as Figure 18 shows.

Table 5. Comparison table of results for individual module.

Improved Modules	YOLOv5s	Ghost Convolution	DenseBlock	SE Module
Number of trainings	100	100	100	100
Recognition rate(mAP)	0.685	0.650	0.713	0.660
Model size(mb)	14.07	13.99	14.43	14.67
Extrapolation time(ms)	4.2	4.3	4.8	4.4

Table 6. Comparison table of results for the synthesis improved module.

Network Structure	YOLOv5s	Dense + Ghost + SE	Dense + Ghost	Ghost + SE	Dense + SE
Number of trainings	100	100	100	100	100
Recognition rate(mAP)	0.685	0.731	0.73	0.753	0.685
Model size(mb)	14.07	14.80	14.36	14.59	15.15
Extrapolation time(ms)	4.2	8.5	5.0	4.5	8.6

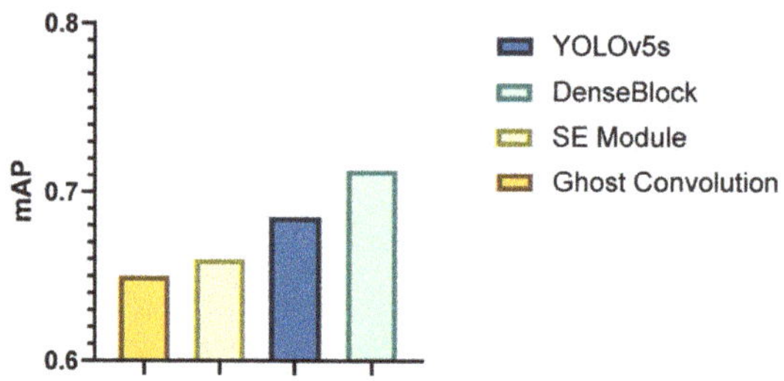

Figure 17. Single-module mAP histogram.

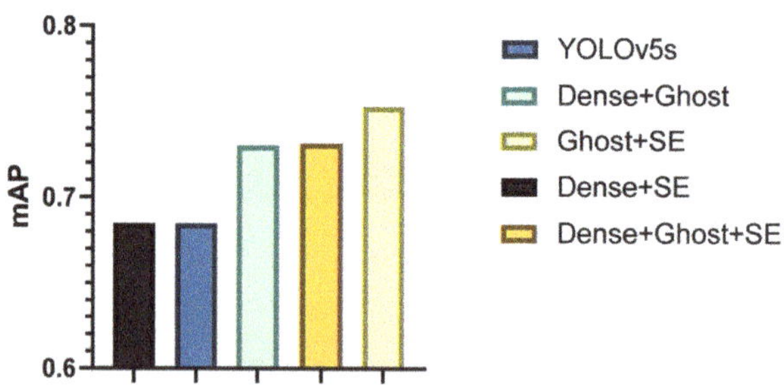

Figure 18. Comprehensive improvement of mAP histogram.

5.1.1. Training Results

Figure 16 shows the comparison between the detection accuracy of the DenseBlock, Ghost convolution and SE modules and the detection accuracy of the original YOLOv5 algorithm.

The characteristics and applicable scenes of each module can be drawn from Figure 19, and from Figure 19a, the confidence loss of the DenseBlock module is significantly lower than that of other modules. That is, the module is more effective in improving the detection accuracy and stability of the target. As can be seen from Figure 19b,c, although the SE module can improve the recognition accuracy, it will lead to a decrease in the recall rate; from Figure 19d, when used alone, the DenseBlock module has the most obvious improvement, but the mAP value of Ghost convolution and SE module does not improve significantly. A combination of these modules and the comprehensive improvement comparison chart is shown in Figure 20.

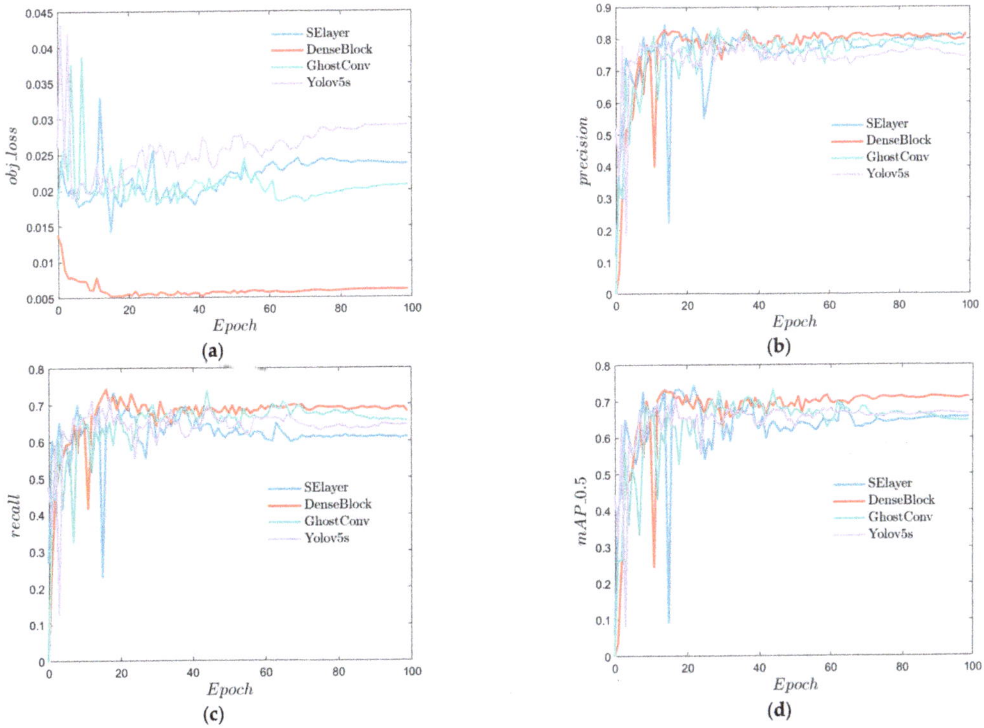

Figure 19. Comparison of results of single-module training. (**a**) Target loss. (**b**) Accuracy rate. (**c**) Recalling rate. (**d**) mAP value.

As can be seen from Figure 20a,b, the target loss value and anchor-frame loss value after the combination of DenseBlock, Ghost Convolution and SE module are the lowest. As can be obtained from Figure 20c, the accuracy of the three module combinations is also the highest. In Figure 20d, although the recall rate after the combination of DenseBlock, Ghost convolution, and SE module is not the highest. It has the smallest fluctuation after 40 epochs and is more stable. As can be seen from Figure 20e, although the mAP value is not significantly improved when using the Ghost convolution and SE module alone, the combined effect is obvious. There is a mutual inhibition effect between the DenseBlock module and the SE module, resulting in no obvious difference between the superimposed effect of the two and the original algorithm. From the analysis of the module principle, SE is a hybrid single-layer, multi-channel information feature used to improve the detection ability. At the same time, the use of the DenseBlock module with multiple feature layers in series makes the feature complexity increase instead of decrease, reducing the detection accuracy. Compared with other improvements, the comprehensive improvement in detection ability has improved the detection stability, while maintaining the lowest target loss value and the best detection effect. However, in some cases where the model detection speed is required to be high, or the size and computing power of the model are limited by the installed equipment, using the Ghost + SE improvement module with similar comprehensive improvement effect may be an option.

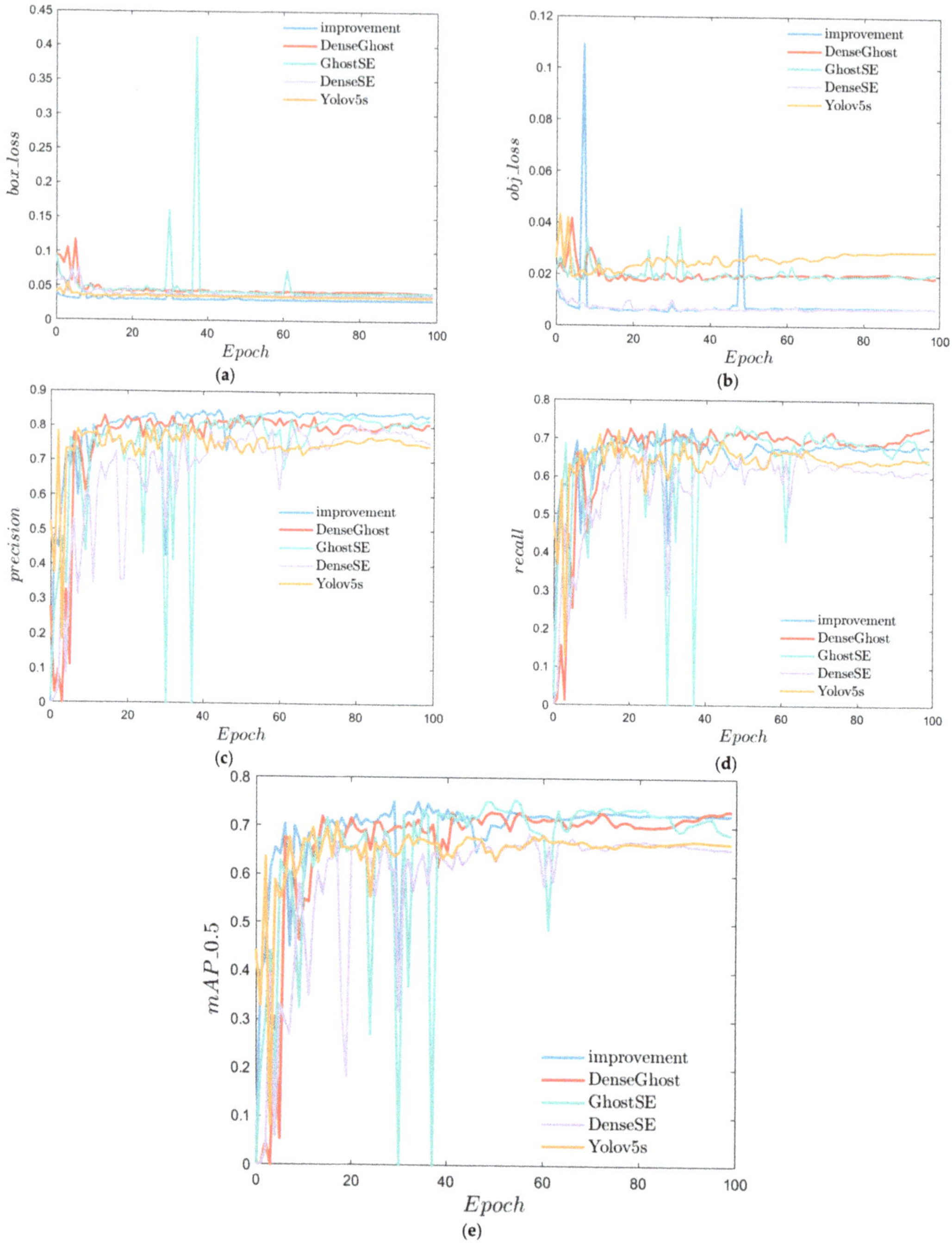

Figure 20. Comprehensive improvement comparison chart. (**a**) Target loss. (**b**) Anchor-frame loss; (**c**) Accuracy rate. (**d**) Recalling rate. (**e**) mAP value.

5.1.2. Testing Results

The results of the improved network for infrared vehicle target detection are shown in Figures 21–25.

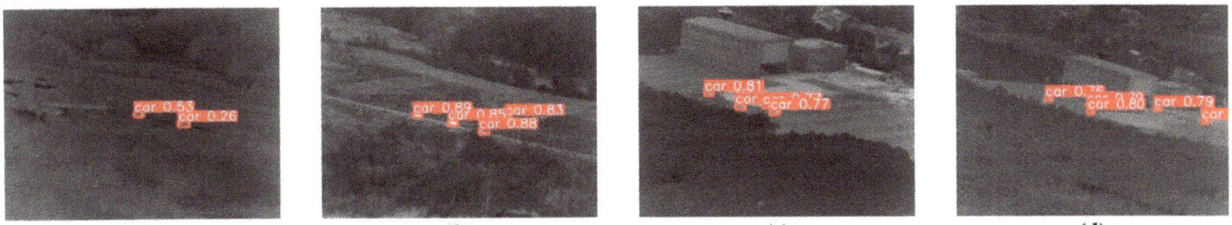

(a) (b) (c) (d)

Figure 21. YOLOv5 detection map. (**a**) Scene 1. (**b**) Scene 2. (**c**) Scene 3. (**d**) Scene 4.

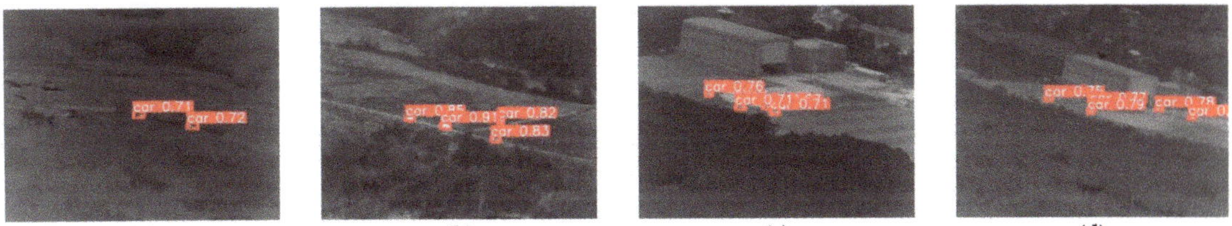

(a) (b) (c) (d)

Figure 22. Dense + Ghost detection diagram. (**a**) Scene 1. (**b**) Scene 2. (**c**) Scene 3. (**d**) Scene 4.

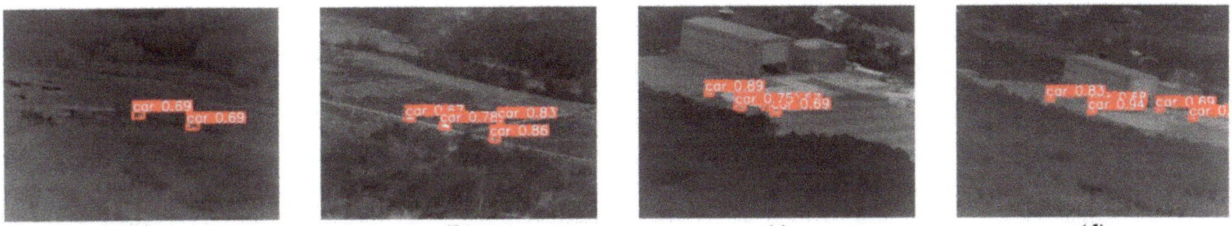

(a) (b) (c) (d)

Figure 23. Dense + SE detection diagram. (**a**) Scene 1. (**b**) Scene 2. (**c**) Scene 3. (**d**) Scene 4.

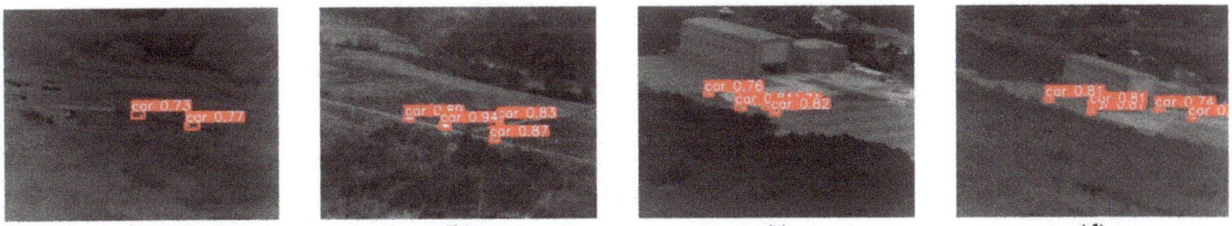

(a) (b) (c) (d)

Figure 24. Dense + Ghost detection diagram. (**a**) Scene 1. (**b**) Scene 2. (**c**) Scene 3. (**d**) Scene 4.

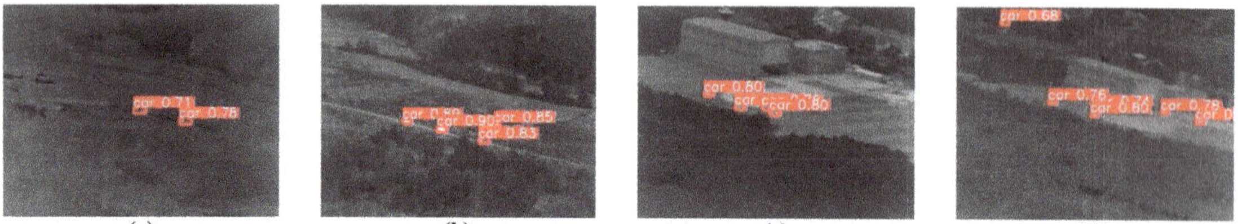

(a) (b) (c) (d)

Figure 25. Dense + Ghost + SE detection diagram. (**a**) Scene 1. (**b**) Scene 2. (**c**) Scene 3. (**d**) Scene 4.

It can be seen from Figures 21–25 that for the two small targets in scene 1, the detection accuracy of Dense + Ghost is improved by 18% and 46%, respectively, compared with the original YOLOv5. Dense + SE is improved by 16% and 43%, respectively, and Dense + Ghost is respectively improved by 18% and 46%. Dense + Ghost is improved by 20% and 51%, and Dense + Ghost + SE is improved by 18% and 52%, respectively. In the objectives of scene 2 and scene 3, the combination of the two modules is improved compared to the original YOLOv5, and the detection effect of the Dense + Ghost + SE combination is not much different from that of the two combinations. At the same time, in scene 4, the Dense + Ghost + SE modules detect the target vehicle that is not detected by other modules. In general, the Dense + Ghost + SE modules combination has better detection performance for small targets, and has a higher probability to detect targets that could not be found in the previous network due to low accuracy.

6. Conclusions

The article analyzes the characteristics of infrared vehicle images, starting from the four improvement modules of DenseBlock, Ghost Convolution, SE Module, and EIOU. The original YOLOv5 network is improved, and experiments are carried out on the effect of each module. The advantages and disadvantages of each module are analyzed, and the two combinations are compared and analyzed, and the following conclusions are drawn:

1. When the module is used alone, the accuracy of DenseBlock and EIOU modules are significantly improved, and the Ghost convolution and SE modules are not significantly improved, which is almost the same as the original network, or even lower.
2. When the module is used in combination, in addition to the combination of DenseBlock module and SE module, the other combinations have obvious improvement effects. When using three modules at the same time, the target loss value is the lowest, the accuracy rate is the highest, and the mAP value is the most stable.
3. For a small target with occlusion, whether it is the original YOLOv5 or the two–two combination module, it has not been detected, and the phenomenon of missed detection has occurred. When using three modules at the same time, the occlusion targets can be effectively detected, and the rate of missed detection can be reduced.
4. When using the improved algorithm in this paper, the insertion-extraction module can be adjusted according to different task requirements. For example, the DenseBlock module can be added to the detection target requiring higher stability. If a higher detection probability is required, the SE module can be added to the neck layer of the improved network. If higher detection speed is required, DenseBlock or SE module can be removed.

Combined with the experimental results and conclusions, the next steps are clarified:

1. Although the missed target is detected, the confidence is not high, and the network needs to be further optimized.
2. In the actual scene, the infrared vehicle target is not only interfered by the background of vegetation, buildings, etc., but also by smoke and electromagnetic interference, resulting in the degradation of the image quality. How to extract the vehicle target in the complex interference environment is a challenge for future work.

Author Contributions: Conceptualization, Y.F. and Q.Q.; methodology, Y.F.; software, Q.Q.; validation, S.H.; Y.L. and J.X.; formal analysis, Y.F.; resources, F.C.; data curation, Q.Q.; writing-original draft preparation, S.H.; writing-review and editing, M.Q.; supervision, M.Q.; funding acquisition, Y.F. All authors have read and agreed to the published version of the manuscript.

Funding: This research was funded by Key Basic Research Projects of the Basic Strengthening Program, grant number 2020-JCJQ-ZD-071.

Data Availability Statement: Not applicable.

Conflicts of Interest: The authors declare no conflict of interest.

References

1. Kim, J.; Hong, S.; Baek, J.; Lee, H. Autonomous vehicle detection system using visible and infrared camera. In Proceedings of the 2012 12th International Conference on Control, Automation and Systems, Jeju, Korea, 17–21 October 2012; pp. 630–634.
2. Chen, D.; Jin, G.; Lu, L.; Tan, L.; Wei, W. Infrared Image Vehicle Detection Based on Haar-like Feature. In Proceedings of the 2018 IEEE 3rd Advanced Information Technology, Electronic and Automation Control Conference (IAEAC), Chongqing, China, 12–14 October 2018; pp. 662–667.
3. Liu, Y.; Su, H.; Zeng, C.; Li, X. A Robust Thermal Infrared Vehicle and Pedestrian Detection Method in Complex Scenes. *Sensors* **2021**, *21*, 1240. [CrossRef] [PubMed]
4. Iwasaki, Y.; Kawata, S.; Nakamiya, T. Vehicle detection even in poor visibility conditions using infrared thermal images and its application to road traffic flow monitoring. In *Emerging Trends in Computing, Informatics, Systems Sciences, and Engineering*; Springer: New York, NY, USA, 2013; pp. 997–1009.
5. Tang, T.; Zhou, S.; Deng, Z.; Zou, H.; Lei, L. Vehicle detection in aerial images based on region convolutional neural networks and hard negative example mining. *Sensors* **2017**, *17*, 336. [CrossRef] [PubMed]
6. Liu, X.; Yang, T.; Li, J. Real-time ground vehicle detection in aerial infrared imagery based on convolutional neural network. *Electronics* **2018**, *7*, 78. [CrossRef]
7. Ouyang, L.; Wang, H. Vehicle target detection in complex scenes based on YOLOv3 algorithm. *IOP Conf. Ser. Mater. Sci. Eng.* **2019**, *569*, 052018. [CrossRef]
8. Li, L.; Yuan, J.; Liu, H.; Cao, L.; Chen, J.; Zhang, Z. Incremental Learning of Infrared Vehicle Detection Method Based on SSD. In Proceedings of the 2020 IEEE 20th International Conference on Communication Technology (ICCT), Nanning, China, 28–31 October 2020; pp. 1423–1426.
9. Mahmood, M.T.; Ahmed, S.R.A.; Ahmed, M.R.A. Detection of vehicle with Infrared images in Road Traffic using YOLO computational mechanism. *IOP Conf. Ser. Mater. Sci. Eng.* **2020**, *928*, 022027. [CrossRef]
10. Zhu, Z.; Liu, Q.; Chen, H.; Zhang, G.; Wang, F.; Huo, J. Infrared Small Vehicle Detection Based on Parallel Fusion Network. *Acta Photonica Sin.* **2022**, *51*, 0210001.
11. Redmon, J.; Divvala, S.; Girshick, R.; Farhadi, A. You Only Look Once: Unified, Real-Time Object Detection. In Proceedings of the 2016 IEEE Conference on Computer Vision and Pattern Recognition (CVPR), Las Vegas, NV, USA, 27–30 June 2016; pp. 779–788. [CrossRef]
12. Zhang, X.; Zhu, X. Vehicle Detection in the aerial infrared images via an improved YOLOv3 network. In Proceedings of the 2019 IEEE 4th International Conference on Signal and Image Processing (ICSIP), Wuxi, China, 19–21 July 2019; pp. 372–376.
13. Li, Z.; Zhou, F. FSSD: Feature fusion single shot multibox detector. *arXiv* **2017**, arXiv:1712.00960.
14. Redmon, J.; Farhadi, A. Yolov3: An incremental improvement. *arXiv* **2018**, arXiv:1804.02767.
15. Han, J.; Liao, Y.; Zhang, J.; Wang, S.; Li, S. Target Fusion Detection of LiDAR and Camera Based on the Improved YOLO Algorithm. *Mathematics* **2018**, *6*, 213. [CrossRef]
16. Deng, Z.; Yang, R.; Lan, R.; Liu, Z.; Luo, X. SE-IYOLOV3: An Accurate Small Scale Face Detector for Outdoor Security. *Mathematics* **2020**, *8*, 93. [CrossRef]
17. Zhang, X.; Zhu, X. Moving vehicle detection in aerial infrared image sequences via fast image registration and improved YOLOv3 network. *Int. J. Remote Sens.* **2020**, *41*, 4312–4335. [CrossRef]
18. Wang, Z.; Wu, L.; Li, T.; Shi, P. A Smoke Detection Model Based on Improved YOLOv5. *Mathematics* **2022**, *10*, 1190. [CrossRef]
19. Kasper-Eulaers, M.; Hahn, N.; Berger, S.; Sebulonsen, T.; Myrland, Ø.; Kummervold, P. Short Communication: Detecting Heavy Goods Vehicles in Rest Areas in Winter Conditions Using YOLOv5. *Algorithms* **2021**, *14*, 114. [CrossRef]
20. Wu, W.; Liu, H.; Li, L.; Long, Y.; Wang, X.; Wang, Z. Application of local fully Convolutional Neural Network combined with YOLO v5 algorithm in small target detection of remote sensing image. *PLoS ONE* **2021**, *16*, e0259283. [CrossRef] [PubMed]
21. The Third "Aerospace Cup" National Innovation and Creativity Competition Preliminary Round, Proposition 2, Track 2, Optical Target Recognition, Preliminary Data Set. Available online: https://www.atrdata.cn/#/customer/match/2cdfe76d-de6c-48f1-abf9-6e8b7ace1ab8/bd3aac0b-4742-438d-abca-b9a84ca76cb3?questionType=model (accessed on 15 March 2022).
22. Jiang, B.; Ma, X.; Lu, Y.; Li, Y.; Feng, L.; Shi, Z. Ship detection in spaceborne infrared images based on Convolutional Neural Networks and synthetic targets. *Infrared Phys. Technol.* **2019**, *97*, 229–234. [CrossRef]

23. Shi, M.; Wang, H. Infrared Dim and Small Target Detection Based on Denoising Autoencoder Network. *Mob. Netw. Appl.* **2020**, *25*, 1469–1483. [CrossRef]
24. Alrasheedi, A.F.; Alnowibet, K.A.; Saxena, A.; Sallam, K.M.; Mohamed, A.W. Chaos Embed Marine Predator (CMPA) Algorithm for Feature Selection. *Mathematics* **2022**, *10*, 1411. [CrossRef]
25. Sharma, A.K.; Saxena, A. A demand side management control strategy using Whale optimization algorithm. *SN Appl. Sci.* **2019**, *1*, 870. [CrossRef]

electronics

Article

Monitoring Tomato Leaf Disease through Convolutional Neural Networks

Antonio Guerrero-Ibañez [1] and Angelica Reyes-Muñoz [2,*]

[1] Faculty of Telematics, University of Colima, Colima 28040, Mexico
[2] Computer Architecture Department, Polytechnic University of Catalonia, 08860 Barcelona, Spain
* Correspondence: angelica.reyes@upc.edu

Abstract: Agriculture plays an essential role in Mexico's economy. The agricultural sector has a 2.5% share of Mexico's gross domestic product. Specifically, tomatoes have become the country's most exported agricultural product. That is why there is an increasing need to improve crop yields. One of the elements that can considerably affect crop productivity is diseases caused by agents such as bacteria, fungi, and viruses. However, the process of disease identification can be costly and, in many cases, time-consuming. Deep learning techniques have begun to be applied in the process of plant disease identification with promising results. In this paper, we propose a model based on convolutional neural networks to identify and classify tomato leaf diseases using a public dataset and complementing it with other photographs taken in the fields of the country. To avoid overfitting, generative adversarial networks were used to generate samples with the same characteristics as the training data. The results show that the proposed model achieves a high performance in the process of detection and classification of diseases in tomato leaves: the accuracy achieved is greater than 99% in both the training dataset and the test dataset.

Keywords: convolutional neural networks; deep learning; disease classification; generative adversarial network; tomato leaf

Citation: Guerrero-Ibañez, A.; Reyes-Muñoz, A. Monitoring Tomato Leaf Disease through Convolutional Neural Networks. *Electronics* **2023**, *12*, 229. https://doi.org/10.3390/electronics12010229

Academic Editors: Taiyong Li, Wu Deng and Jiang Wu

Received: 8 November 2022
Revised: 20 December 2022
Accepted: 20 December 2022
Published: 2 January 2023

1. Introduction

Tomato is one of the most common vegetables grown worldwide and is a high source of income for farmers. The 2020 statistical report of the Food and Agriculture Organization Corporate Statistical Database (FAOSTAT) indicates that world tomato production was 186.821 million tons [1]. In Mexico, the tomato is one of the main crops within the national production, being considered as a basic ingredient both in Mexican cuisine and in general in the cuisine of various parts of the world. According to a report published by Our World in Data in 2020, Mexico is among the top ten countries with the highest production of tomatoes, with a production of 4.1 million tons per year [2]. The Mexican Ministry of Agriculture, Livestock, Rural Development, Fishing and Food (MALRDFF) through the AgriFood and Fisheries Information Service presented the report on Mexico's AgriFood Trade Balance, indicating that tomato is the second most exported agricultural product, with avocado taking the first place. Besides this, tomato production in Mexico has an annual variation of 5.3% from 2011 to 2020 [3]. However, production is affected by different circumstances. The Food and Agriculture Organization (FAO) estimates that crop diseases are responsible for losses ranging from 20 to 40% of total production [4]. Various diseases of the tomato plant can affect the product in terms of quantity and quality, thus decreasing productivity. Diseases can be classified into two main groups [5]. The first group of diseases is related to infectious microorganisms including viruses, bacteria, and fungi. These types of diseases can spread rapidly from plant to plant in the field when environmental conditions are favorable. The second group of diseases is caused by non-infectious chemical or physical factors including adverse environmental factors, physiological or nutritional disorders and herbicide injury. While it is true that non-infectious diseases cannot spread from plant to plant, diseases can spread if the entire plantation is exposed to the same adverse factor [6].

Some special conditions can cause plant diseases. Specifically, there is a conceptual model known as the disease triangle which describes the relationship between three essential factors: the environment, the host and the infectious agent. If any of these three factors is not present, then the triangle is incomplete, and therefore the disease does not occur. There are abiotic factors such as air flow, temperature, humidity, pH, and watering that can significantly affect the plant. The infectious agent is a kind of organism that attacks the plant such as fungi, bacteria, virus, among others. The host is the plant which is affected by a pathogen. When these factors occur simultaneously, disease is produced [7]. Generally, diseases are manifested by symptoms that affect the plant from the bottom up and many of these diseases have a rapid spread process after infection.

Figure 1 shows some of the most common diseases affecting tomato leaves including mosaic virus, yellow leaf curl virus, target spot, two-spotted spider mite, septoria leaf spot, leaf mold, late blight, early blight, and bacterial spot.

Figure 1. Representative images of the most common diseases affecting tomato leaves: (**a**) mosaic virus, (**b**) yellow leaf curl virus, (**c**) target spot, (**d**) two-spotted spider mite, (**e**) septoria leaf spot, (**f**) leaf mold, (**g**) late blight, (**h**) early blight and (**i**) bacterial spot.

Crops require continuous monitoring for early disease detection and thus the ability to apply proper mechanisms to prevent its spread and the loss of production [8].

The traditional methods used for the detection of plant diseases focus on the visual estimation of the disease by experts; studies of morphological characteristics to identify the pathogens; and molecular, serological, and microbiological diagnostic techniques [9]. The visual estimation method for plant disease identification is based on the analysis of characteristic disease symptoms (such as lesions, blight, galls and tumors) or visible signs of a pathogen (uredinospores of Pucciniales, mycelium or conidia of Erysiphales). Visual estimation is very subjective, as it is performed according to the experience of experts, so the accuracy of identification cannot be measured, and it is affected by temporal variation [10]. Microscopic methods focus on pathogen morphology for disease detection. However, these methods are expensive, time-consuming in the detection process and lead to low detection efficiency and poor reliability. In addition, farmers do not have the necessary knowledge to carry out the detection process, and agricultural experts cannot be in the field all the time to carry out proper monitoring.

New innovative techniques need to address the challenges and trends demanded by the new vision of agricultural production that requires higher accuracy levels and near real-time detection.

In recent years, different technologies such as image processing [11,12], pattern recognition [13,14] and computer vision [15,16] have rapidly developed and been applied to agriculture, specifically on automation of disease and pest detection processes. Traditional computer vision models face serious problems due to their complex preprocessing and design of image features that are time-consuming and labor-intensive. In addition, their efficiency is conditioned by the accuracy in the design of feature extraction mechanisms and the learning algorithm [17].

Recently, the problem of plant disease detection has been addressed by deep learning technology, a subset of machine learning that is gaining momentum in disease identification due to the increase in computing power, storage capabilities and the availability of large data sets. Within the deep learning environment, one of the most widely used techniques for image classification, object detection and semantic segmentation are Convolutional Neural Networks (CNN) [18,19]. CNNs are useful for locating patterns in images, objects, and scenes by learning from the data obtained from the image for classification, eliminating the need for manual extraction of the features being searched for. CNN consist of several layers (such as convolutional, pooling and fully connected layers) to learn features from different training data [20,21]. This paper presents an architecture based on CNNs and data augmentation for early disease identification and classification in tomato leaves. The objective of the work is to implement a robust architecture that allows examining the relationship between the images of tomato leaves and the detection of a possible disease and performing a classification task to predict the type of disease with high accuracy levels.

The remainder of this article is organized as follows. Section 2 presents a brief discussion of previous research that has been conducted addressing the problem of disease identification in tomato. Section 3 explains in detail the CNN architecture proposed for tomato leaf disease identification and classification. A discussion of the experimental results obtained is presented in Section 4. Finally, Section 5 closes the paper with conclusions and future direction of the research work.

2. Related works

Plants disease detection has been studied for a long time. With respect to disease identification in tomatoes, much effort has been made using different tools such as classifiers focused on color [22,23], texture [24,25] or shape of tomato leaves [26]. Early efforts focused on support vector machines [27–30], decision trees [31,32] or neural network-based [33–35] classifiers. Visual spectrum images obtained from commercial cameras have been used for disease detection in tomato. The images obtained were processed under laboratory conditions, applying mechanisms such as stepwise multiple linear regression [36] and clustering process [37]. It is worth mentioning that the sample population for both works ranged between 22 and 47 for the first method and included 180 samples for the second experiment.

CNNs have rapidly become one of the preferred methods for disease detection in plants [38–40]. Some works have focused their efforts on identifying features with better quality through the process of eliminating the limitations generated by lighting conditions and uniformity in complex environment situations [41,42]. Some authors have developed real-time models to accelerate the process of disease detection in plants [43,44]. Other authors have created models that contribute to the early detection of plant diseases [45,46]. In [47], the authors make use of images of tomato leaves to discover different types of diseases. The authors apply artificial intelligence algorithms and CNN to perform a classification model to detect five types of diseases obtaining an accuracy of 96.55%. Some works evaluated the performance of deep neural network models applied to tomato leaf disease detection such as in [48], where the authors evaluated the LeNet, VGG16, ResNet and Xception models for the classification of nine types of diseases, determining that the VGG16 model is the one that obtained the best performance with an accuracy of 99.25%. In [49], the authors applied the AlexNet, GoogleNet and LeNet models to solve the same problem, obtaining accuracy results ranging between 94% and 95%. Agarwal et al. [50] developed their own CNN model based on the structure of VGG16 and compared it with different machine learning models (including random forest and decision trees) and deep learning models (VGG16, Inceptionv3 and MobileNet) to perform the classification of the 10 classes, obtaining an accuracy of 98.4%.

Several researches have focused on combining deep learning algorithms with machine learning algorithms to address and improve the accuracy of the classification problem, for example, MobileNetv2 and NASNetMobile that were used to extract features from leaves and those features were combined with classification networks such as random forest, support vector machines and multinomial logistic regression [51]. Other works have applied algorithms such as YOLOv3 [45], Faster R-CNN [52,53] and Mask R-CNN [54,55] to detect disease states in plants.

Some efforts have been made to reduce the computational cost and model size such as Gabor filters [56] and K-nearest neighbors (KNN) [57] that have been implemented to reduce computational costs and overhead generated by deep learning. In [58], the authors reduced the computational cost by using the SqueezeNet architecture and minimizing the number of 3 × 3 filters.

3. Materials and Methods

In this section, we explain in detail the proposed architecture for the detection of diseases in tomato leaves. In general, the proposed architecture takes tomato leaves as input images and the output is a set of labels indicating (1) the type of disease in the image being analyzed or whether the leaf is healthy, (2) the label showing the predicted value obtained by our model, and (3) the prediction percentage.

Figure 2 shows the complete process of the algorithm that we applied for the process of detection and classification of diseases in tomato leaves. The global algorithm is composed of four stages: (a) creation of the experimental dataset, (b) creation of the proposed architecture, (c) distribution of the dataset, and (d) process of training and evaluation of the model.

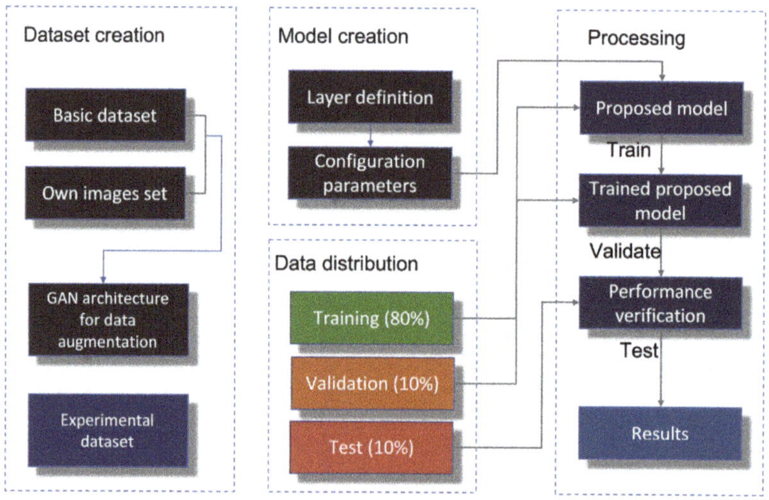

Figure 2. Representation of the proposed architecture for tomato disease detection.

3.1. Dataset Creation

As a first step, we proceeded to create the experimental dataset that would be used for training, validation, and performance evaluation of the proposed architecture. The public dataset available in [59] consists of 11,000 images that were the basis of our dataset. The images represent 10 categories, including nine types of diseases (tomato mosaic virus, target spot, bacterial spot, tomato yellow leaf curl virus, late blight, leaf mold, early blight, two-spotted spider mites, septoria leaf spot) and one category of healthy leaves. The dataset was complemented with 2500 images obtained from different crop fields in Mexico. The total number of images that made up our dataset was 13,500.

One of the problems that datasets face with deep neural network models is that when training the model, overfitting can occur, i.e., a model with high capacity may be able to "memorize" the dataset [60]. A technique known as data augmentation is used to avoid the problem of overfitting. The goal of applying data augmentation is to increase the size of the dataset, and it is widely used in all fields [61]. Commonly, data augmentation is performed by two methods. The first method, known as the traditional method, aims to obtain a new image, which contains the same semantic information but does not have the ability of generalization. These methods include translation, rotation, flip, brightness adjustment, affine transformation, Gaussian noise, etc. The main drawbacks of these methods may be their poor quality and inadequate diversity.

Another method is the use of Generative Adversarial Networks (GANs), which are an approach to generative modeling using deep learning methods, such as CNNs, that aim to generate synthetic samples with the same characteristics as the given training distribution [62]. GAN models mainly consist of two parts, namely the generator and the discriminator [63]. The generator is a model used to generate new plausible examples from the problem domain. The discriminator is a model used to classify examples as real (from the domain) or fake (generated).

To create our experimental dataset, we made use of GAN to avoid the overfitting problem. To build our GAN, we define two separate networks: the generator network and the discriminator network. The first network receives a random noise, and from that number, the network generates images. The second network, the discriminator, defines whether the image it receives as input is "real" or not.

Because the images that complemented the dataset were not balanced for each category, the GAN network generated images that contributed to balance the dataset. The dataset was increased from 13,500 to 15,000 images, distributing the generated images in the different categories to create a balanced dataset.

3.2. Model Creation

Figure 3 shows the proposed CNN architecture for disease detection in tomato. The network has 112 × 112 color images as input, which are normalized to (0, 1) values. The proposed convolutional network has four convolutional layers that use filters whose values were 16, 32, 64, and 128, respectively. These values were assigned in that order since the layers closer to the beginning of the model learn convolutional filters less effectively than the layers closer to the result. In addition, the kernel size, which represents the width and height of the 2D convolution window, was set to a value of 3 × 3. This value was the recommended value for the number of filters to be used. Finally, rectified linear unit (ReLU) was used as the activation model for each convolved node.

After applying the convolutional layer, the maximum clustering layer was applied to down-sample the acquired feature map and condense the most relevant features into patches. This process is repeated for each of the convolutional layers defined in the architecture.

The result of the last MaxPooling layer is passed to a MaxAveragePooling layer to be converted to a column vector and connected to the dense layer of 10 output nodes (which represent the 10 categories) used as softmax activation. Each node represents the probability of each category for the evaluated image. Table 1 shows the information of the layer structure of the proposed model.

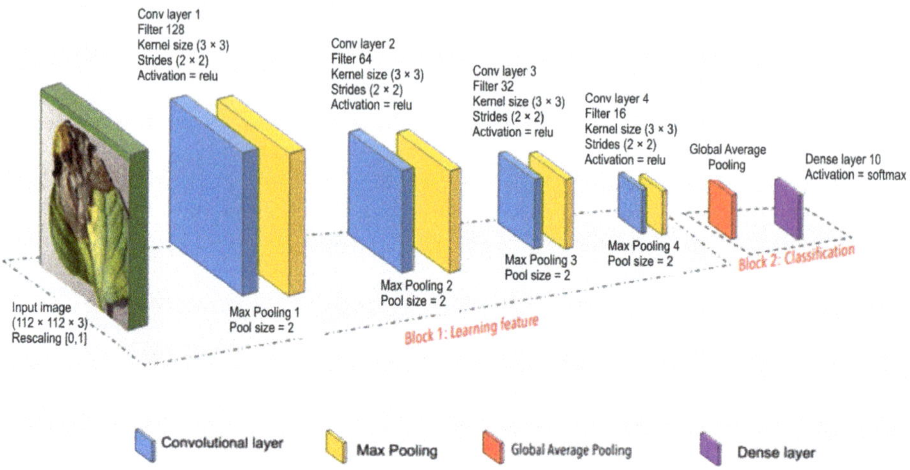

Figure 3. Representation of the proposed algorithm for tomato disease detection.

Table 1. Information on the layers structure of the proposed model.

Layers	Parameters
Conv2D	Filters: 128, kernel size: (3,3), activation: "relu", input shape: (112,112,3)
MaxPool2D	Pool size: (2,2)
Conv2D	Filters: 64, kernel size: (3,3), activation: "relu"
MaxPool2D	Pool size: (2,2)
Conv2D	Filters: 32, kernel size: (3,3), activation: "relu"
MaxPool2D	Pool size: (2,2)
Conv2D	Filters: 16, kernel size: (3,3), activation: "relu"
MaxPool2D	Pool size: (2,2)
Dropout	Rate: 0.2
GlobalAveragePooling2D	
Dense	Units: 10, activation: "softmax"

3.3. Data Distribution

One of the most common strategies to split the dataset into training and validation sets is assigning percentages, for example, 70:30 or 80:20. However, one of the problems that can arise with this strategy is that it is uncertain whether high validation accuracy indicates a good model. When performing this division, it could happen that some information is missing in the data that are not used for training, causing a bias in the results.

We apply a k-fold cross-validation method to evaluate the performance of the model. The k-folds method tries to ensure that all features of the dataset are in the training and validation phases. The k-fold cross-validation method divides the dataset into subsets as k number. Therefore, it repeats the cross-validation method k times. Common values in machine learning are k = 3, k = 5, and k = 10. We use k = 5 to provide good trade-off of low computational cost and low bias in an estimate of model performance.

3.4. Model Creation

For the training process, we use Adam as the optimization algorithm. Adam updates network weights iterative based on training data. The loss function was categorical_crossentropy, one of the most used loss functions for multi-class classification models where there are two or more output labels. The number of epochs for the training and validation process was 200. The steps_per_epoch parameter was 12,000, and for the validation the parameter it was 3000. Table 2 shows a summary of some of the parameters used for the training and validation phase.

Table 2. Training Parameters for the Proposed Model.

Parameter	Value
Optimization algorithm	Adam
Loss function	Categorical cross entropy
Batch size	32
Number of epochs	200
Steps per epoch	12,000
Validation steps	3000
Activation function for conv layer	ReLu

4. Results

In this section, we describe the scenario setup and the results obtained in the performance evaluation process of the proposed model.

4.1. Environmental Setup

Our model was developed in Google Collaboratory, a free Python development environment that runs in the cloud. Google Collaboratory is widely used for the development of machine learning and deep leaning projects. In our project, we use the following libraries: Tensorflow, an open-source library used for numerical computation and automated learning; Keras, a library used for the creation of neural networks; numpy, used for data analysis and mathematical calculations; matplotlib used for graph management and TensorBoard to visually inspect the different runs and graphs.

The model was trained with 200 epochs. We applied early stopping to monitor the performance of the model for the 200 epochs on a held-out validation set during the training to reduce overfitting and to improve the generalization of the neural network. For the evaluation of the model, the validation accuracy scheme allowed early stopping to be activated during the process.

Since our problem is a multi-class classification model, we use the Adam algorithm as the optimizing algorithm. In addition, the cross-entropy categorical loss function was used due to the nature of the multi-class classification environment. During the training process, we implemented checkpoints to save the model with the best validation accuracy, and thus be able to load it later to continue training from the saved state if necessary.

4.2. Evaluation Metrics

To analyze the performance of our model, the following four metrics were considered. The first metric to evaluate was accuracy, which represents the behavior of the model across all classes. Accuracy is calculated as the ratio between the number of correct predictions to the total number of predictions (Equation (1)).

Precision was our second metric, which represents the accuracy of the model in classifying a sample as positive. This parameter is calculated as the ratio of the number of positive samples correctly classified to the total number of samples classified as positive (Equation (2)).

We also analyzed the recall parameter, which measures the ability of the model to detect positive samples and is calculated as the ratio of the number of positive samples correctly classified to the total number of positive samples (Equation (3)).

Finally, we analyzed the *F*1 score parameter. This metric combines the precision and recall measures to obtain a single value. This value is calculated by taking the harmonic mean between precision and recall (Equation (4)).

The following equations were used to calculate accuracy, precision, recall and $F1$ score:

$$Accuracy = \frac{TP + TN}{TP + TN + FP + FN} \quad (1)$$

$$Precision = \frac{TP}{TP + FP} \quad (2)$$

$$Recall = \frac{TP}{TP + FN} \quad (3)$$

$$F1\ Score = 2 \times \frac{(precision \times recall)}{(precision + recall)}. \quad (4)$$

4.3. Results and Discussion

In this section, we analyze the results obtained in the evaluation of the performance of the proposed CNN model in tomato crops. We compare our results with some of the proposed models published in the literature.

4.3.1. Validation of the Proposed Model

The validation of the model was analyzed by applying the k-fold cross-validation procedure to estimate the performance of our algorithm on the tomato images dataset. We define a k-value of 5 to split the dataset. The Scikit-Learn machine learning library was used to implement the k-fold method returning a list of scores calculated for each of our five folds.

Figure 4 shows the results obtained by applying the k-folds method to evaluate the performance of the model. The results show a stability of the model, as it is observed in the different metrics analyzed. There is a very similar behavior with the five folds both in the training phase and in the validation phase, demonstrating that there is no overfitting in the proposed model.

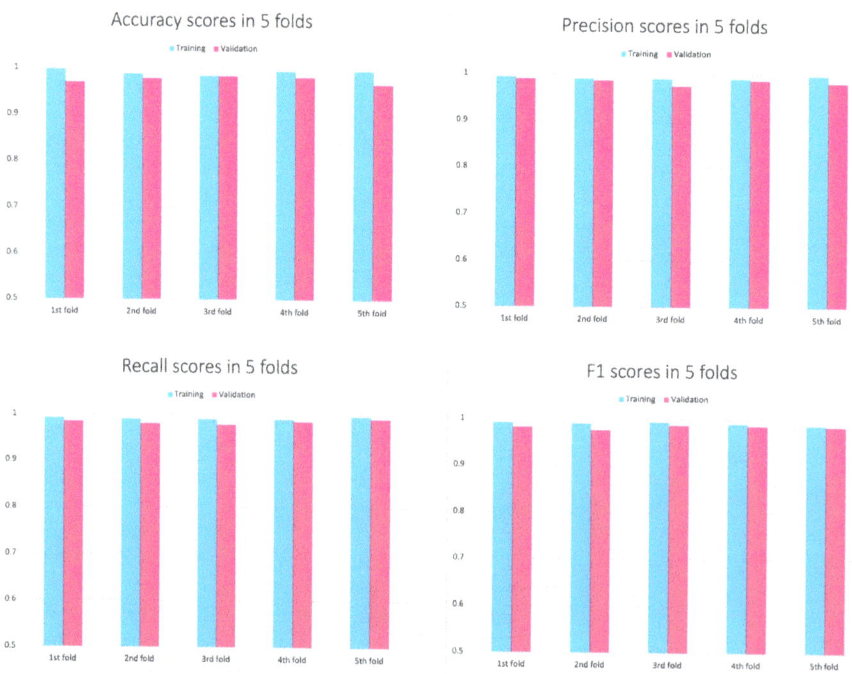

Figure 4. Results obtained for k-folds.

Figure 5 demonstrates the performance of our model in the training and validation stages for identification and classification of tomato leaf diseases. The results achieved a training accuracy of 99.99%. The time used for the training process was 6234 s in the MGPU (Multiple-Graphics Processing Unit) environment. The proposed model achieved a validation accuracy of 99.64% in leaf disease classification.

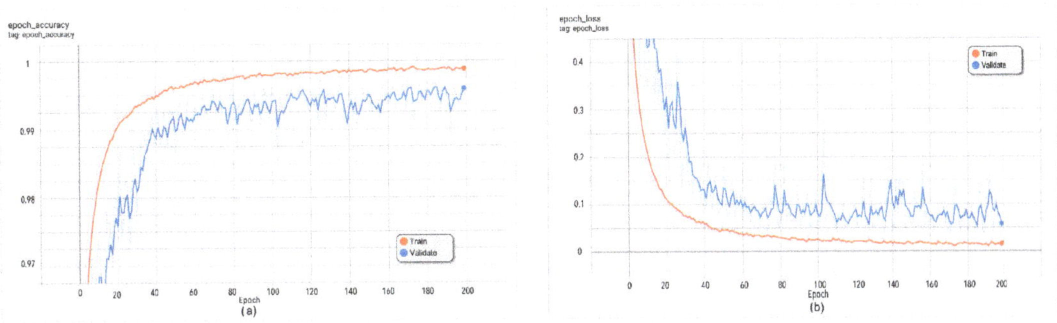

Figure 5. Results obtained of the proposed model during the training and validation phases: (**a**) accuracy and (**b**) loss.

Figure 6 shows the confusion matrix obtained in the evaluation of the proposed model. The confusion matrix shows the true positive (TP), true negative (TN), false positive (FP) and false negative (FN) values obtained for each class evaluated [64].

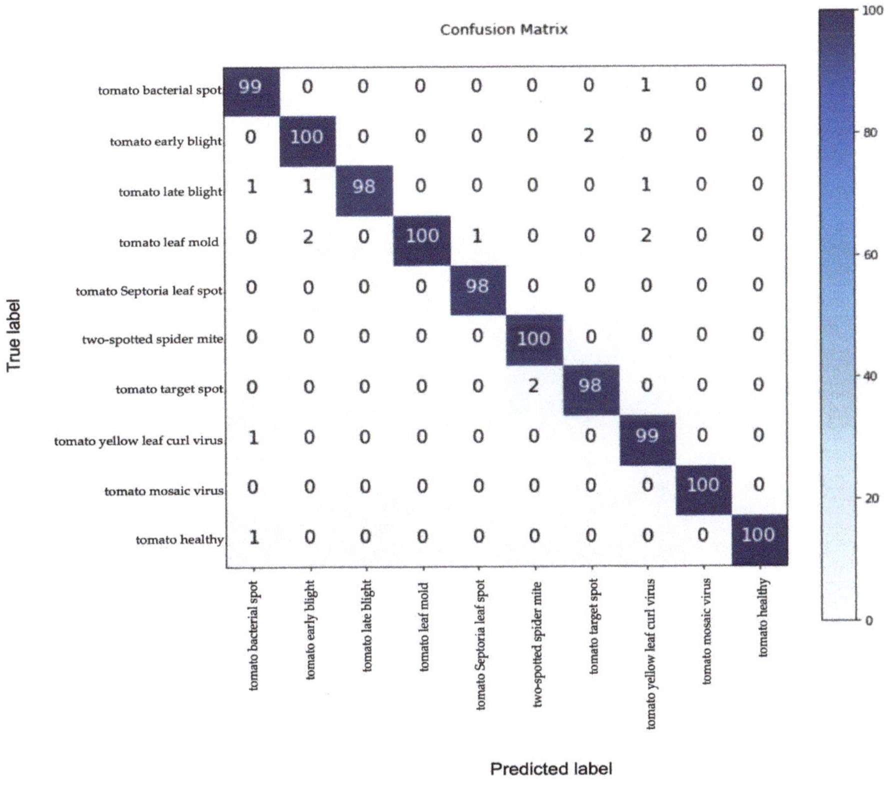

Figure 6. Confusion matrix of the proposed model.

According to the results, which are reflected in the confusion matrix, we can see that the proposed model was able to predict half of the classes that were evaluated using the test dataset with a 100% accuracy. For the rest of the classes, the model reached an accuracy level of at least 98%, thus obtaining better values than those of several of the works proposed in the literature.

Table 3 presents the results obtained in the classification performance of the proposed model on each of the classes defined within the experimental dataset. According to the data reflected in the table, the value obtained for the recall metric is high for each category defined in the dataset; this allows inferring the performance of the proposed model, which is able to correctly classify the corresponding disease with accuracy higher than 98%.

Table 3. Class-wise Performance of the Proposed Model.

Class	Precision	Recall	F1 Score	Support
Tomato bacterial spot	0.99	0.990	0.99	100
Tomato early blight	1	1	0.98	100
Tomato late blight	0.97	0.98	0.97	100
Tomato leaf mold	1	1	0.99	100
Tomato Septoria leaf spot	0.99	0.98	0.97	100
Tomato Two-spotted spider mite	1	1	0.98	100
Tomato target spot	0.99	0.98	0.99	100
Tomato yellow leaf curl virus	0.98	0.98	0.98	100
Tomato mosaic virus	1	1	0.99	100
Tomato healthy	1	1	0.99	100

The architecture and weights obtained from the proposed model were saved as a hierarchical data file to be used during the prediction process. The prediction process uses a dataset with a total of 1350 images. The matplotlib library was used to visualize the prediction result. For each prediction, the image, the true result, and the result of the prediction made with the proposed model were displayed, together with the percentage of accuracy. Figure 7 shows some results of the predictions made by the model.

Figure 7. Sample predicted images using the proposed model.

4.3.2. Comparison of the Model

Finally, our model was compared with other techniques proposed in the literature (Widiyanto et al. [65], Afif Al Mamun et al. [66], Kaur et al. [67], AlexNet [68]; Inception-v3-

Net, ResNet-50 and VGG16Net [69]). Figure 8 presents the results of the comparison and shows that for the accuracy and recall metrics, the proposed model obtained the best results, reaching an accuracy of 99.9%. With respect to the precision metric, the proposed algorithm had a result only lower than the VGG16Net technique, but with a result of 0.99. For the F1 metric, the proposed model had a similar result to that of the VGG16Net technique.

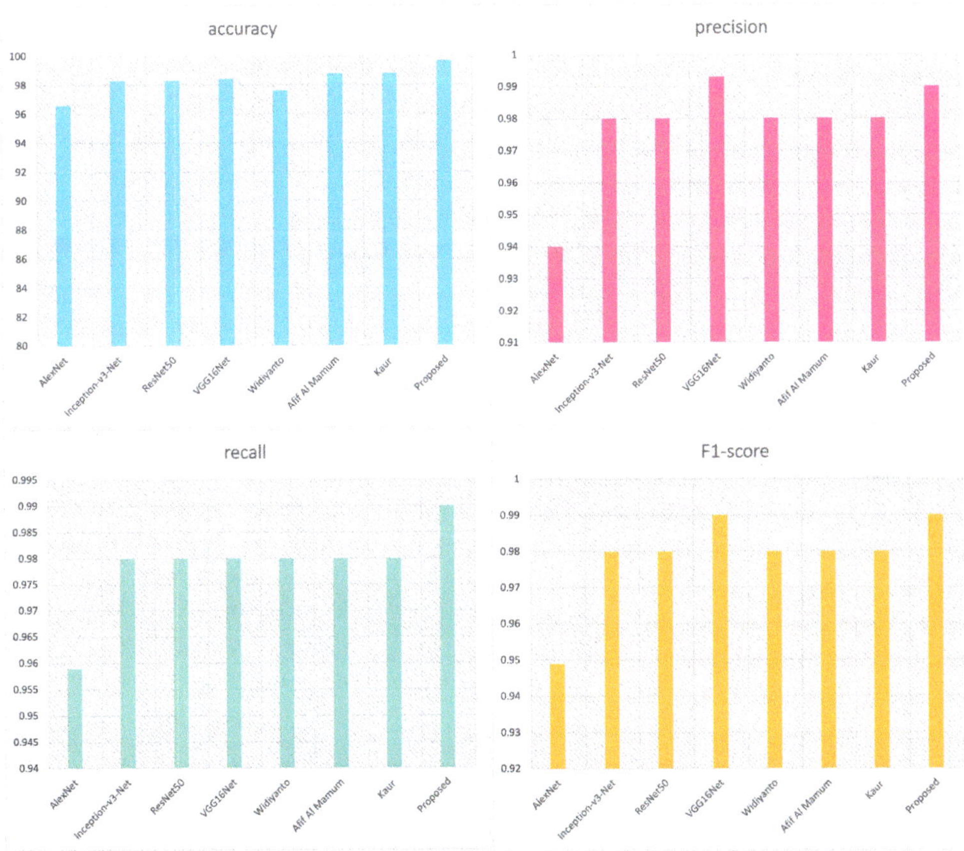

Figure 8. Performance comparison of proposed model and existing models.

In addition, a comparison was made of the complexity of the proposed model and some of the other models included in the comparison (data were not obtained for some of the models used in the comparison). Specifically, the number of trainable parameters and the size of the model were analyzed. The data obtained are shown in Table 4. Finally, Table 5 shows a summary of the performance of the models using the metrics accuracy, precision, recall and F1 score.

Table 4. Complexity comparison of proposed model and existing models.

	ResNet	VGG16Net	Inception-v3-Net	AlexNet	Proposed
Trainable parameters (Millions)	26.7	39.4	24.9	44.7	5.6
Model size (MB)	98	128	92	133	36

Table 5. Class-wise Performance of the Proposed Model.

Reference	Accuracy	Precision	Recall	F1 Score
Widiyanto, et al. (2019)	97.6	0.98	0.98	0.98
Afif Al Mamum et al. (2020)	98.77	0.98	0.98	0.98
Kaur et al. (2019)	98.8	0.98	0.98	0.98
Proposed model	99.64	0.99	0.99	0.99

5. Conclusions

In this research, we propose an architecture based on CNNs to identify and classify nine different types of tomato leaf diseases. The complexity in detecting the type of disease lies in the fact that the leaves deteriorate in a similar way in most of the tomato diseases. It means that it is necessary to develop a deep image analysis to judge the types of tomato leave diseases with a proper accuracy level.

The CNN that we design is a high-performance deep learning network that allows us to have a complex image processing and feature extraction through four modules: the module dataset creation that makes an experimental dataset using public datasets and photographs taken in the fields of the country; model creation that is in charge of parameters configuration and layers definition; data distribution to train, validate and test data; and processing for the optimization and performance verification.

We evaluate the performance of our model via accuracy, precision, recall and the F1-score metrics. The results showed a training accuracy of 99.99% and a validation accuracy of 99.64% in the leaf disease classification. The model correctly classifies the corresponding disease with a precision of 0.99 and an F1 score of 0.99. The recall metric has a value of 0.99 on the classification of the nine tomato diseases that we analyzed.

The resulting confusion matrix describes that our classification model was able to predict half of the classes that were evaluated using the test dataset with a 100% accuracy. For the rest of the classes, the model reached an accuracy level of 98%, thus obtaining better values than those of several of the works proposed in the literature.

Author Contributions: Conceptualization, A.G.-I.; Methodology, A.G.-I. and A.R.-M.; Software A.G.-I.; Validation, A.G.-I. and A.R.-M.; Formal analysis, A.G.-I.; Resources, A.R.-M.; Data curation, A.G.-I.; Writing—review & editing, A.G.-I. and A.R.-M. All authors have read and agreed to the published version of the manuscript.

Funding: This work was partially funded by the State Research Agency of Spain under grant number PID2020-116377RB-C21.

Data Availability Statement: The datasets generated during the current study are available from authors on reasonable request.

Conflicts of Interest: The authors declare no conflict of interest.

References

1. Food and Agriculture Organization of the United Nations. "FAOSTAT" Crops and Livestock Products. Available online: https://www.fao.org/faostat/en/#data/QCL (accessed on 19 October 2022).
2. Los Productos Agropecuarios Más Exportados. 1 July 2022. Available online: https://mundi.io/exportacion/exportacion-productos-agropecuarios-mexico/ (accessed on 2 November 2022).
3. Ritchie, H.; Rosado, P.; Roser, M. Agricultural Production—Crop Production Across the World. 2020. Available online: https://ourworldindata.org/agricultural-production (accessed on 2 December 2022).
4. Food and Agriculture Organization of the United Nations. FAO—News Article: Climate Change Fans Spread of Pests and Threatens Plants and Crops—New FAO Study. Available online: https://www.fao.org/news/story/en/item/1402920/icode/ (accessed on 19 October 2022).
5. Gobalakrishnan, N.; Pradeep, K.; Raman, C.J.; Ali, L.J.; Gopinath, M.P. A Systematic Review on Image Processing and Machine Learning Techniques for Detecting Plant Diseases. In Proceedings of the 2020 International Conference on Communication and Signal Processing (ICCSP), Chennai, India, 28–30 July 2020; pp. 0465–0468. [CrossRef]

6. Damicone, J.; Brandenberger, L. Common Diseases of Tomatoes: Part I. Diseases Caused by Fungi—Oklahoma State University. 2016. Available online: https://extension.okstate.edu/fact-sheets/common-diseases-of-tomatoes-part-i-diseases-caused-by-fungi.html (accessed on 19 October 2022).
7. Ahmad, A.; Saraswat, D.; El Gamal, A. A survey on using deep learning techniques for plant disease diagnosis and recommendations for development of appropriate tools. *Smart Agric. Technol.* **2023**, 3, 100083. [CrossRef]
8. DeChant, C.; Wiesner-Hanks, T.; Chen, S.; Stewart, E.L.; Yosinski, J.; Gore, M.A.; Nelson, R.J.; Lipson, H. Automated Identification of Northern Leaf Blight-Infected Maize Plants from Field Imagery Using Deep Learning. *Phytopathology* **2017**, *107*, 1426–1432. [CrossRef]
9. Bock, C.H.; Poole, G.H.; Parker, P.E.; Gottwald, T.R. Plant Disease Severity Estimated Visually, by Digital Photography and Image Analysis, and by Hyperspectral Imaging. *Crit. Rev. Plant Sci.* **2010**, *29*, 59–107. [CrossRef]
10. Bock, C.H.; Parker, P.E.; Cook, A.Z.; Gottwald, T.R. Visual Rating and the Use of Image Analysis for Assessing Different Symptoms of Citrus Canker on Grapefruit Leaves. *Plant Dis.* **2008**, *92*, 530–541. [CrossRef] [PubMed]
11. Devaraj, A.; Rathan, K.; Jaahnavi, S.; Indira, K. Identification of Plant Disease using Image Processing Technique. In Proceedings of the 2019 International Conference on Communication and Signal Processing (ICCSP), Chennai, India, 4–6 April 2019; pp. 0749–0753. [CrossRef]
12. Mugithe, P.K.; Mudunuri, R.V.; Rajasekar, B.; Karthikeyan, S. Image Processing Technique for Automatic Detection of Plant Diseases and Alerting System in Agricultural Farms. In Proceedings of the 2020 International Conference on Communication and Signal Processing (ICCSP), Chennai, India, 28–30 July 2020; pp. 1603–1607. [CrossRef]
13. Phadikar, S.; Sil, J. Rice disease identification using pattern recognition techniques. In Proceedings of the 2008 11th International Conference on Computer and Information Technology, Khulna, Bangladesh, 24–27 December 2008; pp. 420–423. [CrossRef]
14. Sarayloo, Z.; Asemani, D. Designing a classifier for automatic detection of fungal diseases in wheat plant: By pattern recognition techniques. In Proceedings of the 2015 23rd Iranian Conference on Electrical Engineering, Tehran, Iran, 10–14 May 2015; pp. 1193–1197. [CrossRef]
15. Thangadurai, K.; Padmavathi, K. Computer Visionimage Enhancement for Plant Leaves Disease Detection. In Proceedings of the 2014 World Congress on Computing and Communication Technologies, Trichirappalli, India, 27 February–1 March 2014; pp. 173–175. [CrossRef]
16. Yong, Z.; Tonghui, R.; Changming, L.; Chao, W.; Jiya, T. Research on Recognition Method of Common Corn Diseases Based on Computer Vision. In Proceedings of the 2019 11th International Conference on Intelligent Human-Machine Systems and Cybernetics (IHMSC), Hangzhou, China, 24–25 August 2019; Volume 1, pp. 328–331. [CrossRef]
17. Khirade, S.D.; Patil, A.B. Plant Disease Detection Using Image Processing. In Proceedings of the 2015 International Conference on Computing Communication Control and Automation, Pune, India, 26–27 February 2015; pp. 768–771. [CrossRef]
18. Kamilaris, A.; Prenafeta-Boldú, F.X. Deep learning in agriculture: A survey. *Comput. Electron. Agric.* **2018**, *147*, 70–90. [CrossRef]
19. Li, L.; Zhang, S.; Wang, B. Plant Disease Detection and Classification by Deep Learning—A Review. *IEEE Access* **2021**, *9*, 56683–56698. [CrossRef]
20. Lee, S.H.; Chan, C.S.; Wilkin, P.; Remagnino, P. Deep-plant: Plant identification with convolutional neural networks. In Proceedings of the 2015 IEEE International Conference on Image Processing (ICIP), Quebec City, QC, Canada; 2015; pp. 452–456. [CrossRef]
21. Zhang, Y.; Song, C.; Zhang, D. Deep Learning-Based Object Detection Improvement for Tomato Disease. *IEEE Access* **2020**, *8*, 56607–56614. [CrossRef]
22. Widiyanto, S.; Wardani, D.T.; Pranata, S.W. Image-Based Tomato Maturity Classification and Detection Using Faster R-CNN Method. In Proceedings of the 2021 5th International Symposium on Multidisciplinary Studies and Innovative Technologies (ISMSIT), Ankara, Turkey, 21–23 October 2021; pp. 130–134. [CrossRef]
23. Zhou, X.; Wang, P.; Dai, G.; Yan, J.; Yang, Z. Tomato Fruit Maturity Detection Method Based on YOLOV4 and Statistical Color Model. In Proceedings of the 2021 IEEE 11th Annual International Conference on CYBER Technology in Automation, Control, and Intelligent Systems (CYBER), Jiaxing, China, 27–31 July 2021; pp. 904–908. [CrossRef]
24. Hlaing, C.S.; Zaw, S.M.M. Tomato Plant Diseases Classification Using Statistical Texture Feature and Color Feature. In Proceedings of the 2018 IEEE/ACIS 17th International Conference on Computer and Information Science (ICIS), Singapore, 6–8 June 2018; pp. 439–444. [CrossRef]
25. Lu, J.; Shao, G.; Gao, Y.; Zhang, K.; Wei, Q.; Cheng, J. Effects of water deficit combined with soil texture, soil bulk density and tomato variety on tomato fruit quality: A meta-analysis. *Agric. Water Manag.* **2021**, *243*, 106427. [CrossRef]
26. Kaur, S.; Pandey, S.; Goel, S. Plants Disease Identification and Classification Through Leaf Images: A Survey. *Arch. Comput. Methods Eng.* **2018**, *26*, 507–530. [CrossRef]
27. Bhagat, M.; Kumar, D.; Haque, I.; Munda, H.S.; Bhagat, R. Plant Leaf Disease Classification Using Grid Search Based SVM. In Proceedings of the 2nd International Conference on Data, Engineering and Applications (IDEA), Bhopal, India, 28–29 February 2020; pp. 1–6. [CrossRef]
28. Rani, F.A.P.; Kumar, S.N.; Fred, A.L.; Dyson, C.; Suresh, V.; Jeba, P.S. K-means Clustering and SVM for Plant Leaf Disease Detection and Classification. In Proceedings of the 2019 International Conference on Recent Advances in Energy-efficient Computing and Communication (ICRAECC), Nagercoil, India, 7–8 March 2019; pp. 1–4. [CrossRef]

29. Padol, P.B.; Yadav, A.A. SVM classifier based grape leaf disease detection. In Proceedings of the 2016 Conference on Advances in Signal Processing (CASP), Pune, India, 9–11 June 2016; pp. 175–179. [CrossRef]
30. Mokhtar, U.; Ali, M.A.S.; Hassenian, A.E.; Hefny, H. Tomato leaves diseases detection approach based on Support Vector Machines. In Proceedings of the 2015 11th International Computer Engineering Conference (ICENCO), Cairo, Egypt, 29–30 December 2015; pp. 246–250. [CrossRef]
31. Sabrol, H.; Satish, K. Tomato plant disease classification in digital images using classification tree. In Proceedings of the 2016 International Conference on Communication and Signal Processing (ICCSP), Melmaruvathur, India, 6–8 April 2016; pp. 1242–1246. [CrossRef]
32. Chopda, J.; Raveshiya, H.; Nakum, S.; Nakrani, V. Cotton Crop Disease Detection using Decision Tree Classifier. In Proceedings of the 2018 International Conference on Smart City and Emerging Technology (ICSCET), Mumbai, India, 5 January 2018; pp. 1–5. [CrossRef]
33. Molina, F.; Gil, R.; Bojacá, C.; Gómez, F.; Franco, H. Automatic detection of early blight infection on tomato crops using a color based classification strategy. In Proceedings of the 2014 XIX Symposium on Image, Signal Processing and Artificial Vision, Armenia, Colombia, 17–19 September 2014; pp. 1–5. [CrossRef]
34. Pratheba, R.; Sivasangari, A.; Saraswady, D. Performance analysis of pest detection for agricultural field using clustering techniques. In Proceedings of the 2014 International Conference on Circuits, Power and Computing Technologies [ICCPCT-2014], Nagercoil, India, 20–21 March 2014; pp. 1426–1431. [CrossRef]
35. Shijie, J.; Peiyi, J.; Siping, H.; Haibo, L. Automatic detection of tomato diseases and pests based on leaf images. In Proceedings of the 2017 Chinese Automation Congress (CAC), Jinan, China, 20–22 October 2017; pp. 2510–2537. [CrossRef]
36. Jones, C.; Jones, J.; Lee, W.S. Diagnosis of bacterial spot of tomato using spectral signatures. *Comput. Electron. Agric.* **2010**, *74*, 329–335. [CrossRef]
37. Borges, D.L.; Guedes, S.T.D.M.; Nascimento, A.R.; Melo-Pinto, P. Detecting and grading severity of bacterial spot caused by Xanthomonas spp. in tomato (Solanum lycopersicon) fields using visible spectrum images. *Comput. Electron. Agric.* **2016**, *125*, 149–159. [CrossRef]
38. Lakshmanarao, A.; Babu, M.R.; Kiran, T.S.R. Plant Disease Prediction and classification using Deep Learning ConvNets. In Proceedings of the 2021 International Conference on Artificial Intelligence and Machine Vision (AIMV), Gandhinagar, India, 24–26 September 2021; pp. 1–6. [CrossRef]
39. Militante, S.V.; Gerardo, B.D.; Dionisio, N.V. Plant Leaf Detection and Disease Recognition using Deep Learning. In Proceedings of the 2019 IEEE Eurasia Conference on IOT, Communication and Engineering (ECICE), Yunlin, Taiwan, 3–6 October 2019; pp. 579–582. [CrossRef]
40. Marzougui, F.; Elleuch, M.; Kherallah, M. A Deep CNN Approach for Plant Disease Detection. In Proceedings of the 2020 21st International Arab Conference on Information Technology (ACIT), Giza, Egypt, 28–30 November 2020; pp. 1–6. [CrossRef]
41. Ngugi, L.C.; Abdelwahab, M.; Abo-Zahhad, M. Tomato leaf segmentation algorithms for mobile phone applications using deep learning. *Comput. Electron. Agric.* **2020**, *178*, 105788. [CrossRef]
42. Elhassouny, A.; Smarandache, F. Smart mobile application to recognize tomato leaf diseases using Convolutional Neural Networks. In Proceedings of the 2019 International Conference of Computer Science and Renewable Energies (ICCSRE), Agadir, Morocco, 22–24 July 2019; pp. 1–4. [CrossRef]
43. Mattihalli, C.; Gedefaye, E.; Endalamaw, F.; Necho, A. Real Time Automation of Agriculture Land, by automatically Detecting Plant Leaf Diseases and Auto Medicine. In Proceedings of the 2018 32nd International Conference on Advanced Information Networking and Applications Workshops (WAINA), Krakow, Poland, 16–18 May 2018; pp. 325–330. [CrossRef]
44. Divyashri., P.; Pinto, L.A.; Mary, L.; Manasa., P.; Dass, S. The Real-Time Mobile Application for Identification of Diseases in Coffee Leaves using the CNN Model. In Proceedings of the 2021 Second International Conference on Electronics and Sustainable Communication Systems (ICESC), Coimbatore, India, 4–6 August 2021; pp. 1694–1700. [CrossRef]
45. Liu, J.; Wang, X. Early recognition of tomato gray leaf spot disease based on MobileNetv2-YOLOv3 model. *Plant Methods* **2020**, *16*, 83. [CrossRef] [PubMed]
46. Khasawneh, N.; Faouri, E.; Fraiwan, M. Automatic Detection of Tomato Diseases Using Deep Transfer Learning. *Appl. Sci.* **2022**, *12*, 8467. [CrossRef]
47. Mim, T.T.; Sheikh, M.H.; Shampa, R.A.; Reza, M.S.; Islam, M.S. Leaves Diseases Detection of Tomato Using Image Processing. In Proceedings of the 2019 8th International Conference System Modeling and Advancement in Research Trends (SMART), Moradabad, India, 22–23 November 2019; pp. 244–249. [CrossRef]
48. Kumar, A.; Vani, M. Image Based Tomato Leaf Disease Detection. In Proceedings of the 2019 10th International Conference on Computing, Communication and Networking Technologies (ICCCNT), Kanpur, India, 6–8 July 2019; pp. 1–6. [CrossRef]
49. Tm, P.; Pranathi, A.; SaiAshritha, K.; Chittaragi, N.B.; Koolagudi, S.G. Tomato Leaf Disease Detection Using Convolutional Neural Networks. In Proceedings of the 2018 Eleventh International Conference on Contemporary Computing (IC3), Noida, India, 2–4 August 2018; pp. 1–5. [CrossRef]
50. Agarwal, M.; Gupta, S.K.; Biswas, K. Development of Efficient CNN model for Tomato crop disease identification. Sustain. *Comput. Inform. Syst.* **2020**, *28*, 100407. [CrossRef]
51. Al-Gaashani, M.S.A.M.; Shang, F.; Muthanna, M.S.A.; Khayyat, M.; El-Latif, A.A.A. Tomato leaf disease classification by exploiting transfer learning and feature concatenation. *IET Image Process.* **2022**, *16*, 913–925. [CrossRef]

52. Pathan, S.M.K.; Ali, M.F. Implementation of Faster R-CNN in Paddy Plant Disease Recognition System. In Proceedings of the 2019 3rd International Conference on Electrical, Computer & Telecommunication Engineering (ICECTE), Rajshahi, Bangladesh, 26–28 December 2019; pp. 189–192. [CrossRef]
53. Zhou, G.; Zhang, W.; Chen, A.; He, M.; Ma, X. Rapid Detection of Rice Disease Based on FCM-KM and Faster R-CNN Fusion. *IEEE Access* **2019**, *7*, 143190–143206. [CrossRef]
54. Mu, W.; Jia, Z.; Liu, Y.; Xu, W.; Liu, Y. Image Segmentation Model of Pear Leaf Diseases Based on Mask R-CNN. In Proceedings of the 2022 International Conference on Image Processing and Media Computing (ICIPMC), Xi'an, China, 27–29 May 2022; pp. 41–45. [CrossRef]
55. Wang, Q.; Qi, F.; Sun, M.; Qu, J.; Xue, J. Identification of Tomato Disease Types and Detection of Infected Areas Based on Deep Convolutional Neural Networks and Object Detection Techniques. *Comput. Intell. Neurosci.* **2019**, *2019*, 9142753. [CrossRef]
56. Kirange, D. Machine Learning Approach towards Tomato Leaf Disease Classification. *Int. J. Adv. Trends Comput. Sci. Eng.* **2020**, *9*, 490–495. [CrossRef]
57. Lu, J.; Ehsani, R.; Shi, Y.; De Castro, A.I.; Wang, S. Detection of multi-tomato leaf diseases (late blight, target and bacterial spots) in different stages by using a spectral-based sensor. *Sci. Rep.* **2018**, *8*, 2793. [CrossRef]
58. Durmuş, H.; Güneş, E.O.; Kırcı, M. Disease detection on the leaves of the tomato plants by using deep learning. In Proceedings of the 2017 6th International Conference on Agro-Geoinformatics, Fairfax, VA, USA, 7–10 August 2017; pp. 1–5. [CrossRef]
59. Tomato Leaf Disease Detection. Available online: https://www.kaggle.com/datasets/kaustubhb999/tomatoleaf (accessed on 24 October 2022).
60. Konidaris, F.; Tagaris, T.; Sdraka, M.; Stafylopatis, A. Generative Adversarial Networks as an Advanced Data Augmentation Technique for MRI Data. In Proceedings of the VISIGRAPP, Prague, Czech Republic, 25–27 February 2019.
61. Kukacka, J.; Golkov, V.; Cremers, D. Regularization for Deep Learning: A Taxonomy. *arXiv* **2017**, arXiv:1710.10686.
62. Pandian, J.A.; Kumar, V.D.; Geman, O.; Hnatiuc, M.; Arif, M.; Kanchanadevi, K. Plant Disease Detection Using Deep Convolutional Neural Network. *Appl. Sci.* **2022**, *12*, 6982. [CrossRef]
63. Goodfellow, I.; Pouget-Abadie, J.; Mirza, M.; Xu, B.; Warde-Farley, D.; Ozair, S.; Courville, A.; Bengio, Y. Generative Adversarial Networks. *Commun. ACM* **2020**, *63*, 139–144. [CrossRef]
64. Geetharamani, G.; Arun Pandian, J. Identification of plant leaf diseases using a nine-layer deep convolutional neural network. *Comput. Electr. Eng.* **2019**, *76*, 323–338. [CrossRef]
65. Widiyanto, S.; Fitrianto, R.; Wardani, D.T. Implementation of Convolutional Neural Network Method for Classification of Diseases in Tomato Leaves. In Proceedings of the 2019 Fourth International Conference on Informatics and Computing (ICIC), Semarang, Indonesia, 16–17 October 2019; pp. 1–5. [CrossRef]
66. Mamun, M.A.A.; Karim, D.Z.; Pinku, S.N.; Bushra, T.A. TLNet: A Deep CNN model for Prediction of tomato Leaf Diseases. In Proceedings of the 2020 23rd International Conference on Computer and Information Technology (ICCIT), Dhaka, Bangladesh, 19–21 December 2020; pp. 1–6. [CrossRef]
67. Kaur, M.; Bhatia, R. Development of an Improved Tomato Leaf Disease Detection and Classification Method. In Proceedings of the 2019 IEEE Conference on Information and Communication Technology, Allahabad, India, 6–8 December 2019; pp. 1–5. [CrossRef]
68. Nachtigall, L.; Araujo, R.; Nachtigall, G.R. Classification of apple tree disorders using convolutional neural networks. In Proceedings of the 2016 IEEE 28th International Conference on Tools with Artificial Intelligence (ICTAI), San Jose, CA, USA, 6–8 November 2016; pp. 472–476.
69. Ferentinos, K.P. Deep learning models for plant disease detection and diagnosis. *Comput. Electron. Agric.* **2018**, *145*, 311–318. [CrossRef]

electronics

MDPI

Article

Hemerocallis citrina Baroni Maturity Detection Method Integrating Lightweight Neural Network and Dual Attention Mechanism

Liang Zhang [1,†], Ligang Wu [1,2,*,†] and Yaqing Liu [2,*]

1 College of Mechanical and Electrical Engineering, Shanxi Datong University, Datong 037003, China
2 School of Information Science and Technology, Dalian Maritime University, Dalian 116026, China
* Correspondence: wuligang@sxdtdx.edu.cn (L.W.); liuyaqing@dlmu.edu.cn (Y.L.)
† These authors contributed equally to this work.

Abstract: North of Shanxi, Datong Yunzhou District is the base for the cultivation of *Hemerocallis citrina Baroni*, which is the main production and marketing product driving the local economy. *Hemerocallis citrina Baroni* and other crops' picking rules are different: the picking cycle is shorter, the frequency is higher, and the picking conditions are harsh. Therefore, in order to reduce the difficulty and workload of picking *Hemerocallis citrina Baroni*, this paper proposes the GGSC YOLOv5 algorithm, a *Hemerocallis citrina Baroni* maturity detection method integrating a lightweight neural network and dual attention mechanism, based on a deep learning algorithm. First, Ghost Conv is used to decrease the model complexity and reduce the network layers, number of parameters, and Flops. Subsequently, combining the Ghost Bottleneck micro residual module to reduce the GPU utilization and compress the model size, feature extraction is achieved in a lightweight way. At last, the dual attention mechanism of Squeeze-and-Excitation (SE) and the Convolutional Block Attention Module (CBAM) is introduced to change the tendency of feature extraction and improve detection precision. The experimental results show that the improved GGSC YOLOv5 algorithm reduced the number of parameters and Flops by 63.58% and 68.95%, respectively, and reduced the number of network layers by about 33.12% in terms of model structure. In the case of hardware consumption, GPU utilization is reduced by 44.69%, and the model size was compressed by 63.43%. The detection precision is up to 84.9%, which is an improvement of about 2.55%, and the real-time detection speed increased from 64.16 *FPS* to 96.96 *FPS*, an improvement of about 51.13%.

Keywords: deep learning; lightweight neural networks; attentional mechanisms; *Hemerocallis citrina Baroni*; maturity detection

Citation: Zhang, L.; Wu, L.; Liu, Y. *Hemerocallis citrina Baroni* Maturity Detection Method Integrating Lightweight Neural Network and Dual Attention Mechanism. *Electronics* **2022**, *11*, 2743. https://doi.org/10.3390/electronics11172743

Academic Editor: Rashid Mehmood

Received: 2 August 2022
Accepted: 25 August 2022
Published: 31 August 2022

1. Introduction

In recent years, the policies of agricultural revitalization strategy and agricultural poverty alleviation have achieved many successes. Under the background of a rural revitalization strategy, facing the opportunities and challenges in the process of rapid development of agriculture, the Ministry of Agriculture and Rural Affairs has implemented science and technology to assist agriculture, accelerate the integration of rural industries and digital economy, solve the problems of low efficiency and quality, and actively encourage and promote the efficient development of smart agriculture.

In the process of sowing and growing [1–3], fertilizing and watering [4–6], pest monitoring [7–9], and fruit picking [10,11] of agricultural products [12], smart agriculture plays an irreplaceable role in improving the quality of agricultural products; it makes all the work more convenient and efficient, so smart agriculture has received a wide range of attention from researchers.

At present, the effective combination of artificial intelligence technology and smart agriculture has become a key research topic, whereas computer vision [13] and deep learning technology have become effective measures to promote rural revitalization and agricultural poverty alleviation. Based on deep learning methods and computer vision techniques, in Ref. [14], an accurate quality assessment of different fruits was efficiently accomplished by using the Faster RCNN target detection algorithm. Similarly, in Ref. [15], the authors accomplished tomato ripening detection based on color and size differentiation using a two-stage target detection algorithm, Faster R-CNN, which has an accuracy of 98.7%. In addition, Wu [16] completed the detection of strawberries in the process of strawberry picking by a U^2-Net network image segmentation technique, and applied it to the automated picking process.

Compared with two-stage target detection algorithms, single-stage target detection algorithms are more advantageous, and the YOLO algorithm is a typical representative. With the YOLOv3 target detection algorithm, Zhang et al. [17] precisely located the fruit and handle of a banana, which is convenient for intelligent picking and operation, and the average precision was as high as 88.45%. In Ref. [18], the detection of a palm oil bunch was accomplished using the YOLOv3 algorithm and its application in embedded devices through mobile and the Internet of Things. Wu et al. [19] accomplished the identification and localization of fig fruits in complex environments by using the YOLOv4 target detection algorithm, which distinguishes and discriminates whether the fig fruits are ripe or not. Zhou et al. [20] completed the ripening detection of tomatoes by K-means clustering and noise reduction processing based on the YOLOv4 algorithm, but the detection speed was only 5–6 FPS, which could not meet the needs of real-time detection.

In summary, it can be seen that deep learning methods and computer vision techniques are widely used in smart agriculture in previous studies [21,22]. However, with the improvement and update of the YOLO algorithm, the YOLOv5 algorithm was proposed, but, currently, there are less applications in agriculture related fields. Inspired by existing studies, we applied the YOLOv5 algorithm to the maturity detection process of *Hemerocallis citrina Baroni* because North of Shanxi, Datong Yunzhou District, has been known as the hometown of *Hemerocallis citrina Baroni* since ancient times, and is also the planting base of organic *Hemerocallis citrina Baroni*.

In recent years, the *Hemerocallis citrina Baroni* industry in Yunzhou District has entered the fast track of development. As the leading industry of "one county, one industry", it has brought rapid economic development while promoting rural revitalization. However, at present, the picking of *Hemerocallis citrina Baroni* mainly relies on manual completion, and whether the *Hemerocallis citrina Baroni* is mature or not relies entirely on experience to distinguish. Therefore, the main work and contributions of this paper are as follows:

a. Computer vision technology and deep learning algorithms are applied to the maturity detection of *Hemerocallis citrina Baroni*, and highly accurate maturity detection of whether the *Hemerocallis citrina Baroni* are mature and meet the picking standards, providing ideas for improving the picking method and reducing the cost of picking labor.
b. The lightweight neural network is introduced to reduce the number of network layers and model complexity, compress the model volume, and lay the foundation for the embedded development of picking robots.
c. Combined with the dual attention mechanism, it improves the tendency of feature extraction and enhances the detection precision and real-time detection efficiency.

The remainder of this paper is organized as follows. Section 2 offers YOLOv5 object detection algorithms. In Section 3, the GGSC YOLOv5 network structure and its constituent modules are presented. Section 4 introduces the model training parameters, advantages of the lightweight model, and analysis of experiments results. Finally, conclusions and future work are given in Section 5.

2. YOLOv5 Object Detection Algorithms

Currently, there are two types of deep learning target detection algorithms, one-stage and two-stage. The one-stage target detection algorithm performs feature extraction on the entire image to complete end-to-end training, and the detection process is faster, but less precise. Common algorithms include YOLO, SSD, Retina Net, etc. The two-stage target detection algorithm selectively traverses the entire image by pre-selecting the boxes, which is a slower detection process, but has higher precision. Faster RCNN, Cascade RCNN, Mask RCNN, etc. are common algorithms.

With continuous improvements and updates [23], the YOLOv5 target detection algorithm has improved detection precision and model fluency. As shown in Figure 1, YOLOv5 mainly consists of four parts: input head, backbone, neck, and prediction head. The input head is used as the input of the convolutional neural network, which completes the cropping and data enhancement of the input image, and backbone and neck complete the feature extraction and feature fusion for the detected region, respectively. The prediction head is used as the output to complete the recognition, classification, and localization of the detected objects [24].

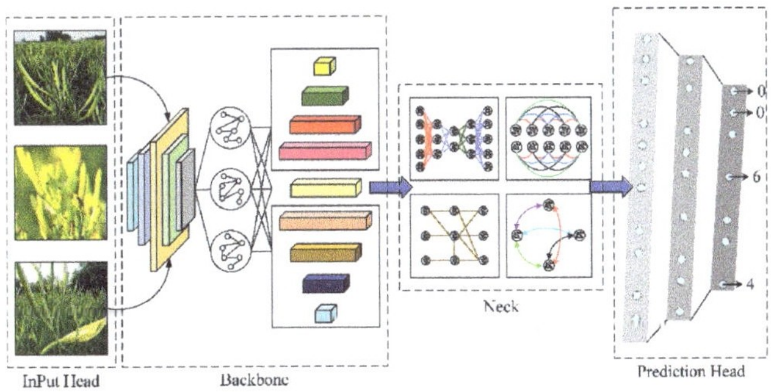

Figure 1. YOLOv5 algorithm process.

In the version 6.0 of the YOLOv5 algorithm, the focus module is replaced by a rectangular convolution with *stride* = 2, the CSP residual module [25] is replaced by the C3 module, and the size of the convolution kernel in the spatial pyramid pooling (SPP) [26] module is unified to 5. However, the problems of the complex model structure, redundant feature extraction during convolution, and large number of parameters and computation of the model still exist, which are not suitable for mobile and embedded devices.

To address the above problems, the GGSC YOLOv5 detection algorithm based on a lightweight and double attention mechanism is proposed in this paper, and applied to *Hemerocallis citrina Baroni* recognition, with the obvious advantages of the lightweight model and excellent detection performance in the detection process.

3. Deep Learning Detection Algorithm GGSC YOLOv5

3.1. Ghost Lightweight Convolution

In limited memory and computational resources, deploying efficient and lightweight neural networks is the future development direction of convolutional neural networks [27]. In the feature extraction process, the traditional convolution traverses the entire input image sequentially, with many similar feature maps generated by adjacent regions during the convolution process. Therefore, traditional convolutional feature extraction is computationally intensive, inefficient, and redundant in terms of information.

As shown in Figure 2, Ghost Conv [28] takes advantage of the redundancy characteristic of the feature map, and first generates m intrinsic feature maps by a few traditional convolutions. Then, the Φ_i cheap linear operation is performed on m intrinsic feature maps, such that each intrinsic feature map produces $s-1$ new feature maps. Lastly, the m intrinsic feature maps and $s-1$ new feature maps are spliced together to complete the lightweight convolution operation.

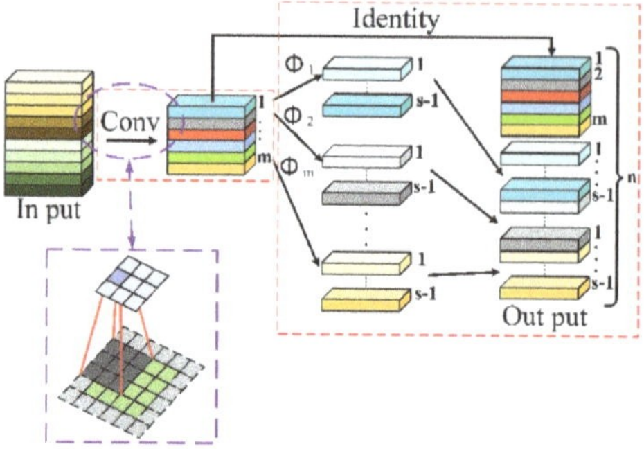

Figure 2. The Ghost Convolution process.

The Ghost Conv process is less computationally intensive and more lightweight than traditional convolution due to the cheap linear operations introduced in the process. Therefore, the theoretical speedup ratio (r_s) and model compression ratio (r_c) of Ghost Conv and traditional convolution are as follows, respectively:

$$r_s = \frac{C_T}{C_G} = \frac{c \times k \times k \times ms \times h' \times w'}{c \times k \times k \times m \times h' \times w' + m \times k \times k \times (s-1) \times h' \times w'} = \frac{c \times s}{c+s-1} \approx s \quad (1)$$

$$r_c = \frac{c \times k \times k \times ms}{c \times k \times k \times m + m \times k \times k \times (s-1)} = \frac{c \times s}{c+s-1} \approx s \quad (2)$$

where $h \times w$ and $h' \times w'$ are the height and width sizes of the input and output images.c is the number of input channels, ms is the number of output channels, and $k \times k$ is the custom convolution kernel size. C_T and C_G are the convolutional computations of traditional convolution and Ghost Conv, respectively.

In summary, the r_s and r_c of Ghost Conv are only $1/s$ of the traditional convolution due to the introduction of cheap linear operations. It can be seen that Ghost Conv has obvious advantages of being lightweight, with a lower number of parameters and computation compared with the traditional convolution.

3.2. Ghost Lightweight Bottleneck

Ghost Bottleneck [28] is a lightweight module consisting of Ghost Conv, Batch Normalization (BN) layers, down sampling, and activation functions. Its design method and model structure are similar to that of the Res Net residual network, which has the features of a simple model structure, easy application, and high operational efficiency. Since the number of channels remains constant before and after feature extraction, the module can be plug-and-play.

The structure of the Ghost Bottleneck model is shown in Figure 3. Ghost Bottleneck mainly consists of two Ghost Conv stacks, where the input image is passed through the first Ghost Conv to increase the number of channels, normalized by the BN layer, and the nonlinear properties of the neural network model are increased by the ReLU activation function. Subsequently, it goes through a second Ghost Conv to reduce the number of channels, thus ensuring that the number of output channels is the same as before the first Ghost Conv operation. Lastly, the output after twice Ghost Conv and the original input after down sampling are spliced and stacked by Add operation, which increases the amount of information of the desired features, while the number of channels remains the same.

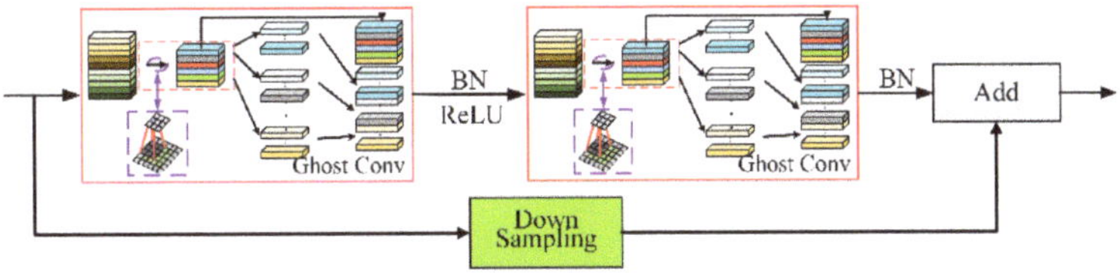

Figure 3. Ghost Bottleneck modules.

The structure of the Ghost Bottleneck model is similar to MobileNetv2. The BN layer is retained after compressing the channels without using the activation function, so the original information of feature extraction is retained to the maximum extent. Compared with other residual modules and cross-stage partial (CSP) network layers, Ghost Bottleneck uses fewer convolutional and BN layers, and the model structure is simpler. Therefore, using Ghost Bottleneck makes the number of model parameters and the Flops calculation lower, the number of network layers less, and the lightweight feature more obvious.

3.3. SE Attentional Mechanisms

The Squeeze-and-Excitation channel attention mechanism module (SE Module) [29] consists of two parts: Squeeze and Excitation. First, the SE Module performs the Squeeze operation on the feature map obtained by convolution to get the global features on the channel. Subsequently, the Excitation operation is performed on the global features, which learns the relationship between each channel and obtains the weight values of different channels. Lastly, the weight values of each channel are multiplied on the original feature map to obtain the final features after performing the SE Module.

The SE Module feature extraction process is shown in Figure 4. During the Squeeze operation, global average pooling is used to obtain global features, and the output z_c is obtained by the compression function F_{sq} according to the compression aggregation strategy. During the Excitation operation, the dimension is first reduced, and then, the dimension is increased. The output s after Excitation is obtained by the excitation function F_{ex}, and the relationship between channels is obtained by the feature capture mechanism of $Sigmoid$ to complete the feature extraction.

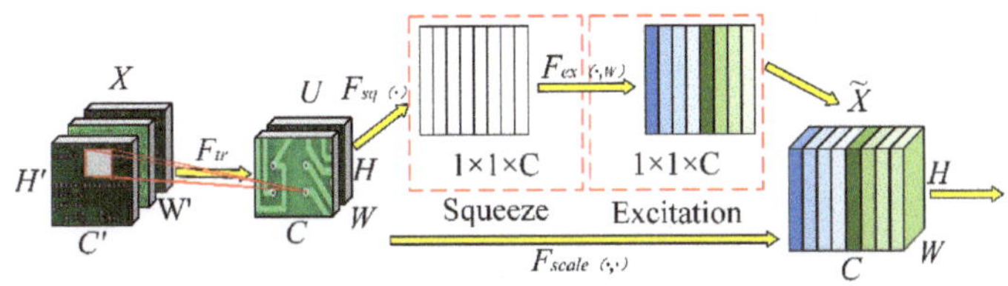

Figure 4. SE attentional mechanisms.

In the SE channel attention mechanism model, the Squeeze-Excitation function can be expressed as $x_c = F_{scale}(u_c, s_c) = s_c u_c$, where u_c represents the c-th feature map in the Squeeze-Excitation process, and s_c represents the weight of the c-th feature map.

The SE channel attention mechanism adaptively [30] accomplishes the adjustment of feature weights during feature extraction, which is more conducive to obtaining the required feature information. Therefore, the SE channel attention mechanism is introduced into the model structure, which can enhance the discrimination ability of the model, and improve the detection accuracy and maturity detection effect.

3.4. CBAM Attentional Mechanisms

The Convolutional Block Attention Module (CBAM) [31] attention mechanism is an efficient feed-forward convolutional attention model, which can perform the propensity extraction of features sequentially in channel and spatial dimensions, and it consists of two sub-modules: Channel Attention Module (CAM) and Spatial Attention Module (SAM).

The CBAM feature extraction process is shown in Figure 5. First, compared with the SE Module, the channel attention mechanism in CBAM adds a parallel maximum pooling layer, which can obtain more comprehensive information. Second, the CAM and SAM modules are used sequentially to make the model recognition and classification more effective. Lastly, since CAM and SAM perform feature inference sequentially along two mutually independent dimensions, the combination of the two modules can enhance the expressive ability of the model.

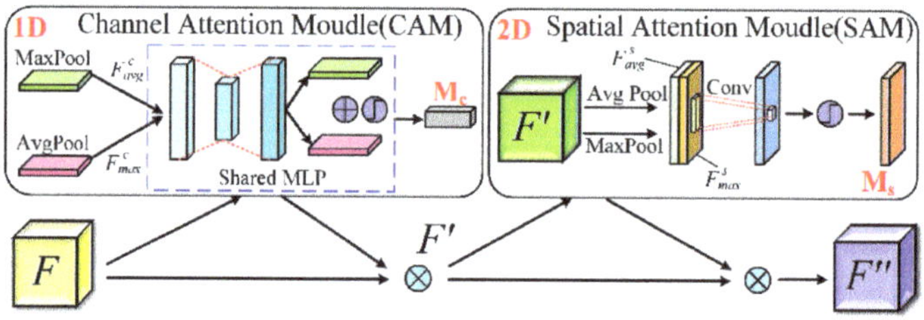

Figure 5. CBAM attentional mechanisms.

In the CAM module, with the input feature map performing maximum pooling and average pooling in parallel, the shared network in the multilayer perceptron (MLP) performs feature extraction based on the maximum pooling feature maps $F_{\max}$ and average pooling feature maps F_{avg} to produce a 1D channel attention map M_c. The CAM convolution calculation can be expressed as:

$$\begin{aligned} M_c(F) &= \sigma[MLP(AvgPool(F)) + MLP(MaxPool(F))] \\ &= \sigma[W_1(W_0(F^c_{avg})) + W_1(W_0(F^c_{\max}))] \end{aligned} \quad (3)$$

where σ denotes the sigmoid function, and W_0 and W_1 denote the weights after pooling and sharing the network, respectively.

In the SAM module, the input feature map performs maximum pooling and average pooling in parallel, and a 2D spatial attention map M_s is generated by traditional convolution. The SAM convolution calculation can be expressed as:

$$\begin{aligned} M_s(F) &= \sigma[f^{k\times k}(AvgPool(F); MaxPool(F))] \\ &= \sigma[f^{k\times k}(F^s_{avg}; F^s_{\max})] \end{aligned} \quad (4)$$

where $f^{k\times k}$ represents a traditional convolution operation with the filter size of $k \times k$.

The CBAM attention mechanism is an end-to-end training model with plug-and-play functionality, and, thus, can be seamlessly fused into any convolutional neural network. Combined with the YOLO algorithm, it can complete feature extraction more efficiently and obtain the required feature information without additional computational cost and operational pressure.

3.5. GGSC YOLOv5 Model Structure

The improved GGSC YOLOv5 algorithm model structure and module parameters are shown in Figure 6, which combines Ghost Conv and Ghost Bottleneck modules to achieve a light weight, and introduces the dual attention mechanism of SE and CBAM to improve detection precision (P) and real-time detection efficiency.

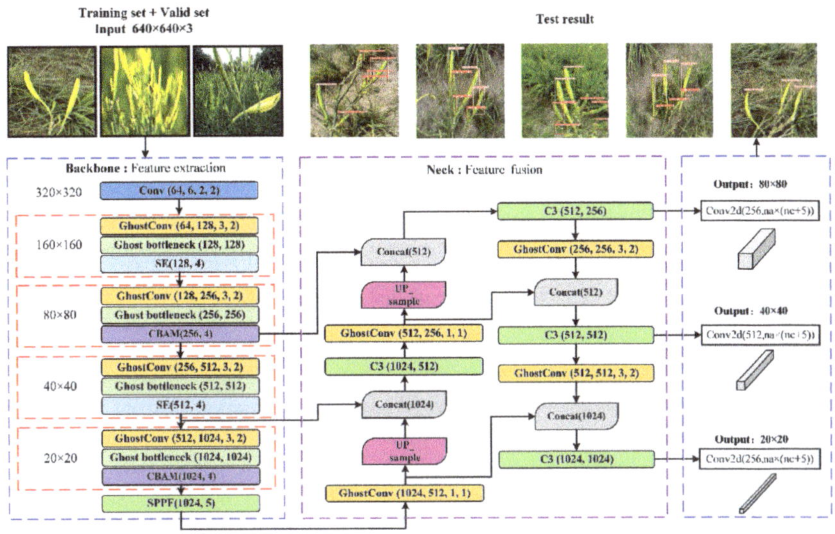

Figure 6. GGSC YOLOv5 model structure.

In the GGSC YOLOv5 algorithm feature extraction network backbone, Ghost Conv and Ghost Bottleneck module sets are used instead of traditional convolution and C3 modules, respectively, which reduces the consumption of memory and hardware resources in the convolution process. The SE and CBAM attention mechanisms are used alternately after each module group (Ghost Conv and Ghost Bottleneck) to enhance the tendency of feature extraction, enabling the underlying fine-grained information and the high-level semantic information to be extracted effectively. In the feature fusion network neck, images with different resolutions are fused by Concatenate and Up-sample, which makes the localization information, classification information, and confidence information of the feature map more accurate.

After the feature extraction and feature fusion, three different tensor are generated at the output prediction head by conventional convolution, Conv2d: $(256, na \times (nc+5))$, $(512, na \times (nc+5))$, and $(1024, na \times (nc+5))$, corresponding to three sizes of output: 80×80, 40×40, and 20×20, where 256, 512, and 1024 denote the number of channels, respectively. $na \times (nc+5)$ represents the relevant parameters of the detected object; the number of anchors for each category and the number of categories of detected objects are denoted by na and nc, respectively. The four localization parameters and one confidence parameter of the anchor are represented by 5.

4. Experiments and Results Analysis

4.1. Model Training

The experiments in this paper were carried out using the Python 3.8.5 environment and CUDA 11.3, under Intel Core i9-10900k@3.7 GHz, NVidia GeForce RTX 3080 10G, and DDR4 3600 MHz dual memory hardware.

For the dataset, 800 images of *Hemerocallis citrina Baroni* were available after photographs on the spot, screening, dataset production, and classification. Among them, 597 images are used as the training set, 148 images are used as the validation set, and 55 images are used as the test set.

In this paper, the original YOLOv5 and the improved GGSC YOLOv5 algorithm use the same parameter settings, the image input is 640×640, the learning rate is 0.01, the cosine annealing hyper-parameter is 0.1, the weight decay coefficient is 0.0005, and the momentum parameter in the gradient descent with momentum is 0.937. A total of 300 epochs and a batch size of 12 are used during training.

The GGSC YOLOv5 training process of the *Hemerocallis citrina Baroni* recognition method based on the lightweight neural network and dual attention mechanism is shown in Algorithm 1.

Algorithm 1: Training process of GGSC YOLOv5.

Determine: Parameters, Anchor, lr, L_{ciou}, IOU_{th}.
InPut: Training dataset, Valid dataset, Label set.
Loading: Train models, Valid models.
Ensure: In Put, Backbone, Neck, OutPut. Algorithm environment.
N iterations of training. i-th iteration training($i \leq N$):
Feature extraction Net:
a: Rectangular convolution
b: i-th iteration($i \leq 2$):
Ghost Conv-Ghost Bottleneck-SE, feature extraction.
Ghost Conv-Ghost Bottleneck-CBAM, feature extraction.
c: Feature fusion.
d: Predicted Head: classification c_i, confidence p_i.
e: Positioning error, category error, confidence error.
f: $\sum loss$.
Val Net:
a: Test effect of model ϖ_i.
b: Calculate P, R and $Val\ loss$.
c: Adjust lr and update strategy.
Save results of the i-th training: weight π_i, and model ϖ_i.
Update: Weight: $\pi_{i+1} \leftarrow \pi_i$, Model: $\varpi_{i+1} \leftarrow \varpi_i$.
Temporary storage model ϖ'.
Plot: Result curve, Save best model ϖ', Output.
End Train

4.2. Model Lightweight Analysis

The cultivation of *Hemerocallis citrina Baroni* has the characteristics of vast area, dense plants, and different growth. Therefore, recognition methods based on computer vision and deep learning are widely used in robotic picking, and the lightweight features are more in line with the practical needs and future development direction of embedded devices.

The GGSC YOLOv5 algorithm takes advantage of the redundancy of the feature maps to reduce the model complexity while improving the efficiency of feature extraction and the relationship between channels.

A comparison of the model parameters of GGSC YOLOv5 and the original YOLOv5 is shown in Figure 7. The improved algorithm has the obvious advantages of being lightweight. In terms of model structure, the number of network layers is reduced from 468 to 313, which is about 33.12% less. The number of parameters and the number of Flops operations decreased significantly, by about 63.58% and 68.95%, respectively. In terms of memory occupation and hardware consumption, the GPU utilization [32] was reduced from 6.9 G to 3.8 G, a reduction of about 44.69%. The volume of the model trained is reduced from 92.7 M to 33.5 M, a compression of about 63.43%. At the same time, the time required for training 300 epochs is reduced by 3.4%.

Neural network algorithms based on computer vision and deep learning have high requirements on the hardware and computing power of microcomputers. The memory and computational resources of the picking robot are limited in the recognition process of *Hemerocallis citrina Baroni*, so the GGSC YOLOv5 algorithm can show its advantages of being lightweight, and can reduce the demand of hardware equipment for the picking device.

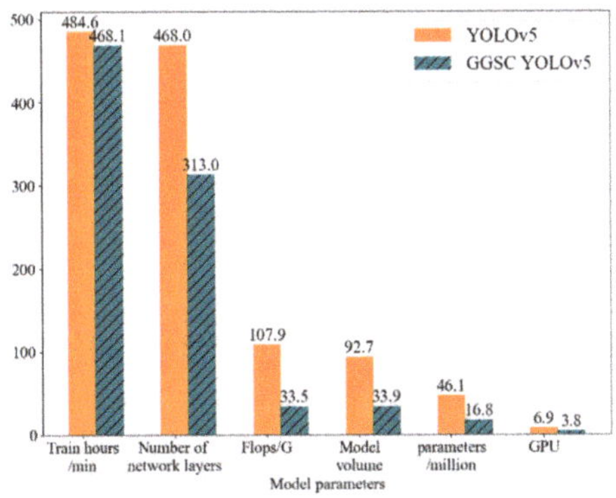

Figure 7. Comparison of the number of model parameters.

4.3. Model Training Process Analysis

The loss function convergence curves during model training are shown in Figure 8. As can be seen from the figure, the loss function curves of the training and validation sets of the training process show an obvious convergence trend, and the convergence speed of the validation set is faster, which proves that the model has excellent learning performance during the training process, so it shows more satisfactory results in the validation process.

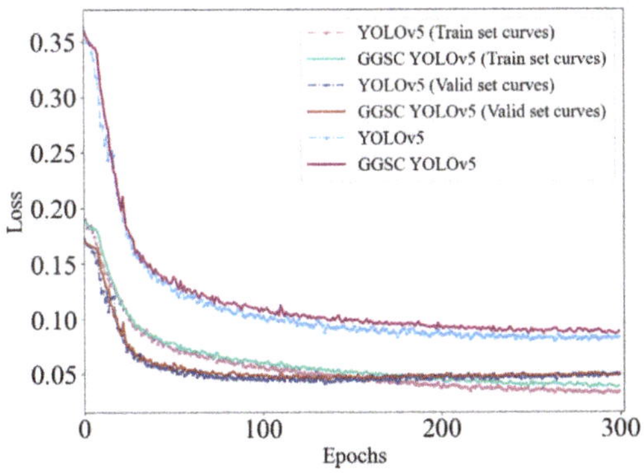

Figure 8. Loss value convergence curve with epoch times.

During the training process of the previous 50 times, the feature extraction is obvious, the learning efficiency is high, and the loss function continues to decline. After 100 iterations of training, the convergence trend of GGSC YOLOv5 and YOLOv5 algorithms is roughly the same, showing a gradual stabilization trend, and the loss function value does not decrease, the model converges, and the detection accuracy tends to be stable.

In the field of deep learning target detection, the reliability of the resulting model can be evaluated by calculating the precision (P), recall (R), and harmonic mean (F_1) based on the number of positive and negative samples.

The $R-P$ curve consists of R and P. It can show the variation trend of model P with R. The area under the $R-P$ curve line can indicate the average precision (AP) of the model, and the larger the area under the $R-P$ curve line, the higher the AP of the model, and the better the comprehensive performance.

The $R-P$ curves of the GGSC YOLOv5 and YOLOv5 algorithms are shown in Figure 9. In the figure, the trend and area under the line are approximately the same for both curves. During the model training, the AP value of GGSC YOLOv5 is 0.884, whereas the AP value of the YOLOv5 algorithm is 0.890, which is a very small difference. However, the model structure of the GGSC YOLOv5 algorithm is simpler, lighter, and requires less hardware devices, memory size, and computer computing power.

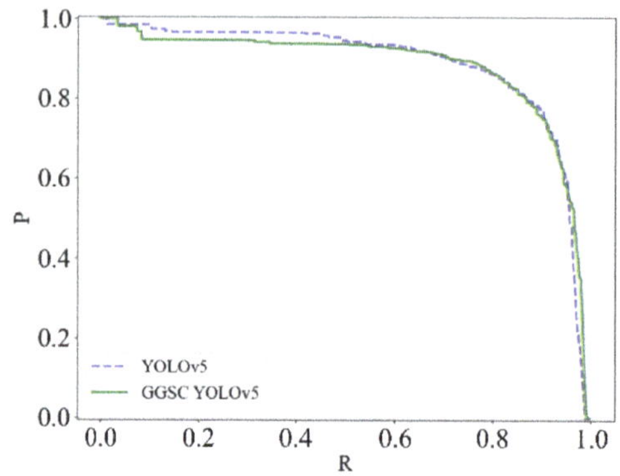

Figure 9. Comparison with $R - P$ curve.

The harmonic mean F_1 is influenced by P and R, which can reflect the comprehensive performance of the model. The value of F_1 is higher, and the model has better equilibrium performance for P and R, and vice versa. P and R indexes in the model can be effectively assessed by F_1, which can determine whether there is a sharp increase in one and a sudden decrease in the other. Therefore, it is an important indicator to assess the reliability and comprehensiveness of the model.

The variation trend of YOLOv5 and GGSC YOLOv5 harmonic mean value curves F_1 with confidence is shown in Figure 10. After combining the lightweight network and the dual attention mechanism, the GGSC YOLOv5 algorithm has the same harmonic mean value as YOLOv5, both of which are 0.84. The results show that the harmonic performance of the improved algorithm is not affected on the basis of achieving lightweight.

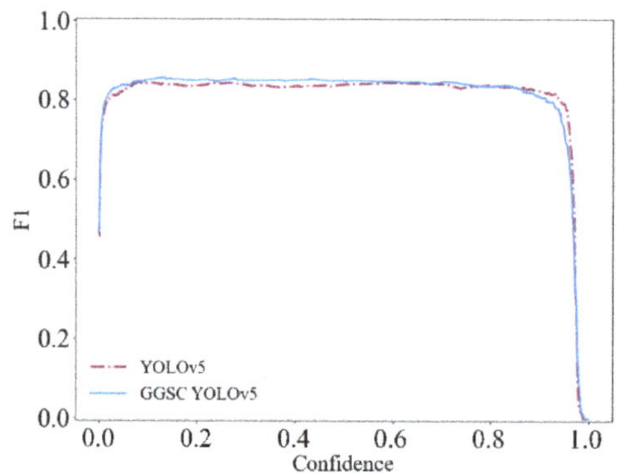

Figure 10. Comparison of the F_1 curves for before and after algorithm improvement.

In the process of identifying whether *Hemerocallis citrina Baroni* is mature, detection precision (P) is a key performance indicator that has a decisive impact on the picking results. In the process of picking *Hemerocallis citrina Baroni*, the higher the P, the more accurate the picking, and the lower the loss, the higher the income, and vice versa.

The precision curves of the YOLOv5 algorithm and the GGSC YOLOv5 algorithm, as well as the curve fitting, are shown in Figure 11. As can be seen from the figures, Figure 11a shows the original data and precision curves of the YOLOv5 and GGSC YOLOv5 algorithms. Figure 11b,c show the original and fitted curves of the YOLOv5 and GGSC

YOLOv5 algorithms, respectively. Figure 11d shows the fitted precision curves of the YOLOv5 and GGSC YOLOv5 algorithms.

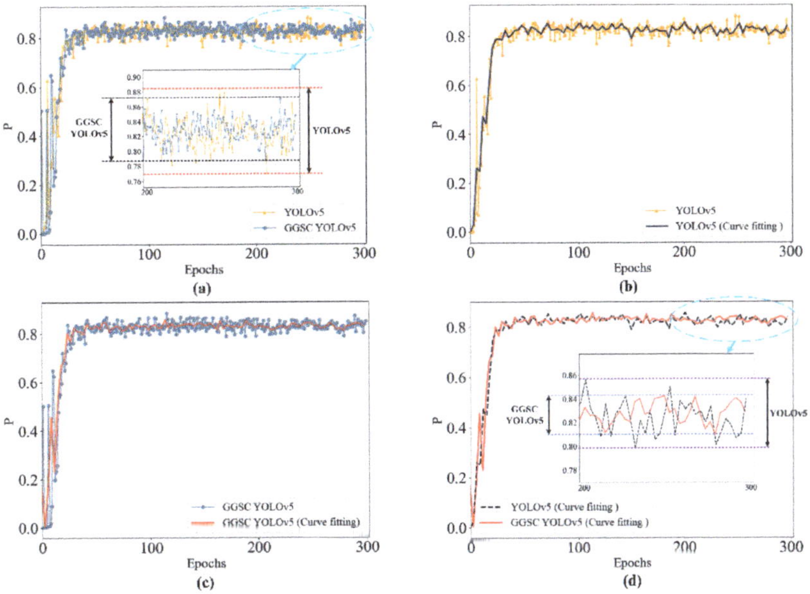

Figure 11. Model precision curve.

In the original precision curve and the fitted precision curve, GGSC YOLOv5 has less fluctuation range and higher precision compared with YOLOv5. During the training process, the final precision of YOLOv5 is 82.36%, whereas the final precision of GGSC YOLOv5 is 84.90%. After the introduction of Ghost Conv, Ghost Bottleneck, and the double attention mechanism, not only is a light weight achieved, but also the detection precision is improved by 2.55%.

The precision and fast picking of *Hemerocallis citrina Baroni* is a prerequisite to ensure picking efficiency. Therefore, for the maturity detection of *Hemerocallis citrina Baroni*, in addition to the detection precision, the real-time detection speed is also a crucial factor.

The real-time detection speed is determined by the number of image frames processed per second (FPS). The more frames processed per second, the faster the real-time detection speed and the better the real-time detection performance of the model, and vice versa. In the maturity detection process of *Hemerocallis citrina Baroni*, the real-time detection speed of the YOLOv5 algorithm is 64.14 FPS, whereas the GGSC YOLOv5 is 96.96 FPS, which exceeds the original algorithm by about 51.13%. It can be seen that based on computer vision technology and deep learning methods, the GGSC YOLOv5 algorithm can complete the recognition of *Hemerocallis citrina Baroni* with high accuracy and efficiency.

In summary, the average precision and harmonic mean performance of GGSC YOLOv5 and YOLOv5 algorithms are approximately the same. However, in model lightweight analysis, the GGSC YOLOv5 algorithm has more prominent advantages, which is in line with the future development direction of neural networks, and can also meet the needs of embedded devices in agricultural production. The experimental results of the training process show that GGSC YOLOv5 has higher detection precision and real-time detection speed, which can effectively improve the picking efficiency and meet the needs of *Hemerocallis citrina Baroni* picking.

Figure 12 compares the maturity detection results of the YOLOv5 algorithm and the improved GGSC YOLOv5 lightweight algorithm for *Hemerocallis citrina Baroni*. It can be seen from Figure 12a that the improved algorithm has higher coverage and detection precision for yellow flower detection, with the same confidence threshold and intersection-over-union ratio threshold. In the multi-plant environment, GGSC YOLOv5 was more effective in detecting the overlap of *Hemerocallis citrina Baroni* fruits, whereas in the single-plant environment, the GGSC YOLOv5 algorithm gave a higher confidence in classification and maturity detection. In contrast, the GGSC YOLOv5 algorithm proposed in this paper has better maturity detection ability, and it can accurately identify highly dense, overlapping, and obscured *Hemerocallis citrina Baroni* fruits.

Figure 12. *Cont.*

Figure 12. Maturity detection results of *Hemerocallis citrina Baroni* by different algorithms. (**a**) Test results of multi-plant and single-plant environment. (**b**) Test results of rainy weather environment.

In crop growing and picking, special environmental factors (e.g., rainy weather) can affect the normal picking work. Therefore, in order to verify the effectiveness of the proposed algorithm in this paper under multiple scenarios, the detection results of different algorithms in rainy weather environments are presented in Figure 12b. The experiments show that special factors such as rain and dew adhesion do not affect the effectiveness of the proposed algorithm, and it shows better maturity detection and detection results than the original algorithm, which shows that the proposed algorithm has good generalization and derivation ability.

5. Conclusions

In this paper, we propose a deep learning target detection algorithm, GGSC YOLOv5, based on a lightweight and dual attention mechanism, and apply it to the picking maturity detection process of *Hemerocallis citrina Baroni*. Ghost Conv and Ghost Bottleneck are used as the backbone networks to complete feature extraction, and reduce the complexity and redundancy of the model itself, and the dual attention mechanisms of SE and CBAM are introduced to increase the tendency of the model feature extraction, and improve the detection precision and real-time detection efficiency. The experimental results show that the proposed algorithm achieves an improvement of detection precision and detection efficiency under the premise of being lightweight, and has strong discrimination and generalization ability, which can be widely applied in a multi-scene environment.

In future research and work, the multi-level classification of *Hemerocallis citrina Baroni* will be carried out. Through the accurate maturity detection of different maturity levels of *Hemerocallis citrina Baroni*, it will be able to play different edible and medicinal roles at different growth stages, and can then be fully exploited to enhance the economic benefits.

Author Contributions: Conceptualization, L.Z., L.W. and Y.L.; methodology, L.Z., L.W. and Y.L.; software, L.Z. and Y.L.; validation, L.W. and Y.L.; investigation, Y.L.; resources, L.W.; data curation, L.Z.; writing—original draft preparation, L.Z., L.W. and Y.L.; writing—review and editing, L.W. and Y.L.; visualization, L.Z.; supervision, L.W. and Y.L.; project administration, Y.L.; funding acquisition, L.W. All authors have read and agreed to the published version of the manuscript.

Funding: This research was funded by the Shanxi Provincial Philosophy and Social Science Planning Project, grant number 2021YY198; and Shanxi Datong University Scientific Research Yun-Gang Special Project (2020YGZX014 and 2021YGZX27).

Acknowledgments: The authors would like to thank the reviewers for their careful reading of our paper and for their valuable suggestions for revision, which make it possible to present our paper better.

Conflicts of Interest: The authors declare no conflict of interest. The funders had no role in the design of the study; in the collection, analyses, or interpretation of data; in the writing of the manuscript, or in the decision to publish the results.

References

1. Lin, Y.D.; Chen, T.T.; Liu, S.Y.; Cai, Y.L.; Shi, H.W.; Zheng, D.; Lan, Y.B.; Yue, X.J.; Zhang, L. Quick and accurate monitoring peanut seedlings emergence rate through UAV video and deep learning. *Comput. Electron. Agric.* **2022**, *197*, 106938. [CrossRef]
2. Perugachi-Diaz, Y.; Tomczak, J.M.; Bhulai, S. Deep learning for white cabbage seedling prediction. *Comput. Electron. Agric.* **2021**, *184*, 106059. [CrossRef]
3. Feng, A.; Zhou, J.; Vories, E.; Sudduth, K.A. Evaluation of cotton emergence using UAV-based imagery and deep learning. *Comput. Electron. Agric.* **2020**, *177*, 105711. [CrossRef]
4. Azimi, S.; Wadhawan, R.; Gandhi, T.K. Intelligent Monitoring of Stress Induced by Water Deficiency in Plants Using Deep Learning. *IEEE Trans. Instrum. Meas.* **2021**, *70*, 1–13. [CrossRef]
5. Patel, A.; Lee, W.S.; Peres, N.A.; Fraisse, C.W. Strawberry plant wetness detection using computer vision and deep learning. *Smart Agric. Technol.* **2021**, *1*, 100013. [CrossRef]
6. Liu, W.; Wu, G.; Ren, F.; Kang, X. DFF-ResNet: An insect pest recognition model based on residual networks. *Big Data Min. Anal.* **2020**, *3*, 300–310. [CrossRef]
7. Wang, K.; Chen, K.; Du, H.; Liu, S.; Xu, J.; Zhao, J.; Chen, H.; Liu, Y.; Liu, Y. New image dataset and new negative sample judgment method for crop pest recognition based on deep learning models. *Ecol. Inf.* **2022**, *69*, 101620. [CrossRef]
8. Jiang, H.; Li, X.; Safara, F. IoT-based Agriculture: Deep Learning in Detecting Apple Fruit Diseases. *Microprocess. Microsyst.* **2021**, *91*, 104321. [CrossRef]
9. Orano, J.F.V.; Maravillas, E.A.; Aliac, C.J.G. Jackfruit Fruit Damage Classification using Convolutional Neural Network. In Proceedings of the 2019 IEEE 11th International Conference on Humanoid, Nanotechnology, Information Technology, Communication and Control, Environment, and Management (HNICEM), Laoag, Philippines, 29 November–1 December 2019; pp. 1–6. [CrossRef]
10. Herman, H.; Cenggoro, T.W.; Susanto, A.; Pardamean, B. Deep Learning for Oil Palm Fruit Ripeness Classification with DenseNet. In Proceedings of the 2021 International Conference on Information Management and Technology (ICIMTech), Jakarta, Indonesia, 19–20 August 2021; pp. 116–119. [CrossRef]
11. Gayathri, S.; Ujwala, T.U.; Vinusha, C.V.; Pauline, N.R.; Tharunika, D.B. Detection of Papaya Ripeness Using Deep Learning Approach. In Proceedings of the 2021 3rd International Conference on Inventive Research in Computing Applications (ICIRCA), Coimbatore, India, 2–4 September 2021; pp. 1755–1758. [CrossRef]
12. Wu, D.; Wu, C. Research on the Time-Dependent Split Delivery Green Vehicle Routing Problem for Fresh Agricultural Products with Multiple Time Windows. *Agriculture* **2022**, *12*, 793. [CrossRef]
13. An, Z.; Wang, X.; Li, B.; Xiang, Z.; Zhang, B. Robust visual tracking for UAVs with dynamic feature weight selection. *Appl. Intell.* **2022**, *14*, 392–407. [CrossRef]

14. Kumar, A.; Joshi, R.C.; Dutta, M.K.; Jonak, M.; Burget, R. Fruit-CNN: An Efficient Deep learning-based Fruit Classification and Quality Assessment for Precision Agriculture. In Proceedings of the 2021 13th International Congress on Ultra-Modern Telecommunications and Control Systems and Workshops (ICUMT), Brno, Czech Republic, 25–27 October 2021; pp. 60–65. [CrossRef]
15. Widiyanto, S.; Wardani, D.T.; Wisnu Pranata, S. Image-Based Tomato Maturity Classification and Detection Using Faster R-CNN Method. In Proceedings of the 2021 5th International Symposium on Multidisciplinary Studies and Innovative Technologies (ISMSIT), Ankara, Turkey, 21–23 October 2021; pp. 130–134. [CrossRef]
16. Wu, H.; Cheng, Y.; Zeng, R.; Li, L. Strawberry Image Segmentation Based on U^ 2-Net and Maturity Calculation. In Proceedings of the 2022 14th International Conference on Advanced Computational Intelligence (ICACI), Wuhan, China, 15–17 July 2022; pp. 74–78. [CrossRef]
17. Zhang, R.; Li, X.; Zhu, L.; Zhong, M.; Gao, Y. Target detection of banana string and fruit stalk based on YOLOv3 deep learning network. In Proceedings of the 2021 IEEE 2nd International Conference on Big Data, Artificial Intelligence and Internet of Things Engineering (ICBAIE), Nanchang, China, 26–28 March 2021; pp. 346–349. [CrossRef]
18. Mohd Basir Selvam, N.A.; Ahmad, Z.; Mohtar, I.A. Real Time Ripe Palm Oil Bunch Detection using YOLO V3 Algorithm. In Proceedings of the 2021 IEEE 19th Student Conference on Research and Development (SCOReD), Kota Kinabalu, Malaysia, 23–25 November 2021; pp. 323–328. [CrossRef]
19. Wu, Y.J.; Yi, Y.; Wang, X.F.; Jian, C. Fig Fruit Recognition Method Based on YOLO v4 Deep Learning. In Proceedings of the 2021 18th International Conference on Electrical Engineering/Electronics, Computer, Telecommunications and Information Technology (ECTI-CON), Chiang Mai, Thailand, 19–22 May 2021; pp. 303–306. [CrossRef]
20. Zhou, X.; Wang, P.; Dai, G.; Yan, J.; Yang, Z. Tomato Fruit Maturity Detection Method Based on YOLOV4 and Statistical Color Model. In Proceedings of the 2021 IEEE 11th Annual International Conference on CYBER Technology in Automation, Control, and Intelligent Systems (CYBER), Jiaxing, China, 27–31 July 2021; pp. 904–908. [CrossRef]
21. Jose, N.T.; Marco, M.; Claudio, F.; Andres, V. Disease and Defect Detection System for Raspberries Based on Convolutional Neural Networks. *Electronics* **2021**, *11*, 11868. [CrossRef]
22. Wang, J.; Wang, L.Q.; Han, Y.L.; Zhang, Y.; Zhou, R.Y. On Combining Deep Snake and Global Saliency for Detection of Orchard Apples. *Electronics* **2021**, *11*, 6269. [CrossRef]
23. Zhou, X.; Ma, H.; Gu, J.; Chen, H.; Deng, W. Parameter adaptation-based ant colony optimization with dynamic hybrid mechanism *Eng. Appl. Artif. Intell.* **2022**, *114*, 105139. [CrossRef]
24. Chen, H.Y.; Miao, F.; Chen, Y.J.; Xiong, Y.J.; Chen, T. A Hyperspectral Image Classification Method Using Multifeature Vectors and Optimized KELM. *IEEE J. Sel. Top. Appl. Earth Obs. Remote Sens.* **2021**, *14*, 2781–2795. [CrossRef]
25. Wang, C.Y.; Mark Liao, H.Y.; Wu, Y.H.; Chen, Y.H.; Hsieh, J.W.; Yeh, I.H. CSPNet: A New Backbone that can Enhance Learning Capability of CNN. In Proceedings of the 2020 IEEE/CVF Conference on Computer Vision and Pattern Recognition Workshops (CVPRW), Seattle, WA, USA, 14–19 June 2020; pp. 1571–1580. [CrossRef]
26. He, K.; Zhang, X.; Ren, S.; Sun, J. Spatial Pyramid Pooling in Deep Convolutional Networks for Visual Recognition. *IEEE Trans. Pattern Anal. Mach. Intell.* **2015**, *37*, 346–361. [CrossRef] [PubMed]
27. Zhao, H.; Liu, J.; Chen, H.; Li, Y.; Xu, J.; Deng, W. Intelligent diagnosis using continuous wavelet transform and gauss convolutional deep belief network. *IEEE Trans. Reliab.* **2022**, *2022*, 1–11. [CrossRef]
28. Han, K.; Wang, Y.; Tian, Q.; Guo, J.; Xu, C. GhostNet: More Features from Cheap Operations. In Proceedings of the 2020 IEEE/CVF Conference on Computer Vision and Pattern Recognition (CVPR), Seattle, WA, USA, 14–19 June 2020; pp. 1577–1586. [CrossRef]
29. Hu, J.; Shen, L.; Albanie, S.; Sun, G.; Wu, E. Squeeze-and-Excitation Networks. *IEEE Trans. Pattern Anal. Mach. Intell.* **2020**, *42*, 2011–2023. [CrossRef] [PubMed]
30. Yao, R.; Guo, C.; Deng, W.; Zhao, H. A novel mathematical morphology spectrum entropy based on scale-adaptive techniques. *ISA Trans.* **2022**, *126*, 691–702. [CrossRef] [PubMed]
31. Woo, S.; Park, J.; Lee, J.Y.; Kweon, I.S. CBAM: Convolutional Block Attention Module. In Proceedings of the 15th European Conference on computer vision (ECCV), Munich, Germany, 8–14 September 2018; Springer: Cham, Switzerland, 2018; pp. 3–19.
32. Deng, W.; Ni, H.; Liu, Y.; Chen, H.; Zhao, H. An adaptive differential evolution algorithm based on belief space and generalized opposition-based learning for resource allocation. *Appl. Soft Comput.* **2022**, *127*, 109419. [CrossRef]

electronics

Article

Abnormal Cockpit Pilot Driving Behavior Detection Using YOLOv4 Fused Attention Mechanism

Nongtian Chen [1,*], Yongzheng Man [2] and Youchao Sun [3]

1 College of Aviation Engineering, Civil Aviation Flight University of China, Guanghan 618307, China
2 College of Civil Aviation Safety Engineering, Civil Aviation Flight University of China, Guanghan 618307, China
3 College of Civil Aviation, Nanjing University of Aeronautics and Astronautics, Nanjing 211106, China
* Correspondence: chennongtian@hotmail.com; Tel.: +86-83-8518-3621

Abstract: The abnormal behavior of cockpit pilots during the manipulation process is an important incentive for flight safety, but the complex cockpit environment limits the detection accuracy, with problems such as false detection, missed detection, and insufficient feature extraction capability. This article proposes a method of abnormal pilot driving behavior detection based on the improved YOLOv4 deep learning algorithm and by integrating an attention mechanism. Firstly, the semantic image features are extracted by running the deep neural network structure to complete the image and video recognition of pilot driving behavior. Secondly, the CBAM attention mechanism is introduced into the neural network to solve the problem of gradient disappearance during training. The CBAM mechanism includes both channel and spatial attention processes, meaning the feature extraction capability of the network can be improved. Finally, the features are extracted through the convolutional neural network to monitor the abnormal driving behavior of pilots and for example verification. The conclusion shows that the deep learning algorithm based on the improved YOLOv4 method is practical and feasible for the monitoring of the abnormal driving behavior of pilots during the flight maneuvering phase. The experimental results show that the improved YOLOv4 recognition rate is significantly higher than the unimproved algorithm, and the calling phase has a mAP of 87.35%, an accuracy of 75.76%, and a recall of 87.36%. The smoking phase has a mAP of 87.35%, an accuracy of 85.54%, and a recall of 85.54%. The conclusion shows that the deep learning algorithm based on the improved YOLOv4 method is practical and feasible for the monitoring of the abnormal driving behavior of pilots in the flight maneuvering phase. This method can quickly and accurately identify the abnormal behavior of pilots, providing an important theoretical reference for abnormal behavior detection and risk management.

Keywords: pilot abnormal behavior; behavior detection; YOLOv4 algorithm; CBAM; flight safety

Citation: Chen, N.; Man, Y.; Sun, Y. Abnormal Cockpit Pilot Driving Behavior Detection Using YOLOv4 Fused Attention Mechanism. *Electronics* **2022**, *11*, 2538. https://doi.org/10.3390/electronics11162538

Academic Editor: George A. Papakostas

Received: 26 July 2022
Accepted: 9 August 2022
Published: 13 August 2022

1. Introduction

Overall, 60% to 80% of flight accidents are caused by human factors. The statistics from the Civil Aviation Safety Annual Report show that in the past 10 years, the proportion of flight accidents caused by pilot and flight crew factors has been as high as 67.16% [1]. With the rapid development of civil aviation, the air transportation volume has increased significantly, and ensuring aviation safety has resulted in higher requirements for civil aviation pilots. According to relevant aviation accident statistics, most of the flight accidents are caused by the abnormal behavior of pilots, and the abnormal behavior of pilots in the cockpit is directly or indirectly related to flight accidents and symptoms. On 10 July 2018, an oxygen mask incident occurred in the airspace of Guangzhou on a flight from Hong Kong to Dalian in China. The investigation results showed that the cause of the incident was that the co-pilot smoked electronic cigarettes in the cockpit (abnormal behavior). An adjacent air conditioning unit was mistakenly shut down, resulting in a lack of oxygen in the cabin, triggering the incident. The Civil Aviation Administration of China issued an advisory circular (AC-121-FS-2018-130) on the flight operation style of civil aviation pilots, regulating pilots during the whole process of flight operations from the pre-flight stage to the flight operation stage, post-flight, and short-stop and station-based stages, driving behavior to improve the professionalism of pilot teams. Possible solutions for how to identify and monitor the abnormal behavior of pilots effectively, prevent the possible consequences of the risk of abnormal behavior of pilots, and explore and establish an effective mechanism to reduce human errors from the perspective of intrinsic safety have attracted the attention of many researchers. Therefore, it is of great practical significance to carry out research on the identification, monitoring, and early warnings of the abnormal driving behavior of pilots to regulate this driving behavior of pilots and ensure the safety of aviation operations.

Abnormal behavior research originated in the 1960s, and first explored the mechanism of abnormal behavior from the perspective of the behavioral environment [2]. Action recognition is an important field in computer vision and has been the subject of extensive research. It is widely used in pedestrian detection [3], robot vision [4], car driver detection [5], intelligent monitoring [6], and worker detection [7]. With the development of information technology, more scholars are using information detection technology to carry out abnormal behavior research. Abnormal behavior identification and detection processes are used to locate and detect abnormal actions; that is, the accurate identification of a certain action. The traditional detection technology has problems such as poor robustness to changing targets and long and redundant detection windows, which limit the improvement of the accuracy and speed during target detection. With the emergence of convolutional neural networks, due to their better representative learning ability, the research on abnormal behavior detection began to develop in the direction of convolutional neural network technology. In 2017, Wang et al. [8] tried to use the depth map method for the first time to identify the hand movements of cockpit pilots and to implement an approach and landing safety analysis. Liu et al. extracted time series of 3D human skeleton key points using Yolov4 and applied a mean shift target tracking algorithm, then converted key points into spatial RGB data and put them into a multi-layer convolution neural network for recognition [9]. Zhou et al. proposed a new framework for behavior recognition [10]. In this framework, we propose an object depth estimation algorithm to compute the 3D spatial location object information and use this information as the input to the action recognition model. At the same time, to obtain more spatiotemporal information and better deal with long-term videos, combined with the attention mechanism, spatiotemporal convolution and attention-based LSTMs (ST-CNN and ATT-LSTM) are proposed. Incorporating deep spatial information into each segment, the model focuses on the extraction of key information, which is crucial for improving the behavior recognition performance. Some scholars have proposed an abnormal target detection method based on the T-TINY-YOLO network model. The YOLO network model is used to train the calibrated abnormal behavior data to achieve end-to-end abnormal behavior classification, thereby achieving abnormal target detection for specific application scenarios [11]. Some scholars have studied the impact of civil aviation pilots' work stress on unsafe behavior based on a correlation analysis and multiple regression analysis [12]. In 2018, Yang et al. used the heads-up display to perform pattern recognition for pilot behavior, and for the first time proposed a behavior recognition framework that included pilot eye movements, head movements, and hand movements [13]. The deep-learning-based anomaly detection reduces human labor and its decision making ability is comparatively reliable, thereby ensuring public safety. Waseem et al. proposed a two-stream neural network in this direction for anomaly detection in surveillance [14]. Qu proposed a future frame prediction framework and a multiple instance learning (MIL) framework by leveraging attention schemes to learn anomalies [15]. Other scholars have used 3D ConvNets to identify anomalies from surveillance videos [16]. Waseem et al. presented an efficient light-weight convolutional neural network (CNN)-based anomaly recognition framework that is functional in surveillance environments with reduced time complexity [17]. One-shot image recognition has been explored for many applications in the computer vision community. One-shot anomaly recognition can be efficiently handled according to the 3D-CNN model [18]. The low reliability during feature and tracking box detection is still a problem in visual object tracking, An et al. proposed a robust tracking method for unmanned aerial vehicles (UAV) using dynamic feature weight selection [19]. Wu designed a road travel time calculation method across time periods. Considering the time-varying vehicle speed, fuel consumption, carbon emissions, and customer time window, the satisfaction measure function and economic cost measure function based on the time window were adopted [20]. Regarding the study by Chen [21], in order to improve the accuracy and generalization ability during hyperspectral image classification, in their paper a feature extraction method combining a principal component analysis (PCA) and local binary pattern (LBP) was developed for hyperspectral images, which provided a new idea for processing hyperspectral images. Zhou [22] proposed an ant colony optimization (ACO) algorithm based on parameter adaptation using a particle swarm optimization (PSO) algorithm with global optimization ability, a fuzzy system with fuzzy reasoning ability, and a 3-Opt algorithm with local search ability, namely PF3SACO. Yao et al. proposed a scale-adaptive mathematical morphological spectral entropy (AMMSE) approach to improve the scale selection. In support of the proposed method, two properties of the mathematical morphological spectra (MMS), namely the non-negativity and monotonic decrease, were demonstrated [23].

In recent years, deep learning has achieved outstanding performance in many fields, such as image processing, speech recognition, and semantic segmentation. Now, the commonly used neural networks include deep Boltzmann machines (DBM), recurrent neural networks (RNNs) [24], and convolutional neural networks (CNNs) [25]. In 2015, Girshick [26] first proposed the R-CNN algorithm for abnormal behavior recognition, which effectively improved the recognition accuracy. The improved algorithms, such as Fast R-CNN and Faster R-CNN, proposed later have higher efficiency and accuracy in abnormal behavior recognition [27,28]. These improved methods improve the speed of the information collection, information processing ability, and transmission speed, and provide important theoretical and technical support for abnormal behavior identification and early warnings. There are many regional models based on deep learning, including SSP [29], SSD [30], and YOLO [31,32].

In short, many scholars have carried out studies on abnormal behavior recognition and have achieved many effective results, but further research is needed on the abnormal behavior recognition algorithms and monitoring effects, especially as research combined with the abnormal behavior of pilots in the civil aviation industry is rare. This paper proposes an abnormal pilot behavior monitoring and identification algorithm based on an improved YOLOv4 (you only look once) approach, adopts a deep-learning-based abnormal behavior target detection algorithm, and introduces a convolutional attention mechanism module (CBAM) for the feature fusion of the backbone network. A convolutional back attention module is used to enhance the perception of the model in the channel and space, and finally to extract the features through the convolutional neural network and monitor and identify the abnormal behavior of pilots in order to provide a reference for the identification of the abnormal behavior of pilots and the norms of pilot behavior.

2. Overview of the Method

2.1. Convolutional Neural Networks

In recent years, convolutional neural networks (CNNs) have made great progress in image and video processing. The CNNs extract the high-level semantic features of images through the deep neural network structure, and complete the recognition and classification of complex images and videos. A convolutional neural network is generally a feed-forward neural network formed by overlapping convolutional layers, pooling layers, and fully connected layers, and the characteristics include local connections, weight sharing, and aggregation. These properties make convolutional neural networks invariant to certain degrees of translation, scaling, and rotation. The role of the convolutional layer is to extract the features of a local area, and the different convolution kernels are equivalent to different feature extractors. The role of the pooling layer is to perform feature selection and reduce the number of features, thereby reducing the number of parameters. The learning rate is a very important parameter in such algorithms. Here, Softmax is selected as the classifier, and the optimization algorithm of the learning rate is the adaptive algorithm Adam. Its calculation formula is:

$$m_t = \beta_1 m_{t-1} + (1 - \beta_1) g_t \tag{1}$$

$$v_t = \beta_2 v_{t-1} + (1 - \beta_2) g_t^2 \tag{2}$$

Here, t is the time, m_t is the first-order moment estimation of the gradient, v_t is the second-order moment estimation of the gradient, and β_1 and β_2 are the exponential decay rated of the moment estimation, ranging from 0 to 1. When calculating the deviation correction, Equation (2) will be used, where $m\hat{}_t$ and $v\hat{}_t$ are the corrections of the sum:

$$m\hat{}_t = \frac{m_t}{1 - \beta_1{}^t} \tag{3}$$

$$v\hat{}_t = \frac{v_t}{1 - \beta_2{}^t} \tag{4}$$

The gradient is updated using Equation (5):

$$\theta_{t+1} = \theta_t - \mu m\hat{}_t / (\sqrt{v\hat{}_t} + \varepsilon) \tag{5}$$

Here, ε is a numerically stable small constant; θ_t represents the gradient to be updated, generally 10^{-8}; and μ is the step size, generally 0.001.

2.2. YOLO

The YOLO algorithm is an object recognition and localization algorithm based on a deep neural network. It is characterized by improving the speed of the deep learning target detection process and meeting the requirements for real-time monitoring to a certain extent. The CNN algorithm convolves the image through the convolutional neural network, but the detection speed is low, which cannot meet the needs of real-time monitoring. The characteristic of the YOLO algorithm is that only one CNN operation is needed for the image, and the corresponding region and position of the regression prediction frame can be obtained in the output layer. The algorithm steps are as follows.

Divide the original image into $S \times S$ grid cells. If an object falls in the grid, then the feature of the grid is the object (if multiple objects fall in this grid, the closest object in the center is the feature of this grid).

(1) The features of each grid need B bounding boxes (Bbox) to return. To ensure accuracy, the B box features corresponding to each grid are also the same.

(2) Each *Bbox* also predicts 5 values: x, y, w, h, and the confidence. The (x, y) is the relative position of the center of the *Bbox* in the corresponding grid, and (w, h) is the length and width of the *Bbox* relative to the full image. The range of these 4 values is [0, 1], and these 4 values can be calculated from the features. The confidence is a numerical measure of how accurate a prediction is. Let us set the reliability as C, where I refers to the intersection ratio between the predicted *Bbox* and the ground-truth box in the image; the probability of containing objects in the corresponding grid of the *Bbox* is P, and the formula is:

$$C = P\,i \tag{6}$$

(3) If the grid corresponding to the *Bbox* contains objects, then P = 1, otherwise it is equal to 0. If there are N prediction categories, plus the confidence of the previous *Bbox*

prediction, the $S \times S$ grid requires o output information. The calculation method is as follows:

$$o = S \times S \times (1 + B + N) \tag{7}$$

For each grid, the confidence that it belongs to each category will also be predicted. Among them, the B boxes can only belong to one category, which corresponds to the first step, and its characteristics are the same.

2.3. YOLOv4

2.3.1. CSPDarkent–53

CSPDarknet-53 is based on the YOLOv4 backbone network, and is modified and improved on the basis of it, finally forming a backbone structure that includes 5 CSP modules. The CSP module divides the feature map of the base layer into two parts; that is, the original stack of residual blocks is split into two different parts on the left and right. The main part is used to continue the original stack of residual blocks, and the other part is similar to the residual edge, which is directly connected to the end after a small amount of processing. They are then merged through a cross-stage hierarchy. Through this processing, the accuracy of the model is also ensured based on reducing the amount of calculation.

2.3.2. Prediction Box Selection

The prediction principle of YOLOv4 is to divide the image into 13 × 13, 26 × 26, and 52 × 52 networks, and each network node is responsible for the prediction of one area. YOLOv4 uses a clustering method to select candidate frames, and the cluster centers are divided into 3 scales of different sizes according to the different sizes used for prediction. The calculation formula for the offset predicted by the network is:

$$b_x = \sigma(t_x) + c_x \tag{8}$$

$$b_y = \sigma(t_y) + c_y \tag{9}$$

$$b_w = p_w e^{t_w} \tag{10}$$

$$b_h = p_h e^{t_w} \tag{11}$$

Here, (c_x, c_y) is the distance from the upper left corner (when the prediction frame is selected, the values of c_x and c_y are 1); (p_w, p_h) are the length and width of the prior frame, respectively; p_w and p_h are determined manually; (t_x, t_y) is the offset of the target center point relative to the upper left corner of the grid, where the prediction point is located; (t_w, t_h) are the width and height of the prediction frame, which are related to p_w and p_h, respectively (see Equations (10) and (11)) and with which the width and height of the Bbox are obtained; σ is the activation function, indicating the probability between [0, 1].

3. The Improved Network Model

3.1. Channel Attention Mechanism

An attention mechanism is a method of processing data, which imitates the human visual system, integrates local visual structures, focuses attention on important points among a lot of information, selects key information, and ignores other unimportant information. A Channel attention block (CAB) can model the dependencies of different channel features, fuse multi-channel feature images, and adaptively adjust their feature weights. The channel attention module rescales the weights of each input channel so that the key region feature channels containing the target object have a greater contribution during convolution. The idea is to enhance the weight of the key channels and reduce the weight of invalid channels. The channel attention can be expressed as Equation (12):

$$M_C(F) = \sigma(MLP(AvgPool(F)) + MLP(MaxPool(F))) \tag{12}$$

In the formula, F represents the feature of the input, where σ represents the activation function, and *AvgPool*() and *MaxPool*() represent the processes of average pooling and maximum pooling, respectively.

3.2. Spatial Attention Mechanism

In the process of behavior recognition, when the pilots perform abnormal behaviors such as calling and smoking, the location features (such as gradients and grayscales) will change drastically. Therefore, the spatial attention mechanism (SAB) can be used in the feature map. By increasing the weights of the key parts, the network can focus more on crucial features and improve the feature extraction ability of the network. The channel attention mechanism is the part that the network one must pay attention to, and the spatial attention mechanism gives the locations of key features. The specific implementation process is shown in Equation (13):

$$M_S(F) = \sigma(f^{7\times 7}([AvgPool(F); MaxPool(F)])) \quad (13)$$

In the formula, F also represents the feature of the input; 7×7 convolution is used for feature extraction, then average pooling and maximum pooling are used for evaluation, and finally normalization is performed according to the activation function σ.

3.3. Attention Mechanism Fusion

This article involves the fusion of the spatial attention mechanism and channel attention mechanism, which will be used for the detection of behavior recognition in the YOLOv4 algorithm. For an input video F, the global information for each feature channel is obtained using the global average pooling and maximum pooling operations, and then the future channel attention vector is obtained through two fully connected layers, which are used to weigh the input feature F channel using M_C(F). In addition, this feature is input to the 3×3 convolution layer and output by the sigmoid function and $M_S(F)$, which gives the feature F'. Its structure is shown in Figure 1. In this paper, the attention mechanism is added to the two effective feature layers extracted from the backbone network, and the attention mechanism is also added to the results after up-sampling. The attention mechanism in this paper can enhance the feature extraction ability of the model based by increasing a small amount of the computation. Due to the large size gap in the image dataset, after the attention mechanism is introduced, the image features can be extracted from multiple scales, which strengthens the model's ability to detect images.

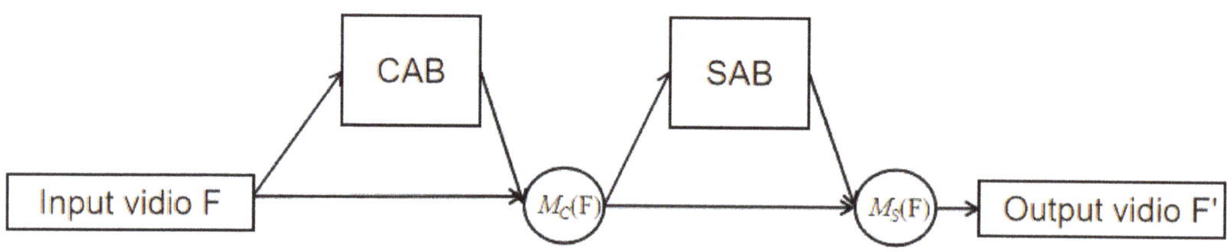

Figure 1. Fusion of CAB and SAB attention mechanisms.

4. Test and Analysis

4.1. Environment Settings

The environmental configuration of this experiment is shown in Table 1, and the hardware configuration of the comparison experiment is the same configuration.

Table 1. Numbers of various types of images.

Operating System	Windows 10
CPU	i7-10750H
RAM	16G
GPU	RTX 2060Ti
Language	Python 3.6
Backbone Network	CSPDarkent–53

The research subjects in this paper are pilots. There is no special public dataset available at present, so the database must be established by itself. The database data in this paper mainly come from the relevant action pictures taken by us, pictures that meet the requirements for the existing datasets, and relevant pictures searched on the Internet. The dataset is prepared according to the deep learning standard dataset format in VOC 2007. The specific steps are: (1) use the labeling tool to classify the abnormal behavior in the image, whereby the category names are calling and smoking; (2) create relevant files according to the standard dataset format and save the files, including pictures, sizes, and coordinates for target detection, then divide the dataset into a training set and test set at a ratio of 9:1. Table 2 shows the numbers of images in the various categories in the dataset.

Table 2. Numbers of various types of images.

Type	Calling	Smoking
Number	500	500

4.2. Detection Process

According to the driving behavior requirements and flight guidelines for civil aircraft pilots, smoking and calling during the flight can be called abnormal pilot behaviors. Therefore, in the identification process, these typical abnormal behaviors are identified to provide a basis for the implementation of abnormal driving behavior monitoring and early warning processes. The process of detecting abnormal pilot behaviors is shown in Figure 2. The main process is as follows.

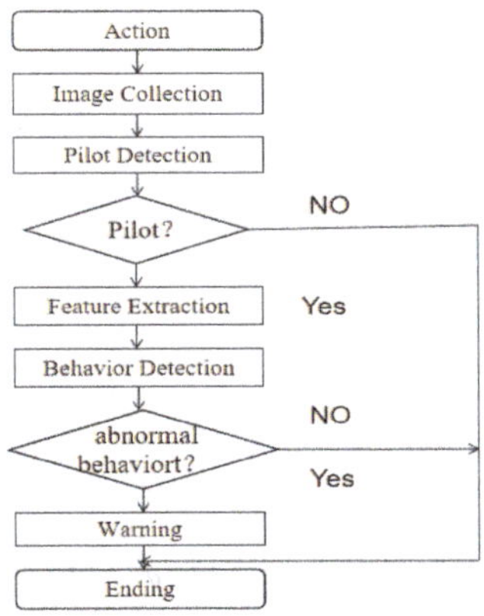

Figure 2. Abnormal behavior detection process.

(1) Taking with a camera in the cockpit of an aircraft simulator, capture images in frame units from the live video stream captured by the camera;
(2) Perform abnormal behavior detection on the pilots. Use the model trained by the deep learning YOLO v4 algorithm to locate the pilot area, and when the area is detected, the abnormal behavior can be identified according to the model;

(3) Monitor the video. When there is abnormal behavior, it will give a warning. After the frame detection ends, enter the next frame.

4.3. Model Training

In the improved YOLOv4 model training, the smaller the loss value of the model structure the better, and the expected value is 0. To achieve the best performance for the model, during training the number of iterations is set to 600, the weight decay coefficient is set to 0.0001, and the learning rate momentum is set to 0.9 to prevent the model from overfitting. The maximum training batch is set to 8, the loss function value drops sharply from 0 to 300 times, and the loss number decreases slowly from 300 to 600 times. After 400 iterations, the loss value tends to stabilize around 0.05, and the model reaches the maximum excellence state. The training loss is shown in Figure 3.

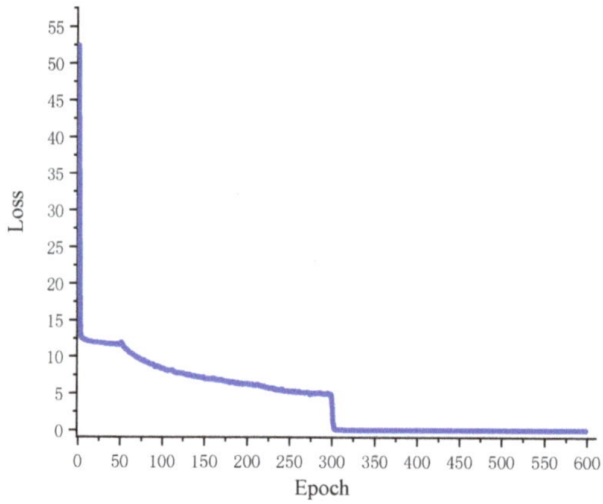

Figure 3. Loss map.

4.4. Evaluation of the Model Performance

It can be seen from the Figure 4 below that for the two types of abnormal behaviors, the recognition rate of the improved YOLOv4 is significantly higher than that of the unimproved algorithm.

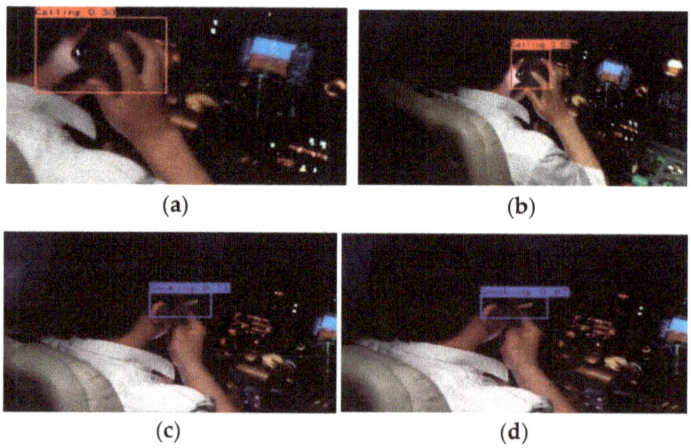

Figure 4. Object detection results using the YOLOv4 method (**a**,**c**) and object detection results using our proposed method (**b**,**d**).

When compared with the original YOLOv4 algorithm, the unified video is input and the darknet backbone network is used for training. In order to make the model converge as soon as possible, this experiment adopts the method of transfer learning. The experimental data are shown in Table 3.

Table 3. Comparison between the original YOLO algorithm and the improved YOLO algorithm used in this article.

	YOLOv4/%	Improved YOLOv4/%
Smoking	71	82
Calling	50	82

In the evaluation of the pilots' abnormal behavior recognition effect, the important parameters are as follows: T_P indicates that abnormal behavior is detected, and there is also abnormal behavior in the actual picture (the number of samples detected by the algorithm); T_N indicates that no abnormal behavior is detected, and the actual picture is not abnormal (the number of correct error samples detected by the algorithm); F_N means that no abnormal behavior is detected, but abnormal behavior is present in the actual graph (the number of samples that the algorithm detects wrong); F_P means that no abnormal behavior is detected, and there is no abnormal behavior in the actual graph (the number of correct samples needed for the algorithm to detect errors); the recall rate (R) is the ratio of the number of abnormal behaviors detected to the total number of abnormal behaviors; the precision rate (P) is the ratio of the number of correctly detected abnormal behaviors to the total number of abnormal behaviors [33–35]. The average precision (AP) measures the accuracy of the model from the two aspects of precision and recall. It is a direct evaluation standard for model accuracy, and it can also be analyzed using the detection effect of a single category.

$$R = \frac{T_P}{T_P + F_N} \tag{14}$$

$$P = \frac{T_P}{T_P + F_P} \tag{15}$$

The abnormal behavior recognition results obtained according to Equations (14) and (15) are shown in Table 4.

Table 4. Evaluation indicators of behavior detection.

Abnormal Behavior	P/%	R/%	mAP/%
Smoking	85.54	85.54	89.23
Calling	75.76	87.36	87.35

5. Conclusions

An abnormal pilot behavior monitoring method based on the improved YOLO v4 algorithm was proposed. The method was verified by collecting abnormal behavior recognition datasets. The recognition rate was improved compared to the original basis. The CSPDarkent-53 framework was used to train the recognition model, which enhanced the method. The robustness of the training model was 85.54% for docking calls and smoking recognition. This method expands the training set through data augmentation, thereby achieving high-accuracy recognition with less training data. The algorithm performance needs to be further improved in later research. The next step is to explore the implantation of the algorithm into the camera terminal for practical applications.

The deep learning algorithm based on the improved YOLOv4 abnormal driving behavior monitoring algorithm can effectively identify the abnormal driving behavior of pilots. The attention mechanisms (CAB and SAB) were introduced to enhance the model's perception in channels and spaces. The image semantic features are extracted based on the deep neural network structure, and the image and video recognition and classification of the pilots' driving behavior are then completed.

In the next step, we will continue to improve the network so that the network is not limited to feature extraction in the spatial domain, and we will also add some information in the time domain so as to further improve the generalization ability of the model.

Author Contributions: Conceptualization, N.C. and Y.S.; formal analysis, investigation, writing of the original draft, N.C. and Y.M. All authors have read and agreed to the published version of the manuscript.

Funding: This research was funded by the National Natural Science Foundation of China, grant number U2033202; the Key R&D Program of the Sichuan Provincial Department of Science and Technology (2022YFG0213); and the Safety Capability Fund Project of the Civil Aviation Administration of China (ASSA2022/17).

Data Availability Statement: The data used to support the findings of this study are included within the article.

Acknowledgments: Written informed consent has been obtained from the patients to publish this paper.

Conflicts of Interest: The authors declare no conflict of interest.

References

1. *Annual Report on Aviation Safety in China, 2018*; Civil Aviation Administration of China: Beijing, China, 2019.
2. Xiao, W.; Liu, H.; Ma, Z.; Chen, W. Attention-based deep neural network for driver behavior recognition. *Future Gener. Comput. Syst.* **2022**, *132*, 152–161. [CrossRef]
3. Zhang, S.; Benenson, R.; Omran, M.; Hosang, J.; Schiele, B. How far are we from solving pedestrian detection? In Proceedings of the IEEE Conference on Computer Vision and Pattern Recognition, Las Vegas, NV, USA, 27–30 June 2016; Volume 12, pp. 1259–1267.
4. Senicic, M.; Matijevic, M.; Nikitovic, M. Teaching the methods of object detection by robot vision. In Proceedings of the IEEE International Convention on Information and Communication Technology, Kansas City, MO, USA, 20–24 May 2018; Volume 7, pp. 558–563.
5. Nemcová, A.; Svozilová, V.; Bucsuházy, K.; Smíšek, R.; Mezl, M.; Hesko, B.; Belák, M.; Bilík, M.; Maxera, P.; Seitl, M.; et al. Multimodal features for detection of driver stress and fatigue. *IEEE Trans. Intell. Transp. Syst.* **2021**, *22*, 3214–3233. [CrossRef]
6. Popescu, D.; Stoican, F.; Stamatescu, G.; Ichim, L.; Dragana, C. Advanced UAV-WSN system for intelligent monitoring in precision agriculture. *Sensors* **2020**, *20*, 817. [CrossRef] [PubMed]
7. Mneymneh, B.E.; Abbas, M.; Khoury, H. Vision-based framework for intelligent monitoring of hardhat wearing on construction sites. *J. Comput. Civ. Eng.* **2019**, *33*, 04018066. [CrossRef]
8. Wang, T.; Fu, S.; Huang, D.; Cao, J. Pilot action identification in the cockpit. *Electron. Opt. Control.* **2017**, *24*, 90–94.
9. Liu, Y.; Zhang, S.; Li, Z.; Zhang, Y. Abnormal Behavior Recognition Based on Key Points of Human Skeleton. *IFAC-PapersOnLine* **2020**, *53*, 441–445. [CrossRef]
10. Zhou, K.; Hui, B.; Wang, J.; Wang, C.; Wu, T. A study on attention-based LSTM for abnormal behavior recognition with variable pooling. *Image Vis. Comput.* **2021**, *108*, 104–120. [CrossRef]
11. Ji, H.; Zeng, X.; Li, H.; Ding, W.; Nie, X.; Zhang, Y.; Xiao, Z. Human abnormal behavior detection method based on T-TINY-YOLO. In Proceedings of the 5th International Conference on Multimedia and Image Processing, Nanjing, China, 10–12 January 2020; pp. 1–5.
12. Li, L.; Cheng, J. Research on the relationship between work stress and unsafe behaviors of civil aviation pilots. *Ind. Saf. Environ. Prot.* **2019**, *45*, 46–49.
13. Yang, K.; Wang, H. Pilots use head-up display behavior pattern recognition. *Sci. Technol. Eng.* **2018**, *18*, 226–231.
14. Ullah, W.; Ullah, A.; Hussain, T.; Muhammad, K.; Heidari, A.A.; Del Ser, J.; Baik, S.W.; De Albuquerque, V.H.C. Artificial Intelligence of Things-assisted two-stream neural network for anomaly detection in surveillance Big Video Data. *Future Gener. Comput. Syst.* **2022**, *129*, 286–297. [CrossRef]
15. Li, Q.; Yang, R.; Xiao, F.; Bhanu, B.; Zhang, F. Attention-based anomaly detection in multi-view surveillance videos. *Knowl.-Based Syst.* **2022**, *252*, 109348. [CrossRef]
16. Maqsood, R.; Bajwa, U.; Saleem, G.; Raza, R.H.; Anwar, M.W. Anomaly recognition from surveillance videos using 3D convolution neural network. *Multimed. Tools Appl.* **2021**, *80*, 18693–18716. [CrossRef]
17. Ullah, W.; Ullah, A.; Hussain, T.; Khan, Z.A.; Baik, S.W. An efficient anomaly recognition framework using an attention residual LSTM in surveillance videos. *Sensors* **2021**, *21*, 2811. [CrossRef]

18. Ullah, A.; Muhammad, K.; Haydarov, K.; Haq, I.U.; Lee, M.; Baik, S.W. One-shot learning for surveillance anomaly recognition using siamese 3d. In Proceedings of the International Joint Conference on Neural Networks, Glasgow, UK, 19–24 July 2020; pp. 1–8.
19. An, Z.; Wang, X.; Li, B.; Xiang, Z.; Zhang, B. Robust visual tracking for UAVs with dynamic feature weight selection. *Appl. Intell.* **2022**, 1–14. [CrossRef]
20. Wu, D.; Wu, C. Research on the Time-Dependent Split Delivery Green Vehicle Routing Problem for Fresh Agricultural Products with Multiple Time Windows. *Agriculture* **2022**, *12*, 793. [CrossRef]
21. Chen, H.; Miao, F.; Chen, Y.; Xiong, Y.; Chen, T. A hyperspectral image classification method using multifeatured vectors and optimized KELM. *IEEE J. Sel. Top. Appl. Earth Obs. Remote Sens.* **2021**, *14*, 2781–2795. [CrossRef]
22. Zhou, X.; Ma, H.; Gu, J.; Chen, H.; Deng, W. Parameter adaptation-based ant colony optimization with dynamic hybrid mechanism. *Eng. Appl. Artif. Intell.* **2022**, *114*, 105139. [CrossRef]
23. Yao, R.; Guo, C.; Deng, W.; Zhao, H. A novel mathematical morphology spectrum entropy based on scale-adaptive techniques. *ISA Trans.* **2022**, *126*, 691–702. [CrossRef]
24. Sherstinsky, A. Fundamentals of recurrent neural network (RNN) and long short-term memory (LSTM) network. *Phys. D Nonlinear Phenom.* **2020**, *404*, 132306. [CrossRef]
25. Kattenborn, T.; Leitloff, J.; Schiefer, F.; Hinz, S. Review on Convolutional Neural Networks (CNN) in vegetation remote sensing. *ISPRS J. Photogramm. Remote Sens.* **2021**, *173*, 24–49. [CrossRef]
26. Girshick, R. Fast R-CNN. Computer Science. In Proceedings of the 2015 IEEE International Conference on Computer Vision (ICCV), Santiago, Chile, 7–13 December 2015.
27. Ren, S.; He, K.; Girshick, R.; Sun, J. Faster r-cnn: Towards real-time object detection with region proposal networks. *Adv. Neural Inf. Processing Syst.* **2015**, *28*, 1137–1149. [CrossRef] [PubMed]
28. Deng, W.; Ni, H.; Liu, Y.; Chen, H.; Zhao, H. An adaptive differential evolution algorithm based on belief space and generalized opposition-based learning for resource allocation. *Appl. Soft Comput.* **2022**, *127*, 109419. [CrossRef]
29. Huang, Z.; Huang, L.; Gong, Y.; Huang, C.; Wang, X. Mask scoring r-cnn. In Proceedings of the IEEE Conference on Computer Vision and Pattern Recognition, Long Beach, CA, USA, 15–20 June 2019; pp. 6409–6418.
30. He, K.; Zhang, X.; Ren, S.; Sun, J. Spatial Pyramid Pooling in Deep Convolutional Networks for Visual Recognition. *IEEE Trans. Pattern Anal. Mach. Intell.* **2015**, *37*, 1904–1916. [CrossRef]
31. Liu, W.; Anguelov, D.; Erhan, D.; Szegedy, C.; Reed, S.; Fu, C.Y.; Berg, A.C. SSD: Single shot multibox detector. In *Lecture Notes in Computer Science*; Springer: Cham, Swizerland, 2016; Volume 9905, pp. 21–37.
32. Redmon, J.; Divvala, S.; Girshick, R.; Farhadi, A. You Only Look Once: Unified, Real-Time Object Detection. In Proceedings of the 2016 IEEE Conference on Computer Vision and Pattern Recognition, Las Vegas, NV, USA, 27–30 June 2016; pp. 779–788.
33. Chen, B.; Wang, X.; Bao, Q.; Jia, B.; Li, X.; Wang, Y. An Unsafe Behavior Detection Method Based on Improved YOLO Framework. *Electronics* **2022**, *11*, 1912. [CrossRef]
34. Kumar, T.; Rajmohan, R.; Pavithra, M.; Ajagbe, S.A.; Hodhod, R.; Gaber, T. Automatic face mask detection system in public transportation in smart cities using IoT and deep learning. *Electronics* **2022**, *11*, 904. [CrossRef]
35. Wahyutama, A.; Hwang, M. YOLO-Based Object Detection for Separate Collection of Recyclables and Capacity Monitoring of Trash Bins. *Electronics* **2022**, *11*, 1323. [CrossRef]

electronics

Article

Quantum Dynamic Optimization Algorithm for Neural Architecture Search on Image Classification

Jin Jin [1], Qian Zhang [2], Jia He [3,*] and Hongnian Yu [4]

1 School of Software Engineering, Chengdu University of Information Technology, Chengdu 610225, China
2 Active Network (Chengdu) Co., Ltd., Chengdu 610021, China
3 School of Computer Science, Chengdu University of Information Technology, Chengdu 610225, China
4 School of Computing, School of Engineering and the Built Environment, Edinburgh Napier University, Edinburgh 16140, UK
* Correspondence: hejia@cuit.edu.cn

Abstract: Deep neural networks have proven to be effective in solving computer vision and natural language processing problems. To fully leverage its power, manually designed network templates, i.e., Residual Networks, are introduced to deal with various vision and natural language tasks. These hand-crafted neural networks rely on a large number of parameters, which are both data-dependent and laborious. On the other hand, architectures suitable for specific tasks have also grown exponentially with their size and topology, which prohibits brute force search. To address these challenges, this paper proposes a quantum dynamic optimization algorithm to find the optimal structure for a candidate network using Quantum Dynamic Neural Architecture Search (QDNAS). Specifically, the proposed quantum dynamics optimization algorithm is used to search for meaningful architectures for vision tasks and dedicated rules to express and explore the search space. The proposed quantum dynamics optimization algorithm treats the iterative evolution process of the optimization over time as a quantum dynamic process. The tunneling effect and potential barrier estimation in quantum mechanics can effectively promote the evolution of the optimization algorithm to the global optimum. Extensive experiments on four benchmarks demonstrate the effectiveness of QDNAS, which is consistently better than all baseline methods in image classification tasks. Furthermore, an in-depth analysis is conducted on the searchable networks that provide inspiration for the design of other image classification networks.

Keywords: quantum dynamics; global optimization; neural architecture search; image classification

Citation: Jin, J.; Zhang, Q.; He, J.; Yu, H. Quantum Dynamic Optimization Algorithm for Neural Architecture Search on Image Classification. *Electronics* **2022**, *11*, 3969. https://doi.org/10.3390/electronics11233969

Academic Editor: Dimitris Apostolou

Received: 4 November 2022
Accepted: 23 November 2022
Published: 30 November 2022

1. Introduction

Deep learning (DL) methods have shown great potential for such applications as computer vision and natural language processing [1]. Image classification is one of the four major tasks of computer vision. Given an input image, the image classification task aims to determine the category of the image [2].

To effectively deal with a classification task, multiple network architectures, i.e., ResNet [3], DensNet [4], and SENet [5], have been proposed. These new architectures have heuristic significance for designing neural networks, such as the residual module in ResNet, which has now become the basic module in many network architectures.

However, designing DL algorithms requires designers to have rich experiences. It is a challenging task to design neural network architectures due to the fact that little prior knowledge on architecture design is available and the designed structures are problem-dependent. In that case, the ability to automatically generate the correct network architecture for any given task has become a new requirement [6,7]. One way to generate these architectures is to use evolutionary algorithms (EA) [8]. Traditional topological neuroevolution research is the exploration of early neural network architecture searches [9,10]. EA uses neural networks to simplify search, weighting, structured search, and multi-objective search [11,12].

Google Research showed that Regular Evolutionary Algorithms (REA) [13] work well in neural network architecture.

Recent research on the neural network architecture search problem has brought a new trend in evolutionary neural network architecture search. However, there are two main challenges: (1) most well-designed new algorithms could not be used for neural network architecture search [14–16]; and (2) each search algorithm is only experimented on for a specific search space during the search process and is not verified on other, well-known search spaces [17,18].

The quantum dynamics optimization algorithm (QDO) is an iterative optimization algorithm [19] constructed by simulating the optimization process of the quantum dynamics equation. In the quantum dynamics optimization algorithm, the evolutionary process of the optimization algorithm over time is transformed into a quantum dynamics process. In the quantum dynamics optimization algorithm, the modulus of the wave function represents the distribution of the solution. Therefore, the evolution process is the evolution process of the optimization algorithm solution. The modulus of the quantum wave function can be obtained from the ensemble theory in physics, where the probability distribution represents the probability distribution of the quantum particles in a given state. The tunneling effect, potential barrier estimation, and other theories in quantum mechanics can effectively facilitate the optimization process of the optimization algorithm.

Here we explore an application of quantum dynamic optimization algorithms for a neural architecture search (NAS) problem. In the neural network architecture search problem, novelty search facilitates the discovery of excellent architectures [20]. The quantum dynamics optimization algorithm can effectively jump out of the local optimum and find the global optimum by using the tunnel effect. It is a well-designed intelligent optimization algorithm. The potential barrier estimation in quantum mechanics can make reasonable use of the information on non-optimal solutions in the process of algorithm optimization, thereby increasing the diversity of solutions. In the neural network architecture search problem, some non-optimal architectures may evolve into optimal architectures after iteration. The properties of these two aspects of the quantum dynamics optimization algorithm suggest that it may be a better solution to the neural network architecture search problem.

The proposed method is shown in Figure 1. Quantum dynamics optimization algorithms are competitive optimizations proposed in [19]. Recent research primarily focuses on improving quantum dynamics optimization algorithms [21]. By introducing different mechanisms, the optimization performance of the algorithm is further improved by improving the performance of the algorithm. Unlike previous studies, the method in [19] does not improve the performance of the algorithm for specific optimization tasks. Instead, it uses the most basic quantum dynamic optimization algorithm (QDO) to explore its application in neural network architecture research.

The NAS method relies on a search strategy to determine the next architecture to be evaluated, and a performance evaluation strategy to evaluate its performance [8]. This article will focus on search strategies. To evaluate the performance of the search algorithm more comprehensively, we use table-based NAS benchmarks as the benchmark dataset [22–24].

The contributions of this work can be summarized as follows:

- In addition to conventional evolutionary algorithms, for the first time, this paper applies a quantum heuristic optimization algorithm as a search algorithm for a neural network architecture search problem. We transform the applicability of quantum dynamics optimization algorithms from traditional optimization problems to neural network architecture search problems. The designed algorithm does not depend on specific data and is a general neural network architecture search algorithm.
- Reduce the problem search space by defining reasonable discretization encoding methods, and quantum heuristic rules. The use of the quantum tunneling effect and barrier estimation principle makes the proposed algorithm more competitive with general evolutionary methods.

- Conduct extensive experiments on NAS-Benchmark to demonstrate the effectiveness of the proposed models.

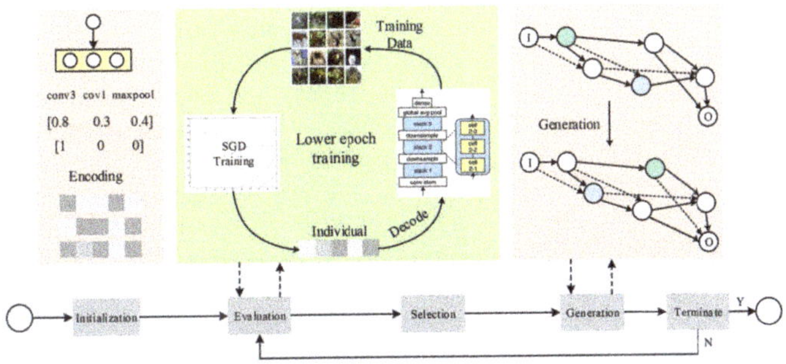

Figure 1. Pipeline of the QDO-NAS.

We first describe the quantum dynamic optimization algorithm (QDO; Section 2), then describe how to apply QDO to NAS (Section 3), and then Section 4 verifies the effectiveness of the search algorithm proposed for table-based benchmarks, such as NAS-Bench-101 [22], NAS-Bench-1Shot1 [23], NAS-Bench-201 [24], and NSATs-Bench [25].

2. Quantum Dynamic Optimization

The quantum dynamics optimization algorithm is an iterative optimization algorithm [19], in which the evolution of the optimization algorithm is transformed over time into a quantum dynamic process. The theories such as the tunneling effect and potential barrier estimation in quantum mechanics can effectively promote the optimization process of optimization algorithms.

According to the basic iterative operation of the optimization algorithm under the quantum dynamics model, the basic iterative process can be obtained in Algorithm 1.

Algorithm 1: Pseudocode of QDO.

Randomly generate k copies of free particle in the domain $[d_{min}, d_{max}]$, $\sigma_s = d_{max} - d_{min}$
while *(stop condition is not satisfied)* **do**
  Initialize $Ac = 0$
  while $(\sigma < \sigma_k)$ **do**
    for $i=1$ *to* k **do**
      Generate $x'[i] \sim N(x[i], \sigma^2)$
      if $(f(x'[i]) \leq f(x[i]))$ **then**
        $x[i] = x'[i]$, update the ith particle
      else
        $x[i] = x'[i]$, update the ith particle according to $T \propto e^{-\frac{\Delta x \sqrt{\Delta f}}{\sigma}}$
      end
    end
    $Ac = Ac + 1$
    Calculate the σ_k for k copies
  end
  $x_{\text{worse}}[i] = x_{\text{aver}}[i]$
  $\sigma = \sigma/2$
end
Output: $x_{best}[i]$

All operations of this basic iterative process are obtained by using the theoretical platform of the quantum dynamics of the optimization algorithm and the approximation and estimation of the objective function. The specific steps of QDO are as follows.

1. Generate k sampled individuals in the domain $[d_{min}, d_{max}]$.
2. The probability evolution of the location distribution of k sampled individuals can be considered as the evolution of the particle wave function modulus. The larger the value of k, the closer to the probability distribution of the wave function modulus. The initial mean square error σ takes the length of the domain. When the initial mean square error is large, the algorithm is not sensitive to the initial position of the sampled individual.
3. Generate new solutions with a normal distribution $x'[i] \sim N(x[i], \sigma^2)$, if the new solution $f(x'[i]) \leq f(x[i])$; that is, the new solution is better than the old solution, then the new solution is directly accepted; if the new solution is worse than the old solution, it can be considered from the physical image that the particle is blocked by the potential barrier, and the difference solution is accepted according to the probability that the barrier penetrates the transmission coefficient T.
4. This iterative process is repeated until the mean square error of the $x[i]$ positions of the k sampled individuals is less than or equal to the mean square error of the current normal sampling.
5. Replacing the worst position with the mean of the sampled individuals $x[i]$, $x_{\text{worse}}[i] = x_{\text{aver}}[i]$ reduces the mean square error of normal sampling and enters a smaller scale to perform the same iterative process.
6. If the algorithm meets the set maximum function evolution times $maxFE$, the entire iterative process ends, and the optimal solution $x_{\text{best}}[i]$ among the current k sampled individuals $x[i]$ is output.

3. Proposed Method

3.1. NAS Problem Black Box Modeling

The principle of NAS is to give a set of candidate neural network structures called the search space and use a certain strategy. During the search for the optimal network structure, the pros and cons of the neural network structure are measured via the performance of some indicators, such as accuracy and speed degree to measure, called performance evaluation.

In the NAS problem, the form of the fitness function is unknown; it belongs to the black-box optimization problem [26]. It has the characteristics of nonlinearity and non-convexity, and intelligent optimization algorithms have natural advantages for solving such problems.

In the neural network architecture search problem, the search space represents and defines the variables of the optimization problem; that is, it is the basic components of the problem that need to be optimized, such as convolution size, stride, what kind of pooling, and the number of layers of the network.

The search strategy specifies the algorithm used to search for the optimal architecture. These algorithms include: random search [27], Bayesian optimization [28], evolutionary algorithms [26], reinforcement learning [29], and gradient-based algorithms [30]. Among them, Google's reinforcement learning search method was an earlier exploration in 2017. This paper made architecture search more popular [31], and later research institutions, such as Uber, OpenAI, and Deepmind, began to apply evolutionary algorithms to this field. NAS has become a key application of evolutionary computing, and many domestic companies have also begun the same attempt.

Formally, NAS can be modeled as a black-box optimization problem, as shown in Equation (1):

$$\begin{cases} \arg\min_A = \mathcal{L}(A, \mathcal{D}_{\text{train}}, \mathcal{D}_{\text{fitness}}) \\ \text{s.t. } A \in \mathcal{A} \end{cases} \tag{1}$$

where $\mathcal{A}$ represents the search space of the potential neural architecture, and $\mathcal{L}(\cdot)$ measures the fitness evaluation $\mathcal{D}_{fitness}$ on the dataset D_{train}. $\mathcal{L}(\cdot)$ is usually non-convex and non-differentiable. *s.t.* is the abbreviation of subject to (such that), which means to be bound. In principle, NAS is a complex optimization problem with a series of challenges, such as complex constraints, discrete representations, two-layer structures, a high computational cost, and multiple conflicting criteria. A NAS algorithm refers to an optimization algorithm specially designed to efficiently and efficiently solve the problem represented by Equation (1). The following section will explore the application of the quantum dynamics optimization algorithm (QDO) in neural network architecture search.

3.2. QDNAS

Recent NAS methods and benchmarks parameterize the unit structure of deep neural networks into directed graphs. The realization of the unit structure can be seen as assigning related operations from a set of choices or values, such as selecting the predecessor and successor of a node in a directed graph or an operator that selects a node.

The selection of the candidate unit structure belongs to the discrete optimization problem. It can be seen from the basic iterative process of QDO that the basic operation of QDO is Gaussian sampling in continuous space.

We discretize it, that is, set a function as Equation (2). For example, the value obtained by sampling [cov3,cov1,maxpool] is [0.8,0.3,0.4], then the discretized value is [1,0,0].

$$f(x) = \begin{cases} 1, & x \geqslant 0.5 \\ 0, & \text{else} \end{cases} \tag{2}$$

The algorithm involves the problem of replacing the difference solution with the mean value, which is explained here with the solution search matrix of NAS-Bench-101. When NAS-Bench-101 searches, the adjacency matrix is used to encode the network architecture; that is, the sampled particles are the adjacency matrix. Suppose the two sampled particles are $x_1 = \begin{bmatrix} 0.3 & 0.2 & 0.4 \\ 0.1 & 0.6 & 0.3 \\ 0.3 & 0.7 & 0.2 \end{bmatrix}$ and $x_2 = \begin{bmatrix} 0.2 & 0.8 & 0.3 \\ 0.9 & 0.1 & 0.4 \\ 0.6 & 0.2 & 0.1 \end{bmatrix}$, then $x_{aver} = \begin{bmatrix} 0.25 & 0.5 & 0.35 \\ 0.5 & 0.35 & 0.35 \\ 0.45 & 0.45 & 0.5 \end{bmatrix}$. The final architectural adjacency matrix obtained by the function $discrete(x)$ is $X = \begin{bmatrix} 0 & 1 & 0 \\ 1 & 0 & 0 \\ 0 & 0 & 0 \end{bmatrix}$ QDNAS is shown in Algorithm 2. Figure 1 shows the framework of the algorithm. To demonstrate the performance of the framework, several state-of-the-art NAS methods are compared in the simulation experiments section.

The specific steps of QDONAS are:

1. Initialize the population, specifying the dataset D to use.
2. Randomly sample the architecture in the search space and assign it to a queue pop_i.
3. The particles are discretized according to Equation (2).
4. Generate new particle according to $POP_i' = regularized(POP_i + \sigma N(0,1))$.
5. If $f(POP_i) < f(POP_i')$, then POP_i' is assigned to POP_i. Otherwise, the poor solution is accepted with a certain probability. In this part, the probability is 0.1. This probability is selected on the basis of many trials.
6. Replace the worst position with the mean of the sampled individuals $pop_{worst} = pop_{aver}$, and discretize the sampled individuals again.
7. Keep repeating lines 2 to 12 in QDNAS until the maximum number of iterations is reached.

QDO is a sampling-based method, but the difference from random sampling is that QDO can effectively use the information from the previous generation of individuals. QDO introduces a Gaussian distribution in the sampling process. The probability of a Gaussian distribution in the range of σ is 65.26%, and the probability of falling into the range of 3σ is 99.74%. In other words, the particles will move to the vicinity of the better solution with a small step length, which ensures the mining of the algorithm. At the same time, in order to ensure the diversity of the population, the difference is accepted with a certain probability to ensure the diversity of the population. At the end of the iteration of each group, a certain perturbation mechanism is introduced through mean replacement to avoid premature stagnation of the algorithm.

Algorithm 2: Pseudocode of QDNAS.

Input: f:NAS problem with evaluation,D:The reference dataset,k:Population size
Output: Best architecture.
while *(stop condition is not satisfied)* **do**
    for $i = 1$ *to* k **do**
        $pop_i \leftarrow random\_configuration()$
        $POP_i \leftarrow discretized\_architecture()$
        $f_i \leftarrow evaluate_{a}rchitecture(POP_i)$
        Generate POP_i' for POP_i to $POP_i' = regularized(POP_i + \sigma N(0,1))$
        if ($f(POP_i') < f(POP_i)$) **then**
            $POP_i = POP_i'$, update the ith particle
        else
            $POP_i = POP_i'$, update the ith particle according to T
        end
    end
end
POP_{worst}=POP_{aver}
Output:POP_{best}

The pipeline of our method is shown in Figure 1. Initialization is performed first, the initial population is uniformly sampled, and the initial population is discretized. That is, discretization is performed with 0.5 as the threshold. Each individual obtains an initial structure through decoding. We evaluate these structures and record the evaluation results as the fitness value of the individual. We choose the better individual as the next generation and accept the difference with a certain probability. We generate new individuals with a Gaussian distribution around the current individual. We judge whether the termination condition is met; if it is met, the loop ends; if it is not met, the loop will continue.

4. Experiments

We verified the performance of QDNAS in four recent NAS benchmark tests, NAS-Bench-101, NATs-Bench, NAS-Bench-1shot1, and NAS-Bench-201. Different articles use different hyperparameters/data enhancement/regularization/etc. when retraining the searched network structure. Using NAS-Bench can make a fair comparison of each NAS algorithm.

For the image classification task, this paper chooses the default dataset Cifar-10 of NAS-Bench. The CIFAR-10 dataset has a total of 6×10^4 color images, and the size of these images is 32×32, divided into 10 non-overlapping classes. During an architecture search, the training dataset uses CIFAR-10, and the final search network is a network suitable for image classification.

The benchmark test algorithm is Random Search (RS) [27], Tree-Structured Parzen Estimator (TPE) [8], and Regularized Evolution Algorithm (REA) [32]. The experimental parameters are set to NP = 40 and the transmission coefficient is 0.1. Among these algorithms, REA is the preferred benchmarking algorithm, first because REA and QDO are both heuristic algorithms and secondly, because REA has demonstrated excellent performance in past work. For each algorithm, we conduct 500 independent experiments and record the mean performance of the immediate validation regret.

4.1. Nas-Bench-101

The NAS-Bench-101 dataset contains 423k samples, mapping the model structure to the corresponding index (run time and accuracy) traverses the entire search space, making it possible to perform complex analysis on the entire search space.

NAS-Bench-101: The dataset table contains the CNN structure and corresponding training/evaluation indicators using Cell coding. The dataset is Cifar-10 (40k training/10k verification/10k test). Each model was repeatedly trained and evaluated three times under four types of Epochs $E_{\text{stop}} \in \left\{ \frac{E_{\text{max}}}{3^3}, \frac{E_{\text{max}}}{3^2}, \frac{E_{\text{max}}}{3^1}, E_{\text{max}} \right\} = \{4, 12, 36, 108\}$. The indicators used in NASBench101 are: training accuracy, validation accuracy, testing accuracy, number of parameters, and training time.

Figures 2 and 3 show the performance of the search algorithm QDO. Figure 2 shows the trajectory of test accuracy and verification accuracy in 10 tests. Red represents the verification accuracy, and blue represents the test accuracy. It can be seen from the figure that for Random search, the curve is more scattered, which means that the results of each run are quite different, indicating that the randomness is strong. Regarding the regular evolutionary algorithm, this problem has been improved to a certain extent, but it still has a certain degree of randomness. The QDO algorithm verification accuracy rate is relatively concentrated, indicating that the algorithm is robust. However, only two test accuracy rates have large deviations. Furthermore, in the visualization of Figure 2, the comparison of the three can be seen.

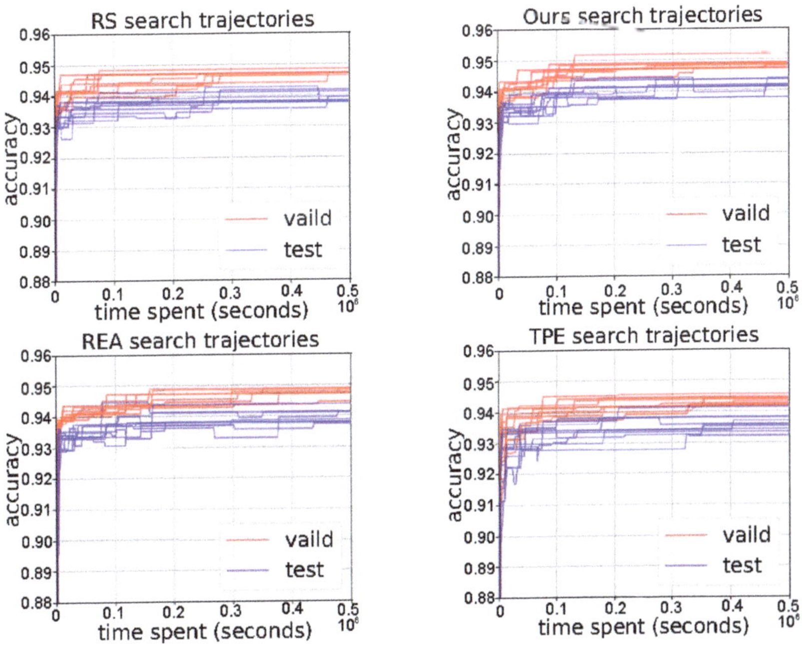

Figure 2. Search trajectories of Random search, REA, and QDO on NAS-Bench-101.

4.2. Nas-Bench-201

NAS-Bench-201 has trained more than 15,000 neural networks on three datasets (CIFAR-10, CIFAR-100, and ImageNet-16-120) based on different random number seeds and different hyperparameters many times. It provides the training and testing time after each training epoch, the loss function and accuracy of the model in the training set/validation set/test set, model parameters after training, model size, model calculation amount, and other important information. With NAS-Bench-201, every NAS algorithm can be compared fairly. Different articles use different hyperparameters/data enhancement/regulations/etc.

when retraining the searched network structure. Using the NAS-Bench-201 API, each researcher can fairly compare the searched network structure.

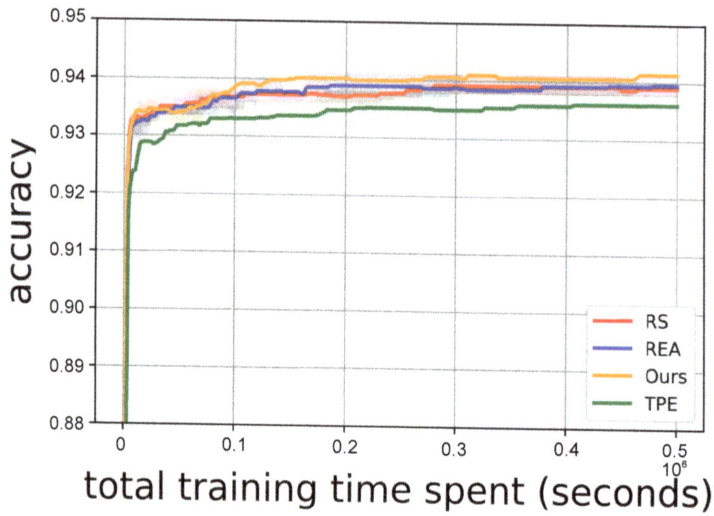

Figure 3. Comparison of the mean test accuracy along with error bars on NAS-Bench-101.

Figures 4 and 5 show the comparative performance of the algorithms. From the comparative performance analysis of the four algorithms, it can be seen that in 10 test experiments, the random search algorithm is more random, and the accuracy of each search changes greatly.

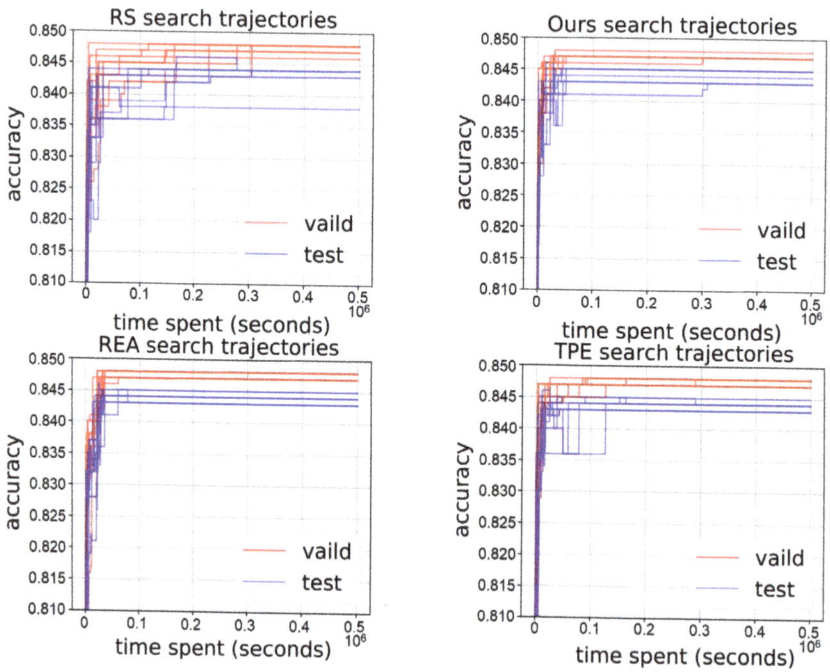

Figure 4. Search trajectories of Random search, REA, and QDO on NAS-Bench-201.

Figure 6 shows the instant validation regret after 500 independent runs. From the results, we can see that for Cifar10, we conclude that even though TPE is better than other algorithms at the beginning it is much slower when approaching the global optimum. The test regrets of DE and RE are almost the same, while RS has shown excellent convergence performance after recovering from the misleading early assessment, and its convergence speed is faster than other algorithms.

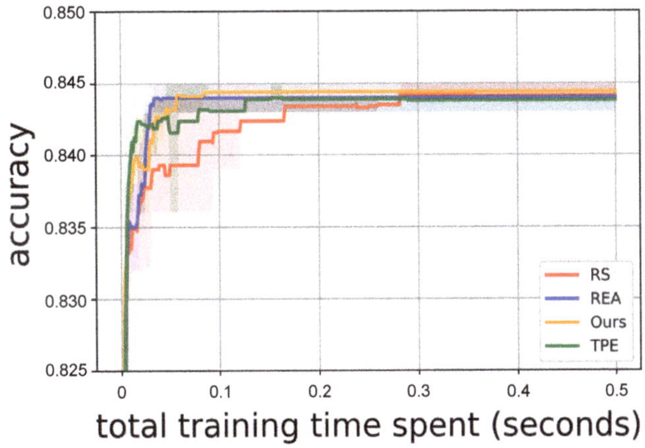

Figure 5. Comparison of the mean test accuracy along with error bars on NAS-Bench-201.

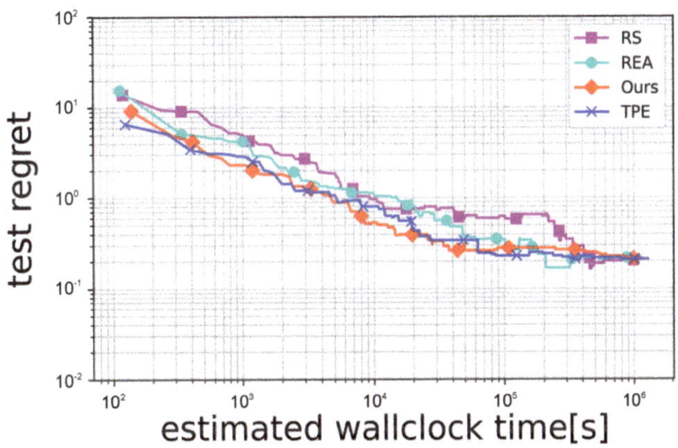

Figure 6. A comparison of the mean test regret performance of 500 independent runs as a function of estimated training time for NAS-Bench-201 on Cifar10

4.3. *Nas-Bench-1shot1*

NAS-Bench-1shot1 modifies the cell-level topology based on NAS-Bench-101 while keeping the network-level topology unchanged. NAS-Bench-1shot1 makes the NAS approach more practical. It defines three search spaces that are convenient for the weight-sharing algorithm to use: search space 1, search space 2, and search space 3. The number of schemas available for searching are 6240, 29160, and 363648.

It can be seen from Figure 7 that RS has better performance in the initial search stage, the reason may be that a better architecture is randomly searched, and when the iteration time is around the point of 2500, REA and QDO are better due to the algorithm itself having a better search mechanism, so it quickly locks in a better search area. When the time is 2700, QDO shows an overwhelming advantage, and the accuracy of the searched architecture is higher. As the iteration progresses, the performance of several algorithms on the NAS-Bench-1Shot1 test set gradually tends to be the same.

Figure 8 shows the immediate test regret after 500 independent runs. It can be seen from the results that both RS and REA performed better in the initial stage, but the QDO algorithm performed better in the later stage, and TPE performed better in the middle stage, but there was premature stagnation. The performance of the QDO algorithm is average in the early stage, but there is a rapid convergence in the later stage. The REA algorithm outperforms other algorithms in the later architecture search.

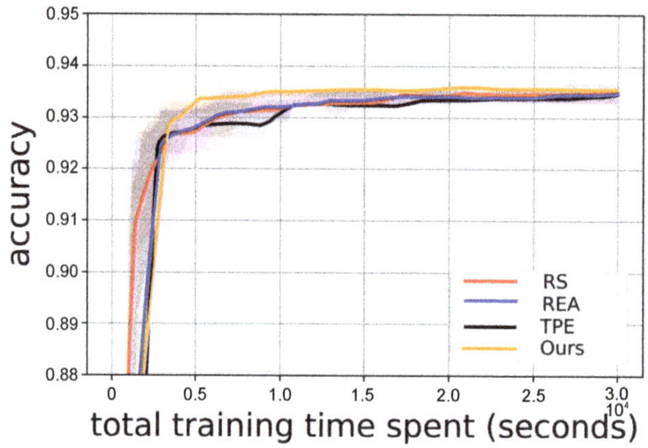

Figure 7. A comparison of the mean test regret performance of 500 independent runs as a function of estimated training time for NAS-Bench-1Shot1 on Cifar10

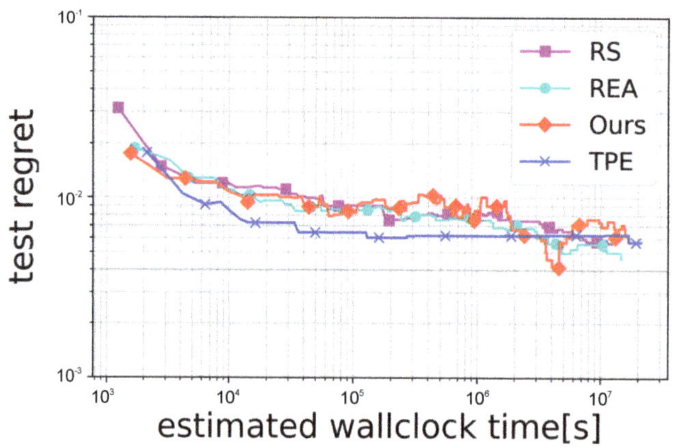

Figure 8. Search trajectories of Random search, REA, and QDO on NAS-Bench-1Shot1.

4.4. NATs-Bench

NATs-Bench is based on NAS-Bench201, which expands the NAS-Bench201 dataset into three, namely CIFAR10, CIFAR100, and ImageNet-16-120. NATS Bench includes 15,625 candidate neurons in the three datasets. Among them, the topological search space S_t is applicable to all NAS methods and the size of the search space S_s complements the lack of architecture size analysis. The average convergence curves of the four algorithms on the NATs-Bench test set are shown in Figure 9. From the visual analysis of the average convergence curve, it is known that QDO and REA have better robustness.

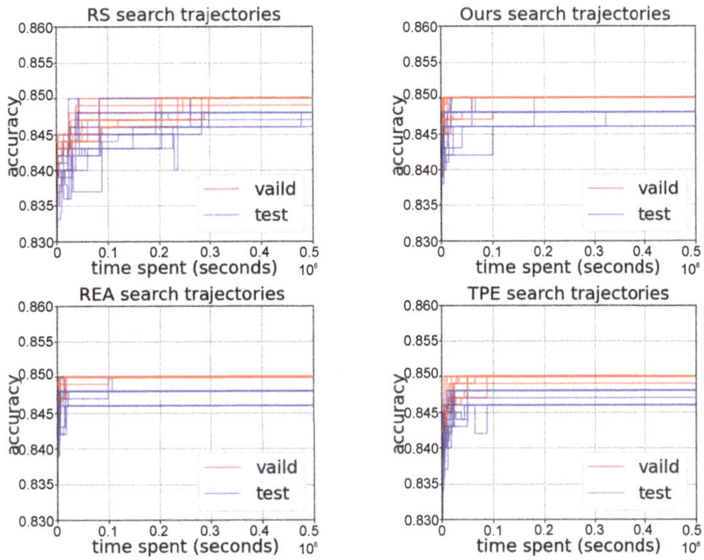

Figure 9. Comparison of the mean test accuracy along with error bars.

4.5. Results Discussion

We record the statistical results of the experimental data of the QDO algorithm, as shown in the table, in which Table 1 records the experiments of the benchmark algorithm for the NAS-Bench-101, NAS-Bench-201, and NAS-Bench-1Shot1 test sets results. The bold words in the table indicate the top ranking. From the experimental results, on the Cifar10 classification dataset; the optimal architecture searched by the QDO algorithm on NAS-Bench-101 is 0.003 higher than the accuracy rate of RS and REA, while in NAS, excellent results were also obtained on Bench-210 and NAS-Bench-1Shot1. REA is a baseline algorithm proposed by the Google AI research team. It is proven that QDO is competitive in architecture search problems.

Table 1. Statistical experimental results for NAS-Bench on the Cifar10 dataset.

	NAS-101	NAS-201	NAS-1Shot1
RS	0.940	0.939	0.946
TPE	0.933	0.936	0.947
RE	0.940	0.940	0.947
Ours	**0.943**	**0.942**	0.947

In addition to the optimization performance, robustness is also an important factor. Whether the algorithm is sensitive to randomness during training and searching is also a measure of whether the NAS algorithm is good. Since REA and QDO performed better in the previous experiments, this part of the experiment only compares the REA and QDO algorithms. Figure 10 is the empirical cumulative distribution of the final test regret after 500 runs of REA and QDO. Based on the robust performance ratios of REA and DE on different test sets in the figure, it can be seen that the robustness of the QDO algorithm on NAS-Bench-101 is significantly better than that of the REA algorithm, while on the other three datasets, the two algorithms' robustness differs little.

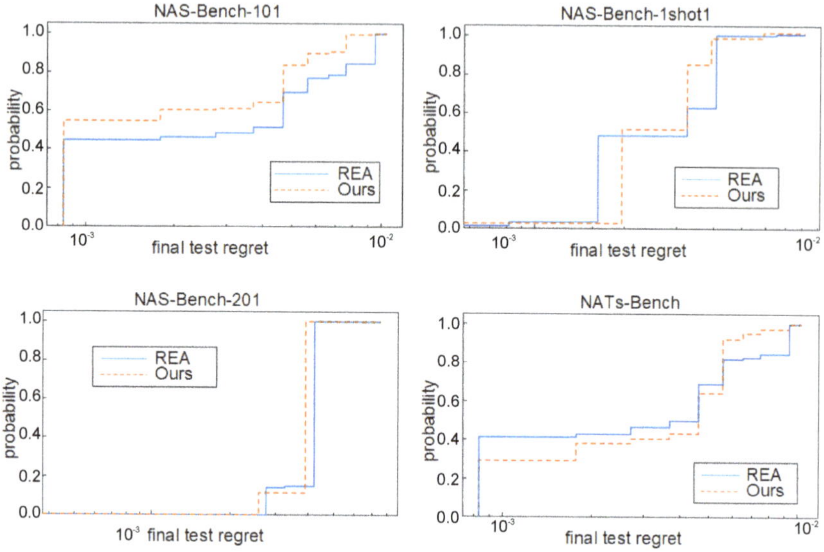

Figure 10. Empirical cumulative distribution of the final test regret after 500 runs of REA and QDNAS.

5. Conclusions

We proved that the quantum dynamics optimization algorithm can be used for a neural network architecture search. The quantum dynamics optimization algorithm is a sampling-based algorithm. Due to the quantum tunneling effect, it has advantages in dealing with mixed data types and high-dimensional optimization problems. Therefore, QDO may be a good candidate for NAS, which may help discover novel but unknown architectures. Since the quantum dynamics optimization algorithm has a natural parallelism, we will explore the parallel implementation of the algorithm in the architecture search in the future.

First, we performed classification recognition on the CIFAR-10 image classification dataset. It should be noted here that by adjusting the core size and number of channels of the convolutional and pooling layers, the algorithm can be easily applied to other fields.

Author Contributions: Conceptualization, Q.Z. and H.Y.; methodology, J.J.; formal analysis, J.H. All authors have read and agreed to the published version of the manuscript.

Funding: Project of Sichuan Science and Technology Department (2021Z005).

Data Availability Statement: Not applicable.

Acknowledgments: Thanks to Sichuan Intelligent Tolerance Design and Testing Engineering Research Center.

Conflicts of Interest: The authors declare no conflict of interest.

References

1. LeCun, Y.; Bengio, Y.; Hinton, G. Deep learning. *Nature* **2015**, *521*, 436–444. [CrossRef] [PubMed]
2. Yin, D.; Gontijo Lopes, R.; Shlens, J.; Cubuk, E.D.; Gilmer, J. A fourier perspective on model robustness in computer vision. *Adv. Neural Inf. Process. Syst.* **2019**, *32*, 13276–13286.
3. He, K.; Zhang, X.; Ren, S.; Sun, J. Deep residual learning for image recognition. In Proceedings of the IEEE Conference on Computer Vision and Pattern Recognition, Las Vegas, NV, USA, 27–30 June 2016; pp. 770–778.
4. Huang, G.; Liu, Z.; Van Der Maaten, L.; Weinberger, K.Q. Densely connected convolutional networks. In Proceedings of the IEEE Conference on Computer Vision and Pattern Recognition, Honolulu, HI, USA, 21–26 July 2017; pp. 4700–4708.
5. Hu, J.; Shen, L.; Sun, G. Squeeze-and-excitation networks. In Proceedings of the IEEE Conference on Computer Vision and Pattern Recognition, Salt Lake City, UT, USA, 18–23 June 2018; pp. 7132–7141.
6. Guo, Y.; Luo, Y.; He, Z.; Huang, J.; Chen, J. Hierarchical neural architecture search for single image super-resolution. *IEEE Signal Process. Lett.* **2020**, *27*, 1255–1259. [CrossRef]

7. Wang, Y.; Liu, Y.; Dai, W.; Li, C.; Zou, J.; Xiong, H. Learning Latent Architectural Distribution in Differentiable Neural Architecture Search via Variational Information Maximization. In Proceedings of the IEEE/CVF International Conference on Computer Vision, Montreal, QC, Canada, 10–17 October 2021; pp. 12312–12321.
8. Elsken, T.; Metzen, J.H.; Hutter, F. Neural architecture search: A survey. *J. Mach. Learn. Res.* **2019**, *20*, 1997–2017.
9. Sun, Y.; Xue, B.; Zhang, M.; Yen, G.G. Evolving deep convolutional neural networks for image classification. *IEEE Trans. Evol. Comput.* **2019**, *24*, 394–407. [CrossRef]
10. Stanley, K.O.; Miikkulainen, R. Evolving neural networks through augmenting topologies. *Evol. Comput.* **2002**, *10*, 99–127. [CrossRef] [PubMed]
11. Sun, J.D.; Yao, C.; Liu, J.; Liu, W.; Yu, Z.K. GNAS-U 2 Net: A New Optic Cup and Optic Disc Segmentation Architecture With Genetic Neural Architecture Search. *IEEE Signal Process. Lett.* **2022**, *29*, 697–701. [CrossRef]
12. Gong, M.; Liu, J.; Qin, A.K.; Zhao, K.; Tan, K.C. Evolving deep neural networks via cooperative coevolution with backpropagation. *IEEE Trans. Neural Netw. Learn. Syst.* **2020**, *32*, 420–434. [CrossRef]
13. Real, E.; Moore, S.; Selle, A.; Saxena, S.; Suematsu, Y.L.; Tan, J.; Le, Q.V.; Kurakin, A. Large-scale evolution of image classifiers. In Proceedings of the International Conference on Machine Learning. PMLR, Sydney, NSW, Australia, 6–11 August 2017; pp. 2902–2911.
14. Niu, R.; Li, H.; Zhang, Y.; Kang, Y. Neural Architecture Search Based on Particle Swarm Optimization. In Proceedings of the 2019 3rd International Conference on Data Science and Business Analytics (ICDSBA), Istanbul, Turkey, 11–12 October 2019; pp. 319–324.
15. Xie, L.; Yuille, A. Genetic cnn. In Proceedings of the IEEE International Conference on Computer Vision, Venice, Italy, 22–29 October 2017; pp. 1379–1388.
16. Junior, F.E.F.; Yen, G.G. Particle swarm optimization of deep neural networks architectures for image classification. *Swarm Evol. Comput.* **2019**, *49*, 62–74. [CrossRef]
17. Sun, Y.; Xue, B.; Zhang, M.; Yen, G.G.; Lv, J. Automatically designing CNN architectures using the genetic algorithm for image classification. *IEEE Trans. Cybern.* **2020**, *50*, 3840–3854. [CrossRef] [PubMed]
18. Xue, Y.; Wang, Y.; Liang, J.; Slowik, A. A self-adaptive mutation neural architecture search algorithm based on blocks. *IEEE Comput. Intell. Mag.* **2021**, *16*, 67–78. [CrossRef]
19. Wang, P.; Xin, G.; Jiao, Y. Quantum Dynamics Interpretation of Black-box Optimization. *arXiv***2021**, arXiv:2106.13927.
20. Zhang, M.; Li, H.; Pan, S.; Liu, T.; Su, S.W. One-Shot Neural Architecture Search via Novelty Driven Sampling. In Proceedings of the IJCAI, Yokohama, Japan, 11–17 July 2020; pp. 3188–3194.
21. Jin, J.; Wang, P. Multiscale Quantum Harmonic Oscillator Algorithm with Guiding Information for Single Objective Optimization. *Swarm Evol. Comput.* **2021**, *65*, 100916. [CrossRef]
22. Ying, C.; Klein, A.; Christiansen, E.; Real, E.; Murphy, K.; Hutter, F. Nas-bench-101: Towards reproducible neural architecture search. In Proceedings of the International Conference on Machine Learning. PMLR, Long Beach, CA, USA, 9–15 June 2019; pp. 7105–7114.
23. Zela, A.; Siems, J.; Hutter, F. Nas-bench-1shot1: Benchmarking and dissecting one-shot neural architecture search. *arXiv* **2020**, arXiv:2001.10422.
24. Dong, X.; Yang, Y. Nas-bench-201: Extending the scope of reproducible neural architecture search. *arXiv* **2020**, arXiv:2001.00326.
25. Dong, X.; Liu, L.; Musial, K.; Gabrys, B. Nats-bench: Benchmarking nas algorithms for architecture topology and size. *IEEE Trans. Pattern Anal. Mach. Intell.* **2021**, *44*, 3634–3646. [CrossRef] [PubMed]
26. Liu, Y.; Sun, Y.; Xue, B.; Zhang, M.; Yen, G.G.; Tan, K.C. A survey on evolutionary neural architecture search. *IEEE Trans. Neural Netw. Learn. Syst.* **2021**. 1–21. [CrossRef] [PubMed]
27. Li, L.; Talwalkar, A. Random search and reproducibility for neural architecture search. In Proceedings of the Uncertainty in Artificial Intelligence. PMLR, virtual online, 3–6 August 2020; pp. 367–377.
28. Kandasamy, K.; Neiswanger, W.; Schneider, J.; Poczos, B.; Xing, E.P. Neural architecture search with bayesian optimisation and optimal transport. In Proceedings of the 32nd International Conference on Neural Information Processing Systems, Montréal, Canada, 3–8 December 2018; pp. 2020–2029
29. Chen, Y.; Meng, G.; Zhang, Q.; Xiang, S.; Huang, C.; Mu, L.; Wang, X. Renas: Reinforced evolutionary neural architecture search. In Proceedings of the IEEE/CVF Conference on Computer Vision and Pattern Recognition, Long Beach, CA, USA, 15–20 June 2019; pp. 4787–4796.
30. Santra, S.; Hsieh, J.W.; Lin, C.F. Gradient descent effects on differential neural architecture search: A survey. *IEEE Access* **2021**, *9*, 89602–89618. [CrossRef]
31. Zoph, B.; Le, Q.V. Neural architecture search with reinforcement learning. *arXiv* **2016**, arXiv:1611.01578.
32. Real, E.; Aggarwal, A.; Huang, Y.; Le, Q.V. Regularized evolution for image classifier architecture search. In Proceedings of the AAAI Conference on Artificial Intelligence, Honolulu, HI, USA, 27 January–1 February 2019; Volume 33, pp. 4780–4789.

electronics

MDPI

Article

A Lightweight Border Patrol Object Detection Network for Edge Devices

Lei Yue, Haifeng Ling *, Jianhu Yuan and Linyuan Bai

Field Engineering College, Army Engineering University of PLA, Nanjing 210022, China
* Correspondence: haifeng_ling@aeu.edu.cn; Tel.: +86-181-8498-2962

Abstract: Border patrol object detection is an important basis for obtaining information about the border patrol area and for analyzing and determining the mission situation. Border Patrol Staffing is now equipped with medium to close range UAVs and portable reconnaissance equipment to carry out its tasks. In this paper, we designed a detection algorithm TP-ODA for the border patrol object detection task in order to improve the UAV and portable reconnaissance equipment for the task of border patrol object detection, which is mostly performed in embedded devices with limited computing power and the detection frame imbalance problem is improved; finally, the PDOEM structure is designed in the neck network to optimize the feature fusion module of the algorithm. In order to verify the improvement effect of the algorithm in this paper, the Border Patrol object dataset BDP is constructed. The experiments show that, compared to the baseline model, the TP-ODA algorithm improves mAP by 2.9%, reduces GFLOPs by 65.19%, reduces model volume by 63.83% and improves FPS by 8.47%. The model comparison experiments were then combined with the requirements of the border patrol tasks, and it was concluded that the TP-ODA model is more suitable for UAV and portable reconnaissance equipment to carry and can better fulfill the task of border patrol object detection.

Keywords: object detection; deep learning; computer vision; border patrol

Citation: Yue, L.; Ling, H.; Yuan, J.; Bai, L. A Lightweight Border Patrol Object Detection Network for Edge Devices. *Electronics* **2022**, *11*, 3828. https://doi.org/10.3390/electronics11223828

Academic Editor: Taiyong Li

Received: 22 October 2022
Accepted: 15 November 2022
Published: 21 November 2022

1. Introduction

In recent years, illegal acts such as drug trafficking, smuggling, border crossing and smuggling have been prohibited in border areas, and the workload of border patrol tasks has only increased. Considering the problem of limited patrol force, the relevant management departments have equipped border patrol staffing with drones or handheld portable reconnaissance equipment [1], which has greatly improved the management capability of the border, while reducing the risk of border patrol and solving many of the existing problems of traditional border patrol [2]. However, the use of UAV (Unmanned Aerial Vehicle) platforms and portable reconnaissance equipment for border patrol missions has also raised some issues that need to be further addressed, the most important of which is the ability of the patrol reconnaissance equipment to detect border patrol objects. Most of the existing UAV and reconnaissance equipment are equipped with high-definition optical cameras, which can acquire objects at different distances, but at the same time will generate a large amount of image video data. However, the computing power of edge devices is generally insufficient, so it is important to develop a border patrol detection model that can be easily deployed on edge devices such as UAV platforms and handheld portable reconnaissance terminals. The traditional method of border patrol reconnaissance is mainly through close reconnaissance, or the use of long-range photographic equipment to capture images and video data of suspicious areas, and then use the communication transmission equipment carried to transmit the data to the rear for analysis and judgment. However, subject to technical problems, the detection field of view is limited, inefficient and ineffective, which is a very prominent problem. With the continuous development of computer technology, faster, more accurate, more efficient object detection technology has emerged.

In recent years, deep learning has been greatly developed. Whether deep learning can be used in the field of object detection is also being studied by scholars. An important turning point in the field of object detection occurred when AlexNet [3] was proposed. As a result, the scope of object detection application research has been expanded. Thus far, deep learning has been widely used in various fields of computer vision, which has important research significance and application value in national security, military [4], transportation [5], medical [6] and life.

After the emergence of Alex Net, Ross B Girshick et al. [7] proposed R-CNN in 2014, and then the R-CNN algorithm underwent the evolution of Fast R-CNN and Faster R-CNN. Compared to the traditional detection algorithm, the performance has been greatly improved. Since then, more and more detection algorithms based on convolutional neural networks have been proposed, such as MSCNN [8], M2Det [9], EfficientNet [10], etc., and the accuracy and detection speed are constantly improving.

According to different network design paradigms, we classify existing object detection algorithms into one-stage detection algorithms and two-stage detection algorithms. The above detection algorithm is a two-stage detection algorithm, which has a high detection accuracy, but a slow detection speed, and is not applicable to the problem of border patrol object detection proposed in this paper. In order to solve this problem, this paper uses the representative one-stage detection algorithm YOLOv5 [11] as the baseline model, which is the representative one-stage detection algorithm of the YOLO series. Compared to the YOLOv1-4 [12–15] detection algorithm and the two-stage detection algorithm, the most prominent features of the YOLOv5 detection algorithm are its fast detection speed and high detection accuracy, which can meet the requirements of real-time.

In this study, a border patrol object detection algorithm, TP-ODA, was designed for the carriage of UAV platforms or portable border patrol reconnaissance equipment. As the most widely used detection algorithm of the current YOLO series, the YOLOv5 detection algorithm has made a good balance between detection accuracy and detection speed, but there are still many redundant parameters in its network, which need to be further improved. We therefore propose a lightweight and less resource intensive border patrol object detection algorithm. First, the Ghost structure is improved based on the lightweight attention module and is combined with the benchmark network to rebuild the feature extraction network. Then, the bounding box loss function of the benchmark algorithm was modified to solve the problem of sample detection box imbalance. Finally, a depth-separable convolution was introduced, and the neck network was reconstructed, while the feature fusion module PDOEM (Patrol Duty Object Detection Efficient Modules) was designed to optimize the feature fusion structure of the algorithm. The experiments were conducted on our self-built border patrol task dataset BDP (Border Defense Patrol), which was prepared for this study. The results show that the TP-ODA (Typical Border Patrol-Object Detection Algorithm) network reduces many parameters and reduces the size, which is very suitable for border patrol object detection tasks. Compared to previous studies, the main contributions of this paper are as follows.

1. In order to improve the feature extraction capability of the network for different dimensions and improve the performance degradation of the model after compression, we proposed a lightweight feature extraction structure BP-Sim, which takes into account the functions of the original feature extraction structure and reduces the occupation of computing resources. Aiming at the unbalance problem of the sample detection frame of the benchmark model, the EIOU loss function is introduced to further improve the detection accuracy of the model.

2. In order to further compress the volume of the model and reduce the resource occupation, we designed the feature fusion module PDOEM to improve the fusion ability of the model to the deep feature information. Combined with the depth-separable convolution, the neck feature fusion network of the model was reconstructed.

3. To address the problem of the confidentiality of the information involved in the border patrol domain and the existing public datasets that cannot be well used for border patrol detection tasks, the border patrol task dataset BDP is constructed to train and evaluate the performance of the object detection model.

The rest of this paper is structured as follows. Section 2 describes some of the most important related works. Section 3 describes the proposed the object detection network. Section 4 describes the experimental preparation. The experimental results and analysis are described in Section 5. Finally, a summary and outlook are given in Section 6.

2. Related Work

At present, series detection algorithms are widely used, and many scholars have undertaken a lot of research work in common detection fields. In medicine, the detection algorithm is used to detect breast tumors [16] and to fight against COVID-19 [17,18]. In the field of agriculture, it is used to detect plant diseases [19] and pests and for crop production [20]. In industrial applications, it is used to detect defects on the surface of steel strips [21]. In the transportation field, it is used to solve road congestion [22] and road failure problems [23].

Many scholars have also done a lot of research in the field of military object detection [24,25]. As our border patrol object detection task is not only a military object detection task, with the complexity of security maintenance, border patrol, reconnaissance and duty operation tasks, the border patrol object detection algorithm is required to have a certain generalization detection performance, but also the ability to detect military objects. Guangdi Zheng et al. [26] used the YOLOv3 algorithm for the detection of low-resolution infrared objects present on the terrestrial battlefield and trained the model with the aid of visible samples. Hui Peng et al. [4] used the YOLO detection algorithm to detect five common military weapons in order to obtain a fuller sense of the battlefield situation. Xingkui Zhu et al. [27] proposed TPH-YOLOv5 based on the YOLOv5x network, combining the transformer and CBAM, and used a larger network to detect small objects in UAV aerial photography. M. Krišto et al. [28] used the YOLOv3 detection model to detect abnormal behaviors in border areas and found the case of sneaking around objects and illegal border crossings in a timely manner.

From the above study, it can be concluded that the YOLO series detection algorithm generalizes well and the detectability can basically meet the needs of various fields. However, based on our research, we believe that the existing detection algorithms for detecting border patrol objects still need to address two aspects:

1. Most studies have improved the detection accuracy of military-type objects in complex environments and UAV aerial images, but the model resource consumption has increased accordingly, which poses a serious limitation for embedded devices with limited computing power.

2. Border patrol object detection differs from traditional image detection in that the data obtained during border patrol has obvious peculiarities because of the various forms of data collection. The first is that the border patrol objects have strong regional restrictions and can only be collected in special areas, and the second is that most of the border patrol objects are in the state of obscuration and camouflage, so the quality of the collected images is not high, so for the object detection model, objects with a camouflage nature and tiny objects in aerial images are difficult to detect.

In order to apply large neural network models to UAV platforms and portable reconnaissance equipment, we have conducted an in-depth study of network model parameter reduction. Lightweight detection networks have gained more attention because they can reduce the resource footprint of the model and speed up detection by reducing a small amount of detection accuracy. The core idea of the detection algorithm compression is to reduce the computational complexity and spatial complexity of the model by modifying the way the network is constructed while ensuring the model accuracy as much as possible, so that the neural network detection algorithm can be deployed in UAVs with limited computational performance, embedded edge devices such as portable reconnaissance devices, thus establishing a link from academic research to practical applications.

Currently, there are two main types of model volume optimizations. One type is compression of the model, using methods such as knowledge distillation and model de-branching to reduce the number of parameters and unnecessary computational consumption of the model, which has a limited scope for model compression and a large impact on accuracy. We therefore chose another type of light-weighting method for optimization. This class of optimization method mainly introduces the idea of lightweight networks in the structure of benchmark models, such as SqueezeNet [29], MobileNet series [30,31], ShuffleNet series [32,33], Xception [34], etc. By using different convolution methods and structures, the models are made lighter. Currently, it is common to use lightweight networks to optimize benchmark models for object detection tasks in common scenarios. A common approach is to use lightweight backbone networks in large detection models, such as Youchen Fan et al. [35] used YOLOv3 and improved with GhostNet to have good detection results when detecting infrared images of vehicles. Minghua Zhang et al. [36] proposed light-weighting using MobileNetV2 and depth-separable convolution for detecting underwater objects; J. Feng et al. [37] used MobileNet as the backbone network to modify the original model for detecting rail defects. Tianhao Wu et al. [38] adapted the network structure of YOLOv5 and designed the YOLOv5-Ghost algorithm for use on a CARLA vehicle and a distance detection system in a virtual environment. The aforementioned study significantly reduced the model resource consumption, but the detection accuracy was not high.

While research work on lightweight networks has great application, there has been little research in the area of border patrol object detection. In response to the current situation, we have designed a border patrol object detection model that is less resource intensive and more efficient in detection.

3. Method

The basic framework of the YOLOv5 detection model mainly includes Input, Backbone, Neck, Prediction and four other parts. Input part: Mainly adjust the image to 640 × 640 ratio, and zoom, enhance and other processing. The Backbone module uses the Darknet-53 network to facilitate the training of the model and the extraction of multiple scale features. The Neck module draws on the function of fusing multi-scale feature information completed by FPN [39] and PANet [40]. This part can fuse the feature information of different depths so as to reduce the loss of semantic information due to feature extraction, so that the model training can obtain more training information, which is conducive to the improvement in algorithm performance. The Prediction part is composed of three detection heads, which are used to predict the feature map and to obtain the position and category of the detected object in the image.

3.1. The Improved Network Structure

In order to make the model less resource intensive, we compared various lightweight networks and finally chose GhostNet to optimize the backbone network. In order to improve the feature representation of the detected object, we embed the lightweight attention mechanism module SimAM into the GhostNet network and design the BP-Sim (Border Patrol-SimAM module) structure to optimize the feature extraction network, which further reduces the parameters of the model while improving the accuracy. In addition, in order to improve the feature fusion performance of the model, the PDOEM feature fusion module is designed and combined with depth separable convolution to reconstruct the feature fusion structure, and finally, the EIOU loss function is introduced to optimize the design for the problem that the loss function in the original benchmark model has the problem of sample detection box imbalance leading to the decrease in detection accuracy and the slowdown of the model convergence.

In Figure 1, the images are input into the backbone network, and feature extraction and slicing are first performed using ordinary convolution, and then the processed images are input into GhostConv and BP-Sim structures, and the feature images after the above operations are divided into multiple levels and passed to the Neck for concat operation. In the Neck structure, the feature information is extracted using depth-separable convolution, then the feature map is resized after upsampling and connected with the feature information of the backbone part, and finally the feature map obtained from the concat operation is input to the PDOEM module for information mining.

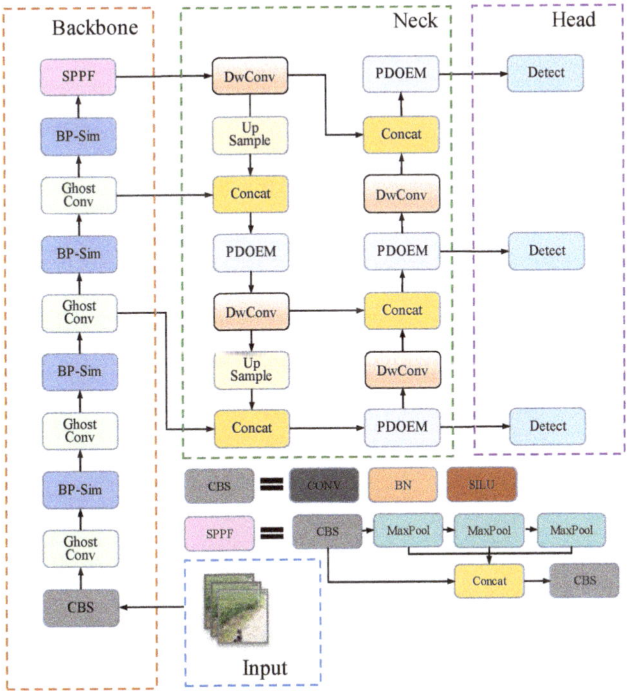

Figure 1. TP-ODA border patrol object detection network architecture.

3.2. Lightweight Network Design Module

Border patrol missions using UAVs or portable reconnaissance devices require not only the accurate detection of suspicious objects in the border area, but also requires minimizing the resource consumption of the network to meet the edge device load requirements. Next, we optimize the design of the backbone part of the benchmark network.

The common convolution operation is to apply the convolution kernel to the local image, slide the high latitude and low latitude in the local image, then form the correspondence in space and complete the convolution, and obtain the convolution kernel after many repetitions. The above operation enables the model to achieve better accuracy through multiple training, but it also requires many convolution operations, which has an enormous consumption of computational resources. Due to this problem, some lightweight networks remove some redundant features by removing some of the redundant feature information while reducing the model performance to achieve the effect of streamlining the model. However, some scholars' research proves that the redundant feature information also exists in the redundant features contributing to the model's comprehensive understanding of image data, which becomes an important part of the model performance improvement. As shown in Figure 2, it is with this in mind that, instead of trying to eliminate redundant feature maps, GhostNet uses cheaper computation to obtain redundant feature maps.

According to our previous research on the lightweight network, it is concluded that GhostNet [41,42] is more prominent in terms of comprehensive performance. Therefore, we will carry out further optimization of the detection model's resource footprint in conjunction with GhostNet.

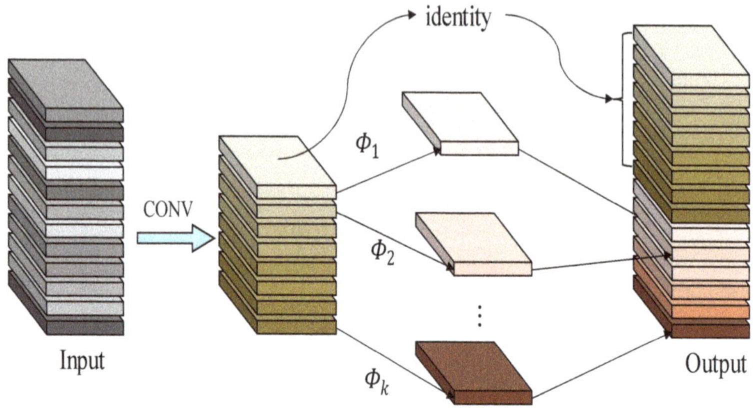

Figure 2. Ghost module structure description.

The backbone of the benchmark network uses many traditional convolutional neural networks, which are mainly used to extract image features. These networks contain a large number of parameters that occupy a large amount of computational resources and memory. Therefore, influenced by GhostNet idea, we use the Ghost convolutional network to replace part of traditional convolutional networks in the backbone network.

3.3. *Feature Information Extraction Module*

In the actual border patrol environment, which contains multiple types of environments such as desert, snow, jungle, and grass, the use of UAV platforms or other reconnaissance equipment for detection can lead to low image quality, blurred object backgrounds, and loss of feature information due to the harsh natural environment. The presence of these factors greatly increases the detection difficulty of the network. Studies in recent years have concluded that the use of attention mechanism modules can enhance the network's ability to extract image feature information. Therefore, to improve the model's ability to extract effective feature information and not to increase the model's excessive number of parameters and computational effort, we designed the BP-Sim and PDOEM modules in the network.

The improvement steps for the backbone network are: Considering that the backbone network is not sufficient for processing image information with different dimensions of feature semantic information, especially in the case of border patrol image data, which are mostly blurred images, top view captured images and diverse scales. We first optimized the feature extraction structure. Considering that the direct use of the lightweight network in the optimization process of the benchmark network would lead to a reduction in the detection accuracy of the model, and that the original bottleneck connection network contained a large number of parameters, we redesigned the bottleneck structure of the model by modifying the bottleneck network on the basis of the original C3 structure, removing part of the regular convolutional network from this structure and the BP-Sim network is obtained by replacing the regular convolutional module with a lighter convolutional module and embedding the SimAM [43,44] attention mechanism. The network exploits the sensitivity of the attention mechanism with useful information to improve the network's ability to mine feature information. The BP-Sim bottleneck structure feature extraction structure is shown in Figure 3.

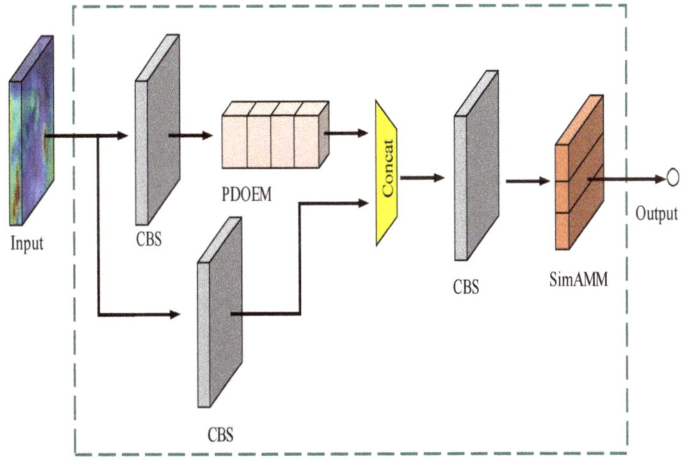

Figure 3. BP-Sim bottleneck structure feature extraction structure.

In Figure 3, the feature image first goes through traditional convolution to obtain one input edge of concat operation; in the other input, the feature map is extracted using traditional convolution, and while going through PDOEM for dimensionality reduction and enhancement, difficult feature information mining is performed with the help of the attention mechanism in this module, and the obtained feature information is connected with another edge of the feature extraction; finally, the connected feature map is extracted and information is mined again.

The existing attention module is commonly used to improve the output results of each layer. This kind of operation usually generates one-dimensional or two-dimensional weights along the channel or spatial dimension and treats the positions in the space or channel equally, which will lead to the limitation of the model's cue discrimination ability. In order to realize the effect brought by the attention mechanism to the model, SimAM referred to the idea of spatial inhibition in neuroscience and gave higher priority to the neurons with obvious spatial inhibition effects.

$$e_t(w_t, b_t, y, x_i) = (y_t - \hat{t})^2 + \frac{1}{M-1}\sum_{i-1}^{M-1}(y_0 - \hat{x}_i)^2 \tag{1}$$

where t and x_i denote the object neuron and the input feature $X \in R^{C\times H\times W}$ other neurons in the same channel $\hat{t} = w_t t + b_t$, respectively, $\hat{x}_i = w_t x_i + b_i$ is t and x_i linear transformation. w_t and b_t are linearly varying weights and biases, i is the spatial dimension index, M is the number of channel neurons, and y_0 and y_t are two different values. For the convenience of use and operation, the binary label is used for the above, and a regularization term is added to the energy function formula to obtain the final energy function formula. According to the principle that each channel has M energy functions, the analytical solution Formula (4) is obtained:

$$e_t(w_t, b_t, y, x_i) = \frac{1}{M-1}\sum_{i=1}^{M-1}(-1 - (w_t x_i + b_t))^2 + (1 - (w_t t + b_t))^2 + \lambda {w_t}^2 \tag{2}$$

$$w_t = \frac{2(t - u_t)}{(t - u_t)^2 + 2\sigma_t^2 + 2\lambda} \tag{3}$$

$$b_t = -\frac{1}{2}(t + \mu_t)w_t \tag{4}$$

Including the $\mu_t = \frac{1}{M-1}\sum_{i=1}^{M-1} x_i$ and $\sigma_t^2 = \frac{1}{M-1}\sum_{i}^{M-1}(x_i - \mu_t)^2$ is the mean and variance of all neurons except t. The minimum energy Equation (5) is obtained:

$$e_t^* = \frac{4(\hat{\sigma}^2 + \lambda)}{(t - \hat{\mu})^2 + 2\hat{\sigma}^2 + 2\lambda} \tag{5}$$

According to Equation (5), the lower the energy, the more different the neuron is from the surrounding neurons. Therefore, the importance of each neuron can be obtained by $1/(e_t^*)$. SimAM uses the operation of scaling instead of adding the feature refinement, and the refinement process of the whole module is shown in Equation (6).

$$\widetilde{X} = sigmoid\left(\frac{1}{E}\right) \tag{6}$$

3.4. *Improvement of Feature Fusion Module*

As the baseline network uses more common convolutional modules, and the traditional convolutional modules are large in size and have a large number of parameters. Therefore, we modified the backbone part of the baseline model and replaced the general convolutional module in the backbone with the GhostConv module, which reduces the number of parameters of the model with little reduction in accuracy. Inspired by this idea, we also replaced the basic convolutional module in the neck network with GhostConv, but the training results were not very good. In response to the experimental results, we considered that the model also needs to capture useful feature information and suppress noise information when performing feature fusion, so we kept part of the general convolution in the Neck network and replaced the original convolution module with the depth separable convolution, and connected SimAM after the DBS convolution module, and finally built the PDOEM feature fusion module, as shown in Figure 4 We use the PDOEM module to replace some of the normal convolutional modules in the Neck part in order to improve the situation of the inadequate extraction of high-level feature information and waste of computational resources when the network is fused with features, and because the addition of the attention module does not use too many computational resources, it is important for the model compression design and overall performance improvement.

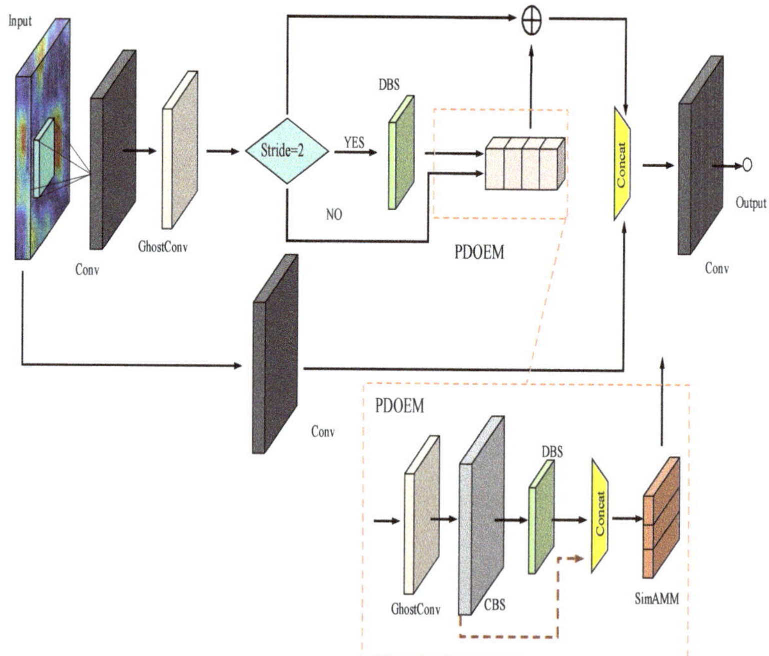

Figure 4. Structure design of the PDOEM feature information extraction module.

3.5. Loss Function Improvement

The loss function in the YOLO family of models is mainly composed of three parts: Bounding Box loss function, object confidence loss function and class loss function. In the YOLOv5 model, CIOU open is used to calculate the loss of the bounding box by default. CIOU is based on DIOU [45] with the addition of the influence factor αv. Where α denotes the weight parameter and v is used to measure the consistency of the aspect ratio, taking the αv influence factor into account can further consider the relationship between the prediction frame and the real frame, improve the regression accuracy when the real frame and the prediction frame IOU are larger or even included, and enhance the suppression of the model loss function. The effect improves the suppression of the model loss function, and finally improves the model convergence accuracy.

$$L_{CIOU} = 1 - IOU + \frac{\rho^2(b, b^{gt})}{c^2} + \alpha v \tag{7}$$

$$\alpha = \frac{v}{1 - IOU + v} \tag{8}$$

$$v = \frac{4}{\pi}\left(arctan\frac{w^{gt}}{h^{gt}} - arctan\frac{w}{h}\right)^2 \tag{9}$$

However, as reflected by v in Equations (8) and (9), the aspect ratio difference of the CIOU loss function cannot reflect the real aspect difference and confidence value, which hinders the similarity optimization of the model and reduces the convergence speed of the model. Therefore, in the study by Zhang et al., based on the CIOU loss function, the aspect ratio of the model was decomposed and the EIOU [46] loss was refined. The EIOU loss function is defined, as shown in Equation (10):

$$L_{EIOU} = L_{IOU} + L_{dis} + L_{asp} = 1 - IOU + \frac{\rho^2(b, b^{gt})}{c^2} + \frac{\rho^2(w, w^{gt})}{C_w^2} + \frac{\rho^2(h, h^{gt})}{C_h^2} \tag{10}$$

$$L_{Focal-EIOU} = IOU\gamma L_{EIOU} \tag{11}$$

This loss function consists of three parts: Overlap loss, center distance loss, width and height loss. Where C_w and C_h represent the width and height of the minimum bounding box. The EIOU loss function retains the advantages of CIOU loss function, and at the same time, considering the situation that the gradient is too large to affect the training accuracy caused by the imbalance problem of the Bounding box samples, the idea of Focal loss is introduced on the basis of the EIOU loss function, and the Focal EIOU loss function is proposed after the combination. The definition is shown in Equation (11). The $IOU = |A \cap B| / |A \cup B|$ and γ in the formula represent the coefficients that control the degree of outlier suppression. Focal EIOU loss function separates the low quality and high-quality anchor boxes from classifying the training samples.

4. Experiment Preparation

In this section, the border patrol dataset BDP used in the experiments, the experimental environment configuration, and the model performance evaluation metrics are introduced.

4.1. Introduction to the Dataset

Due to the confidentiality of the information involved in the field, the image information related to border patrol is relatively scarce, so this paper creates the BDP dataset by offline collection, online collection of public video information, and network images. The BDP dataset has a total of more than 2600 samples, containing a total number of 11,000 labeled boxes, involving different tasks, different natural scenes of pedestrians, soldiers on duty, vehicles, camouflage vehicles, trucks and other common objects at the border. Some of the sample images of the dataset are shown in Figure 5. Due to the various methods of data collection, involving aerial photography, overhead cameras and some portable photographic devices, the dataset has various scales and complex image backgrounds, and some of the model objects are obscured, blurred, and individual features are difficult to be extracted completely. We normalized the dataset and then used the image annotation software LabelImg for annotation. The dataset is divided into the training set, test set and validation set in the ratio of 8:1: 1 for training and performance testing of the model.

Figure 5. Sample images from the BDP dataset.

4.2. Introduction to Experimental Environment

The experimental platform for the experiments in this paper were performed on a workstation on Ubuntu20.04. The GPU is NVIDIA TITAN V 12 G. The neural network is built with Pytorch1.10 as the basic framework and programmed with Python language, and the specific parameters are shown in Table 1.

Table 1. Experimental parameter configuration.

Parameter	Disposition
CPU	Intel(R) Xeon(R) Gold 5118 × 2 CPU @2.29 GHz
GPU	NVIDIA TITAN V 12 G
Systems	Ubuntu 20.04
CUDA	11.3

4.3. Evaluation Indicators

In order to verify the comprehensive performance of the TP-ODA algorithm, this paper mainly selects mAP@0.5 (the average AP of all categories when the IOU is set to 0.5), mAP@0.5:0.95 (the average mAP under different IOU thresholds), FPS, GFLOPs, the number of parameters and the model size to evaluate the model performance.

The mAP value is the average value of all AP values, which can be used to evaluate the detection effect of the algorithm for multi-class objects. AP represents the result of evaluating the detection results of each class, which is related to the precision value and recall value of the model. The specific definition is as follows

$$AP = \int_0^1 PdR \tag{12}$$

$$mAP = \frac{1}{N}\sum_{i=O}^{N} AP_i \tag{13}$$

TP, FP, and FN represent the number of correct detections, false detections, and missed detections, respectively. TP represents the number of instances that themselves belong to this class of objects and can be accurately detected by the model. In contrast, FP represents the number of instances that do not belong to this class of objects themselves, but are misjudged as such objects due to insufficient model performance. Here, true positive (TP) is the number of positive samples predicted to be positive, false positive (FP) is the number of samples predicted to be positive but is actually negative, and false negative (FN) is the number of samples predicted to be negative but is actually positive.

$$Precision = \frac{TP}{TP + FP} \times 100\% \tag{14}$$

$$Recall = \frac{TP}{TP + FN} \times 100\% \tag{15}$$

The size of the model is the size of the model stored after the final model training. The detection speed of the detection model is measured by the number of images per second (FPS) denoting the number of images that can be processed per second, and T denoting the time it takes to process an image. The average FPS detection time includes the inference time of the model, the average detection processing time, and the non-maximum suppression processing time.

$$FPS = \frac{1}{T} \tag{16}$$

5. Experimental Process

For the application scenario of the UAV border patrol detection, which is the focus of the paper, improving the detection speed of the model, reducing the parameters and computation of the model, and reducing the consumption of memory resources of the model are the main requirements for model selection while maintaining the detection accuracy of the model.

5.1. Implementation Details

Model training process: To prevent overfitting and skipping the optimal solution, the momentum factor is set to 0.937, and the stochastic gradient descent method is used to adjust the parameters. The batchsize is set to 32. Epochs were trained for 300 rounds, with an initial learning rate of 0.01 for the first 200 rounds and a weight decay of 0.0005 for the last 100 rounds. The overlap coefficient of the Mixup was set to 0.7. When the loss function and accuracy are gradually stable, the optimal weight of the algorithm is obtained. In the image preprocessing process, the image size is resized to 640×640 before being input into the network for training.

The YOLOv5 model includes a variety of different structures depending on the depth and width of the network. In this paper, some YOLOv5 models with different depths and widths are selected for experiments. As the detection objects in the VisDrone2019 [47] dataset involve common objects, such as vehicles and pedestrians, and the characteristics of small and dense objects are similar to the characteristics of a part of the objects on patrol, we first use the Visdrone2019 dataset to carry out the baseline model selection experiment. The training process does not load the pre-training weights, a batchsize of 16, epochs are iterated 300 times, and the other parameters are selected as the default parameters of the algorithm for training. The model after training is tested on the test dataset in the Visdrone2019 dataset, and the relevant parameters are shown in Table 2.

Table 2. Baseline training results for different structures (Visdrone-2019 dataset).

Method	P	R	mAP@0.5	mAP@0.5:0.95	FPS	GFLOPs	Model Size (MB)	Parameter (M)
YOLOv5s	0.368	0.314	0.269	0.139	79	15.9	14.4	7.03
YOLOv5m	0.434	0.332	0.311	0.169	74	48.1	42.2	20.9
YOLOv5l	0.44	0.355	0.325	0.181	60	107.9	92.9	46.2
YOLOv5x	0.459	0.37	0.341	0.193	48	204.2	173.2	86.2

As can be seen from Table 2, the YOLOv5x model has the highest detection accuracy, but the slowest detection speed, the largest amount of model calculation and parameters, and the largest memory occupation. The YOLOv5s model has the smallest memory, the smallest amount of calculation and the smallest number of parameters, but the detection accuracy and the detection accuracy are low. The accuracy difference between the YOLOv5x model and YOLOv5x model is 7.2%, but the model occupies a large amount of memory, calculation and the number of parameters, and the model detection speed is increased by 64.58%. Therefore, the YOLOv5 model has the advantages of fast detection speed, small overall model size and high detection accuracy, which meets the needs of the patrol duty object detection studied in this paper. At the same time, considering the real-time requirements of the task and the limited computing resources of the edge devices to be carried out in the future. Therefore, this paper chooses the YOLOv5s model as the baseline model, analyzes the existing and possible future problems of the actual task, makes objected improvements to the baseline model, and proposes a detection algorithm TP-ODA that is more suitable for patrol duty detection tasks.

5.2. Ablation Experiments

We use the model after improving the loss function for training and detection on the BDP dataset. Table 3 represents the improved experimental results. From the experimental results, we know that the detection performance of the baseline detection algorithm on the BDP dataset is good. Compared to the baseline model, the mAP of the improved loss function detection algorithm is improved by 2.1% and the FPS is improved by 8.3%. From the experimental results, it is clear that the improvement in the loss function has more practical significance for the border patrol detection task proposed in this paper.

Table 3. Loss function improvement case parameters on the BDP dataset (batch = 32).

Baseline Method	All	FPS	GFLOPs	Model Size (MB)	Parameter (M)
L	0.559	78	**107.8**	92.9	46.1
L + EIOU	0.571	81	107.8	92.9	46.1
S	0.532	108	15.8	14.1	7.0
S + EIOU	0.553	117	15.8	14.1	7.0

To verify the effectiveness of the other improvement modules used in this paper for the algorithm, we conducted ablation experiments on the BDP dataset. To ensure the fairness of the model evaluation, we set the same parameters for each variable.

The experimental procedure and the resulting relevant parameters are shown in Tables 4–6. To test the performance of the algorithm for detecting images of different scale sizes, the detected images are adjusted to the sizes of 640 and 1024 in this thesis and input to the model for detection. However, according to the actual computational capacity of the edge devices, the number of images input to the network in a single pass is adjusted in the experiments, and the batchsize is set to 1, which means that only 1 image is input to the model for detection at a time, so as to mimic the situation that the UAV platform or other patrol reconnaissance devices have a limited number of images to process in a single pass due to less computational resources. The comprehensive experimental results show that the TP-ODA proposed in this chapter has better performance for the UAV border patrol object detection task. The specific experimental detection results are as follows.

Table 4. The results of ablation experiments performed by the improved module. Batchsize = 32, image size = 640.

Method	Loss Function	Ghost Module	BP-Sim	PDOEM	mAP @0.5/	mAP@ 0.5:0.95	FPS	Model Size/MB	Parameter (M)	GFLOPs
Baseline					0.532	0.227	108	14.1	7.01	15.8
Model 1	√				0.553	0.231	117	14.1	7.01	15.8
Model 2	√	√			0.528	0.223	118	7.6	3.6	8.1
Model 3	√	√	√		0.554	0.245	121	6.1	2.9	6.6
Model 4	√	√	√	√	0.561	0.249	118	5.1	2.4	5.5

Table 5. Batchsize = 1, image size = 640.

Method	Loss Function	Ghost Module	BP-Sim	PDOEM	mAP @0.5/	mAP@ 0.5:0.95	FPS	Model Size/MB	Parameter (M)	GFLOPs
Baseline					0.531	0.23	72	14.1	7.01	15.8
Model 1	√				0.554	0.221	50	14.1	7.01	15.8
Model 2	√	√			0.538	0.223	58	7.6	3.6	8.1
Model 3	√	√	√		0.558	0.247	71	6.1	2.9	6.6
Model 4	√	√	√	√	0.566	0.251	82	5.1	2.4	5.5

Table 6. Batchsize = 1, image size = 1024.

Method	Loss Function	Ghost Module	BP-Sim	PDOEM	mAP @0.5/	mAP@ 0.5:0.95	FPS	Model Size/MB	Parameter (M)	GFLOPs
Baseline					0.476	0.189	103	14.1	7.01	15.8
Model 1	√				0.505	0.202	120	14.1	7.01	15.8
Model 2	√	√			0.467	0.186	117	7.6	3.6	8.1
Model 3	√	√	√		0.48	0.187	120	6.1	2.9	6.6
Model 4	√	√	√	√	0.514	0.211	116	5.1	2.4	5.5

Model 1 mainly improves the imbalance problem of the detection box sample of the model. As can be seen from the three groups of experimental data in Tables 4–6, the detection accuracy and detection speed of the model are improved. Based on Model 1, Model 2 is designed for lightweight, and inspired by the idea of GhostNet, the ordinary convolutional neural network is optimized. The experimental results show that, after Model 2 was replaced with a module that consumes less computational resources, the detection accuracy in the three sets of experiments was reduced by 2.5%, 1.6% and 2.8%, respectively, but the number of model parameters and computational effort were reduced substantially, including a 46.1% reduction in model volume, a 48.64% reduction in the number of parameters, a 48.73% reduction in GFLPOS, and a 3.7% increase in detection speed.

Considering the patrol task that the improved algorithm will use, and aiming at the complex and diverse detection background, we build Model 3 based on Model 2, mainly by adding a lightweight feature information extraction module BP-Sim in the network. The purpose is to enhance the effective information expression ability of the detection object in the complex patrol task environment, and to have better sensitivity to the useful features of each dimension of the border patrol image. The experimental results show that the detection accuracy of Model 3 is improved by 1.8%, 2.0% and 1.3%, the model size is reduced by 19.74%, the number of parameters is reduced by 19.44%, and the GFLOPs is reduced by 18.52%. In the comparison of detection speed, Model 3 is increased by 2.54%, 22.41% and 2.56%, respectively.

To address the impact of noise information when fusing features and the large size of the neck network of the benchmark model, this study adds the feature fusion module PDOEM to the neck network on the basis of Model 3. From the results of the three sets of experiments, it can be seen that the detection accuracy of the model was improved by 0.7%, 0.8%, and 3.4%, respectively, and the model volume was reduced by 16.39%, the parameter volume is reduced by 17.24%, and the GFLOPs was reduced by 16.67%. In terms of detection speed, except for the 2nd group of experiments in which the detection speed of the model increased by 15.49%, the other two groups of experiments decreased by 2.57% and 3.33%, but still belonged to the model with high detection efficiency.

5.3. Model Comparison Experiment

In order to illustrate the performance of the improved algorithm in this paper, we selected some images from the border patrol detection dataset for detection. The main characteristics of the selected graphics are: Highly similar detection environment, blurred object background, diverse number of objects, etc. The selected objects are mainly vehicle objects and soldiers on duty commonly found in border patrol. In addition, this section selects representative detection models from various types of detection models for comparison experiments. The experimental comparison results are shown in Figures 6–8 and Table 7.

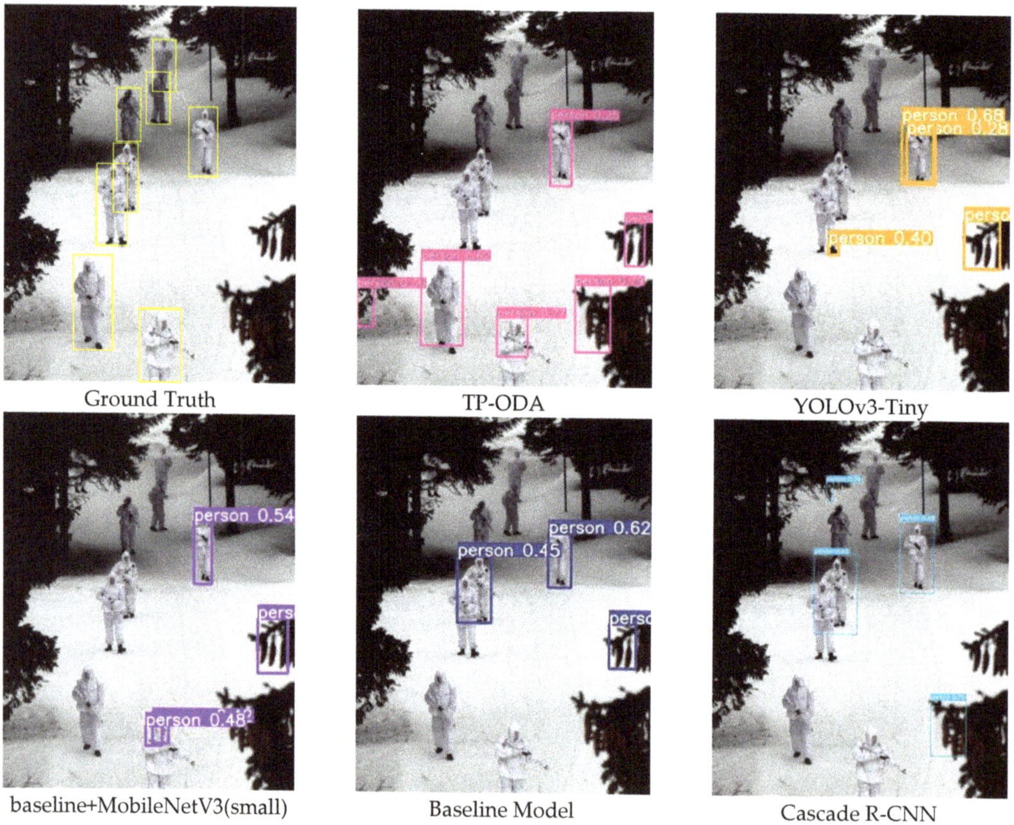

Figure 6. Snow scene visualization detection results.

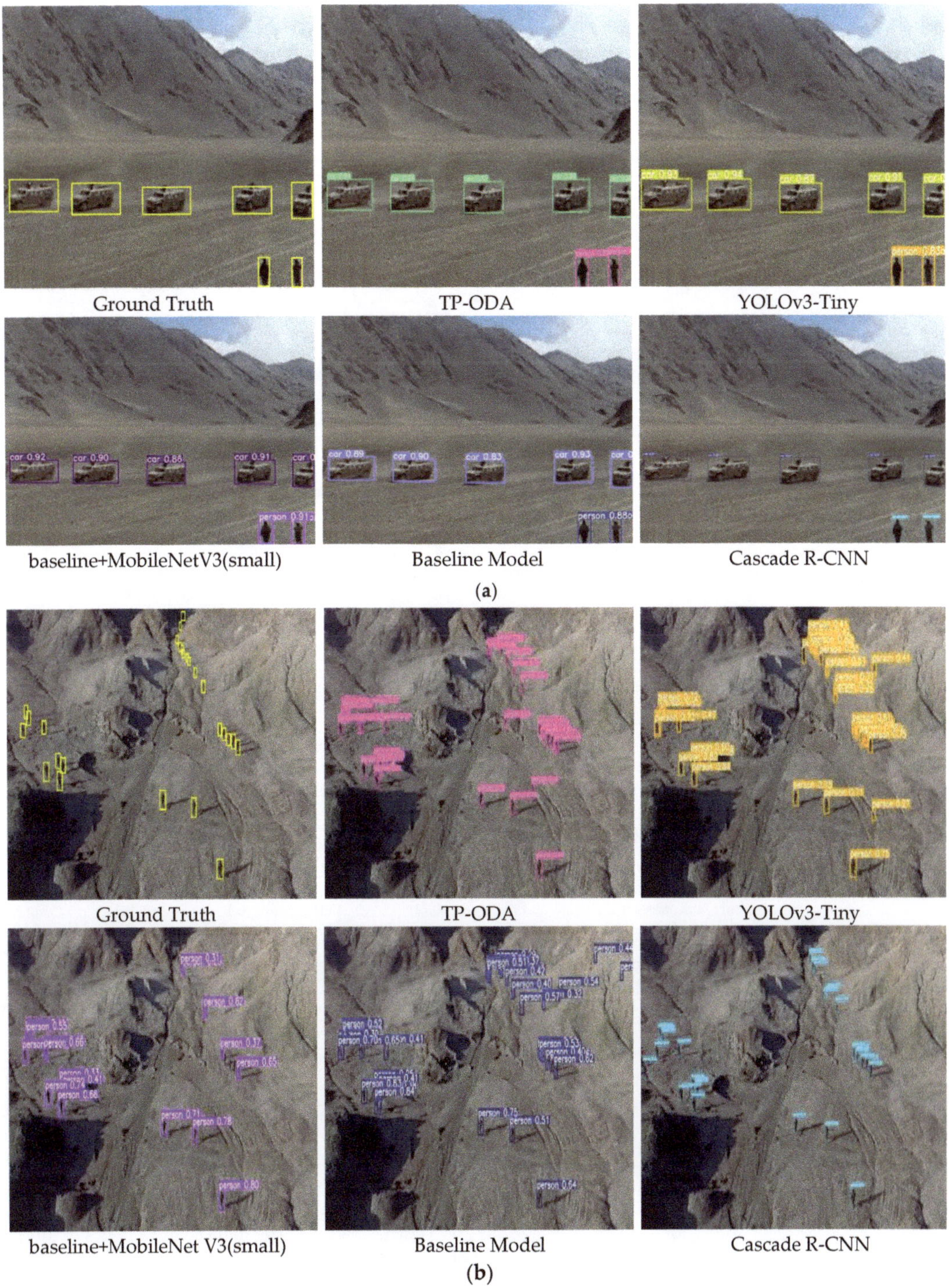

Figure 7. Desert background visualization detection results. (**a**) Low-altitude horizontal view. (**b**) Overhead view.

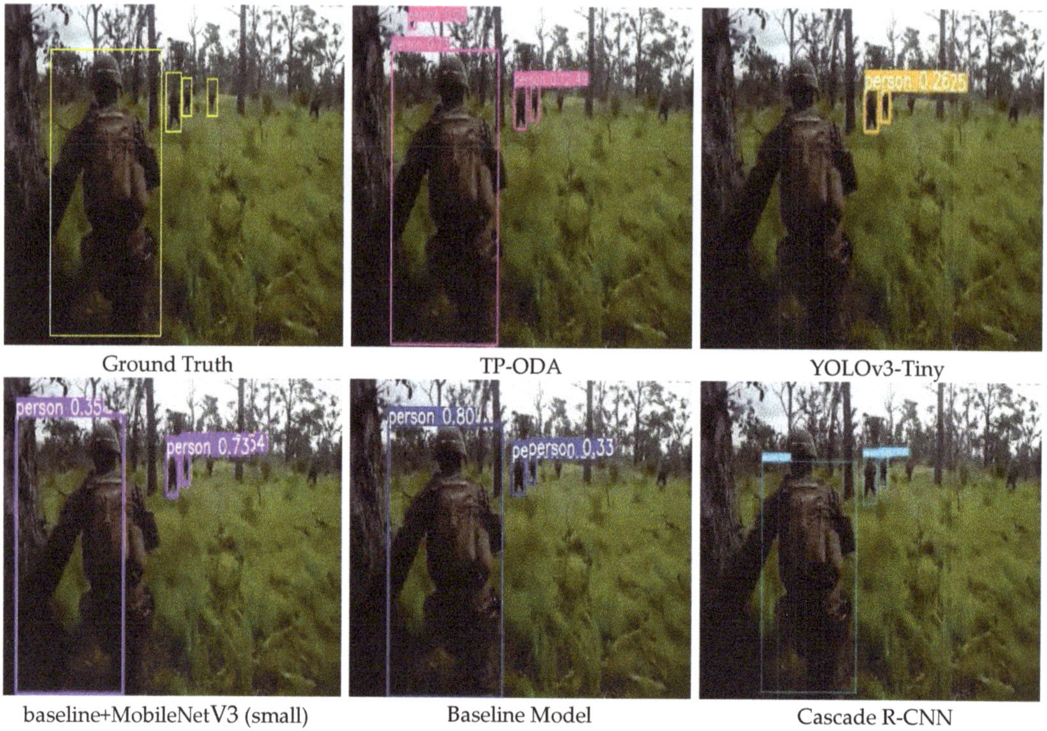

Figure 8. Jungle background visualization detection results.

Table 7. The TP-ODA model was compared to the other models.

Method Baseline	mAP @0.5	mAP @0.5:0.95	FPS	Model Size/(MB)	Parameter (M)	GFLOPs
TP-ODA	0.561	0.249	117	5.1	2.4	5.5
Baseline	0.532	0.227	108	14.1	7.01	15.8
+MobileNeV3(small)	0.53	0.221	121	7.2	3.5	6.1
+EfficientNet	0.517	0.218	112	7.7	3.7	7.6
+ShuffleNet v2	0.497	0.21	**133**	6.1	3.5	**3.1**
YOLOv3-tiny	0.505	0.205	100	16.6	8.6	12.9
Cascade R-CNN	**0.585**	**0.255**	11	165.0	68.9	234.4

The detected environment in Figure 6 is a snowy scene, and the detected objects have a high similarity to the detection background, which is very challenging for the model. From the results, it can be seen that all the detections have missed and false detections. The Cascade R-CNN algorithm and the TP-ODA algorithm both detect three objects, and the benchmark model detects two objects, but also three object false detections, and the Cascade R-CNN only has one object. The experimental results show that the improved algorithm in this chapter is slightly less accurate than the Cascade R-CNN and better than the benchmark algorithm and other detection algorithms on this class of object detection task.

Figure 7 shows two sets of detected objects against a desert background, involving detection categories of soldiers and vehicles on duty. The main characteristics of this group of images are the large number of objects and the small size of the objects. From the results of the two sets of experiments, it can be concluded that all the detection algorithms can detect the vehicle objects and the algorithms have good overall performance, but when detecting pedestrian objects in this type of scene, the YOLOv3-Tiny and Baseline+MobileNetV3 detection algorithms show different degrees of missed detection, and the baseline model and TP-ODA show false detection, with the baseline. The Cascade R-CNN detection algorithm does not show false detections or missed detections, but the TP-ODA algorithm has a higher confidence value in the detection results, which is closer to the real frame.

Figure 8 shows the detected objects in the jungle environment, which are mainly characterized by the different scales of the objects to be detected, and the fuzzy and complex detection backgrounds. All five sets of experimental results failed to detect all the objects, among which the YOLOv3-Tiny detection algorithm had more missed detections, and only two objects were detected in both sets of data. the Baseline model and TP-ODA detected three objects, which was better than the other models. While the TP-ODA algorithm showed one false detection case, the detection results were closer to the true value.

Table 6 indicates that the results of the TP-ODA model with other models for comparison experiments. In the experimental results, the detection algorithm in this paper guarantees the detection speed and detection accuracy, and the number of parameters and computation volume of the model are significantly reduced, and the accuracy is improved by 2.9%, the model parameter volume is reduced by 65.76%, the model volume is reduced by 63.83%, and the computation volume is reduced by 65.19% compared to the benchmark model. In the detection speed comparison experiments, the model with ShuffleNet v2 for light processing has the fastest inference speed with a FPS of 133, which exceeds the detection speed of the benchmark model by 23.14% and that of TP-ODA by 13.67%, but the model computation and the number of parameters are higher than those of the TP-ODA algorithm by more than two-fifths and the model volume is larger. In terms of detection accuracy, the two-stage network shows a stronger advantage, with the accuracy value exceeding that of the TP-ODA algorithm by 2.24%, but the comprehensive performance of the algorithm in this paper is more advantageous in completing the border patrol detection task in terms of the comprehensive model size, detection accuracy and detection speed.

6. Conclusions

In this study, we designed a lightweight detection network for detecting border patrol objects for use with the UAV platforms and portable reconnaissance equipment often used by border patrols. In order to be better used on edge devices, we used the YOLOv5 detection algorithm as the benchmark model and took the reduction of network size and the consumption of computational resources as the starting point. We proposed the TP-ODA detection network in three aspects: Volume compression of the model, improving the semantic information representation of object features and optimizing the loss function of the model, and verify through experiments that the improvement module has a positive effect on the improvement of the model. Synthesizing the improvement work in this paper, the following conclusions can be drawn: We used stacking to reconstruct the backbone network using the lightweight module, reducing the resource consumption by nearly one-third, while using BP-Sim to further optimize the feature extraction function of the network and enhance the detection capability of the model for border patrol hard-to-detect images. Then, we used the EIOU loss function to improve the problem of the detection frame sample imbalance leading to accuracy degradation and convergence slowdown; finally, we designed the feature fusion module PDOEM for the problem of the large size of the neck network feature fusion structure, which further compresses the model while reducing the impact of noise information on the model feature fusion and further enhances the difficult sample feature information mining capability.

This paper verifies, through ablation experiments, that the introduced method and designed module have good effects on algorithm performance improvement, and further verifies that the TP-ODA detection algorithm has better detection performance in the border patrol detection task by comparing it with other lightweight algorithms and common detection algorithms and meets the requirements of the border patrol detection task for real-time and accuracy.

Combining the experimental results and conclusions of this paper, the next research directions are also clarified as follows.

1. The border patrol detection task is an all-weather task, and the next step of the model performance improvement needs to consider training in a richer and more diverse task environment.

2. The improved model will be mounted into resource-constrained edge devices to test the detection performance of the algorithm in reality, and to be able to find the problems with the model in such a way to further improve the algorithm performance.

Author Contributions: Conceptualization, H.L. and L.Y.; methodology, L.Y. and L.B.; software, L.Y.; validation, H.L. and J.Y.; formal analysis, L.B.; investigation, H.L.; resources, L.Y.; data curation, L.Y., L.B.; writing—original draft preparation, L.Y.; writing—review and editing, H.L.; visualization, L.B.; supervision H.L.; project administration, J.Y. All authors have read and agreed to the published version of the manuscript.

Funding: This research was supported by the Military Graduate Student Fund (KYGYJWXX22XX).

Institutional Review Board Statement: Not applicable.

Informed Consent Statement: Informed consent was obtained from all subjects involved in the study.

Data Availability Statement: Not applicable.

Conflicts of Interest: The authors declare no conflict of interest.

References

1. Pedrozo, S. Swiss Military Drones and the Border Space: A Critical Study of the Surveillance Exercised by Border Guards. *Geogr. Helv.* **2017**, *72*, 97–107. [CrossRef]
2. Abushahma, R.I.H.; Ali, M.A.M.; Rahman, N.A.A.; Al-Sanjary, O.I. Comparative Features of Unmanned Aerial Vehicle (UAV) for Border Protection of Libya: A Review. In Proceedings of the IEEE 2019 IEEE 15th International Colloquium on Signal Processing & Its Applications (CSPA), Penang, Malaysia, 8–9 March 2019; pp. 114–119.
3. Alom, M.Z.; Taha, T.M.; Yakopcic, C.; Westberg, S.; Sidike, P.; Nasrin, M.S.; Essen, B.C.V.; Awwal, A.A.S.; Asari, V.K. The History Began from AlexNet: A Comprehensive Survey on Deep Learning Approaches. *arXiv* **2018**, arXiv:1803.01164.
4. Peng, H.; Zhang, Y.; Yang, S.; Song, B. Battlefield Image Situational Awareness Application Based on Deep Learning. *IEEE Intell. Syst.* **2020**, *35*, 36–43. [CrossRef]
5. Buch, N.; Velastin, S.A.; Orwell, J. A Review of Computer Vision Techniques for the Analysis of Urban Traffic. *IEEE Trans. Intell. Transp. Syst.* **2011**, *12*, 20. [CrossRef]
6. Esteva, A.; Chou, K.; Yeung, S.; Naik, N.; Madani, A.; Mottaghi, A.; Liu, Y.; Topol, E.J.; Dean, J.; Socher, R. Deep Learning-Enabled Medical Computer Vision. *NPJ Digit. Med.* **2021**, *4*, 5. [CrossRef] [PubMed]
7. Girshick, R.B.; Donahue, J.; Darrell, T.; Malik, J. Rich Feature Hierarchies for Accurate Object Detection and Semantic Segmentation. In Proceedings of the 2014 IEEE Conference on Computer Vision and Pattern Recognition, Columbus, OH, USA, 23–28 June 2014; pp. 580–587.
8. Jiang, G.; He, H.; Yan, J.; Xie, P. Multiscale Convolutional Neural Networks for Fault Diagnosis of Wind Turbine Gearbox. *IEEE Trans. Ind. Electron.* **2019**, *66*, 3196–3207. [CrossRef]
9. Zhao, Q.; Sheng, T.; Wang, Y.; Tang, Z.; Chen, Y.; Cai, L.; Ling, H. M2Det: A Single-Shot Object Detector Based on Multi-Level Feature Pyramid Network. *arXiv* **2019**, arXiv:1811.04533. [CrossRef]
10. Tan, M.; Le, Q.V. EfficientNet: Rethinking Model Scaling for Convolutional Neural Networks. *arXiv* **2019**, arXiv:1905.11946.
11. Ultralytics. YOLOv5. Available online: https://github.com/ultralytics/yolov5 (accessed on 5 December 2021).
12. Redmon, J.; Divvala, S.K.; Girshick, R.B.; Farhadi, A. You Only Look Once: Unified, Real-Time Object Detection. In Proceedings of the 2016 IEEE Conference on Computer Vision and Pattern Recognition (CVPR), Las Vegas, NV, USA, 27–30 June 2016; pp. 779–788.
13. Redmon, J.; Farhadi, A. YOLO9000: Better, Faster, Stronger. In Proceedings of the 2017 IEEE Conference on Computer Vision and Pattern Recognition (CVPR), Honolulu, HI, USA, 21–27 July 2017; pp. 6517–6525.
14. Redmon, J.; Farhadi, A. YOLOv3: An Incremental Improvement. *arXiv* **2018**, arXiv:1804.02767.
15. Bochkovskiy, A.; Wang, C.-Y.; Liao, H.-Y.M. YOLOv4: Optimal Speed and Accuracy of Object Detection. *arXiv* **2020**, arXiv:2004.10934.
16. Mohiyuddin, A.; Basharat, A.; Ghani, U.; Peter, V.; Abbas, S.; Naeem, O.B.; Rizwan, M. Breast Tumor Detection and Classification in Mammogram Images Using Modified YOLOv5 Network. *Comput. Math. Methods Med.* **2022**, *2022*, 1–16. [CrossRef] [PubMed]
17. Walia, I.S.; Kumar, D.; Sharma, K.; Hemanth, J.D.; Popescu, D.E. An Integrated Approach for Monitoring Social Distancing and Face Mask Detection Using Stacked ResNet-50 and YOLOv5. *Electronics* **2021**, *10*, 2996. [CrossRef]
18. Loey, M.; Manogaran, G.; Taha, M.H.N.; Khalifa, N.E.M. Fighting against COVID-19: A Novel Deep Learning Model Based on YOLO-v2 with ResNet-50 for Medical Face Mask Detection. *Sustain. Cities Soc.* **2020**, *65*, 102600. [CrossRef] [PubMed]

19. Liu, J.; Wang, X. Tomato Diseases and Pests Detection Based on Improved Yolo V3 Convolutional Neural Network. *Front. Plant Sci.* **2020**, *11*, 898. [CrossRef]
20. Chen, W.; Lu, S.; Liu, B.; Chen, M.; Li, G.; Qian, T. CitrusYOLO: A Algorithm for Citrus Detection under Orchard Environment Based on YOLOv4. *Multim. Tools Appl.* **2022**, *81*, 31363–31389. [CrossRef]
21. Kou, X.; Liu, S.; Cheng, K.I.-C.; Qian, Y. Development of a YOLO-V3-Based Model for Detecting Defects on Steel Strip Surface. *Measurement* **2021**, *182*, 109454. [CrossRef]
22. Al-qaness, M.A.A.; Abbasi, A.A.; Fan, H.; Ibrahim, R.A.; Alsamhi, S.H.; Hawbani, A. An Improved YOLO-Based Road Traffic Monitoring System. *Computing* **2021**, *103*, 211–230. [CrossRef]
23. Du, Y.; Pan, N.; Xu, Z.; Deng, F.; Shen, Y.; Kang, H. Pavement Distress Detection and Classification Based on YOLO Network. *Int. J. Pavement Eng.* **2020**, *22*, 1659–1672. [CrossRef]
24. Liu, Y.; Wang, C.; Zhou, Y. Camouflaged People Detection Based on a Semi-Supervised Search Identification Network. *Def. Technol.* **2021**, *in press*. [CrossRef]
25. Fang, Z.; Zhang, X.; Deng, X.; Cao, T.; Zheng, C. Camouflage People Detection via Strong Semantic Dilation Network. In Proceedings of the ACM TURC 2019: ACM Turing Celebration Conference—China, Chengdu China, 17–19 May 2019; pp. 1–7.
26. Zheng, G.; Wu, X.; Hu, Y.; Liu, X. Object Detection for Low-Resolution Infrared Image in Land Battlefield Based on Deep Learning. In Proceedings of the IEEE 2019 Chinese Control Conference (CCC), Guangzhou, China, 27–30 July 2019; pp. 8649–8652.
27. Zhu, X.; Lyu, S.; Wang, X.; Zhao, Q. TPH-YOLOv5: Improved YOLOv5 Based on Transformer Prediction Head for Object Detection on Drone-Captured Scenarios. In Proceedings of the 2021 IEEE/CVF International Conference on Computer Vision Workshops (ICCVW), Montreal, BC, Canada, 11–17 October 2021; pp. 2778–2788.
28. Kristo, M.; Ivasic-Kos, M.; Pobar, M. Thermal Object Detection in Difficult Weather Conditions Using YOLO. *IEEE Access* **2020**, *8*, 125459–125476. [CrossRef]
29. Iandola, F.N.; Moskewicz, M.W.; Ashraf, K.; Han, S.; Dally, W.J.; Keutzer, K. SqueezeNet: AlexNet-Level Accuracy with 50x Fewer Parameters and <1 MB Model Size. *arXiv* **2016**, arXiv:1602.07360.
30. Sandler, M.; Howard, A.G.; Zhu, M.; Zhmoginov, A.; Chen, L.-C. MobileNetV2: Inverted Residuals and Linear Bottlenecks. In Proceedings of the 2018 IEEE/CVF Conference on Computer Vision and Pattern Recognition, Salt Lake City, UT, USA, 18–23 June 2018; pp. 4510–4520.
31. Howard, A.G.; Zhu, M.; Chen, B.; Kalenichenko, D.; Wang, W.; Weyand, T.; Andreetto, M.; Adam, H. MobileNets: Efficient Convolutional Neural Networks for Mobile Vision Applications. *arXiv* **2017**, arXiv:1704.04861.
32. Zhang, X.; Zhou, X.; Lin, M.; Sun, J. ShuffleNet: An Extremely Efficient Convolutional Neural Network for Mobile Devices. In Proceedings of the 2018 IEEE/CVF Conference on Computer Vision and Pattern Recognition, Salt Lake City, UT, USA, 18–23 June 2018; pp. 6848–6856.
33. Ma, N.; Zhang, X.; Zheng, H.; Sun, J. ShuffleNet V2: Practical Guidelines for Efficient CNN Architecture Design. In Proceedings of the ECCV, Munich, Germany, 8–14 September 2018.
34. Chollet, F. Xception: Deep Learning with Depthwise Separable Convolutions. In Proceedings of the 2017 IEEE Conference on Computer Vision and Pattern Recognition (CVPR), Honolulu, HI, USA, 21–27 July 2017; pp. 1800–1807.
35. Fan, Y.; Qiu, Q.; Hou, S.; Li, Y.; Xie, J.; Qin, M.; Chu, F. Application of Improved YOLOv5 in Aerial Photographing Infrared Vehicle Detection. *Electronics* **2022**, *20*, 2344. [CrossRef]
36. Zhang, M.; Xu, S.; Song, W.; He, Q.; Wei, Q. Lightweight Underwater Object Detection Based on YOLO v4 and Multi-Scale Attentional Feature Fusion. *Remote. Sens.* **2021**, *13*, 4706. [CrossRef]
37. Feng, J.H.; Yuan, H.; Hu, Y.Q.; Lin, J.; Liu, S.; Luo, X. Research on Deep Learning Method for Rail Surface Defect Detection. *IET Electr. Syst. Transp.* **2020**, *10*, 436–442. [CrossRef]
38. Wu, T.-H.; Wang, T.-W.; Liu, Y.-Q. Real-Time Vehicle and Distance Detection Based on Improved Yolo v5 Network. In Proceedings of the 2021 3rd World Symposium on Artificial Intelligence (WSAI), Guangzhou, China, 18–20 June 2021; pp. 24–28.
39. Lin, T.-Y.; Dollár, P.; Girshick, R.B.; He, K.; Hariharan, B.; Belongie, S.J. Feature Pyramid Networks for Object Detection. In Proceedings of the 2017 IEEE Conference on Computer Vision and Pattern Recognition (CVPR), Honolulu, HI, USA, 21–27 July 2017; pp. 936–944.
40. Wang, K.; Liew, J.H.; Zou, Y.; Zhou, D.; Feng, J. PANet: Few-Shot Image Semantic Segmentation With Prototype Alignment. In Proceedings of the 2019 IEEE/CVF International Conference on Computer Vision (ICCV), Seoul, Republic of Korea, 27 October–2 November 2019; pp. 9196–9205.
41. Han, K.; Wang, Y.; Tian, Q.; Guo, J.; Xu, C.; Xu, C. GhostNet: More Features From Cheap Operations. In Proceedings of the 2020 IEEE/CVF Conference on Computer Vision and Pattern Recognition (CVPR), Seattle, WA, USA, 13–19 June 2020; pp. 1577–1586.
42. Kong, L.; Wang, J.; Zhao, P. YOLO-G: A Lightweight Network Model for Improving the Performance of Military Targets Detection. *IEEE Access* **2022**, *10*, 55546–55564. [CrossRef]
43. Yang, L.; Zhang, R.-Y.; Li, L.; Xie, X. SimAM: A Simple, Parameter-Free Attention Module for Convolutional Neural Networks. In Proceedings of the 38th International Conference on Machine Learning, Virtual, 18–24 July 2021; p. 12.
44. Zhu, D.; Qi, R.; Hu, P.; Su, Q.; Qin, X.; Li, Z. YOLO-Rip: A Modified Lightweight Network for Rip Currents Detection. *Front. Mar. Sci.* **2022**, *9*. [CrossRef]
45. Zheng, Z.; Wang, P.; Liu, W.; Li, J.; Ye, R.; Ren, D. Distance-IoU Loss: Faster and Better Learning for Bounding Box Regression. In Proceedings of the AAAI Conference on Artificial Intelligence, New York, NY, USA, 7–12 February 2020.

46. Zhang, Y.-F.; Ren, W.; Zhang, Z.; Jia, Z.; Wang, L.; Tan, T. Focal and Efficient IOU Loss for Accurate Bounding Box Regression. *Neurocomputing* **2022**, *506*, 146–157. [CrossRef]
47. Wen, L.; Zhu, P.F.; Du, D.; Bian, X.; Ling, H.; Hu, Q.; Liu, C.; Cheng, H.; Liu, X.; Ma, W.; et al. VisDrone-SOT2019: The Vision Meets Drone Single Object Tracking Challenge Results. In Proceedings of the 2019 IEEE/CVF International Conference on Computer Vision Workshop (ICCVW), Seoul, Republic of Korea, 27–28 October 2019; pp. 199–212.

electronics

Article

Innovative Hyperspectral Image Classification Approach Using Optimized CNN and ELM

Ansheng Ye [1,2], Xiangbing Zhou [3,*] and Fang Miao [1]

1 Key Lab of Earth Exploration & Information Techniques of Ministry Education, Chengdu University of Technology, Chengdu 610059, China; yas@cdu.edu.cn (A.Y.); miaof@cdut.edu.cn (F.M.)
2 School of Computer Science, Chengdu University, Chengdu 610106, China
3 School of Information and Engineering, Sichuan Tourism University, Chengdu 610100, China
* Correspondence: zhouxb@uestc.edu.cn

Abstract: In order to effectively extract features and improve classification accuracy for hyperspectral remote sensing images (HRSIs), the advantages of enhanced particle swarm optimization (PSO) algorithm, convolutional neural network (CNN), and extreme learning machine (ELM) are fully utilized to propose an innovative classification method of HRSIs (IPCEHRIC) in this paper. In the IPCEHRIC, an enhanced PSO algorithm (CWLPSO) is developed by improving learning factor and inertia weight to improve the global optimization performance, which is employed to optimize the parameters of the CNN in order to construct an optimized CNN model for effectively extracting the deep features of HRSIs. Then, a feature matrix is constructed and the ELM with strong generalization ability and fast learning ability is employed to realize the accurate classification of HRSIs. Pavia University data and actual HRSIs after Jiuzhaigou M7.0 earthquake are applied to test and prove the effectiveness of the IPCEHRIC. The experiment results show that the optimized CNN can effectively extract the deep features from HRSIs, and the IPCEHRIC can accurately classify the HRSIs after Jiuzhaigou M7.0 earthquake to obtain the villages, bareland, grassland, trees, water, and rocks. Therefore, the IPCEHRIC takes on stronger generalization, faster learning ability, and higher classification accuracy.

Keywords: hyperspectral image classification; CNN; ELM; PSO; deep feature

Citation: Ye, A.; Zhou, X.; Miao, F. Innovative Hyperspectral Image Classification Approach Using Optimized CNN and ELM. *Electronics* **2022**, *11*, 775. https://doi.org/10.3390/electronics11050775

Academic Editor: Byung Cheol Song

Received: 21 January 2022
Accepted: 1 March 2022
Published: 2 March 2022

1. Introduction

Remote sensing image (RSI) classification is to divide the image into several regions by using specific rule or algorithm according to the spectral features, geometric texture features, or other features [1–3]. Each region is a set of ground and objects with the same characteristics, or a lot of RSIs are divided into several sets through some methods, and each set represents a kind of ground or object category. It is a very important basic problem and plays a very important position in the field of RSIs [4–6]. Therefore, the research on remote sensing image classification method has become an important direction, which has very important theoretical significance and practical application value.

In recent years, many classification methods of RSIs have been proposed, which can be divided into two categories of manual visual interpretation and computer classification [7]. The manual visual interpretation is the most traditional classification method, which has large workload, low efficiency, and requires rich professional knowledge and interpretation experiences [8–10]. With the rapid development of computer techniques, the automatic classification method of RSIs replaces the manual visual interpretation classification method. The more complex computer technology uses the spectral brightness value of pixels and the spatial relationship between pixels and their surrounding pixels to realize pixel classification. Tran et al. [11] presented a sub-pixel and per-pixel classification method to analyze the impact of land cover heterogeneity. Khodadadzadeh et al. [12] presented a new hyperspectral spectral-spatial classifier. Li et al. [13] presented a novel classification method of RSIs based on the probabilistic fusion of pixel-level and superpixel-level classifiers. Li et al. [14] presented a novel pixel-pair method. Mei et al. [15] presented a novel pixel-level perceptual subspace learning method. Pan et al. [16] presented a new central pixel selection strategy based on gradient information to realize texture image classification. Bey et al. [17] presented a new land cover assessment methodology. Yan et al. [18] presented a triple counter domain adaptation approach for learning domain invariant classifier. Li et al. [19] presented a novel multi-view active learning approach based on sub-pixel and super-pixel. Ma and Chang [20] presented a novel mixed pixel classification approach.

The single pixel spectral classification method can obtain the hyperspectral spectral-spatial classification results, but they still exist at low classification accuracy and high time complexity. The signal processing method on computer has the characteristics of large amount of calculation and can obtain high classification accuracy. However, the high-resolution RSIs have high spatial resolution and complexity. It is very difficult to classify high-resolution RSIs by using traditional classification methods. Therefore, it is urgent to deeply study a fast classification approach that can be effectively applied to high-resolution RSIs [21,22]. As a field of artificial intelligence, deep learning has attracted extensive attention, and has gradually become one of the important technologies to promote the development of artificial intelligence. Therefore, many scholars have applied deep learning to remote sensing image classification and proposed many features extraction and classification methods. Romero et al. [23] presented a sparse feature unsupervised learning approach based on greedy hierarchical unsupervised pretraining method. Sharma et al. [24] presented a new deep patch-based CNN. Maggiori et al. [25] presented a dense pixel-level classification model. Wang et al. [26] presented a HRSI classification method using principal component analysis (PCA) and guided filtering, deep learning architecture. Ji et al. [27] presented a novel three-dimensional CNN to automatically classify crops. Ben et al. [28] presented 3-D deep learning approach. Xu et al. [29] presented a novel RSI classification model using generative adversarial network. Tao et al. [30] presented a novel reinforced deep neural network (DNN) with depth and width. Liang et al. [31] presented a new RSI classification approach using stacked denoising autoencoder. Li et al. [32] presented a novel region-wise depth feature extraction model. Li et al. [33] presented an adaptive multiscale deep fusion residual network. Yuan et al. [34] presented a classification approach based on rearranged local features. Zhang et al. [35] presented a new dense network with multi-scales. Zhang et al. [36] presented a new feature aggregation model based on 3-D CNN. Chen et al. [37] presented a novel deep Boltzmann machine based on the conjugate gradient update algorithm. Xiong et al. [38] presented a novel deep multi-feature fusion network based on two different deep architecture branches. Tong et al. [39] presented a channel-attention-based DenseNet network. Zhu et al. [40] presented a new deep network with dual-branch attention fusion. Raza et al. [41] presented a four-layer classification network based on visual attention mechanisms. Li et al. [42] presented a classification approach by combining generative adversarial network (GAN), CNN with long short-term memory. Gu et al. [43] presented a pseudo labeled sample generation method. Guo et al. [44] presented a novel self-supervised gated self-attention GAN. Li et al. [45] presented a novel locally preserving deep cross embedded classification network. Lei et al. [46] presented a novel deep convolutional capsule network using spectral-spatial features. Cui et al. [47] presented a dual-channel deep learning recognition model. Peng et al. [48] presented an efficient search framework to discover optimal network architectures. Guo et al. [49] presented a novel semi-supervised scene classification method using GAN. Dong et al. [50] presented a pixel cluster CNN. Li et al. [51] presented a new RSI classification approach using error-tolerant deep learning. Li et al. [52] presented a gated recursive neural network. Dong et al. [53] explored the potential of the reference-based super-resolution method. Wu et al. [54] presented a self-paced dynamic infinite mixture model. Karadal et al. [55] presented automated classification of remote sensing images based on multileveled MobileNetV2 and DWT. Ma et al. [56] presented a novel adaptive hybrid fusion network for multiresolution remote sensing images classification.

Cai et al. [57] presented a novel cross-attention mechanism and graph convolution integration algorithm. Zhang et al. [58] presented a convolutional neural architecture for remote sensing image scene classification. Hilal et al. [59] presented a new deep transfer learning-based fusion model for remote-sensing image classification. Li et al. [60] presented a multi-scale fully convolutional network to exploit discriminative representations. In addition, some new optimization algorithms are proposed [61–72], which can optimize the parameters of classification models.

Because the CNN has good feature extraction ability, these classification methods based on CNN have obtained better classification effects. It has attracted extensive attention and has been widely applied in RSIs. However, the structure and parameter selection of the CNN seriously affect its learning accuracy. Therefore, the enhanced PSO algorithm with global optimization ability is employed to optimize and determine the parameters of the CNN to obtain the optimized parameter values for constructing an optimized CNN, which is applied to effectively extract the multi-layer features of HRSIs to form a multi-feature fusion matrix. Then, the ELM is employed to realize the classification of HRSIs. The effectiveness is verified by typical data set and actual HRSIs after Jiuzhaigou M7.0 earthquake.

The main contributions of this paper are described as follows.

(1) For the slow convergence and low accuracy of the PSO, an enhanced PSO based on fusing multi-strategy (CWLPSO) is proposed by adding new acceleration factor strategy and inertia weight linear decreasing strategy.
(2) For the difficultly determining the parameters of the CNN, an optimized CNN model using CWLPSO is developed to effectively extract the deep features of HRSIs.
(3) The ELM with strong generalization ability, fast learning ability, and the constructed feature vector are combined to realize the accurate classification of HRSIs.
(4) An innovative classification method of HRSIs based on CWLPSO, CNN, and ELM, namely, IPCEHRIC is proposed.

2. Basic Methods

2.1. CNN

The CNN is a feedforward neural network, which includes convolution calculation and representative algorithm. It has the representation learning ability and can classify the input information according to its hierarchical structure. The CNN includes input layer, hidden layer, and output layer, which is shown in Figure 1.

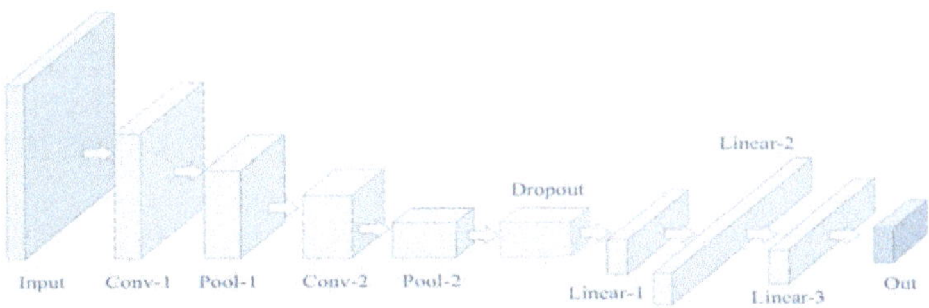

Figure 1. The structure of the CNN.

The structure of the CNN is described in detail as follows.

Input layer. It can deal with multidimensional data, and the input features need to be standardized.

Hidden layer. It includes convolution operation, pooling operation, and full connection layer. The convolution layer is used to extract features from input data through the convolution operation of multiple convolution cores to obtain and construct the feature map. The pooling layer is to select features and filter information from the feature map to retain important features, and preset the pooling function. The full connection layer is equivalent to the hidden layer in the network. The output is obtained.

Convolution kernel. When the convolution kernel works, it will regularly scan input features, multiply and sum the input features, and superimpose the deviation. The output of the $l+1$ layer is described as follow.

$$Z^{l+1}(i,j) = [Z^l \otimes w^{l+1}](i,j) + b = \sum_{k=1}^{K_l} \sum_{x=1}^{f} \sum_{y=1}^{f} [Z_k^l(s_0 i + x,\ s_0 j + y) w_k^{l+1}(x,y)] + b$$
$$(i,j) \in \{0,1,\ldots,L_{l+1}\} \quad L_{l+1} = \frac{L_l + 2p - f}{s_0} + 1 \tag{1}$$

where, b is the offset, Z^l and Z^{l+1} represents the convolution input and output of the $l+1$ layer, L_{l+1} is the size of Z_{l+1}. In here, it is assumed that the length and width of the characteristic graph are the same. $Z(i,j)$ corresponds the pixels of the feature map, K is the number of channels, f, s_0 and p are the convolution layer parameters, which correspond to the kernel size, convolution step size and number of filling layers. Especially, when the kernel is f = 1, the step size is s_0 = 1, and when a filled unit convolution kernel is not included, the cross-correlation calculation is equivalent to matrix multiplication, and a fully connected network is established between the convolution layers.

$$Z^{l+1} = \sum_{k=1}^{K_l} \sum_{x=1}^{L} \sum_{y=1}^{L} (Z_{i,j}^l w_k^{l+1}) + b = w_{l+1}^T Z_{l+1} + b,\ L^{l+1} = L \tag{2}$$

Output layer. The output layer is the same, and the output result is obtained.

2.2. PSO

The PSO is an intelligent algorithm, which was proposed by Eberhart and Kennedy in 1995 [73]. At first, it was to study the predation behavior of birds. Inspired by this, it carried out modeling research on bird activities. In PSO, the update formula of the particle velocity and position are described as follows.

$$v_{m+1} = \omega v_m + c_1 r_1 (pbest_m - x_m) + c_2 r_2 (gbest_m - x_m) \tag{3}$$

$$x_{m+1} = x_m + v_{m+1} \tag{4}$$

where, v_{m+1} represents the velocity of particles, ω is the inertia weight factor, c_1 and c_2 are learning factors, ω, c_1, and c_2 are usually preseted in advance. r_1 and r_2 represent a random number, $pbest_m$ is the optimal value of individual, $gbest_m$ is the optimal value of swarm. The function used to evaluate the fitness value of particles is called fitness function, i.e., objective function. In most cases, the fitness value is smaller, the particle is better. The optimal value of the individual and the optimal value of swarm are generally updated by the following formula.

$$pbest_{m+1} = \begin{cases} x_{m+1}, f(x_{m+1}) < f(pbest_m) \\ pbest_m, otherwise \end{cases} \tag{5}$$

$$gbest_{m+1} = \begin{cases} pbest_{m+1}, f(pbest_{m+1}) < f(gbest_{m+1}) \\ gbest_{m+1}, otherwise \end{cases} \tag{6}$$

If the value of x_{m+1} is smaller than the value of the individual extreme value, then $pbest_{m+1}$ is equal to x_{m+1}. On the contrary, the individual extreme value is not updated. If the value of $gbest_{m+1}$ is greater than the value of the individual extreme value, then $gbest_{m+1}$ is equal to $gbest_{m+1}$.

2.3. ELM

The ELM is one of the commonly used neural network models in machine learning. Its essence is a machine learning method based on single-hidden layer feed forward network (SLFN). Compared with back propagation (BP) neural network model that uses gradient descent algorithm to update the weight in the field of machine learning, the ELM can randomly generate the threshold value. It has low computational complexity and less time-consuming. In the classification and regression problems, the structure of the ELM model is generally divided into the input layer, hidden and output layers. The specific structure is shown in Figure 2.

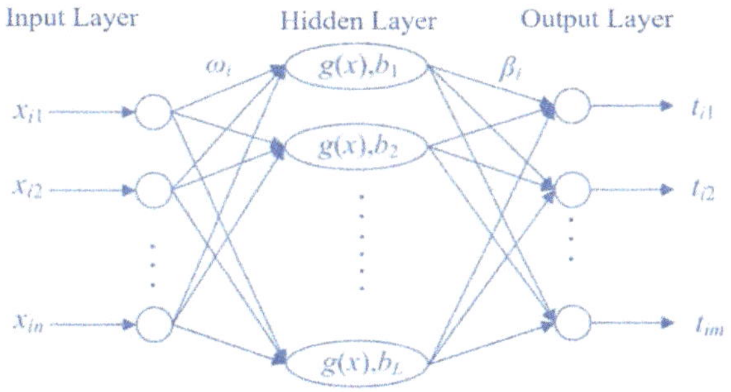

Figure 2. The structure of ELM.

3. Improved Learning Factor and Inertia Weight

Although many researchers have proposed some effective researches and improvements on the shortcomings of PSO, the PSO still has the problems of slow convergence, high time complexity, and low accuracy. Therefore, the acceleration factor strategy and the inertia weight linear decreasing strategy are introduced to propose an enhanced PSO(CWLPSO) in this paper. That is, aiming at the slow convergence speed, a fast convergence strategy with small deviation angle of particle speed and position is adopted to accelerate the convergence of particles. Aiming at the poor search ability, a new improvement strategy of learning factor is proposed in here. That is, different c_1 and c_2 values are selected in order to improve the local search ability of particles in the early stage, enhance the optimization ability of particle swarm and strengthen the overall search ability of particles in the later stage. Aiming at the premature in the later stage, a new linear decreasing strategy of inertia weight is adopted to linearly reduce the inertia weight from the maximum value to the minimum value, so as to avoid the premature and the oscillation in the later stage of the algorithm.

3.1. Improve Learning Factors

The learning factors c_1 and c_2 in the PSO represent the function of the particle itself and the remaining particles removed from the particle itself on the motion route of moving particles. At the same time, they also represent the information exchange between particles, which result in different motion trajectories of particles. Therefore, an improvement strategy of learning factor is designed to improve the local search ability of particles, enhance the optimization ability of particle swarm and strengthen the overall search ability of particles in here. That is, in the early stage of the algorithm, the c_1 value is larger and the c_2 value is smaller, so that the particles can enhance the ability of self-cognition and weaken the swarm cognition of the particles. However, in the later stage of the algorithm, the c_1 value decreases and the c_2 value increases, it can improve the search ability by increasing the c_1 value in the early stage, the proportion of particle swarm will be strengthened in the later stage, so that more particles can learn from the swarm optimum. At the same time, the fewer particles can learn from individual optimum, which is conducive to enhancing the optimization ability, and strengthening the overall search ability of particles. The improved strategy of learning factor is described as follows.

$$c_1 = c_{1max} + (c_{1max} - c_{1min}) * \left(\frac{i}{k}\right) \tag{7}$$

$$c_2 = c_{2min} + (c_{2max} - c_{2min}) * \left(\frac{i}{k}\right) \tag{8}$$

where, c_{1max} and c_{1min} represent the maximum and minimum values of learning factor c_1.c_{2min} and c_{2max} represent the maximum and minimum values of learning factor c_2, i represents the current iterations, and k represents the maximum iterations.

3.2. *Linear Decreasing of Inertia Weight*

Inertia weight plays an important role in PSO. Generally, the inertia weight is generally set to a fixed value between 0.6 and 0.9. The improper selection of inertia weight will cause errors. If the inertia weight is larger, on the one hand, it will help to jump out from the local minimum point and facilitate the global search, on the other hand, it will weaken the local search ability. Therefore, for the premature in the later stage of the algorithm, a new linear decreasing strategy of inertia weight is developed. That is, the inertia weight is linearly reduced from ω_{max} to ω_{min}, which is described as follows.

$$\omega = \omega_{max} - \left(\frac{i * (\omega_{max} - \omega_{min})}{k}\right) \tag{9}$$

where, ω is inertia weight, ω_{max} is maximum value of inertia weight, ω_{min} is minimum value of inertia weight, i is current iteration, and k is maximum iterations.

4. Optimize CNN Using CWLPSO

4.1. *Optimized Idea for CNN*

The CNN with combining weight sharing and local area connection reduces the complexity of the model and the values of parameters. However, the selection of the number of filters, activation function, and learning rate of the CNN seriously affects the learning accuracy. The parameters of the CNN are trained by the steepest gradient descent method, which has a great impact on the learning performance. The proposed CWLPSO has the characteristics of global search ability, population diversity, and fast convergence. Therefore, the CWLPSO is employed to optimize the parameters of the CNN, and an optimized CNN model based on the CWLPSO algorithm is developed in this paper. That is, each particle is a network structure of the CNN. After the CNN calculates the error between the expected value and the actual value, each particle considered the number of filters, activation function learning rate, initial weight, and initial offset of the CNN as particle dimensions. The obtained test error is taken as the fitness function value, the optimal CNN model is selected through the iteration of the CWLPSO.

4.2. *Model of Optimized CNN*

The optimization process of the CNN using CWLPSO is shown in Figure 3.

The specific optimization process of the CNN using CWLPSO are described as follows.

Step 1. Initialize the parameters of the CNN, which include the number of nodes in hidden layer, the learning rate, and so on.

Step 2. Initialize the parameters of the CWLPSO, which include the number of the population, the maximum number of iterations, and the initial learning factor and inertia weight, and so on.

Step 3. Construct the optimization objective function.

Step 4. Calculate the individual fitness values in the population in order to obtain the initial fitness values of the population.

Step 5. Determine whether the end condition is met. If the end condition is met, then the optimal individual is regarded as the optimal parameter value of the CNN and loop Step 7. Otherwise execute Step 6.

Step 6. The velocity and position are updated, then the learning factor and the weight factor is updated. Then return to Step 4.

Step 7. Obtain the optimal parameter values of the CNN and an optimized CNN model is output.

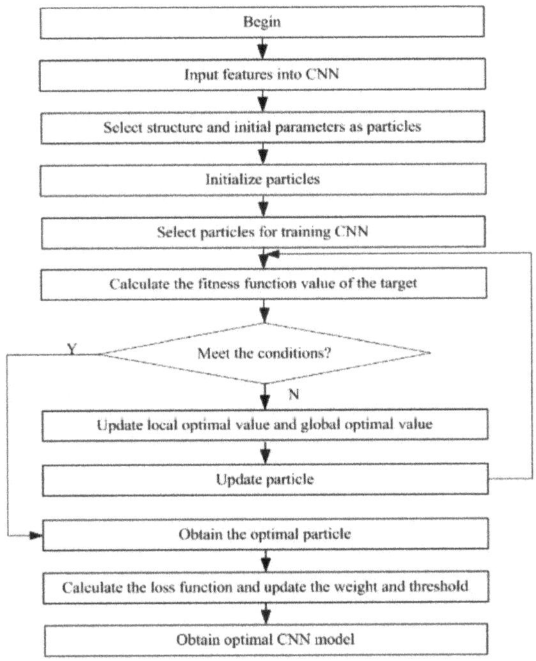

Figure 3. The optimization process of the CNN using CWLPSO.

5. An Innovative Classification Method of HRSIs Using Optimized CNN and ELM

Classification accuracy is important indicators to evaluate the classification model for HRSIs. Therefore, the effective feature extraction of HRSIs is the key factor for affecting classification accuracy. As a deep learning method, the CNN can effectively mine the multi-layer representation feature information. Different levels of representation correspond to different feature attributes of the recognition object. For example, the shallow network mainly represents the texture, edge and other local information of the recognition object, while the deep network represents the more abstract semantics, structure, and other global information. This feature matrix composes of the multi-layer feature attributes of the HRSIs. As a fast machine learning algorithm, the weight parameters of the ELM and the offset parameters on the hidden layer do not need to be adjusted repeatedly through iteration, which can reduce the amount of calculation and shorten the training time. Therefore, in order to make full use of the feature extraction ability of the optimized CNN, the comprehensiveness of multi-layer features and the fast-training speed of the ELM, an innovative classification model of HRSIs based on combining the optimized CNN and ELM, namely IPCEHRIC is developed to improve the robustness and classification effect of the model. The classification process of HRSIs is shown in Figure 4.

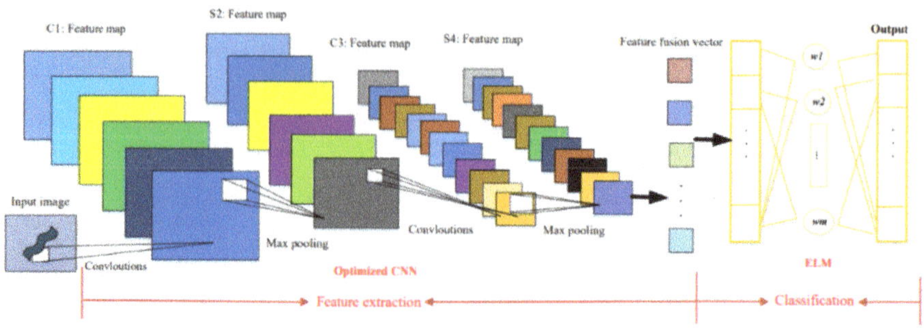

Figure 4. The innovative classification model of HRSIs.

The classification process of the IPCEHRIC is described as follows.

(1) Preprocess HRSIs

Some preprocessed methods, such as whitening processing, normalization processing, gray transformation, image smoothing, interpolation method, and so on are used to eliminate irrelevant information in hyperspectral remote sensing images, restore useful real information, enhance the detectability of relevant information, and simplify the data to the greatest extent, including image denoising, enhancement, smoothing, and sharpening, so as to improve the reliability of feature extraction, image matching, and so on.

(2) Optimize parameters of CNN

The CWLPSO with global optimization capability is employed to optimize and determine the parameters of the CNN, including the number of filters, activation function, learning rate, initial weight, and initial bias as particle dimension. The optimized parameter values are obtained, and an optimized CNN model is constructed.

(3) Extract features

The optimized CNN is essentially a multi-layer perceptron, which is mainly characterized by its local connection and weight sharing mode. When the input data are images, the alternated convolution layer and maximum pool layer by layer are used to automatically complete the feature extraction.

(4) Construct feature matrix

The extracted local features are input into the full connection layer of the first layer in order to form the global features. These images are taken from different feature ranges. Then these extracted features are selected to construct a feature matrix in order to provide feature matrix for the classifier.

(5) Establish ELM classifier

The feature matrix is taken as the input of the ELM, elmtrain( ) function and training sets are created to train the ELM. Then, the trained parameters and elmpredict( ) function are used to test the test set, and finally the classification results are obtained.

6. Experiment Verification and Result Analysis

6.1. Experimental Environment and Parameter Setting

The experimental environment is Intel i7-11700 HQ_CPU_@ _ 2.5GHz, 16G RAM with Windows 10, and the programming language is Matlab 2018b. The IPCEHRIC network structure consists of two convolution layers, two pooling layers and an ELM classifier. The nonlinear activation function of CNN is RELU function, and the ELM classifier uses Sigmoid function. The initial parameters of CWLPSO are $c_{1max} = 2.0$, $c_{1min} = 0.5$, $\omega_{max} = 0.9$, $\otimes$, maximum number of iterations K = 200. The initial parameters of the CNN are the number of convolution kernels (6), and the size of convolution kernels (1 * 3). The initial parameters of the ELM are σ = 0.1, regularity coefficient C = 0.5.

6.2. Pavia University Data

6.2.1. Data Description

Pavia University data set is a hyperspectral remote sensing image data set collected from the University of Pavia in northern Italy by using the airborne reflection optical spectrum imager of Germany. The size of image is 610 × 340, including 42,776 pixels and 9 types of features through excluding a large number of backgrounds. Basic information of Pavia University data is shown in Table 1. A total of 20% of the samples are randomly selected as the training set and 80% of the samples are used as the test set. The number of samples for training and test is shown in Table 2, and the describing of the HRSIs is shown in Figure 5.

Table 1. Basic information of Pavia University data.

Data	Pavia University
Collection location	Northern Italy
Acquisition equipment	ROSIS
Spectral coverage (μm)	0.43–0.86
Data size (pixel)	610 × 340
Spatial resolution (m)	1.3
Number of bands	115
Number of bands after denoising	103
Sample size	42,776
Number of categories	9

Table 2. The number of samples in Pavia University.

Types	Class	Training Samples	Test Samples	Samples
1	Asphalt	1326	5305	6631
2	Meadows	3722	14,927	18,649
3	Gravel	418	1681	2099
4	Trees	612	2452	3064
5	Painted metal sheets	268	1077	1345
6	Bare Soil	1004	4025	5029
7	Bitumen	266	1064	1330
8	Self-Blocking Bricks	736	2946	3682
9	Shadows	188	759	947
	Total	8540	34,236	42,776

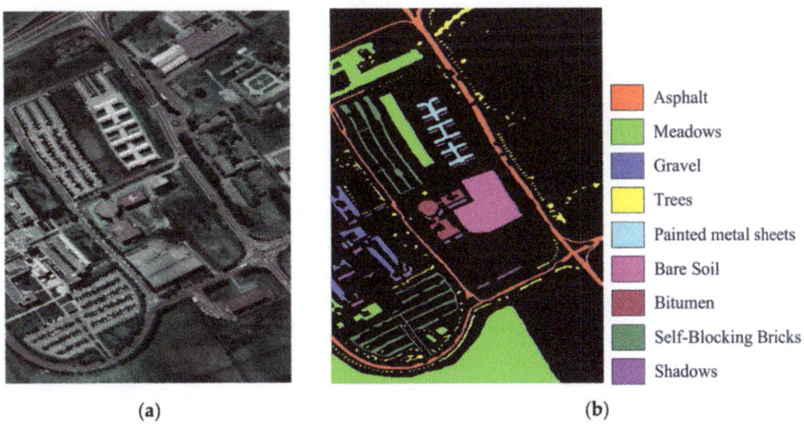

Figure 5. The HRSIs of Pavia University. (**a**) False color composite of HRSI. (**b**) Surface observations.

6.2.2. Experimental Results and Analysis

To verify the effectiveness of the IPCEHRIC, the CNN, local binary pattern (LBP) and CNN (LBP-CNN), CNN and ELM (CNN-ELM), LBP, CNN and ELM (LBP-CNN-ELM), LBP, PCA, CNN and ELM (LBP-PCA-CNN-ELM) are selected in here. The experiment results of the Pavia university data are shown in Table 3. The overall accuracy (OA), average accuracy (AA), and standard deviation (STD) of classification results are calculated for each algorithm.

Table 3. The experiment results of the Pavia University data (%).

Types	Class	CNN	LBP-CNN	CNN-ELM	LBP-CNN-ELM	LBP-PCA-CNN-ELM	IPCEHRIC
1	Asphalt	90.00	89.64	94.72	95.72	99.92	99.96
2	Meadows	89.99	89.79	93.00	95.00	99.12	99.67
3	Gravel	89.70	91.63	99.94	99.94	100.00	100.00
4	Trees	88.90	87.10	89.15	94.17	96.88	99.84
5	Painted metal sheets	86.00	89.91	92.26	96.68	99.72	100.00
6	Bare Soil	88.15	89.90	95.00	96.27	100.00	100.00
7	Bitumen	90.45	92.00	94.15	96.15	99.15	99.82
8	Self-Blocking Bricks	89.83	91.86	93.25	95.01	99.66	100.00
9	Shadows	87.50	93.87	90.90	97.74	97.94	99.15
	OA (%)	85.67	88.75	92.63	95.64	98.95	99.21
	AA (%)	88.95	90.63	93.60	96.30	99.15	99.83
	STD	1.467	1.939	3.022	1.722	1.075	0.279

It can be seen from Table 3 that the IPCEHRIC method obtains the classification accuracies of OA and AA are 99.21 and 99.83%, which are best classification results among the CNN, LBP-CNN, CNN-ELM, LBP-CNN-ELM, LBP-PCA-CNN-ELM, and IPCEHRIC methods. The STD of the IPCEHRIC is 0.279, which is also the least STD among these methods. Among other comparison methods, the LBP-PCA-CNN-ELM method obtains the classification accuracies of OA and AA as 98.95 and 99.15%. While the CNN-ELM method obtains the classification accuracies for OA and AA of 92.63 and 93.60%. Compared with the CNN-ELM, the classification accuracies of OA and AA of the IPCEHRIC are improved by 6.58 and 6.23% than those of the CNN-ELM. This shows that the feature extraction ability of the optimized CNN is better than that of the CNN, which explains the global optimization ability of the CWLPSO algorithm. Therefore, the classification performance of the IPCEHRIC method is significantly better than those of the CNN, LBP-CNN, CNN-ELM, LBP-CNN-ELM, and LBP-PCA-CNN-ELM. The experiment results show that the IPCEHRIC method has higher classification accuracy than other comparison methods. The IPCEHRIC is an effective classification method for HRSIs.

6.3. Actual HRSI after Jiuzhaigou M7.0 Earthquake

6.3.1. Description of HRSI after Jiuzhaigou 7.0 Earthquake

Jiuzhaigou is located in Zhangzha Town, Jiuzhaigou County, Sichuan Province. It is located in the transition zone. It is more than 400 km away from Chengdu. It is a mountain valley with a depth of more than 50 km, with a total area of 64,297 hm^2 and a forest coverage rate of more than 80%. The hyperspectral remote sensing image after Jiuzhaigou M7.0 earthquake on 8 August 2017 is shown in Figure 6.

The HRSI after Jiuzhaigou M7.0 earthquake is saved as *. mat file, which determined the coordinates of different areas by manual frame drawing. Then a matrix consistent with the size of the picture is constructed. The corresponding positions of the matrix with different numbers is marked according to the coordinates of different areas, so as to mark different labels on different areas of the picture, save and generate *.mat file with labels. A data set containing four types of samples is made, which include villages, water, grassland, and trees in the HRSIs after Jiuzhaigou M7.0 earthquake. The number of samples and four types are shown in Table 4.

Figure 6. The HRSI after Jiuzhaigou M7.0 earthquake.

Table 4. The number of samples and four types.

Types	Class	Samples
1	Villages	12,575
2	Water	14,953
3	Grassland	38,790
4	Trees	39,159
Total		105,477

According to the gray value of pixels, the color function is used to set the threshold. The different areas of HRSIs after Jiuzhaigou M7.0 earthquake are marked by different colors. A matrix consistent with the image size is constructed, and the different areas are marked with color. A data set with six types of samples is made, which include the villages, bareland, grassland, trees, water, and rocks in the HRSIs after Jiuzhaigou M7.0 earthquake. The number of samples and six types are shown in Table 5.

Table 5. The number of samples and six types.

Types	Class	Samples
1	Villages	1608
2	Bareland	25
3	Grassland	376,651
4	Trees	110,409
5	Water	5558
6	Rocks	2469
Total		495,087

6.3.2. Experimental Results and Analysis

To prove the ability of the IPCEHRIC to solve practical engineering problems, the hyperspectral remote sensing images after Jiuzhaigou M7.0 earthquake is used for the experimental comparison and analysis. Similarly, the CNN, LBP-CNN, CNN-ELM, LBP-CNN-ELM, and LBP-PCA-CNN-ELM are selected to compare in here. Each algorithm is executed ten times independently. The classification results of HRSI after Jiuzhaigou 7.0

earthquake for four types are shown in Tables 6 and 7. The classification results of HRSI after Jiuzhaigou M7.0 earthquake for six types are shown in Tables 8 and 9.

Table 6. The classification results of HRSIs for 10 times for four types (%).

Times	CNN	LBP-CNN	CNN-ELM	LBP-CNN-ELM	LBP-PCA-CNN-ELM	IPCEHRIC
1	41.47	36.68	69.80	64.38	65.67	89.76
2	41.80	36.68	75.84	64.25	65.16	88.96
3	41.75	36.68	75.98	64.16	65.33	89.99
4	41.73	36.68	75.38	64.40	65.14	89.26
5	41.85	37.02	61.45	64.47	65.47	89.76
6	41.70	37.02	75.80	64.12	65.56	90.58
7	41.86	37.02	74.04	63.83	65.40	91.64
8	41.77	37.02	60.02	64.44	65.19	92.12
9	41.78	36.68	74.80	64.38	65.49	90.99
10	41.76	37.02	75.46	64.10	65.81	89.94
AA (%)	41.75	36.85	71.86	64.25	65.42	90.30
STD	0.109	0.179	6.145	0.201	0.223	1.019

Table 7. The classification results of HRSIs for four types (%).

Types	Class	CNN	LBP-CNN	CNN-ELM	LBP-CNN-ELM	LBP-PCA-CNN-ELM	IPCEHRIC
1	Villages	50.47	46.76	79.16	74.64	75.70	92.46
2	Water	41.80	35.43	78.37	70.47	73.28	90.73
3	Grassland	39.26	33.58	73.78	63.19	72.45	89.15
4	Trees	40.73	36.29	76.12	69.24	75.42	91.48
OA (%)		41.75	36.85	71.86	64.25	65.42	90.30
AA (%)		43.07	38.02	76.86	69.39	74.21	90.96
STD		5.046	5.939	2.422	4.733	1.597	1.396

Table 8. The classification results of HRSIs for 10 times for six types (%).

Times	CNN	LBP-CNN	CNN-ELM	LBP-CNN-ELM	LBP-PCA-CNN-ELM	IPCEHRIC
1	79.77	79.83	99.21	85.12	85.12	99.99
2	79.78	79.85	99.78	84.80	84.14	100.0
3	79.84	79.84	99.99	84.14	84.01	99.98
4	79.78	79.86	99.26	84.80	84.22	99.78
5	79.86	79.84	99.99	85.12	85.46	100.0
6	79.87	79.81	99.21	84.76	86.13	99.77
7	79.88	79.59	99.98	85.46	84.57	100.0
8	79.87	79.59	99.27	86.08	85.12	99.98
9	79.86	79.80	99.98	85.46	86.02	100.0
10	79.86	79.84	99.77	86.43	84.80	99.99
AA (%)	79.84	79.79	99.64	85.22	84.96	99.95
STD	0.043	0.104	0.360	0.672	0.753	0.092

Table 9. The classification results of HRSIs for six types (%).

Types	Class	CNN	LBP-CNN	CNN-ELM	LBP-CNN-ELM	LBP-PCA-CNN-ELM	IPCEHRIC
1	Villages	82.34	85.46	99.46	87.45	90.35	99.98
2	Bareland	86.05	86.04	99.64	89.62	93.46	100.0
3	Grassland	79.98	85.32	99.06	87.17	90.67	100.0
4	Trees	78.46	84.14	99.31	86.43	89.86	99.81
5	Water	83.49	87.25	99.78	87.69	90.34	100.0
6	Rocks	82.16	85.68	99.34	88.03	92.05	99.85
OA (%)		79.84	79.79	99.64	85.22	84.96	99.95
AA (%)		82.08	85.65	99.43	87.73	91.12	99.94
STD		2.658	1.013	0.256	1.072	1.367	0.086

As can be seen from Tables 6–9 that the IPCEHRIC obtains the classification accuracies of AA are 90.30% for four types and 99.95% for six types, respectively, which are best classification results among the CNN, LBP-CNN, CNN-ELM, LBP-CNN-ELM, LBP-PCA-CNN-ELM, and IPCEHRIC methods. The STD of the IPCEHRIC is 1.396 for four types and 0.086 for six types, which are also the least STD among these methods. Among other comparison methods, for four types of the samples, the overall classification effect of these methods is not ideal. Especially, the classification accuracies of the CNN and LBP-CNN are very unsatisfactory. For six types of the samples, the overall classification effect of these methods is better. Especially, the classification accuracies of the CNN-ELM are ideal among CNN, LBP-CNN, CNN-ELM, LBP-CNN-ELM, and LBP-PCA-CNN-ELM. Compared with the CNN-ELM, the classification accuracy of AA of the IPCEHRIC method are improved by 18.44 and 0.31%, which indicate that the optimized CNN has better feature extraction ability and classification performance, and the CWLPSO has better global optimization ability. Therefore, the experiment results show that the classification accuracy of the IPCEHRIC is better than that of other comparison methods. The CWLPSO can optimize and determine the parameters of the CNN in order to construct an optimized CNN model, which can effectively extract the deep features of HRSIs after Jiuzhaigou 7.0 earthquake, so as to obtain a better classification result. It can effectively classify the HRSIs after Jiuzhaigou 7.0 earthquake to obtain the villages, bareland, grassland, trees, water, and rocks in HRSIs after Jiuzhaigou 7.0 earthquake.

The HRSIs after Jiuzhaigou 7.0 earthquake are divided into four types and six types. The classification effects of HRSIs are shown in Figure 7.

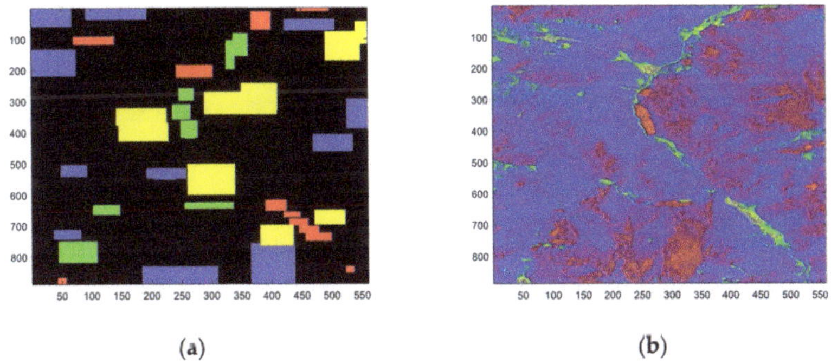

Figure 7. The classification effects of HRSIs after Jiuzhaigou M7.0 earthquake. (**a**) Four types. (**b**) Six types.

As can be seen from Figure 7, the classification effects of six types by using the IPCEHRIC for the HRSIs after Jiuzhaigou M7.0 earthquake is ideal. For actual HRSIs, the IPCEHRIC method has higher classification accuracy, and it is an effective classification method for actual HRSIs.

7. Conclusions

In this paper, an innovative hyperspectral remote sensing image classification method based on combining CWLPSO, CNN, and ELM, namely IPCEHRIC is proposed to obtain the accurate classification results. The CWLPSO with fusing multi-strategy is proposed to optimize the parameters of the CNN. Then the deep features are extracted from HRSIs, which are input into the ELM to realize the accurate classification of HRSIs. Pavia University data and actual HRSIs after Jiuzhaigou 7.0 earthquake are selected to verify the effectiveness of the IPCEHRIC. The experiment results show that the IPCEHRIC obtains the classification accuracies of 99.21% for Pavia University data, 90.30 and 99.95% for actual HRSIs after Jiuzhaigou 7.0 earthquake. The classification results of the IPCEHRIC are better than those of the CNN, LBP-CNN, CNN-ELM, LBP-CNN-ELM, and LBP-PCA-CNN-ELM methods. Compared with the CNN-ELM, the classification accuracies of the IPCEHRIC are improved by 6.58, 21.44, and 0.31%, respectively. This shows that the CWLPSO algorithm can effectively optimize the parameters and obtain reasonable parameter values for CNN to improve the feature extraction ability. Therefore, the IPCEHRIC has certain advantages on classification effect of the HRSIs. Especially, the IPCEHRIC can obtain accurate classification accuracy for actual HRSIs after Jiuzhaigou M7.0 earthquake. It can effectively classify the villages, bareland, grassland, trees, water, and rocks in the HRSIs after Jiuzhaigou M7.0 earthquake and achieve good classification result.

Author Contributions: Conceptualization, A.Y. and X.Z.; Methodology, A.Y.; Software, X.Z.; Validation, F.M. and X.Z.; Resources, F.M.; Writing—original draft preparation, A.Y.; Writing—review and editing, X.Z.; Visualization, X.Z.; Project administration, F.M.; Funding acquisition, X.Z. All authors have read and agreed to the published version of the manuscript.

Funding: This research was funded by the Sichuan Science and Technology Program, grant number 2019ZYZF0169, 2019YFG0307, 2021YFS0407; the A Ba Achievements Transformation Program, grant number R21CGZH0001; the Chengdu Science and technology planning project, grant number 2021-YF05-00933-SN.

Institutional Review Board Statement: Not applicable.

Informed Consent Statement: Not applicable.

Data Availability Statement: Not applicable.

Acknowledgments: The authors would like to acknowledge the UCI Machine Learning Repository.

Conflicts of Interest: The authors declare no conflict of interest.

References

1. Dumke, I.; Ludvigsen, M.; Ellefmo, S.L. Underwater hyperspectral imaging using a stationary platform in the Trans-Atlantic Geotraverse hydrothermal field. *IEEE Trans. Geosci. Remote Sens.* **2019**, *57*, 2947–2962. [CrossRef]
2. Chen, H.; Miao, F.; Chen, Y.; Xiong, Y.; Chen, T. A hyperspectral image classification method using multifeature vectors and optimized KELM. *IEEE J. Sel. Top. Appl. Earth Observ. Remote Sens.* **2021**, *14*, 2781–2795. [CrossRef]
3. Ma, K.Y.; Chang, C.I. Iterative training sampling coupled with active learning for semisupervised spectral–spatial hyperspectral image classification. *IEEE Trans. Geosci. Remote Sens.* **2021**, *59*, 8672–8692. [CrossRef]
4. Chen, Y.; Xiao, Z.; Chen, G. Detection of oasis soil composition and analysis of environmental parameters based on hyperspectral image and GIS. *Arab. J. Geosci.* **2021**, *14*, 1050. [CrossRef]
5. Shimoni, M.; Haelterman, R.; Perneel, C. Hypersectral imaging for military and security applications: Combining myriad processing and sensing techniques. *IEEE Geosci. Remote Sens. Mag.* **2019**, *7*, 101–117. [CrossRef]
6. Luo, X.; Shen, Z.; Xue, R. Unsupervised band selection method based on importance-assisted column subset selection. *IEEE Access* **2018**, *7*, 517–527. [CrossRef]
7. Chang, C.I.; Kuo, Y.M.; Chen, S. Self-mutual information-based band selection for hyperspectral image classification. *IEEE Trans. Geosci. Remote Sens.* **2021**, *59*, 5979–5997. [CrossRef]
8. Lin, Z.; Yan, L. A support vector machine classifier based on a new kernel function model for hyperspectral data. *Mapp. Sci. Remote Sens.* **2015**, *53*, 85–101. [CrossRef]
9. Kang, X.; Xiang, X.; Li, S. PCA-based edge-preserving features for hyperspectral image classification. *IEEE Trans. Geosci. Remote Sens.* **2017**, *55*, 7140–7151. [CrossRef]
10. Yuan, H.; Tang, Y.Y.; Lu, Y. Spectral-spatial classification of hyperspectral image based on discriminant analysis. *IEEE J. Sel. Top. Appl. Earth Observ. Remote Sens.* **2014**, *7*, 2035–2043. [CrossRef]
11. Tran, T.V.; Julian, J.P.; Beurs, K.M. Land cover heterogeneity effects on sub-pixel and per-pixel classifications. *ISPRS Int. J. Geo-Inf.* **2014**, *3*, 540–553. [CrossRef]
12. Khodadadzadeh, M.; Li, J.; Plaza, A.; Ghassemian, H.; Bioucas-Dias, J.M.; Li, X. Spectral-spatial classification of hyperspectral data using local and global probabilities for mixed pixel characterization. *IEEE Trans. Geosci. Remote Sens.* **2014**, *52*, 6298–6314. [CrossRef]
13. Li, S.T.; Lu, T.; Fang, L.Y.; Jia, X.P.; Benediktsson, J.A. Probabilistic fusion of pixel-level and superpixel-level hyperspectral image classification. *IEEE Trans. Geosci. Remote Sens.* **2016**, *54*, 7416–7430. [CrossRef]
14. Li, W.; Wu, G.D.; Zhang, F.; Du, Q.; Hyperspectral, A. Image classification using deep pixel-pair features. *IEEE Trans. Geosci. Remote Sens.* **2017**, *55*, 844–853. [CrossRef]
15. Mei, J.; Wang, Y.B.; Zhang, L.Q.; Zhang, B.; Liu, S.H.; Zhu, P.P.; Ren, Y.C. PSASL: Pixel-level and superpixel-level aware subspace learning for hyperspectral image classification. *IEEE Trans. Geosci. Remote Sens.* **2019**, *57*, 4278–4293. [CrossRef]
16. Pan, Z.B.; Wu, X.Q.; Li, Z.Y. Central pixel selection strategy based on local gray-value distribution by using gradient information to enhance LBP for texture classification. *Expert Syst. Appl.* **2019**, *120*, 319–334. [CrossRef]
17. Bey, A.; Jetimane, J.; Lisboa, S.N.; Ribeiro, N.; Sitoe, A.; Meyfroidt, P. Mapping smallholder and large-scale cropland dynamics with a flexible classification system and pixel-based composites in an emerging frontier of Mozambique. *Remote Sens. Environ.* **2020**, *239*, 111611. [CrossRef]
18. Yan, L.; Fan, B.; Liu, H.M.; Huo, C.L.; Xiang, S.M.; Pan, C.H. Triplet adversarial domain adaptation for pixel-level classification of VHR remote sensing images. *IEEE Trans. Geosci. Remote Sens.* **2020**, *58*, 3558–3573. [CrossRef]
19. Li, Y.; Lu, T.; Li, S.T. Subpixel-pixel-superpixel-based multiview active learning for hyperspectral images classification. *IEEE Trans. Geosci. Remote Sens.* **2020**, *58*, 4976–4988. [CrossRef]
20. Ma, K.Y.; Chang, C.I. Kernel-based constrained energy minimization for hyperspectral mixed pixel classification. *IEEE Trans. Geosci. Remote Sens.* **2021**, *60*, 5510723. [CrossRef]
21. Chen, Y.; Lin, Z.; Zhao, X. Deep learning-based classification of hyperspectral data. *IEEE J. Sel. Top. Appl. Earth Observ. Remote Sens.* **2014**, *7*, 2094–2107. [CrossRef]

22. Liu, L.; Wang, Y.; Peng, J. Latent relationship guided stacked sparse autoencoder for hyperspectral imagery classification. *IEEE Trans. Geosci. Remote Sens.* **2020**, *58*, 3711–3725. [CrossRef]
23. Romero, A.; Gatta, C.; Camps-Valls, G. Unsupervised deep feature extraction for remote sensing image classification. *IEEE Trans. Geosci. Remote Sens.* **2016**, *54*, 1349–1362. [CrossRef]
24. Sharma, A.; Liu, X.W.; Yang, X.J.; Shi, D. A patch-based convolutional neural network for remote sensing image classification. *Neural Netw.* **2017**, *95*, 19–28. [CrossRef] [PubMed]
25. Maggiori, E.; Tarabalka, Y.; Charpiat, G.; Alliez, P. Convolutional neural networks for large-scale remote-sensing image classification. *IEEE Trans. Geosci. Remote Sens.* **2017**, *55*, 645–657. [CrossRef]
26. Wang, L.Z.; Zhang, J.B.; Liu, P.; Choo, K.K.; Huang, F. Spectral-spatial multi-feature-based deep learning for hyperspectral remote sensing image classification. *Appl. Soft Comput.* **2017**, *21*, 213–221. [CrossRef]
27. Ji, S.P.; Zhang, C.; Xu, A.J.; Shi, Y.; Duan, Y.L. 3D convolutional neural networks for crop classification with multi-temporal remote sensing images. *Remote Sens.* **2018**, *10*, 75. [CrossRef]
28. Ben, H.A.; Benoit, A.; Lambert, P.; Ben, A.C. 3-D deep learning approach for remote sensing image classification. *IEEE Trans. Geosci. Remote Sens.* **2018**, *56*, 4420–4434.
29. Xu, S.H.; Mu, X.D.; Chai, D.; Zhang, X.M. Remote sensing image scene classification based on generative adversarial networks. *Remote Sens. Lett.* **2018**, *9*, 617–626. [CrossRef]
30. Tao, Y.T.; Xu, M.Z.; Lu, Z.Y.; Zhong, Y.F. DenseNet-based depth-width double reinforced deep learning neural network for high-resolution remote sensing image per-pixel classification. *Remote Sens.* **2018**, *10*, 779. [CrossRef]
31. Liang, P.; Shi, W.Z.; Zhang, X.K. Remote sensing image classification based on stacked denoising autoencoder. *Remote Sens.* **2018**, *10*, 16. [CrossRef]
32. Li, P.; Ren, P.; Zhang, X.Y.; Wang, Q.; Zhu, X.B.; Wang, L. Region-wise deep feature representation for remote sensing images. *Remote Sens.* **2018**, *10*, 871. [CrossRef]
33. Li, G.; Li, L.L.; Zhu, H.; Liu, X.; Jiao, L.C. Adaptive multiscale deep fusion residual network for remote sensing image classification. *IEEE Trans. Geosci. Remote Sens.* **2019**, *57*, 8506–8521. [CrossRef]
34. Yuan, Y.; Fang, J.; Lu, X.Q.; Feng, Y.C. Remote sensing image scene classification using rearranged local features. *IEEE Trans. Geosci. Remote Sens.* **2019**, *57*, 1779–1792. [CrossRef]
35. Zhang, C.J.; Li, G.D.; Du, S.H. Multi-scale dense networks for hyperspectral remote sensing image classification. *IEEE Trans. Geosci. Remote Sens.* **2019**, *57*, 9201–9222. [CrossRef]
36. Zhang, C.J.; Li, G.D.; Lei, R.M.; Du, S.H.; Zhang, X.Y.; Zheng, H.; Wu, Z.F. Deep feature aggregation network for hyperspectral remote sensing image classification. *IEEE J. Sel. Top. Appl. Earth Observ. Remote Sens.* **2020**, *13*, 5314–5325. [CrossRef]
37. Chen, C.; Ma, Y.; Ren, G.B. Hyperspectral classification using deep belief networks based on conjugate gradient update and pixel-centric spectral block features. *IEEE J. Sel. Top. Appl. Earth Observ. Remote Sens.* **2020**, *13*, 4060–4069. [CrossRef]
38. Xiong, W.; Xiong, Z.Y.; Cui, Y.Q.; Lv, Y.F. Deep multi-feature fusion network for remote sensing images. *Remote Sens. Lett.* **2020**, *11*, 563–571. [CrossRef]
39. Tong, W.; Chen, W.T.; Han, W.; Li, X.J.; Wang, L.Z. Channel-attention-based densenet network for remote sensing image scene classification. *IEEE J. Sel. Top. Appl. Earth Observ. Remote Sens.* **2020**, *13*, 4121–4132. [CrossRef]
40. Zhu, H.; Ma, W.P.; Li, L.L.; Jiao, L.C.; Yang, S.Y.; Hou, B. A dual-branch attention fusion deep network for multiresolution remote-sensing image classification. *Inf. Fusion* **2020**, *58*, 116–131. [CrossRef]
41. Raza, A.; Huo, H.; Sirajuddin, S.; Fang, T. Diverse capsules network combining multiconvolutional layers for remote sensing image scene classification. *IEEE J. Sel. Top. Appl. Earth Observ. Remote Sens.* **2020**, *13*, 5297–5313. [CrossRef]
42. Li, J.T.; Shen, Y.L.; Yang, C. An adversarial generative network for crop classification from remote sensing timeseries images. *Remote Sens.* **2021**, *13*, 65. [CrossRef]
43. Gu, S.W.; Zhang, R.; Luo, H.X.; Li, M.Y.; Feng, H.M.; Tang, X.G. Improved SinGAN integrated with an attentional mechanism for remote sensing image classification. *Remote Sens.* **2021**, *13*, 1713. [CrossRef]
44. Guo, D.E.; Xia, Y.; Luo, X.B. Self-supervised GANs with similarity loss for remote sensing image scene classification. *IEEE J. Sel. Top. Appl. Earth Observ. Remote Sens.* **2021**, *14*, 2508–2521. [CrossRef]
45. Li, Y.S.; Zhu, Z.H.; Yu, J.G.; Zhang, Y.J. Learning deep cross-modal embedding networks for zero-shot remote sensing image scene classification. *IEEE Trans. Geosci. Remote Sens.* **2021**, *59*, 10590–10603. [CrossRef]
46. Lei, R.M.; Zhang, C.J.; Liu, W.C.; Zhang, L.; Zhang, X.Y.; Yang, Y.C.; Huang, J.W.; Li, Z.X.; Zhou, Z.Y. Hyperspectral remote sensing image classification using deep convolutional capsule network. *IEEE J. Sel. Top. Appl. Earth Observ. Remote Sens.* **2021**, *14*, 8297–8315. [CrossRef]
47. Cui, X.P.; Zou, C.; Wang, Z.S. Remote sensing image recognition based on dual-channel deep learning network. *Multimed. Tools Appl.* **2021**, *80*, 27683–27699. [CrossRef]
48. Peng, C.; Li, Y.Y.; Jiao, L.C.; Shang, R.H. Efficient convolutional neural architecture search for remote sensing image scene classification. *IEEE Trans. Geosci. Remote Sens.* **2021**, *59*, 6092–6105. [CrossRef]
49. Guo, D.G.; Xia, Y.; Luo, X.B. GAN-based semisupervised scene classification of remote sensing image. *IEEE Geosci. Remote Sens. Lett.* **2021**, *18*, 2067–2071. [CrossRef]

50. Dong, S.X.; Quan, Y.H.; Feng, W.; Dauphin, G.; Gao, L.R.; Xing, M.D. A pixel cluster CNN and spectral-spatial fusion algorithm for hyperspectral image classification with small-size training samples. *IEEE J. Sel. Top. Appl. Earth Observ. Remote Sens.* **2021**, *14*, 4101–4114. [CrossRef]
51. Li, Y.S.; Zhang, Y.J.; Zhu, Z.H. Error-tolerant deep learning for remote sensing image scene classification. *IEEE Trans. Cybern.* **2021**, *51*, 1756–1768. [CrossRef] [PubMed]
52. Li, B.Y.; Guo, Y.L.; Yang, J.G.; Wang, L.G.; Wang, Y.Q.; An, W. Gated recurrent multiattention network for VHR remote sensing image classification. *IEEE Trans. Geosci. Remote Sens.* **2021**, *60*, 5606113. [CrossRef]
53. Dong, R.M.; Zhang, L.X.; Fu, H.H. RRSGAN: Reference-based super-resolution for remote sensing image. *IEEE Trans. Geosci. Remote Sens.* **2022**, *60*, 5601117. [CrossRef]
54. Wu, E.Q.; Zhou, M.; Hu, D.; Zhu, L.; Tang, Z.; Qiu, X.Y.; Deng, P.Y.; Zhu, L.M.; Ren, H. Self-paced dynamic infinite mixture model for fatigue evaluation of pilots' brains. *IEEE Trans. Cybern.* **2021**. [CrossRef] [PubMed]
55. Karadal, C.H.; Kaya, M.C.; Tuncer, T.; Dogan, S.; Acharya, U.R. Automated classification of remote sensing images using multileveled MobileNetV2 and DWT techniques. *Expert Syst. Appl.* **2021**, *185*, 115659. [CrossRef]
56. Ma, W.P.; Shen, J.C.; Zhu, H.; Zhang, J.; Zhao, J.L.; Hou, B.; Jiao, L.C. A novel adaptive hybrid fusion network for multiresolution remote sensing images classification. *IEEE Trans. Geosci. Remote Sens.* **2022**, *60*, 5400617. [CrossRef]
57. Cai, W.W.; Wei, Z.G. Remote sensing image classification based on a cross-attention mechanism and graph convolution. *IEEE Geosci. Remote Sens. Lett.* **2022**, *19*, 80002005. [CrossRef]
58. Zhang, Z.; Liu, S.H.; Zhang, Y.; Chen, W.B. RS-DARTS: A convolutional neural architecture search for remote sensing image scene classification. *Remote Sens.* **2022**, *14*, 141. [CrossRef]
59. Hilal, A.M.; Al-Wesabi, F.N.; Alzahrani, K.J.; Al Duhayyim, M.; Hamza, M.A.; Rizwanullah, M.; Diaz, V.G. Deep transfer learning based fusion model for environmental remote sensing image classification model. *J. Remote Sens.* **2022**. [CrossRef]
60. Li, R.; Zheng, S.Y.; Duan, C.X.; Wang, L.B.; Zhang, C. Land cover classification from remote sensing images based on multi-scale fully convolutional network. *GEO Spat. Inf. Sci.* **2022**. [CrossRef]
61. Wei, Y.Y.; Zhou, Y.Q.; Luo, Q.F.; Deng, W. Optimal reactive power dispatch using an improved slime mould algorithm. *Energy Rep.* **2021**, *7*, 8742–8759. [CrossRef]
62. Deng, W.; Zhang, X.; Zhou, Y.; Liu, Y.; Zhou, X.; Chen, H.; Zhao, H. An enhanced fast non-dominated solution sorting genetic algorithm for multi-objective problems. *Inf. Sci.* **2022**, *585*, 441–453. [CrossRef]
63. Li, T.; Qian, Z.; Deng, W.; Zhang, D.; Lu, H.; Wang, S. Forecasting crude oil prices based on variational mode decomposition and random sparse Bayesian learning. *Appl. Soft Comput.* **2021**, *113*, 108032. [CrossRef]
64. Deng, W.; Xu, J.; Zhao, H.; Song, Y. A novel gate resource allocation method using improved PSO-based QEA. *IEEE Trans. Intell. Transp. Syst.* **2020**. [CrossRef]
65. Cui, H.; Guan, Y.; Chen, H. Rolling element fault diagnosis based on VMD and sensitivity MCKD. *IEEE Access* **2021**, *9*, 120297–120308. [CrossRef]
66. Wang, X.; Wang, H.; Du, C.; Fan, X.; Cui, L.; Chen, H.; Deng, F.; Tong, Q.; He, M.; Yang, M.; et al. Custom-molded offloading footwear effectively prevents recurrence and amputation, and lowers mortality rates in high-risk diabetic foot patients: A multicenter, prospective observational study. *Diabetes Metab. Syndr. Obes.* **2022**, *15*, 103–109.
67. Deng, W.; Shang, S.; Cai, X.; Zhao, H.; Zhou, Y.; Chen, H.; Deng, W. Quantum differential evolution with cooperative coevolution framework and hybrid mutation strategy for large scale optimization. *Knowl. Based Syst.* **2021**, *224*, 107080. [CrossRef]
68. Deng, W.; Xu, J.; Gao, X.; Zhao, H. An enhanced MSIQDE algorithm with novel multiple strategies for global optimization problems. *IEEE Trans. Syst. Man Cybern. Syst.* **2020**, *52*, 1578–1587. [CrossRef]
69. Zhang, Z.H.; Min, F.; Chen, G.S.; Shen, S.P.; Wen, Z.C.; Zhou, X.B. Tri-partition state alphabet-based sequential pattern for multivariate time series. *Cogn. Comput.* **2021**. [CrossRef]
70. Ran, X.; Zhou, X.; Lei, M.; Tepsan, W.; Deng, W. A novel k-means clustering algorithm with a noise algorithm for capturing urban hotspots. *Appl. Sci.* **2021**, *11*, 11202. [CrossRef]
71. Chen, H.; Zhang, Q.; Luo, J. An enhanced Bacterial Foraging Optimization and its application for training kernel extreme learning machine. *Appl. Soft Comput.* **2020**, *86*, 105884. [CrossRef]
72. Cui, H.; Guan, Y.; Chen, H.; Deng, W. A novel advancing signal processing method based on coupled multi-stable stochastic resonance for fault detection. *Appl. Sci.* **2021**, *11*, 5385. [CrossRef]
73. Kennedy, J.; Eberhart, R. Particle swarm optimization. *IEEE Int. Conf. Neural Netw. Perth* **1995**, *4*, 1942–1948.

electronics

Article

A Novel Color Image Encryption Algorithm Using Coupled Map Lattice with Polymorphic Mapping

Penghe Huang [1], Dongyan Li [1], Yu Wang [2], Huimin Zhao [3,4,*] and Wu Deng [3,*]

1 Software Technology Institute, Dalian Jiaotong University, Dalian 116028, China
2 School of Cyber Science and Engineering, Southeast University, Nanjing 211189, China
3 School of Electronic Information and Automation, Civil Aviation University of China, Tianjin 300300, China
4 Traction Power State Key Laboratory, Southwest Jiaotong University, Chengdu 610032, China
* Correspondence: hmzhao@cauc.edu.cn (H.Z.); wdeng@cauc.edu.cn (W.D.)

Abstract: Some typical security algorithms such as SHA, MD4, MD5, etc. have been cracked in recent years. However, these algorithms have some shortcomings. Therefore, the traditional one-dimensional-mapping coupled lattice is improved by using the idea of polymorphism in this paper, and a polymorphic mapping–coupled map lattice with information entropy is developed for encrypting color images. Firstly, we extend a diffusion matrix with the original 4×4 matrix into an $n \times n$ matrix. Then, the Huffman idea is employed to propose a new pixel-level substitution method, which is applied to replace the grey degree value. We employ the idea of polymorphism and select f(x) in the spatiotemporal chaotic system. The pseudo-random sequence is more diversified and the sequence is homogenized. Finally, three plaintext color images of $256 \times 256 \times 3$, "Lena", "Peppers" and "Mandrill", are selected in order to prove the effectiveness of the proposed algorithm. The experimental results show that the proposed algorithm has a large key space, better sensitivity to keys and plaintext images, and a better encryption effect.

Keywords: coupled map lattice; polymorphic mapping; color image; hash function; pixel level

Citation: Huang, P.; Li, D.; Wang, Y.; Zhao, H.; Deng, W. A Novel Color Image Encryption Algorithm Using Coupled Map Lattice with Polymorphic Mapping. *Electronics* **2022**, *11*, 3436. https://doi.org/10.3390/electronics11213436

Academic Editor: Stefanos Kollias

Received: 13 September 2022
Accepted: 20 October 2022
Published: 24 October 2022

1. Introduction

In recent years, with the popularity of computers, multimedia messages have been transported through the network, causing more attention to be paid to information security. The hash algorithm is a traditional method used to encrypt passwords. When a password is created in clear text, it is run through a hash algorithm to produce the password text stored in the file system. The U.S. standard of a hash function is SHA-1 (Secure Hash Algorithm 1) with 160 bits of output length [1]. It is difficult to be sure of the security of a hash function with 160 bits of output length, and it was cracked in 2017. Additionally, other hash algorithms such as MD5 (Message-Digest Algorithm 5), MD4, and RIPEMD (RACE Integrity Primitives Evaluation Message Digest) have also been cracked [2]. Recently, encryption algorithms with higher security have become a research hotspot. Chaos encryption is a relatively new encryption idea developed in recent years, and spatiotemporal chaos is the best among them. Chaos in nonlinear science refers to a deterministic but unpredictable motion state [3–6]. Chaos has the characteristics of sensitivity to initial conditions, pseudo-randomness, and ergodicity, which makes chaos closely related to cryptography. In recent years, the security, complexity, and speed of image encryption algorithms based on chaos theory have become a research hotspot [7–15]. In addition, some algorithms are also proposed for image processing, image encryption, model optimization, function solutions, fault diagnosis, data security, etc. [16–28].

The spatiotemporal chaos model derives from the classical natural fluid-mechanics model, and the spatiotemporal chaos model has many advantages. For example, the effect of the pseudo-random sequence generated by the coupled lattice is better than the low-dimensional chaotic model, and the coupled lattice's iterative efficiency is better than that of the low-dimensional chaotic model. However, it is found in this study that the local chaotic mapping in the previous method of coupled lattice mapping only chooses a kind of chaotic mapping, and in order to avoid the periodic window and other problems, the local chaotic mapping parameter range is also smaller. In 2004, a new encryption theory, the idea of the Polymorphic Cipher (PMC), was proposed by Roelgen, and the sequence cipher was able to generate a new dynamic [29]. Because polymorphic cryptography belongs to the self-compiled class of encryption algorithms, when an attacker attacks the system [30–32], the parameters produced by the attacker can be started from the compiler. Because most self-compiled systems are composed of unidirectional functions and are unreadable, they can be reassembled according to attack parameters and unidirectional functions. They can resist differential attacks and brute force attacks. Therefore, based on the idea of polymorphism proposed by Roelgen, this paper increases the local chaotic map to 4, and achieves the goal of polymorphism. Experiments show that the polymorphic coupled lattice map generates better pseudo-random sequences. The keystream generator is the key to the sequence cipher.

On the other hand, there are two typical links in the chaotic cryptosystem, namely scrambling and diffusion. The combination of scrambling and diffusion improves the security of cryptosystems [30,32], but there are still some drawbacks, and some cryptosystems that conform to this rule have been cracked. The main reason is that the chaotic dynamic performance is not fully considered when designing the algorithm. Coupled mapping lattice (CML)-based spatiotemporal chaotic systems are applied to chaotic cryptography to overcome these shortcomings. Coupled lattices have better chaotic dynamics, including more parameters, larger key spaces, and longer periods. Some encryption algorithms based on coupled lattices are not related to plaintext images [33–35]. The output ciphertext image relies only on the key, which has been shown to be insecure and not resistant to chosen plaintext/ciphertext attacks [36]. In this paper, we use the idea of polymorphism to improve the traditional one-dimensional-mapping coupled lattice, and construct a selective chaotic map. It can make one-dimensional coupled map lattices produce various pseudo-random sequences based on different chaotic maps. Additionally, the key space is larger than the traditional one-dimensional coupled map lattices. Moreover, the uneven distribution of chaotic sequences in one-dimensional coupled lattices is rearranged to produce homogeneous sequences, and the encryption effect is better.

The experimental results and security analysis showed that the algorithm based on the CML with polymorphic mapping can achieve the goal of polymorphism, improve the traditional one-dimensional-mapping coupled lattice, and construct a selective chaotic map.

The structure of this paper is as follows. In Section 2, we briefly discuss some basic knowledge of polymorphic spatiotemporal chaotic systems and random ergodicity, including extension of the T diffusion matrix, the polymorphic CML, and the replacement of the pixel value. In Section 3, the algorithm proposed in this paper is described in detail, including key generation, and the encryption and decryption processes of the algorithm. Section 4 shows the experimental results. A detailed security analysis of the algorithm is given in Section 5. Finally, the characteristics and shortcomings of the algorithm are summarized.

2. Polymorphic Spatiotemporal Chaotic Systems and Random Ergodicity

2.1. Extension of T Diffusion Matrix

The reversible matrix $\boldsymbol{T}$ is only in a 4×4 format in Ref. [37], but the diffusion effect is significant in matrix $\boldsymbol{T}$. Therefore, this paper extends the matrix $\boldsymbol{T}$ to the $N \times N$ effect, and the effect of the original matrix is the same. Furthermore, when the matrix $\boldsymbol{T}$ is extended to a new reversible diffusion matrix, it maintains the reversible property and good diffusion effect of the original matrix.

Suppose $\boldsymbol{P}$ is the clear matrix, and the diffusion formula is

$$T \times P = P'. \tag{1}$$

P' is a plaintext matrix after diffusion.

The original matrix is

$$T = \begin{bmatrix} n & 1 & 1 & 1 \\ n+1 & 1 & 2 & 2 \\ n+2 & 1 & 2 & 3 \\ n+3 & 1 & 2 & 3 \end{bmatrix}, \tag{2}$$

where n is an arbitrary value for the control variable.

After extension matrix $\boldsymbol{T}$:

$$T = \begin{bmatrix} n & 1 & 1 & 1 & \dots & 1 \\ n+1 & 1 & 2 & 2 & \dots & 2 \\ n+2 & 1 & 2 & 3 & \dots & 3 \\ \vdots & \vdots & \vdots & \vdots & \ddots & \vdots \\ n+i & 1 & 2 & 3 & \dots & i-1 \end{bmatrix} \tag{3}$$

where $i(i > 3)$ is the size of the number of rows generated. n can take any value for the control variable, which guarantees the invertibility of the matrix. To ensure the effectiveness of the implementation, a format of 4×4 and above is recommended.

2.2. Polymorphic CML

Only one kind of chaotic mapping is selected in the one-dimensional coupled image lattice; the result is very good, but the chaotic sequence is not uniform, and the number of iterations needs to be abandoned. Moreover, if the parameter selection of the chaotic map is not good, it will easily lead to the phenomenon of the periodic window. So, this paper multiplies the pseudorandom sequence and the matrix $\boldsymbol{T}$ mentioned in the previous paper, and diffuses a coupled image lattice to the uniform state because of the reversible property of the matrix $\boldsymbol{T}$. So, in this paper, we add four chaotic maps to the traditional chaotic map.

A coupled image lattice is a simple model that can describe the complex dynamics of closed systems and is defined as Equation (4).

$$x_{n+1}(i) = (1-\varepsilon)f(x_n) + \frac{\varepsilon}{2}[f(x_n(i+1) + f(x_n(i-1))], \varepsilon \in (0,1) \tag{4}$$

when $i = 1, 2, ..., L$, L represents the size of the lattice. In this paper, when the CML system is used at the pixel level, it is set to 10; n represents the evolution time, $\varepsilon \in (0,1)$ is the coupling coefficient, and the $x_n(L) = x_n(0)$ edge conditions are satisfied. ε and $x_0(1)$ and $x_0(2)$ are used as keys. When $f(x)$ is a chaotic map, the dynamical characteristics of the system are also chaotic. In the course of the study, $f(x) = 4x(x-1)$ and $\varepsilon = 0.5$, and 10 grids were selected and iterated 10,000 times. There was a phenomenon of two-level differentiation. When different coefficients were selected, the histogram data showed different degrees of two-level differentiation. As shown in Figure 1.

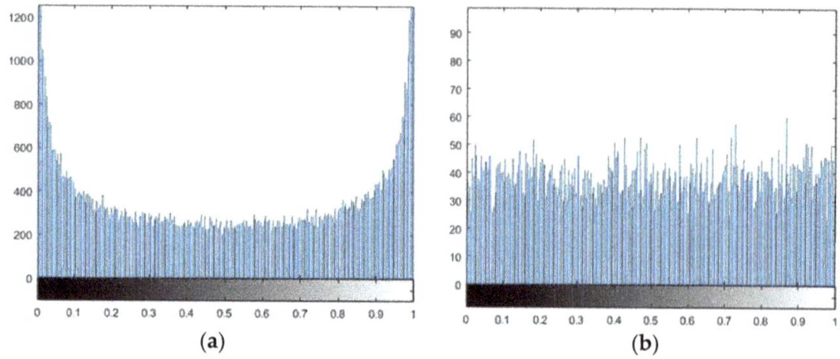

Figure 1. Pseudorandom sequence distribution. (a) The original CML chaotic sequence is distributed. (b) After the T matrix is completed, the CML chaotic sequence is distributed.

Chaotic maps are used to generate chaotic sequences, which are random sequences generated by simple deterministic systems. Therefore, using the idea of polymorphism, let $f(x)$ choose between them to increase the diversity of random sequences and increase the security of the algorithm. The chaotic map $f(x)$ is defined as

$$f(x) = a_0[\mu x_n(1 - x_n)] + a_1[bx_n - ax_n^3] + a_2[x_n/\alpha] + a_3[(1 - x_n)/(1 - \alpha)]\mathrm{mod}1 \quad (5)$$

The chaotic mappings and their corresponding parameter ranges are shown in Table 1. In this paper, when designing $f(x)$, we use a simple chaotic map to design 15 alternative mappings that can increase the change in the pseudo-random sequence. When $a_0a_1a_2a_3 = 1111$ represents all the chaotic maps, it is all selected and discarded as the state of $a_0a_1a_2a_3 = 0000$.

Table 1. Chaotic mappings and parameter ranges.

$f(x)$	Parameter Range
$x_{n+1} = \mu x_n(1 - x_n)$	$\mu \in (0,4), x_{n+1} \in [0,1]$
$x_{n+1} = bx_n - ax_n^3$	$\mathrm{b} \in [0,3], x_n \in [-c,c]$
$x_{n+1} = x_n/\alpha$	$0 \le x_n \le 0.5, \alpha \in (0,1)$
$x_{n+1} = (1 - x_n)/(1 - \alpha)$	$0.5 \le x_n \le 1, \alpha \in (0,1)$

2.3. Use of the Probability Replacement of the Pixel Value

Suppose that for a plaintext image P of size $M \times N$, M is the number of rows and N is the number of columns. If we divide them evenly into n parts, calculate the frequency of every pixel value in every small image and sort them, and then, use the generated pixel values to replace the others, the pixel values can be replaced [38]. Additionally, because the pixel value is 8 bit, the higher pixel values have more information than the originals. When filling 0 to expand the digit, we need to fill 0 on the left side [39–41]. However, considering the complexity of key management, we recommend $n \le 16$ as follows:

Step 1: Judge the parity property of the $M \times N$ image; if it is an odd number, i chooses 1; if it is even, i chooses 2; the number of part n should satisfy

$$n = \frac{M \times N}{(2i)^2 \times (2i - 1)^2}, \ i \in \{1,2\}. \quad (6)$$

Step 2: Calculate the frequency of the pixel value in every plaintext image, and construct a Huffman tree based on the frequency of the pixel value by using the Huffman encoding rule.

Step 3: Because the obtained Huffman code does not satisfy 8 bits, it needs to be extended. Because the information of high pixel values is more than that of low pixel values, it needs to fill in 0 on the left when the number 0 is extended, but not on the right side. The effects are shown in Figures 2 and 3.

Figure 2. RGB-channel image of Lena: (a) original plaintext, R-channel image; (b) original plaintext, G-channel image; (c) plaintext, B-channel image.

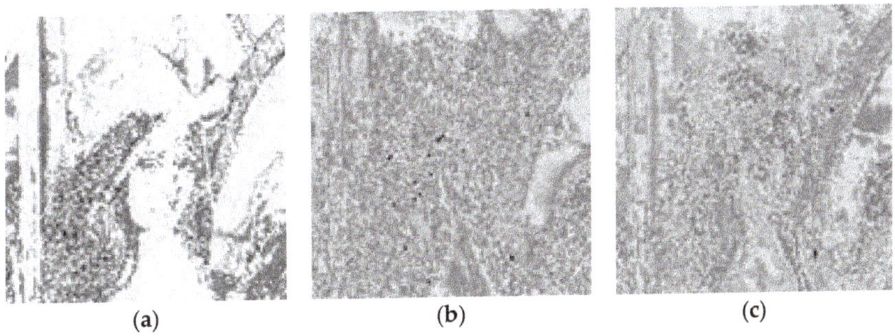

Figure 3. RGB-channel image after the replacement of Lena pixel value: (a) replacement, after R-channel image; (b) replacement, G-channel image; (c) replacement, B-channel image.

3. Image Encryption Algorithm Based on CML with Polymorphic Mapping

3.1. Key Generation

The key parts of the cryptosystem include the control parameter ε of the CML system; the initial values $x_0(1)'$ and $x_0(2)'$; the selected $a_0a_1a_2a_3$; the initial parameters of the chaotic mapping x_0, x_1; the n, i; and the RSA 1024-bit keys in the diffusion matrix T.

3.2. Encryption Algorithm Process

For a plaintext image P of size $M \times N$, the encryption process is as follows:

Step 1: Given the initial key, the SHA-256 algorithm is used to transform the sequence into binary encoding. After the sequence $Hash = [h_1, h_2, h_3, \ldots, h_{256}]$, the first 8 bits is selected, the first 4 bits are judged and the latter 4 bits are selected.

Step 2: According to the initial key, $Hash = [h_1, h_2, h_3, \ldots, h_{256}]$ is obtained, $H_1 = [h_9, h_{10}, h_{11}, h_{12}, h_{13}]$ is selected to obtain the selection sequence of $a_0a_1a_2a_3$, and the specific $f(x)$ is selected.

Step 3: From the key $H_2 = [h_9, h_{10}, h_{11}, h_{12}, h_{13}]$, the sequence is transformed into the CML's initial parameter and coupling coefficient. i is any value and the coupling coefficient is calculated by Equation (7).

$$\varepsilon = [(H \times n \times 10^{-2})\text{mod}i]\text{mod}1 \tag{7}$$

Step 4: The initial values generated by $x_0(1)$ and $x_0(2)$ are replaced by pixel values. The formula is as follows.

$$\begin{cases} x_0(1) = [\text{Haffman}(i_1) \times 0.123]\text{mod}1 \\ x_0(2) = [\text{Haffman}(i_2) \times 0.234]\text{mod}1 \end{cases} \quad (8)$$

Step 5: Disposal. The scrambling processing is performed using the function $sort(\cdot)$. If A is the vector to be sorted, $[B,\ index] = sort(A)$, where B is the sorted vector A, and $index$ is the index of each item in B corresponding to vector A.

Step 6: The n,i of the diffusion matrix $\boldsymbol{T}$ is selected from the rule of probability substitution.

Step 7: The two value sequences seq_i are determined; the formula is as follows:

$$seq_i = \begin{cases} 1, & x_i > 0.5 \\ 0, & x_i \le 0.5 \end{cases} \quad (9)$$

Step 8: The bit-OR operation is performed at the end of this pixel level encryption process. Figure 4 describes the process of the encryption algorithm.

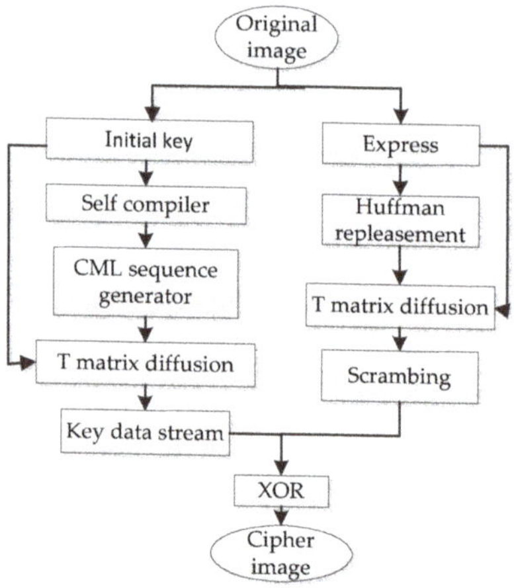

Figure 4. Flowchart of the image encryption algorithm based on CML with polymorphic mapping.

3.3. Decryption Process of Algorithm

The Huffman code used in the encryption process is irreversible. In the process of encryption, the probability of each pixel value in the image is completely destroyed. So, in the process of decryption, the Huffman code is processed separately. This paper uses the traditional RSA scheme to deal with the problem. The other steps are the inverse of the encryption process [42,43].

4. Experimental Results

Here, we choose three plaintext color images of 256 × 256 × 3, "Lena", "Peppers" and "Mandrill", to simulate the algorithm in this paper. We choose the initial key (hash, diffusion matrix) to verify the effect, and the selected matrix size is 8 × 8 sequences, where $n = 2$; the local parameters are related to the initial key. The experimental results are the same as the expected experimental results. All three plaintext images of 256 × 256 × 3 can be encrypted, and intuitively, no clear plaintext information appears in the image. Figures 5–7 are the simulation results.

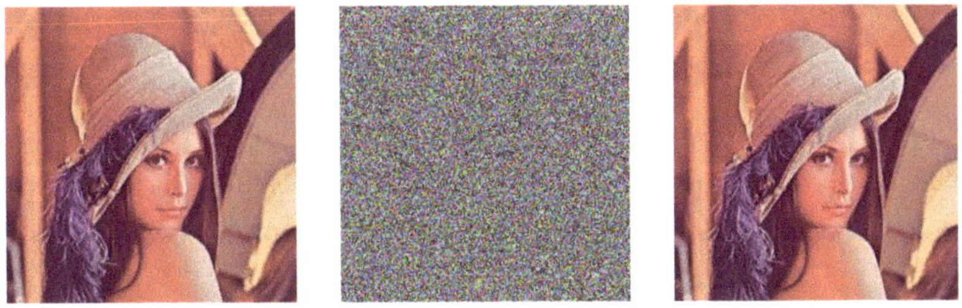

(a) (b) (c)

Figure 5. Lena encryption process. (**a**) Original image Lena; (**b**) encrypted image Lena; (**c**) decrypted image Lena.

(a) (b) (c)

Figure 6. Mandrill encryption process. (**a**) Original image Mandrill; (**b**) encrypted image Mandrill; (**c**) decrypted image Mandrill.

(a) (b) (c)

Figure 7. Peppers encryption process. (**a**) Original image Peppers; (**b**) encrypted image Peppers; (**c**) decrypted image Peppers.

5. Security Analysis

In this section, we will conduct a theoretical analysis and numerical simulations of a violent attack, statistical attack, differential attack, chosen plaintext attack, etc., and compare the results with Refs. [31,32,44].

5.1. Key-Space Analysis

A large enough key space can resist violent attacks and improve the security of encryption algorithms. The key space includes all keys used in the scrambling and diffusion processes. Valid keys for this algorithm are as follows:

The initial key is:

$$Hash = [5312fb609f60384731fcfcb95deef3602239bf61f865a07bd8e08d818d22e9fa].$$

Since the initial key used in this paper is generated by the hash function of the SHA-256 algorithm, there are a total of 256 bits, and there are 256 cases of probability replacement of the pixel values in this paper, as well as the n, i part of the diffusion matrix T and the public key, plus the secret 1024 bits in the RSA algorithm. So, if the computing precision of the computer is 10^{-14}, the algorithm key space designed in this paper is $2^{256} \times 256 \times 4 \times 2^{10} \approx 2^{276}$, far greater than that of the password system; thus, this algorithm can resist the a violent attack.

5.2. Statistical Analysis

Image histogram analysis and adjacent-pixel correlation are two very important statistical properties of image encryption algorithms, and can reflect the algorithm's ability to resist statistical attacks.

5.2.1. Histogram Analysis

Figures 8 and 9 give the histograms of the RGB channels of the plaintext and ciphertext images of "Lena". Comparing their histograms, it can be found that the histogram distribution of encrypted plaintext images is more uniform than before encryption. Therefore, an image encrypted by this algorithm makes a statistical analysis attack difficult.

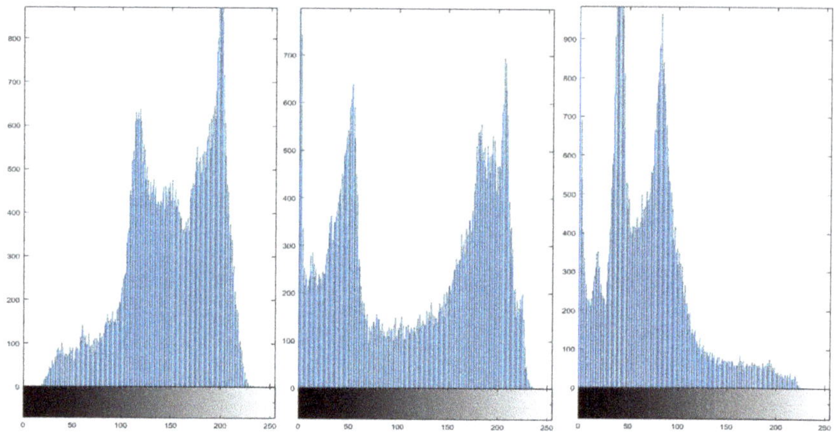

(a) (b) (c)

Figure 8. RGB histogram of Lena. (**a**) Lena R-channel pixel histogram; (**b**) Lena G-channel pixel histogram; (**c**) Lena B-channel pixel histogram.

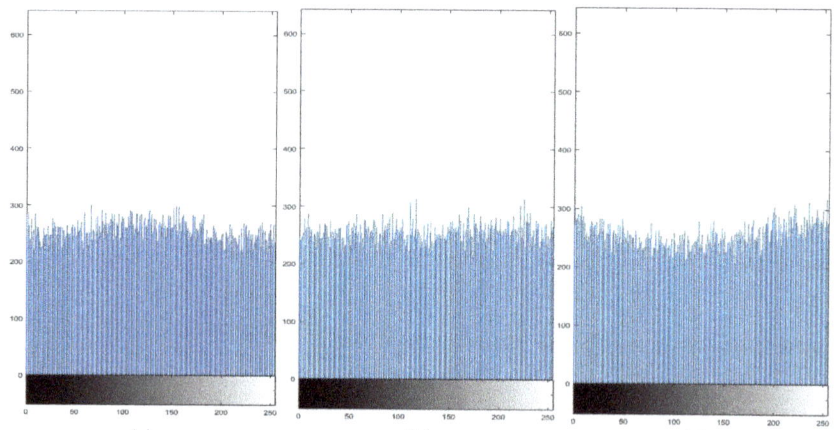

(a) (b) (c)

Figure 9. RGB histogram of Lena after encryption. (**a**) encrypted, Lena's R-channel pixel histogram; (**b**) encrypted, G-channel pixel histogram; (**c**) encrypted, B-channel pixel histogram.

5.2.2. Adjacent Pixel Correlation

To resist statistical attacks, the correlation between adjacent pixels must be effectively reduced [31,32,44]. Using Equation (10) to calculate the correlation between the adjacent pixels of the plaintext and the ciphertext images, we randomly select 10,000 pairs of pixels in the plaintext and the ciphertext images of the "Lena" R channel, as shown in Figure 10. The correlations between the adjacent pixels of the horizontal, vertical, and diagonal lines of these pixels are also tested, as shown in Table 2.

$$r_{xy} = \frac{\mathrm{cov}(x,y)}{\sqrt{D(x)}\sqrt{D(y)}} \tag{10}$$

when,

$$\mathrm{cov}(x,y) = \frac{1}{N}\sum_{i=1}^{N}(x_i - E(x))(y_i - E(y)), D(x) = \frac{1}{N}\sum_{i=1}^{N}(x_i - E(x))^2, E(x) = \frac{1}{N}\sum_{i=1}^{N}x_i. \tag{11}$$

Table 2. Correlation between adjacent pixels of plaintext and ciphertext images.

Lena	Plaintext			Ciphertext		
	R	G	B	R	G	B
Horizontal	0.988	0.983	0.955	0.002	0.009	0.001
Vertical	0.974	0.951	0.935	0.032	−0.002	0.052
Diagonal	0.974	0.950	0.921	0.002	0.019	0.025

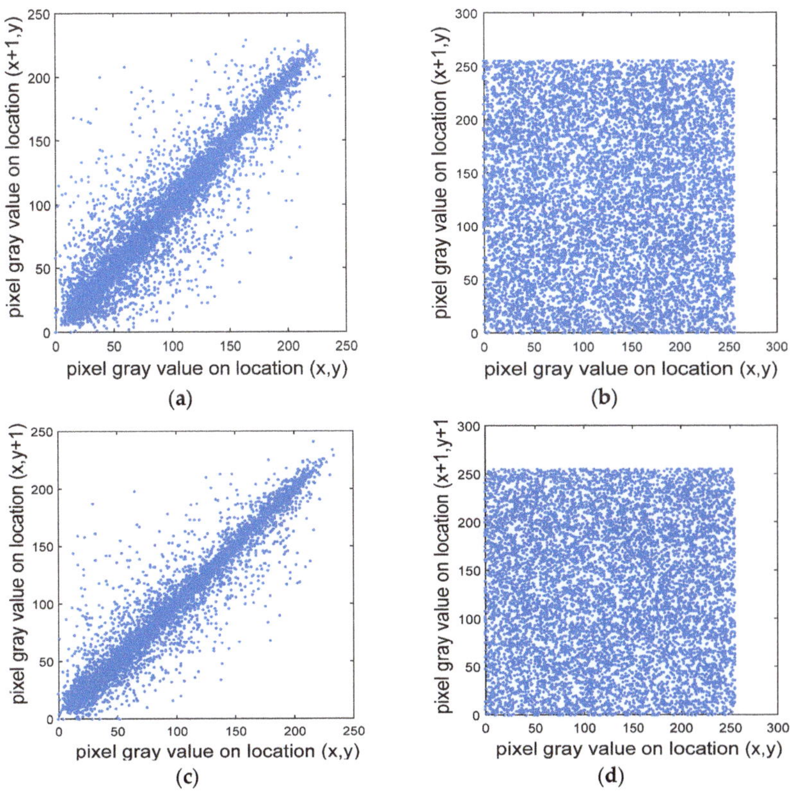

Figure 10. *Cont.*

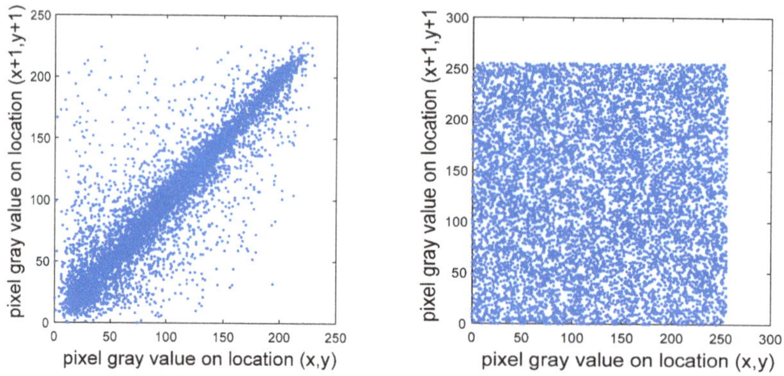

Figure 10. The level of adjacent-pixel correlation. (**a**) Lena plaintext; (**b**) Lena ciphertext horizontal correlation; (**c**) Lena plaintext vertical correlation; (**d**) Lena clear text vertical correlation; (**e**) Lena plaintext diagonal correlation; (**f**) Lena plaintext diagonal correlation.

5.2.3. Information Entropy

Entropy is an index used to measure uncertainty [31,32,44], that is, the probability of discrete random events. The more chaotic the system is, the higher the information entropy, and vice versa. The value can be calculated by Equation (12).

$$H(s) = \sum_{i=0}^{2^L-1} p(s_i) \log_2 \frac{1}{p(s_i)} \tag{12}$$

Here $p(s_i)$ is the probability that s_i occurs. The information entropy of the encrypted ciphertext image should be closer to 8. The results in Table 3 show that the encrypted information of the ciphertext image is not easy to leak, and it can better resist statistical attacks.

Table 3. Information entropy of clear and ciphertext images.

Lena	Information Entropy of Plaintext Image			Information Entropy of Ciphertext Image		
	R	G	B	R	G	B
Our paper	7.253	7.594	6.688	7.990	7.925	7.904
Ref. [31]	-	-	-	7.883	7.875	7.570
Ref. [32]	-	-	-	7.880	7.854	7.953
Ref. [44]	-	-	-	7.950	7.968	7.882

5.2.4. Resistance to Differential Attacks

A differential attack analyzes the data of the image based on the ciphertext before modification and after modification, and obtains the key by making small changes to the text. Here, the number-of-pixels change rate (NPCR) and the unified mean change intensity (UACI) are calculated [31,45]. The larger the value of NPCR, the more sensitive the encryption algorithm is to changes in the original plaintext. The larger the UACI value, the greater of the average change intensity. The number-of-pixels change rate is calculated as follows:

$$NPCR = \frac{\sum_{i,j} D(i,j)}{W \times H} \times 100 \tag{13}$$

$$UACI = \frac{1}{W \times H} [\sum_{i,j} \frac{|c_1(i,j) - c_2(i,j)|}{255}] \times 100 \tag{14}$$

where W and H denote the width and height of the image, respectively. Additionally, c_1 and c_2 denote two ciphertext images after the original plaintext is changed by one pixel. If $c_1(i,j) \neq c_2(i,j)$, then $D(i,j) = 1$; otherwise, $D(i,j) = 0$. The experimental results in Table 4 show that, in general, NPCR is close to 99.6049% and UACI is close to 33.4635% [32,46]. The results show the advantages of the algorithm in resisting differential attacks.

Table 4. NPCR and UACI values for ciphertext images.

Lena	NPCR/%			UACI/%		
	R	G	B	R	G	B
Ref. [31]	41.96	41.96	41.96	33.25	33.25	33.25
Ref. [32]	86.68	86.68	86.68	32.51	32.43	32.43
Ref. [44]	94.68	95.68	98.68	33.46	34.50	35.49
Our paper	98.44	98.42	98.44	33.38	33.28	33.38

5.2.5. Robustness Analysis

We use a noise attack and block attack to test the robustness of the algorithm [32,44]. Compared with other common noise types, salt-and-pepper noise has a greater direct impact on the ciphertext image. Therefore, this experiment considers the effect on the plaintext image after adding salt-and-pepper noise to the algorithm. We add different strengths of salt-and-pepper noise to the plaintext and use the same key for encryption and decryption. Figure 11 shows the encrypted image with three noise values of 0.02, 0.12, and 0.2, respectively.

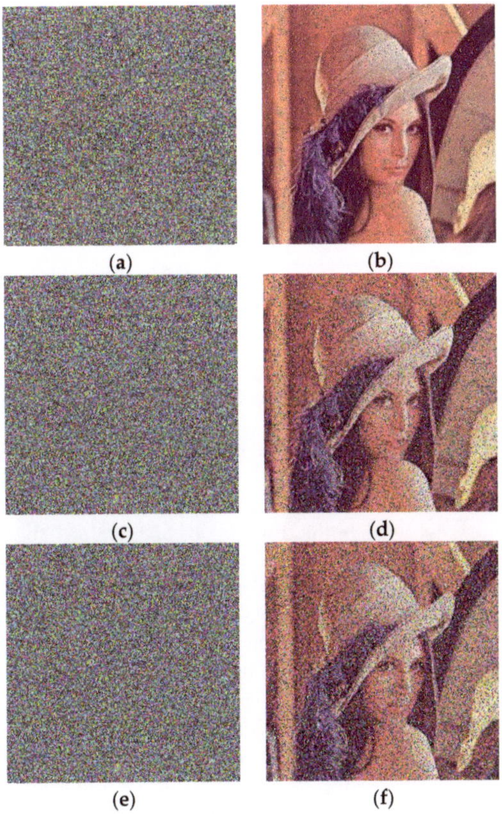

Figure 11. Noise experiment adding noise intensity of (**a**) 0.02, (**b**) 0.12, and (**c**) 0.2 after the encrypted image, and (**d–f**) is their corresponding decryption image.

5.2.6. Sensitivity Analysis

The main part of this article contains the initial key part, the n, i parameter in the matrix T, and the cryptographic part of the RSA algorithm. The RSA algorithm is a traditional encryption method, so it has not been tested in sensitivity tests. This paper only changed the initial key

$$Hash = [5312fb609f60384731fcfcb95deef3602239bf61f865a07bd8e08d818d22e9fb],$$

and the n,i parameters in the matrix $\boldsymbol{T}$, and the n,i of the replacement of the pixel values under the premise that the RSA was not cracked. The experimental results in Figure 12 show that the encryption scheme of the CML color image based on the polymorphism principle designed in this paper has good sensitivity to keys.

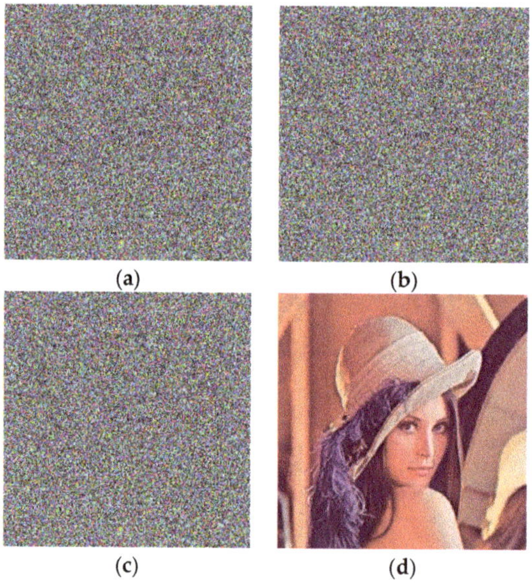

Figure 12. Sensitivity experiment. (**a**) Changes the hash value; (**b**) changes the n of the $\boldsymbol{T}$ matrix and decrypts the image of the n,i; (**c**) changes the decryption image of the Huffman replacement rule; and (**d**) the correct key decryption.

5.2.7. Complexity Analysis

The time-consuming nature of the calculation of floating-point data in an encryption algorithm was considered in this section. In generating the chaotic sequence using the CML system, $\Theta(L \times M \times N)$ iterations of floating-point data were performed. When using the $\boldsymbol{T}$ matrix to perform the diffusion of pixel values, the corresponding computational complexity was $\Theta(n \times M \times N)$ operations of floating-point data. It is less efficient when using Matlab R2016b to calculate and select CML sequences, but it can still meet the needs of real cryptosystems. This article mainly considers the security of the image encryption algorithm, so this does not violate the original intention of this article.

The running environment of this algorithm is 8.00 GB RAM, Intel(R) Core(TM) Intel(R) Xeon(R) CPU at 2.67 GHz, the operating system is Windows 8, and the simulation software is Matlab R2016b. We know that Matlab is an excellent piece of simulation software, but its efficiency is low; the algorithm's running speed and programming language, CPU, memory size, operating system, etc. have certain purposes. It is less efficient when using Matlab to calculate and select CML sequences, but it can still meet the needs of real cryptosystems. The purpose of this article is to propose a more secure image encryption algorithm, so this is not contrary to the original intention of this article.

6. Conclusions

A new polymorphic coupled map lattice based on information entropy is developed for encrypting color images in this paper. Firstly, we extend a diffusion matrix with the original 4×4 matrix into an $n \times n$ matrix. Then, the Huffman idea is employed to propose a new pixel-level substitution method, which is applied to replace the grey degree value. We employ the idea of polymorphism and select f(x) in the spatiotemporal chaotic system. The pseudo-random sequence is more diversified and the sequence is homogenized. Three plaintext color images of $256 \times 256 \times 3$, "Lena", "Peppers" and "Mandrill", are selected in order to prove the effectiveness of the proposed algorithm. The results show the advantages of the algorithm in resisting differential attacks. An encrypted image with three noise values of 0.02, 0.12, and 0.2 is obtained. The security of the image encryption algorithm does not violate our original intention. Therefore, the results of brute-force attacks, statistical attacks, and plaintext attacks show that the algorithm has good security. In addition, in our study, the mixed model gradually replaced the single CML model, and showed better results in resisting various typical attacks [47]. Therefore, the hybrid model of the genetic algorithm and CML will be further studied.

Author Contributions: Conceptualization, P.H. and D.L.; methodology, P.H. and Y.W.; software, Y.W. and D.L.; validation, D.L.; formal analysis, P.H.; investigation, P.H. and H.Z.; resources, D.L. and H.Z.; data curation, D.L.; writing—original draft preparation, P.H.; writing—review and editing, W.D.; visualization, W.D.; supervision, D.L.; project administration, W.D.; funding acquisition, W.D. All authors have read and agreed to the published version of the manuscript.

Funding: This research was funded by the National Natural Science Foundation of China under grant number 61771087, and the Research Foundation for the Civil Aviation University of China under grant numbers 3122022PT02 and 2020KYQD123.

Informed Consent Statement: Not applicable.

Conflicts of Interest: The authors declare no conflict of interest.

References

1. Stinson, D. *Cryptography: Theory and Practice*, 2nd ed.; CRC Press Public House of Electronic Industry: Boca Raton, FL, USA, 2002.
2. Long, S. A comparative analysis of the application of hashing encryption algorithms for MD5, SHA-1, and SHA-512. *J. Phys. Conf. Ser.* **2019**, *1314*, 012210. [CrossRef]
3. Wang X, Y.; Zhang, H.L. A novel image encryption algorithm based on genetic recombination and hyper-chaotic systems. *Nonlinear Dyn.* **2016**, *83*, 333–346. [CrossRef]
4. Wang, X.; Zhang, M. An image encryption algorithm based on new chaos and diffusion values of a truth table. *Inf. Sci.* **2021**, *579*, 128–149. [CrossRef]
5. Li, Z.; Peng, C.; Tan, W.; Li, L. A novel chaos-based color image encryption scheme using bit-level permutation. *Symmetry* **2020**, *12*, 1497. [CrossRef]
6. Zarebnia, M.; Parvaz, R. Image encryption algorithm by fractional based chaotic system and framelet transform. *Chaos Solitons Fractals* **2021**, *152*, 111402. [CrossRef]
7. Wu, D.; Wu, C. Research on the time-dependent split delivery green vehicle routing problem for fresh agricultural products with multiple time windows. *Agriculture* **2022**, *12*, 793. [CrossRef]
8. Li, X.; Zhao, H.; Yu, L.; Chen, H.; Deng, W.; Deng, W. Feature extraction using parameterized multisynchrosqueezing transform. *IEEE Sens. J.* **2022**, *2*, 14263–14272. [CrossRef]
9. Zhou, X.B.; Ma, H.J.; Gu, J.G.; Chen, H.L.; Deng, W. Parameter adaptation-based ant colony optimization with dynamic hybrid mechanism. *Eng. Appl. Artif. Intell.* **2022**, *114*, 105139. [CrossRef]
10. Li, T.Y.; Shi, J.Y.; Deng, W.; Hu, Z.D. Pyramid particle swarm optimization with novel strategies of competition and cooperation. *Appl. Soft Comput.* **2022**, *121*, 108731. [CrossRef]
11. Chen, H.Y.; Miao, F.; Chen, Y.J.; Xiong, Y.J.; Chen, T. A hyperspectral image classification method using multifeature vectors and optimized KELM. *IEEE J. Sel. Top. Appl. Earth Obs. Remote Sens.* **2021**, *14*, 2781–2795. [CrossRef]
12. Yao, R.; Guo, C.; Deng, W.; Zhao, H.M. A novel mathematical morphology spectrum entropy based on scale-adaptive techniques. *ISA Trans.* **2022**, *126*, 691–702. [CrossRef] [PubMed]
13. Zhao, H.M.; Liu, J.; Chen, H.Y.; Chen, J.; Li, Y.; Xu, J.J.; Deng, W. Intelligent diagnosis using continuous wavelet transform and gauss convolutional deep belief network. *IEEE Trans. Reliab.* **2022**, 1–11. [CrossRef]
14. Wei, Y.Y.; Zhou, Y.Q.; Luo, Q.F.; Deng, W. Optimal reactive power dispatch using an improved slime Mould algorithm. *Energy Rep.* **2021**, *7*, 8742–8759. [CrossRef]

15. Deng, W.; Ni, H.C.; Liu, Y.; Chen, H.L.; Zhao, H.M. An adaptive differential evolution algorithm based on belief space and generalized opposition-based learning for resource allocation. *Appl. Soft Comput.* **2022**, *127*, 109419. [CrossRef]
16. Chen, H.Y.; Fang, M.; Xu, S. Hyperspectral remote sensing image classification with CNN based on quantum genetic-optimized sparse representation. *IEEE Access* **2020**, *8*, 99900–99909. [CrossRef]
17. Deng, W.; Zhang, L.; Zhou, X.; Zhou, Y.; Sun, Y.; Zhu, W.; Chen, H.; Deng, W.; Chen, H.; Zhao, H. Multi-strategy particle swarm and ant colony hybrid optimization for airport taxiway planning problem. *Inf. Sci.* **2022**, *612*, 576–593. [CrossRef]
18. Song, Y.; Cai, X.; Zhou, X.; Zhang, B.; Chen, H.; Li, Y.; Deng, W.; Deng, W. Dynamic hybrid mechanism-based differential evolution algorithm and its application. *Expert Syst. Appl.* **2023**, *213*, 118834. [CrossRef]
19. Zhang, Z.; Huang, W.G.; Liao, Y.; Song, Z.; Shi, J.; Jiang, X.; Shen, C.; Zhu, Z. Bearing fault diagnosis via generalized logarithm sparse regularization. *Mech. Syst. Signal Process.* **2022**, *167*, 108576. [CrossRef]
20. Li, N.; Huang, W.G.; Guo, W.J.; Gao, G.Q.; Zhu, Z. Multiple enhanced sparse decomposition for gearbox compound fault diagnosis. *IEEE Trans. Instrum. Meas.* **2020**, *69*, 770–781. [CrossRef]
21. Xu, G.; Bai, H.; Xing, J.; Luo, T.; Xiong, N.N. SG-PBFT: A secure and highly efficient distributed blockchain PBFT consensus algorithm for intelligent Internet of vehicles. *J. Parallel Distrib. Comput.* **2022**, *164*, 1–11. [CrossRef]
22. Zheng, J.J.; Yuan, Y.; Zou, L.; Deng, W.; Guo, C.; Zhao, H. Study on a novel fault diagnosis method based on VMD and BLM. *Symmetry* **2019**, *11*, 747. [CrossRef]
23. Wu, X.; Wang, Z.C.; Wu, T.H.; Bao, X.G. Solving the family traveling salesperson problem in the adleman–lipton model based on DNA computing. *IEEE Trans. NanoBioscience* **2021**, *21*, 75–85. [CrossRef] [PubMed]
24. Cao, H.; Shao, H.; Zhong, X.; Deng, Q.; Yang, X.; Xuan, J. Unsupervised domain-share CNN for machine fault transfer diagnosis from steady speeds to time-varying speeds. *J. Manuf. Syst.* **2022**, *62*, 186–198. [CrossRef]
25. Zhou, Y.; Zhang, J.; Yang, X.; Ling, Y. Optimal reactive power dispatch using water wave optimization algorithm. *Oper. Res.* **2020**, *20*, 2537–2553. [CrossRef]
26. Xu, G.; Dong, W.; Xing, J.; Lei, W.; Liu, J. Delay-CJ: A novel cryptojacking covert attack method based on delayed strategy and its detection. *Digit. Commun. Netw.* 2022, *in press*. [CrossRef]
27. Li, X.; Shao, H.; Lu, S.; Xiang, J.; Cai, B. Highly-efficient fault diagnosis of rotating machinery under time-varying speeds using LSISMM and small infrared thermal images. *IEEE Trans. Syst. Man Cybern. Syst.* **2022**, *30*, 135–142. [CrossRef]
28. Ren, Z.; Han, X.; Yu, X.; Skjetne, R.; Johan, B.; Leira, S.; Zhu, M. Data-driven simultaneous identification of the 6DOF dynamic model and wave load for a ship in waves. *Mech. Syst. Signal Process.* **2023**, *184*, 109422. [CrossRef]
29. Roellgen, C.B. Polymorphic cipher theory. 2004. Available online: http://www.ciphers.de/products/polymorphic_cipher_theory.html (accessed on 12 September 2022).
30. Mackowski, D.W.; Mishchenko, M.I. Calculation of the T matrix and the scattering matrix for ensembles of spheres. *J. Opt. Soc. Am. A* **1996**, *13*, 2266–2278. [CrossRef]
31. Behnia, S.; Akhshani, A.; Mahmodi, H.; Akhavan, A.J.C.S. A novel algorithm for image encryption based on mixture of chaotic maps. *Chaos Soliton & Fract.* **2008**, *35*, 408–419.
32. Hussain, I.; Shah, T.; Gondal, M.A. Image encryption algorithm based on PGL(2,GF(28)) S-boxes and TD-ERCS chaotic sequence. *Nonlinear Dynam.* **2012**, *70*, 181–187. [CrossRef]
33. Hussain, I.; Shah, T.; Gondal, M.A. An efficient image encryption algorithm based on S8 S-box transformation and NCA map. *Opt. Commun.* **2012**, *285*, 4887–4890. [CrossRef]
34. Zhu, Z.L.; Zhang, W.; Wong, K.W.; Yu, H. A chaos-based symmetric image encryption scheme using a bit-level permutation. *Inf. Sci. Int. J.* **2011**, *181*, 1171–1186. [CrossRef]
35. Hussain, I.; Gondal, M.A. An extended image encryption using chaotic coupled map and S-box transformation. *Nonlinear Dynam.* **2014**, *76*, 1355–1363. [CrossRef]
36. Baptista, M.S. Cryptography with chaos. *Phys. Lett. A* **1998**, *240*, 50–54. [CrossRef]
37. Jain, A.; Rajpal, N. A robust image encryption algorithm resistant to attacks using DNA and chaotic logistic maps. *Multimed. Tools Appl.* **2016**, *75*, 5455–5472. [CrossRef]
38. Rehman, A.U.; Liao, X.; Kulsoom, A.; Abbas, S. Selective encryption for gray images based on chaos and DNA complementary rules. *Multimed. Tools Appl.* **2015**, *74*, 4655–4677. [CrossRef]
39. Huang, X.; Ye, G. An image encryption algorithm based on hyper-chaos and DNA sequence. *Multimed. Tools Appl.* **2014**, *72*, 57–70. [CrossRef]
40. Bakhshandeh, A.; Eslami, Z. An authenticated image encryption scheme based on chaotic maps and memory cellular automata. *Opt. Lasers Eng.* **2013**, *51*, 665–673. [CrossRef]
41. Xu, L.; Li, Z.; Li, J.; Hua, W. A novel bit-level image encryption algorithm based on chaotic maps. *Opt. Lasers Eng.* **2016**, *78*, 17–25. [CrossRef]
42. Zhang, Q.; Guo, L.; Wei, X. Image encryption using DNA addition combining with chaotic maps. *Math. Comput. Model.* **2010**, *52*, 2028–2035. [CrossRef]
43. Liu, H.; Wang, X.Y.; Kadir, A. Image encryption using DNA complementary rule and chaotic maps. *Appl. Soft Comput.* **2012**, *12*, 1457–1466. [CrossRef]
44. Rhouma, R.; Belghith, S. Cryptanalysis of a spatiotemporal chaotic image/video cryptosystem. *Phys. Lett. A* **2008**, *372*, 5790–5794. [CrossRef]

45. Akhshani, A.; Behnia, S.; Akhavan, A.; AbuHassana, H.; Hassana, H. A novel scheme for image encryption based on 2D piecewise chaotic maps. *Opt. Commun.* **2010**, *283*, 3259–3266. [CrossRef]
46. Hussain, I.; Shah, T.; Gondal, M.A. Image encryption algorithm based on total shuffling scheme and chaotic S-box transformation. *J. Vib. Control.* **2014**, *20*, 2133–2136. [CrossRef]
47. Nematzadeh, H.; Enayatifar, R.; Motameni, H. Medical image encryption using a hybrid model of modified genetic algorithm and coupled map lattices. *Opt. Lasers Eng.* **2018**, *110*, 24–32. [CrossRef]

electronics

MDPI

Article

One-Dimensional Quadratic Chaotic System and Splicing Model for Image Encryption

Chen Chen [1], Donglin Zhu [2], Xiao Wang [3,*] and Lijun Zeng [1]

[1] Nanhang Jincheng College, Nanjing 211156, China
[2] College of Mathematics and Computer Science, Zhejiang Normal University, Jinhua 321004, China
[3] Xingzhi College, Zhejiang Normal University, Jinhua 321000, China
* Correspondence: tianzhu213@zjnu.cn

Abstract: Digital image transmission plays a very significant role in information transmission, so it is very important to protect the security of image transmission. Based on the analysis of existing image encryption algorithms, this article proposes a new digital image encryption algorithm based on the splicing model and 1D secondary chaotic system. Step one is the algorithm of this article divides the plain image into four sub-parts by using quaternary coding, and these four sub-parts can be coded separately. Only by acquiring all the sub-parts at one time can the attacker recover the useful plain image. Therefore, the algorithm has high security. Additionally, the image encryption scheme in this article used a 1D quadratic chaotic system, which makes the key space big enough to resist exhaustive attacks. The experimental data show that the image encryption algorithm has high security and a good encryption effect.

Keywords: 1D quadratic chaotic system; image encryption; splicing model; DNA coding

Citation: Chen, C.; Zhu, D.; Wang, X.; Zeng, L. One-Dimensional Quadratic Chaotic System and Splicing Model for Image Encryption. *Electronics* **2023**, *12*, 1325. https://doi.org/10.3390/electronics12061325

Academic Editor: Gwanggil Jeon

Received: 13 February 2023
Revised: 1 March 2023
Accepted: 9 March 2023
Published: 10 March 2023

1. Introduction

With the development of technologies such as artificial intelligence and 5G and the internet of things, we have entered the times of big data information. However, due to the sharing and openness of computer networks, information security is facing great challenges. Most of the information in the network is carried by images, so it is very necessary to protect information security. Meanwhile, researchers have adopted a series of digital image encryption schemes [1–5]. Some researchers put forward the image encryption algorithm based on DNA computing and chaotic system, which protects its safe transmission of images in the network to some extent [6–14]. Reference [1] put forward an image encryption algorithm based on a one-dimensional composite chaotic mapping system, which is composed of logistic mapping and tent mapping. The algorithm has high complexity and insufficient key space. Reference [2] put forward an image encryption method based on diffusion (JPD) and joint permutation, which determines which pixels will be permuted and diffused by hyperchaotic sequence. Reference [7] put forward an image encryption algorithm based on one-dimensional fractional chaotic mapping, which uses chaotic mapping to design parallel DNA coding to encrypt images. The algorithm has a greater key space. References [15,16] put forward image encryption algorithms based on a logistic chaotic system and a sine mapping system, respectively. Although its scheme is simple, it adopts a low-dimensional logical chaotic system, and the number of parameters is small, which leads to less key space. In addition, the mapping is easy to predict, and the ability to resist exhaustive attacks is poor. The author of reference [17] proposed an encryption algorithm based on quaternary separation of the original image and hyperchaotic system, which has a good anti-attack ability, but the calculation speed is not fast enough, and the key space is not large enough. Reference [18] encrypts the image by generating chaotic sequence and bit cross-diffusion through iterative logical mapping, which has a larger key space. Therefore, the choice of a chaotic system is very significant, which will affect the whole image encryption scheme.

In order to ensure that its key space is exceptionally large and the calculation time is appropriate, the paper uses 1D quadratic chaotic mapping to encrypt digital images. Because 1D quadratic mapping has three adjustable parameters, it will obtain a larger key space, and its calculation speed is faster than that of a high-dimensional chaotic system.

According to different digital image encryption methods, encryption technologies of the image are roughly split into image encryption technology based on matrix transformation, chaos, frequency domain, SCAN language and DNA computing, etc. At present, the most popular encryption technology is based on DNA computing, which has the characteristics of low embodied energy, high concurrency, and high storage density and can meet the space and speed requirements of DNA sequences. Therefore, encryption methods are widely used in the field of information hiding, which is based on DNA computing [8–12]. Reference [8] put forward an encryption algorithm of DNA coding and sequence based on constructing short DNA chains and long DNA chains. Reference [18] put forward a way to modify the pixels of original images by DNA encoding. Reference [9] put forward a novel image encryption algorithm, which is based on an intertwining logistic map and DNA coding. However, these experiments only use DNA bases as operating objects and require harsh laboratory environments and expensive experimental equipment. At present, the laboratory cannot always meet such requirements. Therefore, the image encryption methods which combine DNA computing with a chaotic system were introduced by countless researchers. In recent years, some researchers have abandoned the disadvantages of traditional DNA encryption algorithms using complex biological operations and used the idea of DNA subsequence operations to scramble and spread pixel values. However, there is no perfect encryption algorithm for image information encryption, which has its advantages and disadvantages. Decryption technology is also constantly improving, so digital image encryption needs further research.

According to the existing digital image encryption algorithm, this paper puts forward the following improvement measures:

1. Based on the quaternary coding theory, the plaintext image is edited into four sub-parts, and each sub-part can be coded by different coding rules, which makes it more difficult for attackers to crack the original image.
2. There are three key parameters of a 1D quadratic chaotic map, which are significantly expanded compared with the traditional 1D chaotic map in parameter space. This algorithm uses a 1D quadratic chaotic map to encrypt the original image. Large key space makes the encryption algorithm more robust.
3. Using DNA sequence XOR operation to diffuse the pixel value of the digital image. In the process of digital image encryption, the mosaic model is introduced, which makes it difficult for image attackers to recover the original image.

2. Relevant Knowledge

2.1. D Quadratic Chaotic Map

The general 1D quadratic chaotic formula is defined as follows $f(x) = mx^2 + nx + k$ when $m \neq 0$ and

$$k = \frac{2a - a^2 + n^2 - 2n}{4m} \tag{1}$$

where $a \in (3.5699, 4]$ this map will be chaotic. Equation (1) can be solved in reverse, and its solution is:

$$\begin{cases} a_1 = 1 - \sqrt{(n-1)^2 - 4mk} \\ a_2 = 1 + \sqrt{(n-1)^2 - 4mk} \end{cases} \tag{2}$$

For a_2, we should make:

$$3.5699 < a_2 \leq 4 \tag{3}$$

By Equation (3), can obtain:

$$6.6 < (n-1)^2 - 4mk \leq 9 \tag{4}$$

If $(n-1)^2 - 4mk = 9$, map p will be full. $p(x) = mx^2 + nx + k$ is chaotic when condition (4) holds; the 1D quadratic function is chaotic because it is topologically conjugate with logical chaotic mapping [19]. The 1D quadratic function is chaotic because it is topologically conjugate with logical chaotic mapping [19]. The values of three adjustable parameters k, n and m of a 1D quadratic chaotic map need to meet the limitations of Equation (4). In the implementation of the encryption algorithm, we usually determine the values of n and k at random first and then determine the value range of the third parameter, m, by Formula (4).

Generally, the low-dimensional chaotic map having a lesser key space will lead to difficulty in resisting exhaustive attacks, while 1D quadratic map contains three adjustable parameters, so the encryption algorithm using the 1D quadratic map has a larger key space.

2.2. The Splicing Model

The splicing model was proposed by Tom Head [20]. The basic theory of splicing model details as below:

Suppose there is an abstract alphabet M and two strings $x = x_1k_1k_2x_2$, $y = y_1k_3k_4y_2$, which is composed of symbols of M. The primary splicing operation refers to the conversion of $(x_1k_1k_2x_2, y_1k_3k_4y_2)$ to $(x_1k_1k_4y_2, y_1k_3k_2x_2)$ under the premise of rule $r = k_1\#k_2\$k_3\#k_4$. Figure 1 shows the conversion process.

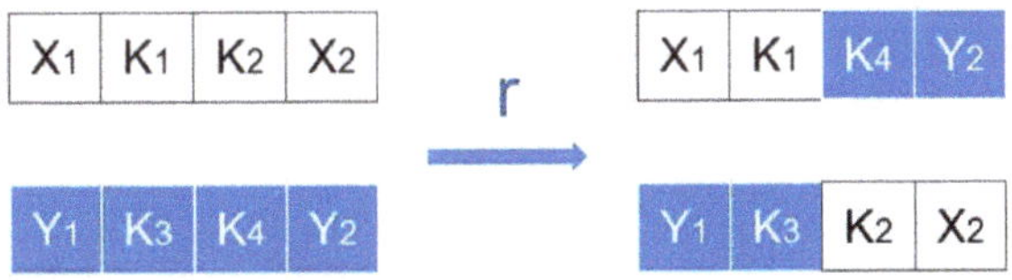

Figure 1. The splicing operation.

2.3. DNA Computing

2.3.1. DNA Encoding and Decoding

In the realm of number theory, a positive integer W can be replaced by H integers smaller than it. Defined as follows:

$$\begin{cases} m_1 = W\text{mod}h; \\ m_2 = (W/h)\text{mod}h; \\ m_3 = (W/h^2)\text{mod}h; \\ \ldots \\ m_N = (W/h^{H-1})\text{mod}h \end{cases} \tag{5}$$

where h is a positive integer smaller than W. These performed calculations are reversible, and the value of W can be found according to Equation (6).

$$W = ((((W/h^H) \times h + m_H) \times h + m_{H-1}) \ldots) \times h + m_1 \tag{6}$$

We divide the plaintext image into four sub-parts by using the quaternary principle; each sub-part is coded separately, and each sub-part is transformed independently on the internet, so the encrypted image of no sub-part is incomplete. Therefore, the information interceptor cannot obtain the original image without any DNA sequence matrix, which increases the difficulty for attackers to crack the original image information and improves the security of the original image information.

For example, let us assume that the first W of the original image, according to Formula (6), is 125, and we choose $h = 4$. In this way, after four modular operations, the value of W is zero. Four position integers $m_1 = 1$, $m_2 = 3$, $m_3 = 3$, $m_4 = 1$ are the results of expression (5), so the value of each sub-section is m_1, m_2, m_3, m_4 individually, and the value of W can be found in reverse according to Formula (6) that $W = 125 = (((0 \times 4 + 1) \times 4 + 3) \times 4 + 3) \times 4 + 1$.

Through the calculation of Formula (5), a grayscale image can get four sub-segments with pixel values of 0, 1, 2 and 3. These four sub-fragments can be expressed by four nucleic acid bases, which are adenine, cytosine, guanine, and thymine, respectively. Among them, adenine is represented by A, cytosine by C, guanine by G and thymine by T. In this paper, Table 1 provides 24 coding schemes. Therefore, by using quaternary and DNA coding, the plaintext image can be divided into four sub-parts, and the grayscale image can be turned into four DNA sequence matrices. These four DNA sequence matrices are got by DNA coding using DNA coding rules. Therefore, using the quaternary image encryption method changes the statistical characteristic of the plain image information.

Table 1. Maping rules.

	0	**1**	**2**	**3**
(1)	C	G	T	A
(2)	C	T	G	A
(3)	G	C	T	A
(4)	G	T	C	A
(5)	T	G	C	A
(6)	T	C	G	A
(7)	A	G	T	C
(8)	T	G	A	C
(9)	G	T	A	C
(10)	G	A	T	C
(11)	T	A	G	C
(12)	A	T	G	C
(13)	C	T	A	G
(14)	C	A	T	G
(15)	T	A	C	G
(16)	A	C	T	G
(17)	A	T	C	G
(18)	T	C	A	G
(19)	G	A	C	T
(20)	G	C	A	T
(21)	C	G	A	T
(22)	C	A	G	T
(23)	A	C	G	T
(24)	A	G	C	T

Figure 2 shows the DNA coding. The DNA coding process is as follows:

A sub-image with a size of 5×5 is obtained from the pixel values of the plaintext image from (208.1) to (212.5).

The second step is to perform four modulo-4 operations on the pixel values, respectively, with the result of the first operation as the pixel value of the first sub-image, the result of the second operation as the pixel value of the second sub-image, and so on until all four sub-images are generated. Finally, according to the coding method, the four sub-image matrices are coded to obtain four DNA sequence matrices. Similarly, the gray values of other parts of the original image can be coded in the same way [17].

In the process of encryption, four DNA sequence matrices are encoded by different rules, so in the process of decoding, four DNA matrices should be decoded by specific rules. Therefore, in order to obtain the original image information, the attacker needs to have four matrix sequences at the same time, all of which are indispensable.

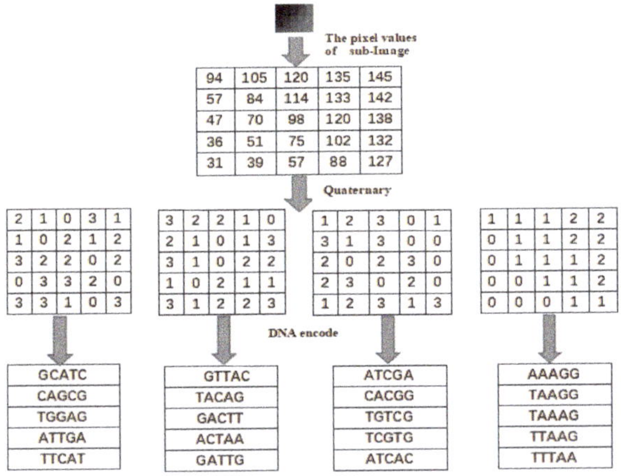

Figure 2. DNA encoding.

2.3.2. XOR Operation for DNA Sequence

Traditional digital calculation methods cannot meet the requirements of calculation. Researchers have put forward some biological calculation methods, such as DNA sequence XOR operation, DNA sequence subtraction operation, and DNA sequence addition operation. The exclusive-or operation of DNA sequence is adopted, which is put forward on the basis of traditional modulo-two operation. In Table 2, the operation rules of DNA sequence XOR operation are listed.

Table 2. XOR rules.

XOR	G	T	C	A
G	C	A	G	T
C	G	T	C	A
T	A	C	T	G
A	T	G	A	C

3. Image Encryption Scheme

There are three stages in the encryption process: (1) transforming a plaintext image into four DNA sequence matrices; (2) generating a 1D quadratic chaotic sequence and diffusing four matrix pixel values through DNA XOR operation; and (3) a mosaic model is introduced and the four matrices are combined into an image matrix.

3.1. *The Basic Theory Introduction*

This subsection will introduce the concrete flow of the image encryption scheme based on a 1D quadratic chaotic system and splicing model. First of all, the plain image is encoded into four sub-regions with pixel values of 0, 1, 2, and 3 by using quaternary. Then, the replaced four sub-images are coded into four DNA sequence matrices by DNA coding rules. During the second step, we use the XOR operation of DNA sequences and chaotic sequences produced by the 1D secondary chaotic system to diffuse pixel values. Ultimately, the pixel values are diffused again through the splicing model, these matrices are combined into one image matrix by using the quaternary system, and in the final stages of the image encryption scheme, the encrypted digital holograph is obtained. Figure 3 displays the process and steps of the encryption scheme described above.

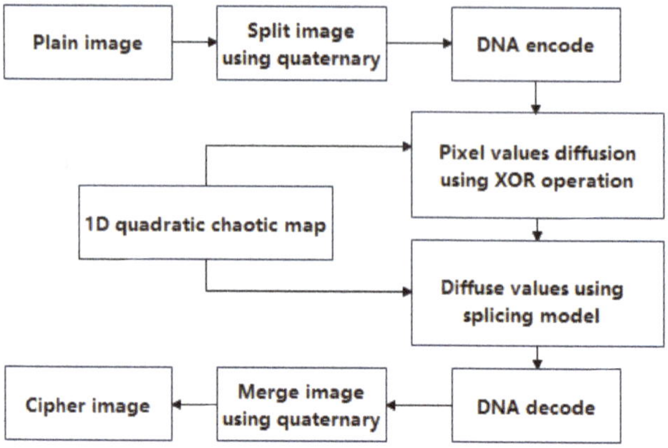

Figure 3. Encryption Process.

3.2. The Generation of Secret Key

By the following operations, the key can be obtained:

1. Read the original image M, which size is M × H.
2. The statistical data can be obtained by the following calculation:

$$p = 10 \times \sum_{i=1}^{M} \sum_{j=1}^{H} a_{ij} / MH \text{mod} 1 \tag{7}$$

3. Set $k = -9/8, n = 1$, and assign values to four parameters m_1, m_2, m_3, and m_4. Randomly select the parameters m_1, m_2, and m_3 in the chaotic region, and $m_4 = (m_1 + m_2 + m_3 + m'_4)/4$ in which m'_4 is randomly picked out in the chaotic region.
4. Additionally, four chaotic sequences are generated $\{x_i\}$, $\{y_i\}$, $\{s_i\}$, and $\{t_i\}$, according to

$$x_{i+1} = f(x_i) = m x_i{}^2 + x_i - 9/8 \tag{8}$$

through using four initial conditions and four sets of parameters $x_0 + p/10$, $y_0 + p/10$, $s_0 + p/10$, and $(x_0 + y_0 + s_0 + t_0)/4 + p/10$, where x_0, y_0, s_0, and t_0 all these parameters are randomly selected in the chaotic region.

We selected the parameters m_1, m_2, m_3, and m'_4, initial keys x_0, y_0, s_0, and t_0 as the secret keys.

3.3. Encryption Process

From Figure 3 above, the specific encryption algorithm process is as follows:

1. Divide the plain image M(m, h) into four sub-images according to the operation Formula (5), and convert them into four sub-matrices HA, HB, HC, HD of size (m, h).
2. According to the coding rules of DNA sequence rules (2), (8), (13), and (19) in Table 1, encode the matrices HA, HB, HC, HD into four DNA sequence matrices FA, FB, FC, FD, respectively.
3. Generate four chaotic sequences $\{x_i\}$, $\{y_i\}$, $\{s_i\}$, and $\{t_i\}$, which are the consequences of the 1D quadratic chaotic system under the condition that initial values are $x_0 + p/10$, $y_0 + p/10$, $s_0 + p/10$, and $(x_0 + y_0 + s_0 + t_0)/4 + p/10$.
4. Scrambling the DNA sequence matrices FA, FB, FC, FD is based on the following formula:

$$\begin{aligned} FA(v,k) &= FA(fx(v), fy(k)); \\ FB(v,k) &= FB(fy(v), fz(k)); \\ FC(v,k) &= FC(fz(v), fq(k)); \\ FD(v,k) &= FD(fq(v), fx(k)); \end{aligned} \tag{9}$$

In which $v = 1,2,\ldots m$, $k = 1,2,\ldots h$, $FA(v,k)$, $FB(v,k)$, $FC(v,k)$, and $FD(v,k)$ are DNA sequence matrices. The values at the (v,k) positions of FA, FB, FC, FD can be scrambled to obtain a new DNA sequence matrix NA, NB, NC, ND.

5. Diffuse the pixel values via chaotic sequences and DNA sequence XOR operation. In addition, we obtain SA, SB, SC, SD, which are the DNA sequence matrices.
6. Taking a column of DNA sequence matrix SA, SB, SC, SD as a subsequence, four one-dimensional arrays QA, QB, QC, QD can be obtained, and then the arrays QA, QB, QC, QD are scrambled by using the idea of the splicing model, following these steps:

   If $x(v) + y(v) < 1$, implement the following formula:

$$QA\{v\} \leftrightarrow QB\{v\} \tag{10}$$

   If $z(v) + q(v) < 1$, implement the following formula:

$$QC\{v\} \leftrightarrow QD\{v\} \tag{11}$$

   The value range of v is an integer from 1 to m; the value of k is an integer from 1 to n.

7. Decoding the DNA sequence matrices QA, QB, QC, QD according to the second DNA decoding rule (6), (11), (18), and (24) in Table 1 can obtain four matrices OA, OB, OC, OD.
8. The matrices of these values are reorganized using Equation (6). Lastly, we got the encrypted digital holograph.

The procedure of decrypt image is the reserve order of encrypt image. In the other words, the encrypted image is complemented as the contrary operations of encryption algorithm, and the only change is that the secret image is used in Step 2 among the decryption algorithm.

4. Experiment and Analysis

4.1. Exhaustive Attacks

4.1.1. Analysis of Key Space

It is very significant for the robustness of the image encryption scheme that the capacity of key space. If the capacity of the key space is small, it cannot resist the exhaustive attack. The key space represents the total number of selectable keys in the image password. In the image encryption algorithm, eight adjustable parameters, including the parameters m_1, m_2, m_3, m_4' and initial key x_0, y_0, s_0, t_0 are chosen as secret keys. Presume that the maximum calculation accuracy is 10^{-x}. According to the value ranges of the eight adjustable parameters, the image encryption algorithm's key space is calculated as follows $(10^{x-1} \times 0.53)^4 \times (10^{x-1} \times 0.85)^4 = 10^{8x-10} \times 4.12$. If the operational precision $x = 14$, the capacity of the key space is $4.12 \times 10^{102} \approx 2^{340}$. The calculation results turn out that the key space of the scheme is big enough to effectively resist exhaustive attacks. In Table 3, the key space size of our scheme is compared with that of other documents.

Table 3. Key space.

	Our Method	In Ref. [1]	In Ref. [17]	In Ref. [19]	In Ref. [21]
Key space	2^{340}	2^{209}	2^{100}	2^{340}	2^{261}

4.1.2. Key Sensitivity

Obviously, the encryption method using a 1D quadratic chaotic system put forward in the dissertation is sensitive to all initial keys, under the condition that we cannot obtain the plain image result from a small modification to input conditions. Figure 4 demonstrates the conclusions of the key sensitivity test, and decrypted digital holography under only 10^{-14} inappreciable difference in its secret keys m_1, m_2, m_3, m_4', $s_0, t_0 x_0$, and y_0, respectively.

We can sum up that the original image information can be extracted only if the secret keys are consistent. The decrypted digital holograph cannot reflect the true information of the plaintext image if any small change in the primary key values. Therefore, our scheme has a greater level of security and can withstand exhaustive attack efficiency.

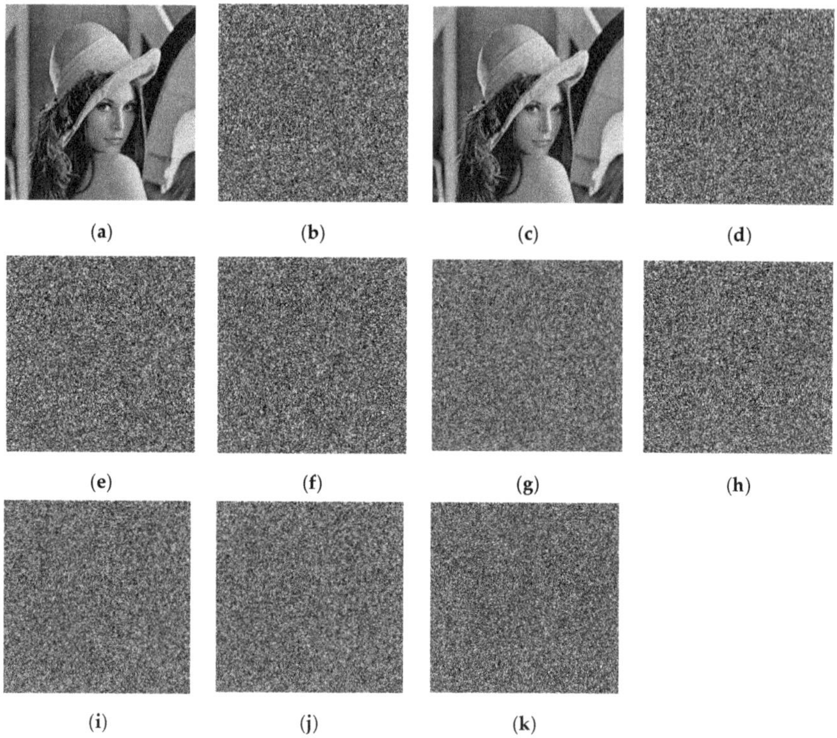

Figure 4. (**a**) "Lena" image; (**b**) cipher image (initial encryption key); (**c**) decrypted image (initial encryption key); (**d**) $m_1 + 10^{-14}$; (**e**) $m_2 + 10^{-14}$; (**f**) $m_3 + 10^{-14}$; (**g**) $m'_4 + 10^{-14}$; (**h**) $x_0 + 10^{-14}$; (**i**) $y_0 + 10^{-14}$; (**j**) $s_0 + 10^{-14}$; (**k**) $t_0 + 10^{-14}$. Key sensitivity test: (**d–k**) Decrypted image with the wrong key.

4.2. Statistical Attacks

4.2.1. Gray Histogram

A gray histogram describes each pixel value's frequency in a gray image. Typically, original image pixel values are concentrated on some specific gray values, and encrypted pixel values of the image are evenly distributed on all gray values. The gray histogram of the original image and encrypted digital holograph are demonstrated in Figure 5. From the figure, the distribution of pixel values in the original image is uneven, mainly concentrated on several gray values. However, the pixel values of the encrypted digital holograph are relatively evenly distributed on all gray values. The image encryption system has influenced and changed the distribution of pixel values. The algorithm with a high sense of resisting statistical attacks, which ensures the security of images in the process of transmission.

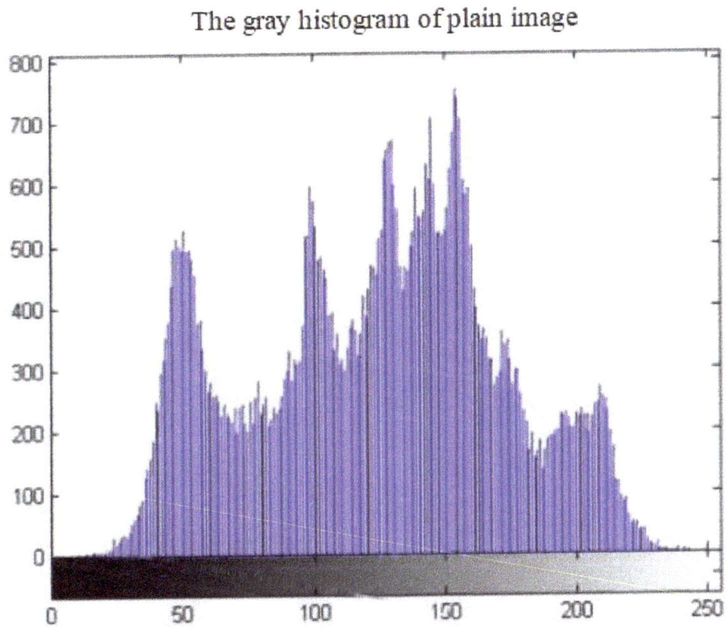

(**a**) The gray histogram (original image)

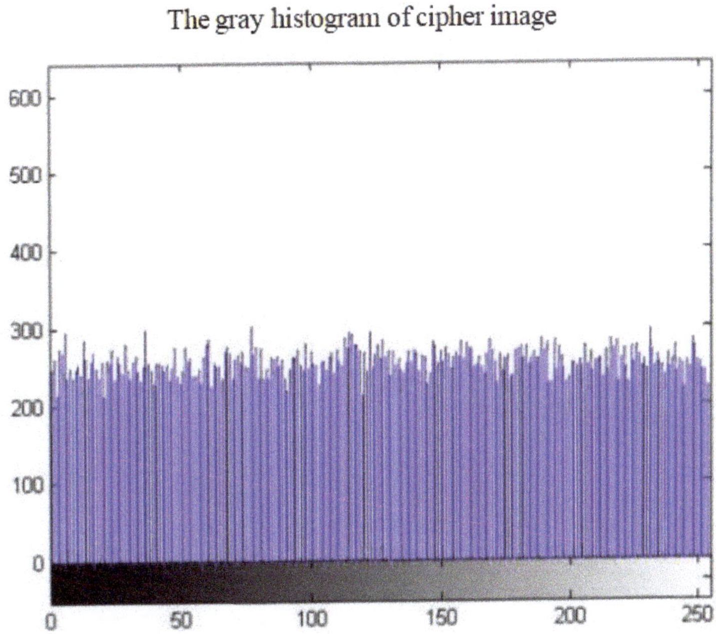

(**b**) The gray histogram (encrypted holograph)

Figure 5. Gray histogram analysis.

4.2.2. Correlation Coefficient Analysis

The quality of scrambling and diffusion of the image encryption system can be expressed by calculating the relationship between adjacent pixels of the encrypted digital holograph. The greater the degree of encryption scrambling and diffusion, the smaller the correlation coefficient of the neighboring pixels of the encrypted digital holograph, indicating that the relationship of the adjacent pixels of the encrypted digital holograph is weaker. If the calculated values of neighboring pixels in the original image show a linear distribution, the correlation between neighboring pixels will be strong. The distribution of neighboring pixel values of the encrypted image should be irregular, and the correlation between neighboring pixels should be weak. When the correlation coefficient of the encrypted digital holograph is close to zero, it shows that the encryption scheme has good robustness.

The correlation coefficient r_{st} of neighboring pixels of the image may be calculated by the subsequent formula.

$$P(s) = \frac{1}{H}\sum_{k=1}^{H} s_k \tag{12}$$

$$Q(s) = \frac{1}{H}\sum_{k=1}^{H} (s_k - E(s))^2 \tag{13}$$

$$\mathrm{cov}(s,t) = \frac{1}{H}\sum_{k=1}^{H} (s_k - E(s))(t_k - E(t)) \tag{14}$$

$$r_{st} = \frac{\mathrm{cov}(s,t)}{\sqrt{Q(s) \times Q(t)}} \tag{15}$$

The pixel values of two adjacent pixels in the image are denoted by s and t, respectively, and $\mathrm{cov}(s,t)$ is covariance, $P(s)$ is mean, $Q(s)$ is variance.

First, 1000 pairs of adjacent pixels were selected from the original image, and the correlation was calculated in horizontal, vertical, and diagonal directions. Similarly, 1000 pairs of adjacent pixels were selected in the same position in the encrypted image, and the correlation was calculated in horizontal, vertical, and diagonal directions again. Figure 6 correlation coefficient analysis demonstrates the relationship between the two horizontally adjacent pixels in the original image and in the encrypted digital holograph is very different. From Figure 6a, the pertinence of two horizontally adjacent pixels is strong. From Figure 6b, the pertinence of two horizontally adjacent pixels is weak.

From Table 4 below, it can be concluded that the correlation coefficient between two neighboring pixels of the encrypted image with the original image of "lenna.bmp" is close to 0, and the relationship between neighboring pixels of the image is weak. By comparing the correlation between neighboring pixels of the original image and the encrypted image, the following conclusions can be drawn. In the encryption algorithm, a 1D quadratic chaotic system was used to generate the key and scramble the image, and the mosaic model was introduced to participate in scrambling the image. The correlation between adjacent pixel values of the scrambled image was very low. It showed that the encryption algorithm can effectively resist statistical attacks.

Table 4. Correlations coefficients.

Direction	Lenna	Cipher Image	In Ref. [1]	In Ref. [5]	In Ref. [13]	In Ref. [17]
Horizontal	0.9277	0.0015	−0.0062	−0.0020	−0.0119	0.0015
Vertical	0.9168	−0.0021	−0.0001	−0.0065	−0.0087	0.0018
Diagonal	0.8871	−0.0020	0.0018	0.0087	−0.0045	0.0018

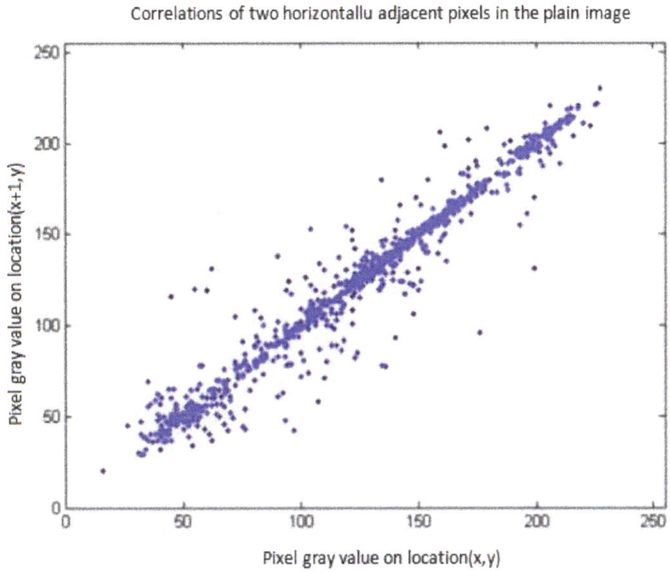

(**a**) lenna.bmp

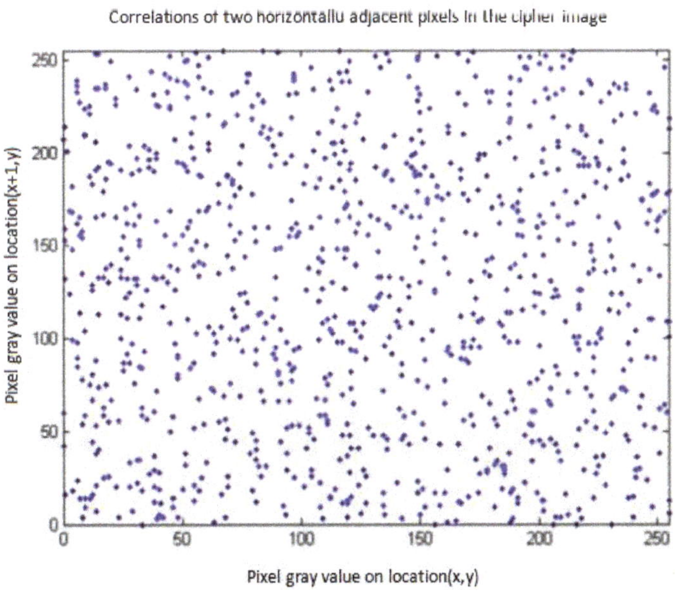

(**b**) Cipher image

Figure 6. Correlation coefficient analysis.

4.3. Differential Attacks

The calculation results of the following formula can measure the ability of the encryption algorithm to resist differential cryptanalysis. The change rate of image pixel number (the number of pixels change rate, NPCR) is calculated by the Formula (16), and the even average change intensity of the image (the unified average changing intensity, UACI) is calculated by the Formula (17). The magnitude of these values reflects the ability of the encryption algorithm to resist differential cryptanalysis. The larger these two values are, the more sensitive the image encryption algorithm is to small changes in gray images.

$$NPCR = \frac{\sum_{s=1}^{H}\sum_{t=1}^{U} C(s,t)}{H \times U} \times 100\% \tag{16}$$

$$UACI = \frac{\sum_{s=1}^{H}\sum_{t=1}^{U} |T_1(s,t) - T_2(s,t)|}{H \times U \times 255} \times 100\% \tag{17}$$

where H, U are the size of cipher image, $T_1(s,t)$ represents the pixel value of one ciphertext image at (s,t) position, and $T_2(s,t)$ represents the pixel value of another ciphertext image at the same position. $C(s,t)$ is determined as

$$C(s,t) = \begin{cases} 0, & if \quad T_1(s,t) = T_2(s,t); \\ 1, & if \quad T_1(s,t) \neq T_2(s,t); \end{cases} \tag{18}$$

NPCR and UACI analysis of the 256 × 256 Lena image and Baboon image were carried out by existing methods. The values in Table 5 show the approximate theoretical values. It can be concluded that the image encryption algorithm based on the 1D quadratic chaotic system and splicing model excellent in resisting differential cryptanalysis.

Table 5. UACI and NPCR of our innovate algorithm and other algorithms.

	UACI	NPCR
Lena	33.4685%	99.6092%
Baboon	33.4687%	99.6089%
In Ref. [1] (Lena)	33.48%	99.61%
In Ref. [5] (Lena)	33.4477%	99.6063%
In Ref. [14] (Lena)	33.4645%	99.6096%
In Ref. [17] (Lena)	33.505%	99.571%
In Ref. [21] (Lena)	34.61%	99.65%

4.4. Information Entropy

In information theory, information entropy refers to the average amount of information received, and it can also represent the unpredictability and uncertainty of image information. Information entropy is also an index to measure the quality of the image encryption scheme. If the information entropy is close to 8, it indicates that the image encryption algorithm is excellent. If the entropy of an image encryption algorithm is far less than 8, the encryption scheme has certain security problems. The information entropy of an encrypted digital hologram can be calculated according to the Formula (19).

$$P(X) = -\sum_{i=0}^{m} Q(x_i) \log_2 Q(x_i) \tag{19}$$

x_i is the value of the ith position of the grayscale image, the $Q(x_i)$ is the frequency of x_i's appearance, and m is the size of the grayscale [22]. The following Table 6 shows the information entropy values of the encrypted digital holograph in this thesis and those of encrypted images under other algorithms.

Table 6. The entropy analysis.

Images	In Ref. [1]	In Ref. [8]	In Ref. [17]	In Ref. [21]	Our Method
Lena	7.9978	7.9971	7.9971	7.9975	7.9994
Baboon	7.9974	7.9973	/	/	7.9991

By comparing the values in Table 6, indicated that the encryption algorithm proposed in this thesis is very competitive. According to our encryption algorithm, the information entropy of encrypted digital holography is 7.9994 and 7.9991, respectively, which shows that the algorithm is excellent because the value is infinite and close to the theoretical value of 8.

4.5. Encryption Speed Test

In the proposed algorithm, the plaintext image is divided into four matrices, which can be encrypted at the same time, and four cycles are parallel, so the total number of cycles is $1/4(M+N)$, and this algorithm's time complexity is chiefly expressed as $O(1/4(M+N))$. The number of cycles of the traditional image encryption algorithm with a single pixel as the processing unit is equal to the number of pixels, and the number of cycles is $(M \times N)$. Therefore, the time complexity of this kind of encryption algorithm is $O(M \times N)$. This algorithm significantly improves the encryption speed compared with the encryption algorithm in references. In this thesis, the experimental diagram Lena was decrypted in the experimental environment, and its running time is shown in Table 7. The actual running efficiency of the encryption algorithm was influenced by many factors, such as running environment and programming skills, so the specific running time of the algorithm was not compared, but the time complexity of the algorithm was compared.

Table 7. Encryption speed test.

	In Ref. [2]	In Ref. [14]	Our Method
Time complexity	$O(6N^2)$	$O(M \times N)$	$O(1/4(M+N))$

5. Conclusions

This article presents the digital image encryption system based on a 1D quadratic chaotic system and splicing model. Firstly, the plaintext image was divided into four sub-parts by using the quaternary principle, and each sub-part was coded separately. If an attacker wants to obtain the original image, he must have all the sub-parts at the same time, which increases the difficulty for the attacker to crack the image. In addition, the encryption system encrypted the image using 1D quadratic chaotic mapping, which not only increased the key space of the algorithm but also improved the randomness. Finally, the mosaic model was introduced in the process of digital image encryption to ensure the security of the algorithm. Security analysis and experimental results show that the encryption scheme is not only highly secure, but also resistant to various attacks from the outside world, for instance, statistical attacks, exhaustive attacks, and score-checking attacks and has good robustness.

Author Contributions: Data curation, formal analysis, C.C.; software, validation, C.C. and L.Z.; supervision, D.Z.; writing—review and editing, C.C. and X.W. All authors have read and agreed to the published version of the manuscript.

Funding: This work was supported in part by the National Natural Science Foundation of China under Grant numbers 62272418 and 62002046.

Institutional Review Board Statement: Not applicable.

Informed Consent Statement: Not applicable.

Data Availability Statement: Dataset used in this study may be available on demand.

Conflicts of Interest: The authors declare no conflict of interest.

References

1. Zhu, S.; Deng, X.; Zhang, W. A New One-Dimensional Compound Chaotic System and Its Application in High-Speed Image Encryption. *Appl. Sci.* **2021**, *11*, 11206. [CrossRef]
2. Li, T.Y.; Shi, J.Y.; Zhang, D.H. Color image encryption based on joint permutation and diffusion. *J. Electron. Imaging* **2021**, *30*, 013008. Available online: https://www.spiedigitallibrary.org/journals/journal-of-electronic-imaging/volume-30/issue-1/013008/Color-image-encryption-based-on-joint-permutation-and-diffusion/10.1117/1.JEI.30.1.013008.full?SSO=1 (accessed on 12 January 2022). [CrossRef]
3. Zhu, D.; Huang, Z.; Liao, S. Improved Bare Bones Particle Swarm Optimiztion for DNA Squence Dsign. *IEEE Trans. NanoBiosci.* **2022**, *35*. Available online: https://ieeexplore.ieee.org/document/9943286 (accessed on 9 December 2022).
4. Li, Z.; Peng, C.; Tan, W.; Li, L. A novel chaos-based color image encryption scheme using bit-level permutation. *Symmetry* **2020**, *12*, 1497. Available online: https://www.mdpi.com/2073-8994/12/9/1497 (accessed on 9 February 2022). [CrossRef]
5. Geng, S.T.; Tao, W.; Wang, S.D. A novel image encryption algorithm based on chaotic sequences and cross-diffusion of bits. *IEEE Photon.* **2021**, *13*, 6276–6281. Available online: https://ieeexplore.ieee.org/stamp/stamp.jsp?arnumber=9291442 (accessed on 9 February 2022).
6. Li, T.; Yang, M.; Wu, J.; Jing, X. A Novel Image Encryption Algorithm Based on a Fractional-order Hyperchaotic Sysetem and DNA Computing. *Complexity* **2017**, *2017*, 9010251. Available online: https://www.hindawi.com/journals/complexity/2017/9010251/ (accessed on 13 February 2022).
7. Zhu, S.; Deng, X.; Zhang, W.; Zhu, C. Image Encryption Scheme Based on Newly Designed Chaotic Map and Parallel DNA Coding. *Mathematics* **2023**, *11*, 231. Available online: https://www.mdpi.com/2227-7390/11/1/231 (accessed on 1 February 2023). [CrossRef]
8. Zou, C.; Wang, X.; Zhou, C. A novel image encryption algorithm based on DNA strand exchange and diffusion. *Elsevier Appl. Math. Comput.* **2022**, *430*, 127291. Available online: https://www.sciencedirect.com/science/article/abs/pii/S0096300322003654 (accessed on 13 December 2022). [CrossRef]
9. Dua, M.; Wesanekar, A.; Gupta, V. Differential evolution optimization of intertwining logistic map-DNA based image encryption technique. *J. Amb. Intell. Human. Comput.* **2020**, *11*, 3771–3786. Available online: http://link.springer.com/article/10.1007/12652-019-01580-z (accessed on 9 February 2022). [CrossRef]
10. Soni, R.; Johar, A.; Soni, V. An Encryption and Decryption Algorithm for Image Based on DNA. In Proceedings of the 2013 International Conference on Communication Systems and Network Technologies, Gwalior, India, 6–8 April 2013; Volume 12, pp. 478–481. Available online: https://ieeexplore.ieee.org/abstract/document/6524442 (accessed on 9 February 2022).
11. Gupta, S.; Jain, A. Efficient Image Encryption Algorithm Using DNA Approach. In Proceedings of the 2015 2nd International Conference on Computing for Sustainable Global Development, INDIACom, New Delhi, India, 11–13 March 2015; Volume 8, pp. 726–731. Available online: https://ieeexplore.ieee.org/abstract/document/7100345 (accessed on 13 February 2022).
12. Som, S.; Kotal, A.; Chatterjee, A.; Dey, S.; Palit, S. A Colour Image Encryption Based on DNA Coding and Chaotic Sequences. *ICETACS* **2013**, *112*, 108–114. Available online: https://ieeexplore.ieee.org/document/6691405 (accessed on 13 February 2022).
13. Liu, Q.; Liu, L.F. Color image encryption algorithm based on DNA coding and double chaos system. *IEEE Access* **2020**, *35*, 3581–3596. Available online: https://ieeexplore.ieee.org/document/9082588 (accessed on 6 March 2022). [CrossRef]
14. Zhang, Q.Y.; Han, J.T.; Ye, Y.T. Multi-image encryption algorithm based on image hash, bit-plane decomposition and dynamic DNA coding. *IET Image Proc.* **2020**, *68*, 726–731. Available online: https://ietresearch.onlinelibrary.wiley.com/doi/full/10.1049/ipr2.12069 (accessed on 6 March 2022). [CrossRef]
15. Matthews, R. On the derivation of a Chaotic encryption algorithm. *Cryptologia* **1989**, *13*, 29–42. Available online: https://www.tandfonline.com/doi/abs/10.1080/0161-11899186374 (accessed on 13 February 2022). [CrossRef]
16. Belazi, A.; El-Latif, A. A simple yet efficient S-box method based on chaotic sine map. *Optik* **2017**, *130*, 1438–1444. Available online: https://www.sciencedirect.com/science/article/abs/pii/S0030402616314887 (accessed on 6 March 2022). [CrossRef]
17. Niu, H.; Zhou, C.; Wang, B.; Zheng, X.; Zhou, S. Splicing Model and Hyper- Chaotic System for Image Encryption. *J. Electr. Eng.* **2016**, *67*, 78–86. Available online: https://sciendo.com/article/10.1515/jee-2016-0012 (accessed on 13 February 2022). [CrossRef]
18. Zhu, X.S.; Liu, H.; Liang, Y.R. Image encryption based on Kronecker product over fifinite fifields and DNA operation. *Optik* **2020**, *224*, 164725. Available online: https://www.sciencedirect.com/science/article/abs/pii/S0030402620305611 (accessed on 6 March 2022). [CrossRef]
19. Liu, L.F.; Wang, J. A cluster of 1D quadratic chaotic map and its applications in image Encryption. *Math. Comput. Simul.* **2022**, *204*, 89–114. Available online: https://www.sciencedirect.com/science/article/abs/pii/S0378475422003329 (accessed on 6 March 2022). [CrossRef]
20. Tom, H. Splicing and Regularity. *Bull. Math. Biol.* **1987**, *49*, 737. Available online: https://www.sciencedirect.com/science/article/abs/pii/S0092824087900188 (accessed on 13 February 2022). [CrossRef]

21. Zheng, J.; Hu, H.P. A symmetric image encryption scheme based on hybrid analog-digital chaotic system and parameter selection mechanism. *Multimed. Tools Appl.* **2021**, *27*, 176–191. Available online: https://link.springer.com/article/10.1007/s11042-021-10751-0 (accessed on 13 February 2022). [CrossRef]
22. Kamarposhti, M.S.; Mohammad, D.; Rahim, M.; Yaghobi, M.I. Using 3-cell chaotic map for image encryption based on biological operations. *Nonlinear Dyn.* **2014**, *7*, 407–416. Available online: https://link.springer.com/article/10.1007/s11071-013-0819-6 (accessed on 13 February 2022). [CrossRef]

electronics MDPI

Article

Design Science Research Framework for Performance Analysis Using Machine Learning Techniques

Mihaela Muntean * and Florin Daniel Militaru

Business Information Systems Department, Faculty of Economics and Business Administration, West University of Timisoara, 300223 Timisoara, Romania
* Correspondence: mihaela.muntean@e-uvt.ro

Abstract: We propose a methodological framework based on design science research for the design and development of data and information artifacts in data analysis projects, particularly managerial performance analysis. Design science research methodology is an artifact-centric creation and evaluation approach. Artifacts are used to solve real-life business problems. These are key elements of the proposed approach. Starting from the main current approaches of design science research, we propose a framework that contains artifact engineering aspects for a class of problems, namely data analysis using machine learning techniques. Several classification algorithms were applied to previously labelled datasets through clustering. The datasets contain values for eight competencies that define a manager's profile. These values were obtained through a 360 feedback evaluation. A set of metrics for evaluating the performance of the classifiers was introduced, and a general algorithm was described. Our initiative has a predominant practical relevance but also ensures a theoretical contribution to the domain of study. The proposed framework can be applied to any problem involving data analysis using machine learning techniques.

Keywords: design science research; performance analysis; machine learning; classification algorithms; clustering algorithms

Citation: Muntean, M.; Militaru, F.D. Design Science Research Framework for Performance Analysis Using Machine Learning Techniques. *Electronics* **2022**, *11*, 2504. https://doi.org/10.3390/electronics11162504

Academic Editors: Taiyong Li, Wu Deng and Jiang Wu

Received: 20 July 2022
Accepted: 8 August 2022
Published: 11 August 2022

1. Introduction

Design science research is a research paradigm with well-established conceptualizations applicable in engineering and, more recently, in the field of information systems.

According to Pfeffers et al. [1], design science research (DSR) is important in disciplines oriented towards the creation of successful artifacts. In data analysis, key artifacts are the "useful data artifacts" (UDA) and data-related information artifacts [2]. UDAs are "nonrandom subsets or derivative digital products of a data source, created by an intelligent agent (human or software) after performing a task on the data source", e.g., a labelled dataset or train and test dataset, while information artifacts refer to the objectives of the solution and requirements for final data visualizations or data specifications. Based on the importance of data/information artifacts in data analysis, we propose the design and development of a DSR process in this field of investigation.

Performance measurement is "the process of collecting, analyzing, and/or reporting information regarding the performance of an individual, group, organization, system, or component" [3]. According to Stroet [4], performance measuring is influenced by the usage of machine learning (ML) techniques "in a way that it becomes more accurate through the use of more current and accurately collected data, performance data are gathered easier, is done more continuous, is less biased and done with a more proactive attitude than before ML was implemented in the process". Managers and employees are frequently evaluated using 360-degree feedback. In general, 360 feedback focuses on behaviors and competencies more than basic skills, job requirements, and performance objectives. Therefore, the 360 feedback is incorporated into a larger performance management process and it is clearly "communicated on how the 360 feedback will be used". Because 360-feedback is time-consuming, the use of machine learning techniques for analyzing performance data determines the fluidization of the entire process, and the evaluation results are obtained in real time [4,5].

The process is a priori reviewed with staff members, and is started by collecting confidential information from managers' colleagues and sending the evaluation form to be completed by the employees [6]. Data are automatically collected and integrated into a single dataset. Further, mean values for each competence for all evaluated managers are calculated. The resulting dataset is subjected to analysis using machine learning algorithms.

The paper develops a theoretical applied research discourse based on:

- a methodological framework using design science research (DSR) for data analysis with machine learning techniques, such as classification algorithms;
- a theoretical approach to classification evaluation metrics;
- a set of competencies for evaluating the managers' performance using 360 feedback;
- an approach to apply the methodological framework to a performance related dataset.

2. Materials and Methods

2.1. Machine Learning Techniques—Clustering and Classification

The study of machine learning (ML) led to the development of many methods depending on the purpose, data representation, and learning strategy. Depending on the experience gained during the learning, we distinguish supervised, unsupervised, or semi-supervised learning methods. In addition, learning can occur through reinforcement or through "learning" [7]. According to El Bouchefry and De Souza [8], "ML algorithms are programs of data-driven inference tools that offer an automated means of recognizing patterns in high-dimensional data". Supervised algorithms search for inherited structures in a dataset, whereas unsupervised algorithms provide the correct labels or function values.

Both clustering and classification algorithms are proven to be successful in different analyses [9]. Classification algorithms require labelled datasets to perform the learning process. We proposed applying classification algorithms to datasets that were previously subjected to a clustering process. According to Alapati and Sindhu [10], the accuracy of a classifier can be improved by applying a classification algorithm to clustered data. We propose the following phases for the prediction analysis and performance prediction (Figure 1).

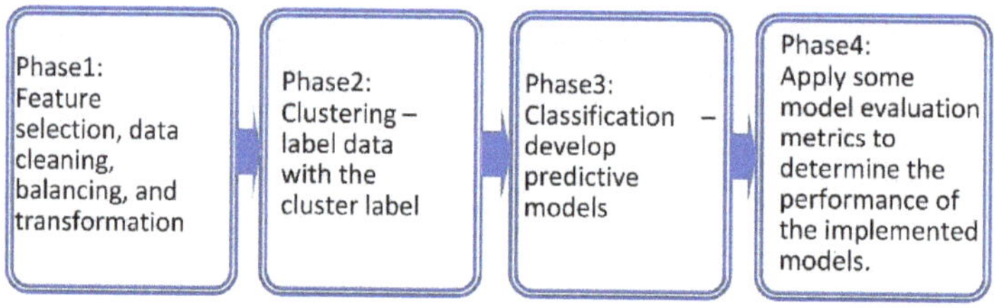

Figure 1. Prediction analysis phases.

To evaluate the quality of the classification, the performance of the classifier was analyzed, regardless of whether it may be, with the help of the following measures: sensitivity, specificity, accuracy, and F1 score [11,12].

2.1.1. Feature Selection

Not all attributes or features are important for a specific learning task. The challenging task in feature selection is to obtain an optimal subset of relevant and non-redundant features, which will provide an optimal solution without increasing the complexity of the modelling task [13]. According to Dash and Koot [14], for clustering tasks, it is not so obvious which features are to be selected: some of the features may be redundant, some are irrelevant, and others may be "weakly relevant". In the context of classification, feature selection techniques can be categorized as filter methods (ANOVA, Pearson correlation, and variance thresholding), wrapper methods (forward, backward, and stepwise selection), embedded methods (LASSO, RIDGE, and decision tree), and hybrid methods [15]. All feature selection methods help reduce the dimensionality of the data and the number of variables, while preserving the variance of the data.

2.1.2. Clustering

Clustering is an unsupervised learning problem that involves finding a structure in a collection of unlabelled data. A cluster is "a collection of objects that are similar between them and dissimilar to objects belonging to other clusters" [16]. Clustering algorithms can be classified as hierarchical, partitioning, density, grid, or model-based (Figure 2). According to Witten, Frank, Hall, and Pal [17], a cluster contains instances that bear a stronger resemblance to each other than to other instances.

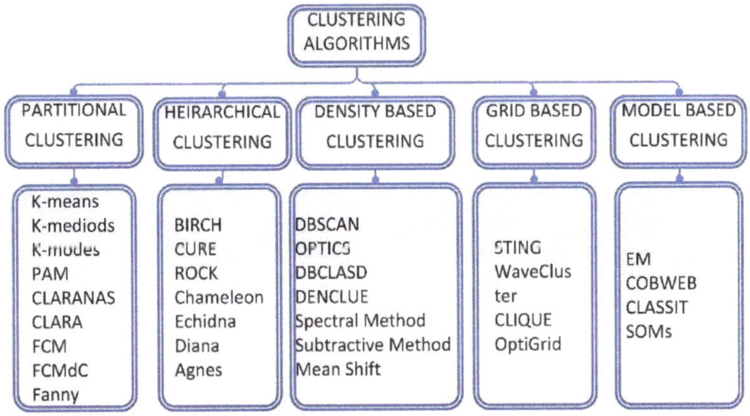

Figure 2. Clustering algorithms [18].

Partitional clustering algorithms divide datasets into mutually disjointed partitions. Data points are assigned to K clusters using an iterative process [19]. The partitional clustering techniques start with randomly chosen clustering, and then optimize the clustering according to the accuracy measurements. Owing to its simplicity and low time complexity, the K-means algorithm is commonly used for mining data and labeling them with cluster labels [20]. This requires pre-defining the number of clusters K, and the optimal K value is determined a priori [21]. Determining the optimal number of clusters is fundamental for clustering. According to Loukas [22], the optimal number of clusters depends on the method used for measuring similarities and the parameters used for partitioning (the elbow method, silhouette analysis, and gap statistics method).

Hierarchical clustering can be divided into two types: agglomerative (bottom-up) and divisive (top-down) clustering. Data objects (instances) are organized into a tree of clusters called a dendrogram. Each intermediate level can be viewed as combining two clusters from the next lower level (bottom-up) or splitting a cluster from the next higher level (top-down) [23]. Frequently applied in the construction of taxonomies, hierarchical clustering requires considerable computational and storage resources for deploying the dendrogram. Unfortunately, once a merge or split step is performed, it cannot be undone. Therefore, it is recommended to integrate hierarchical clustering with other techniques for multi-phase clustering.

Density-based clustering algorithms identify distinctive clusters in the data based on the idea that "a cluster in a data space is a contiguous region of high point density", separated from other such clusters by contiguous regions of low point density [24]. The algorithms detect areas where points are concentrated, and where they are separated by areas that are empty or sparse.

Grid-based approaches are popular for mining clusters in large multidimensional spaces, in which clusters are regarded as denser regions than their surroundings. Such an algorithm is concerned not with data points but with the value space that surrounds them [25].

Finally, model-based clustering assumes that data are generated by a model, and attempts to recover the original model from the data.

2.1.3. Classification

Classification algorithms are supervised learning techniques that are used to identify the category (class) of new data. The classification involves the following processing phases (Figure 3).

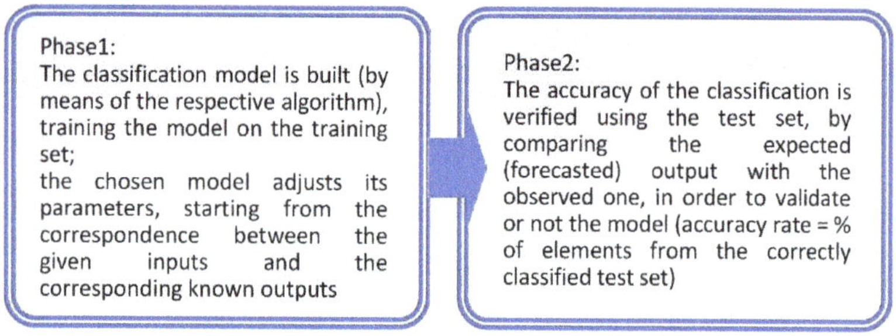

Figure 3. Classification process.

Among the most well-known models (methods) used for classification, we can mention the following [26]: decision trees, Bayesian classifiers, neural networks, k-nearest neighbor classifiers, statistical analysis, genetic algorithms, rough sets, rule-based classifiers, memory-based reasoning, support vector machines (SVMs), and boosting algorithms.

Binary classification (Figure 4) refers to classification tasks that have only two class labels (k-nearest neighbors, decision trees, support vector machines, and naive Bayes), whereas multiclass classification refers to classification tasks that have more than two class labels (k-nearest neighbors, decision trees, naive Bayes, random forest, and gradient boosting).

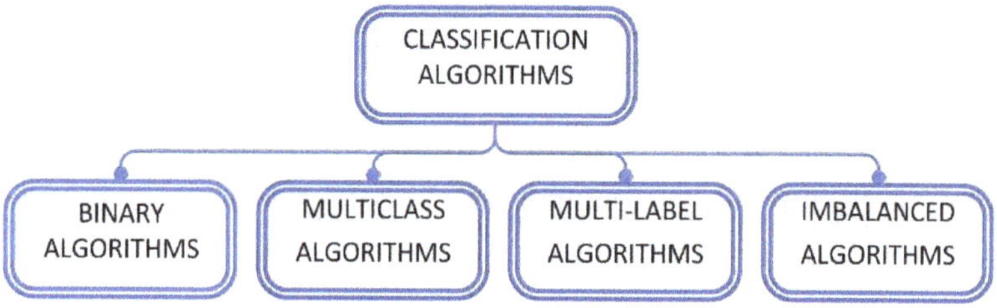

Figure 4. Classification algorithms.

A multi-label classifier can predict one or more labels for each data instance (multi-label decision trees, multi-label random forests, and multi-label gradient boosting). Unbalanced classification processes determine the classification of an unequal number of instances into classes (cost-sensitive logistic regression, cost-sensitive decision trees, and cost-sensitive support vector machines).

According to [27], it is necessary to first identify business needs and then map them to the corresponding machine learning tasks (Figure 5). After establishing the business requirements, the requirements for the machine learning algorithm were established. Characteristics, such as the accuracy of the algorithm, training time, linearity, number of parameters, and number of features influence the classifier selection [5]. The accuracy reflects the effectiveness of a model, that is, the proportion of true results in all cases. The training time varies from one classifier to another. Many machine learning algorithms use linearity. The parameters are the values that determine the algorithm behavior, and a large number of features substantially influence the training time [28]. Classification performance can be improved by mixed approaches [29].

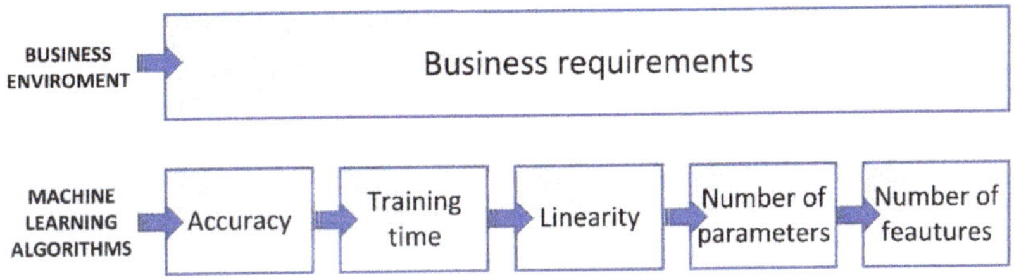

Figure 5. Criteria for selecting machine learning algorithms [28].

2.2. Design Science Research

According to Nunamaker et al. [30], research is "represented by its objectives and methods, whereby the objectives require a methodological approach to integrate theory building, system development, and experimentation". On a theoretical scale (Figure 6), the degree of theoretical importance is represented on one side versus the practical relevance on the other side [31].

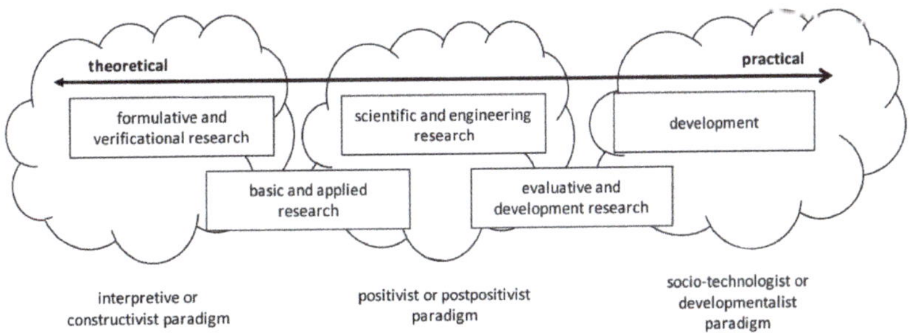

Figure 6. Research paradigm [30].

Research in information systems and data science implies an interdisciplinary research process that fits more than one paradigm [31].

Design science research (DSR) is a paradigm that is accepted in disciplines, such as engineering. This research paradigm is extended to information systems and data science [32]. As asserted by Hevner et al. [32], guidelines for design science research include methodological choices for the DSR process.

Several research methodologies were developed to support the DSR process [33]. The main methodologies are the systems development research methodology (SDRM) [30], DSR process model (DSRPM) [34], design science research methodology (DSRM) [7], action design research (ASR) [35], soft design science methodology (SDSM) [36], and participatory action design research (PADR) [37].

According to Nunamaker et al. [33], SDRM is a five-step research process that includes the following steps: constructing a conceptual framework, developing a system architecture, analyzing and designing the system, building the (prototype) system, and observing and evaluating the system.

In their "Design Research in Information Systems", Vaishnavi and Kuechler [34] explain the process steps of design research. By pointing out the importance of artifacts, the DSR process includes the following steps: awareness of the problem, suggestion, development, evaluation, and conclusion.

Peffers et al. [1] proposed a six-step design science research methodology: identifying the problem and motivation, defining the objectives of a solution, design and development, demonstration, evaluation, and communication. DSR methodology is "an artifact-centric creation and evaluation approach" [1,34]. The research methodology implies the design cycle of "artifacts of practical value to either the research or professional audience" [38,39]. Artifacts are systems, applications, methods, data models, data sets, and others "that could contribute to the efficacy of information systems and business analysis in organizations" [40].

ADR methodology combines action research with DSR [33]. It includes four phases: problem formulation, building intervention and evaluation, reflection and learning, and formalization of learning [35].

The eight activities of SDSM are: learning about a specific problem, inspiring and creating the general problem and general requirements, intuiting the general solution, general evaluation, designing specific solution for specific problem, specific evaluation, constructing specific solution, and post evaluation [33,36].

The PADR methodology is recommended for developing solutions to problems involving large heterogeneous groups of stakeholders [33,37]. It consists of the following steps: diagnosis and problem formulation, action planning, action taking: design, impact evaluation, and reflection and learning.

Based on the DSRM and DSRPM, we recommend the methodological framework shown in Figure 7 for performing data analysis.

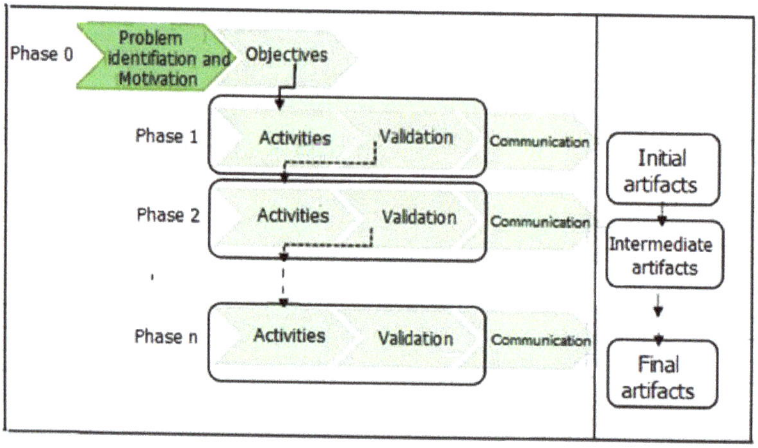

Figure 7. Design science research framework.

The activities shown in Figure 7 indicate the design and development of the artifacts. Furthermore, the artifacts are evaluated, and after validation, they are e communicated and processed in the next phase [41]. Artifact evaluation provides a better interpretation of the problem and feedback to improve the quality of designed artifacts [42].

Owing to its focus on developing information artifacts, DSR is a research approach with a predominant practical relevance. Artifacts are designed and developed in order to improve business activities, processes, or to support decisions. Therefore, the targeted business beneficiaries of the artifacts are involved in their testing and validation [31].

3. Methods

3.1. Artifacts Development in Design Science Research

"Current design science research method does not have a systematic methodological process to follow in order to produce artifacts" [43]. In general, the following research methods, techniques and tools are used for artifact design and development (Table 1).

Table 1. DSR process. Research methods, techniques and tools.

Phase	Activity	Research Methods, Techniques and Tools
Phase 0	Problem identification and Motivation Objectives establishment	Brainstorming, systems review and analysis, literature study, interviews, focus group
Phase 1 Phase n	Activities	Literature review, system analysis, field and case study. artifacts engineering
	Validation	Simulation, informed argument, controlled experiment, case study, field study
	Communication	Communication framework

We propose an approach to prediction analysis (Figure 1) in a DSR framework (Figure 7) using appropriate research methods, techniques and tools (Table 1).

Artifact engineering using machine learning techniques implies a set of activities and tasks that are highlighted in Table 2. The initial, intermediate, and final artifacts were established for each phase.

Table 2. DSR process. Using machine learning techniques.

Phase	Activity/Task	Artifact (Output)
Ph0. Phase 0	A01. Problem identification and Motivation A02. Objectives establishment	O01. Objectives of the solution O02. Requirements for final data visualizations
Ph1. Phase 1	A11. Understanding the data Task1. Establishing the data that are necessary for the business analysis, Task2. Identifying the issues that affect the data quality A12. Validation of the information artifacts A13. Communicating the result	O11. Data specifications
Ph2. Phase 2	A21. Data set design and development Task1. Accessing data sources and retrieving data Task2. Features selection Task3. Data cleaning Task4. Data transforming A22. Validation of the information artifact A23. Communication of the result	O21. Dataset
Ph3. Phase 3	A31. Data modelling through clustering Task1. Choosing the clustering algorithm depending on analysis objectives, type of data, size of the dataset, cleanliness of the data Task2. Performing the clustering A32. Validation of the information artifact A33. Communication of the result	O31. Labelled dataset
Ph4. Phase 4	A41. Data modelling through classification Task1. Establishing the training (80%) and test dataset (20%) Task2. Train the model for different classification algorithms Task3. Test the models Task4. Evaluate each model and select the best classification algorithm A42. Validation of the information artefacts A43. Communication of the result	O41. Training dataset O42. Test dataset O43. Classification data models O44. Results of the testing process O45. Evaluation metrics results
Ph5. Phase 5	A51. Prediction with the best classification algorithm	O51. Class labels and scores for the new dataset

Our proposal establishes all necessary processing to perform data analysis in general, and performance analysis in particular.

Data analysis is part of a larger business process, such as the process of evaluating performance, and is meant to add value to a business [7]. Data analysis takes primary information from the information flows and returns the information artifacts to the information flows in the corporate environment. As part of the performance management process, the proposed framework is closely linked to process elements downstream and upstream. This implies a scalable deployment approach containing the following stages: top management involvement, proper planning and scoping, introducing the data analysis in terms of a business case, implementing the DSR process, and maintaining a solid data governance program.

3.2. Metrics for Evaluating Classification Models

Classification algorithms are widely used to make predictions and meaningful decisions [42]. Once a classification algorithm produces a model, it is evaluated with respect to certain criteria such as accuracy, ROC curve, or F1 score [44].

According to the prediction approach (Figure 1), classification represents the third phase after feature selection and clustering [11]. A classification model is constructed by applying a classifier to the training dataset (80% of the data). Furthermore, classification accuracy was verified using a test set (20% of the data) by comparing the forecasted output (class label) with the observed output (cluster label provided by the clustering algorithm). Building acceptable classification models implies, despite accuracy and justifiability, that the model should be in line with the existing domain knowledge [45].

According to Choi et al. [46], six evaluation metrics are recommended to evaluate multilevel classification: accuracy, precision, recall, F1-score, receiver operating characteristic curve, and AUC. A greater number of indicators are used in specific contexts, such as software fault predictions [47]. In addition, the classification performance was measured using G-mean, J-coefficient, error rate, and balance. A review of evaluation metrics for data classification evaluations presented a set of suitable indicators for obtaining the optimal classifier: accuracy, error rate, sensitivity, specificity, precision, recall, F-score, geometric mean, average accuracy, average error rate, average precision, average recall, and average F-score [48].

In all approaches, the basic metrics are true positive (TP), true negative (TN), false positive (FP), and false negative (FN) [49]. A true positive is a predicted outcome that is similar to the actual class (cluster label). A false positive result occurs when the classifier labels (or categorizes) a data instance that it should not contain. A true negative result occurs when the classifier does not correctly label (or categorize) the output. A false negative result occurs when the classifier does not label a data instance but should have. Based on these considerations, we introduced the following metrics to evaluate the classification performance (Table 3).

Table 3. Classification model evaluation metrics (adapted from [49]).

No.	Metric Name	Metric Description	
0	true positive (TP), true negative (TN), false positive (FP), false negative (FN)		
1	Confusion matrix (CM)	is a summary of the prediction results; the number of correct and incorrect predictions is summarized with count values and broken down by class	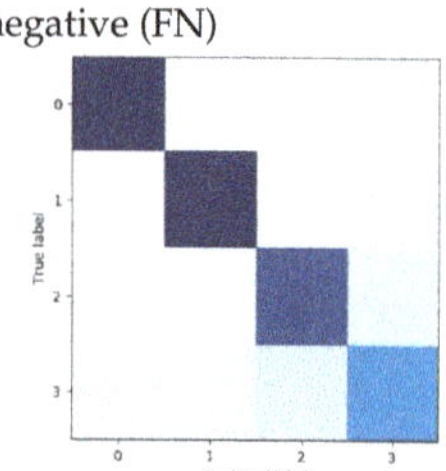

Table 3. *Cont.*

No.	Metric Name	Metric Description	
2	Precision (P)	is the number of correct classes returned by the classification model.	$Precision = \frac{TP}{TP+FP}$
3	Recall (R)	is the ability of a model to find all relevant cases within a dataset, and is the number of true positives divided by the number of true positives plus the number of false negatives	$Recall = \frac{TP}{TP+FN}$
4	F1 score	is the harmonic mean of precision and recall	$F1 = 2 \times \frac{Precision \times Recall}{Precision + Recall}$ $F1 = \frac{2TP}{2TP+FP+FN}$
5	Receiver operating characteristic (ROC) curve	the ROC curve shows how the recall vs. precision relationship changes as we vary the threshold for identifying a positive data point in our model	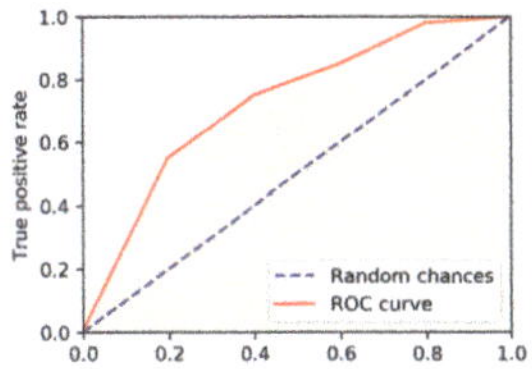
6	Area under the ROC curve (AOC)	is the measure of the ability of a classifier to (AOC) distinguish between classes and is used as a summary of the ROC curve.	
7	Accuracy (A)	the number of correct predictions made as a ratio of all predictions made.	$Accuracy = \frac{TP+TN}{TP+FP+FN+TN}$
8	Classification report	is used to measure the quality of predictions from a classification algorithm; the following measures are displayed: precision, recall, F1, and support scores.	precision recall f1-score support 0 0.73 0.79 0.76 146 1 0.58 0.49 0.53 85 accuracy 0.68 231 macro avg 0.65 0.64 0.64 231 weighted avg 0.67 0.68 0.67 231

Accuracy is widely used to evaluate the classification performance. Additionally, in the case of imbalanced datasets, the F1-score and metrics presented in Table 3 were used [49].

3.3. General Algorithm for Determining the Classification Model Evaluation Metrics

Let DS be a labelled dataset with N instances and different NC class labels. During the training phase, a classification model was generated, and predicted class labels were added during the testing phase (1).

$$Y_{Class(j)}, Y_{Predicted_{Class(j)}} \in \left\{ class_{label(i)} \right\}, \quad i = 1, NC; \; j = 1, N \tag{1}$$

Metrics TP(i), TN(i), FP(i), FN(i), Precision(i), Recall(i), Accuracy(i) and f1(i) were calculated for each class_label(i) according to Pseudocode 1.

The classification report was assembled, and the global metrics of precision, recall, accuracy, and F1 for the classification algorithm were determined, as indicated in Pseudocodes 2.

We recommend using MS Power BI to perform the data analysis. It is used in business and industry sectors as an integral part of the technological and information systems framework. In a self-service manner, business users can integrate data from a variety of sources, perform advanced analysis, and design dashboards for process tracking and decision support. Automated machine learning (AutoML) for dataflows enables business analysts to train, validate, and invoke machine learning models directly in MS Power BI. Pycaret, an open source, low-code machine learning library in Python, accessible from MS Power BI offers support for automated machine learning workflow.

Pseudocode 1

```
FOR i IN 1..NC DO
  FOR j IN 1..N DO
    IF Y_Predicted_Class(j) = Y_Class(j)  AND Y_Predicted_Class(j) =class_label(i)
    TP(i) = TP(i) + 1;
    IF Y_Predicted_Class(j)<>Y_Class(j)  AND Y_Class(j)<>class_label(i)
    FN(i) = FN(i) + 1;
    IF Y_Predicted_Class(j)<>class_label(i)  AND Y_Class(j)<>class_label(i)
    TN(i) = TN(i) + 1;
    IF Y_Predicted_Class(j) = class_label(i)  AND Y_Class(j)<>class_label(i)
    FP(i) = FP(i) + 1;
  IF TP(i) + FP(i)<>0
    Precision(i) = TP(i)/((TP(i) + FP(i)));
  ELSE
    Precision(i) = 0;
  IF TP(i) + FN(i)<>0
    Recall(i) = TP(i)/((TP(i) + FN(i)));
  ELSE
    Recall(i) = 0;
  IF TP(i) + TN(i) + FP(i) + FN(i)<>0
    Accuracy(i) = ((TP(i) + TN(i)))/((TP(i) + TN(i) + FP(i) + FN(i)));
  ELSE
    Accuracy(i) = 0;
  IF Precision(i) + Recall(i)<>0
    f1(i) = 2*(Precision(i)*Recall(i))/((Precision(i) + Recall(i)));
  ELSE
    f1(i) = 0;
```

Pseudocode 2

```
FOR i IN 1..NC DO
  classification_report(i,1) = Precision(i);
  classification_report(i,2) = Recall(i);
  classification_report(i,3) = Accuracy(i);
  classification_report(i,4) = f1(i);
  Global_precision = Global_precision + Precision(i);
  Global_recall = Global_recall + Recall(i);
  Global_accuracy = Global_accuracy + Accuracy(i);
  Global_f1 = Global_f1 + f1(i);
Global_precision = Global_precision/NC;
Global_recall = Global_recall/NC;
Global_accuracy = Global_accuracy/NC;
Global_f1 = Global_f1/NC;
```

4. Analysis and Results

Right from the beginning, the objectives of our theoretical applied discourse were established. Objective one aims to the introduction of a methodological framework using design science research for data analysis. Based on relevant references on DSR [1,2,31–37], we propose a multi-phase framework (Figure 7). Further, the development of artifacts was systematized by establishing activities and tasks specific to each phase within the DSR framework (Table 2). Concrete specifications regarding the use of machine learning algorithms are formulated.

The second objective refers to the unitary approach of metrics for evaluating the performance of classification algorithms. The main evaluation metrics were briefly presented (Table 3) and a general algorithm for determining the classification model evaluation metrics was proposed (Pseudocodes 1 and 2).

The next two objectives, mentioned in the introductory chapter, aim at the application of the theoretical considerations for performance analysis.

The analysis regarding the "managerial capacity" of decision makers was performed using the DSR framework, in compliance with the phases listed in Table 2. A 360-degree evaluation form was chosen as the investigation tool and means of data collection [50]. The following competencies are evaluated: decision making ability, conflict management, relationship management, employee motivation, influence and negotiation, strategic thinking, results orientation, and last but not least planning and organization. Each competence was based on four statements, each of which was assessed by assigning a score on a scale of one to five. The resulting competency scores are in a range from 4 to 20 points (Appendix A).

The dataset centralizes the scores obtained by various managers and contains 195 final instances (Figure 8). Eight competencies (decision making ability, conflict management, relationship management, employee motivation, influence and negotiation, strategic thinking, result orientation, planning, and organization) were selected for data analysis using machine learning techniques, such as clustering and classification.

ID_Manager	Decision makeing	Conflict management	Relationship management	Employee motivation	Influence and negotiation	Strategic thinking	Results orientation	Planning and oraganization	Region	Industry sector
188	15	19	12	10	4	7	12	17	3	3
189	19	12	7	11	5	4	9	15	1	4
190	14	11	20	16	9	16	14	16	4	5
191	18	20	17	10	5	19	18	20	3	3
192	12	18	12	8	12	15	17	10	2	5
193	10	13	16	9	16	13	14	14	5	5
194	15	11	5	7	13	6	9	10	1	3
195	18	19	14	16	20	14	9	18	1	6

Figure 8. O21. Dataset. Partial data.

The dataset contained unlabeled data and required further annotation. This was achieved by modelling the data through clustering. PyCaret's clustering module is an unsupervised machine learning module that groups of a set of objects such that those in the same group (called a cluster) are more similar to each other than to those in other groups. Clustering was performed using the K-means algorithm (Script 1, Figure 9).

Script 1

```
from pycaret.clustering import *
dataset = get_clusters(dataset, num_clusters = 4, ignore_features = ['ID_Manager',
'Industry_sector', 'Region'])
```

The classification module is "a supervised machine learning module that is used for classifying elements into groups. The goal is to predict discrete and unordered categorical class labels" [26]. We used various classification algorithms (Table 2) and calculated evaluation metrics for each algorithm. The models were saved as pkl files. (Script 2).

ID_Manager	Decision makeing	Conflict management	Relationship management	Employee motivation	Influence and negotiation	Strategic thinking	Results orientation	Planning and oraganization	Region	Industry sector	Cluster
182	7	18	11	6	11	15	14	10	3	5	Cluster 1
183	7	19	8	7	14	18	9	4	5	1	Cluster 0
184	8	17	5	8	16	20	8	7	5	4	Cluster 1
185	7	16	14	14	18	12	19	4	4	5	Cluster 1
186	13	13	4	18	13	6	15	7	5	4	Cluster 3
187	19	20	14	12	16	17	16	7	4	5	Cluster 1
188	15	19	12	10	4	7	12	17	3	3	Cluster 0
189	19	12	7	11	5	4	9	15	1	4	Cluster 0
190	14	11	20	16	9	16	14	16	4	5	Cluster 2
191	18	20	17	10	5	19	18	20	3	3	Cluster 3
192	12	18	12	8	12	15	17	10	2	5	Cluster 0
193	10	13	16	9	16	13	14	14	5	5	Cluster 1
194	15	11	5	7	13	6	9	10	1	3	Cluster 0
195	18	19	14	16	20	14	9	18	1	6	Cluster 2

Figure 9. O31. Labelled dataset. Partial data.

Script 2

```
clf1 = setup(df, target = 'Cluster', silent = True, ignore_features = ['ID_Manager',
'Industry_sector','Region'])
# train multiple models
algorithms = ['knn','dt','catboost','nb','rbfsvm','lr','gpc','mlp','rf','qda','ada','gbc','lda','et',
'xgboost','lightgbm','svm','ridge']
models = [create_model(i) for i in algorithms]
final_models = [finalize_model(models[i]) for i in range(len(algorithms))]
for x in range(len(algorithms)):
save_model(final_models[x], 'D:/'+ algorithms [x])
```

After training different classification algorithms, the models were tested (Script 3). The predicted class labels are associated with each instance of the test dataset (Figure 10).

Script 3

```
algorithms = ['knn','dt','catboost','nb','rbfsvm','lr','gpc','mlp','rf','qda','ada','gbc','lda','et',
'xgboost','lightgbm','svm','ridge']
from pycaret.classification import *
for i in range(len(algorithms)):
clasificator = load_model('D:/' + algorithms[i])
dataset = predict_model(clasificator, data = dataset)
dataset.rename(columns = {'Label':'Label_' + algorithms[i],'Score': 'Score_' + algorithms[i]},
inplace = True)
```

The evaluation metrics were calculated for each classification model according to the previously described "general algorithm for determining the classification model evaluation metrics" (Pseudocodes 1 and 2).

We created, trained, and deployed a machine leaning model for each classification algorithm available in PyCaret library. The following algorithms, which are listed in alphabetical order, were applied: adaboost (ada), cat booster classifier (catboost), decision tree (dt), extra tree classifier (et), extreme gradient boosting (xgboost), gaussian process classifier (gpc), gradient boosting classifier (gbc), light gradient boosting (lightgbm), linear disc analysis (lda), logistic regression (lr), k nearest neighbor (knn), multi level perceptron (mlp), naives bayes (nb), random forest (rf), ridge classifier (ridge), support vector machine (svm and rbfsvm), and quadratic disc analysis (qda) [26]. They are representative for all classification algorithm categories (Figure 4).

ID_Manager	Cluster	Label_ada	Score_ada	Label_catboost	Score_catboost	Label_dt	Score_dt	Label_knn	Score_knn	Label_svm	Label_rf	Score_rf
157	Cluster 0	Cluster 0	0.4129	Cluster 0	0.8155	Cluster 0	1.0	Cluster 0	0.6	Cluster 2	Cluster 0	0.66
158	Cluster 0	Cluster 0	0.4192	Cluster 0	0.9672	Cluster 0	1.0	Cluster 0	0.8	Cluster 0	Cluster 0	0.84
159	Cluster 0	Cluster 0	0.443	Cluster 0	0.9712	Cluster 0	1.0	Cluster 0	0.8	Cluster 0	Cluster 0	0.7
160	Cluster 3	Cluster 3	0.8683	Cluster 3	0.9997	Cluster 3	1.0	Cluster 3	0.8	Cluster 3	Cluster 3	0.89
161	Cluster 0	Cluster 3	0.3897	Cluster 0	0.9457	Cluster 0	1.0	Cluster 0	0.4	Cluster 3	Cluster 0	0.41
162	Cluster 0	Cluster 0	0.461	Cluster 0	0.9352	Cluster 0	1.0	Cluster 0	0.8	Cluster 0	Cluster 0	0.51
163	Cluster 2	Cluster 0	0.4192	Cluster 0	0.9463	Cluster 0	1.0	Cluster 2	0.6	Cluster 0	Cluster 0	0.68
164	Cluster 1	Cluster 1	0.6242	Cluster 1	0.8416	Cluster 1	1.0	Cluster 1	0.4	Cluster 0	Cluster 1	0.61
165	Cluster 1	Cluster 1	0.3063	Cluster 1	0.9053	Cluster 1	1.0	Cluster 0	0.6	Cluster 1	Cluster 1	0.6
166	Cluster 2	Cluster 2	0.5816	Cluster 2	0.9976	Cluster 2	1.0	Cluster 2	0.8	Cluster 2	Cluster 2	0.8
167	Cluster 2	Cluster 0	0.4785	Cluster 0	0.9649	Cluster 0	1.0	Cluster 0	0.8	Cluster 0	Cluster 0	0.79
168	Cluster 0	Cluster 0	0.4192	Cluster 0	0.9752	Cluster 0	1.0	Cluster 3	0.6	Cluster 0	Cluster 0	0.74
169	Cluster 0	Cluster 2	0.4031	Cluster 0	0.7355	Cluster 1	1.0	Cluster 0	0.4	Cluster 0	Cluster 0	0.5
170	Cluster 0	Cluster 0	0.4192	Cluster 0	0.9841	Cluster 0	1.0	Cluster 0	0.8	Cluster 0	Cluster 0	0.73
171	Cluster 0	Cluster 0	0.4192	Cluster 0	0.9353	Cluster 0	1.0	Cluster 0	0.8	Cluster 0	Cluster 0	0.61
172	Cluster 0	Cluster 3	0.5149	Cluster 3	0.6472	Cluster 3	1.0	Cluster 0	0.6	Cluster 3	Cluster 0	0.57

Figure 10. O44. Results of the testing process.

The automated machine learning (AutoML) workflow implemented by Scripts 2 and 3 generated the data artifacts specified in Table 2.

Further, the evaluation metrics for each algorithm were processed, namely precision, recall, accuracy, and f1 metric (Figure 11). Script 4 is part of the Auto ML approach.

Script 4

```
algorithms = ['knn','dt','catboost','nb','rbfsvm','lr','gpc','mlp','rf','qda','ada','gbc','lda','et',
'xgboost','lightgbm','svm','ridge']
final_models = [finalize_model(models[i]) for i in range(len(al))]
for x in range(len(al)):
save_model(final_models[x], 'D:/[x])
best = compare_models(include = al)
results = pull()
print(results)
```

algorithm	precision	recall	accuracy	f1
catboost	0.883333333333333	0.822222222222222	0.923076923076923	0.829465709728868
et	0.853174603174603	0.769444444444444	0.91025641025641	0.761808367071525
lr	0.793434343434343	0.733333333333333	0.91025641025641	0.746637426900585
xgboost	0.829365079365079	0.747222222222222	0.897435897435898	0.764139369402527
gbc	0.782142857142857	0.730555555555556	0.897435897435897	0.741239316239316
rf	0.79040404040404	0.711111111111111	0.897435897435897	0.73969298245614
mlp	0.782738095238095	0.761111111111111	0.897435897435897	0.752752639517345
dt	0.708333333333333	0.691666666666667	0.871794871794872	0.683080808080808
ridge	0.688354037267081	0.644444444444444	0.858974358974359	0.654375896700144
svm	0.6625	0.641666666666667	0.858974358974359	0.647368421052632
knn	0.657738095238095	0.572222222222222	0.833333333333333	0.598646125116713
ada	0.627083333333333	0.594444444444444	0.833333333333333	0.605927698032961
gpc	0.803030303030303	0.458333333333333	0.794871794871795	0.47192513368984
nb	0.460227272727273	0.413888888888889	0.782051282051282	0.413293650793651
rbfsvm	0.115384615384615	0.25	0.730769230769231	0.157894736842105
lightgbm	0.115384615384615	0.25	0.730769230769231	0.157894736842105
lda	0.388888888888889	0.397222222222222	0.717948717948718	0.387967914438503
qda	0.312333864965444	0.394444444444444	0.705128205128205	0.346203346203346

Figure 11. O45. Evaluation metrics synthesis.

According to the values obtained for accuracy, as well as for the other metrics, the CatBoost algorithm proved to be the best performant classification algorithm in our analysis. Therefore, this will be investigated further (Figure 12). CatBoost is an algorithm for gradient boosting of decision trees. According to Pramoditha [51], CatBoost is one of the best machine learning models for tabular heterogeneous datasets.

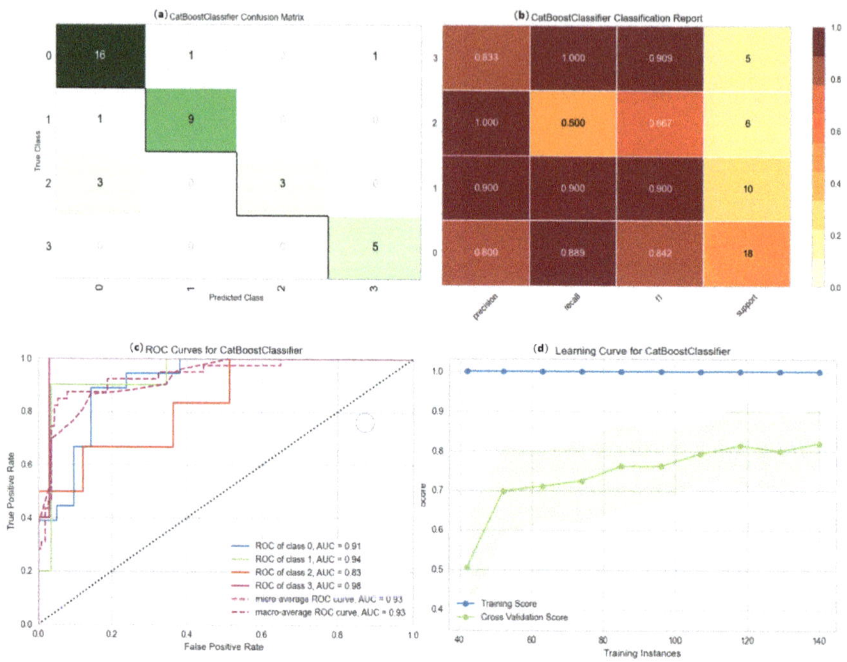

Figure 12. CatBoost algorithm. Evaluation metrics.

The confusion matrix contains the values of true positive (TP), true negative (TN), false positive (FP), and false negative (FN) calculated for each class (Figure 12a). We observed that most instances were correctly labelled. Most instances that were incorrectly labelled belonged to class 0 (cluster 0).

The classification report presents the main classification matrices, namely, precision, recall, and F1 score for each class (Figure 12b). We can notice that:

- The algorithm has a significant ability to label instances correctly, particularly in classes 1 and 2.
- For classes 0, 1, and 3, the algorithm had a high capacity to find all instances; however, it correctly labelled only half of the instances in class 2.
- The values of f1 for classes 0, 1, and 3 are appropriate and approximately equal to an average of 0.9; however, for class 2, f1 is only 0.667. Although the precision of the classification of the instances in class 2 was 1, the algorithm identified only three out of the six instances of class 2.

The graph of the ROC curve shows that the classification model can place the instances in a single class (Figure 12c). The graph shows that the instances of classes 0, 1, and 3 are approximately equal to the algorithm average of 0.93, indicating that these classes are well-separated. The only class for which a lower score was obtained was class 2, which had a score of 0.83. However, even for this class, the model provides a good measure of the separability.

The learning curve for the CatBoost classifier indicated that increasing the number of instances in the training set led to an increase in the validation score (Figure 12d). The training score maintains a value of one, which indicates that the model perfectly integrates each newly added instance.

According to Huilgol [52], accuracy is used when true positives and true negatives are decisive in the analysis, whereas the F1-score is used when false negatives and false positives are the most important. Furthermore, the accuracy can be used when the class distribution is similar, whereas the F1-score is a better metric when dealing with imbalanced classes.

The use of machine learning techniques for performance analysis makes a significant contribution when operating with large datasets [27]. We identified concrete applications of our proposal, namely: the application of the procedure within a multinational company or in statistical research studies on companies.

The Power BI application integrates the data obtained through 360-feeback and performs the analysis. The results are available to the management boards and research coordinators.

DSR is applied in various business and industrial engineering areas [53]. The literature indicates different approaches to designing artifacts [31–41]. Our proposal comes to offer a framework for data analysis using machine learning techniques. The theoretical discourse was applied to a performance analysis.

5. Conclusions

DSR opens new research perspectives in information systems and data analysis. We managed to complete an artifact design-centric approach adapted for data analysis. The proposed DSR framework describes a multi-phase process containing activities and tasks that allow the design, development, testing, validation, and communication of the considered data and information artifacts.

Artifacts engineering is performed using machine learning techniques. We recommend the use of AutoML to automate the iterative tasks of machine learning model development. Mainly based on classification algorithms, the workflow also provides for the evaluation of the applied algorithms.

The proposed design science research was applied in a managerial performance evaluation project. Further steps are necessary to define a secure connection to the operational HR database, where performance data are stored. In this sense, we are concerned to respect all internal regulations and data governance prescriptions.

Author Contributions: Conceptualization, M.M.; methodology, M.M.; software, F.D.M.; validation, M.M. and F.D.M.; writing—review and editing, M.M. and F.D.M. All authors have read and agreed to the published version of the manuscript.

Funding: This research received no external funding.

Acknowledgments: This study was conducted by Muntean Mihaela, associate member of the East European Center for Research in Economics and Business (ECREB) at the Faculty of Economics and Business Administration, West University of Timisoara. Florin Daniel Militaru, contributed to the completion of the paper, with the results of research undertaken within the Business Information Systems Department.

Conflicts of Interest: The authors declare no conflict of interest.

Appendix A

Table A1. The 360 feedback form for measuring a manager's performance [51].

Competence	Statements	Evaluation Scale				
Decision making capacity	Assess the implications of a strategic or potentially risky decision and the impact it may have on the organization	1	2	3	4	5
	Make good decisions based on a mix of analysis, intuition, experience and logic.	1	2	3	4	5
	Most of the solutions and suggestions offered by him/her prove to be correct and precise in time.	1	2	3	4	5
	It takes less popular measures when the situation demands it or implements decisions even if it does not have the consent of all its subordinates.	1	2	3	4	5
Conflict management	Manage issues firmly, directly and in a timely manner.	1	2	3	4	5
	He/she is an active listener, able to understand the source of conflicts and to suggest proper solutions.	1	2	3	4	5
	It easily reaches armistices and agreements, with the involvement of a minimum number of third parties.	1	2	3	4	5
	He/she finds ways out of difficult situations and manages to value the disputes.	1	2	3	4	5
Relationships management	It relates well to all categories of people, regardless of the hierarchical level, both inside and outside the company.	1	2	3	4	5
	Communication with colleagues is clear and efficient.	1	2	3	4	5
	Provides current, direct, complete, actionable, positive and/or corrective feedback to others.	1	2	3	4	5
	Encourages open dialogue within the team.	1	2	3	4	5
Employee motivation	Maintains a constant dialogue with the team members for whom he is responsible for the quality and quantity of work and results	1	2	3	4	5
	Appreciate the extra effort and communicate its recognition	1	2	3	4	5
	He/she is actively concerned with the development of the staff for whose performance he/she is responsible	1	2	3	4	5
	Request input from each person in the team, support visibility and invest in the right people with authority	1	2	3	4	5
Influence and negotiation	He convinces others and gains their support	1	2	3	4	5
	Use convincing arguments and ideas	1	2	3	4	5
	He/she tends to negotiate whenever he/she has the opportunity	1	2	3	4	5
	It is not discouraged by arguments against its objectives	1	2	3	4	5
Strategic thinking	Is capable to formulate new strategies and competitive plans.	1	2	3	4	5
	Can accurately anticipate future consequences and trends.	1	2	3	4	5
	Can draw up a realistic and motivating strategic plan.	1	2	3	4	5
	Think long-term, corroborating information and market trends, anticipating possible developments and alternative action plans.	1	2	3	4	5
Results orientation	He focuses his efforts on priority tasks, reserving time for other activities as well.	1	2	3	4	5
	Shows a passion for business, reflected in a "can-do" attitude	1	2	3	4	5
	Helps others manage priorities by focusing on critical activities for success.	1	2	3	4	5
	Do not get lost in irrelevant details by quickly finding the shortest path to the result.	1	2	3	4	5
Planning and organization	Can organize people, activities and resources to finish projects successfully.	1	2	3	4	5
	Can coordinate multiple activities at once to accomplish one goal.	1	2	3	4	5
	He plans his activity ahead of time and sets realistic deadlines.	1	2	3	4	5
	He is a systematic and well-organized person who sets clear priorities.	1	2	3	4	5

References

1. Peffers, K.; Tuunanen, T.; Rothenberger, M.A.; Chatterjee, S. A design science research methodology for information systems research. *J. Manag. Inf. Syst.* **2007**, *24*, 45–77. [CrossRef]
2. Paquette, J. A Brief Introduction to Useful Data Artifacts—And the Next Generation of Data Analysis Systems. Medium, 2021. Available online: https://medium.com/tag-bio/a-brief-introduction-to-useful-data-artifacts-and-the-next-generation-of-data-analysis-systems-1f42ef91ce92 (accessed on 30 December 2021).
3. Behn, R.D. Why measure performance? different purposes require different measures. *Public Adm. Rev.* **2003**, *63*, 586–606. [CrossRef]
4. Stroet, H. AI in Performance Management: What Are the Effects for Line Managers? Bachelor's Thesis, University of Twente, Enschede, The Netherlands, 2020.
5. Bhardwaj, G.; Singh, S.V.; Kumar, V. An empirical study of artificial intelligence and its impact on human resource functions. In Proceedings of the 2020 International Conference on Computation, Automation and Knowledge Management (ICCAKM), Dubai, United Arab Emirates, 9–10 January 2020. [CrossRef]
6. Eight-Step Guide to Performance Evaluations for Managers—The Management Center. 2021. Available online: https://www.managementcenter.org/article/eight-step-guide-to-performance-evaluations-for-managers/ (accessed on 30 December 2021).
7. Attaran, M.; Deb, P. Machine learning: The new 'big thing' for competitive advantage. *Int. J. Knowl. Eng. Data Min.* **2018**, *5*, 277–305. [CrossRef]
8. El Bouchefry, K.; de Souza, R.S. Learning in big data: Introduction to machine learning. In *Knowledge Discovery in Big Data from Astronomy and Earth Observation*; Elsevier: Amsterdam, The Netherlands, 2020; pp. 225–249. [CrossRef]
9. Chakraborty, T. EC3: Combining clustering and classification for Ensemble Learning. In Proceedings of the 2017 IEEE International Conference on Data Mining (ICDM), New Orleans, LA, USA, 18–21 November 2017. [CrossRef]
10. Alapati, Y.K.; Sindhu, K. Combining Clustering with Classification: A Technique to Improve Classification Accuracy. *Int. J. Comput. Sci. Eng.* **2016**, *5*, 336–338.
11. Bertsimas, D.; Dunn, J. Optimal classification trees. *Mach. Learn.* **2017**, *106*, 1039–1082. [CrossRef]
12. Durcevic, S. 10 Top Business Intelligence and Analytics Trends for 2020. Information Management. 2019. Available online: https://www.information-management.com/opinion/10-top-business-intelligence-and-analytics-trends-for-2020 (accessed on 20 March 2022).
13. Walowe Mwadulo, M. A review on feature selection methods for classification tasks. *Int. J. Comput. Appl. Technol. Res.* **2016**, *5*, 395–402. [CrossRef]
14. Dash, M.; Koot, P.W. Feature selection for clustering. In *Encyclopedia of Database Systems*; Springer: Boston, MA, USA, 2009; pp. 1119–1125. [CrossRef]
15. Rong, M.; Gong, D.; Gao, X. Feature selection and its use in big data: Challenges, methods, and Trends. *IEEE Access* **2019**, *7*, 19709–19725. [CrossRef]
16. Madhulatha, T.S. An overview on clustering methods. *IOSR J. Eng.* **2012**, *2*, 719–725. [CrossRef]
17. Witten, I.H.; Frank, E.; Hall, M.A. *Data Mining: Practical Machine Learning Tools and Techniques*; Elsevier: Amsterdam, The Netherlands, 2011.
18. Ghosal, A.; Nandy, A.; Das, A.K.; Goswami, S.; Panday, M. A short review on different clustering techniques and their applications. In *Advances in Intelligent Systems and Computing*; Springer: Singapore, 2019; Volume 937, pp. 69–83. [CrossRef]
19. Celebi, M.E.; Kingravi, H.A. Linear, deterministic, and order-invariant initialization methods for the K-means clustering algorithm. In *Partitional Clustering Algorithms*; Springer: Cham, Switzerland, 2014; pp. 79–98. [CrossRef]
20. Sinaga, K.P.; Yang, M.-S. Unsupervised K-means clustering algorithm. *IEEE Access* **2020**, *8*, 80716–80727. [CrossRef]
21. Papas, D.; Tjortjis, C. Combining clustering and classification for Software Quality Evaluation. In *Artificial Intelligence: Methods and Applications*; Springer: Cham, Switzerland, 2014; pp. 273–286. [CrossRef]
22. Loukas, S. K-Means Clustering: How It Works & Finding the Optimum Number of Clusters in the Data. Medium, 2020. Available online: https://towardsdatascience.com/k-means-clustering-how-it-works-finding-the-optimum-number-of-clusters-in-the-data-13d18739255c (accessed on 30 December 2021).
23. Rani, Y.; Harish, R. A study of hierarchical clustering algorithm. *Int. J. Inf. Comput. Technol.* **2013**, 3, 1115–1122.
24. Webb, G.I.; Fürnkranz, J.; Fürnkranz, J.; Fürnkranz, J.; Hinton, G.; Sammut, C.; Sander, J.; Vlachos, M.; Teh, Y.W.; Yang, Y.; et al. Density-based clustering. In *Encyclopedia of Machine Learning*; Springer: Boston, MA, USA, 2011; pp. 270–273. [CrossRef]
25. Grabusts, P.; Borisov, A. Using grid-clustering methods in data classification. In Proceedings of the International Conference on Parallel Computing in Electrical Engineering, Warsaw, Poland, 25 September 2002. [CrossRef]
26. Duda, R.O.; Hart, P.E. *Pattern Classification and Scene Analysis*; Wiley: New York, NY, USA, 1973; Volume 3.
27. Narula, G. Machine Learning Algorithms for Business Applications—Complete Guide. Emerj, 2021. Available online: https://emerj.com/ai-sector-overviews/machine-learning-algorithms-for-business-applications-complete-guide/ (accessed on 30 December 2021).
28. How to Select a Machine Learning Algorithm—Azure Machine Learning. 2021. Available online: https://docs.microsoft.com/en-us/azure/machine-learning/how-to-select-algorithms (accessed on 30 December 2021).
29. Zhao, L.; Lee, S.; Jeong, S.P. Decision tree application to classification problems with boosting algorithm. *Electronics* **2021**, *10*, 1903. [CrossRef]

30. Nunamaker, J.F.; Chen, M.; Purdin, T.D.M. Systems development in information systems research. *J. Manag. Inf. Syst.* **1990**, *7*, 89–106. [CrossRef]
31. Weber, S. Design Science Research: Paradigm or Approach? AMCIS 2010 Proceedings. 2010. Available online: https://aisel.aisnet.org/amcis2010/214/ (accessed on 2 March 2022).
32. Hevner, A.; March, S.; Park, J.; Ram, S. Design science in information systems research. *MIS Q. Manag. Inf. Syst.* **2004**, *28*, 75–105. [CrossRef]
33. Venable, J.R.; Heje, J.P.; Baskerville, R.L. Choosing a Desing Science Research Methodology. ACIS 2017 Proceedings. 2017. Available online: https://aisel.aisnet.org/acis2017/112 (accessed on 2 March 2022).
34. Vaishnavi, V.; Kuechler, W.; Petter, S. (Eds.) *Design Science Research in Information Systems*; Association for Information Systems: Atlanta, GA, USA, 2004. Available online: http://www.desrist.org/design-research-in-information-systems/ (accessed on 2 March 2022).
35. Sein, M.K.; Henfridsson, O.; Purao, S.; Rossi, M.; Lindgren, R. Action design research. *MIS Q. Manag. Inf. Syst.* **2011**, *35*, 37–56. [CrossRef]
36. Baskerville, R.; Pries-Heje, J.; Venable, J. Soft design science methodology. In Proceedings of the 4th International Conference on Design Science Research in Information Systems and Technology—DESRIST'09, Philadelphia, PA, USA, 6–8 May 2009. [CrossRef]
37. Bilandzic, M.; Venable, J. Towards participatory action design research: Adapting Action Research and Design Science Research Methods for Urban Informatics. *J. Community Inform.* **2011**, *7*. [CrossRef]
38. Ahmed, M.; Sundaram, D. Design Science Research Methodology: An Artefact-Centric Creation and Evaluation Approach. In Proceedings of the Australasian Conference on Information Systems (ACIS), Sydney, Australia, 30 November–2 December 2011.
39. Herselman, M.; Botha, A. Evaluating an artifact in Design Science Research. In Proceedings of the 2015 Annual Research Conference on South African Institute of Computer Scientists and Information Technologists—SAICSIT'15, Stellenbosch, South Africa, 28–30 September 2015. [CrossRef]
40. Peffers, K.; Tuunanen, T.; Niehaves, B. Design science research genres: Introduction to the special issue on exemplars and criteria for applicable Design Science Research. *Eur. J. Inf. Syst.* **2018**, *27*, 129–139. [CrossRef]
41. Muntean, M.; Dănăiaţă, D.; Hurbean, L.; Jude, C. A Business Intelligence & Analytics framework for clean and affordable energy data analysis. *Sustainability* **2021**, *13*, 638. [CrossRef]
42. Elragal, A.; Haddara, M. Design science research: Evaluation in the lens of Big Data Analytics. *Systems* **2019**, *7*, 27. [CrossRef]
43. Achampong, E.K.; Dzidonu, C. Methodological Framework for Artefact Design and Development in Design Science Research. *J. Adv. Sci. Technol. Res.* **2017**, *4*, 1–8. Available online: https://www.researchgate.net/publication/329775397_Methodological_Framework_for_Artefact_Design_and_Development_in_Design_Science_Research (accessed on 30 December 2021).
44. Chicco, D.; Jurman, G. The advantages of the Matthews correlation coefficient (MCC) over F1 score and accuracy in binary classification evaluation. *BMC Genom.* **2020**, *21*, 6. [CrossRef]
45. Martens, D.; Baesens, B. Building acceptable classification models. In *Annals of Information Systems*; Springer: Boston, MA, USA, 2009; pp. 53–74. [CrossRef]
46. Choi, J.-G.; Ko, I.; Kim, J.; Jeon, Y.; Han, S. Machine Learning Framework for multi-level classification of company revenue. *IEEE Access* **2021**, *9*, 96739–96750. [CrossRef]
47. Muhammad, R.; Nadeem, A.; Azam Sindhu, M. Vovel metrics—Novel coupling metrics for improved software fault prediction. *PeerJ Comput. Sci.* **2021**, *7*, e590. [CrossRef] [PubMed]
48. Hossin, M.; Sulaiman, M.N. A review on evaluation metrics for Data Classification Evaluations. *Int. J. Data Min. Knowl. Manag. Process* **2015**, *5*, 2. [CrossRef]
49. Vujovic, Ž.Đ. Classification Model Evaluation Metrics. *Int. J. Adv. Comput. Sci. Appl.* **2021**, *12*, 6. [CrossRef]
50. Apt360. Chestionar Pentru Evaluarea Managerilor Din Prima Linie. 2019. Available online: https://www.evaluare360.ro/wp-content/uploads/2019/01/Chestionar-angajati-Manageri-prima-line2019.pdf (accessed on 30 December 2021).
51. Pramoditha, R. 5 Cute Features of CatBoost. Towardsdatascience, 2021. Available online: https://towardsdatascience.com/5-cute-features-of-catboost-61532c260f69 (accessed on 30 December 2021).
52. Huilgol, P. Accuracy vs. F1-Score. Medium, 2021. Available online: https://medium.com/analytics-vidhya/accuracy-vs-f1-score-6258237beca2 (accessed on 30 December 2021).
53. Goecks, L.S.; De Souza, M.; Librelato, T.P.; Trento, L.R. Design Science Research in practice: Review of applications in Industrial Engineering. *Gest. Prod.* **2021**, *28*, e5811. [CrossRef]

electronics

MDPI

Article

KNN-Based Consensus Algorithm for Better Service Level Agreement in Blockchain as a Service (BaaS) Systems

Qingxiao Zheng [1], Lingfeng Wang [1], Jin He [1,*] and Taiyong Li [2,*]

1 Industry College of Blockchain, Chengdu University of Information Technology, Chengdu 610225, China
2 School of Computing and Artificial Intelligence, Southwestern University of Finance and Economics, Chengdu 611130, China
* Correspondence: hejin@cuit.edu.cn (J.H.); litaiyong@gmail.com (T.L.)

Abstract: With services in cloud manufacturing expanding, cloud manufacturers increasingly use service level agreements (SLAs) to guarantee business processing cooperation between CSPs and CSCs (cloud service providers and cloud service consumers). Although blockchain and smart contract technologies are critical innovations in cloud computing, consensus algorithms in Blockchain as a Service (BaaS) systems often overlook the importance of SLAs. In fact, SLAs play a crucial role in establishing clear commitments between a service provider and a customer. There are currently no effective consensus algorithms that can monitor the SLA and provide service level priority. To address this issue, we propose a novel KNN-based consensus algorithm that classifies transactions based on their priority. Any factor that impacts the priority of the transaction can be used to calculate the distance in the KNN algorithm, including the SLA definition, the smart contract type, the CSC type, and the account type. This paper demonstrates the full functionality of the enhanced consensus algorithm. With this new method, the CSP in BaaS systems can provide improved services to the CSC. Experimental results obtained by adopting the enhanced consensus algorithm show that the SLA is better satisfied in the BaaS systems.

Keywords: BaaS system; blockchain consensus algorithm; KNN; service level agreement; transaction priority

Citation: Zheng, Q.; Wang, L.; He, J.; Li, T. KNN-Based Consensus Algorithm for Better Service Level Agreement in Blockchain as a Service (BaaS) Systems. *Electronics* **2023**, *12*, 1429. https://doi.org/10.3390/electronics12061429

Academic Editor: Ping-Feng Pai

Received: 4 February 2023
Revised: 28 February 2023
Accepted: 14 March 2023
Published: 16 March 2023

1. Introduction

Blockchain, business analytics, and the Internet of Things (IoT) are the emerging industry trends to which scholars and practitioners have paid much attention in recent years. The state-of-the-art research related to these technologies has been summarized by Zhang and Chen [1]. Blockchain as a Service (BaaS) is a new technology that combines cloud computing and blockchain technology. As a third-party service, BaaS provides customers with the ability to create and manage blockchain-based networks through cloud technology. It is a relatively new technology trend that provides third-party services within the blockchain technology domain. Blockchain applications are more than just cryptocurrency transactions. They have expanded to encompass all types of secure transactions. As a result, hosting services are increasingly in demand. Blockchain technology has been used to provide services to more customers as a service model through the cloud. This model works similarly to SaaS, PaaS, and IaaS models, which support the usage of cloud-based applications and storage. Blockchain technology is complex, and much effort is required to build, maintain, and monitor a blockchain system when applied. In order to increase the accessibility of the blockchain and distributed ledgers, we need to leverage blockchain with lower costs and less overhead, especially for businesses. BaaS is a promising technical option that can meet these goals [2]. However, critical issues in current public blockchain systems prevent them from being used as a generic platform for different services and applications. Bitcoin can handle about 5.5 transactions per second (TPS), and Ethereum can process about 20 TPS, which is far below the mainstream payment systems. There is no silver bullet that solves all of these problems due to the Trilemma, as mentioned by the founder of Ethereum, Vitalik Buterin: public blockchain systems can have only two of the following three properties: decentralization, scalability, and security [3].

Most previous studies have not solved the scalability issue well. It is difficult for cloud service providers (CSPs) to guarantee effective SLA with cloud service consumers (CSCs). There are some studies that achieve better data query/sharing services based on blockchain service, such as BlockShare [4], Verifiable Query Layer (VQL) [5], and vChain+ [6], but these services cannot solve the SLA issue. To address the SLA issue in the BaaS environment, this paper proposes a novel KNN-based consensus algorithm by classifying the transactions with priority. Any factor that impacts the priority of the transaction can be used to calculate the distance in the KNN algorithm. Such factors include the SLA definition, the smart contract type, the CSC type, and the account type.

This paper has three main contributions: (1) A simple supervised learning method, KNN, is used to build a consensus algorithm for the first time. (2) With the realization of the full functionality of the enhanced consensus algorithm, the CSP in the BaaS systems can provide improved services to the CSC. (3) Experimental results demonstrate that the SLA is better satisfied in the BaaS systems. The transaction with higher priority that arrives later is executed early.

We have organized the rest of the paper as follows. Section 2 provides a review of related work. Section 3 depicts the problem studied in this paper. Section 4 describes preliminaries, such as BaaS, cloud computing SLA in BaaS, and the KNN algorithm. The proposed KNN-based consensus algorithm is detailed in Section 5. Section 6 reports and analyzes the experimental results. Section 7 concludes this paper.

2. Related Work

2.1. Evolution of Consensus Algorithms

In decentralization, any node in a blockchain can submit a transaction to be stored in the system, so it is important that there are processes that can ensure that each node reaches a consensus to accept or reject the submitted transactions. These processes are essentially considered consensus algorithms.

PoW is the first consensus protocol used in blockchain. It works with Bitcoin and Ethereum, among others. In each round of consensus, PoW uses computational power competition to decide which node can pack recent transactions into a new block. PoW guarantees eventual consistency based on the major distributed nodes with high computational power in reaching a consensus. It is a probabilistic-finality consensus protocol [7].

PoS was created to overcome shortages that occur when PoW consumes too much computational power. In each round of consensus, PoS considers not only the computational power but also the stake held when deciding which node can pack recent transactions into a new block. The difference between PoS and PoW is the importance of the amount of stake (coins) and of how many times the nonce is adjusted. PoS is also a probabilistic-finality consensus protocol [7].

Raft reaches a consensus by an elected leader. A node in a blockchain system with Raft is either a leader or a follower and can be a candidate in an election scenario when a current leader is unavailable. The Raft leader has the responsibility of logging replications to the followers, and it periodically notifies the followers of its alive state by sending a heartbeat message. Raft implements a consensus based on the leader schema. The whole blockchain system has only one elected leader, which has full responsibility for managing logged replications to followers.

PBFT provides a practical Byzantine state machine replication that tolerates the Byzantine Generals' Problem caused by malicious nodes. It assumes that these malicious nodes have independent failures and send manipulated messages. Distributed nodes in a blockchain system with PBFT are appointed as leaders, in turn, and others are appointed as backup nodes. All nodes in the blockchain system assume that all honest nodes will make an agreement by using predefined rules when communicating with each other.

The above consensus algorithms are the main types of consensus algorithms used in the blockchain system. They have different decentralization and transaction throughput capabilities, and these consensus algorithms have their own application scenarios based on the requirement of decentralization and performance grades.

The data structure of the transaction in most blockchains is simple. It includes a receiver address, transaction amount, etc. In a typical blockchain system, such as Bitcoin, the receiver address is located in the "Locking-Script" field of a transaction output, and the transaction amount is located in the "Amount" field of a transaction, as shown in Tables 1 and 2. The blockchain node verifies the validity and effectiveness of the transaction, while transactions are not classified or processed with priority in the consensus procedure since there is no field in the transaction data structure to describe the transaction priority or type [8]. There is an opportunity for optimization by classifying and processing transactions with priority. The method introduced in this paper uses a strategy that ensures that transactions with higher priority can be processed in a timely manner.

Table 1. The structure of a transaction in Bitcoin.

Size	Field	Description
4 bytes	Version	Specifies which rules this transaction follows
1–9 bytes (VarInt)	Input Counter	How many inputs are included
Variable	Inputs	One or more transaction inputs
1–9 bytes (VarInt)	Output Counter	How many outputs are included
Variable	Outputs	One or more transaction outputs
4 bytes	Locktime	A Unix timestamp or block number

Table 2. The structure of a transaction output in Bitcoin.

Size	Field	Description
8 bytes	Amount	Bitcoin value in satoshis
1–9 bytes (VarInt)	Locking-Script Size	Locking-Script length in bytes, to follow
Variable	Locking-Script	A script defining the conditions needed to spend the output

2.2. *QoS Assurance*

Previous studies show that most of the recently developed public blockchain systems focus on increasing transaction throughput to improve scalability. Even if the existing consortium blockchain TPS is improved compared with public blockchains, the efficiency of the consensus algorithm is still low, and its fault tolerance is still poor [9].

Blockchain technology plays an important role in supporting Service Level Agreements (SLAs) that guarantee quality of service (QoS) standards for various service providers. Meanwhile, although smart contracts are applied in traditional cloud providers, SLAs are rarely used to provide improved service [10].

The blockchain data are used in the BaaS system to provide a range of operational services, such as search queries and task submission on the blockchain [11]. Driven by BaaS, the content of a cloud service becomes more abundant, and the CSC increases its requirements for QoS [12,13]. In order to solve the QoS assurance problem between the CSP and CSC, one method is proposed to support the cloud computing service level agreement. The purpose of this agreement is to create a healthy environment for operations on the network so that the CSC can enjoy not only the service promised verbally by the CSP but also a service that is regulated and fully protected [14].

Existing research recognizes the critical role played by the service provider [15], but it lacks a valid method that enhances the consensus algorithm with improved QoS assurance. Since smart contracts stand on the application layer, providing QoS assurance for smart contracts is relatively inefficient and is not the best choice; it is better to put this assurance in the kernel module of the BaaS system for all transactions.

According to the above studies, the existing consensus algorithms cannot provide effective support for SLAs between a CSP and a CSC. It is important to provide QoS assurance in a consensus algorithm, and how various transactions are classified is key in supporting QoS. As the KNN is one of the simplest classification methods, it was chosen here for classifying transactions. The main aim of a KNN is to find k training samples that are closest to the new sample and assign the majority label of the k samples to the new sample. Despite its simplicity, the KNN has been successful in solving a wide range of regression and classification problems, including handwritten characters and image recognition scenarios. As a non-parametric approach, it often succeeds in classification situations where the decision boundary is highly irregular [16].

In this paper, we introduce a KNN-based consensus algorithm for improved service level agreements in BaaS systems. Even with the efficiency or poor fault tolerance in BaaS systems, the QoS assurance between the CSC and the CSP is better achieved with the enhanced consensus algorithm.

3. Problem Definition

Performance and scalability are always key non-functional requirements in application systems, and such application systems generally achieve extremely high transaction throughput. China's central bank digital currency, DCEP, for example, has about 220,000 TPS. While blockchain systems or BaaS achieve a lower transaction throughput, Bitcoin has 5.5 TPS, and Ethereum has 20 TPS on average. The CSP in BaaS can only provide a similar transaction throughput performance; it cannot meet the requirements of the CSC in the SLA due to the limitation of throughput [17].

The two major challenges of blockchain, scalability and throughput issues, have been studied and improved extensively as the below methods.

Consortium blockchain does not use high-power consensus algorithms such as PoW. They consume much effort and have a complicated consensus process. Hyperledger Fabric is a typical consortium blockchain that uses a Raft or PBFT consensus algorithm [18] to reach a consensus faster than a public blockchain that uses PoW or PoS. It can achieve higher throughput than a public blockchain, and its throughput is 3500 TPS on average [17].

The Ethereum community scheduled a scaling method that performs sharding to improve Ethereum's scalability and capacity. It splits Ethereum data horizontally to spread the load. After Ethereum upgrades to 2.0 with sharding, it is expected to reach 100,000 TPS [17].

NeuChain utilized an ordering-free consensus that makes ordering implicit through deterministic execution to markedly improve the throughput of the blockchain system. The distributed experimental results show that NeuChain can achieve 47.2–64.1X throughput improvement over HyperLedger Fabric [19].

Some hardware methods to improve blockchain performance have been proposed. For instance, a FPGA-based NoSQL caching system with high performance was proposed to improve the throughput and scalability of the blockchain system, and this can increase the throughput to about 10,000 TPS when a cache hit occurs [20].

Except for the above performance optimization for consensus algorithms, some proposals for the optimization of other aspects related to the blockchain system and the blockchain-based framework have also been researched. For some special scenarios, such as confidential transactions, the SymmeProof method, used to reduce the transmission cost, was proposed, and it can improve communication efficiency and indirectly improve the transaction throughput for special types of transactions [21]. A mechanism where full nodes can be compensated fairly for their full blockchain data storage and where clients can query blockchain data effectively was constructed [22]. LineageChain provides an innovative method to support efficient provenance and history data query processing [23]. The secure performance of the blockchain-based federated learning framework has been proposed to be optimized [24].

Due to the need to establish trust between completely anonymous entities, a time-consuming mining-based consensus mechanism was used. Thus, it takes a long time to achieve transaction finality and results in much lower transaction throughput. The limitation of throughput can be increased by using the methods mentioned above. However, compared to traditional e-business application systems that do not adopt blockchain technology, the optimized blockchain still presents a gap between throughput performance and the requirements of e-business scenarios. Although some of the methods mentioned above can improve the throughput of the blockchain system to different degrees, they generally cannot be applied for most scenarios.

Considering the existing studies on blockchain performance optimization, the throughput of a blockchain system cannot reach the same magnitude as traditional e-business application systems. Therefore, another approach where the CSP of BaaS provides an SLA that meets the CSC's requirements is needed.

4. Preliminaries

4.1. BaaS

Blockchain as a Service (BaaS) is a service provided by third parties that create and manage cloud-based networks for customers building their own blockchain applications. The decentralization of blockchain, the pervasiveness of IoT, and the high computing power of cloud computing are combined in BaaS, while the transparency and openness of the system are ensured. The main functional behaviors of blockchain, such as off-chain and on-chain synchronization, node validity, consensus, and forking, are managed by BaaS. The CSC can fully outsource the technical overhead to the CSP [25].

BaaS inherits blockchain's challenges, synchronization mechanism, transaction throughput, storage space, network congestion, accessibility, and cost issues, among others. As discussed in Section 3, the transaction throughput of the blockchain system cannot be improved to match the magnitude of traditional e-business application systems. BaaS also has a transaction throughput issue that cannot be completely resolved. This paper depicts a method to optimize SLAs for key transactions when transaction throughput cannot be further promoted in BaaS.

4.2. Cloud Computing SLAs in BaaS

BaaS is introduced as an important part of the cloud service platform of several giant enterprises that can provide a trustworthy decentralization service, such as the Alibaba-built BaaS system on Kubernetes, the IBM-built BaaS system on Bluemix, and the Microsoft-built BaaS system on the Microsoft Azure cloud platform [2].

An SLA formally defines the relationship between two or more parties in BaaS, one of which is the CSC and one of which is the CSP. It specifies what CSCs can be served by a CSP, the obligations that both the CSC and the CSP shall fulfill, the objectives of the service related to performance, availability, and security, and the processes that guarantee compliance with SLAs. In general, an SLA includes the following typical components: the type of service to be provided; the desired performance level of the service; the reporting process that occurs when the service is unstable or unavailable; the time frame for responding and issuing a problem resolution; the schema for monitoring and reporting the service level; the consequences that result when the CSP does not fulfill its promises; termination clauses; and constraints of service. The SLA is used to evaluate the QoS provided by the CSP in BaaS, as in IaaS, PaaS, and SaaS.

4.3. KNN

As a typical supervised learning method in machine learning, the k-nearest neighbors algorithm (KNN) has shown its advantages for both classification and prediction [26–28]. It is a supervised learning classifier and is used to classify or predict the grouping of an individual data point according to the distance between different feature vectors. KNN has two main phases: (1) the training phase, in which feature vectors are stored and labels of the training samples are classified, and (2) the classification phase, in which an unlabeled vector is classified by assigning the most frequent label among the k training samples that are nearest to that vector. Although it can be used in either regression or classification, it is typically used as a classification algorithm, as in this paper.

The parameter k of the KNN has an extraordinary impact on the classification result, and the data impact the best choice of k. In general, a larger k reduces the effect of noise on classification, but it is then less distinct among class boundaries. Cross-validation is used when assigning different k values to different test samples in previous solutions. A kTree method that learns different optimal k values for different tests of individual data points is proposed in the training stage during kNN classification [29].

Although KNN was developed by Joseph Hodges and Evelyn Fix in 1951 [30], due to its simple implementation and relatively excellent performance, it, along with its improved methods, has been widely used in the applications of several industries in the last three years, including cancer diagnosis in medicine [31], gas-bearing reservoir prediction in geophysics [32], and antenna optimization and design in the electronic industry [33]. This paper applies the KNN to classify the priority of the transaction in BaaS, and transactions are executed with different priorities based on priority classification. It should be noted that the KNN can be replaced by other classification models in practice.

5. KNN-Based Consensus Algorithm

We propose a KNN-based consensus algorithm in this paper, and we describe this algorithm in three subsections: the Priority-Queue-Enabled Transaction Pool, Attribute Selection in the KNN-Based Consensus Algorithm, and Transactions Classified to a Different Priority Queue by Adopting the KNN-Based Consensus Algorithm. The first subsection introduces how the existing consensus algorithm puts newly received transactions into the transaction pool and points out that our optimization aim is to classify newly received transactions and determine their priority according to the classification results. The second subsection explains how classification attributes are selected, and the third subsection introduces how transactions are classified and how the transaction pool is filled according to the classified priority queue.

5.1. *The Priority-Queue-Enabled Transaction Pool*

The blockchain system only allows a limited number of transactions since a block can only contain a limited number of transactions. Transactions that exceed the limit of arrival should not be included in the block. For example, in a four-node consortium blockchain system, a Practical Byzantine Fault Tolerance algorithm [34] (also known as PBFT) is applied, where multiple clients connect to the blockchain system, and the transaction count limit of the transaction pool is set to 1000. As the leader node, Node4 picks the transactions to be sent to the Transaction Pool. However, once it reaches the pool limit, the transactions that arrive later will not be sent to the transaction pool. The existing PBFT consensus protocol can be illustrated in Figure 1.

A new priority-queue-enabled transaction pool is introduced in this paper. Based on the attributes, received transactions should be classified into queues of different priorities, and each queue is a first-in first-out queue. The transactions that arrive later are cached once the queue is full. The details on how the attributes are selected and how the incoming transactions are classified will be presented in the following subsections.

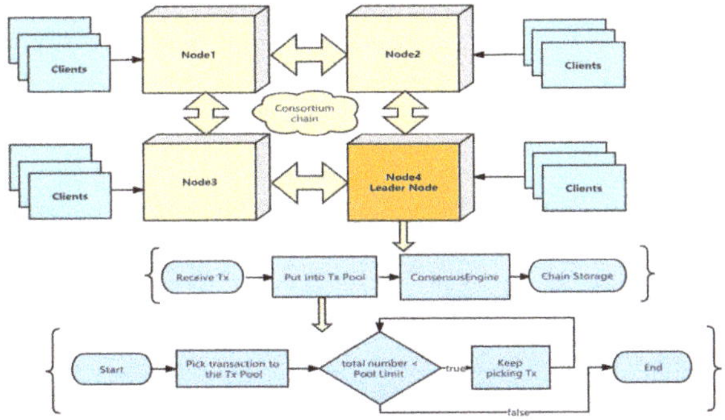

Figure 1. The PBFT consensus protocol.

5.2. *Attribute Selection in the KNN-Based Consensus Algorithm*

The account type, the CSC (cloud service consumer) type, and the contract type i chosen as the attributes to be used in the KNN algorithm to calculate the distance.

(1) Account type. There are different kinds of roles in the blockchain system. Roles based on access control should usually be defined in the system, such as chain administrators, system administrators, and ordinary accounts. Chain administrators have access control permissions, that is, grant permissions. System administrators need to manage permissions related to system functions, and each permission should be granted independently, including contract deployment, user table creation, node management, and system parameter modification. Chain administrators can authorize other accounts to be chain administrators or system administrators or authorize ordinary accounts to write table lists. Table 3 lists the permissions related to the roles.

Table 3. Permissions related to the roles.

Permission Type
Permission of chain administrators
Permission of system administrators
Permission to deploy contracts
Permission to create user tables
Permission to manage nodes
Permission to modify system parameters
Permission to write user tables

(2) CSC type. Transactions from different CSCs have varying priorities. Since CSC clients run similar CSCs, they should have similar priorities. However, if a CSC client experiences limitations in terms of CPU, memory, storage, or network resources, it may need to adjust the CSC priority accordingly.

(3) Contract type. Besides contracts relating to chain management, there are many smart contracts. Some of them handle time-critical applications, some of them handle applications that are not so critical but are urgent, and some of them do not care about the timing. In the consortium Blockchain FISCO BCOS [35], for example, there are many contract types in the consortium Blockchain system, as shown in Tables 4 and 5.

Table 4. Address range in FISCO BCOS [35].

Address Use	Address Range
Ethereum precompiled	0x0001–0x0008
Reserve	0x0008–0x0fff
Precompiled contracts in FISCO BCOS	0x1000–0x1006
Reserved in FISCO BCOS	0x1007–0x5000
Interval assigned by user	0x5001–0xffff
Reserved for CRUD	0x10000+
Used by Solidity	Other address

Table 5. Precompiled contracts in FISCO BCOS [35].

Address	Feature
0x1000	managing system parameters
0x1001	contract of the table factory
0x1002	implementing CRUD operations
0x1003	managing consensus nodes
0x1004	Contract Name Services
0x1005	managing storage table authorities
0x1006	configuring parallel contracts

The chain management contracts have higher priority when we send transactions to the Tx pool. For those fundamental contracts, such as the table factory and CRUD operations, we cannot determine the priority that depends on the CSC's request. The contract type can be used to calculate the priority.

5.3. Transactions Classified to a Different Priority Queue by Adopting the KNN-Based Consensus Algorithm

A KNN-based consensus algorithm is proposed to select the transactions for the queue. When a CSC is registered, we can obtain the key properties of the transactions in this CSC, which may impact its transaction priority, such as the SLA type, the contract type, the account type, the CSC type, the CPU type, the memory size, the storage type, and the network bandwidth.

Classification is an important task in machine learning. The KNN algorithm is simple and accurate and is used for regression models and pattern classification [36]. The term "non-parametric" is used when there are no parameters, or there is a fixed number of parameters, regardless of data size. The size of the training dataset determines the parameters, although no assumptions need to be made about the underlying data distribution. Therefore, KNN is probably the best choice for any classification study that involves little or no prior knowledge of the data distribution. KNN is also a lazy learning method, which means it stores all training data and waits to generate test data without creating a learning model [37]. This is the reason why the KNN algorithm was chosen to optimize the consensus algorithm.

The KNN algorithm classifies as follows: there is an existing set of sample data or a training set. All of these data have been labeled, and we know the class of each piece of data. When a new piece of data has no label, we compare that new piece of data with every existing piece of data. We then take the nearest neighbors and check their labels. We look at the k data that are most similar to the known dataset, which is what k represents. Finally, we perform a majority vote on the similar k data, and the label of the winning vote is selected as the new class to be assigned to the new piece of data. The detailed steps to calculate distance and determine the k value are listed below:

(1) The distance calculation and normalization procedure is as follows: We can use

$$d(p,q) = \sqrt{\sum_{i=0}^{n}(p_i - q_i)^2}$$

to calculate the Euclidean Distance between input data and existing data. Which term in the above equation makes the most difference? It must be the one with the largest magnitude. To reduce the impacts of the magnitude, we need to normalize the sample data to give all factors an equivalent weight. In this paper, every attribute is scaled from 0 to 1, which can be formulated as

$$newValue = (oldValue - min)/(max - min).$$

(2) The KNN algorithm does not need a training procedure. However, the selection of k is important for accuracy. Basically, k should be an integer between 1 and 20. We divided the sample data into two portions: 90% of them was used for the known set, while the remaining 10% was for testing. We increase k successively and calculate the accuracy. The k value that achieves the highest accuracy is chosen for classifying the transactions from the incoming client in the final algorithm. Algorithm 1 describes the procedure by which the incoming transaction is classified into different priority queues, while Algorithm 2 details how transactions are collected and sent to T_x Pool. In the system, there are N queues starting from Q_1 to Q_N, where Q_N and Q_1 have the highest and lowest priority, respectively. n is in 1 N, and $Q_n.size$ denotes the number of transactions in Q_n.

Algorithm 1 Transaction classification.

Input: T_x: The incoming transaction; $Q_1 \ldots Q_N$: The priority Q list from Q_1 to Q_N; *sample*: The sample dataset;
Output: updated Q_n

```
Initial trainingData = prepareLabeledDataset(sample);    ▷ Prepare training dataset
n = classify(Tx,trainingData,k)    ▷ Classify Tx with given training data and k value,n means which Qn it belongs to
if (Qn.size < MAX_SIZE) then
    PUSH Tx to Qn;    ▷ PUSH Tx to corresponding Qn once the Q isn't full
else
    Save Tx to memory pool;    ▷ Otherwise temporarily save the Tx to memory pool
end if
return Qn
```

Algorithm 2 Prepare the T_x Pool.

Input: $Q_1 \ldots Q_N$: The priority queue list from Q_1 to Q_N; *POOL_LIMIT*: The limit on the number of transactions that can be accommodated in the *T_x_Pool*;
Output: T_x_*Pool*

```
Set Tx_Pool to empty, j = N; ▷ j = N ⟶ start to pick up priority queue item from QN to Q1
while (Tx_Pool.size < POOL_LIMIT) and (j > 0) do    ▷ Keep filling the pool in order of priority of the queue items until j ≤ 0 or Tx_Pool is full
    Fill Tx_Pool with Qj
    j − −    ▷ Move to the next lower priority queue item
end while
Save remaining queue items to memory pool if the Tx_Pool is full;
return Tx_Pool
```

With the KNN algorithm, the consensus algorithm can be optimized with SLA assurance. Any transaction that is classified with higher priority can be handled earlier. Figure 2 shows the data flow through which transactions are selected and sent to the transaction pool.

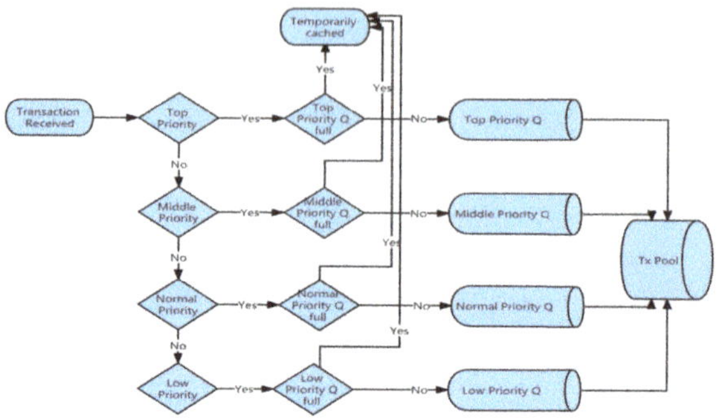

Figure 2. Data flow through which transactions are selected and sent to the transaction pool.

Figure 3 provides the overall framework of the enhanced KNN-enabled consensus algorithm. As shown in the figure, the newly added KNN-enabled transaction classification module is a new concept in the BaaS system. Any CSP can integrate this module into its BaaS framework when it wants to provide a guaranteed SLA to the CSC. Some minor changes are required when preparing the transaction pool, which picks up transactions in the order of priority. If the SLA of one transaction is 1 s, 2000 transactions come in within 1 s, the TPS of the BaaS system is 1000, and the CSP receives this transaction with a sequence number 1100. Only transactions with a sequence number smaller than 1000 can be handled, so this transaction cannot meet the requirements of the SLA. With this KNN-enabled consensus algorithm, since it has a higher priority, it can be sent to the transaction pool with a smaller sequence number (e.g., 100), and it can be handled earlier within the SLA.

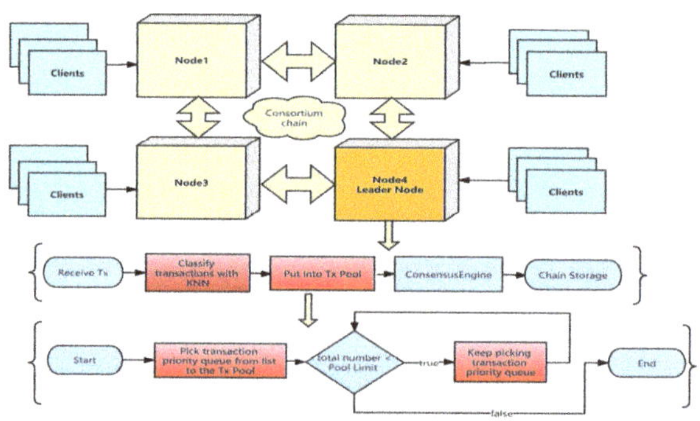

Figure 3. The KNN-enabled PBFT consensus algorithm.

6. Simulation Experiments and Analysis

6.1. TPS Limit Measured from the Existing BaaS System

To evaluate the proposed consensus algorithm, we ran a performance test on the well-known FISCO-BCOS Consortium Blockchain system. The flowchart of the test process is shown in Figure 4.

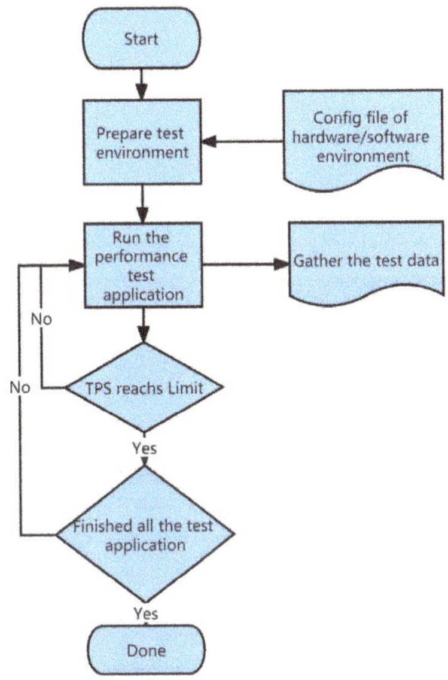

Figure 4. Performance test flow chart.

We deployed a Cloud Virtual Machine standard type S3 on the Tencent Cloud to simulate the BaaS system. Tables 6 and 7 list the details of the hardware and software environment, respectively.

Table 6. Virtual machine hardware configuration.

Hardware Type	Hardware Configuration
CPU	Intel Xeon Cascade Lake 8255C (2.5 GHz)
vCPU	4 Core
Memory (GB)	4 GB
System disk type	High-performance cloud disk
System disk size (GB)	50 GB
Bandwidth	100 Mbps

Table 7. Virtual machine software configuration.

Software Type	Software Version
FISCO BCOS	V2.7.2
OS	CentOS 7.6 64 bit
WeBase	V1.5.1
JAVA	jdk1.8.0
IDE	IntelliJ IDEA 2022.2.2 (Ultimate Edition)

A JAVA performance testing application [38] was used to measure the TPS on the BaaS simulation system. It started at 1000 transactions and set the TPS limit from 10 to 100 with a step of 10. Figure 5 shows the Actual TPS/TPS Limit results. The TPS limit setting is the maximum number of transactions that the testing application is allowed to send, and the actual TPS is the actual number of transactions that the testing application sends. If the actual TPS is smaller than the TPS limit setting, then the testing application has reached the maximum TPS supported by the BaaS system.

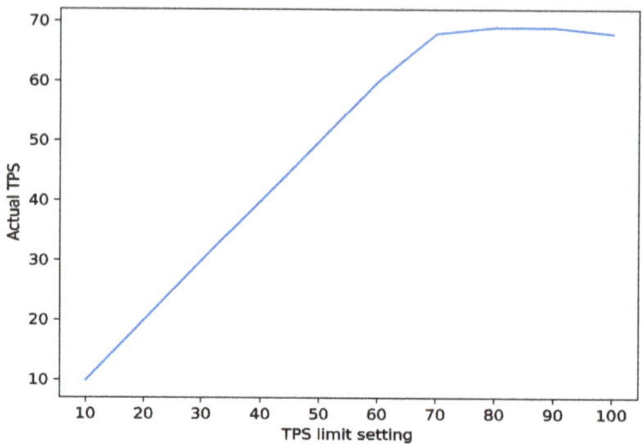

Figure 5. TPS limit of the performance evaluation system.

6.2. Simulation Experiments with the Existing Consensus Algorithm

We considered transaction type, account type, and CSC type as the input features of the KNN. The normalized values are from 0 to 1, where 0 is the highest priority and 1 is the lowest priority.

Without a KNN-based consensus algorithm, the transactions should be handled in a FIFO way. In this way, the transaction that arrived early will be served early. We generated 1000 transactions with different transaction types, account types, CSC types, and arrival times. Table 8 shows part of the transaction data. Algorithm 3 illustrates how the transaction pool picks up the transactions in a FIFO way. Correspondingly, Figure 6 shows a scatter diagram of the handled transactions.

Table 8. Samples of transactions.

Transaction Type	Account Type	CSC Type	Arriving Time (ms)
0.503	0.536	0.592	506
0.014	0.399	0.454	539
0.655	0.547	0.991	52
0.328	0.095	0.579	152
0.271	0.810	0.502	822
0.734	0.675	0.667	391
...	...	...	...
0.608	0.726	0.806	84
0.472	0.610	0.017	935
0.410	0.239	0.538	257
0.870	0.715	0.617	23

Algorithm 3 Transactions selected with the FIFO method.

Input: T_x_Table: A 2D array as the transaction table, one row presents one T_x;
Output: Q: An FIFO Q with all transactions;
Set Q to empty;
while ($Q.size < T_x\_MAX$) and (T_x_Table has unused transaction) **do** ▷ Keep picking up T_x fromT_x_Table until Q is full or T_x_Table has no unused transaction
  Get a new T_x from T_x_Table
  Insert T_x to Q
end while
return Q

Figure 6 shows that, with the FIFO method, transactions that arrive earlier will be handled earlier, even if it has a lower priority classified by their attributes. The transaction start time is irrelevant to its priority, and the higher priority transaction will not be handled earlier.

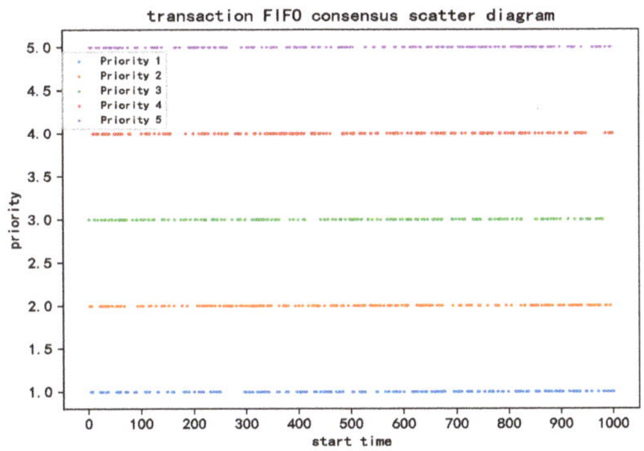

Figure 6. FIFO way transaction scatter diagram.

6.3. Simulation Experiments with the KNN-Enabled Enhanced Consensus Algorithm

6.3.1. First Round of k Value Selection

In this paper, Algorithms 1 and 2 are used to classify 1000 transactions into 5 priority queues using the KNN algorithm, in which the number of nearest neighbors, k is an important parameter. First, we apply Algorithm 4 to the selection of the best k value. It adopts the KNN classification using Scikit-learn in python. Generally, the dataset is split into a training set and a testing set. We then run the KNN classifier with different k values. The accuracy score is used to check the accuracy of our KNN model and the k value. The k value with the highest accuracy score should be selected as the best k value for handling unknown incoming transactions and checking its target priority.

Algorithm 4 KNN k value selection with a single training/test set split.

Input: *Training_Data*: The prepared transaction training data; *Training_Target*: The target priority of the prepared transaction training data; k: The k value used in KNeighborsClassifier function;

Output: *accuracy_score*: The accuracy score of the given k value;

1: $Xtrain, Xtest, ytrain, ytest = train\_test\_split(Training\_Data, Training\_Target, test\_size = 0.3)$; ▷ The dataset is split into a training set and a testing set, and the testing set size is 30 percent
2: *knn* = neighbors.KNeighborsClassifier(k, *weights* = "distance") ▷ Prepare the KNN classifier by using the Scikit-learn module with the specified k value
3: *knn*.fit($Xtrain, ytrain$) ▷ Fit the KNN classifier with the split training set
4: *yprediction* = knn.predict($Xtest$) ▷ Predict the target priority of the split testing set
5: $accuracy\_score = accuracy\_score(ytest, yprediction)$ ▷ Check the accuracy of the target priority of the testing set
6: **return** *accuracy_score*

We plotted the accuracy of different k values ranging from 1 to 20 in Figure 7. It demonstrates that $k = 12$ can achieve the highest accuracy among all k values. Therefore, we used $k = 12$ for KNN in all remaining experiments. A flowchart for classifying all 1000 transactions into the 5 priority queues once k is fixed is shown in Figure 8.

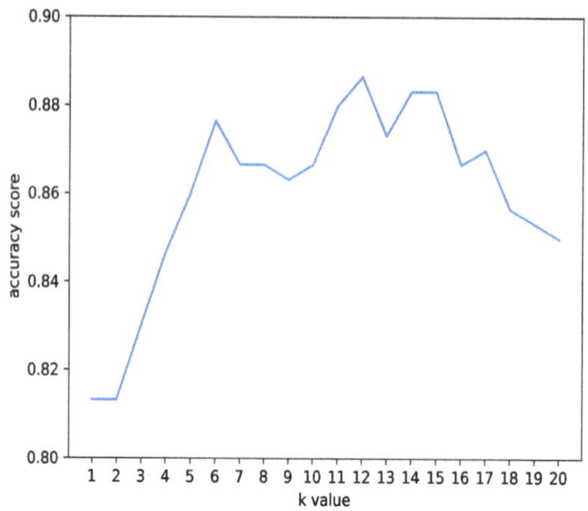

Figure 7. *k* value selection.

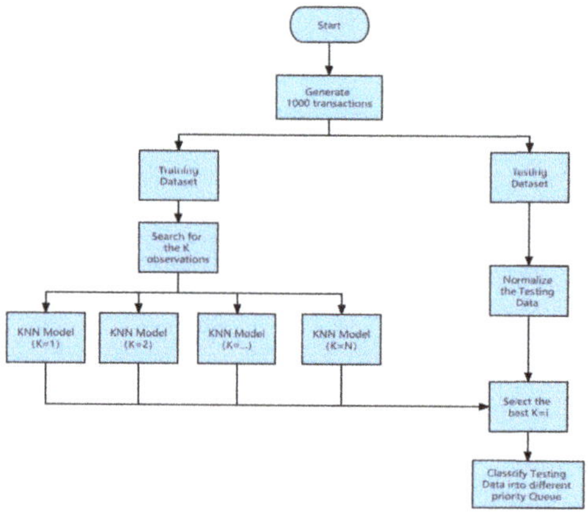

Figure 8. Transaction classification.

6.3.2. Choose a *k* Value in K-Fold cross-Validation

In Section 6.3.1, we presented the initial method for selecting the best value of *k*. However, is this the optimal *k* value? In the first round's KNN *k* value, we only use a single training/test set split. The test set will only include a small portion of randomly selected data. In this scenario, the test set may not accurately represent "new unseen data", which could lead to an overestimation of performance if it is used alone (due to potentially significant variability in the test results). By using cross-validation, all available data can be used for testing purposes, thereby ensuring that "bad" observations also play a role during the testing process. The train–test split and k-fold cross-validation are both examples of resampling methods in statistics. Resampling methods involve taking a sample from a dataset and using it to estimate unknown quantities. These techniques are particularly useful in machine learning and data analysis when a limited amount of data is available for model training and evaluation. The *k* value generated by using only one training/test set split will change due to the selection of the training/test set. We must use k-fold cross-validation to eliminate this effect, so we use the following algorithm to ensure that we consider all of the elements in the dataset. We finally obtain a *k* value of 4, as shown in Figure 9 below.

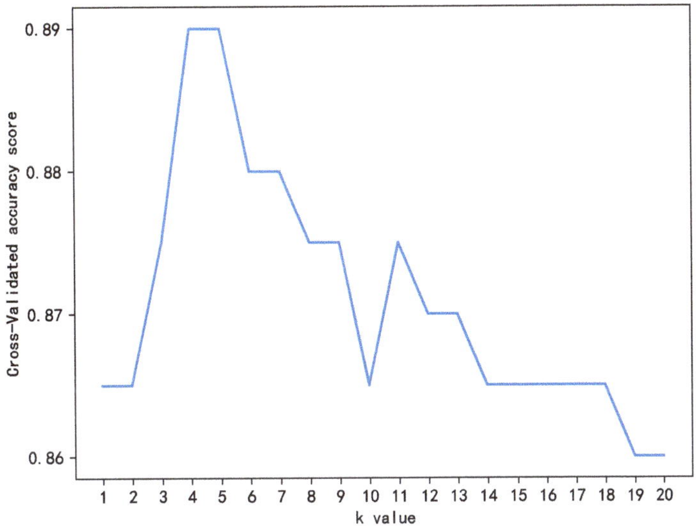

Figure 9. Cross-validated accuracy score scatter diagram.

6.3.3. Performance Optimization and Evaluation

When we obtain an optimal k value, we use 1000 transactions as the training set, and their layout is shown in Figure 10. In the figure, different categories of data in the training set data sometimes overlap (meaning that the categories of this part of the data are blurred). This part of the data will cause some model overfitting. Based on the learning curve in Figure 11, we know that there are still opportunities to optimize performance. One idea is to directly remove this part of the overlapping data, which is referred to as a clipping method.

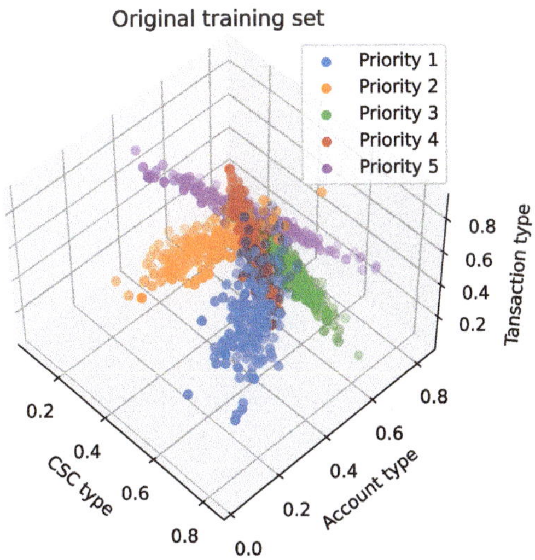

Figure 10. Original training set scatter diagram.

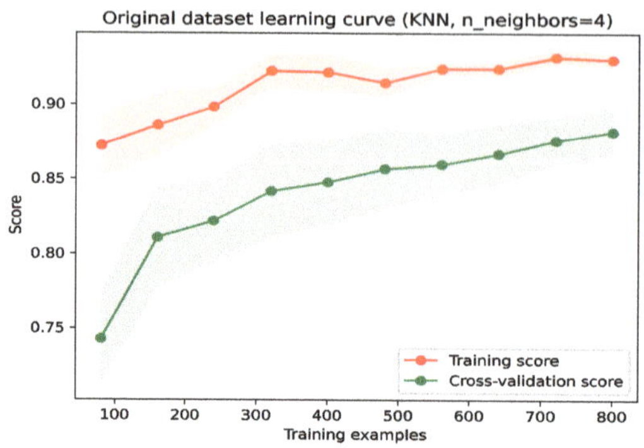

Figure 11. Original training set learning curve scatter diagram.

The clipping method randomly divides the training set, D, into two parts. One part is used as a new training set, and the other part is used as a test set. Based on the new training set, the KNN method is used to classify the test set, and the misclassified samples are removed from the entire training set. Since the division of the training set D is randomly divided, it is difficult to ensure that the samples in the overlapping part of the data will be eliminated in the first clip. After obtaining the new training set, the above operations can be repeated, and clearer class boundaries can be obtained. We can obtain its layout image (Figure 12) and learning curve (Figure 13), as shown below. Compared with the original training set, we achieved improved performance with a smaller size.

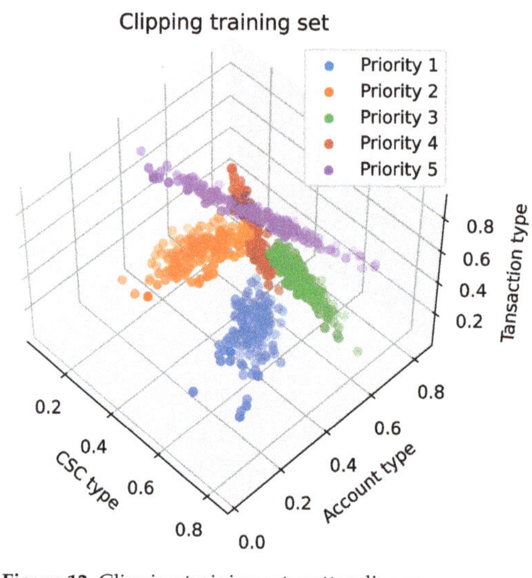

Figure 12. Clipping training set scatter diagram.

By observing the learning curve optimized by the clipping method, it can be seen that when the number of samples is around 300, it already has a good fitting performance. At the same time, as shown by the layout of samples in Figure 12, there are a large number of samples in the center of each class, indicating that we can reduce the size of the training set by compressing the KNN training set. The compressing method is used when a large number of samples of the same type are concentrated in the center of the cluster, and these concentrated samples have little effect on classification, so these samples can be discarded. The training set is divided into two parts in this method. The first part is a store that contains a portion of the samples, and the second part is a grab bag that contains the remaining samples. The store is used for the training set of the KNN model, and the grabbag is used for the test set. The misclassified samples are moved from the grab bag to the store. The store continues to be used with increased samples, and the grab bag with decreased samples is used to train and test the KNN model again until all samples in the grab bag are correctly classified or until the number of samples in the grab bag is 0. After compression, the store keeps a portion of the randomly selected samples at initialization as well as the misclassified samples in each subsequent cycle. Since the clipping method removes all outliers, these selected misclassified samples are concentrated at the edge of the cluster and are considered correct samples with a large classification effect. The final training set is smaller. We can see its layout in Figure 14. The learning curve in Figure 15 shows that the training set still has a similar accuracy to that of the clipping training set.

Each transaction will be executed with its priority, and arrival time is only used when the transactions have the same priority. If two transactions have the same priority, the transaction that arrived earlier will be executed earlier. Table 9 describes the priority and new start time of each transaction based on its attributes.

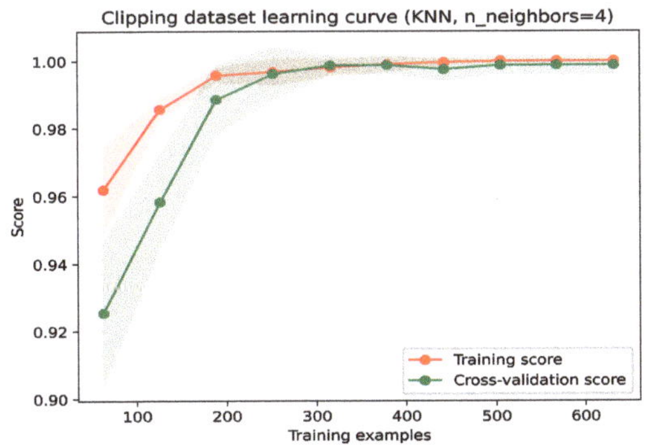

Figure 13. Clipping training set learning curve scatter diagram.

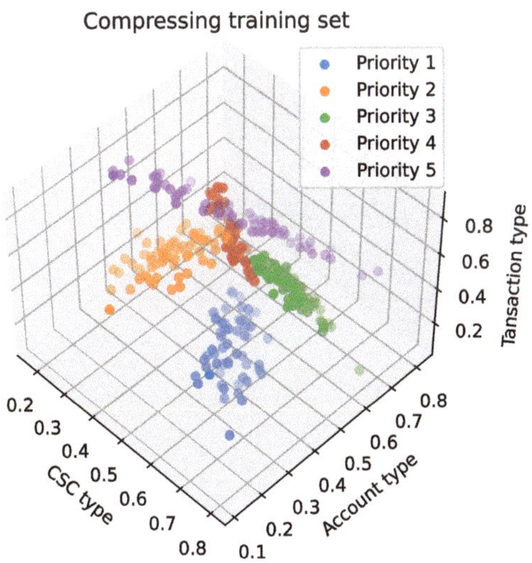

Figure 14. Compressing training set scatter diagram.

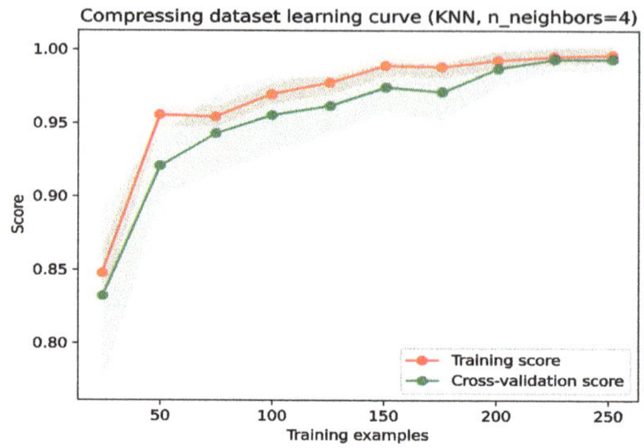

Figure 15. Compressing training set learning curve scatter diagram.

Table 9. Transactions with priority and new start time.

Transaction Type	Account Type	CSC Type	Arrival Time (ms)	Priority	Start Time (ms)
0.077	0.132	0.058	396	1	1
0.133	0.029	0.226	330	1	2
0.031	0.007	0.213	500	1	3
0.208	0.017	0.174	132	1	4
0.116	0.069	0.007	326	1	5
…	…	…	…	…	…
0.067	0.306	0.009	499	1	12
0.034	0.143	0.545	331	2	13
0.496	0.066	0.171	393	2	14
0.248	0.371	0.122	788	2	15
…	…	…	…	…	…

With the proposed KNN consensus algorithm, the scatter diagram of the transactions is shown in Figure 16, where 1 is the highest priority, and 5 is the lowest priority. Differently from the start time that only relates to the arrival time in the FIFO method, as shown in Figure 6, the start time with the KNN-based consensus algorithm relates to the priority of the transaction, which introduces the QoS method to the consensus algorithm and helps to better achieve SLA requirements and provide BaaS users an improved experience.

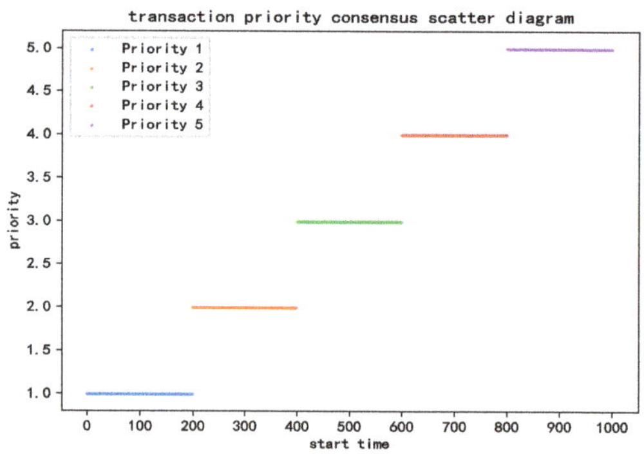

Figure 16. Prioritized transaction scatter diagram.

When a new transaction needs to be added to the transaction pool, it needs to be classified by the KNN algorithm. The prediction is only determined by the number of sample points in the training set, which is a constant value once the training set is finalized. The time complexity of this algorithm is O(1), and the space complexity is also O(1), which is irrelevant to the number of transactions in the transaction pool. After adopting the clipping and compressing algorithms, the number of samples in the training set is greatly reduced while ensuring a good fitting performance. The given example reduces the number of samples from 1000 to 200+. Algorithm 5 describes how a new transaction is added to the priority queue with the new compressed training set.

Algorithm 5 New transaction classification.

Input: T_x: The new transaction; $Q_1 \ldots Q_N$: The priority Q list from Q_1 to Q_N; k: The best k in K-fold Cross-Validation *Traning_data*: The compressed training data; *Traning_target*: The compressed training target;
Output: Updated Q_n

Initial $KNN\_clf = neighbors.KNeighborsClassifier(n\_neighbors = k)$; ▷ Prepare KNeighborsClassifier with the best k value
$KNN\_clf.fit(Traning\_data, Traning\_target)$ ▷ Fit the compressed training data and target to the KNeighborsClassifier
$predict = KNN\_clf.predict(T_x)$ ▷ Predict the class of new incoming T_x
if ($predict == n$) and ($Q_n.size < MAX\_SIZE$) **then**
  PUSH T_x to Q_n; ▷ PUSH T_x to Q_n once the Q isn't full
else
  Save T_x to memory pool; ▷ Otherwise temporarily save the T_x to memory pool
end if
return Q_n

Compared with existing blockchain consensus algorithms, the proposed KNN-based consensus algorithm guarantees that higher priority transactions are executed earlier. Table 9 shows that the CSC type is important for calculating the priority. If a CSC has a short SLA requirement, its CSC type should be assigned with a high priority. This helps to deliver services to the CSCs within the SLA limitation in the BaaS system. Considering the transaction with attributes {0.034, 0.143, 0.545} in Table 9 as an example, its arrival time is 331. Without the proposed KNN-based optimization consensus algorithm, the transaction pool assigns it with a sequence number of 331. If the SLA of this transaction has a short duration, the transaction may miss the SLA. With the KNN-based consensus algorithm, however, it should be classified into a higher target priority queue. In this way, the transaction pool assigns it with a sequence number of 13 and, therefore, is more likely to satisfy the SLA.

7. Conclusions

Most existing consensus algorithms do not consider the priority. If a high-priority transaction comes late, it needs to wait until other, lower-priority transactions are handled. Due to the TPS limitation, it is difficult to meet SLA requirements in the BaaS system. This paper proposes a KNN-based consensus algorithm to enhance the SLA handling in the BaaS system. With the KNN-based consensus algorithm, each transaction is handled based on its priority. The transactions that arrive late but have high priority can be handled early. In this way, the BaaS system can better satisfy the SLA between the CSP and the CSC. The proposed KNN-based blockchain consensus algorithm is a common solution, and we only choose three attributes for classification. The experimental results illustrate the advantages of the proposed algorithm. In the future, we will consider more attributes for classification and try using other classification methods that can outperform the KNN.

Author Contributions: Conceptualization, Q.Z., L.W. and J.H.; formal analysis, Q.Z., L.W. and J.H.; investigation, Q.Z. and L.W.; methodology, Q.Z., L.W. and T.L.; project administration, L.W. and J.H.; resources, J.H. and T.L.; software, Q.Z., T.L. and L.W.; supervision, J.H.; validation, Q.Z. and T.L.; writing—original draft preparation, Q.Z., L.W. and J.H.; writing—review and editing, Q.Z., L.W. and T.L. All authors have read and agreed to the published version of the manuscript.

Funding: This work was supported by the Ministry of Education of Humanities and Social Science Project (grant no. 19YJAZH047), the Ministry of Science and Technology Innovation Method Work

Special Project (grant no.2017IM030100), Sichuan Provincial Higher Education Talent Training Quality and Teaching Reform Project (grant no. JG2021-995), Sichuan Provincial Higher Education Talent Training Quality and Teaching Reform Project (grant no. JG2021-1016), and the Social Practice Research for Teachers of Southwestern University of Finance and Economics (grant no. 2022JSSHSJ11).

Data Availability Statement: All the data in this paper are publicly available. Please contact the corresponding author to obtain them.

Conflicts of Interest: The authors declare that they have no conflict of interest.

References

1. Zhang, C.; Chen, Y. A review of research relevant to the emerging industry trends: Industry 4.0, IoT, blockchain, and business analytics. *J. Ind. Integr. Manag.* **2020**, *5*, 165–180. [CrossRef]
2. Song, J.; Zhang, P.; Alkubati, M.; Bao, Y.; Yu, G. Research advances on blockchain-as-a-service: Architectures, applications and challenges. *Digit. Commun. Netw.* **2021**, *8*, 466–475. [CrossRef]
3. Buterin, V. A next-generation smart contract and decentralized application platform. *White Pap.* **2014**, *3*, 1–36. Available online: https://ethereum.org/en/whitepaper/#a-next-generation-smart-contract-and-decentralized-application-platform (accessed on 28 January 2023).
4. Peng, Z.; Xu, J.; Hu, H.; Chen, L. BlockShare: A Blockchain Empowered System for Privacy-Preserving Verifiable Data Sharing. *IEEE Data Eng. Bull.* **2022**, *45*, 14–24.
5. Wu, H.; Peng, Z.; Guo, S.; Yang, Y.; Xiao, B. VQL: Efficient and Verifiable Cloud Query Services for Blockchain Systems. *IEEE Trans. Parallel Distrib. Syst.* **2022**, *33*, 1393–1406. [CrossRef]
6. Wang, H.; Xu, C.; Zhang, C.; Xu, J.; Peng, Z.; Pei, J. vChain+: Optimizing Verifiable Blockchain Boolean Range Queries. In Proceedings of the 2022 IEEE 38th International Conference on Data Engineering (ICDE), Kuala Lumpur, Malaysia, 9–12 May 2022; pp. 1927–1940. [CrossRef]
7. Sayeed, S.; Marco-Gisbert, H. Assessing Blockchain Consensus and Security Mechanisms against the 51% Attack. *Appl. Sci.* **2019**, *9*, 1788. [CrossRef]
8. Akcora, C.G.; Gel, Y.R.; Kantarcioglu, M. Blockchain networks: Data structures of Bitcoin, Monero, Zcash, Ethereum, Ripple, and Iota. *Wiley Interdiscip. Rev. Data Min. Knowl. Discov.* **2022**, *12*, e1436. [CrossRef] [PubMed]
9. Du, M.; Chen, Q.; Ma, X. MBFT: A New Consensus Algorithm for Consortium Blockchain. *IEEE Access* **2020**, *8*, 87665–87675. [CrossRef]
10. Li, D.; Deng, L.; Cai, Z.; Souri, A. Blockchain as a service models in the Internet of Things management: Systematic review. *Trans. Emerg. Telecommun. Technol.* **2022**, *33*, e4139. [CrossRef]
11. Samaniego, M.; Jamsrandorj, U.; Deters, R. Blockchain as a Service for IoT. In Proceedings of the 2016 IEEE International Conference on Internet of Things (iThings) and IEEE Green Computing and Communications (GreenCom) and IEEE Cyber, Physical and Social Computing (CPSCom) and IEEE Smart Data (SmartData), Chengdu, China, 15–18 December 2016; pp. 433–436. [CrossRef]
12. Ardagna, D.; Casale, G.; Ciavotta, M.; Pérez, J.F.; Wang, W. Quality-of-service in cloud computing: Modeling techniques and their applications. *J. Internet Serv. Appl.* **2014**, *5*, 1–17. [CrossRef]
13. Viriyasitavat, W.; Da Xu, L.; Bi, Z.; Hoonsopon, D.; Charoenruk, N. Managing qos of internet-of-things services using blockchain. *IEEE Trans. Comput. Soc. Syst.* **2019**, *6*, 1357–1368. [CrossRef]
14. Tan, W.; Zhu, H.; Tan, J.; Zhao, Y.; Xu, L.D.; Guo, K. A novel service level agreement model using blockchain and smart contract for cloud manufacturing in industry 4.0. *Enterp. Inf. Syst.* **2022**, *16*, 1939426. [CrossRef]
15. Rashid, A.; Chaturvedi, A. Cloud computing characteristics and services: A brief review. *Int. J. Comput. Sci. Eng.* **2019**, *7*, 421–426. [CrossRef]
16. Pedregosa, F.; Varoquaux, G.; Gramfort, A.; Michel, V.; Thirion, B.; Grisel, O.; Blondel, M.; Prettenhofer, P.; Weiss, R.; Dubourg, V.; et al. Scikit-learn: Machine Learning in Python. *J. Mach. Learn. Res.* **2011**, *12*, 2825–2830.
17. Kshetri, N. The Economics of Central Bank Digital Currency [Computing's Economics]. *Computer* **2021**, *54*, 53–58. [CrossRef]
18. Yang, G.; Lee, K.; Lee, K.; Yoo, Y.; Lee, H.; Yoo, C. Resource Analysis of Blockchain Consensus Algorithms in Hyperledger Fabric. *IEEE Access* **2022**, *10*, 74902–74920. [CrossRef]
19. Peng, Z.; Zhang, Y.; Xu, Q.; Liu, H.; Gao, Y.; Li, X.; Yu, G. NeuChain: A Fast Permissioned Blockchain System with Deterministic Ordering. *Proc. VLDB Endow.* **2022**, *15*, 2585–2598. [CrossRef]
20. Sanka, A.I.; Cheung, R.C. Efficient High Performance FPGA based NoSQL Caching System for Blockchain Scalability and Throughput Improvement. In Proceedings of the 2018 26th International Conference on Systems Engineering (ICSEng), Sydney, NSW, Australia, 18–20 December 2018; pp. 1–8. [CrossRef]
21. Gao, S.; Peng, Z.; Tan, F.; Zheng, Y.; Xiao, B. SymmeProof: Compact Zero-Knowledge Argument for Blockchain Confidential Transactions. *IEEE Trans. Dependable Secur. Comput.* **2022**, 1. [CrossRef]
22. Cai, C.; Xu, L.; Zhou, A.; Wang, C. Toward a Secure, Rich, and Fair Query Service for Light Clients on Public Blockchains. *IEEE Trans. Dependable Secur. Comput.* **2022**, *19*, 3640–3655. [CrossRef]

23. Ruan, P.; Chen, G.; Dinh, T.T.A.; Lin, Q.; Ooi, B.C.; Zhang, M. Fine-Grained, Secure and Efficient Data Provenance on Blockchain Systems. *Proc. VLDB Endow.* **2019**, *12*, 975–988. [CrossRef]
24. Peng, Z.; Xu, J.; Chu, X.; Gao, S.; Yao, Y.; Gu, R.; Tang, Y. Vfchain: Enabling verifiable and auditable federated learning via blockchain systems. *IEEE Trans. Netw. Sci. Eng.* **2021**, *9*, 173–186. [CrossRef]
25. Onik, M.M.H.; Miraz, M.H. Performance Analytical Comparison of Blockchain-as-a-Service (BaaS) Platforms. In *Emerging Technologies in Computing*; Miraz, M.H., Excell, P.S., Ware, A., Soomro, S., Ali, M., Eds.; Springer International Publishing: Cham, Switzerland, 2019; pp. 3–18.
26. Samet, H. K-nearest neighbor finding using MaxNearestDist. *IEEE Trans. Pattern Anal. Mach. Intell.* **2007**, *30*, 243–252. [CrossRef]
27. Martínez, F.; Frías, M.P.; Pérez-Godoy, M.D.; Rivera, A.J. Dealing with seasonality by narrowing the training set in time series forecasting with kNN. *Expert Syst. Appl.* **2018**, *103*, 38–48. [CrossRef]
28. Li, T.; Qian, Z.; Deng, W.; Zhang, D.; Lu, H.; Wang, S. Forecasting crude oil prices based on variational mode decomposition and random sparse Bayesian learning. *Appl. Soft Comput.* **2021**, *113*, 108032. [CrossRef]
29. Zhang, S.; Li, X.; Zong, M.; Zhu, X.; Wang, R. Efficient kNN Classification With Different Numbers of Nearest Neighbors. *IEEE Trans. Neural Netw. Learn. Syst.* **2018**, *29*, 1774–1785. [CrossRef] [PubMed]
30. Fix, E. *Discriminatory Analysis: Nonparametric Discrimination, Consistency Properties*; USAF School of Aviation Medicine: Dayton, OH, USA, 1985; Volume 1.
31. Mahfouz, M.A.; Shoukry, A.; Ismail, M.A. EKNN: Ensemble classifier incorporating connectivity and density into kNN with application to cancer diagnosis. *Artif. Intell. Med.* **2021**, *111*, 101985. [CrossRef] [PubMed]
32. Song, Z.H.; Sang, W.J.; Yuan, S.Y.; Wang, S.X. Gas-Bearing Reservoir Prediction Using k-nearest neighbor Based on Nonlinear Directional Dimension Reduction. *Appl. Geophys.* **2022**, 1–11. [CrossRef]
33. Cui, L.; Zhang, Y.; Zhang, R.; Liu, Q.H. A Modified Efficient KNN Method for Antenna Optimization and Design. *IEEE Trans. Antennas Propag.* **2020**, *68*, 6858–6866. [CrossRef]
34. Castro, M.; Liskov, B. Practical byzantine fault tolerance. In Proceedings of the Third Symposium on Operating Systems Design and Implementation, New Orleans, LA, USA, 22–25 February 1999; Volume 99, pp. 173–186.
35. FISCO BCOS Platform. Available online: https://github.com/fisco-bcos (accessed on 28 January 2023).
36. Abu Alfeilat, H.A.; Hassanat, A.B.; Lasassmeh, O.; Tarawneh, A.S.; Alhasanat, M.B.; Eyal Salman, H.S.; Prasath, V.S. Effects of distance measure choice on k-nearest neighbor classifier performance: A review. *Big Data* **2019**, *7*, 221–248. [CrossRef]
37. Wettschereck, D.; Aha, D.W.; Mohri, T. A review and empirical evaluation of feature weighting methods for a class of lazy learning algorithms. *Artif. Intell. Rev.* **1997**, *11*, 273–314. [CrossRef]
38. FISCO BCOS Performance Demo Program. Available online: https://github.com/FISCO-BCOS/java-sdk-demo/blob/main/src/main/java/org/fisco/bcos/sdk/demo/perf/PerformanceOk.java (accessed on 28 January 2023).

electronics

Article

Peak Shaving and Frequency Regulation Coordinated Output Optimization Based on Improving Economy of Energy Storage

Daobing Liu [1,2], Zitong Jin [1,2], Huayue Chen [3,*], Hongji Cao [1,2], Ye Yuan [1,2], Yu Fan [1,2] and Yingjie Song [4]

1 College of Electrical Engineering and New Energy, China Three Gorges University, Yichang 443002, China; liudb@ctgu.edu.cn (D.L.); 201908520721024@ctgu.edu.cn (Z.J.); 201908520721145@ctgu.edu.cn (H.C.); 202008580021232@ctgu.edu.cn (Y.Y.); 202008580021033@ctgu.edu.cn (Y.F.)
2 Hubei Provincial Key Laboratory for Operation and Control of Cascade Hydropower Station, China Three Gorges University, Yichang 443002, China
3 School of Computer Science, China West Normal University, Nanchong 637002, China
4 College of Computer Science and Technology, Shandong Technology and Business University, Yantai 264005, China; songyj@sdtbu.edu.cn
* Correspondence: sunnyxiaoyue20@cwnu.edu.cn

Abstract: In this paper, a peak shaving and frequency regulation coordinated output strategy based on the existing energy storage is proposed to improve the economic problem of energy storage development and increase the economic benefits of energy storage in industrial parks. In the proposed strategy, the profit and cost models of peak shaving and frequency regulation are first established. Second, the benefits brought by the output of energy storage, degradation cost and operation and maintenance costs are considered to establish an economic optimization model, which is used to realize the division of peak shaving and frequency regulation capacity of energy storage based on peak shaving and frequency regulation output optimization. Finally, the intra-day model predictive control method is employed for rolling optimization. An intra-day peak shaving and frequency regulation coordinated output optimization strategy of energy storage is proposed. Through the example simulation, the experiment results show that the electricity cost of the whole day is reduced by 10.96% by using the coordinated output strategy of peak shaving and frequency regulation. The obtained further comparative analysis results and the life cycle economic analysis show that the profit brought by the proposed coordinated output optimization strategy is greater than that for separate peak shaving or frequency modulation of energy storage under the same capacity.

Keywords: energy storage; model predictive control; peak shaving and frequency regulation; output optimization

Citation: Liu, D.; Jin, Z.; Chen, H.; Cao, H.; Yuan, Y.; Fan, Y.; Song, Y. Peak Shaving and Frequency Regulation Coordinated Output Optimization Based on Improving Economy of Energy Storage. *Electronics* **2022**, *11*, 29. https://doi.org/10.3390/electronics11010029

Academic Editor: Jonghoon Kim

Received: 23 November 2021
Accepted: 21 December 2021
Published: 22 December 2021

1. Introduction

Under the goal of "carbon neutralization", energy storage has become the focus of development because of its rapid charging and discharging characteristics. On the power generation side, energy storage can be connected to make the power grid more "friendly" towards new energy sources such as wind power and photovoltaic [1–4]. On the user side, energy storage can cut the peaks and fill the valleys, improving users' power consumption habits and reducing peak power consumption. According to the "14th five-year plan", China's energy storage will reach more than 30 million kilowatts in 2025. Compared with 2020, the scale of the energy storage market will expand nearly tenfold, and local policies and market mechanisms will be better, which means that the application of energy storage in various scenarios needs to be further improved. With the increase in energy storage reserve capacity on the user side, making good use of this energy storage capacity can increase the system stability and the economy of energy storage on the user side [5–9].

At present, China mainly implements two-part electricity price and timeshare electrovalence policies for industrial users, hoping that industrial users can change their electricity consumption habits, but industrial production habits are difficult to change [10–13]. Therefore, the industrial users can be equipped with energy storage systems to reduce the maximum demand of users, according to the policy, and adopt the strategy of low charge and high discharge according to the time-of-use electricity pricing to charge during low electricity price periods, discharging in the high electricity price periods and peak load periods. This method not only improves the power consumption habits of users, but also obtains economic benefits by using the peak valley electricity price difference and the maximum demand electricity charge difference [14–18]. However, the energy storage battery needs more deep discharge when participating in the peak shaving on the user side, which will produce a large battery degradation effect, limiting the economy of peak shaving.

Therefore, the economic benefits of user-side energy storage participating in frequency regulation can improve the economy of user equipped energy storage. At present, China's small capacity energy storage power stations cannot be allowed to compete for frequency regulation services, but the establishment of auxiliary service markets such as frequency regulation and standby is conducive to guiding investment to improve the flexibility of power systems [19–25]. With the improvement of energy storage service market mechanisms, the future frequency regulation service market will certainly expand to individual participation, so the energy storage on the user side can not only achieve low absorption and high amplification, but it can also participate in the frequency regulation service market to obtain revenue [26–29], which will encourage industrial users to actively equip energy storage batteries and reduce peak power consumption.

In other countries, the frequency regulation market such as Pennsylvania–New Jersey–Maryland (PJM) in the United States is relatively mature. In this market, the energy storage devices represented by batteries and aircraft turbines has been introduced into the frequency regulation service market [30]. Salles et al. [31] used the battery energy storage systems in an Italian shopping mall to shave the peak consumption and get benefit from it. It has been proven that the strategy including peak shaving can increase the economy on the user side. However, this ignores that energy storage can also generate benefits by participating in frequency regulation services. The energy on the user side is used to participate in the frequency regulation service in the power market to obtain income [32–36]. They make the energy on the user side follow the frequency regulation signals in the PJM market for equivalent output, similar to energy storage. Shi et al. [37] used the battery storage system for peak shaving and frequency regulation through the joint optimization framework on the user side. Based on the degradation effect of energy storage batteries, it was found that the joint optimization has super linear gain compared with energy storage for frequency regulation or peak shaving alone, but this method is only used in the day-ahead planning stage, and simply follows the frequency regulation signal during the day's frequency regulation real-time output. It fails to achieve real-time optimization, and the peak shaving model only considers the peak cost in the electricity price, not the difference of timeshare electrovalence. Based on the prediction model, Liu et al. [38] proposed a model predictive control (MPC) intra-day rolling optimization frequency regulation model. The model considers the degradation effect, but it does not consider the operation and maintenance cost of the battery, and while it achieves intra-day optimization, it does not consider the day-ahead bidding capacity of the energy storage.

On the basis of this research, this paper puts forward a strategy for day-ahead peak shaving and frequency regulation planning and a frequency regulation rolling optimization output strategy for user-side intra-day energy storage. The strategy considers the degradation cost and operation and maintenance cost of energy storage. By solving the economic optimal model of peak shaving and frequency regulation coordinated output a day ahead, the division of peak shaving and frequency regulation capacity of energy storage is obtained, and a real-time output strategy of energy storage is obtained by MPC intra-day rolling optimization. Finally, through the 24-h economic analysis of the strategy proposed in this paper and the economic analysis of the whole life cycle, it can be concluded that the economic benefit of energy storage participating in peak shaving and frequency regulation coordinated output is much higher than that of energy storage batteries participating in peak shaving or frequency regulation under the same capacity. Through simulation, it is demonstrated that energy storage participating in peak shaving can reduce the battery degradation cost when energy storage is used for frequency regulation by reducing the number of battery cycles, thereby increasing the service life of energy storage batteries.

The main contributions of this work are described as follows:

- A peak shaving and frequency regulation coordinated output strategy based on the existing energy storage participating is proposed to improve the economic problem of energy storage development and increase the economic benefits of energy storage on the industrial park.
- The profit and cost models of peak shaving and frequency regulation are established.
- The benefits brought by the output of energy storage, degradation cost and operation and maintenance costs are considered to establish an economic optimization model.
- The intra-day model predictive control method is employed for rolling optimization.

2. Establishment of the Peak Shaving Model

In order to promote staggered peak power consumption, the industrial peak valley electricity price of a city in China is shown in Table 1. For industrial parks with two-part electricity pricing, the electricity charge includes the electricity charge and the basic electricity charge [39]. The electricity charge is calculated according to the amount of electricity consumption, and the basic electricity charge is calculated according to the maximum demand. For convenience of expression, function $f_1(x, y)$ is defined as follows.

$$f_1(x,y) = \begin{cases} x-y & (x > y) \\ 0 & (x \leq y) \end{cases} \tag{1}$$

where, x and y are mathematical variables.

Table 1. The peak valley price.

Type	Time Interval	Electricity Price/(Yuan/kW h) $^{-1}$
Valley	23:00–07:00	0.414
Normal	07:00–08:00; 11:00–18:00	0.782
Peak	08:00–11:00; 18:00–23:00	1.149

On this basis, the total electricity charge for industrial users is calculated as follows.

$$M_{\text{elec}} = C_x \cdot S_o + C_{x1} \cdot f_1(\max(\mathbf{s}), S_o) + \sum_{t=1}^{T} C_{\text{elec}}(t) \cdot s(t) \cdot t_s \tag{2}$$

where, C_x is the demand price when the actual maximum demand is within the maximum contract limit, S_o is the maximum contract limit, and C_{x1} is the demand price of the excess part when the actual maximum demand exceeds the maximum contract limit. According to the regulations, $C_{x1}/C_x = 2$. Let $\mathbf{s} = [s(1), s(2), \ldots, s(T)]$ be the vector of power demand. $C_{\text{elec}}(t)$ is the hourly electricity price, $s(t)$ is the power demand of the industrial park, t_s is the data time step, and T is the amount of data.

Typical daily load curve of industrial park is shown in Figure 1.

According to the daily load curve and electricity price table, the power demand of the industrial park is large when the electricity price is high, but the power demand is small when the electricity price is low, so the power consumption cost is high.

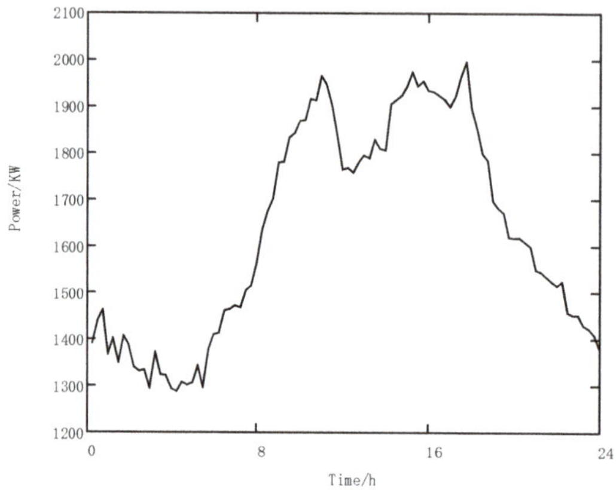

Figure 1. Daily load curve of industrial park.

2.1. Income from Energy Storage Participating in Peak Shaving

By providing energy storage to reduce the maximum demand by charging in the low electricity price period and discharging in the peak electricity price period users can reduce the total electricity charge. The difference between the electricity charge without energy storage and peak shaving by energy storage is the income from participating in user side peak shaving, which is expressed as follows:

$$\begin{aligned} M_{\text{peak.1}} &= C_{x1} \cdot [f_1(\max(\mathbf{s}), S_o) - f_1(\max(\mathbf{s}-\mathbf{b}), S_o)] + \\ &\sum_{t=1}^{T} C_{\text{elec}}(t) \cdot s(t) \cdot t_s - \sum_{t=1}^{T} C_{\text{elec}}(t) \cdot [s(t) - b(t)] \cdot t_s \end{aligned} \tag{3}$$

where, $b(t)$ is the output of energy storage at each time. $\mathbf{b} = [b(1), b(2), \ldots, b(T)]$ is the vector of energy storage actions. This formula expresses the saving of electricity cost after energy storage participates in peak shaving, but the energy storage itself will deteriorate during charging, and daily maintenance is required to ensure the normal operation of energy storage.

2.2. Energy Storage Output Cost

In the operation of battery energy storage, the operation cost is a key problem that must be considered, and the degradation cost comes from the degradation of the battery under repeated charge and discharge cycles [40–42]. Different batteries show different degradation characteristics. Lithium-ion batteries are a widely used form of battery energy storage. Therefore, the degradation cost model in this paper is mainly based on lithium-ion batteries.

In this paper, the rain flow cycle counting method is used to calculate the degradation cost of the energy storage battery. According to paper [43], the relationship between the depth of discharge (DOD) of an energy storage battery and its cycle life is described as follows:

$$N_{\max} = -1302D_{OD}^5 + 4427D_{OD}^3 - 8925D_{OD} + 10{,}500 \tag{4}$$

where, $N_{\max}$ is the cycle life (Times) of the battery, and D_{OD} is the discharge depth of the energy storage battery.

Using the rain flow counting method, the SOC change of the energy storage battery can be obtained according to the energy storage output $b(t)$ of each cycle, and then the cycle output times and output depth of the energy storage in each output cycle can be obtained. In a certain cycle, the energy storage has conducted n cycle output, and the corresponding discharge cycle depth is set as $D_{OD}(1)$, $D_{OD}(2)$, ... , $D_{OD}(n)$. Then, the battery life decay rate in a certain energy storage output cycle is given as follows [40]:

$$\gamma = \sum_{i=1}^{n} \frac{1}{N_{\max}(D_{OD}(i))} \cdot 100\% \tag{5}$$

where γ is the decay rate of battery life, and $N_{\max}$ ($D_{OD}(i)$) is the maximum number of discharge cycles corresponding to $D_{OD}(i)$. Therefore, the degradation cost generated after one cycle of output of the energy storage battery is expressed as follows.

$$f(b) = \gamma \cdot (C_S P_r + C_B E_r) \tag{6}$$

where C_S is the unit power cost of the PCS, that is, the unit power cost of the energy storage converter; P_r is the rated configuration power of the energy storage; C_B is the unit capacity cost and E_r is the energy storage capacity.

Energy storage operation and maintenance cost refers to a series of costs such as battery maintenance, repair and inspection to ensure the normal use of energy storage battery within the specified service life [44], which is related to the charging and discharging power and battery capacity of energy storage.

$$g(b(t)) = C_{\text{POM}} \sum_{t=1}^{T} b(t) + C_{\text{BOM}} \sum_{t=1}^{T} b(t) \cdot t_s \tag{7}$$

where C_{POM} is the unit power operation and maintenance cost; C_{BOM} is the operation and maintenance cost per unit capacity, that is the operation and maintenance cost corresponding to absorbing/releasing 1 MWh of energy.

2.3. Model Establishment

According to the benefits and costs described above, the daily energy storage output planning model aiming at the lowest total electricity charge in the industrial park is established as follows:

$$M_{\text{peak}} = \min_{b_1(t)} C_{x1} \cdot f_1(\max(\mathbf{s} - \mathbf{b}_1), S_o) + \sum_{t=1}^{T} f(b_1(t)) + \sum_{t=1}^{T} g(b_1(t)) + \sum_{t=1}^{T} C_{\text{elec}}(t) \cdot [s(t) - b_1(t)] \cdot t_s \tag{8}$$

where $b_1(t)$ is the variable, meaning the output of energy storage for peak shaving at each time, $\mathbf{b_1} = [b_1(1), b_1(2), \ldots, b_1(T)]$ is the vector of battery actions for peak shaving.

The constraints are:

(1) SOC constraint of energy storage battery

$$\text{SOC}_{\min} - \text{SOC}_1 \le \frac{\sum_{\tau=1}^{t} b_1(\tau) \cdot t_s}{E_1} \le \text{SOC}_{\max} - \text{SOC}_1 \tag{9}$$

where $\text{SOC}_{\max}$ and $\text{SOC}_{\min}$ respectively represent the maximum and minimum state of charge in the discharge area of the energy storage battery, SOC_1 represents the SOC of the energy storage battery at the initial time, and E_1 represents the peak shaving capacity of energy storage.

(2) Same constraint as initial state

$$\sum_{t=1}^{T} b_1(t) = 0 \tag{10}$$

Each optimization process is a cycle. During this cycle, the SOC of the energy storage battery shall be consistent, so as to facilitate the optimization and output of multiple cycles.

(3) Maximum power constraint of energy storage charge and discharge

$$0 \leq b_{i1}(t) \leq P_r \tag{11}$$

$$0 \leq b_{o1}(t) \leq P_r \tag{12}$$

where $b_{i1}(t)$ represents the charging power of the battery during peak shaving, and $b_{o1}(t)$ represents the discharge power of the battery during peak shaving.

3. Optimization Model of Energy Storage Battery Participating in Frequency Regulation

The energy storage battery has good response speed and climbing ability, so it can adapt to flexible frequency regulation signals. In this paper, the Reg_D frequency regulation signal of the American PJM market is used as the frequency regulation action instruction of energy storage battery. Figure 2 shows a one-hour Reg_D frequency regulation signal, which is expressed in normalized form and ranges from [−1,1]. In this frequency regulation market, each energy storage power station willing to participate in frequency regulation service needs to submit bidding application and bidding capacity the day before the frequency regulation day. After winning the bid, the energy storage battery needs to output according to the frequency regulation signal. At the same time, the frequency regulation market will compensate the capacity of the winning energy storage. However, the energy storage unit that fails to comply with the regulations will also be adjusted and punished.

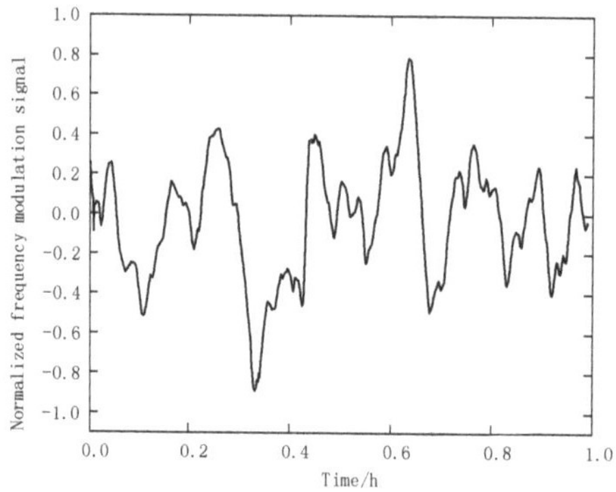

Figure 2. The signal of the Reg_D.

The power that battery energy storage needs to respond to in the process of frequency regulation P_{need} is described as follows.

$$P_{\text{need}} = r(t) \cdot C_J \tag{13}$$

where $r(t)$ is the Rrg_D frequency regulation real-time signal and C_J is the bid-marked capacity.

When participating in the frequency regulation service market, the mileage of the energy storage battery following the frequency regulation signal determines the benefits brought by the energy storage. Deeper following of the signal will give more frequency regulation mileage benefits and reduce the penalty caused by insufficient output. However, deeper following means a larger span of energy storage output, which will also bring more degradation, operation and maintenance costs. Therefore, a frequency regulation optimization model with the most economical energy storage battery is established.

(1) Traditional objective function

$$M_r = \min_{C, b_2(t)} \sum_{t=1}^{T} f(b_2(t)) + \sum_{t=1}^{T} g(b_2(t)) + c_{\text{mis}} \cdot \sum_{t=1}^{T} |b_2(t) - C \cdot r(t)| - c_t \cdot T \cdot C - R_b \tag{14}$$

where C and $b_2(t)$ are the variables, C is the bidding capacity, $b_2(t)$ is output of energy storage for frequency regulation at each time. c_{mis} is the penalty coefficient, which represents the penalty amount required for every 1 MW·h of deviation between the energy storage output and the frequency regulation signal and c_t is the frequency regulation compensation coefficient, which represents the compensation amount for each 1 MW energy storage successfully bid by the grid service market every hour. R_b is mileage compensation, and its calculation method is described as follow.

$$R_b = K \cdot c_{bp} \cdot r_b \tag{15}$$

where, K is the frequency regulation performance index, c_{bp} is the frequency regulation mileage price, r_b is the frequency regulation mileage in a certain frequency regulation stage. The calculation method was according to reference [45,46].

(2) Constraints

$$C \geq 0 \tag{16}$$

$$C < P_2^{\max} \tag{17}$$

$$\text{SOC}_{\min} - \text{SOC}_1 \leq \frac{\sum_{\tau=1}^{t} b_1(\tau) \cdot t_s}{E_1} \leq \text{SOC}_{\max} - \text{SOC}_1 \tag{18}$$

$$\sum_{t=1}^{T} b_2(t) = 0 \tag{19}$$

$$0 \leq \max(b_{i2}(t)) < P_2^{\max} \tag{20}$$

$$0 \leq \max(b_{o2}(t)) \leq P_2^{\max} \tag{21}$$

where, ${P_2}^{\max}$ is the maximum power of frequency regulation of energy storage.

(3) Improve the objective function

In the constraint condition (18), the sum of the energy storage battery's output for frequency regulation in an optimization cycle is 0. Therefore, in a frequency regulation optimization cycle, the output of the energy storage battery cannot impact the electricity charge, but it will have an impact on the basic electricity charge. This will fluctuate the total electricity charge after the energy storage battery participates in frequency regulation, so the objective function is improved as follows:

$$M_r = \min_{c,b_2(t)} \sum_{t=1}^{T} f(b_2(t)) + C_{x1} \cdot f_1(\max(\mathbf{s} - \mathbf{b_2}), S_o) + \sum_{t=1}^{T} g(b_2(t)) + c_{\text{mis}} \cdot |b_2(t) - C \cdot r(t)| - c_t \cdot T \cdot C - R_b \tag{22}$$

where $\mathbf{b_2} = [b_2(1), b_2(2), \ldots, b_2(T)]$ is the vector of battery actions for frequency regulation.

4. Energy Storage Frequency Regulation and Peak Shaving Output Planning

4.1. Joint Optimization of Frequency Regulation and Peak Shaving

Based on the storage battery's fixed capacity, an optimization strategy is proposed for the joint output of frequency regulation and peak shaving.

In this strategy, first the energy storage battery capacity is optimized by day ahead allocation to obtain the optimal economic peak shaving and frequency regulation capacity allocation, then the day ahead peak shaving planning and the optimal bidding capacity of energy storage frequency regulation are obtained. The MPC model is used to optimize the intra-day energy storage frequency regulation output, then the total intra-day energy storage output is obtained.

By predicting the user's required load and Reg_D of the next day, according to the peak and valley electricity charge, the maximum contract limit and other parameters, we take the energy storage output and peak shaving and frequency regulation capacity as variables and optimize them with the goal of optimizing the economy of energy storage and peak shaving output at the same time, so as to obtain the optimal allocation of energy storage frequency regulation and peak shaving capacity. The model is as follows:

Objective function is described as follows.

$$\begin{aligned}\underset{C,b_1(t),b_2(t),E_1,E_2}{M_{\text{both}}} &= \min C_{x1}\cdot f_1(\max(\mathbf{s}-\mathbf{b}_1-\mathbf{b}_2),S_o)+\sum_{t=1}^{T}C_{\text{elec}}(t)\cdot[s(t)-b_1(t)]\cdot t_s+\sum_{t=1}^{T}f(b_1(t)+b_2(t))\\&+\sum_{t=1}^{T}g(b_1(t)+b_2(t))+c_{\text{mis}}\cdot\sum_{t=1}^{T}|b_2(t)-C\cdot r(t)|-R_b-c_t\cdot T\cdot C\end{aligned} \tag{23}$$

Constraints:

$$C \geq 0 \tag{24}$$

$$C \leq P^{\max}-\max(b_1(t)) \tag{25}$$

$$\text{SOC}_{\min}-\text{SOC}_1 \leq \frac{\sum_{\tau=1}^{t} b_1(\tau)\cdot t_s}{E_1} \leq \text{SOC}_{\max}-\text{SOC}_1 \tag{26}$$

$$\text{SOC}_{\min}-\text{SOC}_1 \leq \frac{\sum_{\tau=1}^{t} b_2(\tau)\cdot t_s}{E_2} \leq \text{SOC}_{\max}-\text{SOC}_1 \tag{27}$$

$$E_1+E_2=E_o \tag{28}$$

$$-P_o \leq b_1(t)+b_2(t) \leq P_o \tag{29}$$

$$\sum_{t=1}^{T} b_1(t)=0 \tag{30}$$

$$\sum_{t=1}^{T} b_2(t)=0 \tag{31}$$

where E_1 is the occupied capacity of peak shaving, E_2 is the occupied capacity of frequency regulation, E_o is the rated capacity of the energy storage battery and P_o is the rated power of energy storage battery. Using this model, the capacity E_1 and E_2 of peak shaving and frequency regulation can be optimized. We can bring the obtained E_1 and E_2 into the peak shaving and frequency regulation models to obtain the planned energy storage peak shaving output $b_1(t)$, the maximum peak shaving output $\max(b_1(t))$, and the energy storage frequency regulation bidding capacity C. These optimization results will affect the parameter setting of intra-day frequency regulation optimization.

4.2. Intra-Day Optimization of Frequency Regulation and Peak Shaving

Since the frequency regulation signal changes according to the real-time state of the system, the frequency regulation output needs to respond to the real-time signal. C, obtained from the above optimization results, shall be used for bidding. If the bidding is successful, intra-day rolling output optimization of energy storage shall be carried out according to the actual Reg_D signal intra-day, and intra-day peak adjustment shall be carried out according to the day-ahead planning. The flow chart of day ahead and day ahead optimization is shown in Figure 3.

During intra-day frequency regulation optimization, rolling optimization is carried out according to MPC framework to obtain real-time output. MPC can consider future information, so it can solve the problem of short sightedness of model optimization [47]. In the output strategy of this paper, MPC is divided into the following steps: (1)At the current time, t, of receiving frequency regulation signal, based on the current state of energy storage battery, the prediction model is used to predict the frequency regulation signal in [t, t + 1, ..., t + n]. (2) According to the predicted frequency regulation signals [r (t), r (t + 1), ..., r (t + n)] into the frequency regulation optimization model, the economic optimal solution [b_2(t), b_2(t + 1), . . . , b_2(t + n)] is obtained. (3) b_2(t) is used as the energy storage frequency regulation output at time t, and the initial energy storage state at time t + 1 is obtained. (4) Repeat the above steps at time t + 1. Finally, the optimal economic frequency regulation output of the energy storage battery considering mileage income, degradation effect and operation and maintenance cost is obtained.

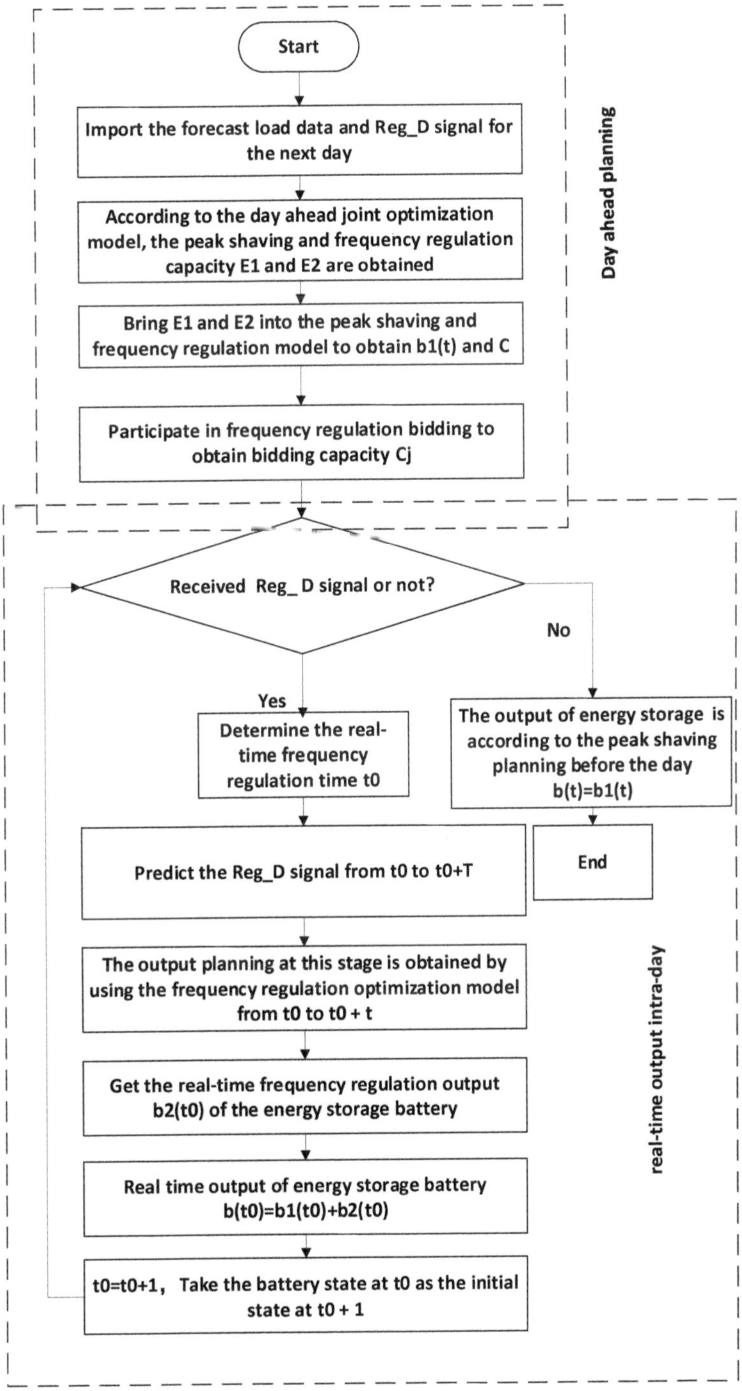

Figure 3. The output plan flow chart of peak shaving and frequency regulation.

5. Life Cycle Economic Analysis Model

5.1. Life Cycle Cost Calculation Model

The economics and investability of the strategy proposed in this paper can be seen through the economic analysis of its whole lifetime. Therefore, the cost and benefit model of energy storage participating in peak and frequency regulation on the user side is established in this section. The life cycle cost of energy storage refers to all direct, indirect, derived or non-derived costs that occur or may occur during the entire life cycle of energy storage participating in peak and frequency regulation. The life cycle cost model of the energy storage system in this paper includes its investment cost, operation and maintenance costs, scrap processing cost, and penalties caused by frequency regulation. Since the degradation cost is calculated from the investment cost of the energy storage battery, so the degradation cost is not considered during the full life cycle. the present value method is used to convert the cost of the entire life cycle into the present value of the investment cost at the initial moment of investment as follows:

$$C_T = C_1 + C_2 + C_3 + C_4 \tag{32}$$

where C_T is the present value of total cost, C_1 is the present value of the initial investment cost of energy storage throughout the life cycle, C_2 is the present value of energy storage operation and maintenance cost, C_3 is the present value of energy storage disposal cost and C_4 is the present value of the penalty cost of energy storage participating in frequency regulation. The expressions are as follows.

$$C_1 = C_S P_r + \sum_{k=0}^{n} C_B E_r (1+r)^{-[kT_{\text{LCC}}/(n+1)]} \tag{33}$$

$$C_2 = C_{\text{POM}} P_r \{[(1+r)^{T_{\text{LCC}}} - 1]/[r(1+r)^{T_{\text{LCC}}}]\} + \sum_{t=1}^{T_{\text{LCC}}} C_{\text{BOM}} W (1+r)^{-t} \tag{34}$$

$$C_3 = C_{\text{Pscr}} P_r (1+r)^{-T_{\text{LCC}}} + \sum_{k=1}^{n+1} C_{\text{Escr}} E_r (1+r)^{-k \cdot T_{\text{life}}} \tag{35}$$

$$C_4 = \sum_{t=1}^{T_{\text{LCC}}} F_c (1+r)^{-t} \tag{36}$$

where n is number of replacements, T_{LCC} is the number of years considered in the whole life cycle and $n = T_{\text{LCC}}/T_{\text{life}}$. T_{life} is the cycle life of the energy storage. r is the discount rate, W is the annual charge and discharge capacity of energy storage, C_{Pscr} is the unit power scrap disposal cost, C_{Escr} is the disposal cost per unit capacity and F_c is the annual frequency regulation penalty for energy storage.

5.2. Life Cycle Benefit Calculation Model

The benefits of energy storage participating in user-side peaking and frequency regulation come from the electricity price difference of peaking, frequency regulation capacity compensation and frequency regulation mileage compensation. It is expressed as the following formula.

$$R_T = R_1 + R_2 \tag{37}$$

where R_1 is the present value of peak shaving income and R_2 is the frequency regulation revenue present value.

$$R_1 = \sum_{t=1}^{T_{\text{LCC}}} M_p (1+r)^{-t} \tag{38}$$

$$R_2 = \sum_{t=1}^{T_{\text{LCC}}} M_t (1+r)^{-t} \tag{39}$$

where M_p is the peak shaving annual revenue and M_t is the frequency regulation annual income.

6. Example Analysis

6.1. Parameter Setting

In order to verify the effectiveness of the scheme in improving the economy of energy storage on the user side, the actual Reg_D signal and industrial park load are used to simulate and verify. The experimental model is optimized by the CVX software package in MATLAB, which is a general software package to solve convex optimization problems. the parameters appearing in the model are assigned values as shown in Table 2. Because the frequency regulation signal adopts the Reg_D of the PJM market in the United States, the currency unit in this paper is the US dollar. The frequency tariff is converted from one month to a single day price. Because the research focus of this paper is not to determine the optimal value of the user's maximum contract limit, and the optimal value of the contract limit is a long-term fixed value and cannot be changed every day, so the maximum contract limit is specified as the determined value in the experimental process of this paper.

Table 2. The numerical table of the parameters.

Parameter Name	Numerical Value
Demand electricity price C_x (\$/MW)	215
Maximum contractual limit S_o (MW)	0.185
Valley electricity price (\$/MW·h)	59.14
Normal electricity price (\$/MW·h)	111.71
Pick electricity price (\$/MW·h)	164.14
Unit power cost C_S (\$/MW)	257,000
Energy storage battery unit capacity cost C_B (\$/MW·h)	384,000
Unit power operation and maintenance cost C_{POM} (\$/MW)	10,000
Unit capacity operation and maintenance cost C_{BOM} (\$/MW·h)	10
The number of years considered in the whole life cycle T_{LCC} (year)	20
Unit power scrap cost C_{Pscr} (\$/MW)	1000
Unit capacity scrap cost C_{Escr} (\$/MW·h)	1000
Penalty coefficient c_{mis} (\$/MW)	500
Frequency regulation compensation coefficient c_t (\$/MW)	30
Mileage compensation coefficient c_{bp} (\$/MW)	4
Discount rate r (%)	6

The parameters of energy storage battery used in this paper are shown in Table 3.

Table 3. The parameters of the energy storage battery.

Parameter Name	Numerical Value
Maximum charge/discharge power P^{max} (MW)	1
Upper limit of stored energy (MWH)	1
Initial value of state of charge	0.5
Maximum state of charge (S^{max})	0.8
Minimum state of charge (S^{min})	0.2
Lithium battery cycle life N	800

6.2. Peak Shaving and Frequency Regulation Day-Ahead Optimization

In this paper, a long short-term memory (LSTM) network is used to predict the load and frequency regulation signal. Because the time steps of peak shaving and frequency regulation are different, peak shaving needs to optimize the electricity price and load demand of the whole day as a reference, so the optimization step is hour level, while the step size of Reg_D signal is 2 s, which is too different from the peak shaving time step. If they are optimized for 24 h, there will be up to 43,200 frequency regulation signals, which undoubtedly increases the optimization complexity. Therefore, in the day-ahead capacity planning stage in this paper, the load data is divided into 2 s from the original steps of 15 min, so four data in an hour are divided into 1800 data to match the frequency regulation steps, so Equation (18) can be solved to get E_1 and E_2. This process is repeated 24 times to obtain 24 groups of E_1 and E_2 per day, and the average value is taken for the final peak shaving and frequency regulation capacity allocation.

According to the capacity planning model of peak shaving and frequency regulation and the parameters given above, an energy storage battery with a maximum power of 1 MW and capacity of 1 MW·h was used to carry out the day-ahead peak shaving and frequency regulation planning on the user side. The obtained results are E_1 = 0.8 MW·h and E_2 = 0.2 MW·h. Then, we bring E_1 into the peak shaving model shown in Equation (8) and so get the power curve required by the user after peak shaving is shown in Figure 4. The energy storage output and SOC changes are shown in Figures 5 and 6. The maximum output power of energy storage peak regulation is P_1^{max} = 0.13 MW. According to Figure 4, the energy storage battery charges in the night when the electricity price is low, and the energy storage discharges in the morning and afternoon when the electricity price is high, so as to reduce the power demand of users in the time when the electricity price is high. Maximum demand from industrial users is reduced based on maximum contract quotas.

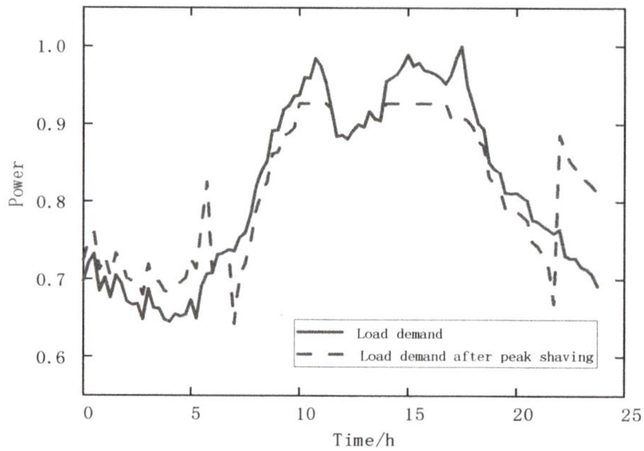

Figure 4. The result of day-ahead peak shaving.

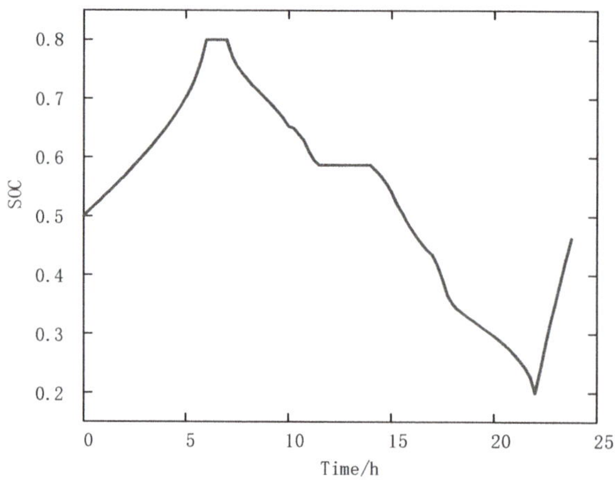

Figure 5. The change curve of the battery's peak-shaving SOC.

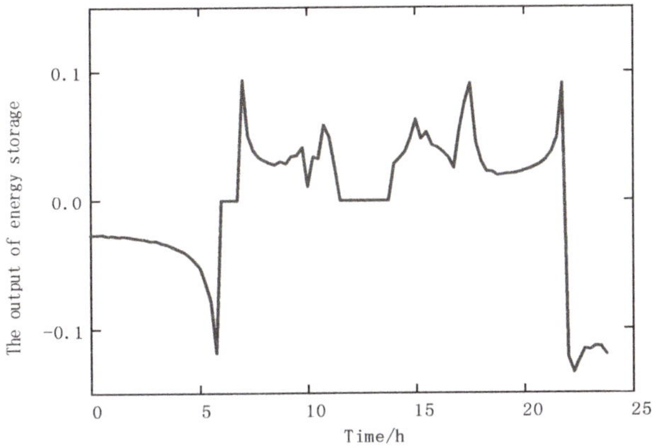

Figure 6. The change curve of the battery's peak-shaving output.

According to the calculation rule of the user electricity charge, the 24 h electricity charge without energy storage battery is $2487, of which the demand electricity charge is $230. After adding the output of the energy storage battery, the electricity charge for 24 h is $2446, including the demand electricity charge of $199 and degradation cost and operation and maintenance cost of $52. Therefore, the energy storage power station is equipped with energy storage battery for peak shaving, which has limited savings on electricity charges. This is because if the energy storage output is small and the peak shaving is small, it has little impact on electricity charges. When the energy storage output is large, although the electricity charges are reduced, the degradation costs and operation and maintenance costs of the energy storage will also increase, resulting in no significant savings in total electricity charges.

Taking E_2 = 0.2 MWH and P_2^{max} = 0.87 MW into the frequency regulation model, the optimal power C = 0.87 MW. The variation results of energy storage frequency regulation output and SOC are shown in Figures 7 and 8.

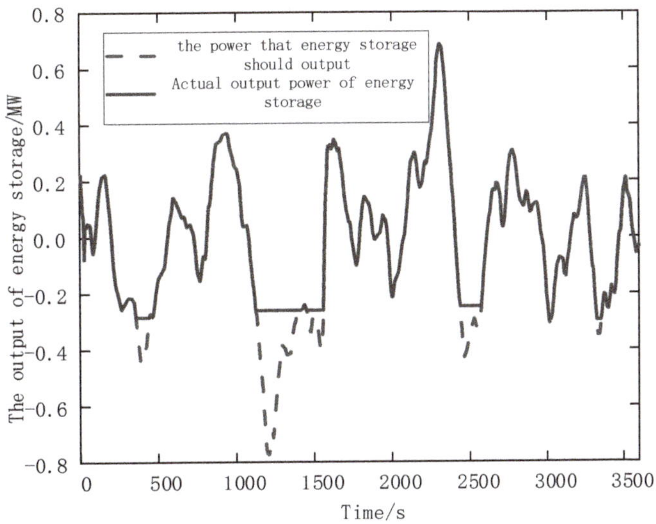

Figure 7. The output of day-ahead frequency regulation.

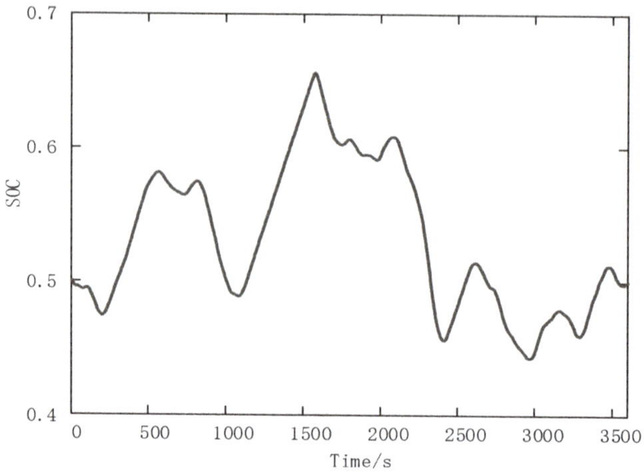

Figure 8. The change curve of battery's SOC of frequency regulation.

It can be seen from the Figure 7 that the energy storage battery tracks the Reg_D signal and sends output most of the time. When large-scale output of the energy storage is required, the model will take into account the degradation effect of the energy storage battery, operation and maintenance costs and power demand at the user side, so that the energy storage battery only responds to some frequency regulation commands and reduces the output depth. In this hour, the electricity charge of the industrial park is \$57.37. Participating in the service market through frequency regulation, the optimized electricity charge is \$37.60, including degradation cost and operation and maintenance cost of \$9.12. Thus, the user-side energy storage battery can participate in the market frequency regulation auxiliary service, which can effectively reduce the user's electricity charge.

6.3. Intra-Day Real-Time Optimization

According to the LSTM that predicts the frequency regulation signal, the MPC model described in Section 4 is used for rolling training on the frequency regulation day in which the energy storage with power of 0.87 MW and capacity of 0.2 MWH is used. The 24-h frequency regulation output is shown in Figure 9.

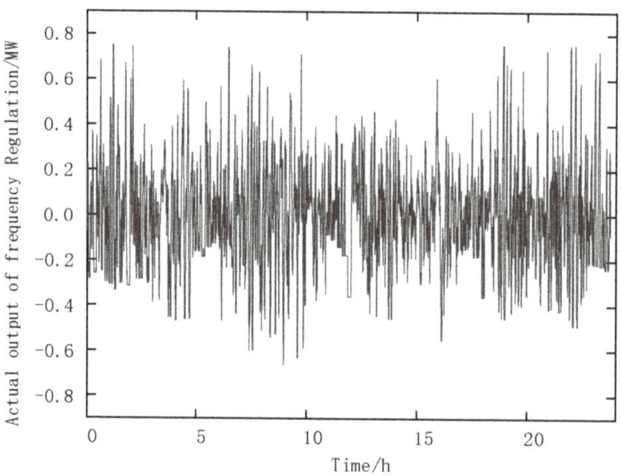

Figure 9. The output of storage frequency regulation in real time during 24 h.

Since the total output of the energy storage battery in a day is equal to the sum of the frequency regulation output and the peak shaving output, we can take any continuous two hours in a day to observe, and the actual total output of energy storage is shown in Figure 10.

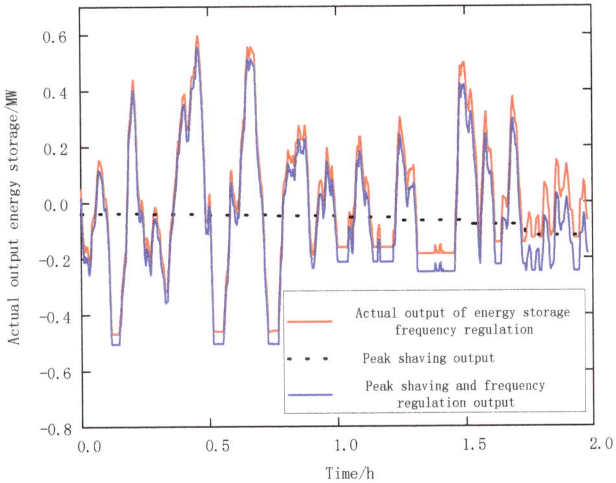

Figure 10. The intra-day total output of the storage.

6.4. Economic Analysis

This section compares the coordinated output of peak shaving and frequency regulation of energy storage with the economic benefits obtained by peak shaving or frequency regulation alone for a whole day. As shown in Figure 11, the impact of using 1 MW and 1 MWH energy storage batteries for one day under different schemes on the electricity charge of the industrial park is shown respectively.

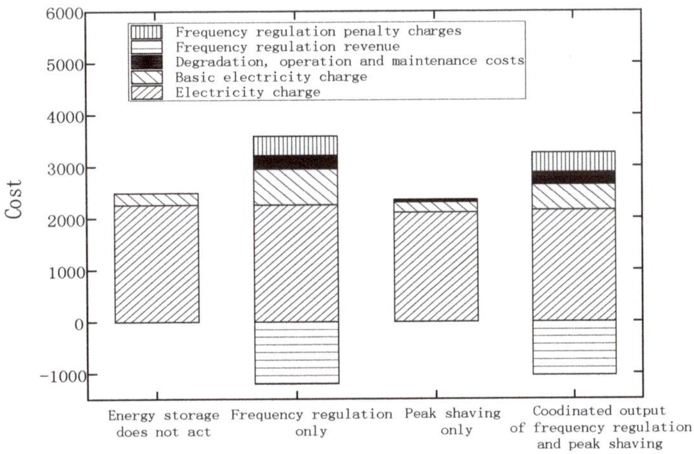

Figure 11. The economy contrasts of different schemes.

It can be seen from the Figure 11 that the 24 h electricity charge of users obtained through the strategy in this paper is reduced by 10.96% compared with the output without energy storage, 5.8% compared with the output of peak shaving only for energy storage, and 3.6% compared with the output of frequency regulation only for energy storage. The benefit brought by the combined output of energy storage peak shaving and frequency regulation is better than that of the frequency regulation service or peak shaving alone with batteries of the same capacity and power.

This is due to the Reg_D frequency regulation signal frequently crossing the zero value, and the SOC of the battery can be recovered by following the signal. Therefore, there is little demand for capacity during energy storage frequency regulation. Although the profit obtained by using 1 MW bidding capacity is greater than that obtained by using 0.87 MW capacity for frequency regulation, it will also increase the degradation cost of energy storage battery each time following the signal. If 0.87 MW power is used for frequency regulation and 0.13 MW power is used for peak shaving, the benefit of frequency regulation is less than that of 1 MW power frequency regulation, but the cost of degradation benefit is lower, and the benefit of peak shaving will be obtained. Therefore, the optimal economic results of frequency regulation and peak shaving will be obtained. The degradation costs incurred by adopting various schemes are shown in Table 4.

Table 4. Degradation costs of different schemes.

Scheme	Degradation Cost
1 MW, 1 MW·h Frequency regulation	\$261
1 MW, 1 MW·h Peak shaving	\$51.4
0.87 MW, 0.2 MW·h Frequency regulation individually	\$200
0.13 MW, 0.8 MW·h Peak shaving individually	\$52
1 MW, 1 MW·h Combined output of frequency regulation and peak shaving	\$236

It can be seen from Table 4 that the sum of degradation cost generated by 0.87 MW, 0.2 MW·h frequency regulation and 0.13 MW, 0.8 MW·h peak shaving alone is \$16 more than that generated by 1 MW, 1 MW·h combined frequency regulation with peak shaving. The reason for the cost reduction is that in the process of joint output, the frequency regulation signal has quite a lot of time, which is contrary to the peak shaving of energy storage, thus reducing the discharge depth of the storage battery and reducing the cost of degradation.

6.5. Economic Analysis Based on Life Cycle

According to the 24 h energy storage peak shaving and frequency modulation output, the SOC change of the energy storage battery during the day is shown in Figure 12.

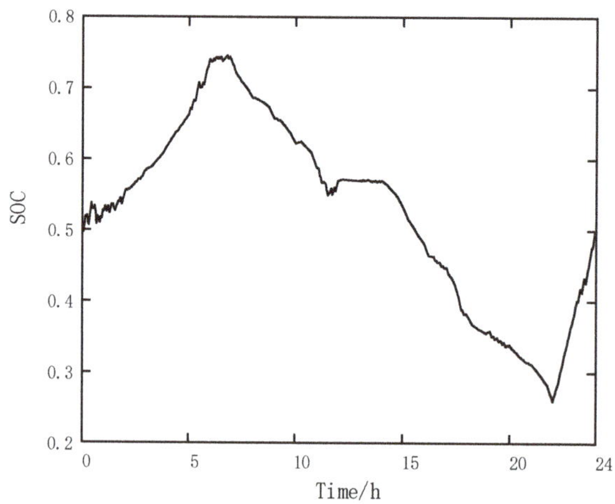

Figure 12. Change curve of SOC in 24 h.

According to the SOC change data over 24 h, the cycle times and cycle depth of peak shaving and frequency regulation of energy storage in a day can be obtained by using the rain flow counting method, as shown in the Figure 13. The discharge cycle is composed of peak shaving deep cycle and several frequency regulation shallow cycles. A total of 109 cycles are carried out in 24 h, and the sum of cycle depths is 0.7171. According to the average cycle life of lithium battery, the operation life under the strategy proposed in this paper is 3 years.

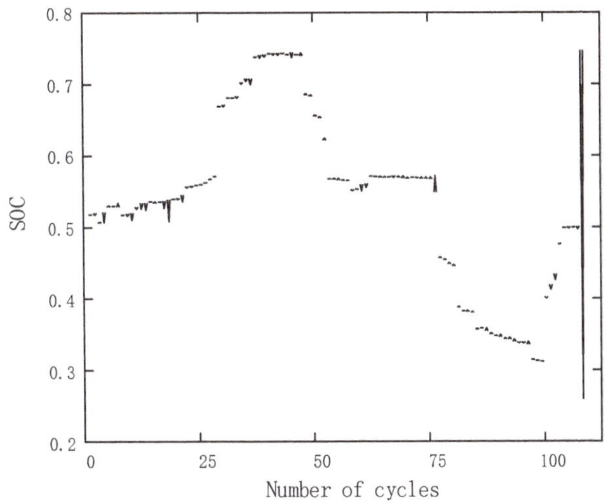

Figure 13. The result of rain flow.

In order to reflect the profitability of the strategy proposed, the rate of return on investment is introduced for evaluation. The rate of return on investment can be calculated by the ratio of the average annual net income over the entire life cycle of the system to the total investment amount, and the greater the rate of return on investment, the better the profitability of the project, which can be calculated as follows.

$$R_{\text{inv}} = \frac{N_B}{K} \times 100\% \tag{40}$$

where, K is the annual average investment of the project, namely $K = C_T/T_{\text{LCC}}$. N_B is the average annual net income during the life cycle of the system.

According to the life of the energy storage battery, the economic analysis of the whole life cycle is carried out, and the costs and benefits are shown in Table 5.

Table 5. Economic analysis result of life cycle.

Parameter	Numerical Value
Present value of the initial investment cost C_1 ($)	1.98×10^6
Present value of energy storage operation and maintenance cost C_2 ($)	3.12×10^5
Present value of energy storage disposal cost C_3 ($)	4.00×10^3
Present value of the penalty cost C_4 ($)	1.31×10^6
Present value of peak shaving income R_1 ($)	1.49×10^5
Present value of frequency regulation income R_2 ($)	7.56×10^6
$R_T - C_T$ ($)	4.09×10^6
Rate of return on investment R_{inv} (%)	5.66

For comparison, using the same method, frequency regulation of 1 MW, 1 MW·h energy storage batteries and peak shaving of 1 MW, 1 MW·h energy storage batteries are used to perform a full-life economic analysis. The results are shown in Tables 6 and 7.

Table 6. Life cycle analysis of 1 MW, 1 MW·h energy storage for peak shaving.

Parameter	Numerical Value
Present value of the initial investment cost C_1 ($)	8.55×10^5
Present value of energy storage operation and maintenance cost C_2 ($)	7.31×10^4
Present value of energy storage disposal cost C_3 ($)	1.33×10^3
The cycle life of the energy storage T_{life} (year)	8
Present value of peak shaving income R_1 ($)	3.57×10^5
$R_T - C_T$ ($)	-5.72×10^5
Rate of return on investment R_{inv} (%)	−3.08

Table 7. Life cycle analysis of 1 MW, 1 MW·h energy storage for frequency regulation.

Parameter	Numerical Value
Present value of the initial investment cost C_1 ($)	5.15×10^6
Present value of energy storage operation and maintenance cost C_2 ($)	3.37×10^5
Present value of energy storage disposal cost C_3 ($)	1.21×10^4
Present value of the penalty cost C_4 ($)	1.09×10^6
The cycle life of the energy storage T_{life} (year)	1
Present value of frequency regulation income R_2 ($)	9.61×10^6
$R_T - C_T$ ($)	3.02×10^6
Rate of return on investment R_{inv} (%)	2.29

Available from Tables 5–7, the life cycle analysis results are different. For T_{life}, peak shaving only is the largest, followed by peak shaving and frequency regulation coordination output, and only frequency regulation is the smallest. This is because the SOC changes are different for different output strategies. When energy storage performs frequency modulation only, it needs to constantly switch between the two states of charging and discharging for tracking the Reg_D signal. Therefore, the corresponding number of energy storage cycles per day will increase. The cycle life of the energy storage battery is fixed, and when the number of cycles is reached, the battery needs to be replaced so greater battery replacement costs will be incurred. Therefore, the C_1 is the largest when the energy storage battery only participates in frequency regulation service. Similarly, T_{life} is the largest when energy storage participates peak shaving only, so the number of battery replacement required in the whole life cycle is the smallest. Therefore, C_1 is the smallest when energy storage only shaves the peaks, and the characteristics of C_3 are the same as C_1. Although the investment cost of energy storage only peak shaving is low, the income from energy storage only peak shaving is too small, such that the net income in the whole life cycle of energy storage only peak shaving is negative, and the rate of return on investment is also negative. Although the frequency regulation gain of the energy storage battery is very high when it is used only, the service life of the energy storage is too short due to long-term multiple cycles. By comparison, under the operation of the strategy proposed in this paper, the energy storage battery can reduce the output cycle required for participating in frequency regulation through peak regulation output (as shown in the Figure 12). At the same time, the problem of low peak shaving income is compensated by the high income of frequency regulation services, so the income and life of energy storage batteries coexist, which has a higher investment value.

7. Conclusions

In order to improve the economy and investability of energy storage on the user side, this paper puts forward the peak shaving and frequency regulation coordinated output strategy in which the industrial park energy storage battery participates in the system frequency regulation service while peak shaving to obtain additional income.

The strategy divides the peak shaving and frequency regulation capacity of energy storage and obtains the output of peak shaving plan day ahead. The real-time output with optimal economy is obtained through MPC rolling optimization intra-day, and the degradation effect and operation and maintenance cost are considered while the maximum frequency regulation capacity compensation and mileage compensation are obtained, so as to improve the total revenue of the industrial park energy storage intra-day. Finally, the whole life cycle economic analysis of the strategy proposed in this paper shows that the peak shaving and frequency regulation coordinated output on the user side has a larger rate of return. When China's frequency regulation service market is better in the future, this strategy provides a new idea for industrial park energy storage to improve its economy.

Author Contributions: Methodology, Z.J.; software, Y.Y.; validation, Y.Y. and Y.F.; formal analysis, Y.S. and Y.F.; investigation, Y.F.; resources, D.L.; writing—original draft preparation, Z.J.; writing—review and editing, H.C. (Huayue Chen) and Y.S.; visualization, H.C. (Hongji Cao); funding acquisition, Y.S. and D.L. All authors have read and agreed to the published version of the manuscript.

Funding: This research was jointly funded by the Yantai Key Research and Development Program, grant number 2020YT06000970; the Wealth Management Characteristic Construction Project of Shandong Technology and Business University, grant number 2019ZBKY019.

Institutional Review Board Statement: Not applicable.

Informed Consent Statement: Not applicable.

Data Availability Statement: Not applicable.

Conflicts of Interest: The authors declare no conflict of interest.

References

1. Sun, Z.; Tian, H.; Wang, W. Research on economy of echelon utilization battery energy storage system for user-side peak load shifting. *Acta Energ. Sol. Sin.* **2021**, *42*, 95–100.
2. Ye, J. *Research on Technical Adaptability and Benefit Evaluation of Energy Storage System in Typical Application Scenarios*; North China Electric Power University: Beijing, China, 2020.
3. Zhang, Z.H.; Min, F.; Chen, G.S.; Shen, S.P.; Wen, Z.C.; Zhou, X.B. Tri-partition state alphabet-based sequential pattern for multivariate time series. *Cogn. Comput.* **2021**. [CrossRef]
4. Ran, X.; Zhou, X.; Lei, M.; Tepsan, W.; Deng, W. A novel k-means clustering algorithm with a noise algorithm for capturing urban hotspots. *Appl. Sci.* **2021**, *11*, 11202. [CrossRef]
5. Chen, H.; Zhang, Q.; Luo, J. An enhanced Bacterial Foraging Optimization and its application for training kernel extreme learning machine. *Appl. Soft Comput.* **2020**, *86*, 105884. [CrossRef]
6. Cui, H.; Guan, Y.; Chen, H.; Deng, W. A novel advancing signal processing method based on coupled multi-stable stochastic resonance for fault detection. *Appl. Sci.* **2021**, *11*, 5385. [CrossRef]
7. Zhao, H.; Li, D.; Deng, W.; Yang, X. Research on vibration suppression method of alternating current motor based on fractional order control strategy. *Proc. Inst. Mech. Eng. Part E J. Process. Mech. Eng.* **2017**, *231*, 786–799. [CrossRef]
8. Salles, M.B.C.; Gadotti, T.N.; Aziz, M.J.; Hogan, W.W. Potential revenue and breakeven of energy storage systems in PJM energy markets. *Environ. Sci. Pollut. Res. Int.* **2021**, *28*, 12357–12368. [CrossRef]
9. He, G.; Chen, Q.; Kang, C.; Pinson, P.; Xia, Q. Optimal bidding strategy of battery storage in power markets considering performance-based regulation and battery cycle life. *IEEE Trans. Smart Grid* **2016**, *7*, 2359–2367. [CrossRef]
10. Zheng, J.; Yuan, Y.; Zou, L.; Deng, W.; Guo, C.; Zhao, H. Study on a novel fault diagnosis method based on VMD and BLM. *Symmetry* **2019**, *11*, 747. [CrossRef]
11. Zhou, Y.; Zhang, J.; Yang, X.; Ling, Y. Optimal reactive power dispatch using water wave optimization algorithm. *Oper. Res.* **2020**, *20*, 2537–2553. [CrossRef]
12. Zhong, K.; Zhou, G.; Deng, W.; Zhou, Y.; Luo, Q. MOMPA: Multi-objective marine predator algorithm. *Comput. Methods Appl. Mech. Eng.* **2021**, *385*, 114029. [CrossRef]
13. Wei, Y.Y.; Zhou, Y.Q.; Luo, Q.F.; Deng, W. Optimal reactive power dispatch using an improved slime mould algorithm. *Energy Rep.* **2021**, *7*, 8742–8759. [CrossRef]
14. Deng, W.; Zhang, X.X.; Zhou, Y.Q.; Liu, Y.; Zhou, X.B.; Chen, H.L.; Zhao, H.M. An enhanced fast non-dominated solution sorting genetic algorithm for multi-objective problems. *Inform. Sci.* **2022**, *585*, 441–453. [CrossRef]

15. Li, T.Y.; Qian, Z.J.; Deng, W.; Zhang, D.Z.; Lu, H.H.; Wang, S.H. Forecasting crude oil prices based on variational mode decomposition and random sparse Bayesian learning. *Appl. Soft Comput.* **2021**, *113*, 108032. [CrossRef]
16. Kulpa, J.; Kamiński, P.; Stecuła, K.; Prostański, D.; Matusiak, P.; Kowol, D.; Kopacz, M.; Olczak, P. Technical and economic aspects of electric energy storage in a mine shaft-budryk case study. *Energies* **2021**, *14*, 7337. [CrossRef]
17. Deng, W.; Xu, J.; Zhao, H.; Song, Y. A novel gate resource allocation method using improved PSO-based QEA. *IEEE Tran. Intell. Transp. Syst.* **2020**, 1–9. [CrossRef]
18. Xue, J.; Ye, J.; Xu, Q. Interactive package and diversified business mode of renewable energy accommodation-n with client distributed energy storage. *Power Syst. Technol.* **2020**, *44*, 1310–1316.
19. Cui, H.; Guan, Y.; Chen, H. Rolling element fault diagnosis based on VMD and sensitivity MCKD. *IEEE Access* **2021**, *9*, 120297–120308. [CrossRef]
20. Deng, W.; Shang, S.; Cai, X.; Zhao, H.; Zhou, Y.; Chen, H.; Deng, W. Quantum differential evolution with cooperative coevolution framework and hybrid mutation strategy for large scale optimization. *Knowl. -Based Syst.* **2021**, *224*, 107080. [CrossRef]
21. Cheng, B.; Powell, W.B. Co-optimizing battery storage for the frequency regulation and energy arbitrage using multi-scale dynamic programming. *IEEE Trans. Smart Grid* **2018**, *9*, 1997–2005. [CrossRef]
22. Senchilo, N.D.; Ustinov, D.A. Method for determining the optimal capacity of energy storage systems with a long-term forecast of power consumption. *Energies* **2021**, *14*, 7098. [CrossRef]
23. Chau, T.K.; Yu, S.S.; Fernando, T.; Iu, H.H. Demand-side regulation provision from industrial loads integrated with solar pv panels and energy storage system for ancillary services. *IEEE Trans. Ind. Inform.* **2018**, *14*, 5038–5049. [CrossRef]
24. Deng, W.; Xu, J.; Gao, X.; Zhao, H. An enhanced MSIQDE algorithm with novel multiple strategies for global optimization problems. *IEEE Trans. Syst. Man. Cybern. Syst.* **2020**. [CrossRef]
25. Meng, L.; Zafar, J.; Khadem, S.K.; Collinson, A.; Murchie, K.C.; Coffele, F.; Burt, G.M. Fast frequency response from energy storage system a review of grid standards, projects and technical issues. *IEEE Trans. Smart Grid* **2020**, *11*, 1566–1581. [CrossRef]
26. Lepszy, S. Analysis of the storage capacity and charging and discharging power in energy storage systems based on historical data on the day-ahead energy market in Poland. *Energy* **2020**, *213*, 118750. [CrossRef]
27. Liu, Z.; Feng, D.; Wu, F.; Zhou, Y.; Fang, C. Contract demand decision for electricity users with stochastic photovoltaic generation. *Proc. CSEE* **2020**, *40*, 1865–1873.
28. Zhao, H.M.; Liu, H.D.; Jin, Y.; Dang, X.J.; Deng, W. Feature extraction for data-driven remaining useful life prediction of rolling bearings. *IEEE Trans. Instrum. Meas.* **2021**, *70*, 3511910. [CrossRef]
29. Arias, N.B.; López, J.C.; Hashemi, S.; Franco, J.F.; Rider, M.J. Multi-objective sizing of battery energy storage systems for stackable grid applications. *IEEE Trans. Smart Grid* **2021**, *12*, 2708–2721. [CrossRef]
30. Walawalkar, R.; Apt, J.; Mancini, R. Economics of electric energy storage for energy arbitrage and regulation in New York. *Energy Policy* **2007**, *35*, 2558–2568. [CrossRef]
31. Salles, M.B.C.; Huang, J.; Aziz, M.J.; Hogan, W.W. Potential arbitrage revenue of energy storage systems in PJM. *Energies* **2017**, *10*, 1100. [CrossRef]
32. Barchi, G.; Pierro, M.; Moser, D. Predictive energy control strategy for peak shaving and shifting using BESS and PV generation applied to the retail sector. *Electronics* **2019**, *8*, 526. [CrossRef]
33. Wang, Y.; Pei, C.; Li, Q.; Li, J.; Pan, D.; Gao, C. Flow shop providing frequency regulation service in electricity market. *Energies* **2020**, *13*, 1767. [CrossRef]
34. Karpilow, A.; Henze, G.; Beamer, W. Assessment of commercial building lighting as a frequency regulation resource. *Energies* **2020**, *13*, 613. [CrossRef]
35. Cai, J. Optimal Building Thermal Load Scheduling for Simultaneous Participation in Energy and Frequency Regulation Markets. *Energies* **2021**, *14*, 1593. [CrossRef]
36. Vatandoust, B.; Ahmadian, A.; Golkar, M.A.; Elkamel, A.; Almansoori, A.; Ghaljehei, M. Risk-averse optimal bidding of electric vehicles and energy storage aggregator in day-ahead frequency regulation market. *IEEE Trans. Power Syst.* **2019**, *34*, 2036–2047. [CrossRef]
37. Shi, Y.; Xu, B.; Wang, D.; Zhang, B. Using battery storage for peak shaving and frequency regulation: Joint optimization for superlinear gains. *IEEE Trans. Power Syst.* **2018**, *33*, 2882–2894. [CrossRef]
38. Liu, Q.; Liu, M.; Lu, W. Control method for battery energy storage participating in frequency regulation market considering degradation cost. *Power Syst. Technol.* **2021**, *45*, 3043–3051.
39. Li, J.; Zhang, J.; Mu, G.; Ge, Y.; Yan, G.; Shi, S. Day ahead optimal scheduling strategy of peak regulation for energy storage considering peak and valley characteristics of load. *Electr. Power Autom. Equip.* **2020**, *40*, 128–133.
40. Xu, B.; Zhao, J.; Zheng, T.; Litvinov, E.; Kirschen, D.S. Factoring the cycle aging cost of batteries participating in electricity markets. *IEEE Trans. Power Syst.* **2018**, *33*, 2248–2259. [CrossRef]
41. Cao, J.; Harrold, D.; Fan, Z.; Morstyn, T.; Healey, D.; Li, K. Deep reinforcement learning-based energy storage arbitrage with accurate lithium-ion battery degradation model. *IEEE Trans. Smart Grid* **2020**, *11*, 4513–4521. [CrossRef]
42. Li, X.; Ma, R.; Wang, S.; Zhang, S.; Li, P.; Fang, C. Operation control strategy for energy storage station after considering battery life in commercial park. *High Volt. Eng.* **2020**, *46*, 62–70.
43. Li, W.C.; Tong, Y.B.; Zhang, W.G. Energy storage capacity allocation method of electric vehicle charging station considering battery life. *Adv. Technol. Electr. Eng. Energy* **2019**, *39*, 55–63.

44. Huang, J.; Li, X.; Chang, M.; Li, S.; Liu, W. Capacity allocation method considering the energy storage battery participating in the primary frequency modulation technology economic mode. *Trans. China Electrotech. Soc.* **2017**, *32*, 112–121.
45. Chen, Y.; Leonard, R.; Keyser, M.; Gardner, J. Development of performance-based two-part regulating reserve compensation on miso energy and ancillary service market. *IEEE Trans. Power Syst.* **2015**, *30*, 142–155. [CrossRef]
46. Papalexopoulos, A.D.; Andrianesis, P.E. Performance-based pricing of frequency regulation in electricity markets. *IEEE Trans. Power Syst.* **2014**, *29*, 441–449. [CrossRef]
47. Batiyah, S.; Zohrabi, N.; Abdelwahed, S.; Sharma, R. An MPC-based power management of a PV/battery system in an islanded DC microgrid. In Proceedings of the 2018 IEEE Transportation Electrification Conference and Expo (ITEC), Long Beach, CA, USA, 13–15 June 2018; pp. 231–236.

electronics

MDPI

Article

Machine Learning-Driven Approach for a COVID-19 Warning System

Mushtaq Hussain [1], Akhtarul Islam [2], Jamshid Ali Turi [3], Said Nabi [1,*], Monia Hamdi [4], Habib Hamam [5,6,7,8], Muhammad Ibrahim [9,10,*], Mehmet Akif Cifci [11,12] and Tayyaba Sehar [1]

1 Department of Computer Science and Information Technology, Virtual University of Pakistan, Lahore 54000, Pakistan
2 Statistics Discipline, Science, Engineering and Technology (SET) School, Khulna University, Khulna 9208, Bangladesh
3 Department of Management Information System, University of Tabuk, Tabuk 47512, Saudi Arabia
4 Department of Information Technology, College of Computer and Information Sciences, Princess Nourah bint Abdulrahman University, P.O. Box 84428, Riyadh 11671, Saudi Arabia
5 Faculty of Engineering, Moncton University, Moncton, NB E1A3E9, Canada
6 International Institute of Technology and Management, Commune d'Akanda, BP, Libreville 1989, Gabon
7 Department of Electrical and Electronic Engineering Science, School of Electrical Engineering, University of Johannesburg, Johannesburg 2006, South Africa
8 Spectrum of Knowledge Production & Skills Development, Sfax 3027, Tunisia
9 Department of Computer Engineering, Jeju National University, Jeju 63014, Republic of Korea
10 Department of Information Technology, University of Haripur, Haripur 22620, Pakistan
11 Department of Computer Engineering, Bandirma Onyedi Eylul University, Balikesir 10200, Turkey
12 Informatics, Klaipeda State University of Applied Sciences, LT-91274 Klaipeda, Lithuania
* Correspondence: said.nabi@vu.edu.pk (S.N.); ibrahimmayar@uoh.edu.pk (M.I.)

Abstract: The emergency of the pandemic and the absence of treatment have motivated researchers in all the fields to deal with the pandemic situation. In the field of computer science, major contributions include the development of methods for the diagnosis, detection, and prediction of COVID-19 cases. Since the emergence of information technology, data science and machine learning have become the most widely used techniques to detect, diagnose, and predict the positive cases of COVID-19. This paper presents the prediction of confirmed cases of COVID-19 and its mortality rate and then a COVID-19 warning system is proposed based on the machine learning time series model. We have used the date and country-wise confirmed, detected, recovered, and death cases features for training of the model based on the COVID-19 dataset. Finally, we compared the performance of time series models on the current study dataset, and we observed that PROPHET and Auto-Regressive (AR) models predicted the COVID-19 positive cases with a low error rate. Moreover, death cases are positively correlated with the confirmed detected cases, mainly based on different regions' populations. The proposed forecasting system, driven by machine learning approaches, will help the health departments of underdeveloped countries to monitor the deaths and confirm detected cases of COVID-19. It will also help make futuristic decisions on testing and developing more health facilities, mostly to avoid spreading diseases.

Keywords: time series; forecasting; COVID-19; machine learning; warning system; PROPHET; health

Citation: Hussain, M.; Islam, A.; Turi, J.A.; Nabi, S.; Hamdi, M.; Hamam, H.; Ibrahim, M.; Cifci, M.A.; Sehar, T. Machine Learning-Driven Approach for a COVID-19 Warning System. *Electronics* **2022**, *11*, 3875. https://doi.org/10.3390/electronics11233875

Academic Editors: Taiyong Li, Wu Deng, Jiang Wu and Juan M. Corchado

Received: 19 August 2022
Accepted: 14 November 2022
Published: 23 November 2022

1. Introduction

The introduction should briefly place the study in a broad context and highlight why it is important. Coronaviruses (termed as CoVs) are a group of viruses that infect birds and mammals. They also cause widespread diseases, such as Severe Acute Respiratory Syndrome Coronavirus (SARS-CoV), Middle East Respiratory Syndrome, Coronavirus (MERS-CoV), and the 2019 Novel Coronavirus (2019-nCoV, also known as COVID-19) [1]. The COVID-19 outbreak started in Wuhan Province, China, in late December 2019, and patients died due to organ dysfunction syndrome [2,3]. The Chinese government reported that the causative pathogen was a coronavirus identified by genomic sequencing and electron microscopy. The virus originated in bats and was eventually transmitted to humans via an intermediate host (probably the raccoon dog) [4].

In many instances, the major symptoms of COVID-19, were fever, cough, and shortness of breath, resembling those of seasonal influenza [5]. Since it was first recognized, COVID-19 has spread exponentially across the world. According to world meters, as of the 5 October 2022, 11:13 GMT, the COVID-19 pandemic has affected 228 countries and territories worldwide and two international conveyances with 624,430,759 confirmed 6,553,537 deaths, 604,462,445 recovered cases, and 13,414,777 active cases. Even after the substantial efforts made by scientists and scholars worldwide, COVID-19 has no standard cure method through vaccines [6]. Nonetheless, some of the patients of the COVID-19 pandemic are recovering with the aid and the proper administration of antibiotic medications. Right now, the world needs a speedy solution to tackle the further spread of COVID-19. The emergence of COVID-19 infection has forced researchers from various disciplines to explore this novel virus. Machine learning is a branch of AI that essentially focuses on the production of systems that can learn from trained examples and improve without being explicitly programmed [7]. Machine Learning has played a significant part in many fields, e.g., medical care [8], medical informatics [9], and agriculture [10]. Moreover, different ML models have optimization problems and mathematical techniques [11] that can be used to solve these problems. Similarly, ML algorithms have been used to understand and detect COVID-19, which has alleviated the enormous strain on healthcare systems while offering the most effective diagnostic and prognostic tools for COVID-19 pandemic patients.

The COVID-19 pandemic has seriously affected population health across the globe. The forecasting of COVID-19 research efforts has become critical and, with the advancement of computers and software technology, AI has played a vital role in the healthcare system in the detection and clinical diagnosis of diseases. Much research has focused on the treatment, prediction, as well as the formulation of COVID-19 [12].

A variety of ML techniques have been used to predict the mortality risk of COVID-19 patients. Pourhomayoun et al. [13] have used a support vector machine (SVM), artificial neural network (ANN), random forest (RF), decision tree, logistic regression, and K-nearest neighbor to detect the mortality risk of patients due to COVID-19 infection.

Researchers have also focused on modeling, predicting, and forecasting the spread of COVID-19 based on the time-series recorded data of COVID-19. Sarkar et al. [14] proposed the SARII mathematical model to forecast the dynamic transmission of COVID-19, which was an extended version of the SEIR model. The proposed model is based on six dynamics behaviors, i.e., susceptible, asymptomatic, recovered, infected, and quarantined. An alternative version of the SEIR model was proposed by Abbasi et al. [15], named SQEIAR, which considered the two parameters, quarantined individuals and asymptomatic individuals, to describe COVID-19. Similarly, Ribeiro et al. [16] used applied regression models ARIMA, cubist regression, random forest, SVR, rigid regression, and stacking-ensemble learning for the forecasting of COVID-19 cases in Brazil. According to the obtained results, researchers observed that SVM regression and stacking-ensemble are better in forecasting. Apart from linearity, many researchers have used nonlinearity structures to predict COVID-19 cases. Peng et.al [17], used the SVR with a Gaussian kernel and claimed the better prediction of COVID-19 cases.

Various ML algorithms and deep learning techniques have been utilized in the literature to compute COVID-19. Different methodologies, including long short-term memory (LSTM), ARIMA, and JNARNN, were built using ML and deep learning [18,19]. However, this research did not analyze the performance model's link between positive cases and input features. This study explores the performance time series of the ML model on the COVID-19 dataset and identifies the characteristics most closely associated with positive COVID-19 cases. The prognosis of death and verified detection cases (of COVID-19) is a weekly concern for numerous nations. The current dataset displayed daily confirmations and death cases in various nations; however, such a dataset was not ordered weekly, and not all the observations of the existing dataset were available (many attributes were missing), which must be fixed.

For this research, we utilized the COVID-19 virus dataset which is available online for research purposes. In this dataset, the COVID-19 observations, such as confirmed cases, death cases, and recovery cases, are organized by date for many U.S. states. In addition, the dataset comprises data from 10 March 2020–29 March 2020.

In the subsequent section, the relevant literature will be presented. Materials and methods will be discussed in the following section. Section 4 will next exhibit the experiment described in this paper. The final section will present issues, difficulties, and conclusions.

Consequently, a new COVID-19 warning system can be constructed using an ML technique, for instance, by comparing the performance of the "Time Series ML Algorithm" to the "Statistical Time Series Model." This would aid healthcare professionals and physicians in diagnosing COVID-19 pandemic patients and recommending recent anti-bodies medication (for recovery). Additionally, implementing the time series ML algorithm (to avoid the pandemic) would limit the spread of the COVID-19 pandemic in situations where human-to-human interaction is inevitable.

The main objective of this study is twofold. First, to estimate the weekly-confirmed instances of COVID-19 and potential deaths using patient history data in different nations; second, to create a warning system that can evaluate the performance of various ML time series models with statistical time series models. Since the present pandemic data (of COVID-19) is only available in abundance, the investigation of the following research issues constitutes a significant contribution to the body of research.

1. What are the appropriate time series models for predicting patients infected with the COVID-19 virus?
2. What will the number of death cases and possible confirmed detected cases in the coming weeks be, based on various given features in the form of data at input, such as date, country, detected cases, and deaths?

2. Related Work

False positives are often observed in research when the literature of papers and methodologies are considered. As a result, it is essential to develop methods that make becoming faster and gaining more accurate results easier while simultaneously reducing human-induced errors. This section of the study examines the procedures and methodologies of the literature.

To help policymakers manage the disease and related emerging situations, the authors [6] devised a COVID-19 pandemic prediction tool. This tool was based on data from patients from India to keep track of infected cases. They assumed that control strategies, such as quarantines and lockdowns, would prevail. Their results suggested that India could experience the end of the pandemic by March 2021. The model was developed on the basis of least-square fitting of the novel coronavirus behavior and is based on real-world data for a particular time, but the least-square technique was unable to address the overfitting issue.

Ganiny et al. [19], based on the Indian perspective, employed an autoregressive integrated moving average (ARIMA) model that utilizes the past trajectory and forecasts the future evolution of COVID-19. Their model predicted the number of infected cases, active cases, recoveries, and deaths due to the pandemic. They suggest some robust control strategies to mitigate the spread of COVID-19.

Wadhwa et al. [20] predicted recovery, death, and active cases of COVID-19 patients by applying a linear regression technique from Indian records. Their model predicted the extension of lockdown based on empirical results. They applied graphical tools to showcase the predicted results more comprehensively.

Saima et al. [21] studied the trends of COVID-19 in the eastern Mediterranean regions using a statistical method. Their analysis revealed that Iran was the worst affected country, followed by Saudi Arabia and Pakistan. The United Arab Emirates and Saudi Arabia had the lowest fatality rates, while Pakistan and Lebanon had moderate fatalities. They suggest following strict recommendations, based on epidemiological principles, to reduce COVID-19 cases.

Yadav et al. [22] utilized ML tools to analyze the transmission and growth rates of COVID-19 patients across various countries. They further correlated the weather conditions and the COVID-19 cases and predicted the pandemic's end time frame. They exploited support vector machine algorithms (SVM) for these tasks.

The model demonstrated a high accuracy of 98% and proved its efficacy compared to recent forecasting models.

Ricardo, M. A. V., et al. [23] applied reduced-space Gaussian process regression, related to chaotic dynamical systems, to forecast COVID-19-related deaths from 82 days' data. Empirical results asserted that Gaussian mean-field models were able to be employed to gather information regarding the pandemic's spread, recovery, and fatality rates. They also devised a reduced-space Gaussian process regression model to estimate when saturation would be achieved in the USA (regarding the pandemic).

Hamzah et al. [24] also introduced a predictive model based on the Corona Tracker (an online platform for reliable analysis, and statistics, of COVID-19) to forecast COVID-19-related cases, recoveries, and deaths. They exploited susceptible exposed infectious recovered (SEIR) modeling to keep track and predict COVID-19 outbreaks.

Moreover, they classified and analyzed the queried news into positive and negative categories based on the people's sentiments. Furthermore, they tried to understand the economic and political impacts of COVID-19. Overall, they observed that more negative articles exist in the given domain than positive ones.

Mahajan et al. [25] utilized a compartmental epidemic model (SIPHERD) to predict COVID-19 active, confirmed, and death cases in India. Their results show that social-distancing measures, increasing daily tests, and strict lockdown significantly impacted the reduction of COVID-19.

Moreover, the authors [26] employed the SEIR model to extract the epidemic curve from the epidemiological data of COVID-19. They also applied an AI framework to forecast the disease. Their model was trained using 2003 SARS data. They predicted that the epidemic peak would gradually rise and then fall in China. Their dynamic model demonstrated its efficacy in forecasting COVID-19 epidemic sizes and peaks.

Shahid et al. [27] also presented a COVID-19 time series prediction model by employing LSTM, bidirectional long short-term memory (Bi-LSTM), support vector regression (SVR), and autoregressive integrated moving average model (ARIMA) techniques. They evaluated their model using the R square score, root mean square error (RMSE), and mean absolute error indices (MAEI). Their results suggest that the Bi-LSTM model is the best-suited model for such pandemic predictions, especially for better management and planning.

According to Xue et al. [28], in 2020, COVID-19 still needed to be completely understood. The authors believe that scientists and doctors were struggling to find COVID-19 instances. COVID-19 tests include viral tests to determine whether the patients are infected and antibody tests to determine if the patients have been infected before. The paper aims to reduce the false positive rate.

Various ML algorithms and deep learning techniques have been utilized in the literature to compute COVID-19. Different methodologies, including LSTM, ARIMA, and JNARNN, were built using ML and deep learning [29,30]. However, this research did not analyze the performance model's link between positive cases and input features. This study explores the performance time series of the ML model on the COVID-19 dataset and identifies the characteristics most closely associated with positive COVID-19 cases. The prognosis of death and the verified detection cases (of COVID-19) is a weekly concern for numerous nations.

Mansour et al. [17] provide a unique unsupervised DL-based variational autoencoder model for COVID-19 identification and classification. They utilized the Adagrad approach to modify the Inception v4 model hyperparameters to improve the classification performance.

Accordingly, an intelligent COVID-19 positive cases detection system was developed in this study using ML time series algorithms. The main task of the proposed architecture is to provide an efficient method for predicting COVID-19 patient-positive cases. Based on the performance of the proposed system, the health department can find daily positive cases in different areas of the country.

3. Materials and Methods

We studied COVID-19 data from many countries across the globe, which are freely accessible online for research purposes. The forecasting system of the current study, driven by machine learning approaches, will help the health departments of underdeveloped countries to monitor the death and confirmed cases of COVID-19. It will also help make futuristic decisions on testing and developing more health facilities, mostly to avoid spreading diseases. Using an ML technique, for instance, by comparing the performance of the "Time Series ML Algorithm" to the "Statistical Time Series Model", would aid healthcare professionals and physicians in diagnosing COVID-19 pandemic patients and recommending recent anti-bodies medication (for recovery).

The dataset contains information regarding the COVID-19 virus in various countries and cities. In addition, the dataset contains daily record data for many countries/cities. The dataset includes records for other countries and cities beginning on 11 March 2020 and ending on 29 March 2020. Using the time series models depicted in Figure 1, we followed the steps below to explore COVID-19 predictions for the subsequent week.

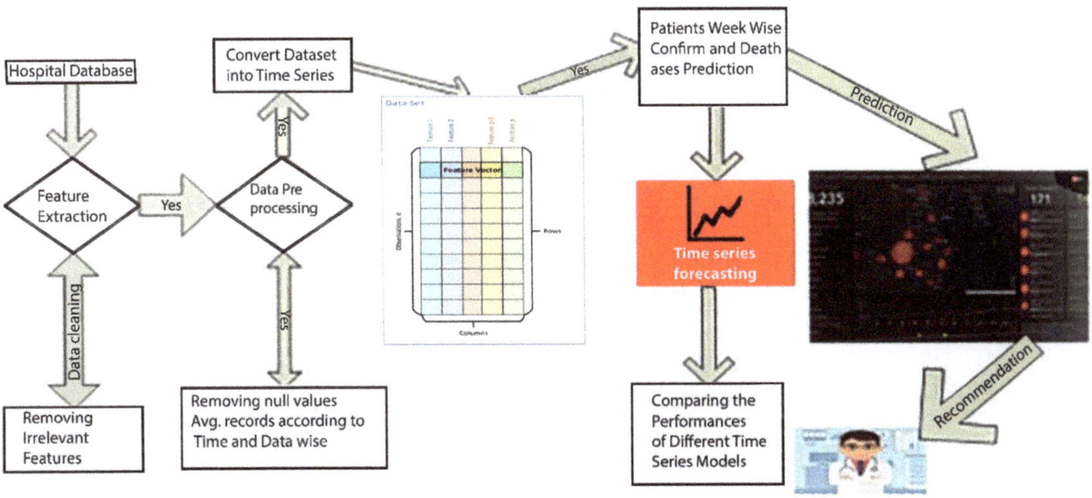

Figure 1. Proposed framework.

Figure 1 depicts the data collection process, followed by the feature extraction procedure. If there are irrelevant characteristics, they are eliminated. Following this, we transfer the data into the preprocessing procedures, eliminating null values and transforming the data into a time series. The whole dataset is disseminated to the Week Wise section, where the death and survival instances are verified. Either it is moved to time series forecasting to be compared with several models or the prediction is moved to an expert.

Data description: Table 1 provides specifics on the features extracted from the dataset. It describes how to exclude the pertinent features/attributes of the dataset to construct time-series prediction models for the COVID-19 cases of different counties, including laboratory-confirmed cases, recovered cases, and deaths in the following week or hours.

Extracted features include date, state, country, confirmed cases, recovered cases, deaths, and population.

Table 1. Attribute description of the COVID-19 dataset.

Attribute	Description
Date	The date on which the data was recorded
State	State from where the COVID-19 patients belong
County	Count y from where COVID-19 patients belong
Confirmed	Number of confirmed COVID-19 patients
Recovered	Number of recovered COVID-19 patients
Deaths	Number of deceased COVID-19 patients
Population	The total population in the state

Data preprocessing: Before applying time series forecasting models, we removed missing values by applying the median imputation method. We checked the dataset's stationarity property, which is a relevant feature of time series and that confirms the data suitability, specifically for the time series-related issues. Moreover, many time series models only work on stationary data. Stationarity data has a constant up and down movement, and it also has a constant mean and variance. Since the data used in this study is not stationary because the status of the time series of COVID-19 dataset statistics and the properties that changed over time, i.e., MSE and RMSE, we applied Python differencing techniques (to convert the data into the stationary format). For that reason, we subtracted the current value from the next value. Then we used a partial autocorrelation function (PACF) plot to check for stationary properties in the dataset.

Machine learning techniques: In order to apply time series forecasting models to predict next week and hours of confirmed detected and death cases (based on ML), there are multiplication classification techniques for the datasets (such as PROPHET, auto regression, ARIMA, and LSTM). The technique that we use (in this study) consists of different ways to extract and classify features that help predict futuristic issues. The details of the current study ML model are below.

3.1. PROPHET

Without a high level of expertise, the prediction is difficult for ML researchers because it often needs more skills than they possess in terms of programming language. PROPHET is the Facebook data ML technique, which is open source and available in Python and R languages. Researchers can use this tool without any programming skills. It is an algorithm that is used to build a forecasting model for time series data based on an additive approach. The algorithm was first introduced in 2017, and unlike the traditional time series technique, PROPHET tries to fit additive regression (called curve fitting) [31]. PROPHET is very robust within missing data, handles outliers very well, and is best with time series, strong seasonal effects, and several seasons of historical data [32].

3.2. Autoregressive Model (AR)

The automotive regressive (AR) model predicts the next timestamp value by applying regression and previous values. The analysis of nature, economics, stock markets, and other time series-based systems frequently employs the AR model. AR models provide a number of advantages over other time series models, such as their ability to operate on continuous variables. The AR model predicts the next timestamp value by regressing and using previous values. The AR model is commonly used in analyzing nature, economics, stock markets, and other time series-based processes. AR models have some advantages over other time series models; for example, they work on continuous values.

3.3. Auto Regressive Integrated Moving Average (ARIMA)

We used the auto regressive integrated moving average (ARIMA) model as our dataset was non-stationary, whereas the integration part was "Stationized" the time series.

3.4. Long Short-Term Memory

LSTM is used to solve the learning models for recurrent neural networks to provide promising results on many tasks, such as constructing prediction and language models [33]. It solves challenging tasks (large time-lags) that recurrent network algorithms [34] have never solved. LSTM is used to solve the learning models for recurrent neural networks to produce promising results on a variety of tasks, including building prediction and language models [33]. It solves complex tasks (long time lags) that have never been solved by recurrent network algorithms [34].

4. Results and Discussion

In this study, we used time-series ML models rather than other ML models with no time dimension. The time series forecasting model is based on previously observed values [35].

We discovered a positive but weak correlation (r = 0.032) between "Confirmed" and "Recovered" cases in Table 2. Every day, COVID-19 confirmed and recovered patients move in the same positive direction, while the increase in confirmed patients is very high compared to recovered patients of COVID-19. The maximum number of COVID-19 patients is 21,873, whereas the maximum number of recovered patients with COVID-19 is 10. Between deaths and confirmed cases, we found a strong positive correlation (r = 0.796). As COVID-19 is confirmed and detected, the death rate of COVID-19 patients also increased, with the maximum reported death of 281. The variables of "Population" and "Confirmed" also showed a positive but weak correlation (r = 0.154), which means that as the population increases, the confirmed patients also increase.

Table 2. Correlation analysis and descriptive statistics for different variables (descriptive statistics).

	N	Minimum	Maximum	Mean	Std. Deviation	Correlation with "Confirmed" (r)	*p*-Value
Confirmed	16,585	0.00	21,873.00	22.79	323.776	1	<0.001 ***
Recovered	16,585	0.00	10	0.008	0.1625	0.032	<0.001 ***
Deaths	16,585	0.00	281	0.311	4.0454	0.796	<0.001 ***
Population	16,452	88	39,512,223.0	387,142.67	1,997,001.27	0.154	<0.001 ***

Note: *** $p < 0.001$.

Table 3 demonstrates that when the number of confirmed cases grows, the death rate will increase by 0.010 times. We determined that a one-unit (100,000) increase in population contributes 0.016 times to the death factor for the "Population" variable. Here, R^2 equals 0.640, which shows the amount of a dependent variable's variance explained by the independent variables in a regression model. The model's inputs can explain approximately 64 percent of the observed variation. We also benefited from the same-day confirmed cases to predict death, though, for one day of COVID-19 cases, it can be concluded that 2.2% died while 75.9% recovered and 21.9% were still in isolation or being treated at the last follow-up.

Table 3. Multiple linear regression estimation considering "Deaths" as a dependent variable.

	Unstandardized Coefficients		*p*-Value	95.0% Confidence Interval for B	
	B	Std. Error		Lower Bound	Upper Bound
(Constant)	0.032	0.019	0.094	−0.006	0.07
Confirmed	0.01	0.000	0.000	0.01	0.01
Population	0.016	0.001	0.000	0.014	0.018

Note: B (Biases) is a training parameter that needs to be optimized during the training process. *p*-value is the probability value corresponding to the likelihood of gaining a data value.

Non-stationary data represent that the mean and the standard deviation are not constant for given data during the time curve described by [35]. With the help of data visualization, we can understand the pattern, trend, and correlation between the variables for COVID-19 predictions, based on the time series-based ML approach.

Figure 2 shows how the confirmed COVID-19 patients and the dead COVID-19 patients, from 10 March 2020–28 March 2020. The figure also shows a sharp increase in both confirmed and unconfirmed deaths of COVID-19 patients during this time. As the number of confirmed COVID-19 patients rises, so does the death rate among these patients. The disturbing fact is that the death toll on March 26 exceeded one thousand.

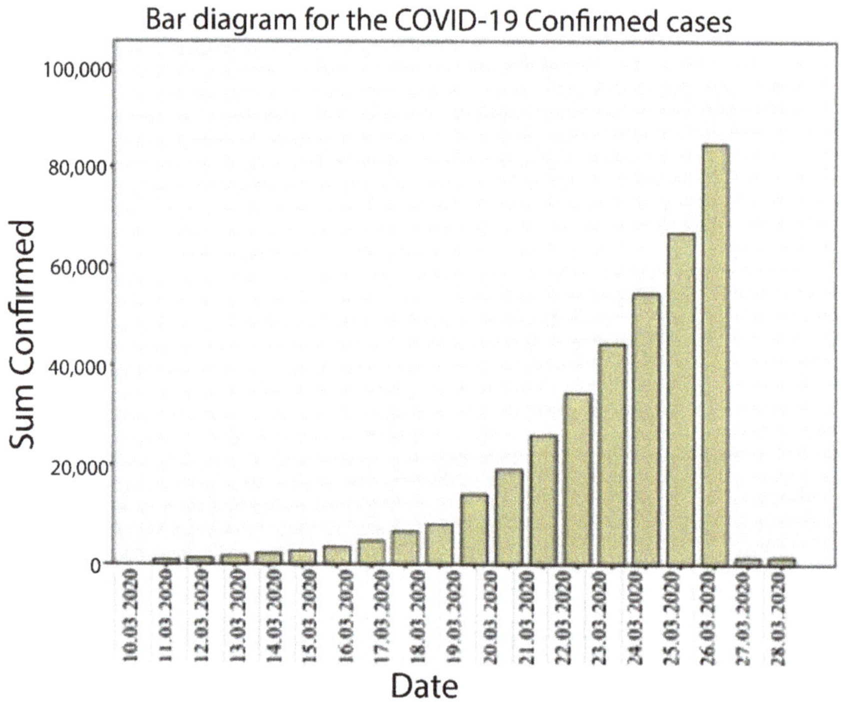

Figure 2. Bar diagram shows the daily confirmed cases of COVID-19 in current study dataset.

Figure 3 depicts the confirmed COVID-19 patients and deceased COVID-19 patients between March 10, 2020, and March 28, 2020. The figure also shows a sharp increase in confirmed and unconfirmed COVID-19 patient deaths over this period. The COVID-19 epidemic affects all sectors of the population but disproportionately negatively impacts the most disadvantaged social groups.

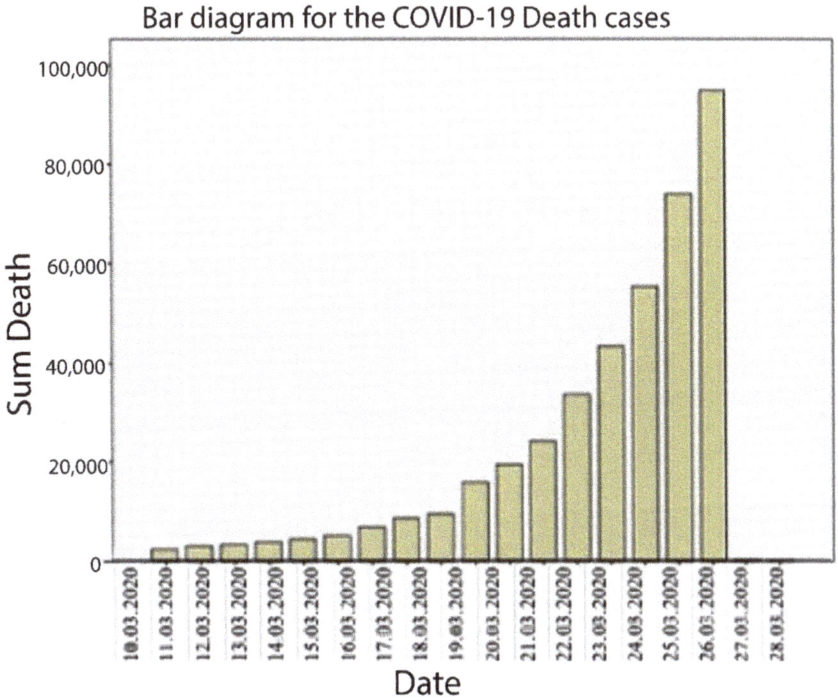

Figure 3. Bar diagram shows the daily death cases due to COVID-19 in current study dataset.

Figure 4 illustrates the daily confirmed cases during COVID-19. In addition, it implies that the rate of confirmed patients jumped dramatically after 21 March 2020. From the beginning of 11 March 2020–21 March 2020, as confirmed in Figure 4, the number of COVID-19-confirmed cases increased gradually. From 21 March 2020, there was an alarming increase in the number of confirmed COVID-19 cases. After 23 March 2020, the number of confirmed COVID-19 cases surged. Globally, there was an increase in the world of confirmed COVID-19 cases. Figures 3 and 4 demonstrate a quick decline in the amount of confirmed and fatal cases due to the lack of data in some date-specific datasets.

4.1. Design of the Predictive Models and Experimental Setup

In this study, for the COVID-19 predictions, we used the available data for research purposes online [36–41]. From different countries' data from the date 11 March 2020–29 March 2020, we used different Python modules to visualize and describe the data and then trained ML time series models with 80% of data and tested on 20% of data. The PROPHET is a simple time-series algorithm that gives a quick result during the initial stage of modeling. Therefore, we used a Python module to implement the PROPHET algorithm.

Nevertheless, to implement the PROPHET algorithm in Python, the dataset must have NAN-values (or missing values) in the features column; therefore, we leave some NaN values in the dataset. Next, we changed the date column into a date index. We trim the current study dataset to keep only those rows that fall within the period from 10 March 2020–31 March 2020. Before running the model, we rename the dataset column into two columns that are ds (Date) and y (confirmed cases). For LSTM, we also converted the data into three dimensions because LSTM only works on three-dimensional data.

The LSTM model has been enhanced with four layers: the first two layers have 40 neurons each, the third layer has 25 neurons, and the final layer has one neuron. In addition, the model is employed as an Adam optimizer, which utilizes square errors as loss functions. Before applying the AR model, we checked the stationary properties of the dataset because the AR model only works on stationary data. Since the current study dataset was non-stationary (in nature), we took severe differences and finally obtained the stationary data. Nevertheless, the AR and ARIMA models were trained on default values.

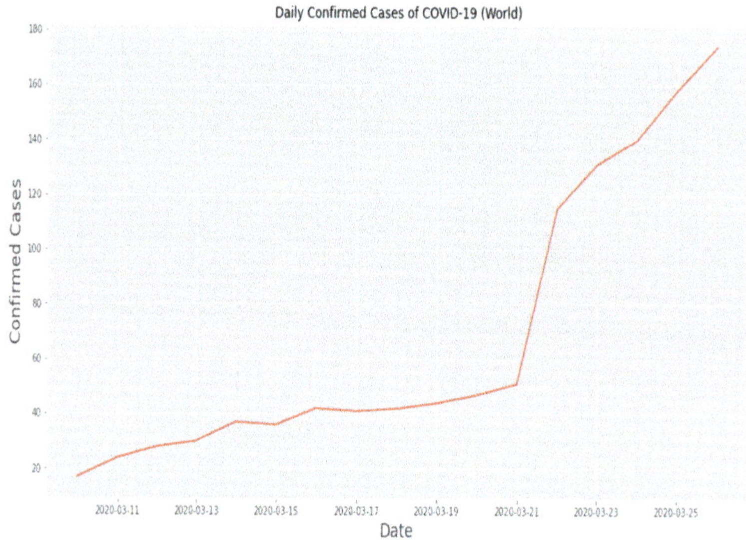

Figure 4. Visualization of daily confirmed cases during COVID-19 disease.

4.2. Performance Measurements

To analyze the performance of time series ML models, we employed root mean square error (RMSE) performance metrics. RMSE is the square root of the mean squared error (MSE), which is converted to RMSE by taking its square root. MSE is measured in square units of the target variable, whereas RMSE is measured in the same units. MSE penalizes greater errors more harshly than the squared loss function from which it is derived, and it penalizes greater errors more harshly due to its structure. It measures the deviation between the value predicted by the ML model and the actual value. We predicted the confirmed identified cases and deaths of individuals needing medical care. Consequently, if the number of confirmed deaths is considerable (according to the provided country's statistics), the health agency can take the necessary instances to reduce COVID-19. This is a time series ML-based problem, and the dataset is freely available for research purposes [36].

We investigated the performance of the AR, PROPHET, ARIMA, and LSTM time series classifiers to identify the optimal ML time series models for forecasting the daily confirmed detected cases in different countries during COVID-19. The input variables were daily confirmed cases and death cases of the patient in different countries. The output variables were next week's confirmed and death cases of the patients.

The PROPHET time series model was trained using 80% COVID-19 training data. Then we tested trained PROPHET classifiers on 20% test data. The model received an RMSE value of 29.07, and the results are shown in Table 4. Whereas Figure 5 shows a comparison between the actual confirmed cases and predicted confirmed cases of PROPHET algorithms.

In Figure 6, y-hat or y-hat represent the estimated or predicted values in predictive models. In a regression or other predictive model, the estimated or anticipated values are referred to as y-hat values.

Table 4. ML time series models' performance on COVID-19 test data.

Models	RMSE (Root Mean Square Error)
PROPHET	29.07
LSTM	130.11
AR model	10.49
ARIMA Model	34.75

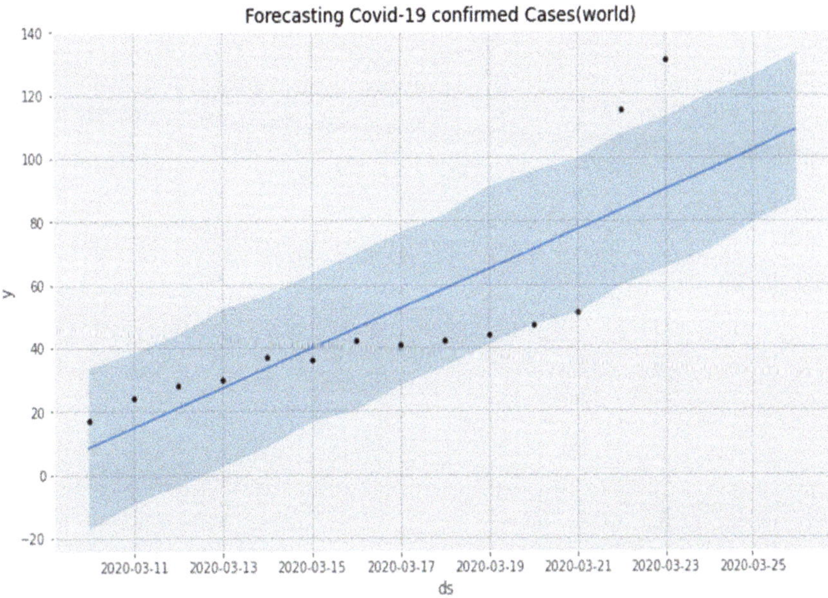

Figure 5. Daily predicted cases of COVID-19 using the PROPHET ML algorithm.

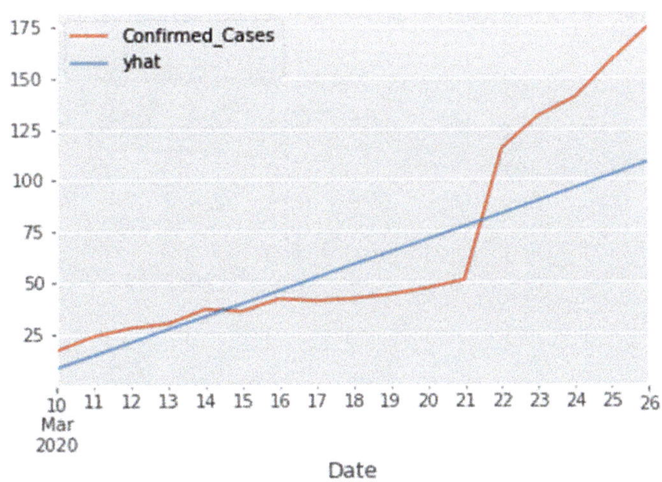

Figure 6. Comparison of actual and predicted cases of COVID-19 using Prophet. The X-axis shows the number of COVID-19 confirmed cases.

Figure 7 depicts PROPHET's forecasts based on test results. The date is represented by ds, and the confirmed instances for the provided dates are represented by y.

	yhat	yhat_lower	yhat_upper	y
ds				
24/03/2020	96.015084	71.563744	119.940583	140
25/03/2020	102.267617	79.103752	126.790716	158
26/03/2020	108.520149	87.350423	132.865530	174

Figure 7. PROPHET forecasting on test data. The ds represents the date and y represents the confirmed cases for the given dates.

Figure 8 shows the AR predicting outcomes. The AR is shown in the middle of the graph, and residuals at various time steps are displayed beside each observation. Both the proven and predicted instances are evident.

Date	Confirmed cases	Predictions
24/03/2020	140	26.088554
25/03/2020	158	27.168270
26/03/2020	174	28.391363

Figure 8. Comparison of the actual confirmed case and the predicted case of the PROPHET algorithms.

Figure 9 represents the predicted value of the test data, whereas the blue line represents the actual value.

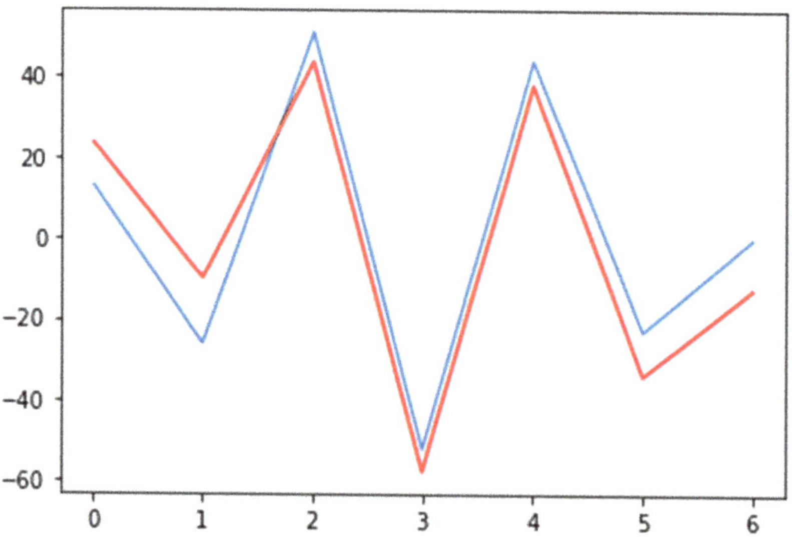

Figure 9. The AR prediction on test data. The red line: predicted value; the blue line: the actual value.

Next, we built the AR and ARIMA models using Python. The Figure 9 shows prediction results of AR model. The AR and ARIMA models (RMSE 10.49 and 34.75, respectively) are shown in Table 4. Additionally, We tuned the AR and ARIMA model parameters using Python because they affect the performance of AR models. Finally, we constructed the time series model LSTM of deep learning using the Python KERAS framework. The performance of LSTM could be better because LSTM requires a considerable amount of data.

Finally, we compared the performance of time series ML models on COVID-19 datasets. The experiment results indicate that developers can integrate the AR and PROPHET time series model into the COVID-19 death warning system and predict confirmed and fatal patient cases in the country (with high performance). Popular algorithms, e.g., the LSTM and ARIMA models, perform well with various real-world issues. Nonetheless, the outcome shows that the performance of these models is inadequate. We found that the size of the confirmed COVID-19 is positively correlated with the level of death caused by COVID-19. Furthermore, from the beginning, the confirmed cases and deaths increased. Regression analysis shows a positive association between the "size of the total population" and the "size of the infected population", along with the number of deaths from COVID-19.

5. Strengths and Limitations

The strength of the study includes, but is not limited to, the performance of PROPHET and AR Time Series algorithms, which is high and can easily integrate with health systems because the PROPHET and AR algorithms are simple and can be easily implemented without programming skills. Moreover, this study also, in the same way, carries certain limitations, which need to be addressed in future research studies. The dataset was limited only to country, state, confirmed cases, recovered cases, deaths, dates, and population as an input related to the COVID-19 disease. At the same time, COVID-19 also depends on other factors, such as age, weather, and even gender, whose inclusion may increase ML models' performance. Similarly, in the current study, the number of patient records was limited and the ML model performance may increase if we increase the number of documents in the dataset.

6. Conclusions

Researchers have encountered various challenges when attempting to construct a warning system that can predict the rapid development and spread of COVID-19. Some issues are hardware resources, DL network architecture repair, and data availability. A massive dataset is required to implement DL methods, such as LSTM, for prediction. The absence of such datasets may result in inaccurate and improper conclusions—consequently, the performance of deep learning architectures declines concerning these warning systems. In addition, there is uncertainty associated with medical datasets. Another problem with the datasets is the lack of phenotypic data, such as gender and age. Moreover, for the prognosis of the disease using computer-assisted early warning systems, several elements (such as infection of neighbor/friend/family member, climatic circumstances, policies to prevent the spread of the disease by countries, and the average age of the community) come into play. The nature of COVID-19 is still largely unclear, so the probability of mutation is a formidable obstacle.

This study examined the performance of time series ML models for predicting patients' confirmed, detected, and death cases over the following week (using a given dataset for research purposes). After training the LSTM, AR, PROPHET, and ARIMA models, we calculated the predictions of confirmed and detected death cases for the next week. The findings predict that PROPHET and AR models have the lowest RMSE error for making predictions concerning the confirmed, detected, and death cases. Furthermore, the present research suggests that we can include PROPHET and AR models in the COVID-19 hospital dashboard. Based on the time series ML technique, it can also predict the medical personnel and government institutions' ability to predict, detect, and confirm COVID-19 death cases in the nation over the next week.

Governments across the globe have adopted various measures to contain the COVID-19 epidemic. Among these measures are the closure of public education and leisure places, such as schools, colleges, universities, movie theaters, retail malls, and parks, and the restriction of face-to-face meetings via obligatory "social distancing". The majority of the global population must adhere to these extraordinary measures. As the number of medical facilities restricted in many developing countries, the exponential development of

COVID-19 cases places a tremendous strain on health professionals and services; it causes a shortage of intensive care facilities in hospitals. The early prediction of this pandemic may assist governments, planning officials, and physicians in addressing the health issue more effectively. Thus, a COVID-19 warning system equipped with AI and ML may provide a great source of assistance.

Author Contributions: M.H. (Mushtaq Hussain): conceptualization, methodology, writing, visualization, and supervision; M.A.C. and A.I.: investigation, data curation, conceptualization, methodology, and writing; J.A.T.: formal analysis, writing, review, and editing; S.N.: formal analysis, visualization, writing, review, editing, and hand journal correspondence; M.H. (Monia Hamdi): formal analysis, review, editing, funding; H.H.: validation, analysis, review, and editing; M.I.: review and editing; and T.S.: writing, revising, and editing. All authors have read and agreed to the published version of the manuscript.

Funding: This research was funded by Princess Nourah bint Abdulrahman University Researchers Supporting Project number (PNURSP2022R125), Princess Nourah bint Abdulrahman University, Riyadh, Saudi Arabia.

Data Availability Statement: The current study data are publicly available online. No participants' personal information (e.g., name or address) was included in this study. The dataset used in the current study is publicly available at: https://doi.org/10.7910/DVN/URHUOV (accessed on 25 September 2022), https://dataverse.harvard.edu/dataset.xhtml?persistentId=doi:10.7910/DVN/URHUOV (accessed on 25 September 2022).

Conflicts of Interest: The authors declare no conflict of interest.

References

1. McIntosh, K.; Perlman, S. Coronaviruses, including severe acute respiratory syndrome (SARS) and Middle East respiratory syndrome (MERS). In *Mandell, Douglas, and Bennett's Principles and Practice of Infectious Diseases*; Bennett, J.E., Dolin, R., Blaser, M.J., Eds.; Elsevier Health Sciences: Amsterdam, The Nederlands, 2015; pp. 1928–1936.e2.
2. Morens, D.M.; Daszak, P.; Taubenberger, J.K. Escaping Pandora's box—Another novel coronavirus. *N. Engl. J. Med.* **2020**, *382*, 1293–1295. [CrossRef] [PubMed]
3. Tu, W.-J.; Cao, J.; Yu, L.; Hu, X.; Liu, Q. Clinicolaboratory study of 25 fatal cases of COVID-19 in Wuhan. *Intensiv. Care Med.* **2020**, *46*, 1117–1120. [CrossRef] [PubMed]
4. Bennett, J.E.; Dolin, R.; Blaser, M.J. *Mandell, Douglas, and Bennett's Principles and Practice of Infectious Diseases*, 9th ed.; Elsevier Health Sciences: Amsterdam, The Netherlands, 2014.
5. Gralinski, L.E.; Menachery, V.D. Return of the Coronavirus: 2019-nCoV. *Viruses* **2020**, *12*, 135. [CrossRef]
6. Sahoo, B.K.; Sapra, B.K. A data driven epidemic model to analyse the lockdown effect and predict the course of COVID-19 progress in India. *Chaos Solitons Fractals* **2020**, *139*, 110034. [CrossRef] [PubMed]
7. Shishvan, O.R.; Zois, D.; Soyata, T. Machine intelligence in healthcare and medical cyber physical systems: A survey. *IEEE Access* **2018**, *6*, 46419–46494. [CrossRef]
8. Chen, C. Ascent of machine learning in medicine. *Nat. Mater.* **2019**, *18*, 407.
9. Swapnarekha, H.; Behera, H.S.; Nayak, J.; Naik, B. Role of intelligent computing in COVID-19 prognosis: A state-of-the-art review. *Chaos Solitons Fractals* **2020**, *138*, 109947. [CrossRef]
10. Deng, W.; Ni, H.; Liu, Y.; Chen, H.; Zhao, H. An adaptive differential evolution algorithm based on belief space and generalized opposition-based learning for resource allocation. *Appl. Soft Comput.* **2022**, *127*, 109419. [CrossRef]
11. Yao, R.; Guo, C.; Deng, W.; Zhao, H. A novel mathematical morphology spectrum entropy based on scale-adaptive techniques. *ISA Trans.* **2022**, *126*, 691–702. [CrossRef]
12. Lalmuanawma, S.; Hussain, J.; Chhakchhuak, L. Applications of machine learning and artificial intelligence for COVID-19 (SARS-CoV-2) pandemic: A review. *Chaos Solitons Fractals* **2020**, *139*, 110059. [CrossRef]
13. Wu, D.; Wu, C. Research on the Time-Dependent Split Delivery Green Vehicle Routing Problem for Fresh Agricultural Products with Multiple Time Windows. *Agriculture* **2022**, *12*, 793. [CrossRef]
14. Sarkar, K.; Khajanchi, S.; Nieto, J.J. Modeling and forecasting the COVID-19 pandemic in India. *Chaos Solitons Fractals* **2020**, *139*, 110049. [CrossRef]
15. Ribeiro, M.H.D.M.; da Silva, R.G.; Mariani, V.C.; dos Santos Coelho, L. Short-term forecasting COVID-19 cumulative confirmed cases: Perspectives for Brazil. *Chaos Solitons Fractals* **2020**, *135*, 109853. [CrossRef]
16. Xue, Y.; Onzo, B.M.; Mansour, R.F.; Su, S.B. Deep Convolutional Neural Network Approach for COVID-19 Detection. *Comput. Syst. Sci. Eng.* **2022**, *42*, 201–211. [CrossRef]
17. Mansour, R.F.; Escorcia-Gutierrez, J.; Gamarra, M.; Gupta, D.; Castillo, O.; Kumar, S. Unsupervised Deep Learning based Variational Autoencoder Model for COVID-19 Diagnosis and Classification. *Pattern Recognit. Lett.* **2021**, *151*, 267–274. [CrossRef]

18. Yan, Z.; Wang, Y.; Yang, M.; Li, Z.; Gong, X.; Wu, D.; Zhang, W.; Wang, Y. Predictive and analysis of COVID-19 cases cumulative total: ARIMA model based on machine learning. *medRxiv* **2022**. [CrossRef]
19. Ganiny, S.; Nisar, O. Mathematical modeling and a month ahead forecast of the coronavirus disease 2019 (COVID-19) pandemic: An Indian scenario. *Model. Earth Syst. Environ.* **2021**, *7*, 29–40. [CrossRef]
20. Wadhwa, P.; Aishwarya; Tripathi, A.; Singh, P.; Diwakar, M.; Kumar, N. Predicting the time period of extension of lockdown due to increase in rate of COVID-19 cases in India using machine learning. *Mater. Today Proc.* **2020**, *37*, 2617–2622. [CrossRef]
21. Dil, S.; Dil, N.; Maken, Z.H. COVID-19 trends and forecast in the Eastern Mediterranean Region with a Particular Focus on Pakistan. *Cureus* **2020**, *12*, e8582. [CrossRef]
22. Yadav, M.; Perumal, M.; Srinivas, M. Analysis on novel coronavirus (COVID-19) using machine learning methods. *Chaos Solitons Fractals* **2020**, *139*, 110050. [CrossRef]
23. Velásquez, R.M.A.; Lara, J.V.M. Forecast and evaluation of COVID-19 spreading in USA with reduced-space Gaussian process regression. *Chaos Solitons Fractals* **2020**, *136*, 109924. [CrossRef] [PubMed]
24. Hamzah, F.B.; Lau, C.; Nazri, H.; Ligot, D.V.; Lee, G.; Tan, C.L.; Bin Mohd Shaib, M.K.; Binti Zaidon, U.H.; Binti Abdullah, A.; Chung, M.H.; et al. CoronaTracker: Worldwide COVID-19 outbreak data analysis and prediction. *Bull. World Health Organ.* **2020**, *1*, 1–32.
25. Mahajan, A.; A Sivadas, N.; Solanki, R. An epidemic model SIPHERD and its application for prediction of the spread of COVID-19 infection in India. *Chaos Solitons Fractals* **2020**, *140*, 110156. [CrossRef] [PubMed]
26. Yang, Z.; Zeng, Z.; Wang, K.; Wong, S.-S.; Liang, W.; Zanin, M.; Liu, P.; Cao, X.; Gao, Z.; Mai, Z.; et al. Modified SEIR and AI prediction of the epidemics trend of COVID-19 in China under public health interventions. *J. Thorac. Dis.* **2020**, *12*, 165–174. [CrossRef]
27. Shahid, F.; Zameer, A.; Muneeb, M. Predictions for COVID-19 with deep learning models of LSTM, GRU and Bi-LSTM. *Chaos Solitons Fractals* **2020**, *140*, 110212. [CrossRef]
28. Cheikhrouhou, O.; Mahmud, R.; Zouari, R.; Ibrahim, M.; Zaguia, A.; Gia, T.N. One-Dimensional CNN Approach for ECG Arrhythmia Analysis in Fog-Cloud Environments. *IEEE Access* **2021**, *9*, 103513–103523. [CrossRef]
29. Kırbaş, İ.; Sözen, A.; Tuncer, A.D.; Kazancıoğlu, F.Ş. Comparative analysis and forecasting of COVID-19 cases in various European countries with ARIMA, NARNN and LSTM approaches. *Chaos Solitons Fractals* **2020**, *138*, 110015. [CrossRef]
30. Zeroual, A.; Harrou, F.; Dairi, A.; Sun, Y. Deep learning methods for forecasting COVID-19 time-Series data: A Comparative study. *Chaos Solitons Fractals* **2020**, *140*, 110121. [CrossRef]
31. Jockers, M.L.; Thalken, R. Introduction to dplyr. In *Text Analysis with R*; Springer: Cham, Switzerland, 2020; pp. 121–132.
32. Liu, S.; Sweeney, C.; Srisarajivakul-Klein, N.; Klinger, A.; Dimitrova, I.; Schaye, V. Evolving oxygenation management reasoning in COVID-19. *Diagnosis* **2020**, *7*, 381–383. [CrossRef]
33. Huang, Z.; Xu, W.; Yu, K. Bidirectional LSTM-CRF models for sequence tagging. *arXiv* **2015**, arXiv:150801991. [CrossRef]
34. Hochreiter, S.; Schmidhuber, J. Long short-term memory. *Neural Comput.* **1997**, *9*, 1735–1780. [CrossRef]
35. Nguyen, D.H.D.; Tran, L.P.; Nguyen, V. Predicting Stock Prices Using Dynamic LSTM Models. In *Applied Informatics*; Florez, H., Leon, M., Diaz-Nafria, J., Belli, S., Eds.; Springer: Cham, Switzerland, 2019.
36. Center for Systems Science and Engineering (CSSE). *Coronavirus COVID-19 Global Cases*; Johns Hopkins University (JHU): Baltimore, MD, USA, 2020.
37. Shoeibi, A.; Khodatars, M.; Alizadehsani, R.; Ghassemi, N.; Jafari, M.; Moridian, P.; Khadem, A.; Sadeghi, D.; Hussain, S.; Zare, A.; et al. Automated detection and forecasting of COVID-19 using deep learning techniques: A review. *arXiv* **2020**, arXiv:200710785. [CrossRef]
38. Cifci, M.A. SegChaNet: A Novel Model for Lung Cancer Segmentation in CT scans. *Appl. Bionics Biomech.* **2022**, *2022*, 1139587. [CrossRef]
39. Alizadehsani, R.; Roshanzamir, M.; Hussain, S.; Khosravi, A.; Koohestani, A.; Zangooei, M.H.; Abdar, M.; Beykikhoshk, A.; Shoeibi, A.; Zare, A.; et al. Handling of uncertainty in medical data using machine learning and probability theory techniques: A review of 30 years (1991–2020). *Ann. Oper. Res.* **2021**, 1–42. [CrossRef]
40. Alizadehsani, R.; Sani, Z.A.; Behjati, M.; Roshanzamir, Z.; Hussain, S.; Abedini, N.; Hasanzadeh, F.; Khosravi, A.; Shoeibi, A.; Roshanzamir, M.; et al. Risk Factors Prediction, Clinical Outcomes and Mortality of COVID-19 Patients. *J. Med. Virol.* **2020**, *93*, 2307–2320. [CrossRef]
41. Cifci, M.A. Derin Öğrenme Metodu ve Ayrık Dalgacık Dönüşümü Kullanarak BT Görüntülerinden Akciğer Kanseri Teşhisi. *Mühendislik Bilimleri Ve Araştırmaları Derg.* **2022**, *4*, 141–154.

electronics

MDPI

Article

An Effective Model of Confidentiality Management of Digital Archives in a Cloud Environment

Jian Xie [1], Shaolong Xuan [1,*], Weijun You [2,*], Zongda Wu [1,*] and Huiling Chen [3]

1 Deparment of Computer Science and Engineering, Shaoxing University, Shaoxing 312000, China
2 Department of Management, Office of Natural Science Foundation of Zhejiang Province, Hangzhou 310006, China
3 College of Computer Science and Artificial Intelligence, Wenzhou University, Wenzhou 325035, China
* Correspondence: zjsxxsl@163.com (S.X.); uweijun@zjinfo.gov.cn (W.Y.); zongda1983@163.com (Z.W.)

Abstract: Aiming at the problem of confidentiality management of digital archives on the cloud, this paper presents an effective solution. The basic idea is to deploy a local server between the cloud and each client of an archive system to run a confidentiality management model of digital archives on the cloud, which includes an archive release model, and an archive search model. (1) The archive release model is used to strictly encrypt each archive file and archive data released by an administrator and generate feature data for the archive data, and then submit them to the cloud for storage to ensure the security of archive-sensitive data. (2) The archive search model is used to transform each query operation defined on the archive data submitted by a searcher, so that it can be correctly executed on feature data on the cloud, to ensure the accuracy and efficiency of archive search. Finally, both theoretical analysis and experimental evaluation demonstrate the good performance of the proposed solution. The result shows that compared with others, our solution has better overall performance in terms of confidentiality, accuracy, efficiency and availability, which can improve the security of archive-sensitive data on the untrusted cloud without compromising the performance of an existing archive management system.

Keywords: cloud; digital archives; confidentiality management; information system

Citation: Xie, J.; Xuan, S.; You, W.; Wu, Z.; Chen, H. An Effective Model of Confidentiality Management of Digital Archives in a Cloud Environment. *Electronics* **2022**, *11*, 2831. https://doi.org/10.3390/electronics11182831

Academic Editor: Irene Moser

Received: 8 May 2022
Accepted: 13 July 2022
Published: 7 September 2022

1. Introduction

In cloud computing, pay-per-use enables an organization to obtain the required sources from the shared pool of configurable computing resources anytime, anywhere and on-demand [1–3], therefore, greatly reducing the organization's expenditure on business operations and archive management and then improving the service efficiency of the organization. To this end, governments and enterprises in various countries have promoted the cloud-first strategy [4–6], i.e., the cloud computing model is given priority in the process of institutional informatization, such that the proportion of archival documents formed and managed on the cloud is becoming higher and higher. Archive management on the cloud has become the general trend [7–9]. However, although storing digital archives on the cloud can reduce the management cost and improve management efficiency, it also results in some negative effects, the most prominent of which is the security of archives on the cloud [10–12]. In a cloud computing environment, the archives of an organization are not stored on a trusted local server but are stored and managed by the cloud server, resulting in each archive and its owner being separated from each other, i.e., making each archive in an uncontrollable area, and in turn posing a serious threat to the security of archival materials [13–15]. Such security threat mainly includes two aspects: (1) external threat, i.e., hackers' attack on the cloud service provider (which has been verified by endless hacking incidents) [16]; and (2) internal threat, i.e., inside jobs from workers of the cloud service provider (driven by interests, it is possible for management workers to maliciously stealing sensitive archival information) [17]. In a word, the security issue of archives on the cloud (i.e., how to ensure the security of sensitive archival data on the untrusted cloud) has become one of the main obstacles to restricting the management of archives on the cloud, which has attracted more and more attention.

1.1. Related Works and Limitations

Aiming at the problem of the security of archives on the cloud, scholars from the field of social sciences conducted more research from the perspective of laws and regulations and believe that the solution of the problem requires governments to formulate relevant laws and regulations for guidance [18,19]. Most of the countries in the world have successively formulated relevant standards and specifications, such as the Guideline for Document Management in Cloud Computing Environment in the United States, Advice on Risk Management of Cloud Computing File Storage in Australia, and Guidelines for Cloud Storage and Digital Permanent Preservation in the United Kingdom. In recent years, China has also intensively promulgated three relevant laws and regulations, i.e., Cybersecurity Law, Data Security Law, and Personal Information Protection Law, which play an important role in ensuring the security of archives on the cloud [20,21]. However, endless incidents of privacy breaches show that confidentiality management of archives on the cloud requires not only laws and regulations, but also the support of technical methods [22–25].

In order to ensure the security of archive data, a digital archive management system uses a variety of technical methods and strategies, such as identity authentication, access control and data encryption. Below, we briefly introduce the technical features of these methods and analyze their application limitations in the confidentiality management of archives on the cloud. (1) Identity authentication is the process of user identity confirmation, to prevent illegal users from accessing system resources illegally [11,26]. Specifically, it can be divided into two categories, i.e., single-factor authentication [27–29] (such as username and password authentication, smart card authentication, dynamic password authentication and biometric authentication) and two-factor authentication [30,31] (which combines two kinds of single-factor authentication to further strengthen the security of identity authentication). (2) Access control is to restrict access to unauthorized resources or the use of unauthorized functions, according to the specific identity of a user [32]. Specifically, it can be divided into discretionary access control (DAC) and mandatory access control (MAC) [33]. Identity authentication and access control have been widely used in operating systems, database systems, file management systems [34], etc. Although the two kinds of technical methods can prevent external users from illegally accessing the sensitive data in a digital archives system, to alleviate the archive security problem to a great extent, they cannot separate the support of the server side (they assume that the server side is trusted), i.e., they only target external illegal attackers of a digital archive system, and cannot prevent the internal staff of the untrusted server side (or the hackers who conquer the server) from accessing the archive-sensitive data [35–37]. However, the cloud is not trustworthy, and it is the main source causing the archive security problem. Therefore, the problem of confidentiality management of archives on the cloud cannot be solved by traditional access control and identity authentication. (3) Data encryption refers to strictly encrypting sensitive data before being stored in an untrusted server, so that even if the encrypted data is leaked, it is difficult to be understood, consequently, ensuring data security [38,39]. Therefore, it is an important means to solve the data security problem in a cloud environment [40–43]. However, there are a large number of query operations defined over archive-sensitive data in a digital archive system (such as querying archives by user names). Once the sensitive data stored on the cloud is strictly encrypted, the ciphertext data would lose many inherent characteristics of the corresponding plaintext data (such as orderliness, similarity and comparability), resulting in most of the original archive query operations in an archive system no longer being performed correctly on the ciphertext data, consequently, damaging the accuracy of archival search [44,45]. In order to solve the ciphertext search problem, we can first transmit all the ciphertext data on the cloud back to a local server, decrypt the ciphertext data, and then perform archive query operations on the decrypted data. However, for such a method, since almost all the process of archive search is completed locally, it not only completely loses the cost-efficiency advantage of archive management on the cloud, but also seriously reduces the efficiency of archive search (it needs huge overhead for network transmission and decryption). Therefore, the problem of confidentiality management of digital archives on the cloud cannot be directly solved by a traditional data encryption method.

In addition, scholars from the field of library science also try to solve the problem of the security of archives on the cloud from the perspective of technical methods [46–48]. However, the methods proposed by them are usually developed based on some original technical methods from a digital archive management system (i.e., identity authentication, access control, data encryption, etc.), so it is difficult to meet the actual needs of confidentiality management of archives on the cloud. For the problem of cloud data security, scholars from the field of information sciences have also conducted in-depth and systematic research and proposed many effective technical methods [49–53]. However, these methods are not specifically proposed for digital archives systems, so they still cannot meet the practical application requirements of confidentiality management of archives on the cloud in terms of availability, effectiveness and security. To sum up, under the existing architecture of a digital archive cloud management platform, it remains to be further discussed and studied how to improve the security of archive-sensitive data on the untrusted cloud without compromising the availability of an archive system and the effectiveness of archive search.

1.2. Contributions

In this paper, we propose an effective solution for confidentiality management of archives on the cloud, which can improve the security of sensitive archive data on the cloud without affecting the efficiency of archive search. Its basic idea is to deploy a local server between the cloud and each client of an archive system to run a confidentiality management model of digital archives on the cloud (specifically, which includes an archive release model and an archive search model), which acts as a layer of middleware between the cloud and the client, to achieve transparency for users and the cloud, and then achieve effective integration with the existing archive management system. Specifically, the contributions of this paper mainly include the following three aspects. (1) Propose a confidentiality release model of archives on the cloud, which is responsible for strictly encrypting the archive files and archive data released by an administrator, generating archive feature data for the archive data, and then submitting them to the cloud for storage to ensure the security of archive data. (2) Propose a confidentiality search model of archives on the cloud, which is responsible for rewriting and transforming the query operations defined on archive data submitted by an inquirer, so that it can be correctly executed on feature data on the cloud (to filter out most of the non-target records) to ensure the accuracy and efficiency of archive search. (3) Both theoretical analysis and experimental evaluation demonstrate the overall performance of the proposed solution, i.e., it can satisfy the actual requirements in terms of data security, query accuracy, and query efficiency. This paper gives a valuable study attempt on confidentiality management of archives on the cloud, which has positive significance for promoting the application and development of cloud computing technology in digital archives management.

2. Problem Statement

2.1. System Framework

In Figure 1, we show the basic framework of a confidentiality management model of digital archives on the cloud adopted in this paper. It can be seen that it mainly includes the following four roles, i.e., archive administrators and their management interfaces (trusted), archive inquirers and their query interfaces (trusted), a local server (trusted) and the cloud server (untrusted). The functions of the four types of roles are briefly described below.

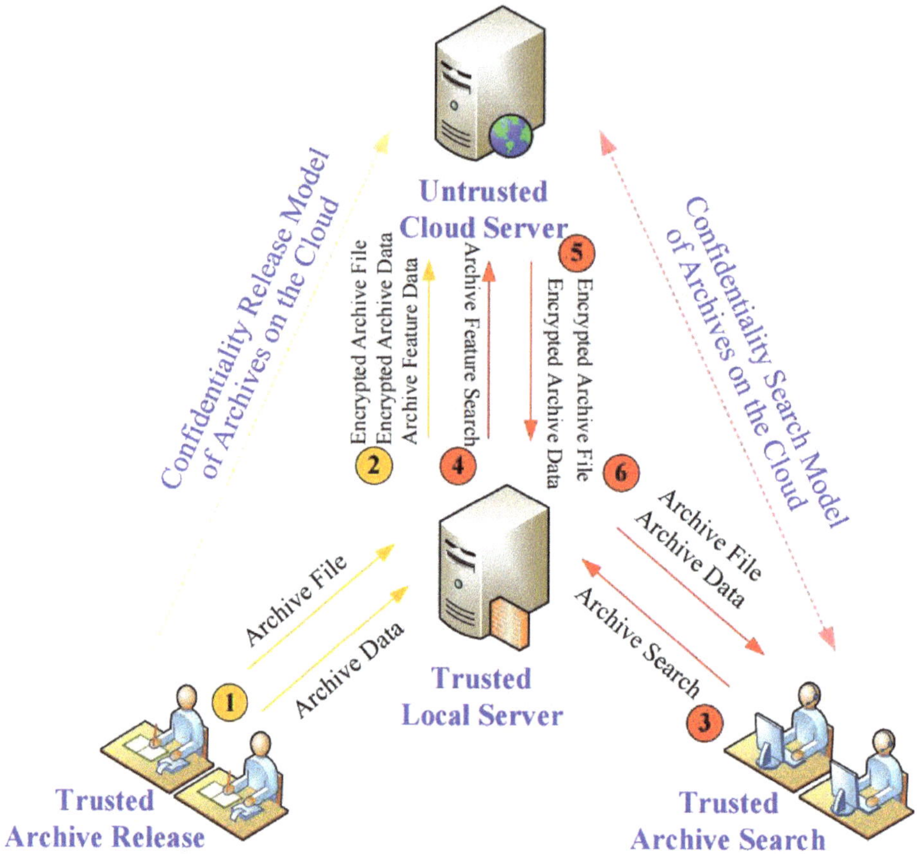

Figure 1. A framework of a confidentiality management model of archives on the cloud.

1. Archive administrator (also known as archive entry clerk): through a trusted archive management interface, who submits digital archive files (electronic scanning pictures) and their corresponding archive data (usually in the form of tables, which are used to record archive description data to facilitate archive search).
2. Archive inquirer: through a trusted archive search interface, who performs archive search operations (i.e., perform related archive query operations defined on archive description data) to obtain target archive files and related materials.
3. Cloud server: which is deployed on the untrusted cloud, is responsible for storing archive files (in the form of ciphertext), archive description data (in the form of ciphertext) and archive feature data submitted by the local server, and is also responsible for executing archive search requests submitted by the local server.
4. Local server: which is deployed on the trusted local, responsible for strictly encrypting the archive files and archive description data submitted by an archive administrator, generating the corresponding archive feature data, and then submitting them to the cloud server for storage, and recording the corresponding encryption key data and setting parameters locally (i.e., responsible for running the confidentiality release model of archives on the cloud). In addition, it is also responsible for rewriting the archive search requests submitted by an archive inquirer, so that they can be correctly executed on the feature data of the cloud (to filter out most non-target records on the cloud) to ensure the accuracy and efficiency of archive search (i.e., responsible for running the confidentiality search model of archives on the cloud).

2.2. Design Goals

In order to satisfy the practical application requirements of confidentiality management of archives on the cloud, a confidentiality management model constructed based on the framework shown in Figure 1 should meet the following three constraints.

1. Ensuring the security of archive data, which includes archive file security, archive data security and feature data security, i.e., from the encrypted archive files, encrypted archive data and feature data submitted by the local server, it is impossible for the cloud server to accurately know the original archive files and sensitive archive data.
2. Ensuring the accuracy of archive search. With the help of archive feature data constructed by the confidential release model of archives on the cloud, each archive query operation defined on the archive data submitted by an archive inquirer can be effectively converted into a feature query operation defined on the feature data (i.e., Step 4 in Figure 1), so that the result returned by the cloud server by executing the feature query operation (i.e., the data returned by Step 5 in Figure 1) contains the real search result to ensure the accuracy of archive search.
3. Ensuring the efficiency of archive search. With the help of feature data, the cloud server can eliminate most of the non-target records on the cloud by executing each feature query operation constructed by the confidential search model of archives on the cloud, so as to reduce the amount of archive data returned to the client (i.e., the data returned by Step 6 in Figure 1), and in turn, ensure the efficiency of archive data search.

3. Proposed Solution

3.1. Archive Confidentiality Model

On the basis of the framework of Figure 1, this paper constructs a confidentiality management model of digital archives on the cloud, which mainly includes two sub-models, i.e., a confidentiality release model of archives on the cloud, and a confidentiality search model of archives on the cloud. Here, the confidentiality release model corresponds to Steps 1 and 2 in Figure 1, i.e., the process of the local server to encrypt archive files and archive data released by an archive administrator, and attach archive feature data, which can be further shown in Figure 2. The description can be divided into the following four steps.

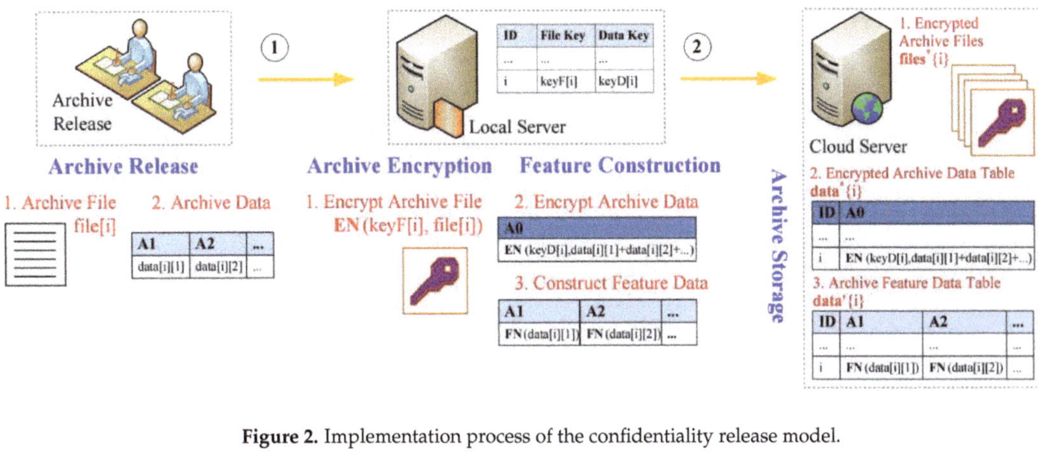

Figure 2. Implementation process of the confidentiality release model.

Step 1.1. Archive Release (executed by an archive administrator). An archive administrator releases an archive file and its corresponding archive description data through an archive management interface. The archive description data is denoted by (data[i][1], data[i][2], . . .), where data[i][j] denotes a sensitive archive data item (i.e., some private data such as name,

ID number, phone number, home address, etc., which cannot be known to the cloud). The archive file is denoted by file$[i]$ (usually which is an electronic scanning picture).

Step 1.2. Archive Encryption (executed by the local server). First, the local server generates an archive file key (denoted by keyF$[i]$) and an archive data key (denoted by keyD$[i]$) randomly. Then, using a traditional encryption algorithm (such as RSA, etc.), the local server strictly encrypts the archive file and archive description data submitted by an archive administrator, so as to obtain an encrypted archive file (denoted by file*) and an encrypted archive data (denoted by data*), which are, respectively, denoted by Equations (1) and (2).

$$\text{data}^*[i] = \mathbf{EN}\,(\text{keyD}[i], \text{data}[i][1] + \text{data}[i][2] + \ldots) \tag{1}$$

$$\text{file}^*[i] = \mathbf{EN}\,(\text{keyF}[i], \text{file}[i]) \tag{2}$$

Finally, the local server submits the encrypted archive file and the encrypted archive data to the cloud server for storage and stores the archive file key and archive data key locally (note that the secret keys are generated dynamically and randomly, and the secret keys of archive files are different from each other, and the keys of archive data are also different from each other).

Step 1.3. Feature Construction (executed by the local server). First, the local server generates the corresponding archive feature data (denoted by data$'$) for archive-sensitive data, which is denoted by Equation (3).

$$(\text{data}'[i][1] = \mathbf{FN}(\text{data}[i][1]), \text{data}'[i][2] = \mathbf{FN}(\text{data}[i][2]), \ldots) \tag{3}$$

Then, the local server submits the feature data to the cloud server for storage. The parameters related to feature construction are stored on the local server (note that the same archive data item uses the same feature parameter, and different items use different feature parameters).

Step 1.4. Archive Storage (executed by the cloud server). The cloud server stores the encrypted archive data and archive feature data in its archive database, as well as the encrypted archive files in its storage devices. Then, it establishes the associations (e.g., using URLs) between the archive data records of the database and the encrypted archive files.

The confidentiality release model of archives on the cloud corresponds to Steps 3 to 6 in Figure 1, i.e., the process of the local server to rewrite and replace each archive query operation defined on the archive description data released by an archive enquirer with a feature query operation defined on the corresponding archive feature data, and the process of decrypting and filtering the archive query result returned by the cloud server. The process can be further described in Figure 3, which can be divided into the following four steps.

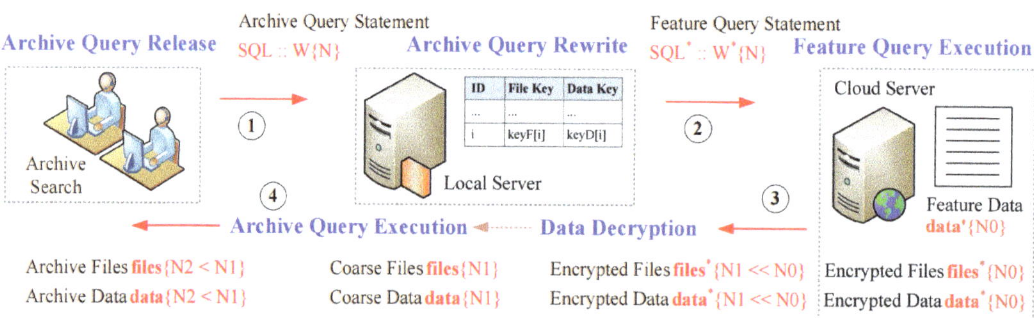

Figure 3. Implementation process of the confidentiality search model.

Step 2.1. Query Release (executed by an archive inquirer). An archive inquirer submits an archive query statement (defined on archive description data) through an archive query interface, to the local server. An archive query statement is mainly composed of a series of basic query conditions defined on archive data items and connected by logical operations. To this end, the basic query conditions of an archive query statement can be denoted by (W[1], W[2],..., W[N]).

Step 2.2. Query Rewrite (executed by the local server). The local server converts each archive query statement defined on archive description data published by an inquirer into a feature query statement defined on the corresponding feature data and then submits it to the cloud server for execution. A feature query statement is mainly composed of a series of basic query conditions defined on feature data and connected by logical operations, which can be denoted by Equation (4).

$$(\mathrm{W}^*[1] = \mathrm{TR}(\mathrm{W}[1]), \mathrm{W}^*[2] = \mathrm{TR}(\mathrm{W}[2]), \ldots, \ \mathrm{W}^*[\mathrm{N}] = \mathrm{TR}(\mathrm{W}[\mathrm{N}])) \tag{4}$$

Step 2.3. Query Execution (executed by the cloud server). The cloud server executes the feature query statement submitted by the local server on the feature dataset data$'\{\mathrm{N}_0\}$ (where N_0 denotes the size of the feature dataset), and then returns the set of encrypted archive data data$^*\{\mathrm{N}_1\}$ ($\mathrm{N}_1 \ll \mathrm{N}_0$) and the set of encrypted archive files files$^*\{\mathrm{N}_1\}$ to the local server.

Step 2.4. Result Decryption (executed by the local server). For the encrypted archive dataset returned by the cloud server, in combination with the associated archive data keys saved by the local server, after decrypting the encrypted data, the local server obtains the corresponding plaintext archive dataset denoted by data $\{\mathrm{N}_1\}$. Second, the local server executes the original archive query statement issued by the archive manager on the plaintext archive dataset to obtain the target archive dataset denoted by data $\{\mathrm{N}_2\}$ ($\mathrm{N}_2 < \mathrm{N}_1$). Let files$^*\{\mathrm{N}_2\}$ denote the set of the encrypted archive files associated with the dataset data $\{\mathrm{N}_2\}$. Finally, the local server decrypts the ciphertext file set in combination with the locally stored file keys to obtain the corresponding plaintext archive file set files $\{\mathrm{N}_2\}$, and return the archive file set files $\{\mathrm{N}_2\}$ and the archive data set data $\{\mathrm{N}_2\}$ to the client.

3.2. Feature Construction and Query Rewriting

From the previous section, we can see that feature construction (Step 1.3) is the key to the confidentiality release model of archives on the cloud, and query rewriting (Step 2.3) is the key to the confidentiality search model of archives on the cloud. Moreover, we can see the query rewriting strategy is closely dependent on the feature construction strategy. To this end, this section first gives a simple and effective strategy for feature construction and then constructs a corresponding strategy for query rewriting accordingly. Note that the types of archive data items are mainly divided into numerical type (e.g., real number, date, etc.) and text type (i.e., character string). The strategy of feature construction proposed in this paper can be applied to the two data types at the same time, whose process mainly includes the following three steps.

Step 3.1. Suppose that an archive-sensitive data item $\mathbf{A}_0$ contains n basic units, respectively, denoted by $\mathbf{A}_1, \mathbf{A}_2, \ldots, \mathbf{A}_n$. For each basic unit $\mathbf{A}_k$, we divide the domain composed of all its possible values into N_k subdomains, denoted by $\mathbb{D}_1^k, \mathbb{D}_2^k, \ldots, \mathbb{D}_{N_k}^k$, which should meet the constraints.

1. None of the subdomains is an empty set, i.e., $\mathbb{D}_i^k \neq \varnothing$;
2. Any two subdomains do not overlap, i.e., $\mathbb{D}_i^k \cap \mathbb{D}_j^k = \varnothing$;
3. The union of all subdomains is equal to the domain itself of the basic unit, i.e., $\cup \mathbb{D}_i^k = \mathbb{D}_k$.

Step 3.2 Assign an identifier to each subdomain $\mathbb{D}_i^k$, denoted by $\varpi_k\left(\mathbb{D}_i^k\right)$. All the identifiers should meet the following constraints.

1. Each identifier itself is selected from the domain $\mathbb{D}_k$ of the basic unit, i.e., $\varpi_k\left(\mathbb{D}_i^k\right) \in \mathbb{D}_k$;

2. All identifiers remain in order, i.e., if $\omega_k\left(\mathbb{D}_i^k\right) \geq \omega_k\left(\mathbb{D}_j^k\right)$, then $\forall \mathrm{a} \in \mathbb{D}_i^k\, \forall \mathrm{b} \in \mathbb{D}_j^k \rightarrow \mathrm{a} \geq \mathrm{b}$;
3. The length of each identifier is equal to that of the maximum value of the domain $\mathbb{D}_k$, i.e., $\left|\omega_k\left(\mathbb{D}_i^k\right)\right| = |\mathbf{max}(\mathbb{D}_k)|$.

Based on the settings of the above two steps, any specific value a_k of the subdomain $\mathbf{A}_k$ can be mapped to an identifier value of the same length with a_k, i.e., a feature mapping function is determined, denoted by $\mathbf{FN}_k(\mathrm{a}_k) = \omega_k\left(\mathbb{D}_i^k\right)$, where $\mathbb{D}_i^k$ is the subdomain which contains a_k. Now, based on the settings of the above two steps, we have actually determined n feature mapping functions for the archive-sensitive data item $\mathbf{A}_0$, which are denoted by $\mathbf{FN}_1, \mathbf{FN}_2, \ldots, \mathbf{FN}_n$ (corresponding to the basic units $\mathbf{A}_1, \mathbf{A}_2, \ldots, \mathbf{A}_n$, respectively).

Step 3.3 For any value a of the archive data item $\mathbf{A}_0$, based on the settings of the above two steps, we assume that the values corresponding to the basic units of $\mathbf{A}_0$ are $\mathrm{a}_1, \mathrm{a}_2, \ldots, \mathrm{a}_n$, respectively, i.e., $\mathrm{a} = \mathrm{a}_1\mathrm{a}_2\ldots\mathrm{a}_n$. Then, based on the functions $\mathbf{FN}_1, \mathbf{FN}_2, \ldots, \mathbf{FN}_n$, it can be mapped to a new feature value (i.e., feature data), denoted by Equation (5).

$$\mathrm{a}' = \mathbf{FN}(\mathrm{a}) = \mathbf{FN}_1(\mathrm{a}_1)\, \mathbf{FN}_2(\mathrm{a}_2) \ldots \mathbf{FN}_n(\mathrm{a}_n) \tag{5}$$

Example 1. *Take the archive-sensitive data item Name as an example to briefly describe the construction process of feature data. Here, we assume that the maximum length of the name field is 8 Chinese characters (i.e., it contains 8 basic units). First, let us consider the first basic unit. Note that there are 20902 common Chinese characters, and their UNICODE codes are between 0X4E00 and 0X9FA5. To this end, we simply divide the value range of Chinese characters into 209 subdomains (so the size of each subdomain is equal to 100) by using an equal-width strategy, respectively, denoted by* $\mathbb{D}_1^1, \mathbb{D}_2^1, \ldots, \mathbb{D}_{209}^1$ *(Step 3.1). Then, we assign an identifier for each subdomain according to the following strategy* $\omega_1(\mathbb{D}_k^1) = k$ *(Step 3.2),*

$$\omega_1\left(\mathbb{D}_1^1\right) = 0\mathrm{X}0001;\ \omega_1\left(\mathbb{D}_2^1\right) = 0\mathrm{X}0002; \ldots;\ \omega_1\left(\mathbb{D}_{209}^1\right) = 0\mathrm{X}00\mathrm{D}1 \tag{6}$$

To simplify the presentation, for the remaining seven basic units of the data item, we apply the same subdomain division and identifier assignment strategies as the first unit, i.e., $\omega_1 = \omega_2 = \ldots = \omega_8$, so we have that $\mathbf{FN}_1 = \mathbf{FN}_2 = \ldots = \mathbf{FN}_8$. Now, for any given specific name, we can generate its corresponding feature data value. For example, for a Chinese name whose UNICODE encode is 0X8BF8 0X845B 0X4EAE, its feature data value after feature construction is as 0X009A 0X008B 0X0002.

Based on the settings of Steps 3.1 and 3.2, we can see that feature data has the same length and format as its corresponding archive data, so feature data generated in Step 3.3 can be directly stored in the field $\mathbf{A}_0$ of archive data tables. So far, after feature mapping, feature data (instead of archive data) is stored in archive-sensitive data item fields of the cloud database. However, this makes the query operations defined on the archive data items issued by an archive inquirer no longer correctly executed in the cloud database. To this end, the purpose of query rewriting is to transform each archive query condition into a feature query condition defined on feature data. Since an archive query statement is mainly composed of a series of basic query condition items connected by logical operators, below, we briefly discuss how to rewrite three kinds of basic archive query condition items (i.e., equivalent query, implication query and range query), and then introduce Algorithm 1 to show how an archive query statement is rewritten.

Algorithm 1 Query Rewriting.

(1) Input: an archive query statement; **(2) Output**: a feature query statement

```
Divide the archive query statement into a series of basic archive query condition items;
FOREACH basic archive query condition item DO
    IF the item is an equivalent condition item THEN
        CALL Conversation 1.1 to convert it into a feature equivalent query condition;
    ELSEIF the item is an implication condition item THEN
        CALL Conversation 1.2 to convert it into a feature implication query condition;
    ELSEIF the item is a range condition item THEN
        CALL Conversation 1.3 to convert it into a feature range query condition;
    END IF
END FOR
RETURN a feature query statement constructed based on the feature query conditions.
```

Conversion 1.1. Equivalent Query Conversion: A basic equivalent query conditional item defined on an archive-sensitive data item $\mathbf{A}_0$ can be generally expressed as $\mathbf{A}_0 = \mathrm{a}_0$, where $\mathrm{a}_0 = \mathrm{a}_1\mathrm{a}_2\ldots\mathrm{a}_n$ represents a constant defined on the archive data item $\mathbf{A}_0$. Then, the archive equivalent query condition item can be converted into a feature equivalent query condition, denoted by Equation (7)

$$\begin{aligned}\mathbf{TR}(\mathbf{A}_0 = \mathrm{a}_0) \Rightarrow \mathbf{A}_0 = \mathbf{FN}(\mathrm{a}_0) \Rightarrow \mathbf{A}_0 \\ = \mathbf{FN}_1(\mathrm{a}_1)\,\mathbf{FN}_2(\mathrm{a}_2)\ldots\mathbf{FN}_n(\mathrm{a}_n)\end{aligned} \tag{7}$$

An implication query conditional item is generally constructed based on the predicate LIKE, whose general syntax can be represented by LIKE<match string>. The matching string can contain a variety of wildcards, among which % (which denotes to match a character string of any length) is the most representative wildcard. Below, we only discuss the conversion of a left direction implication query condition based on the wildcard %.

Conversion 1.2. Implication Query Conversion: A basic implication query condition item defined on an archive-sensitive data item $\mathbf{A}_0$ can be generally expressed as $\mathbf{A}_0$ **LIKE** a_0%. Based on the setting of Step 3.1, we assume that a_0 completely covers k basic units (i.e., $\mathbf{A}_1, \mathbf{A}_2, \ldots, \mathbf{A}_k$) of the data item $\mathbf{A}_0$, and the values corresponding to the k basic units are $\mathrm{a}_1, \mathrm{a}_2, \ldots, \mathrm{a}_k$, respectively. Then, the implication query condition item can be converted into a feature implication query condition, denoted by Equation (8).

$$\mathbf{TR}(\mathbf{A}_0\ \mathbf{LIKE}\ \mathrm{a}_0\%) \Rightarrow \mathbf{A}_0\ \mathbf{LIKE}\ \mathbf{FN}_1(\mathrm{a}_1)\,\mathbf{FN}_2(\mathrm{a}_2)\ldots\mathbf{FN}_k(\mathrm{a}_k)\% \tag{8}$$

Conversion 1.3. Range Query Conversion: A range query condition item defined on an archive-sensitive data item $\mathbf{A}_0$ can be generally expressed as $\mathbf{A}_0 \geq \mathrm{a}_0$. Then, the range query condition item can be converted into a feature range query condition, denoted by (9).

$$\mathbf{TR}(\mathbf{A}_0 \geq \mathrm{a}_0) \Rightarrow \mathbf{A}_0 \geq \mathbf{FN}(\mathrm{a}_0) \Rightarrow \mathbf{A}_0 \geq \mathbf{FN}_1(\mathrm{a}_1)\,\mathbf{FN}_2(\mathrm{a}_2)\ldots\mathbf{FN}_n(\mathrm{a}_n) \tag{9}$$

Example 2. *Take querying the archive-sensitive data item Name as an example to briefly describe the query rewriting process. Assume that an archive inquirer wants to query the digital archive information from the persons named "ZhangSan" or surnamed "Liu". Then, an archive query statement defined on archive data submitted from a query interface can be presented as follows*

SELECT * FROM DATA **WHERE** Name = "ZhangSan" **OR** Name **LIKE** "Liu%"

It can be seen that the statement contains two basic archive query conditional items. Then, the feature query statement generated by the local server after equivalent query transformation and implication query transformation can be presented as follows

SELECT * FROM DATA **WHERE** Name = **TR**("ZhangSan") **OR** Name **LIKE** TR("Liu")%

From Examples 1 and 2, we can see that the query rewriting strategy is closely dependent on the feature construction strategy, but the converted feature query statement can be directly executed by the cloud database, and most of the non-target records can be filtered out on the cloud accordingly, thereby ensuring the accuracy and efficiency of archive search (please refer to the accuracy analysis and efficiency analysis in Section 4 for detail).

4. Analysis and Evaluation

4.1. Security Analysis

Generally, the cloud server is considered to be honest but curious, i.e., although it can follow the protocol specifications related to cloud services, it remains curious about archive files and archive data. In other words, the cloud server is not trusted. To this end, in this section, we theoretically analyze the security of the confidentiality management model of digital archives on the cloud, including archive file security, archive data security, and feature data security, i.e., analyze the possibility that the untrusted cloud server obtains sensitive archive information, according to the encrypted archive files, encrypted archive data and archive feature data submitted by the local server.

Observation 1.1. The confidentiality management model proposed in this paper can effectively ensure the security of digital archives on the cloud. The model adopts a traditional encryption algorithm to strictly encrypt the digital archive files (in the form of images), and the keys are stored in a trusted local server. As a result, the cloud server can obtain neither the keys nor the archive content based on the encrypted archive files.

Observation 1.2. The confidentiality management model proposed in this paper can effectively ensure the security of sensitive data of digital archives on the cloud. The model adopts a traditional encryption algorithm to strictly encrypt archive-sensitive data, and the keys are stored in the trusted local server. As a result, the cloud server can obtain neither the keys, nor the archive-sensitive data based on the encrypted archive data.

Explanation: Observations 1.1 and 1.2 are easy to be explained. The security of traditional encryption algorithms has been proved by a lot of practice, i.e., without knowing the secret key, it is almost impossible for an attacker to directly obtain the plaintext corresponding to the ciphertext. However, the secret key is stored in the trusted local server, which cannot be obtained by the cloud server.

In order to support archive search, the confidentiality management model proposed in this paper introduces feature data for archive data, which inevitably reflects some key characteristics (such as comparability and similarity) of archive data, consequently, leading to some risk of privacy leakage. This risk can be measured by the possibility of an attacker successfully guessing the corresponding archive data based on the feature data.

Observation 1.3. The confidentiality management model proposed in this paper can effectively ensure the security of archive feature data on the cloud. Below, we analyze the probability of the cloud successfully guessing the archival data based on feature data under the worst case. At this time, we assume that the attacker has completely understood the feature construction process of an archive data item and obtained the relevant feature parameters on the local server, i.e., the attacker has mastered the feature function **FN**. Assume that the archive data item contains n basic units and the value range $\mathbb{D}_k$ of each basic unit is divided into N_k subdomains. Now, given any feature data a′, the possibility of the attacker successfully guessing the corresponding plaintext data a can be measured as the Equation (10).

$$\mathbf{PR}(\mathrm{a}\,|\mathrm{a}') = \frac{\text{the size of the domain of a}'}{\text{the size of the domain of a}} = \frac{(N_1 \cdot N_2 \cdot \ldots \cdot N_n)}{|\mathbb{D}_1| \cdot |\mathbb{D}_2| \cdot \ldots \cdot |\mathbb{D}_n|} = \frac{N}{|\mathbb{D}|} \tag{10}$$

It can be seen that the range of the value of N (equal to the accumulation of the numbers of the subdomains of all the basic units) is $[1, |\mathbb{D}|]$, and the feature data security can be controlled by adjusting the value of N. Moreover, it can be seen that when the value of N is smaller (i.e., when each basic unit is roughly divided), the possibility of the attacker obtaining the plaintext would be very small, i.e., even if the cloud server has obtained the feature mapping function, it is difficult to further obtain the archive data according to feature data. Below, the value of $N/|\mathbb{D}|$ is referred to as feature threshold. The larger the feature threshold, the worse the security of feature data, and the smaller the feature threshold, the better the security of feature data. Moreover, the feature threshold value would affect the efficiency of archive search (see Section 4.3 for detail). Based on the above three observations, it can be further concluded that the confidentiality management model of archives on the cloud constructed in this paper can effectively ensure the security of archive files, archive data and feature data, i.e., it has good security.

4.2. Accuracy Analysis

In this section, we analyze the accuracy of the archive confidentiality search model proposed in this paper. In the mode, with the help of feature data, each query operation defined on archive data would be transformed into a feature query operation defined on feature data. In order to ensure the accuracy of archive search, the result returned from the cloud by executing each feature query operation has to contain the exact result corresponding to the archive query operation. To this end, below, we first introduce Observation 2.1 and Observation 2.2 to demonstrate that each feature query condition obtained based on Conversations 1.1 to 1.3 can ensure the accuracy of archive search.

Observation 2.1. Let W denote an implication query condition before conversion, and W^* the feature implication query condition after conversion (Conversion 1.2). Then, for any archive data $a_1a_2\ldots a_n$, if its corresponding feature data $a'_1a'_2\ldots a'_n$ ($a'_k = \mathbf{FN}_k(a_k)$) meets the condition W^*, it certainly meets the condition W.

Explanation: An implication query condition is only targeted for textual data (not for numeric data). Let $b_1b_2\ldots b_m$ denote the text constant associated with the implication query condition W, and $b'_1b'_2\ldots b'_m$ denote its feature data. Because the feature data $a'_1a'_2\ldots a'_n$ meets the feature implication query condition W^* (i.e., it contains $b'_1b'_2\ldots b'_m$), i.e., it exists $k \leq n$, such that $a'_1a'_2\ldots a'_k = b'_1b'_2\ldots b'_m$. Based on Conversation 1.2, we can conclude that the text constant $b_1b_2\ldots b_m$ corresponding to the feature data $b'_1b'_2\ldots b'_m$ is certainly contained in the archive data $a_1a_2\ldots a_n$ corresponding to the feature data $a'_1a'_2\ldots a'_n$, i.e., the archive data $a_1a_2\ldots a_n$ meets the implication query condition W.

Observation 2.2. Let W denote a range query condition before conversion, and W^* the feature range query condition after conversion (Conversion 1.3). Then, for any archive data $a_1a_2\ldots a_n$, if its corresponding feature data $a'_1a'_2\ldots a'_n$ meets the condition W^*, it certainly meets the condition W.

Explanation: A range condition is targeted for both textual data and numeric data. Let b_0 denote the text constant associated with the range query condition W. For any given archive data a_0, it may not be consistent with the length of the constant b_0. In this situation, it can be right-padded with zeros (encode values) for short text data, or left-padded with zeros (integer values) for numeric data, to make both with the same length. Let $a_0 = a_1a_2\ldots a_n$ and $b_0 = b_1b_2\ldots b_n$, and $a'_0 = a'_1a'_2\ldots a'_n$ and $b'_0 = b'_1b'_2\ldots b'_n$ denote the feature data corresponding to a_0 and b_0. Because the feature data $a'_1a'_2\ldots a'_n$ meets the range condition W^*, i.e., $a'_1a'_2\ldots a'_n \geq b'_1b'_2\ldots b'_n$ (it is assumed to be greater than the comparison), we conclude that there certainly exists that $1 \leq k \leq n$ such that $a'_1 = b'_1, a'_2 = b'_2, \ldots, a'_k \geq b'_k$. Based on the constraints of the previous feature construction strategy (Step 3.2), we can further conclude that $a_1 = b_1,\ a_2 = b_2,\ \ldots, a_k \geq b_k$ ($a_1a_2\ldots a_n \geq b_1b_2\ldots b_n$), i.e., the archive data meets the range query condition W.

An equivalence query can be regarded as a special implication query or a special range query, so its accuracy analysis is no longer presented. Note that equivalence query, implication query and range query are three kinds of the most common basic conditions, and other query conditions can be completed directly or indirectly by means of them. Therefore, based on the above observations, we can further conclude that various query operations defined on archive-sensitive data can be converted into feature query operations defined on feature data, and the results returned from the cloud by executing these feature query operations certainly contain the real results corresponding to the original query operations, i.e., the confidentiality management model of archives on the cloud proposed in this paper can effectively ensure the accuracy of archive search.

4.3. Efficiency Evaluation

In this section, we evaluate the efficiency of the archive confidentiality search model through experiments, i.e., whether the feature query can filter out most of the non-target data on the cloud, so as to improve the archive search efficiency. The experiments were run on a randomly generated table of one million digital archive data records. The experiment selects two sensitive archive data items (name and birthday), which are text type and numerical type, respectively. From the archival data (i.e., Name and Birthday), which are text type and numerical type, respectively. From the search process of archive data shown in Figure 3, it can be seen that the search efficiency of feature data depends on the filtering effect of the feature query operations obtained by query transformation on the non-target records on the cloud. To this end, we introduce the following definition to measure search efficiency.

Definition 1. *Let W denote a query condition before transformation, and W^* denote the feature query condition defined on feature data after transformation. Let N_0 denote the number of archive records, N_2 denote the number of records that meet the archive query W, and N_1 denote the number of records that meet the feature query W^*. Then, the search efficiency of feature data can be measured by the filtering effect of the feature query on the non-target records, i.e.,* ***FR*** $(W^*, W) = (N_0 - N_1)/(N_0 - N_2)$.

The efficiency evaluation is divided into three groups of experiments, i.e., range query on numeric data, range query on textual data, and implication query on textual data. (1) The first group of experiments aims to evaluate the efficiency of range query operations on numeric data. The experimental results are shown in Figure 4, where the abscissa represents the feature threshold (see Observation 1.3 for detail), and the ordinate represents the query efficiency. It can be seen that the filtering effect of feature range query operations on the non-target records would become worse as the feature threshold decreases. This is because the decrease in the feature threshold would increase the number of possible plaintexts corresponding to each feature data value, resulting in a decrease in the query efficiency measure. However, even if the feature threshold is smaller (e.g., less than 2^{-12}), each feature range query operation can still filter out most of the non-target records (greater than 0.99), thereby, reducing the scale of the records returned to the client, and in turn, greatly improving the range query efficiency. (2) The second group of experiments aims to evaluate the efficiency of range query operations on textual data, and the experimental results are shown in Figure 5. It can be seen that the change trend of the range query efficiency measure of textual data with respect to the feature threshold is consistent with that of numerical data. (3) The third group of experiments aims to evaluate the implication query efficiency of textual data, and the experimental results are shown in Figure 6. It can be seen that with the decrease in the feature threshold, the filtering effect of feature implication query operations on the non-target records would become worse (the change trend is basically the same as that of textual range query operations); however, compared with textual range query operations, implication feature query operations have a better filtering effect on non-target records (i.e., having greater values for the efficiency measure). This is because the target record set of an implication query is extremely smaller (usually thousands), while the target record set of a range query is extremely larger (usually hundreds of thousands).

From the three groups of experiments mentioned above, we can draw a conclusion that both for implication query conditions or range query conditions, both for textual data and numerical data, by executing the feature query conditions obtained through feature transformation, the cloud can filter out most non-target records (greater than 0.99), thereby reducing the scale of records returned to a client, and in turn, effectively reducing the time overhead of archive search, i.e., feature data has good search efficiency.

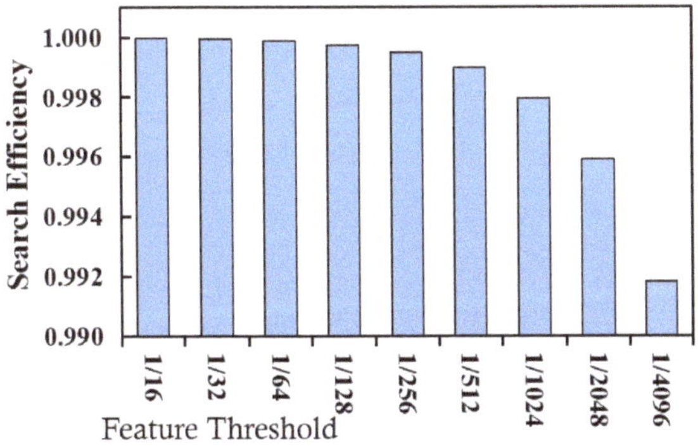

Figure 4. Evaluation results for numeric data range query efficiency.

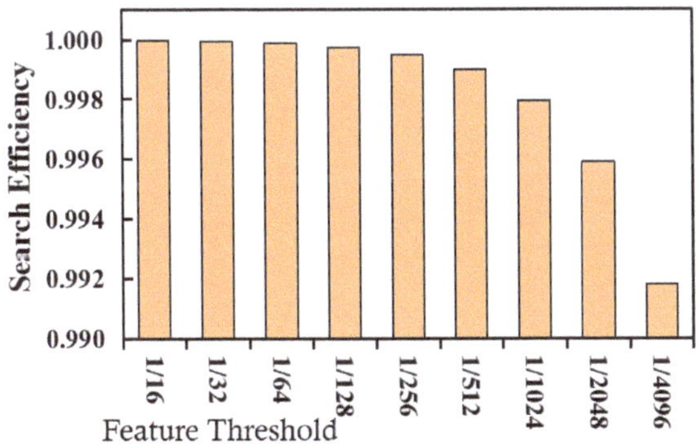

Figure 5. Evaluation results for textual data range query efficiency.

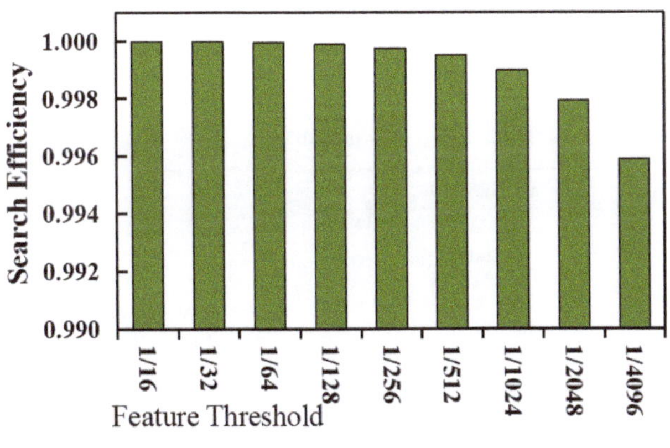

Figure 6. Evaluation result for textual data implication query efficiency.

Finally, Table 1 presents a brief comparison between our proposed solution and other related ones mentioned in Section 1.1. From the table, we see that compared with others, our solution has better overall performance in terms of confidentiality, accuracy, efficiency and availability, which demonstrates again that our solution can well meet the goals presented in Section 2.2. At last, it should be pointed out that although the solution of this paper is targeted at the confidentiality management of digital archives in a cloud environment, it can be transferred to other problems of data confidentiality management as well, such as multimedia data management [54,55], knowledge management [56–60], and series management [61–64].

Table 1. A comparison between our solution and other related ones.

Methods	Confidentiality	Accuracy	Efficiency	Availability
Our solution	Good	Good	Good	Good
Identity Authentication	Not good	Good	Good	Good
Access Control	Not good	Good	Good	Good
Encryption	Good	Good	Good	Not good

5. Conclusions

Aiming at the problem of confidentiality management of digital archives in a cloud environment, this paper constructs an archive release model and an archive search model, whose basic idea is to strictly encrypt all archive files and their corresponding archive data on a trusted local server, before they are submitted to the cloud for storage, to ensure the security of archive data on the untrusted cloud. In order to solve the problem of archive search, the solution also adds additional feature data to the encrypted archive data, so that each query operation defined on archive data can be executed on the cloud, thereby, greatly improving the efficiency of archive data query, and in turn ensuring the effectiveness of archive search. This paper presents a valuable research attempt on the problem of confidentiality management of archives on the cloud. The solution proposed in this paper can effectively balance the security of archive data and the effectiveness of archive search, i.e., it can ensure the security of sensitive archive information on the untrusted cloud, without affecting the efficiency and accuracy of archive search. It has positive significance for promoting the further application and development of cloud computing technology in archives management.

However, the proposal of this paper is not the end of our work. In future work, we will try to further study some problems, e.g., (1) how to simplify the archive release model and the archive search model to reduce the workload of the local server; (2) how to design different feature construction schemes for different archive data types, to improve the efficiency and security; and (3) the practical implementation of the proposed method in a management system of digital archives in a cloud environment.

Finally, Table 2 describes some key symbols used in the paper.

Table 2. Symbols and their meanings.

Symbols	Meanings
$\mathrm{data}[i]$	A sensitive archive data record
$\mathrm{data}^*[i]$	An encrypted archive data record
$\mathrm{data}'[i]$	An archive feature data record
$\mathrm{W}[i]$	A basic query condition defined on archive data
$\mathrm{W}^*[i]$	A basic query condition defined on feature data
$\mathbf{A}_k$	A basic unit of an archive-sensitive data item
$\mathbb{D}_i^k$	A subdomain of the domain of the basic unit $\mathbf{A}_k$
$\varpi_k\left(\mathbb{D}_i^k\right)$	An identifier of the subdomain $\mathbb{D}_i^k$
$\mathbf{FN}_k$	A feature mapping function for the basic unit $\mathbf{A}_k$
a	A value of an archive data item
$\mathbf{FN}(\mathrm{a})$	A feature mapping function for an archive data item
a′	A feature value of an archive data item
TR	A condition conversion function

Author Contributions: Methodology, S.X.; writing—original draft preparation, J.X. software, W.Y.; writing—review and editing, Z.W.; software, H.C. All authors have read and agreed to the published version of the manuscript.

Funding: The work is supported by the key project of Humanities and Social Sciences in Colleges and Universities of Zhejiang Province (No 2021GH017), Humanities and Social Sciences Project of the Ministry of Education of China (No 21YJA870011), Zhejiang Philosophy and Social Science Planning Project (No 22ZJQN45YB) and National Social Science Foundation of China (No 21FTQB019).

Institutional Review Board Statement: Not applicable.

Data Availability Statement: Not applicable.

Conflicts of Interest: The authors declare no conflict of interest.

References

1. Liu, J.; Wang, X.; Shen, S.; Fang, Z.; Yu, S.; Yue, G.; Li, M. Intelligent jamming defense using DNN Stackelberg game in sensor edge cloud. *IEEE Internet Things J.* **2021**, *9*, 4356–4370. [CrossRef]
2. Liu, J.; Wang, X.; Shen, S.; Yue, G.; Yu, S.; Li, M. A bayesian q-learning game for dependable task offloading against ddos attacks in sensor edge cloud. *IEEE Internet Things J.* **2020**, *8*, 7546–7561. [CrossRef]
3. Shen, S.; Huang, L.; Zhou, H.; Yu, S.; Fan, E.; Cao, Q. Multistage signaling game-based optimal detection strategies for suppressing malware diffusion in fog-cloud-based IoT networks. *IEEE Internet Things J.* **2018**, *5*, 1043–1054. [CrossRef]
4. Wu, Z.; Shen, S.; Li, H.; Lu, C.; Xie, J. A basic framework for privacy protection in personalized information retrieval. *J. Organ. End User Comput.* **2021**, *33*, 1–26. [CrossRef]
5. Wu, B.; Zhao, Z.; Cui, Z.; Wu, Z. Secure and efficient adjacency search supporting synonym query on encrypted graph in the cloud. *IEEE Access* **2019**, *7*, 133716–133724. [CrossRef]
6. Wu, Z.; Shen, S.; Li, H.; Zhou, H.; Zou, D. A comprehensive study to the protection of digital library readers' privacy under an untrusted network environment. *Libr. Hi Tech* **2021**. [CrossRef]
7. Wu, Z.; Xie, J.; Lian, X.; Su, X.; Pan, J. A privacy protection approach for xml-based archives management in a cloud environment. *Electron. Libr.* **2019**, *37*, 970–983. [CrossRef]
8. Wu, Z.; Xuan, S.; Xie, J.; Lu, C.; Lin, C. How to ensure the confidentiality of electronic medical records on the cloud: A technical perspective. *Comput. Biol. Med.* **2022**, *147*, 105726. [CrossRef]
9. Wu, Z.; Xu, G.; Lu, C.; Chen, E.; Jiang, F.; Li, G. An effective approach for the protection of privacy text data in the CloudDB. *World Wide Web* **2018**, *21*, 915–938. [CrossRef]
10. Mei, Z.; Zhu, H.; Cui, Z.; Wu, Z.; Peng, G.; Wu, B.; Zhang, C. Executing multi-dimensional range query efficiently and flexibly over outsourced ciphertexts in the cloud. *Inf. Sci.* **2018**, *432*, 79–96. [CrossRef]
11. Wang, T.; Bhuiyan, M.Z.A.; Wang, G.; Li, L.; Wu, J. Preserving balance between privacy and data integrity in edge-assisted Internet of Things. *IEEE Internet Things J.* **2019**, *7*, 2679–2689. [CrossRef]
12. Wu, Z.; Shen, S.; Lu, C.; Li, H.; Su, X. How to protect reader lending privacy under a cloud environment: A technical method. *Libr. Hi Tech* **2021**. [CrossRef]
13. Wu, Z.; Xie, J.; Pan, J.; Su, X. An effective approach for the protection of user privacy in a digital library. *Libri* **2019**, *69*, 315–324. [CrossRef]
14. Wu, Z.; Shen, S.; Lian, X.; Su, X.; Chen, E. A dummy-based user privacy protection approach for text information retrieval. *Knowl. Based Syst.* **2020**, *195*, 105679. [CrossRef]
15. Chen, Z.; Xu, W.; Wang, B.; Yu, H. A blockchain-based preserving and sharing system for medical data privacy. *Future Gener. Comput. Syst.* **2021**, *124*, 338–350. [CrossRef]
16. Cui, Z.; Wu, Z.; Zhou, C.; Gao, G.; Zhao, Z.; Wu, B. An efficient subscription index for publication matching in the cloud. *Knowl.-Based Syst.* **2016**, *110*, 110–120. [CrossRef]
17. Wu, B.; Chen, X.; Wu, Z.; Zhao, Z. Privacy-guarding optimal route finding with support for semantic search on encrypted graph in cloud computing scenario. *Wirel. Commun. Mob. Comput.* **2021**, *2021*, 6617959. [CrossRef]
18. Wu, Z.; Xie, J.; Zheng, C.; Chen, E. A framework for the protection of user behavior preference privacy of digital library. *J. Libr. Sci. China* **2018**, *44*, 72–85.
19. Cheng, Z.; Yue, D.; Shen, S.; Hu, S. Secure frequency control of hybrid power system under DoS attacks via lie algebra. *IEEE Trans. Inf. Forensics Secur.* **2022**, *17*, 1172–1184. [CrossRef]
20. Shen, Y.; Shen, S.; Li, Q.; Wu, Z. Evolutionary privacy-preserving learning strategies for edge-based IoT data sharing schemes. *Digit. Commun. Netw.* **2022**. [CrossRef]
21. Shen, Y.; Shen, S.; Wu, Z.; Yu, S. Signaling game-based availability assessment for edge computing-assisted IoT systems with malware dissemination. *J. Inf. Secur. Appl.* **2022**, *66*, 103140. [CrossRef]
22. Feng, S.; Wu, C.; Zhang, Y.; Olvia, G. WSN deployment and localization using a mobile agent. *Wirel. Pers. Commun.* **2017**, *97*, 4921–4931. [CrossRef]

23. Feng, S.; Shi, H.; Huang, L.; Shen, S.; Yu, S. Unknown hostile environment-oriented autonomous WSN deployment using a mobile robot. *J. Netw. Comput. Appl.* **2021**, *182*, 103053. [CrossRef]
24. Liu, J.; Shen, S.; Yue, G.; Han, R.; Li, H. A stochastic evolutionary coalition game model of secure and dependable virtual service in sensor-cloud. *Appl. Soft Comput.* **2015**, *30*, 123–135. [CrossRef]
25. Li, H.; Zhu, Y.; Wang, J.; Liu, J.; Shen, S.; Gao, H. Consensus of nonlinear second-order multi-agent systems with mixed time-delays and intermittent communications. *Neurocomputing* **2017**, *251*, 115–126. [CrossRef]
26. Abuarqoub, A. D-FAP: Dual-factor authentication protocol for mobile cloud connected devices. *J. Sens. Actuator Netw.* **2019**, *9*, 1. [CrossRef]
27. Wang, S.; Cong, Y.; Zhu, H.; Chen, X.; Qu, L. Multi-scale context-guided deep network for automated lesion segmentation with endoscopy images of gastrointestinal tract. *IEEE J. Biomed. Health Inform.* **2020**, *25*, 514–525. [CrossRef]
28. Fan, C.; Hu, K.; Feng, S.; Ye, J. Heronian mean operators of linguistic neutrosophic multisets and their multiple attribute decision-making methods. *Int. J. Distrib. Sens. Netw.* **2019**, *15*, 1550147719843059. [CrossRef]
29. Cao, Y.; Sun, Y.; Min, J. Hybrid blockchain–based privacy-preserving electronic medical records sharing scheme across medical information control system. *Meas. Control* **2020**, *53*, 1286–1299. [CrossRef]
30. Liu, J.; Yu, J.; Shen, S. Energy-efficient two-layer cooperative defense scheme to secure sensor-clouds. *IEEE Trans. Inf. Forensics Secur.* **2017**, *13*, 408–420. [CrossRef]
31. Li, T.; Wang, H.; He, D.; Yu, J. Blockchain-based privacy-preserving and rewarding private data sharing for IoT. *IEEE Internet Things J.* **2022**. [CrossRef]
32. Lu, C.; Wu, Z.; Liu, M.; Guo, J. A patient privacy protection scheme for medical information system. *J. Med. Syst.* **2013**, *37*, 9982. [CrossRef] [PubMed]
33. Li, Q.; Zhang, Q.; Huang, H.; Zhang, W. Secure, efficient and weighted access control for cloud-assisted industrial IoT. *IEEE Internet Things J.* **2022**. [CrossRef]
34. Li, T.; Wang, H.; He, D.; Yu, J. Synchronized provable data possession based on blockchain for digital twin. *IEEE Trans. Inf. Forensics Secur.* **2022**, *17*, 472–485. [CrossRef]
35. Wu, Z.; Li, G.; Liu, Q.; Xu, G.; Chen, E. Covering the sensitive subjects to protect personal privacy in personalized recommendation. *IEEE Trans. Serv. Comput.* **2018**, *11*, 493–506. [CrossRef]
36. Zhang, S.; Ren, W.; Tan, X.; Wang, Z.; Liu, Y. Semantic-aware dehazing network with adaptive feature fusion. *IEEE Trans. Cybern.* **2021**. [CrossRef]
37. Fu, J.; Wang, N.; Cai, Y. Privacy-preserving in healthcare blockchain systems based on lightweight message sharing. *Sensors* **2020**, *20*, 1898. [CrossRef]
38. Wu, Z.; Shi, J.; Lu, C.; Chen, E. Constructing plausible innocuous pseudo queries to protect user query intention. *Inf. Sci.* **2015**, *325*, 215–226. [CrossRef]
39. Wu, Z.; Xu, G.; Yu, Z.; Zhang, Y. Executing SQL queries over encrypted character strings in the Database-As-Service model. *Knowl.-Based Syst.* **2012**, *35*, 332–348. [CrossRef]
40. Wu, Z.; Shen, S.; Zhou, H.; Li, H.; Xu, G. An effective approach for the protection of user commodity viewing privacy in e-commerce website. *Knowl.-Based Syst.* **2021**, *220*, 106952. [CrossRef]
41. Wu, Z.; Zheng, C.; Xie, J.; Zhou, Z.; Xu, G.; Chen, E. An approach for the protection of users' book browsing preference privacy in a digital library. *Electron. Libr.* **2018**, *36*, 1154–1166. [CrossRef]
42. Zhou, H.; Shen, S.; Liu, J. Malware propagation model in wireless sensor networks under attack defense confrontation. *Comput. Commun.* **2020**, *162*, 51–58. [CrossRef]
43. Zhao, L.; Lin, T.; Zhang, D.; Zhou, K. An ultra-low complexity and high efficiency approach for lossless alpha channel coding. *IEEE Trans. Multimed.* **2019**, *22*, 786–794. [CrossRef]
44. Chen, L. Road vehicle recognition algorithm in safety assistant driving based on artificial intelligence. *Soft Comput.* **2021**, 1–10. [CrossRef]
45. Wu, Z.; Li, R.; Zhou, Z.; Su, X. A user sensitive subject protection approach for book search service. *J. Assoc. Inf. Sci. Technol.* **2020**, *71*, 183–195. [CrossRef]
46. Wu, Z.; Li, G.; Shen, S.; Xu, G. Constructing dummy query sequences to protect location privacy and query privacy in location-based services. *World Wide Web* **2021**, *24*, 25–49. [CrossRef]
47. Wu, Z.; Lu, C.; Zhao, Y.; Xie, J.; Zhou, H.; Su, X. The protection of user preference privacy in personalized information retrieval: Challenges and overviews. *Libri* **2021**. [CrossRef]
48. Wu, Z.; Wang, R.; Li, Q.; Lian, X.; Xu, G.; Chen, E. A location privacy-preserving system based on query range cover-up for location-based services. *IEEE Trans. Veh. Technol.* **2020**, *69*, 5244–5254. [CrossRef]
49. Li, Q.; Cao, Z.; Ding, W.; Li, Q. A multi-objective adaptive evolutionary algorithm to extract communities in networks. *Swarm Evol. Comput.* **2020**, *52*, 100629. [CrossRef]
50. Shen, S.; Zhou, H.; Feng, S.; Huang, L.; Liu, J.; Yu, S. HSIRD: A model for characterizing dynamics of malware diffusion in heterogeneous WSNs. *J. Netw. Comput. Appl.* **2019**, *146*, 102420. [CrossRef]
51. Liu, J.; Wang, X.; Yue, G.; Shen, S. Data sharing in VANETs based on evolutionary fuzzy game. *Future Gener. Comput. Syst.* **2018**, *81*, 141–155. [CrossRef]

52. Shen, S.; Hu, K.; Huang, L.; Li, H.; Han, R. Quantal response equilibrium-based strategies for intrusion detection in WSNs. *Mob. Inf. Syst.* **2015**, *2015*, 179839. [CrossRef]
53. Jiang, G.; Shen, S.; Hu, K.; Huang, L. Evolutionary game-based secrecy rate adaptation in wireless sensor networks. *Int. J. Distrib. Sens. Netw.* **2015**, *11*, 975454. [CrossRef]
54. Zhou, Q.; Zhao, L.; Zhou, K.; Lin, T.; Wang, H. String prediction for 4: 2: 0 format screen content coding and its implementation in AVS3. *IEEE Trans. Multimed.* **2020**, *23*, 3867–3876. [CrossRef]
55. Wu, Z.; Xu, G.; Zhang, Y.; Li, G.; Hu, Z. GMQL: A graphical multimedia query language. *Knowl. Based Syst.* **2012**, *26*, 135–143. [CrossRef]
56. Wu, Z.; Zhu, H.; Li, G.; Cui, Z.; Li, J.; Huang, H.; Chen, E.; Xu, G. An efficient Wikipedia semantic matching approach to text document classification. *Inf. Sci.* **2017**, *393*, 15–28. [CrossRef]
57. Pan, J.; Zhang, C.; Wang, H.; Wu, Z. A comparative study of Chinese named entity recognition with different segment representations. *Appl. Intell.* **2022**, 1–13. [CrossRef]
58. Xu, G.; Wu, Z.; Li, G.; Chen, E. Improving contextual advertising matching by using Wikipedia thesaurus knowledge. *Knowl. Inf. Syst.* **2015**, *43*, 599–631. [CrossRef]
59. Li, Q.; Li, L.; Wang, W.; Zhong, J. A comprehensive exploration of semantic relation extraction via pre-trained CNNs. *Knowl. Based Syst.* **2020**, *194*, 105488. [CrossRef]
60. Xu, G.; Zong, Y.; Jin, P.; Wu, Z. KIPTC: A kernel information propagation tag clustering algorithm. *J. Intell. Inf. Syst.* **2015**, *45*, 95–112. [CrossRef]
61. Yan, W.; Li, G.; Wu, Z.; Wang, S.; Yu, P. Extracting diverse-shapelets for early classification on time series. *World Wide Web* **2020**, *23*, 3055–3081. [CrossRef]
62. Li, Q.; Cao, Z.; Zhong, J.; Li, Q. Graph representation learning with encoding edges. *Neurocomputing* **2019**, *361*, 29–39. [CrossRef]
63. Bai, B.; Li, G.; Wang, S.; Wu, Z.; Yan, W. Time series classification based on multi-feature dictionary representation and ensemble learning. *Expert Syst. Appl.* **2021**, *169*, 114162. [CrossRef]
64. Wu, Z.; Jiang, T.; Su, W. Efficient computation of shortest absent words in a genomic sequence. *Inf. Process. Lett.* **2010**, *110*, 596–601. [CrossRef]

electronics

MDPI

Article

Spatial and Temporal Normalization for Multi-Variate Time Series Prediction Using Machine Learning Algorithms

Alimasi Mongo Providence [1], Chaoyu Yang [2,*], Tshinkobo Bukasa Orphe [1], Anesu Mabaire [1] and George K. Agordzo [3]

[1] School of Economics and Management, Anhui University of Science and Technology, Huainan 232000, China
[2] School of Artificial Intelligence, Anhui University of Science and Technology, Huainan 232000, China
[3] School of Mathematics and Big Data, Anhui University of Science and Technology, Huainan 232000, China
* Correspondence: yangchy@aust.edu.cn

Abstract: Multi-variable time series (MTS) information is a typical type of data inference in the real world. Every instance of MTS is produced via a hybrid dynamical scheme, the dynamics of which are often unknown. The hybrid species of this dynamical service are the outcome of high-frequency and low-frequency external impacts, as well as global and local spatial impacts. These influences impact MTS's future growth; hence, they must be incorporated into time series forecasts. Two types of normalization modules, temporal and spatial normalization, are recommended to accomplish this. Each boosts the original data's local and high-frequency processes distinctly. In addition, all components are easily incorporated into well-known deep learning techniques, such as Wavenet and Transformer. However, existing methodologies have inherent limitations when it comes to isolating the variables produced by each sort of influence from the real data. Consequently, the study encompasses conventional neural networks, such as the multi-layer perceptron (MLP), complex deep learning methods such as LSTM, two recurrent neural networks, support vector machines (SVM), and their application for regression, XGBoost, and others. Extensive experimental work on three datasets shows that the effectiveness of canonical frameworks could be greatly improved by adding more normalization components to how the MTS is used. This would make it as effective as the best MTS designs are currently available. Recurrent models, such as LSTM and RNN, attempt to recognize the temporal variability in the data; however, as a result, their effectiveness might soon decline. Last but not least, it is claimed that training a temporal framework that utilizes recurrence-based methods such as RNN and LSTM approaches is challenging and expensive, while the MLP network structure outperformed other models in terms of time series predictive performance.

Keywords: spatial-temporal systems; neural networks; machine learning; information systems; forecasting; time series

Citation: Providence, A.M.; Yang, C.; Orphe, T.B.; Mabaire, A.; Agordzo, G.K. Spatial and Temporal Normalization for Multi-Variate Time Series Prediction Using Machine Learning Algorithms. *Electronics* **2022**, *11*, 3167. https://doi.org/10.3390/electronics11193167

Academic Editor: Jordi Guitart

Received: 18 July 2022
Accepted: 26 September 2022
Published: 1 October 2022

1. Introduction

A crucial part of many industry sectors is forecasting, which is the method of predicting the present significance of time series data [1]. Forecasting distribution networks and airline requests, finance price levels, power, and traffic or weather systems are just a few examples of its applications. As disputed to univariate (solitary time series data) predicting, multi-variate time series analysis is frequently necessary for large statistics of linked time series data. Suppliers may, for instance, need to forecast the sales and requests of millions of different commodities at tens of thousands of varying locations, resulting in billions of marketing time series data. The statistical modeling research has widely covered multivariate time series prediction, which entails learning from traditional multi-variate data to estimate the prospective qualities of several factors. However, the majority of the studies (also latest) concentrate on linear methods, deal with situations in which only a small number of variables are available, and only occasionally use forecasting horizons greater than another uses. In financial predicting research, particularly Dynamic Factor Models (DFM) [2], which are typically limited to linear methods, will more frequently consider huge difference setups. The big information transformation is currently calling for methods that can handle very multiple amounts of non-linear time series, possibly closely associated or predicated, and forecast their transformation over longer timeframes [3]. The Internet of Things (IoT) devices, of which the key effect is the development of a constant flow of spatiotemporal transmissions predicted to be generated and evaluated, serve as the most obvious source of inspiration [4]. This is already taking place in an increasing number of research and applied fields, including financial services, meteorology, industrial activities, atmospheric engineering, and physical sciences. As a result, performing univariate forecasting instead of multi-variate forecasting is also a common strategy. The most popular statistical forecasting techniques in use today in business include exponential smoothing (ES), auto-regressive AR and ARIMA designs, and more overall state space designs [5]. These techniques utilize a simple mathematical model of historical data and future predictions. Until recently, these techniques consistently surpassed machine learning techniques such as RNNs in large-scale predicting contests. Multi-task univariate projections, which shares deep learning design variables across all sequence, possibly including some series-specific basis functions or parameterized range of fundamental, is a major factor in the latest achievements of deep learning for forecasting. For instance, a hybrid ES-RNN framework [6] simultaneously learns different seasons and level ES variables for every sequence to regulate those gained by the M4 predicting competition. This model forecasts each series using a single cohesive univariate RNN prototype.

In many commercial and industrial implementations, time series predictions are a critical issue. For example, if a public transportation provider can predict that a specific area will experience a supply issue in the coming hours, they could allocate enough capacity to reduce queuing times in that area in advance [7]. As another illustration, a Robo-advisor that can anticipate a prospective financial collapse can help an investor avoid financial loss. Real-world time series frequently exhibit varied dynamics because of the complicated and constantly changing influencing factors. This makes them exceptionally non-stationary. For example, the state of the road, its location, the time of day, and the climate all have an important influence on the amount of circulation that passes over it. The latest season, value, and product all play a role in determining a product's sales within the retail industry [8]. Time series forecasting faces a great deal of difficulty as a result of the diverse interactions. This would then research multi-variate time series prediction in this research, their several variables changing over time. Numerous disciplines, including physical sciences, engineering, weather forecasting, and tracking of human health, are conducting extensive research on time series prediction [9]. We proposed the most appropriate technique to test, evaluate, and verify the most popular forecasting methodologies using a collection of information. It appears that using just a few of the models cannot be a suitable method for simulating the hydrological time series. In such cases, time series modeling and artificial intelligence models might be combined to account for hydrological processes rather than utilizing a single model [10]. It is well recognized that the experimental dataset, the machine learning model, and the use of efficient variables for model creation depending on such a challenge are all very important components in building a reliable machine learning technique [11].

A non-linear multi-variate (or variable) time series' multi-step ahead prediction is recognized to be quite challenging. If a forecasting task is modeled as an autoregressive procedure, this task poses several formidable difficulties for every learning machine, including the high dimensional space of outputs and inputs, cross-sectional and seasonal high dependence (which results in both non-linear multi-variate connections within inputs as well as a nonlinear framework within outputs), and last but not least, the danger of error reproduction [12]. Earlier work has concentrated on a particular sub-problem product's sale within the issue of the one-step-ahead predictive model of sequential multi-variate time series, discussions of the issue of multiple-step-ahead predicting of such a univariate time series, and the latest manuals that individuals take into account linear methods [13].

In a wide range of works, the topic of dimensionality decrease is protected more normally, although without addressing how to expand it to multiple steps predicting.

Extraction of structures and characteristics that characterize the key characteristics of the data is frequently the first step in the analysis and investigation of a time series (and other kinds of analysis). The exploitation of worldwide time series characteristics (including spectral characteristics measured with a conversion, such as the Discrete Cosine or Wavelet Transforms) and utilization of such global characteristics (that characterize time series characteristics as a collective) for archiving are standard techniques in the research [14]. Worldwide, fingerprints of multi-variate time series data could be extracted using correlations, transfer functions, statistical groupings, spectral characteristics, Singular-Value Decomposition (SVD), and related eigen decomposition. Tensor degradation is the equivalent analysis procedure on a tensor that can be utilized to depict the time dynamics of multi-modal information [15]. Costly methods include tensor and matrix degradation processes, probabilistic methods (such as Dynamic Topic Modeling, DTM), and autoregressive incorporated moving average (ARIMA) predicated analysis, which divide a statistical model into informatics, moving average, and autoregressive elements for simulation and prediction. A dependable structure for modeling and learning time series structures is offered by conventional time series forecasting techniques, such as ARIMA and state-space models (SSMs). These techniques, however, have a great implication for the normality of a time series, which poses serious practical challenges if the majority of the influencing factors are not accessible. Deep learning methods have recently advanced to the point where they can now handle complex dynamic nature as a single entity, even in the absence of increased affecting variables. Recurrent neural networks (RNN) [16], long-short term memories (LSTM), Transformer, Wavenet [17], and temporal convolution networks are popular neural structures used on time series information (TCN). The key would be to further modify various components of different kinds from the initial measurement. Interactions that set dynamics apart from the spatial or temporal perspective can then be collected. This research offers two different types of normalization configurations that individually improve its high-frequency and local elements: Spatial Normalization (SN) and Temporal Normalization (TN) [18]. To do this, academics have become interested in applying ML approaches to create models that are more potent with greater accuracy. The shortcomings of traditional modeling methods were widely addressed by ML approaches to solve complicated environmental technical challenges [19]. This paper's contribution is the refinement of further categories of original measuring elements. Connections that set dynamics apart from the temporal or spatial view can then be represented. In this study, two different types of normalization modules are presented that individually enhance the high-frequency and local elements: temporal normalization (TN) and spatial normalization (SN). In particular, the local component makes it easier to separate dynamics from the spatial perspective, and the high-frequency component helps to distinguish dynamics from the temporal view. The system can uniquely fit every cluster of data because of its difference over space and time, specifically those long-tailed samples. The paper also demonstrates how the approach compares to existing state-of-the-art (SOTA) methods that use mutual relationship development to discern between dynamics.

Numerous applications produce and/or use multi-variate temporal data, but experts frequently lack the tools necessary to effectively and methodically look for and understand multi-variate findings. Efficient prediction models for multi-variate time series are crucial because of the incorporation of sensory systems into vital applications, including aviation monitoring, construction energy efficiency, and health management. Time series prediction methods have been expanded from univariate predictions to multi-variate time series predictions to meet this requirement. However, naive adaptations of prediction approaches result in an unwanted rise in the expense of model simulation, and more crucially, a significant decline in prediction accuracy since the extensive models are unable to represent the fundamental correlations between variates. However, research has shown that investigating both geographical and temporal connections might increase predictive performance. These effects also influence how MTS will develop in the future, making it crucial to include them in time series forecasting work. Conventional approaches, however, have inherent limitations in separating the components produced by each type of effect from the source data. Two different normalization components are suggested with machine learning techniques to do this. The local component underlying the original data as well as the improved high-frequency element is separated by the suggested temporal and spatial normalization. Additionally, these modules are simple to include in well-known deep learning architectures like Wavenet and Transformer. Mixtures and original data can be difficult to distinguish using conventional methods. In this way, it incorporates well-known neural networks, such as the multi-layer perceptron (MLP), complex deep learning techniques, such as RNN and LSTM, two recurrent neural networks, support vector machines (SVM), and its application to regression, XGBoost, and others.

2. Related Works

Modern applications, such as climatic elements and requirement predicting, have high-dimensional time series estimation difficulties. In the latter requirement, 50,000 pieces must be predicted. The data are irregular and contain missing values. Modern applications require scalable techniques that can handle noisy data with distortions or missing values. Classical time series techniques often miss these issues. This research gives a basis for data-driven temporal learning and predicting, dubbed temporal regularized matrix factorization (TRMF). Create new regularization methods and scalable matrix factorization techniques for high-dimensional time series analysis with missing values. The proposed TRMF is comprehensive and includes multiple time series assessment methods. Linking autoregressive structural correlations to pattern regularization is needed to better comprehend them.

According to experimental findings, TRMF is superior in terms of scalability and prediction accuracy. Specifically, TRMF creates greater projections on real-world datasets such as Wal-Mart E-commerce data points and is two requirements of magnitude quicker than some other techniques [20].

Using big data and AI, it is possible to predict the citywide audience or traffic intensity and flow. It is a crucial study topic with various applications in urban planning, traffic control, and emergency planning. Combining a big urban region with numerous fine-grained mesh grids can display citywide traffic data in 4D tensors. Several grid-based forecasting systems for citywide groups and traffic use this principle to do reevaluating the intensity and in-out flow forecasting issues and submitting new accumulated human mobility source data from a smartphone application. The data source has many mesh grids, a fine-grained size distribution, and a high user specimen. By developing pyramid structures and a high-dimensional probabilistic model based on Convolutional LSTM, we offer a new deep learning method dubbed Deep Crowd for this enormous crowd collection of data. Last but not least, extensive and rigorous achievement assessments have been carried out to show how superior its suggested Deep Crowd is when compared to various state-of-the-art methodologies [21].

Regional forecasting is crucial for ride-hailing services. Accurate ride-hailing demand forecasting improves vehicle deployment, utilization, wait times, and traffic density. Complex spatiotemporal needs between regions make this job difficult. While non-Euclidean pair-wise correlation coefficients between possibly remote places are also crucial for accurate predicting, typical systems focus on modeling Euclidean interrelations between physically adjacent regions. This paper introduces the spatiotemporal multi-graph convolution network for predicting ride-hailing consumption (ST-MGCN). Non-Euclidean pair-wise relationships between regions are encoded into graphs before explicitly modeling correlation coefficients using a multi-graph transform. Perspective Landscaping recurrent neural networks, which add context-aware limits to re-weight historical observational data, are presented as a way to use global data to build association coefficients. This tests the suggested concept using two large-scale ride-hailing requirement data sources from the true world and finds that it consistently outperforms benchmarks by more than 10% [7].

To address multi-horizon probability forecasting, we use a data-driven technique to predict a time series distribution over upcoming horizons. Observed changes in historical data are vital for predicting long-term time series. Traditional methodologies rely on building a temporal relationship by hand to explore historical regularities, which is unrealistic for predicting long-term series. Instead, they propose learning how to use deep neural systems to display hidden knowledge and generate future predictions. In this study, an end-to-end deep-learning structure for multi-horizon time series prediction is proposed, along with temporal focus procedures to more efficiently capture latent patterns in historical data that are relevant for future prediction. Based on latent pattern properties, several future projections can be made. To accurately demonstrate the future, we also suggest a multi-modal fusion process that combines characteristics from various periods of history. Results from experiments show that the method produces cutting-edge outcomes on two sizable predicting data sources in various fields [22]. Multi-horizon prediction sometimes uses static (time-invariant) confounders, recognized future components, and other external time series only identified in the past. Deep learning techniques abound. "Black-box" designs do not describe how they employ real-world inputs. The Temporal Fusion Transformer (TFT) demands maximum multi-horizon prediction using temporal insights. TFT uses self-attention structures to create long-term connections and factors for learning temporal correlation at different scales and uses elements to pick relevant characteristics and gating pieces to override superfluous components. This improves performance above baselines on a wide range of real data sources and exhibits three TFT use cases [23]. Academic research struggles with hierarchical time series prediction. The precision of each hierarchical system, notably the interrupted time series at the bottom, is explored thoroughly. Hierarchical reunification boosts system productivity. This article provides a hierarchical prediction-to-alignment technique that considers bottom step projections changeable to improve upper hierarchical prediction performance. The bottom stage employed Light GBM for occasional time series and N-BEATS for constant time series. Hierarchical prediction with orientation is a simple but effective bottom-up improvement that adjusts for biases hard to discover at the bottom. It increases the average accuracy of less accurate estimates. The first author developed this study's technique in M5 Predicting Precision tournaments. The business-oriented approach may be effective for strategic business planning [24].

3. Methodology

3.1. Normalization

Since normalization is initially used in deep image processing, now almost all deep learning activities have seen a significant improvement in model performance. Each normalization approach has been future to report a specific gathering of computer vision applications, including group normalization, instance normalization, positional normalization layer normalization, and batch normalization [25]. Instance normalization, which was initially intended for image generation due to its ability to eliminate style data from its images, does have the highest opportunity for research. Researchers have discovered that attribute statistical data could collect an image's design and that after initializing the statistical data, the remaining characteristics are in charge of the image's substance. This ability to deliver an image's material in the fashion of another image, also recognized as extracting features, is made possible by its distinguishable assets. Similar to scale details in the time series is the style data in the image. Another area of research investigates the rationale behind the normalization trick's facilitation of deep neural network learning. One of their key findings is that normalization could improve the evaluation of an attribute space, allowing the framework to retrieve characteristics that are more different.

Figure 1 presents a high-level assessment of the structure used. Along the computation path, a few significant variables and their shapes have been branded at the appropriate locations. The structure normally has a Wavenet-like structure, with the addition of modules for spatial and temporal normalization, collectively referred to as ST-Norm or STN.

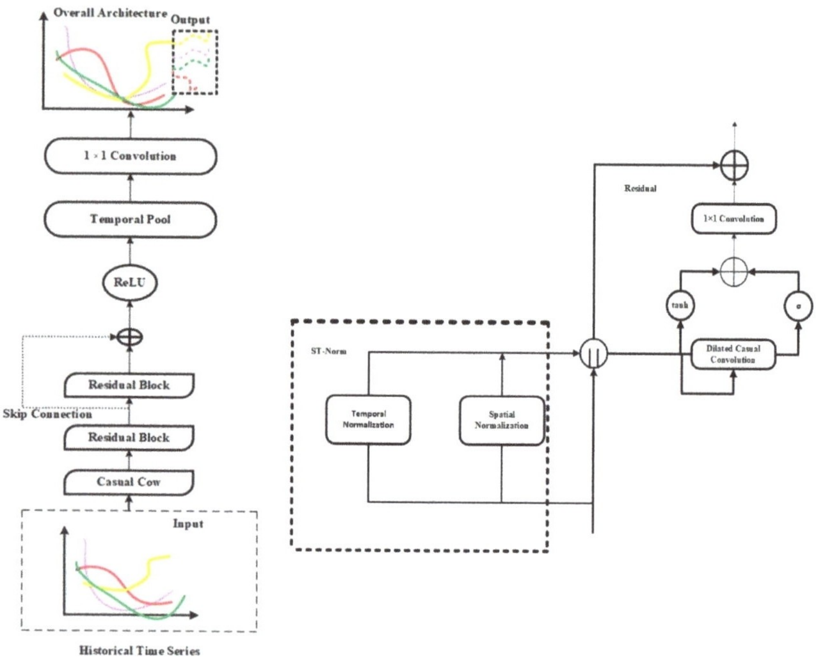

Figure 1. Overall high-level assessment structure.

3.2. *Dilated Causal Convolution*

This segment provides a brief introduction to dilated causal convolution, which applies the filter while skipping beliefs. The causal convolution on component t for the 1D signal $z \in \mathbb{R}^T$ and a filter $f : \{0, \ldots, k-1\} \rightarrow \mathbb{R}$ is described in Equation (1) as follows:

$$F(t) = (c * f)(t) = \sum_{i=0}^{k-1} f(i).c_{t-i} \tag{1}$$

This formula is simple to categorize for multi-dimension signals, but for the sake of brevity, it will not be included it here. To guarantee length continuity, padding (zero as well as recreate) to the dimension of $k-1$ is added to a left tail of a transmitter [26]. To give so every component a larger receptive ground, we combined several causal convolution layers.

Figure 2 shows the structure of dilated causal convolution. Trying to cause an outburst of characteristics when predicting long history is one drawback of using causal convolution, because the diameter of a kernel or its number of layers grows linearly with the dimensions of a receptive sector [27]. The obvious solution to this problem is pooling, but doing so compromises the signal's order details. To achieve this, dilated causal convolution is used, a form that encourages the exponential growth of an approachable pitch. The structured computing method is expressed in Equation (2).

$$F(t) = (c *_d f)(t) = \sum_{i=0}^{k-1} f(i).c_{t-d}.i, \tag{2}$$

In Equation (2), d is its component for dilation. Typically, d grows exponentially with network depth (namely, 2^l at stage l of a system). The variable d which denotes the dilated convolution operator decreases to the $*_d$ a normal convolution controller if d is 1 or (2^0).

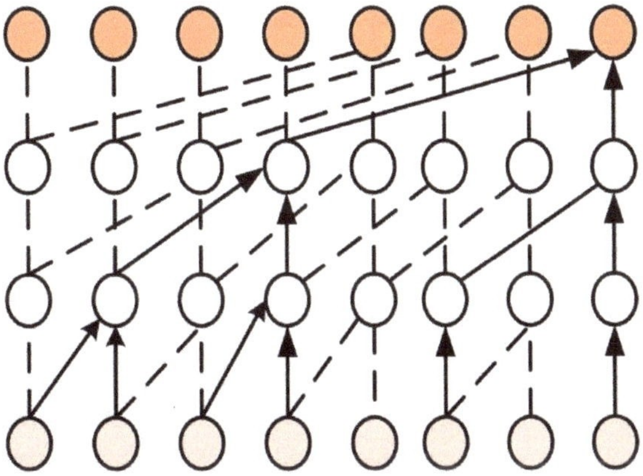

Figure 2. Dilated causal convolution architecture.

3.3. Neural Networks

3.3.1. Multi-Layer Perceptrons (MLPs)

A brief overview of Random Forests, multi-layer perceptrons (MLP), XGBoost, long-short term memory (LSTM) systems, and support vector regressors (SVR) is provided as an introduction to creating the research self-contained. MLP is regarded as an effective technique for capturing interactions among the parameter estimation that are not linear. It is being used effectively in hydrology, particularly time series modeling, hazard identification, and sediment supply [28]. The most popular artificial neural networks (ANN) for the classification of a regression issue are multi-layer perceptrons (MLPs). An input layer, each or many hidden layers, as well as an output layer, make up this category of designs [29]. A three-layer MLP is shown in Figure 3.

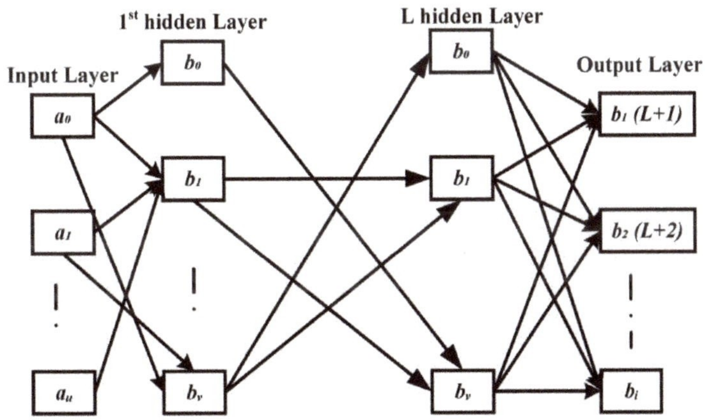

Figure 3. Multi-layer perceptron network graph layer.

A network diagram, for example, can produce the following results in Equation (3)

$$A(t) = \alpha_0 + \sum_{l=1}^{L}\left(\sum_{j=1}^{q}\alpha_{jl}g\left(\beta_{0jl} + \sum_{i=1}^{p}\beta_{ijl}X_{t-i}\right)\right) + \epsilon_t \tag{3}$$

The numbers $L,_{P,q}$ represent the number of hidden sections, inputs $X_t(i = 1, 2, \ldots p)$, and nodes in a specific hidden layer, respectively. The ReLU function $(g(x) = 1/(1 + e^{-a}))$ or the convolution $((e^a - e^{-a})/(e^a + e^{-a}))$ are some examples of activation functions

$ReLU(g(x) = \max(0, x))$. Equation (4) becomes simpler for networks including a single hidden layer:

$$X_t = \alpha_0 + \sum_{j=1}^{q} \alpha_j g\left(\beta_{0j} + \sum_{i=1}^{p} \beta_{ij} X_{t-i}\right) + \epsilon_t \quad (4)$$

3.3.2. LSTM

The LSTM is a design for such a recurrent neural network composed of three gates and two states: Input gate, Output gate, Forget gate, Cell state, and Hidden state.

Figure 4 displays the network's total schematic. In the mentioned reasoning, we will use it as a constant. This Figure 4 includes the hyperbolic tangent $\tanh(a) = (e^a + e^{-a})/(e^a - e^{-a})$ and the sigmoid $\sigma(a) = 1/(1 + e^{-a})$. The elements of the vector are subjected to activation functions [30]. Additionally, the element-wise addition and multiplication processes are denoted by $\odot$ and $\oplus$. The two related matrices are finally concatenated vertically when two lines intersect. The formula for this procedure is $\dagger : A \dagger B = \begin{pmatrix} A \\ B \end{pmatrix}$.

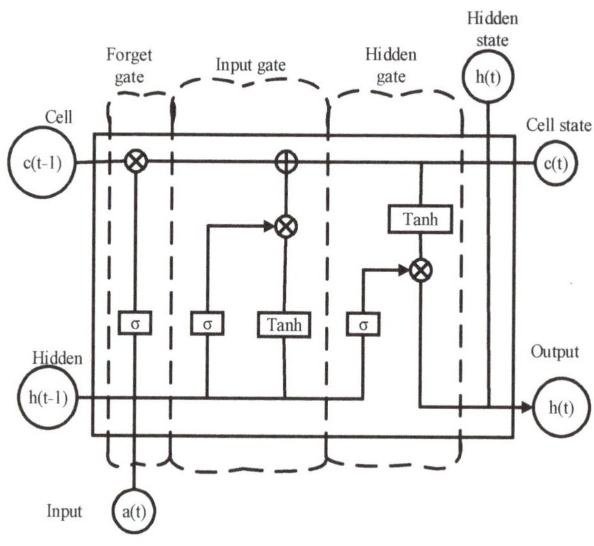

Figure 4. The cell structure of Long-Short Term Memory.

One could demonstrate $h^{\langle r \rangle}$ and $c^{\langle r \rangle}$ as a component of $h^{\langle r-1 \rangle}$, $c^{\langle r-1 \rangle}$, and $a^{\langle r \rangle}$ without using these formulae in Equation (5):

$$\begin{gathered} c^{\langle r \rangle} = \left(c^{\langle r-1 \rangle} \oplus \sigma\left(\ominus_f\left(h^{\langle r-1 \rangle} \dagger a^{\langle r \rangle}\right)\right)\right) \oplus \\ \left(\left(\tanh\left(\ominus_c\left(h^{\langle r-1 \rangle} \dagger x^{\langle r \rangle}\right)\right) \odot \sigma\left(\ominus_i\left(h^{\langle r-1 \rangle} \dagger a^{\langle r \rangle}\right)\right)\right)\right) \\ h^{\langle r \rangle} = \tanh\left(c^{\langle r \rangle}\right) \odot \sigma\left(\ominus_o\left(h^{\langle r-1 \rangle} \dagger a^{\langle r \rangle}\right)\right) \end{gathered} \quad (5)$$

where $a^{\langle r \rangle}$, $c^{\langle r \rangle}$, and $h^{\langle r \rangle}$ stand for the input signal (time series amount at time r), approximate output importance for time r, and cell condition at time r, respectively. The characteristics of an LSTM framework were its matrices $\ominus_f, \ominus_i, \ominus_c, \ominus_o$.

3.3.3. SVM

The ϵ-insensitive loss capability is used by its support vector regression (SVR) algorithm, which was developed. The time series analysis A_t in SVR is transformed nonlinear, Φ, from its input space to such greater dimensional storage, which is denoted in Equations (6) and (7):

$$\Phi = \mathbb{R}^n \rightarrow F \tag{6}$$

$$f(a) = < u,\ \Phi(a) > + y \tag{7}$$

where the linear function f is minimized by a vector of characteristics (also known as weights) called w ^{2} F, and b ^{2} R is continuous. SVR typically selects the insensitive loss function again for minimization procedure instead of more traditional loss functions, such as the mean absolute percentage error (MAPE) and the least mean average error (MAE) [31]. One must reduce the risk formalized function to reduce its weight vector w, and subsequently, the function of f in Equation (8):

$$\begin{array}{c} min1/2\left|\left|w\right|\right|_2 + C\sum_{i=1}^{l}(\xi_i + \xi_i^*) \\ s.t. b_i - < w, \Phi(a_i) > - b \le \epsilon + \xi_i \\ < w, \Phi(a_i) > + y - b_i \le \epsilon + \xi_i^* \end{array} \tag{8}$$

where $\epsilon \geq 0$ represents the separation among the real charge of y and the assumed shape of f. Slack variables $\xi, \xi^* \geq 0$ are added to accommodate errors bigger than that ϵ. When fitting training data, the regularization constant C is utilized to define the trade-off between generalization and precision.

In actuality, the Lagrangian multiplier-based expressions of w and f have been utilized in Equation (9):

$$\begin{array}{c} w = \sum_{i=1}^{t}(\alpha_i - \alpha_i^*)\Phi(a_i) \\ f(a) = \sum_{i=1}^{t}(\alpha_i - \alpha_i^*)K(a_i - a) - y \end{array} \tag{9}$$

where $(\alpha_i - \alpha_i^*) \leq C$ denotes the partial derivative of $\Phi(xi)$ and $\Phi(x)$, also known as the kernel features, and where $K(a_i, a)$. The literature provides more information on support vector machines and how to use them to solve regression issues.

3.3.4. Tree-Based Techniques

(a) Random forest regressor

As ensemble learning methods, random forests (RF) [32] could be used for correlation by averaging many different regression trees $z_n(a, \ominus_m, D_n)$, where $\ominus$ has been the model's parameters variable and D_n would be the training set $(A_1, B_1), \ldots, (A_n, B_n)$. The assessment of the combined trees' regression function in Equation (10):

$$\overline{r}_n(A, D_n) = \mathbb{E}_{\ominus}[r_n(A, \ominus, D_n)] \tag{10}$$

where $\mathbb{E}$ has been the assumption or conditional mean for the specified posterior distribution. To create a forest of nonlinear individual trees, packing and feature stochastic are also used. When predictions are made by a committee rather than by individual trees, the results are more precise. References provide thorough configurations of a random forest classifier.

(a) XGBoost

The gradient tree boosting (GBT) development known as XGBoost (eXtreme Gradient Boosting) is indeed a tree ensemble machine learning technique. The forecast is described as follows at the time (or phase) r in Equation (11):

$$\hat{b}^{(r)} = \sum_{k=1}^{t} f_k(a_i) = \hat{b}_i^{(r-1)} + f_i(a_i) \tag{11}$$

where a_i would be the feature variable, also known as the input observation, which refers to the prior time values within the time series analysis set. Moreover, at time t, $f_i(a_i)$ seems to be the learner, which is typically a regression tree. The XGBoost framework employs a normalized objective function to protect the excessively of its training examples, as shown in Equation (12):

$$O^{(t)} = \sum_{k=1}^{n} l\left(\hat{b}, b\right) + \sum_{k=1}^{t} \Omega(f_i) \tag{12}$$

where t denotes the leaf count, Ω denotes the leaf score, and O denotes the regularization attribute. The leaf node splitting minimum loss value O is represented by the parameter. The research of Chen and Guestrin provides more information on the XGBoost framework and how it was put into practice.

3.4. Temporal Normalization

The goal of temporal normalization (TN) is to smooth out the high frequency of local and global elements of a hybrid signal. High-frequency elements and the low-frequency elements are each summarized separately here, using the two notations that are demonstrated in Equation (13):

$$C^{high}_{i,t} = C^{lh}_{i,t} C^{gh}_{t} \cdot C^{low}_{i,t} = C^{II}_{i,t} C^{gl}_{t}. \tag{13}$$

The reasonable inference that low-frequency element evolving charges are significantly slower than high-frequency element evolving rates is the foundation for the suitability of TN. Technically speaking, every low-frequency element over a period roughly equals a constant [33]. This presumption allows us to pertain TN to time sequence without its need for additional characteristics that characterize the frequency. Many real-world issues where the specialized frequency is unavailable can benefit from this characteristic. To achieve a desirable form for whom unique amounts could be determined from information, begin by expanding $C^{high}_{i,t}$ in Equation (14):

$$\begin{aligned} C^{high}_{i,t} &= \frac{C^{high}_{i,t} - EC^{highl}_{i,t} i}{\sigma C^{high}_{i,t} i + \epsilon} \sigma C^{highl}_{i,t} i + EC^{highl}_{i,t} i \\ &= \frac{C^{high}_{i,t} C^{low}_{i,t} - C^{low}_{i,t} EC^{highl}_{i,t} i}{C^{low}_{i,t} \sigma C^{highl}_{i,t} i + \epsilon} \sigma C^{highl}_{i,t} i + EC^{highl}_{i,t} i \\ &= \frac{C_{i,t} - EC_{i,t} C^{low}_{i,t}, i}{(\pm)\sigma Z_{i,t} Z^{low}_{i,t}, i + \epsilon} \sigma C^{highl}_{i,t} i + EC^{highl}_{i,t} i \\ &= \frac{C_{i,t} - EC_{i,t} C^{low}_{i,t}, i}{\sigma Z_{i,t} Z^{low}_{i,t}, i + \epsilon} (\pm)\sigma C^{highl}_{i,t} i + EC^{highl}_{i,t} i, \end{aligned} \tag{14}$$

where $C_{i,t}$ is perceptible, ϵ is a minor constant to maintain numerical stability, and the high-frequency impact mostly on ith time series over time is represented mostly by vectors $E\ C^{highl}_{i,t} i$ and $(\pm)\sigma\ C^{highl}_{i,t} i$, which can be estimated by a sequence of the learnable feature vector γ^{high}_i and β^{high}_i with a size of d_z, the values of $E\ C_{i,t} C^{low}_{i,t}, i$ and $\sigma\ C_{i,t} C^{low}_{i,t},\ i$ can be calculated by Equation (15):

$$\begin{aligned} EC_{i,t} C^{low}_{i,t}, i &\approx 1/\delta \sum_{t'=1}^{\delta} C^{low}_{i,t-t'+1} C^{low}_{i,t} \\ &\approx 1/\delta \sum_{t'=1}^{\delta} C^{low}_{i,t-t'+1} C^{low}_{i,t-t'+1} \\ &\approx 1/\delta \sum_{t'=1}^{\delta} C^{low}_{i,t-t'+1} - EC_{i,t} C^{low}_{i,t}, i^2 \end{aligned} \tag{15}$$

where δ is a time interval when the low-frequency element roughly stays continuous. For the sake of easiness, make several input appropriate actions in the task equitable. Research can acquire the recognition of the high-frequency element by replacing the estimates of four non-observable parameters in Equation (16):

$$C_{i,t}^{high} = \frac{C_{i,t} - EC_{i,t}C_{i,t}^{low}, i}{\sigma C_{i,t}C_{i,t}^{low}, i + \epsilon}\gamma_i^{high} + \beta_i^{high} \tag{16}$$

Notably, TN and instance normalization (IN) for image data have a special connection in which style acts as a low-frequency element and material as a high-frequency element [33]. The research is novel because it identifies the source of TN within the perspective of MTS and pieces together TN step-by-step of its source.

3.5. Spatial Normalization

Refining local elements, which are made up of the native high-frequency component as well as the regional low-frequency element, is the goal of spatial normalization (SN) [34]. The first step in achieving this goal is to get rid of global elements, which are caused by factors corresponding to the time of day, day of the week, climate, etc. This also presents two notations for enumerating regional and global elements as given in Equation (17):

$$C_t^{global} = C_t^{gh} C_t^{gl}, C_{i,t}^{local} = C_{i,t}^{lh} C_{i,t}^{ll}. \tag{17}$$

The suitability of SN is also predicated on the idea that all time series will be affected similarly by global impacts. Here, it is significant to say that we do not also strictly need global effects to have identical effects on every time series. The specified local element could supplement those impacts which are not equally identified in each time series. This begins by extending $C_{i,t}^{local}$ to a representation where another term can either be delegated with trainable parameters or an approximation based on data in Equation (18):

$$\begin{aligned} C_{i,t}^{local} &= \frac{C_{i,t}^{local} - EC_{i,t}^{local} t}{\sigma C_{i,t}^{locall} t + \epsilon}\sigma C_{i,t}^{local} t + EC_{i,t}^{local} t \\ &= \frac{C_{i,t} - EC_{i,t}C_t^{global}, t}{(\pm)\sigma C_{i,t}C_t^{global}, t + \epsilon}\sigma C_{i,t}^{global} t + EC_{i,t}^{global} t \\ &= \frac{C_{i,t} - EC_{i,t}C_t^{global}, t}{\sigma C_{i,t}C_t^{global}, t + \epsilon}(\pm)\sigma C_{it}^{local} t + EC_{it}^{local} t \end{aligned} \tag{18}$$

where $C_{i,t}$ is easily observable and $(\pm)\sigma\left(C_{i,t}^{local} t\right)$ and $E\left(C_{i,t}^{local} t\right)$ have been estimated by two teachable vectors[2] (γ local and β local), the prediction of $EC_{i,t}C_t^{global}$, t and σ $EC_{i,t}C_t^{global}$, t could be derived from it so information in the following methods in Equation (19):

$$\begin{aligned} EC_{i,t}C_t^{global}, t &= \tfrac{1}{N}\sum_{j=1}^{N} C_{j,t}^{global} C_t^{global} \\ &\tfrac{1}{N}\sum_{j=1}^{N} c_{j,t} \\ \sigma^2 C_{i,t}C_t^{global}, t &= E[\left(C_{i,t} - EC_{i,t}C_t^{global}, t\middle| C_t^{global}, t\right. \\ &= \tfrac{1}{N}\sum_{j=1}^{N} C_{i,t} - EC_{i,t}C_t^{global}, t^2 \end{aligned} \tag{19}$$

This can get the composite illustration of local elements by putting the estimations of four non-observable factors into Equation (18), which reads as follows (Equation (20)):

$$C_{i,t}^{global} = \frac{C_{i,t} - EC_{i,t}C_t^{global}, t}{\sigma C_{i,t}C_t^{global}, t + \epsilon}\gamma local + \beta local \tag{20}$$

In the spatial domain, SN is TN's counterpart because it uses high-frequency elements to represent local elements and low-frequency elements to represent global elements. The model takes into account fine-grained variability by removing the local and high-frequency elements with the actual signal, which is extremely helpful in time series prediction.

3.6. Learning and Forecasting

This designates $C^{(L)} \in \mathbb{R}^{N_l \times T_{in} \times d_z}$ as the result of the final residual block, in which every row of $C^{(L)} \in \mathbb{R}^{T_{in} \times d_z}$ stands for a different variable. Next, temporal accumulation is undertaken for each variable using a temporal pooling block. Depending on the issue being investigated, various pooling processes could be used, including max pooling and mean pooling [35]. In this instance, the pooling result is chosen to represent the entire signal by choosing the vector from the latest time slot. Finally, depending on the recognition acquired by a common fully connected layer, each creates different forecasting for each variable. The goal in the learning stage is to reduce the mean squared error (MSE) among the expected attributes and the standards obtained from the ground truth. Additionally, this goal can be maximized using the Adam optimizer.

4. Result and Discussion

4.1. Data Collection

To verify the efficacy of ST-Norm from various angles, perform comprehensive studies on three common data sources in this section by using Jupyter notebook. Using three real-world datasets, such as PeMSD7, BikeNYC, and Electricity, a framework can be verified, as well as statistics for each dataset and the correlating planned task settings. The principles in every dataset can be simplified to make training easier, and when testing is complete, the principles can be converted back to their original magnitude. In addition to SN and TN, an instance normalization (IN) control unit can be added. The sample size is four, and the batch pattern's input length is sixteen. For every DCC element's kernel size for such Wavenet backbone is set to 2, and its related dilation percentage is 2^i, where i have been the layer's index (counting from 0). Together, these settings allow Wavenet's output to recognize 16 input steps. Each DCC contains 16 hidden channels (d_z). To make the duration of a DCC output equivalent to 16, zero-stuffing can be extended to the left tail of an input. The Adam optimizer has a training set of 0.0001.

4.2. Evaluation Metrics

Then, we use the mean absolute error (MAE), mean squared error (RMSE), and mean absolute error (MSE) proportion to confirm the prototype (MAPE). For every model and every dataset, the test is repeated 10 times, and the average of outcomes is presented. Graph Wavenet has a similar architecture to MTGNN. The primary distinction would be that the former depends on a soft graph with a perpetual probability for every pair of nodes to be attached. This model designs segment-level correlation while capturing long-term correlations in time series information via the use of a consideration method, where its explanations and enquires are produced by the correlational convolution over a specific setting. MTGNN introduces a graph-learning device to create inter-variate connections. The diagram learning method particularly links every center node to its top k closest neighbors in a specified dimensional space. Wavenet is the primary design of MTGNN for sequential simulation.

A graph-learning component is also included in AGCRN to help develop inter-variable relationships. Additionally, it models the sequential connection for every time series using a customized RNN.

LSTNet consists of two parts: an LSTM with a supplemental skip correlation from an overtime component, and a traditional autoregressive design.

Graph Wavenet has a similar structure to MTGNN. The primary distinction would be that the distinction generates a soft chart with a constant likelihood that every pair of base stations is attached.

TCN's architecture is similar to Wavenet's, with the exception that every residual block's nonlinear transition is composed of two rectified linear units (ReLU).

This is also evaluated by using TCN and Inductor function when STN is used similarly before every layer's causal convolution procedure.

4.3. Ablation Study

This creates numerous variations as regards verifying the efficacy of SN and TN. This will also test a combination that includes both STN and a graph-learning module to see if it enhances STN. Since the normal Wavenet backbone is present in all of the variations, Wavenet is left out of the phrase for simplicity.

On each of the three datasets, these variants were tested, and Table 1 presents the overall findings. SN and TN both relate to the improvement. Moreover, STN's achievement slightly improves with a responsive graph-learning device. This shows that STN significantly outperforms and replaces the graph-learning device.

Table 1. Study of ablation.

		STN	*GSTN*	*SN*	*TN*	*Graph*	*Vanilla*
P	*MAE*	2.46	2.48	2.64	2.65	2.98	3.45
	RMSE	5.19	5.22	2.45	4.65	7.35	6.32
	MAPE	19.5	19.3	18.5	18.7	21.5	24.9
B	*MAE*	2.76	2.84	3.04	2.88	3.45	3.95
	RMSE	5.14	5.23	5.33	5.28	5.67	6.54
	MAPE	5.85	5.96	6.34	6.08	7.35	8.34
E	*MAE*	17.6	18.7	20.5	20.87	21.54	22.3
	RMSE	37.5	39.65	39.54	38.32	44.87	43.8
	MAPE	15.3	15.66	14.32	15.32	15.23	15.45

4.4. Model Optimization

To identify any underfitting or overfitting, learning shapes of prediction accuracy upon that train and validation datasets have been used. Each applicant's model's effectiveness was represented as a loss function and plotted against the epochs for both the training and validation sets. One could quickly determine whether the model might result in increased variation (overfit) or bias by correlating and examining the patterns of plotted loss features (underfit). K-fold cross-validation is utilized when a model has several hyperparameters because there were also too several more possible mixtures of hyperparameter attributes. A manual method should, therefore, be avoided because it is time-consuming. This method divides the training set into k smaller sets. Several of the k-folds is processed in the manner described below:

- A structure is trained by utilizing $k - 1$ of folds as training information;
- The architecture was authorized on a residual data portion.

The two stages above were recurring k-times by utilizing another data portion for validations. By computing the mean calculated for the k steps, the effectiveness identified by k-fold cross-validation is obtained. Although this method may be computationally time affordable, it does not necessitate as much information as repairing a specialized validation set. Other methods exist that might be slightly different, but usually adhere to the same fundamentals.

4.5. XGBoost

Several parameters could be given specific values to characterize XGBoost models. They must be chosen towards optimizing the performance of the approach on a particular dataset and in a manner that guards against both underfitting and overfitting as well as unnecessary difficulty. These parameters also include effect, learning algorithm, lambda, alpha, and the number of boosting repetitions. In XGBoost, the amount of shaped consecutive trees is referred to as the number of boosting iterations. The largest number of splits is determined by the tree's maximum depth; a high maximum depth can lead to overfitting. Before growing trees, random subsampling corresponds to a particular ratio of a training dataset within every iteration. The optimization method is stronger by using a learning rate, which essentially lessens the effect of every individual tree and allows future trees to enhance the framework. Increases in the variables make the model extra conservative because they are normalization terms for weight training. The values for each parameter are obtained by applying a grid search algorithm and a 10-fold cross-validation procedure. In total, there were 500 boosting repetitions, 25 maximum tree depths, and 1 (0.8) for the subsample ratio of the training dataset, 0.5 for the subsample ratio of the columns, and $\lambda = 1$, $\alpha = 0.2$, and 0.1 for the learning rate.

4.6. Analysis of Hyperparameter Configurations

We investigated the impact of various hyperparameter configurations in the suggested modules. The percentage of historical steps entered into the model, the dimensions of a DCC kernel, the batch dimensions, and the element of hidden channels (d_Z) are the four hyperparameters that need a manual process set by professionals. The study's findings are shown in Figure 5, which can infer that STN not only improves effectiveness, but also multiplies the achievement stability in various hyperparameter configurations.

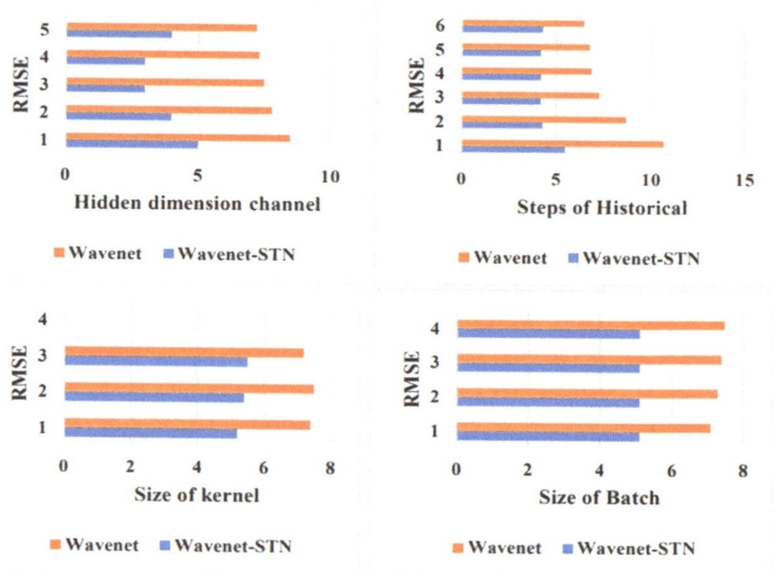

Figure 5. Analysis of hyperparameters.

They can be applied to the raw input data to see if they reduce the problems raised and to show how TN and SN redefine the extracted features. The original quantity is against both the temporally normalized amount and the spatially normalized quantity. There are differences among regions and days within the pairwise connection between the make sure and the temporally normalized quantity as well as among the original measure and the spatially normalized quantity. Insufficient SOTA methods suggest that to improve the local element, various time series should be made to have a mutual relationship with one another. In essence, they highlight the local element of individual time series by contrasting a pair of time series that have similar global elements over time. Multiple connections reflect the individuality of every time series are produced, for instance, by contrasting the three-time series within a single time series (referred to as an anchor). Dissimilar time series might need to be multiplied along various anchors because it is frequently unknown which ones are eligible anchors. These approaches use a graph-learning component to investigate every potential couple of time series to mechanically recognize the anchor for each time series. Here, $O(TN^2)$ refers to the computational complexity. The method's normalization modules, in contrast to other ones that have been suggested in this field, only call for $O(TN)$ operations.

4.7. Model Comparison

The MASE of every method is established for various lead times to demonstrate their effects on various time series predictions. MASE for every model is then calculated for every lead time. Table 2 shows the proposed model's comparison with existing methods. Figure 6 demonstrates the competitive performances of Multi-variate RNN, LSTM, MLP, and SVM for 5–15 min early time series prediction. However, their prediction accuracy declines significantly as lead time lengthens. Given the strong correlation, a simulation model such as MLP can be supplemented by using both temporal and spatial data as input. As a result, predicting accuracy could be increased. Figure 6 demonstrates that MLP performs better than other methods for every lead time. This happens because MLP, which has a deep architecture having nonlinear properties, can accurately capture and recreate the entire target, whereas other approaches cannot do so, owing to their linear characteristics.

Table 2. Performance comparison with MASE.

Leading Time/ Methods	MASE			
	RNN	**LSTM**	**SVM**	**MLP**
5	98.1	85.3	71.3	68.2
10	94.8	80.1	70.3	64.5
15	96.7	90.2	68.6	56.3
25	98.1	91.4	73.9	78.2
30	90.1	86.3	77.3	60.6

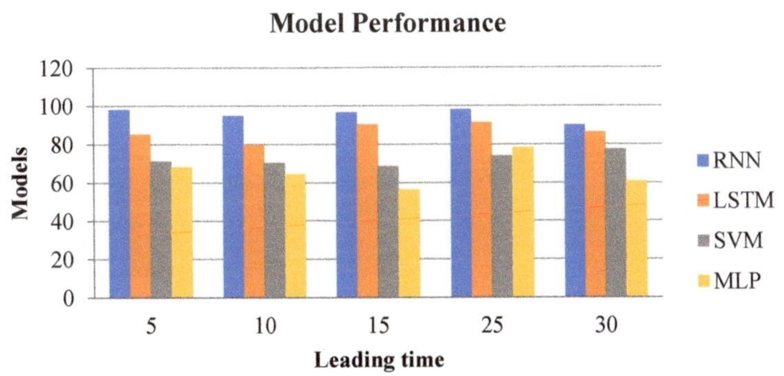

Figure 6. Comparison of the proposed method.

4.8. Computation Time Comparison

Table 3 lists the times for training (on the training set) as well as predicting (on the testing set) for every approach. The training time for the many multi-variate algorithms that have been suggested, including MLP, RNN, LSTM, and SVM, is the total of the training times, as well as the prediction time, determined in the same manner. The persistent method's learning and forecasting times are 0 and are, therefore, not listed. As seen in Table 3, because periodic retraining and refreshing of forecasts are eliminated, the time effectiveness is comparable to other approaches while training spatial-temporal predictions.

The MTS spatial and temporal prediction is taught live by incremental learning, which saves time by avoiding periodic model retraining and refreshing.

Table 3. Time computation comparison.

Leading Method/Methods	RNN		LSTM		SVM		MLP	
Process/(s)	**Train**	**Pred**	**Train**	**Pred**	**Train**	**Pred**	**Train**	**Pred**
5	1545	4.21	22,540	20.09	17,680	0.38	134	0.25
10	1445	3.82	20,020	21.13	16,560	0.33	185	0.22
15	1685	3.45	37,140	22.55	15,580	0.34	182	0.34
20	1510	3.41	25,860	23.42	14,230	0.34	225	0.32
25	1415	3.30	39,770	21.01	17,380	0.36	261	0.37
30	1500	3.29	22,590	24.10	12,730	0.25	228	0.59

5. Conclusions

This paper proposes a novel method for factorizing MTS data. We suggest spatial and temporal normalization following factorization, which improves the local and high-frequency components of the MTS data, respectively. Due to the nonlinear nature of demand variations, current research demonstrates that statistical methods lack predictive value in real-world circumstances. In particular, predictions beyond a few hours may be imprecise when employing these algorithms. As evidenced by the multiple plots in the current edition, these shifts occur after a few hours, particularly when the demand varies drastically. The results of the experiment illustrate the capability and performance of these two components. However, this study has significant limitations, including controlling the modeling process through improved machine learning model parameters and locating appropriate variables for taking the input from the models into consideration. Future studies may take into account the use of various hybrid frameworks with optimization techniques as a unique addition to spatial and temporal prediction.

Author Contributions: Conceptualization, A.M.P. and C.Y.; methodology, A.M.P. and C.Y.; software, G.K.A.; validation, A.M.P. and C.Y.; formal analysis, A.M.P. and C.Y.; investigation, A.M.P.; resources, A.M.P.; data curation, A.M.P.; writing—original draft preparation, A.M.P.; writing—review and editing, A.M.P., T.B.O. and A.M.; visualization, G.K.A.; supervision, C.Y.; funding acquisition, C.Y. All authors have read and agreed to the published version of the manuscript.

Funding: The National Natural Science Foundation of China under Grant No. 61873004.

Conflicts of Interest: The authors declare no conflict of interest.

References

1. Fildes, R.; Nikolopoulos, K.; Crone, S.F.; Syntetos, A.A. Forecasting and operational research: A review. *J. Oper. Res. Soc.* **2008**, *59*, 1150–1172. [CrossRef]
2. Forni, M.; Hallin, M.; Lippi, M.; Reichlin, L. The Generalized Dynamic Factor Model. *J. Am. Stat. Assoc.* **2005**, *100*, 830–840. [CrossRef]
3. Perez-Chacon, R.; Talavera-Llames, R.L.; Martinez-Alvarez, F.; Troncoso, A. Finding Electric Energy Consumption Patterns in Big Time Series Data. In *Distributed Computing and Artificial Intelligence, 13th International Conference*; Omatu, S., Semalat, A., Bocewicz, G., Sitek, P., Nielsen, I.E., García García, J.A., Bajo, J., Eds.; Springer International Publishing: Cham, Switzerland, 2016; Volume [illegible], pp. [illegible]. [CrossRef]
4. Galicia, A.; Torres, J.F.; Martínez-Álvarez, F.; Troncoso, A. [illegible] Forecasting Techniques Applied to Big Electricity Time Series. In *Advances in Computational Intelligence*; Rojas, I., Joya, G., Catala, A., Eds.; Springer International Publishing: Cham, Switzerland, 2017; Volume 10306, pp. 165–175. [CrossRef]
5. Hyndman, R.; Koehler, A.B.; Ord, J.K.; Snyder, R.D. *Forecasting with Exponential Smoothing: The State Space Approach*; Springer Science & Business Media: West Lafayette, IN, USA, 2008.
6. Makridakis, S.; Hyndman, R.J.; Petropoulos, F. Forecasting in social settings: The state of the art. *Int. J. Forecast.* **2020**, *36*, 15–28. [CrossRef]

7. Geng, X.; Li, Y.; Wang, L.; Zhang, L.; Yang, Q.; Ye, J.; Liu, Y. Spatiotemporal Multi-Graph Convolution Network for Ride-Hailing Demand Forecasting. *Proc. Conf. AAAI Artif. Intell.* **2019**, *33*, 3656–3663. [CrossRef]
8. Ding, D.; Zhang, M.; Pan, X.; Yang, M.; He, X. Modeling Extreme Events in Time Series Prediction. In Proceedings of the 25th ACM SIGKDD International Conference on Knowledge Discovery & Data Mining, Anchorage, AK, USA, 4–8 August 2019; pp. 1114–1122. [CrossRef]
9. Piccialli, F.; Giampaolo, F.; Prezioso, E.; Camacho, D.; Acampora, G. Artificial intelligence and healthcare: Forecasting of medical bookings through multi-source time-series fusion. *Inf. Fusion* **2021**, *74*, 1–16. [CrossRef]
10. Fathian, F.; Mehdizadeh, S.; Sales, A.K.; Safari, M.J.S. Hybrid models to improve the monthly river flow prediction: Integrating artificial intelligence and non-linear time series models. *J. Hydrol.* **2019**, *575*, 1200–1213. [CrossRef]
11. Gul, E.; Safari, M.J.S.; Haghighi, A.T.; Mehr, A.D. Sediment transport modeling in non-deposition with clean bed condition using different tree-based algorithms. *PLoS ONE* **2021**, *16*, e0258125. [CrossRef] [PubMed]
12. Bontempi, G.; Ben Taieb, S. Conditionally dependent strategies for multiple-step-ahead prediction in local learning. *Int. J. Forecast.* **2011**, *27*, 689–699. [CrossRef]
13. Ben Taieb, S.; Bontempi, G.; Atiya, A.F.; Sorjamaa, A. A review and comparison of strategies for multi-step ahead time series forecasting based on the NN5 forecasting competition. *Expert Syst. Appl.* **2012**, *39*, 7067–7083. [CrossRef]
14. Kolda, T.; Bader, B.W. Tensor Decompositions and Applications. *SIAM Rev.* **2009**, *51*, 455–500. [CrossRef]
15. De Silva, A.; Hyndman, R.J.; Snyder, R. The vector innovations structural time series framework: A simple approach to multivariate forecasting. *Stat. Model.* **2010**, *10*, 353–374. [CrossRef]
16. Bai, S.; Kolter, J.Z.; Koltun, V. An Empirical Evaluation of Generic Convolutional and Recurrent Networks for Sequence Modeling. *arXiv* **2018**, arXiv:1803.01271. [CrossRef]
17. Oord, A.V.D.; Dieleman, S.; Zen, H.; Simonyan, K.; Vinyals, O.; Graves, A.; Kalchbrenner, N.; Senior, A.; Kavukcuoglu, K. WaveNet: A Generative Model for Raw Audio. *arXiv* **2016**, arXiv:1609.03499. [CrossRef]
18. Liu, C.-T.; Wu, C.-W.; Wang, Y.-C.F.; Chien, S.-Y. Spatially and Temporally Efficient Non-local Attention Network for Video-based Person Re-Identification. *arXiv* **2019**, arXiv:1908.01683. [CrossRef]
19. Safari, M.J.S. Hybridization of multivariate adaptive regression splines and random forest models with an empirical equation for sediment deposition prediction in open channel flow. *J. Hydrol.* **2020**, *590*, 125392. [CrossRef]
20. Yu, H.-F.; Rao, N.; Dhillon, I.S. Temporal Regularized Matrix Factorization for High-dimensional Time Series Prediction. *Adv. Neural Inf. Processing Syst.* **2016**, *29*, 9.
21. Jiang, R.; Cai, Z.; Wang, Z.; Yang, C.; Fan, Z.; Chen, Q.; Tsubouchi, K.; Song, X.; Shibasaki, R. DeepCrowd: A Deep Model for Large-Scale Citywide Crowd Density and Flow Prediction. *IEEE Trans. Knowl. Data Eng.* **2021**. [CrossRef]
22. Fan, C.; Zhang, Y.; Pan, Y.; Li, X.; Zhang, C.; Yuan, R.; Wu, D.; Wang, W.; Pei, J.; Huang, H. Multi-Horizon Time Series Forecasting with Temporal Attention Learning. In Proceedings of the 25th ACM SIGKDD International Conference on Knowledge Discovery & Data Mining, Anchorage, AK, USA, 4–8 August 2019; pp. 2527–2535. [CrossRef]
23. Anderer, M.; Li, F. Hierarchical forecasting with a top-down alignment of independent-level forecasts. *Int. J. Forecast.* **2022**, *38*, 1405–1414. [CrossRef]
24. Lin, S.; Xu, F.; Wang, X.; Yang, W.; Yu, L. Efficient Spatial-Temporal Normalization of SAE Representation for Event Camera. *IEEE Robot. Autom. Lett.* **2020**, *5*, 4265–4272. [CrossRef]
25. Wang, J.-H.; Lin, G.-F.; Chang, M.-J.; Huang, I.-H.; Chen, Y.-R. Real-Time Water-Level Forecasting Using Dilated Causal Convolutional Neural Networks. *Water Resour. Manag.* **2019**, *33*, 3759–3780. [CrossRef]
26. Zhang, X.; You, J. A Gated Dilated Causal Convolution Based Encoder-Decoder for Network Traffic Forecasting. *IEEE Access* **2020**, *8*, 6087–6097. [CrossRef]
27. Zhao, W.; Wu, H.; Yin, G.; Duan, S.-B. Normalization of the temporal effect on the MODIS land surface temperature product using random forest regression. *ISPRS J. Photogramm. Remote Sens.* **2019**, *152*, 109–118. [CrossRef]
28. Botalb, A.; Moinuddin, M.; Al-Saggaf, U.M.; Ali, S.S.A. Contrasting Convolutional Neural Network (CNN) with Multi-Layer Perceptron (MLP) for Big Data Analysis. In Proceedings of the 2018 International Conference on Intelligent and Advanced System (ICIAS), Kuala Lumpur, Malaysia, 13–14 August 2018; pp. 1–5. [CrossRef]
29. Shi, X.; Li, Y.; Yang, Y.; Sun, B.; Qi, F. Multi-models and dual-sampling periods quality prediction with time-dimensional K-means and state transition-LSTM network. *Inf. Sci.* **2021**, *580*, 917–933. [CrossRef]
30. de Myttenaere, A.; Golden, B.; Le Grand, B.; Rossi, F. Mean Absolute Percentage Error for regression models. *Neurocomputing* **2016**, *192*, 38–48. [CrossRef]
31. Aschner, A.; Solomon, S.G.; Landy, M.S.; Heeger, D.J.; Kohn, A. Temporal Contingencies Determine Whether Adaptation Strengthens or Weakens Normalization. *J. Neurosci.* **2018**, *38*, 10129–10142. [CrossRef]
32. Diao, Z.; Wang, X.; Zhang, D.; Liu, Y.; Xie, K.; He, S. Dynamic Spatial-Temporal Graph Convolutional Neural Networks for Traffic Forecasting. In Proceedings of the AAAI Conference on Artificial Intelligence, Honolulu, HI, USA, 27 January–1 February 2019; Volume 33, pp. 890–897. [CrossRef]
33. Makridakis, S.; Spiliotis, E.; Assimakopoulos, V. Statistical and Machine Learning forecasting methods: Concerns and ways forward. *PLoS ONE* **2018**, *13*, e0194889. [CrossRef]

34. Safari, M.J.S.; Arashloo, S.R. Sparse kernel regression technique for self-cleansing channel design. *Adv. Eng. Inform.* **2021**, *47*, 101230. [CrossRef]
35. Mohammadi, B.; Guan, Y.; Moazenzadeh, R.; Safari, M.J.S. Implementation of hybrid particle swarm optimization-differential evolution algorithms coupled with multi-layer perceptron for suspended sediment load estimation. *CATENA* **2021**, *198*, 105024. [CrossRef]

MDPI
St. Alban-Anlage 66
4052 Basel
Switzerland
www.mdpi.com

Electronics Editorial Office
E-mail: electronics@mdpi.com
www.mdpi.com/journal/electronics